Geography of the Physical Environment

The *Geography of the Physical Environment* book series provides a platform for scientific contributions in the field of Physical Geography and its subdisciplines. It publishes a broad portfolio of scientific books covering case studies, theoretical and applied approaches as well as novel developments and techniques in the field. The scope is not limited to a certain spatial scale and can cover local and regional to continental and global facets. Books with strong regional focus should be well illustrated including significant maps and meaningful figures to be potentially used as field guides and standard references for the respective area.

The series appeals to scientists and students in the field of geography as well as regional scientists, landscape planners, policy makers, and everyone interested in wide-ranging aspects of modern Physical Geography. Peer-reviewed research monographs, edited volumes, advance and undergraduate level textbooks, and conference proceedings covering the major topics in Physical Geography are included in the series. Submissions to the Book Series are also invited on the theme 'The Physical Geography of...', with a relevant subtitle of the author's/editor's choice. Please contact the Publisher for further information and to receive a Book Proposal Form.

Sujit Mandal · Ramkrishna Maiti ·
Michael Nones · Heinz R. Beckedahl
Editors

Applied Geomorphology and Contemporary Issues

Editors
Sujit Mandal
Department of Geography
Diamond Harbour Women's University
Diamond Harbour, West Bengal, India

Michael Nones
Department of Hydrology
and Hydrodynamics
Polish Academy of Sciences
Warsaw, Poland

Ramkrishna Maiti
Department of Geography and
Environment Management
Vidyasagar University
Midnapore, West Bengal, India

Heinz R. Beckedahl
Department of Geography, Environment
of Science and Planning
University of Swaziland
Swaziland, Eswatini

ISSN 2366-8865 ISSN 2366-8873 (electronic)
Geography of the Physical Environment
ISBN 978-3-031-04534-9 ISBN 978-3-031-04532-5 (eBook)
https://doi.org/10.1007/978-3-031-04532-5

© The Editor(s) (if applicable) and The Author(s), under exclusive license to Springer Nature Switzerland AG 2022

This work is subject to copyright. All rights are solely and exclusively licensed by the Publisher, whether the whole or part of the material is concerned, specifically the rights of translation, reprinting, reuse of illustrations, recitation, broadcasting, reproduction on microfilms or in any other physical way, and transmission or information storage and retrieval, electronic adaptation, computer software, or by similar or dissimilar methodology now known or hereafter developed.
The use of general descriptive names, registered names, trademarks, service marks, etc. in this publication does not imply, even in the absence of a specific statement, that such names are exempt from the relevant protective laws and regulations and therefore free for general use.
The publisher, the authors, and the editors are safe to assume that the advice and information in this book are believed to be true and accurate at the date of publication. Neither the publisher nor the authors or the editors give a warranty, expressed or implied, with respect to the material contained herein or for any errors or omissions that may have been made. The publisher remains neutral with regard to jurisdictional claims in published maps and institutional affiliations.

Cover image by Sonja Weber, München

This Springer imprint is published by the registered company Springer Nature Switzerland AG
The registered company address is: Gewerbestrasse 11, 6330 Cham, Switzerland

Contents

Part I Climate Change and Rivers Response

1 **Large-Scale Sediment Transport Modelling: Development, Application, and Insights** 3
   Kai Tsuruta and Marwan A. Hassan

2 **An Appraisal to Anthropogeomorphology of the Chel River Basin, Outer Eastern Himalayas and Foreland, West Bengal, India** 19
   Sonam Lama and Ramkrishna Maiti

3 **Channel Migration Vulnerability in the Kaljani River Basin of Eastern India** 53
   Moumita Dutta and Sujit Mandal

4 **Exploring Change of River Morphology and Water Quality in the Stone Mine Areas of Dwarka River Basin, Eastern India** 77
   Indrajit Mandal and Swades Pal

5 **An Attempt to Forecast Seasonal Precipitation in the Comahue River Basins (Argentina) to Increase Productivity Performance in the Region** 97
   Maximiliano Vita Sanchez, Marcela Hebe González, and Alfredo Luis Rolla

6 **Channel Instability in Upper Tidal Regime of Bhagirathi-Hugli River, India** 127
   Chaitali Roy and Sujit Mandal

7 **The Pattern of Extreme Precipitation and River Runoff using Ground Data in Eastern Nepal** 147
   Shakil Regmi and Martin Lindner

8 **Climate Change and its Impact on Catchment Linkage and Connectivity** 167
   Manudeo Singh and Rajiv Sinha

9   Inter-decadal Variability of Precipitation Patterns
    Increasing the Runoff Intensity in Lower Reach of Shilabati
    River Basin, West Bengal 179
    Suparna Chaudhury

Part II Land Degradation, Resource Depletion and Livelihood Challenges

10 Are the Badlands of Tapi Basin in Deccan Trap Region
    of India "Vanishing Landscape?" Badland Dynamics:
    Past, Present and Future! 193
    Veena Joshi and Shreeya Kulkarni

11 Soil Piping: Problems and Prospects 217
    H. R. Beckedahl, J. A. A. Jones, and U. Hardenbicker

12 Role of LU and LC Types on the Spatial Distribution
    of Arsenic-Contaminated Tube Wells of Purbasthali
    I and II Blocks of Burdwan District, West Bengal, India 245
    Sunam Chatterjee, Srimanta Gupta, Bidyut Saha,
    and Biplab Biswas

13 Forecasting the Danger of the Forest Fire Season
    in North-West Patagonia, Argentina 257
    Ezequiel A. Marcuzzi, Marcela Hebe González,
    and María del Carmen Dentoni

14 Quantifying the Spatio-seasonal Water Balance and Land
    Surface Temperature Interface in Chandrabhaga River
    Basin, Eastern India 273
    Susanta Mahato and Swades Pal

15 Application of Ensemble Machine Learning Models
    to Assess the Sub-regional Groundwater Potentiality:
    A GIS-Based Approach 293
    Sunil Saha, Amiya Gayen, and Sk. Mafizul Haque

16 Enhancement of Natural and Technogenic Soils Through
    Sustainable Soil Amelioration Products for a Reduction
    of Aeolian and Fluvial Translocation Processes 309
    Sandra Muenzel and Oswald Blumenstein

17 Assessment of Land Use and Land Cover Change
    in the Purulia District, India Using LANDSAT Data 329
    Pritha Das, Prasenjit Bhunia, and Ramkrishna Maiti

18 Prioritization of Watershed Developmental Plan
    by the Identification of Soil Erosion Prone Areas Using
    USLE and RUSLE Methods for Sahibi Sub-Watershed
    of Rajasthan and Haryana State, India 351
    Ajoy Das, Jagmohan Singh, Madan Thakur,
    and Asim Ratan Ghosh

19 Estimation of Soil Erosion Using Revised Universal
    Soil Loss Equation (RUSLE) Model in Subarnarekha
    River Basin, India 381
    Ujjwal Bhandari and Uttam Mukhopadhyay

20 Land Cover Changes in Green Patches and Its Impact
    on Carbon Sequestration in an Urban System, India 397
    Sunanda Batabyal, Nilanjan Das, Ayan Mondal,
    Rituparna Banerjee, Sohini Gangopadhyay,
    and Sudipto Mandal

21 Review on Sustainable Groundwater Development
    and Management Strategies Associated with the Largest
    Alluvial Multi-aquifer Systems of Indo-Gangetic
    Basin in India 411
    Anadi Gayen

Part III Large Dams and River Systems

22 Predicting the Distribution of Farm Dams in Rural South
    Africa Using GIS and Remote Sensing 427
    Jonathan Tsoka, Jasper Knight, and Elhadi Adam

23 Large Dams, Upstream Responses, and Riverbank Erosion:
    Experience from the Farakka Barrage Operation
    in India... 441
    Tanmoy Sarkar and Mukunda Mishra

**Part IV Climate Change, Geomorphic Hazards and Human
Livelihood**

24 Climate Change and Human Performance: Assessment
    of Physiological Strain in Male Paddy Cultivators
    in Hooghly, West Bengal, India 465
    Ayan Chatterjee, Sandipan Chatterjee, Neepa Banerjee,
    and Shankarashis Mukherjee

25 Study on Climate Change and Its Impact on Coastal
    Habitats with Special Reference to Ecosystem Vulnerability
    of the Odisha Coastline, India 475
    Avijit Bakshi and Ashis Kumar Panigrahi

26 The Millennium Flood of the Upper Ganga Delta,
    West Bengal, India: A Remote Sensing Based Study 499
    Sayantan Das and Sunando Bandyopadhyay

27 Tropical Cyclone: A Natural Disaster with Special
    Reference to *Amphan* 519
    Biplab Biswas and Chandi Rajak

28	**Observed Changes in the Precipitation Regime in the Argentinean Patagonia and Their Geographical Implication**.. Paula B. Martin, Victoria A. Oruezabal, and María E. Castañeda	537
29	**An Assessment of Severe Storms, Their Impacts and Social Vulnerability in Coastal Areas: A Case Study of General Pueyrredon, Argentina** .... Ignacio A. Gatti, Paula B. Martin, Elisabet C. Vargas, Mariana Gasparotto, Barbara E. Prario, Elvira E. Gentile, and Leandro G. Patané	547
30	**Modelling and Mapping Landslide Susceptibility of Darjeeling Himalaya Using Geospatial Technology** ...... Biplab Mandal, Subrata Mondal, and Sujit Mandal	565
31	**Climate Change Induced Coastal Hazards and Community Vulnerability in Indian Sundarban** .... Biraj Kanti Mondal	587
32	**Sea-Level Changes Along Bangladesh Coast: How Much Do We Know About It?**.................... M. Shahidul Islam	611
33	**Assessing Channel Migration, Bank Erosion Vulnerability and Suitable Human Habitation Sites in the Torsa River Basin of Eastern India Using AHP Model and Geospatial Technology**....................................... Sourav Dey and Sujit Mandal	635
34	**Spatiotemporal Assessment of Drought Intensity and Trend Along with Change Point: A Study on Bankura District, West Bengal, India** .... Shrinwantu Raha, Suman Kumar Dey, Madhumita Mondal, and Shasanka Kumar Gayen	655
35	**Landslide Susceptibility Assessment and Management Using Advanced Hybrid Machine Learning Algorithms in Darjeeling Himalaya, India** .... Anik Saha and Sunil Saha	667
36	**Predicting the Landslide Susceptibility in Eastern Sikkim Himalayan Region, India Using Boosted Regression Tree and REPTree Machine Learning Techniques** .... Kanu Mandal, Sunil Saha, and Sujit Mandal	683
37	**An Exploratory Analysis of Mountaineering Risk Estimation Among the Mountaineers in the Indian Himalaya** .... Chinmoy Biswas, Koyel Roy, Rupan Dutta, and Shasanka Kumar Gayen	709

Part I
Climate Change and Rivers Response

Large-Scale Sediment Transport Modelling: Development, Application, and Insights

Kai Tsuruta and Marwan A. Hassan

Abstract

Estimating sediment erosion and yield at the regional scale presents a challenging problem for researchers. While most studies have relied on spatially lumped, empirical estimates based on hydrology, the high spatial variability in sediment dynamics calls into question the efficacy of such methods for numerous study sites. Furthermore, it is unclear how to extend these relationships to future climate or landuse scenarios. In this chapter, we provide an overview of efforts made to adapt a high-resolution sediment model into a large-scale, distributed, mechanistic sediment dynamics model. We then review insights gleaned from the application of this model to the Fraser River Basin in British Columbia, Canada including a new empirical estimate for gross erosion based on sediment yield. Finally, we suggest opportunities for generalizations and avenues for future work.

Keywords

Large-scale sediment dynamics · Mechanistic sediment model · Gross erosion approximation · Glaciated basin geomorphology

1.1 Introduction

A region's sediment dynamics are functions of its hydrology and landscape and therefore likely to be sensitive to future changes in climate or landuse (Walling 2009). For large-scale river systems, these changes may have both local and global consequences for flora and fauna. Nutrient and contaminant transport, water quality, and fish habitat are all influenced by sediment dynamics (Motew et al. 2017; Kerr 1995; Sternecker et al. 2013). Sedimentation process has been shown to play an active role in the global carbon cycle (Walling 2009). In the coastal areas, sediment deposition contributes to coastline retreat and advance (Syvitski et al. 2005) as well as floodplain agriculture and aquaculture (Manh et al. 2015).

Despite the importance of sediment dynamics to terrestrial and aquatic biota, there have been few studies to quantify the degree to which climatic and human factors may affect the entire sequence of sediment erosion, transportation, and deposition (Asselman et al. 2003), in part because very few sediment models adequately

K. Tsuruta (✉)
Pacific Climate Impacts Consortium, University of Victoria, Victoria, BC, Canada
e-mail: kaitsuruta@uvic.ca

M. A. Hassan
Department of Geography, University of British Columbia, Vancouver, BC, Canada

simulated these processes (Praskievicz and Chang 2011). While there is a need for adequate large-scale models and its development which presents numerous challenges for researchers, previous studies [e.g. Pelletier (2012); Smith et al. (2011); Bathurst (2010)] argued that because substantial erosion and deposition can occur over relatively small spatial and temporal scales, sediment models need to be distributed or otherwise account for sub-basin variability. Bathurst (2010) additionally argued that for studies involving climate change, it is best if sediment models are physically-based. To further complicate simulations and analysis, basins typically have limited observations of sediment and/or appropriate hydrological drivers (Tsuruta et al. 2019) and climatic effects on sediment are often difficult to disentangle from more direct anthropogenic effects such as land management and dams (Walling 2009).

Among the large-scale simulations of sediment dynamics, most have relied on empirical or stochastic models. Of these models, many are "lumped" in the sense that they do not account for spatial variability within the basin [e.g. Hovius (1998), Syvitski et al. (2003)]. However, several studies (Pelletier 2012; Smith et al. 2011; Vaughan et al. 2017) have argued for the importance of sub-basin representation and a handful of empirical models have accounted for this variability by using a distributed design (Pelletier 2012) or a stochastic approach (Benda and Dunne 1997a, b; Schmitt et al. 2016, 2018).

Whether they take into account spatial variability or not, empirical and stochastic models are difficult to apply in studies involving climate change because it is unclear if their fundamental relationships will remain valid under the new climate. Hence, it is generally preferred to use mechanistic models because the governing equations are based in relationships that are more likely to be invariant under climate changes. Among distributed, mechanistic sediment transport models, SWAT (Arnold et al. 1998; Arnold and Fohrer 2005) is perhaps the most well-known and commonly used. However, computational and data demands typically restrict high resolution sediment simulations with SWAT to small basins (<50 km^2). While, White et al. (2014) did apply SWAT to model sediment dynamics throughout the entire Mississippi Basin, the study calibrated SWAT against long-term (47 years) observed averages and therefore did not demonstrate that the modelling framework was sensitive to land use and climate change or capable of reproducing any seasonal, year-to-year, or decadal variability (Tsuruta et al. 2018).

Recently, Tsuruta et al. (2018) developed a physically-based, fully distributed large-scale sediment dynamics model. To route both water fluxes and sediment, the authors integrated the large-scale model into the Terrestrial Hydrology Model with Biochemistry (THMB) developed in Coe (2000). The integrated model was then applied to the 230,000 km^2 Fraser River Basin (FRB) in British Columbia, Canada. Adaptation and application utilized sub-grid spatial data and high-quality landscape, sediment, and hydrological observations in a watershed well-suited for study. Tsuruta et al. (2019) then drove the calibrated model with hydrological forcings derived from climate scenarios to assess potential impacts of climate change on sediment dynamics across the FRB. As mentioned above, large-scale climate change studies that make use of a distributed, physically-based sediment model are rare and the relevant model development complex. Here, we provide an overview of the large-scale model in Tsuruta et al. (2018, 2019), highlighting some of its key features, rational for development, and utility. We then review outcomes of the FRB simulations, including previously unpublished results, and suggest future areas of exploration.

1.2 A Regional Scale Suspended Sediment Model

In the following section, the rationale behind the development of the large-scale sediment model has been elaborated (Tsuruta et al. 2018, 2019). Figure 1.1 provides a schematic of the model (a) and displays the way an individual grid cell and its sediment dynamics are conceptualized (b). Tables 1.1, 1.2, and 1.3 provide the key

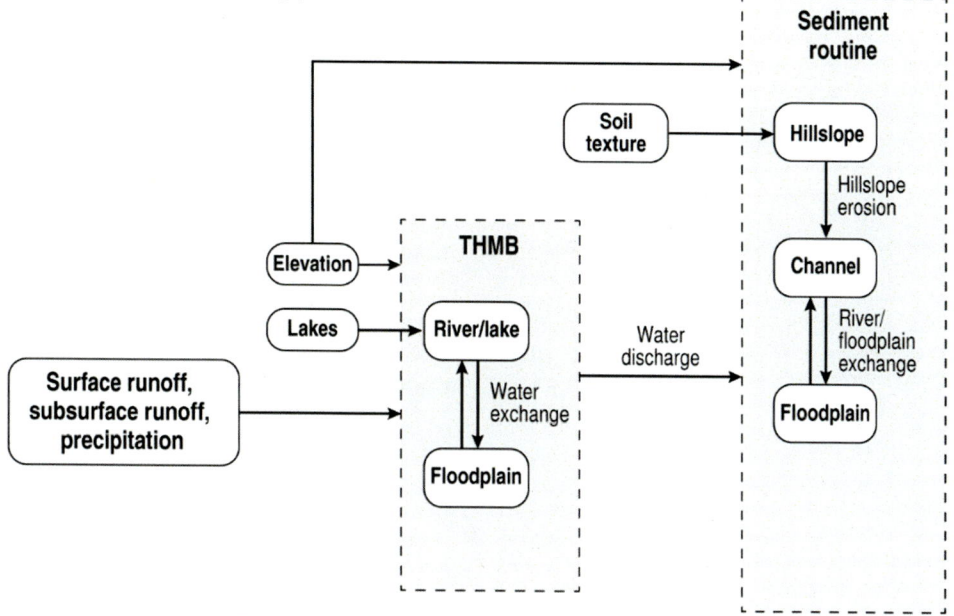
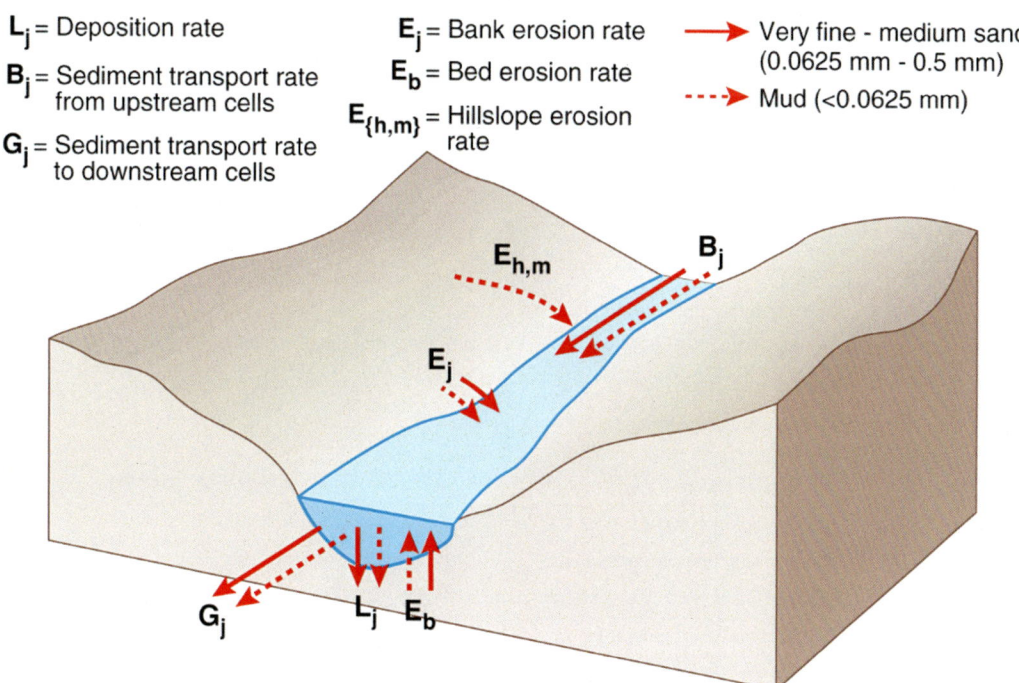

Fig. 1.1 Modelling framework (**a**) and conceptualization of sediment transport model (**b**). Panel (**a**) is modified from Fig. 1 in Tsuruta et al. (2018). Panel (**b**) is modified from Fig. 4 in Tsuruta et al. (2019)

Table 1.1 General transport variables [Modified from Table 1 in Tsuruta et al. (2018)]

Symbol	Units	Description	Equation/value
A	m^2	Grid cell area	
A_r	m^2	Longitudinal river area	$A_r = d \cdot 6.6588 \min\{Q, Q_{bf}\}^{0.4967}$
h	m	Height of river	$h = 0.2307 \min\{Q, Q_{bf}\}^{0.4123}$
h_{bf}	m	Bank-full height of river	$h_{bf} = 0.2307 Q_{bf}^{0.4123}$
V	m^3	Volume of river	$V = A_r h$
ρ_w	kg/m^3	Density of water	$\rho_w = 1000$ kg/m^3
ρ_s	kg/m^3	Density of sediment	$\rho_s = 2650$ kg/m^3
ρ_b	kg/m^3	Bulk density of sediment	$\rho_b = \rho_s(1 - n_p)$
n_p	–	Channel bottom sediment porosity	$n_p = 0.1$
$\beta_t{}^p$	–	Characteristic settling time parameter	$\beta_t = 10$

^{p}Indicates a tuned parameter

Table 1.2 Sand transport variables (Tsuruta et al. 2018)

Symbol	Units	Description	Equation/value
$E_{5,i}$	kg/s	Entrainment rate of D_i at 5% h	$E_{5,i} = v_i \cdot \beta_e(\lambda X_i)^5/(1 + \beta_e/0.3(\lambda X_i)^5) \cdot \rho_s A_r$
E_i	kg/s	Channel erosion rate of D_i	$E_i = F_i E_{5,i}$
L_i	kg/s	Deposition rate of D_i	$L_i = v_i \cdot \overline{C_i}/I[Z_{Ri}] \cdot \rho s A_r$
B_i	kg/s	Transport rate of D_i from upstream	
G_i	kg/s	Transport rate of D_i	$G_i = \max\{E_i + B_i - L_i, 0\}$
X_i	–	Entrainment parameter	$X_i = S^{0.08}(u*sk \cdot Re_{pi}^{0.6}/v_i)(D_i/D50)^{0.2}$
$I(Z_{Ri})$	–	Approximation of Einstein integral	$I(Z_{Ri}) = 1/\sum_{j=0}^{6} c_j Z_{Ri}^j$
Re_{pi}	–	Sediment Reynolds # for D_i	$Re_{pi} = (Rg\,D_i)^{0.5} D_i/v$
v_i	m/s	Vertical settling velocity of D_i	$v_i = g D_i^2 R/(18v)$
$u*sk$	m/s	Shear velocity due to skin friction	$u*sk = (\tau*skg\,D_i(\rho s - \rho w)/\rho w)^{0.5}$
$\tau*sk$	–	Dimensionless bed shear stress due to skin friction	$\tau*sk = 0.05 + 0.7(\tau*Fr^{0.7})^{0.8}$
$\tau*bf$	–	Bank-full channel Shields stress	Gravel: 0.049, sand: 1.86
$\overline{C_i}$	m^3/m^3	Depth-averaged volumetric concentration of ith class	$\overline{C_i} = (E_i + B_i)/(\rho_s V) \cdot (\beta_t 0.05 h/v_i)$
F_i	–	Fraction of bed in ith class	[vf, f, m] = [0.003, 0.01, 0.1]
β_e	–	Wright-Parker constant	$\beta_e = 7.8 \times 10^{-7}$
λ	–	Mixture suppression parameter	$\lambda = (1 - 0.28\sigma_\varphi)$
$\sigma_\varphi{}^p$	–	Grain size standard deviation, phi scale	Sand: 0.6, gravel: 2.6
D_i	mm	ith grain class representative	[vf, f, m] = [0.0088, 0.18, 0.35] mm
D_{50}	m	Median competent grain size predicted by Shields equation	$D50 = h_{bf} S \rho_w/(\tau*bf(\rho s - \rho w))$

^{p}Indicates a tuned parameter

Table 1.3 Mud transport variables and parameters [modified from Table 3 in Tsuruta et al. (2018)]

Symbol	Units	Description	Equation/Value
$E_{h,m}$	kg/s	Hillslope erosion rate of mud	$E_{h,m} = \max\{\beta_h F_m(\tau_h - \tau_{hc})A, 0\}$
E_m	kg/s	Channel erosion rate of mud	$E_m = \max\{e_0 \cdot (\tau_b/\tau_c - 1)A_r, 0\}$
L_m	kg/s	Deposition rate of mud	$L_m = \max\{(1 - \tau_b/\tau_c)v_m C_m A_r, 0\}$
B_m	kg/s	Rate of mud entering from upstream	
G_m	kg/s	Transport rate of mud	$G_m = \max\{E_m + B_m - L_m, 0\}$
τ_h	N/m²	Hillslope shear stress	$\tau_h = gS_h \rho_w R_s \Delta t$
τ_b	N/m²	Channel shear stress	$\tau_b = gS\rho_w h$
τhcp	N/m²	Critical hillslope shear stress	$\tau_{hc} = 0.1$ N/m²
τ_c	N/m²	Critical channel shear stress	$\tau c = k\tau(\rho b - \rho w)^{0.73}$
F_m	–	Fraction of hillslope soil in mud class	From ISRIC data
C_m	kg/m³	Depth-averaged mass concentration of mud in channel	$C_m = E_m(\beta th/v_m(t - 1))/V$
v_m	m/s	Settling velocity of mud	$v_m = 0.08Cj^{1.65}/(Cj^2 + 12.25)^{1.88}$; C_j = median $\{0.15, C_m, 2.11\}$
Hy	–	Hysteresis parameter	
β_rp	–	Rising limb threshold coefficient for hysteresis parameter	$\beta_r = 2.4$
β_fp	–	Falling limb threshold coefficient for hysteresis parameter	$\beta_f = 0.4$
κ_τp	–	Critical channel shear stress parameter	Hy = 0: 0.031, Hy = 1: 0.13
e_0	kg/m²/s	Erosion coefficient rate	$e_0 = 2.0 \times 10^{-4}$ kg/m²/s
β_hp	s/m	Hillslope proportionality constant	$\beta_h = 1.8 \times 10^{-6}$ s/m

pIndicates a tuned parameter

model variables and their governing equations/values. The details of each calculation are provided in Tsuruta et al. (2018).

Broadly, the model operates by generating a mud supply based on the shear stress created by surface runoff. This supply is then provided to the river channel. Within the river channel, both fine sand—which is assumed to be inifinite in supply—and the mud generated by surface erosion can be eroded, deposited, or transported based on water flux and hydraulic geometry, both of which are determined by THMB. THMB conceptualizes the channel as a dynamic rectangular volume and routes the hydrological drivers (precipitation and surface and subsurface runoff) through these volumes. The length of each volume is fixed based on sinuosity and distance from the cell to its downstream neighbor. The height and width are power functions of water flux that are tuned to observations. If the calculated river discharge exceeds bankfull discharge, the river volume ceases to grow and excess water is routed to the floodplain reservoir.

Floodplain sediment dynamics operate on the same principles as in-channel dynamics, but only mud is considered to be available in the floodplains. THMB also has the ability to simulate flow through lakes. It can simulate lake locations internally or users can identify any "lake" grid cells and their corresponding outlet cell based on available landscape maps. THMB then uses water directly upstream of these cells to first fill the lake to the sill height of the outlet prior to allowing any water to move downstream. Erosion of sediment in lakes is considered negligible, while transport and deposition operate on the same principles as in-channel dynamics but using lake flow and geometry.

1.2.1 Input Data

A common challenge for sediment modelling is data availability. In our model, we made an effort to keep data requirements to a minimum, and demonstrate the viability of globally available datasets. Required inputs are as follows: surface and subsurface runoff, rainfall and snowfall, elevation data, soil texture, and lake locations (Fig. 1.1a).

While data quality varies spatially, global elevation data is available [e.g. the Global Multi-resolution Terrain Elevation Map (GMTED); Danielson and Gesch (2011)] as is global-scale soil texture data [e.g. data from the International Soil Reference and Information Centre (ISRIC)] and lake area data [e.g. the Global Water Bodies database; Verpoorter et al. (2014)]. The precipitation and surface and subsurface runoff driving data typically comes in the form of hydrological model outputs. These outputs can be a result of water flux simulations based on historic landuse and climate or any landuse/climate scenarios. Tsuruta et al. (2018, 2019) made use of outputs from the Vertical Infiltration Capacity (VIC) model (Liang et al. 1994, 1996), but distributed results from any appropriate hydrological model can be used. In addition to the required inputs, several types of observational data can be useful in tuning THMB and the sediment model. These observations include river flux, suspended sediment flux, hydraulic geometry, lake sedimentation rates, and grain size distribution. In the absence of available observations, surrogates such as water quality and remote sensing have also been shown to be effective in tuning sediment models (Fagundes et al. 2020).

1.2.2 Major Controls on Sediment

A model should adequately represent the major sediment sources, sinks, and mechanisms for erosion as all three can have substantial variability within a study site. Sediment sources are the portions of the landscape that contribute most of the sediment. They are particularly important in watersheds where the river system generally has sufficient capacity to transport any sediment generated by the landscape to the basin's outlet. Watersheds of this type are referred to as "supply limited" and are characterized by relatively low rates of deposition and a hysteresis between event-scale sediment yield and water-flux caused by the exhaustion of available sediment. In large, undisturbed, heterogeneous basins, sediment sources tend to coincide with areas of high precipitation or runoff, steep hillslopes, and erodible terrain (Wischmeier and Smith 1965, 1978). Mechanisms for sediment generation (i.e. landscape erosion) vary but generally rely on the kinetic energy of rainfall or the shear stress caused by overland flow to detach soil particles. In some basins, mass wasting events (Parker et al. 2011; Pearce and Watson 1986), fires (Warrick and Rubin 2007), and landuse (Manh et al. 2015; Yang et al. 2015) have also been shown to influence sediment supply.

In development of the model, our focus was on watersheds where sediment supply was generated by overland flow. All supply-generating landscape processes were considered to be a function of overland flow and collectively referred to as "hillslope erosion". No distinction was made between these processes nor between any sub-grid features (i.e. hillsides and valleys). To capture the spatial and temporal variability of sediment sources, we estimated hillslope erosion $E_{h,m}$ at each grid cell based on the amount of hillslope shear stress τ_h over some tuned critical value τ_{hc} using the following equation:

$$E_{h,m} = \max\{\beta_h F_m (\tau_h - \tau_{hc}) A, 0\}. \quad (1.1)$$

Here, A is the grid cell area, β_h is a tuned proportionality constant, and F_m is the fraction of soil in the "mud" grain size class (<0.0625 mm). Equation (1.1) is adapted from Patil et al. (2012). The F_m factor was included as a way of accounting for the variance in soil erodibility throughout the basin and to recognize that not all grain-sizes are supply limited. Hillslope shear stress is a function of runoff and hillslope gradient and therefore accounts for the spatial (and in the case of runoff, temporal) variance in these two factors.

At the regional scale, between sediment sources and a watershed's outlet lies a host of different land forms, terrain, vegetation, and climate. In some locations, these properties combine to form sediment sinks—areas where most upstream sediment is deposited. Sinks act to disconnect sediment supply from the outlet and are important to account for in a large-scale model. The major sinks represented in our model are lakes, though THMB also has the capability to represent reservoirs based on site-specific needs. Because THMB explicitly simulates the geometry and water dynamics through lakes, computations of deposition and transport in these features is directly analogous to in-channel dynamics while erosion is considered negligible.

With the lakes, floodplains, and hillslope erosion model in place, the erosion, deposition, and transport routines in the Tsuruta et al. (2018) model should be capable of providing reasonable answers to questions regarding the large-scale dynamics of a simulated basin. Sediment sources can be identified as those locations where hillslope erosion is highest, sinks will be where incoming sediment concentrations are high, but downstream transport is low, and historical and future projected yield at any point within the basin can be produced and analyzed for user-specific purposes.

1.2.3 Example: Fraser River Basin Studies

As an example of the applications of our sediment model we review the results from Tsuruta et al. (2018, 2019) which respectively applied the model to the FRB to study historic sediment dynamics and project future impacts on the basin under various climate scenarios. The FRB is a 230,000 km^2 glaciated watershed in British Columbia, Canada. It contains portions of the Rocky and coastal mountains as well as a low gradient interior plateau (Fig. 1.2).

Runoff in the mountainous sections of the FRB is snowmelt-driven while the interior plateau generally contains hybrid or rain-dominant regimes. Suspended sediment observations along the main channel of the Fraser River suggest the sediment signal is controlled by the snowmelt-driven portions of the basin with the presence of hysteresis indicating the system is supply-limited. The FRB was glaciated during the Pleistocene Epoch leaving behind thick glacial deposits (McLean et al. 1999). These deposits were postglacially incised to form the main branches of the Fraser River and their remobilization acts as the primary supply source for the basin. While mass wasting events and anthropogenic influences are present in the FRB, they do not appear to have a significant impact on sediment flux at the basin scale (Tsuruta et al. 2018). While the delta region of the FRB has undergone significant landuse change as a result of agriculture and urbanization, the basin upstream of Mission—where model simulations end—has seen relatively little anthropogenic influence.

For the FRB simulations, the model was operated at a 1/16° × 1/16° spatial resolution and hourly temporal resolution. Elevation data was provided by GMTED (Danielson and Gesch 2011) and soil texture by ISRIC. Lake areas were identified using British Columbia Geological Survey maps. In both studies, the model was driven by hydrological outputs from VIC simulations performed in Shrestha et al. (2012). Tsuruta et al. (2018) used historic VIC runs while Tsuruta et al. (2019) used VIC output driven by downscaled climate model results from the Coupled Model Intercomparison Project Phase 3 (CMIP3). Based on IPCC climate scenarios A1B, A2, and B1. The radiative forcings of scenarios B1 and A1B are similar to the newer representative concentration pathways (RCP)scenarios 4.5 and 6.0, respectively while scneario A2's radiative forcings are somewhere between the levels of RCP 6.0 and 8.5 (IPCC, 2014).

1.3 Results

Tsuruta et al. (2018, 2019) produced a number of results and projections regarding the current and future sediment dynamics in the FRB. Here, we

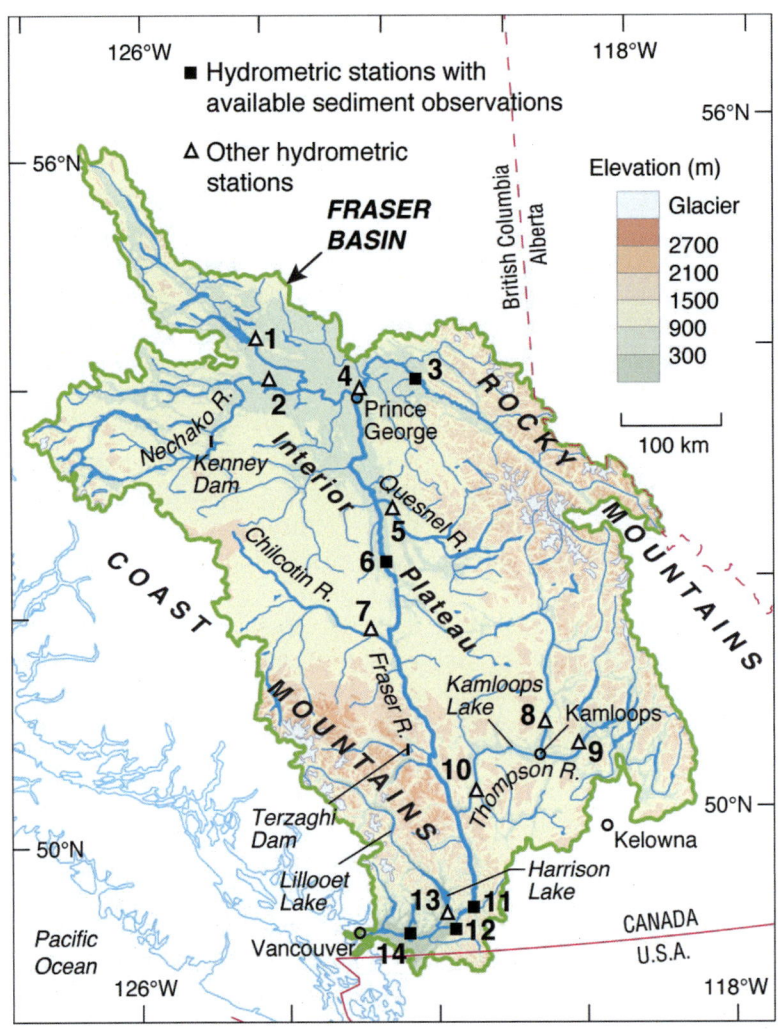

Fig. 1.2 Fraser River Basin (FRB) topography and river system. Numbers 1–14 correspond to the hydrometric stations (Table 1.4) with available observational data

review several outcomes from those studies to demonstrate how our sediment model can be applied and suggest avenues for future studies.

1.3.1 Basin-Wide Sediment Maps

From model simulations, one can produce basin-wide, distributed erosion, deposition, and transport maps demonstrating a basin's seasonal and spatial variation in sediment sources and yield (Fig. 1.3). These maps can be used for a variety of analysis. From Fig. 1.3b, e, one can see hillslope erosion is more substantial in the spring (April–June)—when the basin's snowpack begins to melt—than in the winter season (January–March). Spatially, erosion occurs mainly in the rocky and coastal mountain regions with very little being produced in the interior plateau. Sediment load displays similar patterns (Fig. 1.3a, d). Transport is more widespread and substantial in the spring, and load is largest in the main tributaries throughout the basin as well as sections of the rocky and coastal mountains. Figure 1.3c, f display projected changes to the sediment generation by the end of the century. They indicate that the mountainous regions are also the areas that will see the most change in hillslope erosion under future climate scenarios. Coastal mountain erosion is projected to undergo the largest increase in winter. In spring, erosion is expected to increase in the

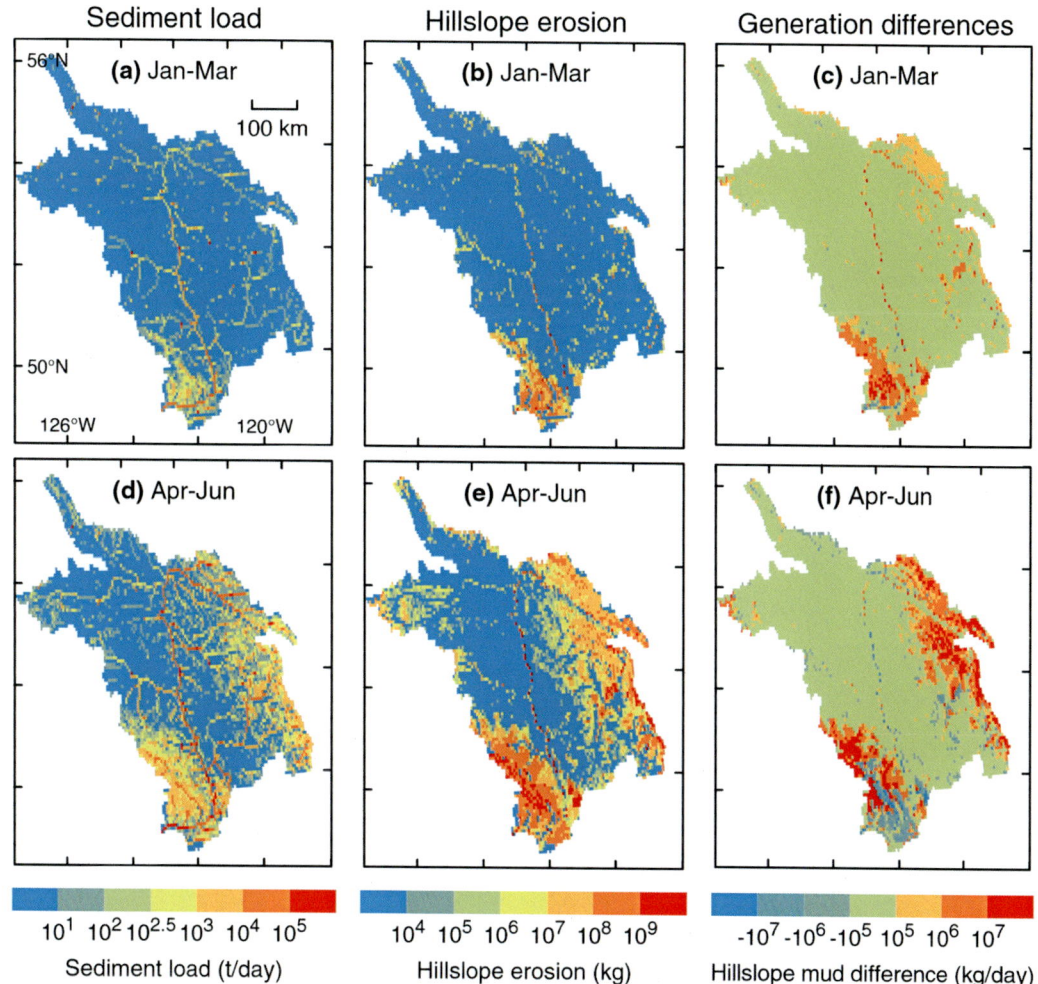

Fig. 1.3 1965–2004 average seasonal (January–March and April–June) simulated sediment load and hillslope erosion, and ensemble mean difference in hillslope mud generation between future (2065–2094) and baseline (1965–1994) periods for A1B scenario. For generation differences, positive values indicate future erosion is larger than baseline. Panels modified from Figs. 6 and 7 in Tsuruta et al. (2018) and Fig. 6 in Tsuruta et al. (2019)

rocky and northern coastal mountains while decreasing in the southern coastal mountains.

1.3.2 Timing of Sediment Load

Model simulations can also be used to perform detailed analysis at any point of particular interest throughout the basin. As an example, Fig. 1.4 from simulations in Tsuruta et al. (2019) displays the simulated seasonal load at two gauging stations under historic forcings as well as under a variety of forcings produced under climate scenarios. Output at Mission, located at the farthest downstream point in the modelled FRB (Fig. 1.2; Table 1.4), is expected to see a shift toward earlier and smaller spring sediment loads in the future while also experiencing a new spike in sediment load from September through January. Load at the outlet of the Chilcotin River, which drains primarily from the interior plateau, is also projected to shift toward an earlier spring peak, but that peak is expected to be larger in the future and no spike between September and January is

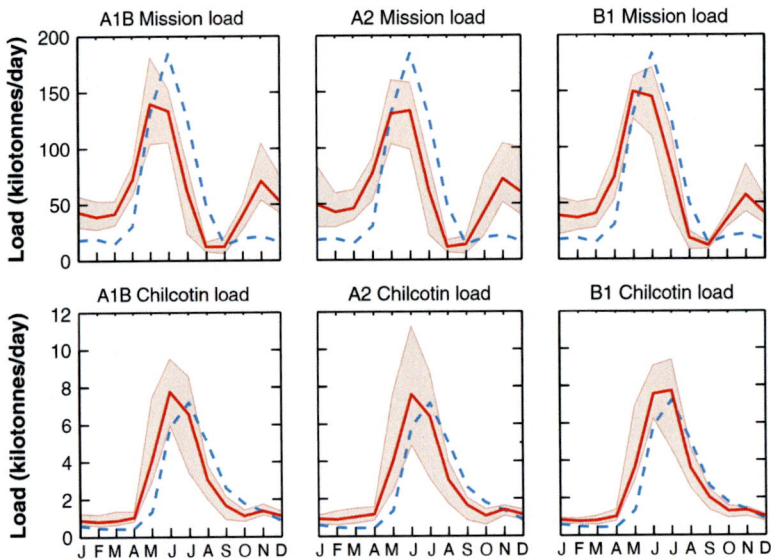

Fig. 1.4 Suspended sediment load at Mission and Chilcotin for baseline (1965–1994) mean (blue line), end-century (2065–2094) ensemble mean (red line), and end-century ensemble range (red region) under scenarios A1B, A2, and B1. Figure modified from Figs. 8 and 10 in Tsuruta et al. (2019)

Table 1.4 Sediment-specific yield (SSY) and delivery ratio (DR) across the FRB. SSY is in units of tonnes/km²/year

Location number	Sub-basin	SSY	DR	Location number	Station name	SSY	DR
1	Stuart	<1	<0.01	8	North Thompson	96	0.15
2	Nechako	1	0.01	9	South Thompson	19	0.02
3[s]	Hansard	161	0.30	10	Thompson	43	0.05
4	Shelley	116	0.27	11[s]	Hope	77	0.13
5	Quesnel	40	0.02	12[s]	Agassiz	77	0.14
6[s]	Marguerite	91	0.19	13[s]	Harrison	9	0.01
7	Chilcotin	44	0.03	14[s]	Mission	73	0.10

[s]Indicates locations with available sediment observations

projected. The physical reasons for these projected changes vary, but are generally related to changes in streamflow and supply and are explored in detail (Tsuruta et al. 2019).

1.3.3 Regional Scale Sources and Sinks

In our view, one of the more interesting results from Tsuruta et al. (2018) is that no major tributary analyzed contributed significant sediment to the main stem of the Fraser River. Surprisingly, this result could have been surmised using only the conceptual model. For tributaries located in the interior plateau (e.g. outlets of the Nechako and Stuart sub-basins) sediment contribution is small because they do not contain sufficient sediment sources relative to the regions of the basin with steep terrain.

Of the analyzed watersheds draining the rocky and coastal mountain areas, all except the North Thompson had lake sinks near their outlet that acted to disconnect the sediment generated from the regions downstream of the confluence. While modelled SSY was relatively large at the outlet of

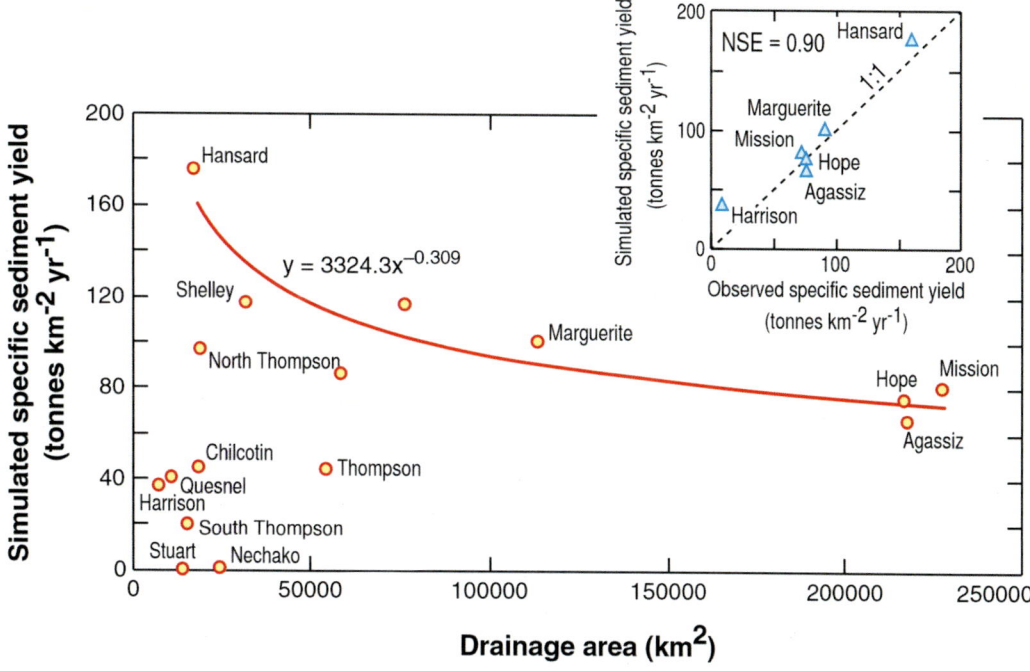

Fig. 1.5 FRB sub-basin specific sediment yield against basin area for 1965–2004 simulated values. Figure modified from Fig. 10 in Tsuruta et al. (2018)

the North Thompson, its drainage area is contained within the Thompson watershed and therefore does not directly contribute to sediment load in the Fraser River. Plotting specific sediment yield (yield per area) against area (Fig. 1.5) displays the dichotomy between yield from the main stem and from that of tributaries (though the North Thompson sub-basin plots relatively high, it drains into the Thompson River prior to meeting the Fraser).

This result clarifies our conceptual model of the FRB and informs any future modelling or prognostication of the basin. As previously stated, because the interior plateau is not considered to contain significant sediment sources, any future changes in that region are not likely to directly affect sediment dynamics downstream. Secondary effects related to changes in the hydrology must be considered, but can be expressed as changes in the driving data. Alternatively, "back of the envelope" calculations could provide researchers with a sense of the magnitude of likely hydrological changes from which related sediment flux changes could be inferred.

The situation is similar in large portions of the rocky and coastal mountains. The Thompson, Chilcotin, and Harrison sub-basins are all disconnected from the main stem in the sense of sediment flux. Combined with the Nechako and Stuart basins located within the interior plateau, these sub-basins account for over 50% of the drainage area of the FRB and do not contribute to sediment along the Fraser River. Any projected future changes in these regions that do not significantly affect their hydrology can be justifiably ignored when projecting related affects to sediment thereby simplifying many geomorphological studies of the future of the FRB.

1.3.4 Erosional Estimates

Analysis from the results of Tsuruta et al. (2018) provides a new estimate for erosion based on yield within the FRB. Since the seminal Walling (1983), understanding the link between erosional processes and sediment yield has been considered a vital study in geomorphology. Yet describing this link mathematically has been challenging in part because basin-scale erosion is difficult to observe. In the FRB, we can instead use distributed simulation results. Table 1.4 shows the delivery ratio (yield per unit of upstream erosion; DR) based on long-term simulations at various points in the FRB. At each of these points, some estimate of long-term specific sediment yield (yield per area; SSY) is also available. Plotting DR against SSY (using Water Survey Canada yield observations where available and Tsuruta et al. (2018) results where no observations are available; Table 1.4) shows a clear relationship between the two (Fig. 1.6). Because the variables and controlling processes are similar, we expect relationships between SSY and SDR to exist in a number of erosional settings. Indeed, using estimates from Trimble (1977), we also see a linear relationship in the U.S. Piedmont (Fig. 1.6).

While a linear relationship between DR and SSY may seem trivial because both variables have yield in their numerator, it does lead to an expression for upstream erosion as a function of yield and area (the relationship can also be inverted to solve for yield):

$$\text{Erosion (kg/year)} = \frac{\text{Yield (kg/year)}}{0.002 \cdot \text{SSY (kg/year/km)} - 0.0225} \quad (1.2)$$

Equation (1.2) provides an estimate of a basin-scale process based on a variable that need to only be observed at a single point and could therefore be valuable for assessing historic changes in basin or sub-basin erosion. Furthermore, it is reasonable to believe Eq. (1.2) may hold in the future. From a sediment transport perspective, the basin is supply not energy limited with disconnection between outlet and source generally coming in the form of sinks (i.e. lakes) that are expected to remain static on the century-scale. Hence, the link between erosion and yield may also remain static, with future climate/hydrological changes that affect erosion being also expressed in the sediment yield.

Ideally, the ratio of yield to erosion can be predicted using simple, fixed variables such as drainage area or relief. However, past studies [e.g. Church and Slaymaker (1989), Piest et al. (1975)] have shown that for many basins the existence of such a predictive formula is unrealistic. In such settings there is value in empirical formulae such as Eq. (1.2). We suggest that Eq. (1.2) is particularly useful as it estimates watershed scale erosion, which is rarely observed because of its high spatial and temporal variability, based on a variable that can be both spatially and temporally lumped.

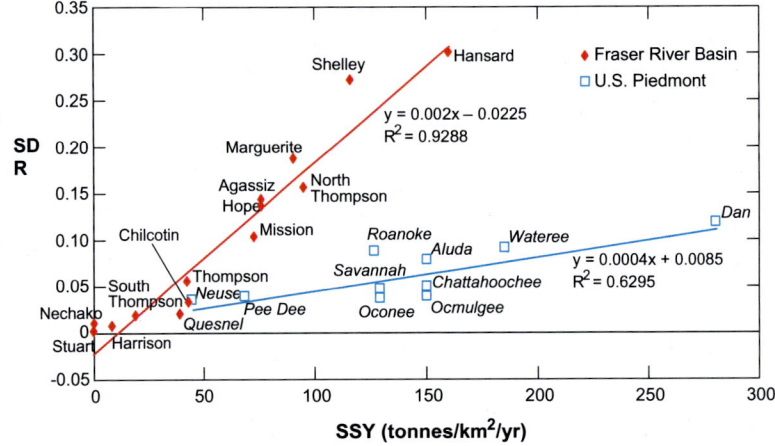

Fig. 1.6 Sediment delivery ratio plotted against specific sediment yield (tonnes/km^2/year) for watersheds within the Fraser River Basin and U.S. Piedmont

1.4 Closing Remarks

In this chapter we have reviewed the sediment model from Tsuruta et al. (2018, 2019) and its application in those studies to investigate the erosion, deposition, and transport of suspended sediment throughout the Fraser River network. The FRB simulations demonstrate that with adequate observational data, physically-based distributed models can reasonably simulate sediment dynamics across a large, diverse landscape.

The results reviewed in this chapter highlight the spatial and temporal variation in modelled sediment dynamics and point to the value of applying a distributed model in large, heterogeneous study sites. Among the more interesting results is the apparent lack of importance of all major tributaries to downstream sediment yield along the Fraser River (Fig. 1.5). Why the watershed has organized in this particular way is an open question. Do other glaciated basins demonstrate this behavior? Can basins with this property be categorized in some other manner? We view these questions as interesting and suggest that their investigation be a focus of future study. We also view the relationship between DR and SSY (Fig. 1.6) as an area for potential further study as it is possible that the linear coefficients between DR and SSY can be predicted based on a watershed's erosional setting, connectivity, and/or geomorphological history. This type of classification of watersheds would be a valuable tool in the study of landscape erosion.

Though not explicitly discussed the Tsuruta et al. (2018, 2019) model could also be used to inform policy decisions regarding landuse, water quality, and wildlife habitat. Its mechanistic nature allows for application under various landuse or climate scenarios and its spatial and temporal resolution can be used to detect changes in sub-basin erosional patterns and seasonal shifts in the sediment signal relevant to pollutant thresholds and salmon spawning cycles.

Despite the utility of large-scale, distributed, physically-based sediment models, there have been relatively few studies to develop and apply such models. As a consequence, there is a dearth of projections related to the impacts of climate and landuse change on regional-scale sediment dynamics. This gap in knowledge points to not only a lack of appropriate models, but also a need to develop better systems for monitoring observations. While well-developed models can feasibly be applied to ungauged catchments, given the relevant infancy of the class of models appropriate for regional studies, we view an increase in observational data as vital for future research. Furthermore, we suggest that sensitivity analysis from modelling exercises such as the ones we have performed can be used to inform researchers what observational data is most relevant.

Acknowledgements Figures were prepared by Eric Leinberger.

References

Arnold JG, Fohrer N (2005) SWAT2000: current capabilities and research opportunities in applied watershed modelling. Hydrol Process 19(3):563–572. https://doi.org/10.1002/hyp.5611

Arnold JG, Srinivasan R, Muttiah RS, Williams JR (1998) Large area hydrologic modeling and assessment part I: model development. J Am Water Resour Assoc 34(1):73–89. https://doi.org/10.1111/j.1752-1688.1998.tb05961.x

Asselman NEM, Middelkoop H, van Dijk PM (2003) The impact of changes in climate and land use on soil erosion, transport and deposition of suspended sediment in the River Rhine. Hydrol Process 17(16):3225–3244. https://doi.org/10.1002/hyp.1384

Bathurst JC (2010) Predicting impacts of land use and climate change on erosion and sediment yield in river basins using SHETRAN. In: Morgan RPC, Nearing MA (eds) Handbook of erosion modelling. Wiley, pp 263–288. https://doi.org/10.1002/9781444328455.ch14

Benda L, Dunne T (1997a) Stochastic forcing of sediment routing and storage in channel networks. Water Resour Res 33(12):2865–2880. https://doi.org/10.1029/97WR02387

Benda L, Dunne T (1997b) Stochastic forcing of sediment supply to channel networks from landsliding and debris flow. Water Resour Res 33(12):2849–2863. https://doi.org/10.1029/97WR02388

Church M, Slaymaker O (1989) Disequilibrium of Holocene sediment yield in glaciated British Columbia. Nature 337:452–454. https://doi.org/10.1038/337452a0

Coe MT (2000) Modeling terrestrial hydrological systems at the continental scale: testing the accuracy of an atmospheric GCM. J Clim 13(4):686–704. https://doi.org/10.1175/1520-0442(2000)013h0686:MTHSATi2.0.CO;2

Danielson JJ, Gesch DB (2011) Global multi-resolution terrain elevation data 2010 (GMTED2010) (Open-File Report Nos. 2011–1073). US Geological Survey, Reston, Virginia

Fagundes HDO, Paiva RCDD, Mainardi Fan F, Costa Buarque D, Csar Fassoni-Andrade A (2020) Sediment modeling of a large-scale basin supported by remote sensing and in-situ observations. CATENA 190: 104535. https://doi.org/10.1016/j.catena.2020.104535

Hovius N (1998) Controls on sediment supply by large rivers. In: Shanley KW, McCabe PJ (eds) Relative role of Eustasy, climate, and tectonism in continental rocks, vol 59. SEPM, Tulsa, Oklahoma, pp 3–16

Kerr SJ (1995) Silt, turbidity and suspended sediments in the aquatic environment: an annotated bibliography and literature review (Southern Region Science and Technology Transfer Unit Technical Report No. TR-008). Ontario Ministry of Natural Resources, Ontario

Liang X, Lettenmaier DP, Wood EF, Burges SJ (1994) A simple hydrologically based model of land surface water and energy fluxes for general circulation models. J Geophys Res Atmosp 99(D7):14415–14428. https://doi.org/10.1029/94JD00483

Liang X, Wood EF, Lettenmaier DP (1996) Surface soil moisture parameterization of the VIC-2L model: evaluation and modification. Global Planet Change 13(1):195–206. https://doi.org/10.1016/0921-8181(95)00046-1

Manh NV, Dung NV, Hung NN, Kummu M, Merz B, Apel H (2015) Future sediment dynamics in the Mekong Delta floodplains: impacts of hydropower development, climate change and sea level rise. Glob Planet Change 127(Suppl C):22–33. https://doi.org/10.1016/j.gloplacha.2015.01.001

McLean DG, Church M, Tassone B (1999) Sediment transport along lower Fraser River: 1. Measurements and hydraulic computations. Water Resour Res 35(8):2533–2548. https://doi.org/10.1029/1999WR900101

Motew M, Chen X, Booth EG, Carpenter SR, Pinkas P, Zipper SC, Loheide SP, Donner SD, Tsuruta K, Vadas PA, Kucharik CJ (2017) The influence of legacy p on lake water quality in a midwestern agricultural watershed. Ecosystems. https://doi.org/10.1007/s10021-017-0125-0

Parker RN, Densmore AL, Rosser NJ, de Michele M, Li Y, Huang R, Whadcoat S, Petley DN (2011) Mass wasting triggered by the 2008 Wenchuan earthquake is greater than orogenic growth. Nat Geosci 4(7):449–452. https://doi.org/10.1038/ngeo1154

Patil S, Sivapalan M, Hassan MA, Ye S, Harman CJ, Xu X (2012) A network model for prediction and diagnosis of sediment dynamics at the watershed scale. J Geophys Res Earth Surf 117(F4):n/a (F00A04). https://doi.org/10.1029/2012JF002400

Pearce AJ, Watson AJ (1986) Effects of earthquake-induced landslides on sediment budget and transport over a 50-yr period. Geology 14(1):52–55. https://doi.org/10.1130/0091-7613(1986)14<52:EOELOS>2.0.CO;2

Pelletier JD (2012) A spatially distributed model for the long-term suspended sediment discharge and delivery ratio of drainage basins. J Geophys Res Earth Surf 117(F2):n/a (F02028). https://doi.org/10.1029/2011JF002129

Piest RF, Kramer LA, Heinemann HG (1975) Sediment movement from loessial watersheds. In: Present and prospective technology for predicting sediment yields and sources (ARS-S-40), pp 130–141

Praskievicz S, Chang H (2011) Impacts of climate change and urban development on water resources in the Tualatin River Basin, Oregon. Ann Assoc Am Geogr 101(2):249–271. https://doi.org/10.1080/00045608.2010.544934

Schmitt RJP, Bizzi S, Castelletti A (2016) Tracking multiple sediment cascades at the river network scale identifies controls and emerging patterns of sediment connectivity. Water Resour Res 52(5):3941–3965. https://doi.org/10.1002/2015WR018097

Schmitt RJP, Bizzi S, Castelletti AF, Kondolf GM (2018) Stochastic modeling of sediment connectivity for reconstructing sand fluxes and origins in the unmonitored Se Kong, Se San, and Sre Pok tributaries of the Mekong River. J Geophys Res Earth Surf 123(1):2–25. https://doi.org/10.1002/2016JF004105

Shrestha RR, Schnorbus MA, Werner AT, Berland AJ (2012) Modelling spatial and temporal variability of hydrologic impacts of climate change in the Fraser River basin, British Columbia, Canada. Hydrol Process 26(12):1840–1860. https://doi.org/10.1002/hyp.9283

Smith SMC, Belmont P, Wilcock PR (2011) Closing the gap between watershed modeling, sediment budgeting, and stream restoration. In: Bennett SJ, Simon A (eds) Stream restoration in dynamic fluvial systems. American Geophysical Union, pp 293–317. https://doi.org/10.1029/2011GM001085

Sternecker K, Wild R, Geist J (2013) Effects of substratum restoration on salmonid habitat quality in a subalpine stream. Environ Biol Fish 96(12):1341–1351. https://doi.org/10.1007/s10641-013-0111-0

Syvitski JP, Peckham SD, Hilberman R, Mulder T (2003) Predicting the terrestrial flux of sediment to the global ocean: a planetary perspective. Sed Geol 162(1):5–24. https://doi.org/10.1016/S0037-0738(03)00232-X

Syvitski JPM, Vörösmarty CJ, Kettner AJ, Green P (2005) Impact of humans on the flux of terrestrial sediment to the global coastal ocean. Science 308(5720):376–380. https://doi.org/10.1126/science.1109454

Trimble SW (1977) The fallacy of stream equilibrium in contemporary denudation studies. Am J Sci 277(7):876–887. https://doi.org/10.2475/ajs.277.7.876

Tsuruta K, Hassan MA, Donner SD, Alila Y (2018) Development and application of a large-scale, physically based, distributed suspended sediment transport model on the Fraser River Basin, British Columbia, Canada. J Geophys Res Earth Surf 123(10):2481–2508. https://doi.org/10.1029/2017JF004578

Tsuruta K, Hassan MA, Donner SD, Alila Y (2019) Modelling the effects of climatic and hydrological regime changes on the sediment dynamics of the Fraser River Basin, British Columbia, Canada. Hydrol Process 33(2):244–260. https://doi.org/10.1002/hyp.13321

Vaughan AA, Belmont P, Hawkins CP, Wilcock P (2017) Near-channel versus watershed controls on sediment rating curves. J Geophys Res Earth Surf 122(10):1901–1923. https://doi.org/10.1002/2016JF004180

Verpoorter C, Kutser T, Seekell DA, Tranvik LJ (2014) A global inventory of lakes based on high-resolution satellite imagery. Geophys Res Lett 41(18):6396–6402. https://doi.org/10.1002/2014GL060641

Walling DE (1983) The sediment delivery problem. J Hydrol 65(1):209–237. https://doi.org/10.1016/0022-1694(83)90217-2

Walling DE (2009) The impact of global change on erosion and sediment transport by rivers: current progress and future challenges. UNESCO, Paris, France

Warrick JA, Rubin DM (2007) Suspended-sediment rating curve response to urbanization and wildfire, Santa Ana River, California. J Geophys Res Earth Surf 112(F2):F02018. https://doi.org/10.1029/2006JF000662

White MJ, Santhi C, Kannan N, Arnold J, Harmel D, Norfleet L, Allen P, DiLuzio M, Wang X, Atwood J, Haney E, Johnson M-V (2014) Nutrient delivery from the Mississippi River to the Gulf of Mexico and the effects of cropland conservation. J Soil Water Conserv 69:26–40. https://doi.org/10.2489/jswc.69.1.26

Wischmeier WH, Smith DD (1965) Predicting rainfall-erosion losses from cropland east of the Rocky Mountains. Agricultural Research Service, U.S. Department of Agriculture in Cooperation with Purdue Agricultural Experiment Station

Wischmeier WH, Smith DD (1978) Predicting rainfall erosion losses: a guide to conservation planning. Department of Agriculture, Science and Education Administration

Yang SL, Xu KH, Milliman JD, Yang HF, Wu CS (2015) Decline of Yangtze River water and sediment discharge: impact from natural and anthropogenic changes. Sci Rep 5(12581). https://doi.org/10.1038/srep1258

An Appraisal to Anthropogeomorphology of the Chel River Basin, Outer Eastern Himalayas and Foreland, West Bengal, India

Sonam Lama and Ramkrishna Maiti

Abstract

The Eastern Himalayas is ecologically fragile and at the same time is also home to some of the world's poorest people. Thus, to assess the nature and extent of resource utilization by the locals and its implications on the geomorphology, the present work attempts to present the prevailing scenario of anthropogenic alteration of the fluvial environment of outer Eastern Himalaya and its immediate foreland through an anthropogeomorphic assessment of a small river basin named the Chel. The assessment was carried out using SRTM DEM, Landsat images, Topographical maps, Google Earth images in ArcGIS 10.1 environment, and MS excel sheets coupled with extensive field surveys. Sediment mining, extensions of embankments, bridge construction, land-use change, extensive cultivation on channel bars, growth of tourism, road widening are the major human activities identified, altering the geomorphology and a few of its individual and combined implications on Chel basin are bed elevation changes, instability of channel bars and thalweg shifting, shifting of pool-riffle sequence, deformation of channel beds, channel planform changes, channel migration, effects on channel geometric and hydraulic parameters has been discussed in detail. Overall, the work reinforces the fact that humans have emerged as one of the most dominant factors among all, changing or altering the natural environment of the earth in this Anthropocene.

Keywords

Anthropogeomorphology · Human alteration · Eastern Himalaya · Sediment mining · Anthropocene

2.1 Introduction

The Anthropocene represents the time since human impacts have become one of the major external forcings on natural processes and humans have interacted with rivers from the time of ancient civilizations (CWC 2000; Sinha et al. 2005). The human imprint on the global environment has now become so large and active that it rivals some of the great forces of nature in its impact on the functioning of the earth system (Steffen et al. 2011). Anthropogeomorphology deals with the human impact on the earth's landforms where human actions transform, correct, and modify natural processes by increasing

S. Lama (✉)
Department of Geography, Darjeeling Government College, Darjeeling, West Bengal 734101, India
e-mail: sonamgeo@yahoo.co.in

R. Maiti
Department of Geography, Vidyasagar University, Medinipur, West Bengal 721102, India

or decreasing their rate of action and by causing the rupture of certain equilibriums that nature will try to reconstitute in different ways (Goudie and Viles 2016). The human–environment interactions are characterized by great diversity, with much feeding back, and many nonlinear processes, thresholds, and time lags (Rudel et al. 2005; DeFries 2008; Lambin and Meyfroidt 2010; An 2012). Human impacts are most frequently related to changing patterns of land use and are of special importance in environmental studies (Dale et al. 1998; Geist and Lambin 2002; Carr et al. 2006). While landforms and soils are subject to the formation, change, and even destruction by natural forces over a geological time scale, changes resulting from human activity usually occur more rapidly and have a strong impact on vegetation, water resources, and soil (Lambin et al. 2003; Vanacker et al. 2003; Geist and Lambin 2006). The industrialization and developmental activities of man during the last few centuries have led to the crossing of the threshold value of balance between man-nature relationships (Ghosh et al. 2016). At times anthropogenic factors dominate and change the channel morphology more significantly than natural events such as floods, drought, and landslides (Petts and Amoros 1996; Rinaldi 2003).

The major river basins of India are considered the important repository of the Anthropocene (Ghosh and Ilahi 2021). It has been found that the fluvial systems are the most sensitive elements of the earth's surface, and any shift in climatic parameters, environmental conditions, and human interferences instigates a rapid response from the fluvial systems (Alila and Mtiraoui 2002; Sridhar 2008; Mujere 2011; Ewemoje and Ewemooje 2011). The Indian subcontinent, which hosts many large and perennial rivers with significant hydrological and geomorphic diversity, is also home to an ancient civilization and is currently one of the most populated and polluted regions on the globe (Sinha et al. 2005; Jain et al. 2016). For example, 61,948 million liters of urban sewage are generated daily in India and more than 38,000 million liters of wastewater go into the major rivers of India (Sengupta, 2018). The estimated polluted riverine length (mainly the rivers of West Bengal, Maharashtra, Assam, Madhya Pradesh, and Gujarat) is 12,363 km, having a BOD (biological oxygen demand) range of 10–30 mg l^{-1}, i.e., severely polluted (Sengupta 2018). On the other side, India has by now about 4500 reservoirs (created by dams on rivers) which have now lost their storage capacity and functionality due to excessive siltation rate (0.475–9.44 mm year^{-1}) (CWC 2019). The anthropogenic disturbances have caused a significant decrease in forest cover from 89 million ha to 63 million ha and an increase in agricultural area from 92 million ha to 140 million ha in India (Jain et al. 2016). Large dams have caused more pronounced dysconnectivity in the sediment fluxes, and as a result, the sediment supply from rivers to oceans has decreased by around 70–80% in most of the Indian river basins (Gupta et al. 2012). Despite this situation of Indian rivers, the impact of anthropogenic forcing on natural geomorphic systems has not been analyzed in detail (Das et al. 2021). The human dimension of geomorphology is a vital area of research that has been largely neglected beyond the local and sub-regional scale (James et al. 2013). Much less has been written about the effectiveness of humans as geomorphic agents in the tropical densely populated countries, like India. The multidisciplinary river studies at modern and historical timescales must be pursued vigorously, for securing the health and future of the Indian rivers (Jain et al. 2016).

The densely settled piedmont zones of young mountains constitute an example of areas where human activity is superimposed on changes induced by natural forces (Tiwari 2000; Liebault and Piegay 2002; Starkel et al. 2008). On a regional scale, the tectonically active Darjeeling-Sikkim Himalayas, with lithology, which is prone to the mass movement, receive the highest annual rainfall along the whole Himalayan front (Starkel 1972; Dhar and Nandargi 2000; Soja and Starkel, 2007). Their margins and piedmont, as a transitional zone down to the lowland plains, is under the strong influence of the adjacent mountains. The nature and extent of the

Himalayas' impact on their piedmont is largely a product of adjustments in the fluxes of water and sediment (Starkel and Basu 2000; Grujic et al. 2006; Ghosh and Carranza 2010). Both are frequently accelerated by various forms of human activity such as agriculture, logging, mineral extraction, and road building at the mountain margin (Froehlich and Starkel 1993; Tiwari 2000).

The present-day economy of the Darjeeling-Sikkim Himalayan margin and piedmont depends mainly on tea estates established in the late nineteenth century (Ray 2002; Prokop and Sarkar 2012). Cultivation of other crops (e.g., rice, millet, areca nut) is largely for local consumption (Bhandari and Kale 2009). A significant part of the region constitutes reserved forests, and tea gardens. Forestry based on commercially valuable trees and tourism is an important contributor to the economy, as well as being an employer of large numbers of people (Government of West Bengal 2008–2009; Madhusudan 2011).

Like other Himalayan margin and piedmont environments, the study area has been experiencing heightened human pressures resulting in environmental degradation from a variety of causal factors, such as deforestation, sediment mining, bridge construction, extensions of embankments, and growth of tourism, etc. In this paper, we have attempted to provide a preliminary account of the anthropogeomorphology of the Chel River basin to highlight the major detrimental types and extent of anthropogenically-induced changes observed in the basin.

2.2 Study Area

With highly dissected northern hilly terrain having high altitudinal variation and gently rolling piedmont surface in the south, Chel River basin has a straddle-like situation between the outer Eastern Himalayan hill surface and sub-Himalayan piedmont North Bengal plains. Chel lies to the left of river Tista after Lish and Gish and unlike Lish and Gish runs parallel to Tista for a considerable length on the piedmont and joins Neora to become Dharala Nadi at 88° 44′ 13″ E, 26° 41′ 45.6″ N which ultimately merges with mighty Tista about 13 km downstream. The study area extends between 26° 41′ 30″ and 27° 5′ 15″ north latitudes and longitudes 88° 37′ 00″ and 88° 45′ 15″ east (Fig. 2.1). The entire watershed covers an area of 321 km^2 with a watershed perimeter of 115.21 km. Elevation ranges from 92 to 2449 m. Geomorphologically, the basin comprises the alluvial plain, piedmont surface, terrace surfaces, lesser Himalayan surfaces, etc. Agriculture in the watershed is largely tea plantation. Chel is one among many other rivers (Gish, Diana, Chamurchi, Rehti, Gabur-Basra, Jainti, etc.) that are dissecting the southern part of Lesser Himalaya with catchment sizes between 50 and 100 km^2 and are in the belt of higher precipitation and form large alluvial fans. "Aggradation follows upstream into the hills, while farther downstream the braided channels change into meandering ones". The basin experiences a tropical monsoonal (Am) type of climate according to Koppen's Climate classification system. The annual total precipitation ranges between 250 and 500 cm, 90% of which happens during four-five months of the high-sun season (May–September) whereas very little precipitation happens during the low-sun season. Mean monthly temperature usually ranges between 18 and 30 °C. (Lama and Maiti 2019a).

The northern mountainous portion of Chel basin lies to the east of Lesser Himalayan Duplex (LHD) and Gish Transverse zone (GTZ), a sinistral strike-slip transverse fault (Mukul et al. 2017). The major portion of the Chel basin south of the mountain front is a part of the Gorubathan recess wherein the absence of Siwalik and Damuda series of rocks and consequent minimal North–South width of the Himalayan arc makes it unique in the Eastern Himalayan belt (Heim and Gansser 1939; Goswami et al. 2013; Srivastava et al. 2017). Further, the Ramgarh Thrust (RT) rather than the Mountain Front Thrust (MFT) defines the mountain front here (Matin and Mukul 2010). The Chel River and its major tributaries namely Manzing and Sukha Khola

Fig. 2.1 **a** India with the state of West Bengal, **b** West Bengal with the location of Chel River Basin, and **c** the elevation character and major drainage lines of the Chel River Basin (elevation derived from SRTM 30 m)

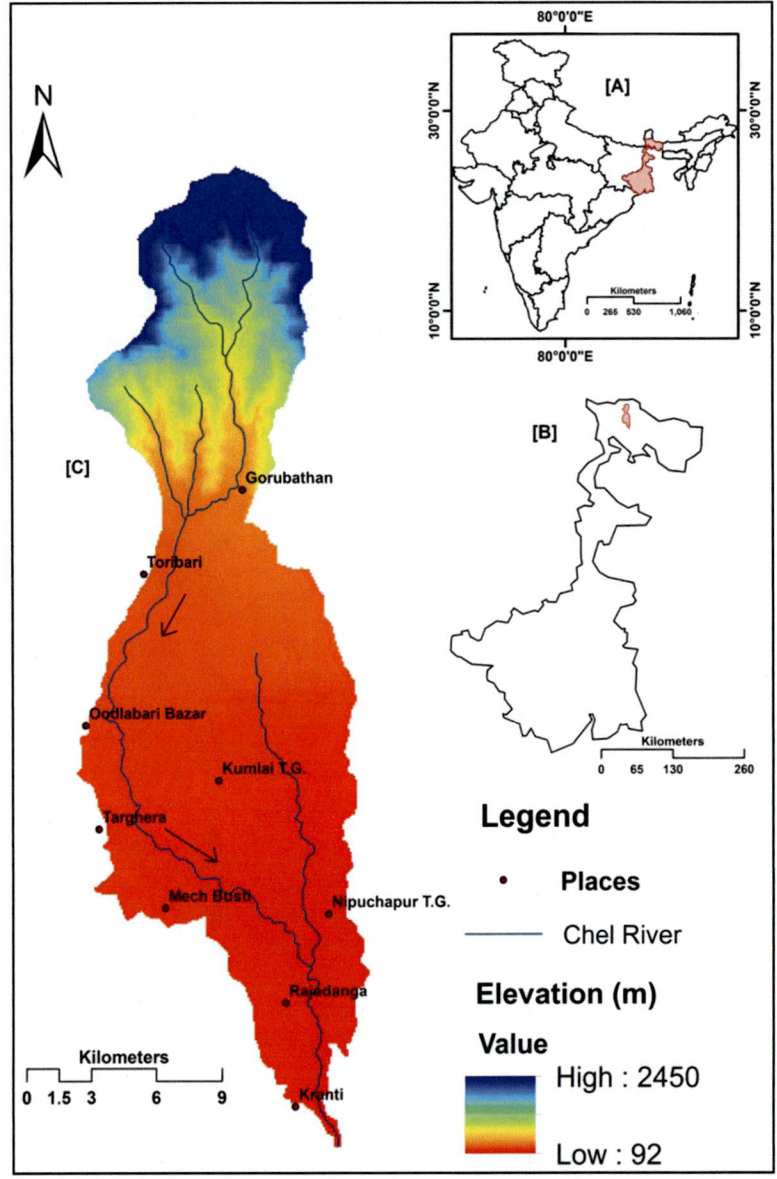

having originated in the higher reaches of outer marginal Himalaya quickly lose much of the surface gradient reaching Putharjhora (300 m approx.) at the tip of the piedmont within 10 km (approx.) from their source. These situations coupled with intense and concentrated rainfall within the basin lay all favorable conditions for a copious amount of sediment aggradation in the piedmont zone. A significant area of the basin is under forests (35.76%) and Tea Gardens (30.11%). As typical of the region, the higher elevated terraces in the Chel basin are stable LULC covered mostly by forests and tea gardens whereas the low-lying areas are cultivated with paddy and seasonal vegetables are much dynamic and frequently get flooded (Table 2.1).

Administratively, the basin spreads over parts of Gorubathan and Malbazar blocks of Kalimpong and Jalpaiguri districts, respectively of West Bengal. These two blocks have

Table 2.1 Hydro-geomorphic characteristics of the Chel River (after Dutt 1966; Starkel et al. 2008; Prokop and Sarkar 2012; Wiejaczka 2016)

River	Mountain catchment (km^2)	Forest area (%)	River gradient in mountains (‰)	River gradient in piedmont (‰)	Maximum discharge (m^3 s^{-1})	Minimum discharge (m^3 s^{-1})	Annual rainfall in the catchment (mm)
Chel	97	52.2	156.6	9.5	184.5	0.06	6096

been considered to get an idea about the demographic aspects of the basin. The municipality of Malbazar and census town of Odlabari lies at the eastern and western edges of the basin boundary, therefore the demographic aspects of these two towns have not been included to avoid getting an inflated figure. The basin is characterized by a higher rate of decadal growth of population 14.12% during 1991–2011 with sizable percentage of SC (16.74) ST (29.05) population coupled with low literacy (67.28%) and a lower percentage of working population (38.87%) (Tables 2.2 and 2.3). Thus, the study area exhibits a very dynamic natural environment superimposed with all traits of a socio-economically backward demographic environment.

Since the Chel River has a flow length of only 16 km in the outer Himalayan mountainous course with 97 km^2 of the catchment area and much of the human activities are concentrated on the piedmont, the study, therefore, has focused much on the piedmont zone of the basin. Further sediment mining has been explained in much detail, for it has emerged as the most widespread and detrimental human activity changing the geomorphology of the region.

2.3 Material and Methods

The study is mostly based on extensive fieldwork conducted during the post-monsoon season (December to February) of 2014–2017. Thalweg shifting and channel bed elevation changes were studied by cross-section measurements at 400 m upstream of the railway bridge and 400 m downstream of the road bridge during 2014, 2015, and 2017. Channel cross profiling was achieved by using a dumpy level and measuring staff considering several change points. The benchmark of 157 m of CWC Chel office located along NH-31C Road Bridge on Chel River at Odlabari has been used for calculation of reduced levels at all the stations. All the cross-profile readings were taken with dumpy level starting from the right bank toward the left in the downstream direction of the river flow. Topographical maps and multi-temporal Landsat data were used to study channel planform dynamics and channel migration in response to the sediment mining and bridge construction. SRTM DEM 2000 (30 m) was compared with ASTER GDEM 2011(30 m) to show the changes in surface terrain both upstream and downstream of the bridges. Extension of embankments was computed from the topographical map of 1970 and Landsat 8 OLI/TIRS 2017. All the Landsat scenes and DEMs were collected from the USGS site (http://earthexplorer.usgs.gov/). One topographical map (1955) 1:250,000 scale U.S. Army corps of Engineers NG 45–8, Series—U502 was acquired from the University of Texas site (https://legacy.lib.utexas.edu/maps/india.html) and two 1:50,000 scale topographical maps (Map no.—78 B/9 and 78B/10) surveyed during 1969–71 were obtained from the Survey of India (SOI). ArcGIS (version 10.1; ESRI, Redlands, CA) software package has been used for the preparation of the GIS database. All the images were processed through ERDAS imagine (v. 9.0) software and then were georeferenced based on Universal Transverse Mercator (UTM) projection system (Northern hemisphere 45 zone and world

Table 2.2 Demographic and socio-economic characteristics of the study area (Census 2011)

Blocks	Population	Sex ratio	Scheduled caste (%)	Scheduled tribe (%)	Literacy (%)	Male literacy (%)	Female literacy (%)	Working population (%)	Working female out of the working population (%)	Main workers (%)
Gorubathan	60,663	953	6.64	23.6	76.88	84.21	69.23	39.79	37.03	23.81
Malbazar	299,556	973	26.84	34.5	57.67	56.72	43.28	37.94	31.63	27.94
Average		963	16.74	29.05	67.28	70.47	56.26	38.87	34.33	25.88

Source Census of India, 2011

geodetic system (WGS 84) manually using GCPs collected during GPS survey.

Channel migration in response to bridge construction was analyzed through channel centerline migration dynamics. This was achieved by drawing transects at 1 km intervals perpendicular to the general trend of the 1994 polygon along the 36 km long study reach of River Chel from near Putharjhora T. E to its confluence with River Neora near Kranti. Thus altogether 35 transects (T1–T35) were drawn and the entire study reach was divided into three smaller reaches namely Reach-A (T1–T11), Reach-B (T11–T25), and Reach-C (T25–T35) (Fig. 2.2). Migration distances of centerlines were computed based on works of Leopold (1973), Gurnell et al. (1994), Yang et al. (1999), and Giardino and Lee (2011). The centerline of the earliest channel (1955) was taken as the origin (Yang et al. 1999) and then the centerline of the later dates was overlaid on the previous ones. The distance between the points of intersection between the centerlines along transects for subsequent years was measured (Chakraborty and Mukhopadhyay 2014). Channel planform dynamics have been attempted through the application of several channel planform indices namely Braiding Index (Brice 1964), Braid-Channel Ratio, and Sinuosity Index (Friend and Sinha's method 1993) for the period 1976–2017 from Landsat images. LULC changes in the study region have been adopted and incorporated from the studies of Prokop and Płoskonka (2014) and Biswas and Banerjee (2018). Secondary data from the Census of India has been largely consulted to understand the socio-economic profile of the basin. All other human activities have been documented by direct observation and in-situ photographs collected during the field works.

2.4 Results

The most detrimental human activities changing the geomorphology of the basin have been discussed in the following section.

Table 2.3 Decadal growth of population in the study area (Census 2011)

Blocks	Decadal growth, 1991–2001 (%)	Decadal growth, 2001–2011 (%)	Average (1991–2011) (%)
Gorubathan	17.03	11.76	14.1
Malbazar	15.39	12.87	14.13
Average	16.21	12.32	14.12

2.4.1 Sediment Mining

Alluvial channels have historically been an attractive source of sand and gravel for a variety of construction activities. The floodplains and terraces are mostly the sites of sediment storage in stream systems and can contain large quantities of boulders, gravel, and sands that can be mined economically (Langer 2003).

The study reach is characterized by highly fluctuating discharge, copious amount of aggradation due to break of slope situation and erodible banks. This altogether has resulted in the development of a braided channel pattern with ever-changing bars. The upper catchment experiences surface runoff and landslides during high-intensity concentrated rainfall months and adds more sediment to the channel downstream. These sediments are available for easy extraction directly from the riverbed and surrounding terraces right from Putharjhora (near mountain front and the tip of break-in slope) up to the confluence of Chel with River Neora. It is not certain when sediment mining began in the Chel basin but through surveys among locals, it was understood that a major peak in the demand for sediments from the Chel basin came during the construction of Teesta Barage at Gazoldoba, in the 1980s. The barrage construction was completed in 1987 but the barrage operation of the water diversion was started at the end of 1997 (Ghosh 2012). Initially, only boulders and large-sized gravels were extracted but with the growth of demand, extraction of other grades of sediments began. In recent years the development of higher connectivity in the form of better and new roads further has increased the demand. Further, the basin is well linked with major growing towns of North Bengal which has made Chel River a great source of sediments. This has attracted an influx of laborers from neighboring areas and far-off places from different regions of North Bengal, Bihar, and even Bangladesh (Gan 2008). Presently the sediments from the Chel River are a major source for growing construction industries in the region, neighboring states, and countries as well. The extraction of sediments is mostly done manually with simple tools throughout the entire stretch of study reach but mechanized extraction to a significant extent was observed up to 10 km upstream and downstream of the Odlabari road and rail bridges (Fig. 2.4a–d). There are two operational local sediment processing units- one at Toribari, 5 km upstream of Odlabari, and another at Odlabari, 1 km downstream of Odlabari Road Bridge (Fig. 2.5a–b). The extracted sediments from the river are brought to these processing units and the materials are processed mechanically with the help of sieves of various sizes. The sieved materials are kept separate, according to their sizes for ready transportation to places of demand. The common methods of sediment extraction practiced are Dry-pit mining, Wet-pit mining, bar skimping or scrapping, and pits on the river terrace or adjacent flood plain.

The types and amount of sediment extraction are seasonal. During the dry season, all grades of sediments ranging from large boulders to fine sand are extracted directly from the riverbed, bars, and surrounding terrace of the entire stretch of the study reach. Generally, large size boulders are collected from the upper section of the Reach-A (Putharjhora-Odlabari), medium-size boulders, gravels, and little proportion of sand are collected from the rest portion of Reach-A and upper portion of Reach-B (Odlabari-Nipuchapur T.G.) whereas sand mining dominates along the rest of Reach-B and entire Reach-C (Nipuchapur T.G.-Kranti) (Fig. 2.3a–d). With

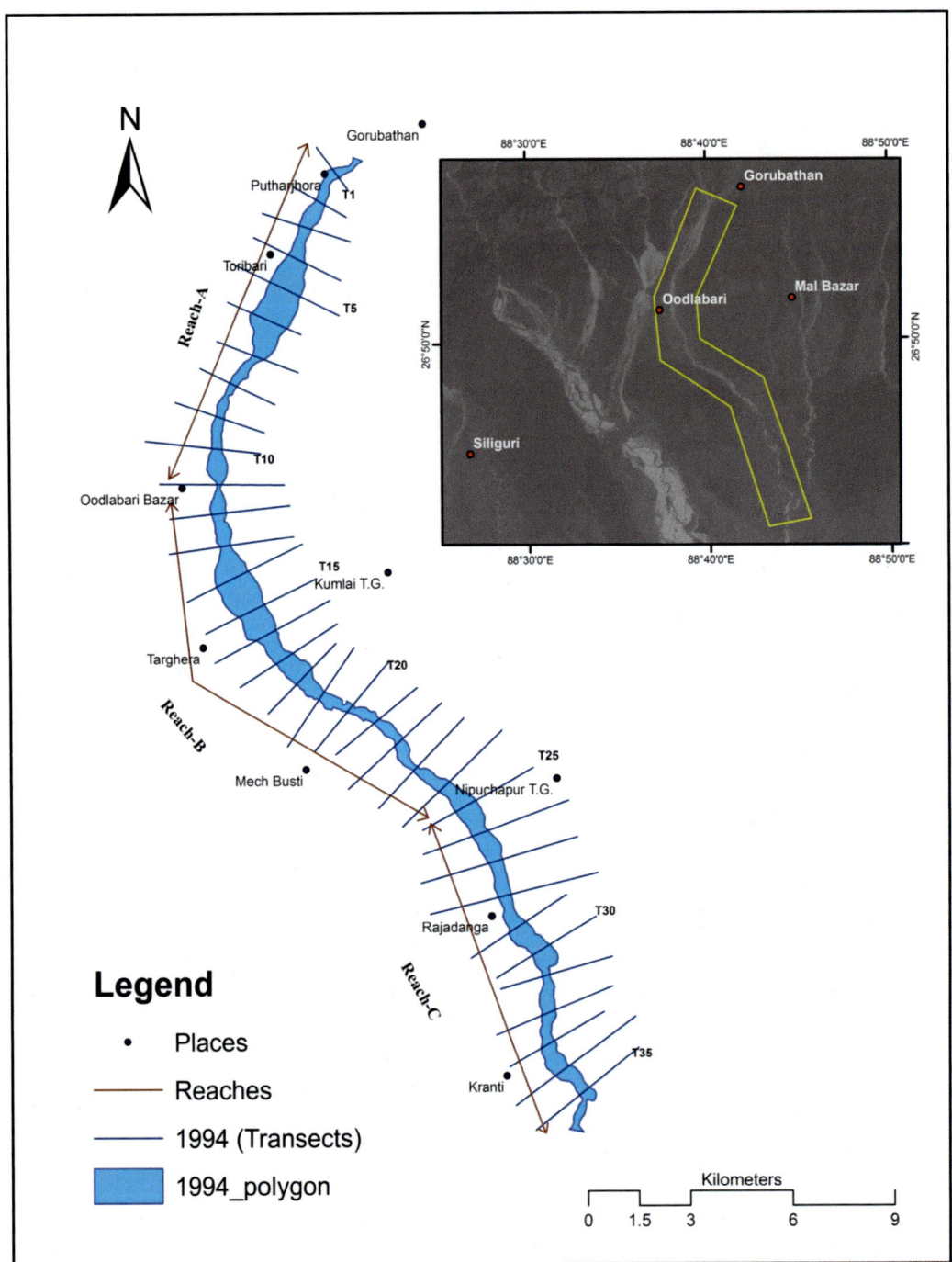

Fig. 2.2 Location of transects at 1 km interval. Channel polygon encompasses all mid-channel bars

the onset of the monsoon season, the extraction activities gradually cease near bank and higher terrace locations as high flow and water depth hinder the in-stream extraction. This season records the lowest extraction volume in the entire year, especially for lower grade sediments and

Fig. 2.3 Field photographs showing Boulder lifting at Toribari, Upper Reach (**a**); mid-channel gravel bar skimping at Odlabari, Middle Reach (**b**); sand mining at Rajadanga, Lower Reach (**c, d**)

sand. As soon as monsoonal winds weaken and flow velocity and depth lowers in the channel by the month of late September and early October, labors lift the larger-sized boulders first that are brought down to greater distance by the higher monsoonal flow. The cycle continues, in the same way, year after year. Various impacts of sediment mining on channel dynamics are discussed below.

2.4.1.1 Thalweg Shifting and Instability of Channel Bars

Thalweg shifting happens when sediment is extracted very close to the river thalweg. The extraction leaves pits of different sizes separated from thalweg by a narrow strip of land. As the water level rises during high flows, the thalweg may capture the pits. Thus, the former off-channel pit transforms into an in-channel pit (Rinaldi et al. 2005).

It has been observed from cross profiles measured at 400 m upstream and 400 m downstream of the Odlabari road and rail bridge, respectively that thalweg points have oscillated to the right and left with time (Fig. 2.6a, b). In the upstream cross profile, the thalweg point migrated near to the right bank in 2015 from its near left bank position in 2014 and it occupied near center position in the year 2017. In the downstream cross profile, the thalweg has consistently moved toward the center from its near left bank location in 2014. The river exhibits a braided channel pattern for most of the lengths of the studied reach except the lower reach (Reach-C). In this multi-channel braided system, there are multiple lower elevation points with very little variation and are separated by low height bars as exhibited by both cross profiles. In such a situation, little disturbance of channel bars separating these multiple flows induced by sediment mining is leading to significant shifting of

Fig. 2.4 Manual way of sediment extraction wherein crushing, and segregation is done on the riverbed only (**a–c**); mechanized sediment extractions collect large volume but crushing and segregation happens with sieves at processing units (**d**)

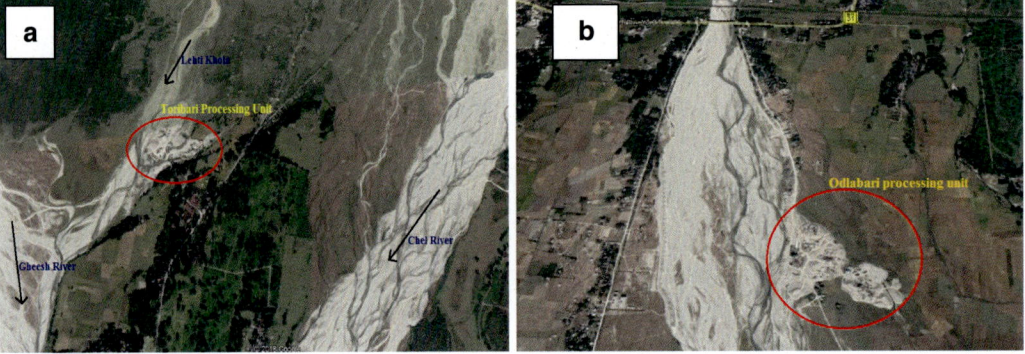

Fig. 2.5 Google Earth images showing sediment processing units. **a** at Toribari and **b** at Odlabari

thalweg points. Thus, due to braided channel pattern and massive sediment mining, thalweg shifting is much more dynamic in the Reach-A and Reach-B.

2.4.1.2 Bed Elevation Change

The drawn cross profiles display a wide variation in terms of shape, configuration, and degree of asymmetry. In terms of bed elevation, the

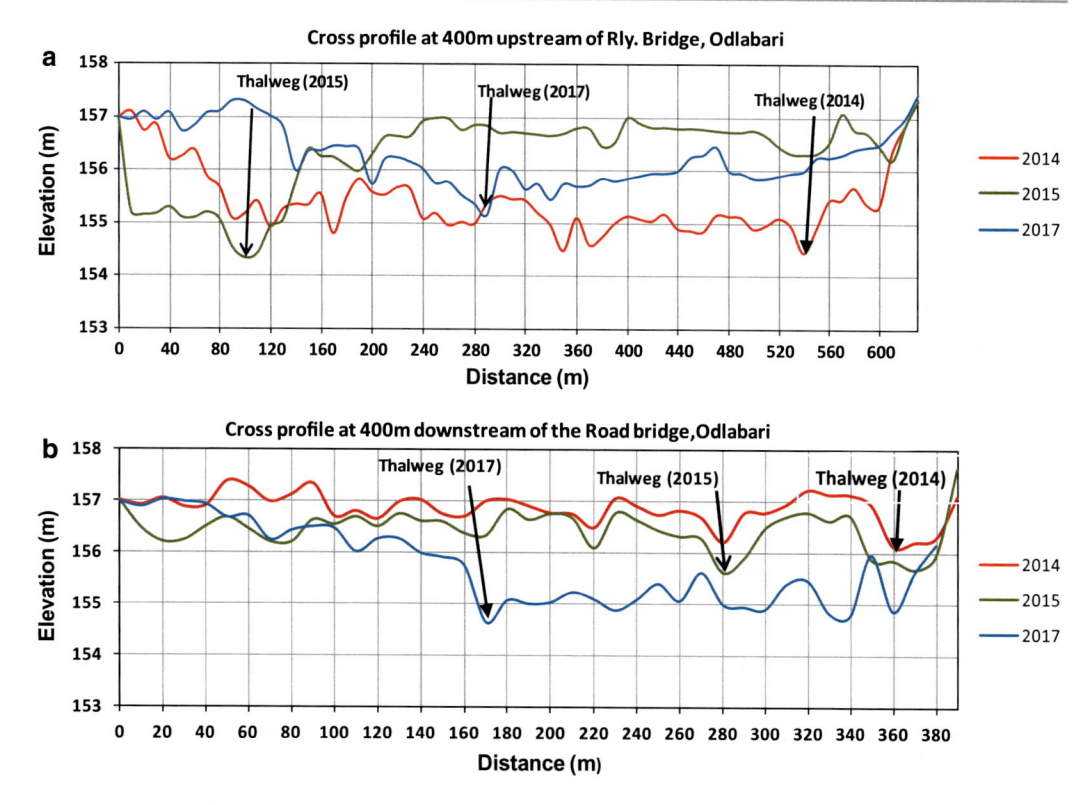

Fig. 2.6 Cross profiles showing bed elevation changes and movement of thalweg during 2014 and 2017 at 400 m upstream of Rail Bridge (**a**) and 400 m downstream of the Road Bridge, Odlabari (**b**)

upstream cross profile exhibits alternate phases of lowering and rise in bed elevation. In the year 2015, its bed elevation has risen by 1.5 m on an average throughout the cross-profile length except near the right bank where the bed elevation has gone below the 2014s bed elevation and the new thalweg point has developed at an elevation of 154.54 m. The 2017 bed elevation falls almost between the bed elevations recorded in 2014 and 2015. It is observable that, close to the right bank, the section experienced deposition of about 2 m and the rest of the cross-profile length experienced erosion by 1 m on average. The fact that this cross profile falls upstream of the twin Odlabari bridges and is nearer to the mountain front explains the bed level rise of 2015. (Fig. 2.6a). Whereas it seems that a huge amount of both manual and mechanized sediment extraction activity operating is the most probable cause for the lowering of bed elevation (Fig. 2.6b). The downstream cross profile on the other hand exhibits consistent lowering, much on the center to left bank section of the cross profile. The consistent lowering of bed elevation in the downstream cross profile can be attributed to the coupled impact of bed scouring caused by the increased flow velocity induced by the twin bridges and huge mechanized sediment extraction being very close to the Odlabari processing unit (Fig. 2.7).

2.4.1.3 Shifting of Pool-Riffle Sequence

Pools are defined as topographic lows and riffles are the topographic highs along the longitudinal profile of a stream. Huge extraction activities, especially the mechanized ones, are creating artificial pools along the studied reach of Chel River (Fig. 2.8b, c). These artificial pools increase the gradient of the upstream channel from the extraction hole, which leads to headward erosion that will tend to move several kilometers upstream and changes the natural

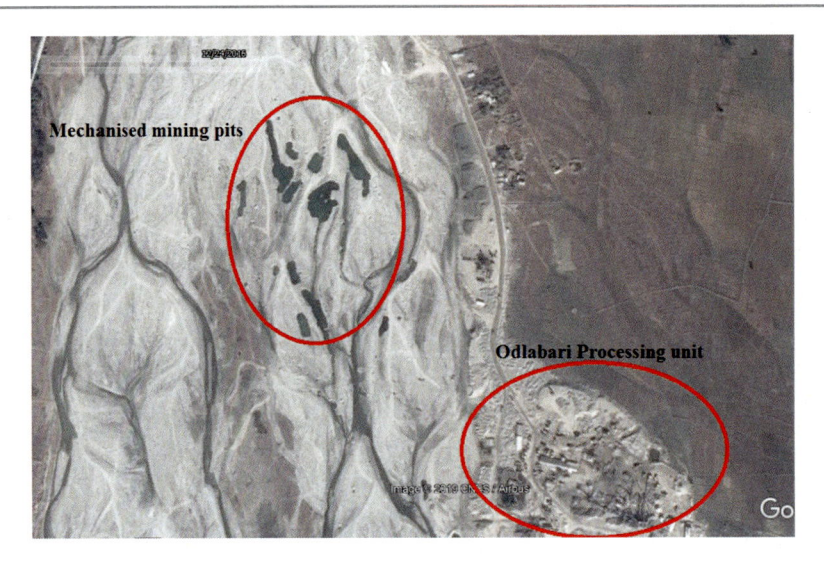

Fig. 2.7 Huge mechanized mining pits near to Odlabari processing unit

Fig. 2.8 Channel bed deformation due to manual sediment mining (**a**); mechanized wet-pit mining creating huge longitudinal depressions (negative topography) (**b**–**c**); dumping of a huge amount of sediments at Odlabari processing unit creating positive topography of sediment mounds (**d**)

sequence of pool and riffles and then change in channel morphology happens (Ghosh et al. 2016).

2.4.1.4 Deformation of Channel Beds

It is been observed that the mining of sediments is deforming the riverbed. The mining pits are creating artificial pools and lows which can be considered as topographic negatives whereas the extracted materials are sorted and stored on riverbeds only which creates temporary topographic positives of mounds ranging anywhere from 1 to 8 m in height. Some of the spoil dumps at the Odlabari processing unit measure 30 m high approximately (Fig. 2.8a, d). These mounds of sediments are stored temporarily from few days to few months. Sometimes it is left till even next season. Some of the excavation pits on the other hand can also last longer if the river cannot fill it in the next high flow season or if the pit remains cut off and thus abandoned from the main channel flow. These artificial highs and lows are creating defaced channel beds and are affecting the river morphology (Fig. 2.8 a–d).

2.4.1.5 Channel Planform Changes

The values of the braid-channel ratio have consistently decreased from its highest mark of 2.27 in 1976 to the lowest value of 0.96 in 2005 and then by 2017, the values in each segment (Reach) became near 1 which is suggestive of the fact that Chel River is gradually transforming itself from a braided channel to a straight one (Table 2.4 and Fig. 2.9). The recent years' clustering of points which otherwise were scattered horizontally elongated further confirms the transition of channel form from braided to straight. Continuous wet-pit mining creates deep elongated trenches on the riverbed and lack of sediments can be partly responsible for this gradual transition. This limits the scope for multi-channel flow.

2.4.1.6 Others Impacts

Due to sediment mining, human pressure is increasing on the channel beds. During daytime, thousands of people tread the riverbed on foot, on bicycles, and on motorbikes. Further sediment transporting vehicles, i.e., pickup trucks, full-size trucks, tractors, tipper, or dumper trucks continuously move on the riverbed. All these movements generally happen across riffles as they are the natural topographic highs. Thus, riverbeds get very compacted across the riffles and along haul ways of the river due to the continuous movement of vehicles. Naturally, the river needs larger stream power to erode these human compacted sections of riverbeds and thus remains largely undisturbed during the entire low flow duration, whereas the river concentrates its energy on a loose section of the bed. The vehicular movements also degrade the growth of vegetation on the riverbeds, banks, and adjacent floodplains, which leads to increased erosion. Further, the surface vibration produced from the continuous vehicular movements loosens weak and high bank materials and thus assists in an accelerated rate of bank erosion. It was found during field surveys that embankments or natural levees were deliberately broken or lowered for easy and short access for men and vehicles to the riverbed (Fig. 2.10). During the high flow time, these points are much more likely to be breached and cause movement of flow toward the low-lying surrounding floodplains. Thus, it also accentuates channel dynamicity. Sediments are generally collected from the base of the banks to get the advantage of easy and quick transportation of sediments which leads to diversion of flow toward the banks and causes bank erosion. Further, the lifting of specific size sediments and leaving behind the others is disrupting the natural distribution and dispersal of sediments both across and along the channel. This is creating an unnatural imbalance in the distribution of critical shear stress (τ_c) both across and along the river channel. Further, though the West Bengal Minor Minerals Rules, 2002; Schedule V (Tamang 2013) prohibits extraction of minerals within a distance of 200 m from any hydraulic structure, reservoir, bridge, canal, road, and other public works or buildings, it was noticed during the field surveys that sediment extraction in the Chel basin is operating flaunting all rules and regulation according to the ease of miners. Sediment

Table 2.4 Reach-wise temporal dynamics in planform

Reach	Year	Lctot (km)	Lcmax (km)	Li (km)	Lr (km)	Sinuosity index (P) = Lcmax/LR	Braiding index, Brice 1964 (B) = 2(?Li)/Lr	Braid-channel ratio, B = Lctot/Lcmax
A	1976	16.18	11.23	8.26	10.45	1.07	1.58	1.44
	1987	14.19	11.6	7.08	10.46	1.11	1.35	1.22
	1994	12	11.83	8.83	10.8	1.1	1.64	1.01
	2005	22.1	11.66	14.13	10.6	1.1	2.67	1.89
	2010	14.23	11.73	4.04	10.7	1.1	0.76	1.21
	2017	13.76	11.8	5.25	10.84	1.1	0.97	1.17
B	1976	26.1	11.5	9.69	12.24	0.94	1.58	2.27
	1987	24.23	11.49	8.73	11.98	0.96	1.46	2.11
	1994	16.01	11.61	8.82	12.04	0.96	1.46	1.38
	2005	22.32	12.32	10.96	12.2	1.01	1.8	1.81
	2010	21.29	11.21	12.67	12.7	0.9	2.01	1.9
	2017	17.5	11.53	7.4	12.23	0.94	1.21	1.52
C	1976	19.71	14.84	6.69	10.9	1.36	1.23	1.33
	1987	15.34	15.34	1.65	11.29	1.36	0.29	1
	1994	13.98	13.98	7.6	10.81	1.29	1.4	0.98
	2005	14.82	14.82	5.75	11.26	1.32	1.02	0.96
	2010	15.51	15.51	4.68	10.52	1.47	0.89	1
	2017	15.46	15.46	6.77	10.46	1.48	1.29	1

Source Computed by authors based on Landsat images (1976–2017)

Fig. 2.9 Temporal variability of sinuosity (B) and braid-channel ratio (P) of River Chel

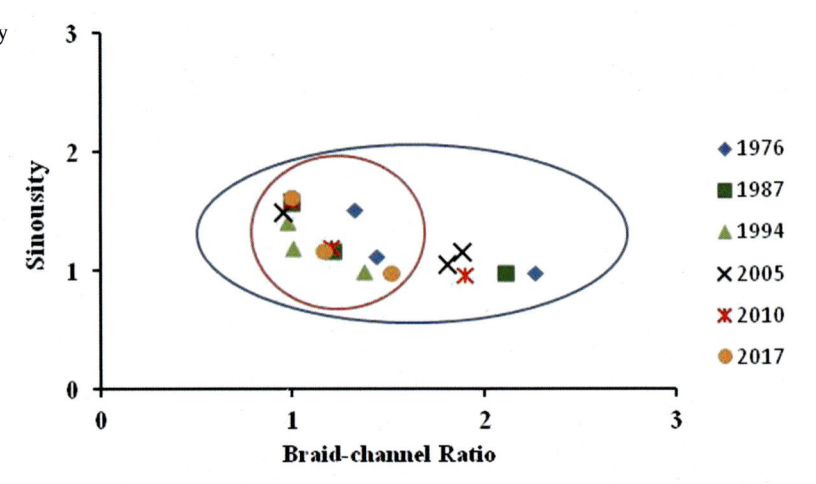

extraction barely 100 m upstream of the rail bridge and even under the road bridge of Odlabari was noticed, most probably to get the comfort of shade provided by bridges on scorching days of March (Fig. 2.11a, b).

2.4.2 Bridge Construction

Human-induced modifications through engineering infrastructures like roads, bridges, dams, and urban landscape when takes place across and

Fig. 2.10 One of the many humans breached embankment points for easy access of men and vehicles into the river

Fig. 2.11 Sediment mining operating barely 100 m upstream of the Odlabari rail bridge (**a**); sediment mining operating under the Odlabari Road Bridge (**b**). Photographs were taken on 21.03.2013

along a river channel, experience severe geomorphologic impacts upon that channel in terms of its transport capacity, sediment supply, riverbed modification, and downstream hazards (Thomas 1956). Several scholars have studied the impact of these infrastructures on the river channel and have established an alteration in the natural state of the channels (Ismail 2009;

Forman and Alexander 1998; Gregory 2006). Bridges, railways, and highways that are built to facilitate surface connectivity affect channel morphology negatively much like the dams. Road-stream crossings especially affect both upstream and downstream of the channel. It also affects the riverine ecology and habitat (Suvendu 2013; Khalifa et al. 2009; Bouska et al. 2011). In the case of the Chel, one rail and one road bridge cross the Chel basin from the middle and thus divides the Chel River into two halves of almost equal lengths. The road bridge is over NH31 connecting Siliguri and Alipurduar which joins the newly built Asian Highway at Telipara, Binnaguri, and the rail bridge is overbroad-gauge line of NEFR (North East Frontier Railway) division connecting Siliguri with Alipurduar. During the construction of bridges, engineers kept the width of bridges minimum for structural efficiency or cost-cutting (Biswas and Banerjee 2018). Chel basin falls within the annual total precipitation range of 250–500 cm much of which is concentrated in the high-sun season from May to June. During 2015 an annual total precipitation of 432.02 cm was recorded at Rangamutee Tea Garden (Rainfall data of Rangamutee T.G.). The concentrated amount of rainfall during monsoon brings down huge sediments and water into the channel but the twin bridges of Odlabari, barely 150 m in across length pose as a great impediment to the natural passage of water and sediments downstream. Very less spacing between the upstream rail bridge and downstream road bridge further aggravates the situation. Presently the channel has been constricted to a greater extent and a bottleneck situation has developed (Fig. 2.12). The morphological and hydrological alterations that the channel is facing upstream and downstream of these twin bridges are discussed below.

2.4.2.1 Surface Elevation Change

The surface relief in both upstream and downstream of rail and road bridges has changed and is observable even within a short period of 2000–2011. The upstream section is experiencing an increment in elevation in response to sedimentation as is evident from the comparison of the Aster GDEM of 2011 (30 m resolution) with the SRTM DEM of 2000 (30 m resolution). The areas within elevation zones of 192–239 m and 239–312 m are growing in size and are increasing downslope which implies sediment increment in the upstream of twin bridges (Fig. 2.13a,

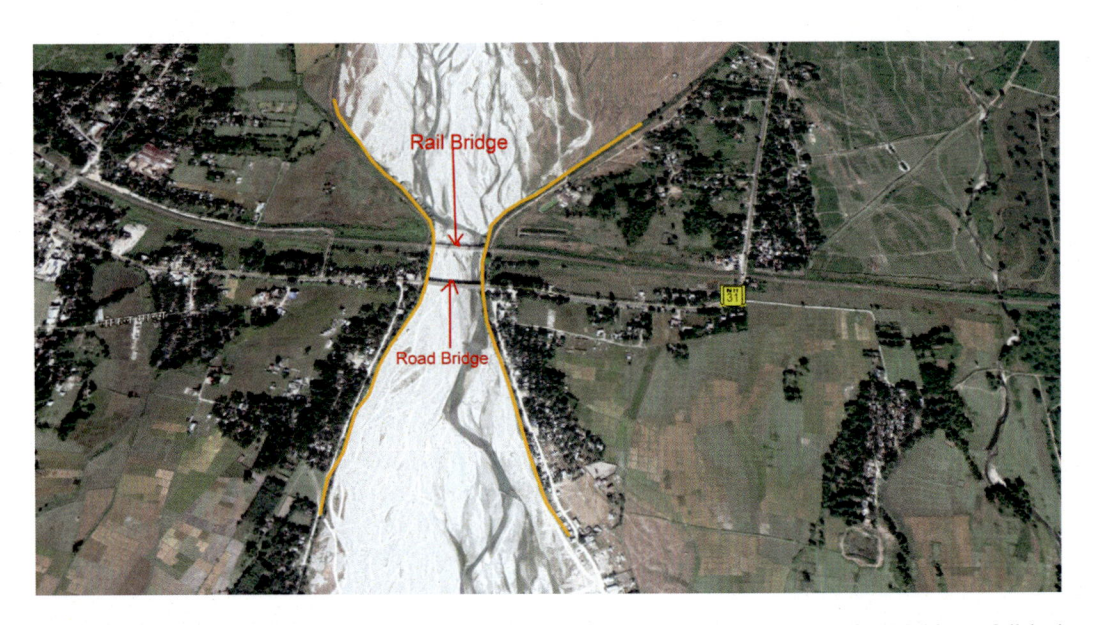

Fig. 2.12 Bottleneck condition of the River Chel (marked with yellow lines) due to Rail and Road Bridge at Odlabari. *Source* Google Earth Image, dated 11.07.2017

b). The bridges hinder the free passage of water and hence the flow velocity gets reduced considerably upstream leading to the deposition of sediments that the river could not carry at that velocity. Whereas the opposite scenario is prevailing in the downstream section of the bridges. This stretch is characterized by bed scouring and bank erosion. Unlike upstream twin bridges, the downstream stretch is experiencing intrusion of low elevation surface zones toward the higher elevation zones in a fragmentary but linear pattern, which suggests bed scouring along these linear zones (Fig. 2.14a, b). It is quite evident from the comparison of DEMs of 2000 and 2011 that the temporary impounding situation occurs during high monsoonal flow in the upstream of the bridges causing an increase in the potential energy of the river which must pass through a constricted section of the bridges increasing downstream flow velocity, i.e., the conversion of potential energy to kinetic energy. This increased velocity translates itself through bed scouring and bank erosion. These findings are well consistent with the bank erosion results wherein the Reach-B (downstream of bridges) accounted for maximum channel migration and consequent bank erosion (Lama and Maiti 2019b).

2.4.2.2 Channel Migration

The general trend of the channel centerline movement is rightward (Fig. 2.16). Out of 35 transects, only nine transects (T12, T13, T14, T15, T21, T27, T29, T33, and T34) show the leftward movement of the centerline. T14 exhibits a maximum total leftward movement of −1081.57 m with a yearly average of −17.44 m whereas T27 demonstrates the minimum leftward centerline movement of −22.98 m with an annual average of −0.37 m. Among the rightward movements, transects T17 record the maximum movement value of 1747.49 m with an annual average movement of 28.19 m whereas T7 exhibit the least movement of 72.2 m only with an annual average movement

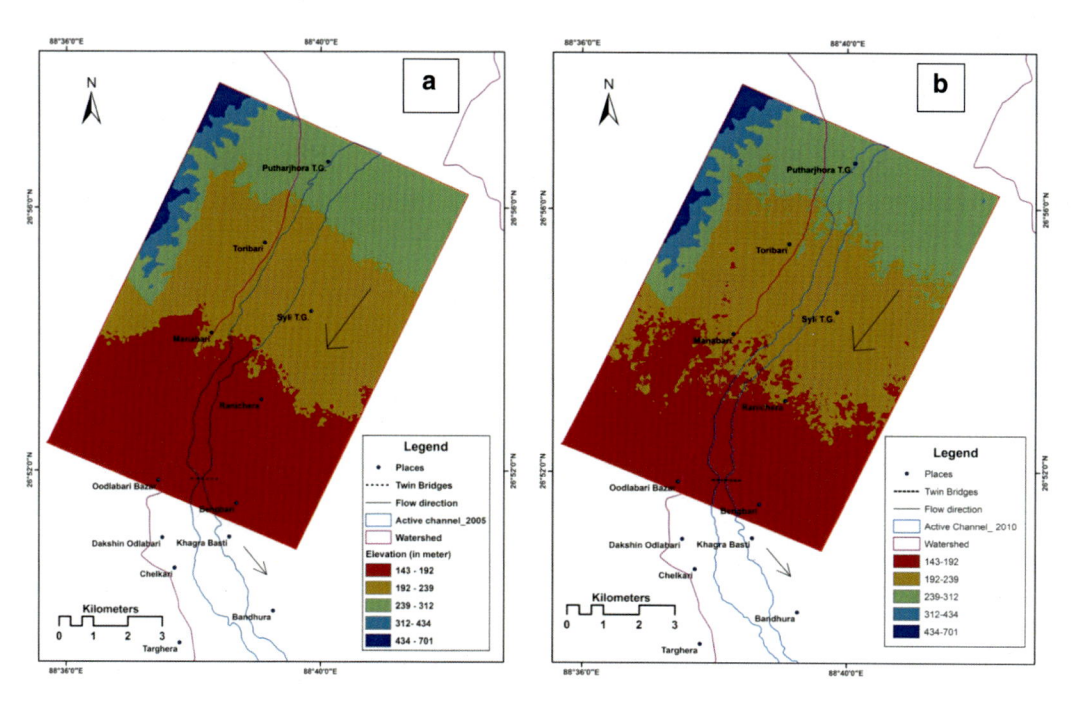

Fig. 2.13 Surface elevation change from Putharjhora to Odlabari bridges in the upstream section in 2000 (**a**) and 2011 (**b**)

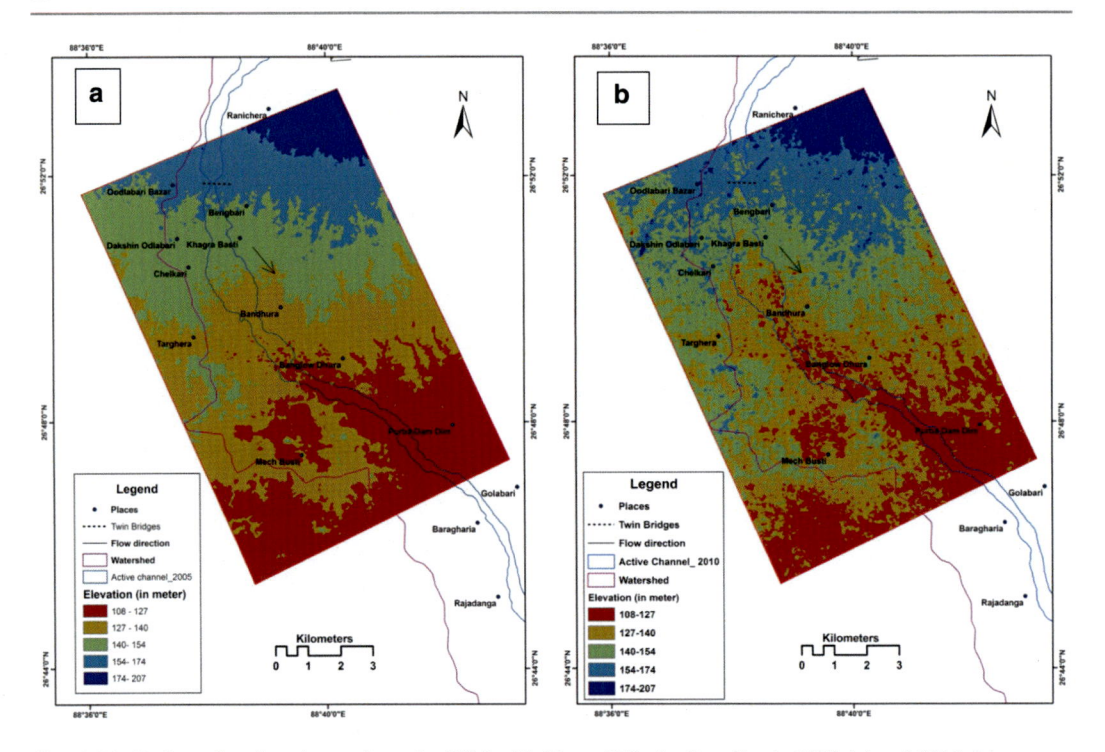

Fig. 2.14 Surface elevation change from the Odlabari bridges till Purba Damdim in 2000 (**a**) and 2011 (**b**)

of 1.16 m (Table 2.5 and Fig. 2.15). Thus, it is evident that the channel section immediately downstream of the twin bridges has experience maximum migration (both rightward and leftward). The results are consistent with the findings of Biswas and Banerjee (2018), wherein based on the six cross-sections drawn along the river, they argue that maximum shifting of the channel has happened across the cross-sections downstream of the twin bridges. Shifting of almost 2.5 km (approx.) has happened since 1913 between downstream bridges to Damdim. Further, they add that the channel is shifting westward in the upstream and eastward, downstream of the twin bridges. Chel River has started sediments deposition soon after the construction of rail bridge and well-formed bars appears by the year 1942. From 1913 to 1942 there was only a rail bridge across its course, but, with the addition of a road bridge soon after 1984, the hydro-morphological impact has amplified significantly. In subsequent years, there have been phases of bar enlargement and modifications (Biswas and Banerjee 2018).

2.4.2.3 Effects on Hydraulic and Geometric Parameters

The bridges have also impacted the hydraulic and geometric parameters of the river. It has been found that the flow velocity is highest in between the bridges. The same is higher in the downstream of the bridges compared to the upstream section. The higher recorded flow velocity in the downstream section compared to the upstream section of the bridges plays a vital role in bed scouring and extensive right bank erosion in the Reach-B (Oodlabari Bazaar-Nipuchapur Tea Garden) section (Lama and Maiti 2019b). The concentration of the highest flow velocity between the bridges is causing scouring and incision of bridge piers threatening their stability (Fig. 2.17a, b). This very process seems responsible for the damage to an old railway bridge and therefore a new bridge has been constructed upstream and is operational presently (Fig. 2.18a). Further, the flood-prone channel width and the width-depth ratio are least in between the bridges whereas it is higher in the

Table 2.5 Statistics of centerline migration dynamics over time

Centerline movement

S. No.	1	2	3	4	5	6	7	8	Total migration (m)	Average migration (m year^{-1})
Time	1955	1970	1976	1987	1994	2005	2010	2017		
Years (accumulative)	1	16	22	33	40	51	56	62		
Transects										
T1	0	278.1	646.92	−283.3	−84.69	112.9	245.05	−444.4	470.58	7.59
T2	0	382	312.15	66.1	−146.86	180	−645.23	−12.6	135.56	2.19
T3	0	517.39	15.66	−85.1	25.6	190.97	−461.4	−44.37	158.75	2.56
T4	0	605.29	−97.42	−167.48	−122.01	679.02	−979.7	218	135.7	2.19
T5	0	528.15	−2.16	−13.37	−232.35	731.77	−916.72	103.6	198.92	3.21
T6	0	343.12	−258.91	35.24	13.34	601.22	−524.6	108.79	318.2	5.13
T7	0	−163.06	103.94	163.71	9.03	377.99	−539.9	120.49	72.2	1.16
T8	0	−236.26	153.54	581.19	−1.66	178.95	−177.73	1.75	499.78	8.06
T9	0	−153.25	164.81	680.6	18.34	−53.87	240.4	−90.24	806.79	13.01
T10	0	−58.76	195.53	576.94	−115.75	280.42	−90.3	−146.19	641.89	10.35
T11	0	−105.99	676.51	−243.14	−8.69	277.62	−16.34	4.55	584.52	9.43
T12	0	−170.51	117.42	115.58	−44.68	24.71	−300.19	167.5	−90.17	−1.45
T13	0	74.21	−1056.5	283.73	6.71	206.37	−246.82	14.04	−718.26	−11.58
T14	0	57.15	−1232.92	837.92	−619.16	876.66	−712.9	−288.32	−1081.57	−17.44
T15	0	−14.65	−1049.97	613.22	−435.05	657.39	−914.9	79.05	−1064.91	−17.18
T16	0	1463.36	−1456.69	135.29	210.69	226.21	−233.25	115.6	461.21	7.44
T17	0	2841.63	−2448.93	547.17	395.84	246.3	−165.66	331.14	1747.49	28.19
T18	0	715.56	−895.61	350.21	502.03	491.02	−528.54	278.5	913.17	14.73
T19	0	341.97	−73.11	173.2	189.02	43.7	−214.32	257.9	718.36	11.59
T20	0	32.18	272.34	57.74	113.5	108.64	−144.64	82	521.76	8.42
T21	0	−125.63	139.5	25.17	−162.59	74.92	48.03	−135.64	−136.24	−2.20
T22	0	340.98	90.78	153.14	−29.44	−15.48	97.2	72.52	709.7	11.45

(continued)

Table 2.5 (continued)

Centerline movement

S. No.	1	2	3	4	5	6	7	8	Total migration (m)	Average migration (m year^{-1})
Time	1955	1970	1976	1987	1994	2005	2010	2017		
Years (accumulative)	1	16	22	33	40	51	56	62		
T23	0	631.7	−756.823	674.75	354.36	−493.71	325.96	−151.16	585.077	9.44
T24	0	642.47	−431.16	540.14	59.95	−46.35	−127.08	137.14	775.11	12.50
T25	0	−43.82	264.9	155.79	120.8	−177.89	−170.43	−17.15	132.2	2.13
T26	0	508.13	103.53	415.39	−5.89	−363.78	35.56	−81	611.94	9.87
T27	0	−75.82	203.62	74.62	−22.17	−142.04	−56.4	−4.79	−22.98	−0.37
T28	0	232.35	84.28	−37.81	−232.8	254.9	−216.74	−16.47	67.71	1.09
T29	0	−108.71	−280.5	658.47	−473.48	448.6	−342.9	42.1	−56.42	−0.91
T30	0	−147.87	292.36	298.58	−386.79	586.42	−478.22	−15.5	148.98	2.40
T31	0	−181.43	190.17	17.78	172.07	15.27	164.45	−120.74	257.57	4.15
T32	0	−122.95	590.57	66.72	−282.31	−0.29	−39.51	56.75	268.98	4.34
T33	0	573.12	−935.25	−32.66	267.82	41.8	−249.24	−12.59	−347	−5.60
T34	0	−783.56	−419.85	−59.39	384.8	−169.8	234.51	−184.63	−997.92	−16.10
T35	0	−6.94	137.52	386.94	152.6	−73.3	31.73	−82.7	545.85	8.80

Source Computed by authors based on Landsat images (1955–2017)

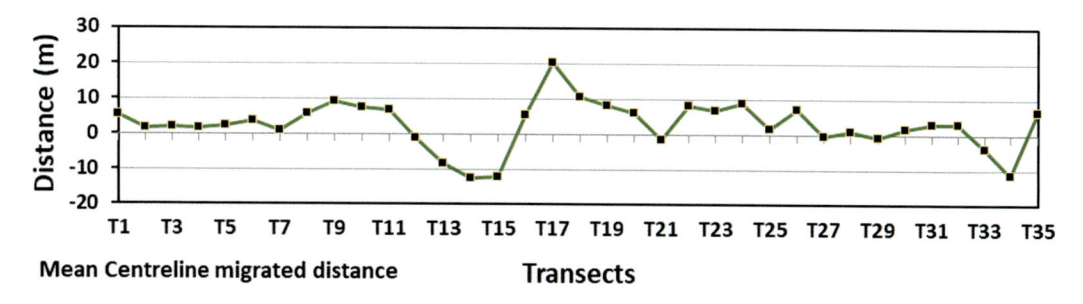

Fig. 2.15 Transect wise mean centerline migrated distance of Chel River (1955–2017)

upstream section compared to the downstream section of the bridges which is suggestive of backwater effect and, consequently, extended flood plain in the upstream section due to obstruction caused by the bridges (Biswas and Banerjee 2018). Thus, the bottleneck condition has enormously increased flood probability both up and downstream with bank erosion and noticeable impact on land uses (Fig. 2.18b).

2.4.3 Extension of Embankments

The embankments (locally known as Bandh) are constricting the active floodplain and thus compelling Chel River to dissipate its energy within the narrow channel bed. During 1970, the total length of the embankment was only 14 km which grew to 41.25 km in 2017 along with 2.53 km of dykes. Dykes have been constructed along severe erosion reaches to strengthen embankments. Thus, there has been three times extension in the length of the embankments and now covers almost 58.1% of the total active channel length of 71 km (Fig. 2.21). The right bank line is almost embanked completely with very little embankment free stretch between Dakshin Odlabari and Targhera. Erection of embankment along the almost entire stretch of the right bank can be explained by the fact that the area beyond the right bank is largely settlement area and cultivable land and the channel is progressively migrating toward the right as seen in (Fig. 2.16). Contrary to this, areas along the left bank are physiographically higher than the right bank area as suggested by the basin asymmetry factor of 40.31 which implies that the basin is tilted toward the right in the downstream flow direction. Thus, the main channel flow reaches near the left bank only at or along with sites of concave channel bends and thus erosion is confined only along these small and fragmented cut banks (Transverse section A–B in Fig. 2.21). But since the main channel flow is mainly concentrated along the left bank, large stable point bars have developed on the inner banks of the channel bend. This development and growth of point bars along the inner banks of left coupled with tiltation toward the right are pushing the main channel more toward the right and thus creating more pressure on the right embankments, especially along stretches where channel bends are near to right banks (Transverse section B–C in Fig. 2.21). Thus, it is been noticed that this has resulted in damage to right embankments along these bends (Fig. 2.19a). Few cases of erosion along the banks without embankments have been noticed which falls on the opposite of banks with embankments (Fig. 2.20b). The area beyond the left bank hosts forests, tea gardens, and agricultural land with no large settlements or towns which altogether explains why much length of the left bank is still free from embankments. Almost the entire left bank upstream of Odlabari bridges has no embankment and similarly the stretch from Bengabari to Purba Damdim the downstream still does not have any embankment. Since the river is flowing confined between higher terraces along the left bank and heavily embanked right bank,

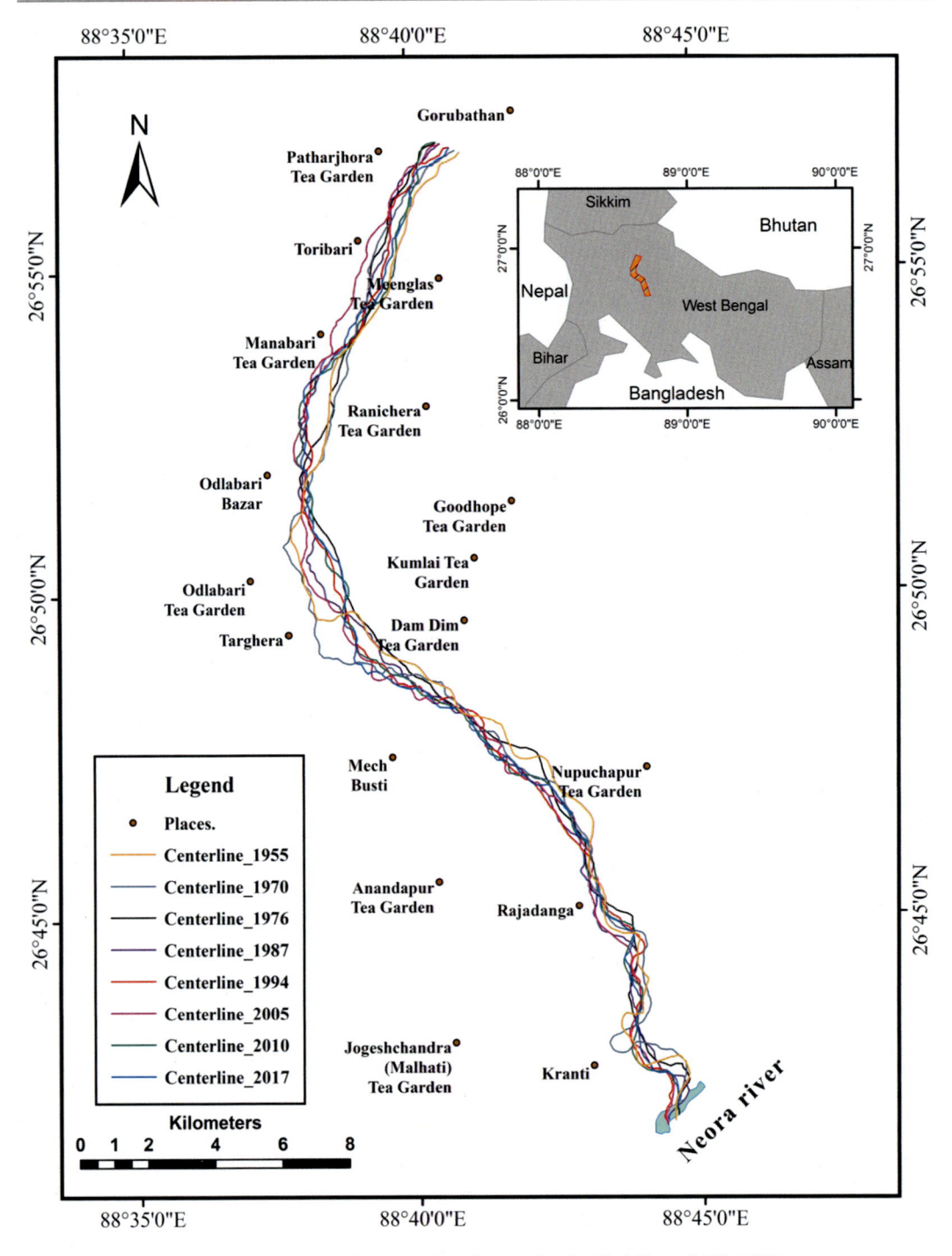

Fig. 2.16 Temporal trends in total centerline migration distance for the Chel River (1955–2017)

unless and until there is very heavy high flow it cannot access and deposit the sediments in the lower flood plains. Due to this the channel bed is rising constantly and thus the difference between channel bed and embankment heights is decreasing progressively (Fig. 2.20a). This

Fig. 2.17 Bed incision at the foot of railway piers (**a**, **b**)

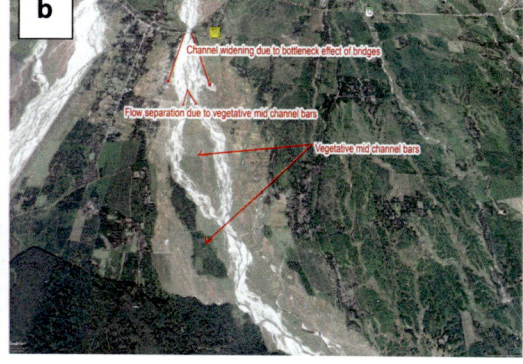

Fig. 2.18 Remnants of old damaged railway piers in between functional Rail and Road Bridge at Odlabari. The number of piers in the new railway bridge has been reduced to only two instead of five in the older one (**a**). Channel widening due to cumulative effects of bottleneck condition and vegetative mid-channel bars in the middle reach of River Chel (b). *Source* Field photograph (**a**), Google Earth Image dated 11.07.2017 (**b**)

Fig. 2.19 Damaged right embankment at transverse section C–D (**a**) (photograph was taken on 21.03.2013). Repairing and strengthening of the embankment by addition of dykes (**b**) (photograph 10.08.2014)

Fig. 2.20 Copious sediment aggradations forming huge gravel bars in the immediate upstream of a railway bridge, Odlabari. Note the least difference in elevation between the riverbed and the embankments (**a**); left embankment with dykes and erosion on the opposite along the mid-channel bars near Rajadanga (**b**)

implies an ever-decreasing threshold level of bank full discharge with a higher probability of flood and channel avulsion at lower discharge in the future.

2.4.4 Change in Land Use

Biswas and Banerjee (2018) have compared the land use of the Chel Basin and surrounding region from 2005 and 2015 classified Landsat images. They found that there have been significant changes in the areas within settlements and tea gardens. The areas within settlement and roads have increased from 5 to 11% and agricultural land from 11 to 30%. Most notably the area with tea gardens has reduced from 25.9% to 10.75%. Forest coverage also has reduced from 37 to 32%. Scrubland has recorded zero in 2015 compared to 10% in 2005 (Fig. 2.22). Altogether it can be said that there has been a decrease in land under natural elements and an increase in land under anthropogenic activities except for tea gardens. The altered land uses harm the channel and it can be seen in the increment of areas within sand and boulder riverbed from 5% in 2000 to 9% during 2015. Much of the areas on both sides of the river below the twin bridges of Odlabari have been converted into settlements and agricultural land and even the channel bars are cultivated intensively seasonally. During monsoon, it is used for paddy cultivation whereas, during the dry winter season, vegetables are cultivated largely.

2.4.5 Intensive Cultivation of Channel Bars

During field surveys, it was noticed that many of the channel bars (both point bars and mid-channel bars) are cultivated. Formation of channel bars begins near Gorubathan as upstream from here the channel flow through the deep constricted valley. Large beetle nut plantations cover much of the mid-channel and point bars from Gorubthan to Putharjhora (Fig. 2.23b). Whereas cultivation of paddy during summer monsoon and vegetables during winter was observed during winter from Toribari till confluence with Neora at Kranti (Fig. 2.23a).

2.4.6 Road Widening

During 2015–2016, the Damdim-Lava road was widened and repaired thoroughly. Fieldworks that coincided during that time revealed the fact that road widening activity affects channel morphology and dynamics mainly in two ways.

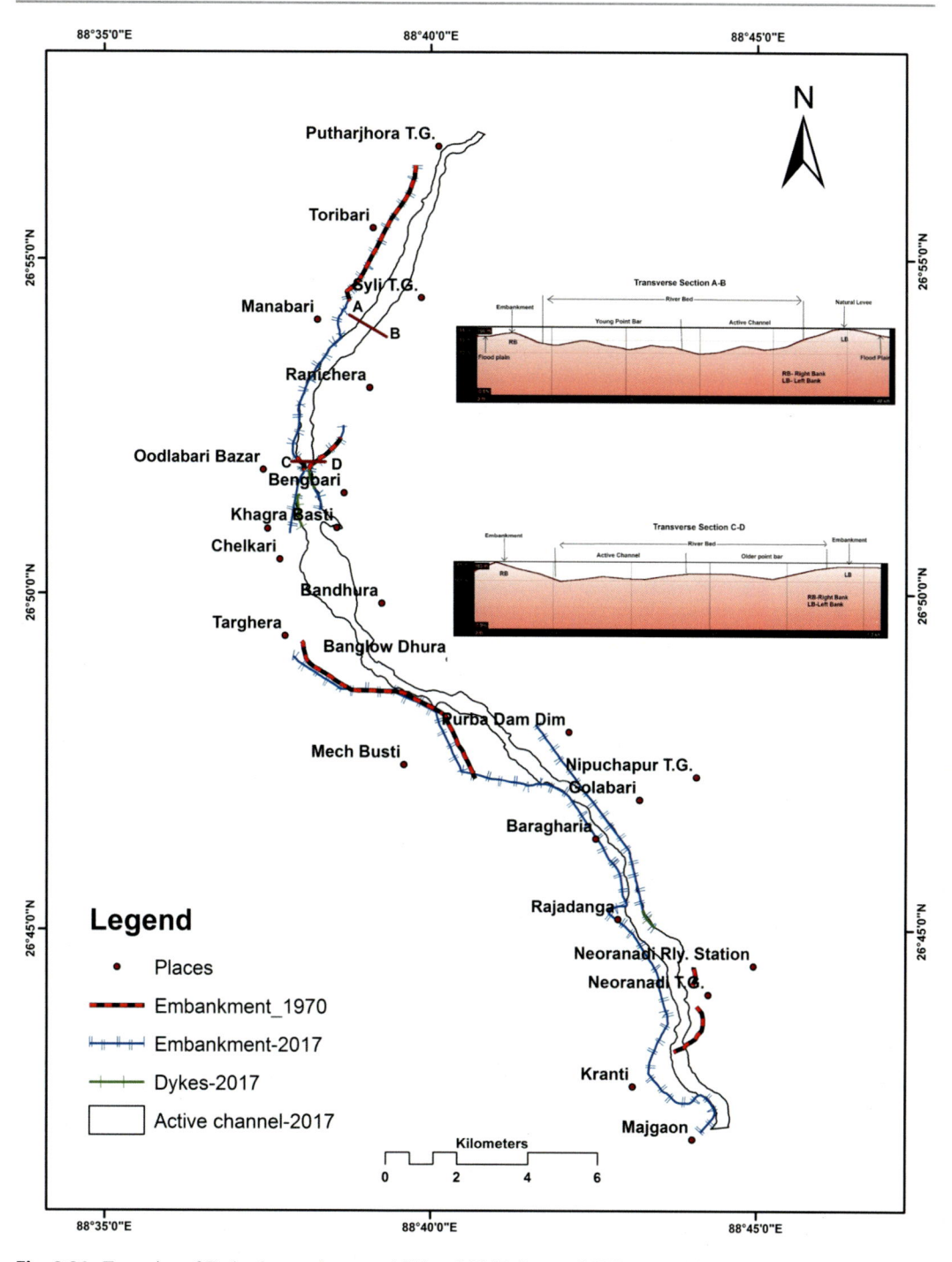

Fig. 2.21 Extension of Embankments between 1970 and 2017. *Source* SOI Topographical Map (1970) and Landsat 8 OLI/TIRS (2017)

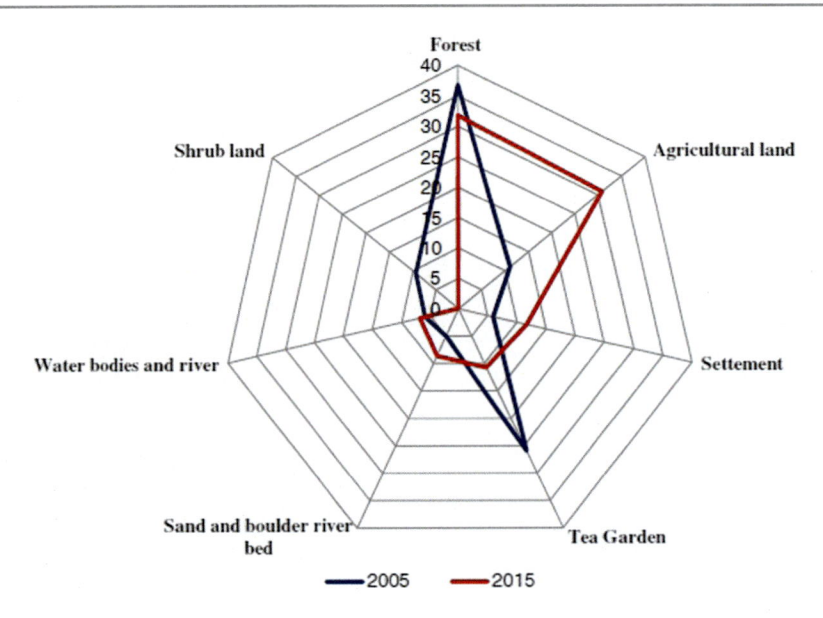

Fig. 2.22 Graphical representation of LULC change (%) during 2005–2015. Adopted from Biswas and Banerjee (2018)

Fig. 2.23 Extensive Paddy field in the foreground being cultivated on the huge left bank point bar at the upstream of Odlabari Rail Bridge (**a**); Betel Nut plantation on mid-channel bar at Gorubathan (**b**)

Firstly, the boulders and loose sediments generated during the slope cutting for road widening were getting dumped into the channel which generally constricts the channel and compels it to divert. Sometimes the dumped sediments completely dam the channel and thus creates temporary channel impoundment upstream until discharge and flow velocity increase enough to breach the newly dammed material (Fig. 2.24c, d). Road widening requires slope cutting above the road which destabilizes the slope and at times triggers landslides which add more sediments to the river (Fig. 2.24a, b). Secondly, it increases the demand for sediments of specific grades for laying and pitching on the road. Further huge quantity of boulders and sand is needed for repair, extension, and construction of new retention walls, culverts, drains, and bridges. The sediments from Chel River were entirely used during the whole widening and repairing phase due to its proximity and thus cost efficiency as well. Thus, it was learned that road widening especially in the upper catchment areas of higher elevation puts pressure on the channel through

Fig. 2.24 Road widening at near Ambiok T.G. (**a**); road widening near Chittong Bridge (**b**); dumping of loose sediments generated during road widening into the channel near Chitong Bridge (**c**); a dumper/tipper truck marked with a red circle can be seen returning after dumping the sediments marked with a blue circle (**d**)

both addition and removal of sediments which individually and combinedly increases channel instability and thus dynamicity.

2.4.7 Growth in Tourism

Due to near saturation and degradation in some aspects, of tourism in Darjeeling and Sikkim, the focus is turning into the vast, still largely unexplored region of Dooars with immense tourism potentiality. The "Dooars" region also serves as a gateway of tourism to Nepal, Bhutan, Assam, and Meghalaya. Thus, the whole region comprising the districts of Darjeeling, Kalimpong, Jalpaiguri, and Coochbehar is gaining huge popularity among tourists looking for a taste of mountains and riverine landscape, forests, and wildlife, lush green tea gardens, adventure, culture, etc. The Chel basin, in particular, falls in the advantageous location as physiographically its upper catchment falls within the lower Himalayas and the middle- and lower part falls within gently rolling North Bengal plains hosting lush green forests and tea gardens. Three popular tourist towns of its basin Odlabari, Gorubathan, and Malbazar are all within a distance of 70 km and 60 km from Bagdogra Airport and New Jalpaiguri Railway Station (NJP), respectively. The basin is in high proximity to many tourist spots of the region such as Neora Valley, Jaldhaka Hydel power projects, Murti, Lataguri, Chapramari Forest, Suntale Khola, Jhalong, Bindu. The Department of Tourism, Govt. of West Bengal promotes tourism in the region with the tagline "Majestic Mountains and Mesmerizing Dooars". Thus, huge

Fig. 2.25 Vacant right lower terrace of Chel River just upstream of Mukti Bridge, Odlabari (Photograph 08.08.2012) (**a**); amenities for tourists and picnickers developed on the same site (13.01.2016) (**b**); close up photograph showing various amenities developed for tourists and picnickers on the same site (13.01.2016) (**c**); billboard erected by Department of Tourism, Govt. of West Bengal along the roadside at Odlabari T.G. near Targh era showing master plan to convert Gazaldoba region into a mega Tourism Hub (Photograph taken 15.01.2016) (**d**)

tourist infrastructures and amenities are growing in the entire basin (Fig. 2.25a–d). Lot more homestays are opening in the rural areas, especially above Gorubathan. Jhandi Eco Hut project in Gorubathan is one such recent and already an established venture. During December and January, people from near and far-off places throng to the Gorubathan area for picnicking. Mukti Bridge over Chel River at Gorubathan has grown into the most popular Picnic destination in the region. Thus, this massive boom in tourism within and in the region of the Chel River basin is affecting the land use and will affect the dynamicity of Chel River.

2.4.8 Surface Coal Mining

During fieldwork, it was noticed that the Chel River basin bears very small and fragmentary pockets of coal seams with a very little volume of very low-quality coal at the confluence zone of Chel, Manzing, and Sukha Khola. Due to little volume and poor grade, commercial mechanized mining has not developed as it is not cost-effective. But because of the diminishing tea industry of the region, some locals especially from Putharjhora T.G. and Gorubathan areas excavate these coals illegally. The coal is collected manually in rice sacks and transported in cycles mostly during the night. They sell it at the rate of Rs. 200–300 mostly to the restaurant and dhaba (local name for roadside informal budget restaurants) owners of Odlabari and Malbazar. This coal mining activity though very small in extent and volume can destabilize the upslope and downslope of the mining hole, pits, or seams. Thus, it may cause landslides of smaller to a larger extent which adds more sediment to the channel and increase its dynamicity. Basu and Ghatowar (1988, 1990) have attributed surface coal mining along with widespread deforestation of mountain catchments and high population density (about 200 inhabitants km^{-2}) to the highest rates of channel widening in the Lish and Gish at the vicinity draining Indian Sikkimese Himalayas and its piedmont.

2.5 Discussion

The study has tried to give an account of the impacts of human actions from socio-economically backward demographic setting on fluvial geomorphology of a very dynamic natural environment. Basin is characterized by a higher rate of decadal growth of population 14.12% during 1991–2011 with a sizable percentage of SC (16.74) ST (29.05) population coupled with low literacy (67.28%) and a lower percentage of working population (38.87%). Sediment mining is happening all along the course of River Chel on the piedmont. Large boulders from Reach-A, gravels from Reach-B, and sand from Reach-C are predominantly extracted. The volume of sediment extraction highly depends on season. Reach-B accounts highest extraction volume of sediment of its nearness to NH-31C and two processing units. Extensive sediment extraction has rendered the Chel River unnatural to a great extent as evident through highly undulating cross profiles suggesting unstable bars leading to frequent thalweg shifting. Wit-pit mechanized sediment extraction along the course is creating elongated deep trench-like topography which creates artificial pools disrupting natural pool-riffle sequence. The riverbed has become deformed with human-induced high and low topography. Chel River is gradually transforming itself from a multi-thread braided channel to a single-thread straight one as suggested by a consistent decrease in the values of the braid-channel ratio from its highest mark of 2.27 in 1976 to the lowest value of 0.96 in 2005 and then by 2017, the values in each segment (Reach) became near to 1. Twin bridges that cross the Chel basin across in West–East direction have constricted the channel to a great extent and a bottleneck situation has developed which seems to have a profound impact on topography, hydraulics, and channel instability upstream and downstream of it. The section upstream of the twin bridges is showing an increment in elevation in response to sedimentation whereas the downstream section is experiencing bed scouring, incision, and bank erosion due to coupled impact of twin bridges and extensive sediment mining (especially mechanized mining). It is been found that the flow velocity is highest in between the bridges. The same is higher in the downstream of the bridges compared to the upstream section. These findings are consistent with the results of Wiejaczka (2016) wherein he accounts that the incision rate of Chel in the foreland is 5 cm $year^{-1}$ and he attributes this trend to the massive bed material extraction from the river. He claims that about 3200 tons of material are extracted every day during the dry season from 2–3 km section of the river just below and above the bridges. The concentration of the highest velocity in between the bridges has translated the energy through scouring and incision of bridge piers. This very process seems responsible for the collapse of an old railway bridge. Twin bridges are also responsible for increasing the instability of the channel as maximum shifting of the channel has been across the transects downstream of the twin bridges.

The length of embankments has grown from 14 km in 1970 to 41.25 km in 2017 along with 2.53 km of dykes. These three times extensions in the length of the embankments have restricted the active floodplain and thus is compelling Chel River to dissipate its energy within the narrow channel bed. Due to this the channel bed is rising constantly and thus the difference between channel bed and embankment heights is decreasing progressively. This also implies an ever-decreasing threshold level of bank full discharge with a higher probability of flood and channel avulsion at lower discharge in the future. Another important human-induced change in channel geomorphology can be LULC change. Changes in land use from forested to agricultural use, in headwater areas, are known to increase discharge and sediment supply, which influences downstream channel morphology in complex ways (Liebault and Piegay 2002; Vanacker et al. 2003; Price and Leigh 2006; Wohl 2006). Widespread deforestation started in the Lish, Gish, and Chel mountain catchments in the mid-nineteenth century and it is continued to the present day, as the result of agriculture, surface coal mining, and settlement expansion (Basu and Ghatowar 1988, 1990; Prokop and Sarkar 2012). Land-use changes are accompanied by parallel enlargement of the landslides which are the most important factor in generating large surface runoff volumes and coarse sediment delivery during rainfall events (Vanacker et al. 2005; Pike et al. 2010). Findings from Biswas and Banerjee (2018) suggest that the areas within settlement and roads have increased from 5 to 11% and agricultural land from 11 to 30%. Most notably the area with tea gardens has reduced from 25.9 to 10.75%. Forest coverage also has reduced from 37 to 32%. Scrubland has recorded zero in 2015 compared to 10% in 2005. There has been a decrease in land under natural elements and an increase in land under anthropogenic activities except for tea gardens. Apart from these major human activities, road widening, surface coal mining intensive cultivation of channel bars, and growth of tourism have also proven significant anthropogenic activities in shaping the geomorphology of the Chel basin.

2.6 Conclusions

The margin and piedmont of Darjeeling-Sikkim Himalaya are gradually shifting from a natural to a human-dominated landscape. Based on the extensive analysis done, it can be concluded that:

1. Sediment mining destabilizes the channel bars separating multi-channels and thus results in migration of thalweg line during little high flow. It is also responsible for lowering and raising the channel beds.
2. Mechanized sediment extraction along wet pits is creating elongated depressions which force water to flow concentrated through these depressions. These pits act as large pools storing much of the water during the low flow time and thus deprive surrounding and downstream reaches by not providing enough flow to these areas.
3. Mechanized wet-pit mining also disrupts the natural pool-riffle sequence and deforms the channel beds. Sediment mining creates both negative and positive topography in the form of mining pits and sediment storing mounds, respectively.
4. Human alteration in the form of twin bridges at Odlabari is disrupting the natural state of hydrological and geometrical properties of the upstream and downstream sections of the bridge. The twin bridges have constricted the channel to such an extent that there is large-scale variation in the elevation surface, velocity, transport capacity, and erosion and deposition the upstream and downstream of the bridges.
5. The embankments have grown in length. It has constricted the flood plain and has disconnected the main channel from its flood plain to a large extent. It reduces the elevation gap between embankments or natural levees and river channel beds due to constant aggradations, the threshold level of bank full discharge, and increasing the likeliness of embankment breaching even in low flow.
6. The areas covering settlement and roads have increased from 5 to 11% and agricultural land from 11 to 30%. Most notably the area with tea gardens has reduced from 25.9 to 10.75%. Forest coverage also has reduced from 37 to 32%. Scrubland has recorded zero in 2015 compared to 10% in 2005. Altogether it can be said that there has been a decrease in land under natural cover and an increase in land

under anthropogenic activities except for tea gardens. The altered land uses harm the channel and it can be seen in the increment of areas within sand and boulder riverbed from 5% in 2000 to 9% during 2015.

7. With less literacy and a higher percentage of SC and ST population, the basin portrays social backwardness. Less than 40% work participation and a higher percentage of marginal workers within and decadal population growth rate (1991–2011) of 14.12% further implies higher exploitation of existing natural resources which is getting reflected by an increase in areas under agriculture, settlement, and sand and boulder riverbed whereas a decrease in area under forest and water bodies and river. A substantial size of the population is engaged in low-wage menial works, which includes sediment mining on the aggradation zone of piedmont after Tea Garden and cultivation. Presently the dwindling tea industry of the region is also forcing the local population to get engaged in the sediment mining process. The same reason is driving locals to coal mining in the nearby Lish and Gish basins.

River Chel is not yet a regulated river in terms of damming or any other major engineering infrastructure but due to the higher decadal growth of human population and socio-economically backward characteristics of the region, the basin is observing the heightened impact of other types of human activities. For the ease of description and understanding, the anthropogenic activities and their impacts have been presented separately but we need to understand that impacts of these activities are all entangled ones and it is difficult to discern the impact of one human activity from another. So, many of the impacts witnessed in the basin are coupled impacts of many human activities. The presented results are of great importance in the context of limiting and maintaining an optimum scale of human impact on the basin for receiving sustained ecological services. Lastly, the findings and scenario presented in this paper can be considered as a typical representation of human-induced changes in fluvial geomorphology along the margin and piedmont (Terai and Dooars) belt of Darjeeling-Sikkim-Bhutan Himalaya.

Acknowledgements The first author would like to acknowledge the financial assistance received in the form of Minor Research Project from the University Grants Commission (UGC), through letter no. F.PHW-135/15-16 (ERO) dated: 02/12/2016. First author would also like to thank friends Mr. Lenin Rai and Miss Asha Lama from Gorubathan and Malbazaar, respectively for taking care of his lodging, fooding, and introducing the study area during the initial phase of the study. Several students from Darjeeling Govt. College, Adarsh Chamling, Supratim Bhatta, Asif Subba, and Junaid Akhtar helped during the field surveys and their help is thankfully acknowledged. We also would like to acknowledge handling reviewer cum editor, Prof. Michael Nones for thorough review and constructive suggestions which greatly enhanced the manuscript.

References

Alila Y, Mtiraoui A (2002) Implications of heterogeneous flood-frequency distributions on traditional stream-discharge prediction techniques. Hydrol Process 16:1065–1082

An L (2012) Modeling human decisions in coupled human and natural systems: review of agent-based models. Ecol Model 229:25–36

Basu SR, Ghatowar L (1988) Landslides and soil-erosion in the Gish drainage basin of the Darjeeling Himalaya and their bearing on North Bengal floods. Studia Geomorphol Carpatho-Balcan 22:105–122

Basu SR, Ghatowar L (1990) The impact of landslides on fluvial processes in the Lish basin of the Darjeeling Himalayas. Geogr Pol 59:77–87

Bhandari L, Kale S (2009) Indian states at a glance 2008–09: West Bengal: performance, facts and figures. Indicus Analytics, New Delhi

Biswas M, Banerjee P (2018) Bridge construction and river channel morphology—a comprehensive study of flow behavior and sediment size alteration of the River Chel, India. Arab J Geosci 11:467

Bouska WW, Keane T, Paukert CP (2011) The effects of road crossings on prairie stream habitat and function. J Freshw Ecol, 499–506

Brice JC (1964) Channel patterns and terraces of the Loup Rivers in Nebraska. Geological Survey Professional Paper, 422-D, Washington DC, D2–D41

Carr DL, Suter L, Barbieri A (2006) Population dynamics and tropical deforestation: state of the debate and conceptual challenges. Popul Environ 27:89–113

Chakraborty S, Mukhopadhyay S (2014) An assessment on the nature of channel migration of River Diana of the sub-Himalayan West Bengal using field and GIS techniques. Arab J Geosci 8(8):5649–5661

Central Water Commission—CWC (2000) Integrated hydrological data book. Central Water Commission of India, New Delhi

Central Water Commission—CWC (2019) National register of large dams. Ministry of Water Resources, Government of India, CWC, New Delhi

Das BC, Ghosh S, Islam A, Roy S (2021) An appraisal to anthropogeomorphology of the Bhagirathi-Hooghly River system. Concepts, ideas, and issues. In: Das BC, Ghosh S, Islam A, Roy S (eds) Anthropogeomorphology of Bhagirathi-Hooghly River system in India. CRC Press, Taylor & Francis Group, pp 213–252

Dale VH, King AW, Washington-Allen RA, McCord RA (1998) Assessing landuse impacts on natural resources. Environ Manag 22:203–211

DeFries R (2008) Terrestrial vegetation in the coupled human-earth system: contributions of remote sensing. Annu Rev Environ Resour 33:369–390

Dhar ON, Nandargi S (2000) A study of floods in the Brahmaputra basin in India. Int J Climatol 20:771–778

Dutt GN (1996) Landslides and soil erosion in the Kalimpong subdivision, Darjeeling district and their bearing on the North Bengal flood. Bull Geol Surv India B 15:62–69

Ewemoje TA, Ewemooje MA (2011) Best distribution and plotting positions of daily maximum flood estimation at Ona River in Ogun-Oshun River Basin, Nigeria. Agric Eng Int J 13(3):1–13

Friend PF, Sinha R (1993) Braiding and meandering parameters. In Best JL, Bristow CS (eds) Braided rivers. Geological Society Special Publications No. 75, Washington, pp 105–111

Forman RT, Alexander LE (1998) Roads and their major ecological effects. Annu Rev Ecol Syst 29:207–231

Froehlich W, Starkel L (1993) The effects of deforestation on slope and channel evolution in the tectonically active Darjeeling Himalaya. Earth Surf Process Landform 18:285–290

Gan B (2008) Child Workers in the Stone Crushing Family. North Bengal Anthropol 1(1):66

Geist HJ, Lambin EF (2002) Proximate causes and underlying driving forces of tropical deforestation. Bioscience 52:143–150

Geist HJ, Lambin EF (2006) Land use and land cover change: local processes, global impacts. In: The IGBP book series. Springer, Berlin

Ghosh K (2012) Changing pattern of channel planform after Gazaldoba Barrage construction: Teesta River, West Bengal. M.Phil. thesis. Jawaharlal Nehru University, New Delhi

Ghosh S, Carranza EJM (2010) Spatial analysis of mutual fault/fracture and slope controls on rock sliding in Darjeeling Himalaya, India. Geomorphology 122:1–24

Ghosh S, Ilahi RA (2021) Responses of fluvial forms and processes to human actions in the Damodar River Basin. In: Das BC, Ghosh S, Islam A, Roy S (eds) Anthropogeomorphology of Bhagirathi-Hooghly River system in India. CRC Press, Taylor & Francis Group, pp 213–252

Ghosh PK, Bandopadhyay S, Jana NC, Mukhopadhyay R (2016) Sand quarrying activities in an alluvial reach of Damodar River, Eastern India: towards a geomorphic assessment. Int J River Basin Manag 14(4):477–489

Goswami CC, Mukhopadhyay D, Poddar BC (2013) Geomorphology in relation to tectonics: a case study from the eastern Himalayan foothills of West Bengal, India. Quatern Int 298:80–92

Goudie AS, Viles HA (2016) Geomorphology in the anthropocene. Cambridge University Press, Cambridge

Government of West Bengal (2008–2009) State forest report West Bengal. Directorate of Forests, Kolkata

Giardino JR, Lee AA (2011) Rates of channel migration on the Brazos River. Report submitted to Texas Water Development Board

Gregory K (2006) The human role in changing river channels. Geomorphology 79:172–191

Grujic D, Coutand J, Bookhagen B, Bonnet S, Blythe A, Duncan C (2006) Climatic forcing of erosion, landscape, and tectonics in the Bhutan Himalaya. Geology 34:801–804

Gupta H, Kao S, Dai M (2012) The role of mega dam in reducing sediment fluxes: a case study of large Asian Rivers. J Hydrol 464–465:447–458

Gurnell AM, Downward SR, Jones R (1994) Channel planform change on the River Dee meanders, 1876–1992. Regul Rivers Res Manage 9:187–204

Heim A, Gansser A (1939) Central Himalaya geological observations of Swiss expedition, 1936. Mem Soc Helv Sci Nat 73:1–245

Ismail S (2009) Evaluation of local scour around bridge piers (River Nile bridges as case study). In: Thirteenth international water technology conference, vol 13. IWTC, Hurghada, Egypt, pp 1249–1260

Jain V, Sinha R, Singh LP, Tandon SK (2016) River systems in India; the Anthropocene context. Proc Indian Natl Sci Acad 82(3):747–761

James LA, Harden CP, Claugue JJ (2013) Geomorphology of human disturbances, climate change and hazards. In: Shroder J (ed) Treatise on geomorphology, vol 13, pp 1–13. Elsevier, San Diego

Khalifa AA, Kheireldin KA, Eltahan AH (2009) Scour around bridge piers applying stream power approach. In: Thirteenth international water technology conference, vol 13. IWTC, Hurghada, Egypt, pp 1261–1280

Lama S, Maiti R (2019a) Morphometric analysis of Chel river basin, West Bengal, India, using geographic information system. Earth Sci India 12:1–23

Lama S, Maiti R (2019b) Bank erosion and accretion along the Putharjhora-Kranti reaches of the Chel River, piedmont Sikkim Himalaya from 1955 to 2017. Earth Sci India 12(III):158–171

Lambin EF, Meyfroidt P (2010) Land use transitions: socio-ecological feedback versus socio-economic change. Land Use Policy 27:108–118

Lambin EF, Geist HJ, Lepers E (2003) Dynamics of land-use and land-cover change in tropical regions. Annu Rev Environ Resour 28:205–241

Langer HW (2003) A general overview of the technology of in-stream mining of sand and gravel resources associated potential environmental impacts, and methods to control potential impacts, open file report. U.S. Department of the Interior, U.S. Geological Survey, pp 4–12

Leopold LB (1973) River channel change with time: an example: address as retiring president of the Geological Society of America, Minneapolis, Minnesota, November 1972. Geol Soc Am Bull 84:1845–1860

Liebault F, Piegay H (2002) Causes of 20th century channel narrowing in mountain and piedmont rivers of southeastern France. Earth Surf Process Landform 27:425–444

Madhusudan K (2011) Nature tourism development and impact assessment in peripheral areas—a study of North Bengal (India). South Asian J Tour Herit 4:90–99

Matin A, Mukul M (2010) Phases of deformation from cross-cutting structural relationships in external thrust sheets: insights from small-scale structures in the Ramgarh thrust sheet, Darjeeling Himalaya, West Bengal. Curr Sci 99:1369–1377

Mujere N (2011) Flood frequency analysis using the Gumbel distribution. Int J Comput Sci Eng 3(7):2774–2778

Mukul M, Srivastava V, Mukul M (2017) Out-of-sequence reactivation of the Munsiari thrust in the Relli River basin, Darjiling Himalaya, India: insights from Shuttle Radar topography mission digital elevation model-based geomorphic indices. Geomorphology 284:229–237

Petts GE, Amoros C (1996) Fluvial hydro systems. Chapman & Hall, London. Rinaldi M (2003) Recent channel adjustments in alluvial rivers of Tuscany, Central Italy. Earth Surf Process Land 28(6):587–608

Pike AS, Scatena FN, Wohl EE (2010) Lithological and fluvial controls on the geomorphology of tropical montane stream channels in Puerto Rico. Earth Surf Process Land 35:1402–1417

Price K, Leigh DS (2006) Morphological and sedimentological responses of streams to human impact in the southern Blue Ridge Mountains, USA. Geomorphology 78:142–160

Prokop P, Płoskonka D (2014) Natural and human impact on the land use and soil properties of the Sikkim Himalayas piedmont in India. J Environ Manag 138:15–23

Prokop P, Sarkar S (2012) Natural and human impact on land use change of the Sikkimese-Bhutanese Himalayan piedmont, India. Quaest Geogr 31(3):63–75

Ray S (2002) Transformations on the Bengal Frontier: Jalpaiguri, 1765–1948. Routledge Curzon, London

Rinaldi M (2003) Recent channel adjustments in alluvial rivers of Tuscany, Central Italy. Earth Surf Proc Land 28(6):587–608

Rinaldi M, Wyzga B, Surian N (2005) Sediment mining in alluvial channels: physical effects and management perspectives. River Res Appl 21:805–828

Rudel T, Coomes O, Moran E, Achard F, Angelsen A, Xu J, Lambin E (2005) Forest transitions: towards a global understanding of land use change. Glob Environ Change 15:23–31

Sengupta S (2018) Cleaning India's polluted rivers. In: Narain S, Bhusan, C, Mahapatra, R, Misra A, Das S (eds) State of India's environment 2018: a down to earth manual, Centre for Science and Environment, New Delhi, pp 52–63

Sinha R, Jain V, Prasad Babu G, Ghosh S (2005) Geomorphic characterization and diversity of the fluvial systems of the Gangetic plains. Geomorphology 70:207–225

Soja R, Starkel L (2007) Extreme rainfalls in Eastern Himalaya and southern slope of Meghalaya Plateau and their geomorphological impacts. Geomorphology 84:170–180

Sridhar A (2008) Fluvial palaeohydrological studies in western India: a synthesis. Earth Sci India 1(1):21–29

Srivastava V, Mukul M, Mukul M (2017) Quaternary deformation in the Gorubathan recess: insights on the structural and landscape evolution in the frontal Darjeeling Himalaya. Quatern Int 462:138–161

Starkel L (1972) The role of catastrophic rainfall in the shaping of the relief of the Lower Himalaya (Darjeeling Hills). Geogr Pol 21:103–147

Starkel L, Basu S (2000) Rains, landslides and floods in the Darjeeling Himalaya. INSA, New Delhi

Starkel L, Sarkar S, Soja R, Prokop P (2008) Present-day evolution of the Sikkimese Bhutanese Piedmont. Prace Geograficzne IGiPZ PAN:219

Steffen W, Grinevald J, Crutzen P, McNeill J (2011) The Anthropocene: conceptual and historical perspectives. Phil Trans R Soc A 369:842–867

Suvendu R (2013) The effect of road crossing on river morphology and riverine aquatic life: a case study in Kunur River basin, West Bengal. Ethiop J Environ Stud Manag 6:835–845

Tamang L (2013) Effects of boulder lifting on the fluvial characteristics of lower Balason Basin in Darjeeling District, West Bengal. Unpublished Ph.D. thesis submitted to the University of North Bengal. Department of Geography and Applied Geography. University of North Bengal

Thomas W Jr (1956) Man's role in changing the face of the earth. University of Chicago Press, Chicago

Tiwari PC (2000) Land-use changes in Himalaya and their impact on the plains ecosystem: need for sustainable land use. Land Use Policy 17:101–111

Vanacker V, Govers G, Barros S, Poesen J, Deckers J (2003) The effect of short term socio-economic and demographic changes on land use dynamics and its corresponding geomorphic response with relation to water erosion in a tropical mountainous catchment. Ecuador Landsc Ecol 18:1–15

Vanacker V, Molina A, Govers G, Van Esch L, Poesen J, Dercon G, Deckers J (2005) River channel response to short-term human-induced change in landscape connectivity in Andean ecosystems. Geomorphology 72:340–353

Wohl E (2006) Human impacts to mountain streams. Geomorphology 79:217–248

Wiejaczka Ł (2016) Riverbeds level changes in the margin and Foreland of the Darjeeling Himalaya during the years with a normal monsoon rainfall. In: Singh RB, Prokop P (eds) Environmental geography of South Asia, advances in geographical and environmental sciences, pp 83–95

Yang X, Damen MCJ, Zuidam RAV (1999) Satellite remote sensing and GIS for the analysis of channel migration changes in the active Yellow River Delta, China. Int J Appl Earth Obs Geoinf 1(2):146–157

Channel Migration Vulnerability in the Kaljani River Basin of Eastern India

Moumita Dutta and Sujit Mandal

Abstract

Channel migration of a river is one of the most common geophysical phenomena in the *Eastern Himalayan foothills* and *Northern plain region*. The meander channel is characterized by the fluvio-geomorphological processes such as avulsion, bank erosion, and channel migration. The present research work focuses on the channel migration vulnerability of Kaljani River in the Northern plain region. For the demarcation of *Channel Migration Zone* (CMZ) *historical migration zone* (HMZ), *avulsion hazard zone* (AHZ), *erosion hazard area* (EHA) and *disconnected migration area* (DMA) were analyzed. HMZs were demarcated based on the raster layers of the Kaljani River courses for the years 1913, 1943, 1955, 1972, 1977, 1990, 2001, 2010, 2017 and 2019 which were then superimposed in ArcGIS 10.3.1 platforms. The AHZ mapping is prepared on the basis of the digitized meandering courses which are prone to cut-off or flow diversion during high discharge beyond CMZ. EHA comprises two components such as *Erosion Setback* (Es) and *Geotechnical Setback* (Gs). For delineating the EHA, the rates of channel migration were measured from 1913 to 2019 along various transects of the Kaljani River and it was summarized at a specific reach scale. DMA is a zone where the bank protection measures are totally controlling the processes of channel migration. The result shows that the Historical Migration Zone covered an area of about 49.33 km^2 during 1913–2019, 27.68 km^2 during 1913–1955, and 40.17 km^2 during 1972–2019. On the basis of HMZ (1913–2019), total of 29 mouzas is vulnerable to channel migration due to bank erosion. Vulnerability scale of the mouzas has been measured on the basis of the frequency of channel appearances by applying the *River Presence Frequency Approach* (RPFA). The incessant migration of the Kaljani River associated with riverbank erosion invited the land loss and wiped out man-made resources in the downstream of Buxa Reserved Forest. Most of the mouzas of Alipurduar-1, Alipurduar-2, Cooch Behar-2, and Tufanganj-1 blocks are more vulnerable to channel migration hazards.

M. Dutta · S. Mandal (✉)
Department of Geography, Diamond Harbour Women's University, South 24 Parganas, Diamond Harbour, West Bengal, India
e-mail: mandalsujit2009@gmail.com

Keywords

Historical migration zone · Channel migration · Vulnerability · Remote sensing · GIS

3.1 Introduction

Channel migration is one of the most significant quasi-natural hazards, which challenges the overall development of socioeconomic status in the foothills of Himalaya (Dey and Mandal 2018b). Alluvial streams are very much dynamic in channel configuration and flow pattern (Schumm 1977; Dey and Mandal 2018a). The River channels are laterally migrating in the floodplain areas within the river valley (Chakraborty and Mukhopadhyay 2014; Bierman and Montgomery 2014). Changes in channel discharges, sediment load, and gentle slope are some of the major factors of channel migration (Macfall et al. 2014). The high stream energy, loose bank and bed materials of the stream made the fluvial system more dynamic. The braided and meander streams are very susceptible to channel migration because of the excessive bank erosion and sedimentation process (Dey and Mandal 2018a). Future channel behavior and vulnerable areas are predicted using Channel Migration Zone. *Channel migration zone* identification and prediction are very much significant for riverbank dwellers because bank migration can damage their lives and properties. The impact of channel migration and bank erosion invites human displacement (Das et al. 2014).

Channel migration is the result of lateral erosion of the stream (Ahmed et al. 2018; Dey and Mandal 2018c). Lateral erosion usually takes place in the lower course of the rivers because of the decreased channel gradient and flow velocity (Ahmed et al. 2018; Dey and Mandal 2018c). The channel migration process is generated by the interaction of different natural and anthropogenic factors such as nature of bank material and channel form (Ahmed et al. 2018; Dey and Mandal 2018c). River course changes are sometimes considered a hazard because it helps to occur in some calamities like flood, avulsion, and bank erosion (Naik et al. 1999; Mani and Patwary 2000). Remote Sensing and GIS is a strong tool to observe the channel courses change at different times compared to other traditional methods like historical maps, cross-sections, sedimentology, etc. (Yang et al. 1999, 2015; Sarkar et al. 2012). There are various methods for estimating *channel* migration zone and vulnerability and some of the methods are well-known and extensively used, i.e., *least squares estimation* and *linear extrapolation method* (Heo et al. 2008), *remote sensing and GIS technology* (Gogoi and Goswami 2013; Nicoll and Hickin 2009), *space-borne techniques and statistical approach* (Laha and Bndyapadhyay 2013), *numerical and graphical methods* (Chakraborty and Mukhopadhyay 2014), *transect method* (Dhari et al. 2014) etc.

The study on channel migration was conducted by Rapp and Abbe (2003), Yang et al. (1999), Thatcher et al. (2009), Das and Saraf (2007), Boyd (2012), Chakraborty and Mukhopadhyay (2014), Das and Pal (2016), Mukherjee and Pal (2017), Boyd and Thatcher (2016), and Dey and Mandal (2018a). The decadal changes in the Kaljani River have developed numerous bends and cut-offs. The Kaljani River is characterized as a meander channel, as well as the course of the Kaljani is predominantly changing on a regular basis due to its meandering nature. During high discharge period, this river transports huge amount of loads into the channel such as bedload and suspended load from upper catchment which leads to aggradation process when the velocity of flow is not sufficient to transport the huge amount of loads. Consequently, the point bar and sidebar are formed along the channel. The concave sides of the meander are more vulnerable to lateral migration due to bank erosion as well as thalweg shifting toward the bank. Flood is a one type of disturbance that changes the channel configuration of the Kaljani River (Maiti 2016). When the frequency of major floods is higher than the recovery time, the channel has flood dominated morphology (Knighton 1998). Every year during monsoon, excessive sedimentation uplift the channel bed of the Kaljani River as well as at the junction of tributaries and subsequently causes the intensity of flood in the lower catchment. A large number of homesteads, houses, farming land, crops, and man-made constructions are being frequently destroyed or damaged during monsoon due to channel migration of the Kaljani. The present study aims to delineate and mapping of the *channel migration zone* (CMZ) vulnerability of the Kaljani River.

3.2 Study Area

The Kaljani River is one of the important river systems in the Eastern Himalayan foothills region. The Kaljani River basin is demarcated by 26° 16′ 22.8″ N to 26° 55′ 37.2″ N latitude and 89° 21′ 46.8″ E to 89° 38′ 38.4″ E longitude and it covers a geographical area of 1306.52 km^2. The study reach is located between the source of the Paro River and the confluence of the Kaljani River with a total length of 54 km (Fig. 3.1).

The basin is dissected by the Kaljani River and its tributaries like Garam, Dima, Nonai, Cheko, Bura Gadadhar, Katajani, etc. Ghargharia has recently met with the mainstream of the Kaljani River as a right bank tributary in 2010 flood, due to the migration of the Torsa River toward the right. In the Kaljani River basin, the Bhutan Himalaya is the main source of the tributaries as well as a number of tributaries originated from the Piedmont and Northern Plain regions of the Himalayan foothills. The entire course of the Kaljani River flows from Greater, Lesser-Shiwalik Himalayas through the Duars to Northern plain region and finally, it meets as an important left-bank tributary to the Torsa River near Deocharai (Plate 3.1a).

The present study area is composed of the Quaternary Sediments of very recent geological formations. The slope direction of the study area in south-eastern, southern, and south-western facing. The *Northern plain region* covers an area of about 488.55 km^2 sharing 37.39% area to the total basin area of the Kaljani River. The Kaljani River basin has been divided into seven

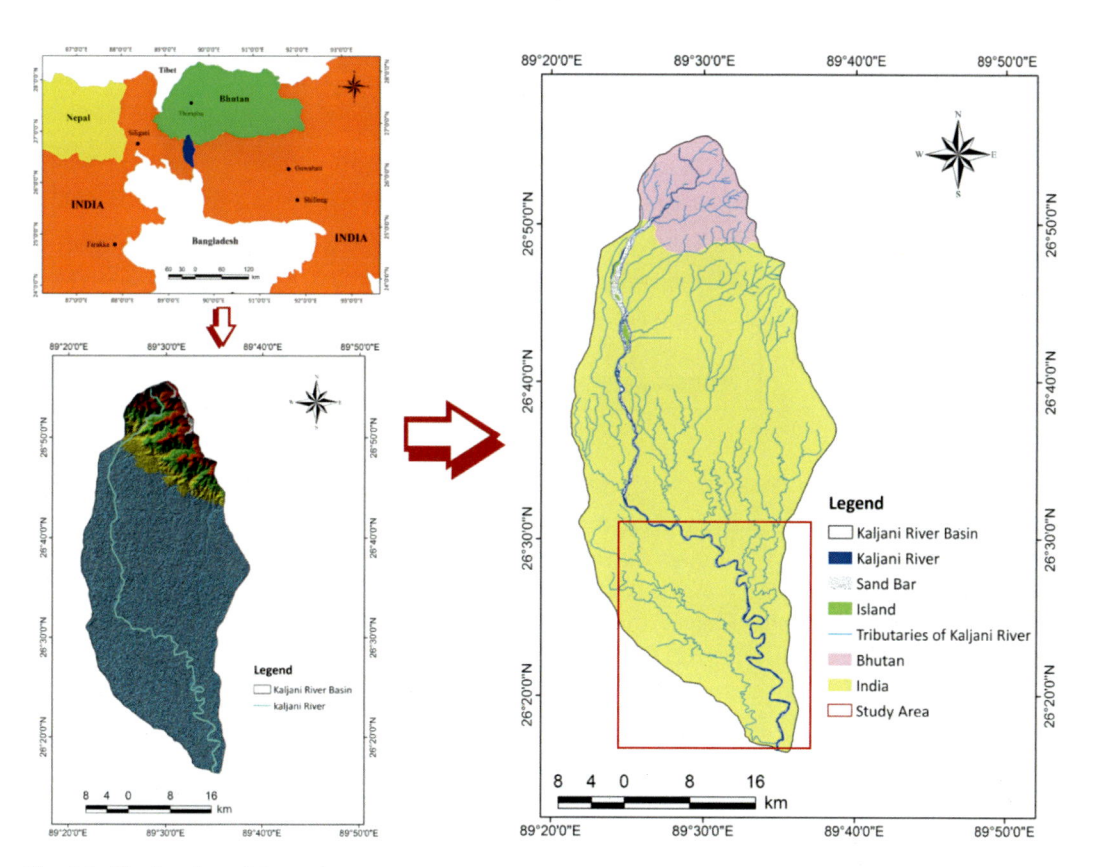

Fig. 3.1 The location of the study area

Fig. 3.2 Changes of the Kaljani River channel during the period of 1913–2019

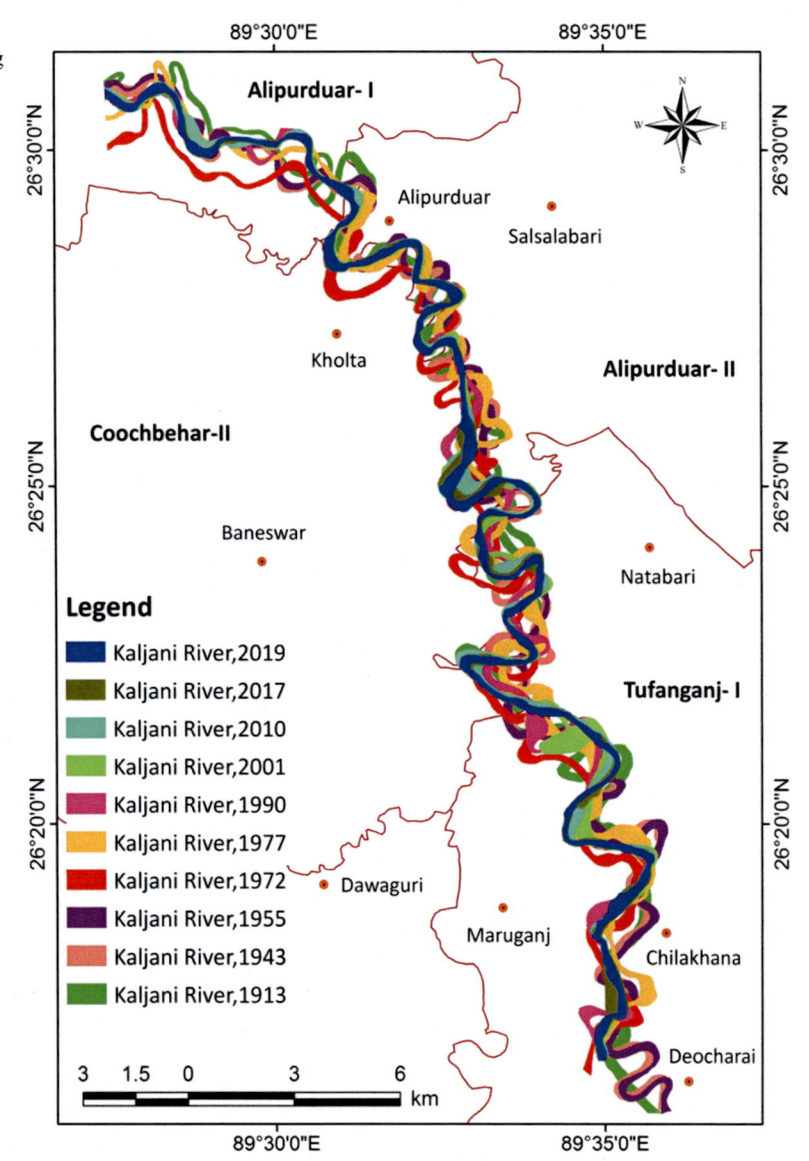

geomorphic units, but in the study area, there are three geomorphic units such as *Lower Terrace, Older Floodplain* and *Younger Active Floodplain*. The floodplain of the Kaljani River is mainly composed of three types of soil. The coarse loamy soil is sharing the highest percentage of area to the total area. Alternatively, the lower part of the study reaches cover with fine to coarse loamy floodplain soil and fine silty to fine loamy soil. The bank and bed materials of the Kaljani River are less cohesive.

3.3 Materials and Methods

3.3.1 Changing Course of the Kaljani River

There are several evidences of the river course changes through *avulsion* in the Duars region of Eastern India. The Kaljani River also changes its course regularly due to avulsion, and bank erosion during high discharge period. The historical

Table 3.1 The details of major data used in this study

S. No.	Types	Publisher	Index/map no.	Spatial coverage	Survey year	Published year	Scale
1	Topographical maps	Survey of India	G-45L (78F)	Jalpaiguri, Rangpur, Goalpara districts and Coochbehar and Bhutan states	1856–92	1913	1:253,440
2	Topographical maps	Survey of India	G-45L (78F)	Goalpara, Jalpaiguri, Rangpur, districts and Paro, Tagana, Thimpu, Wangdu, Phodrang provinces and Coochbehar state	1927–30	1943	1:253,440
3	Edition 2-army map service (AMS) series-U502	Corps of Engineers, U.S. Army	NG 45–8	Jalpaiguri district	1930–33	1955	1:250,000
4	Topographical Maps	Survey of India	78F/6,78 F/7, 78F/11	Jalpaiguri and Koch Bihar districts	1970–71	1977	1:50,000

Table 3.2 The details of Landsat data used in this study

S. No.	Satellite	Sensor	Path	Row	Acquisition date	No. of bands	Spatial resolution (in m)
1	LANDSAT-1	MSS	138	042	23/11/1972	5	60
2	LANDSAT-5	TM	138	042	14/11/1990	7	30
3	LANDSAT-7	ETM+	138	042	20/11/2001	8	30
4	LANDSAT-5	ETM	138	042	05/11/2010	7	30
6	LANDSAT-8	OLI_TIRS	138	042	08/11/2017	11	30
7	LANDSAT-8	OLI_TIRS	138	042	14/11/2019	11	30

Source USGS

maps of different years and photographs have been used to identify the changes in the Kaljani River. The SOI topographical maps, US Army map and USGS Landsat satellite imageries were taken into consideration for detecting the changes in the Kaljani River from 1913 to 2019 (Tables 3.1 and 3.2).

3.3.2 Channel Migration Zone (CMZ) Model

The boundary of the CMZ was demarcated with the help of historical maps, satellite imageries, and intensive field surveys. The Channel Migration Zone was prepared based on the study of Rapp and Abbe (2003), Boyd and Thatcher (2016). A series of components were selected to define the boundaries of CMZ. The delineation of the CMZ is the cumulative product of historical analysis and field experiment (Eqs. 3.1 and 3.2) (Rapp and Abbe 2003; Dey and Mandal 2018a).

Channel Migration Zone (CMZ)
$$= \text{Historical Migration Zone (HMZ)}$$
$$+ \text{Avulsion Hazard Zone (AHZ)}$$
$$+ \text{Erosion Hazard Area (EHA)}$$
$$- \text{Disconnected Migration Area (DMA)}$$
(3.1)

$$EHA = \text{Erosion Setback (Es)} + \text{Geotechnical Setback (Gs)} \quad (3.2)$$

3.3.2.1 Historical Migration Zone (HMZ)

Historical studies are essential to measuring the rates of channel processes by measuring past channel behavior (Dey and Mandal 2018a). Longer record for evaluating channel movement gives high level of accuracy (Rapp and Abbe 2003). The historical migration zone (HMZ) is the summation of all active channels over the chronological period and is demarcated by the outermost extent of channel locations to make a composite zone preparing HMZ (Rapp and Abbe 2003). For the demarcation of the *historical migration zone* (HMZ) of this particular section of the Kaljani River, the polygon layers of the river courses for the years 1913, 1943, 1955, 1972, 1977, 1990, 2001, 2010, 2017 and 2019 have been taken into consideration (Figs. 3.3, 3.4 and 3.5). These polygon layers were sequentially

Fig. 3.3 The historical migration zone map period of 1913–2019

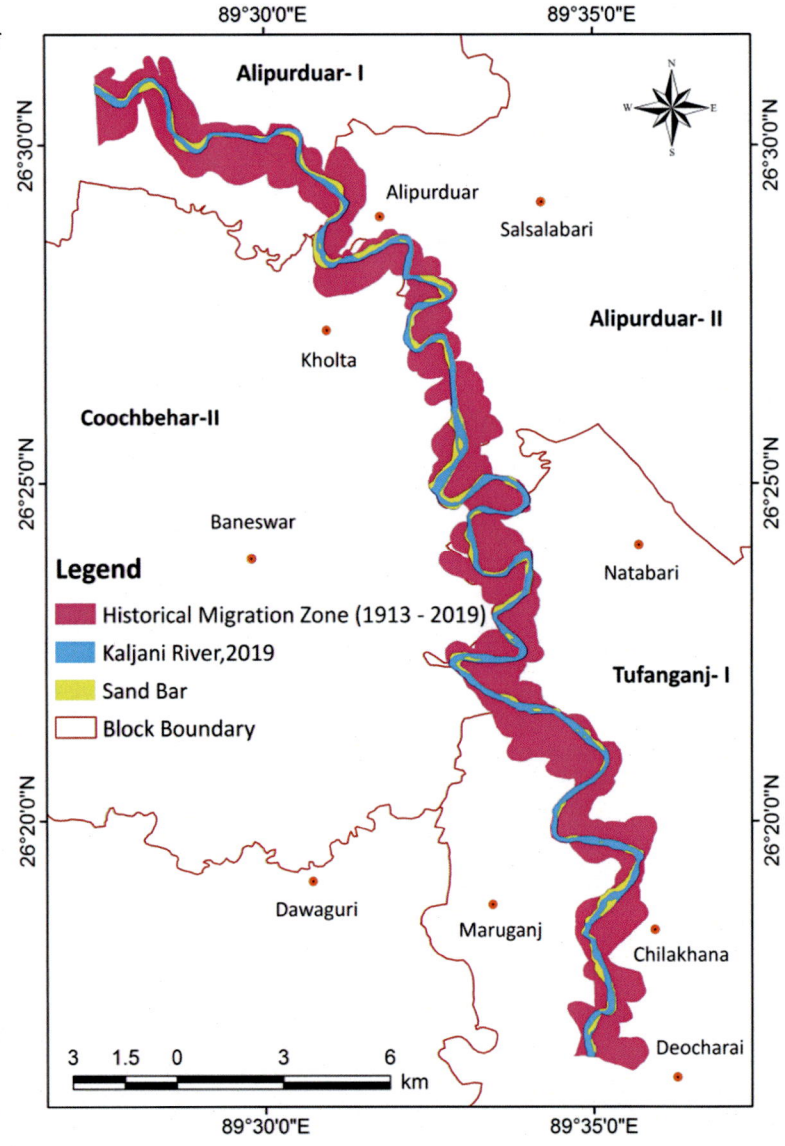

superimposed over each other, and then another single polygon layer has been made. The HMZ has been drawn for three periods, i.e., 1913–2019 (for overall time period), 1913–1955 and 1972–2019.

3.3.2.2 Avulsion Hazard Zone (AHZ)

The *Avulsion Hazard Zone* (AHZ) comprises the areas outside of the HMZ such as secondary channels, cut-off, swales, etc. that are at risk of channel occupation. The main reason behind the delineation of avulsion hazard zones is to predict the possibility of shifting channel locations that may threaten infrastructures. The AHZ mapping is prepared on the basis of the digitized channel bend ways, such as secondary channels, and paleo-channels which are prone to channel occupation outside the HMZ (Rapp and Abbe 2003) (Fig. 3.6).

3.3.2.3 Erosion Hazard Area (EHA)

The *Erosion Hazard Area* (EHA) is defined by the areas outside of the HMZ and AHZ which may be vulnerable to the bank erosion over the timeline of the CMZ (Fig. 3.8). The EHA contains two components, i.e., the Erosion Setback (Es) and the Geotechnical Setback (Gs). The Es is that the area at risk of future bank erosion by the stream flow where the extent of Es is determined by estimating rate of erosion (Tables 3.3 and 3.4). The Gs is defined by the channel and terrace banks that are at risk of mass wasting due to erosion of the bank toe (Rapp and Abbe 2003; Dey and Mandal 2018a). Estimated expectation has been completed to obtain Es on a reach specific scale based on cross-sections for the next 100 years (Thatcher et al. 2009; Washington 2004).

Erosion Setback (Es)

The Es can be estimated by assigning bank conditions and determining the rate of erosion. At first, the Es was calculated on the basis of historical studies to estimated reach wise erosion rates. Next, a percent of time the channel is likely to erode a particular location within the channel's valley bottom (Ce) (Rapp and Abbe 2003).

The Es can be calculated with the following equations (Rapp and Abbe 2003)

$$Ce = Er\left(\frac{Te}{Tr}\right) \quad (3.3)$$

$$Es = T \times Ce \quad (3.4)$$

where Er is the determined bank erosion rate from historical studies, Tr is the average time for channel to reoccupy the same location and Te is the average expected time of the channel to erode at one location.

Geotechnical Setback (Gs)

Once *Erosion Setback* (Es) has been delineated, the *Geotechnical Setb*ack (Gs) is estimated a vertical along the line of Es. Determination of Gs is not essential for vertical embankments composed of well-indurate rock, but it is essential for vertical embankments composed of cracked rock and unconsolidated materials such as alluvium (Rapp and Abbe 2003). In the present study reaches, Gs bear the imprints of the present-day formation of floodplain soil. Most part of the study reaches is categorized by medium to fine sands, fine silt and alluvium soils.

3.3.2.4 Disconnected Migration Area (DMA)

The *Disconnected Migration Area* (DMA) is the part of the CMZ behind a permanently maintained embankment or roads or spurs (Fig. 3.10). It is demarcated by recognizing the zone of the CMZ, where man-made structures physically abolish the processes of channel migration (Rapp and Abbe 2003).

3.3.3 River Presence Frequency Approach (RPFA)

Vulnerability of the channel migration zones of the Kaljani River has been illustrating by superimposing the channel courses for the year 1913, 1943, 1955, 1972, 1977, 1990, 2001, 2010, 2017 and 2019 corresponding to the mouzas. The vulnerability scale has been calculated on the basis of the frequency of channel appearance in the mouzas. This technique is termed as *river*

Table 3.3 Calculation of coefficient of erosion setback and erosion setback (1913–2019)

Cross-section	Channel migration (in meter)		Channel migration in meter/year (Er)		Ce = Er(Te/Tr)		Es = T × Ce	
	Left bank	Right bank	Left bank	Right bank	Left bank	Right bank	Left bank	Right bank
1	137.96	−223.70	1.30	−2.11	0.37	−0.60	36.83	−59.73
2	420.16	−397.74	3.96	−3.75	1.12	−1.06	112.18	−106.20
3	811.09	−900.16	7.65	−8.49	2.17	−2.40	216.56	−240.34
4	249.08	−237.43	2.35	−2.24	0.67	−0.63	66.50	−63.39
5	−914.92	768.24	−8.63	7.25	−2.44	2.05	−244.28	205.12
6	459.06	−411.59	4.33	−3.88	1.23	−1.10	122.57	−109.89
7	668.08	−741.31	6.30	−6.99	1.78	−1.98	178.38	−197.93
8	209.23	−170.23	1.97	−1.61	0.56	−0.45	55.87	−45.45
9	389.87	−205.96	3.68	−1.94	1.04	−0.55	104.09	−54.99
10	262.65	−388.69	2.48	−3.67	0.70	−1.04	70.13	−103.78
11	−14.31	63.96	−0.13	0.60	−0.04	0.17	−3.82	17.08
12	−334.80	390.99	−3.16	3.69	−0.89	1.04	−89.39	104.39
13	543.08	−437.81	5.12	−4.13	1.45	−1.17	145.00	−116.89
14	−763.56	878.96	−7.20	8.29	−2.04	2.35	−203.87	234.68
15	425.28	−115.23	4.01	−1.09	1.14	−0.31	113.55	−30.77
16	−46.73	630.22	−0.44	5.95	−0.12	1.68	−12.48	168.27
17	111.97	−82.03	1.06	−0.77	0.30	−0.22	29.90	−21.90
18	162.56	−342.28	1.53	−3.23	0.43	−0.91	43.40	−91.39
19	1440.19	−1473.81	13.59	−13.90	3.85	−3.94	384.53	−393.51
20	774.80	−662.21	7.31	−6.25	2.07	−1.77	206.87	−176.81
21	−446.58	563.78	−4.21	5.32	−1.19	1.51	−119.24	150.53
22	784.56	−722.90	7.40	−6.82	2.09	−1.93	209.48	−193.01
23	534.28	−496.71	5.04	−4.69	1.43	−1.33	142.65	−132.62
24	−734.48	754.43	−6.93	7.12	−1.96	2.01	−196.10	201.43
25	652.95	−619.55	6.16	−5.84	1.74	−1.65	174.34	−165.42
26	501.55	−424.64	4.73	−4.01	1.34	−1.13	133.91	−113.38
27	637.90	−779.50	6.02	−7.35	1.70	−2.08	170.32	−208.13
28	−17.99	−10.33	−0.17	−0.10	−0.05	−0.03	−4.80	−2.76
29	609.56	−234.41	5.75	−2.21	1.63	−0.63	162.75	−62.59
30	201.78	−127.27	1.90	−1.20	0.54	−0.34	53.88	−33.98

Source USGS, Survey of India and field survey

presence frequency approach (RPFA) and was followed Mukherjee and Pal (2017).

$$RPFA = \sum Rpn/N \quad (3.5)$$

where Rpn is the number of occurrences of the river and N is the total number of years considered for the study.

The mouzas that appear within the channel courses with high frequency are considered as highly vulnerable to channel migration. Frequency approach is a suitable technique to identify channel migration zones (Mukherjee and Pal 2017).

While counting channel migration frequency based on HMZ it is considered that if a channel

Table 3.4 Calculation of coefficient of erosion setback and erosion setback (1972–2019)

Cross-section	Channel migration (in meter)		Channel migration in meter/year (Er)		$Ce = Er(Te/Tr)$		$Es = T \times Ce$	
	Left bank	Right bank	Left bank	Right bank	Left bank	Right bank	Left bank	Right bank
1	−1165.11	1170.64	−24.79	24.91	−12.13	12.19	−1213.10	1218.86
2	−492.35	528.29	−10.48	11.24	−5.13	5.50	−512.64	550.05
3	−444.84	301.11	−9.46	6.41	−4.63	3.14	−463.16	313.52
4	−1028.40	1065.14	−21.88	22.66	−10.71	11.09	−1070.76	1109.02
5	−893.46	1386.62	−19.01	29.50	−9.30	14.44	−930.27	1443.74
6	−370.01	397.31	−7.87	8.45	−3.85	4.14	−385.25	413.68
7	−754.85	575.30	−16.06	12.24	−7.86	5.99	−785.94	599.00
8	−50.78	110.10	−1.08	2.34	−0.53	1.15	−52.87	114.63
9	677.45	−515.18	14.41	−10.96	7.05	−5.36	705.35	−536.40
10	612.26	−209.56	13.03	−4.46	6.37	−2.18	637.48	−218.19
11	−821.01	872.84	−17.47	18.57	−8.55	9.09	−854.83	908.80
12	−633.23	534.84	−13.47	11.38	−6.59	5.57	−659.32	556.87
13	80.33	−40.92	1.71	−0.87	0.84	−0.43	83.64	−42.60
14	−1006.43	801.37	−21.41	17.05	−10.48	8.34	−1047.89	834.38
15	820.20	−709.96	17.45	−15.11	8.54	−7.39	853.99	−739.20
16	−998.66	986.09	−21.25	20.98	−10.40	10.27	−1039.80	1026.71
17	−465.69	586.98	−9.91	12.49	−4.85	6.11	−484.87	611.16
18	68.76	−164.49	1.46	−3.50	0.72	−1.71	71.59	−171.27
19	275.37	−311.65	5.86	−6.63	2.87	−3.24	286.71	−324.49
20	338.59	−303.71	7.20	−6.46	3.53	−3.16	352.53	−316.22
21	−174.52	106.76	−3.71	2.27	−1.82	1.11	−181.70	111.16
22	90.60	−79.86	1.93	−1.70	0.94	−0.83	94.34	−83.15
23	−99.20	66.95	−2.11	1.42	−1.03	0.70	−103.29	69.71
24	−102.46	−54.28	−2.18	−1.15	−1.07	−0.57	−106.68	−56.51
25	453.69	−441.45	9.65	−9.39	4.72	−4.60	472.38	−459.64
26	−313.88	323.46	−6.68	6.88	−3.27	3.37	−326.81	336.78
27	225.78	−440.97	4.80	−9.38	2.35	−4.59	235.09	−459.13
28	−227.57	100.35	−4.84	2.14	−2.37	1.04	−236.95	104.49
29	377.07	−284.41	8.02	−6.05	3.93	−2.96	392.60	−296.12
30	−345.13	194.32	−7.34	4.13	−3.59	2.02	−359.35	202.32

Source USGS and field survey

appeared only one time over the whole period of time in a mouza then the mouza is quite safe. But if a channel appeared in every alternative year then it may be considered as most channel migration vulnerable zones. Following these principles, channel migration vulnerability zones have been classified into high (Above 75%), moderate (50–75%) and low (Below 50%) (Figs. 3.12 and 3.13). If CMZ, AHZ, HMZ and EHA appeared in a mouza, then the mouzas would be located in most channel migration vulnerable zones. The specific ranges, i.e., 75–100%, 50–75%, 25–50%, and 0–25% were adopted to demarcate the severe, high, moderate and low vulnerability zone of CMZ, respectively (Figs. 3.14 and 3.15).

3.4 Results and Discussion

3.4.1 Analysis of Historical Migration Zone (HMZ)

A vast area has been affected by the channel migration in the study reaches of the Kaljani River. The result showed that during the period of 1913–2019 along 54 km reaches of the Kaljani River, 49.33 km^2 area was under the Historical Migration Zone (Fig. 3.3). But, after the engineering construction such as embankments, roads, railways, spars, etc. the river channel started modifying its courses (Fig. 3.10).

During the period of 1913–1955, the overall areal coverage of *the historical migration zone* (HMZ) was 27.68 km^2 (Fig. 3.4). After the construction of the embankments along the left bank of the Kaljani River, the magnitude of channel migration increased due to excessive

Fig. 3.4 The historical migration zone map period of 1913–1955

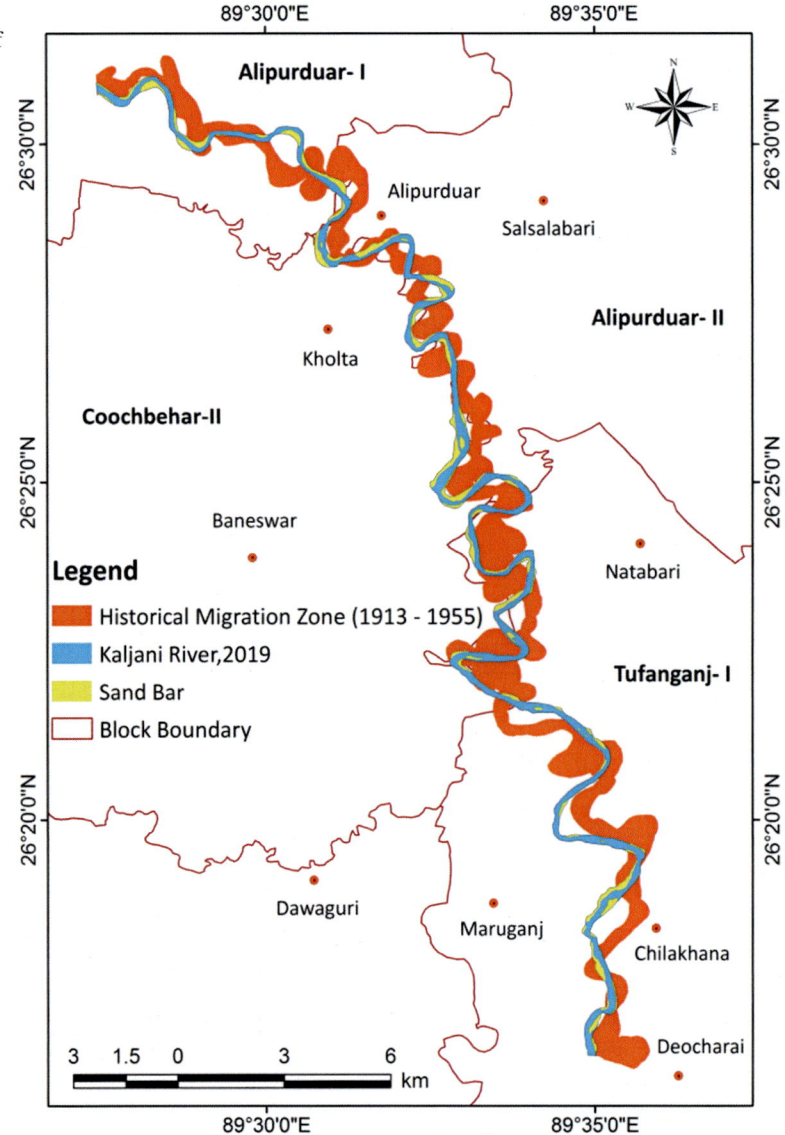

bank erosion toward the right. After the construction of short length of bridges across the natural flow of the channel is obstructed and it becomes turbulent which changes the channel configuration and finally, the channel became more vulnerable to migration. During the period of 1972–2019, the overall areal coverage of *the historical migration zone* (HMZ) was 40.17 km^2 (Fig. 3.5).

3.4.2 Analysis of the Avulsion Hazard Zone (AHZ)

On the basis of the bank stratigraphy analysis, nature of riparian vegetation and location of secondary channels, the AHZ was delineated. In the Kaljani River course, there were many bend ways, non-cohesive and unconsolidated loose bank materials (composed of sand, silt and clay) which

Fig. 3.5 The historical migration zone map period of 1972–2019

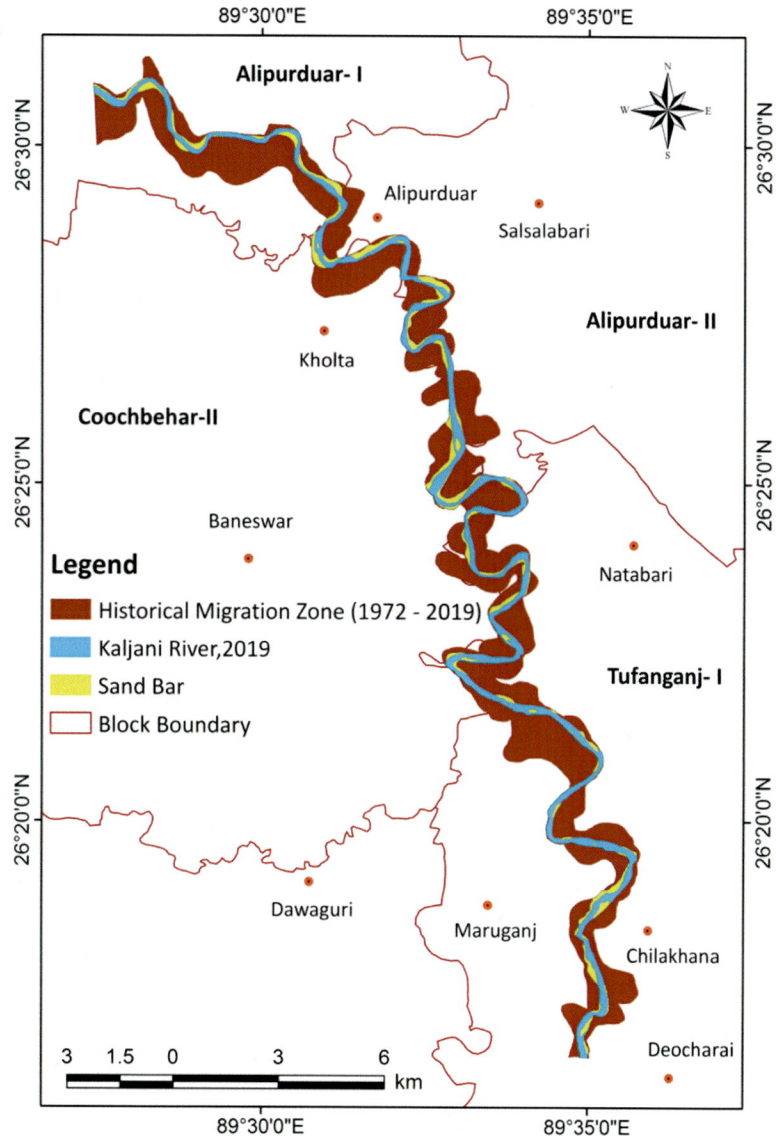

Fig. 3.6 The avulsion hazard zone map

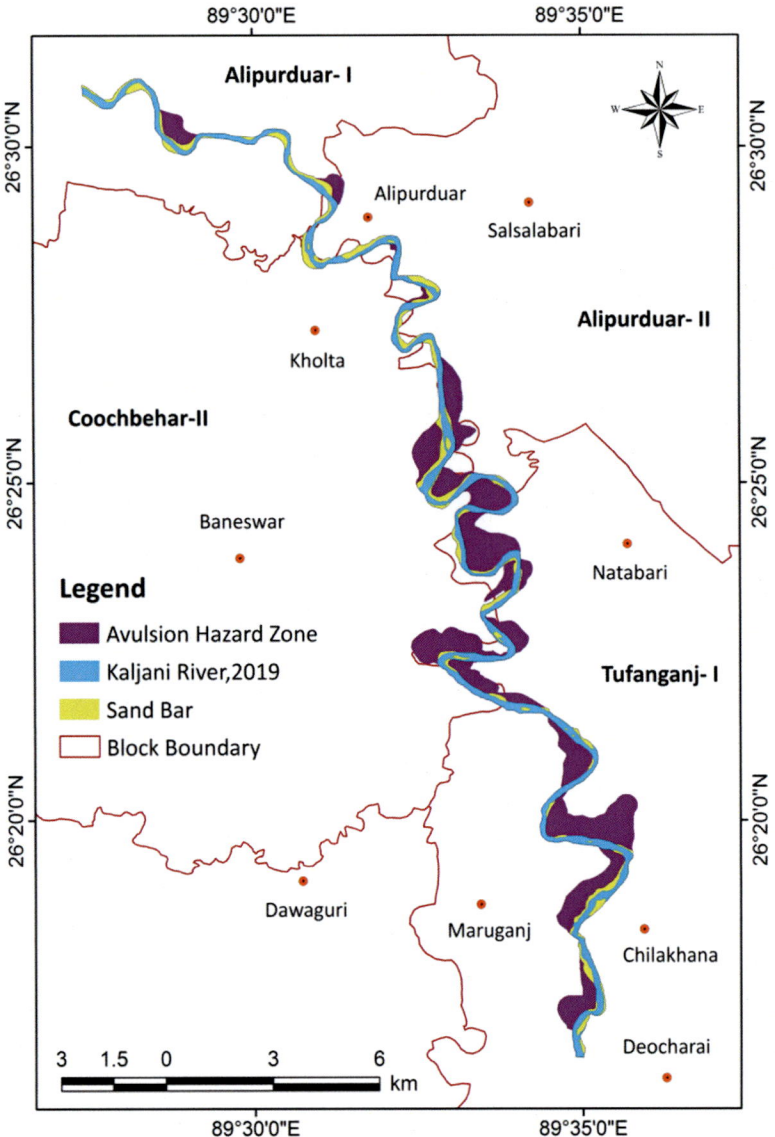

are very much prone to *avulsion hazards* which cause great damage to agriculture and settlement areas. The areal coverage of AHZ is remarkable at all the reaches of the Kaljani River which is covering an area of 12.33 km^2 (Fig. 3.6).

3.4.3 Analysis of Erosion Hazard Area (EHA)

The delineation of the *Erosion Hazard Area* (EHA) is outside of the HMZ and AHZ which may be vulnerable to bank erosion by streamflow that has been initiated by a present fluvial process (Rapp and Abbe 2003; Dey and Mandal 2018a, b, c). To calculate the status of channel migration, 30 cross-sectional studies were being made along the Kaljani River course (Tables 3.3 and 3.4 and Fig. 3.7). On the basis of the *Historical Migration Zone* and *Erosion Setback* (Es), the total length of channel migration and the migration rate have been calculated (Tables 3.3 and 3.4). It was observed that the Maximum cross-sections are attributed to channel migration

Fig. 3.7 Decadal changes of the Kaljani River course in different time periods and the cross-sections

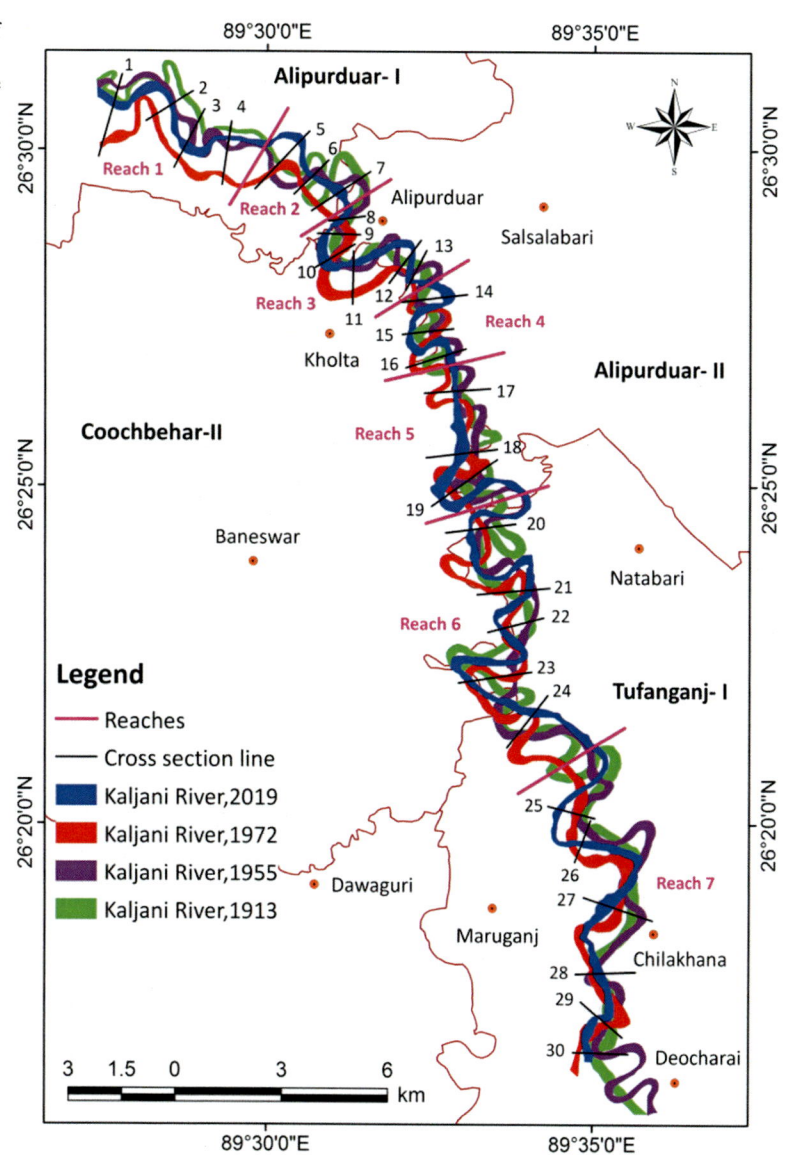

toward the right bank and the total migration has been predicted during the period 1913–2019 (Tables 3.3 and 3.4 and Fig. 3.9).

3.4.3.1 Cross-Sectional Studies and Erosion Hazard Area (EHA)

For the assessment of the *Erosion Setback* and *Geotechnical Setback* of the Kaljani River, 30 cross-sections were selected to obtain data with the help of various measuring instruments (G.P.S. receiver, clinometers, sieve, etc.). (Fig. 3.7).

The *Erosion Hazard Area* maps of the Kaljani River were prepared on the basis of ES and its coefficient in GIS environment (Tables 3.3 and 3.4 and Eqs. (3.3) and (3.4)) (Plate 3.1).

The negative value for both bank sides indicates outward migration of the river and widening of the channel (Tables 3.3 and 3.4) and vice versa. Along the cross-sections 1–5, 6–7, 13, 15, 17–19, 22, 23 and 25–30 the migration tendency of the river is toward the right as of 1913–2019 (Table 3.3 and Fig. 3.9a). On the other hand, along the cross-sections 1–8, 11, 12, 14, 16, 17,

Plate 3.1 a Confluence of Kaljani River near Deocharai. b Under construction of embankment along the Kaljani River, and c Left bank of Kaljani River without protective measures

21, 23, 24, 26, 28 and 30, there is a tendency for the river to migrate toward the left as of 1972–2019 (Table 3.4 and Fig. 3.9b). But at present, these types of migration tendencies are being restricted or controlled by some engineering construction. As a result, the extent of the estimated expected bank line is constricted to its actual extent due to presence of embankment construction at some places in the study reaches (Fig. 3.8).

3.4.4 Delineating the Disconnected Migration Area (DMA)

DMA is a part of CMZ, where channel migration is totally controlled by engineering construction, such as construction of embankments, national or state highways, Railway etc. The aerial coverage of the DMA is 13.5 km^2 (Fig. 3.10). Due to engineering constructions, most of the areas of

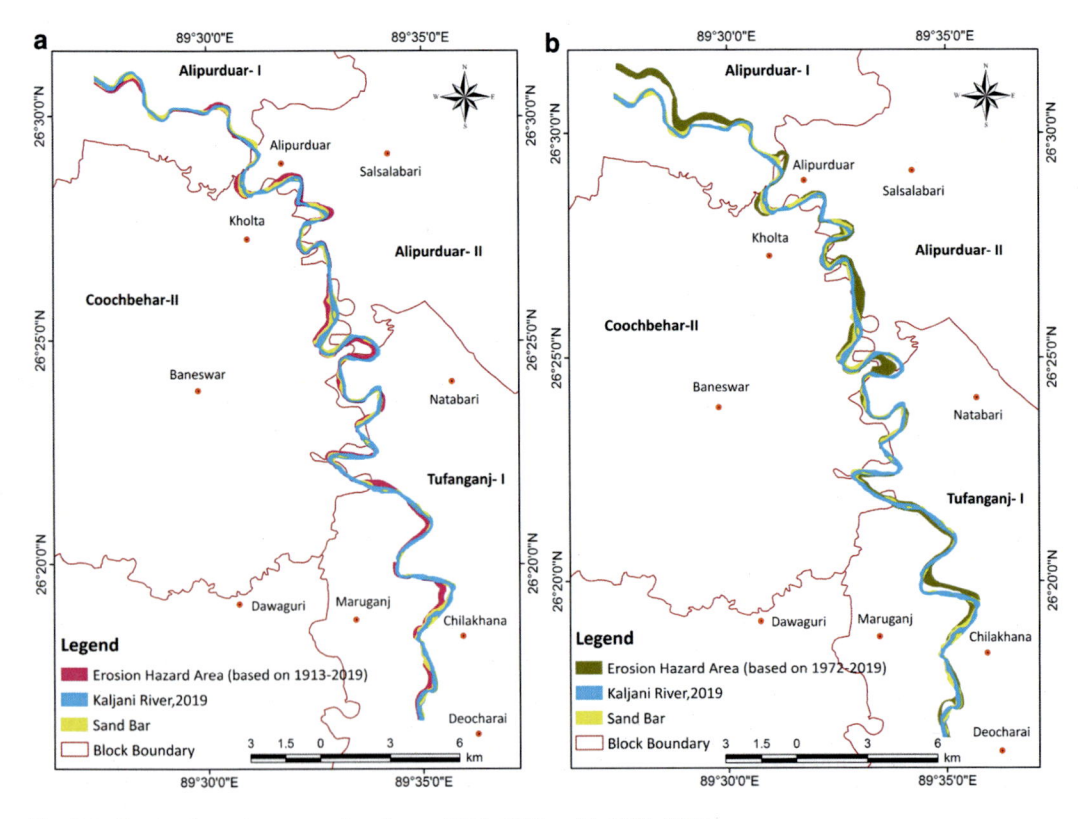

Fig. 3.8 Erosion hazard area map based on **a** 1913–2019 and **b** 1972–2019

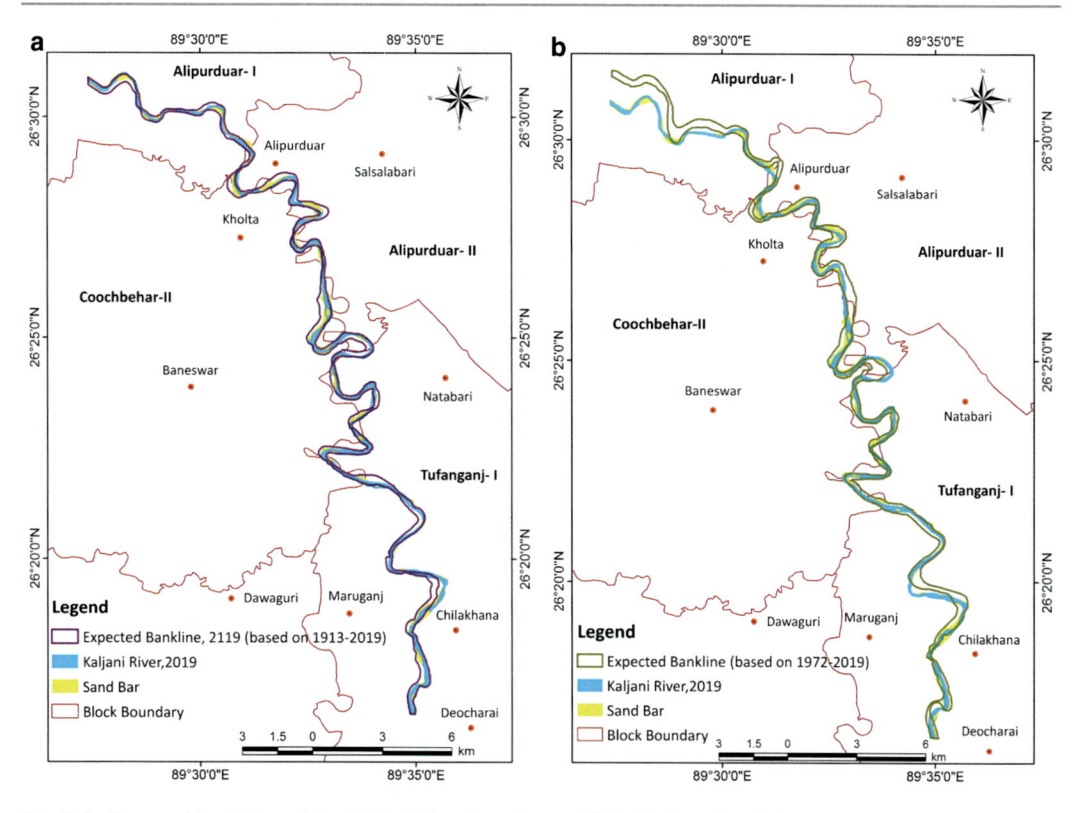

Fig. 3.9 Expected bank line of the Kaljani River based on **a** 1913–2019 and **b** 1972–2019

the both banks are less vulnerable to bank erosion as well as channel migration (Plate 3.1b). Some parts of the Kaljani River have been migrated toward both sides of the channel due to absence of DMA or less protective measures, as well as absence of natural vegetation (Plate 3.1c).

3.4.5 Channel Migration Vulnerability Zone

The *channel migration zone* is the combination of the HMZ, AHZ and EHA (Figs. 3.3, 3.4, 3.5, 3.6, 3.8 and 3.10). The *channel migration zone* map was prepared based on the bank-line migration rate during 1913–2013 and 1972–2072. The areal coverage of the CMZs during the period 1913–2013 is 35.73 km^2 and during 1972–2072 is 35.05 km^2 (Fig. 3.11a, b). Most part of the left bank is protected by DMA (Fig. 3.10) between the confluence of the Garam river and the Nonai River and between the confluence of Bura Gadadhar and Kaljani river which means that the left bank is protected by the engineering construction. Due to channel *avulsion*, the confluence area of the Dima River may be captured by the River Kaljani in future near Alipurduar town (Fig. 3.11a, b).

3.4.5.1 Channel Migration Zone Based on HMZ

Historical channel migration (HMZ) has a direct impact on the bank failure in the Northern plain region. During 1913–2019, total of 29 mouzas, 2 census towns and one municipal area were found vulnerable to channel migration due to bank erosion. The channel migration vulnerability zone has been classified into three categories such as high, moderate and low vulnerable zone (Fig. 3.12). A total of 22 mouzas are categorized as high vulnerability to channel migration. Similarly, Birpara and Sovagang census town, and

Fig. 3.10 Disconnected migration area of the Kaljani River

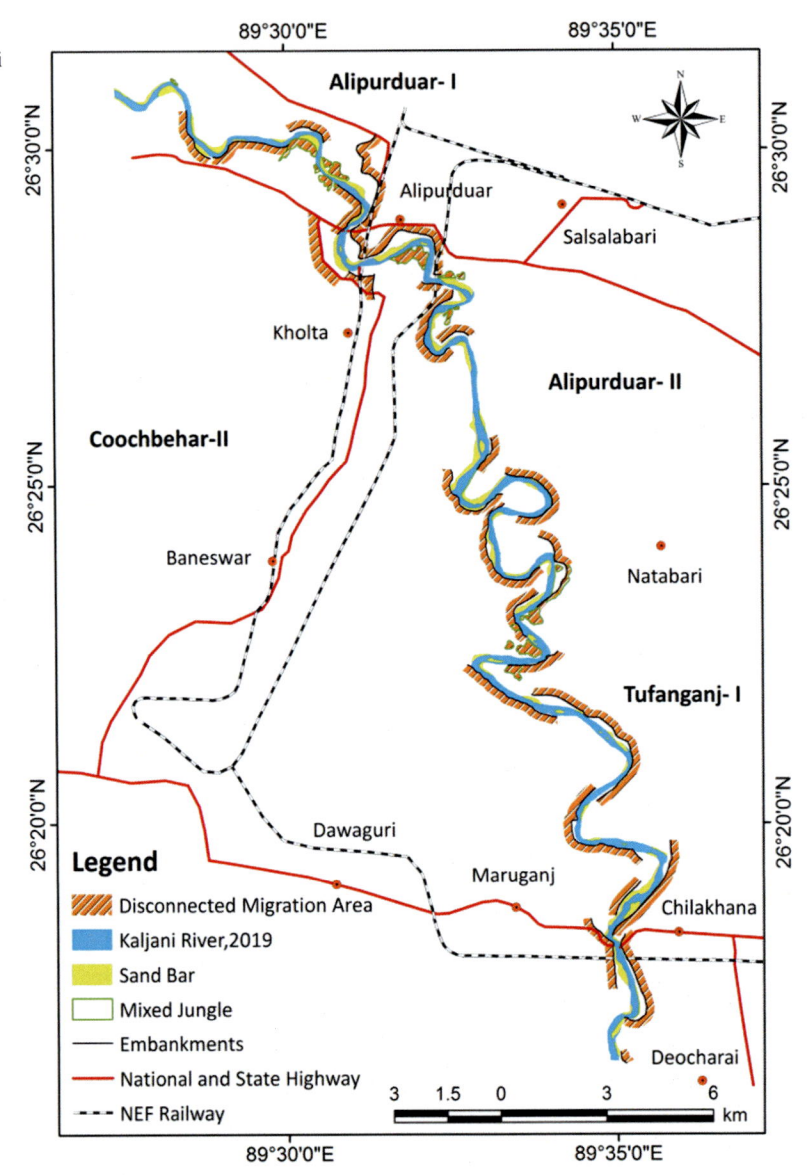

Alipurduar municipality also experience high channel migration vulnerability. Bhelakopa dwitia khanda and Airani chitalia are categorized into moderate vulnerable zone. Five mouzas, i.e., Banchukamari, Chaprarpar, Moamari, Ghogarkuthi pratham khanda and Maradanga are characterized as low vulnerable to channel migration.

During 1972–2019, a total of 26 mouzas, 2 census towns and one municipal area were found vulnerability to bank erosion due to channel migration. The channel migration zone has been classified into three, i.e., high, moderate and low vulnerability zone (Fig. 3.13). Total 23 mouzas are under high channel migration vulnerability.

3.4.5.2 Channel Migration Vulnerability Based on CMZ

Mouza wise appearance of channel is estimated for showing the level of channel migration

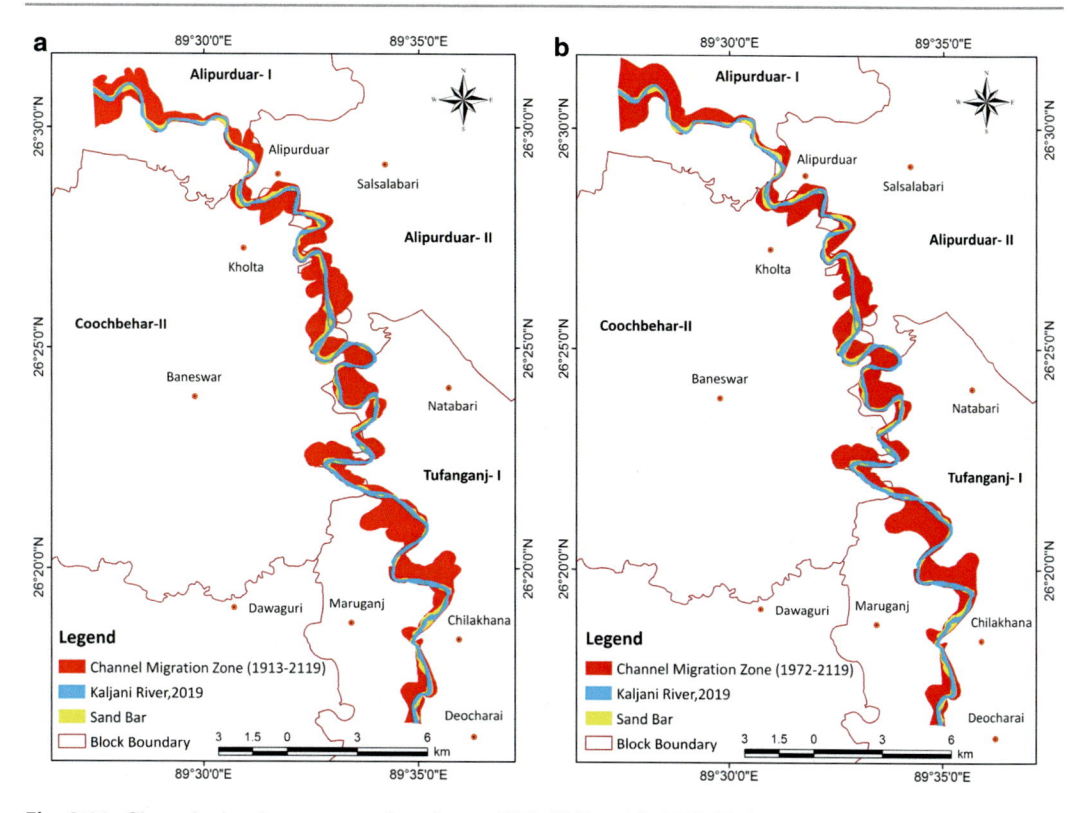

Fig. 3.11 Channel migration zone map based on **a** 1913–2019 and **b** 1972–2019

vulnerability of the river. 29 Mouzas, two Census towns and one municipality were very much susceptible to erosion due to this channel migration. Based on the CMZ, the channel migration vulnerability mouzas were classified into four categories such as, above 75% (sites are affected almost every time in channel migration). The mouzas which are under 50–75%, 25–50% and low below 25% are classified as high, moderate and low vulnerable of channel appearance, respectively (Figs. 3.14 and 3.15).

Based on CMZ (1913–2019), 21 mouzas, two census towns and one municipal area were found as severe channel migration vulnerability where most of the mouzas such as Pararpar, Chandijhar, Kaljani, Ambari, Deocharai, Panisala, Jaigir Chilakhana, etc. are very close to both the river bank sides. Four mouzas, i.e., Kholta, Ghagra, Airani chitalia and Moamari are under high channel migration vulnerability zone. Two mouzas such as Chaprarpar and Banchukamari are experienced with moderate vulnerability. Another two mouzas, i.e., Maradanga and Ghogarkuthi pratham khanda are under low vulnerability (Fig. 3.14).

Based on CMZ (1972–2019), total of 21 mouzas, 2 census towns and one municipal area are under severe channel migration vulnerability. Mouzas of Kholta, Ghagra, Airani chitalia and Moamari experience high channel migration vulnerability. Only Banchukamari mouza is under moderate vulnerability. Chaprarpar and Maradanga mouzas are under low channel migration vulnerability because these are fully protected with DMA (Fig. 3.15).

3.5 Conclusions

The present study is dealt with *channel migration zones* from 1913 to 2019 using two sets of time periods, i.e., 1913–2019 and 1972–2019 for

Fig. 3.12 Vulnerability of mouzas to channel migration of the Kaljani River based on 1913–2019 HMZ

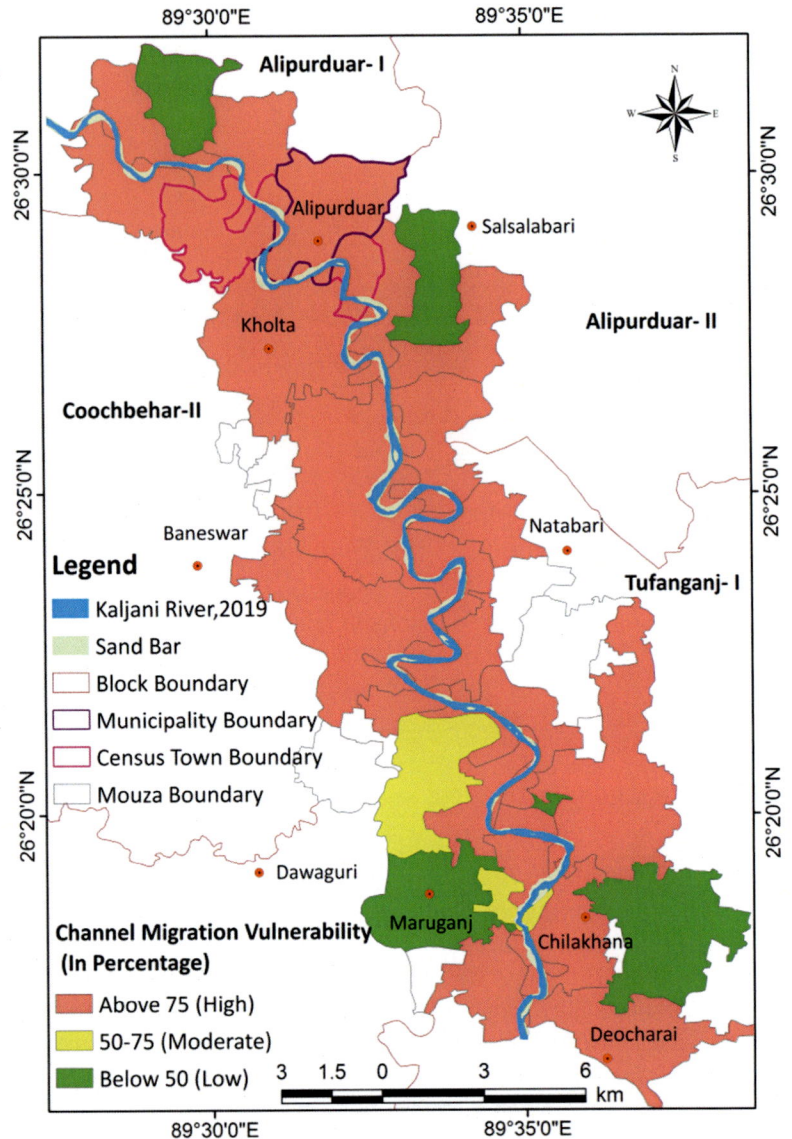

understanding effects of changing the channel migration zone covering the mouzas of Cooch Behar and Alipurduar districts in West Bengal, India. The construction of embankments reduced the predicted CMZs. The both banks of the Kaljani River are highly potential for *avulsion hazards* in the lower reaches from reach-4 to reach-7 in between the confluence of Cheko River and Kaljani River, respectively. The study discovered that the extension of human habitation should be considered away from CMZs. Almost every mouzas are identified highly vulnerable to channel migration and all need adequate protective measures. The present research work may be helpful to administrators, policymakers and local people in introducing proper initiatives to control the financial loss of the river dwellers.

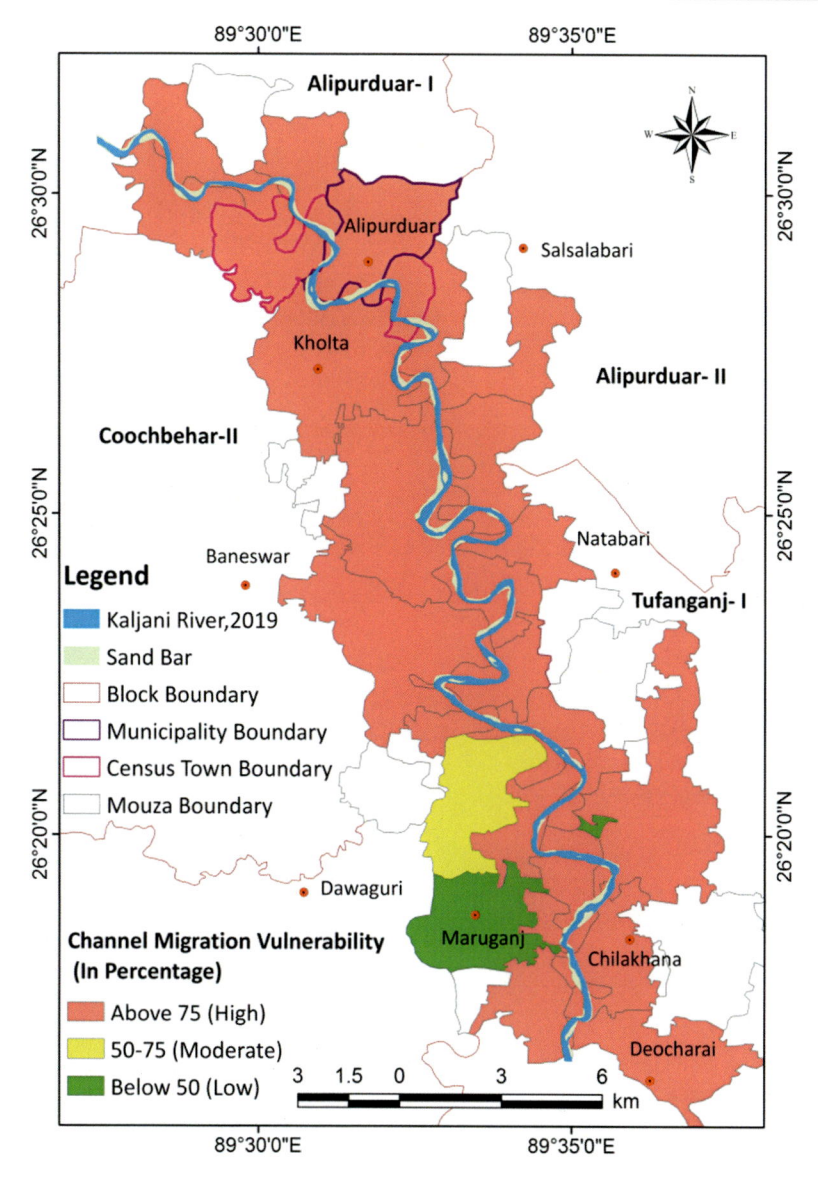

Fig. 3.13 Vulnerability of mouzas to channel migration of the Kaljani River based on 1972–2019 HMZ

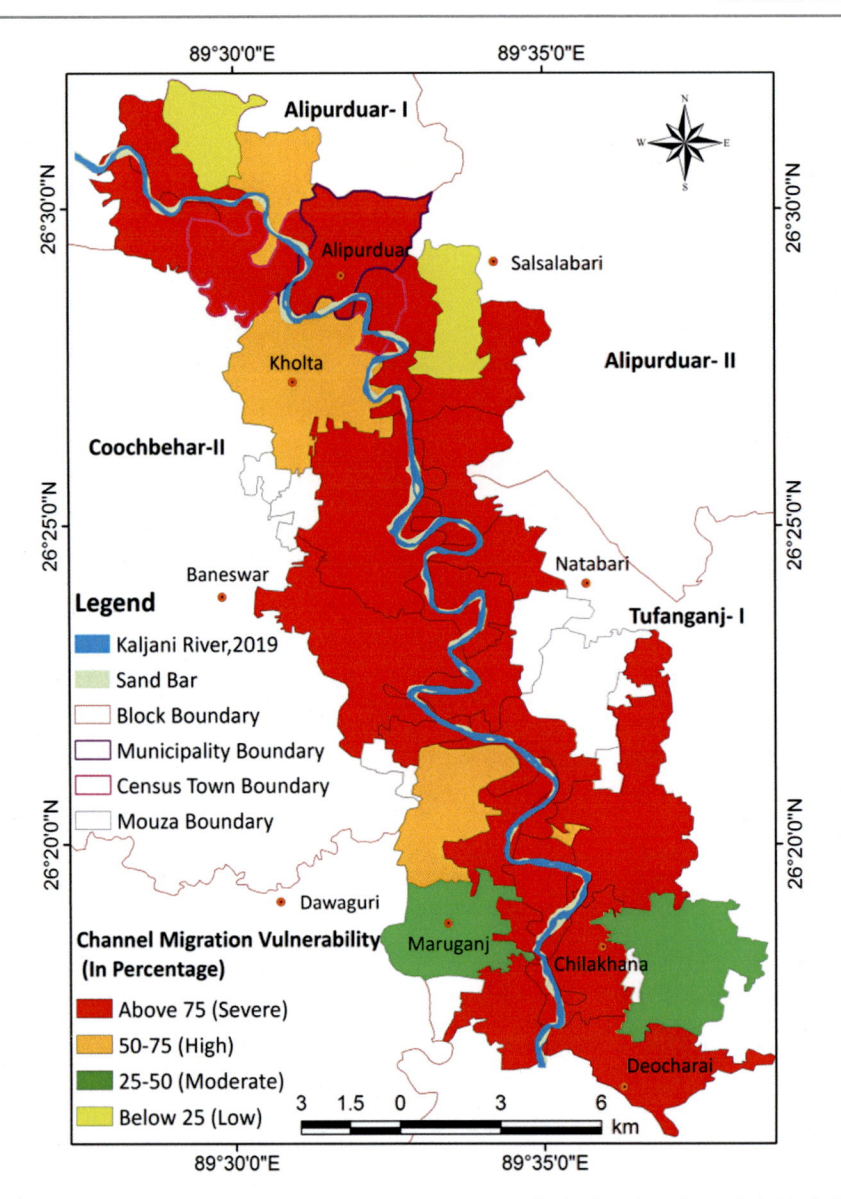

Fig. 3.14 Vulnerability of the mouzas to channel migration of the Kaljani River based on 1913–2019 CMZ

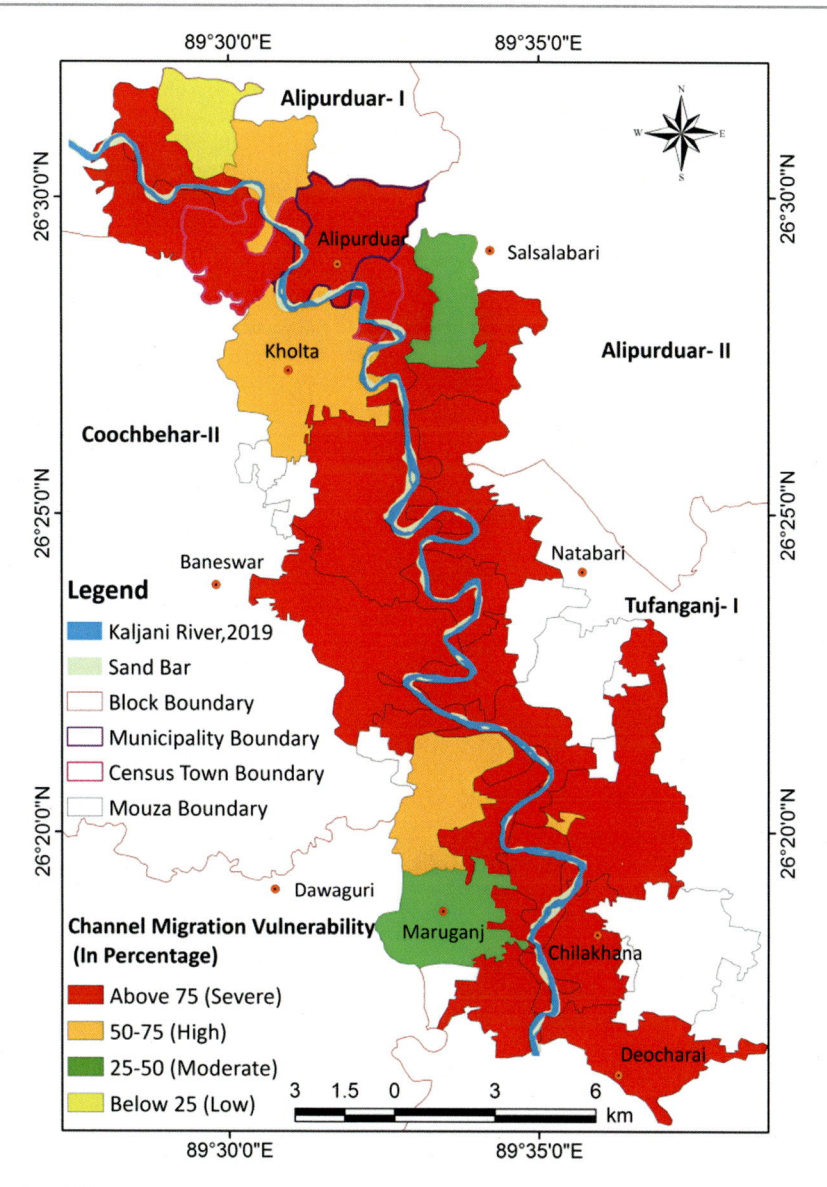

Fig. 3.15 Vulnerability of the mouzas to channel migration of the Kaljani River based on 1972–2019 CMZ

Acknowledgements The authors would like to express special thanks to Mr. Sourav Dey, Dept. of Geography, Darjeeling Govt. College, Darjeeling, India for his kind advice, suggestions and efforts toward the preparation of this paper. The authors are also thankful to the USGS for providing Landsat Imageries and SOI for providing topographical maps.

References

Ahmed I, Das N, Debnath J, Bhowmik M (2018) Erosion induced channel migration and its impact on dwellers in the lower Gumti River, Tripura, India. Spat Inf Res. https://doi.org/10.1007/s41324-018-0196-9

Bierman PR, Montgomery DR (2014) Key concepts in geomorphology. W.H. Freeman, New York, NY

Boyd K (2012) Clark Fork River channel migration zone map development. Phase 1

Boyd K, Thatcher T (2016) Deep creek channel migration zone mapping. Final report

Chakraborty S, Mukhopadhyay S (2014) An assessment on the nature of channel migration of River Diana of the sub-Himalayan West Bengal using field and GIS techniques. Arab J Geosci 8(8):5649–5661. https://doi.org/10.1007/s12517-014-1594-5

Das S, Pal S (2016) Character and cardinality of channel migration of Kalindri River, West Bengal, India. Int Res J Earth Sci 4(1):13–26

Das JD, Saraf AK (2007) Technical note: remote sensing in the mapping of the Brahmaputra/Jamuna river channel patterns and its relation to various landforms and tectonic environment. Int J Remote Sens 28(16):3619–3631

Das TK, Haldar SK, Das GI, Sen S (2014) River bank erosion induced human displacement and its consequences. Living Rev Landsc Res 8(3):1–35. https://doi.org/10.12942/lrlr-2014-3

Dey S, Mandal S (2018a) Channel migration zone mapping of the Torsa River in the Duars and Tal Region of Eastern Himalayan Foothills, India. In: Ghosh A et al (ed) Contemporary research perspectives in geography. Sparrow Publication, Kolkata, pp 269–281

Dey S, Mandal S (2018b) Assessment of channel shifts hazard of the River Torsa in the Eastern Himalayan Foothills, India. Int J Basic Appl Res 8(7):742–758

Dey S, Mandal S (2018c) Assessing channel migration dynamics and vulnerability (1977–2018c) of the Torsa River in the Duars and Tal region of Eastern Himalayan Foothills, West Bengal, India. Spat Inf Res 27(1), 75–86. https://doi.org/10.1007/s41324-018-0213-z

Dhari S, Arya DS, Murumkar AR (2014) Application of remote sensing and GIS in sinuosity and river shifting analysis of the Ganga River in Uttarakhand plains. Appl Geomat. https://doi.org/10.1007/s12518-014-0147-7

Gogoi C, Goswami DC (2013) A study on bank erosion and bank line migration pattern of the Subansiri River in Assam using Remote Sensing and GIS techniques. Int J Eng Sci 2(9):1–6

Heo J, Due TA, Cho HS, Choi SU (2008) Characterization and prediction of meandering channel migration in the GIS environment: a case study of the Sabina River in the USA. Environ Monit Assess 15(2):155–165

Knighton D (1998) Fluvial forms and processes. A New Perspective. Edward Arnold Ltd., London

Laha C, Bndyapadhyay S (2013) Analysis of the changing morphometry of River Ganga, shift monitoring and vulnerability analysis using space-borne techniques: a statistical approach. Int J Sci Publ 3(7)

Macfall J, Robinette P, Welch D (2014) Factors influencing bank geomorphology and erosion of the Haw River, a high order river in North Carolina, since European Settlement. PLoS ONE 9(10):1–12. https://doi.org/10.1371/journal.pone.0110170

Maiti R (2016) Modern approaches to fluvial geomorphology. Primus Books, Delhi

Mani P, Patwary BC (2000) Erosion trends using remote sensing digital data: a case study at Majuli Island. In: Proceedings of brain storming session on water resources problems of North Eastern region, pp 29–35

Mukherjee K, Pal S (2017) Channel migration zone mapping of the River Ganga in the Diara surrounding region of Eastern India. Environ Dev Sustain. https://doi.org/10.1007/s10668-017-9984-y

Naik SD, Chakravorty SK, Bora T, Hussain I (1999) Erosion at Kaziranga National Park, Assam: a study based on multitemporal satellite data. Project report. Space Application Centre (ISRO), Ahmedabad and Brahmaputra Board, Guwahati, pp 70–76

Nicoll TJ, Hickin EJ (2009) Planform geometry and channel migration of confined meandering rivers on the Canadian prairies. Geomorphology 116:37–47

Rapp C, Abbe T (2003) A framework for delineating channel migration zones: Washington State. Department of Ecology and Washington State Department of Transportation. Ecology Final Draft Publication #03-06-027, pp 1–50

Sarkar A, Garg RD, Sharma N (2012) RS–GIS based assessment of river dynamics of Brahmaputra River in India. J Water Resour Prot 4:63–72. https://doi.org/10.4236/jwarp.2012.42008

Schumm SA (1977) The fluvial system. Wiley, New York

Thatcher T, Swindell B, Boyd K (2009) Yellowstone river channel migration zone mapping. Yellowstone River Conservation District Council. ftp://161.7.17.223/Documents/Projects/Channel_Migration_Zones/Yellowstone20090413Report.pdf. Accessed on 12 Sept 2016

Washington DNR (2004) Standard methods for identifying bankfull channel features and channel migration zones. Washington Department of Natural Resources Forest Board Manual, Sections 2, 69

Yang X, Damen CJM, Zuidam AR (1999) Satellite remote sensing and GIS for the analysis of channel migration changes in the active Yellow river Delta, China. Int J Appl Earth Obs Geoinf 1(2):146–157

Yang C, Cai X, Wang X, Yan R, Zhang T, Zhang Q (2015) Remotely sensed trajectory analysis of channel migration in lower Jingjiang reach during the period of 1983–2013. Remote Sens 7:16241–16256. https://doi.org/10.3390/rs71215828

Exploring Change of River Morphology and Water Quality in the Stone Mine Areas of Dwarka River Basin, Eastern India

Indrajit Mandal and Swades Pal

Abstract

In developing countries, growing construction works encourage stone quarry and crushing activities. The middle catchment of east India's Dwarka River basin has a total of 239 quarrying and 982 crushing units, which produces huge stone dust affecting not only air but also river morphology and water quality. The present study aims to ascertain the impact of stone dust on river morphology change and water quality. The study identified growing channel bed aggradations (average: 0.02–0.52 m) due to stone dust. A multi-parametric approach based on machine learning methods like Fuzzy Inference System and Random Forest Algorithm incorporating eleven relevant parameters identified river bed accretion susceptibility due to stone dust. In all the cases 6–17% area is identified as highly susceptible zones. Sediment load is abnormally enhanced exceeding the carrying capacity of the river. River bed mining is identified as the major reason behind the loitering of the thalweg axis of the rivers. Degradation of water quality due to admixing of stone dust is as high as beyond drinkability and irrigability.

Keywords

Stone crushing · Stone dust · River bed accretion · Accretion susceptibility · River bed mining · Thalweg shifting · Water quality

4.1 Introduction

Mining operations consistently lead the developmental process (Xiang et al. 2018; Pal and Mandal 2021). In the twenty-first century, rapid urbanization, industrialization, and many developmental processes are sweeping a high-rise demand for construction materials (Minnullina and Vasiliev 2018). So, control of open-pit mining activity is a greater challenge for good sustainability of the environment (Chen et al. 2015; Lei et al. 2016; Esposito et al. 2017; Xiang et al. 2018). The assessment of the impacts of open-pit mining using geomorphologic knowledge can boost the reasons responsible for the qualitative degradations of our environment and we construct the necessary strategies for the betterment of the future generation (Toy and Hadley 1987; Wilkinson and McElroy 2007; Xu et al. 2019). Rivers are a vital aspect of an open natural system of the earth and it is constantly subject to change and transformation triggered by both natural as well as manmade agents (Nayyeri and Zandi 2018). Mining activity is such a type of

I. Mandal · S. Pal (✉)
Department of Geography, University of Gour Banga, Malda, West Bengal, India
e-mail: swadespal2017@gmail.com

I. Mandal
e-mail: indrajitgeofarakka@gmail.com

activity that changes the natural component like rivers (Festin et al. 2019; Milczarek 2019). On a global scale, river systems have been modified due to increasing sediment and nutrient loads (Xiang et al. 2018). Fluxes of stone dust, fly ash, eroded soil, and fertilizer residues also play a vital role in changing the biogeochemical characters (Holmes et al. 2012; Peña-Ortega et al. 2019) of a river. Change of habitat quality of a river due to changes in channel morphological characters like the roughness of river bed, depth of water, the slope of the bank, sediment load, etc. is also vital concerning the influx of stone dust (Hauer et al. 2013; Costea 2018; Hohensinner et al. 2018). Moreover, direct mining of stone from river beds also causes channel bed modification and changes in flow characters and biological habitat characters of the species living there over (Singh et al. 2016; Wiejaczka et al. 2018).

River water quality modification due to stone quarrying and crushing is one of the important issues in this connection (Nayyeri and Zandi 2018). A huge influx of stone dust can alter the physicochemical composition of the water (Fu et al. 2014, 2016; Qi et al. 2018; Quinn et al. 2018). So the influx of dust in a river or pond water can change physicochemical properties like turbidity, PH level, DO, COD, BOD, etc. (Calle et al. 2017; Barman et al. 2018; Affandi and Ishak 2019). Change of water quality beyond optimum level can change the habitability of species in a river and irrigability in its surrounding agricultural field (Mandal and Pal 2020). The high concentration of total dissolved solids (TDS) above 2250 ppt can make the water non-irrigable as per the Bureau of Indian Standards (BIS 1991). Irrigation with such water can change the soil composition of the agricultural land too (Trujillo-González et al. 2017; Khalid et al. 2018; Pal and Mandal 2019a) which may affect the productivity of the soil. Change in water quality directly affects the livelihood of the fishermen (Samah et al. 2019). Qualitative deterioration of water provides hardship to fish communities which are the mainstay of fishermen's economy (Lyons et al. 2016; Dembowska et al. 2018; Massi et al. 2019; Mariya et al. 2019). Pal et al. (2016) explored that due to qualitative change of water, availability of fish is diminished in Bakreshwar river of India and it also causes shifting of livelihood of the fishermen to other occupations like daily wage laborer, a rickshaw puller etc.

4.2 Study Area

The middle catchment of the Dwarka River basin (3882.71 km^2), a sub-basin of the Mayurakshi River basin, is characterized by Dharwanian sedimentary rocks (Hercynian orogeny, 360–300 million years ago) (Jha and Kapat 2009). The area is therefore extremely rich in stone and related crushing. Bedrock rivers have the potential for direct stone harvesting from the river bed. Almost 239 stone quarrying and 982 crushing centers and so many river bed mining (river bed cutting) centers there, as per the tally of 2018 and these are mostly located in the middle catchment (Pal and Mandal 2017, 2019a, b, 2020) (Fig. 4.1). These are generating a huge volume of stone dust and these are discharging into rivers. River bed stone quarrying (river bed cutting) is also going on at different sites of the rivers. All these can change the hydrological and geo-morphological characteristics of the river and other surface water bodies. In this sense, the present work aims to investigate the impacts of stone quarrying, crushing, and river bed mining on river morphology and also the water quality of the Dwarka river basin of Eastern India.

4.3 Materials and Methods

4.3.1 Methods for the Volume of Stone Dust Estimation on the River Bed

To measure the concentration of stone dust in all six tributaries, eight sites for each tributary have been selected to measure the depth of dust deposition. Direct field digging is done for

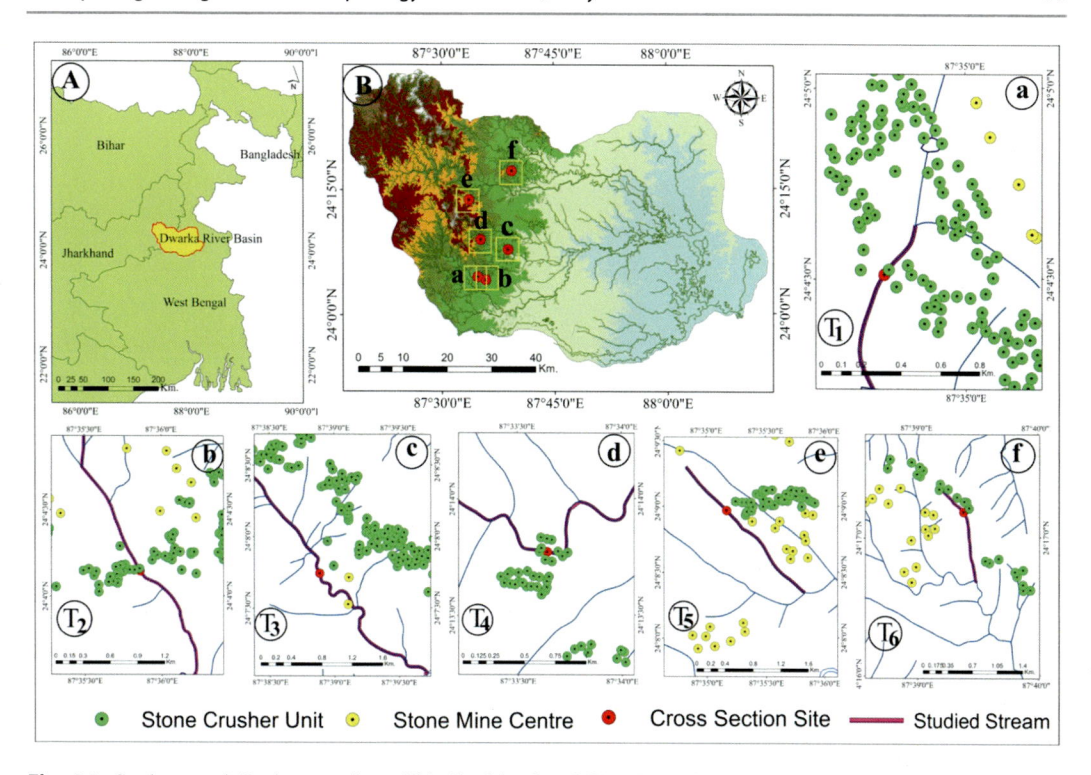

Fig. 4.1 Study area: **A** Basin extension within Jharkhand and West Bengal, **B** showing the field study site within the basin, **a–f** selected tributaries for the study, and cross-section site (T_1 represents tributary 1 and vice versa)

measuring the depth. The average dust depth of all sites is then multiplied by the area of the river for obtaining the total volume of dust deposition in a river.

4.3.2 Methods for Assessing Impacts on Channel Morphology

Cross-sections across the present bed level of some selected six tributary sites used a dumpy level to demonstrate river bed aggradations condition. For obtaining the bed level of the parent rock, each staff site location digging operation has been done and the depth of stone dust is measured. Deducting dust depth from present bed level, bedrock level (profile before dust deposition) is obtained for a comparable illustration of both the profiles (before and after dust deposition). While doing the depth of dust deposition, litho facets of 06 sites have also been assessed.

4.3.2.1 Methods for Assessing the Channel Path Modification Due to Direct Stone Mining from the River Bed

For assessing the impact of direct stone quarrying from river bed on river flow path we have selected some other tributaries of Dwarka river (except the previous six selected) thalweg axis of flow from 2001 to 2017 has been digitized from Google earth image and superimposed for illustrating the change. As the tributaries carry a very less amount of flow, it follows mainly the deepest part of the entire cross-section. It facilitates us to detect the thalweg axis of the concerned period easily without profiling. For making it confirm, the thalweg axis of 2017 is cross-validated with the field. The departure of the digitized thalweg axis from the field-based thalweg axis is <0.6 m in all the cases. For better visibility of the change, only two consecutive reaches of a tributary are shown here.

4.3.3 Assessing Stone Dust Accretion Susceptibility at the Riverbed

4.3.3.1 Data Layer Selection and Preparation for River Bed Accretion Susceptibility

Along with spatial stone dust accretion mapping, stone dust accretion susceptibility of two successive reaches of six selected tributaries has been done based on a multi-parametric approach. Total eleven conditioning parameters like (1) volume of dust heaping (2) direct dust deposition at the river (3) distance of river from dust heaping sites (4) distance from the crusher units (5) river bed slope (6) bed shear stress (7) velocity of flow during pre-monsoon season (8) steepness of bank slope (9) density of water (10) depth of river water (11) velocity of the river in monsoon season have been taken into account for all the selected tributaries. The digital elevation model for the mentioned parameters is prepared in Erdas software. The selection of parameters is based on the field experience and following the predecessors in this field. Jakoyljevic et al. (2009) and Mehdi et al. (2011) also studied these types of susceptibility by using few such parameters in their respective study regions.

Volume of Dust Heaping

Stone dust from the stone crusher units often influxes into the nearby river channel. The field survey was conducted to determine the dust heaping zone and the volume of dust from the distance is estimated at a surface of one square meter per location. Field experience proves that more dust is heaped to the channel near the crushing unit. On this basis, ratings 10, 7.5, 5, 2.5, and 1 are assigned to dust heaping zones, respectively, with extremely high, high, moderate, low, and very low volume. DEM is created in ERDAS remote sensing software based on the rating values of the selected sites.

Direct Dust Deposition at the River

In the present study, the authors identified direct dust deposition point (DDDP) to the river through the field visit. The area of the affected channel is demarcated by direct observation of the field. This analysis is carried out in the light of five integrated buffers from the DDDP (direct dust deposition point) and the rating is given in a specific manner. The highest rating is rated as 10 and granted to the DDDPs immediate peripheral zone (up to 250 m), rating values 7.5, 5, 2.5, and 1 are allocated to the buffer zones 250–500 m, 500–750 m, 750–1000 m and beyond 1000 m. Random points are taken from every buffer area, and every DDDP and DEM are created. Given the subjective but logical approach of assigning a weight to the buffer zone as the deposition rate decays over distance. Priority is given to deciding on buffer radius field experience and perception of people.

Distance from the Crusher Units

This analysis is also carried out considering five dispersed buffers from the units of stone crusher which are the main source of stone dust and rating is given in a specific manner. The highest rating is rated as 10 and granted to the Nearer dust deposition points (NDDP) immediate peripheral zone (up to 250 m), rating values 7.5, 5, 2.5, and 1 are allocated to the buffer zones 250–500 m, 500–750 m, 750–1000 m and beyond 1000 m. DEM is then prepared based on these values.

Bed Shear Stress

The bed shear stress (τ_0) is expressed as a force per unit area of the bed (in N m^{-2}) and increases with flow depth and channel steepness.

$$\tau = yDSw \quad (4.1)$$

where, τ = spatially averaged bed shear stress (N m^2), y = weight density of water (N m^3), D = average water depth (m), Sw = water surface slope.

It is assumed that high dust deposition offers more resistance. Forty sites in each river are selected and shear stress on those sites is computed. The river is reclassified into five rating groups, offering 10, 7.5, 5, 2.5, and 1 at very high, high, moderate, low, and very low-stress levels, based on the range of high and low shear stress values. Based on the rating values taken randomly from different reclassified zones DEM is prepared.

Velocity of Flow During Pre-monsoon and Monsoon Season

The monsoon period and pre-monsoon period flow velocity of the river have been measured. For the selection of velocity measuring sites, random sampling rules have been followed. 40 measuring sites from each studied tributary have been selected for this study and classified into five very high to very low flow velocity zones as done in other layers and finally DEM is prepared for both the seasons.

Steepness of Bank Slope and River Bed Slope

The riverbank slope and bed slope have been measured directly from the field. The water depth and flow of the tributaries are so low that they can be measured. 40 measuring sites are selected from each tributary for this study and plotted into the ArcGIS environment. Abney level is used to measure slope. DEM is prepared based on these values. DEM is then reclassified giving a rating as given in other layers. The maximum rating of 10 is given to the lowest slope class.

Density of Water

The density according to USGS (2018) is only the weight of a specified quantity or amount of material. A common water density measurement unit is gram per milliliter (1 g ml^{-1}) or 1 g per cubic centimeter (1 g cm^{-3}). For analyzing the water density, 40 water samples from each studied tributary has been collected using the hydrometer (it is one of the most basic portable devices for scientific measurement) and classified into low to high flow water density zones as per the measured data and finally, a DEM is prepared.

Depth of River Water

Using simple staff reading depth of water is measured at 40 sites (for each tributary) from the field in post-monsoon season (Mid October to November). Based on these depth values DEM is prepared. Here it is mentioned that a lower depth of river water indicates less volume of water and less volume of water indicates the low stream energy as well as low velocity. So, the decreasing stream energy and velocity influence the probability of dust deposition on the river bed (Hofler et al. 2018; Turowski 2018). Based on this logic, reclassification of the depth layer is done and rating values are assigned. The river is reclassified into five rating classes giving 10, 7.5, 5, 2.5, and 1 at very high, high, moderate, low and very low water depth zones.

Not all parameters affect the susceptibility to dust accretion in the same direction. So for making the data layers unidirectional and standardized in the present case a ten (10) point scale of logical inference is taken into account to assign maximum point rating to a maximum chance of effect in Arc GIS software (v.10.3) by the reclassification method (Pal and Mandal 2017; Pal and Debanshi 2018). Under such a logical assumption, each parameter is categorized into ten classes, and the scale point is allocated. Figure 4.2 displays the used data layers for dust accretion susceptibility.

4.3.3.2 Susceptibility Zoning Using Fuzzy Logic

FIS is a mathematical system, which is formed by the mathematical tool of the multi-value logic base. Zadeh (1965) introduced the theory of fuzzy set and has since been widely used to develop complicated models in various fields.

A fuzzy logic system simulates the whole system to delineate the notion out of inputs spatial data layers of the model. The Mamdani method (Yanar 2003) is the greatest used FIS. This Mamdani method attempts higher performance to appraise people's know-how and factual involvement in classification as well as accountability (Mamdani 1977). This system mainly constitutes four stages (Sami et al. 2014) namely—(i) fuzzification of the data set (ii) evaluation of the rule (iii) inference of the product of fuzzy (iv) operations of the de-fuzzification.

The entire model was accomplished in the ArcGIS environment by applying two tangible tools—(i) membership fuzzy tool (ii) fuzzy

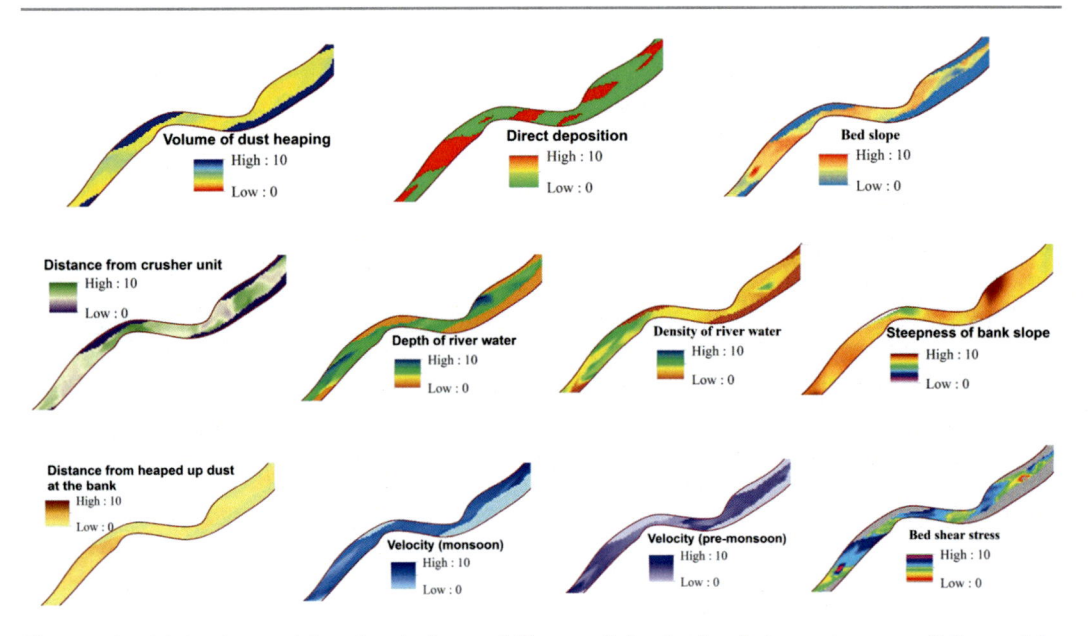

Fig. 4.2 Spatial data layers of the selected tributary 6 (Segment I) for the river bed accretion susceptibility model

overlay tool. First, all data are transformed into fuzzy and subsequently categorized into four categories which, according to the expert judgment, demonstrate high vulnerability to low. Therefore, we used a large membership operation with the help of Eq. 4.2 in the context of a few continuing membership functions.

The large membership function is used to indicate the fact that large values of the input raster have high membership in the fuzzy set.

$$\mu(x) = \frac{1}{1 + \left(\frac{x}{f_j}\right)^{-f_i}} \quad (4.2)$$

where, $\mu(x)$ = fuzzy membership function, f_i = spread (default is 5 in ArcGIS environment for large membership) f_j = midpoint of the range of values, x = degree of a element. The membership value of each component indicates a different degree of support and confidence from 0 to 1 according to this approach (Ercanoglu and Gokceoglu 2002). In Eq. 4.3, the fuzzy set is formulated.

$$a = \{x(fx)\} \quad (4.3)$$

where, a = fuzzy set, x = membership value of the elements, fx = fuzzy membership function.

In this study, the fuzzy 'if–then' method is formulated to make the zonations of river bed accretion susceptibility. The susceptibility maps are categorized into several zones (very high, high, moderate, and low) to explicit the spatial variability of that vulnerability. The following logical structure defines the fuzzy rules, i.e., If the average volume of dust heaping is high or direct dust deposition at the river is high or distance of river from dust heaping sites is less or distance from the crusher units is less or river bed slope is low or bed shear stress is high or velocity of flow during pre-monsoon season is low or steepness of bank slope is high or density of water is high or depth of river water is low and velocity of the river in monsoon season is low then river bed accretion susceptibility will be high.

4.3.3.3 Methods for River Bed Accretion Susceptibility Model Using Random Forest Algorithms (RFA)

The RFA is one of the more versatile data classification algorithms and it can classify a large volume of data using numerous attributes. The algorithm itself is very well suited to parallelization the RAF consists of different attributes.

This is a supervised method of machine learning for classification and regression analysis (Breiman 2001; Youssef et al. 2016; Naghibi et al. 2017; Gayen et al. 2019). A decision tree has to be developed to get the class output and obtain the dependent variable, respectively to classification and regression of the data sets (Kim et al. 2018). This approach consists of several decision trees and integrates them to clarify the spatial connection between the vulnerability of the river bed and the factors of stone dust affecting it (Kim et al. 2018). The advantages of the RFA model in comparison to other methods are (1) it can handle large datasets with high dimensionality as well as avoid the overfitting of the datasets. (2) No assumptions on explanatory and response variables are necessary and (3) it does not require any prior data to transform and rescale the datasets (Youssef et al. 2016; Naghibi et al. 2017). The vector of prediction is indicated with log 2(M + 1), where M indicates the algorithm input number (Kim et al. 2018). Mean squared error can be estimated to test the model's output (Kim et al. 2018).

$$\varepsilon = (V_{observed} - V_{response})^2 \quad (4.4)$$

where ε represents the mean-square error, $V_{observed}$ is the observed data, and $V_{response}$ indicates the variable's outcome.

RF is used to regulate the division in every node by the number of trees and the numbers of predictive variables (Naghibi et al. 2017). The average tree projection is estimated as

$$S = \frac{1}{k} \sum k^{th} v^{response} \quad (4.5)$$

where S marks every forecast of forest and K reflects the trees of the model.

This study moved forward using the four key measures: (1) first preparation and collection of the dataset of river bed accretion susceptibility. Regarding this work, eleven predisposing factors were considered. (2) Preparing data sets for testing and validation by repeated random sampling. (3) River bed accretion susceptibility mapping using RF machine learning algorithm. (4) Finally, validation of the model by the remaining validation datasets. This RF model was implemented in R (version 3.2.4) software using the "randomForest" package. For an RBAS, data representing both sensitive and non-sensitive areas are the main requirements. Conditioning data is typically stored as ArcGIS grid cell layers. The outcome of the decision trees was generated based on the RF model. Finally, the spatial model output was transferred into ArcGIS and reassess into four classes namely high, moderate, low, and very low river bed accretion susceptibility based on natural break statistics. The accuracy of the models was tested. The overall training datasets are usually split by 70% of the samples for training models and others for validation.

Assessing Performance of the Models

A total of 32 field sites from each tributary are visited for measuring the depth of dust accretion over the river bed. Based on these field and model data, Kappa coefficient, area under the curve in Receiver Operating Curve, and correlation coefficient are computed.

For validating the prepared river bed accretion susceptibility models receiver operating characteristics (ROC) curve is prepared. That is invented to appear in the true positive rate (TPR) in respect of false-positive rate (FPR) at assorted threshold contexts. The AUC is known as the fitness rate. It helps to compute the succession and prediction rate. Rasyid et al. (2016) categorized this area under curve into five classes such as (i) excellent (0.90–1.00), (ii) good (0.80–0.90), (iii) satisfactory (0.70–0.80), (iv) poor (0.60–0.70) and (v) fail (0.50–0.60). SPSS statistics software (v.17.0) has been used for preparing the ROC curve as well as the success and prediction rates.

Kappa coefficient (Eq. 4.6) and Overall accuracy (Eq. 4.7) assessment also have been carried out for the derived accretion susceptibility models. Kappa coefficient values range from 0 to 1. As per the recommendation of Monserud and Leemans (1992) that Kappa coefficient values ranges <0.4 represent the poor or very poor agreement, 0.4–0.55 represent the fair agreement, 0.55–0.7 represents good agreement and the 0.7–

0.85 agreement is very good and the >0.85 agreement between models and the ground reality is outstanding.

$$k = \frac{N \sum_{i-1}^{r} Xii - \sum_{i-1}^{r}(xi + {}^*xi + i)}{N^2 - \sum_{i-1}^{r}(xi + {}^*xi + i)} \quad (4.6)$$

where N refers to overall pixel count; r refers to matrix number of rows; Xii equal to the number of rows i observation and columns ii, xi and $x + i$ = marginal totals for both the row.

$$\text{Oac} = \text{Tncs}/100\% \text{ of Tsm} \quad (4.7)$$

where Oac = Overall accuracy, Tncs = complete valid sample number, Tsm = Total sample.

4.3.4 Measuring Sediment Load

Suspended and soluble sediment load is measured at all six selected tributaries. All the cases measuring temporary stations are at the confluence point of the tributaries. The suspended load is measured at different discharge levels and based on these data; the suspended sediment load rating curve is prepared following power regression Eq. 4.8. A rating curve is plotted over a log–log graph. The power regression equation and coefficient of determination can state the direction and degree of change of sediment load in about change of discharge.

$$\log_{10} F(x) = m \log_{10} x + b$$
$$F(x) = x^m \cdot 10^b \quad (4.8)$$

where m is the slope; b is the intercept point on the log plot; $x = \log x$ and $y = \log y$.

4.3.5 Methods for Analyzing Water Quality

40 water samples from various Dwarka river affluents were obtained in different stone mining areas to examine water quality. Physico-chemical properties of water (referred to in Table 4.3) have been tested and the results are compared to the standard state of water quality as defined by BIS (1991) and ISDW (1963). Some water sample parameters such as actual water temperature, pH, and watercolor are gradually measured on-site after water sample collection with the digital thermometer, pH meter, and visual espial. The water sample was collected in glass bottles to analyze the dissolved oxygen and then added dissolved oxygen reagents. We used the azide modification method for the Winkler method here (The technique of azide modification is a standard method for testing the DO). The laboratory test method is used to test the chemical oxygen demand (COD), alkalinity, and total hardness (Th) for analysis. To analyze selected parameters that exceed the threshold limit after Ramesh et al. (2008) we have applied the parameter specified water quality index (PSWQI) with certain necessary modifications. The individual quality ratings (IQRs) of BOD, COD, DO and TDS based on Eq. 4.9 have been done for better understanding.

$$\text{PSWQI} = 100(\text{Vi}/\text{Si}) \quad (4.9)$$

where PSWQI = the selected specific parameter's sub-index, Vi = the tested value of the parameter, Si = standard value.

4.4 Results

4.4.1 Volume of Dust Deposition

Table 4.1 depicts the volume of dust deposition in the selected six tributaries at eight sample sites. The volume of dust differs from 302 to 978 kg m^{-3} in the case of all the tributaries. The mean value varies from 601 to 759 kg m^{-3} and its variation is mainly determined by the nearness of stone crushing units, their total dust production, the velocity of water, etc. Tributary 4 registered maximum dust deposition in its course owing to 489–987 kg m^{-3}.

4.4.2 Channel Bed Accretion

Due to the seasonal rainfall and scanty runoff at the source segment tributaries, stone dust often becomes consolidated over the river bed and pond. Stone dust is often deposited over the river bed and pond due to seasonal rainfall and insufficient drainage in the source segment tributaries. The depth of stone dust varies from 0.02 to 0.52 m according to the January 2018 measurement (Fig. 4.3a–f). At the proximity of the stone crusher units, the dust depth is maximal. The amount of inflow of stone dust to the tributaries is high during the monsoon season due to maximum surface runoff triggered by high rainfall concentration. The average suspended load during the monsoon period is 15,762 g m^{-3}. Due to the rise of discharge during monsoon, a good amount of loose dust is discharged further downstream. A lot of loose dust is released further downstream due to the increase of discharge through monsoons. But such dust heaped the armoring of the river's bed during the non-monsoon seasons.

Field observation of the litho facets (avg. condition) (Fig. 4.4) This difference in composition and their varying width can be explained in terms of size of the produced stone chips, influencing factors of dust deposition in the river bed, and so on.

4.4.3 Simulated Result of River Bed Accretion Susceptibility and Performance Assessment

Figures 4.5 and 4.6, respectively illustrate river bed accretion susceptibility model in selected two reaches of each six tributaries using Fuzzy inference system (FIS) and Random Forest Algorithm (RFA) compiling eleven conditioning parameters. In the case of FIS and RFA very high accretion susceptible zone covers from 5 to 12% in the former model and 4–21% in the case of the latter model. Tributary 5 and 4 have appeared as the most susceptible as per both the applied models. Flowing of those rivers across highly dense crusher units strongly influences such high susceptibility. Figure 4.6 also displays which factor is dominantly responsible for accretion susceptibility. Direct deposition of dust at the river bed, the distance of river from heaped up dust site, distance from crusher unit, bed slope are some dominant conditioning factors for susceptibility.

Table 4.2 shows the estimated K, AUC, and r values of two used models in all the tributaries. K value ranges from 0.81 to 0.93 which means that the relationship between model state and ground reality is good to excellent. While justifying the applicability and interpretability of the applied two models, it can be stated that RFA is more effective in these cases as the average K value (0.91) shows good agreement on the FIS model between model and ground reality (Mean $K = 0.84$). AUC values (>0.8) are also acceptable for both the models but the RFA model (0.9) shows better agreement than the FIS model (0.84). The correlation coefficient between the depth of deposition and model value ranges from 0.71 to 0.82. At the 0.01 confidence level, all of these values are significant. However, the degree of correlation is high in the case of the RFA model.

Table 4.1 Tributary and site-specific volume of dust estimation (kg m^{-3})

Tributary	Volume of dust (kg m^{-3})							
	Site 1	Site 2	Site 3	Site 4	Site 5	Site 6	Site 7	Site 8
Tributary 1 (T_1)	487	964	784	978	648	324	321	302
Tributary 2 (T_2)	845	479	487	475	815	426	654	654
Tributary 3 (T_3)	968	475	963	364	785	463	427	472
Tributary 4 (T_4)	489	889	785	987	624	745	752	804
Tributary 5 (T_5)	897	876	396	874	785	354	365	904
Tributary 6 (T_6)	879	635	478	587	745	748	742	421

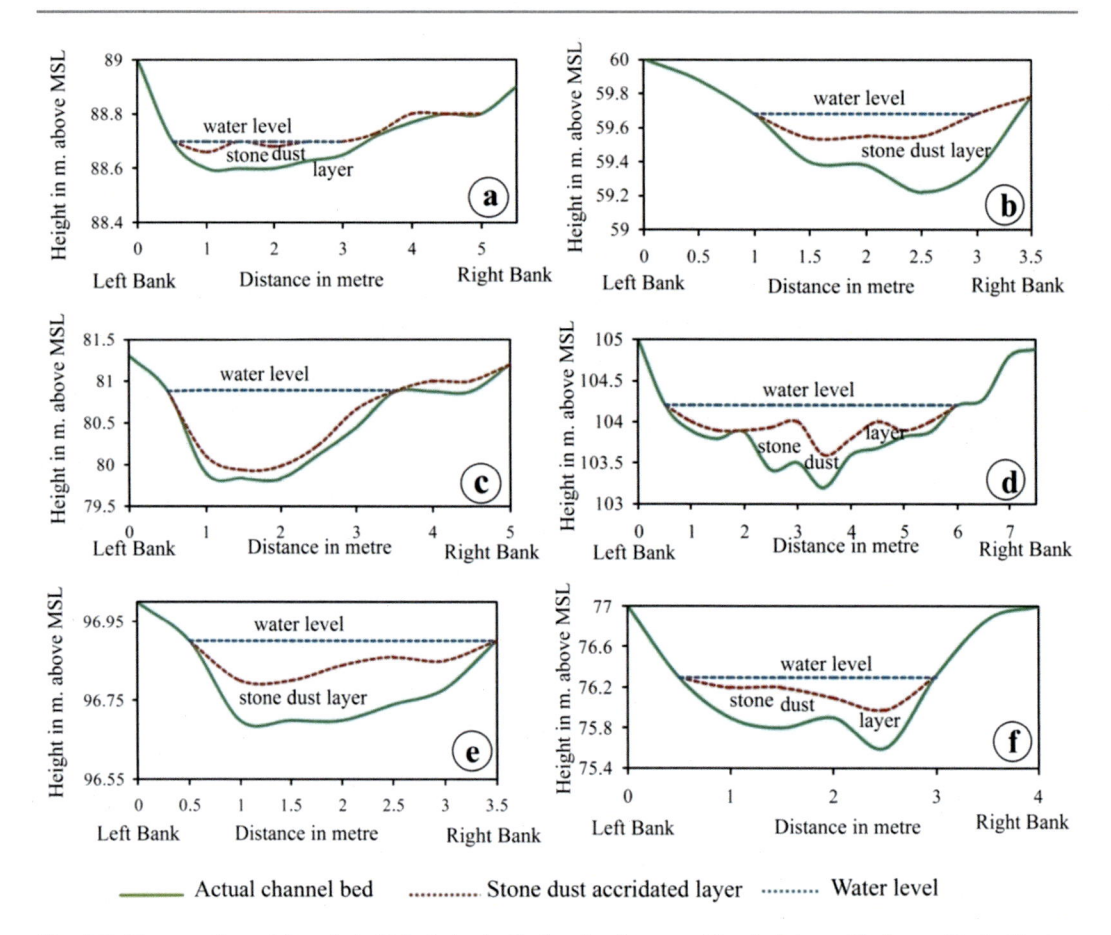

Fig. 4.3 Cross-sections of the selected tributaries (a–f), showing the current level of channel bed accredited with stone dust

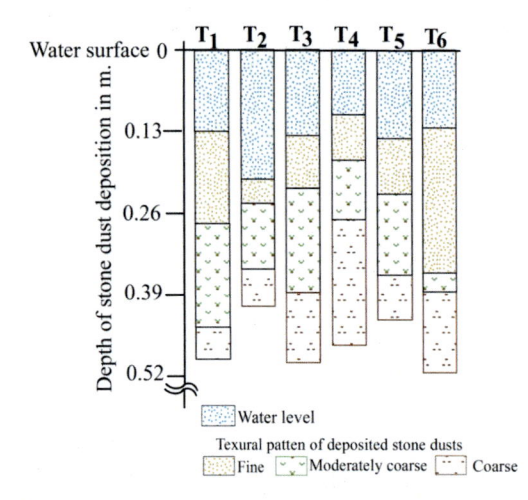

Fig. 4.4 Litho facets of the river bed deposition

4.4.4 Shifting of Flow Paths

Figures 4.7 and 4.8 display the thalweg shifting of a tributary from 2001 to 2017. There is no definite trend of shift due to irregular mining of stone from the river bed. The average yearly shifting is 42 m. Some parts of the selected reaches show a very high rate of annual shifting (>85 m). It mainly depends on where good quality of the stone is found. On a riverbank where there is a good quality of stone for the concerned mining, shifting tendency is toward that side. It causes asymmetric channels and increases the chance of riverbank failure (Das et al. 2013). The average depth of mining in all six tributaries varies from 10 to 80 cm.

Fig. 4.5 River bed accretion susceptibility (RBAS) model using fuzzy inference system (FIS)

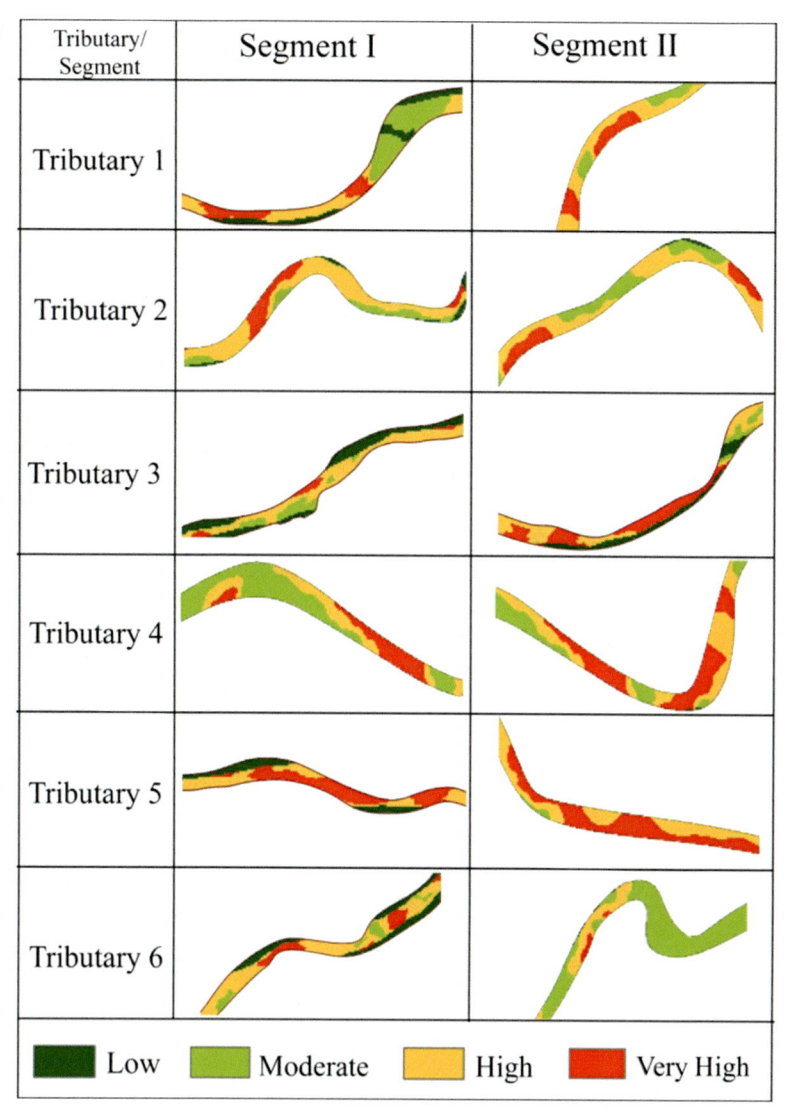

Interestingly, the small-scale mining of stone for grinder, stone pots, statues, etc. causes the formation of small geometric pothole-like features over the river bed. It also causes increasing turbulence in the river flow.

4.4.5 Impact on Sediment Flow

Sediment load in the rivers of the Rarh region is recorded high in most cases. But it is rather high in the areas with the thermal plant producing huge fly ash, stone mining, and crushing units (Pal and Mandal 2019a, b). Such a difference is registered when sediment load is compared with rivers affected by stone dust and free from such a problem. Figure 4.9 depicts the suspended sediment load rating curve describing the influence of discharge on load. The volume of load ranges from 6584 to 79,888 g m^{-3} with a discharge of 1–25.5 m^3 s^{-1} in all six selected affected tributaries. The mean sediment load of the rivers ranges from 7420 to 31,245 g m^{-3}. The highest sediment load is recorded in the tributaries flowing across densely settled crusher units. The composition of load depicts that out of total load >75% is stone dust. While describing the relation between discharge and sediment load, most

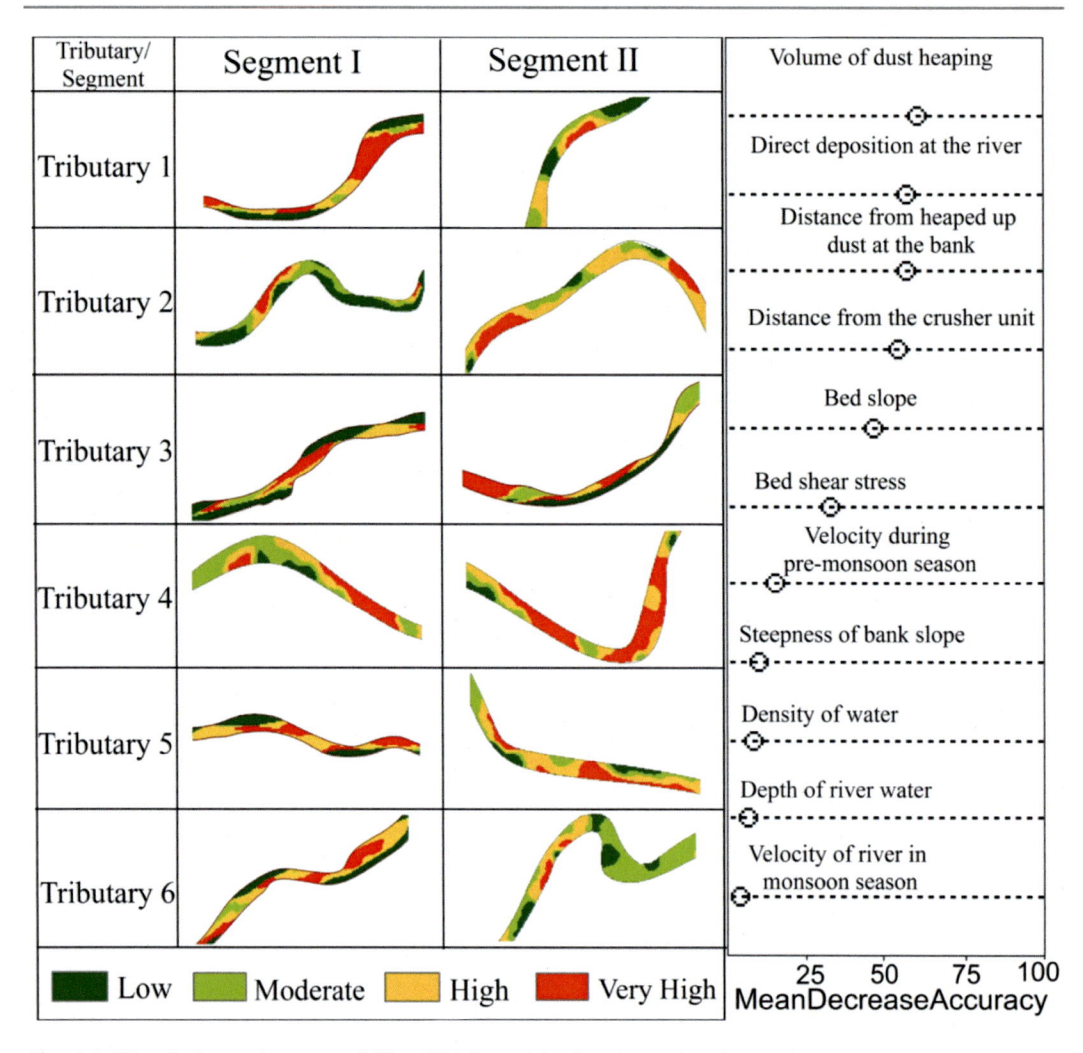

Fig. 4.6 River bed accretion susceptibility (RBAS) model using the random forest algorithm

Table 4.2 Performance level of the applied models for accretion susceptibility

Error estimation	Model	T_1	T_2	T_3	T_4	T_5	T_6
Kappa coefficient	FIS	0.89	0.81	0.92	0.84	0.81	0.82
	RFA	0.92	0.91	0.9	0.89	0.94	0.93
AUC value	FIS	0.821	0.806	0.802	0.874	0.852	0.866
	RFA	0.922	0.905	0.879	0.887	0.912	0.886
Correlation coefficient	FIS						
	RFA	0.745	0.82	0.745	0.654	0.712	0.802

of the tributaries recorded breakpoints indicating decreasing carrying capacity and thus deposition of sediment over the river bed. From the regression graphs for each tributary (Fig. 4.9), it is observed that trend is positive and the coefficient of determination (R^2) value is 0.6–0.75 with a p-value < 0.05. These statistics demonstrate the controlling power of discharge

Fig. 4.7 Shifting of thalweg lines (2001–2017) (Reach 1)

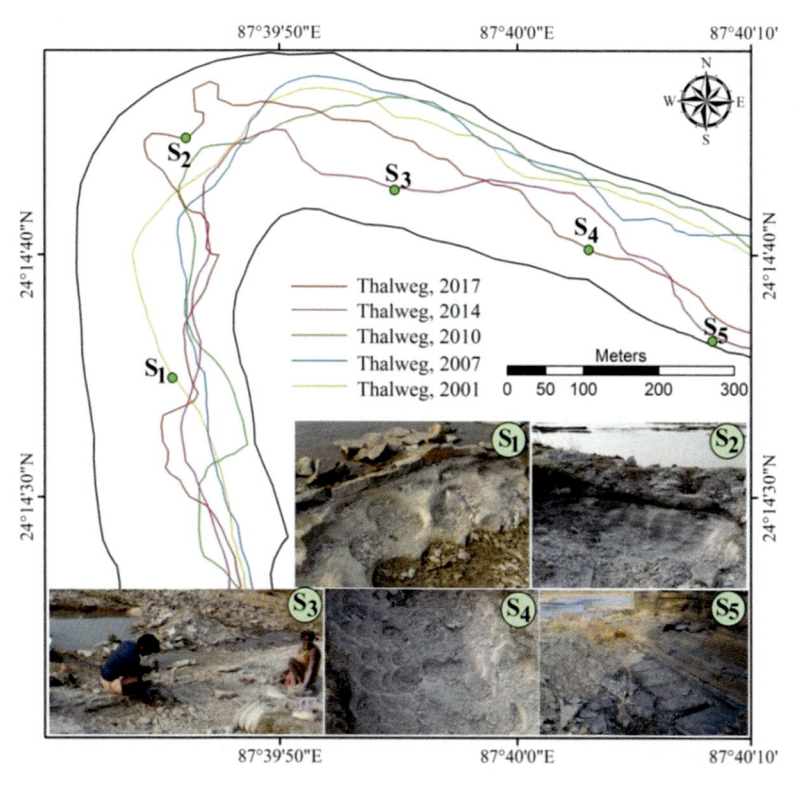

Fig. 4.8 Shifting of thalweg lines (2001–2017) (Reach 2)

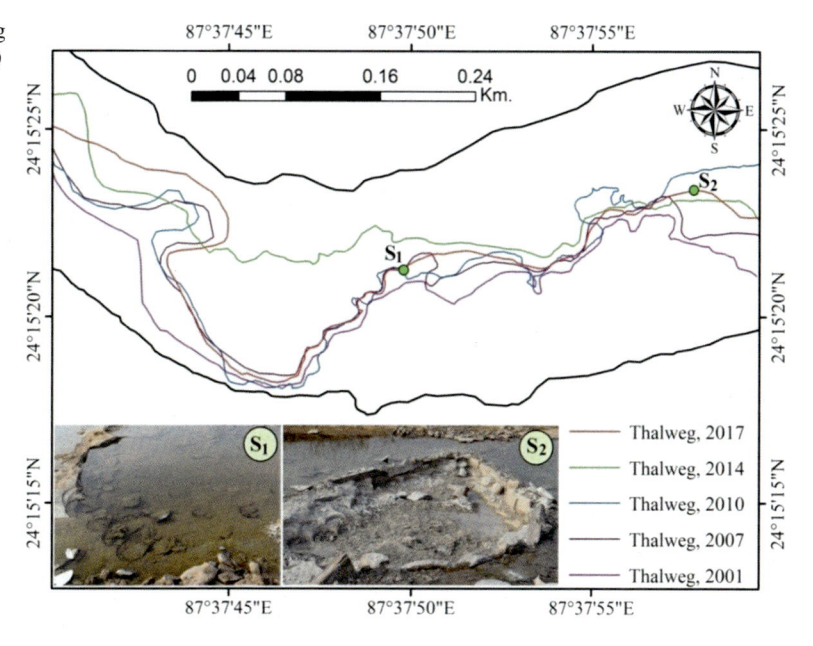

on sediment load is significant but not absolute. If the R^2 value would be near 1 or 1 it could be stated that discharge is the absolute determinant of sediment load. But as the computed value is not 1, the influences of other factors could be anticipated.

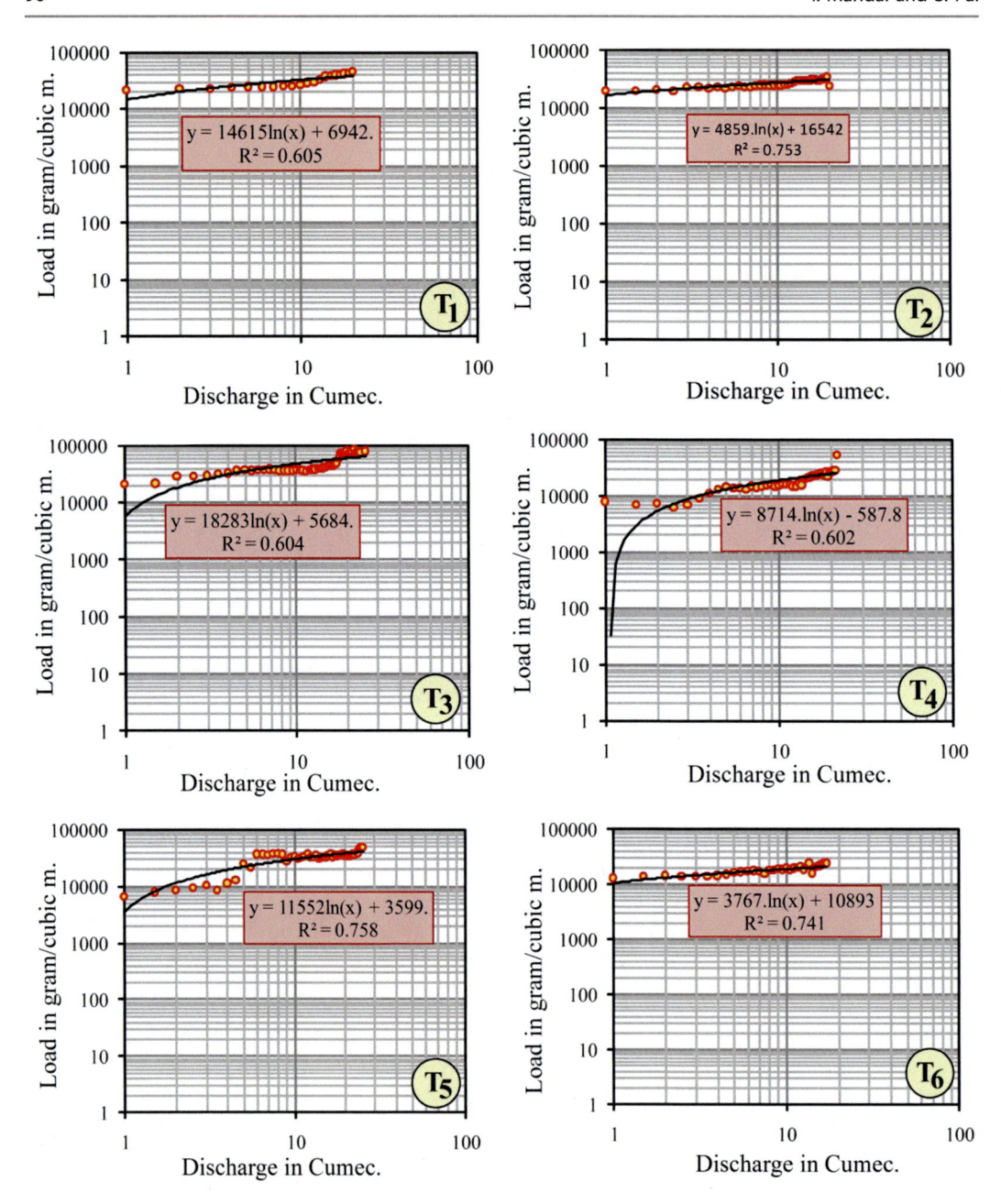

Fig. 4.9 Suspended and dissolved sediment load—(T_1) for selected tributary 1, (T_2) for tributary 2, (T_3) for the selected tributary 3, (T_4) for tributary 4, (T_5) for tributary 5, and (T_6) for the studied tributary 6

4.4.6 Impact on Ambient Water Quality

Table 4.3 shows that the selected tributaries' average water quality condition exceeds the permissible drinking and irrigation threshold set by BIS (1991). For example, the allowable pH for drinkable water is between 6.5 and 8.5 and for irrigation is 6–8.5 but for three tributaries it is observed that pH is >8.5. In all cases, BOD and COD excessively exceed the permissible limit (Table 4.3). Excessive water BOD means that the

Table 4.3 Tributary wise water quality status (physico-chemical properties) with standard permissible limits (BIS 1991)

Physico-chemical parameters	Selected tributaries						Standard permissible limit (mg/L)
	Tributary 1 (T_1)	Tributary 2 (T_2)	Tributary 3 (T_3)	Tributary 4 (T_4)	Tributary 5 (T_5)	Tributary 6 (T_6)	
PH	8.1	8.9	8.4	8.8	8.1	8.6	6.5–8.5
DO (ppm)	3.11	3.23	2.98	2.99	2.89	3.17	4.0–6.0
TDS (mg/l)	2344	2421	2398	2414	2143	1948	500
COD (mg/l)	12,745.4	13,149.4	12,632.8	12,879.4	12,360.8	12,543.2	10
BOD (mg/l)	41.6	47.8	30.3	41.3	35.7	41.8	3
Transparency (cm)	3.7	4.4	4.1	5.9	4.9	5.7	–

river and pond water are oxygen-deficient and that there is an aerobic condition there. The development of aquatic life is very detrimental to this situation. This prevents the cycle of decomposition of organic waste. The total dissolved solids (TDS) measured in the water for all levels are 4–5 times the normal level for the water quality. Water quality is 9–16 times higher in cases of biological oxygen demand (BOD) and chemical oxygen demand (COD) is 1308–1378 times enhanced than the standard limit.

4.5 Discussion

Stone crusher units specifically produce micro stone chips that can yield more dust. Micro-level stone dust particles spread over the larger part of the surrounding crusher center. According to the central pollution control board of India (CPCB 2009), the real dust production area is fairly small it is about 0.5–1 m². But the dust spreads more than 10–15 times larger and definite emission at near about 3–8 m height. Stone dust admixing to the surrounding river water contaminates the water quality (Singh et al. 2016; Okafor and Egbe 2017; Van Duc and Kennedy 2018; Pal and Mandal 2019a, b). The excessive influx of dust can change the river morphology as found in the present case. Increasing sediment load, and river bed aggradations are the direct footprint observed in the tributaries of the Dwarka River basin. Tributaries flowing across densely located crusher units are more affected by this problem. As most of the tributaries carry very scant water or remain dry during the non-monsoon period, rivers get the ambiance of dust deposition (Faershtein et al. 2016; Pal and Mandal 2017, 2019a, b). Moreover, during monsoon time discharge level of the river is not so high to carry out all the dust deposited over the river bed. Moreover, during this time a huge amount of dust was additionally admixed into the river through surface runoff. So, dust remainder within the river is a very common fact and it causes the change in river morphology (Mugade and Sapkale 2015; Ali et al. 2017). If gradually the rivers lose their carrying capacity in this way in future dust may affect wider surfaces specifically agricultural land. Analysis of river bed accretion susceptibility can give a picture of susceptible areas of dust deposition. An advanced simulation model with acceptable accuracy can guide what measures could be taken where for discharging stone dust for the sustenance of the river. Direct mining of stone from river beds creates manmade micro-level pools over there which changes the flow resistance and flow characters (Padmalal et al. 2008; Sreebha and Padmalal 2011; Preciso et al. 2012; Brunier et al. 2014; Wang et al. 2018). The laminar character of flow in most of the cases has converted into the turbulent flow (Lajeunesse et al. 2010; Martín-Vide et al. 2010; Bhuiyan et al. 2014). Excessive sediment load contributed from crusher units is majorly responsible for aggradation of river bed and change of channel morphology. Original bed topography is armored with thick stone dust. People harvest stone from the river bed in a very irregular manner based on the quality of stone and accessibility of harvesting (Padmalal et al. 2008; Kamboj et al. 2017; Sadeghi et al. 2018; Ciszewski 2019). It controls the flow path of the river. Sometimes mining from riverside is caused for riverbank failure and severe change of thalweg line. In this thalweg, a shift is given priority since river bed mining is concerned (Omoti et al. 2016; Kamboj and Kamboj 2019). Overall river shifting is not so high in the studied segment. Mossa and James (2013) reported river shifting due to stone and sand mining from river bed and soil from the river bank. It often causes loss of agricultural land settlement areas. The present study area is sparsely settled and there is very little chance of engulfing the settlement area.

Water quality change is another major concern as reported in the present study area. The enhanced degree of BOD, COD, turbidity level, and PH all can adversely affect the habitat quality of the river and species diversity and safety (Mandal and Pal 2021). Armoring river bed with stone dust creates an artificial habitat platform where species have to be re-adapted. It strongly hampers the functionalities of the primary produces (Ashraf et al. 2011; Monjezi et al. 2013; Brunier et al. 2014; Mudenda 2018). Pal et al. (2016) identified the lowering rate of fish production and breaching of some fish species. Increasing turbidity level can influence to enhance the temperature of the water which withstands against the growth of species.

Fly ash and stone dust-induced water quality change is a very common issue addressed by many scholars (Saha and Padhy 2011; Divya et al. 2012; Ozcan et al. 2012; Pal et al. 2016; Pal and Mandal 2017). But the change of river morphology like the change of sediment load, simulation models for identifying river bed accretion sites due to stone dust deposition, and shifting of thalweg due to river bed mining are some new dimensions of this work.

Few limitations of this work could be pointed out. Gauge data relating to discharge and load for a long time across different rivers can provide a strong database for analysis. Instead of only six tributaries as sample study, more number of samples from different mining and crushing landscape could provide a more comprehensive result.

4.6 Conclusions

Stone dust is often a major cause of channel bed accretion in the tributaries of the river flowing across stone quarrying and crushing units. Admixing of dust in water is responsible for increasing sediment load in water and, when it is beyond carrying capacity, it precipitates in the river bed. Assessing potential areas susceptible to dust accretion in rivers identified 6–17% areas, which represent very high susceptible zones. Direct riverbed stone mining for small-scale industries is caused for frequent diversion of the flow axis of the rivers. River water contamination due to stone dust is well identified and the level of contamination is often beyond irrigability and drinkability. Moreover, such contamination may also affect the normal ecological functioning of river water. The continuous deposition may convert the small tributaries or parts of it into the land. These examples denote the change in fluvio-geomorphic setup of the river. Considering the serviceability of the river, crusher units should be more cautious about discharging stone dust. The ancillary dust refinery industry may solve this problem to some extent as the refine dust may be used for some other industries like fertilizer, cement, etc.

Funding NA.

Conflict of Interest None.

References

Affandi FA, Ishak MY (2019) Impacts of suspended sediment and metal pollution from mining activities on riverine fish population—a review. Environ Sci Pollut Res 1–13

Ali ANA, Ariffin J, Razi MAM, Jazuri A (2017) Environmental degradation: a review on the potential impact of river morphology. In: MATEC web of conferences, vol 103, p 04001

Ashraf MA, Maah MJ, Yusoff I, Wajid A, Mahmood K (2011) Sand mining effects, causes and concerns: a case study from Bestari Jaya, Selangor, Peninsular Malaysia. Sci Res Essays 6(6):1216–1231

Barman B, Kumar B, Sarma AK (2018) Turbulent flow structures and geomorphic characteristics of a mining affected alluvial channel. Earth Surf Proc Land 43 (9):1811–1824

Bhuiyan AA, Amin MR, Karim R, Islam AKMS (2014) Plate fin and tube heat exchanger modeling: effects of performance parameters for turbulent flow regime. Int J Autom Mech Eng 9(1):1768–1781

BIS (1991) Indian standard drinking water-specification. Bureau of Indian Standard, New Delhi. Accessed Sept 2016. https://law.resource.org/pub/in/bis/S06/is.10500.1991.pdf

Breiman L (2001) Random forests. Mach Learn 45(1):5–32

Brunier G, Anthony EJ, Goichot M, Provansal M, Dussouillez P (2014) Recent morphological changes in the Mekong and Bassac river channels, Mekong delta: the marked impact of river-bed mining and implications for delta destabilisation. Geomorphology 224:177–191

Calle M, Alho P, Benito G (2017) Channel dynamics and geomorphic resilience in an ephemeral Mediterranean river affected by gravel mining. Geomorphology 285:333–346

Chen J, Li K, Chang KJ, Sofia G, Tarolli P (2015) Open-pit mining geomorphic feature characterisation. Int J Appl Earth Obs Geoinf 42:76–86

Ciszewski D (2019) The past and prognosis of mining cessation impact on river sediment pollution. J Soils Sediments 19(1):393–402

Costea M (2018) Impact of floodplain gravel mining on landforms and processes: a study case in Orlat gravel pit (Romania). Environ Earth Sci 77(4):119

CPCB (2009) Comprehensive industry document stone crushers. Central Pollution Control Board, Government of India. Series: COINDS/78/2007-08, 1.1–8.21, www.cpcb.nic.in

Das P, Let S, Pal S (2013) Use of asymmetry indices and stability indices for assessing channel dynamics: a study on Kuya river, eastern India. J Eng Comput Appl Sci 2(1):24–31

Dembowska EA, Mieszczankin T, Napiórkowski P (2018) Changes of the phytoplankton community as symptoms of deterioration of water quality in a shallow lake. Environ Monit Assess 190(2):95

Divya CM, Divya S, Ratheesh K, Volga R (2012) Environmental issues in stone crushers. Pudong, Shanghai, China. Accessed 2016 Aug 15. https://businessimpactenvironment.wordpress.com/2012/01/08/environmental-issues-in-stonecrushers/

Ercanoglu M, Gokceoglu C (2002) Assessment of landslide susceptibility for a landslide-prone area (north of Yenice, NW Turkey) by fuzzy approach. Environ Geol 41(6):720–730

Esposito G, Mastrorocco G, Salvini R, Oliveti M, Starita P (2017) Application of UAV photogrammetry for the multi-temporal estimation of surface extent and volumetric excavation in the Sa Pigada Bianca open-pit mine, Sardinia, Italy. Environ Earth Sci 76(3):103

Faershtein G, Porat N, Avni Y, Matmon A (2016) Aggradation–incision transition in arid environments at the end of the Pleistocene: an example from the Negev Highlands, southern Israel. Geomorphology 253:289–304

Festin ES, Tigabu M, Chileshe MN, Syampungani S, Odén PC (2019) Progresses in restoration of post-mining landscape in Africa. J For Res 30(2):381–396

Fu X, Wang SX, Cheng Z, Xing J, Zhao B, Wang JD, Hao JM (2014) Source, transport and impacts of a heavy dust event in the Yangtze River Delta, China, in 2011. Atmos Chem Phys 14(3):1239–1254

Fu X, Cheng Z, Wang S, Hua Y, Xing J, Hao J (2016) Local and regional contributions to fine particle pollution in winter of the Yangtze River Delta, China. Aerosol Air Qual Res 16:1067–1080

Gayen A, Pourghasemi HR, Saha S, Keesstra S, Bai S (2019) Gully erosion susceptibility assessment and management of hazard-prone areas in India using different machine learning algorithms. Sci Total Environ 668:124–138

Hauer C, Schober B, Habersack H (2013) Impact analysis of river morphology and roughness variability on hydropeaking based on numerical modelling. Hydrol Process 27(15):2209–2224

Hofler S, Piberhofer B, Gumpinger C, Hauer C (2018) Status, sources, and composition of fine sediments in upper Austrian streams. J Appl Water Eng Res 6(4):283-297

Hohensinner S, Hauer C, Muhar S (2018) River morphology, channelization, and habitat restoration. In: Riverine ecosystem management. Springer, Cham, pp 41–65

Holmes RM, Mc Clelland JW, Peterson BJ, Tank SE, Bulygina E, Eglinton TI, Staples R (2012) Seasonal and annual fluxes of nutrients and organic matter from large rivers to the Arctic Ocean and surrounding seas. Estuaries Coasts 35(2):369–382

https://www.cpcb.nic.in/National_Air_Quality_Standards.php.

ISDW (1963) Indian standard methods of test for aggregates for concrete IS: 2386 (Part III). Bureau of Indian Standards

Jakovljevic B, Paunovic K, Belojevic G (2009) Road-traffic noise and factors influencing noise annoyance in an urban population. Environ Int 35:552–556. https://doi.org/10.1016/j.envint.2008.10.001

Jha VC, Kapat S (2009) Rill and gully erosion risk of lateritic terrain in South-Western Birbhum District, West Bengal, India. J Eng Comput Appl Sci 2(1):24–31

Kamboj N, Kamboj V (2019) Riverbed mining as a threat to in-stream agricultural flood-plain and biodiversity of Ganges River, India

Kamboj V, Kamboj N, Sharma S (2017) Environmental impact of river bed mining—a review. Int J Sci Res Rev 7(1):504–519

Khalid S, Shahid M, Bibi I, Sarwar T, Shah A, Niazi N (2018) A review of environmental contamination and health risk assessment of wastewater use for crop irrigation with a focus on low and high-income countries. Int J Environ Res Public Health 15(5):895

Kim JC, Lee S, Jung HS, Lee S (2018) Landslide susceptibility mapping using random forest and boosted tree models in Pyeong-Chang, Korea. Geocarto Int 33(9):1000–1015

Lajeunesse E, Malverti L, Charru F (2010) Bed load transport in turbulent flow at the grain scale: experiments and modeling. J Geophys Res Earth Surf 115 (F4)

Lei K, Pan H, Lin C (2016) A landscape approach towards ecological restoration and sustainable development of mining areas. Ecol Eng 90:320–325

Lyons C, Carothers C, Reedy K (2016) Means, meanings, and contexts: a framework for integrating detailed ethnographic data into assessments of fishing community vulnerability. Mar Policy 74:341–350

Mamdani EH (1977) Application of fuzzy logic to approximate reasoning using linguistic synthesis. IEEE Trans Comput 26(12):1182–1191

Mandal I, Pal S (2020) COVID-19 pandemic persuaded lockdown effects on environment over stone quarrying and crushing areas. Sci Total Environ 732:139281

Mandal I, Pal S (2021) Assessing the impact of ecological insecurity on ecosystem service value in stone quarrying and crushing dominated areas. Environ Devel Sustain pp 1–25

Mariya A, Kumar C, Masood M, Kumar N (2019) The pristine nature of river Ganges: its qualitative deterioration and suggestive restoration strategies. Environ Monit Assess 191(9):542

Martín-Vide JP, Ferrer-Boix C, Ollero A (2010) Incision due to gravel mining: modeling a case study from the Gállego River, Spain. Geomorphology 117(3–4):261–271

Massi L, Maselli F, Rossano C, Gambineri S, Chatzinikolaou E, Dailianis T, Lazzara L (2019) Reflectance spectra classification for the rapid assessment of water ecological quality in Mediterranean ports. Oceanologia 61(4):445–459

Mehdi MR, Kim M, Seong JC, Arsalan MH (2011) Spatio-temporal patterns of road trafc noise pollution in Karachi, Pakistan. Environ Int 37:97–104. https://doi.org/10.1016/j.envint.2010.08.003

Milczarek W (2019) Application of a small baseline subset time series method with atmospheric correction in monitoring results of mining activity on ground surface and in detecting induced seismic events. Remote Sens 11(9):1008

Minnullina A, Vasiliev V (2018) Determining the supply of material resources for high-rise construction: scenario approach. In: E3S web of conferences 33. EDP Sciences

Monjezi M, Hasanipanah M, Khandelwal M (2013) Evaluation and prediction of blast-induced ground vibration at Shur River Dam, Iran, by artificial neural network. Neural Comput Appl 22(7–8):1637–1643

Monserud RA, Leemans R (1992) Comparing global vegetation maps with the Kappa statistic. Ecol Model 62:275–293

Mossa J, James LA (2013) Impacts of mining on geomorphic systems. In: Treatise on geomorphology, vol 13, pp 74–95. Geomorphology of human disturbances, climate change, and natural hazards. Academic Press, San Diego, CA

Mudenda L (2018) Assessment of water pollution arising from copper mining in Zambia: a case study of Munkulungwe stream in Ndola, Copperbelt province. Doctoral dissertation, University of Cape Town

Mugade UR, Sapkale JB (2015) Influence of aggradation and degradation on river channels: a review. Int J Eng Tech Res 3(6):209–212

Naghibi SA, Ahmadi K, Daneshi A (2017) Application of support vector machine, random forest, and genetic algorithm optimized random forest models in groundwater potential mapping. Water Resour Manage 31 (9):2761–2775

Nayyeri H, Zandi S (2018) Evaluation of the effect of river style framework on water quality: application of geomorphological factors. Environ Earth Sci 77 (9):343

Okafor FO, Egbe EA (2017) Models for predicting compressive strength and water absorption of laterite-quarry dust cement block using mixture experiment. Niger J Technol 36(2):366–372

Omoti K, Kitetu J, Keriko JM (2016) An assessment of the impacts of gypsum mining on water quality in Kajiado County, Kenya. Kabarak J Res Innov 4 (1):91–104

Ozcan O, Musaoglu N, Seker DZ (2012) Environmental impact analysis of quarrying activities established on and near a river bed by using remotely sensed data. Fresenius Environ Bull 21(11):3147–3153

Padmalal D, Maya K, Sreebha S, Sreeja R (2008) Environmental effects of river sand mining: a case from the river catchments of Vembanad lake, Southwest Coast of India. Environ Geol 54(4):879–889

Pal S, Debanshi S (2018) Influences of soil erosion susceptibility toward overloading vulnerability of the gully head bundhs in Mayurakshi River basin of eastern Chottanagpur Plateau. Environ Dev Sustain 20 (4):1739–1775

Pal S, Mandal I (2017) Impacts of stone mining and crushing on stream characters and vegetation health of Dwarka River basin of Jharkhand and West Bengal, Eastern India. J Environ Geogr 10(1–2):11–21

Pal S, Mandal I (2019a) Impacts of stone mining and crushing on environmental health in Dwarka river basin. Geocarto Int 1–29

Pal S, Mandal I (2019b) Impact of aggregate quarrying and crushing on socio-ecological components of Chottanagpur plateau fringe area of India. Environ Earth Sci 78(23):661

Pal S, Mandal I (2021) Noise vulnerability of stone mining and crushing in Dwarka river basin of Eastern India. Environ Dev Sustain 1–22

Pal S, Mahato S, Sarkar S (2016) Impact of fly ash on channel morphology and ambient water quality of Chandrabhaga River of Eastern India. Environ Earth Sci 75:1268

Peña-Ortega M, Del Rio-Salas R, Valencia-Sauceda J, Mendívil-Quijada H, Minjarez-Osorio C, Molina-Freaner F, Moreno-Rodríguez V (2019) Environmental assessment and historic erosion calculation of abandoned mine tailings from a semi-arid zone of northwestern Mexico: insights from geochemistry and unmanned aerial vehicles. Environ Sci Pollut Res 26 (25):26203–26215

Preciso E, Salemi E, Billi P (2012) Land use changes, torrent control works and sediment mining: effects on channel morphology and sediment flux, case study of the Reno River (Northern Italy). Hydrol Process 26 (8):1134–1148

Qi J, Liu X, Yao X, Zhang R, Chen X, Lin X, Liu R (2018) The concentration, source and deposition flux of ammonium and nitrate in atmospheric particles during dust events at a coastal site in northern China. Atmos Chem Phys 18(2):571–586

Quinn R, Avis O, Decker M, Parker A, Cairncross S (2018) An assessment of the microbiological water quality of sand dams in southeastern Kenya. Water 10 (6):708

Ramesh V, Korwar GR, Mandal UK, Prasad JV, Sharma KL, Yezzu SR, Kandula V (2008) Influence of fly ash mixtures on early tree growth and physicochemical properties of soil in semi-arid tropical Alfisols. Agrofor Syst 73(1):13–22

Rasyid AR, Bhandary NP, Yatabe R (2016) Performance of frequency ratio and logistic regression model in creating GIS based landslides susceptibility map at Lompobattang Mountain, Indonesia. Geoenviron Disasters 3(1):19

Sadeghi SH, Gharemahmudli S, Kheirfam H, Darvishan AK, Harchegani MK, Saeidi P, Vafakhah M (2018) Effects of type, level and time of sand and gravel mining on particle size distributions of suspended sediment. Int Soil Water Conserv Res 6(2):184–193

Saha DC, Padhy PK (2011) Effects of stone crushing industry on Shorea robusta and Madhuca indica foliage in Lalpahari forest. Atmos Pollut Res 2(4):463–476

Samah A, Shaffril HAM, Hamzah A, Samah B (2019) Factors affecting small-scale fishermen's adaptation toward the impacts of climate change: reflections from Malaysian fishers. SAGE Open 9(3):2158244019 864204

Sami M, Shiekhdavoodi MJ, Pazhohanniya M, Pazhohanniya F (2014) Environmental comprehensive assessment of agricultural systems at the farm level using fuzzy logic: a case study in cane farms in Iran. Environ Model Softw 58:95–108

Singh R, Rishi MS, Sidhu N (2016) An overview of environmental impacts of riverbed mining in Himalayan terrain of Himachal Pradesh. J Appl Geochem 18(4):473

Sreebha S, Padmalal D (2011) Environmental impact assessment of sand mining from the small catchment rivers in the southwestern coast of India: a case study. Environ Manage 47(1):130–140

Toy TJ, Hadley RF (1987) Geomorphology and reclamation of disturbed lands. United States. https://www.osti.gov/biblio/5769696

Trujillo-González J, Mahecha-Pulido J, Torres-Mora M, Brevik E, Keesstra S, Jiménez-Ballesta R (2017) Impact of potentially contaminated river water on agricultural irrigated soils in an equatorial climate. Agriculture 7(7):52

Turowski JM (2018) Alluvial cover controlling the width, slope and sinuosity of bedrock channels. Earth Surf Dyn 6(1):29-48

USGS (2018) Water Density, The USGS Water Science School, U.S. Department of the Interior. https://www.usgs.gov/special-topic/water-science-school/science/water-density?qt-science_center_objects=0#qt-science_center_objects. Retrived on 15th Sept 2019 at 01:43 AM

Van Duc B, Kennedy O (2018) Adsorbed complex and laboratory geotechnics of Quarry Dust (QD) stabilized lateritic soils. Environ Technol Innov 10:355–363

Wang JM, Yang XG, Zhou HW, Lin X, Jiang R, Lv EQ (2018) Bed morphology around various solid and flexible grade control structures in an unstable gravel-bed river. Water 10(7):822

Wiejaczka Ł, Tamang L, Piróg D, Prokop P (2018) Socioenvironmental issues of river bed material extraction in the Himalayan piedmont (India). Environ Earth Sci 77(20):718

Wilkinson BH, McElroy BJ (2007) The impact of humans on continental erosion and sedimentation. Geol Soc Am Bull 119(1–2):140–156

Xiang J, Chen J, Sofia G, Tian Y, Tarolli P (2018) Open-pit mine geomorphic changes analysis using multi-temporal UAV survey. Environ Earth Sci 77(6):220

Xu J, Zhao H, Yin P, Bu N, Li G (2019) Impact of underground coal mining on regional landscape pattern change based on life cycle: a case study in Peixian, China. Pol J Environ Stud 28(6)

Yanar TA (2003) The enhancement of the cell-based GIS analysis with fuzzy processing capabilities. MSs thesis. The Middle East Technical University

Youssef AM, Pourghasemi HR, Pourtaghi ZS, Al-Katheeri MM (2016) Landslide susceptibility mapping using random forest, boosted regression tree, classification and regression tree, and general linear models and comparison of their performance at Wadi Tayyah Basin, Asir Region, Saudi Arabia. Landslides 13 (5):839–856

Zadeh LA (1965) Fuzzy sets. Inf Control 8(3):338–353

An Attempt to Forecast Seasonal Precipitation in the Comahue River Basins (Argentina) to Increase Productivity Performance in the Region

Maximiliano Vita Sanchez, Marcela Hebe González, and Alfredo Luis Rolla

Abstract

Precipitation decreased in the last century in the southern Argentinian Andes mountains, especially in winter, and this trend is expected to continue in the future, generating a significant negative impact on the local and national economy. This fact is relevant since the Comahue river basin, compressed by the sub-basins of the Negro, Neuquén and Limay rivers, is characterized by fruit-horticultural production and by the presence of hydroelectric dams. As the river flows are largely influenced by the interannual variability of autumn and winter rainfall, the availability of a good seasonal forecast provides an efficient tool to minimize risks. The study addresses the analysis of the main climatic forcing of precipitation to design some statistical models to forecast autumn rainfall. Using monthly precipitation data for the period 1981–2017 from several national institutions and data of atmospheric variables from the NCEP/NCAR reanalysis, correlation maps have been used to define the best set of predictors. The multiple linear regression methodology was used to build predictive models. The main conclusion is that the South Atlantic Ocean High, the sea surface temperature of the Pacific and Indian Oceans and the polar jet are the most relevant predictors and the designed models have good skill.

Keywords

Seasonal forecast · Precipitation · Statistical models

M. V. Sanchez (✉) · M. H. González
Departamento de Ciencias de la Atmósfera y los Océanos, Facultad de Ciencias Exactas y Naturales, Universidad de Buenos Aires, Buenos Aires, Argentina
e-mail: mvitasanchez@smn.gob.ar
URL: http://perspectiva.at.fcen.uba.ar/

M. V. Sanchez
Dirección de Meteorología Aeronaútica, Servicio Meteorológico Nacional, Buenos Aires, Argentina

M. H. González · A. L. Rolla
CONICET, Centro de Investigaciones del Mar y la Atmósfera (CIMA), Universidad de Buenos Aires, Buenos Aires, Argentina

M. H. González · A. L. Rolla
CNRS, IRD, CONICET, Instituto Franco-Argentino para el Estudio del Clima y sus Impactos (UMI 3351 IFAECI), UBA, Buenos Aires, Argentina

5.1 Introduction

The Comahue river basin is located in north-western Patagonia in Argentina. It drains a total area of 140,000 km^2, almost entirely covering the provinces of Neuquén and Río Negro. The basin is made up of three sub-basins according to the Sub-secretary of Hydric Resources (SsRH): sub-basin of the Limay River (SRL), sub-basin of the Neuquen river (SRN) and sub-basin of the Negro River (SRNe). The climate in the basin varies longitudinally in a very diverse way. On the Andes Mountains, to the west of the SRN and SRL, a temperate and very humid climate is observed with an average annual rainfall that has a range of 500–1000 mm y^{-1} (Garreaud et al. 2013) with a very strong seasonal cycle with rainfall predominantly in the winter semester. The high accumulated rainfall is due to the orographic component is relevant upstream of the mountain (Roe 2005). Towards the east, there are stepped plains and precipitation decrease sharply to about 500 mm y^{-1}. This reduction in precipitation also leads to a rapid disappearance of vegetation up to the Atlantic coast (Carrasco et al. 2002). Due to the fact that in Patagonia the mountain range is lower (of the order of 1500 m) than in the North of Argentina (3000 m), baroclinic waves and their associated precipitating systems can enter the continent generating that rainfall zonal gradient.

Over time, precipitation in Argentina has changed. These results were obtained by several authors (Carril et al. 1997; Barros et al. 2000; Liebmann et al. 2004; Saurral et al. 2016). Some authors observed a negative trend in precipitation in the Comahue basin and the Patagonian Andes, specifically in the high mountain area (González and Vera 2010). This fact is relevant since the Comahue region is characterized not only by fruit and vegetable production but also by the presence of hydroelectric dams distributed along the Limay and Neuquén rivers, providing about 20% of the energy used in the country (see Fig. 5.1). Hydroelectric plants such as Piedra del Águila, El Chocón, Planicie Banderita (Cerros Colorados) and Alicurá are located there and they must be efficiently operated. All the socioeconomic activities mentioned above are directly affected by the interannual variability of rainfall in the basin (Scarpati et al. 2014; González et al. 2020).

There are many relationships between precipitation and some atmospheric forcings, focused on the El Niño Southern Oscillation (ENSO) and the Indian Ocean Dipole (IOD) (Saji et al. 1999), both being Rossby wave train generators that move south to mid-latitudes and move through the Pacific Ocean and enter Argentinian Patagonia (taking advantage of the low altitude of the Andes in that area) with associated precipitation. Another significant oscillation is the Antarctic Oscillation which has an annular pattern also known as the Southern Annular Mode (SAM) (Thompson and Wallace 2000). Previous work demonstrated the influence of SAM on the interannual variability of precipitation in South America (Silvestri and Vera 2003; Gillett et al. 2006; Reboita et al. 2009). The influence that these atmospheric forcings have on precipitation in Patagonia was studied by several authors (González and Herrera 2014; González et al. 2017) and in particular in the Comahue region methodologies for statistical prediction of precipitation were developed (González and Cariaga 2010; González and Vera 2010; González and Domínguez 2012).

It is important to mention that in the North Patagonian basin, winter precipitation is the main source of the Neuquén and Limay rivers flow, snow activity contributes to the flows but it is more important in the summer and spring seasons (Ostertag et al. 2014). In autumn, when the height of the rivers in the basin is close to the maximum allowed, it is convenient to use part of the water to generate electricity. If this is not done, there is a risk that the crowning level will be exceeded, the integrity of the dams will be compromised and the population located downstream will be endangered. On the other hand, when the water is well below the maximum allowed level in autumn, there is the possibility of not generating the necessary energy, especially if the winter (rainy season) is drier than normal. Therefore, autumn precipitation prediction provides a tool that can complement the efficient operation of dams.

Also, there is another important problem to take into account: the Third Communication of the Argentine Republic to the United Nations Framework Convention on Climate Change

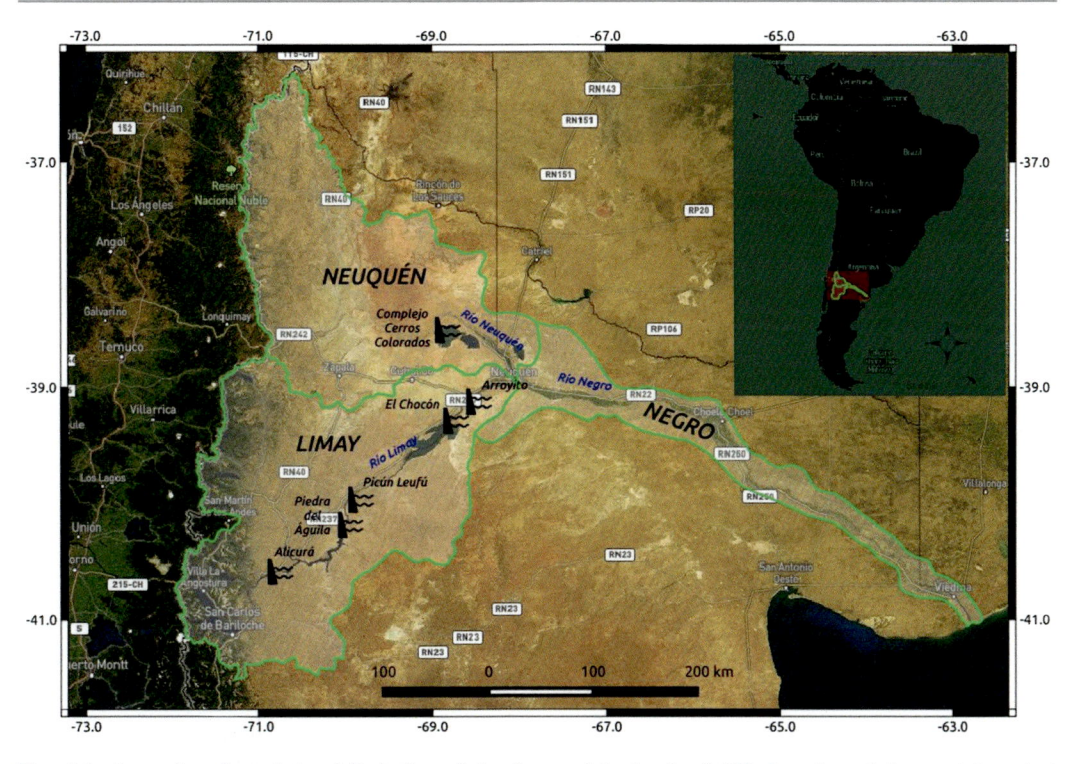

Fig. 5.1 Area of study and the delimitation of the three sub-basins by SsRH. Location of the most important hydroelectric dams in the region

(3CNUCC 2014) and the latest IPCC report (2013) asserted that the area of the Patagonian Andes not only has suffered a decrease in precipitation, but also an increase in extreme precipitation events (droughts and excesses) and a reduction in the average annual flow of up to 30% in the last 20 years. The report also details the possible future scenarios, showing a reduction of 0–10% in annual precipitation in the near future (by 2040), and of 10–20% in the distant future (by 2100).

The main objectives of this work are: to study the behaviour of the Comahue basins, to determine the relationship between the autumn rainfall (March to May, MAM) and the atmospheric and oceanic climate forcings, to define predictors in each of the Comahue sub-basins, to build statistical prediction models of MAM precipitation and to determine the efficiency of the selected models.

5.2 Data and Methodology

In the present work, data from various sources have been used. Data from the Argentine National Geographic Institute (IGN) and The National Aeronautics and Space Administration (NASA) have been used for the topographic analysis. The Shuttle Radar Topographic Mission (SRTM) (NASA JPL 2014) was available to visualize the Digital Elevation Model (DEM) of the basin in tiles of $1° \times 1°$ with a resolution of 1 arcsecond (approximately 30 m).

Accumulated precipitation data from 26 meteorological stations distributed in the Comahue basin for the period 1981–2017 from the Argentine National Meteorological Service (SMN), the SsRH, the Territorial Authority of Comahue basin (AIC) and the National Institute of Agricultural Technology (INTA) have also been used to achieve greater spatial coverage observed in Fig. 5.2 and the source and location of the stations are detailed in Table 5.1. It is necessary to mention the scarce station density in the SRNe where data from only 2 stations was available. The record from the 1981–2010 period has been used to select the predictors and train the models, whilst the 2011–2017 period has been used for verification.

Data used in this study presented less than 10% of missing data which was completed with the monthly average. The outlier or inconsistent values were detected and not considered. To do that, inconsistent data was compared with nearby stations and an outlier value was considered when it exceeded above 4 times the standard deviation of the monthly mean.

To find circulation patterns that affect rainfall in the Comahue basin, a correlation field between MAM precipitation anomalies and some meteorological variables in the previous month (February) were calculated. The variables, obtained from the NCEP-NCAR reanalysis (Kalnay et al. 1996), were sea surface temperature (SST), geopotential height at different levels (g850, g500, g250), zonal (u250, u500, u850) and meridional components of the wind (v500, v850) in middle and lower levels, sea level pressure (SLP), outgoing longwave radiation (OLR) and precipitable water in the column of the atmosphere. Correlations with an absolute value greater than 0.37 were considered statistically significant with 95% confidence, using a Normal test. The correlation fields between precipitation and the previously mentioned variables were used to define predictors in those areas that presented a significant correlation. The predictors were defined as the average of the variable in those significant areas and then a set of predictors independent of each other were considered to build some statistical models using the linear multiple regression (LMR) method.

Linear multiple regression (LMR) is a statistical technique that uses several explanatory variables to predict the outcome of a variable response. The objective of the LMR is to model the linear relationship between the explanatory variables (independent) and the response variable

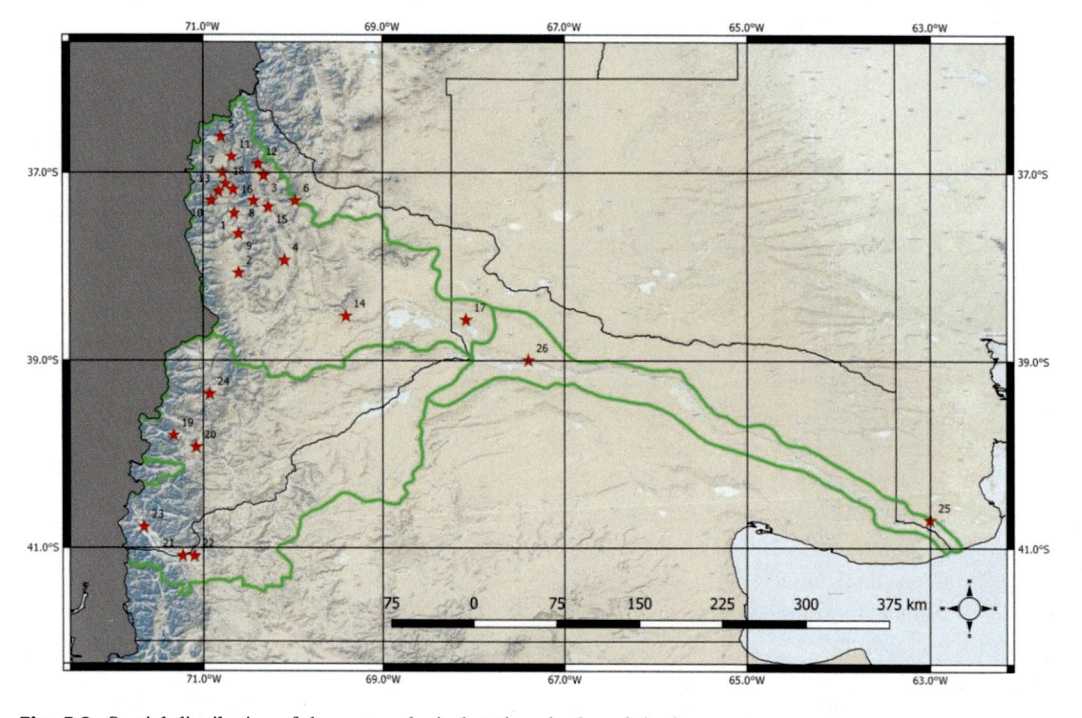

Fig. 5.2 Spatial distribution of the meteorological stations in the sub-basins

Table 5.1 Source and location of the meteorological stations used

Id	Station name	N° Est.	Longitude	Latitude	Source
1	EL CHOLAR	2011	−70.7	−37.4	SsRH
2	LONCOPUE	2014	−70.6	−38.1	SsRH
3	TRICAO MALAL	2079	−70.3	−37.0	SsRH
4	CHORRIACA	2089	−70.1	−37.9	SsRH
5	PICHI NEUQUEN	2091	−70.8	−36.6	SsRH
6	AUQUINCO	2092	−70.0	−37.3	SsRH
7	LAS OVEJAS	2076	−70.8	−37.0	SsRH
8	EL ALAMITO	2072	−70.4	−37.3	SsRH
9	EL HUECU	2078	−70.6	−37.7	SsRH
10	CHOCHOY MALLIN	2085	−70.9	−37.3	SsRH
11	VARVARCO	2077	−70.7	−36.8	SsRH
12	CAJON CURILEUVU	2071	−70.4	−36.9	SsRH
13	LOS MICHES	2080	−70.8	−37.2	SsRH
14	PASO DE INDIOS	2004	−69.4	−38.5	SsRH
15	CHOS MALAL	2013	−70.3	−37.4	SsRH
16	ANDACOLLO	2036	−70.7	−37.2	SsRH
17	NEQUEN	87715	−68.1	−38.6	SMN
18	LOS CARRIZOS	S/N	−70.8	−37.1	AIC
19	MALLEO	2032	−71.3	−39.8	SsRH
20	JUNIN DE LOS ANDES	2040	−71.1	−39.9	SsRH
21	LOS COHIHUES	1838	−71.2	−41.1	SSRH
22	BARILOCHE	87765	−71.1	−41.1	SMN
23	VILLA ANGOSTURA	S/N	−71.7	−40.8	AIC
24	RAHUE	S/N	−70.9	−39.4	AIC
25	VIEDMA	87791	−63.0	−40.7	SMN
26	INTA-ALTO VALLE	1440	−67.4	−39.0	INTA

(dependent). Here, the explanatory variables will be the significant predictors for each sub-basin, whilst the response variable will be the precipitation anomalies for MAM in each sub-basin.

In the LMR it is assumed that the regression function that relates the dependent variable with the independent variables is linear.

$$\widehat{Y}_l = \beta_0 + \beta_1 X_1 + \beta_2 X_2 + \cdots + \beta_k X_k \quad (5.1)$$

where k is the number of predictors used to model the variable. The regression coefficients are chosen so that the sum of squares between y_i and $\widehat{y}_l$ is minimal (SSE) as can be seen in Eq. 5.2. This difference is called residual, so in the process of choosing the parameters, the residual variance $\widehat{y}_l$ will be minimized.

$$\begin{aligned} \text{SSE} &= \sum_{i=1}^{n} e_i^2 = \sum_{i=1}^{n} (\widehat{y}_l - y_i)^2 \\ &= \sum_{i=1}^{n} \big([b_0 + b_1 x_{1,i} + \cdots + b_k x_{k,i}] - y_i\big) \end{aligned}$$
(5.2)

When using this method, some assumptions are made: linearity between the dependent variable and the other independent variables, homoscedasticity, that is, all the disturbances have the same variance, the random disturbances are independent of each other, the random disturbance has a normal distribution and the predictors are obtained without measurement error.

One possibility to obtain the models for each of the sub-basins would be to consider all the predictors found, however, this has great possibilities of producing overfitting (Wilks 2010), the algorithm fits the training data set so well that the noise and singularities of the training data are memorized but performance decreases when testing with an unknown data set. Overfitting the training data leads to a deterioration of the generalizability properties of the model and results in unreliable performance when applied to novel measurements. The amount of data used for the learning process is critical in this context. Small data sets are more prone to overfitting than large data sets (Santos et al. 2009).

Therefore, to minimize this phenomenon, several sets of linearly independent predictors were considered using the correlation matrix between the different predictors for each sub-basin. The Pearson coefficient lower than 0.37 in modulus determines the independence amongst predictors with a 95% confidence level applying a T-student test. This also avoids potential multicollinearity problems (Hyndman and Athanasopoulos 2018). Another factor that was taken into account was the fact that although it is possible to find models with all the sets of predictors, it does not mean that the one with the greatest number of predictors is the best model. Various techniques have been studied to find out the number of predictors required to obtain good performance models. The forward stepwise technique (James et al. 2013) consists of adding the predictors that improve the regression until the difference produced is imperceptible. Another criterion for the selection of predictors is the backward stepwise method that works in the opposite direction. The initial model uses all the predictors and in each step their amount decreases, being removed one by one until residuals are minimized. Another technique was used in this work, which consists of not ruling out all possible combinations but maintaining them and then evaluating which the best combination for each set was.

Different parameters were used to select the best models (Hyndman and Athanasopoulos 2018), such as the adjusted coefficient of determination (Adj R^2), defined in Eq. 5.3. These coefficients were used to minimize the problem of adding predictors which tends to increase the value of the coefficient determination even if they are irrelevant.

$$\text{Adj} R^2 = 1 - (1 - R^2)\frac{n-1}{n-k-1} \quad (5.3)$$

where n is the size of the training sample and k is the number of predictors. The value of Adj R^2 represents the percentage of precipitation variance that is explained by the model.

Another parameter considered was the Akaike Information Criterion (AIC)

$$\text{AIC} = n \log\left(\frac{\text{SSE}}{n}\right) + 2(k+2) \quad (5.4)$$

This coefficient measures the goodness of fit based on the verisimilitude of the model and the complexity based on the number of parameters. The complexity in the AIC is given by k, which is the number of model parameters. The $2k$ penalty of the AIC is equivalent to doing the cross-validation of the model leaving one item of data out (leave one out-cross validation). Since the AIC tends to overestimate the number of predictors needed when n is small, Hurvich and Tsai developed a similar coefficient for (Hurvich and Tsai 1989) trying to solve the aforementioned problem, it is called the Information Criterion of Akaike corrected (AICc) given by Eq. (5.5).

$$\text{AICc} = \text{AIC} + \frac{2(K+2)(K+3)}{n-k-3} \quad (5.5)$$

This coefficient manages to smooth out the bias problem and tends to provide more adequate models for which they only modified the penalty term.

Another parameter used to measure the efficiency of the models was the Schwarz Bayesian Information Criterion (BIC). It is also based on maximum likelihood and it is defined as:

$$\text{BIC} = n \log\left(\frac{\text{SSE}}{n}\right) + (k+2) \log n \quad (5.6)$$

Here the complexity measure incorporates both k and $\ln(n)$. This makes the indicator independent of the sample size and makes it penalize complexity more than the AIC.

As a result, the BIC generally selects the most abstract, simplest model and the one that makes predictions in less detail, whilst the AIC finds a more complex and pragmatic model that makes predictions in greater detail but within our data.

Therefore, the models that have low AIC, AICc and BIC values will be the ones that best model the training series.

For the quantitative analysis, various techniques were used to evaluate the efficiency of the models. New parameters were calculated and different diagrams were used to carry out the analysis. The Taylor diagram (Taylor 2001) provides a concise statistical summary of the efficacy measures in terms of correlation (R^2), standard deviation (SDEV) and the mean square error (RMSD) given by:

$$\text{RMSD} = \sqrt{\frac{\sum_{i=1}^{n}(\widehat{y}_i - y_i)^2}{n}} \quad (5.7)$$

This diagram is especially useful for evaluating complex models. It can be used to summarize the relative merits of a collection of different models. The reference point, in the case of this work, corresponds to the precipitation anomalies for each of the sub-basins in the study period.

The Receiver Operating Characteristic (ROC) curve was built to determine the best models. For this, it is necessary to calculate the Probability of Detection (POD) and the Probability of False Alarm (POFD) (Eqs. 5.8 and 5.9). The observed and forecasted values were categorized into three classes: below normal (less than the first tercile of observed values), normal (between first and second tercile) and above normal (greater than the second tercile). POD and POFD are derived from a contingency table (Table 5.2) constructed for each event (below normal, normal and above normal) as follow:

$$\text{POD} = \frac{\text{True positive}}{\text{True positive} + \text{False negative}} \quad (5.8)$$

$$\text{POFD} = \frac{\text{False positive}}{\text{True negative} + \text{False positive}} \quad (5.9)$$

5.3 Results and Discussion

5.3.1 Morphological Analysis of the Basins

The climate of the northern Patagonian region varies from the high mountains in the west to the semi-desert steppe in the east near the Atlantic coast. In the DEM (Fig. 5.3) it is possible to distinguish the height differences that predominate over the basin. The Neuquén River has an approximate length of 420 km with an altitude of approximately 2000 m. Throughout its mountain range, it receives the contribution of a large number of tributaries. It is in this region where the river has the maximum slope, since after crossing the town of Andacollo (1000 m.s.m) the slope begins to descend at a slower rate. The Limay river has its source in the Nahuel Huapi lake at an altitude of approximately 700 m, and it is fed by other numerous lakes. It has an approximate length of 500 km and a drop of 500 m. When it meets the Neuquén River, the Negro River is formed, which is the longest of the three rivers in the Comahue basin system, with a length of 635 km. It is an allochthonous channel as it has no tributaries and the water comes from more humid places. It flows into the Atlantic Ocean and it is the one with the least slope in the region.

Several morphological parameters were calculated for each sub-basin: the area, the perimeter, the length of the main channels (Lc), the form factor (Kf) (Horton 1932) and the Gravelius coefficient of compactness or Gravelius index (Kc) (Gravelius 1914) (Table 5.3).

Hypsometric curves (HC) (Strahler 1952) (Fig. 5.4) allow us to know the mass distribution in the basin. It is a graph of normalized height concerning the maximum height of the basin as a

Table 5.2 Contingency table of calculation of ROC curve

Contingency table		Observation		
		Yes	No	Total
Prediction	Yes	True positive	False positive	Prediction yes
	No	False negative	True negative	Prediction no
	Total	Observation yes	Observation no	Total

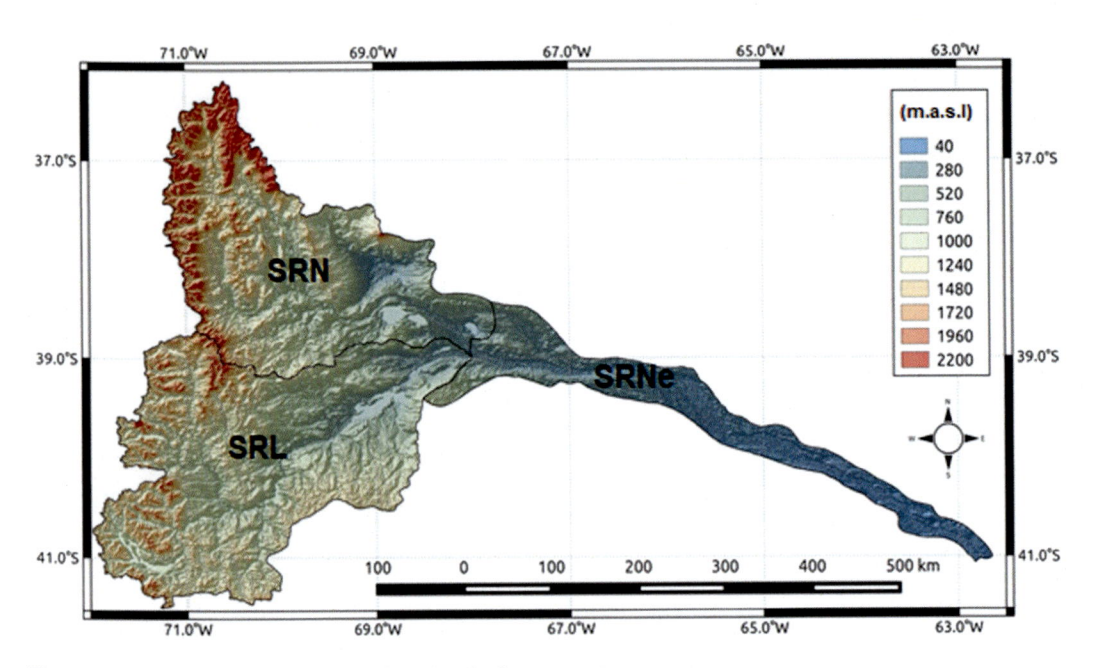

Fig. 5.3 Topographic elevation of the Comahue basins

Table 5.3 Main morphometric parameters of the Comahue basin

Basin	Area (km²)	Perimeter (km)	Lc (km)	Kc	Kf	HI
SRN	50,831.44	1366.64	420.00	1.70	0.29	0.52
SRL	63,682.07	1544.18	380.00	1.71	0.44	0.53
SRNe	20,956.76	1414.82	550.00	2.74	0.07	0.47

function of the accumulated area as a proportion of the total area of the basin. In this way, it is possible to compare the curves of the sub-basins with a theoretical curve. The area under the curve (HI) indicates the stage of development (Llamas 1993). If HI is greater than 0.6 the basin is in a phase of "youth" imbalance and potentially erosive sediment production predominates, if HI is between 0.352 and 0.6 the basin is in a phase of "maturity" equilibrium, whilst values lower than 0.352 indicate "old" basins that have suffered greater erosion (Strahler 1952). The values obtained in the Comahue region show that the HIs all correspond to a stage of maturity, although for SRNe it is in a stage of maturity—old age.

Regarding the morphometric parameters, Kc (Eq. 5.10) indicates the circularity of the basin and it is calculated as the relationship between the perimeter (P) of the basin and that of a circle with the same area (A).

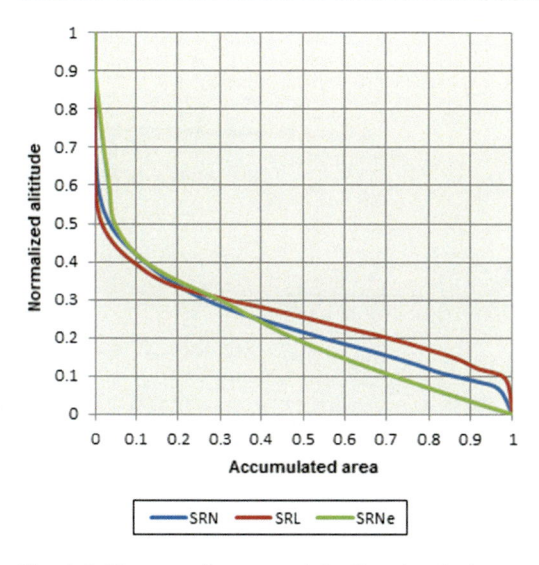

Fig. 5.4 Hypsometric curves of the Comahue basins

$$\text{Kc} = \frac{P}{2\sqrt{\pi A}} \quad (5.10)$$

Therefore, SRN and SRL are oblong basins with Kc values of 1.7 and 1.71, respectively meanwhile SRNe is a rectangular one as Kc is greater than 1.75. To measure how long, the basin is, the Kf (Eq. 5.11) is calculated as the relation between the area and the length of the main channel (Lc) squared.

$$\text{Kf} = \frac{A}{L_c^2} \quad (5.11)$$

It can be seen that the smaller the Kf, the more elongated the sub-basin will be. This is the case of SRNe, with a Kf value of 0.07 For SRN and SRL, the Kf value indicates that they are neither elongated nor widened sub-basins, although SRN tends to be more elongated than SRL.

In general terms, the sub-basins that are more widened have a greater susceptibility to generating floods, since the travel time of the water through the sub-basin is shorter than in elongated sub-basins. Therefore, those that are elongated will have a longer concentration time, contributing to the fact that the flood peaks are less problematic in the event of storms or rains that are concentrated in the sub-basin (Smith and Stopp 1978; Bell 1999).

5.3.2 Climatological Analysis of the Limay and the Neuquen River Basins

Monthly mean precipitation for each of the stations in SRL and SRN for 1981–2010 period was built (Figs. 5.5 and 5.6 respectively). It can be seen that the maximum precipitation occurs in the autumn and winter months and summer is the dry season when the accumulated monthly precipitation does not reach 25 mm.

Precipitation regimes vary considerably depending on their geographic location (Reboita et al. 2010; Garreaud et al. 2013; González et al. 2015) showing a west–east gradient. A greater difference is observed amongst the SRL stations, whilst in SRN the behaviour is more homogeneous. The whole area is dominated by air masses that come from the Pacific Ocean, the presence of the Pacific High, located at 30° S approximately and further south the westerlies are present.

The seasonal movements of the low- and high-pressure systems dominate the precipitation regime throughout the year. In particular, the westerlies and the Pacific High intensify in winter, favouring the entry of precipitating systems from the west. Besides, the SRN is also affected by the entry of air masses from the Atlantic Ocean. The orographic precipitation component in the high mountains also favours precipitation in stations such as Villa La Angostura, as can be seen in Fig. 5.5, registering higher accumulations than the rest of the stations.

The mean areal precipitation in MAM (autumn) was calculated for each sub-basin (Fig. 5.7), showing a decrease of precipitation, as other authors have already recorded (González and Vera 2010). However, non-significant trends were found at 95% confidence by applying a Mann–Kendall test (Mann 1945; Kendall 1975; Gilbert 1987). The MAM precipitation anomalies were calculated (Table 5.4) showing that SRL has a higher precipitation range than Neuquén. Regarding the median, SRL registered more humid events than normal meanwhile this is not the case for SRN and the standard deviation in SRL is higher than in SRN.

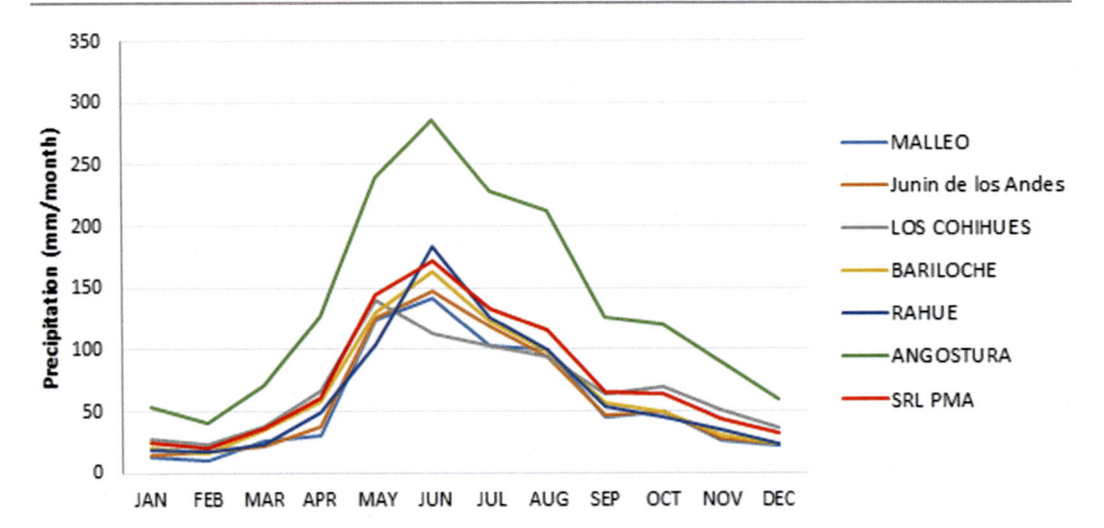

Fig. 5.5 Annual cycle of monthly precipitation in SRL stations

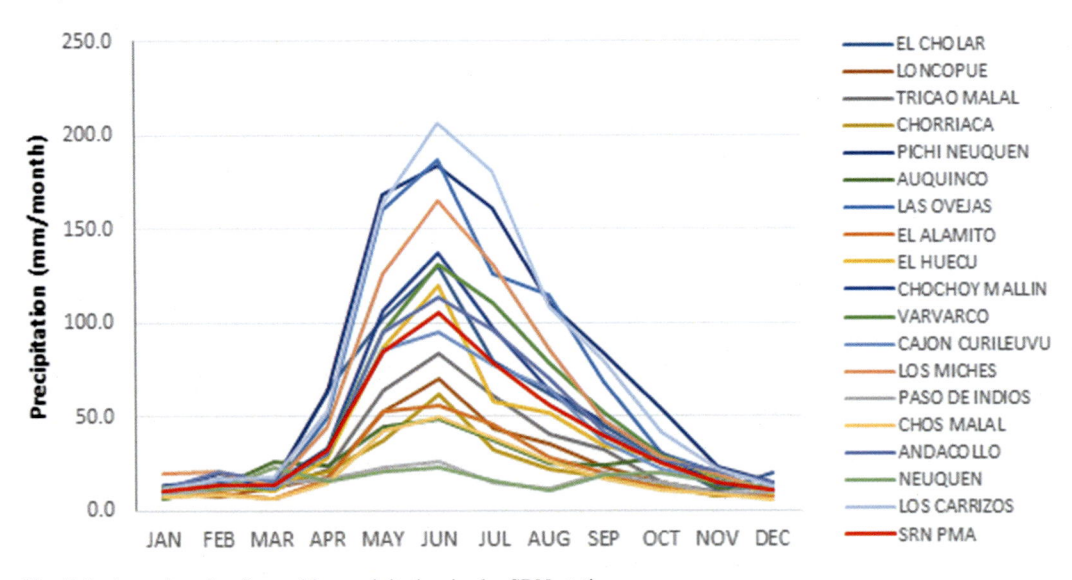

Fig. 5.6 Annual cycle of monthly precipitation in the SRN stations

Therefore, although the sub-basins are similar in the morphological sense, the variability of precipitation shows differences for the period studied.

5.3.3 Climatological Analysis of the Negro River Basin

Only two stations are located in SRNe: Viedma and INTA—Alto Valle (Fig. 5.8). The former is located near the Atlantic coast, with maximum accumulated monthly precipitation in March and in October, both do not exceed 55 mm. This region is affected by the entry of air masses from the Atlantic Ocean due to the influence of the Atlantic High (AAO) (Garbarini et al. 2019) causing precipitation to be more evenly distributed on a seasonal scale, compared to SRN and SRL.

The latter, INTA_Alto Valle station, is close to the source of the Río Negro,

Fig. 5.7 MAM precipitation anomalies for the period 1981–2010 in SRN (**a**) and SRL (**b**)

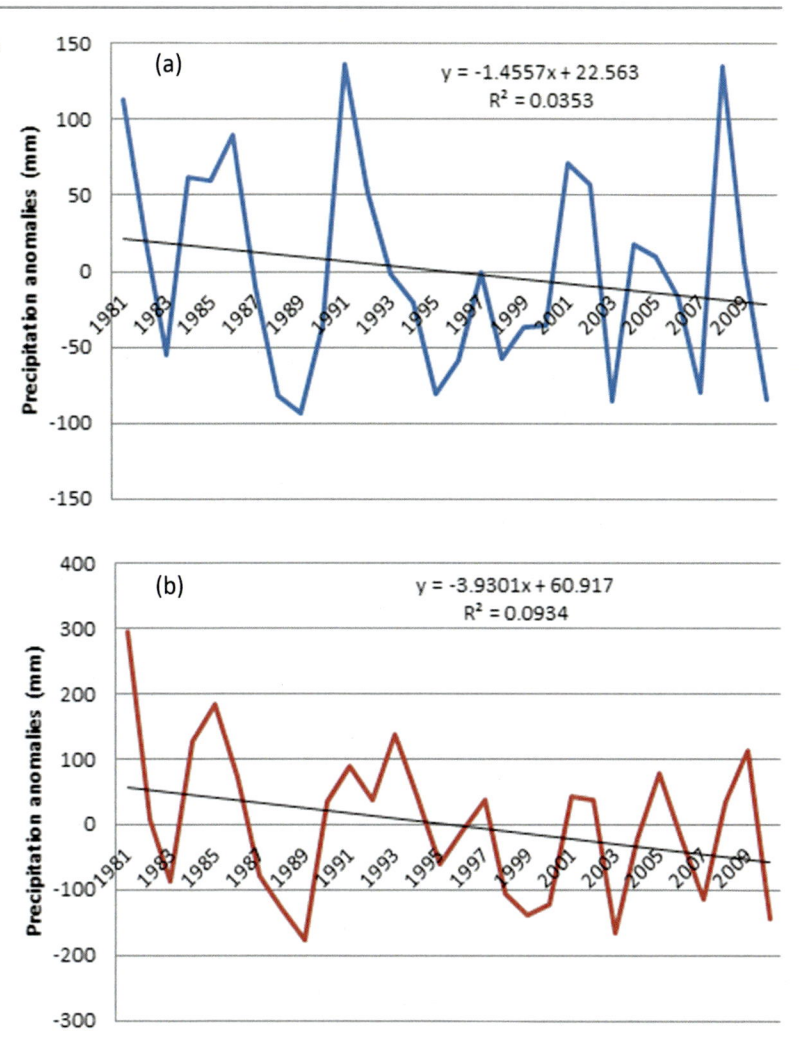

(a) $y = -1.4557x + 22.563$; $R^2 = 0.0353$

(b) $y = -3.9301x + 60.917$; $R^2 = 0.0934$

Table 5.4 Main statistical values of autumn precipitation anomalies for SRN and SRL in the period 1981–2010

Statistical values (mm)	SRN	SRL
Minimum	−93.27	−177.18
Maximum	136.61	294.95
Median	−6.24	20.87
Standard deviation ($n - 1$)	68.17	113.21

influenced by the distance to the Andean mountain range (Garreaud et al. 2013; González et al. 2015), precipitation is maximum in March and reaches only 30 mm. In both cases, precipitation presents an annual cycle with maximums in autumn and summer and it is lower than those registered in SRL and SRN. As for SRL and SRN, MAM precipitation anomalies in SRNe show a non-significant negative linear trend with 95% confidence (Fig. 5.9).

The precipitation anomaly series, averaged over each sub-basin, were calculated and a box plot was built to compare them. The maximum and minimum values of precipitation anomalies, the 1st quartile, the 3rd quartile and the median are represented in Fig. 5.10. In terms of symmetry, SRL is the most asymmetric in contrast to SRN

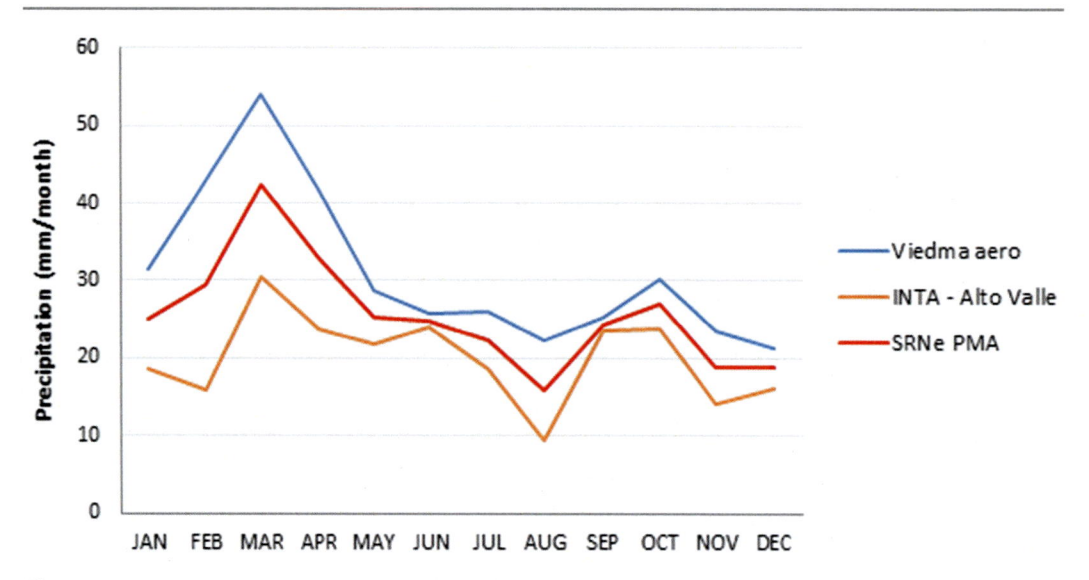

Fig. 5.8 Annual cycle of monthly precipitation for SRNe stations

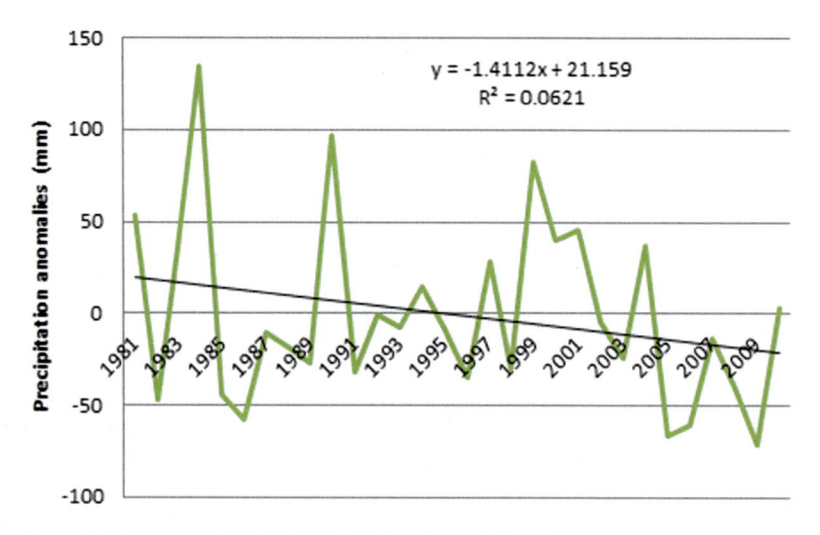

Fig. 5.9 MAM precipitation anomalies for the period 1981–2010 in SRNe

and SRNe. Greater variability is observed in SRL and SRN whilst SRNe presents lower values.

5.3.4 Relationship Between MAM Precipitation and Previous Climate Patterns

To find the predictors to be used as input in the generation of the statistical models, the correlation fields between the MAM precipitation anomalies for each sub-basin and several atmospheric variables one month before (February) have been calculated. Only the most relevant fields are shown in this work (Figs. 5.11, 5.12 and 5.13).

In general, a great similarity can be identified between the lagged correlation fields for the SRN and SRL, whilst for the SRNe no such similarity is appreciated. Western (eastern) zonal wind anomalies at 250 hPa in southern Argentina (40° S–50° S) in February are related to rainy (dry) seasons in SRN and SRL (Figs. 5.11a and

Fig. 5.10 Box diagram for the Comahue basins. The boxes are limited by the 1st and 3rd quartiles respectively, the horizontal bar is the median, the vertical bars are the maximum and minimum values of the series

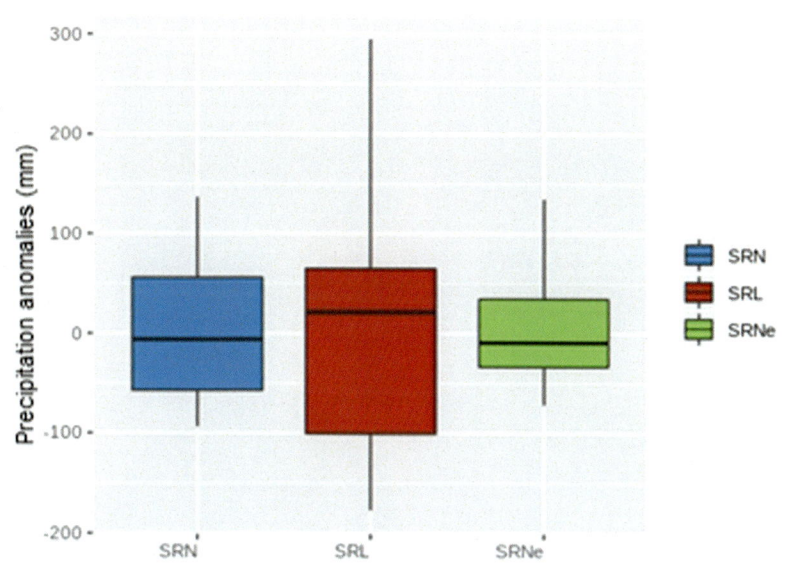

5.12a), the same happens at lower levels (Figs. 5.11b and 5.12b). In SRN and SRL, it was found that positive southern wind anomalies at low levels on the southern coast of Chile are related to rainiest autumns, since the entry of humid air from the south is favoured in that region where the altitude of the mountain range is lower (Figs. 5.11c and 5.12c). Meanwhile, an opposite pattern is observed in SRNe, associated with the circulation in the Atlantic Ocean (Fig. 5.13c). However, these configurations are shifted towards the north in SRNe (Fig. 5.13a, b). This result agrees with Berman et al. (2012) who showed the influence of southern hemisphere circulation on seasonal precipitation in Patagonia. They found a significant correlation between westerly winds in western Patagonia, whilst the opposite behaviour was detected in eastern Patagonia. Zonal (Fig. 5.13b) and meridional (Fig. 5.13c) wind anomalies over the South Atlantic Ocean, strongly related to the intensification (weakening) of the Atlantic High, are associated with positive (negative) MAM precipitation anomalies in SRNe. Some authors have pointed out that when the Atlantic High is displaced towards the east of its mean position, positive rainfall anomalies occur over Patagonia (Mayr et al. 2007; Garbarini et al. 2019).

Positive 500 hPa geopotential height anomalies covering Argentina (30 °S) are associated with positive precipitation anomalies in SRL and SRN (Figs. 5.11d and 5.12d). Negative correlations are also observed in the south of the country, this pattern implies an intensified (weakened) west flow on SRN and SRL, associated with positive (negative) precipitation anomalies. The structure shown in Figs. 5.11d and 5.12d resembles the "South Pacific American Pattern" (PSA1) (Mo and Higgins 1998), showing a Rossby wave trend over the Pacific Ocean which extends from Australia to southern South America. On the other hand, a different pattern is observed in the SRNe: the centre of negative correlation further north, covering a larger area of the country (Fig. 5.13d) and weakened subpolar lows over the Pacific Ocean indicate a weakening (intensification) of the westerly winds, associated with rainy (dry) autumn. This pattern resembles the SAM negative phase (Thompson and Wallace 2000; Marshall 2003). Some authors studied the influence of SAM over precipitation in South America (Reboita et al. 2009; González and Vera 2010) and they pointed out that a positive (negative) phase of SAM is associated with negative (positive) rainfall anomalies in autumn over Patagonia.

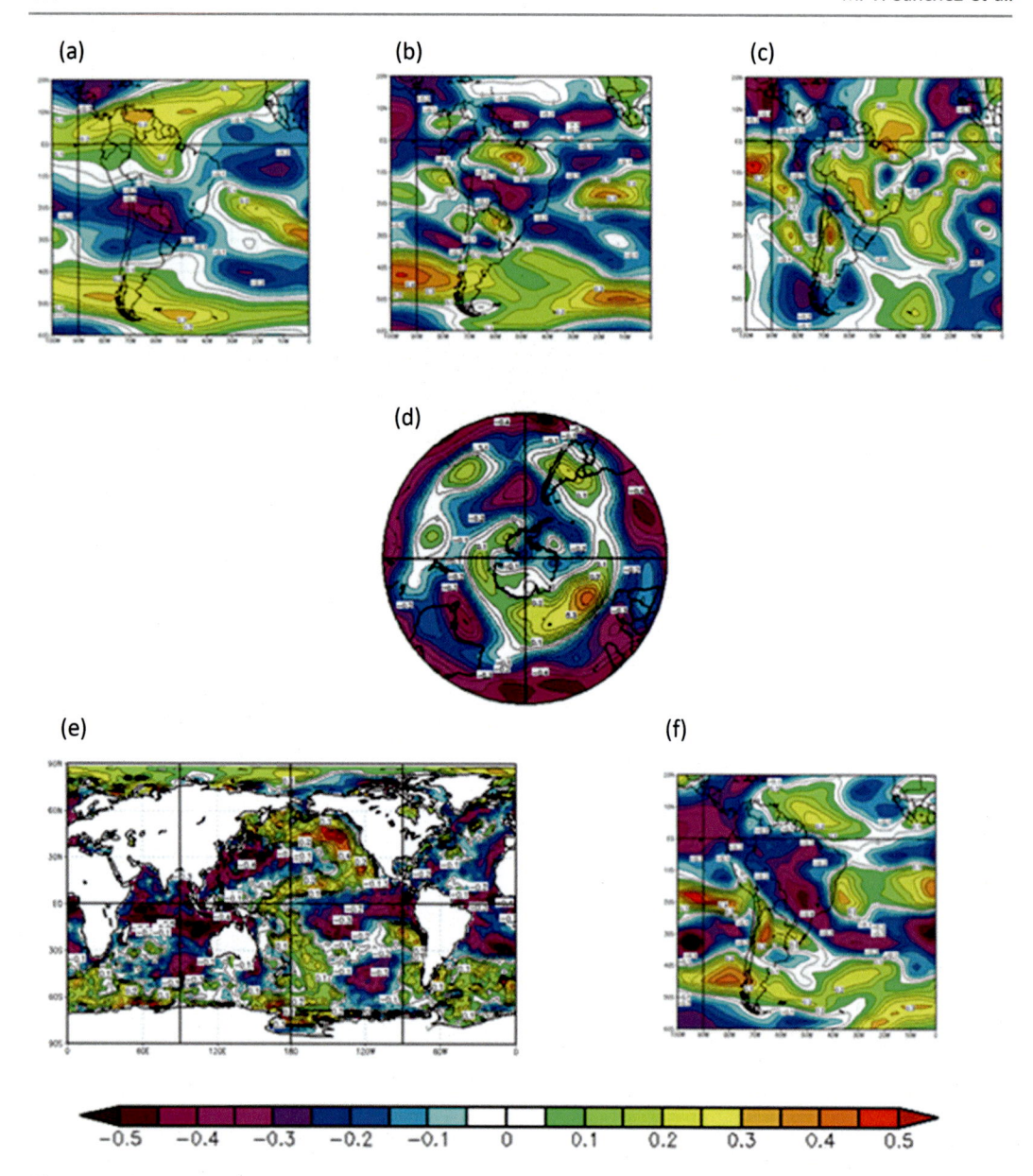

Fig. 5.11 Correlation fields between MAM precipitation anomalies in SRL and climate variables in the previous February: **a** Zonal wind at 250 hPa; **b** zonal wind at 850 hPa; **c** meridional wind at 850 hPa; **d** geopotential height at 500 hPa; **e** sea surface temperature; **f** precipitable water

ENSO is one of the most relevant modes of interannual variability in South America and especially in Argentina. This oscillation affects precipitation in distant regions through atmospheric teleconnections (Diaz and Markgraf 2009; Garreaud et al. 2008). Figures 5.11e and 5.12e show the correlation field between MAM precipitation anomalies in SRL and SRN and SST in February. The cold (warm) ENSO phase in February is associated with the rainiest (the driest) autumns, agreeing with the results found by Garreaud (2009). The signal is opposite in the case of SRNe (Fig. 5.13e) indicating that a warm ENSO phase (Niño) and a cold ENSO phase (Niña) are linked

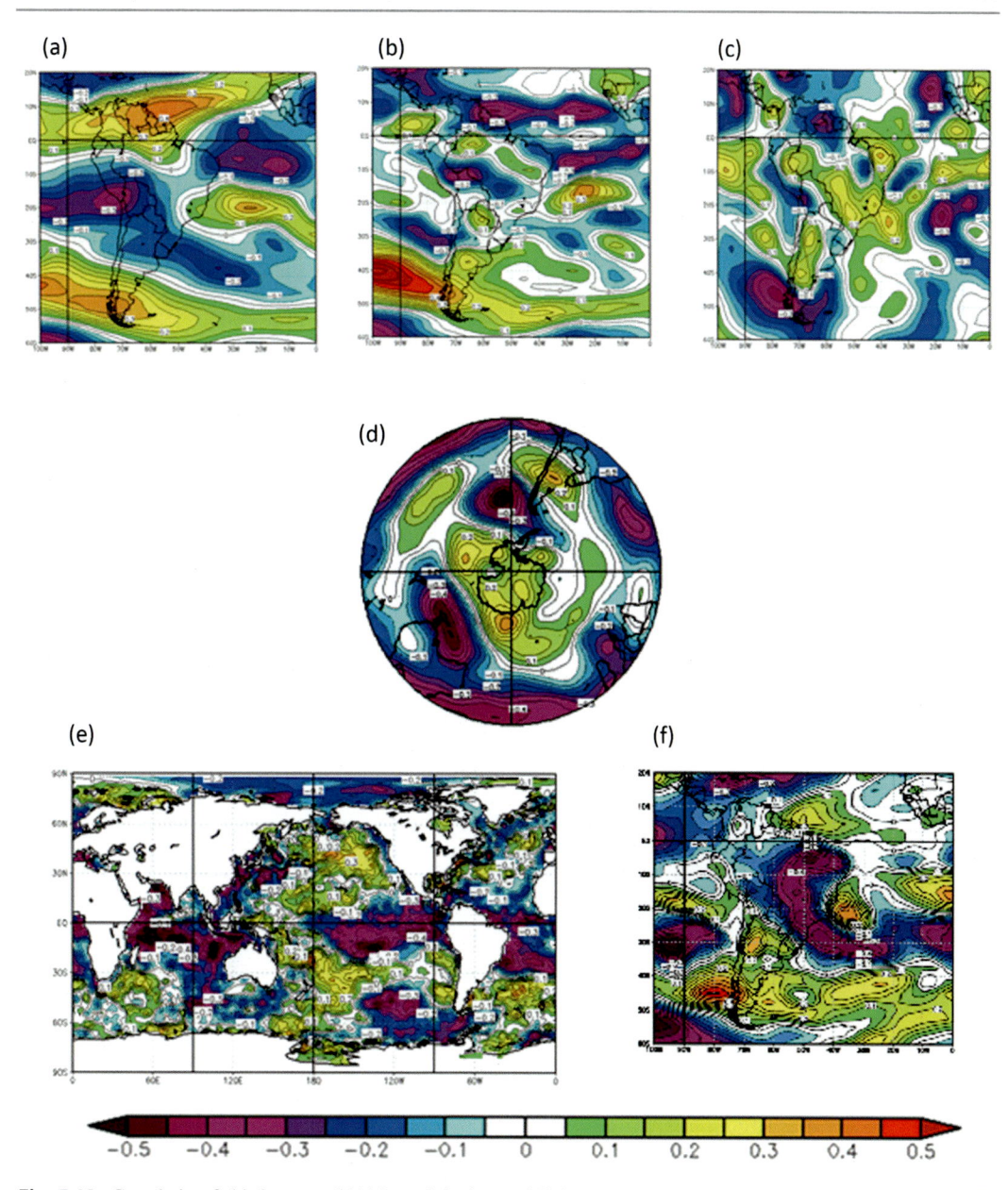

Fig. 5.12 Correlation fields between MAM precipitation anomalies in SRN and climate variables in the previous February: **a** Zonal wind at 250 hPa; **b** zonal wind at 850 hPa; **c** meridional wind at 850 hPa; **d** geopotential height at 500 hPa; **e** sea surface temperature; **f** precipitable water

with positive and negative rainfall anomalies, respectively. This result highlights the difference in the response between the sub-basins. SST anomalies over the Indian Ocean are an important climate forcing too, as they activate Rossby wave trains that travel southward from the tropics to mid-latitudes (Mo and Paegle 2001; Saji et al. 1999) defined the Indian Ocean Dipole (DOI) as two centres of opposite SST anomalies in the northeast and southwest of the Indian ocean. Furthermore, Taschetto and Ambrizzi (2012) defined the generalized cooling or warming of the whole ocean. Figures 5.11e and 5.12e show that a cooling (warming) of the Indian Ocean is related to positive

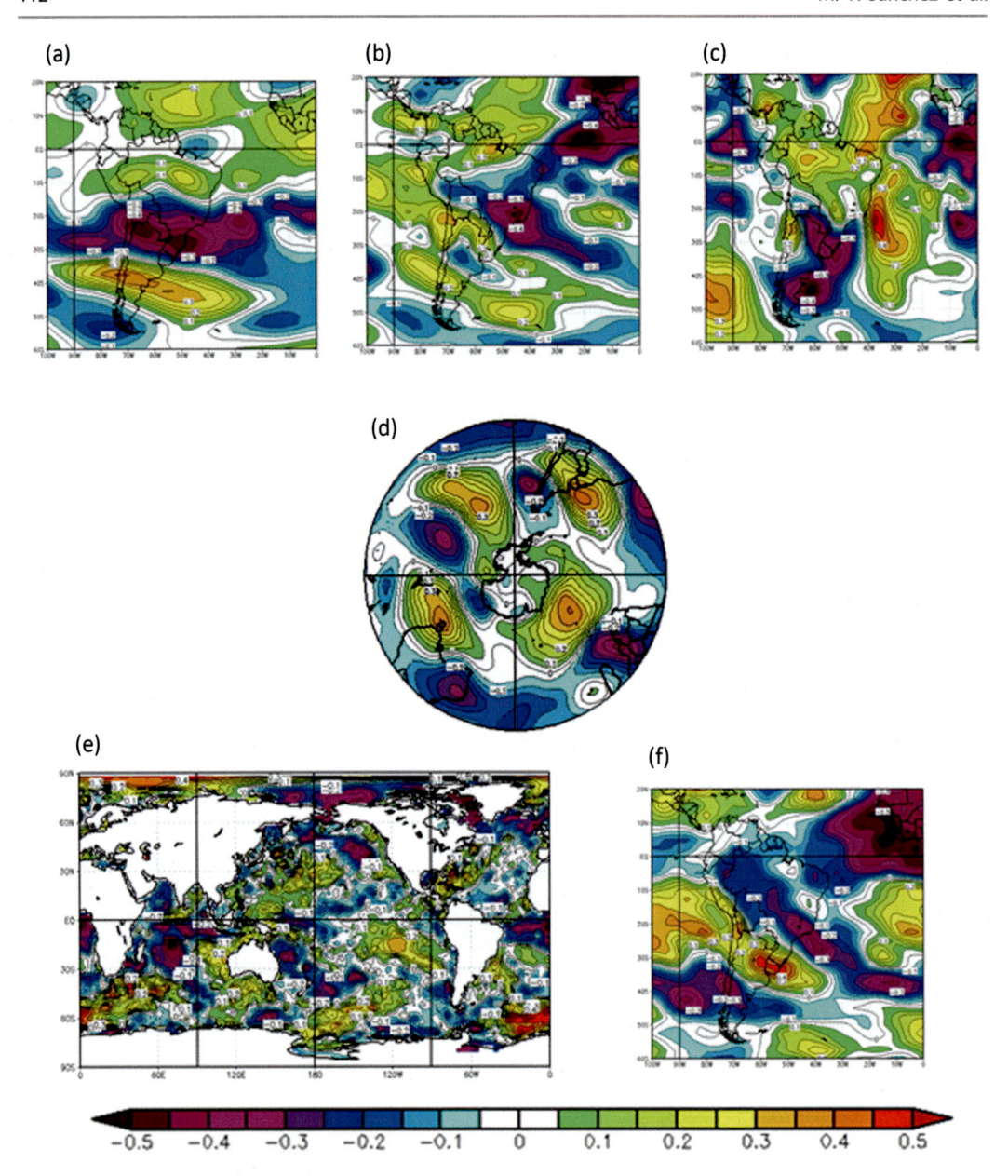

Fig. 5.13 Correlation fields between MAM precipitation anomalies in SRNe and climate variables in the previous February: **a** Zonal wind at 250 hPa; **b** zonal wind at 850 hPa; **c** meridional wind at 850 hPa; **d** geopotential height at 500 hPa; **e** sea surface temperature; **f** precipitable water

(negative) anomalies of MAM precipitation in SRN and SRL. However, the negative (positive) DOI phase in February increases (decreases) precipitation in SRNe.

Nnamchi et al. (2017) investigated the influence of the atmospheric circulation over the tropical Atlantic Ocean on precipitation in summer on the Guinea coast, and they defined the South Atlantic Dipole Index (SAODI). Its positive (negative) phase is anomalous warming (cooling) of the Atlantic Ocean off the coast of Guinea and cooling (warming) off the coast of Brazil. No relationship is observed with MAM rainfall in the case of the SRL and SRN (Figs. 5.11e and 5.12e) but the negative (positive) phase of the SAODI is related to higher (lower) precipitation in SRNe (Fig. 5.13e). However, a dipole with latitudinal distribution is detected for SRN and SRL further south of the SAODI region. This pattern indicates that a cooling of the South Atlantic Ocean north of 30° S and warming south of 30° S, are related to positive MAM anomalies in both sub-basins. Another dipole can be observed in the north Pacific Ocean, with positive correlations on the Pacific coast of North America and negative correlations on the coast of the Asian continent. This pattern favours MAM rainfall in SRL and SRN. This fact can be related to the Pacific Decadal Oscillation (PDO). Kayano and Andreoli (2007) found that a warm PDO phase and a cold ENSO phase are associated with precipitation in South America. The correlation fields between APP anomalies and MAM rainfall in SRL and SRN (Figs. 5.11f and 5.12f) show that a higher moisture content over the basins is related to higher autumn precipitation. The same pattern is observed in SRNe, although slightly shifted to the north, showing the importance of the moist air advection from the Atlantic High (Fig. 5.13f).

5.3.5 Predictors Definition

The method applied to choose the predictors was to define rectangular boxes inside the regions where the correlation resulted statistically significant that is, the absolute value of the Pearson correlation coefficient (R^2) is greater than or equal to 0.37, using a two-tailed Student's t-test with 95% confidence. The predictors were defined as the average of each variable in the boxes for each sub-basin and were detailed in Tables 5.5, 5.6 and 5.7. The name of each predictor is constituted by the variable name and the atmospheric level (if applicable). The predictors were also numbered if there is more than one predictor for the same variable.

Table 5.5 details the 12 predictors defined in SRL: two SST forcing in the northern hemisphere associated with PDO, and as they show a dipolar structure, the difference between them was also considered as a predictor (tsm_3). Similarly, 16 predictors were defined in SRN: the low-level zonal flow in the south of the country is a good predictor as well as the geopotential anomalies south of Australia, the SST in the Indian Ocean and the ENSO area in the Pacific Ocean (Table 5.6). Precipitation in both sub-basins has similar predictors. Finally, 16 predictors were defined in SRNe (Table 5.7): geopotential anomalies in South Australia and sea surface temperature in the Indian Ocean. It is important to mention that most of the possible predictors found are located in regions close to the SRNe as, for example, the precipitable water over the basin.

5.3.6 Design and Goodness of Fit of Statistical Regression Models

Sets of independent predictors were generated to build different multiple linear regression models using the forward stepwise technique. Only those models whose Adj R^2 was greater than or equal to 0.45, which implies that they explain more than 45% of the autumn precipitation, and with minimum AIC, AICc and BIC, were considered. Table 5.8 shows the efficiency coefficients of the selected models. Using the independent predictor sets, 10 models for SRL, 22 models for SRN and 7 models for SRNe were generated. It is important to point out that the models with a greater number of predictors have higher Adj R^2 reaching values of 0.70 in SRN and SRL, whilst Adj R^2 varies from 0.46 to 0.52 in SRNe, despite the increase in the number of predictors. To evaluate the goodness of fit of the models, a qualitative and quantitative verification was carried out using categorized data. The observed and predicted MAM precipitation was categorized into three classes: below normal, normal and above normal events, as was detailed in the Methodology section. The objective was to verify if the models can forecast the observed category. The

Table 5.5 Definition of predictors for MAM precipitation anomalies in SRL

	Variable	Id	Box limits	R
1	app	1	[43° S–47° S; 85° W–75° W]	0.4
2		2	[27° S–33° S; 67° W–65° W]	0.44
3	geo850	1	[30° S–40° S; 140° E–160° E]	−0.51
4	slp	1	[30° S–40° S; 140° E–160° E]	−0.57
5		2	[15° N–5° N; 170° E–150° W]	−0.49
6	tsm	1	[10° S–17° S; 60° E–120° E]	−0.65
7		2	[40° N–50° N; 150° W–140° W]	0.52
8		3	[25° N–35° N; 140° E–160° E]	−0.54
9		4	[Diferencia tsm_2 − tsm_3]	−0.6
10		5	[22° S–30° S; 20° W–0° W]	−0.49
11	u500	1	[17° S–20° S; 30° W–20° W]	0.42
12	u850	1	[40° S–47° S; 100° W–80° W]	0.41

R is the Pearson coefficient between predictors in February and precipitation anomalies in autumn

Table 5.6 Definition of predictors for MAM precipitation anomalies in SRN

	Variable	Id	Box limits	R
1	app	1	[45° S–47° S; 85° W–75° W]	0.46
2		2	[4° S–8° S; 55° W–43° W]	−0.47
3	geo250	1	[45° S–57° S; 105° W–90° W]	−0.52
4	geo500	1	[45° S–57° S; 105° W–90° W]	−0.52
5		2	[37° S–50° S; 135° E–165° E]	−0.51
6	ge850	1	[45° S–57° S; 105° W–90° W]	−0.49
7		2	[30° S–45° S; 135° E–165° E]	−0.59
8	slp	1	[45° S–57° S; 105° W–90° W]	−0.47
9		2	[30° S–45° S; 135° E–165° E]	−0.61
10	tsm	1	[5° S–10° S; 110° W–85° W]	−0.4
11		2	[2° S–15° S; 50° E–90° E]	−0.56
12	u250	1	[45° S–50° S; 80° W–70° W]	0.37
13	u500	1	[45° S–50° S; 80° W–70° W]	0.39
14		2	[15° S–20° S; 32° W–21° W]	0.41
15	u850	1	[40° S–45° S; 100° W–80° W]	0.53
16	v250	1	[35° S–45° S; 85° W–77° W]	−0.4

R is the Pearson coefficient between predictors in February and precipitation anomalies in autumn

quality of the forecast was classified in: "GOOD (G)" when the predicted and the observed categories are equal; "REGULAR (R)" when the observed differs from the predicted in one category and "BAD (B)" when the observed differs in 2 categories from the predicted.

Figure 5.14 shows a similar efficiency for all models in SRL. All models exceed 60% of cases correctly classified (G), and the R forecasts do not reach 40% of the cases. It should be mentioned that none of the forecasts differed in two categories from the observed one, therefore none was a B forecast. Figure 5.15 shows that some models for SRNe have a B prognosis, although all models predict well more than 50% of cases. Something similar to what happens in SRL can be observed

Table 5.7 Definition of predictors for MAM precipitation anomalies in SRNe

	Variable	Id	Límites	R
1	geo850	1	[27° S–35° S; 50° W–40° W]	0.37
2	geo500	1	[33° S–37° S; 55° W–40° W]	0.43
3		2	[40° S–45° S; 145° E–155° E]	0.39
4	geo250	1	[33° S–40° S; 60° W–40° W]	0.44
5		2	[38° S–48° S; 144° E–160° E]	0.45
6	tsm	1	[12° S–22° S; 65° E–75° E]	−0.49
7	u850	1	[20° S–25° S; 60° W–40° W]	−0.46
8	u500	1	[15° S–20° S; 60° W–50° W]	−0.48
9		2	[20° S–30° S; 50° W–40° W]	−0.45
10		3	[35° S–37° S; 80° W–70° W]	0.39
11	u250	1	[23° S–30° S; 65° W–40° W]	−0.49
12		2	[37° S–40° S; 75° W–65° W]	0.39
13	v850	1	[40° S–47° S; 65° W–57° W]	−0.5
14		2	[20° S–30° S; 40° W–35° W]	0.48
15	v500	1	[35° S–45° S; 67° W–57° W]	−0.46
16	app	1	[30° S–35° S; 60° W–50° W]	0.46

R is the Pearson coefficient between predictors in February and precipitation anomalies in autumn

with the SRN models (Fig. 5.16), most of the models do not register B forecasts except for the SRN_1, SRN_3, SRN_4 and SRN_5 models. These models have the least Adj R^2 (Table 5.8). It is also worth mentioning that SRN_5 is the only model with more R forecast than G ones. Most models predict well more than 60% of cases.

Figure 5.17 is the Taylor diagram for SRL, all the models have a similar efficiency and the correlation values range from 0.75 to 0.85. The models SRL_1, SRL_2, SRL_3 and SRL_4 are further away from the reference series, have the greatest mean square error and a standard deviation further from the observed one.

Figure 5.18 shows the same analysis for SRNe. It can be pointed out that all the models are similarly positioned to each other but two of them are the closest to the reference series: SRNe_1 and SRNe_2. They have a correlation value of 0.8, which means that they are highly correlated with observed MAM precipitation anomalies.

They also have a lower mean square error than other models and a low standard deviation.

Figure 5.19 shows the diagram for SRN. It is observed that the SRN_5 model has a low correlation value, a high squared error and it is very far from the reference series. The rest of the models show good performance. The SRN_21 and SRN_22 models have a correlation value of 0.9, a standard deviation similar to the reference one and a minimum root mean square error. It should be noted that these models are the ones with the highest number of predictors.

Figure 5.20a shows ROC space for SRL. All the models are above the diagonal line, indicating that the efficiency of the models exceeds random. It can also be pointed out that the above normal and below normal events are better forecasts than normal events. Figure 5.20b shows ROC space for SRNe. The models have less ability to detect normal rainfall than for above and below cases too. Even more, most of the models for the normal event in

Table 5.8 Models derived using different sets of independent predictors

Model	Equation	AIC	AICc	BIC	Adj R^2
SRL					
SRL_1	preMAM ~ −6.54(e−13) − 1.55(e+02)tsm_1 + 4.76(e+01)dip	266.70	268.30	272.31	0.52
SRL_2	preMAM ~ −7.05(e−13) + 3.04(e+01)u500_1 − 7.04(e+01)tsm_3 − 1.43(e+02)tsm_1	265.08	267.58	272.09	0.56
SRL_3	preMAM ~ −5.70(e−13) + 2.55(e+01)u500_1 + 6.66(e+01)tsm_2 − 1.66(e+02)tsm_1	263.47	265.97	270.48	0.58
SRL_4	preMAM ~ −26(e−13) + 2.73(e+01)u500_1 − 1.44(e+02)tsm_1 + 4.44(e+01)dip	262.18	264.68	269.19	0.60
SRL_5	preMAM ~ −4.95(e−13) + 2.46(e+01)u500_1 + 6.73(e+01)tsm_2 − 1.42(e+02)tsm_1 + 1.26(e+01)app_2	260.11	263.76	268.51	0.63
SRL_6	preMAM ~ −5.56(e−13) + 2.66(e+01)u500_1 − 1.24(e+02)tsm_1 + 1.13(e+01)app_2 + 4.24(e+01)dip	259.78	263.43	268.19	0.64
SRL_7	preMAM ~ −4.13(e−13) + 8.61(e+00)u850_1 + 2.64(e+01)u500_1 + 6.46(e+01)tsm_2 − 1.10(e+02)tsm_1 + 1.38(e+01)app_2	260.09	265.18	269.90	0.64
SRL_8	preMAM ~ 2.73(e−14) + 1.58(e+01)u500_1 + 5.95(e+01)tsm_2 − 1.25(e+02)tsm_1 − 1.43(e+00)g850_1 + 1.43(e+01)app_2	259.85	264.94	269.66	0.65
SRL_9	preMAM ~ −1.17(e−12) + 1.67(e+01)u500_1 + 5.95(e+01)tsm_2 − 1.21(e+02)tsm_1 + −1.40(e+01)slp_1 + 1.34(e+01)app_2	259.55	264.64	269.36	0.65
SRL_10	preMAM ~ −1.25(e−12) + 1.81(e+01)u500_1 − 1.04(e+02)tsm_1 − 1.45(e+01)slp_1 + 1.22(e+01)app_2 + 3.80(e+01)dip	258.97	264.06	268.78	0.66
SRN					
SRN_1	preMAM ~ 3.27(e−14) + 1.82(e+01)u850_1 + 2.80(e+01)u500_2	237.20	238.80	242.80	0.50
SRN_2	preMAM ~ −8.05(e−13) + 3.77(e+00)v850_1 + 3.26(e+01)u500_2 + −8.35(e+00)g250_1	237.30	239.80	244.31	0.52
SRN_3	preMAM ~ −2.27(e−14) + 1.27(e+01)u850_1 + 2.63(e+01)u500_2 + −5.54(e+01)tsm_2	235.37	237.87	242.37	0.55
SRN_4	preMAM ~ −5.20(e−13) + 2.57(e+01)u500_2 − 6.22(e+01)tsm_2 − 4.84(e+00)g250_1	234.84	237.34	241.85	0.55
SRN_5	preMAM ~ 3.02(e−13) + 9.42(e+02)tsm_2 − 22(e−01)g500_2 + 1.60(e+01)app_1	234.21	236.71	241.22	0.56
SRN_6	preMAM ~ 5.00(e−13) + 4.12(e+00)v850_1 − 8.31(e+01)tsm_2 − 9.12(e−01)g500_2 + 2.10(e+01)app_1	233.27	236.92	241.68	0.59
SRN_7	preMAM ~ 2.915(e−13) + 9.67(e+00)u850_1 + 2.12(e+01)u500_2 − 5.91(e+01)tsm_2 − 5.34(e−01)g500_2	232.77	236.42	241.18	0.59
SRN_8	preMAM ~ 3.56(e−13) + 1.98(e+01)u500_2) + 5.10(e+00)u500_1 − 7.24(e+01) tsm_2 − 6.73(e−01)g500_2	232.27	235.92	240.67	0.60
SRN_9	preMAM ~ 4.84(e−13) + 1.15(e+01) u850_1 + 2.62(e+01)u500_2 + −6.99(e+01)tsm_2 + 1.38(e+00)g850_2 − 9.79(e−01)g500_2	233.00	238.09	242.81	0.60
SRN_10	preMAM ~ 3.59(e−13) + 1.56(e+01)u850_1 + 2.34(e+01)u500_2 − 6.00(e01)tsm_2 + 5.66(e−01)g850_1 − 6.25(e−01)g500_2	232.99	238.08	242.80	0.60

(continued)

Table 5.8 (continued)

Model	Equation	AIC	AICc	BIC	Adj R^2
SRN_11	preMAM $\sim$ 1.56(e−13) + 9.31(e+00) u850_1 + 3.61(e+01) u500_2 + 2.59(e+00)g850_1 − 1.93(e+00)g500_1 + v1.26(e+01) app_2	232.73	237.82	242.54	0.61
SRN_12	preMAM $\sim$ 5.72(e−13) + 2.49(e+01)u500_2 + 6.08(e+00) u500_1 − 8.65(e+01)tsm_2 + 1.46(e+00)g850_2 − 1.17(e+00) g500_2	232.26	237.35	242.07	0.61
SRN_13	preMAM $\sim$ 2.44(e−13) + 2.63(e+01) u500_2 + 3.03(e+00) g850_1 − 2.31(e+00)g500_1 − 1.96(e+01)app_2 + 1.44(e+01) app_1	230.66	235.75	240.47	0.63
SRN_14	preMAM $\sim$ 2.07(e−13) + 2.94(e+00) v850_1 + 2.84(e+01) u500_2 + 3.26(e+00)g850_1 − 2.58(e+00)g500_1 − 1.70(e+01) app_2 + 1.76(e+01)app_1	230.23	237.09	241.44	0.65
SRN_15	preMAM $\sim$ 2.09(e−13) + 3.95(e+00)v500_1 + 2.77(e+01) u500_2 + 3.29(e+00)g850_1 − 2.67(e+00)g500_1 − 1.74(e+01) app_2 + 1.88(e+01)app_1	229.83	236.69	241.04	0.65
SRN_16	preMAM $\sim$ 1.07(e−13) u500_2 − 3.84(e+01)tsm_1 + 3.81(e+00)g850_1 − 2.58(e+00)g500_1 − 1.86(e+01)app_2 + 2.08(e+01)app_1	226.64	233.50	237.85	0.69
SRN_17	preMAM $\sim$ 2.93(e−13) + 1.68(e+01)u500_2 − 4.29(e+01) tsm_1 + 3.45(e+00)g850_1 − 3.08(e−01)g500_2 − 2.18(e+00) g500_1 − 1.90(e+01)app_2 + 2.11(e+01)app_1	227.20	2326	239.81	0.69
SRN_18	preMAM $\sim$ 2.92(e−12) − 5.90(e+01)tsm_1 + 6.50(e+00) g850_1 − 7.87(e−01)g500_2 − 6.89(e+00)g500_1 + 2.53(e+01) g250_1 − 2.00(e+01)app_2 + 3.16(e+01)app_1	226.63	235.63	239.24	0.69
SRN_19	preMAM $\sim$ 2.18(e−12) + 1.16(e+01)u500_2 − 4.99(e+01) tsm_1 + 5.96(e+00)g850_1 − 5.52(e−01)g500_2 − 5.96(e+00) g500_1 + 1.88(e+01)g250_1 − 1.92(e+01)app_2 + 2.56(e+01) app_1	226.67	238.24	240.68	0.70
SRN_20	preMAM $\sim$ 3.78(e−12) − 7.08(e+01)tsm_1 + 1.45(e+00) g850_2 + 7.09(e+00)g850_1 − 1.40(e+00)g500_2 − 7.95(e+00) g500_1 + 3.13(e+01)g250_1 − 2.24(e+01)app_2 + 3.47(e+01) app_1	226.11	237.69	240.12	0.70
SRN_21	preMAM $\sim$ 3.03(e−12) + 1.27(e+01)u500_2 − 24(e+01) tsm_1 + 1.57(e+00)g850_2 + 6.54(e+00)g850_1 − 1.19(e+00) g500_2 − 7.01(e+00)g500_1 + 2.46(e+01)g250_1 − 2.18(e+01) app_2 + 2.84(e+01)app_1	225.54	240.21	240.95	0.71
SRN_22	preMAM $\sim$ 3.25(e−12) + 3.35(e+00)v500_1 + 1.38(e+01) u500_2 − 5.28(e+01)tsm_1 + 1.71(e+00)g850_2 + 6.52(e+00) g850_1 − 1.36(e+00)g500_2 − 7.32(e+00)g500_1 + 2.57(e+01) g250_1 − 2.06(e+01)app_2 + 3.07(e+01)app_1	225.51	243.86	242.32	0.72
SRNe					
SRNe_1	preMAM $\sim$ −0.71(e+00) + 1.85(e+01)v500 − 1.08(e+01) v850_1 + 4.19(e+01)v850_2 + 1.13(e+01)u250_1 − 2.52(e+01) u500_2 + 8.49(e+00)u500_3 + 4.00(e+00)geo250_2 − 1.69(e+01)tsm_1	221.74	233.32	235.75	0.52
SRNe_2	preMAM $\sim$ −0.71(e+00) + 2.14(e+01)v500 − 1.43(e+01) v850_1 + 4.48(e+01)v850_2 + 1.08(e+01)u250_1 − 2.54(e+01) u500_2 + 8.92(e+00)u500_3 + 4.42(e+00)geo250_2	221.42	230.42	234.03	0.52

(continued)

Table 5.8 (continued)

Model	Equation	AIC	AICc	BIC	Adj R^2
SRNe_3	preMAM $\sim$ $-0.71(e+00) - 9.07(e+00)$v850_1 $+ 3.36(e+01)$ v850_2 $- 1.25(e+01)$u500_2 $- 9.26(e+00)$geo250_1 $+ 4.93(e+00)$geo250_2	221.50	226.59	231.31	0.49
SRNe_4	preMAM $\sim$ $-0.71(e+00) - 5.84(e+00)$u500_2 $+ 2.23(e+00)$geo250_2 $+ 3.34(e+01)$tsm_1	227.55	230.05	234.55	0.49
SRNe_5	preMAM $\sim$ $-0.71(e+00) + 3.65(e+01)$v850_2 $- 1.41(e+01)$ u500_29.26(e+00)geo250_1 $+ 4.93(e+00)$geo250_2	220.79	224.45	229.20	0.49
SRNe_6	preMAM $\sim$ $-0.71(e+00) - 1.26(e+01)$v850_1 $+ 2.16(e+01)$ v850_2 $+ 0.41(e+00)$geo500_2 $- 2.40(e+01)$tsm_1	222.41	226.06	230.82	0.46
SRNe_7	preMAM $\sim$ $-0.71(e+00) - 1.82(e+01)$v850_1 $+ 1.79(e+01)$ v850_2 $- 6.46(e+00)$u500_1 $+ 0.35(e+00)$geo500_2	222.65	226.31	231.06	0.46

"preMAM" is the MAM precipitation anomaly

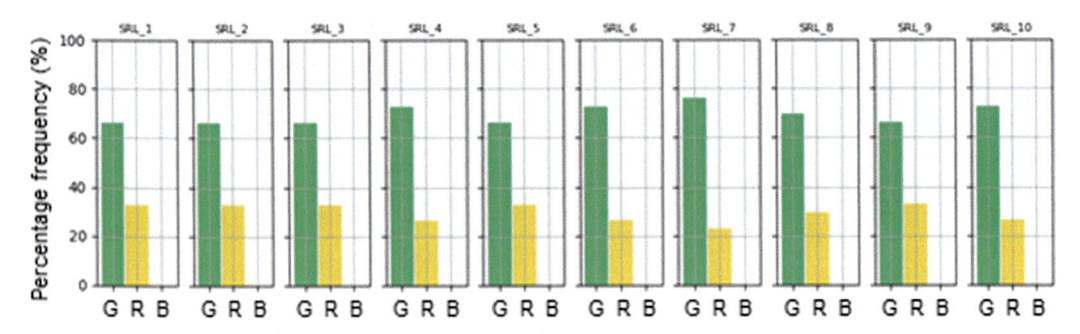

Fig. 5.14 Percentage frequency of the forecast cases for the SRL. The GOOD (G) forecasts are shown in green, the REGULAR (R) ones in yellow and the BAD (B) ones in red

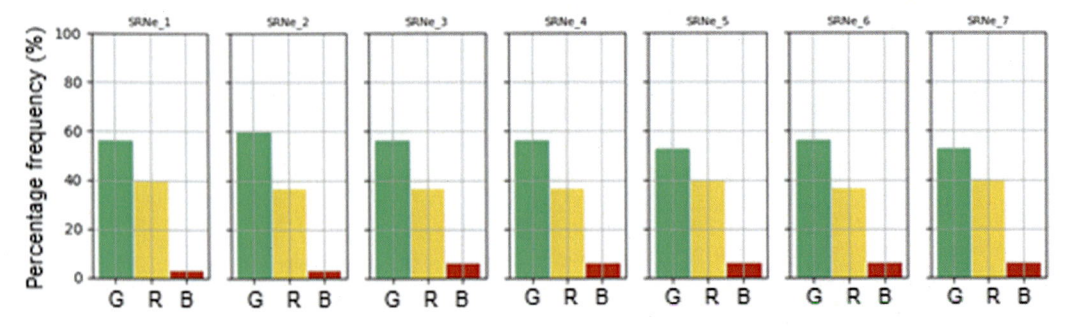

Fig. 5.15 Percentage frequency of the forecast cases for the SRNe. The GOOD (G) forecasts are shown in green, the REGULAR (R) ones in yellow and the BAD (B) ones in red

this sub-basin are very close to the diagonal line. Therefore, SRNe_7 is the best model for SRNe. Finally, observing the ROC space for the SRN models (Fig. 5.20c), it was found that the SRN_21, SRN_20 and SRN_9 models are closer to the upper left corner indicating that they are the best ones for this sub-basin.

To test the efficiency of the models, the forecast was made for a verification period (2011–2017) not included in the period used for

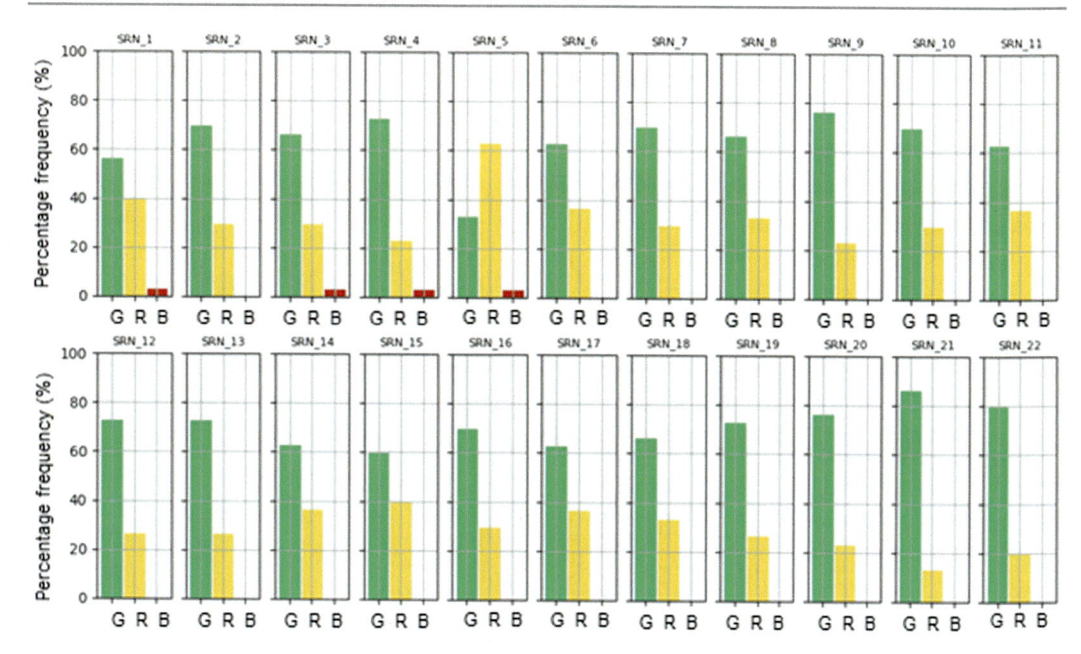

Fig. 5.16 Percentage frequency of the forecast cases for the SRN. The GOOD (G) forecasts are shown in green, the REGULAR (R) ones in yellow and the BAD (B) ones in red

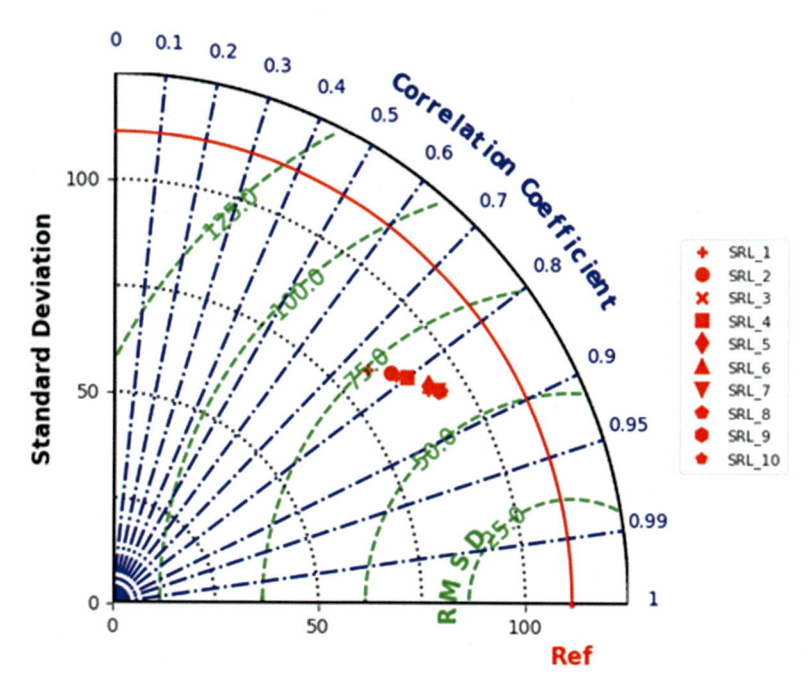

Fig. 5.17 Taylor diagram for MAM precipitation models in SRL. The black dotted curves represent the standard deviation, the green dashed lines represent the mean square error, and the blue lines made up of dots and lines the correlation coefficient. The point of reference is indicated as Ref in red

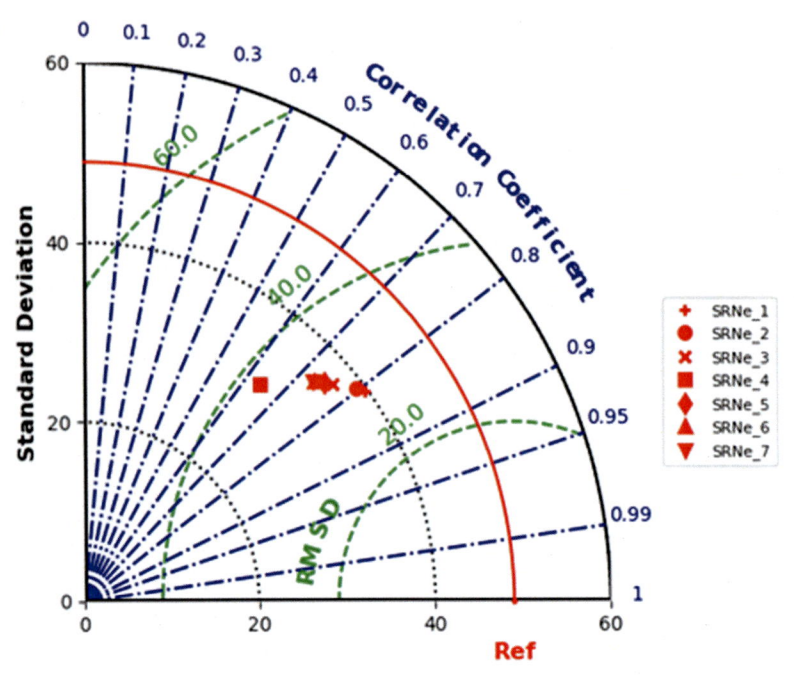

Fig. 5.18 Taylor diagram for MAM precipitation models in SRNe. The black dotted curves represent the standard deviation, the green dashed lines represent the mean square error, and the blue lines made up of dots and lines the correlation coefficient. The point of reference is indicated as Ref in red

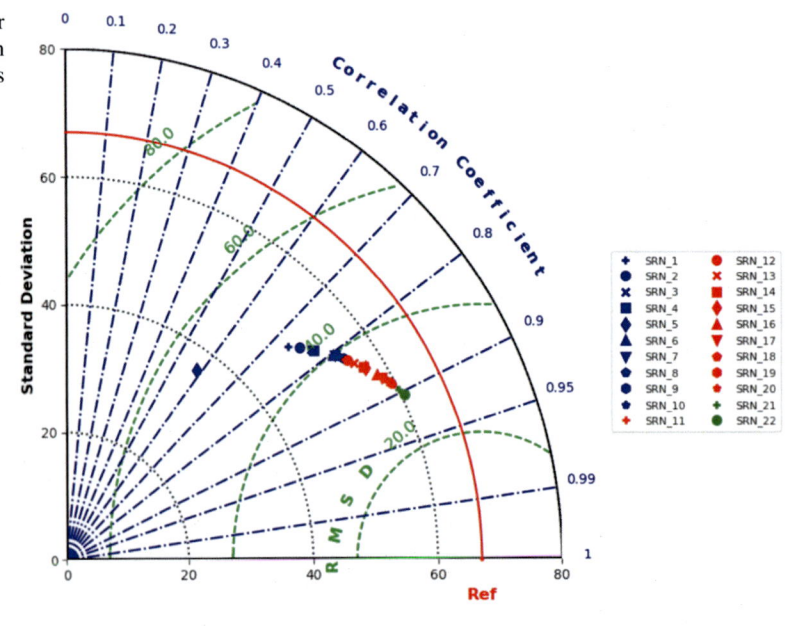

Fig. 5.19 Taylor diagram for MAM precipitation models in SRN. The black dotted curves represent the standard deviation, the green dashed lines represent the mean square error, and the blue lines made up of dots and lines the correlation coefficient. The point of reference is indicated as Ref in red

training. It is not intended to draw statistical conclusions on the performance of the models since the verification sample is very small but simply serves as an application example.

To summarize the results, the time series of the precipitation anomalies for the training period plus the new verification period (1981–2017) were plotted (Figs. 5.21, 5.23 and 5.25) for SRL, SRNe

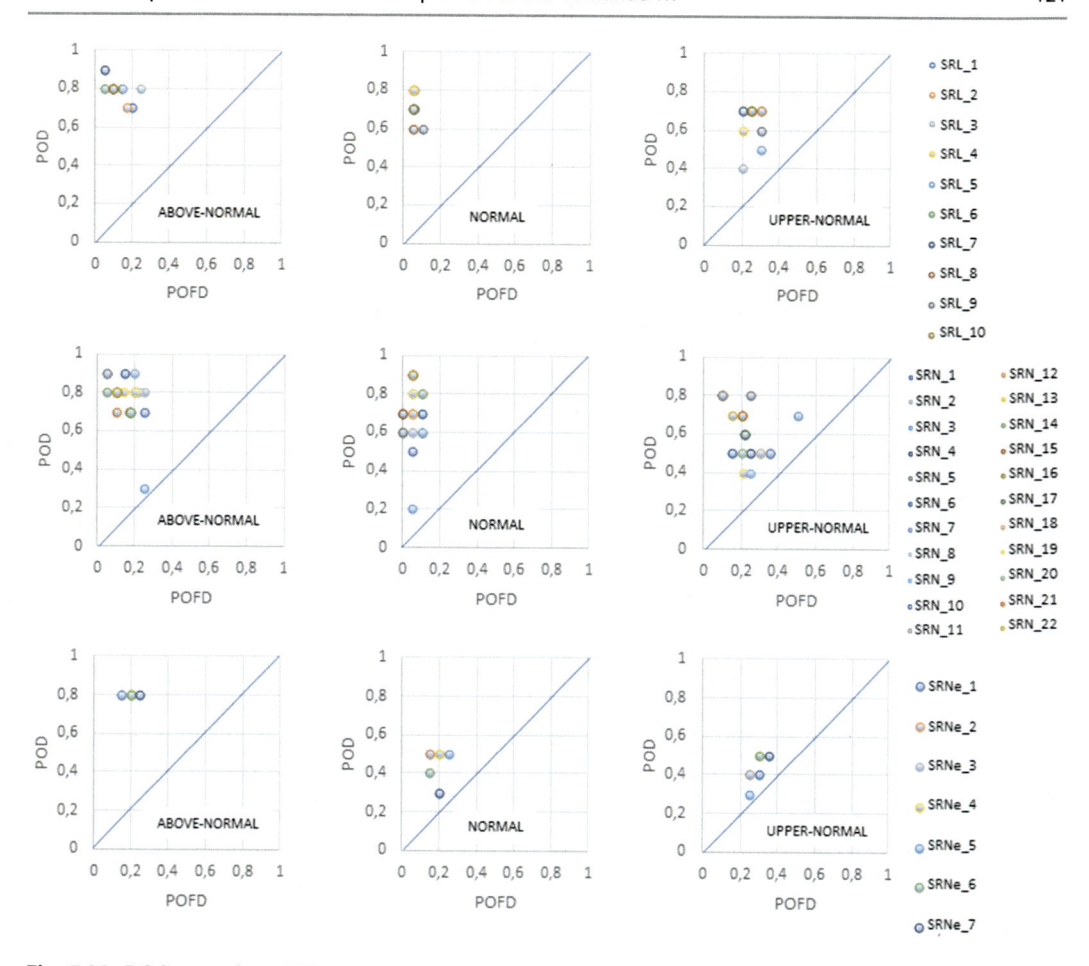

Fig. 5.20 ROC space for **a** SRL (upper panel), **b** SRNe (middle panel) and **c** SRN (lower panel)

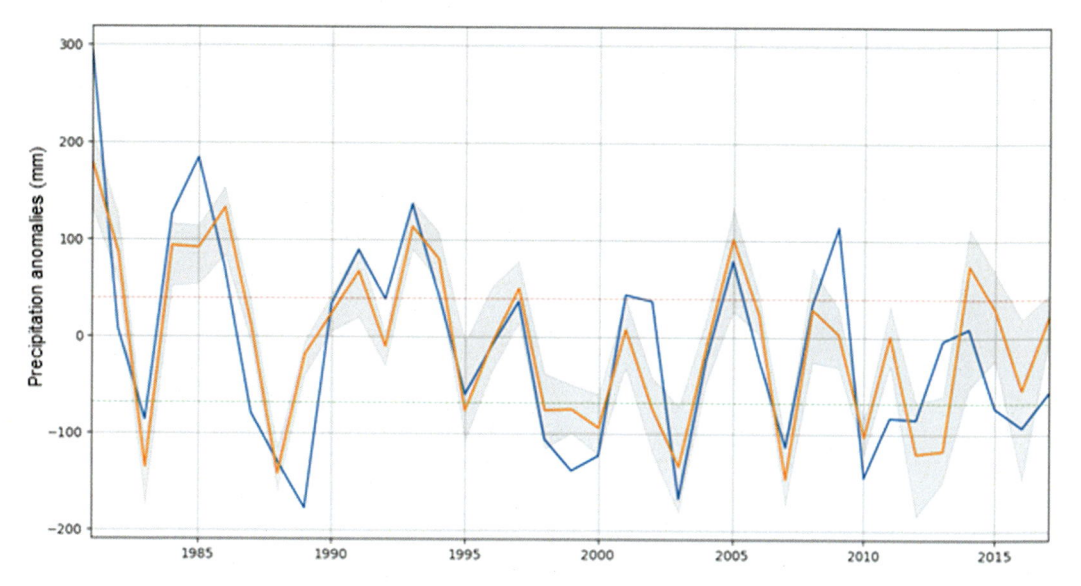

Fig. 5.21 Series of MAM precipitation anomalies in SRL, for the training (1981–2010) and the verification (2011–2017) period. The observed value is represented in blue, the average of the model forecasts in orange and the range in shading. Dot horizontal lines indicate the first and second tercile of the observed values

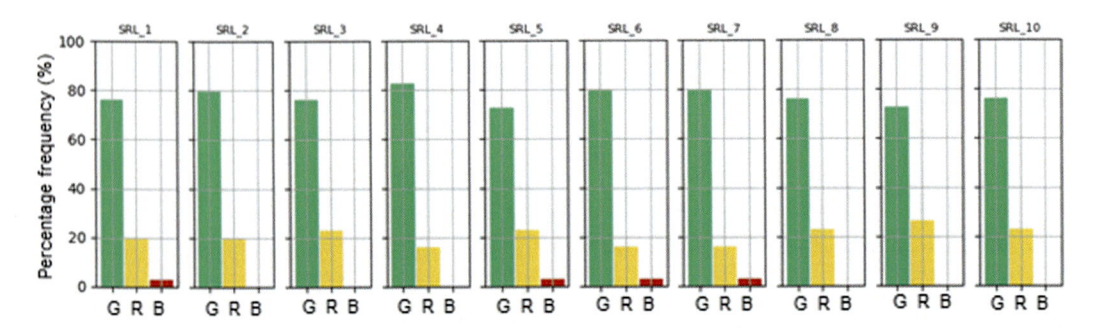

Fig. 5.22 Percentage frequency of the forecast cases for the SRL. The GOOD (G) forecasts are shown in green, the REGULAR (R) ones in yellow and the BAD (B) ones in red for (1981–2017) period

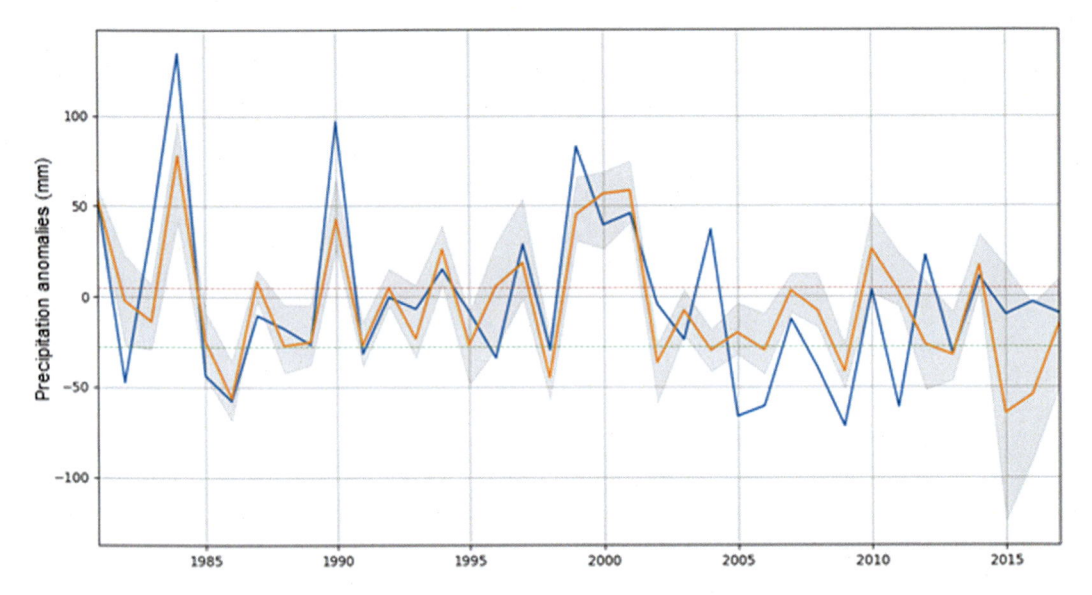

Fig. 5.23 Series of MAM precipitation anomalies in SRNe, for the training (1981–2010) and the verification (2011–2017) period. The observed value is represented in blue, the average of the model forecasts in orange and the range in shading. Dot horizontal lines indicate the first and second tercile of the observed values

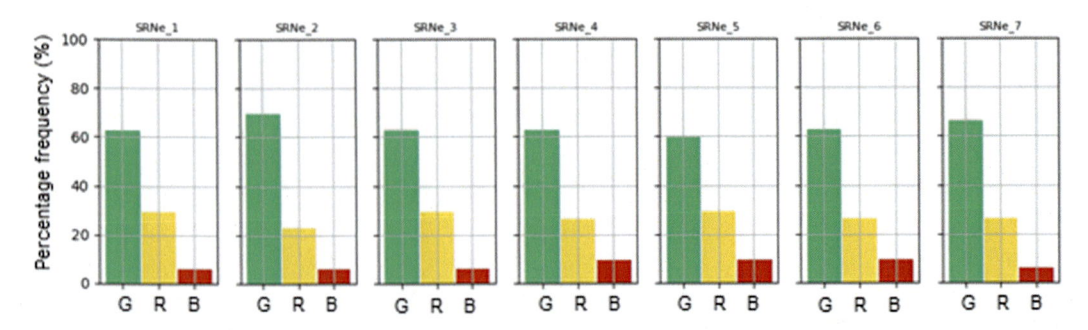

Fig. 5.24 Percentage frequency of the forecast cases for the SRNe. The GOOD (G) forecasts are shown in green, the REGULAR (R) ones in yellow and the BAD (B) ones in red for (1981–2017) period

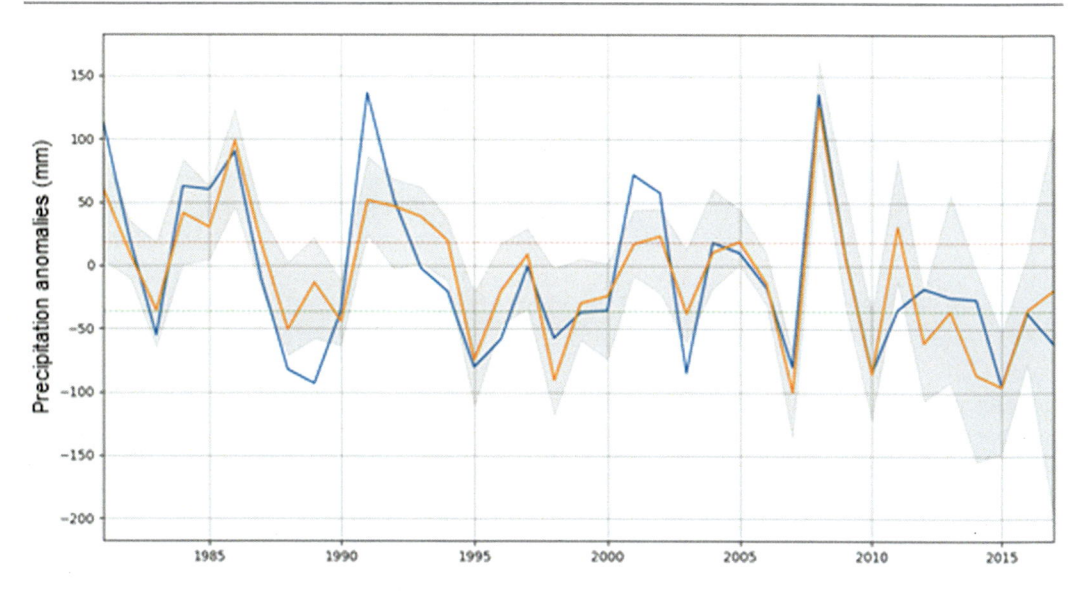

5 An Attempt to Forecast Seasonal Precipitation in the Comahue …

Fig. 5.25 Series of MAM precipitation anomalies in SRN, for the training (1981–2010) and the verification (2011–2017) period. The observed value is represented in blue, the average of the model forecasts in orange and the range in shading. Dot horizontal lines indicate the first and second tercile of the observed values

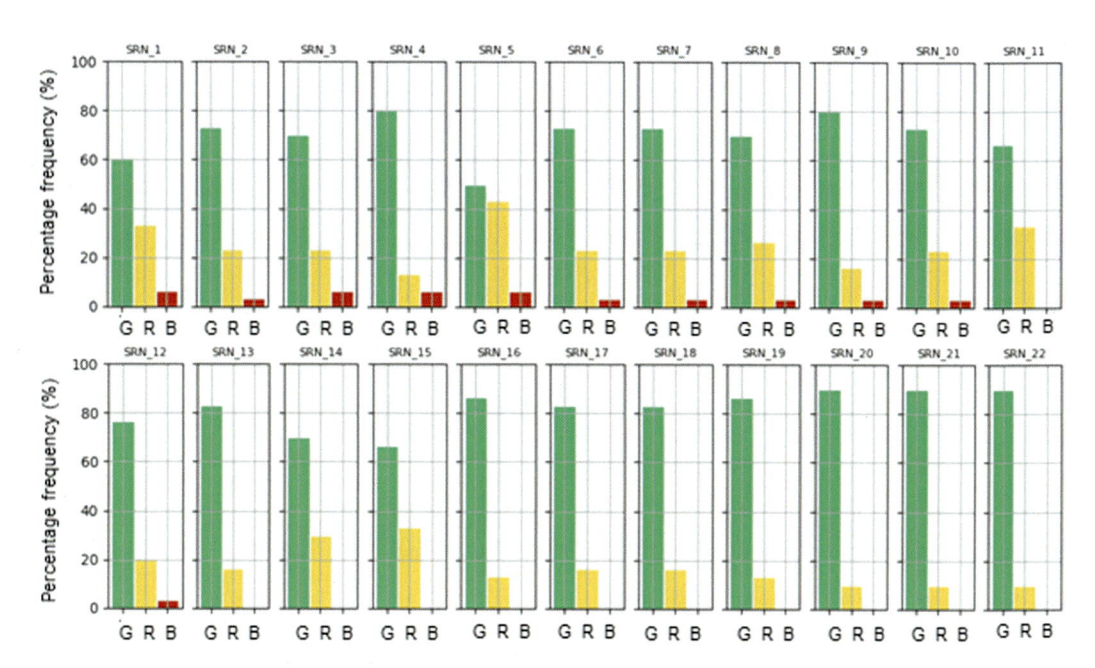

Fig. 5.26 Percentage frequency of the forecast cases for the SRN. The GOOD (G) forecasts are shown in green, the REGULAR (R) ones in yellow and the BAD (B) ones in red but for (1981–2017) period

and SRN respectively. These series were calculated as an average of the forecast derived from all the models in each sub-basin (ensemble mean). The range was also plotted to detect the variability of the ensemble mean. The efficiency analysis for the category forecast performed for the training period (1981–2010) was repeated for the extended period 1981–2017 (Figs. 5.22, 5.24 and 5.26).

In the case of the SRL (Fig. 5.21), the models efficiently forecasted MAM precipitation anomalies in the 2011–2017 period, although most of the events were overestimated. Figure 5.22 shows that 70% of the cases are well estimated by all the models. Only SRL_1, SRL_5, SRL_6 and SRL_7 wrongly forecasted MAM precipitation in 2015. The performance was worse in the case of SRNe in the period 2011–2017 (Fig. 5.23). Something remarkable is that the range of forecasts is maximum in 2015. However, the categories were well predicted in more than 60% of the years by all the models in the entire period 1981–2017 and all the models had less than 10% of the badly predicted cases (Fig. 5.24).

The forecasting models had a reasonable efficiency in the case of SRN (Fig. 5.25), where a high range of variability can be observed in 2017. The well-predicted categories exceeded 60% of the cases, except in the case of the SRN5 model that only reached 50%. Eleven of the twenty-two models had only a small percentage of poorly predicted cases (Fig. 5.26).

5.4 Conclusion

The predictability of autumn precipitation in the three sub-basins of the Comahue region in Argentina was studied. The whole region was described from the climatological and morphological point of view. Predictors of precipitation in the previous February were defined for each sub-basin using correlation methodologies. Precipitation in SRL was found to be more sensitive to SST whilst in the other two sub-basins a greater relationship with atmospheric circulation was observed. In general, a similar behaviour was observed in both Andean basins whilst SRNe showed different behaviour. Precipitation in SRN and SRL was strongly influenced by the temperatures of the sea surface of the Indian Ocean and the ENSO zone in the Pacific Ocean. The intensity of the anticyclone in the Atlantic Ocean also influenced precipitation and the PDO had a special influence in the case of SRL. Once the statistical models had been generated, several of them were discarded because they did not meet the minimum conditions of certain statistical parameters used to evaluate efficiency. The SRN_5 statistical model was not efficient to categorize the precipitation anomalies in SRN. The SRN_20, SRN_21 and SRN_22 models were the best models not only because of the value of the statistics that evaluate efficiency but also because they presented few badly categorized cases in 1981–2017. The models were most efficient in SRL, the best being SRL_8, SRL_9 and SRL_10. Greater efficiency of the SRL and SRN models was observed compared to the efficiency in SRNe. It should be noted that the efficiency of the models is better for forecasting categories than in the case of the accumulated autumn precipitation values. Also, the average of the forecast derived from different models showed to be representative of the precipitation. The results are encouraging and it is possible that if non-linear methodologies such as generalized additive models or neural networks are used, the results will improve significantly.

Acknowledgements Rainfall data was provided by the National Meteorological Service of Argentina (SMN), the Sub-secretary of Hydric Resources (SsRH), the Territorial Authority of Comahue basin (AIC) and the National Institute of Agricultural Technology (INTA). Maps from Argentine National Geographic Institute (IGN) and data from The National Aeronautics and Space Administration (NASA), Shuttle Radar Topographic Mission (SRTM) were used. Images were provided by the NOAA/ESRL Physical Sciences Laboratory, Boulder Colorado from their Web site at http://psl.noaa.gov/. This research was supported by 2020–2022 UBACyT 20020190100090BA, 2018-2020 UBACyT 20620170100012BA and UBACYT 2017-2019 20020160100009BA projects.

References

Barros V, Castañeda ME, Doyle M (2000) Recent precipitation trends in southern South America East of the Andes: an indication of climatic variability. Southern hemisphere paleo- and neoclimates. Springer, Berlin

Bell F (1999) Geological hazards. Their assessment, avoidance and mitigation. Taylor & Francis Group, New York, USA

Berman AL, Silvestre G, Compagnucci MR (2012) Eastern Patagonia seasonal precipitation: influence of southern hemisphere circulation and links with subtropical South American precipitation. J Clim 25:6781–6795

Carrasco JF, Casassa G, Rivera A (2002) Meteorological and climatological aspects of the southern Patagonia icefield. In: Casassa G, Sepúlveda FV, Sinclair RM (eds) The Patagonian icefields. Series of the Centro de Estudios Científicos. Springer, Boston, MA

Carril A, Doyle M, Barros V, Nuñez M (1997) Impacts of climate change on the oases of the Argentinean Cordillera. Clim Res 9:121–129

Diaz HF, Markgraf V (2009) El Nino: historical and paleoclimatic aspects of the southern oscillation. Cambridge University Press, Cambridge

Garbarini EM, González MH, Rolla LA (2019) The influence of Atlantic High on seasonal rainfall in Argentina. Int J Climatol 39:4688–4702

Garreaud RD (2009) The Andes climate and weather. Adv Geosci 22:3–11

Garreaud RD, Vuille M, Campagnucci R, Marengo J (2008) Present-day South American climate. Paleogeogr Palaeoclimatol Palaeoecol 281(3–4):180–195

Garreaud R, Lopez P, Minvielle M, Rojas M (2013) Large-scale control on the Patagonian climate. J Clim 26(1):215–230

Gilbert RO (1987) Statistical methods for environmental pollution monitoring. Van Nostrand Reinhold Co, New York

Gillett NP, Kell T, Jones P (2006) Regional climate impacts of the southern annular mode. Geophys Res Lett 33(23):L23704

González MH, Cariaga ML (2010) Estimating winter and spring rainfall in the Comahue region (Argentine) using statistical techniques. Adv Environ Res 11(5):103–118

González MH, Domínguez D (2012) Statistical prediction of wet and dry periods in the Comahue region (Argentina). Atmos Clim Sci 2:23–31

González MH, Herrera N (2014) Statistical prediction of Winter rainfall in Patagonia (Argentina). Horiz Earth Sci Res 11(7):221–238

González MH, Vera CS (2010) On the interannual wintertime rainfall variability in the Southern Andes. Int J Climatol 30:643–657

González M, Garbarini E, Romero P (2015) Rainfall patterns and the relation to atmospheric circulation in northern Patagonia (Argentine). Adv Environ Res 41(6):85–100

González MH, Garbarini EM, Rolla AL, Eslamian S (2017) Meteorological drought indices: rainfall prediction in Argentina. In: Handbook of drought and water scarcity: principle of drought and water scarcity, vol 1, no 29, pp 540–567

González MH, Losano F, Eslamian S (2020) Rainwater harvesting reduction impact on hydro-electric energy in Argentina. In: Eslamian S (ed) Handbook of water harvesting and conservation. Wiley, NY, USA, 1100 p. ISBN: 978-1-119-47895-9

Gravelius H (1914) Grundrifi der gesamten Gewcisserkunde. Band I: Flufikunde (Compendium of hydrology, vol I. Rivers, in German). Goschen, Berlin

Horton R (1932) Drainage basin characteristics. Trans Am Geophys Union 13:350–361

Hurvich C, Tsai C (1989) Regression and time series model selection in small samples. Biometrika 76:297–307

Hyndman RJ, Athanasopoulos G (2018) Forecasting: principles and practice, 2nd edn. OTexts, Melbourne, Australia. OTexts.com/fpp2. Accessed on May 2018

IPCC (2013) Climate Change 2013: the physical science basis. In: Stocker TF, Qin D, Plattner G-K, Tignor M, Allen SK, Boschung J, Nauels A, Xia Y, Bex V, Midgley PM (eds) Contribution of Working Group I to the fifth assessment report of the Intergovernmental Panel on Climate Change. Cambridge University Press, Cambridge, 1535 p

James G, Witten D, Hastie T, Tibshirani R (2013) An introduction to statistical learning with applications in R. Springer, Berlin

Kalnay E et al (1996) The NCEP/NCAR 40-year reanalysis project. Bull Am Meteor Soc 77:437–471

Kayano MT, Andreoli RV (2007) Relations of South American summer rainfall interannual variations with the Pacific decadal oscillation. Int J Climatol 27:531–540

Kendall MG (1975) Rank correlation methods, 4th edn. Charles Griffin, London

Liebmann B, Vera CS, Carvalho LMV, Camilloni IA, Hoerling MP, Allured D, Barros VR, Báez J, Bidegain M (2004) An observed trend in central South American precipitation. J Clim 17(22):4357–4367

Llamas MR (1993) Hidrología General. Edición española, Servicio Editorial Universidad del País Vasco, 635 p

Mann H (1945) Nonparametric tests against trend. Econometrica 13(3):245–259

Marshall GJ (2003) Trends in the southern annular mode from observations and reanalyses. J Clim 16:4134–4143

Mayr GJ et al (2007) Gap flows: results from the mesoscale alpine programme. Q J R Meteorol Soc 133:881–896

Mo KC, Higgins RW (1998) The Pacific-South American modes and tropical convection during the southern hemisphere winter. Mon Weather Rev 126(6):1581–1596

Mo KC, Paegle JN (2001) The Pacific-South American modes and their downstream effects. Int J Climatol 21:1211–1229. https://doi.org/10.1002/joc.685

NASA JPL (2014) NASA shuttle radar topography mission combined image data set. NASA EOSDIS Land Processes DAAC

Nnamchi HC, Kucharski F, Keenlyside NS, Farneti R (2017) Analogous seasonal evolution of the South Atlantic SST dipole indices. Atmos Sci Lett 18:396–402

Ostertag G, Frassetto F, Solorza N, Salcedo A (2014) Determinación y estado del manto nival de las cuencas del Limay y Neuquén a través de la aplicación de teledetección y SIG. Boletín Geográfico (31):27–41

Reboita MS, Ambrizzi T, Da Rocha RP (2009) Relationship between the southern annular mode and southern hemisphere atmospheric systems. Rev Bras Meteorologia 24(1):48–55

Roe GH (2005) Orographic precipitation. Annu Rev Earth Planet Sci 33:645–671

Saji NH, Goswami BN, Vinayachandran PN, Yamagata T (1999) A dipole mode in the tropical Indian Ocean. Nature 401:360–363

Santos E, Sabourin R, Maupin P (2009) Overfitting cautious selection of classifier ensembles with genetic algorithms. Inf Fusion 10:150–162

Saurral R, Camilloni IA, Barros VR (2016) Low-frequency variability and trends in centennial precipitation stations in southern South America. Int J Climatol 37(4):1774–1793

Scarpati OE, Kruse E, Gonzalez MH, Ismael A, Vich J, Capriolo AD, Caffera RM (2014) Updating the hydrological knowledge: a case study. In: Handbook of engineering hydrology: fundamentals and applications, vol 1, no 23, pp 445–459

Silvestri GE, Vera CS (2003) Antarctic oscillation signal on precipitation anomalies over southeastern South America. Geophys Res Lett 30(21):2115

Smith D, Stopp P (1978) The river basin: an introduction to the study of hydrology. Cambridge University Press, Cambridge

Strahler A (1952) Dynamic basis of geomorphology. Geol Soc Am Bull 63:923–938

Taschetto AS, Ambrizzi T (2012) Can Indian Ocean SST anomalies influence South American rainfall? Clim Dyn 38:1615–1628

Taylor KE (2001) Summarizing multiple aspects of model performance in a single diagram. J Geophys Res 106(D7):7183–7192

Thompson DWJ, Wallace JM (2000) Annular modes in the extratropical circulation. Part I: Month-to-month variability. J Clim 13(5):1000–1016

Wilks DS (2010) Statistical methods in the atmospheric sciences. Elsevier, Amsterdam

https://unfccc.int/documents/67499

Channel Instability in Upper Tidal Regime of Bhagirathi-Hugli River, India

Chaitali Roy and Sujit Mandal

Abstract

Riverbank erosion is a deltaic process that aggravates the channel instability and increases the sediment load in the river. Thus, it can be considered as the reason for sluggish flow of water which further could be a cause of river meander. River Bhagirathi-Hugli has imprints of dynamicity in its deltaic course in South Bengal. The river has experienced several twists and turns within the floodplain area and severe bank erosion due to channel instability. Vertical bank stratification or sedimentary facies analysis in various locations in the study area has been done and plays a major role as a determining factor of channel stability within the floodplain of Bengal basin. This paper attempts to assess the bank erosion pattern and channel instability with the help of multimetric fluvial tool called BEHI model. The Mouza level observation has been done to determine the BEHI scores. More than 20 locations were observed and studied for primary data collection.

Keywords

BEHI model · Bank erosion · Instability · Floodplain · Facies · Stratification

6.1 Introduction

Riverbank erosion is a natural process in the active floodplain region that causes instability of riparian zones. It is the detachment of bank materials or collapsing of river banks as an impact of hydraulic action of streams. Water is the most active agent than wind in the subtropical regions. Sometimes it turns into and hazard and remarkably increases due to intense geotechnical and hydraulic processes (Rosgen 2001). It is hypothesized by Julien (2012) that the process of riverbank erosion and sedimentation were active since palaeo geological time and is represented in the fluvial landforms of the earth. The eroded materials get deposited in the channel and form riffles at the low-velocity zone. Thus the massive bank erosion causes channel instability. In the deltaic flow, river meanders are the common observations. Simply the process of meander growth on plain land causes alternative bank erosion and deposition (Brice 1964; Friend and Sinha 1993; Leopold and Wolman 1957, 1960; Mueller 1968; Rosgen 1994; Schumm 1963, 1985; Williams 1986). According to Simon et al. (1996) and Throne (1999) eroding bank,

C. Roy · S. Mandal (✉)
Department of Geography, Diamond Harbour Women's University, South 24 Parganas, Diamond Harbour, West Bengal 743368, India
e-mail: mandalsujit2009@gmail.com

materials are the source of sediment in the stream, which can have adverse physical and biological consequences. River Ganga Padma is facing serious problems like bank erosion and channel avulsion. River Padma used to flow through the present channel of Bhagirathi-Hugli, after a catastrophe the flow had diverted (Parua 2009). Farakka Barrage was constructed in order to save the Kolkata Port by supplying more water through the channel Hugli. Geomorphic instability of such alluvial river means the fluvial system is still not in equilibrium state and continuously trying to be adjusted within the deltaic floodplain (Ghosh and Mistri 2012).

For proper assessment and management, it is necessary to build a quantitative prediction of stream bank erosion and channel shifting so that the land could respond better to this environmental loss (Darby and Thorne 1993). All the fluvial parameters like width, depth, bank height, bankfull height, sediment characteristics, bank lithology, bank angle, sinuosity, etc. are important to measure stream bank erosion potentiality or instability. Thorne (1982), Simon and Thorne (1996), Darby and Throne (1994, 1996), Thorne (1999), and Simon et al. (1999) had done predictions on streambank stability, bank failures, and its types. Another type of land loss model was prepared by Lohnes (1991) by combing an empirical model with Lohnes and Handy (1968) model of channel stability to quantify the land stability.

The sedimentary deposits in the floodplain reveal the depositional condition under fluvial dominance. Horizontal or layered deposits of sediments or alluvium with specific mineralogical composition make the primary sedentary profile called facies. Swiss geologist Amanz Gressly in 1838 introduced the term facies in modern stratigraphic study (Cross and Homewood 1997). Vertical facies profile, sedimentary contacts, grain size, and color are the important parameters to understand the condition of riverbanks (Cant and Walker 1976, 1978). Channel bank composition and subsurface hydrology determine the compactness and cohesiveness of the bank. Bank made up of hard rock or nonporous media is resistant to erosion, but banks which are composed of fine to medium-grained particles (from 0.340 to 0.028 mm) are vulnerable (Reineck and Singh 1980). It is very much needed to examine the floodplain composition, floodplain evolution, alluvial chronology, sediment supply, sediment depth, grain size, and stratigraphic correlation of infills for scientific study (e.g., Allen 1965; Bridge et al. 1986; Erskine et al. 1992). Fluvial depositional patterns in the old age stage and layering of alluvium were explained by Miall (1985, 2006, 2014). This paper aims to analyze the stream bank condition along both sides of river Bhagirathi-Hugli and assess the nature of stream bank erosion of Bhagirathi-Hugli as well as identify most unstable zones along the river.

The entire Bengal basin is fed by the water of Bhagirathi itself and by its tributaries. It has the largest course in India among all the rivers, almost 1569 miles from its source to its mouth. After crossing the Rajmahal Hill at Jharkhand, river Ganga enters Bengal. It flows through the districts of South Bengal and finally empties into Bay of Bengal. A stretch of river Bhagirathi is selected for the detailed and intensive study to understand the hydromorphological characters of the aforesaid river. From Nabadwip to Kalyani the river is flowing as an inter-district boundary among Nadia, Burdwan, and Hoogly (Fig. 6.1). The latitudinal and longitudinal extension is 88° 22′ 45.63′ E to 88° 24′ 37.69″ E and 23° 24′ 45.46″ N to 22° 59′ 58.97″ N, respectively. The entire flow path within this region is almost 90 km long and the average width is almost 400 m. It has a broad flat floodplain, which is about 10–15 km wide. The stretch of the river covers four administrative blocks of Nadia district on the left side namely Nabadwip, Santipur, Ranaghat, and Chakdah. Purbasthali, Kalna I, Kalna II and Balagarh, and Chinsurah Murga are the administrative blocks of Burdwan, Hoogly, and Howrah districts, respectively.

6.2 Materials and Methods

Sedimentary bank facies is defined as a layering of sediment mass or rock with definite characteristics. The study of stratigraphic alluvium

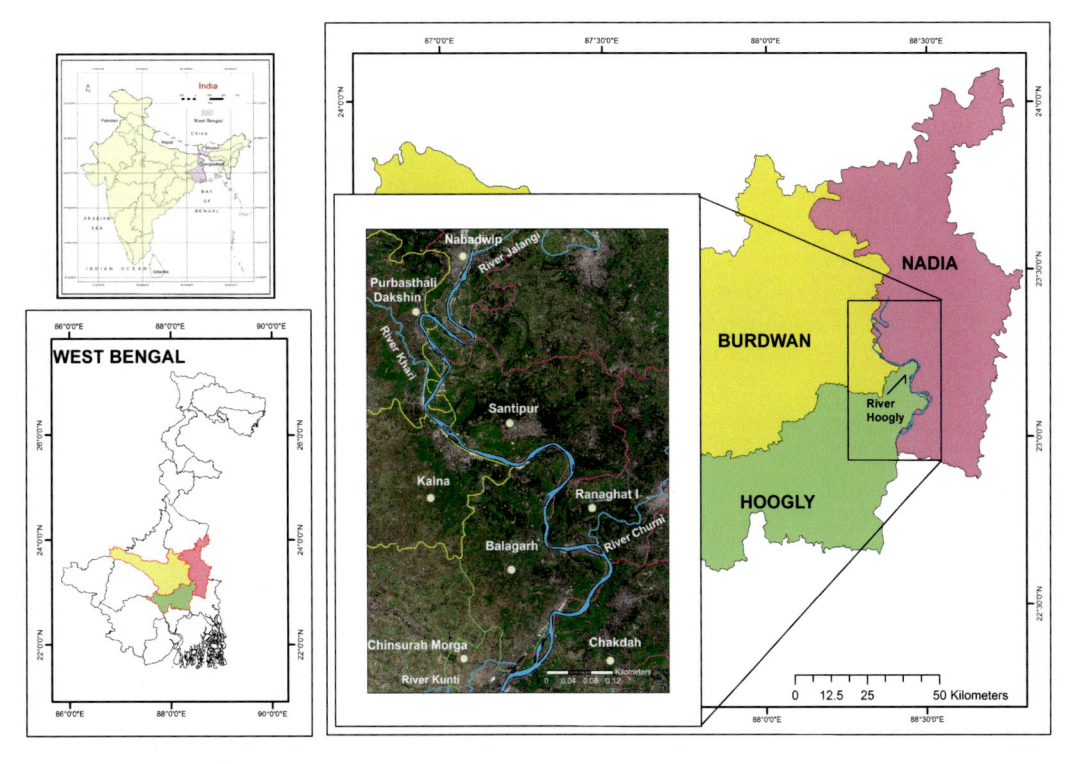

Fig. 6.1 Location of the study area indicating Bhagirathi-Hugli river course along with the floodplain extended over the CD blocks of Nadia, Burdwan, Howrah

deposits on the floodplain of different geological times is done by field-based method. In the post-monsoon season when the subsurfaces litho structures are free from moisture is the ideal time for sedimentary layer detection. Both left and right bank facies analysed at different location within the study area. Measuring tape, staff, and GPS were used to measure the sedimentary layers during field. Different micro geomorphic features like distinct and indistinct layers, laminations, and bioturbation are also identified from vertical exposed bank strata. Soil samples were collected from different layers to know the textural class. Collected field data is then plotted as vertical layers in SedLog 3.1 software.

Bank Erosion Hazard Index (BEHI) is a method proposed by Rosgen (2001). The entire study is done on the basis of site-specific field observation and collected data from Central Water Commission. Instruments that were used for collecting data are GPS, graduated staff, abney level, and digital current meter.

This empirically derived process needs various parameters to be evaluated and weighted. Single weighted values sum up to get overall weighted points and the categorical index is also suggested by Rosgen. The original BEHI method (After Rosgen 2001) will be further modified as modified BEHI. Here for the proper investigation and appropriate analysis.

$$\text{Ratio of bank height to bankfull height} = \frac{\text{Bank height (BH)}}{\text{Bankfull height (BFH)}} \quad (6.1)$$

$$\text{Root depth (RDH)} = \frac{\text{Average plant root depth}}{\text{Bank height (BH)}} \quad (6.2)$$

Root density (RD): The proportion of bank held by plant roots is the root density, expressed in percentage. More root density means more protection for the bank. Eye estimation method had been applied to determine the root density of the bank.

Surface protection (SP): It is the coverage of bank surface by vegetation is expressed in percentage. Rosgen incorporated this factor into his method as surface protection.

Bank angle (BA): Bank angle is the angular difference between bank top and bank toe or water surface. If the bank angle is less acute then the bank has moderate slope and if the bank is vertical (90°) then the bank forms a sharp cliff. Erosional activities of a stream depend upon this bank angle which is measured by abney level instrument from field.

Total of 22 points has been identified for sample study along the course of the river in study area. The overall combined score of all metrics is categorized according to BEHI scale. Very low to very high bank erosion-prone zone is classified under this method. Further for visual representation, geospatial tools are used. Known BEHI scores of different points are plotted along with the map by linear interpolation method and by Inverse Distance Weighting (IDW) technique. Where Σn is the number of observations for interpolation, denominator numbers and d stands for distance between the points (Chang 2010).

$$Z_p = \frac{\sum_{i=1}^{n}\left(\frac{z_i}{d_i^p}\right)}{\sum_{i=1}^{n}\left(\frac{1}{d_i^p}\right)} \quad (6.3)$$

Landsat images of 1972 and 2016 have been used to calculate the eroded area within the span of 44 years. Mouza maps that are collected from the office of Land Reforms, Govt. of West Bengal georeferenced and digitized in ArcGIS 10.3.1 software. A thematic map has been prepared to show the mouzas experienced bank erosion between 1972 and 2019. To identify vulnerable zones and instability along the channel Bhagirathi-Hugli distinct bank erosion zones have been marked on map.

$$\hat{Z}(S_0) = \sum_{i=1}^{N} \lambda_i Z(S_i) \quad (6.4)$$

where $Z(S_i)$ is the calculated value obtained from the thematic map and S_0 is the prediction location, λ is denoted for weight for the measured value and these weighted values are generally based on the lateral distance points which are measured, and prediction location.

6.3 Results and Discussion

6.3.1 Sedimentary River Bank Facies Analysis

The entire study area is a part of GBD, especially mature delta, where the river has deposited alluvium during every flood year. The exposed bank facies tell the lithological setup and depositional manner of the floodplain in study area. Mixed type of sedimentation pattern can be observed as the floodplain is the product of left, right bank tributaries and itself Bhagirathi-Hugli river. According to the genetic classification of floodplain (Nanson and Croke 1992; Schumm 1963; Wilson and Goodbred 2015), Bhagirathi-Hugli floodplain may be formed by three geological stages (1) up to Early Holocene—braid channel accretion (2) up to Late Holocene—lateral point bar accretion (3) Holocene to recent—overbank vertical accretion. There is also the incident of oblique accretion, counter-point accretion, and point bar accretion happened to build delta (Nanson and Croke 1992). According to Guchhait et al. (2016) floodplains like GBD is consist of older and newer alluvium, meander scrolls, cut bank erosion, point bar deposition, back swamps, etc. which are also carrying the imprints of various depositional patterns under the changing role of energy from meandering floodplain to lateral migration scrolled floodplain. The naked vertical sedimentary profiles are examined on the river banks at five different locations. The sedimentary lithosections (Facies 1–5) show the architecture made during different geological ages. In the Holocene period, fine-grained lithofacies were formed (between 3.6 ka and 5.4 ka BP) within Ganga Brahmaputra basin, on the other hand, Bhagirathi-Hugli floodplain had formed between 1 and 1.5 ka BP (Singh et al. 1998). Sedimentary structures are the visible litho features formed under different environmental situations. In the study area, majorly physical sedimentary structure (formed due to transportation and deposition of sediments at the interface of water and sediment) and biogenic sedimentary structure (formed due to interruption within the structure by burrowing organism) were found. Physical structures like horizontal planner lamination, trough cross-bedding, planar cross-bedding, and current ripple cross-bedding have been identified during fieldwork. The structures are coded by alphabets like A, B, C, D (A Horizontal planner stratification, B Trough cross-bedding, C Planar cross-bedding, D Current ripple cross-bedding).

6.3.1.1 River Bank Facies Analysis at Tegharipara

Tegharipara is the place located within the reach 1 of the study area just above the meander belt of Satkulta first study of bank facies has been done, which is nearly 1.8 m high. Six distinct layers have been identified on the right bank of river Bhagirathi. Bottom-most layer (0.6 m) is composed of very fine to fine brown sticky clay or silt. The compactness of silt materials holds the bank from failure as the cohesiveness is greater in this zone. Horizontal planar lamination is traced in this layer proving the paleo flow (Fig. 6.2). Vegetation and bioturbation at the bottom-most part of cane be seen in the photograph. Above it, there is 0.4 m thick layer of brown silt which was formed due to the paleo flood event or bankfull discharge deposits mud within the older path of the channel (Ghosh and Guchhait 2014; Kale 1999; Kale et al. 2010; Kale and Rajaguru 1987; Kochel and Baker 1988; Rajaguru et al. 2011). A very thin pocket layer of fine sand is present in the middle of the bank profile. A distinct 0.3 m thick micaceous sand layer makes the bank unstable but not in a greater way. Horizontal burrows are making the bank materials lose as a result after the threshold limit the bank will collapse. Minor Cross-bedding structure is there in the sand layer. The top-most layer is formed and becomes hard due to presence of biota and humus.

6.3.1.2 Bank Facies Analysis at Dhatrigram

Dhatrigram-Piarinagar is the area of confluence of river Hugli and Khari. The area is facing a serious problems like bank erosion. Vertical bank facies have been observed and examined in order to understand the sedimentary structure. The total bank facies have been drawn from the vertical bank exposure on the right side of the channel. The floodplain on the right side is a part of Damodar sub-basin and forms by the combined alluvium deposits of multiple rivers like Bhariathi, Khari, and Damodar. The entire bank is almost 3 m high from the water level and has prominent layers in display. This sedimentary profile has large sand layers at the bottom (1.2 m) with the mineral composition of mica, quartz. It also shows horizontal planner lamination. The sand composition of this layer makes the bank vulnerable as it is more prone to toe erosion and thus cantilever bank failure occurs. Overburden layers are mainly made up of very fine sand and fine clay minerals carried out by the western rivers like Khari and Damodar. The brown sticky dirty clay has formed a clear distinction from the sand layers below. It is composed of illite, quartz, feldspar, kaolinite, and mottles (Pal and Mukherjee 2008, 2010). Through cross-bedding, planar cross-bedding micro sedimentary structures are the result of cumulative effect of successive events (Fig. 6.3). There is some pocket layer of micaceous sand that signifies varied depositional character in different time periods.

6.3.1.3 Bank Facies Analysis at Santipur

Third bank lithofacies were examined near Santipur on the left side. 3.6 m high vertical bank almost 80% of the bank is composed of loose fine sand materials with cross-bedding structure. The bottom layer (2.5 m) is composed of micaceous sand with silica concentration. Planar cross-stratification shows the unidirectional flow of water in the past days. Deposition of finer

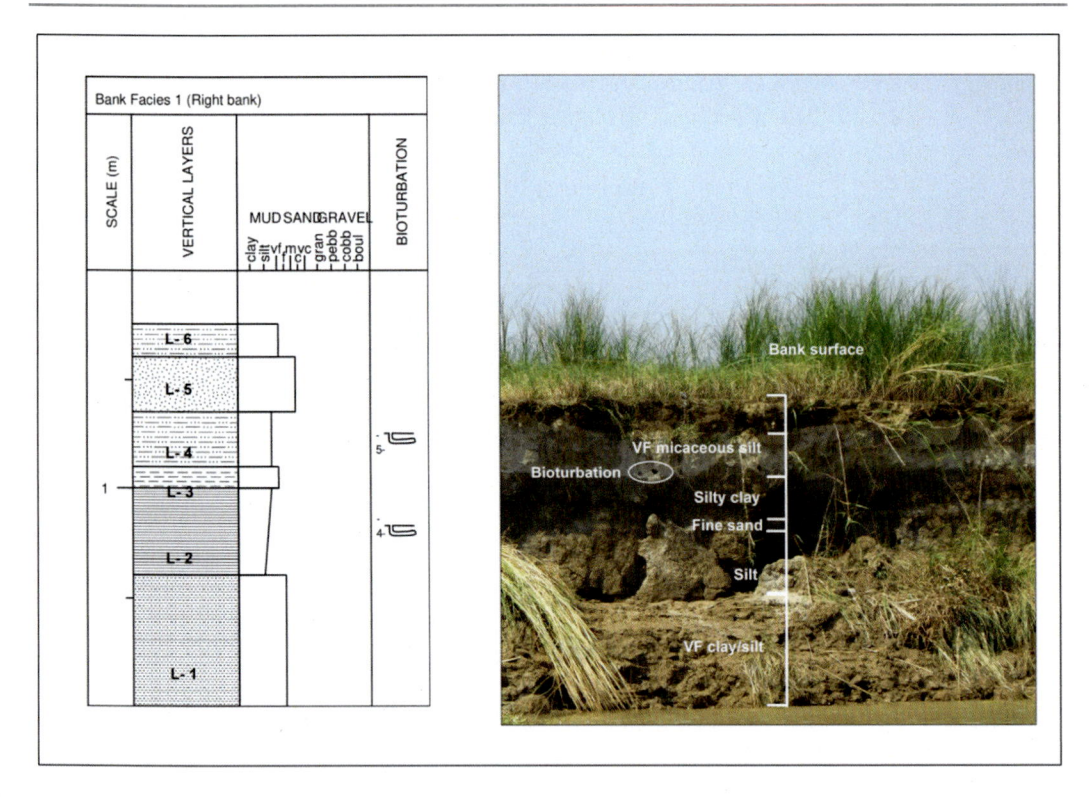

Fig. 6.2 Vertical river bank facies with grain size distribution at Tegharipara

materials at the bottom makes the bank prone to bank erosion. The silty loamy layer is 0.5 m thick and composed of clay particles showing through cross-bedding show the channel dominance in the past. The top-most layer has dripping subsurface hydrological conditions (Fig. 6.4).

6.3.1.4 Bank Facies Analysis at Chak Noapara

Left bank of river Hugli at Chak Noapara is studied for facies analysis. It is one of the erosive banks of the study area. Surface of the bank is occupied by agricultural fields and vegetation through bank erosion occurs due to its sedimentary structure. The bank is only 1.5 m high but is mostly composed of fine-grained sand and clay. Very fine silt and clay are inevitably found at the bottom-most part of the bank and show current ripple cross-bedding structure, which means the flux of flow and its asymmetric profiles show the paleocurrent direction. Layer 2 is dominated by no cohesive sand materials, whereas layer 3 has high concentration of silt and clay (80–90%). A prominent bright white sand layer is the layer 4 which is also got disturbed by horizontal burrows (bioturbation). The planner cross-stratification indicates the inclination of beds toward the flow direction. The top-most layer is the humus layer or the solum layer which has become sticky clayey layer due to presence of vegetation on the top (Fig. 6.5).

6.3.1.5 Bank Facies Analysis at Chandra

The last examined bank facies are located on the right side of the river at Chandra opposite Chak Noapara where the river took a bend toward Balagarh. The total bank height is 4 m under the marshy and stagnant water condition at base. Mud cracks are present in the bottom layer (2.5 m) due to excessive dryness of clay particles present in this layer. The alternate layers of clay and sand depict the various fluvial conditions. Sedimentary structure could not be identified

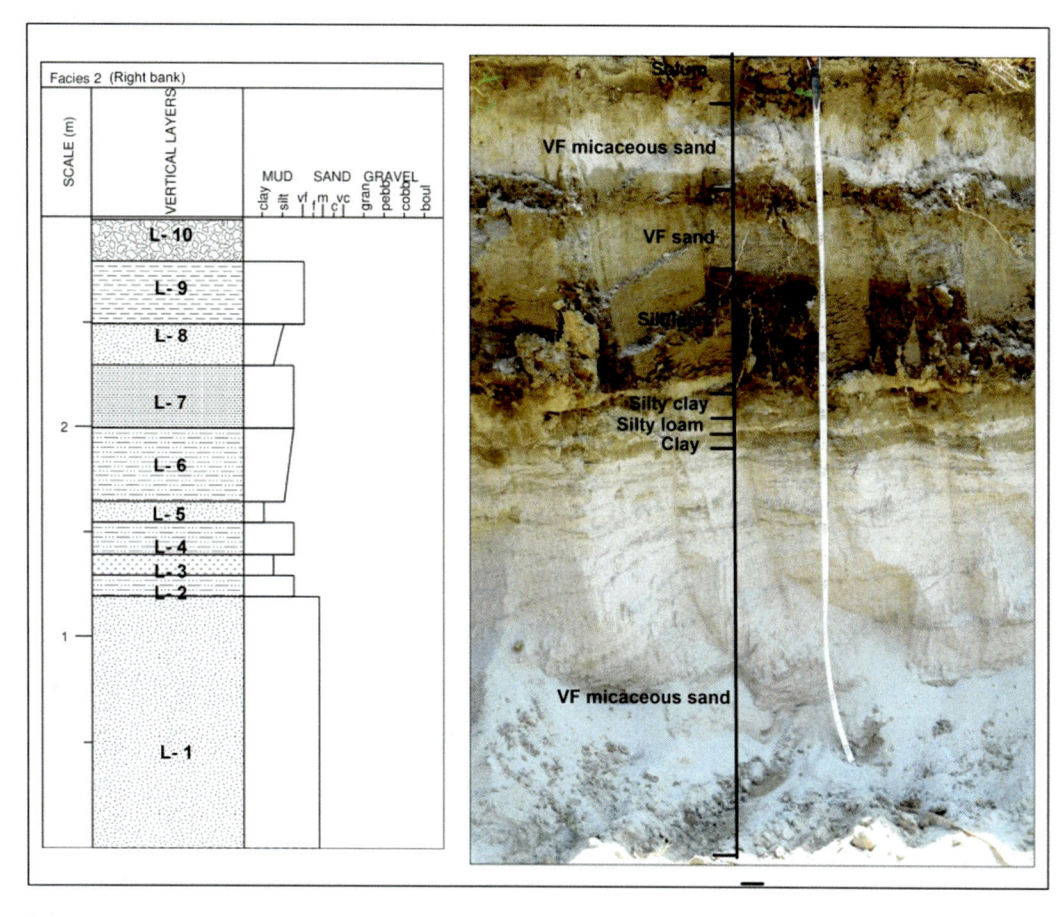

Fig. 6.3 Vertical river bank facies profile at Dhatrigram

because of the contraction of the layer. Rest of the layers above has cross-bedding structure and bioturbation. Silty loam at the top layer and the plant roots make the bank more stable than the others (Fig. 6.6). Here absence of sand layers unlike the previous profiles makes the bank more resistant to bank erosion.

6.3.2 Bank Erosion Hazard Index and Channel Instability

The study area (Nabadwip to Kalyani) is facing moderate to severe bank erosion. For the evaluation of the banks' random locations identified during field visits. Bank height, bankfull height, root depth ratio, surface cover, and bank angle has been measured have been measured. The field investigation was done in the month of November and January. The combined value is showing the BEHI score and depending on that risk of bank erosion can be assessed. If BEHI scores more than 30 then the bank is not so stable and it will be extremely unstable if it crosses the number 40 (Rosgen and Silvey 1996).

According to the survey, result left bank near Jalangi confluence has a low category of bank erosion but suddenly it has increased just above the Gouranga setu (it is a cross-river construction near Nabadwip having four pillars at a distance of 113 m from one another) and the value of BEHI is 34.30 (sl no. 2) which is high bank erosion-prone zone (Table 6.1). In the Nabadwip reach other to meandering points at Saguna and Bhandartikuri (sl no. 3) is the zone of extreme (44) and very high (39.35) bank erosion (sl no.

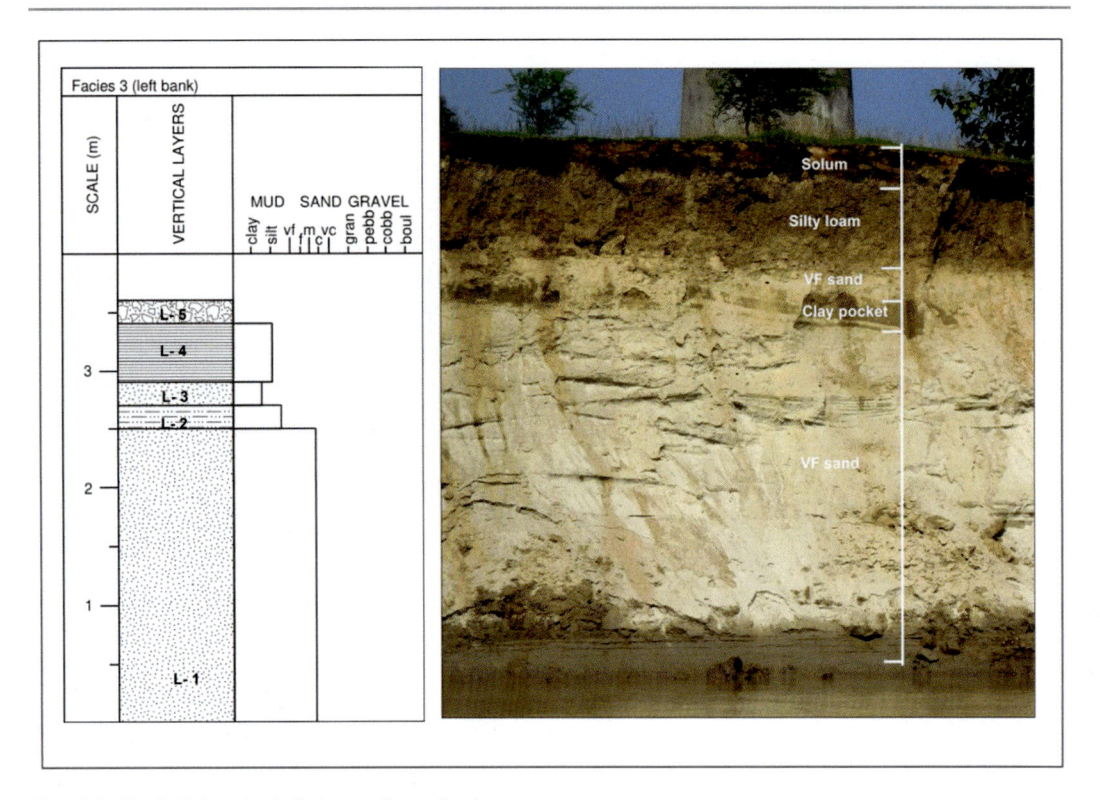

Fig. 6.4 Vertical river bank facies profile at Santipur

4). The bank angle is also high and the surface vegetation coverage is less than 30%. The velocity and channel depth both are high in this region. Bank composition and approaching thalweg make these banks more prone to bank erosion. Fifth point near Bholadanga has been taken for stability assessment, which can be stated as stable. At downstream sl no. 6 is showing the maximum BEHI number, i.e., 40.95 as it has 92° bank angle, 25% surface protection, 13% root density, 14% root depth height and 2.20 is the ratio of bank height to bankfull height (Table 6.1). A cut-off has occurred beside Krisnadebpur and makes the flow straight nowadays and BEHI score is 13.75 (sl no. 13) for both the banks proving that bank erosion hazard probability is low and the banks are stable almost. At Kalna (sl no. 9) though the right bank of the river has 75° slope but the surface protection and bank monitoring make it less erosion-prone where BEHI value is 17.25 (moderate). Near Guptipara there is a bend in river Hugli which has changed the flow direction from southeast to northeast, also due to very less surface coverage and less root density the bank cannot withstand against the chaotic flow of water. BEHI score is 38.9 which is susceptible to bank erosion. Other banks are more or less stable due to their low bank height, low velocity, and more vegetative cover. Even the bank just at Churni confluence has moderate to high bank erosion with 26.7 index value (sl no. 16).

Balagari Char (left bank), Durllabhpur (left bank), Rukeshpur near Baneswarpur (right bank), Char Jirat (left bank), Sija (right bank), Ganga Manoharpur (right bank) are the bank observing sites in Jirat reach. Among them, all near Baneswarpur right bank of Bhagirathi is the most erosion-prone zone (sl no. 19). Velocity and depth of the channel are very high at this point which has been shown in the previous chapter. Bank angle of almost 90°, minimal surface coverage, high vertical difference between bank height and bankfull height trigger the process of

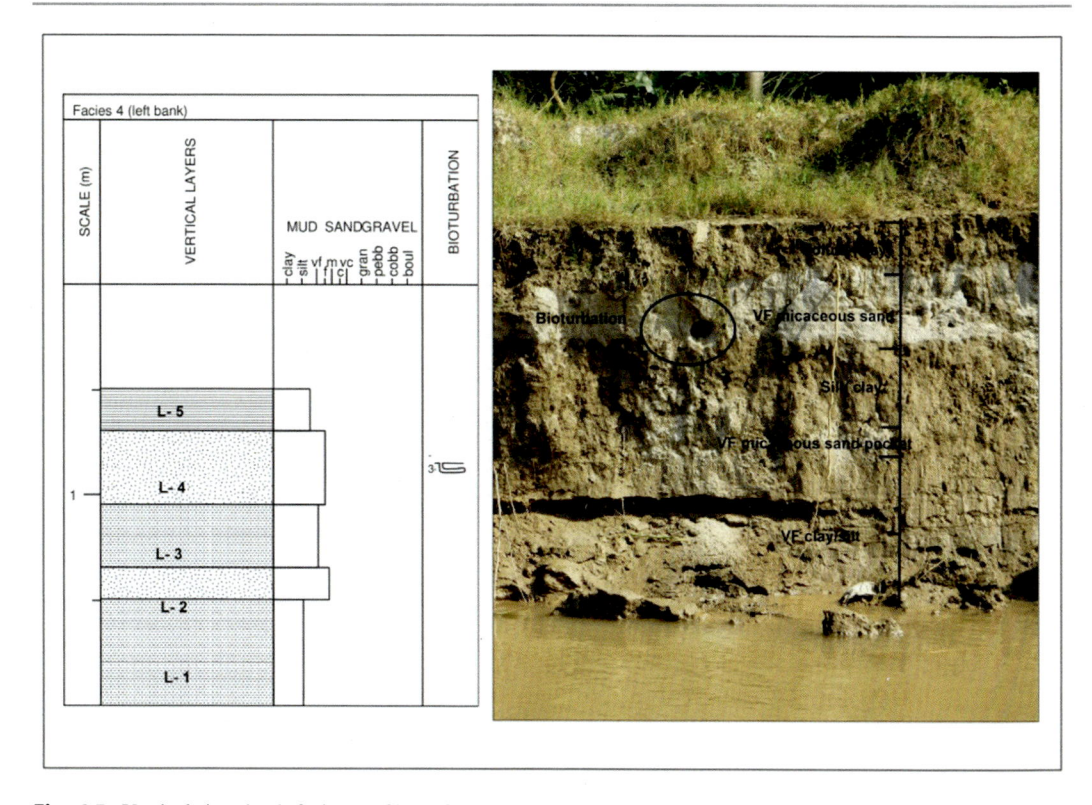

Fig. 6.5 Vertical river bank facies profile at Chak Noapara

bank erosion and channel widening (Table 6.1). The opposite left bank is also susceptible to bank erosion (BEHI score—39.35). Other banks are of least concern in terms of bank erosion as they indicate low index value, thus low bank erosion rate (Fig. 6.7).

6.3.3 Reach Scale Bank Erosion and Land Loss

River Bhagirathi-Hoogy is in its estuarine flow from Murshidabad. After opening of Feeder canal, the main Bhagirathi river has avulsed from its former path in many places. 2.64 km long Farakka Barrage was constructed to divert 40,000 cusec of water into Bhagirathi. This has undoubtedly increased the water supply but the southern districts were badly affected. The upper tidal reach of Hugli river is considered from Nabadwip to Khamargachi. It mainly covers three districts Nadia, Burdwan, and Hugli. Riverbank erosion and deposition have the direct impact on the floodplain of Bhagirathi-Hugli. Riverbank erosion left the imprints of channel instability on the floodplain and reshape the channel morphology reach (Mondal and Satpati 2012). The incidents of bank erosion at different vulnerable and near vulnerable locations during the high monsoonal discharge time increase the sediment with bulk suspended sediment supply within the channel. Velocity, depth, and bank configuration is the combination which is responsible for bank erosion.

From the images of 1972 and 2019, a temporal change has been detected and along with that, the eroded land between the older and newer course of Bhagirathi-Hugli has also been calculated (Fig. 6.8).

Since 1972 the floodplain of Bharirathi-Hugli has undergone many changes like avulsion and cut-off, and mid-channel bar development. Many mouzas got affected due to bank erosion. In Nabadwip reach Bankar dhopali and Satkulta is

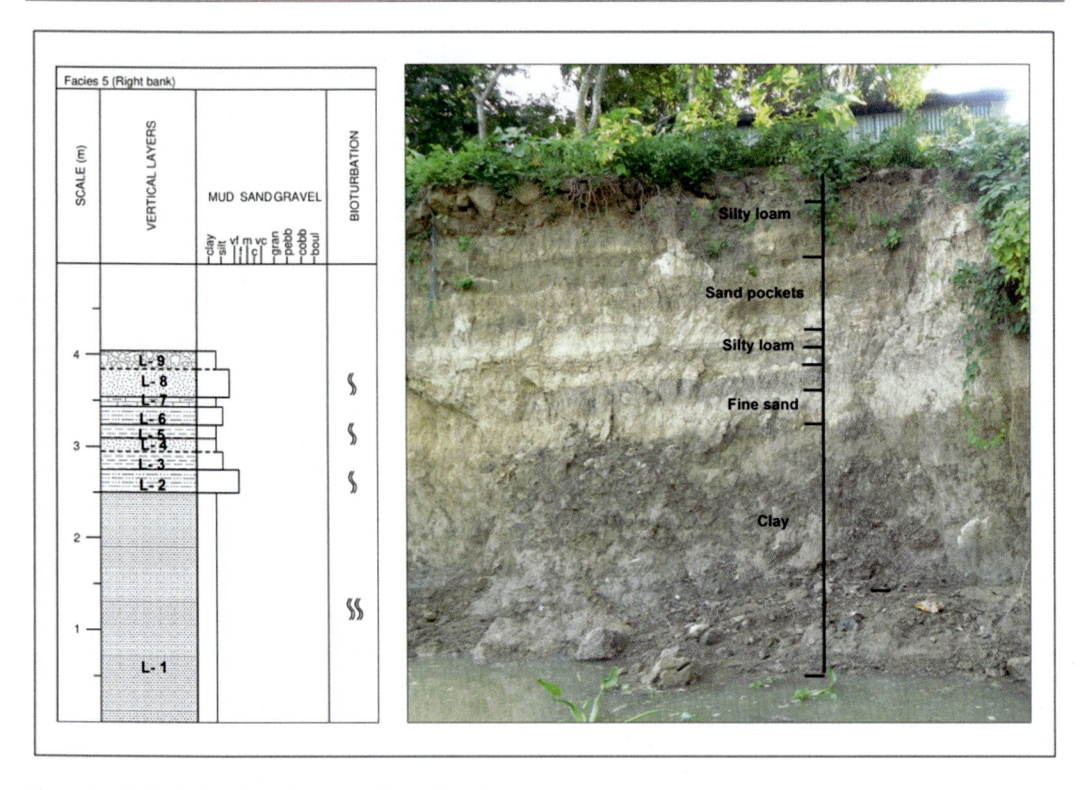

Fig. 6.6 Vertical river bank facies profile at Chandra

showing maximum eroded land, i.e., 1.53 and 1.29 km², respectively. Parmedia, Godkhali, Tegharipara, Bhaluka and Bhandartikuri also lost their land at a moderate scale. Amount of land loss is greater in Kalna Dhatrigram reach because a cut-off occurred after 1955 opposite to Kalna and the river was adjusting its flow within a straighter course. Char Malatipur, Nrisangsapur has the maximum eroded land in this reach (0.51, 0.58). Natunmath Bholadanga, Media, and Krisnadebpur are queued after Char malatipur and Nrisangsapur in terms of land loss (Table 6.2; Fig. 6.9).

Balagarh reach has two consecutive meander bars but did not pose any serious land loss issue in the recent past. Here only Santipur, Phulia, Beharia, and Brittir Char had experienced land loss (not more than 0.5 km²). But Rasulpur Char mouza above Sabujdwip sq.km land 1.5 km² area as gone through land loss, but presently the incident of bank erosion is not so frequent in this region. Par Niamatpur Mouza at Churni confluence had lost its 0.44 km² land lost due to bank erosion and shifting of channel. After confluence, Char Jirat is the Mouza which has highest degree of land loss followed by Tarinipur and Char Jajira. Char Jirat had lost 3.18 km² land in last 44 years.

Vulnerable zones have been identified (2 km buffer zone) by the technique of simple kriging of shifting of meander axis from 1972 to 2016. It is based on the old paths and cut-offs that are left behind on the floodplain. It can be seen from the map (Fig. 6.10) that mostly the meandering part of reach 1 and reach 4 are vulnerable that the other reaches. The meander axis is shifting on a large scale at these places. Due to the kinetic energy that the river achieved after the Churni outfall and reduce in width of the river high rate of meander axis shifting occurred. Lower part of the river (Tribeni) is affected by tidal water thus the banks are almost stable Below Baneswarpur. On the other hand Eocene Hinge line passes near to Pairadanga, which has a minor effect on

Table 6.1 Location-wise BEHI parameters and compound scores

No.	Place	Location	BH/BFH	Score	RDH (%)	Score	RD (%)	Score	SP (%)	Score	BA (°)	Score	Total
1	Swarup ganj	23° 24′ 45.46″ N 88° 22′ 45.63″ E	1.12	2.95	41	4.95	75	2.95	70	2.95	70	4.95	18.75
2	Gadkhali	23° 23′ 12.41″ N 88° 22′ 20.71″ E	2.42	8.5	20	6.95	25	6.95	35	4.95	82	6.95	34.30
3	Satkulta	23° 22′ 53.31″ N 88° 19′ 47.49″ E	4.67	10	6	8.5	10	8.5	10	8.5	100	8.5	44.00
4	Bhandarkuri	23° 21′ 03.39″ N 88° 19′ 36.94″ E	3.8	10	16	6.95	10	8.5	25	6.95	90	6.95	39.35
5	Notun math	23° 18′ 08.89″ N 88° 21′ 28.93″ E	1.15	2.95	22	6.95	40	4.95	50	4.95	56	2.95	22.75
6	Piari nagar	23° 15′ 41.40″ N 88° 19′ 52.42″ E	2.20	8.5	14	8.5	13	8.5	25	6.95	92	8.5	40.95
7	Krishna debpur	23° 14′ 55.28″ N 88° 21′ 09.61″ E	1.005	1.45	60	2.95	35	4.95	90	1.45	40	2.95	13.75
8	Nrishangsapur	23° 15′ 41.40″ N 88° 19′ 52.42″ E	1.183	1.45	62	2.95	45	4.95	60	2.95	29	2.95	13.75
9	Kalna	23° 13′ 33.33″ N 88° 22.04.94″ E	1.508	4.95	96	1.45	60	2.95	75	2.95	75	4.95	17.25
11	Char sultanpur	23° 13′ 00″ N 88° 26′ 00″ E	4.68	10	14	8.5	15	6.95	10	8.5	80	4.95	38.9
12	Chenra char	23° 13′ 07.82″ N 88° 28.34.61″ E	1.86	6.95	22	6.95	20	6.95	25	6.95	65	4.95	32.75
13	Chak Noapara	23° 07′ 46.40″ N 88° 27′ 14.41 E	1.492	4.95	32	4.95	60	2.95	75	2.95	60	2.95	18.75

(continued)

Table 6.1 (continued)

No.	Place	Location	BH/BFH	Score	RDH (%)	Score	RD (%)	Score	SP (%)	Score	BA (°)	Score	Total
14	Chak Noapara	23° 07′ 52.24″ N 88° 27′ 16.28″ E	2.50	8.5	18	6.95	20	6.95	30	4.95	95	8.5	35.85
15	Chandra	23° 07′ 06.42″ N 88° 30′ 47.45″ E	1.117	1.45	50	2.95	35	4.95	40	4.95	50	2.95	17.25
16	Parnia matpur	23° 03′ 59.56″ N 88° 29.12.22″ E	1.56	6.95	15	6.95	30	4.95	70	2.95	80	4.95	26.7
17	Balagari char	23° 03′ 59.56″ N 88° 30′ 47.45″ E	1.10	1.45	51	2.95	45	4.95	60	2.95	40	2.95	15.25
18	Dullabh pur	23° 04′ 41.53″ N 88° 29′ 36.68″ E	1.00	1.45	34	4.95	50	4.95	70	2.95	10	1.45	15.75
19	Rukesh Pur	23° 01′ 11.29″ N 88° 26′ 49.84″ E	4.62	10	5	10	8	8.5	10	8.5	90	6.95	43.95
20	Charjirat	23° 03′ 46.04″ N 88° 27′ 16.40″ E	3.74	10	10	8.5	15	6.95	20	6.95	90	6.95	39.35
21	Sija	23° 02′ 51.91″ N 88° 27′ 30.66″ E	1.42	4.95	40	4.95	40	4.95	80	1.45	45	2.95	19.25
22	Gangamanoharpur	23° 00′ 25.00″ N 88° 02′ 31.52″ E	1.50	4.95	60	2.95	82	1.45	90	1.45	60	2.95	13.75

Table 6.2 Mouza wise land loss estimation for the period of 1972–2019

Nabadwip reach			Dhatrigram Kalna reach			Balagarh reach			Jirat reach		
Ref No.	Mouza	Erosion (km^2)	Ref No.	Mouza	Erosion (km^2)	Ref No.	Mouza	Erosion (km^2)	Ref No.	Mouza	Erosion (km^2)
3	Parmedia	0.17	14	Manik Nagar	0.17	23	Santipur	0.32	62	Durllabhpur	0.6
2	Godkhali	0.3	15	Bholadanga	0.005	24	Beharia	0.5	63	Suksagar	0.03
4	Bankar Dhopadi	1.53	16	Natunmath Bholadanga	0.39	29	Phulia	0.5	67	Baneswarpur	0.16
5	Mohisunra	0.33	17	Chak Bholadanga	0.39	30	Brittir Char	0.42	68	Moktarpur	0.16
6	Teghari	0.39	18	Media	0.26	32	Nilnagar	0.2	69	Sija	0.01
11	Satkulta	1.29	19	Char Malatipur	0.51	34	Pumila	0.12	63	Suksagar	0.37
13	Bhaluka	0.44	20	Gayespur	0.15	34	Malipota	0.6	71	Dumurdahadham	0.8
46	Bhandartikuri	0.5	21	Nrisingsapur	0.58	49	Char Sultanpur	0.7	117	Char Jirat	3.18
47	Ekdala	0.14	23	Char Haripur	0.3	54	Char Rampur	0.1	121	Tarinipur	0.55
48	Paranpur	0.1	37	Gram Kalna	1.1	55	Rasulpur Char	1.5	125	Teligacha	0.35
–	–	–	38	Piarinagar	0.12	115	Raninagar	0.35	127	Char Madhusudanpur	0.2
–	–	–	39	Krisnadebpur	0.26	76	Jhau Mahal	0.18	129	Ganga Manoharpur	0.09
–	–	–	42	Kalna	0.05	56	Chak Noapara	0.03	130	Char Jajira	0.58
–	–	–	48	Paranpur	0.1	84	Par Niamatpur	0.44	131	Sarati	0.09

Fig. 6.7 Bank erosion hazard map of 2 km buffer zone of the river Bhagirathi-Hugli

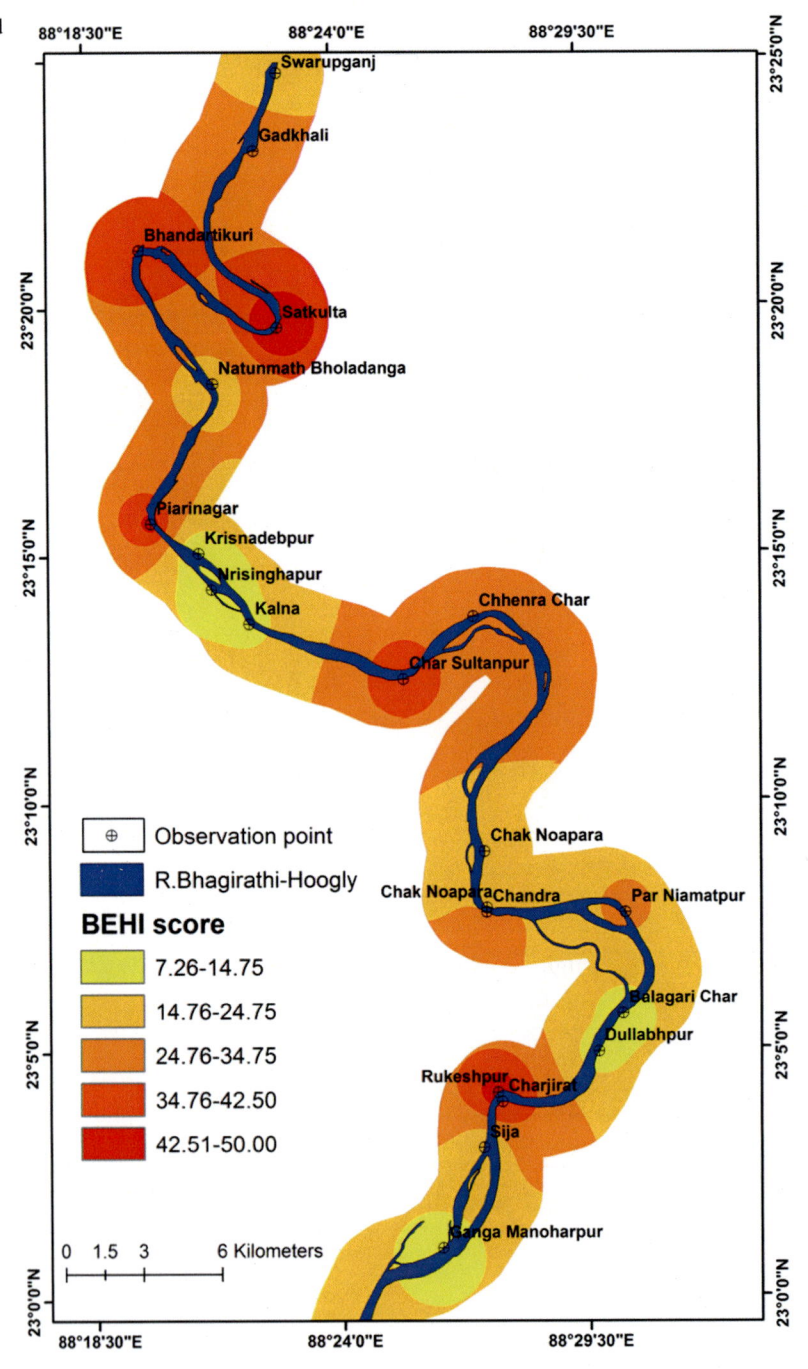

Fig. 6.8 Conditions of river banks. **a** Right bank at Krisnadebpur. **b** Right bank at Guptipara brickfield. **c** Left bank at Santipur. **d** Left bank between Par Niamatpur and Balagari Char

channel avulsion. High kinetic energy of river, hydrodynamics, and tidal influences are the triggering factor developing vulnerable zones. The middle part of the river within the study area is stable after the cut-off at Kalna.

6.4 Conclusion

Channel shifting and bank erosion is an important dynamic aspects of geomorphological evolution and habitat characteristics in large alluvial rivers. The lower segment or deltaic flow of river Ganga is highly engaged in delta building process. The river carries tons of alluvium every year and deposits at the mouth of Bay of Bengal. The river is migrating toward west at downstream of Farakka Barrage. From the Bank, erosion is a common phenomenon as well as threat to socio-economic life of the floodplain dwellers. Application of BEHI method and land loss estimation is somewhat co-related. Those areas that face massive scale bank erosion are also places of land loss. In the study area lower part of Nabadwip reach is unstable than the others. Piarinagar near Khari confluence is prone to bank erosion and also land loss. Bank composition (90% silt) makes the bank fragile and less resistant against bank erosion. Baneshwarpur of Jirat reach is also showing high instability because of the extreme velocity and depth over this region. Krisnadebpur, Kalna, Balagarh, Par Niamatpur has high stability of bank as the course of the river is straight in this region so as the flow.

Fig. 6.9 Mouza wise land loss estimation due to channel shifting (1972–2019)

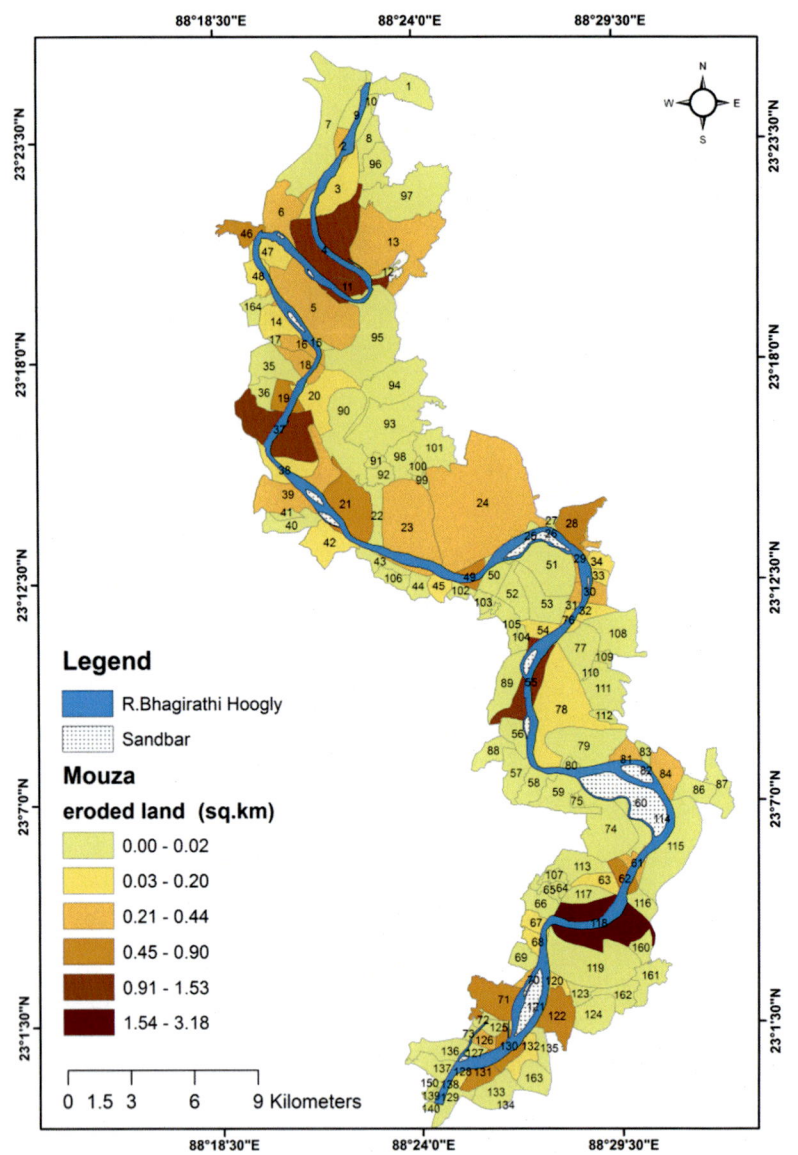

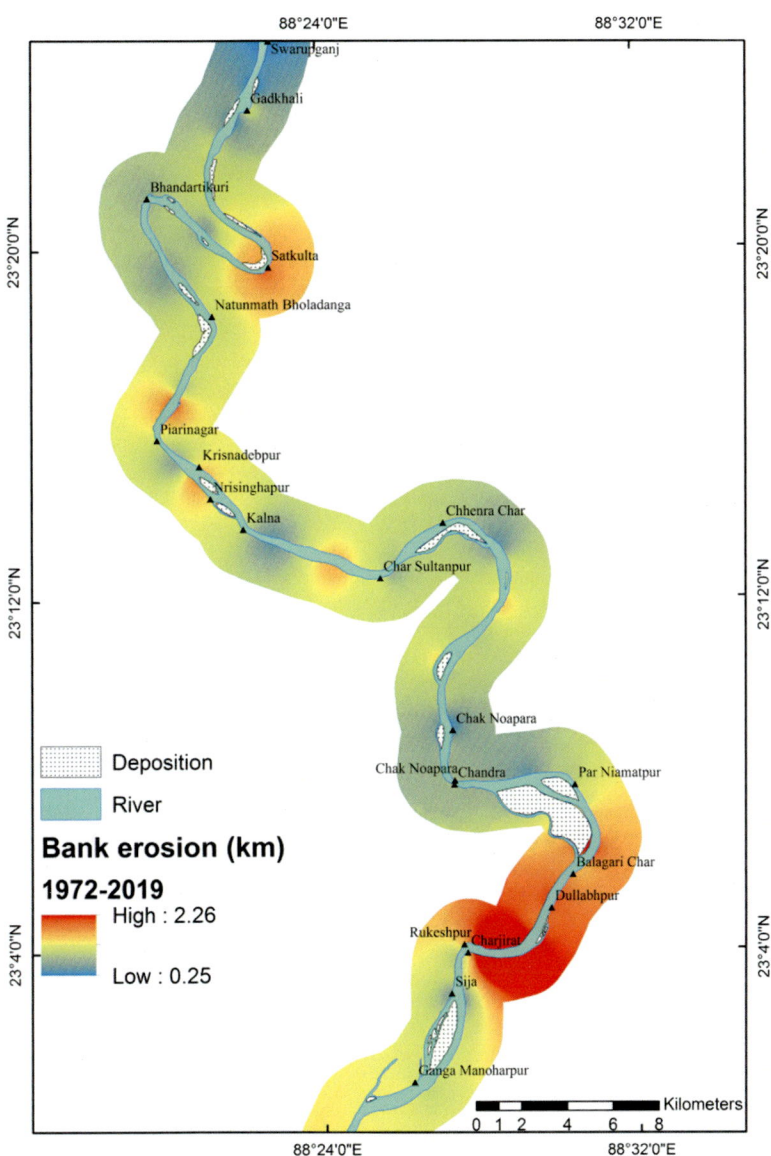

Fig. 6.10 Channel instability assessment on both the banks of Bhagirathi-Hugli river

References

Allen J (1965) A review of the origin and characteristics of recent alluvial sediments. Sedimentology 5:89–191

Brice JC (1964) Channel patterns and terraces of the Loup rivers in Nebraska. Geological survey professional paper 422-D, D1–D41

Bridge JS, Smith ND, Trent F, Gabel SL, Bernstein P (1986) Sedimentology and morphology of a low-sinuosity river, Calamus River, Nebraska Sand Hills. Sedimentology 33:851–870

Cant DJ, Walker RG (1976) Development of a braided-fluvial facies model for the Devonian Battery Point Sandstone, Québec

Cant DJ, Walker G (1978) Fluvial process and facies sequence in the sandy braided South Saskatchewan River, Canada, pp 625–648

Chang K (2010) Introduction to geographic information systems. Tata McGraw Hill Education Private Limited, New Delhi

Cross TA, Homewood PW (1997) Aman's Gressly's role in founding modern stratigraphy. Geol Soc Am Bull 109(12):1617–1630. http://doi.org/10.1130/0016-7606(1997)109<1617:agsrif>2.3.co;2

Darby S, Thorne C (1993) Approaches to modeling width adjustment in curved alluvial channels

Darby SE, Thorne CR (1994) Prediction of tension crack location and riverbank erosion hazards along destabilized channels. Earth Surf Process Land 19:233–245. https://doi.org/10.1002/esp.3290190304

Darby S, Thorne C (1996) Bank stability analysis for the upper missouri river. Nottingham Univ (UNITED KINGDOM)

Erskine W, McFadden C, Bishop P (1992) Alluvial cutoffs as indicators of former channel conditions. Earth Surf Process Land 17:23–37

Friend PF, Sinha R (1993) Braiding and meandering parameters. In: Best JL, Bristow CS (eds) Braided rivers. Special publication 75, Geological Society of London, London, pp 105–111

Ghosh S, Guchhait SK (2014) Palaeoenvironmental significance of fluvial facies and archives of Late Quaternary deposits in the floodplain of Damodar River, India. Arab J Geosci 7:4145–4161

Ghosh S, Mistri B (2012) Hydrogeomorphic significance of sinuosity index in relation to river instability: a case study of Damodar River, West Bengal, India. Int J Adv Earth Sci 1:49–57

Guchhait SK, Islam A, Ghosh S, Das BC, Maji NK (2016) Role of hydrological regime and floodplain sediments in channel instability of the Bhagirathi River, Ganga-Brahmaputra Delta, India. Phys Geogr. https://doi.org/10.1080/02723646.2016.1230986

Julien PY (2012) Erosion and sedimentation, 2nd edn. Cambridge University Press, United Kingdom

Kale VS (1999) Late Holocene temporal patterns of palaeofloods in central and western India. Man Environ 24:109–115

Kale VS, Rajaguru SN (1987) Late Quaternary alluvial history of north-western Deccan upland region. Nature 325:621–614

Kale VS, Achyuthan H, Jaiswal MK, Sengupta S (2010) Palaeoflood records from upper Kaveri River, southern India: evidence for discrete floods during Holocene. Geochronometria 37:49–55

Kochel RC, Baker VR (1988) Paleoflood analysis using slackwater deposits. In: Baker VR, Kochel RC, Patton PC (eds) Flood geomorphology. Wiley, New York, NY, pp 383–422

Leopold LB, Wolman MG (1957) River channel patterns: braided, meandering and straight. Geological survey professional paper 282-B, pp 39–85

Leopold LB, Wolman MG (1960) River meanders. Geol Soc Am Bull 71:769–794

Leopold LB, Wolman MG, Miller JP (1992) Fluvial processes in geomorphology. Dover, New York, NY

Lohnes R (1991) A method for estimating land loss associated with stream channel degradation. Eng Geol 31:115–130

Lohnes R, Handy RL (1968) Slope angles in friable loess. J Geol 76:247–258

Miall AD (1985) Architectural-element analysis: a new method of facies analysis applied to fluvial deposits. Earth Sci Rev 22:261–308

Miall AD (2006) The geology of fluvial deposits. Springer, New York, NY

Miall AD (2014) Fluvial depositional systems. Springer, New York, NY

Mondal M, Satpati LN (2012) Morphometric setting and nature of bank erosion of the Ichamati River in Swarupnagar and Baduria blocks 24-Parganas (N) West Bengal. Indian J Spat Sci 3(2):35–41

Mueller JE (1968) An introduction to the hydraulic and topographic sinuosity indexes. Ann Assoc Am Geogr 58:371–385

Nanson GC, Croke JC (1992) A genetic classification of floodplains. Geomorphology 4:459-486

Pal T, Mukherjee PK (2008) "Orange sand"—a geological solution for arsenic pollution in Bengal Delta. Curr Sci 94:31–33

Pal T, Mukherjee PK (2010) Search for groundwater arsenic in Pleistocene sequence of the Damodar River flood plain, West Bengal. Indian J Geosci 64:109–112

Parua PK (2009) The Ganga—water use in the Indian subcontinent. Springer, Dordrecht

Rajaguru SN, Deotare BC, Gangopadhyay K, Sain MK, Panja S (2011) Potential geoarchaeological sites for luminescence dating in the Ganga Bhagirathi-Hugli Delta, West Bengal, India. Geochronometria 38:282–291

Reineck HE, Singh IB (1980) Depositional sedimentary environments. Springer, New York, 549 p

Rosgen DL (1994) A classification of natural rivers. CATENA 22:169–199

Rosgen DL (2001) A practical method of computing streambank erosion rate. In: Proceedings of the 7th Federal Interagency Sedimentations conference, vol 2, pp 9–15, March 25, 2001, Reno, N.V.

Rosgen DL, Silvey HL (1996) Applied river morphology. Wildland Hydrology Books, Fort Collins CO

Schumm SA (1963) A tentative classification of alluvial river channels. U. S. geological survey circular 477, pp 1–10

Schumm SA (1985) Patterns of alluvial rivers. Annu Rev Earth Planet Sci 13:5–27

Simon A et al (1996) Channel adjustment of an unstable coarse grained alluvial stream: opposing trends of boundary and critical shear stress. Earth Surf Proc Land 21:155–180

Simon A, Thorne CR (1996) Channel adjustment of an unstable coarsegrained stream: opposing trends of boundary and critical shear stress, and the applicability of extremal hypotheses. Earth Surface Proc Landforms 21(2):155–180

Simon et al (1999) Stream bank mechanics and role of bank and near bank processes. In: Darby S, Simon A (eds) Incised River channels. Wiley, New York, pp 123–152

Singh LP, Parkash B, Singhvi AK (1998) Evolution of the lower Gangetic Plain landforms and soils in West Bengal, India. CATENA 33:75–104

Thorne C (1982) In: Hey RD, Bathurst JC, Thorne CR (eds) Processes and mechanism of river bank erosion in gravel bed rivers. Wiley, Chichester, pp 227–271

Thorne C (1999) Bank processes and channel evolution in North Central Mississippi. In: Darby S, Simon A (eds) Incised river channels. Wiley, New York, pp 97–121

Williams GP (1986) River meanders and channel size. J Hydrol 88:147–164

Wilson CA, Goodbred SL (2015) Construction and maintenance of the Ganges-Brahmaputra-Meghna delta: linking process, morphology, and stratigraphy. Ann Rev Marine Sci 7:67–88

The Pattern of Extreme Precipitation and River Runoff using Ground Data in Eastern Nepal

Shakil Regmi and Martin Lindner

Abstract

The topography and positioning of the Himalaya modulate the atmospheric circulation in South Asia leading to summer monsoon events in Nepal. The spatial distribution of monsoonal rainfall is an essential water source for the regional livelihood. However, the occurrence of short-lived, high-intensity precipitation and river runoff results in socio-economic damages. In this study, we rely on 40 years of daily data from 53 precipitation and 35 years of daily data from 14 river runoff ground stations of Eastern Nepal and the Koshi River basin to derive extreme thresholds, amounts and days at the 99th percentile and power-law fit of the precipitation magnitude-frequency relation. We observe extreme precipitation amounts related to the topography relative to days of occurrence and thresholds with the elevation. Likewise, events of extreme river runoff show a strong relationship with the catchment area of the river rather than the topography of the region. Furthermore, precipitation stations below 1 km elevation exhibit a strong power-law relation with magnitude-frequency considerations. Stations at lower elevations generally have lower values of power-law exponents than high elevation areas. This suggests a fundamentally different, yet intense, behavior of the rainfall and flood distribution with an increased occurrence of extreme events in the low elevation regions. These findings are important for hazard assessment and mitigation strategies.

Keywords

Himalaya · Nepal · Koshi River · Monsoon · Extreme precipitation · Extreme river runoff

7.1 Introduction

The hydrogeomorphology of the Himalaya region is dominated by the Indian Summer Monsoon (ISM) (Hannah et al. 2005; Bookhagen and Burbank 2010). The annual precipitation amount usually exceeds 80% during the ISM, which in turn significantly influences the river runoff of the region (Shrestha 2000; Wulf et al. 2010; Panthi et al. 2015). More than a billion people rely on this water budget as the primary source of water for their livelihood (Nandargi and Dhar 2011; Subash and Gangwar 2014). The

S. Regmi (✉)
Department of Forestry and Environmental Technology, South-Eastern Finland University of Applied Sciences, 50100 Mikkeli, Finland
e-mail: shakil.regmi@xamk.fi

M. Lindner
Department of Geography Didactics, Martin-Luther-University Halle-Wittenberg, 06120 Halle (Saale), Germany

top 10% of precipitation, which is a mixture of high-intensity and low-frequency events, are a common occurrence of the ISM and are typically considered as extreme precipitation events (Bookhagen 2010; Malik et al. 2012; Karki et al. 2017). Ongoing discussions of global warming and its impact on the intensity-duration and frequency of precipitation add to the intensifying condition of both wet and dry extreme weather events (Eriksson et al. 2009; Sheikh et al. 2014; Panday et al. 2015; Norris et al. 2020). The extreme precipitation occurs due to many interrelated factors, such as synoptic-scale rainstorms, sudden cloudbursts, high winds, and snowstorms (Barros et al. 2004; Nandargi and Dhar 2011). As a result, it is responsible for the control of mass transport of water and sediment discharge and has a first-order control on the high rate of denudation too (Thiede et al. 2004; Gabet et al. 2008; Bookhagen 2010; Wulf et al. 2012). Against this background, the vulnerability of the Himalayan region in terms of natural factors and socio-economic factors is vivid. However, the understanding of the pattern of extreme events of precipitation and river runoff here is still in its infancy when compared to the South Asian region as a whole (e.g., Barros et al. 2004; Haylock and Goodess 2004; Anders et al. 2006; Malik et al. 2012, 2016; Karki et al. 2017; Ashcroft et al. 2019).

Most meteorological extreme events are associated with floods and landslides, if not both (Bookhagen 2010; Wulf et al. 2010; Sheikh et al. 2014). The 2008 sediment influx in the Koshi River of the eastern plains of Nepal leading to the dam failure, and consequent displacement of more than fifty-thousand people, is one of the most studied recent examples of extreme river runoff (Reddy et al. 2008). Likewise, the most recent event of inundation in July 2019 in central and eastern Nepal due to torrential rainfall, again displaced around ninety-thousand people (Reliefweb 2019). Specifically, the Koshi River floods every ISM season between May to October, causing stress on livelihoods and the environment resulting in human fatalities, erosion of agricultural land, and damaging watershed habitats (Reddy et al. 2008). Recent studies from Malik et al. (2016) and Karki et al. (2017) show relevance of these events as the trend of extreme precipitation is increasing in the plains of Nepal and most parts of India. Likewise, Shrestha et al. (2017) report a statistically insignificant increasing trend of the precipitation intensity over the Koshi River basin from 1975 to 2010.

Contrary to the study of trends, the pattern of extreme environmental events for both precipitation and river runoff is underexplored. Few previous studies try to explain the nature of precipitation and river runoff in the Himalaya range but are limited either on the representation of up-to-date data or are juxtaposed with generalized findings. For example, Nandargi and Dhar (2011) and Panthi et al. (2015) report extreme precipitation on a daily basis and note high precipitation amounts and days on the mid-hills compared to the high Himalaya. Similarly, Bookhagen et al. (2005) find two times higher extreme precipitation events in the high Himalaya than in the plains. The increasing trend and fluctuating pattern of extreme precipitations impact the runoff of a river within the catchment area (Wulf et al. 2015). Furthermore, the climate variability in the Himalaya in regards to variability of the elevation is erratic (Nandargi and Dhar 2011). For example, the central Himalaya varies from 60 m to more than 8000 m from south to north within a distance of approximately 120 km (Dahal and Hasegawa 2008). Similarly, the temperature decreases by 6 °C per 1 km of elevation (Mani 1981). This results in a magnified response to climate change with sudden cloudbursts, or snowstorms, leading to flash floods and landslides.

Wulf et al. (2015) report that rainfall, snowmelt and glacier melt are responsible for 55%, 35%, and 10% of the river discharge, respectively in the western Himalaya, whereas Nepal (2012) found 72% of precipitation events responsible for the runoff of the Koshi River basin. Importantly, Gaume and Payrastre (2018) show that the rainfall intensity per time variable is directly related to the river discharge. In addition, there are further intricacies in the Himalayan region, such as the steep gradients of the orogeny, that have steady erosion rates related to the precipitation (Gabet et al. 2008). Likewise, the high denudation rate of the Himalayan orogeny plays a fundamental role in sediment transport and is the dominant source of total suspended sediment in the downstream river environment (Bookhagen et al. 2005; Wasson et al. 2008). The suspended sediment is a critical aspect because its deposition changes the course of river movement, which in turn affects the livelihoods. For example, the alluvial fan of the Koshi River has shifted horizontally by about 115 km during the last two centuries (Sinha et al. 2008).

The impact of intense precipitation over a short period triggers erosion, flash floods, denudation, and sediment runoff (Ives and Misserli 1989; Gabet et al. 2008). To better describe hydro-meteorological extreme events, (and thereby contribute to the current knowledge of Himalayan hydrogeomorphology), we focus this study on the spatiotemporal pattern of extreme environmental events. The major local factor providing high order control over the precipitation has evolved from elevation to topographical relief in the Himalaya region (Barros and Lang 2003; Barros 2004; Anders et al. 2006; Bookhagen and Burbank 2006, 2010; Bookhagen and Strecker 2008). The topography varies gradually from a one-step rise to a two-step rise from eastern Himalaya, east of Nepal, to the west (Bookhagen and Burbank 2006). In this regard, Nepal (central Himalaya) represents topographical characteristics of both the eastern and western Himalaya. In addition, the impact of precipitation and river runoff induced calamities are a common occurrence in the Koshi River basin of eastern Nepal. Hence this study focuses on the eastern part of Nepal which has a one-step rise in topography (cf. Fig. 7.1).

We use 53 ground stations and 14 stations for monitoring precipitation and river runoff, from eastern Nepal and the Koshi River basin, respectively, to analyse the spatiotemporal pattern of extreme environmental events in eastern Nepal. The precipitation stations have daily precipitation data of 40 years from 1975 to 2014, and river runoff stations have daily river runoff data of 35 years from 1974 to 2008. The indices developed by the Expert Team on Climate Change Detection Monitoring and Indices (ETCCDMI) were modified and used in this study (e.g., Croitoru et al. 2016). Indices of extreme environmental events are percentile-based, which are site-specific and allow the comparison between different regions or stations (Haylock and Goodess, 2004; Krishnamurthy et al. 2009; Bookhagen 2010). The elevation of the precipitation stations ranges from approximately 70–2700 m, while that of river runoff ranges from approximately 140–1500 m. The inclusion of high elevation stations in this study provides additional insight into the nature of extreme environmental events over eastern Nepal (e.g., Archer and Fowler 2004; Nandargi and Dhar 2011; Wulf et al. 2012; Karki et al. 2017).

7.2 Geographic Setting of the Study Area

7.2.1 Precipitation System

The Himalayan range extends for ∼2400 km and is a natural barrier between the mid-latitudes and the tropics (cf. Fig. 7.1a). The Tibetan plateau sits north of the Himalayan range and its elevated landmass (average elevation ∼4 km) experiences rapid warming compared to the surrounding oceans and creates a pressure difference originating from the nearby Bay of Bengal (Flohn 1957; Webster 1986). The weather and pressure anomaly is part of the InterTropical Convergence Zone (ITCZ) which is the drifting process of the east–west oriented precipitation belt from the southern hemisphere during winter to the northern hemisphere during summer (Goswami 2005; Bookhagen et al. 2005).

The large-scale cyclonic vortex from the north of the Bay of Bengal to mainland India forms the Indian Summer Monsoon (ISM) trough and derives the inter-annual variability based on the anomaly of sea-surface temperature (Bookhagen et al. 2005; Ding and Sikka 2006). The ISM has an active and a break period; the active period leads to excessive storms, while break periods lead to draughts (Goswami 2005).

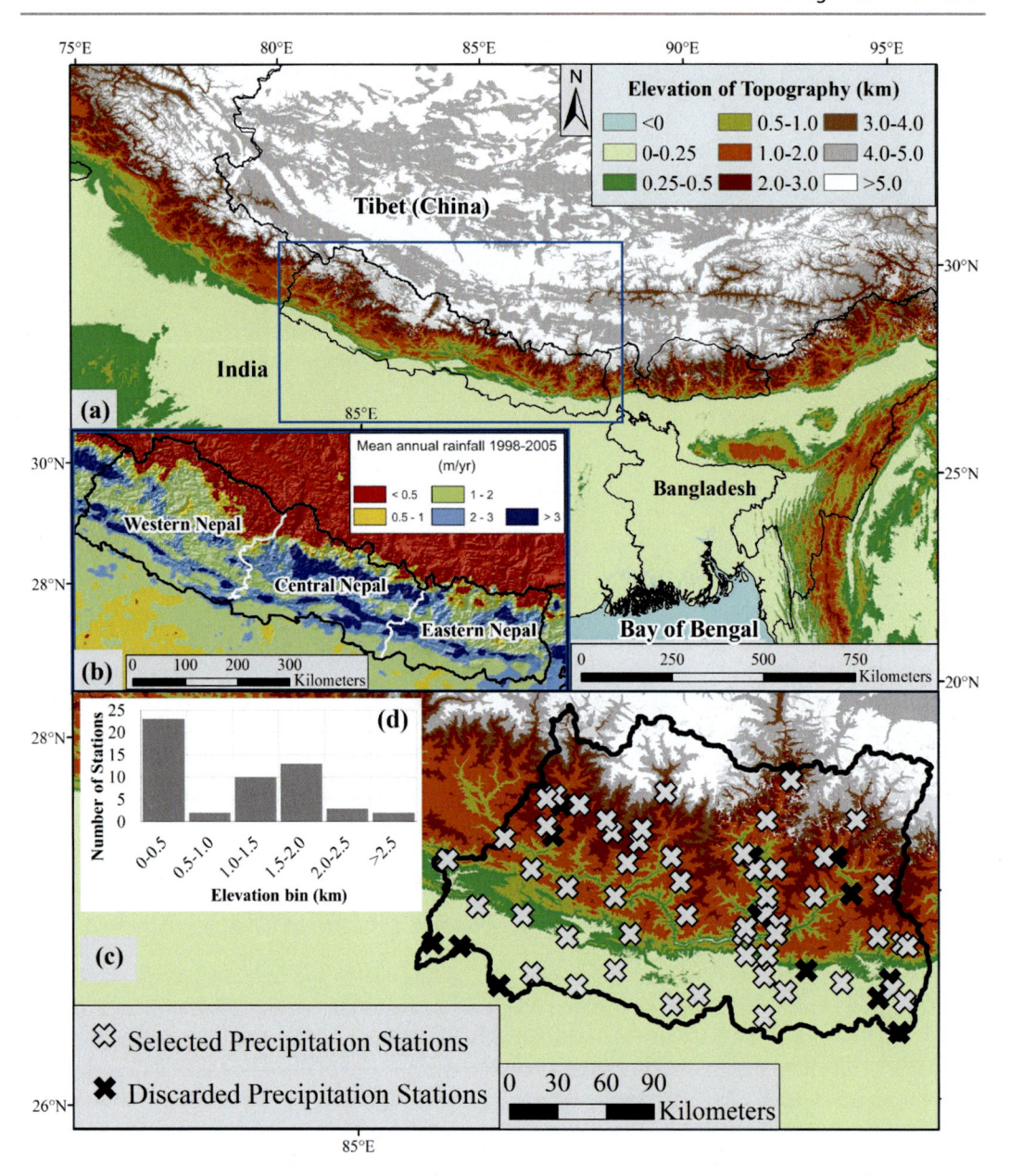

Fig. 7.1 a—Digital topography of the Himalayan arc obtained from the Shuttle Radar Topography Mission (SRTM) with nominal 90 m spatial resolution through the online portal of United States Geological Survey (USGS 2012). The location of Nepal is outlined in the blue box. **b**—Inset shows the average amount of annual rainfall calculated from the Tropical Rainfall Monitoring Mission with a spatial resolution of $\sim 5 \times 5$ km in Nepal, from January 1998 to December 2005 (adapted from Bookhagen and Burbank 2006) including the classification of the topographic swath profile of Eastern Nepal used in this study. **c**—The selected 53 precipitation stations out of the total 67 ground stations from Eastern Nepal. **d**—Frequency of the number of selected stations per elevation bins

The ISM depression enters Nepal through the eastern side and exits through the west, leading to relatively less precipitation in the west. The moisture rises with the topography and decreases pressure and temperature, with an increase in volume leading to the precipitation distribution.

Specifically, this circulation organizes shallow convection at the foothills of south-facing slopes to bring heavy summer monsoon precipitation (Barros and Lang 2003; Bookhagen et al. 2005). Therefore, the positioning and elevation of the Himalaya is essential for the existence of the ISM. In general, the elevation ranges of the Nepal Himalaya are divided into: Terai or Plains (<330 m), Siwalik (200–1500 m), Lower Himalaya (500–3000 m), Higher Himalaya (>2000 m) and Trans Himalaya (2500–4500 m) (Dahal and Hasegawa 2008).

The observable similarity in Fig. 7.1a and b confirms the relationship between the intricate orogeny's relief and precipitation. The relief increases with elevation until the South Tibetan Detachment System divide, after which the Tibetan plateau starts, and here the topography is not intricate but generally uniform. Likewise, there is evidence of strong relief with intense precipitation, but the same relationship with extreme precipitation is yet to be verified (Bookhagen and Burbank 2006, 2010). In general, the precipitation is locally caused by atmospheric and orographic interaction in this region and varies with varying topographical structures.

7.2.2 The Koshi River Basin System

The catchment area of the Koshi River Basin (KRB) starts from the Trans Himalaya till the plains of southern Nepal before mixing with the Ganges river basin of India (Singh et al. 1993). The Koshi River travels ∼200 km with a steepness variation of 55–75 cm/km in the north and 6 cm/km in the south (Sinha et al. 2008). Overall, the elevation of the Koshi River basin ranges from ∼20 m to >8000 m (Devkota and Gyawali 2015). There are various major rivers in the basin namely, Indrawati, Likhu Khola, Dudh Koshi, Tamor and Sun Koshi which lie in Nepal while the Bhote Koshi, Tama Koshi, and Arun lie in both Tibet and Nepal (Nepal 2012). Out of the total basin area, 46% lies in Nepal while the remaining lie in Tibet (Nepal 2016). The confluence of all these rivers forms the Sapta Koshi River, which flows to India and mixes in with the Ganges river.

As the river flows in the high gradient Himalaya, the rate of denudation is also high. Interestingly, the Koshi River basin represents 5% of the Ganges river system but contributes about 25% of the total sediment transport to the Ganges. The total area of the basin is about 57,700 km^2, (see Fig. 7.2 for the geographical location of the Koshi River basin. (Nepal 2012)).

The Koshi River basin on the plains of Nepal is four times wetter than the northern parts (Sharma et al. 2000). Therefore, the Koshi River is known to bring floods in the lowland area of Nepal and India during the monsoon. The significant difference in discharge during the dry and monsoon seasons increases the vulnerability of the river to flooding.

7.3 Data and Method

7.3.1 Data

The precipitation data from 67 ground stations were collected from the Department of Hydrology and Meteorology of Nepal (2016a). The data consist of daily (24 h) observed precipitation (starting from 9 am the previous day) in illimetres (mm). The Department of Hydrology and Meteorology (DHM) uses a US-standard 8-inch (20.32 cm) diameter manual precipitation gauge. The snowfall is measured as a standard rainfall measurement by melting the snow to calculate snow water equivalent. To include stations with uniform length of record, stations with data from 1975 to 2014 were included, while stations with data gaps leading to less than 40 years of data were removed leading to 53 stations, as seen in Fig. 7.1c. Outliers in the data were controlled by the DHM, while the remaining outliers were kept so as not to miss the extreme events (cf. Bohlinger and Sorteberg 2018). The small amount of missing daily data (less than 2%) were left as blank and did not significantly affect the results.

The river runoff data were also collected from the Department of Hydrology and Meteorology (2016b), which uses the cableway system to measure the daily (24 h) runoff in cubic meters

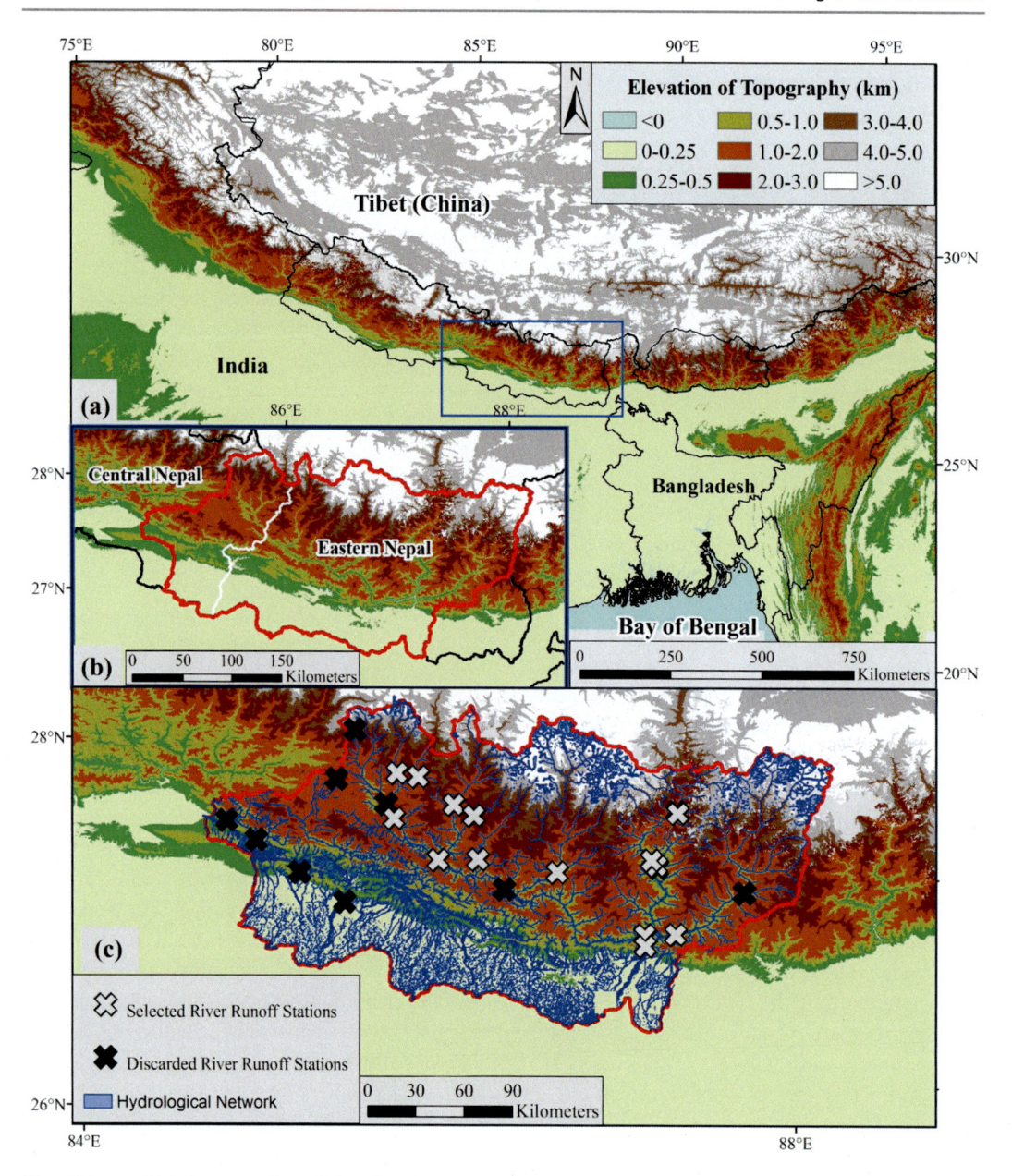

Fig. 7.2 a—Digital topography of the Himalayan arc obtained from the Shuttle Radar Topography Mission (SRTM) with nominal 90 m spatial resolution through the online portal of United States Geological Survey (USGS 2012) and location of Eastern Nepal is outlined by the blue box. **b**—Inset shows the location of the Koshi River basin in Nepal with the border shown in red. **c**—Selected 14 river runoff stations out of the total 23 ground stations from the Koshi River basin

per second (m^3/s). The Koshi River basin is located in 27 districts of Nepal (Hussain et al. 2018). The DHM had data of 23 river runoff stations from the KRB districts. Out of these 23 stations, the 14 final stations for analysis were selected based on the location within a tributary of the main Koshi River known as the Sapta Koshi River, and the amount of daily data available, as seen in Fig. 7.2. The Digital Elevation Model (DEM) of Nepal was obtained from the Shuttle Radar Topography Mission (SRTM), nominal 90 m spatial resolution through the online portal of United States Geological Survey (USGS 2012).

7.3.2 Method of Analysis

The swath profile of east Nepal was divided by an approximately equal distance of longitudes while considering the state district boundaries. Likewise, the study area of KRB was obtained from its corresponding districts. Importantly, the precipitation and KRB study area differ due to the difference of topographical composition between eastern and central Nepal (cf. Fig. 7.1) while the KRB resides in parts of central Nepal too. The topographic swath profile of eastern Nepal was calculated for comparative analysis with the precipitation indices from the SRTM DEM of Nepal. Likewise, the topographic relief was calculated by taking the difference between maximum and minimum elevation within the 5 km radius area from the precipitation stations. The station statistics for precipitation and river runoff were calculated based on the indices presented in Table 7.1.

The indices were calculated for three different time periods, High Precipitation Months (May to October), Low Precipitation Months (November to April), and all months. The calculation was performed per time series of each time period to analyse the respective patterns of indices. The findings of each station are presented graphically or spatially with the categories of elevation, relief and elevation bins of 0.5 km. The extreme threshold for both precipitation and river runoff was the 99th percentile, i.e., events above the 99th percentile from the time series. For spatial comparison of patterns, we also calculate the normalized values of extreme events from the median of the time series (50th percentile).

Furthermore, we conduct a magnitude-frequency analysis of precipitation to estimate high-magnitude and low-frequency events. Here, magnitude is precipitation per day and each event is the sum of at least three or more continuous days of precipitation (>1 mm) including a possible one-day gap of no precipitation. For example, if there are three continuous days of precipitation and the fourth day has no precipitation while the fifth day has precipitation, then this would be one event where magnitude is the sum of total precipitation in these five days divided by number of days. We perform this relation test in different elevation bins and estimate the power-law exponent and its fit using a maximum likelihood method (Clauset et al. 2009). Clauset et al. recommend estimating a minimum value (x_{min}) above which a power-law holds, while the slope of the regression is denoted by α. The scaling parameter or power-law exponent (α) was calculated using Eq. 7.1.

Table 7.1 List of indices, adapted from the Expert Team on Climate Change Detection Monitoring and Indices (ETCCDMI) (Croitoru et al. 2016), was used in this study

Index	Name	Description
R0.1	Total precipitation days	Number of days with >0.1 mm precipitation
PRCPTOT	Total precipitation amount	Total amount of cumulated precipitation
R99	Extreme precipitation threshold	99th percentile threshold value of precipitation
R99d	Extreme precipitation days	Number of days above R99
R99p	Extreme precipitation amount	Total precipitation amount above R99
RROTOT	Total river runoff amount	Total amount of cumulated river runoff
River99	Extreme river runoff threshold	99th percentile threshold value of river runoff
River99p	Extreme river runoff amount	Total river runoff amount above River99
River99d	Extreme river runoff days	Number of days above River99

$$\alpha = 1 + n \left[\sum_{i=1}^{n} \ln \frac{x_i}{x_{\min}} \right]^{-1} \quad (7.1)$$

In Eq. 7.1, x_i, $i = 1, ..., n$, are the observed values of x where $xi \geq x_{\min}$. Here, x is the dataset and $x_{\min}$ is the lower boundary of the power-law behavior. The value which minimizes the Kolmogorov-Smirnov statistic is the best estimate of the $x_{\min}$ calculated using Eq. 7.2, because a too low or high value will either bias the regression or will exclude genuine data points (Shatnawi and Althebyan 2013). The power-law behavior describes the likelihood of high-magnitude, low-frequency events: a low power-law exponent suggests a higher occurrence and magnitude of rare events (Wheeldon and Counsell 2003; Clauset et al. 2009; Wulf et al. 2010; Tang et al. 2020).

$$D = \max|S(x) - P(x)| \quad (7.2)$$

Here D denotes the Kolmogorov–Smirnov statistic. $S(x)$ is the Cumulative Distribution Function (CDF) of the $x_{\min}$ values, and $P(x)$ is the CDF for the best-fitted model when $x \geq x_{\min}$. After estimating α and $x_{\min}$ the p-value was calculated by using the Kolmogorov-Smirnov test by estimating the goodness-of-fit value by applying one thousand iterations between the data and power-law (cf. Clauset et al. 2009). The data are described by a power-law behaviour when the p-value was greater than 0.1.

7.4 Results

7.4.1 Extreme Precipitation and Its Relationship with Topography

The precipitation events of High Precipitation Months (HPM) are identical to those of the annual event. More than 90% and 85% of total precipitation amount and total precipitation days, respectively occur during the HPM. Likewise, out of the annual total precipitation amount and total precipitation days, ∼8% and ∼1% are extreme precipitation amounts and days, respectively. Observations from Figs. 7.3 to 7.4 show that the proportion of extreme precipitation amount is higher in the plains and first rise of topography (around 1 km elevation) than the total precipitation amount during HPM. The pattern of extreme precipitation amount also follows this observation.

Interestingly, the threshold of extreme precipitation also shows two peaks, one at the plains and one at the first rise of topography. However, in general, the threshold value shows a decreasing tendency with elevation while extreme precipitation days increase approximately two-fold with increasing elevation from the plains to the high Himalaya.

The impact of westerly movement, which is the winter precipitation events or winter monsoon, can be seen in the events of Low Precipitation Months (LPM) in Fig. 7.3. Winter monsoon is the result of low-pressure system causing moisture transport to the subtropics of Asia, moving from the west to east (i.e., westerly), which gets captured by the Himalayan orography, leading to precipitation (Barros 2004; Dimri et al. 2006; Cannon et al. 2017). Since the westerly enters from the high Himalaya region, the events of extreme precipitation are also concentrated in the high elevation regions during the LPM. Even in this scenario, the extreme precipitation amount is showing hints of relationship with the topographic gradient band. Moreover, it is important to note that the westerly enter Nepal from the western side and thus its impact in eastern Nepal is comparatively weaker than the HPM when the onset enters from east and southern Nepal.

We find that relief significantly increases with elevation in eastern Nepal. Likewise, the precipitation indices also show a similar pattern with relief, alike topography and elevation. It is interesting to compare the relationship of elevation and relief height with that of extreme precipitation amount in the plains. The plains receive the first onset of monsoon; therefore, it is plausible that the amount of extreme precipitation is higher here. In addition to that, relief height is higher in the plains when the amount of

Extreme Precipitation Indices at 99th Percentile

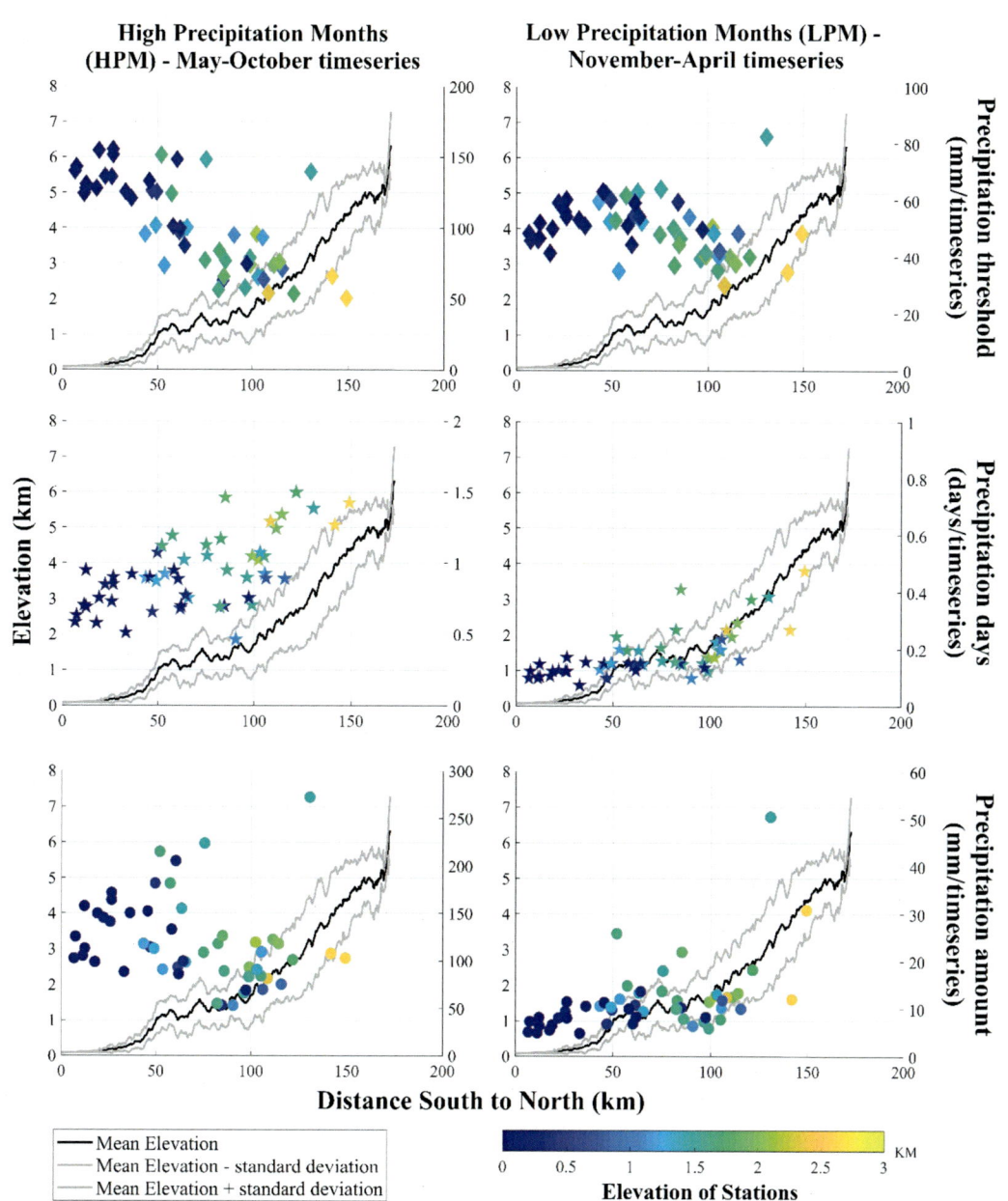

Fig. 7.3 South-to-north topographic profile of Eastern Nepal showing the 99th percentile values during High Precipitation Months (HPM, i.e., May–October) and Low Precipitation Months (LPM, i.e., November–April). The circles represent extreme precipitation amount, the stars represent extreme precipitation days, and the diamond represents extreme precipitation threshold. The color variation of each station represents its elevation, as indicated

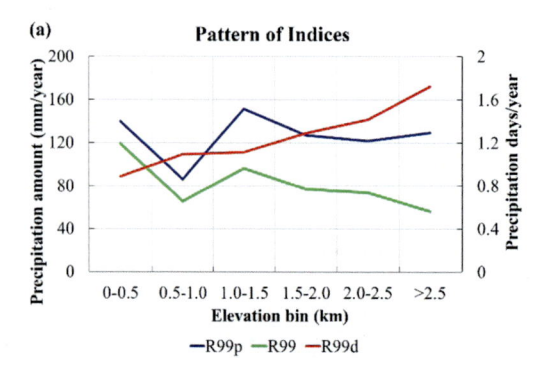

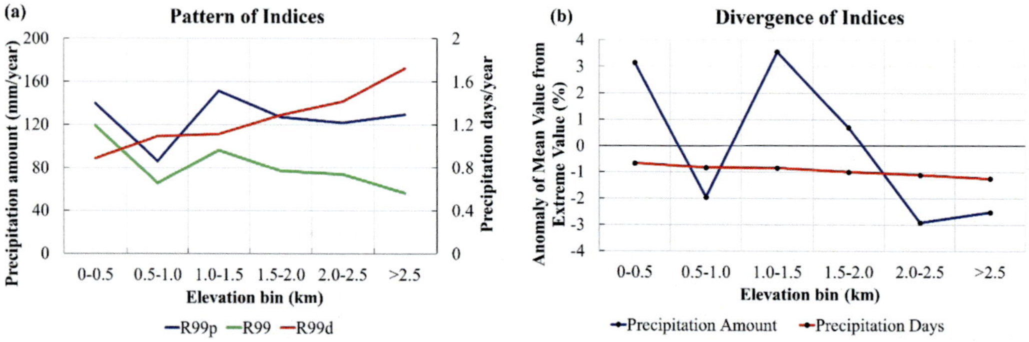

Fig. 7.4 Pattern of extreme precipitation indices (99th percentile) and its comparison with total precipitation events. **a** Mean values of extreme precipitation indices of stations for different elevation bins from less than 0.5 km to more than 2.5 km with a bin size of 0.5 km in Eastern Nepal during all months. **b** Percentage anomaly of divergence (subtracted percentage of mean event from the percentage of extreme event) of extreme amount of precipitation (R99p) from mean amount of precipitation (PRCPTOT); and divergence of extreme number of precipitation days (R99d) from mean number of precipitation days (R0.1). Positive anomaly values refer to higher percentage of extreme events compared to that of mean of total events. This shows that the plains and first rise of topography have more R99p amount compared to higher elevations where the pattern of divergence is lower and negative. In contrast, the anomaly of percentage is close to zero for R99d which is similar to that of R0.1 in Eastern Nepal

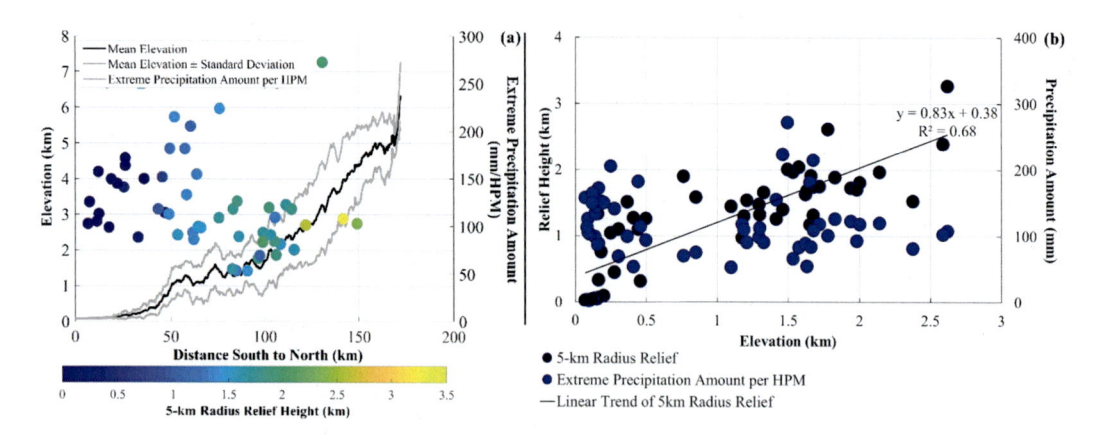

Fig. 7.5 Relation between 5-km radius relief calculated for each station location and its relationship with the values of extreme precipitation amount during High Precipitation Months (HPM, i.e., May–October) in Eastern Nepal at 99th percentile. **a** South-to-north topographic profile of Eastern Nepal showing the 99th percentile extreme precipitation amount when compared to relief shows similar pattern to that of topographic elevation. The color variation of each station represents its 5-km radius relief height. **b** The relationship between relief height and elevation in black circles, and relationship between extreme precipitation amount and elevation in blue circles. Note the high extreme precipitation amount values and high relief height values below the 0.5 km elevation bin, representing a relationship between relief and extreme precipitation amount where there is no topographic step rise to control the precipitation

extreme precipitation is high (cf. Fig. 7.5). The relationship of high relief with high precipitation amount is an established fact. Thus, the high extreme precipitation values of the plains in addition to the topographic one-step rise should be noted in regards to the relationship of extreme precipitation amount with the topographic structure of the Himalaya.

7.4.2 Spatial Distribution and Magnitude-Frequency Relation of Precipitation

The spatial distribution analysis supports the observation of topographic impact on precipitation. The total and, more significantly, the extreme precipitation amount show first and second peak on the plain and first rise of topography, respectively. However, the spatial relationship over the longitudinal difference (west to east) is not visible in eastern Nepal. Usually, the precipitation intensity changes from west to east due to arrival of onset from the east. Similarly, the total and extreme precipitation days also support the previous observations of higher values in higher elevation regions of eastern Nepal (cf. Fig. 7.6). The variance of extreme precipitation threshold (99th percentile) from the median value (50th percentile), when higher, suggests higher values of extreme precipitation. The spatial observation from Fig. 7.7 supports the initial findings of high values in plains and first rise of topography.

Interestingly, the proportion of variance is higher below 1 km elevation compared to higher elevations. The variance of southern slope fronts increases to approximately double that of the higher Himalaya.

The power-law fit of magnitude-frequency relationship of precipitation shows an occurrence of higher-magnitude low-frequency events, which can be attributed to extreme precipitation values (usually >99th percentile). Interestingly, stations from all elevation bins fit the power-law with p-values higher than 0.1 (cf. Table 7.2). However, it is important to note the change of scaling exponents (α) from lower to higher elevation bins. A lower scaling exponent suggests the occurrence of rare and yet high-magnitude events. Moreover, the variance of x_{min} is also critical between different elevation zones.

The comparative lower values of α in lower elevation support the previous findings of high

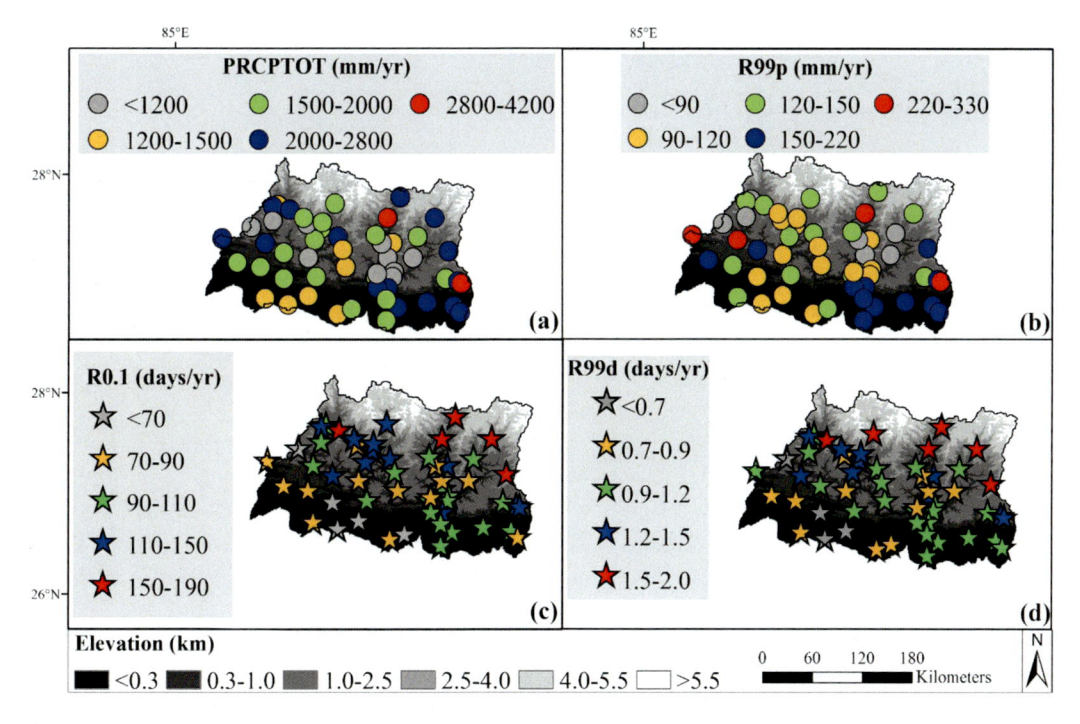

Fig. 7.6 Spatial distribution of annual mean and extreme precipitation events in Nepal. **a** Total mean precipitation amount. **b** Extreme precipitation amount (99th percentile threshold). **c** Number of total mean precipitation days. **d** Number of extreme precipitation days. The circles indicate precipitation amount and stars represent number of precipitation days

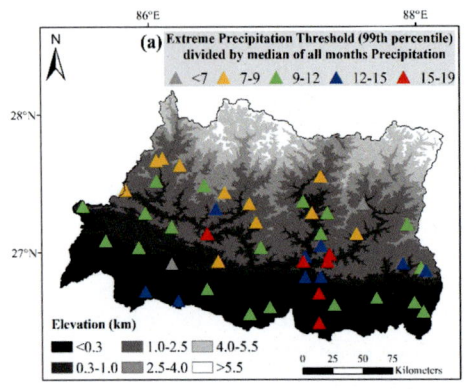

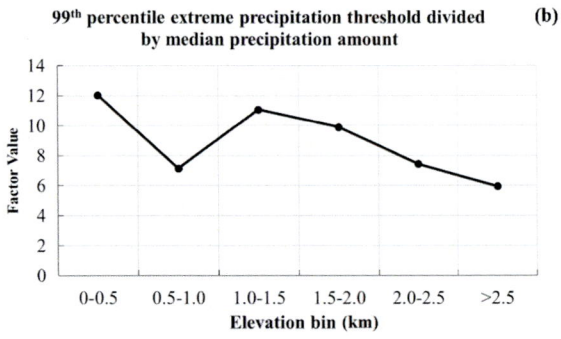

Fig. 7.7 Spatial distribution of extreme precipitation in Eastern Nepal elaborated by showing the 99th percentile extreme precipitation threshold (99th percentile) divided by the median (50th percentile precipitation threshold) during all months. **a** Over the digital elevation model of Eastern Nepal. At the mountain front at the first topographic rise, the 99th percentile is more than ten times higher than the median precipitation. In the higher elevation zones, this relationship is below ten. **b** The same spatial distribution of the extreme precipitation threshold relationship against elevation bins

Table 7.2 The power-law relationship for magnitude-frequency of precipitation for the stations from elevation bins using a maximum likelihood estimator (cf. Clauset et al. 2009)

Elevation Bins (km)	Number of stations in bin	Scaling exponent (α)	α error	x_{min}	x_{min} error	p-value
<0.5	23	5.44	0.50	53.96	8.29	**0.74**
0.5–1.0	2	4.92	0.50	19.67	1.96	**0.96**
1.0–1.5	10	4.99	0.38	38.63	4.21	**0.56**
1.5–2.0	13	5.24	0.34	29.99	3.44	**0.25**
2.0–2.5	3	6.31	0.78	26.63	3.16	**0.77**
>2.5	2	6.76	1.15	21.55	2.32	**0.83**

Bold p-values indicate a statistically significant power-law behaviour

extreme precipitation in plains and first rise of topography. However, the power-law fit of the 0.5–1.0 km elevation bin (which always showed a low amount of extreme precipitation) suggests the threshold analysis cannot capture the status of extreme events completely. Therefore, the impact of extreme precipitation can be felt throughout eastern Nepal as suggested in Table 7.2.

7.4.3 Pattern of Extreme River Runoff

The river morphology in the alpine landscape of the Himalaya decreases its catchment area with the increase of elevation. The river valleys of high elevation are situated within the finely separated ridges, leading to a decreased area for river catchment.

On the contrary, the plains are spread horizontally with wider and larger areas for water catchment in the river. Therefore, the larger and low elevation rivers have a large catchment area, while high elevation rivers have small catchment areas.

We can observe from Fig. 7.8 that the threshold of extreme river runoff and its amount are higher in the larger river catchment areas or low elevation rivers of eastern Nepal. Likewise, the days of extreme river runoff do not show any

Extreme River Runoff Indices at 99th Percentile

High Precipitation Months (HPM) - May-October timeseries

Low Precipitation Months (LPM) - May-October timeseries

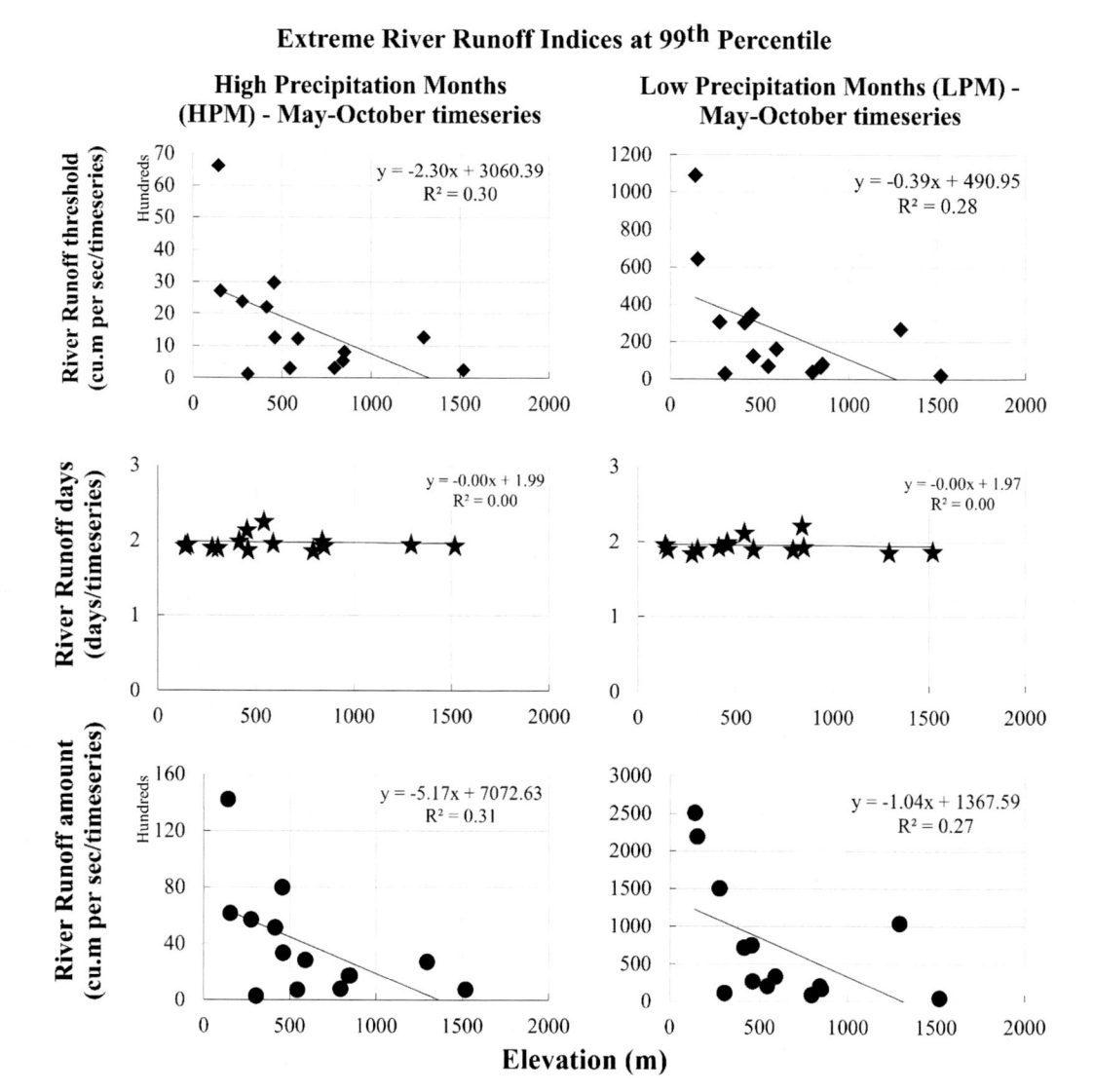

Fig. 7.8 Variance of extreme river runoff threshold, in diamond shape, extreme river runoff days, in star shape, and extreme amount of river runoff, in circle shape, of eastern Nepal, when compared with the elevation of river runoff stations during High Precipitation Months and Low Precipitation Months, however, with similar pattern between both time periods at the 99th percentile. The linear trend provides the relationship of the indices' value against the elevation of stations

comprehensive relationship with the geomorphological structure of the region.

However, the interesting fact in this observation is the range of value between all the stations regardless of elevation or catchment area. The small range shows that the frequency of extreme events is similar throughout the KRB rivers. In general, it indicates similar flood frequency, even with different observed intensities, i.e., the lower intensity river stations also possess a risk of similar flood days as the higher intensity stations.

The aforementioned findings of extreme river runoff are further supported by the observations of Fig. 7.9. The total and extreme river runoff amount, as well as threshold, show a similar pattern in regards to elevation and catchment

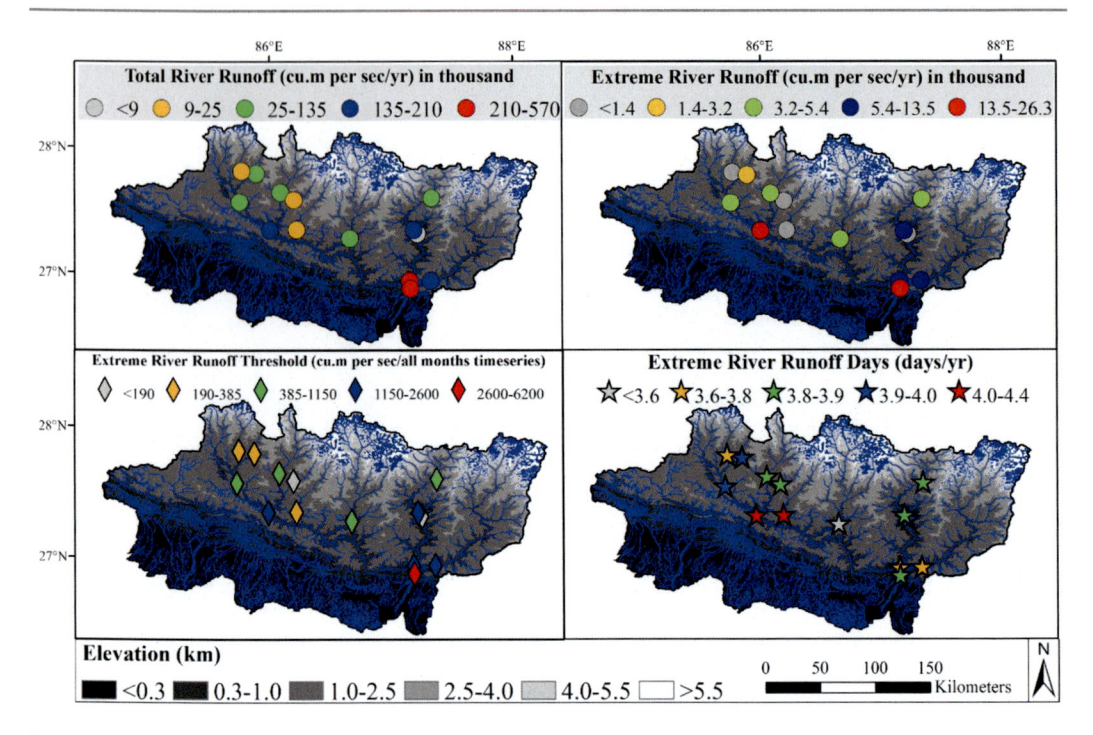

Fig. 7.9 The spatial distribution of the values of total and extreme river runoff parameters in the river runoff stations of the Koshi River Basin network of Eastern Nepal. The circle shapes represent total and extreme river runoff amount, diamonds represent extreme river runoff threshold, and stars represent extreme river runoff days at the 99th percentile

area, decreasing from lower to higher elevation. Likewise, the pattern of extreme river runoff days is almost uniform throughout the KRB.

7.5 Discussion

Our study attempts to decipher the spatiotemporal pattern of extreme precipitation and river runoff in eastern Nepal using ground stations. Previous studies have attempted the same but only in smaller regions while most of them have focused on the trend of events rather than on patterns (Burbank et al. 2003; Barros and Lang 2003; Archer and Fowler 2004; Barros et al. 2004; Anders et al. 2006; Bookhagen and Burbank 2006; Bookhagen 2010; Malik et al. 2012; Panthi, et al. 2015; Norris et al. 2020). Moreover, our study provides a new perspective when compared to analyses using remotely sensed or interpolated data which have limited quality resolutions coupled with variance of under or over-represented data due to the presence of the ubiquitous and intricate ridges along with the topographical structure of the Himalaya (Barros et al. 2006; Anders et al. 2006; Andermann et al. 2011; Yatagai et al. 2012; Cannon et al. 2017).

We focus on eastern Nepal which has a one-step rise of topography at around 1 km elevation. The stepped rise of topography has been associated with bands of precipitation (Bookhagen and Burbank 2006; Anders et al. 2006; Bookhagen 2010; Nandargi and Dhar 2011). Our findings from HPM support this relationship both for total and extreme precipitation amount. However, we find that the southern plains also have an extreme precipitation peak when compared with the total amount of precipitation. Interestingly, this rise is coupled with the high values of 5 km radius relief from the stations. Furthermore, the high reliefs are associated with high values of precipitation amount (Bookhagen and Burbank 2006), but our study hints that the same relationship may hold for high values of extreme precipitation amount. Except for this observation, the pattern exhibited by extreme precipitation amount is similar when compared between topographic elevation and relief height of the station.

Likewise, we also studied the percentile-based threshold of extreme precipitation which has received limited attention in previous studies. It is important to note that threshold variance provides interesting insight about the pattern of extreme precipitation. Since the threshold is objective and site-specific it classifies the intensity for each region without compromising its level of impact (Krishnamurthy et al. 2009; Bookhagen 2010). The extreme precipitation threshold value at the 99th percentile shows the tendency of a decrease with the increment of elevation. Nonetheless, the pattern of threshold in eastern Nepal also shows a peak in the first rise of the topography but, as mentioned earlier, it is comparatively less than the lower elevation zones and decreases gradually with the increase of elevation.

Although the pattern of extreme precipitation amount during LPM shows similar characteristics to that of HPM, the pattern of extreme precipitation threshold is different between these two time periods. The LPM is dominated by westerly flow movement which enters from high Himalaya in the west, while the HPM is dominated by the ISM onset which enters from eastern and southern Nepal. Therefore, during the LPM the threshold values are almost uniformly higher till the mid-high elevation zones. Furthermore, the plains have lower topographical disturbance and high temperature which exhibit stronger convective events and, as the warmer air can store larger amounts of precipitation, the high values of extreme precipitation threshold in the plains during HPM is further plausible (Shrestha and Aryal 2011).

We also find that the number of total and extreme precipitation days increases with the elevation in eastern Nepal during HPM, while the value of precipitation days is significantly lower during the LPM. The extreme precipitation days increase by more than double from lower to higher elevations. The combination of these three findings together leads to noteworthy observations: the lower elevation receives high amounts of extreme precipitation with high threshold in comparatively fewer days, suggesting the formation and impact of torrential precipitation. On the other hand, the high elevation regions experience comparatively less intense extreme precipitation in a high number of extreme precipitation days. The gradients of the Himalaya are fragile and consist of comparatively new geological structures on the Earth (Burbank et al. 2003; Hannah et al. 2005). These findings support the ever-occurring erosion in hills and the floods in the plains (Chaliseand& Khanal 2002; Burbank et al. 2003; Hannah et al. 2005; Dahal and Hasagawa 2008; Nepal 2016).

Our finding of magnitude-frequency relationships in regards to power-law ehaviour also supports these aforementioned findings. In addition, the power-law analysis also suggests that threshold analysis may not be entirely sufficient to understand the occurrence of rare but high-magnitude precipitation events. The stations of each of the elevation bins from less than 0.5 km to more than 2.5 km with a bin size of 0.5 km follow a power-law ehaviour for the magnitude-frequency relationship of the precipitation. Moreover, the scaling exponents are lower at lower elevations, suggesting the occurrence of rare events in these zones too. By contrast, the threshold analysis failed to suggest high extreme precipitation events in certain elevation bins of lower elevation regions. Therefore, it is plausible to state that the impact of rare but high-magnitude events of precipitation are felt throughout the entire region of eastern Nepal, and being more intense in the lower elevation zones.

Spatial analyses also support the topographic and temporal findings of extreme precipitation. The extreme precipitation peaks in the plains and first rise of the topography, while threshold and days of extreme precipitation show contrasting ehaviour in regards to the range of elevation. Furthermore, we found that the extreme precipitation threshold is ten times higher than the median of the precipitation threshold below the 1 km elevation, while that is less than ten times above the 1 km elevation region.

The comparative findings of river runoff are also unprecedented in the sense that the area and quantity of water runoff in the river are independent of the risk of flooding days in eastern Nepal's Koshi River basin. The threshold and amount of extreme river runoff increase with decreasing elevation in the KRB, i.e., the higher the catchment area of the river, the higher the threshold and amount of extreme river runoff. These findings are in line with the studies of Milliman (2001), and Gaume and Payrastre (2018) as they point toward the dependent relationship between the catchment area of the river and the river discharge. On the other hand, the days of extreme river runoff show no relationship with the area and quantity of river runoff and are almost uniform throughout the KRB. Therefore, the frequency of extreme river runoff days which signifies flooding days is almost the same for the largest and smallest rivers of the KRB.

7.6 Conclusion

We used 40 years of daily precipitation data from 53 ground stations, and 35 years of daily river runoff data from 14 ground stations of eastern Nepal and the Koshi River basin, respectively. The stations ranged from 70 to 2700 m, and 140 m to 1500 m for precipitation and river runoff stations, respectively. The spatiotemporal nature of stations was analyzed by adapting the indices set by the Expert Team on Climate Change Detection Monitoring and Indices (ETCCDMI) based on a 99th percentile threshold of the time series. We also calculated a power-law relationship of magnitude-frequency of precipitation to comprehend high-magnitude low-frequency events. The key conclusions are:.

First, we observed a peak of extreme precipitation value with the rise of topography in eastern Nepal, i.e., around 1 km elevation during high precipitation months (May–October). In addition to that, we also observed a peak of extreme precipitation in the plains of eastern Nepal which has a high 5 km radius relief height. The total precipitation amount also showed a similar trend but found that the extreme precipitation amount is higher in the plains with a high relief height and in the first rise of the topography.

Secondly, the variation of the 99th percentile extreme precipitation threshold value decreases with increasing elevation, while the number of extreme precipitation days increases with elevation. Likewise, during the low precipitation months, most indices show higher values in mid-high elevation regions, while these values are finite when compared to that of high precipitation months. The power-law behavior of magnitude-frequency of precipitation shows that lower elevation regions have a higher probability of experiencing high-magnitude low-frequency events.

Finally, we found that the extreme amount and threshold of river runoff (discharge) increases with the increase of catchment area and a decrease of elevation. However, the number of extreme river runoff days is uniform throughout the Koshi River basin, which shows no relationship with the elevation or catchment area of the river. Therefore, the largest and smallest river in the Koshi River basin has similar probabilities of experiencing flooding days.

Acknowledgements This chapter is part of the corresponding author's Ph.D. research project at Martin Luther University Halle—Wittenberg. Thus, the corresponding author thanks Bodo Bookhagen for his supervision during the course of this project. The Ph.D. project was funded by the Graduate Funding Commission of Martin Luther University Halle-Wittenberg under the Graduate Promotion Act of the state of Saxony-Anhalt. The data acquisition from the Department of Hydrology and Meteorology of Nepal was funded by the Institute of Geosciences, University of Potsdam.

References

Andermann C, Bonnet S, Gloaguen R (2011) Evaluation of precipitation data sets along the Himalayan front. Geochem Geophys Geosyst 12(7):Q07023. https://doi.org/10.1029/2011GC003513

Anders AM, Roe GH, Haller B, Montogomery DR, Finnegam NJ, Putkonen J (2006) Spatial patterns of precipitation and topography in the Himalayas. Geol Soc Am Spec Pap 398:39–53. https://doi.org/10.1130/2006.2398(03)

Archer DR, Fowler HJ (2004) Spatial and temporal variations in precipitation in the Upper Indus Basin, global teleconnections and hydrological implications. Hydrol Earth Syst 8(1):47–61. https://doi.org/10.5194/hess-8-47-2004

Ashcroft L, Karoly DJ, Dowdy AJ (2019) Historical extreme rainfall events in Southeastern Australia. Weather Clim Extremes 25:100210. https://doi.org/10.1016/j.wace.2019.100210

Barros AP (2004) On the space-time patterns of precipitation in the himalayan range: a synthesis. In: 6th international GAME conference. Kyoto. Retrieved 12 Aug 2016, from http://www.hyarc.nagoya-u.ac.jp/game/6thconf/html/abs_html/pdfs/T8APB19Oct04100318.pdf

Barros AP, Lang TJ (2003) Monitoring the monsoon in the Himalayas: observations in Central Nepal, June 2001. Mon Weather Rev 131:1408–1427. https://doi.org/10.1175/1520-0493(2003)131%3c1408:MTMITH%3e2.0.CO;2

Barros AP, Chiao S, Lang TJ, Burbank D, Putkonen J (2006) From weather to climate—seasonal and interannual variability of storms and implications for erosion processes in the Himalaya. In Willett SD, Hovius N, Brandon MT, Fisher D (eds) Tectonics, climate, and landscape evolution. Geological Society of America, pp 17–38. https://doi.org/10.1130/2006.2398(02)

Barros AP, Kim G, Williams E, Nesbitt SW (2004) Probing orographic controls in the Himalayas during the monsoon using satellite imagery. Nat Hazard 4:29–51. https://doi.org/10.5194/nhess-4-29-2004

Bohlinger P, Sorteberg A (2018) A comprehensive view on trends in extreme precipitation in Nepal and their spatial distribution. Int J Climatol 38(4):1833–1845. https://doi.org/10.1002/joc.5299

Bookhagen B (2010) Appearance of extreme monsoonal rainfall events and their impact on erosion in the Himalaya. Geomat Nat Haz Risk 1(1):37–50. https://doi.org/10.1080/19475701003625737

Bookhagen B, Burbank DW (2006) Topography, relief, and TRMM-derived rainfall variations along the Himalaya. Geophys Res Lett 33:1–5. https://doi.org/10.1029/2006GL026037

Bookhagen B, Burbank DW (2010) Toward a complete Himalayan hydrological budget: spatiotemporal distribution of snowmelt and rainfall and their impact on river discharge. J Geophys Res Earth Surf 115(F3). https://doi.org/10.1029/2009JF001426

Bookhagen B, Strecker MR (2008) Orographic barrier, high-resolution TRMM rainfall, and relief variations along the eastern Andes. Geophys Res Lett 35 (L06403). https://doi.org/10.1029/2007GL032011

Bookhagen B, Thiede RC, Strecker MR (2005) Abnormal monsoon years and their control on erosion and sediment flux in the high, arid northwest Himalaya. Earth Planet Sci Lett 231:131–146. https://doi.org/10.1016/j.epsl.2004.11.014

Burbank DW, Blythe AE, Putkonen J, Pratt-Sitaula B, Gabet E, Oskin M, Barros A, Ojha TP (2003) Decoupling of erosion and precipitation in the Himalayas. Nature 426:652–655. https://doi.org/10.1038/nature02187

Cannon F, Carvalho LM, Jones C, Norris J, Bookhagen B, Kiladis GN (2017) Effects of topographic smoothing on the simulation of winter precipitation in High Mountain Asia. J Geophys Res Atmos 122:1456–1474. https://doi.org/10.1002/2016JD026038

Chalise SR, Khanal NR (2002). Recent extreme weather events in the Nepal Himalaya. The extreme of the extremes: extraordinary flood. Reykjavik: IADS. Retrieved 11 June 2019, from http://hydrologie.org/redbooks/a271/iahs_271_141.pdf

Clauset A, Shalizi CR, Newman ME (2009) Power-Law Distributions in Empirical Data. Soc Ind Appl Math 51(4):661–703. https://doi.org/10.1137/070710111

Croitoru AE, Piticar A, Burada DC (2016) Changes in precipitation extremes in Romania. Quatern Int 415:325–335. https://doi.org/10.1016/j.quaint.2015.07.028

Dahal RK, Hasagawa S (2008) Representative rainfall thresholds for landslides in the Nepal Himalaya. Geomorphology 100(3–4):429–443. https://doi.org/10.1016/j.geomorph.2008.01.014

Department of Hydrology and Meteorology of Nepal (2016a) Meteorological station. Retrieved from Station Network: http://www.dhm.gov.np/meteorological-station/

Department of Hydrology and Meteorology of Nepal (2016b) Hydrological station. Retrieved from Station Network: http://www.dhm.gov.np/hydrological-station/

Devkota LP, Gyawali DR (2015) Impacts of climate change on hydrological regime and water resources management of the Koshi River Basin, Nepal. J Hydrol Reg Stud 4:502–515

Dimri AP, Mohanty UC, Azadi M, Rathore LS (2006) Numerical study of western disturbances over western Himalayas using mesoscale model. MAUSAM, 57(4), pp 579–590. Retrieved November 17, 2018, from https://metnet.imd.gov.in/mausamdocs/15742_F.pdf

Ding Y, Sikka DR (2006) The Asian monsoon. In: Synoptic systems and weather. Springer, Berlin, pp 131–201. https://doi.org/10.1007/3-540-37722-0_4

Eriksson M, Jianchu X, Shrestha AB, Vaidya RA, Nepal S, Sandstroem K (2009) The changing Himalayas—impact of climate change on water resources and livelihoods in the Greater Himalayas. International Centre for Integrated Mountain Development, Kathmandu. Retrieved 3 Sept 2020, from http://lib.icimod.org/record/26471/files/attachment_593.pdf

Flohn H (1957) Large-scale aspects of the "summer monsoon" in south and East Asia. J Meteorol Soc Jpn 75:180–186. https://doi.org/10.2151/jmsj1923.35A.0_180

Gabet EJ, Burbank DW, Pratt-Sitaula B, Putkonen J, Bookhagen B (2008) Modern erosion rates in the High Himalayas of Nepal. Earth Planet Sci Lett 267:482–494. https://doi.org/10.1016/j.epsl.2007.11.059

Gaume E, Payrastre O (2018) Flood hydrology processes and their variabilities. In: Vinet F (ed) Floods, vol 1. ISTE Press - Elsevier, London, pp 115–127

Goswami BN (2005) South Asian monsoon. In: Lau WK, Waliser DE (eds) Intraseasonal variability in the atmosphere-ocean climate system. Springer, Berlin, pp 19–61. https://doi.org/10.1007/3-540-27250-X_2

Hannah DM, Kansakar SR, Gerrard AJ, Rees G (2005) Flow regimes of Himalayan rivers of Nepal: nature and spatial patterns. J Hydrol 308:18–32. https://doi.org/10.1016/j.jhydrol.2004.10.018

Haylock MR, Goodess CM (2004) Interannual variability of European extreme winter rainfall and links with mean large-scale circulation. Int J Climatol 24(6):759–776. https://doi.org/10.1002/joc.1033

Hussain A, Rasul G, Mahapatra B, Wahid S, Tuladhar S (2018) Climate change-induced hazards and local adaptations in agriculture: a study from Koshi River Basin Nepal. Nat Hazards 91(3):1365–1383

Ives JD, Messerli B (1989) The Himalayan dilemma reconciling development and conservation. Routledge, London

Karki R, Hasson S, Schickhoff U, Scholten T, Bohner J (2017) Rising precipitation extremes across Nepal. Climate, 4, 2–25.https://doi.org/10.3390/cli5010004

Krishnamurthy CK, Lall U, Kwon HH (2009) Changing frequency and intensity of rainfall extremes over India from 1951 to 2003. J Clim 22(18):4737–4746. https://doi.org/10.1175/2009JCLI2896.1

Malik N, Bookhagen B, Marwan N (2012) Analysis of spatial and temportal extreme monsoonal rainfall over South Asia using complex networks. Clim Dyn 39(3–4):971–987. https://doi.org/10.1007/s00382-011-1156-4

Malik N, Bookhagen B, Mucha PJ (2016) Spatiotemporal patterns and trends of Indian monsoonal rainfall extremes. Geophys Res Lett 43.https://doi.org/10.1002/2016GL067841

Mani A (1981) The climate of the Himalaya. In: Lall JS, Modi AD (eds) The Himalaya: aspects of changes. Oxford University Press, New Delhi, pp 3–15

Milliman JD (2001) River inputs. In: Steele JH (ed) Encyclopedia of ocean sciences, 2nd edn. Elsevier Ltd, pp 754–761

Nandargi S, Dhar ON (2011) Extreme rainfall events over the Himalayas between 1871 and 2007. Hydrol Sci J 56(6):930–945. https://doi.org/10.1080/02626667.2011.595373

Nepal S (2012) Evaluating upstream-downstream linkages of hydrological dynamics in the Himalayan region (Doctoral Dissertation). Friedrich-Schiller-University, Jena, Germany

Nepal S (2016) Impacts of climate change on the hydrological regime of the Koshi river basin in the Himalayan region. J Hydro-Environ Res 10:76–89. https://doi.org/10.1016/j.jher.2015.12.001

Norris J, Carvalho LM, Jones C, Cannon F (2020) Warming and drying over the central Himalaya caused by an amplification of local mountain circulation. Clim Atmos Sci 3(1):1–11. https://doi.org/10.1038/s41612-019-0105-5

Panday PK, Thibeault J, Frey KE (2015) Changing temperature and precipitation extremes in the Hindu Kush-Himalayan region: an analysis of CMIP3 and CMIP5 simulations and projections. Int J Climatol 35(10):3058–3077. https://doi.org/10.1002/joc.4192

Panthi J, Dahal P, Shrestha ML, Aryal S, Krakauer NY, Pradhanang SM, Lakhankar T, Jha AK, Sharma M, Karki R (2015). Spatial and temporal variability of rainfall in the Gandaki River Basin of Nepal Himalaya. Climate 3(1):210–226.https://doi.org/10.3390/cli3010210

Reddy DV, Kumar D, Saha D, Mandal MK (2008) The 18 August 2008 Kosi river breach: an evaluation. Curr Sci 95(12):1668–1669. Retrieved 27 Sept 2018, from http://www.indiaenvironmentportal.org.in/files/The%2018%20August%202008%20Kosi%20river%20breach.pdf

Reliefweb (2019, July 11) Nepal: Floods and Landslides —Jul 2019. Retrieved 2 Dec 2019, from Reliefweb: https://reliefweb.int/disaster/fl-2019-000083-npl

Sharma KP, Moore B, Vorosmarty CJ (2000) Anthropogenic, climatic and hydrological trend in the Kosi basin, Himalaya. Clim Change 47:141–165

Shatnawi R, Althebyan Q (2013) An empirical study of the effect of power law distribution on the interpretation of OO metrics. ISRN Softw Eng 198937. https://doi.org/10.1155/2013/198937

Sheikh MM, Manzoor N, Ashraf J, Adnan M, Collins D, Hameed S, Manton MJ, Ahmed AU, Baidya SK, Borgaonkar HP, Islam N (2014) Trends in extreme daily rainfall and temperature indices over South Asia. Int J Climatol 35(7):1625–1637.https://doi.org/10.1002/joc.4081

Shrestha AB, Aryal R (2011) Climate change in Nepal and its impact on Himalayan glaciers. Reg Environ Change 11:65–77. https://doi.org/10.1007/s10113-010-0174-9

Shrestha AB, Bajracharya SR, Sharma AR, Duo C, Kulkarni A (2017) Observed trends and changes in daily temperature and precipitation extremes over the Koshi river basin 1975–2010. Int J Climatol 37(2):1066–1083. https://doi.org/10.1002/joc.4761

Shrestha ML (2000) Interannual variation of summer monsoon rainfall over Nepal and its relation to Southern Oscillation Index. Meteorol Atmos Phys 75:21–28. https://doi.org/10.1007/s007030070012

Singh H, Parkash B, Gohain K (1993) Facies analysis of the Kosi megafan deposits. Sed Geol 85:87–113

Sinha R, Bapalu GV, Singh LK, Rath B (2008) Flood risk analysis in the Kosi river basin, north Bihar using multi-parametric approach of Analytical Hierarchy Process (AHP). J Indian Soc Remote Sens 36:335–349. https://doi.org/10.1007/s12524-008-0034-y

Subash N, Gangwar B (2014) Statistical analysis of Indian rainfall and rice productivity anomalies over the last decades. Int J Climatol 34(7):2378–2392. https://doi.org/10.1002/joc.3845

Tang H, McGuire LA, Kean JW, Smith JB (2020) The impact of sediment supply on the initiation and magnitude of runoff-generated debris flows. Geophys Res Lett. https://doi.org/10.1029/2020GL087643

Thiede RC, Bookhagen B, Arrowsmith JR, Sobel ER, Strecker MR (2004) Climate control on rapid exhumation along the Southern Himalayan Front. Earth Planet Sci Lett 222:791–806. https://doi.org/10.1016/j.epsl.2004.03.015

USGS (2012, October 1) EarthExplorer. Retrieved 14 Nov 2016, from https://earthexplorer.usgs.gov/

Wasson RJ, Juyal N, Jaiswal M, McCullochd M, Sarinb MM, Jaine V, Srivastava P, Singhvi AK (2008) The mountain-lowland debate: deforestation and sediment transport in the upper Ganga catchment. J Environ Manage 88:53–61

Webster PJ (1986) The variable and interactive monsoon. In: Fein, Stephens (eds) Monsoons. John Wiley Company, pp 269–330. Retrieved 26 June 2017, from https://ci.nii.ac.jp/naid/10006234015/#cit

Wheeldon R, Counsell S (2003) Power law distributions in class relationships. In: 3rd IEEE international conference in source code analysis and manipulation. IEEE, Amsterdam, pp 45–54. https://doi.org/10.1109/SCAM.2003.1238030

Wulf H, Bookhagen B, Scherler D (2010) Seasonal precipitation gradients and their impact on fluvial sediment flux in the Northwest Himalaya. Geomorphology 112(1–2):13–21. https://doi.org/10.1016/j.geomorph.2009.12.003

Wulf H, Bookhagen B, Scherler D (2012) Climatic and geologic controls on suspended sediment flux in the Sutlej River Valley, western Himalaya. Hydrol Earth Syst Sci 16:2193–2217. https://doi.org/10.5194/hess-16-2193-2012

Wulf H, Bookhagen B, Scherler D (2015) Differentiating between rain, snow, and glacier contributions to river discharge in the western Himalaya using remote-sensing data and distributed hydrological modeling. Adv Water Resour 7(36):152–169

Yatagai A, Kamiguchi K, Arakawa O, Hamada A, Yasutomi N, Kitoh A (2012) Constructing a long-term daily gridded precipitation dataset for asia based on a dense network of rain gauges. Bull Am Meteor Soc 93(9):1401–1415. https://doi.org/10.1175/BAMS-D-11-00122.1

Climate Change and its Impact on Catchment Linkage and Connectivity

Manudeo Singh and Rajiv Sinha

Abstract

Geomorphic connectivity among different landscape compartments is a result of physical linkage and material flux exchange. Physical linkage is the structural component, and material flux is the functional component of connectivity. Both components are interlinked and define catchments processes and responses. Terrain characteristics such as slope and topography control define structural connectivity, whereas functional connectivity is defined by processes and stimuli such as rainfall events, land cover dynamics, and tectonics. There exists feedback between structural and functional connectivity, and they actively modify each other to keep the geomorphic system in an equilibrium state. Under the climate change scenario, dynamics of the processes such as rainfall and land cover are changing rapidly, which in turn affects the catchment connectivity and linkages. The present study first introduces the concept of geomorphic connectivity in a comprehensive manner and then discusses the impact of climate change on catchment structures and processes in a theoretical framework.

Keywords

Hydro-geomorphic connectivity · Catchment processes · Floodplain-channel connectivity · Hill slope-channel connectivity · Climate change · Connectivity

8.1 Introduction

Among all environmental externalities, global warming is the most prominent one, making climate change the ultimate challenge for world economies (Nordhaus 2019). Climate change at a historical time scale is manifested in several extreme events the frequency of which is expected to increase even further (IPCC 2014). The impact of climate change on the basin linkage and connectivity and ultimately on sediment and hydrological connectivity depends on individual basin characteristics. Usually, topographic factors such as catchment morphometry and catchment land-cover types are the first-order controls that characterize the period and magnitude of the runoffs, and sediment generation and transportation. Therefore, to understand the impact of climate change on catchment linkage and connectivity, it is important to understand how connectivity-defining factors are being modified by climate change.

M. Singh (✉) · R. Sinha
Department of Earth Sciences, Indian Institute of Technology Kanpur, Kanpur 208016, India
e-mail: manudeo@iitk.ac.in

Two major factors that impact the catchment processes and hence connectivity at various scales are land-use change and climate change (Wada et al. 2011; Sinha et al. 2020). Land-use patterns can also get altered by climate change (Sinha et al. 2020) which may imply further alterations in the catchment processes such as evapotranspiration, soil erosion, and runoff generation (Chawla and Mujumdar 2015; Bussi et al. 2016; Op de Hipt et al. 2019; Sinha et al. 2020), amplifying the impact of climate change. For example, a study in the Willamette River Basin, Oregon, USA reported that the regional climate could be significantly warmer in the twenty-first century which might potentially change the dominant vegetation cover type of the basin (Turner et al. 2015). This study predicted that the currently present needle leaf type forest would convert into a mixture of needle leaf and broad leaf type forest. Moreover, the forest might get fragmented. Such climate change-induced alterations in the forest cover type may radically modify the hydrological cycle of the basin (Turner et al. 2015) and the hydrological connectivity.

In various studies involving a comparison between climate change and land-use land-cover (LULC) change as the major control on catchment processes, the impact of climate change is found to be much more significant than the LULC change. For example, Kim et al. (2013) studied the impact of LULC and climate change on the streamflow in the Hoeya River Basin of South Korea. They concluded that among LULC change and climate change, the former has less impact on the stream flows than the latter. However, the impact of LULC change was also significant on the streamflow. Similarly, in Be River catchment of Vietnam, the influence of climate change was observed to be stronger than the influence of LULC change on the hydrological processes (Khoi and Suetsugi 2014).

The present chapter first discusses the concept of connectivity in hydro-geomorphic systems and then presents the ways in which various elements of hydro-geomorphic connectivity are getting impacted by climate change. In particular, possible impacts of climate change on hydrological connectivity and sediment connectivity have been discussed in detail.

8.2 Concept of Connectivity in Hydro-geomorphic Systems

In hydro-geomorphic systems, connectivity is defined as "the efficiency of transfer of materials between system components" (Wohl et al. 2019, pp. 2). Here, water, nutrients, and sediments are materials in the geomorphic system. The catchments, sub-catchments, water bodies, etc. are system components or landscape units. The presence or absence of linkage (or, connections) among different system components defines the degree of interaction (e.g., streamflow rates, sedimentation, ecological functions) in the system (Singh et al. 2021). Further, such connections vary in space and time (Harvey 2002), rendering connectivity to a spatio-temporal dynamic phenomenon.

In hydro-geomorphic systems, there are three types of connectivity i.e. landscape connectivity, hydrological connectivity, and sediment connectivity (Wohl et al. 2019). Further, there are two components—structural and functional connectivity, inherent to all three types of hydro-geomorphic connectivity (Turnbull et al. 2008; Wainwright et al. 2011). All types of hydro-geomorphic connectivity operate in four dimensions: three spatial (longitudinal, lateral, vertical) and one temporal dimension (Ward 1989; Jain and Tandon 2010). Because of its multi-dimensional property, connectivity can be used to understand the inter-and intra-scale hydro-geomorphic (e.g., sediment transport) processes (Bracken et al. 2015). Hydro-geomorphic connectivity gets actively and significantly altered by various anthropogenic factors such as drainage reorganization, land-cover changes, topography alterations (Pringle 2003; Hooke 2006; Fryirs 2013; Singh et al. 2017; Singh and Sinha 2019). Recently, the connectivity concept has been used to evaluate the impacts of climate and land-use change on the navigation of sediments (López-Vicente et al. 2013; Lane et al. 2017) and water

(Smith et al. 2010) within and among various geomorphic units. Therefore, climate change phenomena can impact hydro-geomorphic connectivity in a major way. However, before understanding such impacts and possible measures to minimize them, it is necessary to understand the hydro-geomorphic connectivity and its elements.

Various researchers have used the connectivity concept to understand and evaluate hydrological and sedimentary processes (Brierley et al. 2006, Turnbull et al. 2008, Lexartza-Artza and Wainwright 2011, Jain et al. 2012, Fryirs and Gore 2013, Gomez-Velez and Harvey 2014, Bracken et al. 2015, Lisenby and Fryirs 2017) at different scales and settings. A recent and comprehensive study by Singh et al. (2021) defined a connectivity framework consisting of three basic elements of geomorphic connectivity—connectivity types, connectivity components, and connectivity dimensions (Fig. 8.1). This framework presents the interrelationships and feedbacks among different connectivity elements.

"Hydrologic connectivity is the water-mediated transport of matter, energy, and organisms within or between elements of the hydrologic cycle" (Freeman et al. 2007, p. 1). Further, "the connected transfer of sediment from a source to a sink in a system via sediment detachment and sediment transport, controlled by how the sediment moves between all geomorphic zones in a landscape" (Bracken et al. 2015, p. 177) is defined as sediment connectivity. Recently, Heckmann et al. (2018) presented a working definition of connectivity by encapsulating both sediment and hydrological connectivity: "…we define hydrological and sediment connectivity as the degree to which a system facilitates the transfer of water and sediment through itself, through coupling relationships between its components. In this view, connectivity becomes an emergent property of the system state, reflecting the continuity and strength of runoff and sediment pathways at a given point in time" (Heckmann et al. 2018, pp. 3). Although water and sediment can freely navigate from one

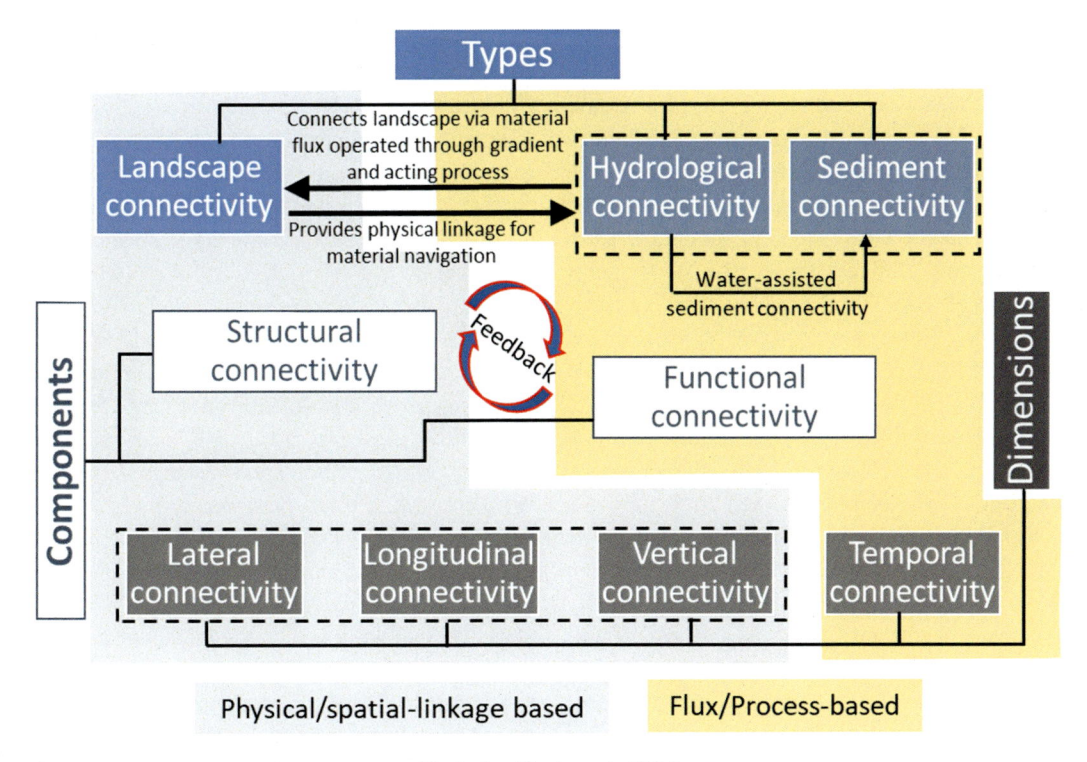

Fig. 8.1 The connectivity framework. Modified after Singh et al. (2021)

landscape to other, the landscapes themselves are static features and they get connected only because of the exchange of water and sediment (Singh et al. 2021). Hence, sediment and hydrological connectivity are the main drivers of landscape connectivity. Further, in many cases, sediment transport is strictly water-controlled and in such scenarios, hydrological connectivity can act as a proxy for sediment connectivity as well. Therefore, in such cases, just by evaluating hydrological connectivity, the other two hydro-geomorphic connectivity can be easily comprehended.

All types of hydro-geomorphic connectivity have two components—structural and functional. The spatial patterns (Turnbull et al. 2008) and physical linkages present at various scales in landscape units define the structural component (Keesstra et al. 2018; Turnbull et al. 2018). Therefore, catchment linkage is the structural component of catchment connectivity. Channel network and Hydrological Response Units (HRUs) are typical examples of structural components. The functional component is the result of the interaction between the structural components and acting processes (Turnbull et al. 2008; Wainwright et al. 2011; Bracken et al. 2015). Sediment and water discharge are typical examples of the functional component of hydro-geomorphic connectivity. On temporal scales, the structural component gets modified by the functional component, thereby, forming a feedback system between these two components (Singh et al. 2021).

Hydro-geomorphic connectivity operates on four dimensions comprising three spatial dimensions i.e. lateral, longitudinal, vertical, and one temporal dimension (Ward 1989; Jain and Tandon 2010). However, all three spatial dimensions are interrelated, and they follow a conservation law, i.e., overall connectivity among different spatial dimensions at a given place remains the same and the connectivity increment in one dimension is a result of connectivity decrement in some other dimension at that place. For example, poor vertical connectivity (percolation) results in runoff generation (strong horizontal connectivity). At a large scale, a decrease in one dimension of connectivity can translate into a decrease in another dimension. For example, reduced vertical connectivity in the floodplain can result in reduced base flow to the channel and therefore, reduced lateral connectivity between floodplain and channel. Further, a reduced lateral hydrological connectivity might result in a reduction of the channel's longitudinal hydrological connectivity. All connectivity types, components, and spatial dimensions also vary with time, and this defines temporal connectivity. The temporal dimension of connectivity induces dynamics in hydro-geomorphic connectivity, i.e., connectivity changes with time. This is where climate change issues become important, and therefore, a conceptual understanding of the impacts of climate change on hydro-geomorphic connectivity needs to be developed.

8.3 Climate Change and Connectivity

Different elements of connectivity are expected to get impacted differently because of climate change. For example, many times, the impact of climate change on hydrological connectivity and sediment connectivity might not be equal. For example, Azari et al. (2016) constructed three climate change scenarios in Northern Iran and found that for the years 2040–2069, stream flows will increase to 5.8%, 2.5%, and 9.5% annually, whereas the sediment yield will increase to 47.7%, 44.5%, and 35.9% respectively. Similar results were observed in Be River catchment of Vietnam where due to climate change, annual streamflow increased by 26.3%, and sediment load by 31.7% (Khoi and Suetsugi 2014). In Northern Iran, a disproportionately higher increment in sediment load than the streamflow was attributed to the power function relationship between streamflow and sediment yield (Azari et al. 2016). Consequently, the impact of frequent and large-intensity floods in northern Iran would result in greater sediment yield than the streamflow. In the case of Vietnam, the disparity among streamflow and sediment yield could be related to the conversion of forest covers into agricultural lands (Khoi and Suetsugi 2014).

Depending upon the present and future conditions and geographical locations, some elements of connectivity can be negatively impacted, and others can be positively impacted. Further, the impact of climate change might not be unidirectional and static. Different regions of the Earth might observe the dynamic impacts of climate change in different periods. For example, a study on the Nile River Basin investigating the impacts of climate change under 2007 IPCC scenarios and found that in the early twenty-first century, due to increased precipitation, the Nile River might observe increased flows, but in the mid- and late twenty-first century, the river might observe decrement in the streamflow induced by a decline in precipitation and increased evaporation (Beyene et al. 2010).

Structural properties of a landscape (i.e., structural connectivity) and dynamics of hydrometeorological processes (such as duration-intensity-frequency of rainfall) control the landscape response and hence, the functional connectivity (Singh et al. 2021). Climate variabilities are expected to translate into extreme river flows and thereby, increased flood risks (Fang et al. 2018). An alteration in structural component influences the functional component (Vanacker et al. 2005; Turnbull et al. 2008; Wainwright et al. 2011; Bracken et al. 2015; Singh and Sinha 2019). Therefore, feedback exists between structural and functional components of hydrogeomorphic connectivity (Fig. 8.2). This feedback is controlled by various terrain and process parameters. It has been demonstrated that any alteration in the spatial pattern of land use in catchments results in a variation in the functional component of connectivity (Vanacker et al. 2005; Singh et al. 2021). Climatic conditions also strongly control the structural connectivity. For example, increasing rainfall usually positively correlates with an increase in vegetation (Jordan et al. 2014). An increased vegetation cover enhances the impedance to the surface water flows, thereby, decreasing the functional connectivity (Singh and Sinha 2019). Such impacts of vegetation cover are more pronounced in catchments with flat terrain. Changing climate can also impact the structural and functional connectivity at a very large scale. For example, the Ganga plains are characterized by two geomorphologically distinctive systems, East Ganga Plains (EGP) and West Ganga Plains (WGP). The EGP is drained by rivers with high sediment flux and low stream power systems, whereas the rivers draining the WGP show high stream power and low sediment flux systems (Roy and Sinha 2017, 2018). Accordingly, an incised topography is a characteristic of WGP, whereas aggradational landforms are the distinctive features of the EGP. In event of climate change, the differential sensitivity of these two distinctive systems is likely to change the stream power and sediment flux relationships, resulting in severe catastrophes.

In the temporal domain, the impact of the functional component on the structural component has been documented. For example, in the case of sediment connectivity, sediment flux is known to modify the morphology of landscapes (Bracken et al. 2015). Such modifications change the structural framework, and ultimately the physical linkage of the landscapes (Turnbull et al. 2008; Bracken et al. 2013). In the following sections, the impact of climate change on the two most important catchment connectivity i.e., hydrological connectivity and sediment connectivity has been discussed in detail.

8.3.1 Impact on Hydrological Connectivity

The impact of climate change on the streamflow and therefore, the hydrological connectivity of basins has been an area of active research (e.g., Arnell 1999; Mimikou et al. 1999; Middelkoop et al. 2001; Chang et al. 2002; Smith et al. 2010, 2013). These studies indicate that there is a conclusive correlation between climate change and variability in stream flows. For example, Arnell (1999), Middelkoop et al. (2001), and Chang et al. (2002) predicted that the regional hydrology of the snow-dominated regions will

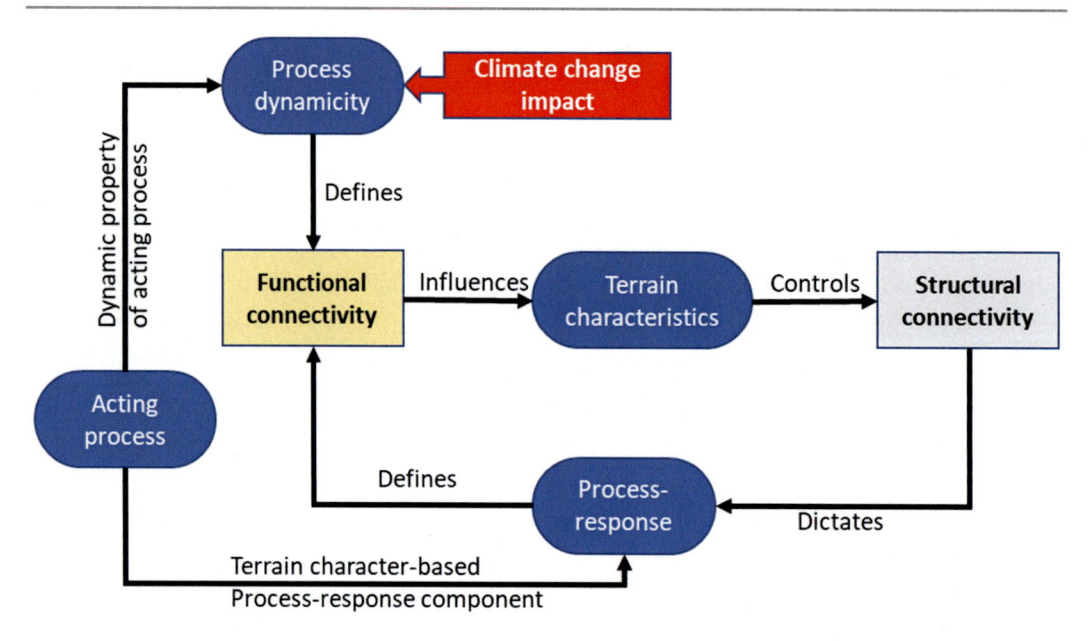

Fig. 8.2 Feedbacks among structural and functional components of hydro-geomorphic connectivity and their controls. Climate change is expected to impact the process dynamics (e.g., rainfall variabilities). Modified after Singh et al. (2021)

observe significant variations under global warming scenarios and the month of maximum runoff are expected to shift. Since hydrological connectivity is expected to be conclusively impacted by climate change, this connectivity can effectively be used to evaluate the climate change-induced modifications to the material fluxes at the catchment scale (Smith et al. 2010). The dynamic aspect of hydrological connectivity is controlled by factors such as "rainfall characteristics, flow path length, and integration, the spatial distribution of areas of low/high abstraction potential and the routing velocity of overland flows" (Smith et al. 2010). It is important to understand how each of such factors will be influenced by climate change and to what degree. The spatio-temporal variations of precipitation (Wainwright and Parsons 2002) and intensity-duration-amount (Bracken et al. 2008) are defining factors of hydrological connections (Smith et al. 2010). Understanding such factors is even more crucial for large and highly populated basins such as Ganga in which monsoonal discharge is expected to increase under climate change scenarios (Whitehead et al. 2018).

A case study by Molina-Navarro et al. (2016) on the Guadalupe River basin, Mexico presents the potential drastic impacts of climate change on basin hydrology which will ultimately impact the basin-scale hydrological connectivity severely. Based on short (2010–2039) and long term (2070–2099) climate change simulations, the authors concluded that the Guadalupe River basin might observe around 45% reduction in runoff in the short-term, and up to 60% reduction in the long-term. Decreased precipitation and increased evapotranspiration are the main factors behind such reductions. Further, it is expected that the aquifer recharge can decrease up to −74%, which will drastically alter the flow of groundwater as well as the base flows.

In regions with large negative water budgets such as the semiarid regions of the world, subsurface flows are typically non-existent and overland flows dominate the outflows (Smith et al. 2010). Even the overland flows generate isolated patches with the occasional connection among such patches (Smith et al. 2010). Such regions are most prone to the changing climate because a reduction in precipitation will diminish the connectivity among the isolated patches even further. Also, an increase in precipitation will impact the region negatively by the generation of flood events. This is because of a sudden increase of connectivity in the horizontal domain without any change in the vertical domain since vertical connectivity such as percolation is a function of lithology which is unlikely to change in pace with precipitation change.

8.3.2 Impact on Sediment Connectivity

Sediment transport and soil erosion processes are expected to be significantly influenced by climate change (Bussi et al. 2016). For example, a study by Jordan et al. (2014) showed that higher rainfall translated into higher total suspended solid (TSS) contribution to surface water systems. The major climate-related stressors that are expected to impact catchment scale sedimentation processes are changes in temperature and precipitation and their subsequent impacts on the vegetation cover and land use (Nearing et al. 2004). These stressors can potentially impact sediment connectivity by altering sediment production and sediment transport (Mullan et al. 2012). For example, a study on the Ganga River estimated that in comparison to the present scenario, the sediment load in this river might increase by 10–40% by mid-century and by 35–79% by the end of the century under changing climate scenarios (Khan et al. 2018).

Extreme precipitation and river discharge strongly control the sediment transportation and hence sediment connectivity within and in-between landscapes. For example, sediment transport in many catchments in Himalayas depends on the precipitation intensity, which in turn dictates the channel morphology and processes. Several Himalayan rivers such as Kosi are highly avulsive (Sinha et al. 2013, 2014) and the high rate of sediment production in its catchment is one of the primary causal factors for its avulsive nature. Climate change may have very significant implications in river basins such as the Kosi draining through Nepal and India where significant spatial variability in sediment connectivity across the basin has been noted; it has also been demonstrated that sediment flux in different sub-basins is controlled by variable slope distribution and land-use/land-cover that are strongly related to the structural connectivity (Mishra et al. 2019). Excessive siltation forms a central problem in the Kosi basin, and it is necessary to understand the implications of siltation on river processes and associated flood risk under climate change scenarios; this would require a comprehensive analysis of sediment connectivity under a modified hydrological regime induced by climate change. This should then lead to developing effective sediment management plans for protection of infrastructure and human lives not just in the Kosi basin but in several other basins across the world.

In addition to the natural hazards, high sediment flux because of modified sediment connectivity is likely to impact the hydroelectric power projects in several basins in a major way. The designed power production capacity of the hydroelectric power plants depends on the discharge of the river and the head at the turbine. The extent of erosion and deposition will change because of spatio-temporal variability in meteorological conditions, and to the hydrological and geomorphological characteristics of the basin. Landslides triggered by hillslope erosion, levee breach, and channel avulsion may result in partial or total abandonment of hydroelectric projects. A recent example is a major landslide in the upstream reaches of the Bhote Koshi (Jure landslide in 2014); this created a large dam upstream, and a small hydroelectric power station downstream became defunct. Further, sediment-extruding mechanisms may be required during higher sediment transport so that the channel is not filled, and sediment does not enter the turbine partially or fully.

In catchments where water-assisted transportation is the prime mode of sediment transfer, sediment connectivity will also decrease with the decrease in hydrological connectivity. However, in arid regions, where eolian transportation is the dominant process of sediment connectivity, a decrease in precipitation will result in higher entrainment of eolian dust (Reynolds et al. 2007) and will enhance sediment transportation. Therefore, the sensitivity of sedimentary processes to climate change is a function of two factors—sediment supply and transport capacity (East and Sankey 2020). In a supply-limited system, a momentary sedimentary response might be observed to an acting forcing, i.e., increased storm activity. Once the supply gets exhausted, the sedimentary response will cease to exist unless more sediments are produced due to weathering (Heimsath et al. 2012).

Perhaps the impact of climate change on the sediment connectivity of a catchment can be best understood in a glacial system. Due to global warming, the glaciers are receding, resulting in an increased extent of paraglacial zones in glaciated regions of catchments (Lane et al. 2017). It is expected that in such events of glacial recession, sediment connectivity will increase in paraglacial regions. This is because (a) deglaciation will expose the underlying sediment, increasing the probability of connectivity between such sediments and stream channels, (b) deglaciation of hillslopes will significantly increase the hillslope-channel connectivity, (c) stream-based sediment transport is faster than glacier-based sediment transport (Lane et al. 2017), (d) lateral migration is easier in proglacial streams than the sub-glacial stream, and therefore, former can gather more sediment than the latter, and (d) glacial debuttressing of valley sidewalls will result into a decrease in the base level of upstream catchments of such valleys— leading to headward erosion in sidewall tributaries (Schiefer and Gilbert 2007). In such deglaciated catchments, with high upslope and downslope sediment connectivity, and increased sediment supply can change the structural connectivity by transforming the river channels to braided systems (East and Sankey 2020). To propagate the impact of climate change-induced deglaciation downstream, a continuous sediment supply will be required. With the cessation of sediment supply, the propagation of climate change response, e.g., braiding of river channel will also cease. Therefore, it could be interesting to evaluate the sensitivity of propagation of climate change impacts in such deglaciated catchment systems.

Recently, a method has been proposed by Zanandrea et al. (2021) to estimate the hydro-sedimentological connectivity index (IHC) of basins as a factor of topography, surface roughness, precipitation, and runoff. It is an enhanced version of the original IC method proposed by Borselli et al. (2008) and Cavalli et al. (2013). The inclusion of precipitation and runoff factors in this method renders it best suitable for the estimation of hydro-geomorphic connectivity under changing climate scenarios. The impact of LULC change can also be implemented in this method by replacing the terrain-derived surface roughness factor with NDVI derived C-factor previously implemented by Singh et al. (2017) and Singh and Sinha (2019). The IHC can estimate both, sediment as well as hydrological connectivity of catchments.

8.4 Conclusions and Outlook

The gravity of the effects of global warming and hence the climate change on the ecology and economy must be realized by humans across the world (Nordhaus 2019). There are a number of factors on which climate change vulnerabilities depend, and such factors vary among communities and places (Panthi et al. 2016). Because of climate change, the water security of the world is under threat (Ravazzani et al. 2015). Therefore, the impact of climate change and global warming on the available water resources for human consumption and for ecological services should always be considered in any water resource management work (Qi et al. 2009), especially since climate change affects every aspect of the hydrological cycle and can potentially affect the water resources by inducing changes in water quantity, subjecting water resources to extreme events, changing water quality. Various aspects of hydrological processes such as peak flow and runoff amount (Prowse et al. 2006), humidity and precipitation (Wang et al. 2008) might get actively altered by climate change.

While it is certain that water resources in basins across the world are going to be impacted by climate change, there are possibilities to mitigate these measures. This can be done by identifying the manageable and unmanageable stressors of water resources (climate change is one of the unmanageable stressors) and trying to reduce the impact of the unmanageable stressors by adjusting the manageable stressors. For example, in the Seyhan River Basin of Turkey, it was found that if water demands can be kept in check, the climate change-induced water scarcity can be mitigated efficiently (Fujihara et al. 2008).

It has been observed that the temporal variabilities in the intensities of precipitation are necessary to understand catchment responses and hydrological connectivity (Wainwright and Parsons 2002; Smith et al. 2010). Therefore, better forecasting of future climate can account for the probable changes in connectivity at catchment scales. An understanding of how different elements of connectivity will get impacted by changing climate can be utilized to mitigate the adverse impacts of climate change.

Understanding the impact of climate change on the catchment linkage and connectivity is most crucial for countries such as India which hosts very diverse geomorphic systems. Various basin-scale projects such as river linking projects and inland waterways projects have been planned in India. The river linkage project is expected to mitigate the flood-drought duality of the country (Misra et al. 2007; Shah and Amarasinghe 2016; Higgins et al. 2018). For such colossal undertakings, the assessment of the catchment response under changing climate is a challenge and should not be ignored.

References

Arnell NW (1999) The effect of climate change on hydrological regimes in Europe: a continental perspective. Glob Environ Chang 9:5–23

Azari M, Moradi HR, Saghafian B, Faramarzi M (2016) Climate change impacts on streamflow and sediment yield in the North of Iran. Hydrol Sci J 61:123–133

Beyene T, Lettenmaier DP, Kabat P (2010) Hydrologic impacts of climate change on the Nile River Basin: implications of the 2007 IPCC scenarios. Clim Change 100:433–461

Borselli L, Cassi P, Torri D (2008) Prolegomena to sediment and flow connectivity in the landscape: a GIS and field numerical assessment. CATENA 75:268–277

Bracken L, Cox N, Shannon J (2008) The relationship between rainfall inputs and flood generation in south–east Spain. Hydrol Proc Int J 22:683–696

Bracken LJ, Turnbull L, Wainwright J, Bogaart P (2015) Sediment connectivity: a framework for understanding sediment transfer at multiple scales. Earth Surf Proc Land 40:177–188

Bracken LJ, Wainwright J, Ali GA, Tetzlaff D, Smith MW, Reaney SM, Roy AG (2013) Concepts of hydrological connectivity: research approaches, pathways and future agendas. Earth-Sci Rev 119:17–34

Brierley G, Fryirs K, Jain V (2006) Landscape connectivity: the geographic basis of geomorphic applications. Area 38:165–174

Bussi G, Dadson SJ, Prudhomme C, Whitehead PG (2016) Modelling the future impacts of climate and land-use change on suspended sediment transport in the River Thames (UK). J Hydrol 542:357–372

Cavalli M, Trevisani S, Comiti F, Marchi L (2013) Geomorphometric assessment of spatial sediment connectivity in small Alpine catchments. Geomorphology 188:31–41

Chang H, Knight CG, Staneva MP, Kostov D (2002) Water resource impacts of climate change in southwestern Bulgaria. GeoJournal 57:159–168

Chawla I, Mujumdar PP (2015) Isolating the impacts of land use and climate change on streamflow. Hydrol Earth Syst Sci 19:3633–3651

East AE, Sankey JB (2020) Geomorphic and sedimentary effects of modern climate change: current and anticipated future conditions in the Western United States. Rev Geophys 58:e2019RG000692

Fang J, Kong F, Fang J, Zhao L (2018) Observed changes in hydrological extremes and flood disaster in Yangtze River Basin: spatial–temporal variability and climate change impacts. Nat Hazards 93:89–107

Freeman MC, Pringle CM, Jackson CR (2007) Hydrologic connectivity and the contribution of stream headwaters to ecological integrity at regional scales1. Wiley Online Library, pp 5–14

Fryirs K (2013) (Dis)Connectivity in catchment sediment cascades: a fresh look at the sediment delivery problem. Earth Surf Proc Land 38:30–46

Fryirs K, Gore D (2013) Sediment tracing in the upper Hunter catchment using elemental and mineralogical compositions: Implications for catchment-scale suspended sediment (dis)connectivity and management. Geomorphology 193:112–121

Fujihara Y, Tanaka K, Watanabe T, Nagano T, Kojiri T (2008) Assessing the impacts of climate change on the water resources of the Seyhan River Basin in Turkey: Use of dynamically downscaled data for hydrologic simulations. J Hydrol 353:33–48

Gomez-Velez JD, Harvey JW (2014) A hydrogeomorphic river network model predicts where and why hyporheic exchange is important in large basins. Geophys Res Lett 41:6403–6412

Harvey AM (2002) Effective timescales of coupling within fluvial systems. Geomorphology 44:175–201

Heckmann T, Cavalli M, Cerdan O, Foerster S, Javaux M, Lode E, Smetanová A, Vericat D, Brardinoni F (2018) Indices of sediment connectivity: opportunities, challenges and limitations. Earth Sci Rev 187:77–108

Heimsath AM, DiBiase RA, Whipple KX (2012) Soil production limits and the transition to bedrock-dominated landscapes. Nat Geosci 5:210–214

Higgins S, Overeem I, Rogers K, Kalina E (2018) River linking in India: downstream impacts on water discharge and suspended sediment transport to deltas. Elem Sci Anth 6

Hooke JM (2006) Human impacts on fluvial systems in the Mediterranean region. Geomorphology 79:311–335

IPCC (2014) Climate change 2013: the physical science basis: Working Group I contribution to the Fifth assessment report of the Intergovernmental Panel on Climate Change. Cambridge university press

Jain V, Tandon S, Sinha R (2012) Application of modern geomorphic concepts for understanding the spatio-temporal complexity of the large Ganga river dispersal system. Curr Sci (Bangalore) 103:1300–1319

Jain V, Tandon SK (2010) Conceptual assessment of (dis)connectivity and its application to the Ganga River dispersal system. Geomorphology 118:349–358

Jordan YC, Ghulam A, Hartling S (2014) Traits of surface water pollution under climate and land use changes: a remote sensing and hydrological modeling approach. Earth Sci Rev 128:181–195

Keesstra S, Nunes JP, Saco P, Parsons T, Poeppl R, Masselink R, Cerdà A (2018) The way forward: can connectivity be useful to design better measuring and modelling schemes for water and sediment dynamics? Sci Total Environ 644:1557–1572

Khan S, Sinha R, Whitehead P, Sarkar S, Jin L, Futter MN (2018) Flows and sediment dynamics in the Ganga River under present and future climate scenarios. Hydrol Sci J 63:763–782

Khoi DN, Suetsugi T (2014) Impact of climate and land-use changes on hydrological processes and sediment yield-a case study of the Be River catchment Vietnam. Hydrol Sci J 59:1095–1108

Kim J, Choi J, Choi C, Park S (2013) Impacts of changes in climate and land use/land cover under IPCC RCP scenarios on streamflow in the Hoeya River Basin, Korea. Sci Total Environ 452–453:181–195

Lane SN, Bakker M, Gabbud C, Micheletti N, Saugy J-N (2017) Sediment export, transient landscape response and catchment-scale connectivity following rapid climate warming and Alpine glacier recession. Geomorphology 277:210–227

Lexartza-Artza I, Wainwright J (2011) Making connections: changing sediment sources and sinks in an upland catchment. Earth Surf Proc Land 36:1090–1104

Lisenby PE, Fryirs KA (2017) Sedimentologically significant tributaries: catchment-scale controls on sediment (dis)connectivity in the Lockyer Valley, SEQ, Australia. Earth Surf Proc Land 42:1493–1504

López-Vicente M, Poesen J, Navas A, Gaspar L (2013) Predicting runoff and sediment connectivity and soil erosion by water for different land use scenarios in the Spanish Pre-Pyrenees. Catena 102:62–73

Middelkoop H, Daamen K, Gellens D, Grabs W, Kwadijk JC, Lang H, Parmet BW, Schädler B, Schulla J, Wilke K (2001) Impact of climate change on hydrological regimes and water resources management in the Rhine basin. Clim Change 49:105–128

Mimikou M, Kanellopoulou S, Baltas E (1999) Human implication of changes in the hydrological regime due to climate change in Northern Greece. Glob Environ Chang 9:139–156

Mishra K, Sinha R, Jain V, Nepal S, Uddin K (2019) Towards the assessment of sediment connectivity in a large Himalayan river basin. Sci Total Environ 661:251–265

Misra AK, Saxena A, Yaduvanshi M, Mishra A, Bhadauriya Y, Thakur A (2007) Proposed river-linking project of India: a boon or bane to nature. Environ Geol 51:1361–1376

Molina-Navarro E, Hallack-Alegría M, Martínez-Pérez S, Ramírez-Hernández J, Mungaray-Moctezuma A, Sastre-Merlín A (2016) Hydrological modeling and climate change impacts in an agricultural semiarid region. Case study: Guadalupe River basin, Mexico. Agric Water Manage 175:29–42

Mullan D, Favis-Mortlock D, Fealy R (2012) Addressing key limitations associated with modelling soil erosion under the impacts of future climate change. Agric Forest Meteorol 156:18–30

Nearing MA, Pruski FF, O'Neal MR (2004) Expected climate change impacts on soil erosion rates: a review. J Soil Water Conserv 59:43–50

Nordhaus W (2019) Climate change: the ultimate challenge for economics. American Econ Rev 109:1991–2014

Op de Hipt F, Diekkrüger B, Steup G, Yira Y, Hoffmann T, Rode M, Näschen K (2019) Modeling the effect of land use and climate change on water resources and soil erosion in a tropical West African catch-ment (Dano, Burkina Faso) using SHETRAN. Sci Total Environ 653:431–445

Panthi J, Aryal S, Dahal P, Bhandari P, Krakauer NY, Pandey VP (2016) Livelihood vulnerability approach to assessing climate change impacts on mixed agro-livestock smallholders around the Gandaki River Basin in Nepal. Reg Environ Change 16:1121–1132

Pringle C (2003) What is hydrologic connectivity and why is it ecologically important? Hydrol Process 17:2685–2689

Prowse T, Beltaos S, Gardner J, Gibson J, Granger R, Leconte R, Peters D, Pietroniro A, Romolo L, Toth B (2006) Climate change, flow regulation and land-use effects on the hydrology of the Peace-Athabasca-Slave system; Findings from the Northern Rivers Ecosystem Initiative. Environ Monit Assess 113:167–197

Qi S, Sun G, Wang Y, McNulty S, Myers JM (2009) Streamflow response to climate and landuse changes in a coastal watershed in North Carolina. Trans ASABE 52:739–749

Ravazzani G, Barbero S, Salandin A, Senatore A, Mancini M (2015) An integrated hydrological model for assessing climate change impacts on water resources of the Upper Po River Basin. Water Resour Manage 29:1193–1215

Reynolds RL, Yount JC, Reheis M, Goldstein H, Chavez P Jr, Fulton R, Whitney J, Fuller C, Forester RM (2007) Dust emission from wet and dry playas in the Mojave Desert, USA. Earth Surf Proc Landforms J British Geomorphol Res Group 32:1811–1827

Roy NG, Sinha R (2017) Linking hydrology and sediment dynamics of large alluvial rivers to landscape diversity in the Ganga dispersal system, India. Earth Surf Proc Land 42:1078–1091

Roy NG, Sinha R (2018) Integrating channel form and processes in the Gangetic plains rivers: implications for geomorphic diversity. Geomorphology 302:46–61

Schiefer E, Gilbert R (2007) Reconstructing morphometric change in a proglacial landscape using historical aerial photography and automated DEM generation. Geomorphology 88:167–178

Shah T, Amarasinghe UA (2016) River linking project: a solution or problem to india's water woes? Indian Water Policy at the Crossroads: Resources, Technology and Reforms. Springer, pp 109–130

Singh M, Sinha R (2019) Evaluating dynamic hydrological connectivity of a floodplain wetland in North Bihar, India using geostatistical methods. Sci Total Environ 651:2473–2488

Singh M, Sinha R, Tandon S (2021) Geomorphic connectivity and its application for understanding landscape complexities: a focus on the hydro-geomorphic systems of India. Earth Surf Proc Land 46:110–130

Singh M, Tandon SK, Sinha R (2017) Assessment of connectivity in a water-stressed wetland (Kaabar Tal) of Kosi-Gandak interfan, north Bihar Plains, India. Earth Surf Proc Land 42:1982–1996

Sinha R, Gaurav K, Chandra S, Tandon SK (2013) Exploring the channel connectivity structure of the August 2008 avulsion belt of the Kosi River, India: application to fl ood risk assessment. Geology 41:1099–1102

Sinha R, Sripriyanka K, Jain V, Mukul M (2014) Avulsion threshold and planform dynamics of the Kosi River in north Bihar (India) and Nepal: a GIS framework. Geomorphology 216:157–170

Sinha RK, Eldho T, Subimal G (2020) Assessing the impacts of land cover and climate on runoff and sediment yield of a river basin. Hydrol Sci J 65:2097–2115

Smith MW, Bracken LJ, Cox NJ (2010) Toward a dynamic representation of hydrological connectivity at the hillslope scale in semiarid areas. Water Resour Res 46

Smith T, Marshall L, McGlynn B, Jencso K (2013) Using field data to inform and evaluate a new model of catchment hydrologic connectivity. Water Resour Res 49:6834–6846

Turnbull L, Hütt M-T, Ioannides AA, Kininmonth S, Poeppl R, Tockner K, Bracken LJ, Keesstra S, Liu L, Masselink R (2018) Connectivity and complex systems: learning from a multi-disciplinary perspective. Appl Network Sci 3:11

Turnbull L, Wainwright J, Brazier RE (2008) A conceptual framework for understanding semi-arid land degradation: ecohydrological interactions across multiple-space and time scales. Ecohydrology 1:23–34

Turner DP, Conklin DR, Bolte JP (2015) Projected climate change impacts on forest land cover and land use over the Willamette River Basin, Oregon, USA. Clim Change 133:335–348

Vanacker V, Molina A, Govers G, Poesen J, Dercon G, Deckers S (2005) River channel response to short-term human-induced change in landscape connectivity in Andean ecosystems. Geomorphology 72:340–353

Wada Y, Van Beek L, Bierkens MF (2011) Modelling global water stress of the recent past: on the relative importance of trends in water demand and climate variability. Hydrol Earth Syst Sci 15:3785–3808

Wainwright J, Parsons AJ (2002) The effect of temporal variations in rainfall on scale dependency in runoff coefficients. Water Resour Res 38:7-1–7-10

Wainwright J, Turnbull L, Ibrahim TG, Lexartza-Artza I, Thornton SF, Brazier RE (2011) Linking environmental régimes, space and time: interpretations of structural and functional connectivity. Geomorphology 126:387–404

Wang S, Kang S, Zhang L, Li F (2008) Modelling hydrological response to different land-use and climate change scenarios in the Zamu River basin of northwest China. Hydrol Proc Int J 22:2502–2510

Ward J (1989) The four-dimensional nature of lotic ecosystems. J North Am Benthological Soc 2–8

Whitehead PG, Jin L, Macadam I, Janes T, Sarkar S, Rodda HJ, Sinha R, Nicholls RJ (2018) Modelling impacts of climate change and socio-economic change on the Ganga, Brahmaputra, Meghna, Hooghly and

Mahanadi river systems in India and Bangladesh. Sci Total Environ 636:1362–1372

Wohl E, Brierley G, Cadol D, Coulthard TJ, Covino T, Fryirs KA, Grant G, Hilton RG, Lane SN, Magilligan FJ, Meitzen KM, Passalacqua P, Poeppl RE, Rathburn SL, Sklar LS (2019) Connectivity as an emergent property of geomorphic systems. Earth Surf Proc Land 44:4–26

Zanandrea F, Michel GP, Kobiyama M, Censi G, Abatti BH (2021) Spatial-temporal assessment of water and sediment connectivity through a modified connectivity index in a subtropical mountainous catchment. CATENA 204:105380

9

Inter-decadal Variability of Precipitation Patterns Increasing the Runoff Intensity in Lower Reach of Shilabati River Basin, West Bengal

Suparna Chaudhury

Abstract

Spatial and inter-decadal variability of precipitation patterns into different storm periods provides abundant impact on runoff, discharge that create the risk of rain-generated floods in this area. Ghatal subdivision is an administrative subdivision of Paschim Medinipur district in the state of West Bengal, India. This area is largely prone to devastating natural floods on a regular interval because of its shape, geophysical condition and geographical location and it is experiencing with riverine floods mainly by the Shilabati River and its tributaries. Heavy to very heavy rainfall associated with average 5–10 days cyclonic storms and depressions during the monsoon season is important factor for creating the annual flood in this area. The instrumental rainfall records of 20 years (2001–2020) reveal that percentage of average storm rainfall comparing to total annual rainfall has increasing from 63.10 in 2001 to 97.10 in 2020 and the average storm rainfall concentration has also exceed than annual rainfall of study area in few years. The highest storm rainfall over the area was 612.6 mm in the year 2017. As side by side percentage of runoff intensity has also increased from 47.36 in 2001 to 52.70 in 2020 that creates the risk of rain-generated floods in this area. Remote sensing data is used as the basic information input for computing runoff using the Soil Conservation Service (SCS) Runoff Curve Number (RCN) model used by US Department of Soil Conservation Service (1972). This empirical model is used for estimation of runoff intensity. So the floodplain users are coped to very heavy flood risks in future.

Keywords

Cyclonic storm and depressions · Monsoon · Runoff · Runoff curve number · Soil conservation service

9.1 Introduction

Rainfall and runoff are significant constitute for generating the river discharge in a watershed. (Zakwan et al. 2017). River basin morphology such as height, length, slope, shape, soil condition and land use have significant impact for the runoff generation in the river basin. Amongst the various methods, Soil Conservation Services and Curve Number (SCS-CN) technique is one of the unique methods for rainfall runoff modelling (Zakwan 2016). Land use and Land cover information is used to estimate the value of

S. Chaudhury (✉)
S.B.S.S.Mahavidyalaya, Paschim Medinipur,
West Bengal, India
e-mail: suparnachaudhury@sbssmahavidyalaya.ac.in

surface roughness or friction which affects the velocity of the overland flow of water and determine the amount of rainfall that will infiltrate into the soil (Ara 2018 and Zakwan 2016). The data on hydrological soil group, land use, antecedent rainfall, storm duration required to estimate surface runoff in catchments, antecedent soil moisture conditions are the basic catchment characteristics used for curve number calculations (Mockus 1949, Sharma et al. 2001). Based on Geographical Information System and Remote Sensing, land-cover and land use changes are identified (Gangodagamage and Agarwal 2001). The geomorphological factors, land use change affect the runoff volume and the runoff rate significantly through interaction with land uses and soils. (Satheeshkumar et al. 2017). Runoff is the most important hydrological variable used for analysing flood frequency and flood potentiality. Accurate and timely prediction of runoff in a drainage basin is mainly done with equation and models or direct measurement at gauging stations by either using a range of equation and models or direct measurement at gauging stations (Moitra 2008). Soil Conservation Services and Curve Number (SCS-CN) method based on Rainfall and Land use data as inputs and all the three antecedent moisture conditions (AMC-1, AMCII and AMCIII) is used in concerned area. The SCS-CN method is useful for calculating volume of runoff from the land surface meets in the river of streams. Remote sensing data is used as the basic information input for computing runoff using the Soil Conservation Service (SCS) Runoff Curve Number (RCN) model proposed by US Department of Agriculture, 1972. This empirical model is used in the present study for estimation of flood discharge in Shilabati river basin on land use conditions and to evaluate the hydrologic response of these measures on runoff. The soil Conservation Service (1972) developed a method for computing abstractions from storm rainfall. For the storm as a whole, the depth of excess precipitation or direct runoff is always less than or equal to the depth of precipitation, likewise after runoff begins, the additional depth of water retained in the watershed, is less than or equal to some potential maximum retention S. There is some amount of rainfall Ia (Initial abstraction before ponding) for which no runoff will occur, so the potential runoff is (P-Ia) (Chow et al. 1988). The volume of water available for runoff increases because of the increased impervious cover provided by bare surface, urbanisation, concrete streets etc. reduce the amount of infiltration changes the hydraulic efficiency associated with artificial channels, curbing gutters and storm drainage collection systems increase the velocity of flow and the magnitude of flood peaks. The SCS method for rainfall–runoff analysis is applied to determine the increase in the amount of runoff caused by urbanisation, increase of concrete road etc.

9.2 The Study Area

The Shilabati River originates in the extended part of Chotonagpur Plateau from the Hurra P.S. of Purulia district at west and has been extended up to eastern part of Hooghly district at Bandar in West Bengal, where it is joined Darakeshwar to originate the river Rupnarayan Bankura.gov.in. Location (2009). This river catchment is characterised by rocky and undulating tract for most of its upper part. It is extended on the lateritic flats up to the village Shimulia at Garbeta block, where the river has divided into two channels then it follows alluvial low land up to Bandar of Ghatal (en.wikipedia.org. wiki 2012). The whole catchment lies between 22° 30′ N–23° 15′ N latitude and 86° 40′E–87° 55′ E longitude (concerned Toposheets no, 73I/12, 73 J/16, 73 N/1, 73 N/5, 73 N/6, 73 N/9, 73 N/10, 73 N/11, 73 N/2). The lower catchment extends between 22° 30′–22° 55′ N and $87^0 15'$–$87^0 50'$E and faces the fury of flood almost annually where Ghatal subdivision is an administrative subdivision of Paschim Medinipur district in the state of West Bengal, India. It is located in the eastern part of the Shilabati river basin which extends between 22° 30′ 30″ N–22° 50′ 30′ N latitude and 87° 31′ 30″ E–87° 55′ E longitude (concerned Toposheets no 73 N/6, 73 N/9, 73 N/10, 73 N/11). Heavy to very heavy rainfall associated with average 5–10 days cyclonic storms and depressions appears during the monsoon season in this area (Flood Monograph 2007). The study area faces the fury of flood almost annually. The floods which occurred in 1978, 1982, 1985, 1987, 1988, 1993, 2001, 2007, 2015, 2017 and 2020 were devastating and caused serious damages of lives and livelihood of the majority of people The inundation was more extensive and loss of life and properties were very severe.

9.3 Materials and Methods

Rainfall data has been collected (2000–2020) from Central Water Commission, Paschim Medinipur, Subdivisional Agricultural Farm Office at Khirpai, Ghatal, Paschim Medinipur. In the present study, the sub-watersheds are identified and the USDA Soil Conservation Service (SCS) runoff curve number (CN) are selected for the Shilabati (main river), Joypanda, Betal, Donai, Tangai, Kubai and Parang (sub-watersheds) from a digitised land use and land cover map derived from ERDAS IMAGINE 9.0 on IRS- LISS-III, Satellite Imagery (2nd April, 2010) and Resourcesat-1 image (2nd May, 2020), Google Map and concerned toposheets no (73I/12, 73I/16, 73 N/1, 73 N/5, 73 N/9, 73 N/10, 73 N/11, 73 M/4, 73 N/2, 73 N/6 etc.) and it is verified through field check. This was accomplished with the help of standard SCS table of runoff curve number modified for Indian conditions. The four hydrological soil groups (A, B, C, D) are used in determining hydrologic soil cover complexes which are used in this method for estimating runoff from rainfall. The soil properties play an important role in the estimation of runoff from the rainfall and in this concern the properties are represented by the hydrological parameters. Direct runoff produced in that watershed by a given precipitation are estimated using the model SCS-CN is widely used and involves the use of simple empirical formulae. The equation requires the rainfall and watershed coefficient as inputs. The Composite Curve Number is computed from the Runoff Curve Number (RCN) which is a quantitative descriptor of the Land and Soil complex and it is derived on the basis of Soil-Vegetation-Land (SVL) and Antecedent Moisture Condition complex. It takes on values from 0 to 100 (Sharma and Kumar 2002). The AMC value is intended to reflect the effect of infiltration on both the volume and rate of runoff according to the infiltration curve Antecedent Moisture Condition (AMC) refers to the water content present in the soil at a given time. (Suresh 1997). The watershed coefficient called curve number (CN) which is an index that represents the combination of hydrologic soil group and land use and land treatment classes. The Curve Number in AMC III condition (as the total rainfall corresponds to the AMC III Condition) for each land use category is then applied in order to estimate Weighted Curve Number for each watershed with following formula proposed by (Schwab et al. 1993). The Potential Maximum Retention (S) in mm is then calculated with the Eq. 9.3. The Runoff in cubic meter is then calculated with (Eq. 9.2). The expression used in SCS runoff is presented as following equations (Sect. 9.3.1).

9.3.1 Model and Potential Retention Equation

The runoff Eq. (Handbook of Hydrology 1972).

$$Qt = f(Pt, S, Ia) \quad (9.1)$$

Precipitation –Runoff- Potential Retention equation (Handbook of Hydrology 1972).

$$Q = (P - 0.2S)^2 / (P + 0.8S) \quad (9.2)$$

(Handbook of Hydrology 1972).

$$S = (25400/CN)254 \text{ in mm} \quad (9.3)$$

Q Actual direct runoff (mm).
P Total rainfall mm.
S Potential maximum retention mm.

CN = SCS Runoff Curve Number, $= f$ (Soil, land, vegetal cover, antecedent moisture condition etc.)

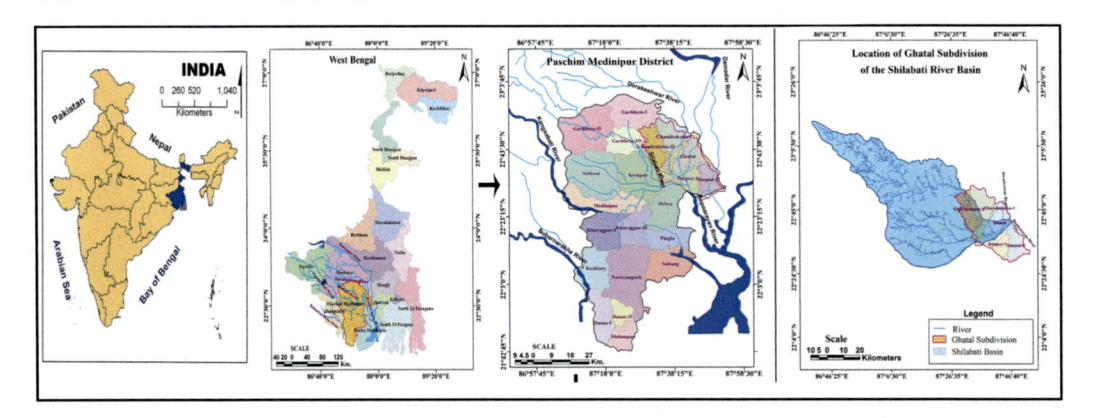

Fig. 9.1 Location of the study area

Fig. 9.2 Flow chart of the methodology of the present study

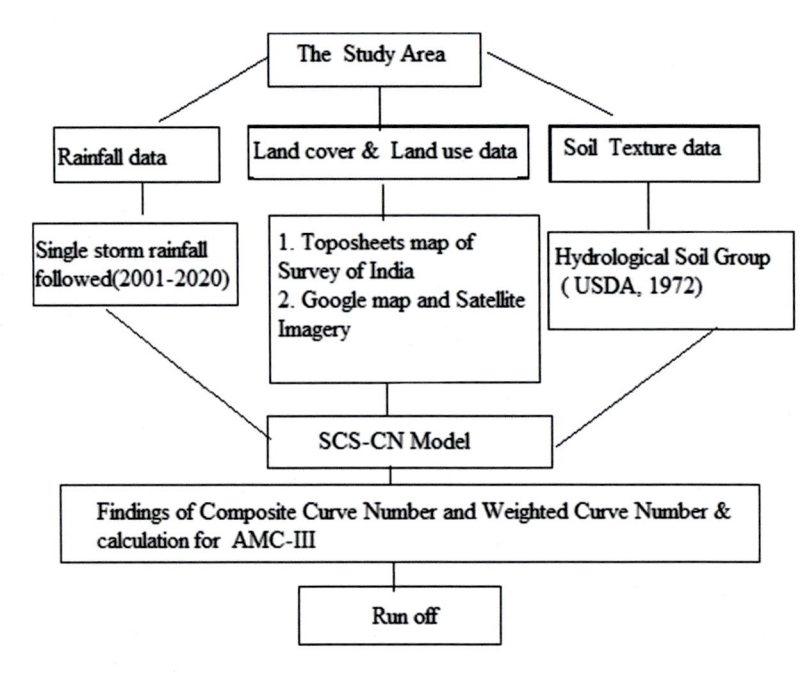

$$\text{Weighted CN} = \frac{CN_{1XA_1} + CN_{2XA_2} \ldots CN_{n*A_n}}{A_1 + A_2 + \ldots A_n} \quad (9.4)$$

CN_1, CN_2… CN_n Curve Number of respective landuse,
A_1 % area under respective landuse.

The runoff is then transferred to discharge in m^3 by multiplying with watersheds area after Schwab et al. (1993).

9.3.2 Rainfall Characteristics in the Study Area (2001–2020)

Intensity of flood in this area varies considerably from year to year. Table 9.1 and Table 9.2 reflects the substantial inter-annual variation in the monsoon rainfall. Sometimes the rainfall is also maintained chiefly by cyclonic storms. Cyclones from the Bay of Bengal and the south–west monsoon current bring very heavy rainfall

Table 9.1 Average rainfall of highest storm period (2001–2020 year)

Year	Highest Storm Duration in each year	Amount of rainfall in mm in highest storm	Average rainfall in mm during storm periods	Average annual rainfall in mm at total catchment area	Percentage
2001	25.9–4.10	220.6	854.80	1354.76	63.10
2002	27.8–6.9	186.6	932.40	1426.52	55.25
2003	23.7–1.8	267.2	1489.00	1549.20	79.55
2004	12.8–23.8	236.8	1403.60	1553.14	95.87
2005	24.7–4.8	215.2	1363.80	1375.04	102.07
2006	8.7–18.7	263.2	1276.60	1686.28	80.87
2007	10.7–20.7	438.2	2355.00	1298.80	98.29
2008	17.7–27.7	128.8	1227.3	2106.00	111.82
2009	16.8–26.8	206.2	1345.4	1780.44	68.93
2010	23.6–3.7	188.8	992.4	1270.10	105.92
2011	7.8–17.8	519.4	2270.02	1001.10	99.13
2012	11.7–21.7	298.4	1484.60	1717.90	132.13
2013	13.1–23.1	325	1903.2	1113.30	133.35
2014	14.8–25.8	266	1296.60	2331.26	81.64
2015	9.7–20.7	491.4	1653.80	1157.16	112.05
2016	1.8–11.8	308.6	1424.20	1449.70	114.08
2017	18.7–28.7	612.6	1748.40	1371.50	103.84
2018	2.7–12.7	172	954.60	1544.80	113.18
2019	12.8–22.8	232	1228.34	1413.30	67.54
2020	15.7–25.7	248.9	1280.80	1319.02	97.10

over the study area (www.meteoprog.es, 2012). The analysis of the available last 20 year's hydrological data provides the vivid picture of rainfall characteristics over this area. Several extreme flood events have occurred in 2001, 2003, 2005, 2007, 2011, 2015, 2017, 2019 etc. Increasing trends of annual peak rainfall during monsoon period at every year increase the tendency of flood. The comparison and analysis of the 20 years' rainfall data display the character of magnitude of floods in this area. The long instrumental rainfall records of 20 years (2001–2020) reveal that the highest storm rainfall over the area was 612.6 mm in the year 2017 which exceed the average annual rainfall at total catchment area (Table 9.1) and the average rainfall of highest storm duration was exceeded (347.43–235.16 = 112.27 mm) before 2010–2020 (Table 9.2). So the flow capacity of this channel had failed to pass the heavy amount of rain water and accumulated in mouth area and creates the water logging situation in lower reach. That is also responsible for intensification of flood usually developed by huge and concentrated rain during a monsoon trough or a single cyclonic storm in this area.

9.3.3 Identification of Drainage Pattern and Hydrological Soil Groups Map of Catchment Area

The Shilabati river catchment comprises the main six tributaries viz; Joypanda, Betal, Donai, Kubai, Tangai and Parang which have also originated almost from the same height from the eastern part of the Chotanagpur plateau and

Table 9.2 Average rainfall of highest storm period in each year at study area (2001–2020)

Year	2001	2002	2003	2004	2005	2006	2007	2008	2009	2010	Total	Average in mm
Total rainfall in mm of highest storm duration at each year	220.6	186.6	267.2	236.8	215.2	263.2	438.2	128.8	206.2	188.8	2351.6	235.16
Year	2011	2012	2013	2014	2015	2016	2017	2018	2019	2020	Total	Average
Total rainfall in mm of highest storm duration at each year	519.4	298.4	325	266	491.4	308.6	612.6	172	232	248.9	3474.3	347.43

follow the same gradient of this river (Fig. 9.3). The orientation and distribution of tributary sub-watersheds show that Donai, Kubai-Tangai and Parang join the parent stream Shilabati at its extreme lower catchment within 9 km reach where the river is extremely incapacitated. So similar discharge condition and similar time of runoff create the accumulation of water that develops flood situation. The soil map of study area was obtained from district planning map of concerned districts (Fig. 9.4). The four hydrological soil groups (A, B, C, D) are identified in determining the hydrologic soil cover complexes which are used in this method for estimating runoff from rainfall. Group **A, B** are sand, loamy sand and sandy loam, silt loam and loam which have low and moderate runoff potentiality but it covered few parts. The transmission rate of these soils is between 0.38 and 0.76 cm /hour and soil group C and D sandy clay loam and clay loam, clay pan or clay layer at or near the surface and shallow soils cover nearly impervious material which are covering most part of the watershed but these soils have moderate to high runoff potentiality because these soils have very slow infiltration rates. These soils have a transmission rate between 0.13 and 0.38 cm/ hour. Basin under study is mostly composed of hydrological soil group D, having high runoff potential. This also contributes high magnitude flood for lower catchment of Shilabati River.

9.3.4 Computation of Curve Number (CN) from Land Use and Hydrological Soil Group

The Curve Number (CN) is computed on the basis of land use and Hydrological soil groups conditions in the catchment area. Eight types of land uses were identified in the Shilabati river basin i.e. fairly dense sal, open scrub, river, wetland, residential area, concrete, metalled and unmetalled road, bare surface, plantation etc. (Fig. 9.5) but it was found that some land use and land cover has been changed and converted between 2010 and 2020 as residential areas,

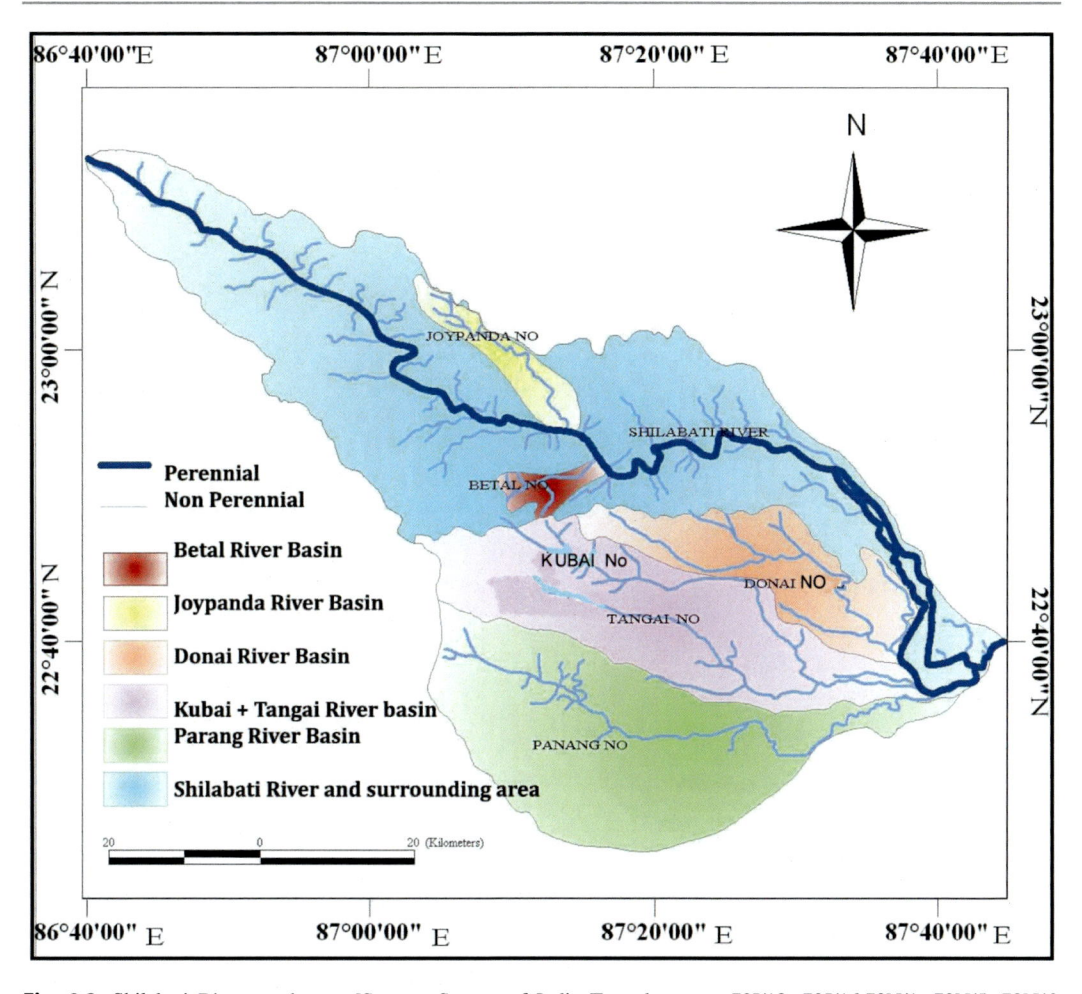

Fig. 9.3 Shilabati River catchment [Source: Survey of India Toposheets no 73I/12, 73J/16,73N/1, 73N/5, 73N/6, 73N/9, 73N/10,73N/11,73N/2], Satellite imagery, Google map (2020)

concrete and metalled roads (Table 9.3 and Fig. 9.6). The Curve Number (CN) and Weighted Curve Number (WCN) of the Shilabati basin and its tributaries were estimated to calculate runoff (Eq. 9.4 and Table 9.4).

9.3.5 Estimation of Runoff Using SCS Curve Number Techniques

Potential retention (S) of rainfall, before 2010 in table no 4 and after 2011–2020 in table no 5 is also calculated on the equation no 3 and Weighted curve number (WCN) reflects in table no 3. Then the Runoff is also estimated on the basis of (eq. no-2) average highest storm rainfall of 235.16 is 385.73 m^3 before 2010 rainfall data and then highest storm rainfall of 347.43 mm is 864.30 m^3 which exceeds (864.30 m^3 − 385.73 m^3 = 478.57 m^3) that increase the runoff discharge after 2010–2020 (reflected in table no 4 and 5) due to variability of precipitation pattern and changes of land use which also increase the flood magnitude basically at lower reach in that study area.

9.4 Result and Discussion

The instrumental rainfall records of 20 years (2001–2020) reveal that inter-decadal variability in percentage of average storm rainfall comparing to total annual rainfall has increasing from

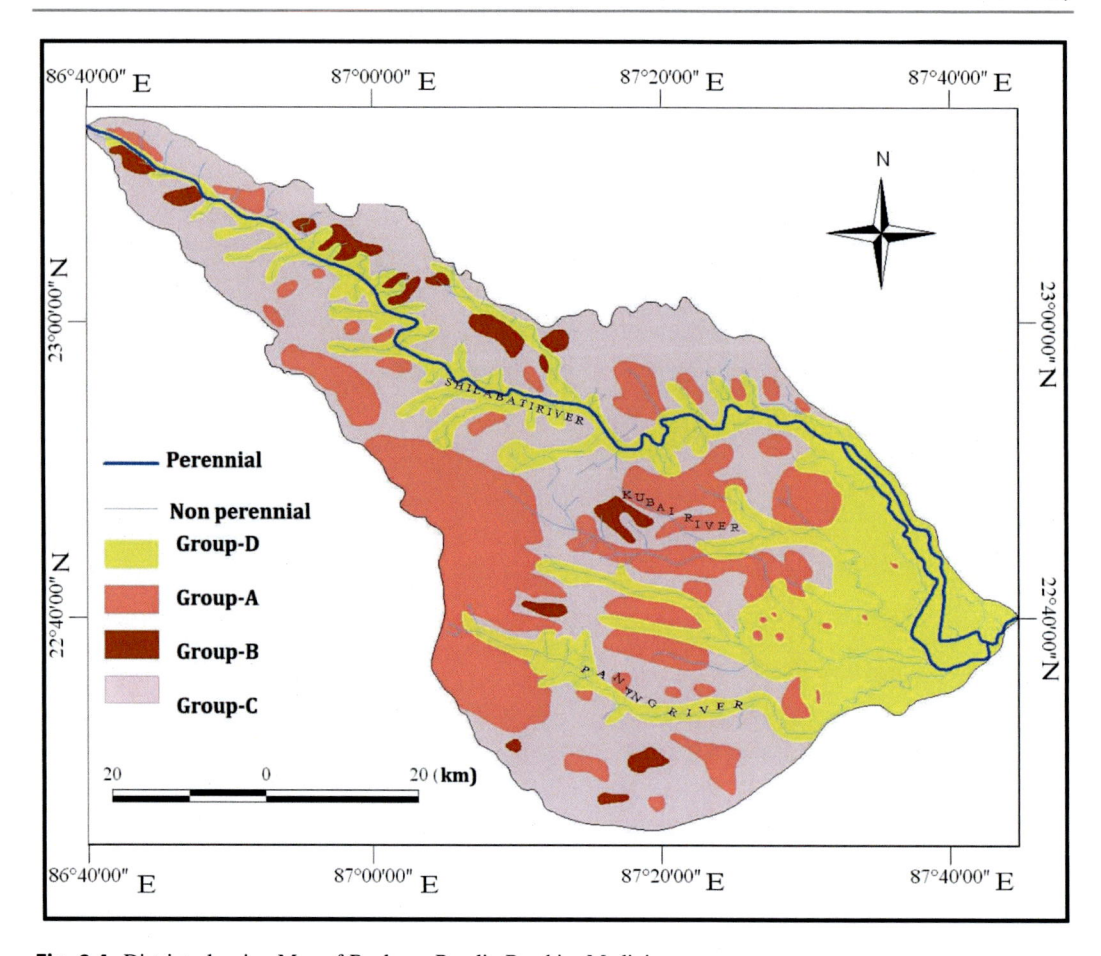

Fig. 9.4 District planning Map of Bankura, Purulia Paschim Medinipur

63.10 in 2001 to 97.10 in 2020 and the average storm rainfall concentration has also exceeded than annual rainfall at study area in few years. The highest storm rainfall over the area was 612.6 mm in the year 2017. After the heavy storm, the low lying depression in the lower reach of this catchment attracts huge flood water and sediments from it upper catchment, sub-tributaries and larger catchments of neighbouring rivers mainly Damodar, Darakeswar and Kangsabati. Most part of the lower catchment lie with a depression of very low gradient and attract huge water from upper catchment. Drainage efficiency of sub-watersheds is estimated in the form of weighted curve number. Shilabati catchment without major tributaries shows lower efficiency of runoff (CN = 63.97) but the maximum drainage efficiency is from the Betal river due to impervious nature of land uses and lower areal coverage. Runoff efficiency is also high for Kubai and Parang (curve number 78.26 and 78.28). The runoff has been calculated on the basis of given weighted curve numbers (WCN) and potential retention (S) of individual river catchment and the cumulative rain during those storm days duration in a particular year is calculated on average of ten years data. The Table 9.5 shows that the input rainfall of 235.16 mm produces the total **385.73 million cubic meter** runoff discharge after interactions on land use and landcover at the Shilabati river catchment and its sub-tributaries catchment. Out of total 385.73 million cubic meter runoff, the tributaries of Donai, Kubai, Tangai and Parang contribute 182.68Mm^3(**47.36%**) which is accumulated in the lower reach of the Shilabati river basin and

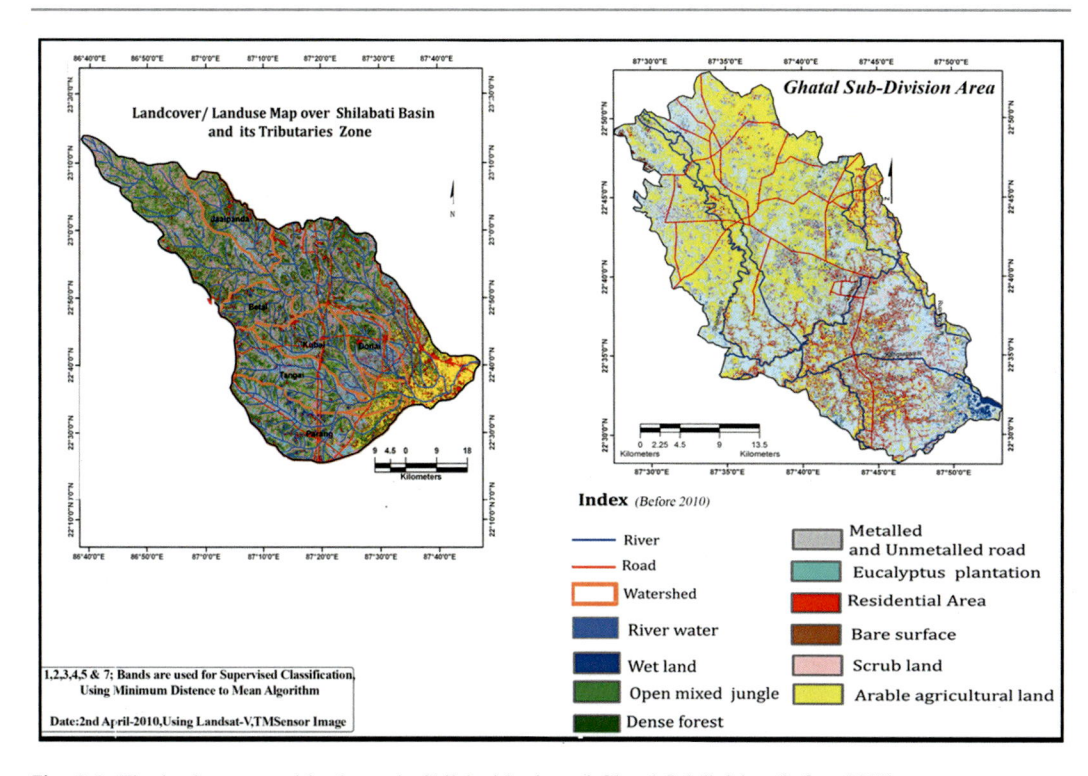

Fig. 9.5 The land covers and land uses in Shilabati basin and Ghatal Subdivision (before 2010)

develops the flood situation. The Table 9.2 also shows that the input rainfall of **347.43 mm** (2011–2020) produces the total 864.30 million cubic meter runoff after interactions on land use and landcover at the Shilabati river catchment and its main tributaries catchment. Out of total **864.30 million cubic meter** runoff, the tributaries of Donai, Kubai, Tangai and Parang contribute 455.50 Mm^3 **(52.70%)** which has been increased **(52.70%-47.30% = 5.4%) almost 6%** from last 2010 in the lower reach of the Shilabati river basin. So the inter-decadal variability of precipitation pattern and land use change accentuates the severity of the flood discharge at main stream and its incapacitated parts of lower reach specially Ghatal Subdivision before 2010 to 2020.

9.5 Conclusion

Inter-decadal variability of rainfall and the maximum runoff efficiency is high at the sub-tributaries mainly Betal, Kubai and Parang river due to impervious nature of land uses and lower areal coverage. The tributaries which have higher drainage efficiency bring more discharge to the main stream at its incapacitated parts. Basin under study is mostly composed of hydrological soil group D, having high runoff potential. This runoff cannot be managed at its lower and sluggish stage, so the trend of increasing runoff will be high risks and contributes regular and high magnitude flood in future at lower reach of Shilabati river specially at Ghatal Subdivision.

Table 9.3 The calculation runoff curve number of Shilabati River by SCS method

Sl No	Catchment area	Soil types	Year	Forest land	River water	Wet land	Bare surface	Open scrub	Plantation	Arable land	Residential area	Concrete, metalled and un-metalled road	Weighted CN
					Composite curve number								
1	Shilabati (Except tributaries)	A, B, C, D	2010	73	66.85	66.23	63.51	64.09	58.7	62.69	60.21	60.01	63.97
			2017	73	66.85	66.23	60.5	60.09	58.7	81.09	69.21	74.08	67.8
2	Joypanda	A, B, C, D	2010	78	83.75	79.68	0	66.42	56.4	84.58	69.27	83.13	75.13
			2017	75	83.75	79.68	0	62.42	56.4	86.64	78.67	86.43	76.1
3	Betal	C, D	2010	88	85.5	80.75	0	88.27	87	89.72	85.35	76.67	85.1
			2017	81	85.5	80.75	0	78.27	87	98.87	94.75	91.82	87.21
4	Tangai + Kubai	A, B, C, D	2010	67	82.03	76.9	85.46	65.87	0	76.48	77.05	88.13	78.26
			2017	67	82.03	78.36	77.46	60.87	0	98.72	97.3	97.24	82.54
5	Donai	A, B, C, D	2010	71	77	83.68	84.36	64.87	0	72.17	75.65	86.17	76.91
			2017	71	77	83.68	74.36	64.87	0	84.17	95.57	95.34	80.29
6	Parang	A, B, C, D	2010	63	83.31	80	83.71	63.72	0	88.9	75.87	86.7	80.48
			2017	55	83.31	80	73.71	63.03	0	95.6	95.87	98.42	78.28

Source Satellite Imagery of concerned area, Topo-sheets and Wiki map of concerned area

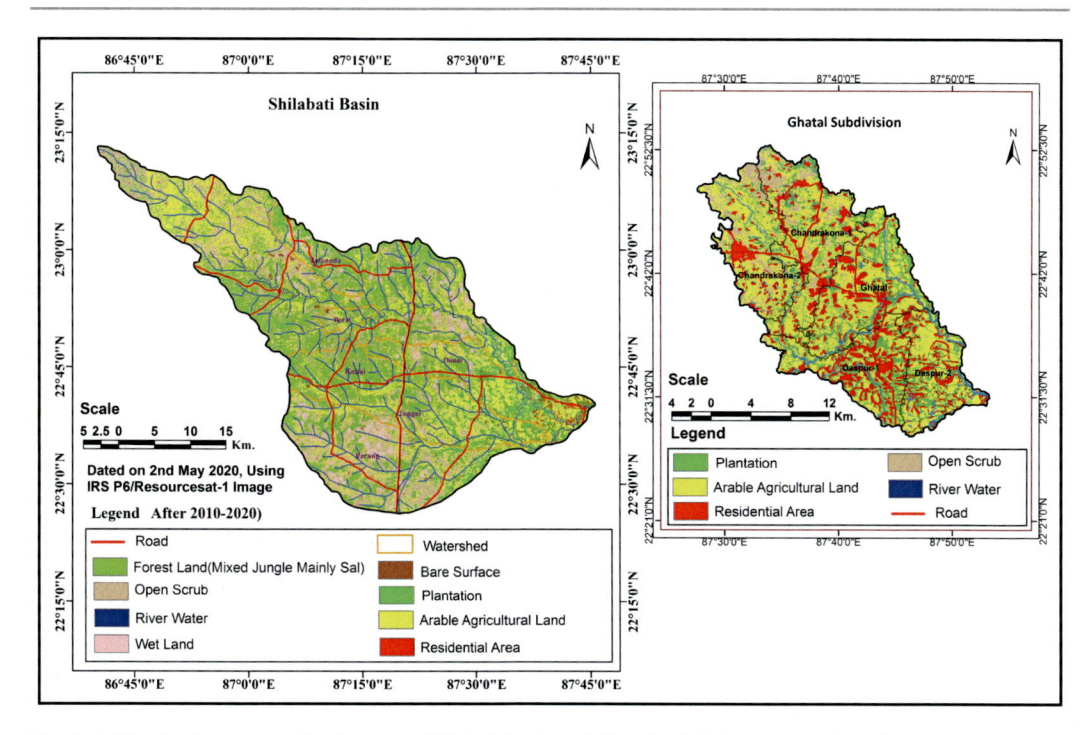

Fig. 9.6 The land covers and land uses in Shilabati basin and Ghatal subdivision area (after 2010)

Table 9.4 Calculation of runoff in (Cubic meter) from the average of highest storm rainfall of 235.16 mm

Name of the main river and Tributaries	Average Rainfall in mm during highest storm period of each year from (2001–2010)	S (potential retention) in mm	Area in km^2	Weighted curve number	Runoff in Mm^3
Shilabati (only the main stream except the mentioned tributaries)	235.16	143.06	1712.5	63.97	173.90
Joypanda	235.16	84.08	125.5	75.13	17.80
Betal	235.16	44.47	63.5	85.10	11.35
Donai	235.16	76.2	239	76.91	35.46
Kubai + Tangai	235.16	70.56	440.8	78.26	67.58
Parang	235.16	70.48	519.2	78.28	79.64
Total runoff in lower catchment (35.46 + 67.58 + 79.64) = 182.68Mm^3 = 47.36%					**385.73**

Bold digits indicate total runoff in storm discharge in whole catchment before 2010

Table 9.5 Calculation of runoff in (Cubic meter) from the average of highest storm rainfall of 347.43 mm

Name of the main river and Tributaries	Average Rainfall in mm during highest storm period of each year from (2011–2020)	S (potential retention) in mm	Area in km^2	Weighted curve number	Runoff in Mm^3
Shilabati (only the main stream except the mentioned tributaries)	347.43	124.30	1712.5	67.80	358.62
Joypanda	347.43	79.77	125.5	76.10	31.33
Betal	347.43	37.25	63.5	87.21	18.85
Donai	347.43	59.93	239	80.91	64.65
Kubai + Tangai	347.43	53.72	440.8	82.54	122.26
Parang	347.43	61.60	519.2	80.48	268.59
Total runoff in lower catchment (64.65 + 122.26 + 268.59) = 455.50 Mm^3 = 52.70%					864. 30

References

Ara Z (2018) Land use classification using remotely sensed images a case study of eastern Sone Canal-Bihar. STIWM-2018, IIT Roorkee

Bankura.gov.in. (2009) Location

Chow VT, Maidment DR, Mays WL (1988) Surface Water, SCS method for Abstractions. Appl Hydrol 147–153

en.wikipedia.org. wiki (2020)

Flood Monograph (2007) Published by Sub divisional Office, Ghatal, Relief Dept

Gangodagamage C, Agarwal SP (2001) Hydrological modeling using remote sensing and GIS, Asian conference on remote sensing 5–9 November 2001

Handbook of hydrology (1972) Soil conservation Department, Ministry of Agriculture, New Delhi

Mockus V (1949) Estimation of total (and Peak Rates of) surface runoff for individual storms. Interim survey report, Grand (Neosho) River Watershed, Exhibit A of Appendix B, USDA, Lincoln, Nebraska

Moitra MM (2008) Crisis of water and its impact on social environment darjiling unpublished thesis, University of Calcutta, pp 154–156

Satheeshkumar S, Venkateswaran S, Kannan R (2017) Rainfall–runoff estimation using SCS–CN and GIS approach in the Pappiredipatti watershed of the Vaniyar sub basin, South India in Model Earth System. Environment 3(24):1–8

Schwab GO, Fangmeier DD, Elliot WJ, Freveret RK (1993) Soil and water conservation engineering. J. Wiley and sons. New York, p 507

Sharma D, Kumar V (2002) Application of SCS model with GIS data base for estimation of runoff in an arid watershed. J Soil Water Conserv 141–145

Sharma T, Kiran PVS, Singh TP, Trivedi AV, Navalgund RR (2001) Hydrologic response of a watershed to land use changes a remote sensing and GIS approach. Int J Remote Sens 22:2018–2095

Suresh R (1997) Soil and water conservation engineering. Standard Publishers Distr, Delhi, pp 48–51

U.S. Soil Conservation Service, USDA (1972) Hydrology, Section 4, SCS, National Engineering Handbook. Washington, D.C.

Zakwan M (2016) Estimation of runoff using optimization technique. Water Energy Int 59(8):42–44

Zakwan M, Muzzammil M, Alam J (2017) Developing stage-discharge relations using optimization techniques, Aquademia: water. Environ Technol 1(2):05

www. meteoprog.es.en. (2012) Ghatal climate

Part II
Land Degradation, Resource Depletion and Livelihood Challenges

Are the Badlands of Tapi Basin in Deccan Trap Region of India "Vanishing Landscape?" Badland Dynamics: Past, Present and Future!

Veena Joshi and Shreeya Kulkarni

Abstract

Soil erosion is one of the most serious hazards human race is facing today. Four million hectares of agricultural lands have been rendered wastelands in India due to rill and gully erosion. Deccan Trap Region of India is characterized by rocky terrain. Sediments are thin and occupy only restricted areas. The study area is a watershed along the Tapi Basin in Maharashtra where alluvial bank deposits are deeply dissected by intricate network of gullies to form badlands. These badlands have been intensively reclaimed for agriculture in the past few decades. Based on various sedimentological and morphological properties, added by rainfall impact analysis, an assessment has been done to evaluate whether such types of land reclamation practices are permanent solution to the problem of land availability and agriculture in these areas. DEMs of three time periods were self-generated using IRS Cartosat I images and changes in the morphometric parameters were detected from them. A field survey was carried out to measure the current gully reactivation in the area. Results indicate that the region is already indicating soil loss beyond the tolerance limit. A land use planning needs to be designed for these areas because the present land use methods are doing more harm than good.

Keywords

Soil erosion · Deccan Trap Region · Badlands · Land reclamation · DEM · Land use planning

10.1 Introduction

Soil erosion is said to occur when the external forces exerted by the agents of erosion exceed the resistance provided by the soil surface on which they act. Gullying and land degradation has been a major research in the field of geomorphology for several years. The problem related to this aspect is investigated in many parts of India. Due to the growing population, there is increasing pressure on every patch of land and they have been brought under various land use practices, especially agriculture. This has led to reclamation of all available land, even those which were earlier unfavourable land for cultivation. Badlands are such type of landscape which are often under the pressure of land reclamation now, wherever they are accessible. Badlands are the areas which are rugged, where vegetation is almost absent and are intensely

V. Joshi (✉) · S. Kulkarni
Department of Geography, Savitribai Phule Pune University, Pune, Maharashtra 411007, India
e-mail: veenaujoshi@gmail.com

dissected. They are mostly fluvial in origin characterized by v-shaped valleys, high drainage densities and short steep slopes, often fingered by gently sloping planar surfaces referred to as pediments (Bryan and Jones 1997).

Land degradation in general and soil erosion in particular are the major concerns of the world today. India has an agrarian economy with vast growing population. Soil degradation directly affects agricultural productivity. The intensity of human interference has increased manifold in the last few decades. Better irrigation facilities and new incentives in agriculture in various parts of India, especially Maharashtra also has triggered this factor. Badlands are dynamic and sensitive landscape and previous studies have shown that disturbing them can induce the land to go into positive feedback mechanism and trigger soil erosion (Joshi and Nagare 2009). There are innumerable research works done on the effects of human interference and change in the landform processes. It is beyond the scope of this paper to include all such studies, however a few classic papers and frequently cited ones and also the studies that are relevant to the present context have been reviewed in the following paragraphs.

The initiation of gullies was the result of unwise utilization of land has been the findings of the study by Wells and Andriamihaja (1993). Removing the natural vegetation has a serious impact on runoff and erosion has been demonstrated by Piegay et al. (2004) and Canton et al. (2004). In India, the study of land cover classification using IRS LISS III image and DEM to understand rugged terrain in Himalayas was carried out by Saha et al. (2005). Gallart et al. (2012) demonstrated high rates of erosion following the reclamation of the badland gullies. Effects of tillage on soil degradation, quality and sustainability were established by Lal in 1993. The physicochemical analysis of clay samples reveals that if organic matter contents can be increased to values above 2%, then remodelled fields tend to stabilize. As a result, some badlands are being irreversibly destroyed and would be better classified as vanishing (Phillips 1998a). The rate of decline of ESP following the reclamation of badlands needs to be ascertained and a mechanism to link the organic matter content and the ESP needs to be established (Phillips 1998b). The undulating surfaces created after reclamation are likely to result in high rates of erosion in the period immediately following reclamation. This results in an active environment, onto which badlands tend to be re-established. Such badlands have been termed as "strained" as opposed to "vanishing" landscape (Rossi and Vos 1993). The role of organic matter in stabilizing and reducing the erosion of dispersive soils has received less attention than that of non-dispersive soils, although a number of such studies do exist, such as, Dong et al. (1983); Gupta et al. (1984) and Muneer and Oades (1989). A historical approach was adopted by Marathianou et al. (2000) to evaluate land use evolution and degradation in Lesvos (Greece). Reynolds and Stafford (2002) edited a volume on global desertification, questioning the role of humans in causing deserts. Zhang et al. (2007) investigated Land use changes and land degradation in China from 1991 to 2001. A Mediterranean case study of the impact of land use/land cover changes on land degradation dynamics was conducted by Bajocco et al. (2012). Matano et al. (2015) investigated the effects of land use change on land degradation and soil properties along Mara River Basin in Kenya, Tanzania. Land degradation and the sustainability of agricultural production in Nigeria was reviewed by Ajayi (2015). Multivariate analysis was employed by Khaledian et al. (2016) to assess and monitor soil degradation during land use change. Badlands forest restoration was designed in Central Spain by Mongil-Manso et al. in 2016. A scientific conceptual framework was made for land degradation neutrality and to bring land in balance by Cowie et al. (2018). Using Earth observations data, Abdel-Kader (2018) assessed and monitored land degradation in the northwest coast region of Egypt. Batunacun et al. (2019) identified the factors driving land degradation in Xilingol, China. Hazbavi et al. (2019) conducted dynamic analysis of watershed-health based on soil erosion. A bibliometric analysis on land degradation that includes the current scenario, development and future directions was proposed by Hualin et al. (2020). Land Degradation Neutrality (LDN) indicators were employed to study land use changes in the subalpine forest ecosystems of Korea by Sangsub et al. (2020).

10.2 Study Area

The site for the proposed work is a part of badlands of the Tapi River in the Jalgaon District of Maharashtra in India. It extends approximately between 75° 32′ 3″ E to 75° 33′ 05″ E and 21° 08′ 33″ N to 21° 09′ 10″ N (Fig. 10.1). The average annual rainfall of the area ranges between 630 and 750 mm. The banks along this stretch of the Tapi River are subject to the processes of gullying and ravination, which have transformed a major part of the landscape into badland terrain. There are many pockets along the river banks where deeply dissected badlands are formed. Land reclamation for the purpose of agriculture of these badlands became a very common activity in the whole region now. Some of these watersheds have been completely altered in the last twenty years including the parts of badlands near the Tapi River.

Figure 10.2 depicts the undisturbed parts of the Tapi Badlands, as well as, freshly activated gullies in a field, which was taken in 2013. Figure 10.3 displays the same spot in 2017, where it is seen that all those previous active gullies were filled up and brought under cultivation.

Careful observation of the field revealed that many erosion concentration points have got generated on the cultivated field. It also appeared that those concentration points more or less followed the previous gully network. Google Earth images of different time periods depicting stages of the reclamation of the badlands followed by series of reactivation of the gully network and refilling, a cyclic pattern of land reclamation and gully reactivation have been depicted in Fig. 10.4. It is clear indication that the field has become completely vulnerable now. This field is just one such example in the whole watershed where the same scenario is occurring everywhere. This had led to the formulation of the research objective, whether the current land use practice is a feasible option which would provide a permanent solution to agriculture in the area under review or otherwise.

10.3 Material and Method

In order to achieve the outlined objective, the following methodologies have been adopted in the study.

- Google Earth images from 2006 till present were obtained for observation. The images were examined to understand the phases of the past gully fillings and reactivation history of the field.
- In order to detect the changes in the morphometry resulting from excessive periodic land reclamation, DEMs of three time periods, such as, 2005, 2010 and 2015 were self-generated from IRS (Indian Remote Sensing) Cartosat I stereo images. Leica Photogrammetry suit (Version 9.2) and Arc GIS 10 were used to create the three DEMs. To study the micro-relief of the field, a detailed survey was conducted using dGPS (differential Global Positioning System) to create DEM of the present field morphology at 5 cm resolution.
- Twenty-five sediment samples were collected from the field to study the sediment properties of the area and to evaluate erodibility of these badland sediments. Textural analysis as well as few chemical parameters that have direct relevance with soil erodibility, such as, pH (power of hydrogen), EC (Electric Conductivity), OC (Organic Carbon/ Content), Na (Sodium), K (Potassium), Ca (Calcium), Mg (Magnesium) were detected from the soil samples. Scour depths of 12 samples were calculated from the silt factors to determine the susceptibility of the topsoil under erosion.
- Morphometric parameters were obtained from the Carto DEMs and indices were calculated that have meaningful relations with the present objective of the study.
- Return periods of the rainfall for 500 years were estimated from 60 years rainfall data of the area, using Anderson–Darling goodness-of-fit test and incorporated in the final synthesis.

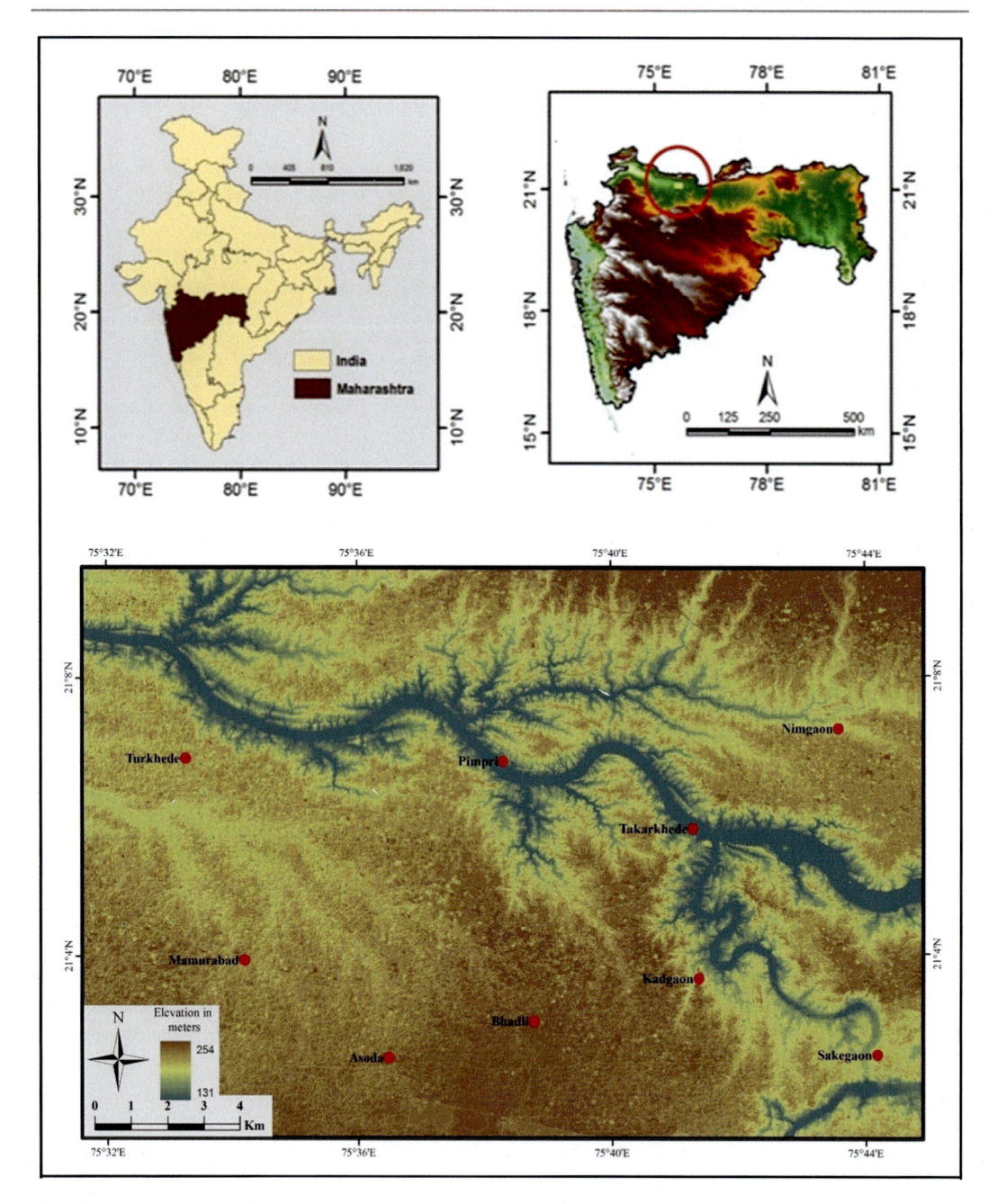

Fig. 10.1 Location map of the study area within Maharashtra state, also showing the DEM of the AOI

Fig. 10.2 The field pictures of the undisturbed parts of the badlands and gully erosion after the clearance of the vegetation and levelling the land for agriculture, within the same watershed

10.4 Results

10.4.1 Comparison of the DEMs

Under the natural conditions, major changes in the hillslope morphology within a time span of 10 years are not expected. But such changes can be possible in an area that is actively interfered by human activities, as in the present study area. Figure 10.5 depicts the flowchart showing the methodology to generate DEM from IRS Cartosat I images and Fig. 10.6 displays the DEMs of the three time periods created at 10 m resolution. The three DEMs were used to assess the changes in the hillslope morphology from 2005 to 2010 and 2015. All the 3 DEMs (12 Dec 2005, 12 Dec 2010, and 10 May 2015) show noteworthy differences in the elevation as well as in the slope.

Remarkable variations can be seen in the relief maps especially between 2005 and 2010 (Fig. 10.7). Relief category between 188 and 254 m increased in the area by 8% between 2005 and 2010. Category between 162 and 177 m decreased by 7% between 2010 and 2015. Changes are more prominent between 2005 and 2010.

Slope of the area also revealed noticeable changes. Area under 14°–29° category increased significantly (5% in five years) from 2005 to 2010 (Fig. 10.8). Area under 14°–29° and 29°–43° categories also increased noticeably (10%, in five years) from 2010 to 2015. Category 0°–14° decreased in the area by 10%. Slope category 43°–57° was absent from 2005 and 2010 DEMs but they start appearing in 2015 DEM. Overall assessment of the slope is that they are visibly changing.

There are a lot of changes in the aspect map also as can be seen in Fig. 10.9. Slope orientation has considerably modified from 2005 to 2010 but the change is not significant between 2010 and 2015. These categories were chosen based on the observation of the main slope aspects of the badland

Fig. 10.3 Scenarios after the badland reclamation are evident in these field pictures. Filling the gullies for cultivation, followed by reactivation of the gully network has been captured in the pictures

terrain in the area. Significant changes in the area can be observed in all the categories. Maximum change can be observed in the category of 71–143 with 15% increase in the area between 2005 and 2010. Category 145–216 and 216–283 decreased by 5% each in the area in the same time. Though the change is slight, pattern remained more or less the same between 2010 and 2015. All the three parameters indicate remarkable changes between 2005 and 2010, implying a major remodelling during this time. Such dramatic changes within such a short time can validate the significant impact of human interference in the region.

Currently, gullies have been filled up and the fields are levelled and put under farming, mainly by cotton and banana. A detailed field observation of the area under review revealed several erosion concentration points and two distinctly activated gullies all across the field.

One of the newly activated gullies oriented towards the main river and another was slopping towards a tributary stream which is situated on one side of the field, guided by an embankment constructed by farmers to reduce erosion of soil from the field. Flutes and pipes were also observed in the field and on the wall of the exposed gullies. In order to conduct a high-resolution mapping of the field, a dGPS survey was conducted for the field and a DEM was created with 5 cm resolution and demonstrated in Fig. 10.10. Erosion concentration points are documented on the DEM. Reactivation of gullies has already began following the path of the old buried gullies, because the energy of the system has not been channelized. These erosion concentration points are targeting the old gully systems because the whole network is controlled by Tapi River. This is a "Downstream controlled" system.

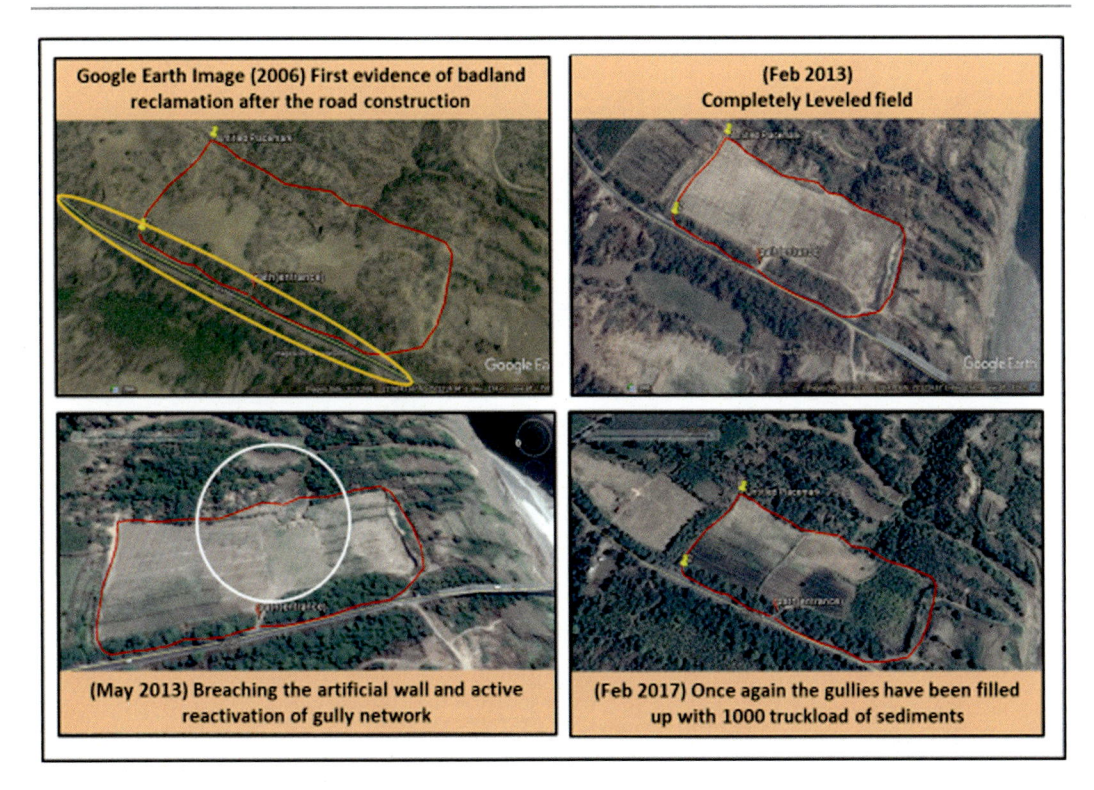

Fig. 10.4 Google Earth images from 2006 till 2016, depicting stages of badland reclamation and gully reactivation in the area (**a** Yellow polygon depicts a road being constructed making the badland watersheds accessible. The reclaimed field is right across the head of an active tributary gully of Tapi River, **b** increase in the reclaimed areas, **c** reactivation of the previous gully after a heavy storm and **d** refilling of the gullies)

10.4.2 Sediment Analysis

Sediments play a crucial role in determining the vulnerability of a region under erosion. Certain sediment properties are known to have a direct influence in the erodibility and a few of such relevant parameters have been detected and the results are presented in the following sections.

10.4.2.1 Power of Hydrogen (pH)

Soil pH is determined by the concentration of hydrogen ions (H^+). It is a measure of the acidity and alkalinity of a soil solution on a scale from 0 to 14. Common acid-forming cations are hydrogen (H^+), aluminium (Al^{3+}) and iron (Fe^{2+} or Fe^{3+}), whereas common base-forming cations include calcium (Ca^{2+}), magnesium (Mg^{2+}), potassium (K^+) and sodium (Na^+) (McCauley et al. 2017). Figure 10.11 demonstrates general pH chart provided by McCauley et al. (2017) and the detected pH values of the sediment samples from the field. pH values range from 7 to 10. Samples ranging between 8 and 10 are sodic in nature McMauley et al. (2017). Following this criterion, the samples are showing a general trend towards sodicity.

10.4.2.2 Organic Matter Content

Organic matter influences the aggregate stability of the soil. Erodible soils contain less than 3.5% of organic matter. Clay particles combine with organic matter form soil aggregates which are usually resistant to erosion. Soil erodibility (K factor) is lessened by larger structural aggregates and rapid soil permeability (Morgan, 1986). The percent organic matter contents of the samples from the field was tested in Zuare Agro, Pune. Figure 10.12 indicates the results of organic content and it reveals that 2/3rd of the samples contain less than 2%. A threshold of 2% is

Fig. 10.5 Flow chart demonstrating steps of the generation of DEM from Cartosat I images, in Leica Photogrammetry Suit 9.2 and Arc GIS 10

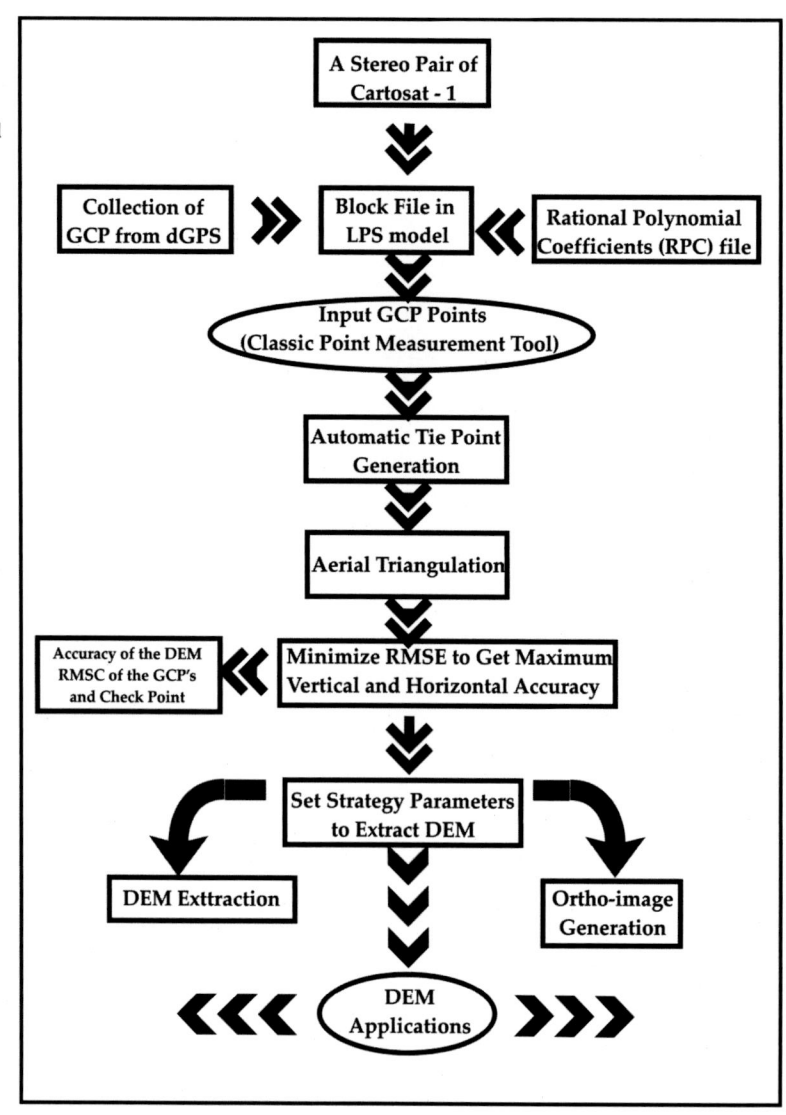

generally accepted for the aggregate stability of the soils.

10.4.2.3 Electrical Conductivity

Soil electrical conductivity (EC) is a measure of the amount of salts in soil (salinity of soil). It is an important indicator of soil-health. EC is correlated to the concentrations of nitrates, potassium, sodium, chloride, sulfate and ammonia but does not provide a direct measurement of salt compounds. Electric conductivity of the samples appears to follow a trend towards changes in textural composition and the topographical position they occupy. EC for the present samples shows great variability and not systematically related to the depth from which the samples were collected. The values of range from 0.16–0.71 mmho/cm. (Fig. 10.12). There is an inverse relationship between EC and soil erodibility. Keeping all other factors constant, if EC is low the rate of removal of soil is low and vice versa.

10.4.2.4 Exchangeable of Sodium Percentage (ESP)

The presence of excessive amount of exchangeable sodium reverses the process of aggregation

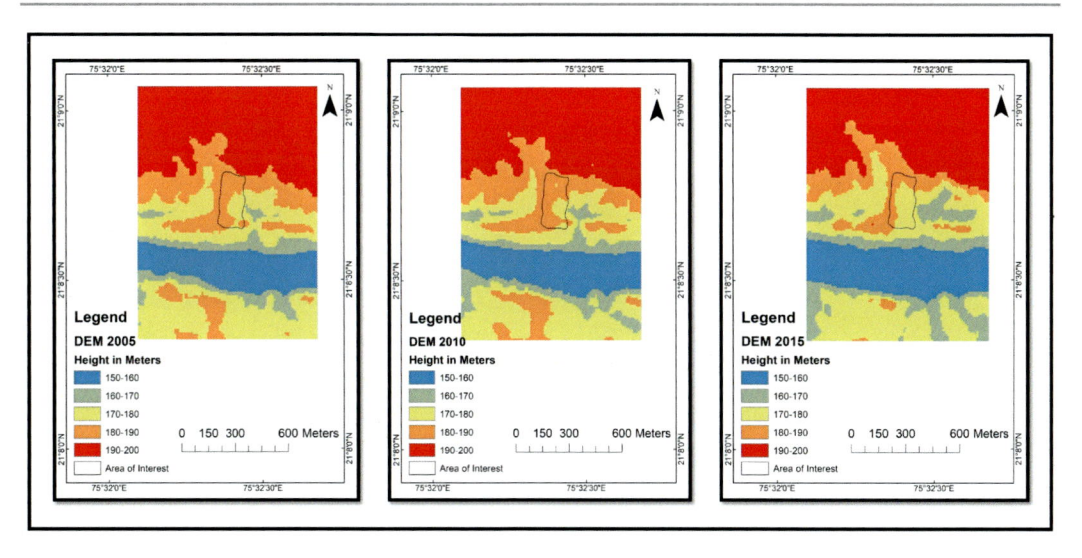

Fig. 10.6 DEMs of 2005, 2010 and 2015, with 10 m resolution

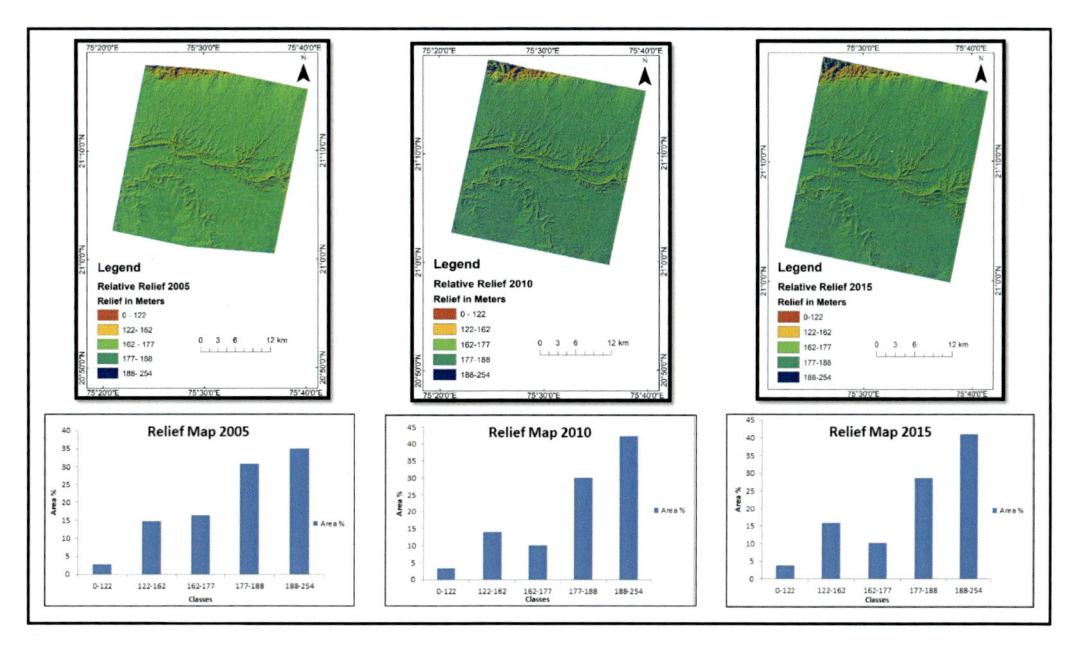

Fig. 10.7 Relief maps and their categories, for the three time periods (2005, 2010 and 2015)

and causes soil aggregates to disperse into their constituent individual soil particle. The sodic soil with few stabilizing agents in topsoil will ultimately succumb to erosion during heavy rain spells via rill or gully erosion. This is especially the case in soil with high silt and clay particle size fractions. Soil sodicity leads to decreased permeability and poor soil drainage over time. The soil with 6% or more content of ESP is known as sodic soil. This is an important parameter because sediments higher than 15% ESP are susceptible to slacking and piping.

The formula used for calculating ESP is as follows;

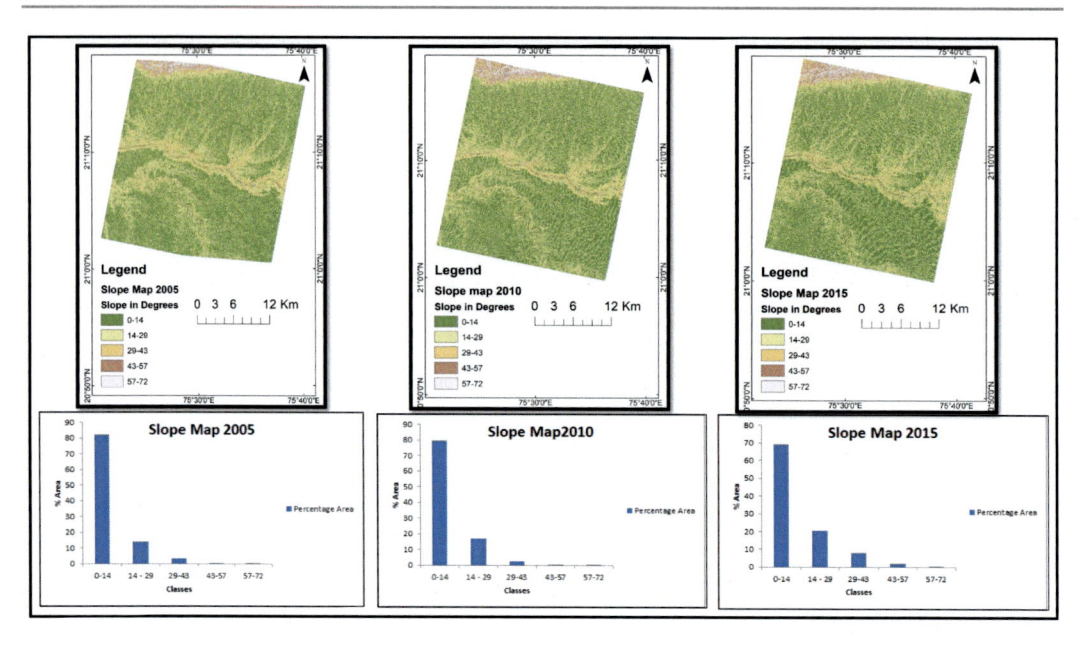

Fig. 10.8 Slope maps and their categories, for the three time periods (2005, 2010 and 2015)

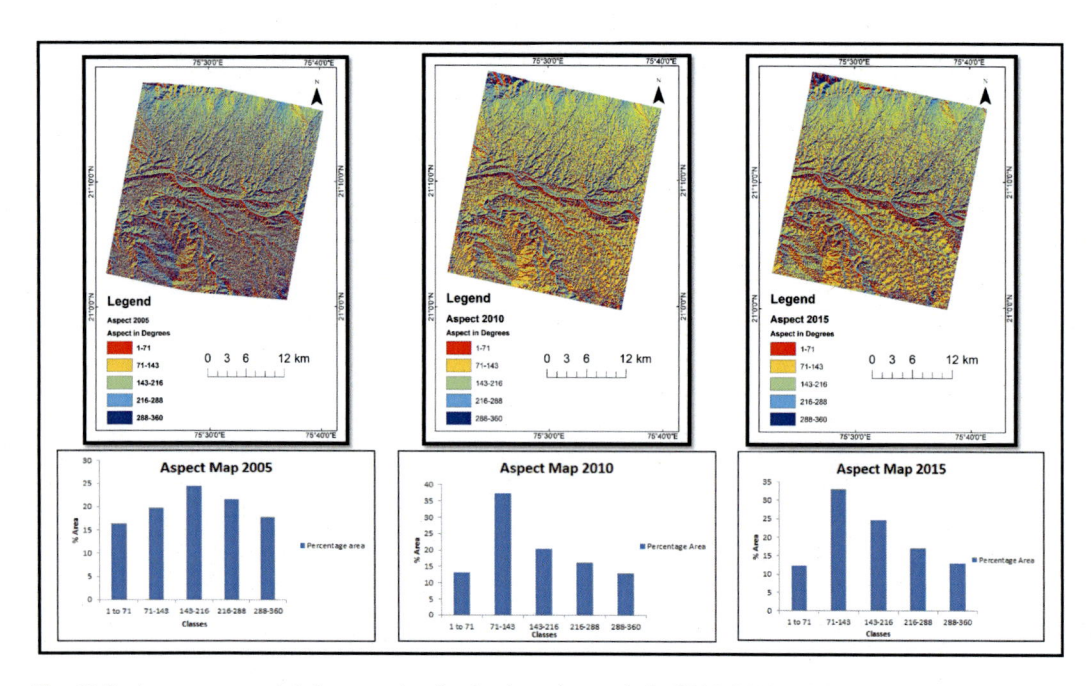

Fig. 10.9 Aspect maps and their categories, for the three time periods (2005, 2010 and 2015)

$$\text{ESP} = \text{Exchangeable } \{(\text{Na})/(\text{Ca} + \text{Mg} + \text{K} + \text{Na})\} * 100 \quad (10.1)$$

Most of the samples ESP range in the value between 5 and 15% (Fig. 10.12). The average value is 10.33%. The high values of ESP indicate that the soils on an average are fairly sodic which measures the erodibility of soil. Some of the detected samples are non-sodic in nature. But on an average, maximum samples fall in the

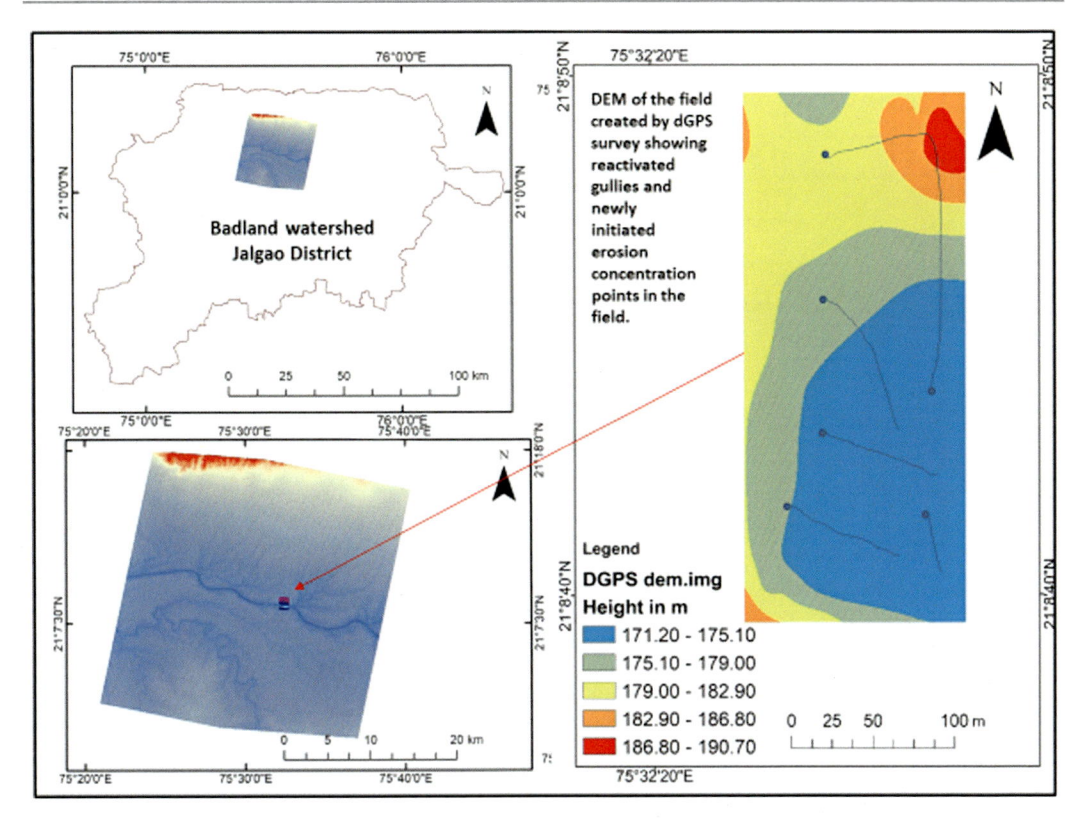

Fig. 10.10 DEM of the field under review created by field survey using dGPS with resolution of 5 cm. Many erosion concentration points can be seen on the DEM, which are following the original gully network

category of moderately sodic class. An association of pH with the general sodicity of the sediments has been depicted in Fig. 10.11 that indicates that samples having pH value between 8 and 10 are sodic in nature. ESP values for the samples indicate that they can be put under the category of moderate to highly erodible category. The pH of 25 samples from the area shows the value between 7 and 9, that also supports the observation.

10.4.2.5 Granulometric Parameter

Textural analysis of the fifteen samples was carried out by using sieving for the coarse sands and sedigraph for silt and clay. It is evident from Fig. 10.12 that silt content is higher than sand and clay in all the samples. Sand allows water to transmit through and hence soil containing higher percentage of sand is not erodible. Clay has the property of forming compact layer as well as shrink and well under wet and dry conditions. They also have strong colloid binding and hence they also do not fall under the category of erodible soils. When soils contain higher percentage of silts, they are known to be highly susceptible to water erosion.

10.4.2.6 *K* Value (Erodibility of Soil)

Some soils have the tendency to erode more than others. This is due to the inherent properties of the soil and not due to any external factor. This is termed as soil erodibility. Soil texture is an important factor whilst studying erodibility. Gravels which require more force to transport are difficult to erode. On the contrary clay particles which are finer are also stable due to their cohesiveness. The most mobile and transportable particles are the very fine sand and silt particles. Organic matter has a big influence on the aggregate stability of the soils. Soils less than

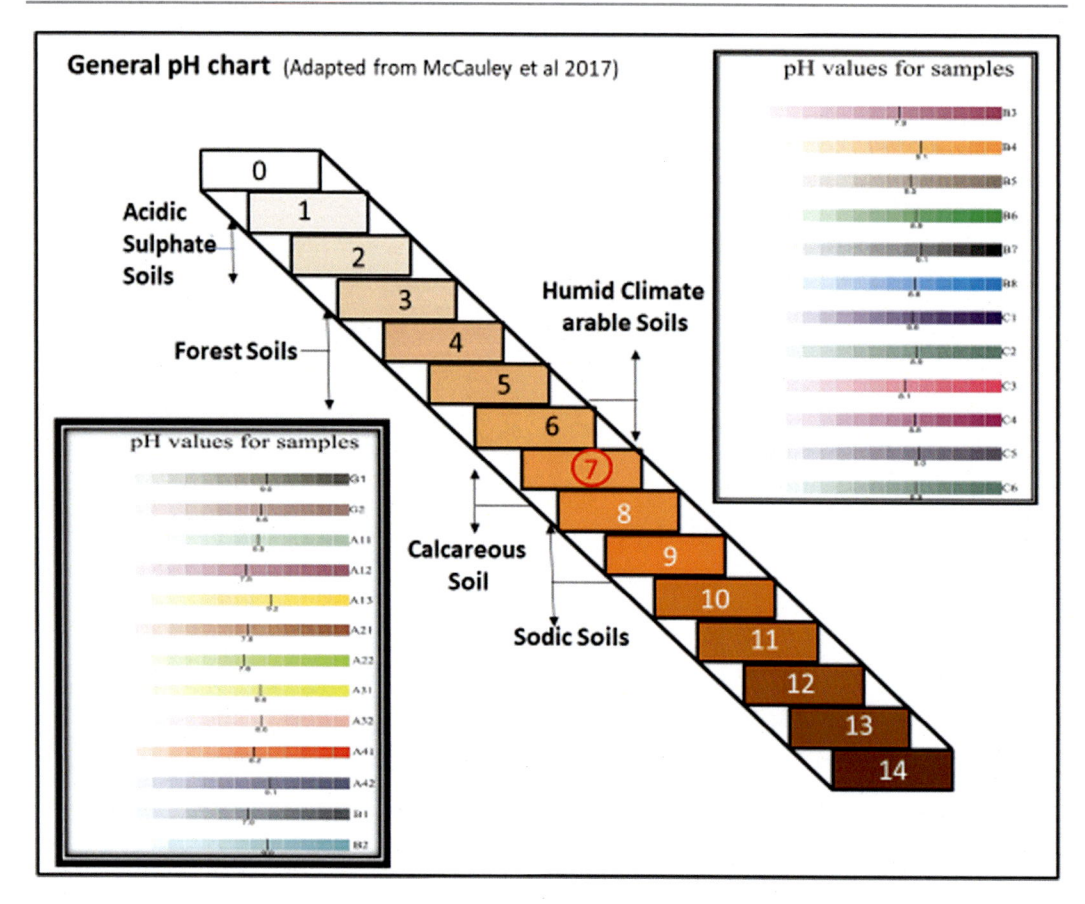

Fig. 10.11 General pH chart in the center (adapted from McCauley et al. 2017) and the pH of the samples from the field. The diagram shows the relation between pH and sodicity of the sediments

3.5% of organic matter are considered erodible. Moreover, clay particles combine with organic matter to form soil aggregates or clods which are usually resistant to erosion. K factor (erodibility) is lessened by larger structural aggregates and rapid soil permeability. However, the erodibility of a soil is the function of complex interactions between the various physical, chemical as well as mineralogical properties of soil. (Morgan 1986). The formula for calculating the K value employed in the study is given below;

$$k = 2.1 \times 10^{-6} * f_p^{1.14}(12 - P_{om}) + 0.0325(S_{stru} - 2) + 0.025(f_{pem} - 3) \quad (10.2)$$

where,

f_p $P_{silt}*(100-P_{clay})$
P_{om} Percent organic matter
S_{stru} Soil structure code used in soil classification
f_{pem} Profile permeability class.

Fifteen samples were selected for calculating the erodibility "K" index. Silt textural group (very fine sand + Silt) forms maximum part of the soils of this region. Numerical codes were assigned to both permeability and soil structure. The soil structure was estimated from field observation and the permeability was detected by testing the soil in Zuare Agro Pvt. Ltd., Pune. The average K value is 0.24, ranging between

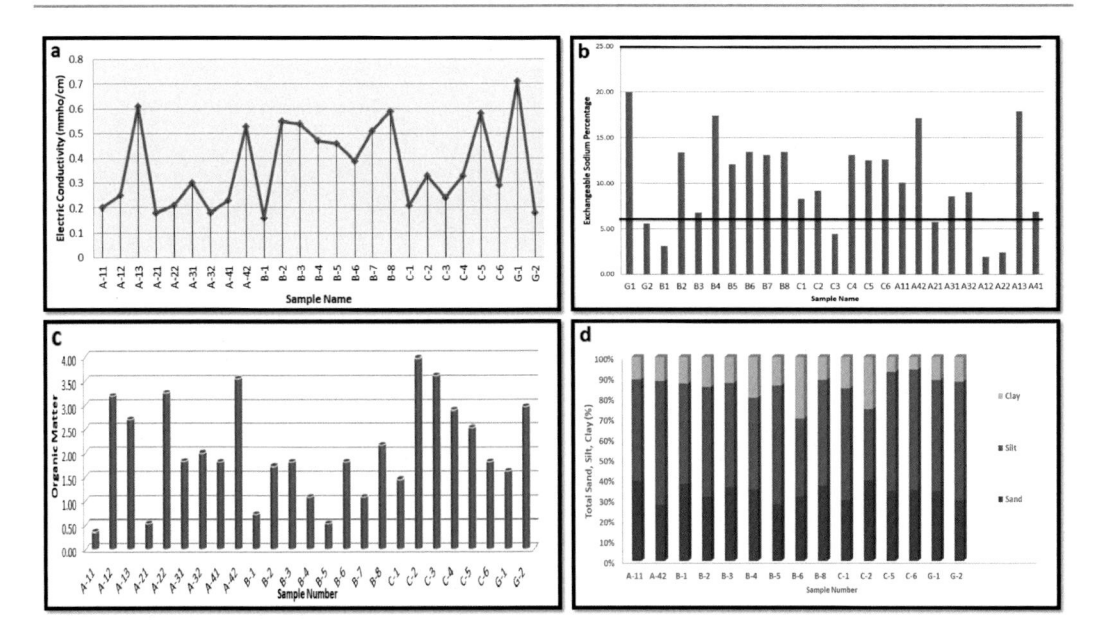

Fig. 10.12 The diagram depicts the results of four sediment parameters of the area (**a** electric conductivity, **b** exchangeable sodium percentage, **c** percentage organic matter, **d** sediment textural classes)

Table 10.1 Soil erodibility (K Index) for the samples

Sample No.	K value	Sample No.	K value
1	0.29	9	0.28
2	0.29	10	0.13
3	0.27	11	0.35
4	0.30	12	0.34
5	0.25	13	0.29
6	0.23	14	0.26
7	0.34	15	0.30
8	0.25		

0.325 and 0.162. Erodibility as indicated by this index is not very high (Table 10.1).

10.4.2.7 Mean Scour Depth

Lacey-Inglis's method of estimating regime depth of flow in loose bed alluvial rivers was first developed by Lacey in 1929 and later by Inglis in 1944 mainly based on the observations of canals in India and neighbouring Pakistan. This technique is therefore used mainly in India for estimation of scour depth around bridge piers in alluvial channels. This technique has been applied in the present study to estimate the susceptibility of the alluvial topsoil forming the badlands. The mean scour depth below the highest flood level (HFL) for natural channels flowing over scourable bed can be calculated theoretically from the following equation,

$$d_{sm} = 1.34 \left(D_b^2 / K_{sf} \right)^{1/3} \quad (10.3)$$

(Source: Indian Road Congress, IRC—5,1998).

Where,

d_{sm} Mean depth of scour in m below the highest flood level

D_b Discharge in cumecs per m width

K_{sf} Silt factor determined for the stratum based on weighted mean diameter of particle in mm.

The silt factors of the samples and estimated scour depth for various silt factors have been displayed in Table 10.2 and the estimated design scour depths for the samples results are outlined in Table 10.3. The mean scour depths of the sediments range in value between 3.06 and 4.39 m. In other words, if the sediments would have been under water, the top 4 m is the portion most vulnerable to erosion under the set of conditions prevailing in the area. These sediments are bank sediments and are directly in contact with water most time of the year, therefore assuming a situation close to the scouring conditions is applicable here.

The calculations are based on D_b = 1.0 cumecs per meter width and severe bend is considered.

10.4.3 Rainfall Factor (R)

Rainfall plays the most crucial parameter whilst studying erosion. Erosion index is a measure of the erosivity or erosive force of a specific rainfall event (Wischmeier and Smith 1965). The rainfall factor "R" of USLE is the number of erosion-index units in a normal year's rain. Wischmeier and Smith (1978) proposed that "R" factor must include the combined effects of moderate, as well as, severe storms on soil loss. The erosivity of rain in other words is the potential ability of rain to cause erosion. It is controlled by the physical characteristics of the rain. Babu et al. (1978) had developed the relationship between average annual erosion index (R factor) and annual rainfall, as well as, seasonal (June–September) rainfall for India based on their analysis of data from 44 stations spread across various rainfall zones of the country. The relationship is thus expressed as;

Table 10.2 Silt factor for different grades of soil and estimated scour depth for various silt factor

Type of bed material	Silt factor	Scour depth (m)	Design scour depth (m)
Fine silt	0.5	2.51	4.39
Silt	0.6	2.37	4.15
Medium sand	1.25	1.86	3.26
Coarse sand	1.5	1.75	3.06

Source (Indian Road Congress) IRC—5 1998

Table 10.3 Scour depth of the samples

Sample No.	Silt factor	Mean scour depth below HFL(m) as per IRC:78	Mean scour depth below HFL (m) assuming D_b^* = 4.2 Cumecs/m
1	0.35	$1.90 \times D_b^{0.67}$	4.97
2	0.55	$1.64 \times D_b^{0.67}$	4.28
3	2.59	$0.98 \times D_b^{0.67}$	2.55
4	2.20	$1.03 \times D_b^{0.67}$	2.69
5	0.35	$1.9 \times D_b^{0.67}$	4.97
6	0.55	$1.64 \times D_b^{0.67}$	4.28
7	0.55	$1.64 \times D_b^{0.67}$	4.28
8	0.55	$1.64 \times D_b^{0.67}$	4.28
9	2.59	$0.98 \times D_b^{0.67}$	2.57
10	0.35	$1.64 \times D_b^{0.67}$	4.97
11	0.35	$1.64 \times D_b^{0.67}$	4.97
12	0.55	$1.64 \times D_b^{0.67}$	4.28

D_b^* Design discharge for foundation per metre with effective linear waterway

$$R = 79 + 0.363P, \quad r = 0.83 \quad (10.4)$$

where,

R Average annual erosion index i.e. R in metric units
P Average annual rainfall in mm.

In the present study, the equation (Eq. 10.4) has been used to compute R (rainfall erosivity) factor. Daily rainfall data (in mm) for the 15-year period from 1998 to 2012 for 15 meteorological stations of Jalgaon District (obtained from the Indian Meteorological Department, Pune) was used to compute the R factor. From the daily rainfall data, the average annual rainfall in Jalgaon was found to be 708.72 mm and the R factor was calculated to be 336.26.

10.4.3.1 Estimated Return Period of the Rainfall

Return period of 60-year rainfall has been calculated for the rainfall of the study area from 1955 to 2015. The highest rainfall during this period was 119.6 mm on 9th Sep. 1960, with a return period of 50 years. Second highest was in 1957 with an amount of 108.7 mm on 22-Jun-1957. Based on these data, estimated return period of 500 years have been calculated employing eight statistical techniques. The techniques used and their parameters have been demonstrated in Fig. 10.13. Out of the eight techniques used for trial estimate, the most suitable three techniques were finally chosen using Anderson–Darling goodness-of-fit test statistic (Table 10.4).

The Anderson–Darling statistic measures how well the data follows a particular distribution. For a given data set and distribution, the better the distribution fits the data, the smaller this statistic will be. Based on this result, the first three ranking techniques, such as, Lognormal (3P), Weibull (3P) and Log-Pearson 3 have been chosen and the final estimates have been made and produced in Fig. 10.13 and Table 10.5. It is interpreted that according to Lognormal (3P) estimate, a rain event of 171 mm/day may occur once in 100 years in these areas, 155 mm/day according to Weibull (3P)'s estimate and 142 mm/day after Log-Pearson 3. The five-hundred-year estimate goes up to 227 mm/day.

The data presented above implies that the chances of a rain event of 227 mm/day occurring in this region are once in 500 years or an event of 194 mm/day is 200 years. If it happens, it will create catastrophe in these fields and everything will get eliminated. Effect will be even more pronounced when the land is disturbed and loosened by the cultivation of crops. Figure 10.2 clearly showed that when undisturbed, these badland slopes are stabilized by vegetation reducing vulnerability. The real threat is when the watershed is disturbed, that progresses towards positive feedback of the system.

10.4.4 Morphometric Parameters

10.4.4.1 Stream Length (SL INDEX) Ratio

For a long time, geomorphology has made use of morphological analysis for the study of evolution and interpretation of landforms. Empirical methods were the major sources for the understanding of morphogenetic processes. From the Carto DEM of 2015, eight badland watersheds have been demarcated as shown in Fig. 10.14. Twenty-four profiles (approximately 3 profiles from each watershed) were drawn from the top of the badland surface to the main river. The semi-log graphs of these badland gullies are presented in Fig. 10.15a, b. All the graphs reveal above-grade condition, indicating a high energy topography and high competence of the badland streams/gullies. The distance between the gullies and the main river is short, hence this also will enhance the rapidity in erosion.

Several physical and mathematical methods were envisaged and applied by many researchers, mainly after 1950 to evaluate the energy of stream network and one amongst such methods one was put forward by Hack (1973) known as Stream Length-Gradient Index (SL Index). This was to determine whether a river would be in geomorphological equilibrium based upon the relationship between the river slope and the areal extent of the watershed.

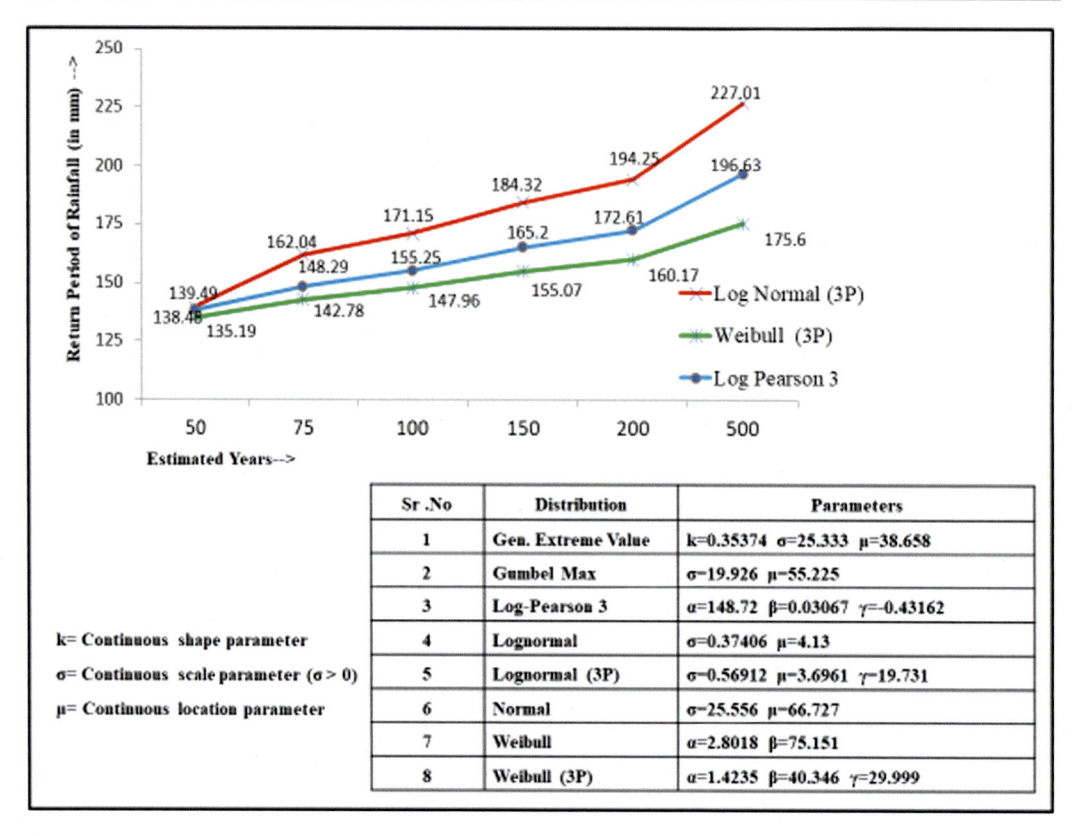

Fig. 10.13 Estimated return period of the rainfall for 500 years, for the three top ranking techniques, such as, Lognormal (3P), Weibull (3P) and Log Pearson 3 (based on Anderson–Darling goodness-of-fit test). Inset is the parameters of the eight statistical techniques used for the trial statistics in Anderson–Darling goodness-of-fit test

Table 10.4 Anderson–Darling goodness-of-fit test

Sr.No	Distribution	Anderson–Darling	
		Statistic	Rank
1	Lognormal (3P)	3.6375	1
2	Weibull (3P)	3.658	2
3	Log-Pearson 3	4.2103	3
4	Lognormal	4.4388	4
5	Gumbel Max	4.7363	5
6	Gen. extreme value	4.7465	6
7	Weibull	5.488	7
8	Normal	6.2589	8

$$SL = h_1 - h_2 / \ln L_2 - \ln L_1 \qquad (10.5)$$

where
1. SL is Hack's Stream Gradient Index.
2. h_1 is the height of first point.
3. h_2 is the height of second point.
4. L_1 is the distance from source to first point.
5. L_2 is the distance from source to second point.

Table 10.5 Estimated return period of rainfall for 500 years

Return period (year)	Probability %	Lognormal (3P) (daily rainfall in mm)	Weibull (3P) (daily rainfall in mm)	Log Pearson 3 (daily rainfall in mm)
50	0.98	139.49	135.19	138.48
75	0.9867	162.04	142.78	148.29
100	0.99	171.15	147.96	155.25
150	0.9933	184.32	155.07	165.2
200	0.995	194.25	160.17	172.61
500	0.998	227.01	175.6	196.63

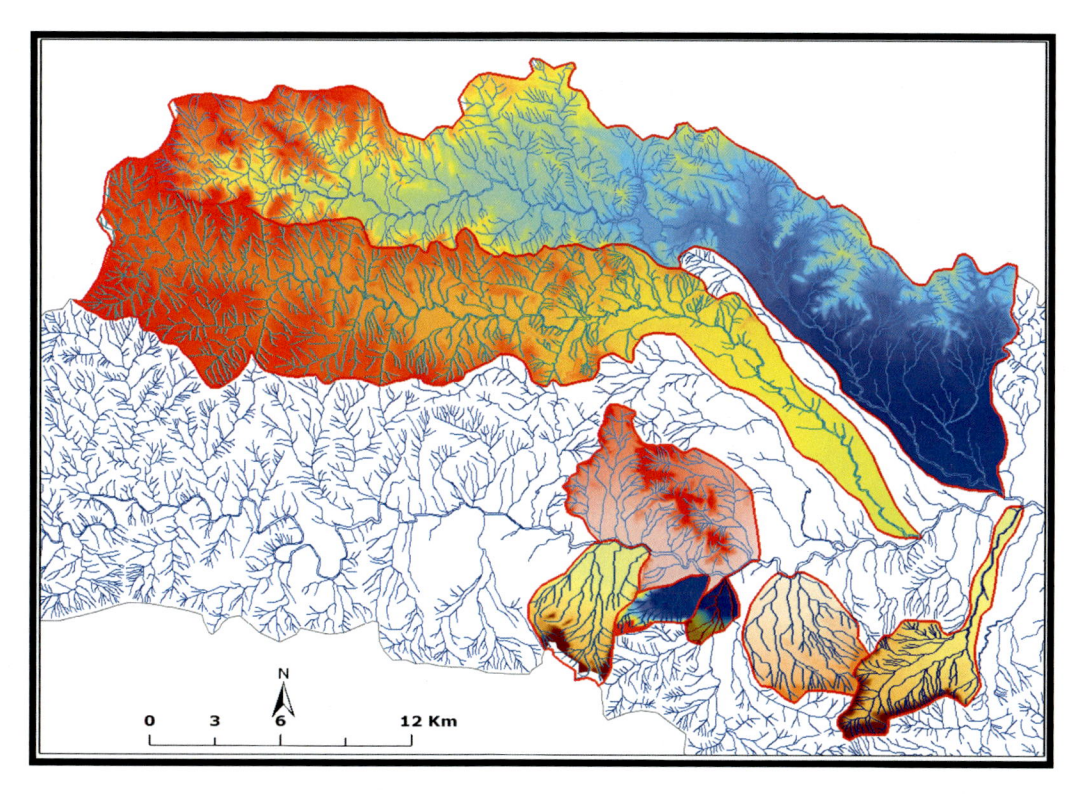

Fig. 10.14 Eight small gully watersheds have been demarcated from 2015 DEM to construct longitudinal profiles and calculate SL Indices

The calculated average SL index value is 211.9072 and the range is between 185 and 240 (Table 10.6). All the profiles are well above the grade level and are actively eroding. Considering the dimensions of the landscape under review, the profile lengths are short and the falls are considerable, overall competence of the streams are high enough to cause rapid erosion in a disturbed watershed.

10.5 Discussion

The present study is a geomorphic evaluation of the feasibility of the present land use pattern to an actively eroding badland and deeply disturbed watershed in a part of the Deccan Trap Region in Maharashtra. Summary of all the indices is depicted in Table 10.7. Together with that,

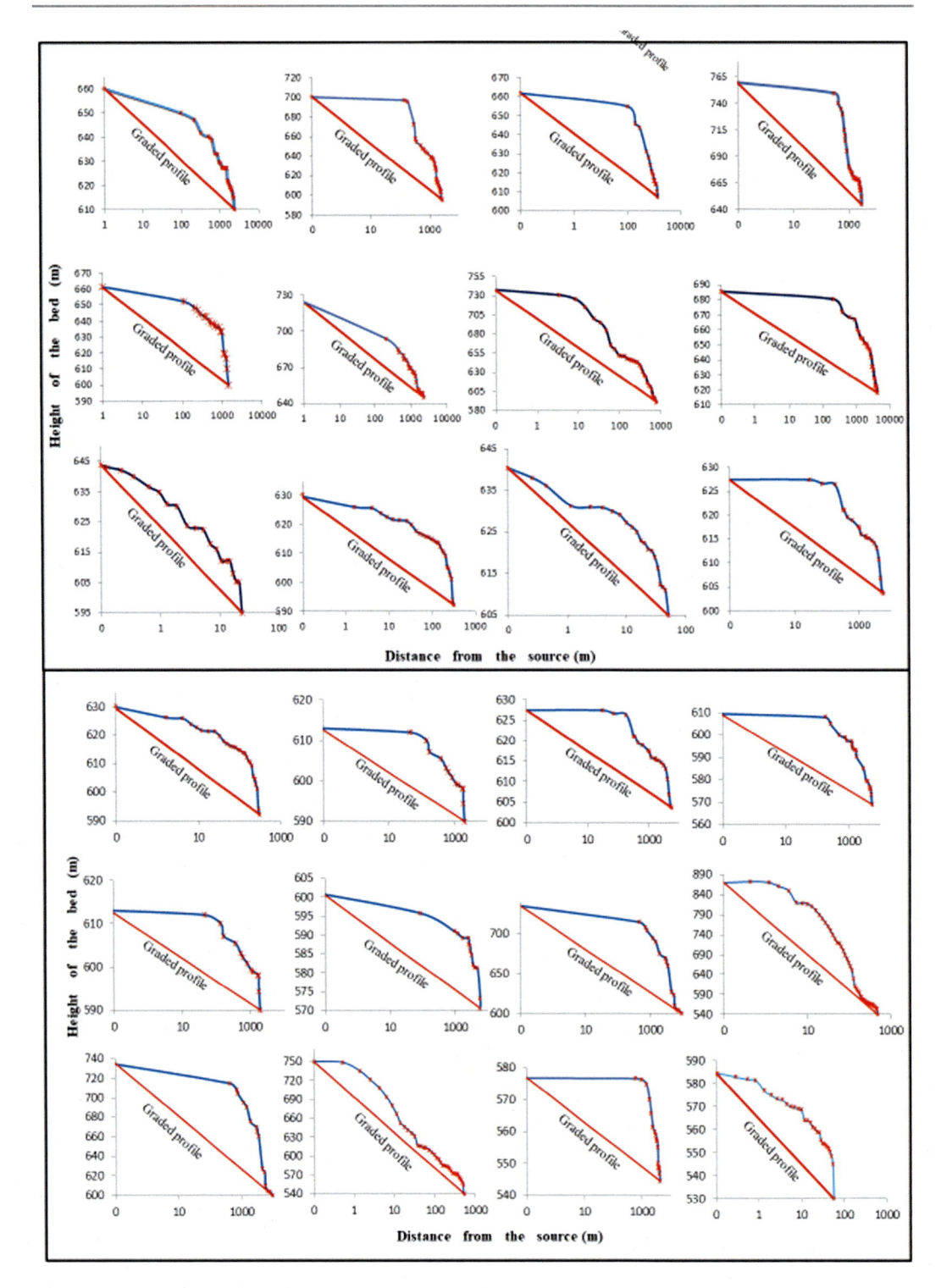

Fig. 10.15 Semi-log profiles of twentyfour badland gullies in the watershed. All the profiles show above-grade condition and high competence

Table 10.6 Hack's (1978) stream gradient index

Number	SL index	Number	SL index
1	240.4334	13	202.5267
2	202.5267	14	205.8139
3	206.629	15	215.8893
4	210.2172	16	216.0215
5	217.5934	17	216.6736
6	216.7907	18	184.6005
7	207.4577	19	220.1867
8	220.1867	20	196.9366
9	196.9366	21	185.6005
10	185.6005	22	212.0038
11	223.0071	23	218.2869
12	240.4334	24	213.3614

Google Earth images were also studied to see changes in the area.

Google Earth images and the DEMs along with the field photographs reveal that the AOI is undergoing a rapid change. There is a steady increase in the reclamation of badland areas and consequences of gully reactivation from 2005 till the present. The biggest change was observed in 2013 when some heavy spells destroyed the artificial wall across the gully and caused widespread gully erosion. Photographs taken from the field in 2013 provided a good evidence of the severity of gully erosion in the area. The gullies were finally filled in 2016 and brought under cultivation once again.

During the recent field visit to this area, reactivation of the gullies and the erosion concentration points were observed all over and also documented in the DEM generated from a field survey. Blocking the gullies and filling them offer just a temporary solution but the energy of the landscape is not diminished. Rainwater will flow with the same vigour and find its way to the river, eroding all the way. It is very clearly seen in the field that several minor rills and two major ones have already formed on the field. With time, it will deepen and once again the old network of gullies will be reactivated, unless a well-engineered diversion is created.

Sediment properties play a crucial role in understanding the erosion scenario of any watershed. Several sediment parameters were detected in the study that are relevant to the objective of the study. The current soil samples contain more percentage of silt and hence more susceptible to erosion. Sediment samples are sodic and calcareous in nature. Sodic soils are prone to a higher erosion rate. Carbon content of the average samples is less than 2%. considering this factor, we can say that soils in this area are not resistant to external force of raindrop impact or runoff. The physicochemical analysis of clay samples shows that if organic matter contents can be increased to values above 2%, then remodelled fields tend to stabilize. As a result, some badlands were being irreversibly destroyed and would be better classified as vanishing (Phillips 1998a, b). The values of electric conductivity are normal ranging from 0.16 to 0.71 mmho/cm. If the EC is low the rate of removal of soil is low and if the rate of removal is high the EC is higher.

When, Na concentrations are high enough to produce sodic soils, erodibility is maximized. Soils with lower electrolyte levels can disperse at lower ESP (Amezketa et al. 2003). Sodium forms a thick layer which will tend to move the clay/soil platelets farther apart and make them more susceptible to dissociation in any environment. The soil with 6% or more content of ESP is known as sodic soil. This is an important parameter because sediments higher than 15% ESP are susceptible to slacking and piping. Most of the samples tend

Table 10.7 Summary of the indices and the results

Parameters	Sub-parameters	Index values	Remarks/interpretations
Sediment	Textural	Silt 60–65%	Erodible
	OM	Avg 1.96%	Less than threshold value of 2 for stability
	ESP	10–15% in majority	Erodible for smectite rich soils
	EC	0.16–0.71 mmho/cm	Variable
	PH	Between 7 and 10	Towards Sodicity
	Silt factor and Scour depth	Between 2.55 m and 4.97 m	In other words, the top 4 m of the sediment is the portion most susceptible to erosion under the set of conditions prevailing in this area
	ESPdepth	Decreasing with dept	Uncertain
	K factor	Avg. 0.24 Between 0.325 and 0.162	Moderately erodible
	Clay	Hydroxy Interlayered Smectites	Shrink-swell, highly erodible (Joshi and Tambe 2008)
Morphometric	Slope Categories	Significantly altered from 2005–2015	Badland reclamation
	Slope aspect	Significantly altered from 2005–2015	
	Relative relief	Significantly altered from 2005–2015	
	Longitudinal profiles	SL Index 211.9 Avg - range between 185 and 240	Well above Grade High energy slopes
Rainfall	Erosivity	R = 336.26	Erosive, especially the first few storms of the monsoon when the badland slopes are bare
	Return period (500)	Lognormal (3P)—227 mm/day Weibull (3P)—175.6 mm/day Log-Pearson 3—196.63 mm/day	Catastrophic Event
dGPS Field survey			Several evidence of reactivation of the earlier filled up gullies

towards sodicity in nature, indicating high erodibility. Overall erodibility based on "K" index show medium to fairly erodible nature of the sediments. Design scour depth of the sediments indicates the susceptibility of the top 4 m of these badland sediments. SL index of 24 samples reveal the above grade high energy system of these badland gullies and streams. Return periods of the high storm events in the region for 50–500 years reveal that 227 mm/day can occur once in 500 years. However stable as of the short term, nothing will stand against a rain event of this dimension in long term.

10.6 Conclusion

On the basis of the data generated during the course of this investigation which have been presented in the previous chapters and summarized in the above paragraphs, it can be concluded that land use is changing very rapidly in this region. During the field visits, there were long interviews with the farmers of the area. Their input became very valuable during the study. First evidence of land filling was vague but it's on record that at least twice the field had been filled in the recent past after constructing an earthen wall across the main tributary gully. The wall breached twice after heavy monsoon rains and the gully network got activated both the times. The last breaching happened during 2013 monsoon. Last filling was performed in 2016 and the field was cultivated again. Around 1000 truckload of soil were dumped to fill these gullies. A rough calculation yielded around 7000 tons of soil that was used to fill these gullies, which is equal to 7000 tons of sediment being eroded from a single field in one monsoon event (qualitative assessment). The same situation is prevailing everywhere within the basin.

Compilation of the entire data generated during the study suggests that the region under review has already crossed the threshold of the tolerance limit of soil erosion mainly as the result of human activities. Though badlands are dynamic landscape, they are stable if they are not disturbed. It is when the natural landscape processes are completely altered by rapid actions of human, that a new set of processes start to develop in such areas to cope with the newly imposed systems. That is what we see in this area. The energy of the system is controlled by the relief and triggered by an erodible soil. Each time, the system goes back to positive feedback mechanism.

From the above discussion it can be concluded that levelling the terrain after clearing the vegetation and filling the natural gullies for farming etc. will not provide a future of agriculture in the area. A soft and hard engineering plan is necessary to divert the channel flow energy. Badlands are complex landscapes and it is not easy to plan diversion channels. Researches need to be conducted that focus on to suggest a more sustainable way of reclamation than what is being practiced now. Sometimes successful operation of badland stabilization has been reported after treating the soil to reduce the threshold of erosion (Phillips 1998a, b), such as, the *calanche* and *biancane* badlands of Italy and Tuscany. This was possible by treating the sodicity of the soil, bringing the ESP below 15% threshold value. Such disappeared badlands are classified as "Vanishing Landscape" in the Red List of Mediterranean landscapes. However, the present study area will never be a vanishing landscape by treating the sediments. A hard/soft engineering design is recommended.

Finally, it is inferred that the area under review, like many other parts of the country is developing rapidly in agriculture. The arable fertile plains are already exhausted hence the further expansion of the activity is by exploiting the badlands, which were once considered unproductive. The only issue that should be seriously considered is whether this development (at the expense of the natural geomorphic feature) is going to be beneficial in the long run or otherwise. By levelling and smoothening the slopes, the natural processes that have long been operating in the area will not suddenly change. The method used by the farmers has already shown degradation in the land. The long-term benefit of the entire activity looks bleak. A land use planning needs to be designed for the gully infested areas because the present crude methods will lead more soil erosion than what was already experiencing in the area.

Acknowledgements The paper is a part of a research project funded by Department of Science and Technology, India and a departmental annual project. The data has been obtained during different projects before finally compiling as one paper. The authors would like to thank the commission for the financial support. Authors would also like to acknowledge the help from many students of the department during the fieldwork, they are Priyanka Hire, Sadashiv Bagul, Kajal Sawkare and Nilesh Susware. We thank NRSC (National Remote Sensing Centre) for making the Cartosat I images available for the study. The Department of Geography, SPPU is duly acknowledged for giving all the facilities to conduct the analysis. Thanks are also due to Zuare Agro, Pune for conducting sediment parameters for the study.

References

Abdel-Kader F (2018) Assessment and monitoring of land degradation in the northwest coast region, Egypt using Earth observations data. Egypt J Remote Sens Space Sci. https://doi.org/10.1016/j.ejrs.2018.02.001

Ajayi A (2015) Land degradation and the sustainability of agricultural production in Nigeria: A review. J Soil Sci Environ Manag 6:234–240

Amezketa E, Aragues R, Carranza R, Urgel B et al (2003) Chemical, spontaneous and mechanical dispersion of clays in arid-zone soils. Span J Agric Res 1:95–107

Babu R, Tejwani KG, Agarwal MP, Bhushan LS et al (1978) Distribution of erosion index and iso-erosion map of India. Indian J Soil Conserv 6(1):1–12

Bajocco S, Angelis AD, Perini L, Ferrara A, Salvati L et al (2012) The impact of land use/land cover changes on land degradation dynamics: a Mediterranean case study. Envron Manag 49:980–989

Batunacun WR, Lakes T, Yunfeng H, Nendel C et al (2019) Identifying drivers of land degradation in Xilingol, China, between 1975 and 2015. Land Use Pol 83:543–559

Bryan RB, Jones JA (1997) The significance of soil piping processes: inventory and prospect. Geomorphology 20:209–218

Canton Y, Barrio GD, Sole-Benet A, Lazaro R et al (2004) Topographic controls on the spatial distribution of ground cover in the Tabernas badlands of SE Spain. CATENA 22:341–365

Cowie AL, Orr BJ, Castillo Sanchez VM, Chasek P, Crossman ND, Erlewein A, Louwagie G, Maron M, Metternicht GI, Minelli S et al (2018) Land in balance: the scientific conceptual framework for land degradation neutrality. Environ Sci Policy 79:25–35

Dong A, Chesters G, Simsiman GV et al (1983) Soil dispersibility. Soil Sci 136(4):208–212

Gallart F, Mariganani M, Perez-Gallego N, Santi E, Maccherini S et al (2012) Thirty years of studies on badlands, from physical to vegetational approaches. A succinct review. CATENA. https://doi.org/10.1016/j.catena.2012.02.008

Gupta RK, Bhumbla DK Abrol IP et al (1984) Effect of sodicity, pH, organic matter and calcium carbonate on the dispersion behaviour of soils. Soil Sci 137(4):245–251

Hack JT (1973): Stream-profile analysis and stream-gradient index: U.S. Geological Survey. Journal Research, 1(4): 421–429

Hualin X, Yanwei Z, Zhilong W, Tiangui L et al (2020) A Bibliometric analysis on land degradation: current status, development, and future directions, Land 9:28. https://doi.org/10.3390/land9010028

Hazbavi Z, Sadeghi SHR, Gholamalifard M et al (2019) Dynamic analysis of soil erosion-based watershed health. Geogr Environ Sustain 3:43–59

Inglis CC (1944) Maxium depth of scour at heads of guide banks, groyens, pier noses and downstream of bridges. Annual Report (Technical), CWPRS, Pune

IRC (1998) Standard specifications and code practice for rod bridges section 1. IRC 5:1998

Joshi VU, Nagare V (2009) Land use and land cover change detection along the Pravara river basin in Maharashtra, using remote sensing and GIS techniques. Acta Geodaetica ET Geophysica Hungarica, AGD Lands Environ 3(2):71–86

Joshi VU, Tambe D (2008) Formation of Hydroxy - Interlayer Smectites (HIS) as an evidence for paleo-climatic changes along the riverine sediments of Pravara River and its tributaries in Maharashtra. Clay Res 27(1–2):1–12

Khaledian Y, Kiani F, Ebrahimi S, Brevik E, Aitkenhead-Peterson J et al (2016) Assessment and monitoring of soil degradation during land use change using multivariate analysis. Land Degrad Dev 28:128–141

Lacey G (1929) Stable channels in alluvium. J Inst Eng 4736:229

Lal R (1993) Tillage effects on soil degradation, soil resilience, soil quality, and sustainability. Soil Tillage Res. 27:1–8

Marathianou M, Kosmas C, Gerontidis S, Detsis V et al (2000) Land-use evolution and degradation in Lesvos (Greece): a historical approach. Land Deg Dev 11(1):63–73

Matano AS, Kanangire CK, Anyona DN, Abuom PO, Gelder FB, Dida GO, Owuor PO, Ofulla AVO et al (2015) Effects of land use change on land degradation reflected by soil properties along Mara River, Kenya and Tanzania. Open J Soil Sci 5:20–38

McCauley A, Jones C, Olson-Rutz K (2017) Soil pH and organic matter. Nutrient management. Module no 8. 1–12

Mongil-Manso J, Navarro-Hevia V, Díaz-Gutiérrez V, Cruz-AlonsoI V, Ramos-Díez I et al (2016) Badlands forest restoration in Central Spain after 50 years under a Mediterranean-continental climate. Ecol Eng 97:313–326

Morgan RPC (1986) Soil erosion and conservation. Longman Group Ltd., Essex

Muneer M, Oades JM (1989) The role of Ca–organic interactions in soil aggregate stability. III. Mechanisms and models. Aust J Soil Res 27:411–423

Phillips C (1998a) a) The badlands of Italy: a vanishing landscape? Appl Geogr 18(3):243–257

Phillips C (1998b) The Crete Senesi, Tuscany: a vanishing landscape? Landsc Urban Plan 41(1):19–26

Piegay H, Walling DE, London N, He Q, Liebault F, Petiot R et al (2004) Contemporary changes in sediment yield in an alpine mountain basin due to afforestation (the upper Drome in France). CATENA 55(2):183–212

Reynolds JF, Stafford SM (eds) (2002) Global desertification: Do humans cause deserts? Dahlem University Press, Berlin

Rossi R, Vos W (1993) Criteria for the identification of a red list of Mediterranean landscapes: three examples in Tuscany. Landsc Urban Plan 24:233–239

Saha AK, Arora MK, Csaplovies E, Gupta RP et al (2005) Land cover classification using IRS LISS III image and DEM in Rugged Terrain; a case study in Himalayas. Geocarto Int 20(2):33–40

Sangsub C, Chan-Beom K, Jeonghwan K, Ah Lim L, Pa K-H, Namin K, Yong SK et al (2020) Land-use changes and practical application of the land degradation neutrality (LDN) indicators: a case study in the subalpine forest ecosystems. Republic Korea Forest Sci Technol 16(1):8–17

Wischmeier WH, Smith DD (1965) Rainfall-erosion losses from cropland east of the rocky mountains: guide for selection of practices for soil and water conservation. USDA Agriculture Research Series, Agriculture Handbook No. 282

Wischmeier WH and Smith DD (1978) Predicting rainfall erosion losses: A guide to conservation planning, USDA Agr. Res. Ser., Agriculture Handbook No. 537.

Wells NA, Andreamiheja B (1993) The initiation and growth of gullies, Madagascar-Are humans to be blamed? Geomorphology 8:1–46

Zhang K, Yu Z, Li X, Zhou W, Zhang D et al (2007) Land use change and land degradation in China from 1991 to 2001. Land Degrad Dev 18:209–219

Soil Piping: Problems and Prospects

H. R. Beckedahl, J. A. A. Jones, and U. Hardenbicker

Abstract

Subsurface soil erosion is an under-reported but widespread erosion phenomenon, and consequently far less researched than surface erosion processes. Piping or tunnel erosion develops in a range of different materials under semiarid through to humid climates. The most severe subsurface erosion has been recorder from loessial soils, colluvial soils and mixed loess colluvium, although it has been reported to a lesser extent in other soils as well. The collapsibility of loess appears to increase with an increase of the clay content, but under higher clay content, collapsibility decreases with the increase in clay content. Especially deeper tunnel erosion in loess can be caused by hydro-consolidation or hydro-collapse after saturation. There is also much evidence from the ex-Gondwana countries suggesting that the sodium content in soils (both in the Sodium Absorption Ratio or SAR and in the Exchangeable Sodium Potential or ESP) may increase the susceptibility of soils to dispersion and so to tunnel erosion, although the exact nature of this relationship is, as yet, poorly understood. Field evidence shows that subsurface erosion is, however, highly significant in that, where it occurs, conventional forms of remediation and rehabilitation are contraindicated. The hydrological impact of piping has also been relatively under-researched compared to its erosional effects, which have been left out of most mathematical erosion models. Research has shown that piping can considerably alter the drainage of hill slopes, contribute significant volumes of storm water to rivers, and affect the ambient water chemistry. Hydrological monitoring in a wide variety of climates and soils shows a range of pipe flow responses as significant contributions to river flow through sources of pipe flow discharge. Further research of both the hydrodynamics and the erosion characteristics of piping will enhance existing hydrological and erosion models.

H. R. Beckedahl (✉)
Department of Geography, Environmental Science and Planning, University of Eswatini, Eswatini, Kingdom of Eswatini
e-mail: hbeckedahl@gmail.com

H. R. Beckedahl
Department of Geography, Geo-informatics and Meteorology, University of Pretoria, Pretoria, South Africa

J. A. A. Jones
Department of Geography and Earth Sciences, Aberystwyth University, Aberystwyth, UK
e-mail: tonyandjenjones@btinternet.com

U. Hardenbicker
Department of Geography and Environmental Studies, University of Regina, Regina, Canada
e-mail: Ulrike.Hardenbicker@uregina.ca

Keywords

Subsurface soil erosion · Piping · Tunnel erosion · Dispersion · Hydrodynamics · Modelling

11.1 Introduction

Piping or tunnel erosion is described as a natural erosion process by which infiltrating water is forming tubular subsurface drainage channels (Dunne 1990; Jones 1981a, b; Verachtert et al. 2010; Wilson 2011; Faulkner 2013). Piping or tunnel erosion was reported and analyzed in many loess sediments and soil developed on loess worldwide (Dunne 1990; Bryan and Jones 1997; Faulkner 2006; Verachtert et al. 2010). A tunnel starts to form where water infiltrates through the surface via soil cracks and concentrated due to changes in permeability. Once the water reaches the zones with a change in soil permeability it moves down slope above the less permeable layer. Though flow carries away fine particles but the movement of water is slow until an outflow forms further downslope. More material can be eroded, subsurface channel ways may build up, and tunnels enlarge rapidly, occasionally in association with hydraulic pressure. If the shear strength of the roof material is exceeded, the enlarged pipes may collapse and gullies may form in a very short period of time. This happens generally during an intensive storm event or after periods of prolonged precipitation (Visser 1969; Beckedahl 1998; Faulkner 2013), but may also follow periods of prolonged desiccation.

Disturbance and changes of the natural vegetation can trigger tunnel gully erosion by increasing the development of soil cracks and increasing surface runoff (Downes 1946; Crouch 1976; Laffan and Cutler 1977). Tunnelling in soil can also result from the dispersion of a sodic soil horizon. Overland flow with low electrolyte concentration is able to enter the soil via desiccation cracks, and disperse the sodic clay subsoils (Crouch 1976; Laffan and Cutler 1977; Jones 1981a, b; Faulkner 2013), or enter through macro pores and so contribute to hill slope quick flow events without meaningful dispersion occurring.

Soil horizons or sediment layers need to have sufficient permeability to enable subsurface run-off and dispersed clays to flow through the macro pores without pore blockages (Wilson et al. 2016). Further rainfall events entrain and erode more material, resulting in both head ward and tail ward linking of cavities forming a continuous tunnel system. Tunnel expansion enables (2016) flowing water to scour the base and undercut sidewalls, causing mass wasting of the gully walls (Laffan and Cutler 1977; Zhu 1997, 2003). Eventually undermining reaches an extent where complete roof collapse occurs and gullies form (Laffan and Cutler 1977). Tunnel erosion processes in loess as a non-dispersive material also result from the liquefaction of non-cohesive soils and sediments with high silt and sand content (Assallay 1998; Rogers et al. 1994). Piping therefore is a continual interplay between the hydrodynamic and the sediment transport process, including dispersion where present. This interplay then is also critical in the environmental aspect and the associated question relating to the rehabilitation of degraded areas that show evidence of subsurface erosion, or piping.

11.2 The Hydrological Impact of Natural Soil Piping

Prior to the beginning of the 1970s, piping was viewed solely as an agent of erosion, most notably as the cause of failure in earth dams (e.g. Terzaghi 1931; Sherard 1953) and as a source of soil erosion in agricultural fields (e.g. Downes 1946; Colclough 1965). The 1960s and 70 s marked a turning point in the understanding of the processes of stream flow generation and of the heterogeneity of hill slope drainage. The term 'hill slope hydrology' was born. The International Association of Hydrological Sciences publication of a volume in its series *Benchmark Papers in Hydrology* devoted to *Stream flow Generation Processes* (Beven 2006) makes this very clear: only three of the 31 papers predate the

1960s. The widespread establishment of research catchments and the involvement of so-called 'forest hydrologists' in America and geographers in Britain focusing attention on the landscape revealed the complexity of processes and the heterogeneity of drainage pathways (e.g. Whipkey 1965; Hewlett and Hibbert 1967; Kirkby and Chorley 1967; Weyman 1970).

The same issue of *Benchmark Papers* acknowledges the ground-breaking paper by Gregory and Walling (1968), who noted the temporary 'out-of-channel' expansion of the stream network during storms along lines that may be termed swales. Jones (1967) discovered just such a network, which he called 'dry thalwegs', while undertaking geomorphological mapping in a small Warwickshire catchment in 1965. Van Meerveld et al. (2020) have recently revisited Bishop et al. (2008) entitled *Aqua Incognita: the unknown headwaters* and added *temporaria* to the title in recognition of such temporary fluctuations. The thin blue line on the map no longer marks the full extent of the drainage network and the process of basin-wide infiltration-excess overland flow envisaged by Horton in 1933 at the very outset of scientific hydrology (also a *Benchmark Paper*) is too simple.

Natural soil pipes fitted into this milleux and extended the range of pathways (Jones 1979). Jones (1981a, b) relates how, sitting on the banks of a small tributary of the River Derwent in Derbyshire having his sandwich lunch in 1968, he noticed holes in the opposite bank that appeared to show signs of having leaked water (Fig. 11.1). Jones (1971) found a close association between clusters of these pipes in the stream bank and percolines—lines of concentrated subsurface seepage (Bunting 1961)—that provide linear extensions of the drainage network and may yield lines of saturation overland flow during storms. Weyman (1970, 1974) found pipes in the Mendip Hills and speculated that they might be a significant source of stream flow, a speculation heavily criticized at the time by the prominent hydrologist Freeze (1974) as 'far from convincing'. Meanwhile, Gilman and Newson (1980) and Jones (1975, 1978) began actually measuring pipe flow in two separate headwater catchments in mid-Wales.

Despite these advances half a century ago, the overwhelming focus of research is still concentrated on erosion rather than hydrology. When Chappell (2010) compiled his excellent review of research in the humid tropics, he found that less than half of the 28 papers reported flow measurements. The considerable amount of field experiments and modelling undertaken by the USDA is understandably focused on soil erosion (e.g. Midgley et al. 2013; Wilson et al. 2016, Xu et al. 2020).

Long-running studies of pipe erosion on loessial soils in Belgium have recently expanded the hydrological aspects. Got (2019) instrumented two catchments, Sippenaeken in the Gueule valley, Eastern Belgium, and Kluisbergen in the Scheldt valley, Western Belgium, mapping pipes using ground penetrating radar (GPR)—a method first applied by Holden et al. (2002)—and monitoring groundwater, pipe flow and isotopes and developing a conceptual model. Verachtert et al. (2010, 2011), Vannoppen et al. (2016, 2017), and Bernatek-Jakiel and Poesen (2018) also studied pipe flow in the Belgian Ardennes, but none of the Belgian studies linked pipe flow to the streams. The same applies to the long history of piping research in Poland (e.g. Bernatek-Jakiel and Poesen 2018).

A considerable amount of research has been undertaken in Japan where pipe flow monitoring began in the early 1980s. Reports reveal significant contributions to overall basin runoff ranging on occasions up to 75% in a number of catchments beginning in the Hachioji basin in the Tama Hills, Honshu, and expanding to Toinotani near Kyoto and the Hitachi Ohta and Hakyuchi basins near Tokyo, and the Jozankei basin in Hokkaido (Yasuhara 1980; Tanaka 1982; Tsukamoto et al. 1982; Ohta et al. 1983; Tanaka et al. 1988; Kitahara 1989, 1994; Kitahara and Nakai 1992; Kitahara et al. 1994; Mizuyama et al. 1994; Sidle et al. 1995; Terajima et al. 1996a, b, 1997, 2000; Uchida et al. 1999, 2001, 2005; Noguchi et al. 1999; Uchida 2000; Koyama and Okumura 2002; Asano and Uchida 2018).

Fig. 11.1 Bank pipes on Burbage Brook, Derbyshire, showing signs of leakage, as seen by a lunchtime observer in 1968

More limited research in Canada has ranged across three different bioclimatic zones from the arid badlands of Alberta and the forests of southern Quebec to the subarctic tundra in the James Bay lowlands of northern Ontario and in the Yukon. In both the tundra and the taiga, the greatest pipe flow activity occurs during the annual snowmelt, reaching 76% in the Lac Laflamme basin (Roberge and Plamondon 1987), rather less in the tundra catchments (Woo and diCenzo 1988; Carey and Woo 2000, 2002). In the badlands, studies were somewhat curtailed by the unexpectedly large discharge overwhelming the instrumentation, but Bryan and Harvey (1985) estimated that pipe flow contributed a third of the catchment runoff. Significant contributions from pipe flow have also been observed in California (Albright 1991; Ziemer and Albright 1987; Swanson et al. 1989; Swanson 1983; Ziemer 1992).

There is also a substantial collection of studies from the tropics, especially in Malaysia in the Danum catchment in Sabah (Bidin 1995; Sayer et al. 2004, 2006; Chappell and Sherlock 2005) and at Bukit Tarek in Peninsular Malaysia (Negishi et al. 2007) and in Singapore (Chappell and Sherlock 2005). In the Western Ghats in southern India, Putty and Prasad (2000) estimated that an average of 25% of stream flow derived from piping, rising to 59% in storm peaks. In Dominica, Walsh et al. (2006) upgraded their estimate to 21%. In the Shanxi Loess Plateau, China, Zhu reports 35–43% derived from pipes (Zhu 1997; Zhu et al. 2002).

In Britain, work in the Institute of Hydrology/Centre for Ecology and Hydrology Plynlimon catchments in Wales has produced very varied estimates ranging from 1 to 61% by Gillman and Newson (1980), but based on just 12 storms and without concurrent stream flow data, to 38% of peak quick flow (Chapman 1994) and just 10% overall by Chapman et al. (1993, 1997). However, Chapman's data also related to just a handful of storm events in a single ephemeral pipe entering a first-order tributary of the Afon Cyff on Plynlimon. More recently, working in the Pennines, Holden has published estimates of 10.5% overall and 30% during stream recession in the deep peat of Little Dodgen Pot Sike and around 29% in the Moor House Nature Reserve, rising to 44% in artificially drained peat (Holden and Burt 2002; Holden et al. 2006).

The first basin-wide study of pipe flow and the first with simultaneous monitoring of stream flow, overland flow and diffuse seepage from bankside seepage zones or percolines began in the Maesnant basin in 1979 (http://www.history-of-hydrology.net/mediawiki/index.php?title=Maesnant_(Wales)_1970s-1990s). In the Maesnant headwater catchment pipes contribute up to 45% of storm flow in the stream via a discontinuous set of pipe networks (Fig. 11.2). The overall network, including ephemerally-flowing and perennially-flowing pipes, has been traced for 750 m up the hillside and the longest integrated, ephemerally-flowing pipe is over 300 m long. This means that the storm flow contributing area is more complex than the previously standard riparian zone abutting the stream bank, and it extends in fingers upslope with frequent interchanges between surface and subsurface flows (Fig. 11.3). It also means that the a/s index (area drained/slope) method of determining contributing areas can be less effective when piping is present (Jones 1986).

Jones (1997a) plotted data from 15 pipe flow gauging sites on Maesnant against data for overland flow and diffuse through flow plotted by Anderson and Burt (1990). The graphs plot peak lag times and peak runoff rates against drainage basin areas; for the Maesnant pipes this meant defining the micro-drainage areas of the pipes using a detailed ground survey. The results showed that pipe flows plot logically between through flow and saturation overland flow (Fig. 11.4). Comparable data from two pipes in the Toinotani basin near Kyoto provided by Uchida et al. (1999) map perfectly into the same space (Jones 2010) and data from Panama reported by Kinner and Stallard (2004) confirm these results. The banks of pipe flow monitoring weirs across the slopes in the Maesnant basin also offer a unique view of how the interaction between basin wetness and storm rainfall affects the patterns of pipe flow over the hillside in terms of the beginning of storm flow and peak flow travel times (Jones 2010, Fig. 11.4). Unfortunately, there has been no other basin-wide intensive monitoring programme elsewhere in the world and it is very difficult to extrapolate the Maesnant results (Jones 1997b). There have, however, been a large number of measurements of pipe flow discharges from individual pipes around the world.

The Maesnant research raised a number of other issues, many of which are still debated in current literature: (1) the degree of connectivity

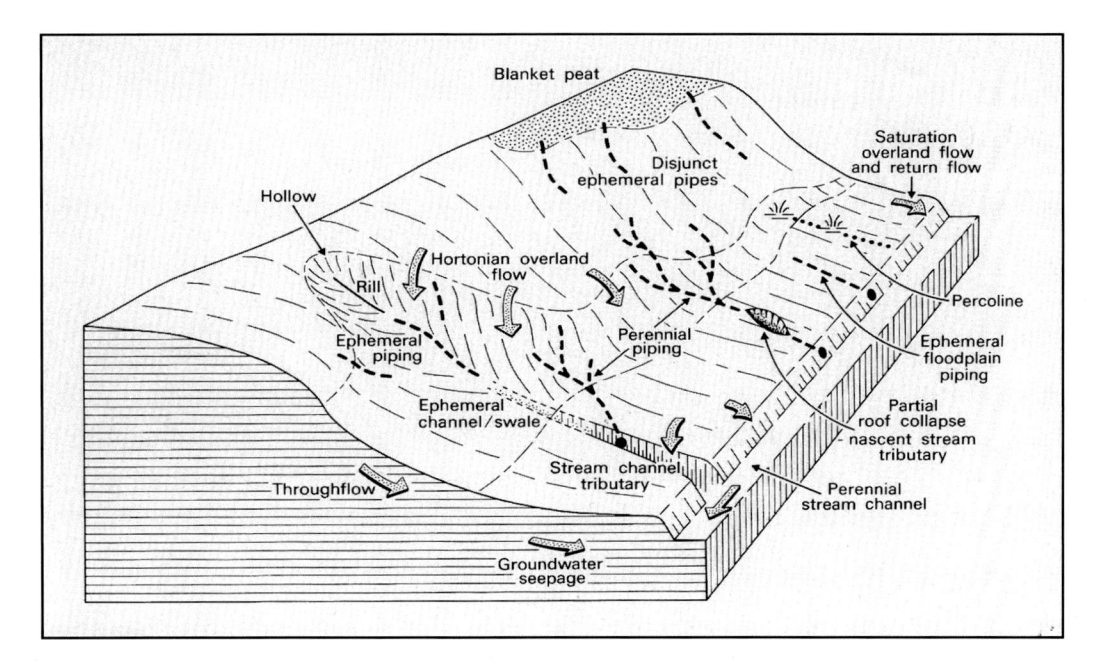

Fig. 11.2 The effects of piping on the pathways of hill slope drainage (after Jones 1981a, b)

Fig. 11.3 The effect of piping on the area of the drainage basin contributing storm runoff to the stream on Maesnant

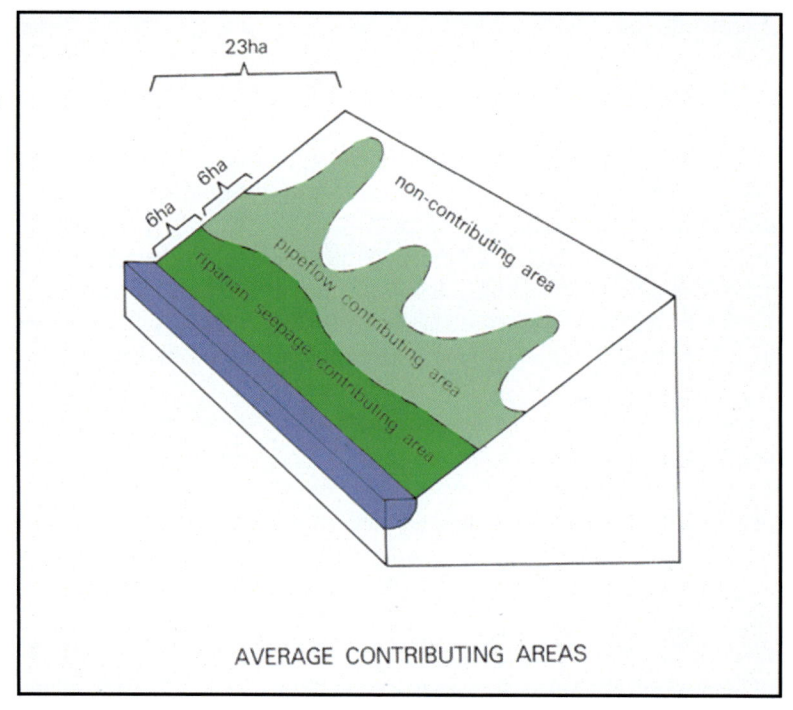

Fig. 11.4 The response of pipeflow lies between throughflow and overland flow. Peak discharge and lag times decline gradually as the micro-drainage areas increase. Pipeflow data from Maesnant, other processes plotted according to the collation by Anderson and Burt (1990)

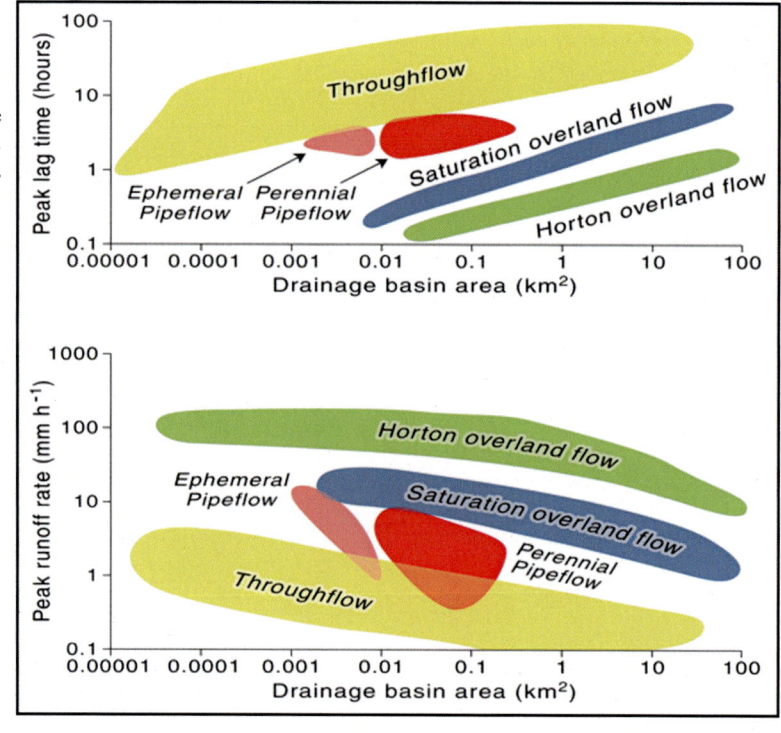

within the hill slope drainage networks; (2) the question of whether pipe flow is discharging 'old' water from previous storms or 'new' water from current rainfall; (3) the extent to which the location and form of pipes are controlled by environmental factors compared with pipes creating the environment; and (4) the implications for environmental management and erosion within catchments.

The question of connectivity is very much a current issue (e.g. McGuire and McDonnell 2010; Got 2019). It is clear from Maesnant that the discontinuities within the ephemeral pipes and between the ephemeral and perennial pipe networks that are apparent on the map are significantly reduced during the more extreme storm flows as water issues from the ends of pipes and continues as saturation overland flow before sinking into another pipe (Fig. 11.5). Elsenbeer and Vertessy (2000) also noted the frequent exchange between surface and subsurface routes that complicate the analytical separation of sources in the La Cuenza basin in the Amazonian rainforest of Peru. And Chappell and Sherlock (2005) observed sinkholes and resurgences along pipe routes on the slopes of the research catchment at the Danum Valley Field Centre in Borneo. In the Institute of Hydrology's Upper Wye catchment, Gilman and Newson (1980) speculated that the ephemeral pipes may be less significant as a source of stream flow because they are not directly connected to the stream but discharge above the riparian bogs. However, where this occurs on Maesnant dye tracing experiments show that it can take just 10 min for the pipe discharge to cross the bogs overland when they are saturated.

The early experiments in the Upper Wye basin, in which Gilman and Newson (1980) used a mechanical pump to stimulate pipe flow artificially showed that the ephemeral pipes are very leaky and suggested that they only produce natural flow if the phreatic surface is raised to the level of the pipe bed. This view has been corroborated by numerous water quality studies in the catchment that show the signature of 'old' groundwater (Sklash et al. 1996; cp. McDonnell 1990). The view from Maesnant is more nuanced: that groundwater is undoubtedly the source of base flow in the perennial pipes and to an extent contributes to storm flow in the ephemeral pipes, but that both types of pipe can also receive 'new' rainfall contributions either directly through sinkholes or cracks above the pipes or indirectly whereby rainwater is delivered to the phreatic surface through macro pores away from the pipes, which is then raised to pipe level. This is confirmed by analyses of pipe flow chemistry that show dilution of the groundwater signature as storms proceed, in the more severe storms and during wetter periods (Hyett 1990).

The implication is that the groundwater element of pipe storm flow is being forced out by a type of piston-flow (Jones 1990, 1997d, 2010). The similarity between pipe flow hydrographs and rainfall hyetographs (Jones and Crane 1982) is clear evidence of either direct rainwater inputs or piston flow or both. Uchida et al. (2005) found a similar response in Panola Mountain, Georgia and in three instrumented catchments in Japan, noting that once significant pipe flow response occurred, the maximum pipe flow rate was sensitive to the measured rainfall intensity. This also means that hydrograph separation into base flow and storm flow is valid, even though some of the storm flow may be old water; hence the validity of calculations of the percentage contribution to stream flow from the pipes. By monitoring rainwater quality simultaneously with pipe flow quality, Hyett also showed that throughputs of sulphate in the pipe were a close match, implying that the pipe water was derived from the current rainfall (Fig. 11.6). Supporting evidence comes from Elsenbeer et al. (1995), who concluded from chemical analyses that quick flow pathways, overland flow and pipe flow, in the Amazonian rainforest at La Cuenca, Peru, were dominated by new water. In the Belgian Ardennes, Verachtart (2011) concluded that pipe flow is a mix of old and new water. Also in Belgium, Got (2019) found that the response of pipe flow to rainfall is driven by a combination of preferential vertical infiltration directly into the pipe network and into the groundwater, which in turn feeds the pipes. He found that seasonal differences in groundwater levels were more important

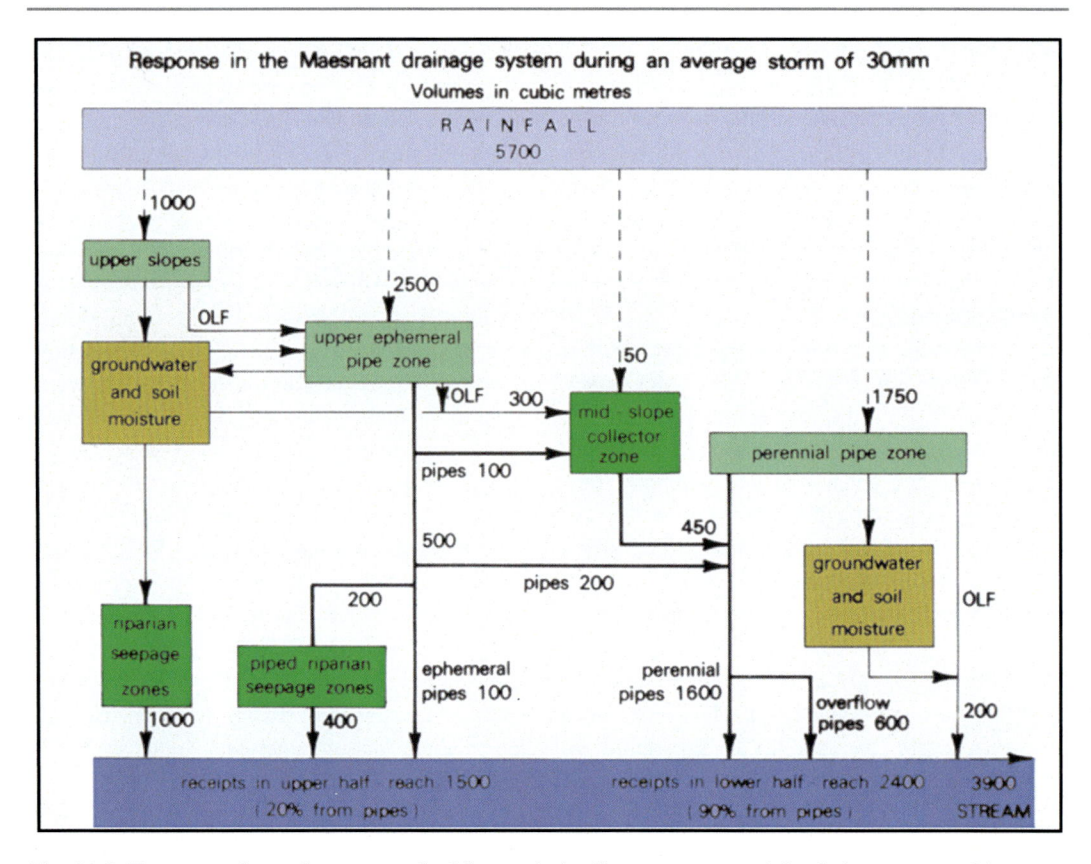

Fig. 11.5 Frequent exchange between overland flow and pipe flow create connectivity during storms on Maesnant

than antecedent wetness. As he states, the level of connectivity between the pipes and the groundwater is critical for response. Uchida et al. (2005) and Vannoppen et al. (2016) cite similar thresholds in response.

The Maesnant catchment sheds light on the question of the degree to which pipes are controlled by their environment or vice versa. The main concentration of ephemeral pipes in the upper mid-slope occurs where the soil cover and the peaty surface horizon are thinnest and more prone to desiccation cracking. On both Maesnant and the Upper Wye, the density of piping following desiccation cracks increased after the severe drought of 1974. But a survey of this area on Maesnant using a double-ring infiltrometer also showed that the general infiltration capacity increased in the vicinity of the pipes (Jones 1990). Since the infiltrometer was never placed directly over a pipe, this implies that the general density of macro pores increases around the pipes and suggests that drawdown by the pipes may be increasing macro pore erosion and opening up more pores. This suggests feedback between the various grades of macro pore (cp. Beven and Germann 1982, 2013).

Lastly, soil piping may not be as severe an agent of erosion in humid temperate climates like the British Isles as it can be in countries like Australia, India and South Africa, but it can nevertheless have important implications for environmental management. Joseph Holden and colleagues have been conducting studies on the impacts of pipe drainage on peat bogs in Northern England for more than two decades (Holden and Burt 2002; Holden et al. 2002). He began by investigating the effect that piping may have on the release of greenhouse gases from the peat due to drying out of the peat and reduction of the carbon sink due to erosion of material

Fig. 11.6 Pipe flow gauging site on Maesnant. Wet grass vegetation follows line of the ephemeral pipe down the hillside

(Holden et al. 2009a, b; Smart et al. 2013; Regensburg et al. 2020). They have explored the efficacy of blocking the pipes but have concluded that this is impractical due to the large number of pipes involved (Regensburg et al. 2020, 2021). They suggest that the best solution is to control gullying, since it is desiccation cracking in the banks of the gullies that is the main cause of pipe development here, especially in those gully banks with south and west facing aspects that receive more sunlight and where there is more active downcutting. They suggest revegetating and reprofiling the gully banks. Regensburg and colleagues also suggest that revegetating bare peat surfaces away from the gullies may reduce desiccation by slowing the velocity of overland flow, improving infiltration and so helping to retain moisture in the peat.

The implications of piping for other aspects of environmental management in the British uplands have also been explored by Jones and colleagues (Jones 2004). One positive effect of pipe drainage is increased biodiversity. Jones et al. (1991) mapped the effect of piping on the patterns of flora in the Maesnant basin, which shows areas of wet grassland around the lines of piping cutting through the more general heath vegetation (Fig. 11.6). The piping has also modified the soil, so that the piping lines are marked by swathes of dry grassland cutting through the more widespread wet grassland and heath vegetation (Fig. 11.7). Meanwhile, a long-term resurvey of stream bank piping in the Burbage Brook Experimental Catchment in Derbyshire, 35 years after Fig. 11.1 was taken, revealed the effect of subsequent afforestation on the opposite side of the basin on the relative intensity of piping. Increased evapotranspiration from the forest seems to have drastically reduced the intensity of piping on that one side and must have had an impact on the overall hydrological regime of the Brook (Jones and Cottrell 2007).

Effects have also been noted on stream water quality. Studies on Maesnant show how pipes can increase the acidity of stream water. In this case, this is especially due to the pipes directing drainage through the peaty O horizon rather than allowing the water to percolate deeper into mineral soil and drift material, making contact with weathering mineral surfaces (Hyett 1990; Jones 1994, 1997c). The increased velocity of pipe flow compared with diffuse seepage also means that acid rainfall is drained through the system faster so that there is less time for chemical buffering and that it can arrive at the stream in time to contribute an acid flush to the hydrograph peak in the stream. The Welsh Acid Rain Programme demonstrated that these acid flushes have a greater ecological impact on stream fauna than average stream water acidity and initiated a system of automatic lime dispensers triggered by the height of the stream water (Gee and Stoner 1989).

Key to the practical application of these research findings will be developing methods of extrapolation. Modelling is still in its infancy and still requires a considerable amount of data that is generally hard to come by or simply not available (Jones 1987, 1988, 1997b). The semi-distributed simulation model developed by Jones and

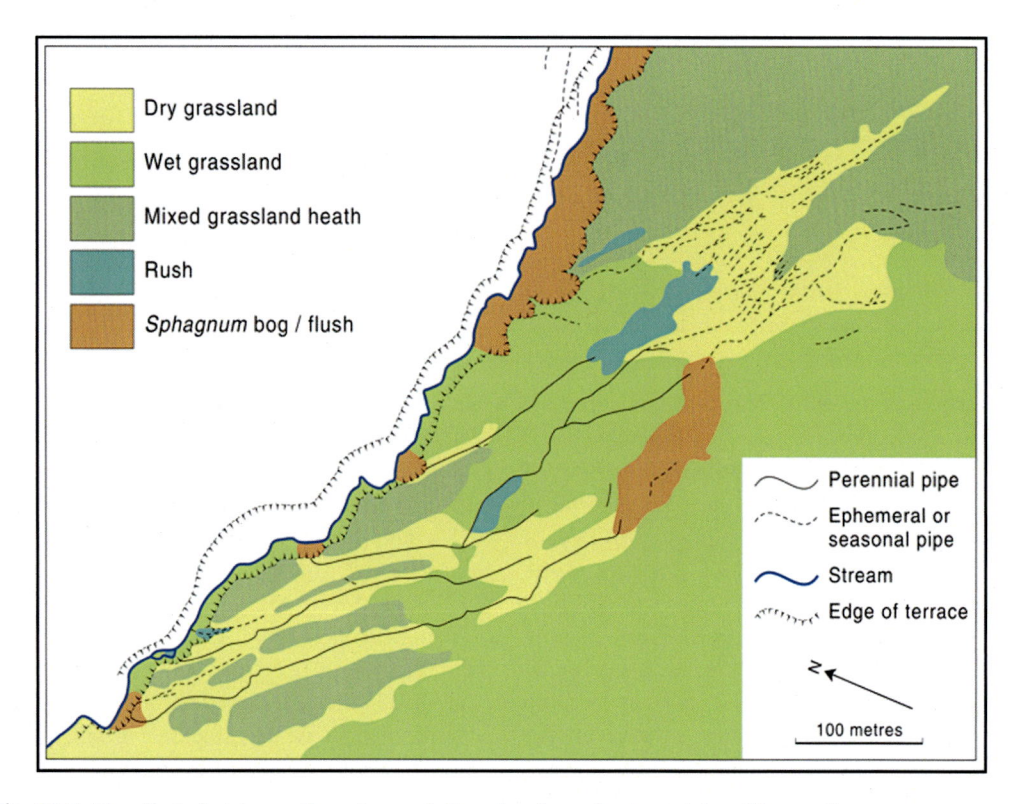

Fig. 11.7 The effect of piping on the pattern and diversity of moorland vegetation (Maesnant)

Connelly (2002) requires data on the level of the phreatic surface around the pipes and still needs a degree of parameterization based on previous measurements. Considerable progress has been made by Nieber and his associates in producing a model of pipe flow that is essentially aimed at soil erosion, with the latest version using actual field data rather than from experiments (Nieber and Warner 1991; Nieber 2001, 2006). The efficacy of this type of modelling does, however, still require verification for countries such as Australia, India and South Africa where, as was alluded previously, soil pipes have on average larger dimensions for a variety of reasons. Progress has also been made by Jeff McDonnell and colleagues building upon field experience in Maimai, New Zealand, and Panola Mountain, Georgia (cp. McGlynn et al. 2002; Tromp-van Meerveld and Weiler 2008). Weiler and McDonnell's (2007) model focuses on short, shallow ephemeral pipes, while McDonnell et al. (2007) attack the wider issues of summarizing or simplifying the heterogeneity of hill slope drainage pathways by seeking an underlying 'set of organizing principles'—still something that is eluding us.

Progress is being made in developing rapid methods of detecting pipe networks. Surface exploration is laborious and Bernatek-Jakiel and Poesen (2018) estimate geomorphological mapping can underestimate the extent of the pipe network by as much as 50%. Analysis of orthophotos can speed surveys, analyzing for linear depressions and roof collapses (Verachtert et al. 2010, 2011, 2013). The topic is ripe for using drones to do the surveying. Jones used 'human drone', i.e. a hang glider, for an initial survey on Maesnant in 1981 (see *History of Hydrology* website). But as piping is a subsurface process that does not necessarily have a surface expression, Ground Penetrating Radar offers a better solution, but is still seriously inefficient (Holden et al. 2002; Got et al. 2014; Got 2019). Dye tracing is also useful for establishing connectivity (Jones and Crane 1982; Wilson et al. 2016).

One potentially encouraging observation comes from Weiler and McDonnell's (2007) experiments with their Hill-vi model. They found that the topology of the network does not need to be known explicitly to efficiently simulate lateral preferential subsurface flow at the hill slope scale, although they were not modelling large, continuous pipe systems. Most crucially, we are still a long way from developing any model that links pipe flow to stream flow, nor one that does justice to the hydrology—stream flow—sediment movement continuum.

11.3 Considerations of Soil Piping on a Loess Soil in Germany: A Case Study

The case study here focuses on recent subsurface erosion and subsurface flow in the eastern Harz foreland in NE-Germany, 20 km west of the city of Halle (Saale), near the town of Langenbogen (Fig. 11.8). The initiation and evolution of subsurface erosion here result from a combination of natural and anthropogenic factors. Anthropogenic factors like deforestation and agriculture and especially the loss of the A-horizon by soil erosion, increase the infiltration into the subsurface layers. The basic requirements for the development of subsurface erosion in the study area are seen to be the significant porosity of the overlaying layers and the weakening of interparticulate bonds between the soil minerals, because physical particle forces control the erosion and dispersion of the material here (Hardenbicker 1998).

Detailed tensiometric recording and soil water measurements indicate a two-component flow system of slow matrix flow and rapid macro pore flow. Since the study area suffers water stress, the rapid macro pore flow is of great importance not only for the subsurface erosion processes but for groundwater pollution as well as resulting in subsurface erosion in the study area. These factors reflect a critical interaction between climatic conditions, soil/regolith characteristics and local hydraulic gradients, and support the context described in the previous section (Hardenbicker 2006).

The most important preconditions for subsurface erosion in this area are thick loess deposits and underlying layers with lower permeability

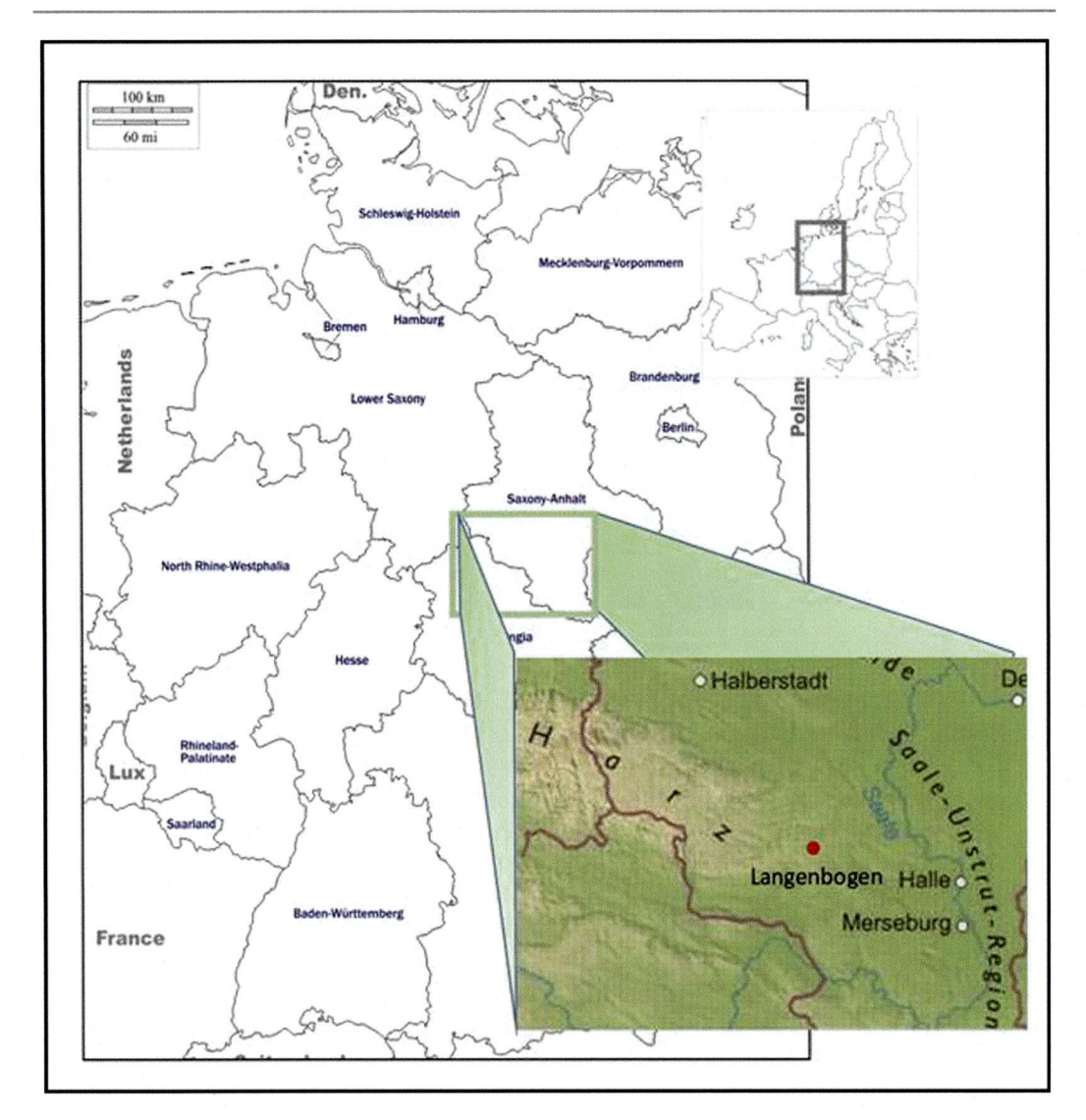

Fig. 11.8 Location map of the Langenbogen case study, Germany

built up by paleosoils, or tertiary weathering zones. Anthropogenic factors like deforestation and agriculture resulting in the truncation of the thick A-horizon by soil erosion—have increased the infiltration into the subsurface layers. Linear hollows, induced by humans since the Middle Ages, have intensified the hydraulic gradient of the subsurface water flow.

The Langenbogen area (Fig. 11.8) is part of the dry loess region of the eastern Harz foreland, situated in the rain shadow area of the Harz Mountains. In the lee of the mountains chernozems developed during the Early Holocene. The mean annual temperature varies between 9.5 and 8.1 °C. The mean annual precipitation ranges from 550 mm/a to less than 450 mm/a with mean annual evapotranspiration exceeding 450 mm. Infiltration is defined in terms of the generalized water budget of this area: 524 mm precipitation, 80 mm runoff and 444 mm evapotranspiration; implying a restricted water flow. Periglacial sediments (loess and solifluidal debris) were accumulated in large areas of the plateaus during the Weichselian glacial (an average of 7 m in vertical thickness) (Hölting and Coldewey 1996). The loess was originally covered with chernozems but now the area is strongly eroded with large quantities of calcic regosols and colluvial soils present.

The loess region of the eastern Harz foreland provides very good conditions for agriculture, although its susceptibility to soil erosion resulting from the properties of the loess and the local relief features is well known. The most widespread form of erosion is sheet erosion by rill and inter-rill flow. The regional soils were transformed by means of truncation or burial, through the effects of erosion and deposition. The intensification of agricultural production by a move towards specialization in cultivation and the use of chemical fertilizers—starting some 30 years ago—coincided with a considerable enlargement of agricultural field-lot sizes.

In the eastern Harz foreland, subsurface erosion—like piping and tunnel erosion—occurs locally and is preferentially located in areas where less permeable paleosoils or saprolite bedrock are underlying loess and associated colluvial deposits.

In the Langenbogen catchment, the subsurface erosion processes occur in a Pleistocene depression that is filled by up to 15 m thick layers of fossil paleosoils, particularly humic zones interbedded with solifluction material (reworked loess) and loess (Fig. 11.9) (Hardenbicker 1998).

Subsurface erosion results from a second drainage network that is connected to the first on the surface, influenced by relatively impermeable layers. Piping does not occur in isolation, and is associated with other forms of erosion like rilling and gullying, mass movement and surface sheet wash. At the same time there is soil erosion at the surface, so the forms built up by the collapsing of soil pipes and calcareous caves can be rapidly filled up by material from surface soil erosion. Consequently it is not easy to distinguish whether the deep gully found in the area is a result of collapsing pipe and cave roofs, or due to linear gully erosion by surface wash as both are triggered by heavy rain. Bachmann et al. (1963) described such roof collapsing in loess and loess deposits after heavy rain and the development of a form 100 m long and 3–5 m deep with varying width near the village Trebitz 20 km north of Halle. Further details were, however, not documented.

11.4 Forms and Processes of Subsurface Erosion

The first visual signs of subsurface erosion are little sinkholes (Fig. 10a and b). Collapse of the pipe roofs often take place in segments resulting in a line of holes following the orientation of the underground pipe, and leaving depressions or gullies on the upper surface. This phenomenon is most common following heavy spring rains and during freeze and thaw cycles in the late winter and early spring. They reach a maximum depth of 2 m and are mainly built up in the loess and closely related to vertical cracks and macro pores within the local soil. Thin humic and clay layers in the 2–3 m thick reworked loess can, however, cause subsurface erosion too. The area between the loess and the underlying reworked loess is particularly affected by this process, so that the roofs of hollow spaces and pipes mainly consist of early Weichselian loess (Hardenbicker 1998).

Next a few pipes at a depth of 4–6 m, where the paleosoils are situated, merge and initiate wide badland like forms by collapsing. In the test plot collapsing of bigger roof sections only took place between 1980 and 1986, (Fig. 10c, d) where loess and reworked loess covers are only a maximum of 1.50 m thick and are underlain by thick sand layers, on which landslips are developing (Fig. 10c). It would appear that the climate of the area is frequently too dry; the loess and the reworked loess are too thick and hence the groundwater table is too deep to get saturated.

The loess is cohesive enough to maintain the walls of soil pipes during the initial phases of formation, but the material is dispersive, allowing the soil to be separated into individual grains which the inflow can entrain and remove (Dunne 1990). This dispersivity is often connected to a high exchangeable sodium percentage (ESP) (Benito et al. 1993; Jones 1981a, b; Laffan

Fig. 11.9 A scanning electro-micrograph of loess from Langenbogen: **a** grains are stabilized by coating of clay and calcium carbonate (enlargement: Calcium carbonate nodules); **b** the single grain structure of reworked loess

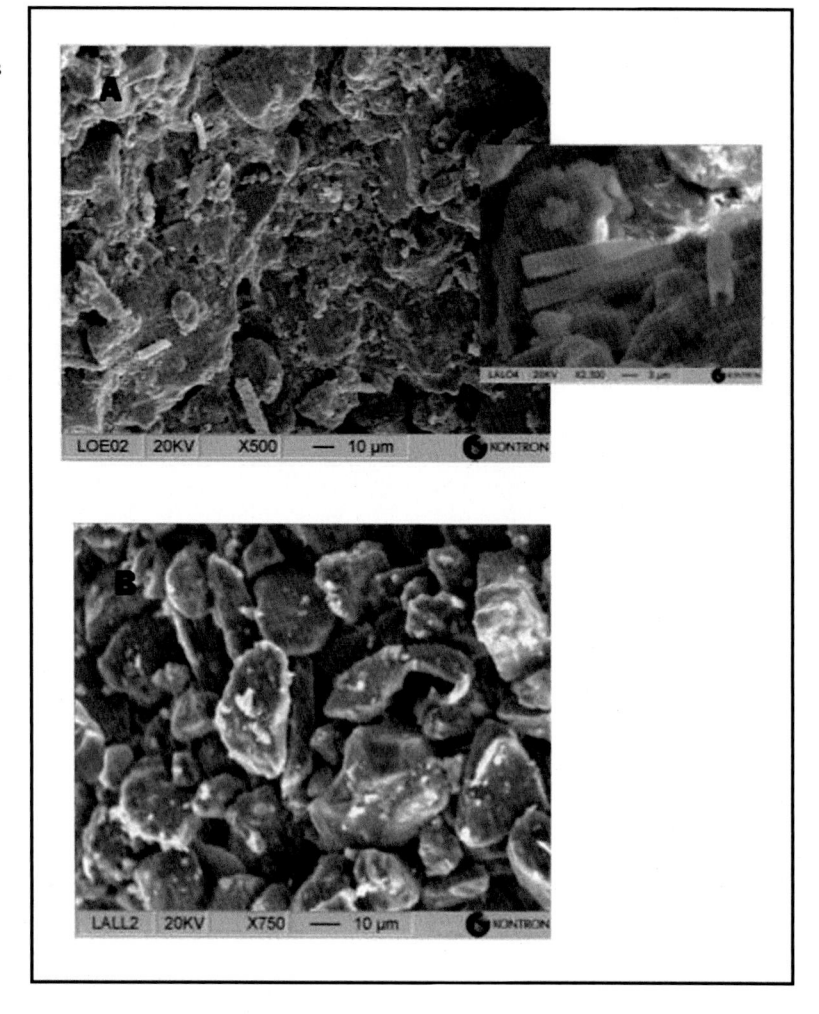

and Cutler 1977). However, loess and reworked loess, silt loam and very fine sandy loam, are relatively low in sodium. Their high silt content may account for their dispersivity, while they contain just enough clay and calcium carbonate for cohesiveness (Table 11.1 and Fig. 11.9).

The grain-size distribution shows a silt content of 50–70%. Figure 9a shows loess in situ forms fairly stable aggregates as a result of secondary formation by clay and calcium carbonate. The vertical saturated conductivity of the loess is more than two times higher than the sandy, non-loessic soil (Scheidig 1934; Hartge and Horn 1992). The relatively high conductivity of the A-Horizon is caused by earth worm activity (Fig. 11.11) (Hardenbicker and Hecht 2000).

Due to the less permeable layers, the infiltrated water is concentrated in the reworked loess and above the saprolite bedrock, resulting in pipe formation in the loess cover and wider collapse forms in the reworked loess below.

11.5 Field Measurements

To clarify the implications of subsurface hydrology for subsurface erosion processes tensiometers were installed in a depth of 1; 2; 3 and 4 m in a test plot (Figs. 11.12 and 11.13) (Hardenbicker and Liermann 2000). From June to September the tensiometer measurements show a wetting of the lower and a drying of the upper soil horizons. First

Fig. 11.10 The landscape at the Langenbogen Loess site. **a** A view into the pipe itself; **b** a field assistant standing in the collapsed section of a soil pipe. **c** smaller collapse structures of pipe roofs; **d** the headwall of the major gully at Langenbogen, and **e** a generalized view of the topography at the Langenbogen site

Table 11.1 Textural values and Atterberg limits for the loess at Langenbogen

	Clay (%)	Silt (%)	Sand (%)	Humus %	CaCO$_3$ %	Wf	Wa	Ip*
Ah ($n = 6$) (min.–max.)	20.3 (18–22)	66.2 (63–70)	13,6 (11–16)	1.3 (0.9–1.5)	7.9 (4.9–11.5)	–	–	–
Loess ($n = 11$) (min.–max.)	13.5 (10–17)	78.0 (70–83)	7.8 (5–10)	0.4 (0.2–0.6)	8.4 (5.0 -10.5)	25.3 (25–26)	19.7 (19–20)	5.4 (4–6)
Reworked loess ($n = 9$) (min.–max.)	14.5 (10–19)	52.1 (39–62)	35.7 (24–55)	0.2 (0.1–0.5)	4.7 (2–5.7)	21.0 (20–24)	16,2 (14–18)	6.2 (5–7)

measurements show that the soil through flow is too shallow to reach the tunnel systems during storm events. Field experiments and investigations indicate that the wetting depth is usually less than 100 cm, but all the tunnel systems are located below this depth. The groundwater is too deep to recharge the tunnel flow.

During heavy rainfall after the harvest on the agriculture field upslope from the test plot (see Fig. 11.14) the water infiltrated the upper soil/sediment (loess) as macro pore flow in the upslope area, and moved as interflow in the reworked loess. The pressure potential in lower horizons (3–4 m) responded to through flow events triggered by the two rainfall events. The tensiometers in the lower half of the profile (3 and 4 m installed in reworked loess) therefore responded while the ones in the upper loess did not. Unsaturated subsurface through flow in the reworked loess and macro pore flow in the upper loess were identified as important pathways operating during these events.

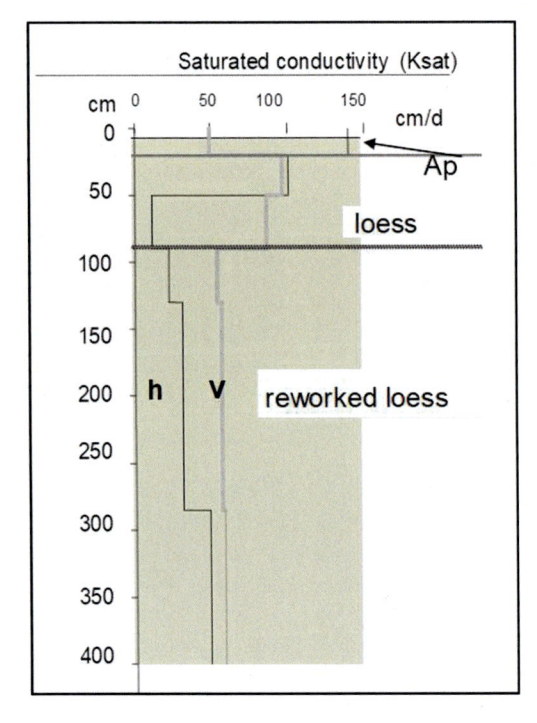

Fig. 11.11 A plot of the saturated conductivity with depth at the Langenbogen site (horizontal = black line, vertical = gray line)

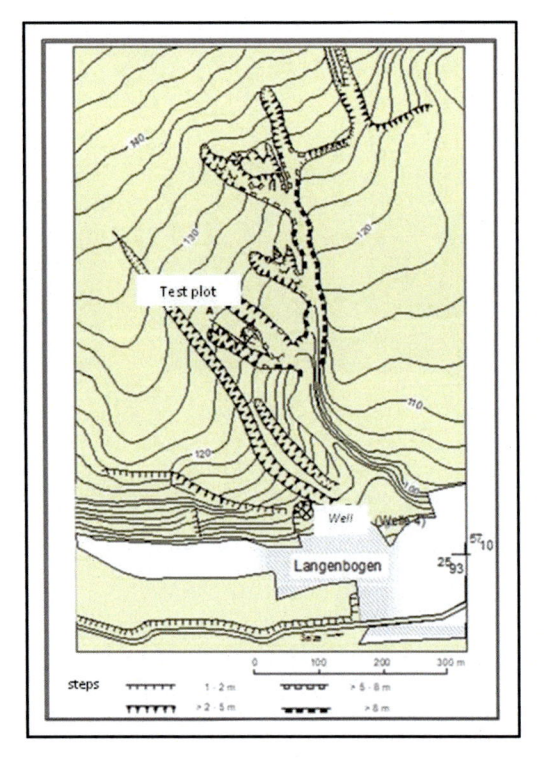

Fig. 11.12 A morphological map of the topographic detail at Langenbogen

Through flow and macro pore flow in loess and in the reworked loess provide only little or no capacity to slow infiltration or to absorb contaminants from contaminated recharge water. As a result, groundwater beneath these well drained soils within the Harz foreland and particularly near areas of subsurface erosion show generally high nitrate concentrations. Shallow wells in the study area show a high nitrate contamination and therefore they cannot be used as potable water. As in other loess areas in the eastern Harz foreland, subsurface erosion is an important additional factor to the well documented surface erosion processes of the local landscape development. As a result of aeolian sedimentation, loess already has a high natural porosity. Sodium is generally absent and dispersion of the clay particles in between the silt and sand grains does not occur on the basis of soil chemistry. Internal subsidence may be due to mechanical removal of loess particles (e.g. suffosion), slaking or collapsing of material on wetting.

Subsurface erosion in this area is also triggered by the difference in water movement between the loess and the underlying, more compacted, reworked loess. Surface waters enter the loess and infiltrate by way of cracks and macro pores. It was shown that water movement in the unsaturated deposits is an important and frequent process in the Langenbogen area.

Numerous other problems, like piping, are directly or indirectly associated with this subsurface erosion process in the eastern Harz foreland. The rapid vertical and horizontal subsurface water flow causes groundwater pollution and drinking water problems in the area, by leaching out nitrate from the arable land.

11.6 Seepage Water Movement Based on the Bromide Tracing

In order to detect lateral slope water transport, the chemical tracer bromide was applied to the test plot at Langenbogen. At the beginning of

Fig. 11.13 Location of the tensiometers at the site, relative to the slope materials

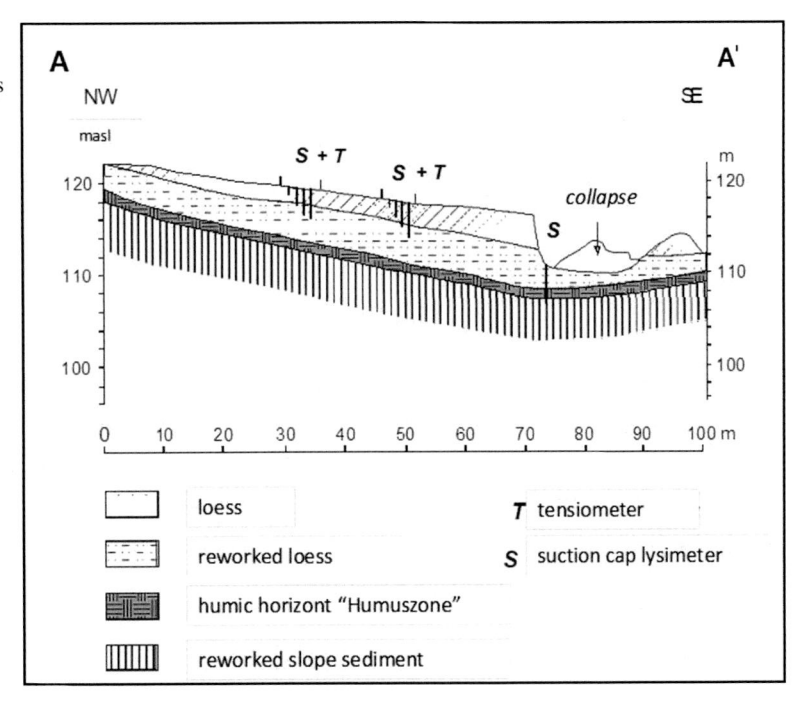

December 1996, 8 g of bromide (mixed with rainwater) were applied to the arable area above the Langenbogen test plot on an area of 10 m 1 m (Fig. 11.15). Shortly after the first bromide application, the temperatures dropped below 0 ° C, and snowfall and rain set in. From December 20, a period of severe frost began, which lasted until January 11, 1997 and was followed by a period with temperatures around 0 °C with precipitation The second bromide tracer application took place on July 9, 1997. The application area was in the measuring field below suction cup group A next to suction cup C2 (Fig. 11.15). The bromide application was repeated on the same area two weeks later on July 22, 1997 (Hardenbicker and Liermann 2000).

On December 13, 1996, the first sampling date after the first bromide application, a bromide concentration of 17.3 mg/l was detected in the seepage water of the suction cup A3 and a bromide concentration of 0.4 mg/l in C2. On January 10, 1997, the seepage water from suction cup C2 had a bromide concentration of 0.3 mg/l and that of suction cup D4 was 9.5 mg/l (Fig. 11.16). From the end of January to the beginning of February 1997, low concentrations of bromide could be measured in the seepage water of the suction cups A3, B4 and D4. Shortly after or simultaneously with the bromide, the potassium and chloride concentrations rose (Figs. 11.16, 11.17 and 11.18). On July 17, 1997, bromide concentrations of 244 mg / l were detected in the seepage water of the suction cup D4. Unlike in January 1997, the nitrate content in the seepage water of the suction cup D4 did not change (Fig. 11.15). Similar high bromide concentrations (280 mg/l) were measured on August 2, 1997 in the seepage water of suction cup C2. At the same time as the bromide, the nitrate values rose (Fig. 11.19).

The relatively high bromide concentrations that were detected shortly after the tracer application on December 13, 1996 in the seepage water from suction cups A3, C2 and D4 indicate an increased lateral seepage water flow, presumably in combination with bypass flow. There was no flow through the entire body of the soil (matrix flow). Only when the field capacity of the topsoil was exceeded on January 10 as a result of the thawing of the ground frost and the onset of precipitation, did extensive seepage water movements become possible causing a

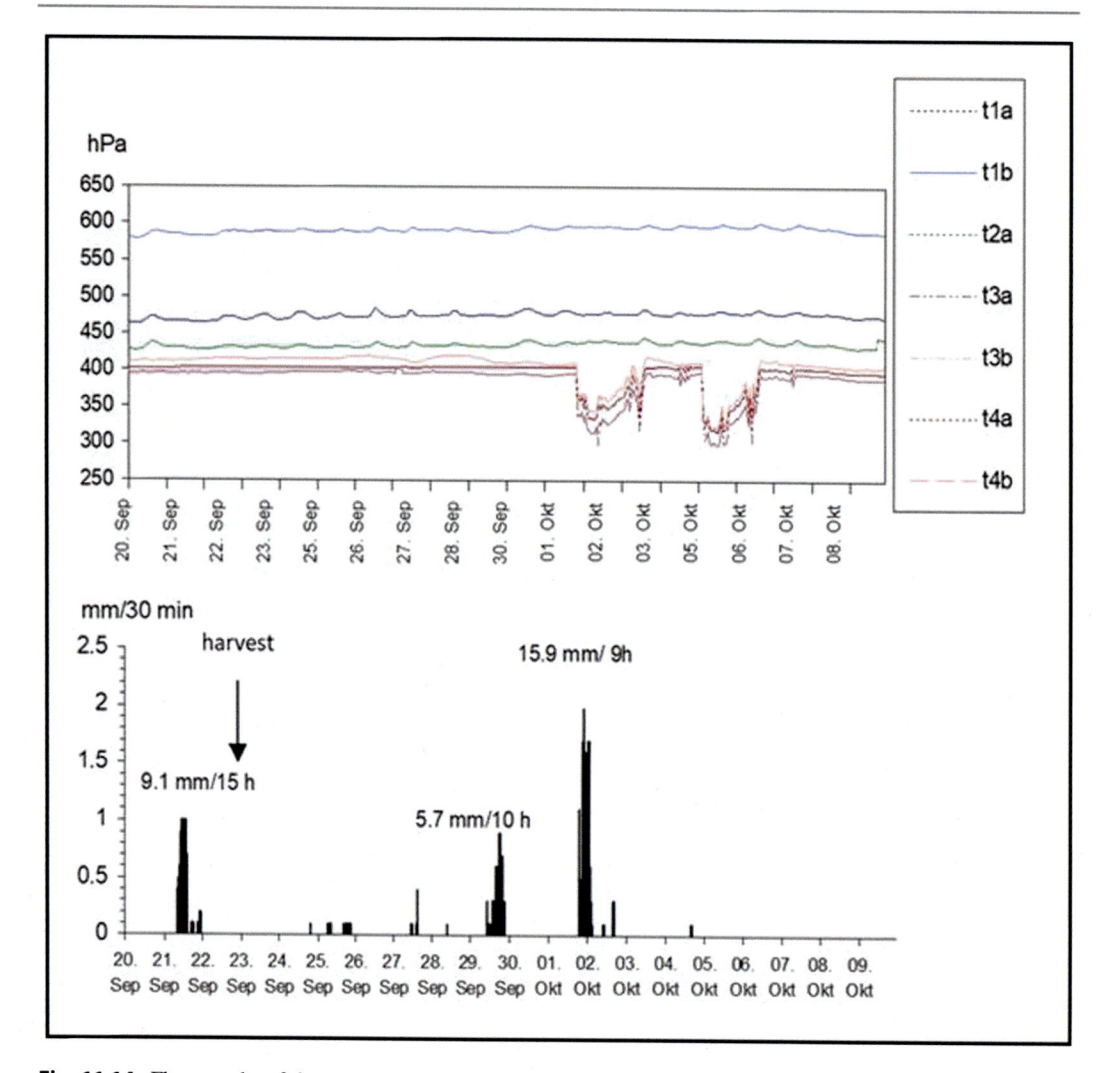

Fig. 11.14 Flow graphs of the pressure potential in relation to storm events

displacement of the old slope water by new. The bromide tracer formed the front of this seepage water movement with a clearly lateral movement component, in the course of which potassium and chloride were also displaced (see Figs. 11.15, 11.16 and 11.18).

In addition to the exchange of slope water by lateral seepage water flows, a preferred flow path must also be available in the Langenbogen test plot. This is the only way to explain the rapid transport of the tracer without any noticeable changes in the ion content of the seepage water. The high concentrations observed are typical of bypass flow and macro pore flow. Since the tracer application took place directly above the suction cup C2 (Fig. 11.20), this bromide shift is mainly due to vertical contamination. Together with the bromide, accumulated nitrate from the 4.4 m thick substrate was shifted under the fallow area below the root zone. In contrast to the bromide concentration in February 1997, the potassium and chloride concentrations in the seepage water did not increase during vertical seepage.

The eastern Harz foreland is one of the driest areas in Germany; with a net negative precipitation after allowing for evaporation. Nevertheless, seepage water investigations have shown that lateral seepage water dominates at a depth of 2 m or more. The results show that in sloping terrain with large hydrographic gradients there is a more frequent exchange of seepage water than a calculation of the vertical seepage would suggest. Presumably, seepage water is temporarily stored underground in hollows or more permeable zones. This seepage water can be mobilized even by small amounts of newly inflowing water.

The subsurface water flow and subsurface erosion processes in the Langenbogen area are clearly not only determined by the topography of the surface. In loess landscapes in particular, the subterranean material flows are controlled by the former topography, which is covered by loess. The figures above illustrate that subsurface topography is more important for lateral leachate transport than the surface topography. In the sloping terrain, not only do increased above-ground and subterranean erosion processes occur, but also the efficient transport of dissolved substances through seepage water, potentially causing groundwater contamination.

11.7 Land Degradation, Subsurface Erosion and Piping

The foregoing sections have highlighted the complexity of the hydrological context, erosion, and monitoring associated with soil pipes. This complexity increases even further when we consider the environmental management implicates of subsurface erosion and soil piping. Beckedahl (1998) identified five different environmental

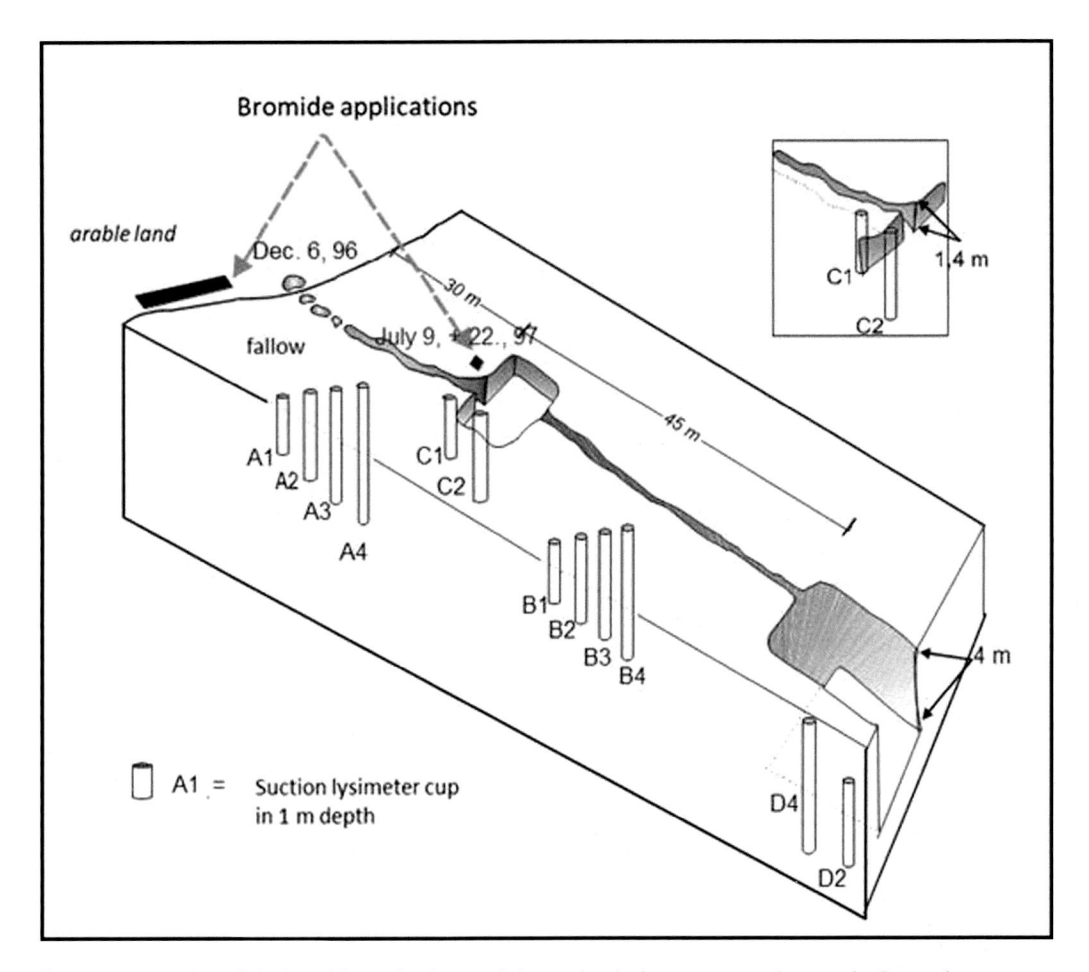

Fig. 11.15 Location of the bromide applications and the suction lysimeter cups at the test plot Langenbogen

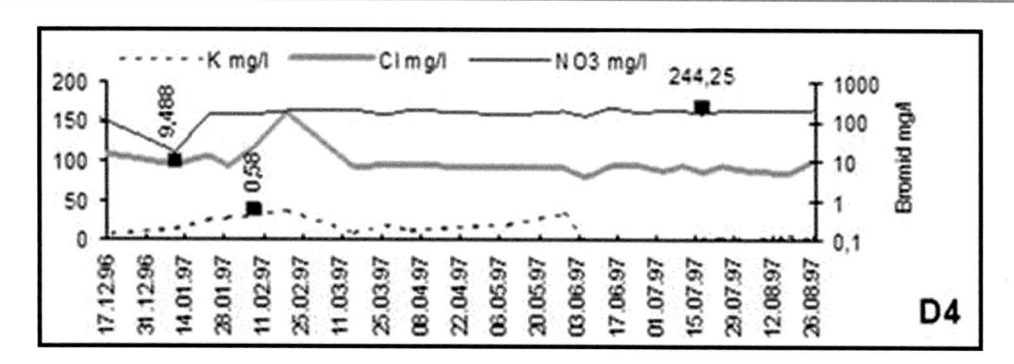

Fig. 11.16 K, Cl, and Br-concentrations (mg/l) in the seepage water of the suction cup D4

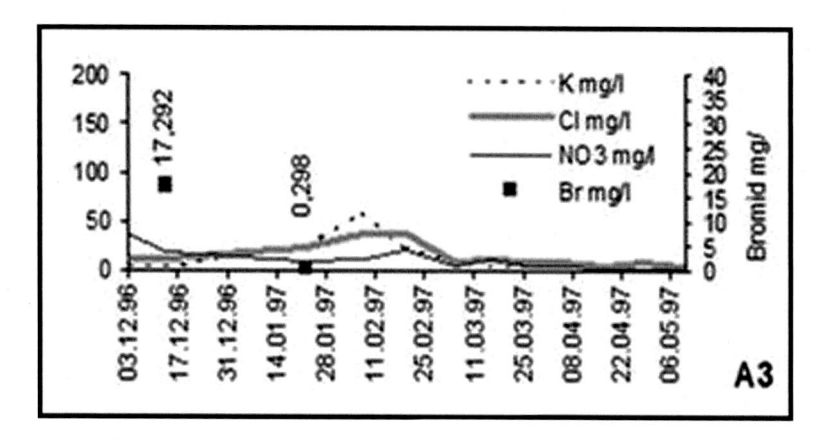

Fig. 11.17 K, Cl, und Br-concentrations (mg/l) in the seepage water of the suction cup A3 (Dec 96–May 97)

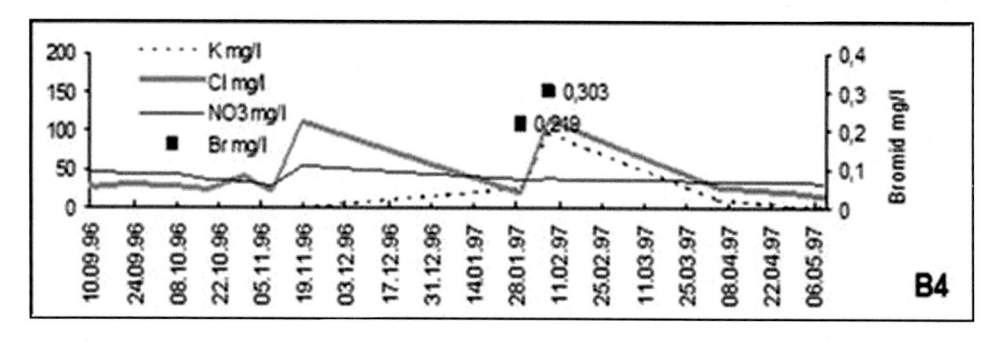

Fig. 11.18 K, Cl, und Br- concentrations (mg/l) in the seepage water of the suction cup B4 (Sept. 96–May 97)

contexts within which piping may develop. Of these the three outlined here (viz. chemistry of the clay mineralogy; hydraulic conditions and suffosion being the most common).

Effective rehabilitation relies on being able to determine and assess the cause of the original degradation (Faulkner 2006; Beckedahl and Gilli 2008; Seutloali et al. 2015; Le Roux and van der Waal 2020) in order to prevent a recurrence of the processes. This is arguably the most important stumbling block, in that much of gully rehabilitation treats all gullies as being essentially the same in origin (Hoffman and Ashwell 2001; Beckedahl 2002; Bork et al. 2003), thereby

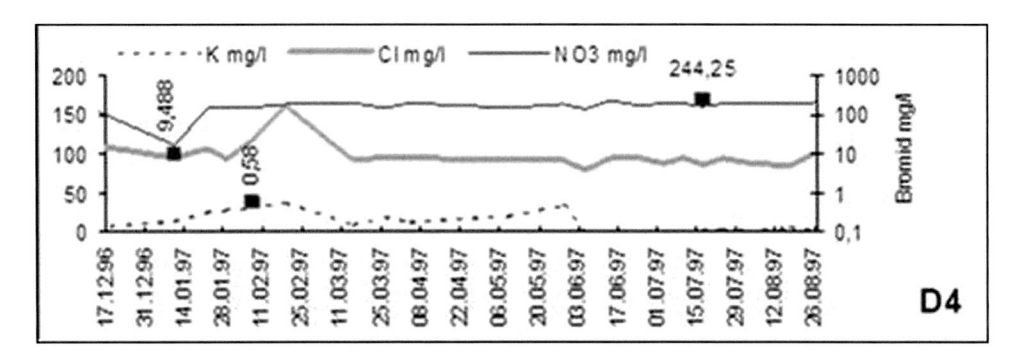

Fig. 11.19 K, Cl, and Br-concentrations (mg/l) in the seepage water of the suction cup D4 (Sept. 96–May 97) K, Cl, and Br-concentrations (mg/l) in the seepage water of the suction cup D4 (Sept. 96–May 97)

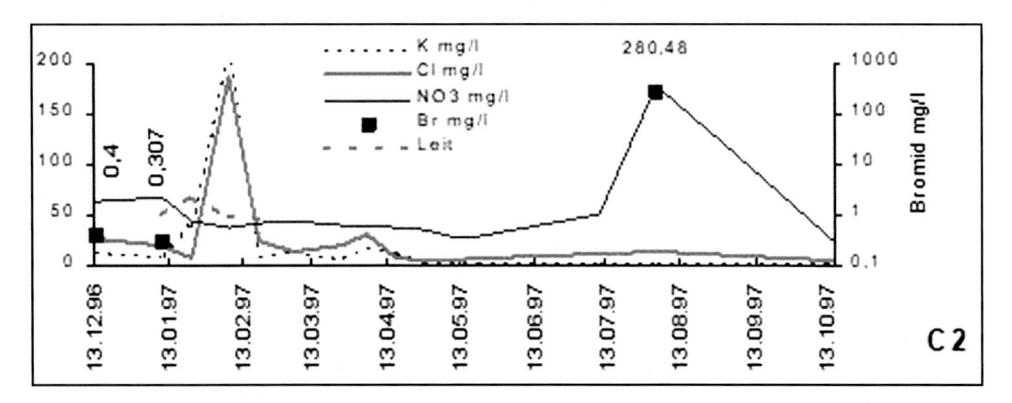

Fig. 11.20 K, Cl, and Br-concentrations (mg/l) in the seepage water of the suction cup C2 (Dec. 96–Oct. 97)

predisposing much of the rehabilitation on dispersive soils to failure, as seen in Fig. 11.21. The challenge associated with installing gabion systems in dispersive soils is that the gabion wall has a temporary damming effect on the water flow down the gully system, as it is intended to have in order to create conditions conducive to low energy water and hence deposition behind the gabion. Unfortunately, the consequence of this is the creation of a positive hydraulic pressure into the gully sidewalls, enhancing the ingress of water into the gully sidewalls. This water will then follow the newly created hydraulic gradient around the gabion wall, and encourage dispersion along the associated percolines and ultimately resulting in a pipe bypassing the gabion (Beckedahl and De Villiers 2001), as shown in Fig. 11.21.

The flow of water along preferred percolines is also argued to be the reason for the short soil pipes existing along gully sidewalls, where water from the surroundings drains down the hydraulic gradient from the surroundings to the pipe or gully bottom (depending on whether or not the roof has already collapsed), instead of the surface flow overtopping the gully sidewall (Beckedahl and Gilli 2010; see also Fig. 11.22).

Under conditions where soils are dispersive as a result of their physical soil properties, or as a result of highly sodic soils with high Exchangeable Sodium Percentages (ESP) and or high Sodium Absorption Ratios (SAR), it is argued that the most effective rehabilitation is through vegetative means (Beckedahl et al. 2020) (Fig. 23a and b).

In this case vetiver grass (*Chrysopogon zizanioides*) was used to stabilize a re-shaped

Fig. 11.21 A failed gabion structure in a gully eroded into a dispersive soil in the Eastern Cape Province, South Africa

gully system in Eswatini. The deep root system of this grass species is being used both to anchor the soil as well as to drain out excess moisture through evapotranspiration. The stabilized soil between the rows of grass are then available for vegetation production. Such systems may, of course, be augmented by planting strategically deep rooted tree varies as well, depending on the

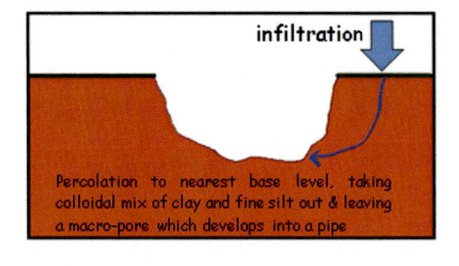

Fig. 11.22 A gully sidewall pipe (left) at Ncise, Eastern Cape Province, South Africa, with a schematic explanation, right

Fig. 11.23 (A) An erosion gully developed in a slightly dispersive soil at Ncganyini, Eswatini. B) The same gully six months later, after it has been reshaped and stabilized using Vetiver grass (*Chrysopogon zizanioides*)

needs of the surrounding communities. The first step would thus be to determine whether the soil associated with a particular gully is dispersive, prior to deciding on a given rehabilitation method.

Acknowledgements J. A. A. Jones would like to thank all those co-workers who kindly updated him by discussing progress and providing recent papers, including Joseph Holden, Jeff McDonnell, Nick Chappell, Jean Poesen, Taro Uchida, Ilja van Meerveld and Tadashi Tanaka. Similarly, the authors wish to thank the many colleagues and students who have contributed to the insights expressed here.

References

Albright JS (1991) Storm hydrograph comparisons of subsurface in a small, forested watershed in northern California. Unpublished M.Sc. thesis, Humboldt State University, Arcata, California, USA

Anderson MG, Burt TP (1990) Subsurface runoff. In: Anderson MG, Burt TP (eds) Process studies in hillslope hydrology. Wiley, Chichester, pp 365–400

Bachmann G, Hoyningen-Huene EV, Reuter F (1963) Über einige Erosionserscheinungen im Löß südlich von Wettin bei Thaldorf (Krs. Hettstedt). Geologie 12 (3):340–348

Beckedahl HR, De Villiers AB (2001) Accelerated erosion by piping in the Eastern Cape Province, South Africa. S Afr Geogr J 82:157–163

Beckedahl HR (2002) Bodenerosion in Africa: ein Überblick. Petermanns Geogr Mitt 146(3):18–25

Beckedahl HR, Gilli A (2008) Problem soils and the rehabilitation of erosion. In: Problem Soils in South Africa—Proceedings of the South African Institute for Engineering and Environmental Geologist's Conference on Problem Soils, 3–4 Nov 2008, Midrand, South Africa, pp 75–79

Beckedahl HR, Gilli A (2010) Adverse soil conditions and the rehabilitation of soil erosion in southern Africa. Geoökodynamik 31:35–44

Beckedahl HR, Mabaso S, Singwane SS, Mamba FS (2020) Community-based gully rehabilitation efforts in the Ezikotheni and Ngcayini Chiefdoms of Eswatini. Report to the Eswatini Environmental Authority, Department of Economic Development and Tourism, Government of Eswatini, Mbabane

Bork H-R, Beckedahl HR, Dahlke C, Geldmacher K, Mieth A, Li Y (2003) Die erdweite Explosion der Bodenerosionsraten im 20.Jh.: Das globale Bodenerosionsdrama – geht unsere Ernährungsgrundlage verloren? Petermanns Geographische Mitteilungen 147 (2):16–23

Benito G, Gutierrez M, Sancho C (1993) The influence of physico-chemical properties on erosion processes in badland areas, Ebro basin, NE-Spain. Z. Geomorph. N. F. 37, Bd. 2:199–214

Bernatek-Jakiel A, Poesen J (2018) Subsurface erosion by soil piping: significance and research needs. Earth Sci Rev 185:1107–1128

Beven KJ (2006) Streamflow generation processes. Benchmark Papers in Hydrology Series No. 1. International Association of Hydrological Sciences, Wallingford, 430pp

Beven KJ, Germann P (1982) Macropores and water flow in soils. Water Resour Res 18(5):1311–1325

Beven KJ, Germann P (2013) Macropores and water flow in soils revisited. Water Resour Res 49:3071–3092

Bidin K (1995) Subsurface flow controls of runoff in a Bornean natural rainforest. Unpublished M.Sc. thesis, University of Manchester, UK

Bryan RB, Harvey LE (1985) Observations on the geomorphic significance of tunnel erosion in a semi-arid ephemeral drainage system. Geographiska Annaler Series A 67:257–273

Bryan RB, Jones JAA (1997) The significance of soil piping processes: inventory and prospect. Geomorphology 20:209–218

Bunting BT (1961) The role of seepage moisture in soil formation, slope development and stream initiation. Am J Sci 259:503–518

Carey SK, Woo M-K (2000) The role of soil pipes as a slope runoff mechanism, Subarctic Yukon, Canada. J Hydrol 233:206–222

Carey SK, Woo M-K (2002) Hydrogeomorphic relations among soil pipes, flow pathways, and soil detachments within a permafrost hillslope. Phys Geogr 23:95–114

Chapman PJ, Reynolds B, Wheater HS (1993) Hydrochemical change along stormwater pathways in a small moorland headwater catchment in mid-Wales, UK. J Hydrol 151:241–265

Chapman PJ, Reynolds B, Wheater HS (1997) Sources and controls of calcium and magnesium in storm runoff: the role of groundwater and ion exchange reactions along water pathways. Hydrol Earth Syst Sci 1:671–685

Chappell NA, Sherlock MD (2005) Contrasting flow pathways within tropical forest slopes of ultisol soils. Earth Surf Proc Land 30:735–753

Colclough JD (1965) Tunnel erosion. Tasman J Agric 36 (1):7–12

Crouch RJ (1976) Field tunnel erosion—a review. J Soil Conserv Serv New South Wales 32(2):98–111

Downes RG (1946) Tunnelling erosion in North-Eastern Victoria. J Commonwealth Sci Ind Res 19(3):283–292

Dunne T (1990) Hydrology, mechanics, and geomorphic implications of erosion by subsurface flow. Special Paper-Geol Soc Am 252:1–28

Elsenbeer H, Lack A, Cassel K (1995) Chemical fingerprints of hydro- logical compartments and pathway characteristics in an Amazonian rainforest catchment. Hydrol Process 14:2367–2381

Elsenbeer H, Vertessy RA (2000) Stormflow generation and flowpath characteristics in an Amazonian rainforest catchment. Hydrol Process 14:2367–2381

Faulkner H (2006) Piping hazard on collapsible and dispersive soils in Europe. In: Boardman J, Poesen J (eds) Soil wrosion in Europe. Wiley, pp 537–562

Faulkner H (2013) Badlands in marl lithologies: a field guide to soil dispersion, subsurface erosion and piping-origin gullies. CATENA 106:42–53

Freeze RA (1974) Stream flow generation. Rev Geophys Space Phys 12(4):627–647

Gee AS, Stoner JH (1989) A review of the causes and effects of acidification of surface waters in Wales and potential mitigating techniques. Arch Environ Contam Toxicol 18:121–130

Gilman K, Newson MD (1980) Soil Pipes and Pipeflow — A Hydrological Study in Upland Wales. Geobooks, Norwich, p 114

Got J-B, Bielders C, Lambot S (2014) Soil piping: detection, hydrological functioning and modelling. In: Lambot S (ed) Proceedings Ph.D. Day ENVITAM, p 7, Louvain-la-Neuve, 05 Mar 2014. http://hdl.handle.net/2078.1/141248

Got J-B (2019) Soil piping: detection, hydrological functioning and modeling. A case study in loess-derived soils in Belgium. Ph.D. thesis, Université Catholique de Louvain, 173pp. http://hdl.handle.net/2078.1/222920

Gregory KJ, Walling DE (1968) The variation of drainage density within a catchment. Bull Int Assoc Sci Hydrol 13(2):61–68

Hardenbicker U (1998) Subterrane Erosion im östlichen Harzvorland. *Zeitschrift für Geomorphologie. Supplementband*, (112):93–103

Hardenbicker U (2006) *Laterale Sickerwasserflüsse als geoökologische und geomorphologische Faktoren in der Landschaftsentwicklung: mit 54 Tabellen*. Inst für Geographie, Universitaet Halle-Wittenberg. ISBN: 386010778X

Hardenbicker U, Hecht C (2000) Bodenphysikalische Prozesse und Ursachen der subterraneo Erosion in Löß und Schwemmlöß im östlichen Harzvorland. *Hercynia-Ökologie und Umwelt in Mitteleuropa*, 33(1):31–41

Hardenbicker U, Liermann R (2000) Erfassung schneller Sickerwasserbewegungern in Löß und Lößderivaten. Jenaer Geographische Schriften Bd. 9:1–11

Hartge KH, Horn R (1992) Die physikalische Untersuchung von Böden. Stuttgart

Hewlett JD, Hibbert AR (1967) Factors affecting the response of small watersheds to precipitation in humid areas. In: Proceedings of the international symposium on forest hydrology (1965), Pennsylvania State University, Pergamon, pp 275–290

Hoffman MT, Ashwell A (2001) Nature divided. Land degradation in South Africa. University of Cape Town Press

Holden J, Smart RP, Chapman PJ, Baird AJ, Billett MF (2009a) The role of natural soil pipes in water and carbon transfer in and from peatlands. In: Baird AJ, Belyea L, Comas X, Reeve A, Slater L (eds) Carbon cycling in Northern Peatlands. American Geophysical Union Monograph, Washington DC, pp 151–164

Holden J, Burt TP, Vilas M (2002) Application of ground-penetrating radar to the identification of subsurface piping in blanket peat. Earth Surf Proc Land 27:235–249

Holden J, Burt TP (2002) Piping and pipeflow in a deep peat catchment. CATENA 48:163–199

Holden J, Evans MG, Burt TP, Horton M (2006) Impact of land drainage on peatland hydrology. J Environ Qual 35:1764–1778

Holden J, Smart RP, Chapman PJ, Baird AJ, Billett MF (2009b) The role of natural soil pipes in water and carbon transfer in and from peatlands. Geophys Monogr Ser 184:1–15

Hölting B, Coldewey WG (1996) Einführung in die allgemeine und angewandte Hydrogeologie. *Enke Verl., Stuttgart*

Hyett GA (1990) The effect of accelerated throughflow on the water yield chemistry under polluted rainfall. Unpublished Ph.D. thesis, University of Wales, Aberystwyth, UK

Jones JAA (1967) Morphology of the Lapworth Valley, Warwickshire. Geogr J 134(2):216–226

Jones JAA (1981a) The nature of soil piping: a review of research, vol 3. Geo Books, Norwich

Jones JAA, Connelly LJ (2002) A semi-distributed simulation model for natural pipeflow. J Hydrol 262 (1–4):28–49

Jones JAA, Cottrell CI (2007) Long-term changes in streambank soil pipes and the effects of afforestation. J Geophys Res 112(F1): F01010-1–F01010-11

Jones JAA, Crane FG (1982) New evidence for rapid interflow contributions to the streamflow hydrograph. Beiträge zur Hydrologie Sonderheft 3:219–232

Jones JAA, Wathern P, Connelly LJ, Richardson JM (1991) Modelling flow in natural soil pipes and its impact on plant ecology in mountain wetlands. In: Nachtnebel P (ed) Hydrological basis of ecologically sound management of soil and groundwater, vol 202. International Association of Hydrological Sciences Publication, pp 131–142

Jones JAA (1971) Soil piping and stream channel initiation. Water Resour Res 7(3):602–610. Reprinted with commentary in Beven KJ (ed) (2006) Streamflow generation processes. Benchmark papers in hydrology, Series No. 1. International Association of Hydrological Sciences Publications, Wallingford, pp 224–232

Jones JAA (1975) Soil piping and the subsurface initiation of stream channel networks. Unpublished Ph.D. thesis, University of Cambridge, Cambridge, UK, p 467

Jones JAA (1978) Soil pipe networks: distribution and discharge. Cambria 5(1):1–21

Jones JAA (1979) Extending the Hewlett model of stream runoff generation. Area 11(2):110–114

Jones JAA (1981b) The nature of soil piping: a review of research. British Geomorphological Research Group Research Monograph 3. GeoBooks, Norwich, 301pp

Jones JAA (1986) Some limitations to the a/s index for predicting basin- wide patterns of soil water drainage. Z Geomorphol 60(Supplement):7–20

Jones JAA (1987) The effects of soil piping on contributing areas and erosion patterns. Earth Surf Proc Land 12(3):229–248

Jones JAA (1988) Modelling pipeflow contributions to stream runoff. Hydrol Process 2:1–17

Jones JAA (1990) Piping effects in humid lands. In: Higgins CG, Coates DR (eds) Groundwater geomorphology: the role of subsurface water in earth-surface processes and landforms. Geological Society of America, Boulder, Special Paper 252, pp. 111–138

Jones JAA (1994) Soil piping and its hydrogeomorphic function. Cuaternario y Geomorfología 8(3–4):77–102

Jones JAA (1997a) Pipeflow contributing areas and runoff response. Hydrol Process 11(1):35–41

Jones JAA (1997b) The role of natural pipeflow in dynamic contributing areas and hillslope erosion: extrapolating from the Maesnant data. Phys Chem Earth 22(3–4):303–308

Jones JAA (1997c) Subsurface flow and subsurface erosion: further evidence on forms and controls. In: Stoddart DR (ed) Process and form in geomorphology. Routledge, London, pp 74–120

Jones JAA (2004) Implications of natural soil piping for basin management in the British uplands. Land Degrad Dev 15(3):325–349

Jones JAA (2010) Soil piping and catchment response. Hydrol Process 24:1548–1566

Kinner DA, Stallard RF (2004) Identifying storm flow pathways in a rainforest catchment using hydrologic and geochemical modelling. J Hydrol 308:67–80

Kirkby MJ, Chorley RJ (1967) Throughflow, overland flow and erosion. Bull Int Assoc Sci Hydrol 12(3):5–21

Kitahara H, Nakai Y (1992) Relationship of pipe flow in to streamflow on a first order watershed. J Jap Forest Soc 74:49–52

Kitahara H, Terajima T, Nakai Y (1994) Ratio of pipe flow to through flow. J Jap Forest Soc 76:10–17

Kitahara H (1989) Characteristics of pipe flow in a subsurface soil layer on a gentle slope (II) Hydraulic properties of pipes. J Jap Forest Soc 70:317–322

Kitahara H (1994) A study on the characteristics of soil pipes influencing water movement in forested slopes. Bull Forest Forest Prod Res Inst 367:63–115

Koyama K, Okumura T (2002) Process of pipeflow runoff with twice increase in discharge for a rainstorm. Trans Jap Geomorphol Union 23:561–584

Laffan MD, Cutler EJB (1977) Landscape, soils, and erosion of a catchment the wither Hills, Marlborough, NZ. J Sci 20:279–289

Le Roux JJ, van der Waal B (2020) Gully susceptibility modelling to support avoided degradation planning. S Afr Geogr J 102(3):406–420. https://doi.org/10.1080/03736245.2020.1786444

McDonnell JJ, Sivapalan M, Vaché K, Dunn S, Grant G, Hagerty R, Hinz C, Hooper R, Kirchner J, Roderick ML, Selker J, Weiler M (2007) Moving beyond heterogeneity and process complexity: a new vision for watershed hydrology. Water Resour Res 43: W07301

McDonnell JJ (1990) A rationale for old water discharge through macropores in a steep, humid catchment. Water Resour Res 26:2821–2832

McGlynn BL, McDonnell JJ, Brammer DD (2002) A review of the evolving perceptual model of hillslope pathways at the Maimai catchment. J Hydrol 257:1–26

McGuire KJ, McDonnell JJ (2010) Hydrological connectivity of hillslopes and streams: characteristic time scales and nonlinearities. Water Resour Res 46: W10543. https://doi.org/10.029/2010WR009341

Midgley TL, Fox GA, Wilson GV, Felice R, Heeren D (2013) In situ soil pipeflow experiments on contrasting streambank soils. Trans ASABE 56(2):479–488

Mizuyama T, Sato I, Kosugi K (1994) Pipe flow and pipe distribution at Tohinotani basin in Ashu experimental forest. Bull Kyoto Univ Forest 66:48–60 (Japanese with English summary)

Nieber JL, Warner GS (1991) Soil pipe contribution to steady subsurface stormflow. Hydrol Process 5 (4):329–344

Nieber JL (2001) The relationship of preferential flow to water quality and its theoretical and experimental quantification. In: Bosch DD, King KW (eds) Preferential flow: water movement and chemical transport in the environment. American Society of Agricultural and Biological Engineers, St. Joseph, pp 1–10

Nieber JL (2006) Lateral preferential flow on hillslopes through pathways formed by biological and mechanical processes. Paper presented at the Biohydrology Conference, Prague. Available at: http://147Ð213Ð145Ð2/biohydrology/abstracts/NieberPlen.pdf

Noguchi S, Tsuboyama Y, Sidle RC, Hosoda I (1999) Morphological characteristics of macropores and the distribution of preferential flow pathways in a forested slope segment. Soil Sci Soc Am J 63:1413–1423

Ohta T, Noguchi H, Tsukamoto Y (1983) Study on the behavior of storm water in a small forested watershed (IV)—role of pipeflow on runoff generation process of storm water. In: Transaction of the 93rd meeting of the Japanese Forestry Society, pp 459–461

Putty MRY, Prasad R (2000) Runoff processes in headwater catchments—an experimental study in Western Ghats, South India. J Hydrol 235:63–71

Regensburg TH, Chapman PJ, Pilkington MG, Chandler DM, Evans MG, Holden J (2020) Controls on the spatial distribution of natural pipe outlets in heavily degraded blanket peat. Geomorphology 367:107322

Regensburg TH, Chapman PJ, Pilkington M, Chandler D, Evans MG, Holden J (2021) Effects of pipe outlet blocking on hydrological functioning in a degraded blanket peatland. Hydrol Process

Rogers CDF, Dijkstra TA, Smalley IJ (1994) Hydroconsolidation and subsidence of loess: studies from China, Russia, North America and Europe: in memory of Jan Sajgalik. Eng Geol 37(2):83–113

Roberge J, Plamondon AP (1987) Snowmelt runoff pathways in a boreal forest hillslope, the role of pipe throughflow. J Hydrol 95:39–54

Sayer AM, Walsh RPD, Bidin K (2004) Pipeflow suspended sediment dynamics and their contribution to stream sediment budgets in small rainforest catchments, Sabah, Malaysia. In: Sidle RC, Tani M, Nik AR, Taddese TA (eds) Forests and water in warm, humid Asia, Proceedings of a IUFRO forest hydrology workshop. Disaster Prevention Research Institute, Uiji, pp 170–173

Sayer AM, Walsh RPD, Bidin K (2006) Pipeflow suspended sediment dynamics and their contribution to stream sediment budgets in small rainforest catchments, Sabah, Malaysia. For Ecol Manage 224:119–130

Seutloali KE, Beckedahl HR, Dube T, Sibanda S (2015) An assessment of gully erosion along major armoured roads in south-eastern region of South Africa: a remote sensing and GIS approach. Geocarto Int 10:1–24. https://doi.org/10.1080/10106049.2015.1047412

Sherard JL (1953) Influence of soil properties and construction methods on performance of homogeneous earth dams. US Department of Interior Bureau of Reclamation, Design and Construction Division Technical Memo No. 645

Scheidig A (1934) Der Löß und seine geotechnischen Eigenschaften. Dresden u, Leipzig

Sidle RC, Kitahara H, Terajima T, Nakai Y (1995) Experimental studies on the effects of pipeflow and throughflow partitioning. J Hydrol 165:207–219

Sklash MG, Beven KJ, Gilman K, Darling WG (1996) Isotope studies of pipeflow at Plynlimon, Wales. Hydrol Process 10(7):921–944

Smart RP, Holden J, Dinsmore K, Baird AJ, Billett MF, Chapman PJ, Grayson R (2013) The dynamics of natural pipe hydrological behaviour in blanket peat. Hydrol Process 27:1523–1534

Swanson ML, Kondolf GM, Boison PJ (1989) An example of rapid gully initiation and extension by subsurface erosion: coastal San Mateo County, California. Geomorphology 2:393–403

Swanson ML (1983) Soil piping and gully erosion along the San Mateo County Coast in central California. In: Proceedings from the second field conference of the American Geomorphological Research Field Group, Chaco Canyon, New Mexico, pp 7–10

Tanaka T, Yasuhara H, Sakai H, Marui A (1988) The Hachioji Experimental Basin study—storm processes and the mechanisms of its generation. J Hydrol 102:139–164

Tanaka T (1982) The role of subsurface water exfiltration in soil erosion processes. Int Assoc Sci Hydrol Publ 137:73–80

Terajima T, Kitahara H, Sakamoto T, Nakai Y, Kitamura K (1996a) Pipe flow significance on subsurface discharge from the valleyhead of a small watershed. J Jap Forest Soc 78:20–28 (in Japanese with English summary)

Terajima T, Sakamoto T, Nakai Y, Kitamura K (1996b) Subsurface discharge and suspended sediment yield interactions in a valley head of a small forested watershed. J Res 1:131–137

Terajima T, Sakamoto T, Nakai Y, Kitamura K (1997) Suspended sediment discharge in subsurface flow from the head hollow of a small forested watershed, Northern Japan. Earth Surf Proc Land 22:987–1000

Terajima T, Sakamoto T, Shirai T (2000) Morphology, structure and flow phases in soil pipes developing in forested hillslopes underlain by a quaternary sand-gravel formation, Hokkaido, Northern Main Island, Japan. Hydrol Process 14:713–726

Terzaghi K (1931) Earth slips and subsidences from underground erosion. Eng News Record 107:90–92

Tromp-van Meerveld I, Weiler M (2008) Hillslope dynamics modeled with increasing complexity. J Hydrol 361(1–2):24–40

Tsukamoto Y, Ohta T, Nogushi H (1982) Hydrological and geomorphological studies of debris slides on forested hillslopes in Japan. Int Assoc Hydrol Sci Publ 137:89–98

Uchida T, Kosugi K, Mizuyama T (1999) Runoff characteristics of pipeflow and effects of pipeflow on rainfall-runoff phenomena in a mountainous watershed. J Hydrol 222:18–36

Uchida T, Kosugi K, Mizuyama T (2001) Effects of pipeflow on hydrological process and its relation to landslide: a review of pipeflow studies in forested headwater catchments. Hydrol Process 15:2151–2174

Uchida T, Tromp-van Meerveld I, McDonnell JJ (2005) The role of lateral pipe flow in hillslope runoff response: an intercomparison of nonlinear hillslope response. J Hydrol 311:117–133

Uchida T (2000) Effects of pipeflow on storm runoff generation processes at forested headwater catchments. Unpublished Ph.D. thesis, Kyoto University, Japan, 118pp

van Meerveld HJ, Sauquet E, Gallart F, Sefton C, Seibert J, Bishop K (2020) Aqua temporaria incognita. Hydrol Process 34:5704–5711

Vannoppen W, Verachtart E, Poesen J (2016) Pipeflow response in loess-derived soils to precipitation and groundwater table fluctuations in a temperate humid climate: Pipeflow response to precipitation and groundwater table fluctuations. Hydrol Process 31(3)

Vannoppen W, Verachtert E, Poesen J (2017) Pipeflow response in loess-derived soils to precipitation and groundwater table fluctuations in a temperate humid climate. Hydrol Process 32:586–596

Verachtart E (2011) Soil piping in a temperate humid climate: the Flemish Ardennes (Belgium). Unpublished D.Sc. thesis, Catholic University of Louvain

Verachtert E, Van Den Eeckhaut M, Martínez-Murillo JF, Nadal-Romero E, Poesen J, Devoldere S, Wijnants N, Deckers J (2013) Impact of soil characteristics and land use on pipe erosion in a temperate humid climate: field studies in Belgium. Geomorphology 192:1–14

Verachtert E, Van Den Eeckhaut M, Poesen J, Deckers J (2010) Factors controlling the spatial distribution of soil piping erosion on loess-derived soils: a case study from central Belgium. Geomorphology 118:339–348

Verachtert E, Maetens W, Van Den Eeckhaut M, Poesen J, Deckers J (2011) Soil loss rates due to piping erosion. Earth Surf Proc Land 36:1715–1725

Weiler M, McDonnell JJ (2007) Conceptualizing lateral preferential flow and flow networks and simulating the effects on gauged and ungauged hillslopes. Water Resour Res 43:W03403. https://doi.org/10.1029/2006WR004867

Weyman DR (1970) Throughflow on hillslopes and its relation to the stream hydrograph. Bull Int Assoc Sci Hydrol 15(2):25–33

Weyman DR (1974) Runoff processes and contributing area and streamflow in a small upland catchment. In: Gregory KJ, Walling DE (eds) Fluvial processes in instrumented watersheds, vol 6. British Geomorphological Research Group Special Publication, London, pp 1433–1443

Whipkey RZ (1965) Subsurface stormflow from forested slopes. Bull Int Assoc Sci Hydrol 10(3):74–85

Wilson GV (2011) Understanding soil-pipe flow and its role in ephemeral gully erosion. Hydrol Process 25 (15):2354–2364

Wilson GV, Rigby JR Jr, Ursic ME, Dabney SM (2016) Soil pipe flow tracer experiments: 1. Connectivity and transport characteristics. Hydrol Process 30(8):1265–1279

Woo M-K, diCenzo P (1988) Pipe flow in James Bay wetlands. Can J Earth Sci 25:625–629

Xu X, Wilson GV, Zheng F, Tang Q (2020) The role of soil pipe and pipeflow in headcut migration processes in loessic soils. Earth Surf Proc Land 45:1749–1763

Yasuhara M (1980) Streamflow generation in a small forested watershed. Unpublished M.Sc. thesis, University of Tsukuba, p 55

Zhu TX, Luk SH, Cai QG (2002) Tunnel erosion and sediment production in the hilly loess region, North China. J Hydrol 257:78–90

Zhu TX (1997) Deep-seated, complex tunnel systems — a hydrological study in a semi-arid catchment, Loess Plateau, China. Geomorphology 20(3–4):255–267

Ziemer RR, Albright JS (1987) Subsurface pipeflow dynamics of north- coastal California swale systems. In: Beschta R, Blinn T, Grant GE, Swanson FJ, Ice GG (eds) Erosion and sedimentation in the Pacific Rim, Proceedings of the Corvallis Symposium. IAHS Publ. no. 165, pp 71–80

Ziemer RR (1992) Effect of logging on subsurface pipeflow and erosion: coastal northern California, USA in Erosion, Debris Flows and Environment in Mountain Regions (Proceedings of the Chengdu Symposium, July 1992), IAHS Publ. no. 209, pp 187–197

Role of LU and LC Types on the Spatial Distribution of Arsenic-Contaminated Tube Wells of Purbasthali I and II Blocks of Burdwan District, West Bengal, India

Sunam Chatterjee, Srimanta Gupta, Bidyut Saha, and Biplab Biswas

Abstract

Earlier works suggest that the floodplain geomorphology has a close association with arsenic (As) contaminated tube wells. However, the association of As-contaminated tube wells and the Land Use Land Cover (LULC) categories is not clear. The focus area of the present study encompasses the Purbasthali I and II block, situated on the west bank of the Bhagirathi River. The study uses the 'average weight score' method to investigate the association between LULC categories with the distribution As-contaminated tube wells. The tube well data has been obtained from the Public Health Engineering Department (PHED), Govt. of West Bengal. IRS, ID: LISS IV (Mx) image of March 2019 is used and updated with recent time Google image of 2019. The contaminated tube wells are overlaid on the LULC map and a buffer circle is configured along each tube well. It is found that they are concentrating within the built-up class. So, within the built-up five major categories are identified, viz. Class-1 denotes a tube well buffer area with the dominance of agricultural land; Class-2 refers to tube well buffer area with the dominance of built-up class; Class-3 represents buffer area with a dominance of water bodies; Class-4 and Class-5 refer to tube well buffer area with a dominance of marshy land and buffer area with a dominance of other LULC classes respectively. Based on the 'weighted on evidence' (WOE) method scores are assigned to the tube wells as per their level of arsenic contamination. +0.5 weight is assigned to mildly contaminated wells (>0.010 to 0.049 mg l^{-1}), and +1.0 weights are assigned to the highly contaminated wells i.e. ≥ 0.05 mg l^{-1}. The average weighted score (C_i) is calculated and normalized with the total weighted value for all contaminated wells. GIS-based WOE model indicates that maximum contaminated wells are situated closely to reframed LULC Class 3 and 4. This signifies that tube wells near the water bodies and marshy lands are prone to arsenic contamination.

Keywords

Arsenic · LU and LC · IRS ID · ILWIS · ArcGIS · Weight score · WOE model

S. Chatterjee · B. Biswas (✉)
Department of Geography, The University of Burdwan, Golapbag, Burdwan, West Bengal 713104, India
e-mail: bbiswas@geo.buruniv.ac.in

S. Gupta
Department of Environmental Science, The University of Burdwan, Golapbag, Burdwan, West Bengal 713104, India

B. Saha
Department of Chemistry, The University of Burdwan, Golapbag, Burdwan, West Bengal 713104, India

12.1 Introduction

Groundwater arsenic contamination is a global environmental problem that affects millions of people around the world (BGS, DPHE 2001; Ravenscroft et al. 2009). It has been reported in different countries such as Bangladesh, India, Pakistan, Nepal, China, Hungary, Vietnam, Thailand, Cambodia, Taiwan, Inner Mongolia, Ghana, Egypt, Japan, Argentina, Mexico, USA, Chile, etc. (Mandal and Suzuki 2002; Ravenscroft et al. 2009). But the situation of Ganga Delta Plain (GDP) is worse (Chetia et al. 2011). The concentration of arsenic in groundwater above 0.01 mg l^{-1} (WHO recommended value) is reported from several places of the Ganga floodplain and the deltaic plain as well (Chakroborti et al. 2018). The shallow aquifer arsenic distribution in GDP varies greatly over short vertical depth and horizontal distances as well (BGS 2001; Neumann et al. 2010). The distributional heterogeneity of dissolved arsenic in the Ganges Delta has puzzled researchers around the world (Neumann et al. 2010). Several pieces of research have been attempted to explain the release process and mobilization mechanism (McArthur et al. 2001; Akai et al. 2004; Islam et al. 2004; Jessen et al. 2008). But, still, scientists are struggling to figure out the micro-level heterogeneity of arsenic contamination. Recent researches believe that arsenic release in earth samples in the absence of oxygen and organic matter plays a vital role in this regard. (McArthur et al. 2004; Reza et al. 2010; Lawati 2012; Tareq et al. 2013). Microbial mediated oxidation of organic carbon is thought to drive the geochemical transformations that release arsenic from sediments. Recharge from ponds carries this degradable organic carbon into the shallow aquifer and groundwater flow, drawn by irrigation pumping, transports pond water to the depth where dissolved arsenic concentrations are greatest (Neumann et al. 2010).

The research in Ganga Delta Plain suggests that human activities like irrigation pumping, excavation of ponds and natural floodplain swamps or standing water bodies (Bills) have influenced the aquifer biogeochemistry and the pattern of arsenic contamination in shallow aquifers (Neumann et al. 2010). It was assumed that the human activities (mode of using the land resource) and surface coverage (natural or anthropogenic) have a profound impact on subsurface or near-surface natural processes of water recharge or discharge (Winter et al. 1998).

The spatial patterns of dissolved arsenic observed at a variety of sites have not been explained yet regarding land use and land cover (LULC) distribution. In the present work, attention has been paid to focus on the spatial association between groundwater arsenic contamination and land use land cover distribution.

12.2 Regional Setting

Physiographically the west bank of the Bhagirathi River is a part of the vast alluvial tract comprising Damodar Fan and Bhagirathi floodplain. Lateritic undulated plain border of the region in the west and gradually stepped into the floodplains of Bhagirathi River. The river itself frames the eastern boundary of the region. River Bhagirathi, one of the major distributaries of River Ganga is draining the region from north to south, following the regional slope. Its meandering nature gives rise to versatility in the landscape. Khari River is another active channel running from the west to meet the River Bhagirathi in the east near *Kalna*. Bhagirathi River and proto-Damodar spill channels jointly form the floodplain. The Holocene sedimentation configures the alluvial stratigraphy of the region. Two major Holocene surfaces are present here: Older Alluvial Plain (OAP) and Younger Deltaic Plain (YDP). The younger deposit consists of sand silt and clay with a fining upward sequence. Groundwater arsenic contamination has been reported from several parts of the Bhagirathi floodplain (Mukherjee et al. 2010).

The Purbasthali I and II blocks are chosen as the study area for present research. Geographically the blocks are located on the western bank of the Bhagirathi River; more precisely on the opposite side of the Bhagirathi–Jalangi confluence. The groundwater lying underneath is severely contaminated with arsenic. The arsenic concentration value ranges between 0.006 and 0.618 mg/l. The people are using arsenic-contaminated groundwater for drinking, domestic and irrigation purposes. Patients with arsenical symptoms are also found from some of the contaminated villages of the blocks. Many researchers have already reported about the presence of arsenic in groundwater in the eastern bank of the Bhagirathi River; but few have focused on the west bank of the river. In the west bank of the Bhagirathi River arsenic contamination has been reported in the groundwater from the district of Purba Barddhaman, Hooghly and Howrah so far. The Purbasthali I and II blocks are the most arsenic-contaminated block in the Purba Barddhaman district.. The spatial extension of the study area varies from 23° 20′ N to 23° 35′ N and 88° 10′ E to 88° 25′ E approximately. Administratively the study area encompasses about 340.91 km^2 (CoI 2011) and comprises about 5.43% of the district population (CoI 2011).

12.3 Material and Methods

Arsenic contamination in the western bank of Bhagirathi floodplains and its spatial association with land use and land cover are studied well by using geospatial technology. Overlay analysis of arsenic-contaminated tube well sources and classified satellite image of 2019 is made. The weight on evidence approach is adopted to analyze the probability of arsenic-contaminated tube wells in different land use and land cover class (Fig. 12.1).

12.3.1 GPS Point Collection

The comprehensive tube well arsenic contamination data are collected from the Public Health Engineering Department (PHED), surveyed in 2006 and have been practiced for the present research work. Its exact coordinate-based location of tube wells is collected through handheld GPS (Garmin etrex-30) and then transferred to the GIS environment (ArcGIS version 10.2) for assessing spatial distribution. The tube well arsenic concentration values are grouped into three major classes, viz. safe tube wells (As concentration value ≤ 0.010 mg l^{-1}), mildly contaminated tube wells (As concentration value 0.011–0.049 mg l^{-1}) and highly contaminated tube wells (As concentration value ≥ 0.050 mg l^{-1}).

12.3.2 Satellite Image Processing and Accuracy Assessment

Remote sensing technology has close compatibility with GIS technology that suits present research objectives. The recent Land Use and Land Cover (LULC) map has been classified from the remotely sensed IRS, ID; LISS-IV (Mx) image of March 2019 and updated with the Google image of 2019. The supervised LU and LC classification technique has been adopted for this purpose. A six-fold classification system is implied for the research. NRC (National Remote Sensing Centre) LULC classification system (Level-1) is used with minor modifications, keeping in mind the study area and to attain the research goal.

12.3.3 Spatial Analysis

In the present work, the prime objective is to explore the spatial association between land use and land cover and tube well locations along with their arsenic contamination value. So for assessing the coexistence between the arsenic-contaminated tube wells and land use land cover units; the co-ordinate locations of tube wells are overlaid over the recently classified image, 2019. GIS analysis found that tube wells are occurring at clusters within the built-up areas or in proximity to it. The majority of them are installed to facilitate the daily needs of the public. To reveal the special association further between the contaminated tube well and land

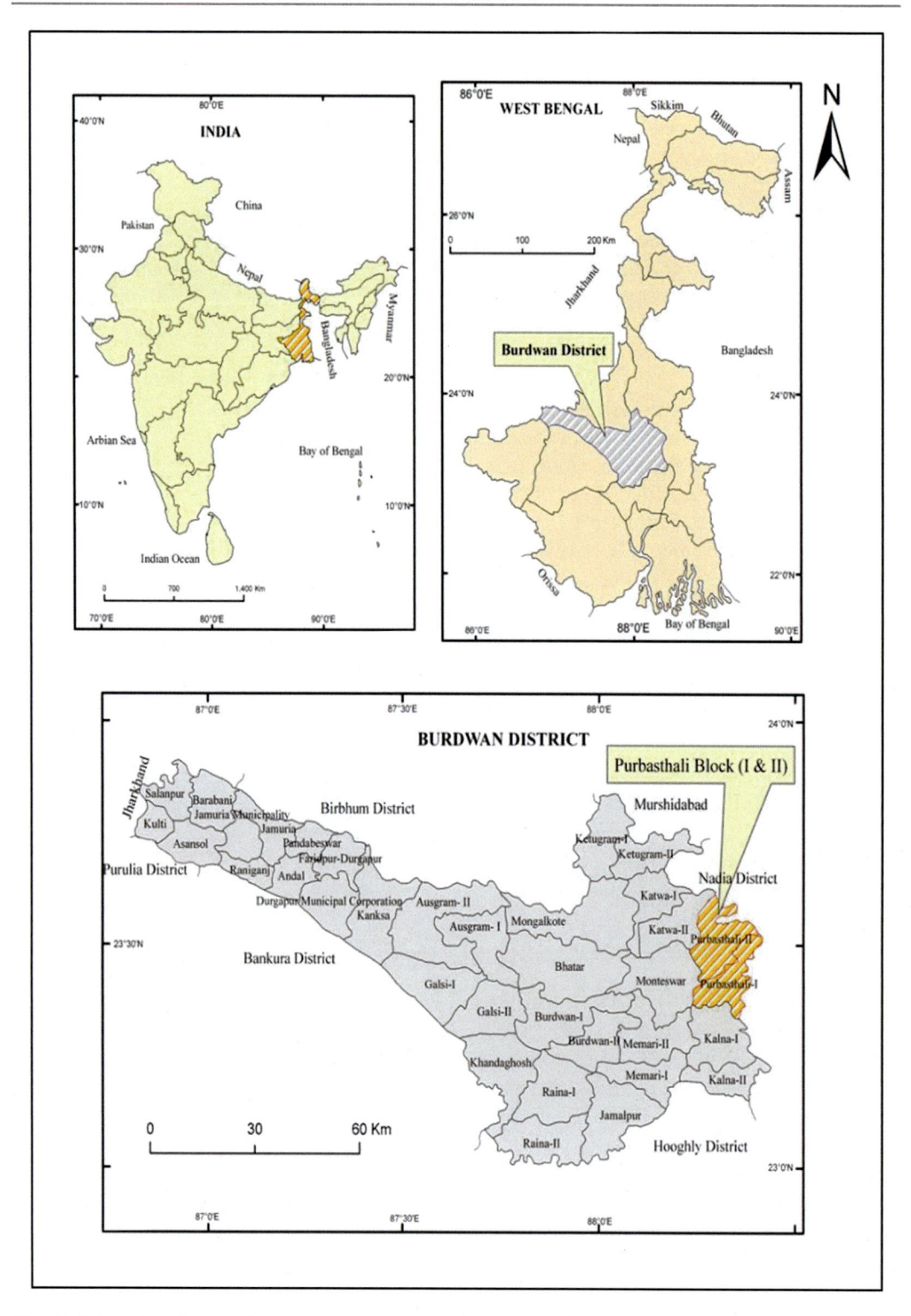

Fig. 12.1 Location of the study area

use, land cover; the researchers are planning to look into the built-up class and assume a 100-m buffer zone around each tube well. During the research, it is observed that the average depth of the tube wells in the study area is 66 m. The nearest three-digit estimated value is 100. So, the 100-m cube has been assumed as a single unit, within which the tube well collects water to supply the people's needs. Therefore, in the present study, a buffer circle of 100 m radius has been imagined centering each tube-well point.

The working hypothesis behind the research is that each buffer area around the tube wells is dominated by one or two land use and land cover class and these have a significant role in the distribution of groundwater arsenic. The dominance of land use, land cover units within the buffer zone is estimated in the GIS environment. The dominance is measured by the maximum area coverage by distinct LU and LC units within the tube well buffers. Depending on the dominance of land use and land cover units five categories of tube well buffers are identified. These are class 1: denotes tube well buffer dominated by agricultural land; class 2: refers to buffer area with a dominance of built-up land; class 3: represents buffer area with significant water bodies; class 4: and class 5: buffer area is giving proxies for marshy land and other land use and land cover class dominance respectively. Based on the weight on evidence approach binary weights are assigned to the mild and highly contaminated tube wells.

Tube wells below WHO limits are treated as safe wells and thus no weight is assigned to them. Mildly contaminated tube wells with arsenic concentration values ranging between 0.011 and 0.049 mg l^{-1} are assigned with +0.5 weights and highly contaminated tube wells (with As concentration 0.05 mg l^{-1}) are assigned with +1.0 weights. A cumulative weight for each buffer class with distinctive land use land cover dominance is calculated by multiplying the mild and high contaminated tube well count with respective weights. Finally, the weighted average score C_i for each buffer class is calculated by using the following algorithm.

$$Ci = \frac{(w1xi + w2yi)}{(w1 + w2)} \quad (12.1)$$

where, the number of, 'mildly contaminated tube wells' for the class C_i will be x_i and the number of 'high contaminated tube wells' for the same class will be y_i. and w_1 and w_2 be the weight of the categories 'mild contaminated tube wells' and 'high contaminated tube wells' respectively.

Finally, the average weighted score (C_i) is calculated and normalized with the total weighted value for all contaminated wells.

12.4 Results

The aqueous distribution of arsenic in the study region tube wells is seemingly yielding water containing low to high levels of arsenic more-or-less randomly, without sketching a simple spatial pattern. So, the situation rose to investigate further regarding the cause of such an intricate distribution pattern. In this regard, it is urgent to judge the role of spatial controls. In the present study, the spatial association between the land use and land cover map and arsenic-contaminated tube wells are made. Before that, it is important to know about recent land use and land cover classification and mapping.

12.4.1 Existing LULC Classification and Mapping

Recent land use and land cover mapping identify six major LU and LC classes, viz. agricultural land, built-up area, marshy land, vegetation cover, water bodies and fallow/current fallow land. Agriculture is the dominant land use class and covers about 54% of the total block administrative area. The built-up area is the second-largest land cover unit and encompasses about 17.38% area. Marshy land is the next dominant land cover class in the poorly drained study region. It covers about 12.46% area. The riverine landscape and its meandering architecture give rise to a significant amount of water bodies and cover about 10.55% area. The natural vegetation patches are severely degraded (Bora 2011). The study reveals about 2.31% of the area belongs to this category. The detailed class wise LU and LC and their areal and spatial distribution are depicted in Table 12.1 and Fig. 12.2, respectively.

12.4.2 Spatial Distribution of Arsenic-Contaminated Tube Wells

The spatial distribution of arsenic-contaminated tube wells in the shallow aquifers of Purbasthali Blocks has been studied by considering $N = 2766$ samples. The tube wells' arsenic concentration values are classified into three categories, viz. safe, mildly contaminated and highly contaminated. The tube wells yielding water with arsenic levels ≤ 0.01 mg l^{-1} are designated as safe; tube wells arsenic concentration value greater than WHO maximum permissible limit, i.e. 0.01 mg/l and less than 0.05 mg l^{-1} are taken into consideration as mild contaminated; while tube wells having dissolved arsenic concentration at or above 0.05 mg l^{-1} are designated as highly contaminated. The analysis shows about 13.34% of the total tube wells are mildly contaminated and 12.90% are highly contaminated while the remaining 73.76% are safe. The detailed class wise frequency distribution of the number of tube-wells with ranges of arsenic content is presented in Fig. 12.3. Finally, the coordinate of tube wells with Arsenic concentration above WHO permissible i.e. 0.010 mg l^{-1} was brought into the GIS environment and represented in Fig. 4a to show the spatial distribution of arsenic-contaminated tube wells.

The regional distribution of arsenic contamination has been mapped by considering the WHO permissible standard. The number of tube wells above WHO maximum permissible limit (i.e. 0.01 mg/l) is counted and the percentage value of arsenic-contaminated wells (≥ 0.01 mg/l) to the total sampled tube wells are calculated and classified into seven classes and illustrated in Fig. 4b. Furthermore, it is important to mention that about 70.85% of the total population and 63.63% of the total area are at risk of arsenic contamination (considering the WHO standards).

12.4.3 Relation Between LULC and Arsenic Contamination

The arsenic distribution map can be represented well at cadastral scale (16 inches of the map represents a 1-mile distance of actual earth surface) but to make the correlative study with LU and LC distribution and tube well arsenic concentration, it is planned to continue the work in an intermediate scale of 1:50,000. In Fig. 12.5, the arsenic-contaminated tube well distribution map and the recent LULC map of 2019 are aligned side by side for understanding the distribution pattern of tube well arsenic contamination and LU and LC types and their underlying relations. It is found that almost all the tube wells ≥ 0.011 mg/L are located although the study area except a narrow strip lying in the western

Table 12.1 Recent land use land covers distribution, 2019

LULC categories	Area in hectare	Area in %
Agriculture	18,562.55	54.45
Natural vegetation	787.5	2.31
Built-up	5926.04	17.383
Marshy	4246.72	12.457
Waterbody	3596.6	10.55
Fallow/current fallow	971.59	2.85
Total	34,091	100

Source IRS-ID, LISS- IV (Mx), Image of March 2019 and Author's calculation

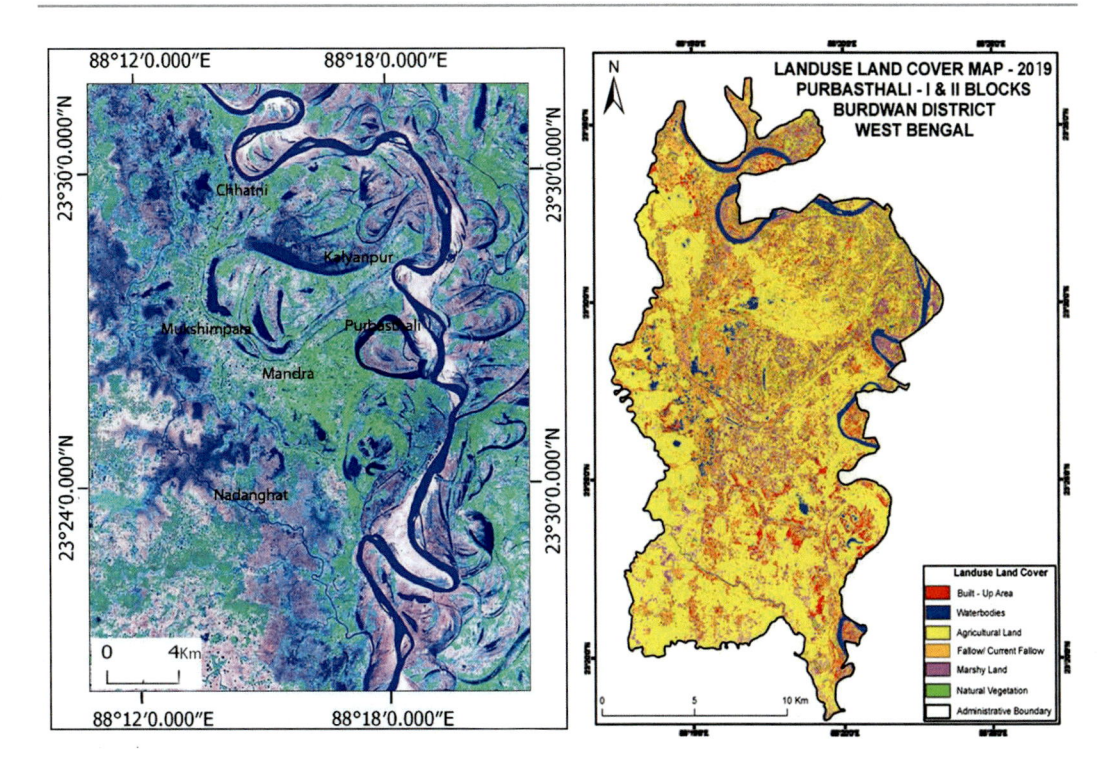

Fig. 12.2 Landscape overview with the LU and LC map

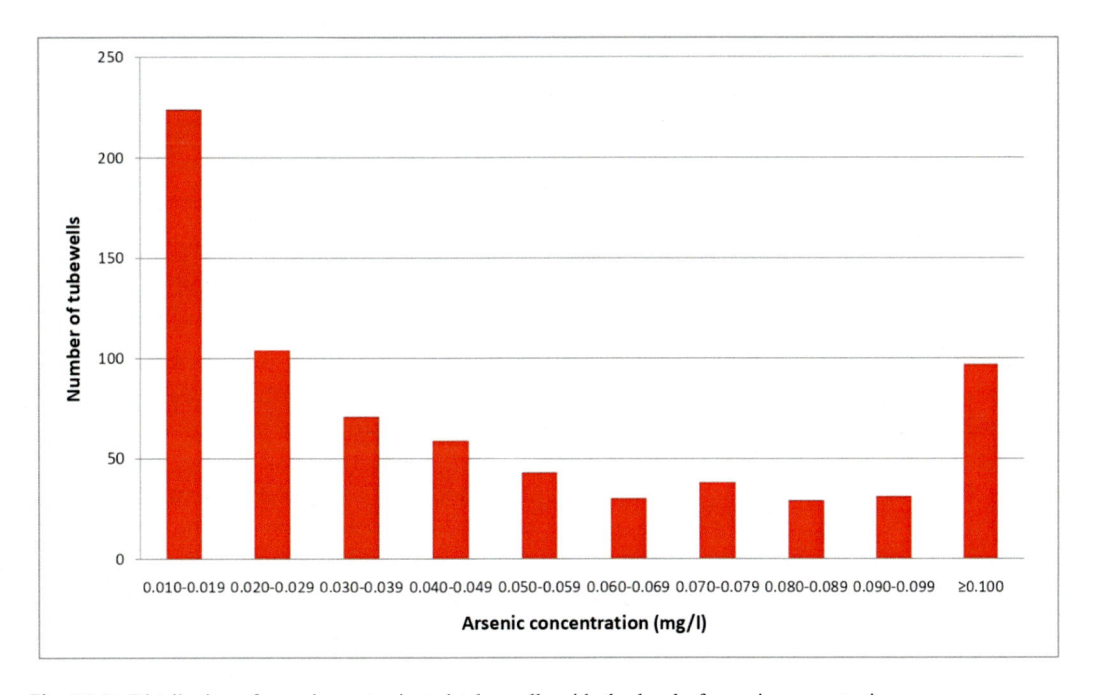

Fig. 12.3 Distribution of arsenic-contaminated tube wells with the level of arsenic concentration

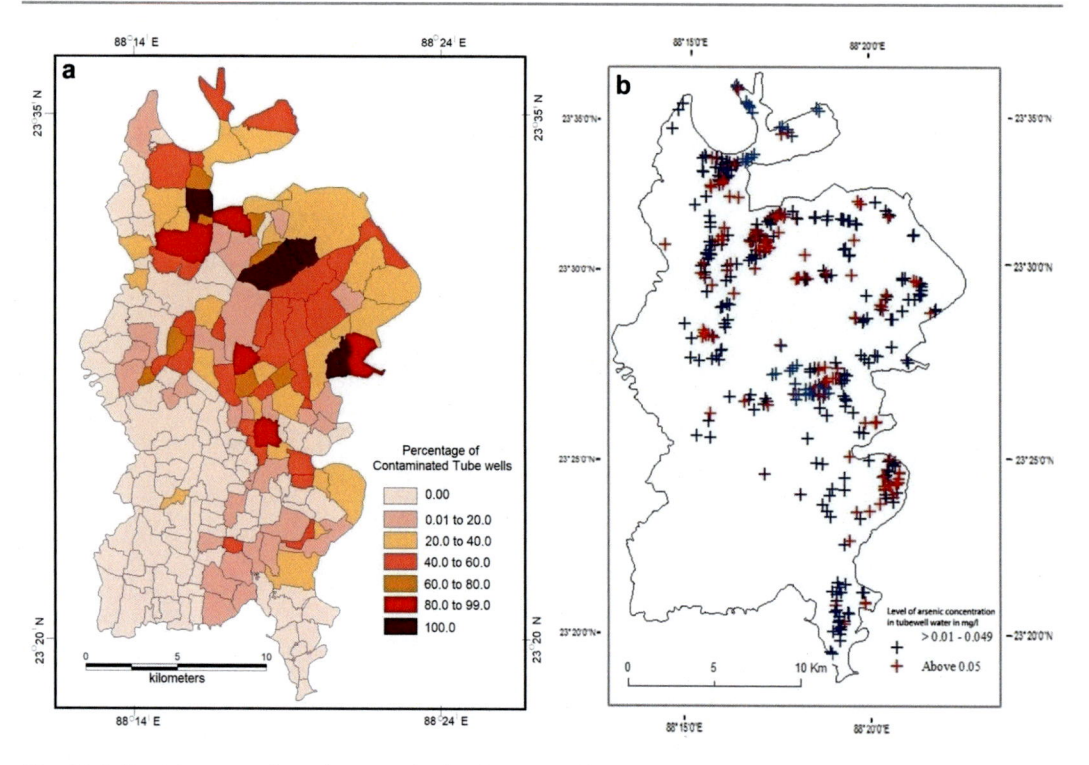

Fig. 12.4 Spatial pattern of arsenic contamination in percentile with GPS-based contaminated tubewell locations

margin of the administrative boundary and south–west corner of the area. The contaminated wells are located near the residential clusters and a few are in agricultural land. Thus the spatial analysis based on tube well buffer zones and dominant LU and LC types is done. It simply calculates the probability of the occurrence of arsenic contamination on reclassified five buffer categories. The cumulative weight scores in each buffer class are calculated by assigning a binary weight scheme. The category wise weighted average score for each category is finally brought into percentage and judged. The maximum value indicates greater intimacy. GIS-based buffer analysis based on WOE approach, along contaminated tube wells indicates that maximum contaminated wells are situated at proximity to reframed categories of buffer circles with the dominance of wetlands, marshy lands and agricultural lands; i.e. Class 1, 3 and 4. The numeric percentages are represented in Table 12.2. The next highest contaminated points belong to the buffer category 5. Buffer category 2 ranks next to category 5 in order of average weighted score. The agricultural land, water bodies and marshy land together carry about 68% weights (Table 12.2; Fig. 12.6).

12.5 Discussion

The distribution of arsenic in groundwater is very irregular. The different land use and land cover units are intimately related to underlying groundwater and are likely to explain many groundwater issues (Lerner and Harris 2009). The present research has shown that the contaminated tube well has an excess with agricultural land, wetlands and Marshy land. In this context, it should be noted that these three land use and land cover units play an important role in aquifer recharge from the surface to the subsurface layers. In different seasons of the year, the fields are flooded sometimes by rainwater or by irrigated water. Wetlands and marshy Lands, on the other hand, are nourished by floodwaters

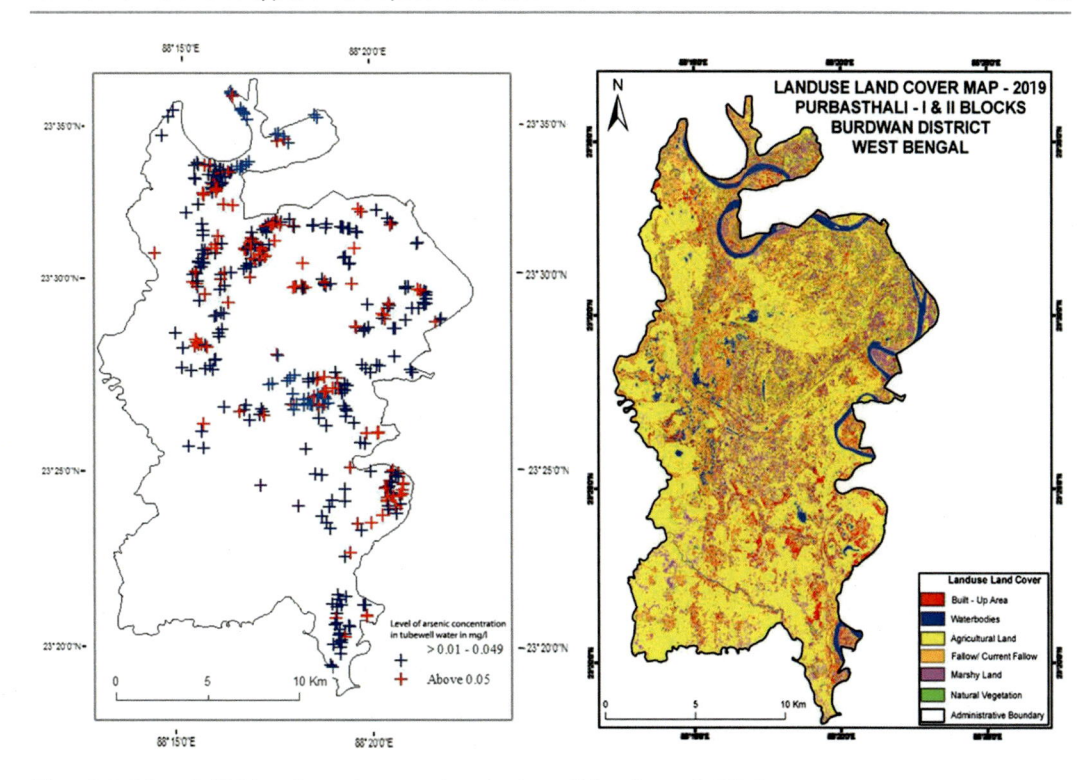

Fig. 12.5 Map of GPS-based arsenic-contaminated tube well location and LULC

Table 12.2 Spatial association analysis using average weight score

LULC categories	No. of mild contaminated tube wells (X_i)	Weighted score ($X_i * W_1$)	No. of high contaminated tube wells (Y_i)	Weighted score ($Y_i * W_2$)	Average weighted score (C_i)	Normalized score in % ($C_i/\sum C_i) * 100$
Class 1	63	31.5	32	32	42.33	10.77
Class 2	68	34	50	50	59	15.01
Class 3	111	55.55	137	137	128.37	32.66
Class 4	84	42	101	101	95.33	24.25
Class 5	72	36	66	66	68	17.31
Total	398	199.05	386	386	393.03	100

Source IRS-ID, LISS- IV (Mx), Image of March 2019, PHED, Govt. of W.B, water testing data (2006) and Author's Calculations

and rainwater. Some of this floodwater evaporates and the rest enters the ground and recharges the groundwater. However, the clay layer above or slightly below the surface plays an important role in this recharging process. This clay layer again controls the percolation amount, rate and water chemistry of water followed (Larsen et al. 2008). Past research has shown that organic carbon dissolved with recharged water is the staple food of microorganisms living in aquifers. These microorganisms arrive in organic carbon-rich areas in search of food and are subsequently released absorbed arsenic from solid plastic grains of sediment to the aquifer. Groundwater usually flows slowly following a hydraulic gradient, the main regulator being the semi-pervasive

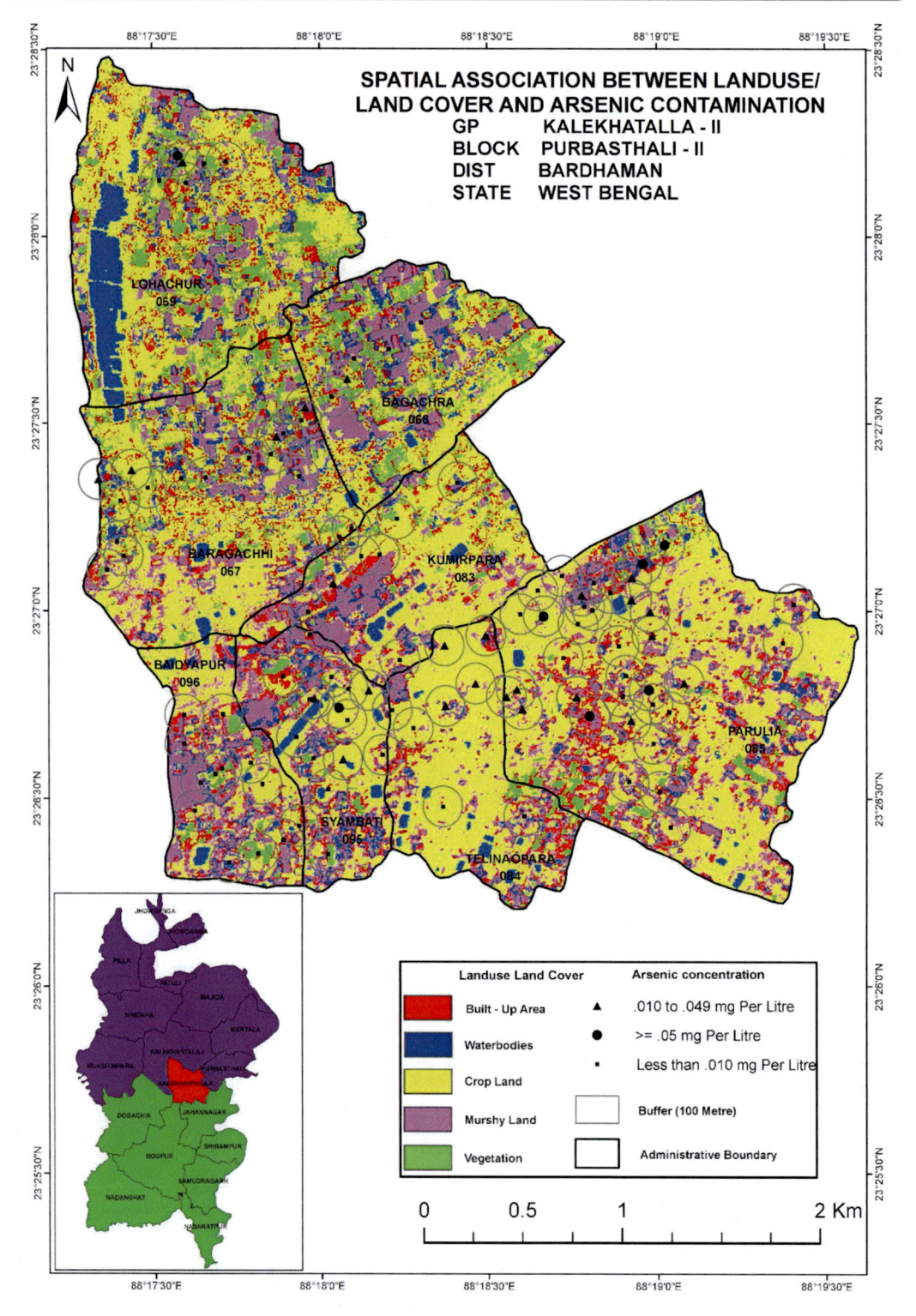

Fig. 12.6 Spatial association analysis between LU and LC and arsenic-contaminated tube wells using buffer zone and average weight score

sediment layer, through which the groundwater flows sub-parallel way along the regional land slope, although in many cases this hydraulic gradient changes due to excessive groundwater extraction for agriculture purposes, especially during paddy cultivation. Most of the area is flood-prone; settlements have developed in the area of ancient levees, while agricultural areas are located close to wetlands. Under normal conditions the hydraulic gradient is following a natural slope; but now due to excessive use of groundwater for agricultural purposes hydraulic gradient changes, it often creates reverse flow. Temporary vacuum induces groundwater recharge laterally and vertically. On the other hand, the supply of organic carbon in the groundwater continues with recharge water and microorganisms arrive in search of food following the hydraulic gradient, the specific biochemical process in the aquifer that separates iron and arsenic from entrapped floodplain sediments in the aquifers (Islam et al. 2004; Postma et al. 2007). This free arsenic comes out with the water of the tube wells and put a handful of arsenic in the whole area. In this context, it should be noted that the floodplain areas are densely populated and have a combination of abundant tube wells which supply drinking water to the people of this densely populated area. In conclusion, the presence of organic carbon and microorganisms in groundwater is the main cause of the arsenic release. But the role of these three special land use types and land cover units in supplying a sufficient amount of recharge water and dissolved organic carbon cannot be denied, although the present study shows how and to what extent organic carbon recharge reaches underground aquifers from these reservoirs. Not researched but there is ample scope for future research on this topic.

12.6 Conclusions

The occurrence of arsenic in the groundwater of the floodplain on the west bank of the Bhagirathi River is controlled by geo-genetic factors as well as the type of land use in the region and the bio physical coverage of the land. The present study proves that there is coexistence between certain fields of land use and land cover and arsenic-contaminated tube wells. It speculates that fine organic particles from the surface may enter the groundwater in the process of recharge. The wetlands, marshes and agricultural fields have a significant influence on groundwater recharge. In this case, the interaction of surface water and groundwater and the process of percolation of surface water through the sedimentary layers and the chemistry of percolating water are very important. In the past, research in the Mekong Basin and the Bengal Basin also supported this. The main source of organic carbon in the aquifer is the in situ organic residues remaining in the aquifer sediments or the tiny organic matter mixed with the floodplain surface water through the recharge process. In this context, it is important to mention that organic carbon and the biochemical reactions of microorganisms release arsenic from sediment into groundwater. Thus, the study seeks to highlight this potential aspect that will help explain the actual pattern of arsenic contamination in floodplains in the coming days.

References

Akai J, Izumi K, Fukuhara H, Masuda H, Nakano S, Yoshimura T (2004) Mineralogical and geomicrobiological investigations on groundwater arsenic enrichment in Bangladesh. Appl Geochem 19:215–230

BGS, DPHE (2001) Arsenic contamination of groundwater in Bangladesh, vol 2. In: Kinniburgh DG, Smedley PL (eds). British Geological Survey, Keyworth

BGS (2001) Arsenic contamination of groundwater in Bangladesh. BGS, Keyworth

Bora U (2011) Floral and faunal diversity of lower middle Ganga. IIT Guwahati, Guwahati

Chakraborti D, Singh SK, Rahman MM, Dutta RN, Mukherjee SC, Pati S, Kar PB (2018) Groundwater arsenic contamination in the Ganga River Basin: a future health danger. Int J Environ Res Public Health 15(2):180

Chetia M, Chatterjee S, Banerjee S, Nath MJ, Singh L, Srivastava RB, Sarma HP (2011) Groundwater arsenic contamination in Brahmaputra river basin: a water quality assessment in Golaghat (Assam), India. Environ Monitor Assess 173(1–4):371–385

CoI (2011) District census hand book. Govt. of India, PCA, New Delhi, India

Islam FS, Gault AG, Boothman C, Polya DA, Charnock JM, Chatterjee DE (2004) Role of metal reducing bacteria in arsenic release from Bengal delta sediments. Nature 430:68–71

Jessen S, Larsen F, Postma D, Viet PH, Nguyen TH, Pham QN et al (2008) Palaeo-hydrogeological control on groundwater arsenic levels in Red River delta, Vietnam. Appl Geochem 23:3116–3126

Larsen F et al (2008) Controlling geological and hydro geological processes in an arsenic contaminated aquifer on the Red River flood plain, Vietnam. Appl Geochem 23:3099–3115

Lawati WA (2012) The role of organics in the mobilization of arsenic in shallow aquifers. The University of Manchester, School of Earth, Atmospheric and Environmental Sciences

Mandal BK, Suzuki KT (2002) Arsenic round the world: a review. Talanta 58:201–235

McArthur JM, Banerjee DM, Hudson-Edwards KA, Mishra R, Purohit R, Ravenscroft P et al (2004) Natural organic matter in sedimentary basins and its relation to arsenic in anoxic groundwater: the example of West Bengal and its worldwide implications. Appl Geochem 19:1255–1293

McArthur JM, Ravenscroft P, Safiulla S, Thirlwall MF (2001) Arsenic in groundwater: testing pollution mechanisms for sedimentary aquifers in Bangladesh. Water Resour Res 37:109–117

Mukherjee PK, Pal T, Chattopadhyay S (2010) Role of geomorphic elements on distribution of arsenic in groundwater—a case study in parts of Murshidabad and Nadia districts, West Bengal. Indian J Geosci 64(1–4):77–86

Neumann RB, Ashfaque KN, Badruzzaman AB, Ashraf Ali M, Shoemaker JK, Harvey CF (2010) Anthropogenic influences on groundwater arsenic concentrations in Bangladesh. Nat Geosci 3:46–52

Postma D, Larsen F, Nguyen TM, Mai TD, Pham HV, Pham QN et al (2007) Arsenic in groundwater of the Red River floodplain, Vietnam: controlling geochemical processes and reactive transport modelling. Geochim Cosmochim Acta 71:5054–5071

Ravenscroft, P, Brammer, H, Richards, K (2009) Arsenic pollution: a global synthesis. Wiley-Blackwell, Oxford

Reza HA, Jean JS, Lee MK, Liu CC, Bundschuh J, Yang HJ et al (2010) Implications of organic matter on arsenic mobilization into groundwater: evidence from north-western (Chapai-Nawabganj), Central (Manikganj) and Eastern (Chandpur) Bangladesh. Water Res 44:5556–5574

Tareq SM, Maruo M, Ohta K (2013) Characteristics and role of groundwater dissolved organic matter on arsenic mobilization and poisoning in Bangladesh. Phys Chem Earth 58–60:77–84

Winter TC, Harvey JW, Franke OL, Alley WM (1998) Ground water and surface water—a single resource. US Geological Survey Circ 1139, Denver, Colorado

Forecasting the Danger of the Forest Fire Season in North-West Patagonia, Argentina

Ezequiel A. Marcuzzi, Marcela Hebe González, and María del Carmen Dentoni

Abstract

An average of 1300 forest fires are officially reported in Argentina annually, triggered either by natural or intentional causes. Once an outbreak has started, topographic and vegetation (fuel) conditions and the weather situation determine the fire behavior. The general evolution of fire danger can be monitored through the Fire Weather Index (FWI), part of the Canadian Forest Fire Danger Rating System (CFFDRS). This indicator was successfully implemented in Argentina since the beginning of 2000. It is based on temperature, relative humidity, wind intensity, and precipitation. Among its advantages, it allows evaluating the possibility of fire ignition, intensity, and spreading. The objective of this work is to achieve a characterization of the danger of the different forest fire seasons in the northern Argentinean Patagonian mountain range, based on Bariloche station data, using the FWI and find a method to forecast the danger. For this purpose, the index was related to atmospheric circulation and sea surface temperature patterns that dominate the region before and during the fire season, using the correlation method. It allowed defining predictors that were used to generate some statistical models, to predict the dangerousness of the future seasons. Sea surface temperatures, anomalies of wind, and the Antarctic Oscillation were the best predictors for this purpose. Some derived statistical models are very useful tools to improve the decision-making regarding the coming fire seasons in this particular area. The incorporation of non-linear techniques seems to improve the efficiency of linear regression models.

Keywords

Forest fires · Prediction · Fire weather index · Argentinean Patagonia · Circulation patterns

E. A. Marcuzzi
Servicio Nacional de Manejo del Fuego, Ministerio de Ambiente y Desarrollo Sostenible de la Nación, Buenos Aires, Argentina

M. H. González (✉)
Facultad de Ciencias Exactas y Naturales, Departamento de Ciencias de la Atmósfera y los Océanos, Universidad de Buenos Aires, Buenos Aires, Argentina
e-mail: gonzalez@cima.fcen.uba.ar

M. H. González
Centro de Investigaciones del Mar y la Atmósfera (CIMA), CONICET—Universidad de Buenos Aires, Buenos Aires, Argentina

M. H. González
CNRS-IRD-CONICET-UBA, Instituto Franco-Argentino para el Estudio del Clima y sus Impactos (UMI 3351 IFAECI), Buenos Aires, Argentina

E. A. Marcuzzi · M. del Carmen Dentoni
UNPSJB—Universidad Nacional de la Patagonia San Juan Bosco, Comodoro Rivadavia, Argentina

13.1 Introduction

An average of 1300 wildfires is reported annually in Argentina from the National Fire Management Service, Ministry of Environment, and Sustainable Development of the Argentine nation (https://www.argentina.gob.ar/ambiente/manejo-del-fuego), which often cause significant material losses, undesired effects on vegetation and risk to people. Most of these fires are caused by people, for reasons that vary from region to region, due to different fuel characteristics, topo reasons as cultural practices or land management policies. Lighting usually causes ignition too in some regions, though in a less amount. Once a fire starts, topography, vegetation fuels, and meteorological conditions are critical factors for the fire behavior, being the weather the most variable in short time and space scales. The weather conditions may produce different fire behaviors under the same topography and fuel conditions, affecting the drying rates of fuels, the spread velocity and direction, the characteristics of the convective columns and even causing the development of pyrocumulus. The behavior of a fire can be monitored through Forest Fire Danger Rating Systems (FFDRS), as the FFDI used in Australia (McArthur 1966, 1967; Gill et al. 1987; Cheney and Sullivan 1997), the National Fire Danger Rating System (NFDRS) (Deeming and Lancaster 1971; Rothermel 1972; Deeming et al. 1977; Fosberg 1978; Goodrick 2002), and CFFDRS developed in Canada (Van Wagner and Pickett 1985; Van Wagner 1987; Forestry Canada Fire Danger Group 1992).

FFDRS are critical for safe and efficient fire management, they are used for prescribed burning planning, resources assignment for preparedness, and definition of strategies and tactics of suppression.

After a revision of existent systems (Dentoni and Muñoz 2012) and a preliminary analysis of feasibility (Taylor 2001), the FWI was adjusted and implemented in Argentina. The process was started in pilot project areas located in the northern Andean region of Patagonia, because of its climate similarities with British Columbia (Taylor 2001) and is nowadays extended to the whole country (Dentoni et al. 2015).

FWI is based on surface meteorological observation of temperature, relative humidity, wind intensity, and precipitation. This is a non-dimensional index that represents the fire intensities as the fire danger increases. It is made up of six standard components. The first three are fuel moisture codes, which track the daily changes in the moisture content of three classes of forest fuels with different drying rates. The last three are fire behavior indices, representing the rate of fire spread, the fuel available for combustion, and the frontal fire intensity. Each of these indicators has different operational applications, such as the possibility of fire ignition, control difficulties, or mop-up requirements (British Columbia, Ministry of Forests, Protection Branch 1978). The analysis of FWI along the year allows for determining the mean characteristics of fire seasons that vary from region to region (Kunst et al. 2003).

In each region, the fire danger conditions and, in consequence, the fire behavior, are the result of different time and space meteorological scale phenomena and that is the reason why to anticipate them is relevant for fire management agencies. Werth (2011), makes a review of critical synoptic patterns for different regions in the United States. Numerous studies have been carried out in other countries that relate global and regional atmospheric patterns with fire season characteristics. Simard et al. (1985), found that fire occurrence in the southern United States decreased during the El Niño-Southern Oscillation (ENSO) warm phase. Skinner et al. (1999) found a relation between burned area and ridges in 500 Hpa and the North American Oscillation pattern (NAO), for different regions of Canada. Williams and Karoly (1999) demonstrated that extreme fire danger in southeast Australia is related to the presence of El Niño. Flannigan et al. (2000) found that sea surface temperatures are good predictors of burned areas in Canada.

According to Hess et al. (2001), 63% of the total burned area occurred during the El Niño phase in Alaska. In addition to this result, Duffy et al. (2005) found that the Pacific Decadal Oscillation (PDO) may be useful to estimate the number of hectares that will be burned in the following season. In Argentina, several studies of meteorological conditions associated with fires have been done (Dentoni et al. 2001; Bianco et al. 2005; Marcuzzi et al. 2006; Marcuzzi and Hoevel 2009; Marcuzzi et al. 2016). Some studies show the influence of large-scale weather patterns on the danger of fires (Kitzberger and Veblen 1997; Veblen et al. 1999, 2003; Kitzberger 2002). In northern Patagonia, between 39° and 42° S, east of the Andes, an important influence of ENSO in the occurrence of fires, is detected. Using this relationship, Kitzberger (2002) designed a fire risk prediction method over the northwest of Patagonia and tested it against the burned area recorded per season in the different national parks of the region between 1950 and 1996. Furthermore, Veblen et al. (1999, 2003), found that at around 40° S, the fires were related to the variation in sea level pressure south of South America, between 50 and 60° S, near the Antarctic Peninsula. Closer in time, Holz and Veblen (2011), related the marks caused by fires in tree rings, located in two areas in southern Chile (west of Chiloé Island and south of the Aysén district), to ENSO, PDO, and the Antarctic Annular Oscillation (AAO), an index that involves a latitudinal pressure gradient as it was defined by Thompson and Wallace (2000). The author found a significant relationship between AAO patterns and the occurrence of fires, especially since the middle of the last century, but not with ENSO. Previously, Silvestri and Vera (2009) and Garreaud et al. (2008), showed that the AAO pattern influences austral spring precipitation, in the north of the Patagonian region in Chile and Argentina. Alessandro (2003) related blocking situations to precipitation in Bariloche.

Despite the aforementioned results, there is still much to investigate about the regional and the global patterns previous to and during fire seasons, that can affect fire danger and lead to critical fire situations. That is why, to advance the knowledge about the processes that cause fire danger, and try to prevent them through the forecast of FWI, the following objectives were proposed to characterize the danger of the different fire seasons in a pilot area, based on the FWI values, to find out atmospheric and oceanic patterns that can be used as predictors of the danger of fires in the coming seasons, and to design statistical models to forecast fire danger and evaluate their efficiency.

13.2 Methodology

The selected study area was the northwest Andean region of Patagonia where FWI was first implemented in Argentina. This area has a high risk of wildfire. The FWI index was calculated using daily data of precipitation (9 am data), temperature, relative humidity, and wind intensity at noon from the National Meteorological Weather Service in Bariloche Aero station, for the period 1980–2019 (Fig. 13.1). The choice of the Bariloche station was due to the fact that it has the most extensive data record in the area. Besides, the mean evolution of the fire danger represents on average the behavior of the closest stations, such as Chapelco and Esquel, as can be derived from Bianchi (2007).

The fire season is considered to begin in September and end in April. The daily FWI values were averaged from 1st September of one year to 30th April of the following year, to calculate the seasonal FWI every year. The initial FWI value of each season is considered to be zero, due to the winter climatic conditions that saturate the organic soil.

Some monthly atmospheric and oceanic variables representative of the local and regional circulation and of the teleconnection process that can modify conditions in remote places were selected to evaluate their relationship with the mean FWI seasonal index. These variables are: 300 (H300), 500 (H500) and 1000 (H1000) Hpa monthly mean geopotential height, sea surface temperature (SST), and the zonal and meridional component of the wind at 850 hpa (U850 and V850, respectively). Data for sea surface temperature and other atmospheric variables were obtained from the National Center of Environmental Prediction

Fig. 13.1 Location of Bariloche, Chapelco, and Esquel weather stations in Argentina (https://www.ign.gob.ar)

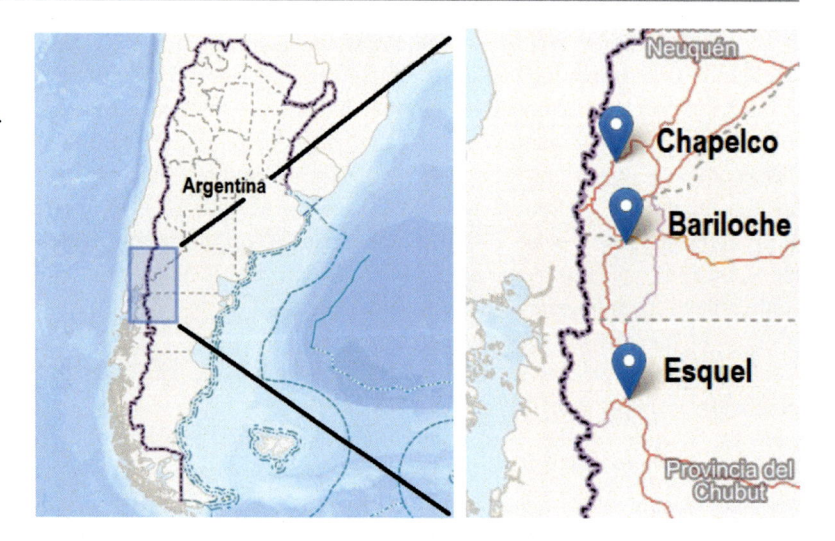

(NCEP) reanalysis (Kalnay et al. 1996). The relationship between the FWI index and the most important circulation patterns that influence the climate of the southern hemisphere: ENSO and AAO were carried out through the values of the respective indices. The SST averaged in the EN3.4 region was used to evaluate the ENSO signal (http://climexp.knmi.nl/data/ihadisst1_nino3.4a.dat). Regarding the AAO index, it is defined as the normalized sea level pressure difference between 40° S and 70° S (Nan and Li 2003), and the data were obtained from http://ljp.lasg.ac.cn/dct/page/65572.

The correlation method was used to determine the relationship between the FWI and the different atmospheric and oceanic variables. FWI of the fire season was correlated with the different variables, calculated in different previous months (May to August) and during the fire season (September to March of the following year). The period 1980–2012 was used to define the predictors and correlation values greater than 0.35 were significant with a 95% confidence level was determined using a normal test. In all cases, a physical explanation was discussed to justify the cause-effect relationship between the variables and the FWI. The correlation fields between the seasonal FWI and the meteorological variable anomalies (SST, H1000, H850, H500, H300, V850, and U850 in previous monthly) were analyzed. These correlation fields made it possible to define predictors as the mean value of the variable in areas with significant correlation with FWI. A set of predictors independent of each other was generated to avoid the problem of multi-correlation (Wilks 1995).

Three different data mining methods were applied to generate statistical forecast models using the same set of independent predictors: Multiple Linear Regression (MLR), Generalized Additive Models (GAM), and Support Vector Regression (SVR).

The Forward Stepwise methodology was used to generate the MLR model. This methodology implies considering each of the predictors one by one and staying only with the combination that provides the best multiple correlation coefficient. The MLR is synthesized in the expression:

$$y = \sum a_i x_i + e \quad (13.1)$$

$$e \sim N(0; \sigma 2) \quad (13.2)$$

where Y is the predictant, x_i is the predictors, a_i are the linear coefficients and e is the error.

Many data in the environmental sciences do not fit simple linear models, so techniques that incorporate non-linear behavior were also used; one of them is GAM (Wood 2006). This method incorporates the possibility that the equation terms can be functions and then uses a smoothing function. In this work, a three-degree spline function was used.

$$y = \sum F(xi) + e \quad (13.3)$$

$$e \sim N(0; \sigma 2) \quad (13.4)$$

The Support Vector Machine (SVM) is a supervised machine learning algorithm that can be employed for both classification and regression purposes (Cortes and Vapnik 1995). It is based on the idea of finding some hyperplane that best divides a dataset into several classes. The hyperplane must have a maximum margin (it is as far as possible from the classes it separates) and must correctly classify as many instances as possible. However, it is not always possible to fulfill both conditions simultaneously, that is why there is a regularization parameter that allows the first condition to be partially "sacrificed" to maximize the number of correctly classified instances (Cortes and Vapnik 1995). SVM is more commonly used in classification problems but it can also be used as a regression method. The SVR uses the same principles as the SVM for classification. Conceptually it works the same as its counterpart for classification, except that the separation hyperplane is now used as a regression surface. The main idea is to minimize error, individualizing the hyperplane which maximizes the margin. The kernel functions transform the data into a higher dimensional feature space to make it possible to perform the linear separation.

The models derived using these three techniques were validated using the cross-validation methodology (Wilks 1995) where all the years except one were used for the construction of the model and the remaining year for the calculation. The process was repeated as many times as there were years to predict for the period 1980–2019. This process allows, on the one hand, to validate the forecast, since the series of observed values can be compared with those predicted with this technique. On the other hand, it allows verifying the stability of the model by observing if the repetition of the methodology always provides the same final set of predictors and if the coefficients of the regression do not differ significantly. This technique is recommended when the data record is short enough to divide into two sub-periods one of training and the other of verification. The goodness of the models was initially quantified through the variance of the FWI that they explain. Therefore, the coefficient of determination (R^2) was calculated. R^2 is defined as the square of the correlation between the actual and the predicted values and represents the proportion of variance in the forecast variable that is explained by the model. It is not a good measure of forecasting skill because adding any variable tends to increase its value even if the variable is irrelevant. Therefore, R^2 adj was designed to overcome this problem:

$$R^2 \text{ adj} = 1 - (1 - R^2)(n-1)/(n-k-1) \quad (13.5)$$

where n is the number of data and k is the number of predictors. Using this statistic, the best model will be the one with the largest R^2 adj. The best model produced by each of the three methodologies has been considered to be analyzed in this work. The selection criteria are based on the highest value of R^2 adj.

Contingency tables were built between the observed and forecast (using cross-validation method) FWI, to test the efficiency of the models. The cases below the first tercile (below normal), those between the first and the second terciles (normal), and those above the second tercile (above normal) were separated. For the categories below and above, the following indices were calculated: the probability of detection (POD), false alarm ratio (FAR), hit rate (H), and area under the curve in a ROC diagram (AUC) (Wilks 1995). The hit rate is the fraction of all the cases when the categorical forecast correctly anticipated the subsequent event. The probability of detection is defined as the fraction of those occasions when the forecasted event occurred where it was also forecasted. The false alarm ratio is the proportion of forecasted events that fail to happen. A ROC (Receiver Operating

Characteristic) curve is a graphical representation of sensitivity (proportion of true positives) versus (1-specificity) (proportion of false positives). Each point in this diagram, together with the origin of coordinates and the maximum value that each variable can take on the coordinate axes, generates a curve. The value of the area under this curve is indicative of the goodness of the model and its best value is unity. A random classification would give a point along the diagonal line, which is also called the non-discrimination line, from the lower left to the upper right corner and this case is associated with the value of 0.5 for AUC.

13.3 Result and Discussion

13.3.1 The Characterization of the Mean FWI Values and the Definition of Its Predictors

Figure 13.2 shows the interannual variability of the seasonal FWI for the period 1980–2019 in Bariloche station. The mean value is 25.6 and the standard deviation is 4.9. Values greater than 26.7 are considered "above normal" meanwhile, while values lower than 23.2 are "below normal"

Table 13.1 Some seasonal FWI statistics

Mean	25.6
Standard deviation	4.9
First tercile	23.2
Second tercile	26.7

(Table 13.1). No significant change in FWI was observed throughout the study period. A linear adjustment of the data showed a trend of -0.03 per year that was not statistically significant using a normal test.

Forecasting models for the mean FWI at the beginning of the period of the greater danger of the fire season in early November will be designed. To determine predictors, the FWI series was correlated with the SST and all the meteorological variables in the previous months: March, April, May, July, August, September, and October. Those areas in the correlation fields between FWI and the variables, where the correlation value was significant with a 95% confidence level, were used to define 23 predictors. Then, only those that were independent of each other were selected. The selected independent predictors are detailed in Table 13.2: V850, U850, SST, and AAO. This fact shows that the dangerousness of the fire season is favored by the SST in August in the subtropical Pacific,

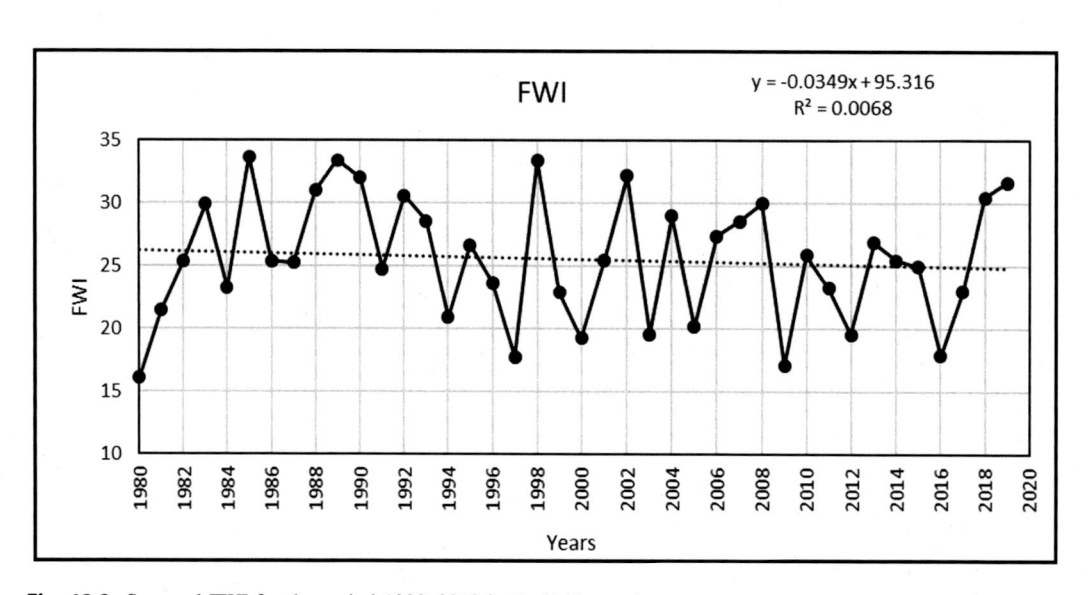

Fig. 13.2 Seasonal FWI for the period 1980–2019 in Bariloche station

Table 13.2 Set of independent predictors for seasonal FWI

Variable	Temporal scale	Spatial scale
V850	July	15–25° W 37–47° S
U850	October	15–30° W 10–20° S
SST	August	100–112° W 20–27° S
AAO	September	–

according to a cold phase pattern of ENSO, by a positive phase of AAO, and by the entry of warm air from the Atlantic to through the anticyclone.

Next, the correlation fields that allowed the definition of the independent predictors and their possible physical explanation will be analyzed. The correlation field between seasonal FWI and V850 in July is shown in Fig. 13.3. The correlations between FWI and V850 show that since July high observed FWI occur when the southern wind has a north component in southern Patagonia (center of negative correlation) and a south component in eastern South Atlantic Ocean. This feature could be related to the presence of anticyclonic anomalies over the Atlantic basin. The pattern does not hold every month but is well observed in July and September.

The correlation field between FWI and U850 in October (Fig. 13.4) shows a dipole over the Pacific Ocean with significant correlations where high values of fire danger would be related to an intensification of the westerlies, the south of 35ºS, and a weakening to the north of 40° S in mid-latitudes. This dipole pattern is maintained in the correlation field from June until November but the correlations are significantly high in August and October. On the other hand, the negative correlation observed off the coast of Brazil in October is marked. This could imply an intensification of the semi-permanent anticyclone of the Atlantic Ocean, which would favor the entry of warm air from the Atlantic to the continent that will later lead to a supply of air from the north in Patagonia.

Figure 13.5 shows the correlation field between FWI and SST in August. Negative correlations can be observed in the south subtropical Pacific Ocean, west of the Chilean coast, indicating a greater danger of fires when the water is colder than normal. This core of correlation becomes significant in June and is maintained until September, in October it loses intensity but in November the relationship becomes significant again. The negative correlations closest to Ecuador seem to be part of the ENSO pattern, indicating that there is a greater danger of fires associated with the cold phase of ENSO (La Niña events).

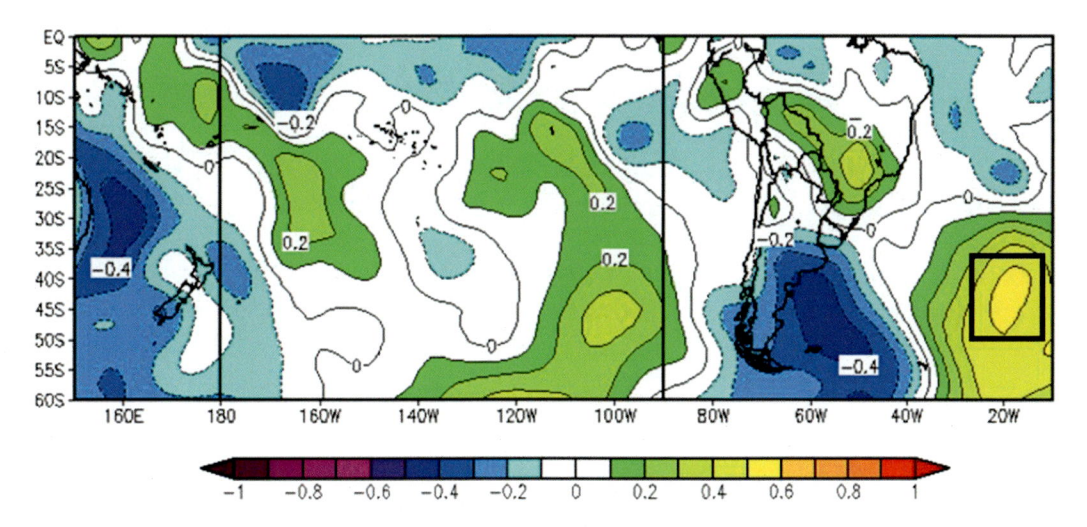

Fig. 13.3 Correlation between seasonal FWI and V850 anomalies in July. The rectangle shows the area where the predictor was defined

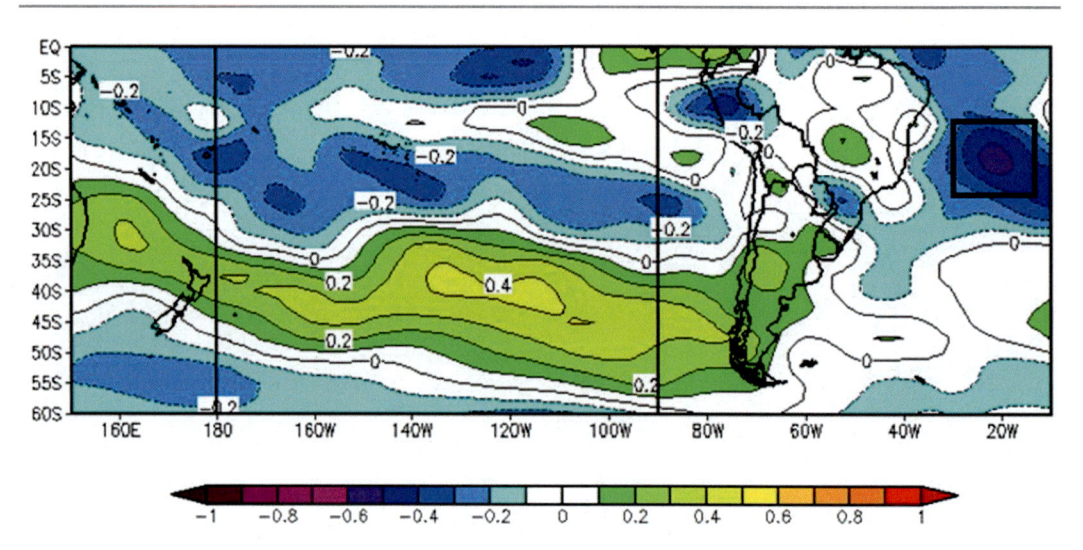

Fig. 13.4 Correlation between seasonal FWI and U850 anomalies in October. The rectangle shows the area where the predictor was defined

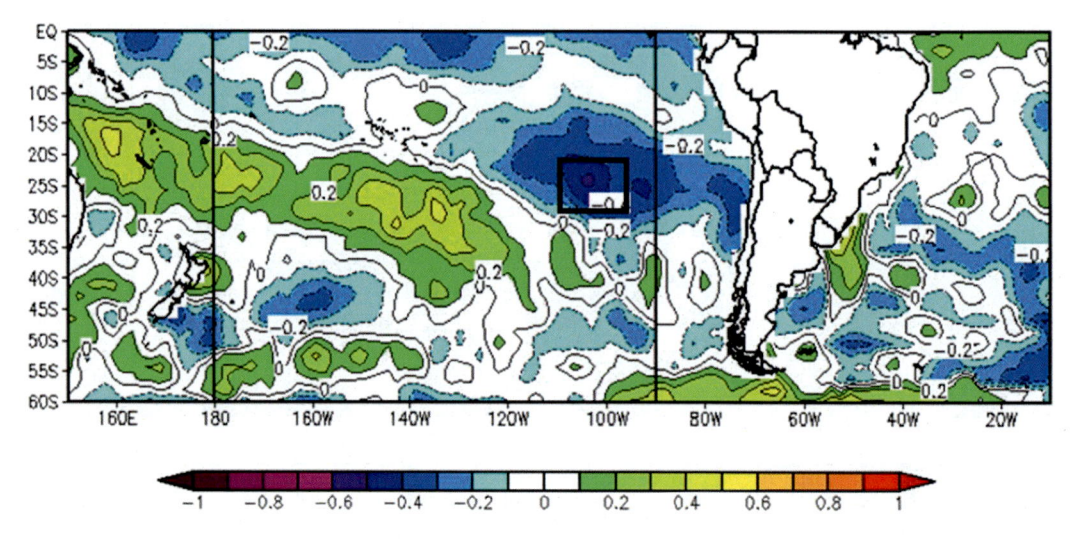

Fig. 13.5 Correlation between seasonal FWI and SST anomalies in August. The rectangle shows the area where the predictor was defined

The correlations between seasonal FWI and the AAO from May to October (Table 13.3) were calculated and it is significant with a 95% confidence level in September. This would indicate that the fire danger is high for the positive AAO phase in September. The AAO is an annular pattern called "Southern Annular Mode" (Thompson and Wallace 2000), its positive phase is defined by negative pressure anomalies at high latitudes combined with cores of positive anomalies at mid-latitudes. This scheme intensifies the westerlies at high latitudes and therefore modifies the trajectory of the systems. Several authors have demonstrated the incidence of this pattern in rainfall in South America. For example, Silvestri and Vera (2003) showed a

Table 13.3 Correlation between seasonal FWI and AAO in previous months

May	June	July	August	September	October
−0.15	0.21	0.3	0.16	*0.42*	−0.12

In italics, correlation significant with 95% confidence level using a normal test

significant relationship with rainfall, particularly in November and December in southeastern South America. Reboita et al. (2009) detected a decrease in frontal activity when the positive phase of the oscillation was present.

To better observe this effect, the correlation field between FWI and H1000 in September is shown (Fig. 13.6). Positive correlation values are observed at mid-latitudes around practically the entire globe and negative correlation values at high latitudes, clearly show the positive phase pattern of AAO. Furthermore, this pattern was observed in the correlation fields from September to October at all levels indicating the strong relationship between FWI and the intensification of both the subpolar low and the subtropical highs and the intensification of the zonal flow in southern Argentina. Figure 13.6 also shows the presence of anticyclonic anomalies over Patagonia associated with a rapid increase o fire danger in spring in the Bariloche area as the precipitation is inhibited and temperatures rise.

13.3.2 The Statistical Forecast Models and Their Efficiency

The set of four independent predictors defined in the previous item was used to apply three statistical methodologies to forecast the FWI at the beginning of November: MLR, GAM, and SVR. The cross-validation technique was used instead of dividing the period into two sub-periods of training and verification to work with records long enough to be statistically significant. This procedure is detailed in the methodology section. Figure 13.7 shows the predicted and observed series of seasonal FWI. The correlation between observed and predicted FWI using the different methodologies is very similar (Table 13.4) and is around the value of 0.5. The model that explained the greatest FWI variance (Fig. 13.8) was SVR (61%) while the other two techniques explained less and a similar percentage (30% GAM and 35% MLR).

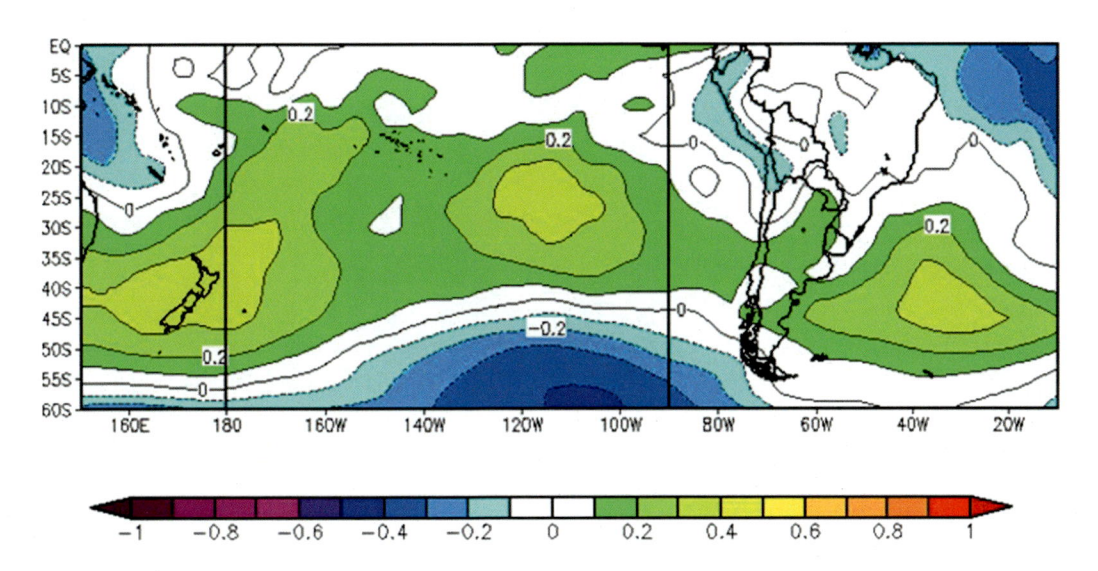

Fig. 13.6 Correlation between seasonal FWI and H1000 anomalies in September

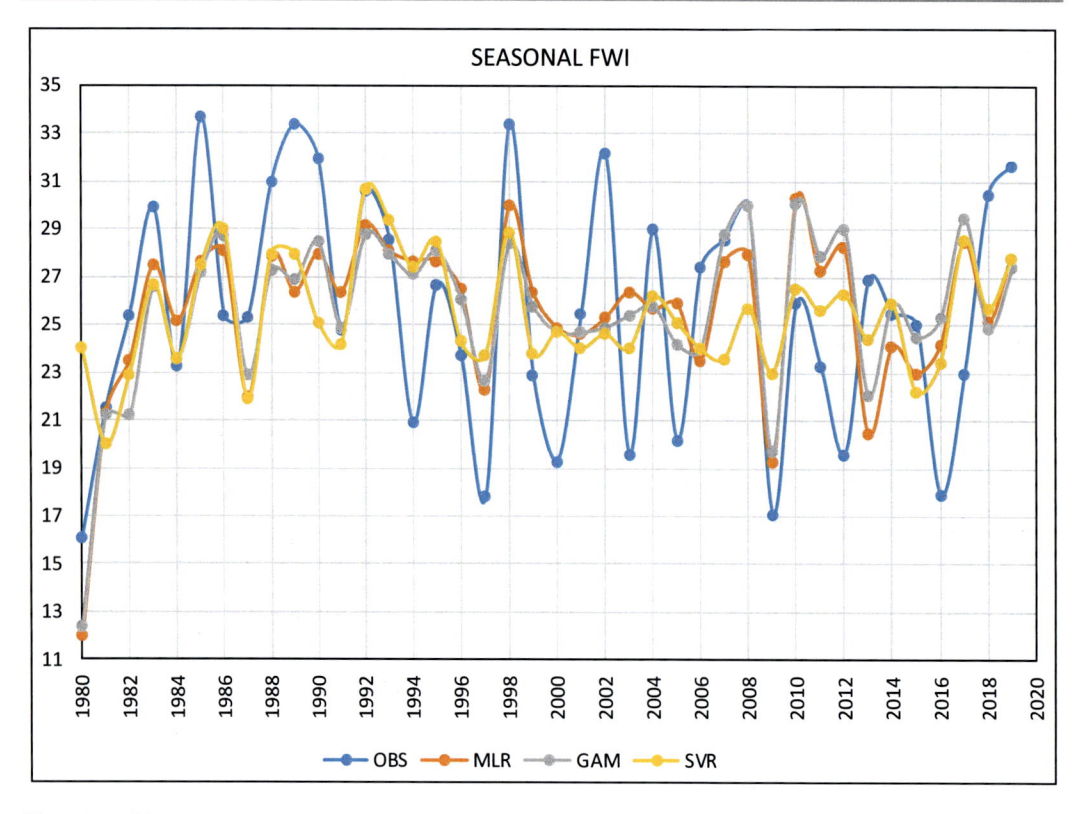

Fig. 13.7 Observed and predicted seasonal FWI using cross-validation technique for MLR, GAM, and SVR methodologies, 1980–2019

Table 13.4 Correlation between observed and predicted FWI and percentage of FWI variance explained by each model

	Correlation with observed values	R^2 adj (%)
MRL	0.51	35
GAM	0.51	30
SVR	0.49	61

To compare the efficiency of the different models, the Hit Rate (Fig. 13.9), POD (Fig. 13.10), FAR (Fig. 13.11), and AUC (Fig. 13.12) coefficients were calculated for the categories *above* normal (values higher than the second tercile) and *below* normal (values lower than the first tercile), detailed in the methodology section. It can be seen that the Hit Rate for the *below* category is close to 72% in the case of MLR and GAM and slightly lower for SVR (68%). Lower values were recorded for the *above* category. The maximum value, in this case, was for GAM (68%) and then SVR (65%) and MLR (62%).

The usefulness of these models is to detect the most dangerous cases *(above)*. The GAM method was the most efficient in detecting the *above* cases (36%). The SVR method detected 21% and the MLR only 14%. The cases *below* were detected in 33% of the cases with GAM and MLR, while the SVR method only detected 21%.

The percentage of the *above* cases in which there were fewer false alarms was using the GAM model (44%). This percentage increased to 50% and 60% in the case of using MLR and SVR, respectively. False alarms for the *below cases* were 43% with MLR and GAM while increasing to 60% using SVR.

In all cases, the AUC coefficient exceeded the value of 50%, indicating that the goodness of fit

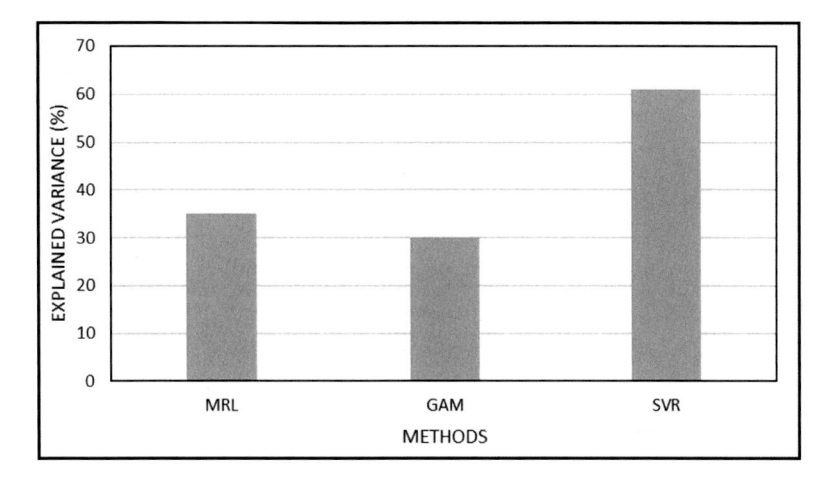

Fig. 13.8 Explained FWI variance using MLR, GAM, and SVR methodologies (1980–2019)

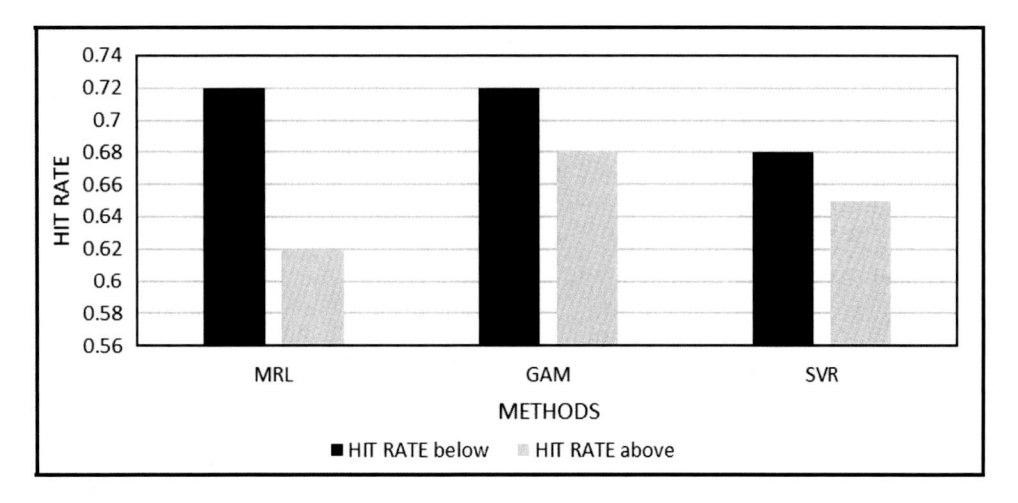

Fig. 13.9 Hit rate for the *above* and *below* categories derived from MLR, GAM, and SVR models

of all the models was better than a random classification. The GAM model recorded an AUC of 61% for the *below* cases and 60% for the *above* cases. In all other cases, the coefficient was lower, although always higher than 50%.

13.4 Conclusions

The seasonal FWI shows an important annual variability but it has not a significant tendency through the studied period. This variable showed to be a good indicator of the seasonal fire danger because it has a significant correlation with the amount of extreme FWI daily values of the fire season. The calculation of the FWI starts in September and it does not take into account previous conditions that can affect the evolution of the fire danger. Various monthly variables were found to be significantly related to the seasonal fire danger and different statistical models could be constructed taking into account the predictors to forecast the seasonal fire danger at the beginning of the fire season.

The results show that the GAM model, which incorporates the possibility of non-linear

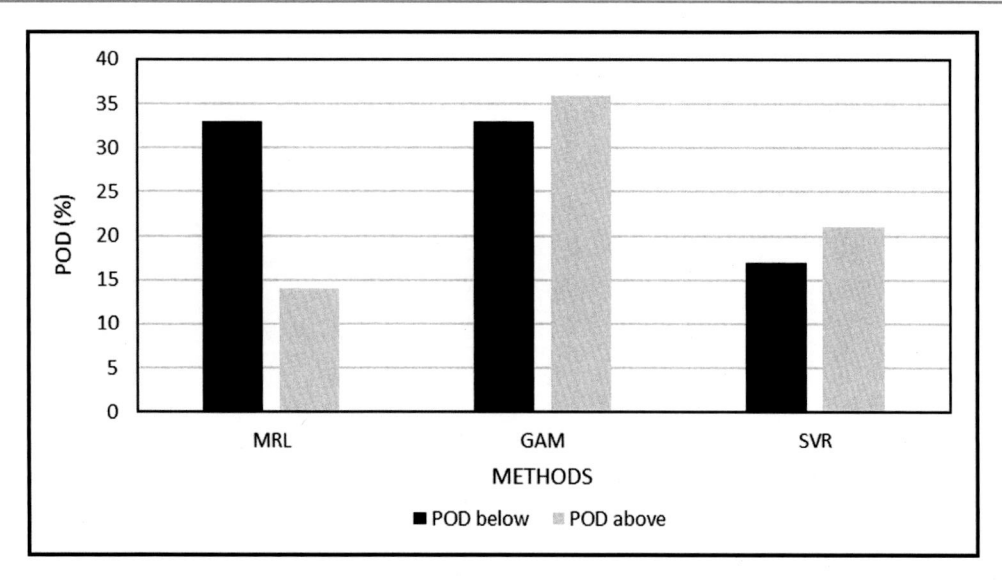

Fig. 13.10 POD for the *above* and *below* categories derived from MLR, GAM, and SVR models

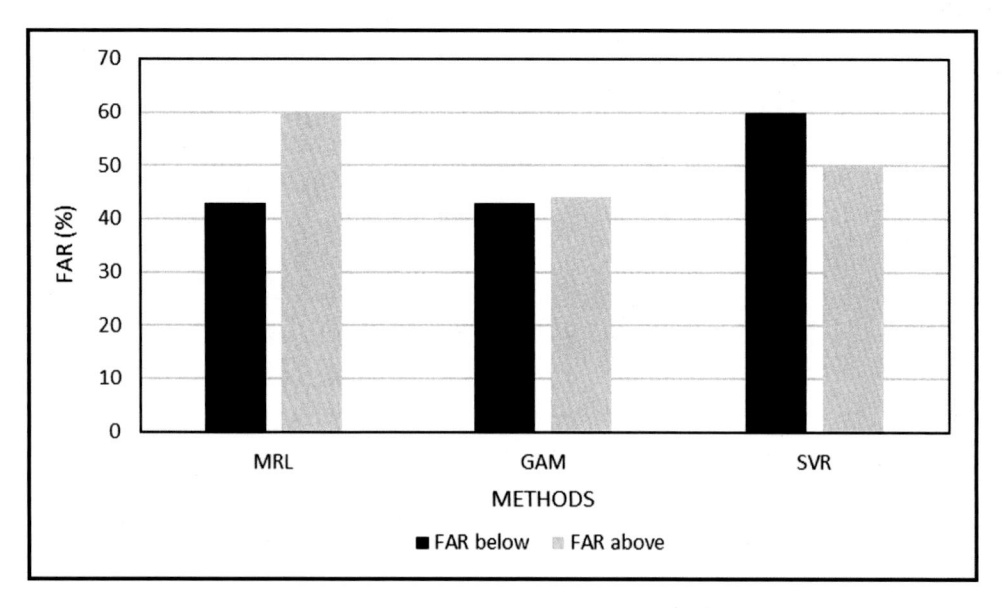

Fig. 13.11 FAR for the *above* and *below* categories derived from MLR, GAM, and SVR models

relationships between seasonal FWI and meteorological variables, has a higher efficiency than the MLR and SVR. GAM model provides the best Hit Rate, POD, FAR, and AUC coefficients for the *above* cases models. The use of the models provides an advantage over chance. However, the efficiency is still low. To face this problem, it is planned to use the neural network methodology in the future. Besides, efficiency could be improved by dividing the record into a verification period and a training period. To forecast each year, the entire previous record would be used for the design of the models. Likewise, the monthly FWI index forecast would also improve the results, instead of forecasting a single seasonal index representative of the entire

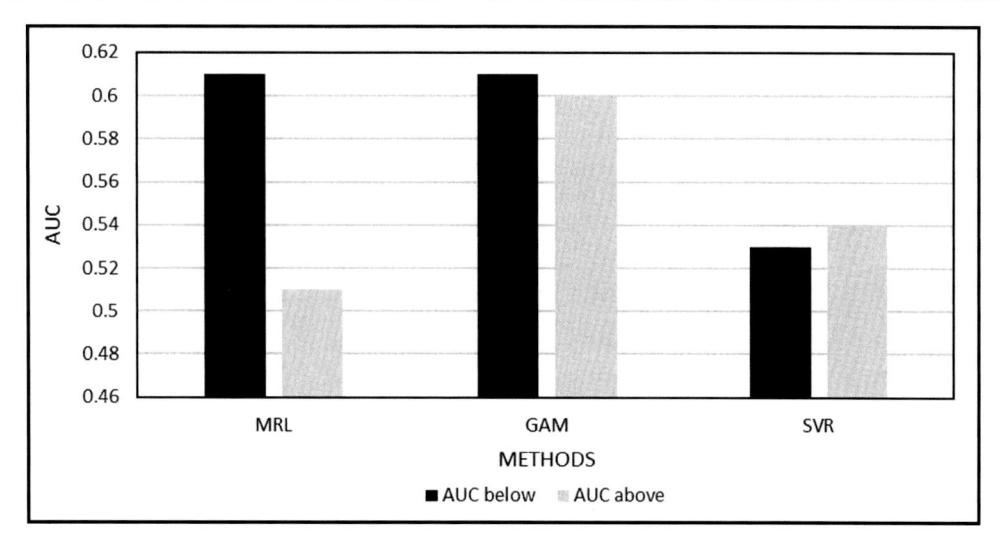

Fig. 13.12 AUC for the *above* and *below* categories derived from MLR, GAM, and SVR models

season. This would allow the incorporation of specific predictors for each month. This will be the subject of future research.

Acknowledgements Meteorological data to calculate were provided by the National Meteorological Service of Argentina. Large scale variables were obtained from the NCEP-NCAR reanalysis. Images were provided by the NOAA/ESRL Physical Sciences Laboratory, Boulder Colorado from their Web site at http://psl.noaa.gov/. This research was supported by UBACYT 2020-2022 20020190100090BA and UBACYT 2017-2019 20020160100009BA projects.

References

Alessandro AP (2003) Influence of blocking on temperature and precipitation in Argentina during the 90's decade. Rev Meterol 28:39–52

Bianchi LA (2007) Métodos para definir clases de peligro de incendio en regiones de argentina basados en un indicador meteorológico. Final project of the Forest Engineering degree, Universidad de la Patagonia San Juan Bosco, Sede Esquel

Bianco J, Di Lucca A, Saurral R, Bertolotti M (2005) Análisis de las condiciones meteorológicas durante los incendios en el centro y sur de la Argentina en mayo de 2005. Publicado en actas del Congremet IX

British Columbia, Ministry of Forests, Protection Branch (1978) Fire weather indices: decision aids for forest operations in British Columbia. 1978. British Columbia Ministry of Forests, Victoria, British Columbia, 40 p

Cheney P, Sullivan A (1997) Grassfires: fuel, weather and fire behaviour. CSIRO Publishing, Collingwood

Cortes C, Vapnik V (1995) Support-vector networks. Mach Learn 20:273–297

Deeming JE, Lancaster JW (1971) Philosophy, implementation—national fire danger rating system. USDA Fire Control Notes v32 i2. 4-8

Deeming JE, Burgan RE, Cohen JD (1977) The National Fire-Danger Rating System—1978. USDA Forest Service General Technical Report INT-39. Intermountain Forest and Range Experiment Station, Ogden, UT, USA

Dentoni MC, Defossé G, Labraga JC, del Valle H (2001) Atmospheric and fuel conditions related to the Puerto Madryn fire of 21 January, 1994. Meteorol Appl 8:361–370

Dentoni MC, Muñoz MM (2012) Informe Técnico No. 1. Sistemas de Evaluación de Peligro de Incendios. Editor: Plan Nacional de Manejo del Fuego. Programa Nacional de Evaluación de Peligro de Incendios y Alerta Temprana

Dentoni MC, Muñoz MM, Marek DS (2015) Fire danger rating as a tool for fire management. Implementation of the CFFDRS: the Argentine experience. In: Alexander ME, Leblon B (eds) Current international perspectives on wildland fires, mankind and the environment. Nova Publishers, pp 101–120

Duffy PA, Walsh JE, Graham JM, Mann DH, Rupp TS (2005) Impacts of large-scale atmospheric-ocean variability on Alaskan fire season severity. Ecol Appl 15:1317–1330

Flannigan M, Todd B, Wotton M, Stocks B, Skinner W, Martell D (2000) Pacific sea surface temperatures and their relation to Area Burned in Canada. In: Third symposium on fire and forest meteorology, 9–14 Jan, Long Beach, CA, by the AMS, Boston, MA

Forestry Canada Fire Danger Group (1992) Development and structure of the Canadian Forest Fire Behaviour Prediction System. Information Report ST-X-3, Forestry Canada, Ottawa, Ontario

Fosberg MA (1978) Weather in wildland fire management: the fire weather index. In: Proceedings, conference on Sierra Nevada Meteorology, American Meteorological Society, South Lake Tahoe, California

Garreaud RD, Vuille M, Compagnucci R, Marengo J (2008) Present-day South American climate. Palaeogeogr Palaeoclimatol Palaeoecol 281(3–4):180–195

Goodrick SL (2002) Modification of the Fosberg Fire Weather Index to include drought. Int J Wildland Fire 11:205–211

Gill AM, Christian KR, Moore PHR, Forrester RH (1987) Bushfire incidence, fire hazard and fuel-reduction burning. Aust J Ecol 12:299–306

Hess JC, Scott CA, Hufford GL, Fleming MD (2001) El Nino and its impact on fire weather conditions in Alaska. Int J Wildland Fire 10:1–13

Holz A, Veblen TT (2011) Wildfire activity in rainforests in western Patagonia linked to the Southern Annular Mode. Int J Wildland Fire 21(2):114–126

Kalnay E, Kanamitsu M, Kistler R, Collins W, Deaven D, Gandin L, Iredell M, Saha S, White G, Woollen J, Zhu I, Chelliah M, Ebisuzaki W, Higgings W, Janowiak J, Mo KC, Ropelewski C, Wang J, Leetmaa A, Reynolds R, Jenne R, Joseph D (1996) The NCEP/NCAR Reanalysis 40 years-project. Bull Amer Meteor Soc 77:437–471

Kitzberger T, Veblen TT (1997) Influences of humans and ENSO on fire history of Austrocedrus chilensis woodlands in northern Patagonia, Argentina. Ecoscience 4:508–520

Kitzberger T (2002) ENSO as a forewarning tool of regional fire occurrence in northern Patagonia, Argentina. Int J Wildland Fire 11:33–39

Kunst C, Bravo S, Panigatti J (2003) El Fuego en los Ecosistemas Argentinos, INTA Santiago del Estero, Argentina 2003, 200 pp

McArthur AG (1966) Weather and grassland fire behaviour. Comm. of Australia Forestry and Timber Bureau Leaflet 100, 23 pp

McArthur AG (1967) Fire behaviour in Eucalypt forests. Department of National Development Forestry and Timber Bureau, Canberra, Leaflet 107

Marcuzzi E, Bianco J, Cerne B, Dentoni MC (2006) Las condiciones atmosféricas durante el incendio del lago Machónico en Diciembre de 1999. Ecofuego 2006. Segunda Reunión. Patagónica y Tercera Reunión Nacional sobre Ecología y Manejo del Fuego. 25 al 28 de abril de 2006, Esquel, Chubut, Argentina

Marcuzzi E, Hoevel R (2009) Condiciones meteorológicas en torno a los incendios ocurridos en el Delta del Paraná durante el mes de Abril de 2008. In: XIII Congreso Latinoamericano e Ibérico de Meteorología (CLIMET XIII) and X Congreso Argentino de Meteorología (CONGREMET X), 05–09 octubre de 2009, Buenos Aires, Argentina

Marcuzzi EA, Nicora MG, Bali JL, Dentoni MC (2016) Análisis de dos situaciones meteorológicas asociadas a incendios por rayos sobre el noroeste de la Patagonia. RALDA 2016. XVI Reunión Argentina y VIII Latinoamericana de Agro-meteorología. 20–23 de Septiembre, Puerto Madryn. Chubut, Argentina, Actas AT4-021, pp 254–255

Nan S, Li J (2003) The relationship between summer precipitation in the Yangtse River Valley and the previous Southern hemisphere Annular Mode. Geophys Res Lett 30(24):2266

Reboita MS, Ambrizzi T, Da Rocha R (2009) Relationship between the Southern Annular Mode and Southern Hemisphere atmospheric systems. Rev Bras Meteorol 24(1):48–55

Rothermel RC (1972) A mathematical model for predicting fire spread in wildland fuels. Res. Pap. INT-115. U.S. Department of Agriculture, Forest Service, Intermountain Forest and Range Experiment Station, Ogden, UT, 40 pp

Silvestri G, Vera C (2003) Antarctic oscillation signal on precipitation anomalies over southeastern South America. Geophys Res Lett 30(21):2115

Silvestri G, Vera CS (2009) Nonstationary impacts of the southern annular mode on southern hemisphere climate. J Climate 22:6142–6148

Simard AJ, Haines DA, Main WA (1985) Relations between El Niño /Southern Oscillation anomalies and wildland fire activity in the United States. Agric for Meteorol 36:93–104

Skinner WR, Stocks BJ, Martell DL, Bonsai B, Shabbar A (1999) The association between circulation anomalies in the mid-troposphere and area burned by wildland fire in Canada. Theor Appl Climatol 63:89–105

Taylor SW (2001) Considerations for applying the Canadian Forest Fire Danger Rating System in Argentina. Victoria, BC Ministry of Forests 26

Thompson DW, Wallace JM (2000) Annular modes in the extratropical circulation. Part I: Month-to-month variability. J Climate 13:1000–1016

Van Wagner CE, Pickett TL (1985) Equations and FORTRAN program for the Canadian forest fire weather index system. Environment Canada, Forestry Service 33, Ottawa, Canada

Van Wagner CE (1987) Development and structure of the Canadian forest fire weather index system. Canadian Forestry Service 35, Ottawa, Canada

Veblen TT, Kitzberger T, Villalba R, Donnegan J (1999) Fire history in northern Patagonia: the roles of humans and climatic variation. Ecol Monogr 69:47–67

Veblen TT, Kitzberger T, Raffaele E, Lorenz DC (2003) Fire history and vegetation changes in northern Patagonia, Argentina. In: Veblen TT, Baker WL, Montenegro G, Swetnam TW (eds) Fire and climatic change in temperate ecosystems of the Western Americas. Springer, New York, pp 265–295

Werth PA (2011) Chapter 3: Critical fire weather patterns. In: Synthesis of knowledge of extreme fire behavior: volume I for fire managers. United States Department of Agriculture Forest Service Pacific Northwest Research Station General Technical Report PNW-GTR-854, pp 25–48

Wilks DS (1995) Statistical methods in the atmospheric sciences (an introduction). International geophysics series. Academic, San Diego, CA, USA, 467 pp

Williams AAJ, Karoly DJ (1999) Extreme fire weather in Australia and the impact of the El Nino-southern Oscillation. Aust Meteorol Mag 48:15–22

Wood S (2006) Generalized additive models: an introduction with R, 2nd edn. CRC Press, Taylors & Francis, 474 pp

Quantifying the Spatio-seasonal Water Balance and Land Surface Temperature Interface in Chandrabhaga River Basin, Eastern India

14

Susanta Mahato and Swades Pal

Abstract

Seasonal water scarcity and thermal uncomfortability in the rural environment are the increasing challenges in the growingly populated Chandrabhaga river basin area of Chottanagpur plateau fringe area. The present work, therefore, intends to investigate spatio-seasonal pattern of water balance and temperature as well as their interlinkages. High degree of spatio-seasonal variation in water balance and the land surface temperature is clearly recognized. Water deficit state is identified in seven Summer and Winter months (maximum 150.59 mm in April) and surplus state in the five monsoon months (310.81 mm in September). Upper and middle catchments are sensitive to water deficit (18.83–150.59 mm) and it withstands against multi-cropping and enhances land surface temperature (LST). LST is high (30–41 °C) in the areas prone to water deficit and it is quite less in the surplus areas. The spatial correlation coefficient between water deficit and LST is −0.2366 and it is 0.1921 between water surplus and LST and these are significant at 0.05 level of significance. It clearly establishes the linkages between water balance state and LST.

Keywords

Water balance · Water deficit or surplus · Spatio-seasonal variation of water balance · Land surface temperature (LST) · Interlinkage between water balance and LST

14.1 Introduction

In the Developing countries, about 90% of freshwater is used for agriculture purposes (Shiklomanov 2000; Perea et al. 2016) and about 70% of people rely on agriculture (Meena and Bisht 2017). In semi-arid regions it causes extra burden on natural storage units (Krol and Bronstert 2007; Yadav and Lal 2018). As consequence of this practice further water deficit, soil moisture deficit, etc. are becoming very usual (Shao et al. 2009; Vero et al. 2014; Zandalinas et al. 2016; Mohanty et al. 2017). Some areas, especially Indian subcontinent suffer from the skewed nature of annual rainfall distribution and evapotranspiration. Thus, growing need but increasing scarcity of such precious and irreplaceable resources is, in fact, an impending concern to them (Wada et al. 2012; Rohde et al. 2015). As the largest consumer of

---

S. Mahato
Special Centre for Disaster Research (SCDR), Jawaharlal Nehru University, New Delhi 110067, India
e-mail: susantamahato@jnu.ac.in

S. Pal (✉)
Department of Geography, University of Gour Banga, Malda 732103, India
e-mail: swadespal2017@gmail.com

fresh water and groundwater in the world, India withdraws 250 km^3 per year (Jha 2013). This amount considers the query how to be managed such growing need for water for agriculture and other sectors. Then how could drinking water be secured for the future? To understand the principal source of water, it is highly essential to consider the rainfall regime. Spatial skewness of rainfall distribution, loss of water through evapotranspiration, runoff, etc. are also essential while dealing with water balance, surplus, or deficit of water over space. Without having any major project for rainwater harvesting, a major portion of rainfall during monsoon months (June to October) is wasted. If some parts of water flow could be arrested seasonal rainfall deficit in non-monsoon seasons to some extent could be managed (Surendran et al. 2015; Mooley and Parthasarathy 1984; Yaduvanshi and Ranade 2017). Moreover, seasonal soil moisture drought is withstanding against the multi-cropping state of many parts of the country (Thober et al. 2015; Zhang et al. 2017) and as a consequence of this either people compel to grow some inferior crops or remain seasonally fallow. In India, large semi-arid tracts including parts of Central India, Western India, and Southern India fall under this tract where this problem is very prominent (Mishra et al. 2014; Sathyanadh et al. 2016). In a word, the scientific study of water balance is highly necessary for effective decision-making of water uses in progressive sectors (Khalaf and Donoghue 2012). For the hydrological cycle and water management, the replenishment rate of groundwater resources plays a key component. Seasonal or perennial water balance deficit may enhance land surface temperature (LST). Assuming this fact the present work started an investigation on the specific relation between water balance deficit and positive temperature anomaly on spatio-seasonal scale. A good number of works highlighted change in land surface temperature in relation to shift of land use/land cover (LU/LC) (Pal and Ziaul 2017; Ziaul and Pal 2016; Jafari and Hasheminasab 2017; Metz et al. 2017) highlighted increase of urban built-up land is caused for enhancing LST. González et al. (2015) and Deo and Şahin (2017) documented that abolishing water bodies, shallowing water bodies, replacing water bodies with other land uses are responsible for rising temperatures. Pal and Ziaul (2017) reported that canopy density is negatively related to LST. But the present interest of study, i.e., role of the water balance on LST is almost unexplored.

In some recent research, remote sensing (RS) and geographical information system (GIS) has contributed a major role in providing information for water resource studies and LST. Some of the eminent scholars successfully explored water balance status, and LST pattern over space using these RS and GIS techniques. Soil moisture balance study by Crosbie et al. (2015), González-Rojí et al. (2018), Hu et al. (2015), water balance study by Bastiaanssen et al. (2005), Hassan-Esfahani et al. (2015), Saiter et al. (2016) are very well known. Spatiotemporal temperature extraction from multi-temporal satellite images is well studied by Lee et al. (2012), Siu and Hart (2013), Jiang and Weng (2017), and Yao et al. (2018) for different urban sectors. But linkages between water balance and LST are still to be investigated. Therefore, the present work focused on calculating water balance by incorporating frequently used parameters like rainfall, evapotranspiration, and runoff on spatio-seasonal scale in the Chandrabhaga river basin of Eastern India and tried to find out the influence of the surplus or deficit water balance on LST.

The Chandrabhaga river basin (area: 119.34 km^2) is located over the Chattanagpur plateau fringe of Rarh tract of Jharkhand and West Bengal (Bagchi and Mukerjee 1983). Geologically 60% of the basin area lies in the upper catchment which is composed of granitic-gneissic rock of Pleistocene age (50 lakh years old) overlain by coarse-grain lateritic soil and 30% of the area at the lower catchment is made with newer alluvium of the Holocene period (Fig. 14.1) (GSI 1985). Coarse-grain lateritic soil with high ferrous and silica content contents insists high range of temperature. Most part of the basin area is dominated by agricultural land with poor qualities of soil fertility and soil moisture (during pre-monsoon 8–11%). The climate of this region is characterized by sub-

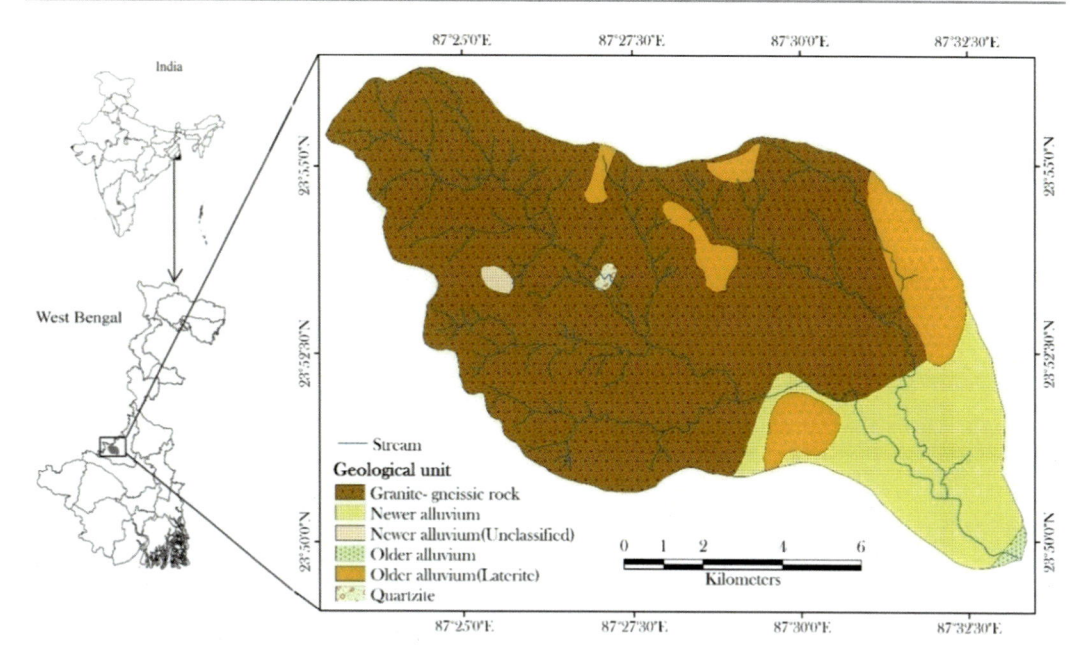

Fig. 14.1 Chandrabhaga river basin showing the geological composition

tropical monsoon with seasonal wet and dry spells of rainfall and cold and hot spells of temperature. The entire year is subdivided into four seasons viz. (1) Winter season (January and February) with low temperature, low humidity level devoid of rain, (2) Pre-monsoon season (March to May) with little rain and high temperature and evaporation, (3) Monsoon season (June to September) with maximum (about 82% of total rain) rain and high temperature and (4) Post-monsoon season (October to November). Average annual rainfall of this basin as gauged by Suri meteorological station is 1444.432 mm. High degree of seasonality of rainfall is reflected by 82% rainfall during the months of June to September (monsoon period).

Monthly spatial rainfall data is extracted from TRMM (Tropical Rainfall Measuring Mission). Although on global scale, the spatial resolution (0.25°) of such multispectral data is considerably high, but for the study of the small-scale hydrological unit, such resolution is still to be considered as coarse (Immerzeel et al. 2009). Duan and Bastiaanssen (2013), Tong et al. (2014) suggested that for downscaling the study, close spacing point data can be created over the study area and inverse distance weighting (IDW) can be adopted for upgrading the spatial scale. Soil texture map of the basin is prepared from the map provided by National Bureau of Soil Survey and Land Use Planning department (India).

14.2 Materials and Methods

LANDSAT OLI/TIRS data (path/row: 139/43; spatial resolution 30 m) for 2016 have been obtained from the US Geological Survey (USGS) Global Visualization Viewer and used for deriving land surface temperature (LST), land use/cover (LuLc), evapotranspiration (ET).

14.2.1 Method for Water Balance Calculation

Simple water balance equation (14.1) can be used for predicting regional and seasonal water deficit and surplus (Gokmen et al. 2013; Crosbie et al. 2015). Precipitation, runoff, evaporation, and recharge components are used for this purpose.

$$P = Q + \mathrm{ET_A} + R_{\mathrm{GW}} \quad (14.1)$$

where, P = precipitation; Q = runoff; $\mathrm{ET_A}$ = Actual Evapotranspiration; R_{GW} = Groundwater recharge.

14.2.2 Method for Surface Runoff Estimation

The Soil Conservation Service-Curve Number (SCS-CN) method (USDA) is used for runoff calculation (Eq. 14.2).

$$Q = \frac{(P - I_a)^2}{P - I_a + S} \quad (14.2)$$

where, Q is actual surface runoff in mm, P is rainfall in mm, I_a is 0.4S/0.3S/0.2S/0.1S (season and climatic region-specific) initial abstraction (mm) or losses of water before runoff begins by soil and vegetation (such as infiltration, or rainfall interception by vegetation), 0.3S is usually used for wet, 0.1S is used dry seasons, S is the potential maximum retention. S can be calculated using Eq. 14.3.

$$S = \frac{25,400}{\mathrm{CN}} - 254 \quad (14.3)$$

Some effective parameters for estimating CN for different spatial units are Land use land cover (LULC), hydrological soil group (HSG), and antecedent moisture condition (AMC). LULC map is prepared from multispectral Landsat OLI image following supervised classification (Maximum Likelihood) technique. Accuracy assessment of the classified image is done based on 231 google reference sites and 131 ground base references. Calculated Kappa coefficient of the classified image is 0.93. Arc-CN-Runoff extension tool of ArcGIS is used for precise quantification of runoff. Land use/land cover thematic layer and polygon of soil groups have been used in ARC-CN runoff.

Relative error (RE) (Eq. 14.4) and Nash–Sutcliffe efficiency (NSE) (Eq. 14.5) (Nash and Sutcliffe 1970), are widely used for evaluating the performance of runoff model (Moriasi et al. 2007). This techniqueis used here for testing the model.

$$\mathrm{RE} = \frac{(Q_i^{\mathrm{cal}} - Q_i^{\mathrm{obs}})}{(Q_i^{\mathrm{obs}})} \times 100\% \quad (14.4)$$

$$\mathrm{NSE} = 1 - \frac{\sum_{i=1}^{n} (Q_i^{\mathrm{cal}} - Q_i^{\mathrm{obs}})^2}{\sum_{i=1}^{n} (Q_i^{\mathrm{cal}} - Q_{\mathrm{mean}}^{\mathrm{obs}})^2} \quad (14.5)$$

where, Q_i^{obs} = runoff of the ith observation, Q_i^{cal} = calculated runoff of the same observation, $Q_{\mathrm{mean}}^{\mathrm{obs}}$ = mean observed runoff.

RE value nearer to 0 specifies high optimality of the model performances. NSE ranges between $-\infty$ and 1, where NSE = 1 is the optimal result. Values of NSE between 0 and 1 are usually assumed as acceptable levels of performance, whereas the value of NSE < 0 points out that the mean observed value is a better predictor than the simulated value, and therefore, is treated as unacceptable (Moriasi et al. 2007). Khatun and Pal (2016) used these testing techniques for assessing the performance level of the runoff model in a similar environment. Actual monthly runoff data is measured at the confluence of Chandrabhaga river at Kaspai for validating models.

14.2.3 Method for Evapotranspiration Estimation

Remote sensing (RS) based Surface Energy Balance Algorithm for Land (SEBAL) is fundamentally used for estimating $\mathrm{ET_A}$. SEBAL utilizes multispectral RS data in corroboration with meteorological data for estimating instantaneous and daily surface energy balance (Eq. 14.6) components (Oliveira et al. 2014). This mechanism produces pixel-to-pixel energy balance information considering the latent heat flux (LE expressed as W m^{-2}). The detail study was made by Bastiaanssen et al. (1998), Silva et al. (2015), and Mahmoud and Alazba (2016).

$$\text{LE} = R_n - G - H \quad (14.6)$$

where, R_n = net radiation (W m^{-2}), G = heat flow in the soil (W m^{-2}), and H = sensible heat flux (W m^{-2}).

The surface albedo (α_s) is worked out from MODIS reflectance data following Trezza et al. (2013) (Eq. 14.7) and the surface emissivity (ε_o) (Eq. 14.8) is articulated based on the LAI (Leaf Area Index) following Allenet al. (2007).

$$\alpha_S = 0.215_{r1} + 0.266_{r2} + 0.242_{r3} + 0.129_{r4} \\ + 0.112_{r6} + 0.036_{r7} \quad (14..7)$$

$$\varepsilon_o = 0.95 + 0.01 \text{LAI} \quad (14.8)$$

where, r_1, r_2, r_3, r_4, r_6, and r_7 represent the reflectance spectral bands of the MODIS data.

The LE conversion is calculated from the evaporative fraction (EF), following Bastiaanssen et al. (1998) as presented in Eqs. (14.9) and (14.10). This function supposes that the EF is steady all through the day, but always it may not be as exhibited by Van Niel et al. (2011) and Van Niel et al. (2012). Ruhoff et al. (2012) rightly portrayed that the instant ET$_A$rate can be acquired as a diurnal average with anticipated inconsistency over outsized scales.

$$\text{EF} = \frac{\text{LE}}{R_n - G} \quad (14.9)$$

$$\text{ET}_{AD} = 0.035 \text{EF} R_{n24h} \quad (14.10)$$

where, R_{n24h} = net radiation of 24 h calculated from daily solar radiation, surface albedo, and atmospheric transmissivity data, as recommended by Bastiaanssen et al. (1998). Sequentially, the daily actual evapotranspiration is combined along with monthly proportion to the potential evapotranspiration (ET$_P$), based on the postulation so as to the share between the ET$_{AD}$ and ET$_P$ remains equivalent right through the month (Morse et al. 2000) (Eq. 14.11).

$$\text{ET}_{AM} = \frac{\text{ET}_{AD}}{\text{ET}_P} \times \text{ET}_{PM} \quad (14.11)$$

where ET$_{AM}$ = monthly actual evapotranspiration and ET$_{PM}$ = monthly cumulative potential evapotranspiration.

For validating ET data derived from satellite data, the empirical formulation can be deployed. Donohue et al. (2010) established that Penman's (1948) equation yields best-fitted result while calculating ET$_P$ using meteorological data. But based on the ground data available here Blaney–Cridle (1945) method is adopted.

$$\text{ET}_{BC} = P(0.46T + 8.13) \quad (14.12)$$

where ET$_{BC}$ potential evapotranspiration is in mm/day for a period of consideration. T = mean daily temperature in °C for a period considered = $(T_{max} + T_{min})/2$; P = mean daily percentage of total annual daytime hours depending on month and latitude.

14.2.4 Method for Estimation of Recharge

Estimation of groundwater recharge is done from the water balance equation. Positive water balance deducing ET and surface runoff from rainfall can be considered as recharge. Deficit water balance indicates the excess use of soil moisture and soil moisture drought. Groundwater table fluctuation data is one of the sensitive indicators of groundwater recharge. It is assumed that commencing rise of the groundwater table or piezometric surface after rainfall signifies groundwater recharge. So, a simple comparison of monthly estimated recharge from water balance equation and monitoring of the degree and direction of groundwater level fluctuation may help to understand the accuracy level of calculation. Some empirical equations formulated by Chaturvedi (1973), Kumar, and Seetapati (2002) can also be used for calculating recharge and validating the water balance-based recharge data. Recharge volume calculated by the Central Ground Water Board (CGWB) is also taken into consideration for validating the same.

$$R = 2.0(p - 15)^{0.4} \quad (14.13)$$

$$R = 1.2(p - 13)^{0.5} \quad (14.14)$$

where R = groundwater recharge from rainfall during monsoon; p = mean rainfall in monsoon (inch).

14.2.5 Method for Land Surface Temperature Assessment

14.2.5.1 Conversion of the Digital Number (DN) to Spectral Radiance (Lλ)

Every object emits thermal electromagnetic energy as its temperature is above absolute zero (K). Following this principle, the signals received by the thermal sensors (ETM+) can be converted to at-sensor radiance. The spectral radiance ($L\lambda$) is calculated using Eq. (14.15) (Landsat Project Science Office 2002).

$$L\lambda = \text{``gain''} * \text{QCAL} + \text{``offset''} \quad (14.15)$$

where "gain" is the slope of the radiance/DN conversion function; DN is the digital number of a given pixel; bias is the intercept of the radiance/DN conversion function. This is also given as:

$$L\lambda = L_{MIN}\lambda + [\{(L_{MAX}\lambda - -L_{MIN}\lambda)/ \\ (\text{QCAL}_{MAX} - \text{QCAL}_{MIN})\} * \text{QCAL}] \quad (14.16)$$

where, $\text{QCAL}_{MIN} = 0$, $\text{QCAL}_{MAX} = 255$ and QCAL = Digital Number (DN of each pixel).

The $L_{MIN}\lambda$ and $L_{MAX}\lambda$ are the spectral radiances for band 6 at digital numbers 0 and 255, respectively. These compute to 3.2 W m^{-2} sr and 12.65 W m^{-2} sr, respectively.

Substitution of the respective values in Eq. (14.16) gives a simpler Eq. (14.17).

$$L\lambda = (0.037059 * \text{DN}) + 3.2 \quad (14.17)$$

14.2.5.2 Conversion of Spectral Radiance (Lλ) to At-Satellite Brightness Temperatures (TB)

Corrections for emissivity (ε) have been applied to the radiant temperatures according to the nature of the land cover. In general, vegetated areas have given a value of 0.95 and non-vegetated areas 0.92 (Nichol 1994). The emissivity corrected surface temperature has been computed following Artis and Carnahan (1982).

$$T_B = \frac{K_2}{\ln\left(\frac{K_1}{L\lambda} + 1\right)} \quad (14.18)$$

where, T_B = At-satellite brightness temperature (K), $L\lambda$ = Spectral Radiance in W m^{-2} sr^{-1} μm^{-1}.

K_1 and K_2 = K_2 and K_1 are two pre-launch calibration constants. (For the Landsat 7 ETM + 6.2 band, these compute to 1282.71 K and 666.09 W m^{-2} sr^{-1} μm, respectively.)

14.2.5.3 Land Surface Temperature (LST)

The obtained temperature values above are referenced to a black body. Therefore, corrections for spectral emissivity (ε) become necessary. These can be done according to the nature of land cover (Snyder et al. 1998) or by deriving corresponding emissivity values from the NDVI values for each pixel. The emissivity corrected land surface temperatures (St) have been computed following Artis and Carnahan (1982).

$$\text{LST} = T_B/[1 + \{(\lambda * TB/\rho) * \ln \varepsilon\}] \quad (14.19)$$

where, LST = Land Surface Temperature (LST) in Kelvin, λ = wavelength of emitted radiance in meters (for which the peak response and the average of the limiting wavelengths (λ = 11.5 μm) (Markham and Barker 1985) is used, $\rho = h*c/\sigma$ (1.438 × 10^{-2} m K), σ = Boltzmann constant (1.38 × 10^{-23} J/K), h = Planck's constant (6.626 × 10^{-34} J s), and c = velocity of light (2.998 × 10^8 m/s) and ε = emissivity (ranges between 0.97 to 0.99). For calculating emissivity of land surface Eq. (14.20) is used.

$$\text{Land surface emissivity } (\varepsilon) = 0.004 * P_v + 0.986 \quad (14.20)$$

where Pv is the proportion of vegetation which can be calculated following Eq. (14.21).

$$P_v = \left(\frac{\text{NDVI}_{j_r} - \text{NDVI}_{\min}}{\text{NDVI}_{\max} - \text{NDVI}_{\min}}\right)^2 \quad (14.21)$$

For convenience, the above-derived LSTs' unit is converted to degree Celsius considering the relation between Kelvin and degree Celsius (0 °C = 273.15 K). Degree Celsius data is rather conventional for interpreting the temperature intensity.

For assessing the influences of deficit or surplus water balance patterns on LST, spatial correlation and regression were made.

14.3 Results and Analysis

14.3.1 Rainfall Analysis

Figure 14.2a–l represents the spatial pattern of rainfall in different months of 2016. The annual precipitation data ranges from 1965.06 to 1991.81 mm with insignificant spatial variation in 2016. In the upper and middle catchments during monsoon times the concentration of rainfall is high. Highest rainfall is recorded in monsoon months specifically in July, August, and September. Location of low-pressure axis orients the distribution of rainfall (Fig. 14.2g–i). More than 80% of rainfall is recorded during monsoon months and other months specifically winter and summer months suffer from rainfall scarcity (Table 14.1).

For validating TRMM data, rain gauge data collected from Suri meteorological station is collected and compared. Result states that there is an identical pattern of average rainfall almost in all months (Fig. 14.3a). Only 0.23–21.65% departure of TRMM data from gauge data is recorded and based on this TRMM data can be accepted as representative. Yong et al. (2010), Melo et al. (2015), Coelho et al. (2017) have also applied similar approach for validating TRMM data in the reference data-scarce state.

14.3.2 Runoff Analysis

Figure 14.4a–l represents the spatial sharing of surface runoff in different months of 2016. The maximum runoff volume is recorded at 332.37 mm and 297.11 mm in the month of August and September, respectively (Fig. 14.4h and i) and this high runoff is triggered by enhanced amount of rainfall in that period. Annual average runoff volume for this study area is 51.77 mm. which is 35.227% of average annual rainfall. Out of total annual average runoff, pre-monsoon, monsoon, post-monsoon and winter seasons, respectively carry 6.20%, 69.30%, 22.75%, and 1.75% to total runoff. This information is quite natural in parity with amount of rainfall that happens during these seasons. Spatially this is usually high in the bare lateritic parts of the basin. In few vegetated patches, high runoff is estimated and it is because of very sparse presence of vegetation (Table 14.2).

For testing, the runoff models of different months' Relative error (RE) and Nash–Sutcliffe efficiency (NSE) indices have been used. Result states that RE value ranges from 0.96 to 27% since January to December indicating very less departure of rainfall between estimated and observed runoff. This information validates the spatial runoff models. Average annual NSE value is 0.782 which is near 1 and therefore runoff model can be treated as responsive.

14.3.3 Evapotranspiration Analysis

Figure 14.5a–l illustrates the monthly spatial pattern of evapotranspiration (ET). Annual evapotranspiration state varies from 960.24 to 1074.15 mm/year spatially. In the lower part of the basin, ET is excessively high in monsoon months and quite less in the middle and upper

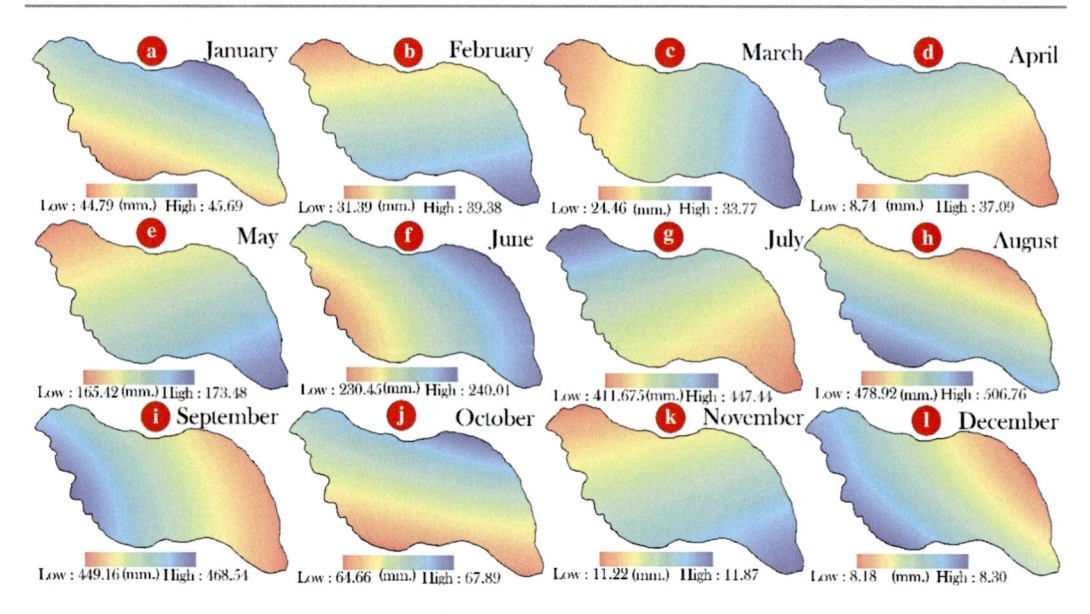

Fig. 14.2 The spatial pattern of rainfall in different months of 2016

part of the basin. Significant seasonal difference in ET is recorded. Highest monthly ET is found in the month of March (138.66 mm) whereas lowest ET (43.73 mm) in the month of December (Table 14.3). Degree of temperature is extremely high during pre-monsoon season, therefore, potential ET is high but actual ET as recorded is not expectedly high. It is because of scarce availability of surface water and soil moisture in this time conditioned by scanty rainfall or no rainfall. In monsoon time, the intensity of temperature is comparatively lower (25–33 °C) than in the pre-monsoon season but due to greater availability of surface water and soil moisture it registers quite greater ET.

Estimated monthly evapotranspiration based on SEBAL method shows a distinct pixel-wise variation in the pattern of ET. The variation of ET for different land covers shows the hourly ET ranged from 0 at bareland and rock outcrops to 0.504 mm per hour in forest areas and the daily ET varies from 0.0 to 12.09 mm per day at the same land cover classes. The mainland covers controlling the ET characteristics is the vegetated land which scatteredly covers about 20.38% of the study area. The least range of variation in the hourly ET values appears in the bare land area indicating absence of vegetation cover. The results show a clear relation between land use/land cover and solar radiation parameters and impact of vegetation cover on the ET values in pixel domain.

Figure 14.3b represents the scatter plot of atmospheric ET and calculated ET for different months of 2016. This regression graph is plotted for showing validity of calculated ET models. Monthly ET data was collected from Suri gauge station for this purpose. Strong coefficient of determination (R^2) value (0.8561) and correlation coefficient (0.785) establish the validity of ET models.

14.3.4 Groundwater Recharge/Deficit

Using water balance method, estimation of monthly groundwater recharge for the year 2016 was done. Average annual recharge is 356.386 mm with significant seasonal and intra-seasonal monthly variation. Highest estimated recharge is recorded in the months of July, August, and September (Table 14.4), and these amounts are

Table 14.1 Average, minimum, and maximum monthly rainfall of Chandrabhaga river basin with descriptive statistics

Months	January	February	March	April	May	June	July	August	September	October	November	December
Min. (mm)	44.79	31.39	24.46	8.74	165.42	230.45	411.68	478.92	449.16	64.67	11.22	8.18
Max. (mm)	45.69	39.38	33.77	34.10	173.49	240.01	447.44	506.77	468.54	67.89	11.87	8.30
Mean (mm)	45.24	35.62	39.30	22.24	169.51	235.82	426.44	492.97	458.28	66.07	11.55	8.24
SD	0.20	1.75	2.35	5.98	1.71	2.28	7.71	6.18	5.03	0.74	0.14	0.03
CV(%)	0.44	4.90	5.97	26.89	1.01	0.97	1.81	1.25	1.10	1.12	1.19	0.33

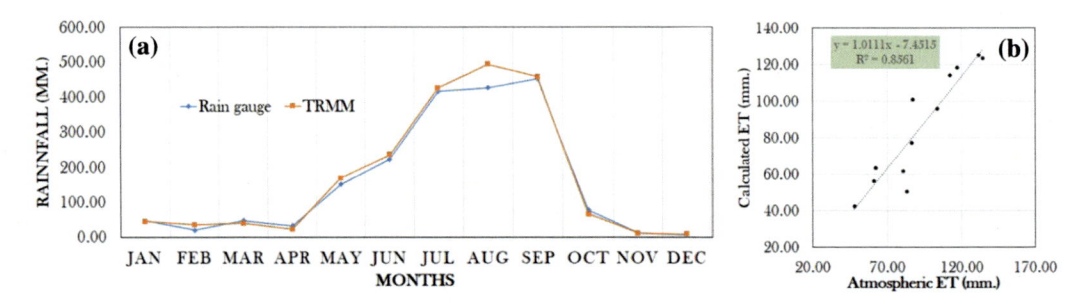

Fig. 14.3 a Comparision of monthly average rainfall data collected from TRMM data and gauge data; **b** regression pattern between monthly average observed and calculated ET

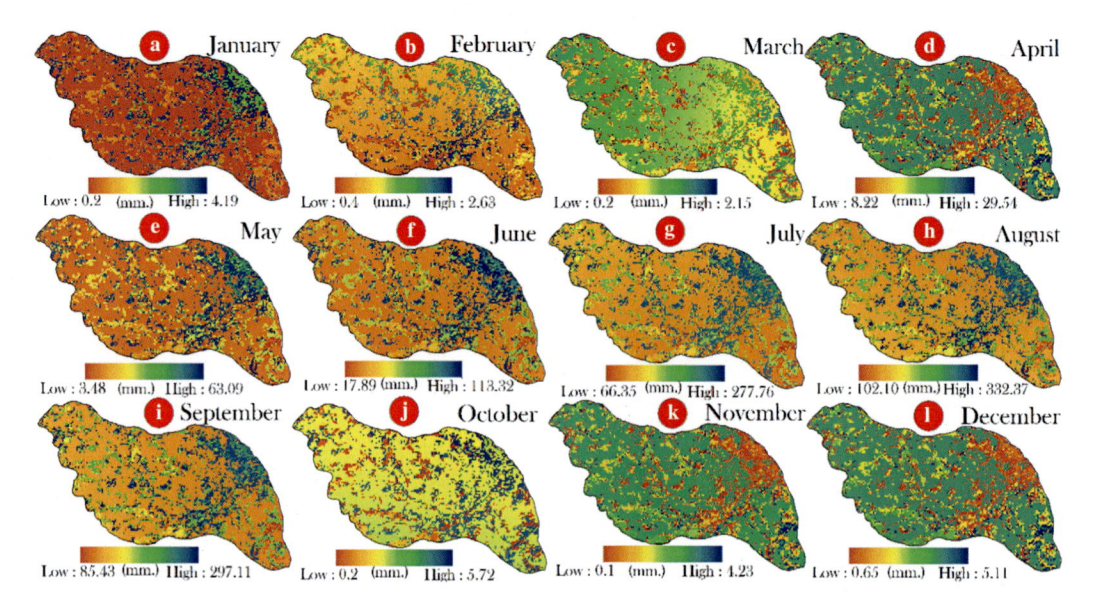

Fig. 14.4 Monthly spatial pattern of estimated surface runoff in Chandrabhaga river basin

44.36, 41.38, and 50.89% of the rainfall, respectively, as extracted from TRMM data. These recharges are the principle source of groundwater in this region. Negative water balance is recorded in winter and summer months indicating seasonal drought. Seasonal soil moisture deficit disrupts agricultural activities in these seasons. Excessive groundwater withdrawal for agriculture and other domestic and commercial uses makes drought states more prominent. Water deficit state is very explicit in the upper and middle catchments of the basin (Fig. 14.6b–d, k, l). Coarser soils with very little moisture retaining capacity, and excess ET are responsible for such stress state of moisture deficit. Poor soil moisture (<7%) in this area also supports this state. Groundwater recharge and rise of water table are well explained by recorded groundwater level data collected from Central Ground Water

Table 14.2 Average, minimum, and maximum monthly surface runoff of Chandrabhaga river basin with descriptive statistics

Months	January	February	March	April	May	June	July	August	September	October	November	December
Min.(mm)	0.20	0.40	0.20	8.22	3.48	17.90	66.35	102.10	85.43	0.20	0.10	0.65
Max. (mm)	4.19	2.63	2.15	29.55	63.09	113.32	277.76	332.37	297.11	5.72	4.23	5.11
Mean (mm)	1.89	1.64	0.78	20.16	21.29	49.61	140.49	187.13	162.31	2.58	2.31	3.11
SD	1.51	0.61	0.38	6.70	19.91	31.55	62.52	69.94	65.94	1.56	1.25	1.41
CV(%)	80.28	37.19	49.17	33.22	93.49	63.60	44.51	37.37	40.63	60.37	54.10	45.19

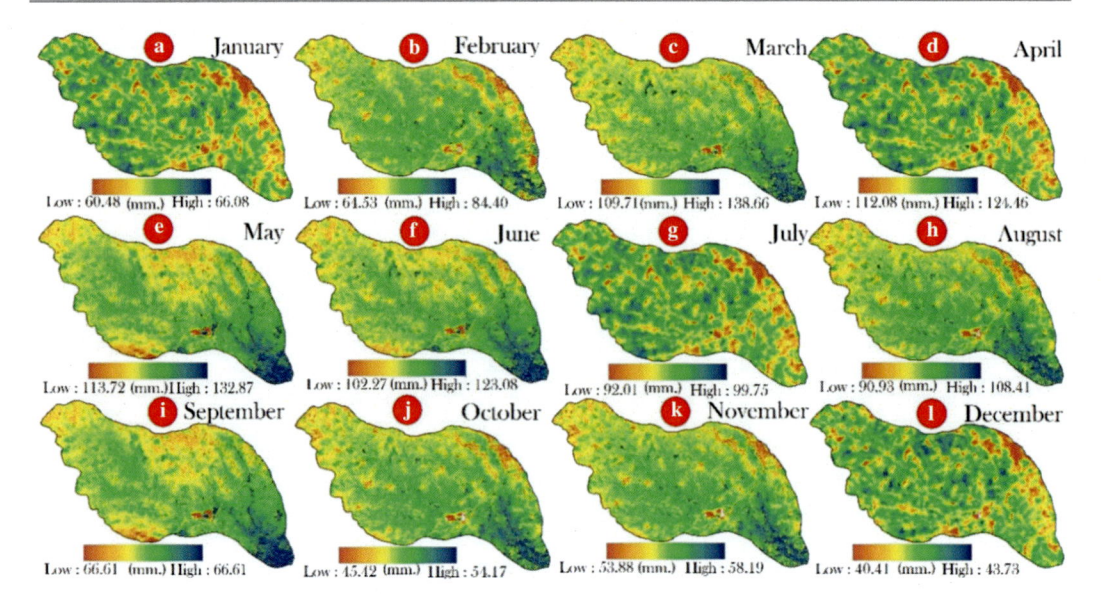

Fig. 14.5 Monthly spatial pattern of estimated ET in Chandrabhaga river basin

Board and State Water Investigation Directorate. Groundwater level during monsoon months is 5 m below ground surface.

14.3.5 Temporal Pattern of Water Balance

Finally, from all these models of water balance components, monthly water balance equations are derived (Table 14.5). January, February, March, April, May, November, and December are appeared as water deficit (negative water balance) months and rest five months are emerged as water surplus (positive water balance) months. Annual water balance states that 15.47% of area falls under water deficit and 84.53% underwater surplus. Calculation of net water balance would result in positive water balance. But, seasonal water deficit cannot be overlooked. Seasonal water deficit not only withstands against agricultural productivity in the densely populated and demanding areas but it can aggravate the temperature condition of the deficit region. Prominent soil cracking, lowering of groundwater level, drying out of ponds and wetlands are some of the evidences for such seasonal water deficit conditions.

14.3.6 Land Surface Temperature

The spatial distribution of LST in different months of 2016 is shown in Fig. 14.7. In winter season, LST ranges from 19.59 to 23.85 °C (mean 22.41 °C), pre-monsoon season temperature ranges from 23.97 to 41.43 °C (mean 35.83 °C) and in monsoon season temperature is confined within 19.74–33.23 °C (mean 26.79 °C) (Table 14.6). High temperature is registered in the upper part of the basin where lateritic bare soil predominates. Temperature regime is not only associated with incoming solar radiation but also strongly linked with land use/cover (Pal and Ziaul 2017; Ziaul and Pal 2016; Fu and Weng 2016; Feng et al. 2014). Green spaces and

Table 14.3 Average, minimum, and maximum monthly ET of Chandrabhaga river basin with descriptive statistics

Months	January	February	March	April	May	June	July	August	September	October	November	December
Min. (mm)	60.48	64.53	109.71	112.08	113.72	102.27	92.01	90.93	57.03	45.42	53.88	40.40
Max. (mm)	66.08	84.40	138.66	124.46	132.87	123.09	99.75	108.41	66.61	54.17	58.20	43.73
Mean (mm)	63.33	77.30	125.17	118.42	123.42	114.30	95.98	101.25	61.89	50.59	56.46	42.43
SD	0.85	1.06	1.32	1.92	1.10	1.14	1.19	1.05	0.55	0.52	0.25	0.44
CV(%)	1.34	1.38	1.05	1.62	0.89	1.00	1.24	1.04	0.89	1.04	0.45	1.04

Table 14.4 Average, minimum, and maximum monthly water balance (surplus or deficit states) of Chandrabhaga river basin with descriptive statistics

Months	January	February	March	April	May	June	July	August	September	October	November	December
Min. (mm)	−23.96	−48.37	−105.23	−150.60	−18.83	7.45	68.60	66.49	105.15	6.59	−49.94	−41.68
Max. (mm)	−15.42	−30.53	−84.03	−119.67	47.71	109.48	268.61	288.29	310.81	19.24	−43.27	−34.67
Mean (mm)	−19.02	−42.37	−96.74	−137.35	24.49	71.49	189.21	204.00	233.24	12.85	−47.25	−38.88
SD	1.40	1.75	2.27	7.79	19.83	31.03	62.29	70.61	66.84	1.78	1.33	1.65
CV(%)	−7.36	−4.12	−2.34	−5.67	80.98	43.40	32.92	34.61	28.66	13.86	−2.81	−4.25

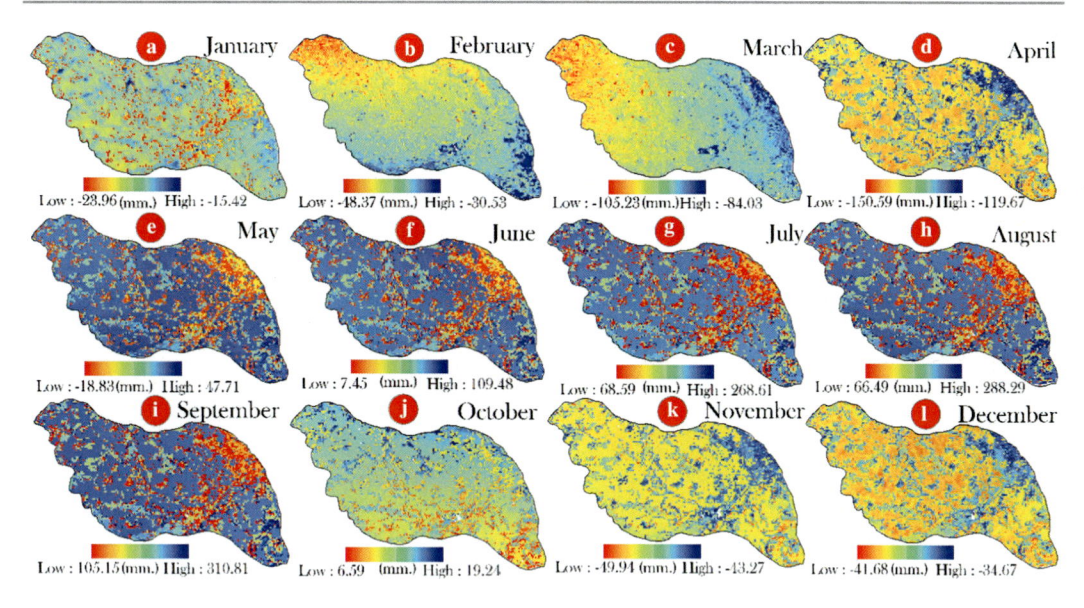

14 Quantifying the Spatio-seasonal Water Balance and Land Surface …

Fig. 14.6 Water deficit and surplus states calculated from water balance equation in different months in 2016

Table 14.5 Mean water balance component values and water balance equation; value within parenthesis indicates the ratio between the hydrological component and ppt

Months	Rainfall	Runoff	ET	Recharge	Water balance equation	Remarks
January	45.24	1.89	24.20	10.12	$P_{45.24} = Q^{0.04} + ET^{1.40} + R^{-0.42}$	Recharge negative
February	35.62	1.64	15.63	14.97	$P_{35.62} = Q^{0.05} + ET^{2.17} + R^{-1.19}$	Recharge negative
March	39.30	0.78	114.95	−86.51	$P_{39.30} = Q^{0.02} + ET^{3.18} + R^{-2.46}$	Recharge negative
April	22.24	20.16	90.82	−110.86	$P_{22.24} = Q^{0.91} + ET^{5.32} + R^{-6.18}$	Recharge negative
May	169.51	21.29	123.00	24.97	$P_{169.51} = Q^{0.13} + ET^{0.73} + R^{0.14}$	Recharge positive
June	235.82	49.61	91.21	110.14	$P_{235.82} = Q^{0.21} + ET^{0.48} + R^{0.30}$	Recharge positive
July	426.44	140.49	203.72	111.11	$P_{426.44} = Q^{0.33} + ET^{0.23} + R^{0.44}$	Recharge positive
August	492.97	187.13	214.00	145.97	$P_{492.97} = Q^{0.38} + ET^{0.21} + R^{0.41}$	Recharge positive
September	458.28	162.31	198.94	138.60	$P_{458.28} = Q^{0.35} + ET^{0.14} + R^{0.51}$	Recharge positive
October	66.07	2.58	43.23	20.20	$P_{66.07} = Q^{0.04} + ET^{0.77} + R^{0.19}$	Recharge positive
November	11.55	2.31	5.28	3.94	$P_{11.55} = Q^{0.20} + ET^{4.89} + R^{-4.09}$	Recharge negative
December	8.24	3.11	2.65	0.92	$P_{8.24} = Q^{0.38} + ET^{5.15} + R^{-4.72}$	Recharge negative

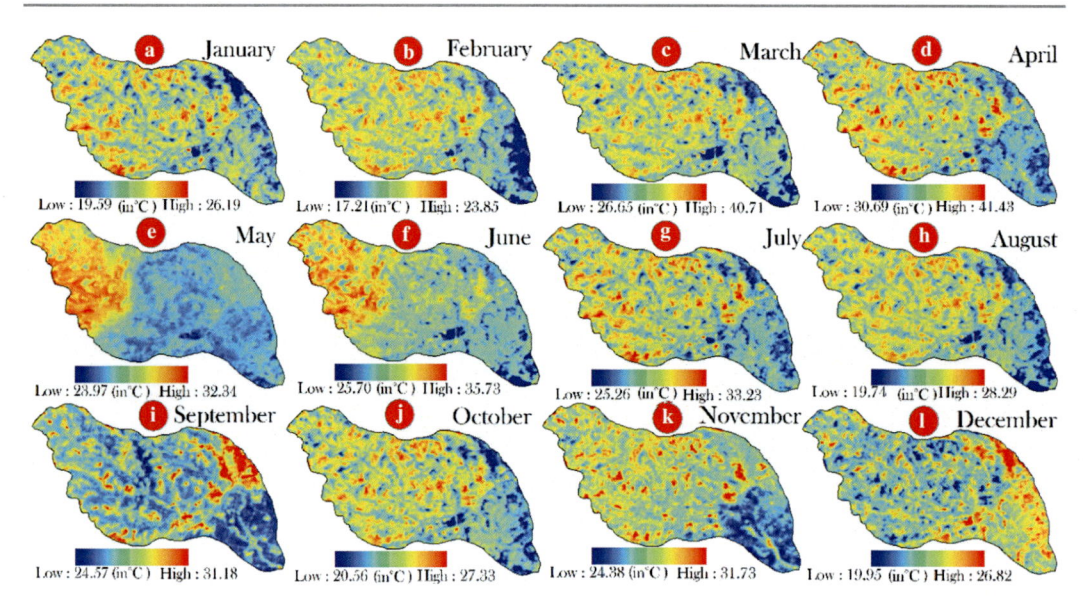

Fig. 14.7 a–l Monthly spatial LST, 2016 in Chandrabhaga river basin

14.3.7 Spatial Association Between LST and Water Deficit

In this section, it is tried to investigate is water deficit area of the study area sensitive to high temperature and reverse in the surplus area. Pixel-based spatial correlation states that although the relation is not so strong but such relation prevails thereon. Correlation coefficient between water deficit intensity and degree of LST is −0.2366 and this value is also significant at 0.05 level of confidence. Water deficit area also experiences strong ET as indicated by significant r value (−0.6748). So, along with other factors, water deficit can also be treated as one of the causes of relatively greater LST in some parts of the basin. On the contrary, when surplus water balance pixels are related to LST of the concerned pixels, it is found that the pixels having greater surplus water balance possess lesser temperature and vice versa (Fig. 14.8).

adjacency of water bodies lead to low-temperature state but bare soil portion of the study area recorded high temperature.

14.4 Conclusion

The present work tried to establish the influence of seasonal and regional water deficit states on enhanced LST and water surplus states on relatively less temperatures. For doing this monthly water balance equations are derived and mapped out spatial scale on the one hand and monthly land surface temperature is estimated and mapped out on the other hand. Finally, spatial associateship is analyzed quantitatively. From the result, it is clearly found that seven months (January to May, November, and December) are identified as water deficit months (up to −150.59 mm), and five months (June to October) are recognized as water surplus months (Highest 310.81 mm). In the deficit months, upper and middle catchments are prone to water deficit. Bare coarse-grain laterite soil also triggers high temperature. LST is also correlation coefficient between water deficit and LST is −0.2366 and it is 0.1785 between water surplus and LST and these are significant at 0.05 level of significance. So, spatial and seasonal water balance state can be treated as one of the reasons for degree of temperature distribution over the study area.

Table 14.6 Average, minimum, and maximum monthly Surface temperature of Chandrabhaga river basin with descriptive statistics

Months	January	February	March	April	May	June	July	August	September	October	November	December
Min. (°C)	19.59	17.21	26.65	30.69	23.97	25.70	25.26	19.74	24.57	20.56	24.38	19.95
Max. (°C)	26.19	23.85	40.71	41.43	32.34	35.73	33.23	28.29	31.18	27.33	31.73	26.82
Mean (°C)	22.56	20.97	35.39	35.54	27.56	31.48	29.05	25.09	25.99	24.39	26.97	23.71
SD	0.96	1.16	1.84	1.79	1.91	1.52	1.33	1.28	0.81	1.06	1.06	0.91
CV(%)	4.25	5.53	5.20	5.03	6.92	4.83	4.58	5.10	3.12	4.35	3.92	3.86

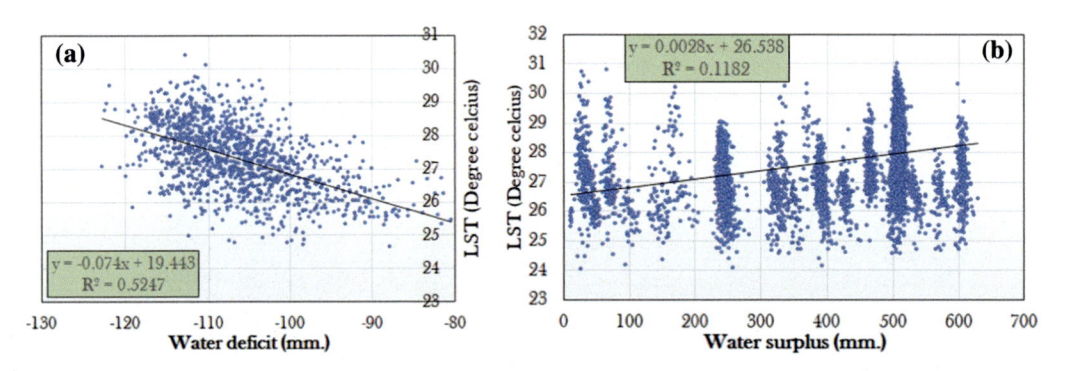

Fig. 14.8 **a**, **b** Correlation between LST and water deficit pixels and water surplus pixels

Funding The first author of the article would like to thank University Grants Commission (UGC Ref. No. 3379/(OBC)(NET-NOV 2017), New Delhi, India for providing financial support as junior research fellowship to conduct the research work presented in this paper.

References

Allen RG, Tasumi M, Morse A, Trezza R, Wright JL, Bastiaanssen W, Kramber W, Lorite I, Robison CW (2007) Satellite-based energy balance for mapping evapotranspiration with internalized calibration (METRIC)—applications. J Irrig Drain Eng 133:395–406. https://doi.org/10.1061/(ASCE)0733-9437(2007)133:4(395)

Artis DA, Carnahan WH (1982) Survey of emissivity variability in thermography of urban areas. Remote Sens Environ 12:313–329

Bagchi K, Mukerjee KN (1983) Diagnostic survey of West Bengal(s). Dept. of Geography, Calcutta University, Pantg Delta & Rarh Bengal

Bastiaanssen WGM, Menenti M, Feddes RA, Holtslag AAM (1998) A remote sensing surface energy balance algorithm for land (SEBAL). 1. Formulation. J Hydrol 212:198–212

Bastiaanssen WGM, Noordman EJM, Pelgrum H, Davids G, Thoreson BP, Allen RG (2005) SEBAL model with remotely sensed data to improve water-resources management under actual field conditions. J Irrig Drain Eng 131(1):85–93

Blaney HF, Criddle WD (1945) Determining water requirements in irrigated areas from climatological data. Processed 17

Chaturvedi RS (1973) A note on the investigation of ground water resources in western districts of Uttar Pradesh. Annual Report, U.P. Irrigation Research Institute, pp 86–122

Coelho VHR, Montenegro S, Almeida CN, Silva BB, Oliveira LM, Gusmão ACV, Freitas ES, Montenegro AAA (2017) Alluvial groundwater recharge estimation in semi-arid environment using remotely sensed data. J Hydrol. https://doi.org/10.1016/j.jhydrol.2017.02.054

Crosbie RS, Davies P, Harrington N, Lamontagne S (2015) Ground trothing groundwater-recharge estimates derived from remotely sensed evapotranspiration: a case in South Australia. Hydrogeol J 23:335–350. https://doi.org/10.1007/s10040-014-1200-7

Deo RC, Şahin M (2017) Forecasting long-term global solar radiation with an ANN algorithm coupled with satellite-derived (MODIS) land surface temperature (LST) for regional locations in Queensland. Renew Sustain Energy Rev 72:828–848

Donohue RJ, McVicar TR, Roderick ML (2010) Assessing the ability of potential evaporation formulations to capture the dynamics in evaporative demand within a changing climate. J Hydrol 386:186–197. https://doi.org/10.1016/j.jhydrol.2010.03.020

Duan Z, Bastiaanssen WGM (2013) First results from Version 7 TRMM 3B43 precipitation product in combination with a new downscaling–calibration procedure. Remote Sens Environ 131:1–13

Feng H, Zhao X, Chen F, Wu L (2014) Using land use change trajectories to quantify the effects of urbanization on urban heat island. Adv Space Res 53(3):463–473

Fu P, Weng Q (2016) A time series analysis of urbanization induced land use and land cover change and its impact on land surface temperature with Landsat imagery. Remote Sens Environ 175:205–214

Gokmen M, Vekerdy Z, Lubczynski MW, Timmermans J (2013) Assessing groundwater storage changes using remote sensing-based evapotranspiration and precipitation at a large semiarid basin scale. J Hydrometeorol 16:129–146. https://doi.org/10.1175/JHM-D-12-0156.1

González GM, Stisen S, Koch J (2015) Retrieval of spatially distributed hydrological properties based on surface temperature rise measured from space for spatial model validation at regional scale

González-Rojí SJ, Sáenz J, Ibarra-Berastegi G, Díaz de Argandoña J (2018) Moisture balance over the Iberian Peninsula according to a regional climate model. The impact of 3DVAR data assimilation. J Geophys Res: Atmos

Hassan-Esfahani L, Torres-Rua A, McKee M (2015) Assessment of optimal irrigation water allocation for pressurized irrigation system using water balance approach, learning machines, and remotely sensed data. Agric Water Manag 153:42–50

Hu P, Liu Q, Heslop D, Roberts AP, Jin C (2015) Soil moisture balance and magnetic enhancement in loess–paleosol sequences from the Tibetan Plateau and Chinese Loess Plateau. Earth Planet Sci Lett 409:120–132

Immerzeel WW, Droogers P, De Jong SM, Bierkens MFP (2009) Large-scale monitoring of snow cover and runoff simulation in Himalayan river basins using remote sensing. Remote Sens Environ 113(1):40–49

Jafari R, Hasheminasab S (2017) Assessing the effects of dam building on land degradation in central Iran with Landsat LST and LULC time series. Environ Monit Assess 189(2):74

Jha AK (2013) Water availability, scarcity and climate change in India: a review. Asian J Water Environ 1(1):50–66

Jiang Y, Weng Q (2017) Estimation of hourly and daily evapotranspiration and soil moisture using downscaled LST over various urban surfaces. Giscience Remote Sens 54(1):95–117

Khalaf A, Donoghue D (2012) Estimating recharge distribution using remote sensing: a case study from the West Bank. J Hydrol 414:354–363

Khatun S, Pal S (2016) Identification of prospective surface water available zones with multi criteria decision approach in Kushkarani river basin of eastern India. Arch Curr Res Int 4(4):1–20

Krol MS, Bronstert A (2007) Regional integrated modelling of climate change impacts on natural resources and resource usage in semi-arid Northeast Brazil. Environ Model Softw 22(2):259–268

Kumar CP, Seetapati PV (2002) Assessment of natural ground water recharge in upper Ganga canal command area. J Appl Hydrol 15(4):13–20

Landsat Project Science Office (2002) Landsat 7 science data user's handbook. Goddard Space Flight Center, NASA, Washington, DC. http://ltpwww.gsfc.nasa.gov/IAS/hand-book/handbook_toc.html. Accessed 10 Sept 2003

Lee TW, Lee JY, Wang ZH (2012) Scaling of the urban heat island intensity using time-dependent energy balance. Urban Climate 2:16–24

Mahmoud SH, Alazba AA (2016) A coupled remote sensing and the Surface Energy Balance based algorithms to estimate actual evapotranspiration over the western and southern regions of South Arabia. J Asian Earth Sci 124:269–283. https://doi.org/10.1016/j.jseaes.2016.05.012

Markham BL, Barker JL (1985) Spectral characterization of the Landsat thematic mapper sensors. Int J Remote Sens 6(5):697–716

Meena AL, Bisht P (2017) Study of variability of rainfall and suitability of farming in sub-humid region: a case study of Jaipur District, Rajasthan, India. Sustain Agri Food Environ Res 5(3)

Melo DCD, Xavier AC, Bianchi T, Oliveira PTS, Scanlon B, Lucas MC, Wendland E (2015) Performance and evaluation of rainfall estimates by TRMM multi-satellite precipitation analysis 3B42V6 and V7 over Brazil. J Geophys Res 120:9426–9436. https://doi.org/10.1002/2015JD023797

Metz M, Andreo V, Neteler M (2017) A new fully gap-free time series of land surface temperature from MODIS LST data. Remote Sens 9(12):1333

Mishra V, Shah R, Thrasher B (2014) Soil moisture droughts under the retrospective and projected climate in India. J Hydrometeorol 15(6):2267–2292

Mohanty BP, Ines AV, Shin Y, Gaur N, Das N, Jana R (2017) A framework for assessing soil moisture deficit and crop water stress at multiple space and time scales under climate change scenarios using model platform, satellite remote sensing, and decision support system. In: Remote sensing of hydrological extremes. Springer, Cham, pp 173–196

Mooley DA, Parthasarathy B (1984) Fluctuations in all India summer monsoon rainfall during 1871–1978. Clim Change 6:287–301

Moriasi DN, Arnold JG, Van Liew MW, Bingner RL, Harmel RD, Veith TL (2007) Model evaluation guidelines for systematic quantification of accuracy in watershed simulations. Trans ASABE 50(3):885–900

Morse A, Allen RG, Tasumi M, Kramber WJ, Trezza R, Wright J (2000) Application of the SEBAL methodology for estimating evapotranspiration and consumptive use of water through remote sensing. University of Idaho, Kimberly, ID, USA, pp 1–220

Nash JE, Sutcliffe JV (1970) River flow forecasting through conceptual models. Part I. A discussion of principles. J Hydrol 10(3):282–290

Nichol JE (1994) A GIS-based approach to microclimate monitoring in Singapore's highrise housing estates. Photogramm Eng Remote Sens 60:1225–1232

Oliveira LMM, Montenegro SMGL, Silva BB, Antonino ACD, Moura AESS (2014) Evapotranspiração real em bacia hidrográfica do Nordeste brasileiro por meio do SEBAL e produtos MODIS. Rev Bras Eng Agríc Ambient 18:1039–1046. https://doi.org/10.1590/1807-1929/agriambi.v18n10p1039-1046

Pal S, Ziaul S (2017) Detection of land use and land cover change and land surface temperature in English Bazar urban centre. Egypt J Remote Sens Space Sci 20(1):125–145

Penman HL (1948) Natural evaporation from open water, bare soil and grass. Proc R Soc London A193:120–145. https://doi.org/10.1098/rspa.1948.0037

Perea RG, Poyato EC, Montesinos P, Morillo JG, Díaz JR (2016) Influence of spatio temporal scales in crop water footprinting and water use management: evidences from sugar beet production in Northern Spain. J Clean Prod 139:1485–1495

Rohde MM, Edmunds WM, Freyberg D, Sharma OP, Sharma A (2015) Estimating aquifer recharge in fractured hard rock: analysis of the methodological challenges and application to obtain a water balance (Jaisamand Lake Basin, India). Hydrogeol J 23(7):1573–1586

Ruhoff AL, Paz AR, Collischonn W, Aragão LEOC, Rocha HR, Malhi YS (2012) A MODIS-based energy balance to estimate evapotranspiration for clear-sky days in Brazilian Tropical Savannas. Remote Sens 4:703–725. https://doi.org/10.3390/rs4030703

Saiter FZ, Eisenlohr PV, Barbosa MR, Thomas WW, Oliveira-Filho AT (2016) From evergreen to deciduous tropical forests: how energy–water balance, temperature, and space influence the tree species composition in a high diversity region. Plant Ecolog Divers 9(1):45–54

Sathyanadh A, Karipot A, Ranalkar M, Prabhakaran T (2016) Evaluation of soil moisture data products over Indian region and analysis of spatio-temporal characteristics with respect to monsoon rainfall. J Hydrol 542:47–62

Shao HB, Chu LY, Jaleel CA, Manivannan P, Panneerselvam R, Shao MA (2009) Understanding water deficit stress-induced changes in the basic metabolism of higher plants–biotechnologically and sustainably improving agriculture and the ecoenvironment in arid regions of the globe. Crit Rev Biotechnol 29(2):131–151

Shiklomanov IA (2000) Appraisal and assessment of world water resources. Water Int 25(1):11–32

Silva BB, Wilcox BP, Silva VPR, Montenegro SMGL, Oliveira LMM (2015) Changes to the energy budget and evapotranspiration following conversion of tropical savannas to agricultural lands in São Paulo State, Brazil. Ecohydrology 8:1272–1283. https://doi.org/10.1002/eco.1580

Siu LW, Hart MA (2013) Quantifying urban heat island intensity in Hong Kong SAR, China. Environ Monit Assess 185(5):4383–4398

Snyder WC, Wan Z, Zhang Y, Feng YZ (1998) Classification-based emissivity for land surface temperature measurement from space. Int J Remote Sens 19(14):2753–2774

Surendran S, Gadgil S, Francis PA, Rajeevan M (2015) Prediction of Indian rainfall during the summer monsoon season on the basis of links with equatorial Pacific and Indian Ocean climate indices. Environ Res Lett 10(9):094004

Thober S, Kumar R, Sheffield J, Mai J, Schäfer D, Samaniego L (2015) Seasonal soil moisture drought prediction over Europe using the North American Multi-Model Ensemble (NMME). J Hydrometeorol 16(6):2329–2344

Tong K, Su F, Yang D, Hao Z (2014) Evaluation of satellite precipitation retrievals and their potential utilities in hydrologic modeling over the Tibetan Plateau. J Hydrol 519:423–437. https://doi.org/10.1016/j.jhydrol.2014.07.044

Trezza R, Allen RG, Tasumi M (2013) Estimation of actual evapotranspiration along the Middle Rio Grande of New Mexico using MODIS and Landsat Imagery with the METRIC Model. Remote Sens 5:5397–5423. https://doi.org/10.3390/rs5105397

Van Niel TG, McVicar TR, Roderick ML, van Dijk AIJM, Renzullo LJ, van Gorsel E (2011) Correcting for systematic error in satellite-derived latent heat flux due to assumptions in temporal scaling: assessment form flux tower observations. J Hydrol 409:140–148. https://doi.org/10.1016/j.jhydrol.2011.08.011

Van Niel TG, McVicar TR, Roderick ML, van Dijk AIJM, Beringer J, Hutley LB, van Gorsel E (2012) Upscaling latent heat flux for thermal remote sensing studies: comparison of alternative approaches and correction of bias. J Hydrol 468–469:35–46. https://doi.org/10.1016/j.jhydrol.2012.08.005

Vero SE, Antille DL, Lalor STJ, Holden NM (2014) Field evaluation of soil moisture deficit thresholds for limits to trafficability with slurry spreading equipment on grassland. Soil Use Manag 30(1):69–77

Wada Y, Beek LP, Sperna Weiland FC, Chao BF, Wu YH, Bierkens MF (2012) Past and future contribution of global groundwater depletion to sea-level rise. Geophys Res Lett 39(9)

Yadav SS, Lal R (2018) Vulnerability of women to climate change in arid and semi-arid regions: the case of India and South Asia. J Arid Environ 149:4–17

Yaduvanshi A, Ranade A (2017) Long-term rainfall variability in the eastern Gangetic plain in relation to global temperature change. Atmos Ocean 55(2):94–109

Yao R, Wang L, Huang X, Niu Y, Chen Y, Niu Z (2018) The influence of different data and method on estimating the surface urban heat island intensity. Ecol Ind 89:45–55

Yong B, Ren L, Hong Y, Wang J, Gourley JJ, Jiang SH, Chen X, Wang W (2010) Hydrologic evaluation of Multisatellite Precipitation Analysis standard precipitation products in basins beyond its inclined latitude band: a case study in Laohahe basin, China. Water Resour Res 46:1–20. https://doi.org/10.1029/2009WR008965

Zandalinas SI, Balfagón D, Arbona V, Gómez-Cadenas A, Inupakutika MA, Mittler R (2016) ABA is required for the accumulation of APX1 and MBF1c during a combination of water deficit and heat stress. J Exp Bot 67(18):5381–5390

Zhang X, Tang Q, Liu X, Leng G, Li Z (2017) Soil moisture drought monitoring and forecasting using satellite and climate model data over Southwestern China. J Hydrometeorol 18(1):5–23

Ziaul S, Pal S (2016) Image based surface temperature extraction and trend detection in an urban area of West Bengal, India. J Environ Geogr 9(3–4):13–25

Application of Ensemble Machine Learning Models to Assess the Sub-regional Groundwater Potentiality: A GIS-Based Approach

Sunil Saha, Amiya Gayen, and Sk. Mafizul Haque

Abstract

Effective data mining models are powerful tools for the prediction and management of sub-regional groundwater resources. In this work, an integrated attempt is employed to assess the groundwater potentiality in C. D. Block of Birbhum District, India using GIS-based novel ensemble machine learning models of Radial Basis Function neural network (RBFnn) in form of RBFnn-Bagging and RBFnn-Dagging. Fourteen hydro-geomorphological factors were used to find the most potential groundwater area. To support the result, observation data of 86 sites were incorporated empirically. Out of these, 70% were randomly split for the training dataset to develop the model and remaining 30% were used for model validation. Results predict excellent groundwater potentiality by the RBFnn-Bagging and RBFnn-Dagging as they covered 17.38% and 13.97% of the study area, respectively. The prediction capacity of newly built models was established with the root mean square error (RMSE), accuracy, precision, and receiver operating characteristic (ROC) curve which shows a satisfactory result as the RMSE values of 0.05 and 0.07 and AUC values of 82.1% and 81.30% are obtained for RBFnn-Bagging and RBFnn-Dagging models respectively. Well-known mean decrease Gini (MDG) from the random forest (RF) algorithm, implemented to determine the relative importance of the factors, reveals that distance from river, pond frequency, aspect, stream junction frequency, elevation, and geomorphology are most useful determinants of groundwater potentiality in the study area. The adopted approach has a wide scope in effective planning and sustainable management of groundwater resources.

Keywords

Groundwater potentiality · Ensemble method · Machine learning models · RBFnn · Root mean square error

15.1 Introduction

Drinking water crisis and groundwater scarcity are major challenges among the various prevailing contemporary issues of the earth. Groundwater is the most important but fast depleting natural resource whose appropriate delineation and management are momentous at this

S. Saha
Department of Geography, University of Gour Banga, Malda, India
e-mail: sunilgeog@ugb.ac.in

A. Gayen (✉) · Sk.M. Haque
Department of Geography, University of Calcutta, Kolkata, India
e-mail: amiya.gayen@midnaporecollege.ac.in

conjuncture. India is the most groundwater-consuming country in the world, which uses nearly 230 km^3 year^{-1} of groundwater (World Bank 2010). According to the World Bank report of 2010, if India does not reduce the use of groundwater, more than 60% of the aquifers will be dried within 20 years. In India, demand for groundwater has been increasing through the green revolution and the pace of industrialization, urbanization, and agricultural practices (Suhag 2016). There are two different types of aquifers in India, i.e., crystalline aquifers (located in peninsular area) and another are alluvial aquifers (developed in the Indo-Gangetic plain). The former is characterized by low permeability and hard rocks and the latter leads in terms of groundwater resources (Suhag 2016). Thus, groundwater quality and potentiality assessment are important tasks at hand for reasons of sustainability and livelihood.

Therefore, most of the aquifers are in critical situations, particularly in semi-arid and arid regions, which may turn into a severe problem. Several researchers have tried to determine aquifer characteristics with the help of sediments beneath, identifying pore space, and fractures in a rock on the earth's surface, which is not adequate for identifying reliable aquifers (Naghibi et al. 2017). Generally, groundwater potentiality assessment including hydro-geological nature of the region especially porosity, aquifer properties, permeability, storage capacity, groundwater recharge, and hydraulic conductivity of the aquifer materials are very pertinent factors. These are broadly dependent on physical variables like geomorphology, geology, rainfall, soil, drainage, and LULC (Saha 2017; Haque et al. 2020). Presently, the unnecessary use of groundwater and unscientific management strategies are affecting the groundwater recharge level (Chaudhry et al. 2019). Therefore, in such circumstances, it is required that an adequate management strategy for groundwater potentiality assessment is framed (Chen et al. 2019). Thus, a groundwater potential map can help to identify the prospect of groundwater yield, which can guide toward proper management of groundwater.

Different popular and well-accepted models have been developed for preparing groundwater potentiality mapping (Corsini et al. 2009; Ozdemir 2011; Lee et al. 2017; Saha 2017; Chen et al. 2019). For example, the analytical hierarchy process (Razandi et al. 2015; Ghosh et al. 2020), the weight of evidence (Tahmassebipoor et al. 2016), and frequency ratio (Guru et al. 2017; Das 2019), fuzzy logic (Mohamed and Elmahdy 2017). Nowadays those models were not applied by researchers because they are unable to solve multi-criteria decision problems. An examination of the literature reveals that the integration of machine learning models has provided better results (Kenda et al. 2018; Chen et al. 2019). So, machine learning models handle data with high dimensionality and provide more perfect results using geographical information systems and remote sensing data (Gayen et al. 2019; Rudin 2019; Haque et al. 2020). Guzman et al. (2015) applied artificial neural network (ANN) and support vector machine (SVM) to predict groundwater potentiality. Guzman et al. (2015) have explained the superiority of the SVM models over ANN models about prediction. Naghibi et al. (2018) also applied some well-accepted machine learning models, i.e., boosted regression tree, classification and regression tree, and random forest for groundwater potentiality prediction. Their study shows that the boosted regression tree model provides a better result with an AUC value of 0.8103. Sajedi-Hosseini et al. (2018) also implemented a few machine learning models for groundwater risk assessment. Thus, the previous research work confirms the prediction capacity of machine learning models to predict groundwater potentiality. The present study has focused on novel ensemble machine learning models of Radial Basis Function neural network (RBFnn)- Bagging (RBFnn-Bagging) and Dagging (RBFnn-Dagging). The primary objective of this research is to prepare a groundwater potentiality map, along with groundwater quality of Md. Bazar Block in Birbhum District, India. Finally, researchers have tried to predict the groundwater controlling efficiency of the applied factors with mean decrease Gini (MDG).

15.2 Materials and Methods

15.2.1 Study Area

The Md. Bazar is a Jharkhand adjacent western block of Birbhum District located in West Bengal, India. It is extended from 87°25′ E to 87°40′ E and 23°55′ N to 24°50′ N (Fig. 15.1). This block was recognized as drought influenced district of West Bengal. This region is formed of gneisses and associated rocks, older alluvium, and older alluvium with lateritic types of aquifer media. The older alluvium has high to moderate yield potentiality but in the cases of older alluvium with laterite rocks, the yield potentiality is limited between 100 and 700 gpd ft^{-2} hydraulic conductivity in the study area (Thapa et al. 2018). This falls under the warm monsoon climate where annual precipitation is approximately 1200 mm and temperature ranges from 6 to 40 °C (Saha 2017). The maximum precipitation occurs from July to September (monsoon period). The long gap of the rainy season and over-increasing pressure of agriculture leads to continuous updraft of groundwater for irrigation which is one of the major issues of this region. The main routes of groundwater recharge in Md. Bazar block is natural and anthropogenic activities such as artificial canals, hydropower dams, and check dams.

15.2.2 Data Used

In the first instance, dug wells locations were collected from the Central Ground Water Board. A total of 85 dug wells and one piezometer were recognized in Md. Bazar Block of Birbhum District and verified using GPS and field survey and considered for a groundwater inventory map (CGWB, 2017). After that, the well and no-well locations were classified into two sets by maintaining 70:30 ratio. 70% of locations were used as training dataset which was applied to predict the GWPMs. At the same time, the unused 30% locations were considered as a validation dataset of the modeling result (Naghibi et al. 2017; Chen et al. 2019).

Fourteen groundwater controlling factors viz., aspect, elevation, curvature, topographical positioning index (TPI), topographical wetness index (TWI), slope, stream junction frequency (SJF), geomorphology, distance to a river, rainfall, pond frequency, land use\land cover (LULC), geology, and soil texture were selected for the development of the GWPMs (Fig. 15.2). Thematic data layers of parameters were prepared using the GIS-spatial analysis tool and the PALSAR Digital Elevation Model (DEM) was taken from the Alaska Satellite facility; LULC map was developed by applying the Sentinal-2 data; rainfall data from Indian Meteorological Department (IMD); soil map from NBSS-LUP; and the geological map was collected from Geological Survey of India (GSI).

15.2.3 Preparing Groundwater Influencing Factors

At first, 12.5 × 12.5 m spatial resolution based PALSAR-DEM data was used to prepare the aspect, elevation, curvature, TWI, and TPI maps (Fig. 15.3a–e). Because these parameters are considered by several researchers (Naghibi et al. 2016, 2017; Chen et al. 2019) to be an essential parameters of the GWPM. Aspect and elevation both are associated with soil moisture, sunlight, temperature, wind, soil development, and precipitation therefore both factors can enhance the rate of groundwater recharge (Golkarian et al. 2018; Gayen et al. 2019). The slope is an important terrain factor that increases the velocity of surface runoff wherein a high slope does not allow infiltration of groundwater (Arabameri et al. 2019). The regional slope angle ranges from 0° to 34.21°. The TWI is applied for measuring the influence of topological conditions on hydro-geomorphic processes. It is the integration of slope and the upstream contributing area per unit orthogonal to the direction of flow (Arabameri et al. 2019). The calculation of TWI is represented in Moore et al. (1991):

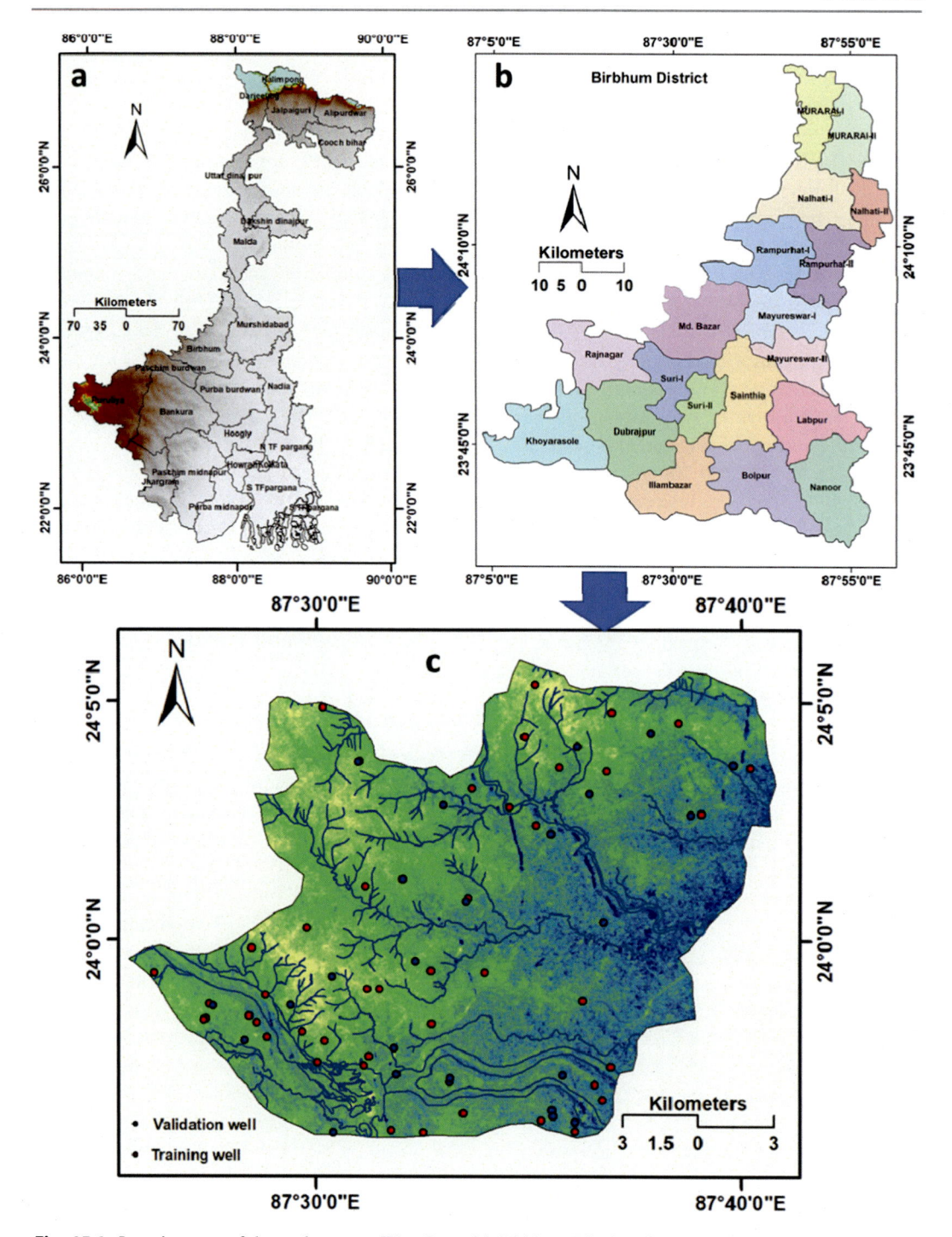

Fig. 15.1 Location map of the study area; **a** West Bengal **b** Birbhum District of West Bengal **c** Md. Bazar block

Fig. 15.2 Flowchart illustrated the applied methodology of groundwater potentiality mapping

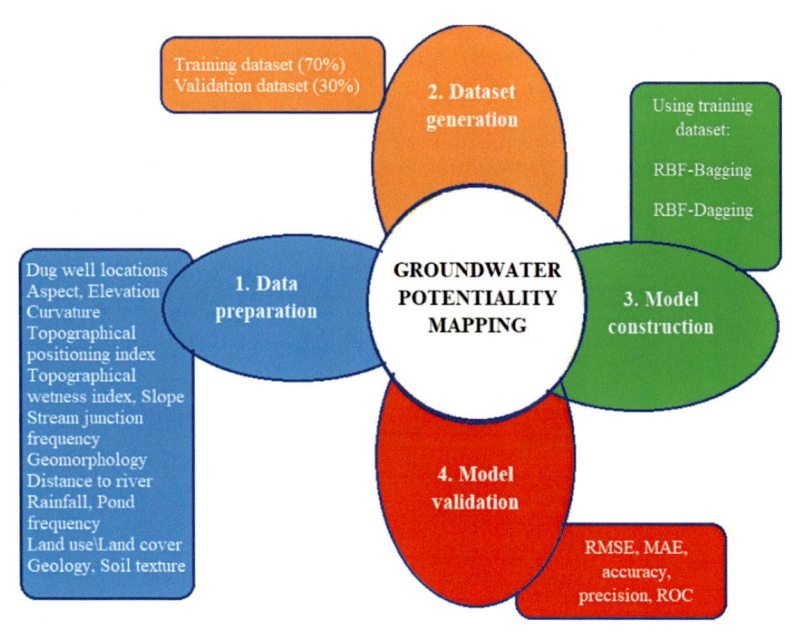

$$\text{TWI} = \ln\left(\frac{As}{\tan\beta}\right) \quad (15.1)$$

where, As denotes cumulative catchment area (m² m⁻¹) and β defines the slope angle.

The TPI and curvature both are exhibited to affect groundwater potentiality (Grohmann and Riccomini 2009; Arabameri et al. 2019). The TPI and curvature maps were developed with the help of PALSAR-DEM data. The TPI has been calculated by using Eq. (15.2).

$$\text{TPI} = Z_0 - \overline{Z} \quad (15.2)$$

$$\overline{Z} = \frac{1}{n_R}\sum_{i \in R} Z_i \quad (15.3)$$

where Z_0 denotes the central point altitude, Z represents the mean altitude within a particular radius (R), and small R defines small ridges and valleys (Weiss 2001). The highest and lowest TPI values within the study area are 0.00 and 1.00.

Pond, drainage, and stream dictate structural characteristics and permeability of an area that influences groundwater storage and movement through a hydraulic gradient (Tien Bui et al. 2017). The distance to river and stream junction frequency maps were developed using the extracted drainage output from the 1:50,000 toposheet maps. Junction indicates the confluence areas of two rivers. Generally, chances of groundwater arability are more in the highest pond frequency areas and confluence zone areas because both are enhancing groundwater recharge processes. The LULC can reflect less susceptibility to groundwater potentiality (Saha 2017). A LULC map of study area was developed using Sentinal-2 data and results were affirmed by applying Cohen's Kappa index with 89.6% Kappa value. The Block is covered by eight LULC classes: reservoir, watercourse, sand cover, settlement, agricultural land, mining area, wasteland, and vegetation cover (Fig. 3l). The duration of the Rainfall and its intensity also play a key role in groundwater recharge (Shekhar and Pandey 2014). Jothibasu and Anbazhagan (2016) noted that rainfall influences GPM accuracy and moving water percolation for that reason spatial distribution of rainfall was taken as a predisposing factor for this study (Fig. 3j).

Soil types are most important predisposing factors for the assessment of the infiltration rate of any region. This study area falls under six major soil types like sandy, clay loam, loamy, sandy loam, sandy clay, and sandy clay loam. Maximum areas of Md. Bazar block is covered by sandy loam and clay loam soil types (Fig. 3n). Generally, potentiality of groundwater infiltration rate is

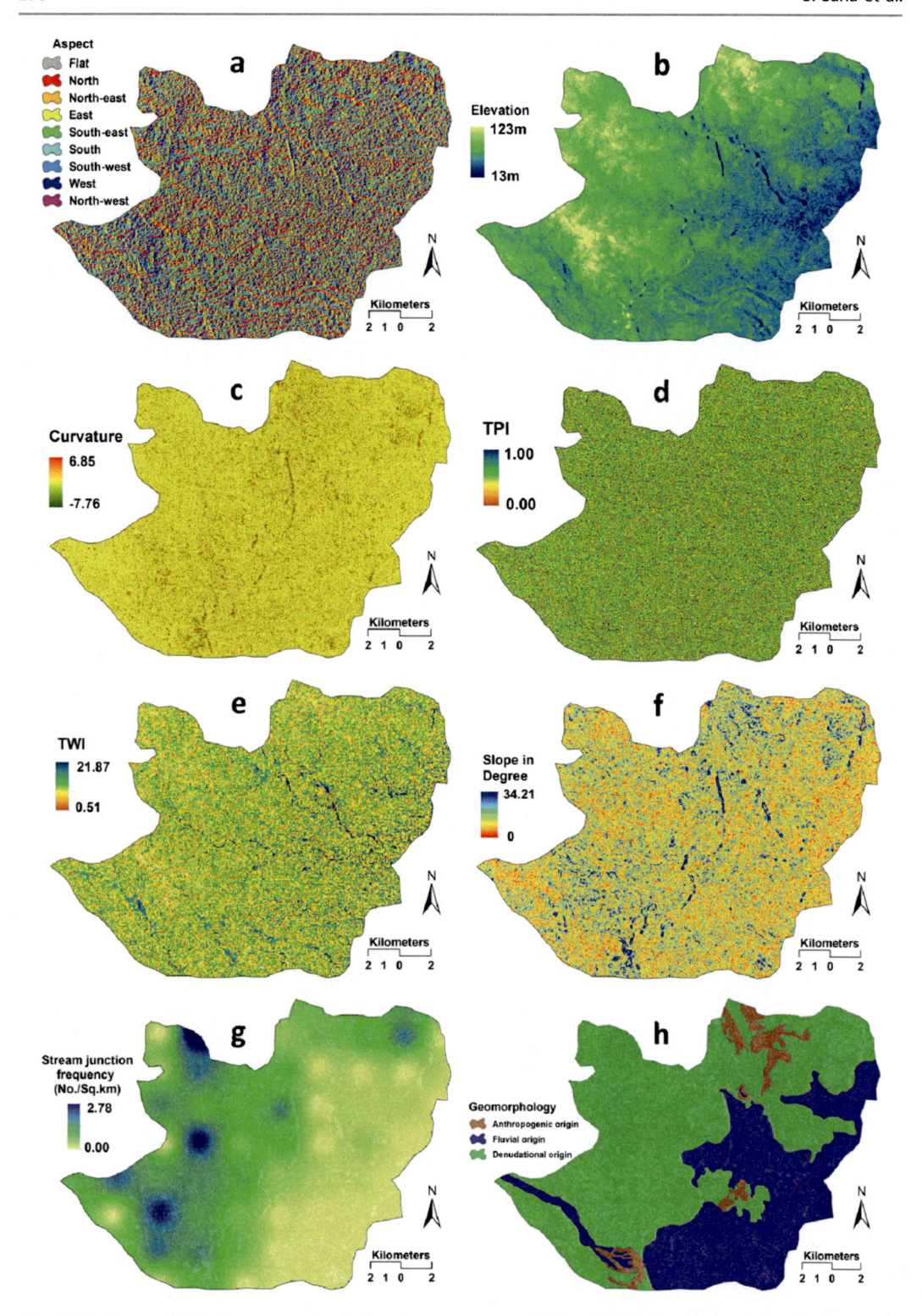

Fig. 15.3 The spatial data layers: **a** aspect, **b** elevation, **c** curvature, **d** TPI, **e** TWI, **f** slope, **g** stream junction frequency, **h** geomorphology, **i** distance to river, **j** rainfall, **k** pond frequency, **l** LULC, **m** geology, **n** soil texture

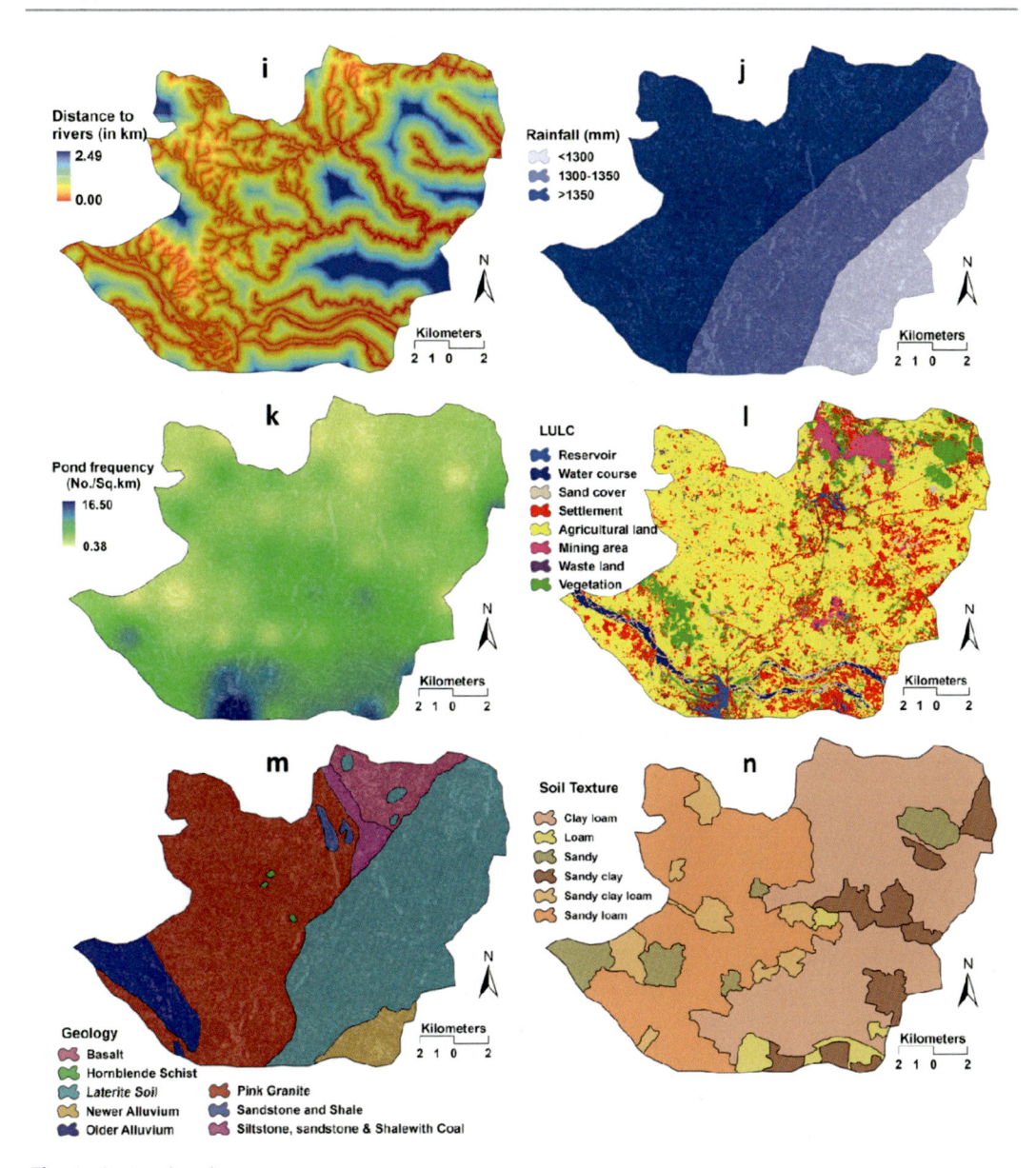

Fig. 15.3 (continued)

higher in sandy regions as compared to loamy or clayey strata. The Md. Bazar block is composed of eight geological formations. The western part is dominated by pink granite whereas rocks belonging to the Vindhyan formation occur to the east (Fig. 3m). The pisolitic and kankar ferruginous concretions are mostly found in the laterite track. Some parts of the block are covered by basaltic rocks and younger alluvium. The block falls under three primary geomorphological regions, i.e., depositional plain, anthropogenic origin, and denudational plain (Fig. 3h).

15.2.4 Machine Learning Ensemble Meta-classifiers Modes for the GWPMs

Novel ensemble models, the RBFnn-Bagging and RBFnn-Dagging, are used for mapping groundwater potentiality in this study. RBFnn originated in the late 1980s is a version of an artificial neural network. In a two-layer neural network, where each hidden unit implements a radial-activated function, RBFs are embedded. A weighted sum of hidden unit outputs is implemented by output units. Although the output is linear, the input into an RBF network is nonlinear. Their exceptional approximation capacities are investigated. RBF networks can model complex mappings due to their nonlinear approximation properties. The RBFnn was used as a base learner in this study. As for the ensemble technique, because of its utility in ensemble estimation, the Bagging and Dagging were applied as the meta-learner.

15.2.4.1 Bagging

The bagging algorithm has introduced by Breiman (1996), is the developer of bootstrapping (Freedman 1981). Several researchers have applied this model to predict susceptibility maps (i.e., flood, landslide, etc.) as this model has excellent performance ability (Hong et al. 2020). The bagging tree is a bagging algorithm comprised of models based on decision trees. This algorithm is selected because it fabricates the decision tree with the help of each produced subset and ultimately, they are assembled within the final model (Hong et al. 2020). It enhances the alignment accuracy by minimizing the inconsistency of the alignment error (Saha et al. 2021; Wu et al. 2020). A bagging classifier is considered a three-step bagging system (Breiman 1996; Yariyan et al. 2020). It is developed as a bootstrap sample through substantive training samples through the displacement approach (Saha et al. 2021). This MLA can promote the success of all arrays of subset by connecting them to the actual feature process for the bagging classification stage; also, this model is not dependent upon the precision of past models (Breiman 1996; Yariyan et al. 2020).

15.2.4.2 Dagging

The Dagging algorithm was introduced by Ting and Witten (1997), using another sampling method to extract a basic classifier. Dagging is very similar to bagging—name is a portmanteau derived from the phrase "disjoint bagging." In dagging, once data is used for classification the subset is "disjointed" (or set aside). In bagging, each subset is not disjointed and the data is returned to the full set to be used again. Dagging is a well-known group-sampling technique using majority votes to combine several classifiers to improve prediction accuracies of basic classifiers (Kotsianti and Kanellopoulos 2007).

15.2.5 Validation of Groundwater Potentiality Models

Models' evaluation and validation is an important steps in prediction work and without validation, the model does not have any scientific significance (Talukdar and Pal 2020; Pal and Mandal 2021). The applied model's prediction capacity was investigated by ROC curve, RMSE, MAE, accuracy, and precision (Chen et al. 2018). The two categories of ROC curve on prediction and success rate, are developed using validation and training datasets, respectively. It is a graphical illustration of model prediction through a diagnostic test (Chen et al. 2019). The area under the curve (AUC) varies from 0.5 to 1.0 and the value close to 1.0 predicts the power of models (Mishra et al. 2020).

Also, error within the predictive models was calculated through RMSE and MAE tests to identify the prediction capacity (Abedinpour et al. 2012). Each error was calculated with the comparison between model values and field observed values (Rahmati et al. 2017). The precision, RMSE, MAE, and AUC have been calculated by using Eqs. (15.4)–(15.7).

$$\text{Precision} = \frac{TP}{TP+FP} \quad (15.4)$$

$$\text{RMSE} = \sqrt{\frac{1}{(N)\sum_{i=1}^{N}(O_i - S_i)^2}} \quad (15.5)$$

$$\text{MAE} = \sqrt{\sum_{i=1}^{n}\frac{(S_i - O_i)}{n}} \quad (15.6)$$

$$\text{AUC} = \frac{\Sigma TP + \Sigma TN}{P + N} \quad (15.7)$$

where TN and TP denote true negative and true positive, FP and FN denote false positive and false negative, O_i and S_i are observed and predicted values, n is the number of observations, P and N are the dug wells location points, and N is the total number of non-dug wells location points.

15.3 Results and Analysis

15.3.1 Groundwater Potentiality Models

At first, two accepted meta classifier based MLAs were developed by applying the training dataset. The constructed models were divided into four classes (i.e., high, very high, moderate, and low) to calculate the groundwater potentiality indices (GWPI) (Chen et al. 2018) (Fig. 4a, b). Actually, the user-defined classification of GWPMs is nearly hard for readers to justify and interpret. Therefore, nature break statistics were most convenient for the arrangement of GWPI following the histogram of data distribution (Chen et al. 2019).

The RBFnn-Bagging produced result shows that low potentiality zone has the maximum area (68.64%), followed by the very high (16.92%), moderate (11.51%), and high (2.92%) in the study area. The corresponding area covered by RBFnn-Dagging mode is 68.03%, 13.70%, 13.59%, and 4.68% for the low, very high, moderate, and high zones, respectively. It is manifest through both models GWPMs; the largest GWP area is found in the southern part of the Md Bazar Block because of the more forest cover and presence of water reservoir (Table 15.1).

15.3.2 Validation and Comparison of Applied Models

For validation and comparing the applied models; RMSE, MAE, accuracy, precision, and ROC were implemented using validation and training data sets (Fig. 5a, b), as they are important aspects to conclude the prediction capacity of applied models (Pal and Mandal 2021).

The results show that the RBFnn-Bagging algorithm has higher AUC values of 0.837 and 0.847, respectively, for the success and prediction rate curves, followed by the RBFnn-Dagging algorithm with an AUC value of 0.793 and 0.829, respectively. So, it is concluded that both models have excellent GWP prediction capacity. The RMSE and MAE values of RBFnn-Bagging and RBFnn-Dagging were calculated for the training phase as 0.237, 0.057, 0.270, and 0.74 and validation phase as 0.039, 0.198, 0.51, and 0.227, respectively (Table 15.2). Also, results of accuracy and precision tests are presented in Table 15.2 for both the applied models. The accuracy and precision values of both models were 0.88, 0.79, 0.83, and 0.75 for RBF-Bagging and RBFnn-Dagging, respectively which indicates that both the models have uniform prediction capacity for assessment of groundwater potentiality.

15.3.3 Significant Factors Identification by MDGs

The significant factors identification is a challenging task because groundwater recharge is impacted by various groundwater controlling factors (Conforti et al. 2010). The mean decrease Gini was applied to evaluate factor's relative importance by using the random forest (RF) algorithm (Breiman 2001). The MDG varies from 14.09 to 286.01. Distance to a river (286.01), pond frequency (229.96), aspect (103.56), stream

Fig. 15.4 The groundwater potentiality maps by RBF-Bagging and RBF-Dagging models

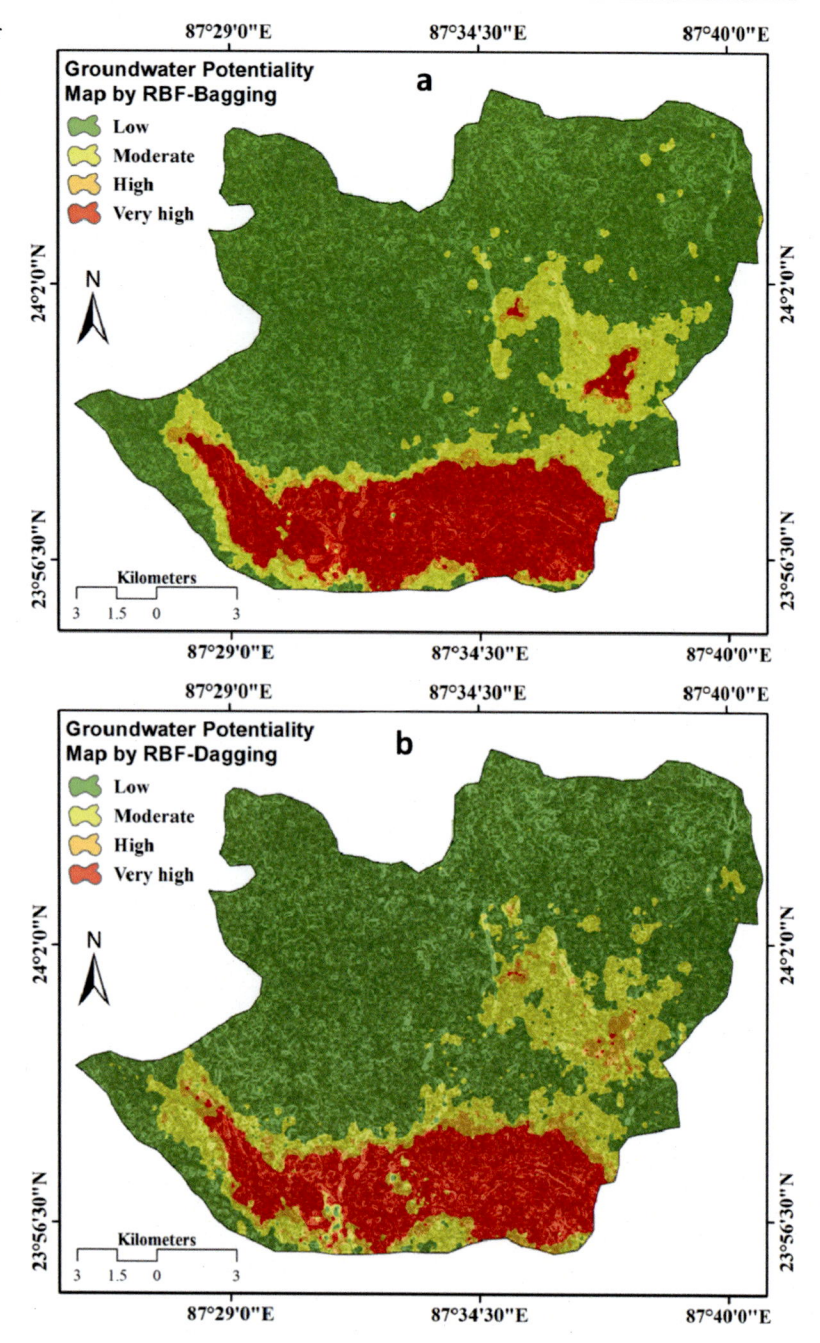

Table 15.1 Areal share under potentiality classes of groundwater potentiality models (area in km²)

GWP classes	RBF-Bagging	RBF-Dagging
Low	68.64	68.03
Moderate	11.51	13.59
High	2.92	4.68
Very high	16.92	13.70

Fig. 15.5 Validation of groundwater potentiality maps applying ROC curve: **a** success rate curve (applying training dataset) and **b** prediction rate curve (applying validation dataset)

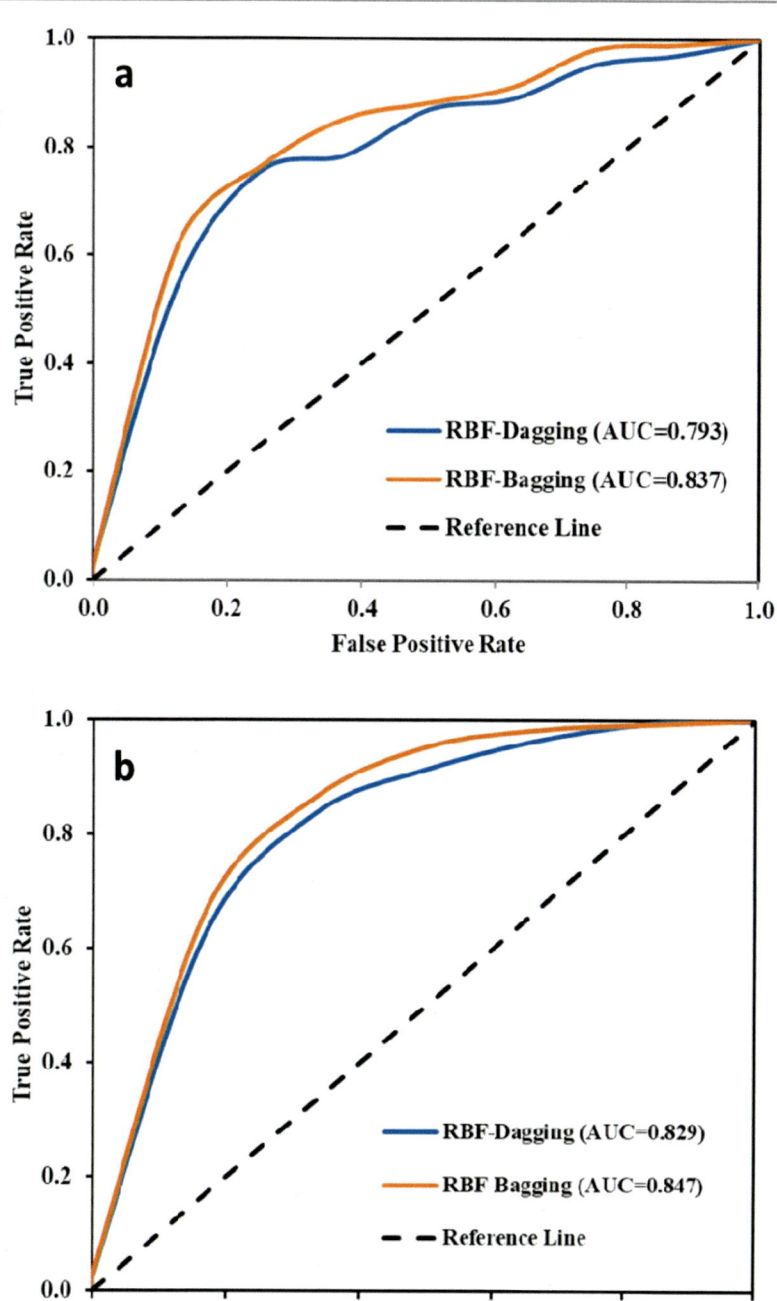

Table 15.2 Estimation of root mean square error (RMSE), MAE, accuracy, and precision for both models

Models	Training dataset				Validation dataset			
	RMSE	MAE	Accuracy	Precision	RMSE	MAE	Accuracy	Precision
RBF-Bagging	0.237	0.057	0.85	0.81	0.039	0.198	0.88	0.83
RBF-Dagging	0.270	0.074	0.71	0.74	0.051	0.227	0.79	0.75

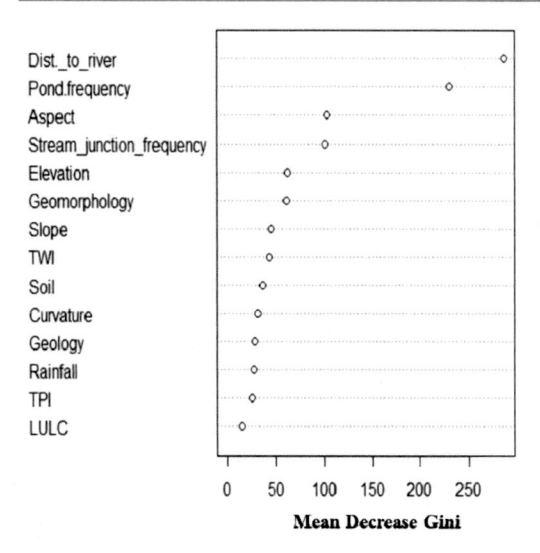

Fig. 15.6 Significant factors identification by mean decrease Gini

Table 15.3 Calculated MDGs values for significant factors identification

Factors	Mean decrease Gini
Curvature	31.45
TPI	24.99
TWI	42.94
Slope	45.32
Distance to river	286.01
Pond frequency	229.96
Aspect	103.56
Elevation	62.06
Stream junction frequency	101.45
Rainfall	26.88
Geology	28.38
Land use/land cover	14.09
Soil types	36.11
Geomorphology	61.10

junction frequency (101.45), elevation (62.06), and geomorphology (61.10) were the most important factors. These were followed in order of influence by the slope (45.32), TWI (42.94), soil types (36.11), curvature (31.45), geology (28.38), rainfall (26.88), TPI (24.99), and LULC (14.09) (Fig. 15.6 and Table 15.3). All the fourteen predisposing factors were subjected to the modelling—purpose because all are contributors to GWP occurrence.

15.4 Discussion

For the groundwater potentiality (GWP) assessment factors like rainfall, land use, slope, elevation, pond frequency, stream junction frequency, distance to a river, TWI, soil texture, geology, geomorphology, curvature, and aspect are used. The elevation and slope are very low in the south-eastern portion of Md. Bazar block. Recharge of groundwater is negatively related to the elevation of study area. Thus, locations that are situated in low elevation areas show high groundwater potentiality at a particular region within the study area rather than being uniformly distributed across it.

In other works (e.g., Corsini et al. 2009; Ozdemir 2011; Rahmati et al. 2017; Naghibi et al. 2017; Chen et al. 2019), similar factors have been used for assessing GWP and the applied relation between the factor used and the wells are also found to be the same. Usually, there is no algorithm with an extreme prediction capacity that works completely as natural processes, and groundwater modeling is a complex and nonlinear process and cannot be based on normal models with a linear structure (Chen et al. 2019). Several researchers have applied MLAs like Bagging and Dagging, in various fields of research, like gully erosion, landslide, flood hazard, and deforestation susceptibility assessment (Chen et al. 2018, 2019; Arabameri et al. 2020; Hong et al. 2020; Pal et al. 2020; Talukdar et al. 2020; Saha et al. 2021). In every case, prediction capacity of the meta classifier ensemble model's results was extremely appreciable. So, the application of machine learning algorithms (MLAs) is not a new thing, but the implication of these machine learning meta-classifiers models for groundwater potentiality assessment is unique.

Previous research work like that by Corsini et al. (2009), Ozdemir (2011), Lee et al. (2017), Naghibi et al. (2018), concluded that MLAs provided adequate results with respect to multivariate and bivariate statistical models. In other studies, like floods, landslides, and assessments of spring potential, the RBFnn-Bagging model has also given good results. In the sense that no overfeeding of data is executed, the RBFnn-Bagging model is the most important. It consists of multiple decision trees with an interaction between predisposing factors and nonlinearity (Hong et al. 2020; Saha et al. 2021). The results also revealed that the processing speed of RBFnn-Bagging is much higher concerning RBFnn-Dagging mode, which means assignment of input factors is very important. As a matter of fact, concerning percentage of the low and high GWP zones, two models displayed a uniform spatial distribution. So, the RBF-Bagging and RBFnn-Dagging models can be applied for hazard vulnerability and susceptibility mappings such as flood, landslide, forest fire, and gully erosion at a local and regional scale.

15.5 Conclusion

Groundwater potential mapping, applying various predisposing factors, is an important aspect of groundwater research. For the accurate experiment of groundwater conditions, several algorithms have been applied around the globe. In this study, a well-accepted methodology was applied to delineate GWP zones in Md. Bazar Block. After critically evaluating the study, fourteen predisposing factors were overlaid with RBF-Bagging and RBFnn-Dagging models. The RBFnn-Bagging and RBFnn-Dagging models identified 16.92 and 13.70% of areas with very high groundwater potentiality and 68.64 and 68.03% of the block with low groundwater potentiality. The results alert that this block may face vulnerable conditions in the future if the government back-steps from introducing various schemes (i.e., rainwater harvesting, dam construction, etc.) and generating awareness among common people. Based on the experiment results the following conclusions can be summarized. First, the RBFnn-Bagging model has better prediction capacity than the RBFnn-Dagging model because the Bagging algorithm can be applied to find out reliable features of the real data. Second, researchers can solve the model overfitting problems by applying the RBF-Bagging model. Third, based on the mean decrease Gini, the most effective factors of groundwater potentiality are the distance to a river, pond frequency, aspect, stream junction frequency, elevation, and geomorphology, respectively. Finally, this proposed approach should be useful for the exploration, development, and management of groundwater. At the outset, it is pertinent that groundwater recharge processes along with their management are taken at the earliest in Md. Bazar Block.

References

Abedinpour M, Sarangi A, Rajput TBS, Singh M, Pathak H, Ahmad T (2012) Performance evaluation of AquaCrop model for maize crop in a semi-arid environment. Agric Water Manag 110:55–66

Arabameri A, Chen W, Blaschke T, Tiefenbacher JP, Pradhan B, Tien Bui D (2020) Gully head-cut distribution modeling using machine learning methods—a case study of NW Iran. Water 12(1):16

Arabameri A, Roy J, Saha S, Blaschke T, Ghorbanzadeh O, Tien Bui D (2019) Application of probabilistic and machine learning models for groundwater potentiality mapping in Damghan Sedimentary Plain, Iran. Remote Sens 11(24):3015

Breiman L (1996) Bagging predictors. Mach Learn 24:123–140

Breiman L (2001) Random forests. Mach Learn 45(1):5–32

Chaudhry AK, Kumar K, Alam MA (2019) Mapping of groundwater potential zones using the fuzzy analytic hierarchy process and geospatial technique. Geocarto Int 1–22

Chen W, Panahi M, Khosravi K, Pourghasemi HR, Rezaie F, Parvinnezhad D (2019) Spatial prediction of groundwater potentiality using ANFIS ensembled with teaching-learning-based and biogeography-based optimization. J Hydrol 572:435–448

Chen W, Shahabi H, Zhang S, Khosravi K, Shirzadi A, Chapi K, Pham BT (2018) Landslide susceptibility modeling based on GIS and novel bagging-based kernel logistic regression. Appl Sci 8(12):2540

Conforti M, Aucelli PP, Robustelli G, Scarciglia F (2010) Geomorphology and GIS analysis for mapping gully erosion susceptibility in the Turbolo stream catchment (Northern Calabria, Italy). Nat Hazards 56(3):881–898

Corsini A, Cervi F, Ronchetti F (2009) Weight of evidence and artificial neural networks for potential groundwater spring mapping: an application to the Mt.

Modino area (Northern Apennines, Italy). Geomorphology 111(1–2):79–87

Das S (2019) Comparison among influencing factor, frequency ratio, and analytical hierarchy process techniques for groundwater potential zonation in Vaitarna basin, Maharashtra, India. Groundw Sustain Dev 8:617–629

Freedman DA (1981) Bootstrapping regression models. Ann Stat 9:1218–1228

Gayen A, Pourghasemi HR, Saha S, Keesstra S, Bai S (2019) Gully erosion susceptibility assessment and management of hazard-prone areas in India using different machine learning algorithms. Sci Total Environ 668:124–138

Ghosh D, Mandal M, Banerjee M, Karmakar M (2020) Impact of hydro-geological environment on availability of groundwater using analytical hierarchy process (AHP) and geospatial techniques: a study from the upper Kangsabati river basin. Groundw Sustain Dev 11:100419

Golkarian A, Naghibi SA, Kalantar B, Pradhan B (2018) Groundwater potential mapping using C5.0, random forest, and multivariate adaptive regression spline models in GIS. Environ Monit Assess 190:149

Grohmann CH, Riccomini C (2009) Comparison of roving-window and search-window techniques for characterising landscape morphometry. Comput Geosci 35:2164–2169

Guru B, Seshan K, Bera S (2017) Frequency ratio model for groundwater potential mapping and its sustainable management in cold desert, India. J King Saud Univ-Sci 29(3):333–347

Guzman SM, Paz JO, Tagert MLM, Mercer A (2015) Artificial neural networks and support vector machines: contrast study for groundwater level prediction. In: Proceedings of the 2015 ASABE annual international meeting

Haque SM et al (2020) Identification of groundwater resource zone in the active tectonic region of Himalaya through earth observatory techniques. Groundw Sustain Dev 10. https://doi.org/10.1016/j.gsd.2020.100337

Hong H, Liu J, Zhu AX (2020) Modeling landslide susceptibility using LogitBoost alternating decision trees and forest by penalizing attributes with the bagging ensemble. Sci Total Environ 718:137231

Jothibasu A, Anbazhagan S (2016) Modeling groundwater probability index in Ponnaiyar River basin of South India using analytic hierarchy process. Model Earth Syst Environ 2:109

Kenda K, Čerin M, Bogataj M, Senožetnik M, Klemen K, Pergar P, Laspidou C, Mladenić D (2018). Groundwater modeling with machine learning techniques: Ljubljana polje Aquifer. Proceedings 2:697

Kotsianti SB, Kanellopoulos D (2007) Combining bagging, boosting and dagging for classification problems. In: International conference on knowledge-based and intelligent information and engineering systems, pp 493–500. Springer

Lee S, Hong SM, Jung HS (2017) GIS-based groundwater potential mapping using artificial neural network and support vector machine models: the case of Boryeong city in Korea. Geocarto Int 1–33

Mishra SV, Gayen A, Haque SM (2020) COVID-19 and urban vulnerability in India. Habitat Int 103:102230

Mohamed MM, Elmahdy SI (2017) Fuzzy logic and multi-criteria methods for groundwater potentiality mapping at Al Fo'ah area, the United Arab Emirates (UAE): an integrated approach. Geocarto Int 32(10):1120–1138

Moore ID, Grayson RB, Ladson AR (1991) Digital terrain modelling: a review of hydrological, geomorphological, and biological applications. Hydrol Process 5(1):3–30

Naghibi SA, Ahmadi K, Daneshi A (2017) Application of support vector machine, random forest, and genetic algorithm optimized random forest model in groundwater potential mapping. Water Resour Manag 31(9):2761–2775

Naghibi SA, Pourghasemi HR, Dixon B (2016) GIS-based groundwater potential mapping using boosted regression tree, classification and regression tree, and random forest machine learning models in Iran. Environ Monit Assess 188(1):1–27

Naghibi SA, Pourghasemi HR, Abbaspour K (2018) A comparison between ten advanced and soft computing models for groundwater qanat potential assessment in Iran using R and GIS. Theor Appl Climatol 131(3):967–984

Ozdemir A (2011) Using a binary logistic regression method and GIS for evaluating and mapping the groundwater spring potential in the Sultan Mountains (Aksehir, Turkey). J Hydrol 405(1):123–136

Pal S, Mandal I (2021) Noise vulnerability of stone mining and crushing in Dwarka river basin of Eastern India. Environ Dev Sustain 1–22

Pal SC, Arabameri A, Blaschke T, Chowdhuri I, Saha A, Chakrabortty R, Lee S, Band S (2020) Ensemble of machine-learning methods for predicting gully erosion susceptibility. Remote Sens 12(22):3675

Rahmati O, Tahmasebipour N, Haghizadeh A, Pourghasemi HR, Feizizadeh B (2017) Evaluating the influence of geo-environmental factors on gully erosion in a semi-arid region of Iran: an integrated framework. Sci Total Environ 579:913–927

Razandi Y, Pourghasemi HR, Neisani NS, Rahmati O (2015) Application of analytical hierarchy process, frequency ratio, and certainty factor models for groundwater potential mapping using GIS. Earth Sci Inf 8(4):867–883

Rudin C (2019) Stop explaining black box machine learning models for high stakes decisions and use interpretable models instead. Nat Mach Intell 1(5):206–215

Saha S (2017) Groundwater potential mapping using analytical hierarchical process: a study on Md. Bazar Block of Birbhum District, West Bengal. Spat Inf Res 25(4):615–626

Saha S, Paul GC, Pradhan B, Abdul Maulud KN, Alamri AM (2021) Integrating multilayer perceptron neural nets with hybrid ensemble classifiers for deforestation probability assessment in Eastern India. Geomat Nat Hazards Risk 12(1):29–62

Sajedi-Hosseini F, Malekian A, Choubin B, Rahmati O, Cipullo S, Coulon F, Pradhan B (2018) A novel machine learning-based approach for the risk assessment of nitrate groundwater contamination. Science Total Environ 644:954–962

Shekhar S, Pandey AC (2014) Delineation of groundwater potential zone in hard rock terrain of India using remote sensing, geographical information system (GIS) and analytic hierarchy process (AHP) techniques. Geocarto Int 30(4):402–421

Suhag R (2016) Overview of ground water in India. PRS Legislative Research ("PRS") standing committee report on Water Resources examined 10

Tahmassebipoor N, Rahmati O, Noormohamadi F, Lee S (2016) Spatial analysis of groundwater potential using weights-of-evidence and evidential belief function models and remote sensing. Arab J Geosci 9(1):79

Talukdar S, Pal S (2020) Wetland habitat vulnerability of lower Punarbhaba river basin of the uplifted Barind region of Indo-Bangladesh. Geocarto Int 35(8):857–886

Talukdar S, Ghose B, Salam R, Mahato S, Pham QB, Linh NTT, Costache R, Avand M (2020) Flood susceptibility modeling in Teesta River basin, Bangladesh using novel ensembles of bagging algorithms. Stoch Env Res Risk Assess 34(12):2277–2300

Thapa R, Gupta S, Guin S, Kaur H (2018) Sensitivity analysis and mapping the potential groundwater vulnerability zones in Birbhum district, India: a comparative approach between vulnerability models. Water Sci 32(1):44–66

Tien Bui D, Bui QT, Ngayen QP, Pradhan B, Nanpak H, Trinh PT (2017) A hybrid artificial intelligence approach using GIS-based neural-fuzzy inference system and particle swarm optimization for forest fire susceptibility modeling at a tropical area. Agric for Meteorol 233:32–44

Ting KM, Witten IH (1997) Stacking bagged and dagged models. Working paper 97/09, University of Waikato, Department of Computer Science, Hamilton, New Zealand

Weiss A (2001) Topographic position and landforms analysis. Poster Presentation, ESRI User Conference, San Diego, CA

World Bank (2010) Deep wells and prudence: towards pragmatic action for addressing groundwater overexploitation in India. 51676, Washington, D.C. http://documents.worldbank.org/curated/en/272661468267911138/Deep-wells-and-prudence-towards-pragmaticaction-for-addressing-groundwater-overexploitation-in-India

Wu Y, Ke Y, Chen Z, Liang S, Zhao H, Hong H (2020) Application of alternating decision tree with AdaBoost and bagging ensembles for landslide susceptibility mapping. CATENA 187:104396

Yariyan P, Janizadeh S, Van Phong T, Nguyen HD, Costache R, Van Le H, Pham BT, Pradhan B, Tiefenbacher JP (2020) Improvement of best first decision trees using bagging and dagging ensembles for flood probability mapping. Water Resour Manag 34(9):3037–3053

Enhancement of Natural and Technogenic Soils Through Sustainable Soil Amelioration Products for a Reduction of Aeolian and Fluvial Translocation Processes

Sandra Muenzel and Oswald Blumenstein

Abstract

The economy of some countries is dominated mostly by mining and agriculture. Enormous amounts of excavated waste material are deposited in huge dams. The tailings substrates in South Africa have been studied and a variety of greenfield revitalization attempts have been made to reduce the consequences or effects of fluvial or aeolian processes. A combination of different soil amendments was used. These have been adapted to the site-specific soil dynamics. There is a visible increase of above-ground and underground biomass. After more than three years, even the extremely acidic gold dumps still had vigorous grass vegetation. The survival of humans and animals depends also on the effective use of the widely varying amounts of precipitation necessary for the growth of crops. With two soil amendments, the storage capacity of these soils for water and nutrients and thus their productivity can be improved. Tests on sandy substrates in Germany with wheat and grass vegetation showed a positive effect on the height growth, the biomass, and the degree of coverage. These amendments have also been used in the Kalahari with maize crops. There was a 50% higher yield of biomass and the corncob compared to NPK fertilizer. It would be possible to stabilize the productivity of crops while at the same time the amount of irrigation can be reduced.

Keywords

Greenfield revitalization · Drought · Spoil dams · Agricultural land use

16.1 Introduction

The range of soil usage is severely limited by their erosive destruction or pollution. Without its filtering, buffering and substance transformation processes, the ecological cycles in the landscape would not be possible. It is known that around 30% of the land surface worldwide is degraded (Nkonya et al. 2016). Every year, 5 to 10 million ha are added (UBA 2015), which is roughly the area of Austria. The costs caused by soil degradation amount to about 300 billion euros annually (Scheub 2016). These figures illustrate the need for action in the application of soil improvement methods. Every euro invested in soil protection today will be a profit of 5 euros in

S. Muenzel (✉)
University of Potsdam, Institute for Environmental Science and Geography, Potsdam, Germany
e-mail: sandra.muenzel@gmx.net

S. Muenzel
Leibniz Institute of Vegetable and Ornamental Crops (IGZ), Großbeeren, Germany

O. Blumenstein
InterEnviroCon GmbH, Potsdam, Germany

the future—half as yield, the other half in the form of better water quality or other ecosystem services (Scheub 2016).

An important cause of the devaluation of soil properties is the extraction of raw materials, such as coal, or the extraction of heavy metals. Both interventions in the landscape not only require a high consumption of land but also contributes to extreme contamination in countries dominated by mineral extraction. South Africa, for example, is also heavily influenced by the extraction of ores. The North-West Province has the world's largest reserves of platinum. The crushed ore is of a fine, sandy grain size. However, since only the platinum group metals are extracted, the substrate still contains large amounts of firmly bound heavy metals, especially chromium and nickel (Münzel 2013). Moreover, about 40% of the world's gold reserves are located in the Witwatersrand mining area in the Gauteng province. In the regions, there are huge tailings dumps with suspended, loose overburden in close proximity to farmland, surface waters, and human settlements. Fluvial and aeolian transport processes can deposit these tailings directly onto these valuable and protected assets. As compact or diffuse deposits, the tailings material thus reaches the surrounding soils, such as oxisols or vertic soil, which results in changes in their soil dynamics (Blumenstein et al. 2010).

On the other hand, in Germany lignite is still one of the most important energy sources. Its share in the gross electricity generation was still around 22.5% in 2018 (Statistisches Bundesamt 2020). The focus of current production is concentrated on the Lusatian mining area, as the second-largest mining area in Germany. A total of 52 million tons of lignite were extracted here in 2019 (Lausitz Energie Bergbau AG 2020). However, such mining is associated with far-reaching consequences and a huge land degradation. The naturally occurring soils are destroyed by the erosion of the hanging (or overlying) tertiary and quaternary cover layers. Hollow depressional forms up to 110 m deep are created, which are refilled or flooded with the dumped substrates after mining has ceased. The dumped areas are characterized by a high nutrient deficiency and low water storage capacity. All of the areas mentioned are extreme sites. They must be re-cultivated to allow forestry or agricultural use again. This is not an easy process due to the properties of the substrate already alluded to.

Degraded land is characterized by some of the following features:

- extremely high or low pH values,
- insufficient buffering against acidity,
- high salinity,
- low or extremely high nutrient storage capacity,
- nutrient deficiencies,
- deficiencies in soil organic matter, or
- instability of the soil structure, increasing susceptibility to erosion.

Furthermore, precipitation in many regions shows an overall decrease with an increased variability. Land users have to apply new technologies to minimize additional negative impacts associated with the proximity of mine deposits. One common practice is to improve the water management, which involves supplemental irrigation. These measures require additional costs and, in the long run, result in changes in soil properties. New tillage techniques could also reduce negative impacts of soil degradation (Aravind et al. 2017; Voutos et al. 2019; von Redwitz et al. 2019).

The best approach to improving soil properties must target the soil directly. The most common approaches include basic amelioration and the addition of fertilizers. However, attempts are also increasingly being made to improve substrate properties by adding soil amelioration products. Substrate melioration is usually followed by planting various tree species. However, woody plants need several years to develop a well-distributed root system. For this reason, grass-herb mixtures are also used, as these roots the soil more quickly and thus provide rapid erosion protection. Examples of soil supplements are auxiliary soils, charcoal, algal lime, primary rock flour, expanded shale, or even plastics such as styromull. However, these substrates have a very one-sided effect, are produced artificially, or are very costly to produce.

As a consequence, the research work of the former "Applied Geoecology" working group at the University of Potsdam, (now IEC GmbH) focuses, among other things, on the development and use of new soil supplements. This is based on the idea of a substrate-specific combination. This is because the diversity of soil properties highlights the difficulties that can arise in treatment options. Therefore, there cannot be a single method or a universally effective soil amelioration product. Before using these combinable amendments, the specific properties and characteristics of the substrates need to be analyzed in relation to the problem that arises. This is especially true for the improvement of soil properties at the extreme sites mentioned above.

This paper presents examples of how the use of newly developed soil amelioration products can affect the cover and growth characteristics of vegetation at various sites and thus can reduce the risk of erosion. In addition to two regions of ore mining in northern South Africa, i.e., Rustenburg and Klerksdorp (Fig. 16.1), re-cultivation areas of lignite mining in northeastern Germany were chosen. Furthermore, the soil amelioration products were used in the southern Kalahari both to cover the soil and simultaneously increase yields in dryland areas.

16.2 Materials and Methods

16.2.1 Soil Amelioration Products

All soil amelioration products presented in this paper were developed, tested, and optimized on a scientific basis over several years (Fig. 16.2). These tests ranged from experiments in the laboratory, container, and greenhouse trials to field experiments on extreme sites. This process was accompanied by extensive data collection on the tested mixture variants in relation to the relevant soil properties and state variables of the plants.

By using this approach three different soil supplements have been developed and tested worldwide (Table 16.1).

Soil amelioration product K1 is a basic, inorganic component that serves to improve the nutrient supply with phosphorus. As a modified source material for the metal industry, it is mainly used in nutrient-poor soils.

Soil amelioration product K3 as an organic component serves as an additional source of nutrients as well as loosening the soil by soil organisms. The water retention capacity of the substrate can be increased by the existing fiber structures and organic matter. The clay-humus complexes form stable soil aggregates. Due to the coarse structure and the presence of soil

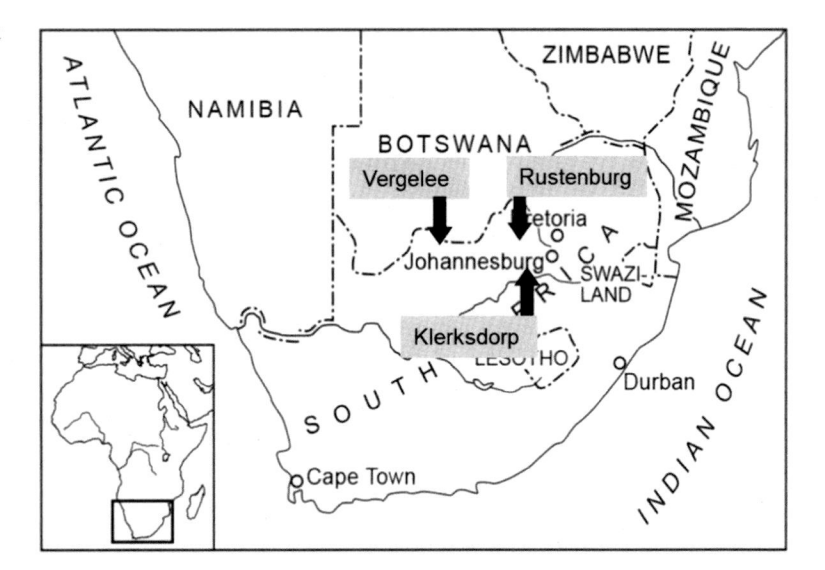

Fig. 16.1 Test locations in Northern South Africa

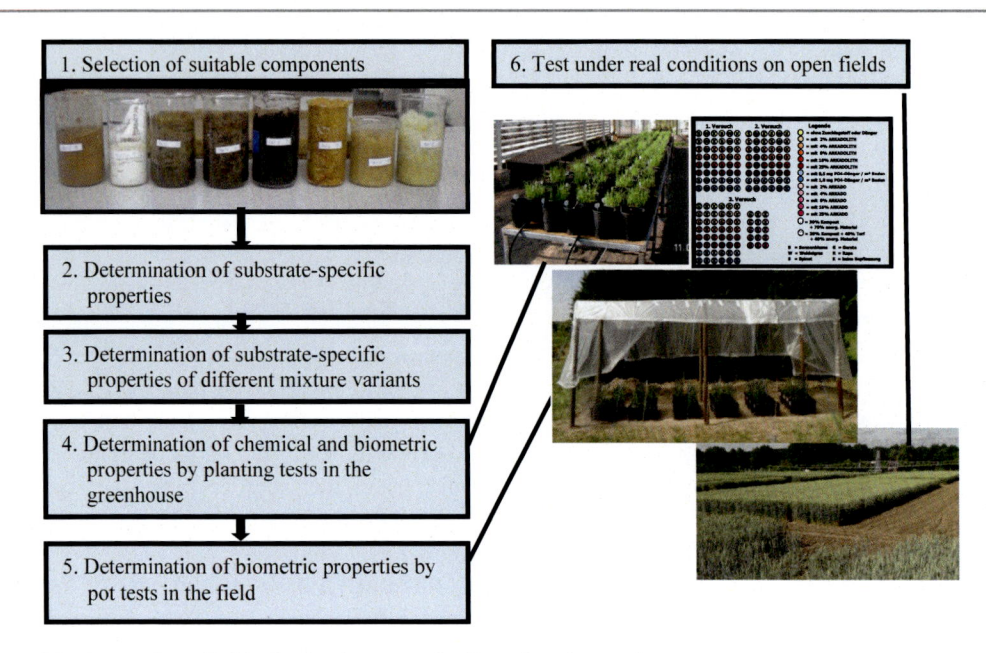

Fig. 16.2 Approach applied in the development of soil amelioration products

Table 16.1 Soil amelioration products developed and applied to sites discussed in the paper; their properties and test areas

	Component K1	Component K3	Component K4
substrate	inorganic	organic	artificial, inorganic
effect	acid buffering; improvement of nutrient (P) supply	increase of organic matter; content increase of CEC	Increase in water storage capacity
source	mineral raw material of the metal industry	woody parts, natural N-sources	polymer with natural nanoparticles
application	acidic and nutrient-poor soils, soil with stagnant moisture	nutrient-poor soils, raw soils	soils in regions with dry periods
test area	platinum and gold tailings in South Africa	platinum and gold tailings in South Africa; lignite coal mining rehabilitation area in Germany; thorn savannah in South Africa	thorn savannah in South Africa

organisms, the soils or substrates are loosened when K3 is added. This in turn leads to a larger pore volume and thus an improvement in gas exchange, percolation, and heat storage. Component K3 is weakly basic. This can counteract soil acidification and thus stabilize the pH value.

Component K4 can be used in regions with a pronounced dry season. It is an artificial polymer with natural nanoparticles and serves to store water and the nutrients it contains. This process of water absorption and storage is reversible. By using it, the onset of a dry stress phase can be delayed and its duration shortened.

The combined use of these components leads to a reduction in the amount of irrigation, fertilizer, and tillage required. In order to prove this, experimental plots were set up on extreme sites worldwide from 2006 onwards. These are located on tailings dumps from platinum and gold ore mining in South Africa, on acid tailings dumps in

Greece, in post-mining landscapes on tipping sand in Germany, and on shifting sand dunes in Chinese semi-deserts (Fig. 16.3).

It should be emphasized that no additional mineral fertilizer and irrigation measures were used in these trials and no additional irrigation took place. The results clearly showed that by adding the ameliorants, a change in pH, loss of ignition, and cation exchange capacity and thus an improvement in soil chemical properties is already achieved within a period of nine months (Münzel and Blumenstein 2012). After seeding, vegetation cover and growth height improved at all investigated sites compared to the untreated areas (Fig. 16.2). The aim is to establish vegetation quickly in order to reduce fluvial and aeolian displacement processes.

16.2.2 Procedure

To identify the improvement of soil properties, biometric parameters were used as these are relevant for minimizing the displacement processes. A control plot and areas with different proportions of soil amelioration products were established in all areas. A comparison of the results should provide information about the different growth successes. On the tailings of the gold and platinum ore mines in South Africa, five test plots of 1 m × 1 m each were established and different amounts of the soil amelioration products K1 and K3 were added depending on the substrate properties (Tables 16.2 and 16.3). Afterward, seeds of a regular seed mixture were sown by hand. The growth height and the degree of cover were determined at intervals of several months. One year after sowing, the above- and below-ground biomass was determined on the tailings of the platinum ore mine, and after three years on the substrate of the gold ore mine. No irrigation or addition of fertilizers took place.

The size of the re-cultivation areas in Germany was 2500 m² each. The LMBV (Lausitzer und Mitteldeutsche Bergbau-Verwaltungsgesellschaft mbH) worked the substrate and added the soil amelioration product K3 (Table 16.4) to a depth of 25 cm (for grass) and 100 cm (for woody plants)

Fig. 16.3 Field trials with K1 and K3 at various extreme locations worldwide

Table 16.2 Added amount of soil amelioration products to test areas of platinum mining

Test area	component K1 [% vol.]	component K3 [% vol.]
Platin 5	3.0	0.0
Platin 6	0.0	3.0
Platin 7	3.0	3.0
Platin 8	1.5	1.5
Platin 9	5.0	5.0

Table 16.3 Added amount of soil amelioration products to test areas of gold mining

Test area	component K1 [% vol.]	component K3 [% vol.]
Gold 1	3.0	3.0
Gold 2	0.0	12.5
Gold 3	12.5	0.0
Gold 4	12.5	12.5
Gold 5	0.0	0.0

Table 16.4 Added amount of soil amelioration products to test areas of Lower Lusatia

Test area	Component K3 [% vol.]	Component E [% vol.]
Lusatia 1	5.0	According to good professional practice
Lusatia 2	5.0	0.0
Lusatia 4	0.0	0.0
Lusatia 5	0.0	According to good professional practice

below-ground level using large agricultural equipment. Three separate plots of 1 m^2 were selected and marked to determine the growth height and the degree of cover of the grasses and herbs in the corresponding areas. The above-ground biomass was determined in an adjacent control plot of 1 m^2. For the recording, 1 m^2 was marked out three times per test plot and these were measured.

In accordance with the experimental design, the growth characteristics of the tree species were also determined on the different substrates. For this monitoring, 24 trees were selected per species and field on the basis of randomization. These were marked for repeated identification. On these selected trees of *Betula pendula*, *Pinus sylvestris*, and *Quercus robur*, the growth height and stem diameter were measured at 50, 86, 156, and 210 days after planting (DAS—Days After Seeding). When measuring selected seedlings for a time series, the nearest individual tree had to be used where a sample specimen had died. Therefore, in the course of the measuring times, apparent contradictions arose in the representation of growth height. Also in this area no irrigation or addition of fertilizers took place.

During the field trials in the Kalahari, six experimental plots of 32 m^2 each were established. After the soil had been plowed, the soil amelioration products were added to a depth of 25 cm below-ground level. For comparison, a trial plot was set up with NPK fertilization according to good professional practice. In each experimental plot, three rows of maize were sown by hand in December 2018 with a total number of 60 grains. Although the experimental plots were fenced, animals used the sowing as a food source. Therefore, thorn bushes were placed around the fencing and in a second series at the beginning of January, two additional rows of 15 seeds each were sown. Half of the test plots were not irrigated. The supplementary irrigation of the other test areas was 121 mm during the trial period, which corresponds to about 50% of the good professional practice there. The plants were harvested in mid-April, about 3 weeks before the usual harvest date.

16.2.3 Biometric Methods

Biometric parameters were used to demonstrate the improvement of soil properties. These include the growth height and stem diameter of woody plants, for grasses and herbs the degree of cover, the number of plant species, and the above-ground and below-ground phytomass. After air-drying, the dry mass could also be determined. In the maize plots, the total number of cobs was also determined. The growth height of the plants was measured using a folding rule. The stem diameter was recorded using a vernier caliper gauge, with the measurement being taken 1 cm above the ground. The degree of cover of the plants was estimated as a percentage of the respective square meters. In order to minimize the subjective influences of this recording, the coverage pattern according to Gehlker (1976) was used. For the determination of the phytomass and dry mass, the grasses and herbaceous plants were harvested on one square meter. The harvested mass was weighed to 0.1 g, using a precision balance. It should be noted that the different water content of the individual plant species could lead to misinterpretations. To obtain comparable results, the dry mass of the harvested vegetation was determined after two weeks of drying in the air.

16.3 Results and Discussion

16.3.1 Re-Greening of Extreme Locations in South Africa

One test site was a tailings dump from platinum ore mining near Rustenburg (Fig. 16.1), whose flushed substrate still contains large amounts of nickel, chromium, and copper (Münzel 2013). Although a pH value of 7.0 could be measured, only a low vegetation cover was present on the dump, which originated from a plantation with acacia trees. Autochthonous grasses were found sporadically.

Another test with the components K1 and K3 took place at gold ore mining dump near Klerksdorp (Fig. 16.3). The greening tests were carried out on the flushing substrate, which was found in the immediate vicinity of the dumps as a deposited sediment package with a layer thickness of more than 50 cm. The pH value of the substrate was very acidic with values around 3.5. This places higher demands on the calculation of the application quantities of the soil amelioration products.

Based on the substrate properties, soil amelioration products of up to 5% by volume were added in platinum ore mining (Table 16.2, Meyer 2012). In gold ore mining, larger additions of up to 12.5% by volume were necessary due to the lower pH values (Table 16.3). The control area was the existing substrate without any addition of soil amelioration products.

The harvested biomass of the individual trial plots. There are clear differences between the trial variants (Fig. 16.4). The lowest growth successes were recorded with the sole use of the soil amelioration product K1 (see Platinum 5, Fig. 16.5). In contrast, additions of 1.5% by volume of the two soil amelioration products resulted in an increased phytomass production of about 100 $g \cdot m^{-2}$ above-ground and 80 g per 0.2 m^3 below-ground. If 3% by volume of each of the two soil amelioration products is added to the platinum tailings substrate (Platinum 7), the biomass increases once again.

On the other hand, the addition of 5% vol. K1 and K3 only lead to a further increase in the below-ground biomass, the above-ground biomass is slightly reduced. The test variant with the sole use of K3 (with 3% vol.) also leads to a very good result (Figs. 16.4 and 16.5). With over 200 $g \cdot m^{-2}$, the above-ground biomass is highest here.

In conclusion, it can be stated that the soil amelioration product K3 primarily promotes above-ground biomass production, whereas the addition of K1 leads to an increase in below-ground biomass production.

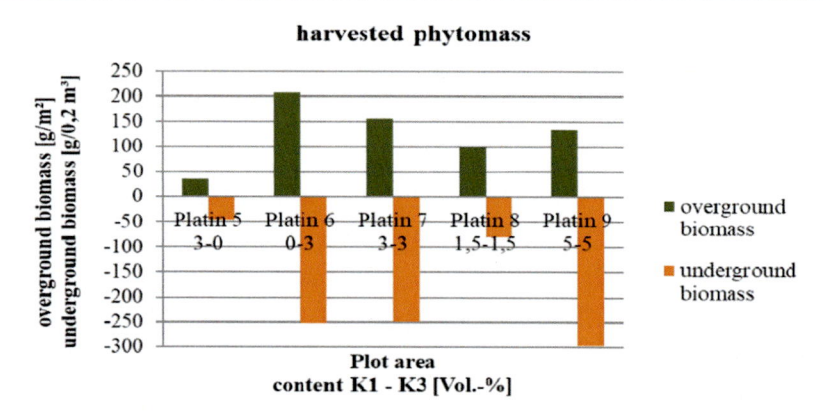

Fig. 16.4 Grass biomass per plot at trial areas of platinum ore mining

Fig. 16.5 Plant plots on tailings of platinum dumps after one year; 5a-Plot with 3% K1 (Platinum 5), 5b- Plot with 3% K3 (Platinum 6), 5c- Plot with 1.5% K1 und 1.5% K3 (Platinum 8), 5d- Plot with 3% K1 und 3% K3 (Platinum 7), 5e-Plot with 5% K1 und 5% K3 (Platinum 9)

An evaluation of the biometric data from field trials with tailings substrate of a gold dump is being studied (Fig. 16.6). It shows a clear dependence of the mixture variants to be used on the objective of vegetation establishment. If the production of above-ground biomass is desired, a relevant proportion of component 3 is necessary. If, on the other hand, intensive rooting is to be achieved, the proportion of component 1 must be increased accordingly. Here even a small amount of soil amelioration product leads to a significantly higher coverage and productivity of the vegetative plant parts, too.

Without the application of soil amelioration products, no vegetation cover existed (Gold 5, Fig. 16.7). The application of K1 alone is already sufficient for grass growth (Gold 3, Fig. 16.6, and Fig. 16.7c). Especially the below-ground biomass is increased by this component. However, growth successes of almost 400 g·m^{-2} can also be recorded in above groundmass.

The soil amelioration product K3 with 12.5% by volume (gold 2) increases, as in the platinum halide substrates, mainly the production of above-ground biomass (up to about 580 g·m^{-2}). It can be seen that the results of this trial are very similar to the variant with 3% vol. incorporation of both components (Gold 1, Fig. 16.6). With regard to the economic efficiency of the measures, the 3% vol. variant is therefore the more sensible one. However, the application of both soil amelioration products at 12.5% vol. each increases the growth of the below-ground biomass once again. The above-ground biomass, on the other hand, decreases again, as can also be seen with the platinum substrate. This leads to the conclusion that the use of larger amounts of component K3 reduces the growth of the above-ground biomass again. In summary, it can be stated that also in this trial the use of 3% by volume leads to the best result in terms of economic efficiency and biomass production. In addition, there is a good basis for reducing aeolian processes.

16.3.2 Reduction of Aeolian Transport Processes in Lower Lusatia

Within the framework of field trials in north-eastern Germany (Fig. 16.8), the effect of component K3, which contributes to an increase in the organic content of the substrates at a rate of 5% by volume, on the growth behavior of grasses, herbs and three different tree species (*Betula pendula, Pinus sylvestris* and *Quercus robur*) was investigated. The initial substrate of the test plots consists of quaternary sands.

Another component consists of residual products containing clay and iron oxide (E), which is provided by the LMBV (Lausitzer und Mitteldeutsche Bergbau-Verwaltungsgesellschaft mbH). The combined effect with K3 was also tested.

The test site consists of partial areas, each 100 m long and 25 m wide (Fig. 16.9). In addition to a control plot without the addition of soil

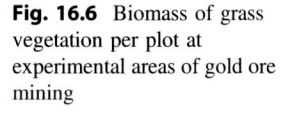

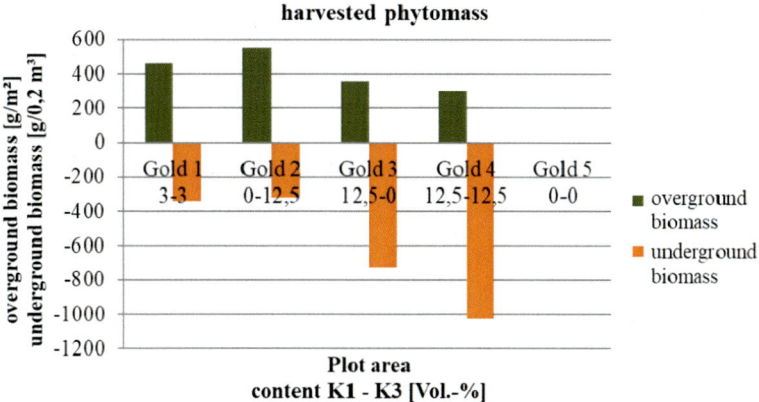

Fig. 16.6 Biomass of grass vegetation per plot at experimental areas of gold ore mining

Fig. 16.7 Plant plots on tailings of gold dumps after 3 years; **a**—Plot with 0% K1 and 0% K3 (Gold 5), **b**—Plot with 12.5% K3 (Gold 2), **c**—Plot wih 12.5% K1 (Gold 3), **d**—Plot with 3% K1 and 3% K3 (Gold 1) and **e**—Plot with 12.5% K1 and 12.5% K3 (Gold 4)

amelioration products, either 5% vol. K3 or component E or a combination of both was applied (Table 16.4). A fence was erected around the trial area to protect against animal predation. The plots were planted at the beginning of the growing season at the end of March 2020.

On 50% of the areas with different substrate compositions (Lusatia 1, 2, 4, 5), a grass and herb mixture (including *Secale multicaule*, *Raphanus sativus*, *Brassica juncea*, *Trifolium*) was sown; on the other half, strips of woody plants were planted at intervals of about 1 m each with annual tree seedlings of *Betula pendula*, *Pinus sylvestris* and *Quercus robur*. In addition to *Secale multicaule*, which forms a strong and extensive root system, various *Trifolium* species served as perennial species, allowing intensive root development up to 1 m deep (Aichele and Schwegler 1996). The species *Raphanus sativus* as well as *Brassica juncea* are often used as green manure in various grass and herb mixtures and are also characterized by being undemanding. They are therefore also suitable for use in dry areas (Bundessortenamt 2020).

The tree species *Betula pendula* are fast-growing and can cope with a lack of nutrients

Fig. 16.8 Sites of use of soil amelioration products in Germany (changed after: wikipedia 2016)

and acidic soil. Therefore, they are considered pioneer woody plants (Lüder 2018). *Quercus robur* is considered a typical tree species of Central Europe. They show good growth performance on nutrient-poor sandy soils, especially in the juvenile stage (Stinglwagner et al. 1998). *Pinus sylvestris* is undemanding and also suitable for nutrient-poor sandy soils (Lüder 2018).

16.3.2.1 Grass-Herbs Mixture

The maximum growth heights of dominant plants of the grass-herb mixture reached 86 days after sowing (Fig. 16.10). The tall species had already reached the maturity phase at this point. The ground-covering *Trifolium* species were included on the basis of the degree of cover to assess the influence of substrate differences.

All plants on the control plot without addition of soil amelioration products (Lusatia 4) did not reach double-digit values in growth height. The addition of component E (Lusatia 5) doubled the height of grass growth (10 cm). Both *Raphanus sativus* and *Brassica juncea* reached 50–60 cm growth height, with a total cover of 20% (Fig. 16.11).

A significant increase in growth height as well as in the degree of cover can be observed when component K3 with 5% by volume is added (Lusatia 1 and 2). The grasses then reach heights of up to 30 cm. While *Brassica juncea* shows a higher growth (up to 90 cm) on plots with a combined application of the soil amelioration products, *Raphanus sativus* dominates in the substrate with sole application of K3 (Lusatia 2). In the further course of the year, all plants of the seed mixture together reached a degree of coverage of 60% in this area. An increase to 75% was only recorded on the plots with additional supplementation of E (Lusatia 1). With regard to the goal of reducing aeolian transport processes, the use of component K3 thus makes a significant contribution (Table 16.5).

When interpreting the occurrence of individual plant species, it can be seen that the on-farm component E primarily promotes the growth of

Fig. 16.9 Copter photo of the individual test areas in Lower Lusatia on 25th June (F. Pustlauck 2020)

Fig. 16.10 Maximum plant height of different plant species on 25th June

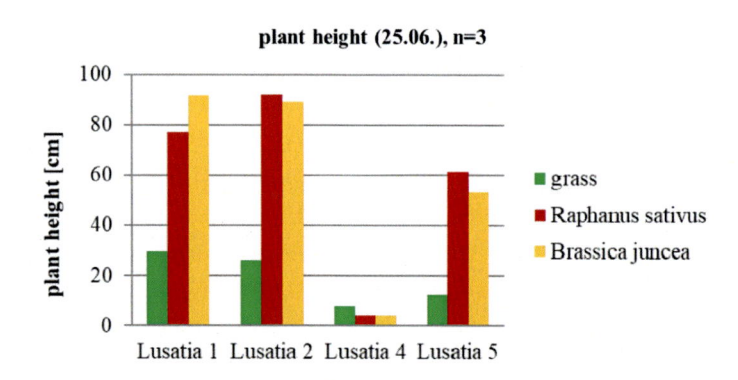

Fig. 16.11 Degree of coverage per square meter (DAS = day after seeding)

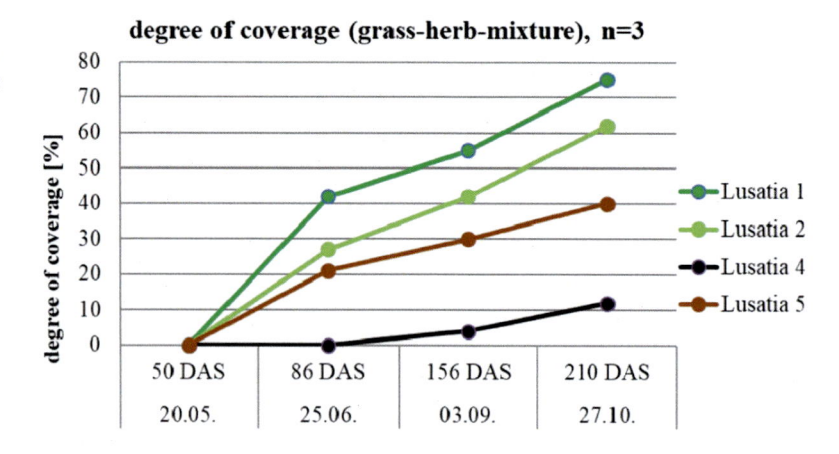

Table 16.5 Number of selected plant species on the three square meters on 25th June

Sum of the plants on three selected 1 m²	Grass	Trifolium	Raphanus sativus	Brassica juncea
Lusatia 1	57	64	19	63
Lusatia 2	31	43	23	54
Lusatia 4	24	8	7	7
Lusatia 5	67	27	20	46

grasses (Fig. 16.10). The use of component K3 did not result in a significant increase in the proportion of grass plants compared to the control or in combination with component E. On the other hand, the use of component K3 resulted in an increase in the proportion of grass plants compared to the control. On the other hand, component K3 mainly promoted *Trifolium* species (Lusatia 1 and 2). This is particularly relevant for a timely soil cover and thus reduction of erosion disposition. Both K3 and component E promote the occurrence of *Raphanus sativus* and *Brassica juncea*. Substrate-specific dominances between the two species are not discernible.

Furthermore, statements of the productivity of the areas are relevant in practice. For this purpose, above-ground phytomass was harvested and weighed for each 1m² area (Fig. 16.12a and b).

The phytomass production of the control plot on pure quaternary sand is less than 50 g·m^{-2}. Any use of soil amelioration products (K3 or E) improves productivity. The highest phytomass (approx. 500 g·m^{-2}) could be detected on plots with the application of a combination of both soil

Fig. 16.12 a Above-ground phytomass of one plot (DAS-days after seeding). **b** Dry matter of one plot (DAS-days after seeding)

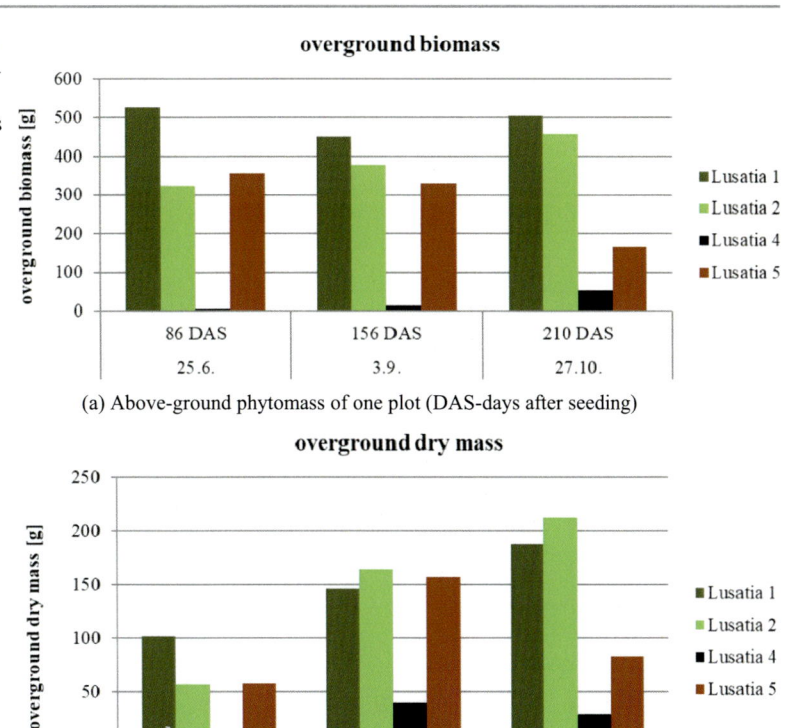

(a) Above-ground phytomass of one plot (DAS-days after seeding)

(b) Dry matter of one plot (DAS-days after seeding)

amelioration products. When assessing the productivity of the dry matter, the area with the sole application of K3 was found to be the most productive. The sole application of the on-farm component E is associated with a low increase in productivity, which is only improved when K3 is added.

16.3.2.2 Woody Plants

The results of the height measurements of the planted *Pinus sylvestris* show that only an average growth of maximum 2 cm (Fig. 16.13a) was achieved in the whole year. *Pinus sylvestris*, which is well adapted to sandy sites, showed good development even on the control plots (Lusatia 4). Plants on plots with the addition of component E show the lowest growth height and the smallest stem diameter. It can be assumed that this has an inhibiting effect on growth. In contrast, *Pinus sylvestris* grows well with the addition of organic matter (K3, Lusatia 1 and 2), which can be seen above all in the increase in stem diameter (Fig. 16.13). The statements give a first trend, but cannot be generalized nor extrapolated.

The planted *Quercus robur* hardly shows any changes in growth height and stem diameter over the course of the year (Fig. 16.14a and b). Only the increase in stem diameter due to component K3 can be considered a reliable statement. *Quercus robur* is subject to slow growth under these site conditions. This is not significantly accelerated by the soil amelioration products, which were used.

As a pioneer plant, *Betula pendula* shows rapid growth during its first vegetation period. A clear correlation is evident between applied soil amelioration products, growth height, and stem diameter. The approximately 20 cm tall annual *Betula pendula* trees grew about 10 cm on the pure quaternary sands of the mining areas (Lusatia 4, Fig. 16.15a). The application of

Fig. 16.13 a Plant height of *Pinus sylvestris* during one vegetation period ($n = 24$). **b** Trunk diameter (below) of *Pinus sylvestris* during one vegetation period ($n = 24$)

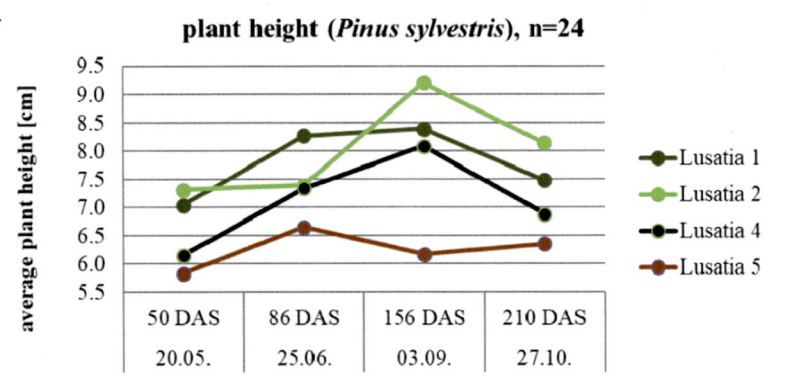

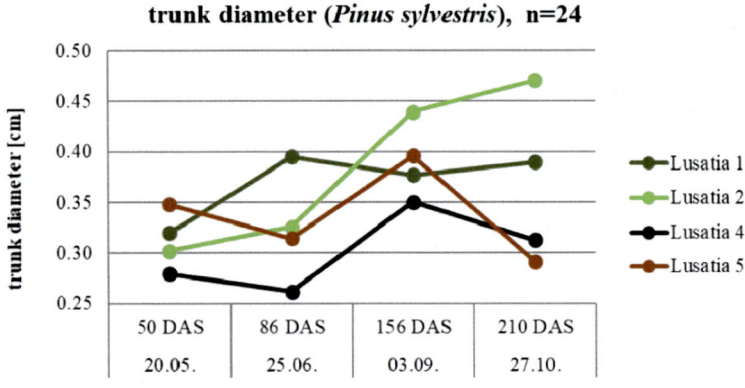

plant height (*Pinus sylvestris*), n=24

(a) Plant height of *Pinus sylvestris* during one vegetation period (n=24)

trunk diameter (*Pinus sylvestris*), n=24

(b) Trunk diameter (below) of *Pinus sylvestris* during one vegetation period (n=24)

component E increases the growth process; the *Betula pendula* reach a growth height of 40 to 48 cm. In the combined application with K3, an inhibitory effect on growth becomes apparent (Lusatia 1). The sole soil improvement by the component K3 leads to the best growth characteristics. Growth heights of 65 cm on average and stem diameters of >1 cm are already achieved after 210 days (Fig. 16.15b). This corresponds to more than a doubling of the measured plant parameters compared to the control plot.

In summary, it can be stated that the growth characteristics of *Betula pendula* as well as *Pinus sylvestris* could be significantly improved by the use of the soil amelioration products E and K3. However, in order to be able to make reliable statements on the effect of the soil amelioration products on the development of *Quercus robur*, a longer observation period is required.

In summary, it can be stated that component K3 is an effective soil amelioration product for vegetation establishment on the rehabilitation sites of Lower Lusatia. In contrast to the pure quaternary sands, the application of component K3 increases the degree of coverage sixfold and the phytomass production. In addition, the growth of the *Trifolium* seed can be promoted in a targeted manner. It is, therefore, possible to achieve protection against aeolian transport after only 5 months.

16.3.3 Reduction of Fluvial Transport Processes and Security of Yields in the Kalahari

In addition to a rapid establishment of vegetation, growth under dry conditions is also important in

Fig. 16.14 a Plant height of *Quercus robur* during one vegetation period (*n* = 24), DAS-days after seeding. **b** Trunk diameter of *Quercus robur* during one vegetation period (*n* = 24), DAS-days after seeding

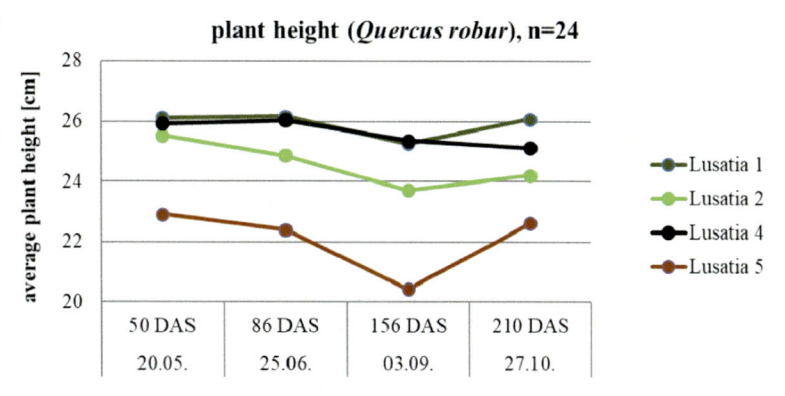

(a) Plant height of *Quercus robur* during one vegetation period (n=24), DAS-days after seeding.

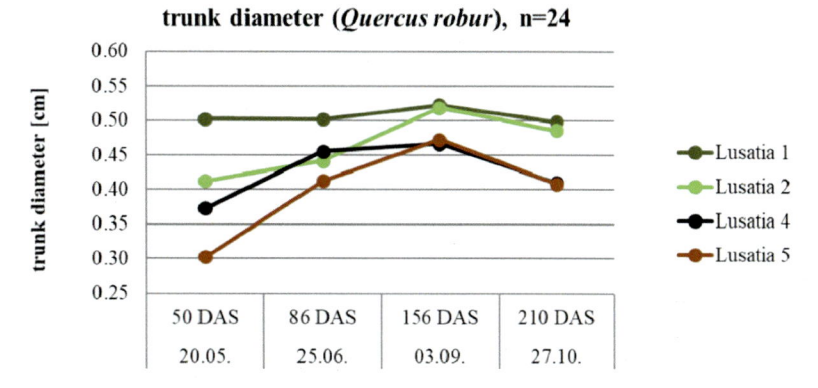

(b) Trunk diameter of *Quercus robur* during one vegetation period (n=24), DAS-days after seeding.

view of the fluctuating rainfall amounts and their temporal distribution.

The field trials in the South African Kalahari are based on results obtained in container trials in Germany (Münzel 2019). Here, the growth of wheat plants on sand substrates was monitored under controlled conditions using the components K3 and K4. The results show a positive effect of the soil amelioration products, which was expressed in height growth, biomass, and cover during the vegetative phase, even under intense drought stress (Münzel 2019).

Based on these results, components K3 and K4 were introduced at the beginning of the 2018 rainy season in areas of a thorn savannah in the southern part of the Kalahari, in the wider surroundings of Vergeleë (North-West Province,). The predominant soil type is an oxisol with an average pH of 5.5. The region is characterized by annual rainfall of around 450 mm. However, these vary greatly annually and mostly fall within 4 months, mostly as heavy rainfall. However, the total precipitation that fell during the study period (January 2019 to 10 April 2019) was only 98 mm. Air temperatures varied between 25 and 39 °C (de Wet 2020, oral communication).

Half of the experimental plots were not irrigated (Kalahari 1, 2, 3). On the other half, the irrigation amounted to 121 mm, which is about 50% of the local good practice (Table 16.6). Due to the low rainfall during the 2018/19 rainy season, even for the region, no yield was recorded on the trial plots without supplementary irrigation. Especially in December, in the initial phase of vegetation development, it was too dry, so that the maize seed dried out on all three trial

plant height (*Betula pendula*), n=24

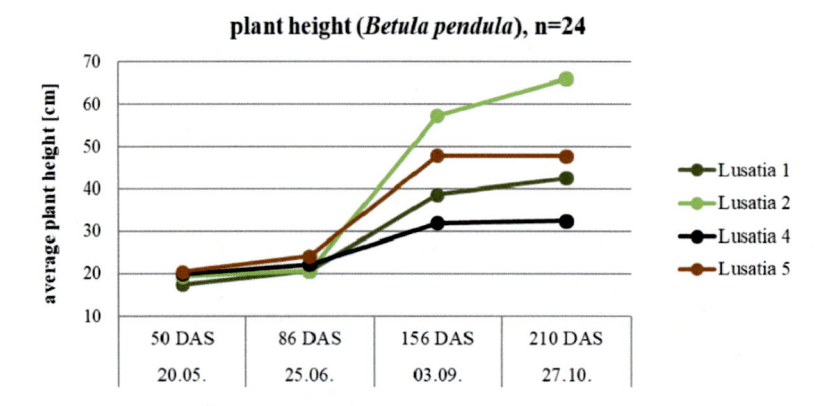

(a) Plant height of *Betula pendula* during one vegetation period (n=24)

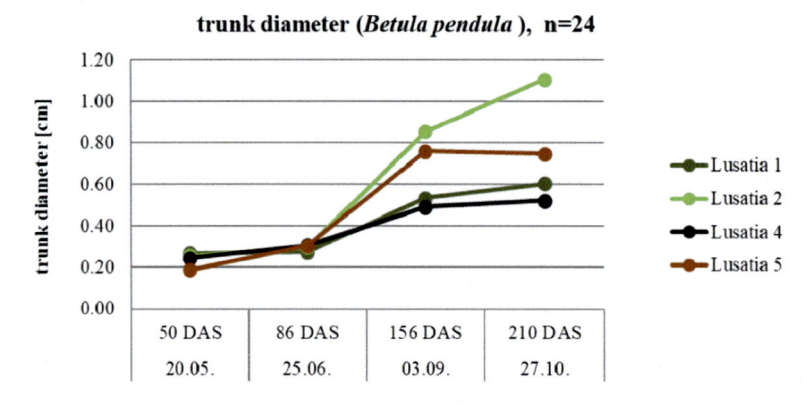

trunk diameter (*Betula pendula*), n=24

(b) Trunk diameter (below) of *Betula pendula* during one vegetation period (n=24)

Fig. 16.15 a Plant height of *Betula pendula* during one vegetation period ($n = 24$). b Trunk diameter (below) of *Betula pendula* during one vegetation period ($n = 24$)

Table 16.6 Added amount of soil amelioration products to test areas in the Kalahari

plot	Component K3 [% vol.]	Component K4 [% vol.]	NPK	Irrigation amount
Kalahari 1	0.0	0.0	According to good professional practice	0.0
Kalahari 2	5.0	0.0	0.0	0.0
Kalahari 3	0.0	0.0	0.0	0.0
Kalahari 4	0.0	0.0	According to good professional practice	50% According to good professional practice
Kalahari 5	5.0	0.0	0.0	50% According to good professional practice
Kalahari 6	5.0	0.3	0.0	50% According to good professional practice

plots. Therefore, only the three irrigated plots (Kalahari 4, 5, and 6) could be evaluated.

Due to the destruction by animals on the field with K3, the evaluation of the first sowing is limited only to the comparison between the NPK variant and the combined use of the two soil amelioration products K3 and K4. A comparison of all variants is only possible for the reseeding (second sowing, Table 16.7).

The growth success, measured by the number of developed plants, indicates an increase in the number of plants due to the application of the soil amelioration products. In the first sowing, 50% of the plants with NPK fertilizer grew successfully, with the use of the components K3 and K4 there were 50% more (Table 16.7). The second sowing also showed similar results. Compared to the NPK fertilization variant, the growth height of the maize plants could be increased by up to 20 cm when mixed with the soil amelioration products K3 and K4. The use of the organic component K3 alone resulted in maize plants that were 5 cm taller on average compared to the NPK variant.

The positive effect of the soil amelioration products was also shown by the total above-ground biomass. Over 100% more yield was achieved compared to the NPK fertilization. The addition of the water-storing component K4 led to a further improvement in harvest success (Fig. 16.16). In terms of the number of harvested cobs, the maize plants with the addition of both components produced the best harvest results (Table 16.7). The number of cobs doubled compared to conventional management with NPK fertilizer.

16.4 Conclusion

The work carried out provides trend statements for the possibilities of using the developed soil amelioration products. Through their application, a greening of mining tailings can be achieved. Under extreme aridity of a thorn savannah, a soil cover with relevant yields can be achieved with a low supplementary irrigation. It is therefore not necessary and economically advisable to resort to mineral fertilizers and intensive irrigation measures. It can be assumed that the applied soil amelioration products can be effective for up to five years. This was confirmed by subsequent on-site inspections in South Africa. This takes ecological and economic sustainability into account. The practical relevance of the results can be found in the savings in operating costs as well as in irrigation and fertilizer quantities and in the overall labor input.

If the soil has lost its natural polyfunctionality or extreme soil properties prevail, one has to specifically determine the composition and dosage of soil amelioration products. As already mentioned at the beginning, there is no single universally effective soil amelioration product or sole method of agrotechnical treatment. The global use of the developed components of soil amelioration products represents an important step for the sustainable use of our vital resource

Table 16.7 Parameter of maize plants as a function of soil improvement

plot	Kalahari 4 (NPK fertilizer)	Kalahari 5 (K3)	Kalahari 6 (K3 and K4)
Number of plants (first/second sowing)	31 von 60/10 von 30	15 von 60/14 von 30	47 von 60/17 von 30
Average height (first/second sowing))	105 cm/61 cm	94 cm/66 cm	126 cm/74 cm
total biomass (first/second sowing)	10.0 kg/0.5 kg	3.1 kg/1.1 kg	26.2 kg/2.4 kg
number of pistons	19	3	40

Fig. 16.16 Maize plants harvesting in the Kalahari (plot with 5% K3 and K4)

soil. In the future, the demands on soil will increase as the speed of change and the degree of interaction increase and new challenges arise, too. Therefore, increased attention should be paid to this resource in order not to "lose the ground under our feet" (Münzel and Blumenstein 2012).

The information presented so far provides the basis for decision-making on measures to optimize site-specific management, re-vegetation of extreme sites, and crop cultivation in drylands. In the future, the number of components should be expanded by further new developments of soil amelioration products. Thus, the synergy of science and practice provides an important contribution to the adaptation of agriculture to contemporary climatic challenges.

Acknowledgements The authors would like to thank the International Bureau of the BMBF (Bundesministerium für Bildung und Forschung) for the financial support of the projects on the mining dumps in South Africa from 2006 to 2012. Many thanks to the farmer Jandré de Wet for providing the area in Kalahari, the support in logistics, and the management of the test areas in 2018 and 2019. The Dr Sandra Münzel would also like to thank the Cooperation Funding of the University of Potsdam (KoUP) for funding travel expenses in 2018 and 2019. In addition, thanks go to the LMBV (Lausitzer und Mitteldeutsche Bergbau-Verwaltungsgesellschaft mbH) for the provision of the lignite rehabilitation areas in Germany. The authors further also gratefully acknowledge the many hardworking hands of field assistants who helped with the measurement and harvest of the test areas.

References

Aichele D, Schwegler HW (1996) Der Kosmos-Pflanzenführer. Bechtermünz Verlag im Weltbild Verlag GmbH, Augsburg

Aravind KR et al. (2017) Task-based agricultural mobile robots in arable farming: a review. Span J Agric Res 15(1):1–16. https://doi.org/10.5424/sjar/2017151-9573

Blumenstein O et al (2010) Ökologische Konsequenzen der Rohstoffsicherung – dargestellt am Beispiel des südafrikanischen Platin- und Goldbergbaus. – Geographie und Schule 183, 24–27

Bundessortenamt (Hrsg., 2020): Getreide, Mais Öl- und Faserpflanzen, Leguminosen, Rüben, Zwischenfrüchte [Elektronische Version]. Beschreibende Sortenliste. de Wet, J. (2020): oral communication on 26.04.2020

Gehlker (1976) Eine Hilfstafel zur Schätzung von Deckungsgrad und Artmächtigkeit, unter https://www.zobodat.at/pdf/Mitt-flori-soz-Arb_NF_19-20_0427-0429.pdf

Lausitz Energie Bergbau AG (Hrsg, 2020) Geschäftsfeld Bergbau. https://www.leag.de/de/geschaeftsfelder/bergbau/ (retrieved 12.04.2020)

Lüder R (2018) Grundlagen der Feldbotanik. Familien und Gattungen einheimischer Pflanzen. Bern: Haupt Verlag; 1. Edition (11. Juni 2018)

Meyer S (2012) Entwicklung und Erprobung von methodischen Grundlagen zur Konzipierung eines Entscheidungshilfesystems für die Begrünung von

Extremstandorten. – CUTEC-Schriftenreihe 75, Papierflieger-Verlag GmbH, Clausthal-Zellerfeld

Münzel S, Blumenstein O (2012) Ökologische und nachhaltige Ergänzungsstoffe für Extremstandorte. Forum Der Geoökologie 23(2):48–51

Münzel S (2013) Aspekte der Evolution von aufgespültem Abraummaterial des Platinerzbergbaus. Dissertationsschrift, University of Potsdam

Münzel S (2019) Wirkung von zwei Bodenergänzungsstoffen auf Trockenstress von Winterweizen-pflanzen – ein Gefäßversuch, unpublished manuscript

Nkonya et al (2016) Economics of land degradation and improvement: an introduction and overview. – In: Nkonya E, Mirzabaev A, von Braun J (eds) Economics of Land Degradation and Improvement – A Global Assessment for Sustainable Development: 1–14: Springer, Cham. https://doi.org/10.1007/978-3-319-19168-3_1

Scheub U (2016) Studie über Bodenerosion weltweit: Eine bodenlose Katastrophe. Taz online, https://taz.de/Studie-ueber-Bodenerosion-weltweit/!5276458/ (retrieved 21.01.2020)

Statistisches Bundesamt (Hrsg.) (2020) Bruttostromerzeugung in Deutschland. unter: https://www.destatis.de/DE/Themen/Branchen-Unternehmen/Energie/Erzeugung/Tablen/bruttostromerzeugung.html (retrieved 16.10.2020)

Stinglwagner G, Haseder I, Erlbeck R (1998) Das Kosmos Wald- und Forstlexikon. Franckh-Kosmos Verlags-GmbH & Co, Stuttgart

UBA (2015) Land Degradation Neutrality: An Evaluation of Methods. Environmental Research of the Federal Ministry for the Environment, Nature Conservation, Building and Nuclear Safety, 62/2015, Project No. 46658 Report No. 002163/E

von Redwitz C et al (2019) Microsegregation in Maize Cropping- a Chance to Improve Farmland Biodiversity. Gesunde Pflanzen 71(2):87–102. https://doi.org/10.1007/s10343-019-00457-7

Voutos Y et al (2019) A Survey on Intelligent Agricultural Information Handling Methodologies. Sustainability 11(12). https://doi.org/10.3390/su11123278

Wikipedia (2016) Location of Lusatia in Central Europe, at https://de.wikipedia.org/wiki/Lausitz (retrieved 31.01.2021)

Assessment of Land Use and Land Cover Change in the Purulia District, India Using LANDSAT Data

Pritha Das, Prasenjit Bhunia, and Ramkrishna Maiti

Abstract

The analysis of land use and land cover change has become necessary and urgent in the field of man–environment relation or resultant global environmental change. The present study analysed temporal and spatial changes of land use and land cover (LULC) in Purulia district covering an area of 6300 km^2 by comparing classified LANDSAT satellite images of 1990 and 2020 coupled by land use transition matrix and Markov Chain model to derive functional information of the spatio-temporal change of the LULC classes. The same analysis was performed at the watershed level. The results show that all selected LULC classes have changed from 1990 to 2020. About 113 km^2 of dense forest (i.e. 21% of the total forest area) has been lost whereas, 452 km^2 of fallow (i.e. 35% of the total fallow land) has been lost because of afforestation and expansion of agriculture. The conversion of dense forest to fallow with vegetation and fallow to fallow with vegetation were the major processes of deforestation and afforestation respectively. The loss of dense forest and gain of fallow with vegetation were lumped with several govt. plantation programmes in the last few years. The transition from fallow to agriculture and from dense forest to fallow with vegetation were the dominant LULC transition processes. The probability of built-up area (98%), fallow with vegetation (96%), and waterbodies (95%) to remain in the same LULC was high. Fallow was noticed as the most disturbed land cover followed by dense forest and agriculture. Future efforts should be made to manage the forest health in this naturally disturbed area where land is sloppy, the soil is infertile, and water is limited. For the proper formulation and implementation of sustainable forest management practices or policies, these findings can be used as primary references.

Keywords

Land use land cover · Transition matrices · Markov chain · Purulia · Watershed

P. Das (✉)
NIGS, Binpur-I, Jhargram, West Bengal, India
e-mail: prithadas.jhargram@gmail.com

P. Bhunia
SACT-I, Santal Bidroho Sardho Satabarshiki Mahavidyalaya, Goaltore, Midnapore, India

R. Maiti
Department of Geography and EM, Vidyasagar University, Midnapore, India
e-mail: ramkrishna@mail.vidyasagar.ac.in

17.1 Introduction

Land cover (LC) and land use (LU) are two major indicators describing the natural and man-made environment concerning both environmental or natural processes, and anthropogenic activities like human settlement and economy. Land covers refer to objects which are mainly naturally originated and subject to change by anthropogenic intervenes like deforestation, cultivation, construction etc. In contrast, land use is a man-made result and refers to the outcome of developmental activities, which means a higher degree of development is coincide with more diversified and complex land uses.

The study of LULC change is one of the main parameters to recognise the environmental modification (Xiao et al. 2006; Basommi et al. 2016) and level of economic development (Currit and Easterling 2009; Najmuddin et al. 2017) at different spatio-temporal scale, such as continent like Africa (Brink and Eva 2009); country like Slovakia (Pazúr and Bolliger 2017) and Mexico (Mas et al. 2004); watershed (Gautam et al. 2003; Allende et al. 2009; Mendoza et al. 2011; Najmuddin et al. 2017); regional (Lambin 1997; Gomez-Mendoza et al. 2006; Wang et al. 2008) or local scale (Lopez et al. 2001; Bayarsaikhan et al. 2009). LULC change is a dynamic process and directly associated with biodiversity loss (Jansen and Gregorio 2002), water and soil quality, runoff and soil erosion rates (Dunjó et al. 2003), local or global food security and poverty (Lambin et al. 2001; Geist and Lambin 2002; Shriar 2002; Carr 2004; Carr et al. 2005; Ewers 2006), human health (Shi et al. 2018), inter and intra-migration (Lopez et al. 2006), environmental hazards (Liu and Shi 2017), etc.

The balance between these could resolve the future biodiversity conservation on every parcel of land over this planet. Similarly, a perfect balance could act as an important active booster not only enhance the economic prosperity of a region, but also resolve the conflicts that arise through man's practices over the environment. Therefore, it is significant to identify the area and quantify the degree of land diversion from land covers to land uses (Lee et al. 1995; Verburg et al. 2009). In recent time tendency to recover, regenerate of land covers such as forest or water bodies through LULC transition have demonstrated. In India, numerous efforts have been given by the Central and State Governments in land conversion procedure and MGNEREGA is the world's largest programme in this circumstance where maximum attentions have been paid in excavation and renovation of water bodies, creation of vegetal cover, and conversion of land for cultivation. Therefore, researchers and policymakers need to realise the interconnections amongst the processes that result in the two-way transition (Mendoza et al. 2011). Such processes and interconnections are so much complex in India that land conflict and man–environment conflict concerning land is an everyday event and often regarded as the main hindrance for any land-related project, makes it time-consuming. India has attracted worldwide attention as the country holds the second populous position of the world, her very fast emerging economy, and increasing share in global trade.

Amongst the land conversions or transitions, deforestation is the most important process of LU change (Lambin 2001), as it is positively related to the other processes of the environment. Many studies have successfully established the close relationship between deforestation and climate change (Malhi et al. 2008; Bonan 2008), loss of biodiversity, increasing CO content and other greenhouse elements (Chakravarty et al. 2012; Barlow et al. 2016), soil erosion and degradation (Lal 1996), flooding (Gentry and Lopez-Parodi 1980), and also the human livelihood (Soltani et al. 2014). For proper policy intervention, it is necessary to understand the deforestation processes and related sub-processes both quantitatively and qualitatively. Deforestation is the conversion of forest land to other land use types (e.g. forest to agricultural land, grassland, built-up area or any other land use types). Thus, in the present study deforestation is defined as dense forest being transformed to other land use types, whilst afforestation is the reverse process. The regulatory factors of deforestation are varied regionally; the most common factors of deforestation are logging for timber, generation of agricultural field, mining and urbanisation, industrialisation, and grazing in developing countries of Asia (Hosonuma et al. 2012). In India, uncontrolled population growth leads to a dramatic increase in food and allied products is considered the main driver of deforestation (Nagdeve 2007; Basnayat 2009). According to the 2019 summary of the Forest Survey of India (FSI), a positive change in forest cover is noticed in India, as it was 21.67% of the total geographical area as of 2019 whereas it was only 19.39% in 1999. But the statistics are not so satisfactory if we considered the states or districts as the study unit. It varies from 3.59% for Haryana, 3.65% for Punjab in 2017 to 86.27% in Mizoram, 79.96% in Arunachal Pradesh. As of 2017, the recorded forest area in West Bengal was 18.68% of the state's geographical area, which was 18.98% in 2015. Types of forest cover are also an important factor to recognise as the mixed and open forest is increasing in the area whereas the decreasing trend is noticed in the case of dense forest.

Current interests of the researchers have been paid for real-time mapping and monitoring the LULC changes at different spatial scales using satellite imageries, conventional aerial photo, digital photograph, topographical sheet, or Google Earth (Shultz et al. 2010; Hegazy and Kaloop 2015; Kibret et al. 2016; Pazúr and Bolliger 2017). Such data or techniques are facing lots of limitations associated with measurement of landscape change during different periods, data quality, data processing techniques etc. (Fuller et al. 2003). The use of satellite imageries improves the quality of LULC assessment since there is a lack of study at different spatial scale. Therefore, multi-temporal analyses of LULC changes should fill the gap to understand the processes and patterns during the historical periods (Mendoza et al. 2011). For the researchers, it is important to realise how these changes over time and what are the factors or driving forces that oversee the rates of LULC change over a particular region and also how these factors vary from region to region. Such understanding would allow us to identify the relationship between LULC processes and socioeconomic variables like population growth (Ningal et al. 2008), migration (Lopez et al. 2006), industrialisation (Currit and Easterling 2009), urbanisation (Lopez et al. 2001).

The watershed is considered as an ideal spatial unit to understand the LULC processes over time as the hydrological processes are easy to identify within this particular spatial unit and other environmental or socioeconomic factors are directly related to these processes. There is no universal or single method for achieving effective watershed management (Naiman et al. 1997; Bhatta et al. 1999; Gautam et al. 2003). Therefore, it could be valuable for watershed management to incorporate explicit watershed information with LULC changes and makes an integrated approach (Mendoza et al. 2011). For that, a watershed can be divided into microscale or sub-watersheds. In the present study, our study area Purulia district of West Bengal is subdivided into five watersheds amongst them Damodar, Silabati, Dwarakeswar, and Kangshabati are belong to Damodar River Basin (DRB) and Subarnarekha belongs to Subarnarekha River Basin (SRB).

This paper analyses the LULC change process over 30 years from 1990 to 2020 in the Purulia district of West Bengal, India as a whole and also at the watershed level. Specifically, the objectives of this paper are:

- To identify the LULC at the 30-year period between 1990 and 2020 using Landsat imageries at district and watershed level.
- To quantify the LULC change through Transition Matrices.
- To predict the future trend of LULC change using Markov Chain at the watershed level.

17.2 The Study Area

Purulia, a district of West Bengal in India is situated in the western side of the state (22°60′N–23°50′N latitudes and 85°75′E–86°65′E longitudes). It belongs to a sub-tropical climate and is characterised by a high rate of evaporation where monsoon is prevalent. The district receives 1100–1500 mm of rainfall annually, with monsoon rains accounting for about 75–80% of the total rainfall. The annual temperature range is high as the average temperature is 2.8 °C in winter to 52 °C in summer approximately. The district is represented by pediplain with some residual hills of the Archean Era that belongs to the peninsular shield of India with an altitude of 150–300 m and covered by the Chhotonagpur Gneissic complex. Based on the difference in physiographic features, the district is sub-divided into three broad micro-physiographic regions namely (a) Damodar–Darkeshwar Upland, (b) Upper Kasai Basin, and (c) Bagmundi–Bundwan Upland. The geology of Purulia mainly composed of granitic terrain consisting of a crystalline basement covered by a very thin layer of soil of haplustalfs subgroup of alfisol group which is mainly rock fragments and weathered materials. This kind of soil is generally infertile. Groundwater potentiality of the district is poor (< 40MCM) to moderate (40–90 MCM), but it is fairly good (> 90 MCM) in the North-Eastern (Raghunathpur-1, 2 and Santuri) and Western (Jhalda-1, 2) parts only. Kangshabati, Kumari, Damodar, Subarnarekha, Dwarakeswar, Silabati are the main rivers of Purulia district. Damodar flows along the northern boundary, Subarnarekha along the southern portion, Kangshabati and Kumari along the middle of the district. Silabati and Dwarakeswar originate along the north-eastern part of the district. Silabati, Dwarakeswar, Kangshabati, Damodar are the sub-watershed of Damodar River Basin (DRB), which is the lifeline of the entire South Bengal including the Purulia district. About 84% of the district's total geographical area covers under the DRB and Kangshabati alone covers 48% of the total geographical area. Subarnarekha belongs to Subarnarekha River Basin (SRB) and serves the southern parts of the district. There are also several small to large dams like Murguma, Pardi, Boronti, Burda, Gopalpur, Saheb Bandh, Moutore, upper and lower dam in Ayodha, Panchet etc., which are mainly used for electricity generation, flood protection, and irrigation. Due to sloppy and undulated topography, a significant amount of water passes as runoff. Instead of sufficient rainfall, Purulia is famous for water scarcity, dryness, and frequent drought events. According to an estimation of the Indian Meteorological Department (IMD), one out of five years is a drought year and all blocks (20) of Purulia are listed under Drought Prone Areas Programme (DPAP) by the Department of Land Reforms, Ministry of Rural Development (GoI). The district lies under the Northern Tropical Dry Deciduous forest (5B/C 1c). Dense forests are found in the hilly parts of Northern and southern (Jhalda1, Jhalda II and Baghmundi, Bandwan, Arsha, and Neturia) blocks whereas, Sal forests mixed with other species like Palash, Kusum, Mahua, Kend, and Neem are very common and extensively distributed throughout the district.

Presently the district of Purulia has four sub-divisions, three municipalities namely Purulia, Raghunathpur and Jhalda amongst them Purulia is the oldest, established in 1876. It is also the district headquarter situated in the Kangshabati watershed mainly. Purulia is the 5th district of West Bengal in terms of area, and 16th in terms of population. According to the 2011 census, the district is the homeland of more than 3 million residents and the number should be 4.7 million after 2051. The percentage share of the population to the total population is highest in the Kangshabati watershed (52.24% in 1951, 51.56% in 2011). But a good increase in this share is observed in the Damodar watershed (23.92% in1951; 25.55% in 2011) where industrialisation is progressing gradually. The principal economic activity undertaken in the study area is agriculture and allied activities and on average 90% of the economically active populations worked on primary sectors (Fig. 17.1).

17.3 Materials and Methods

17.3.1 Materials

LULC was mapped at two different years (1990 and 2020), based on widely and easily available

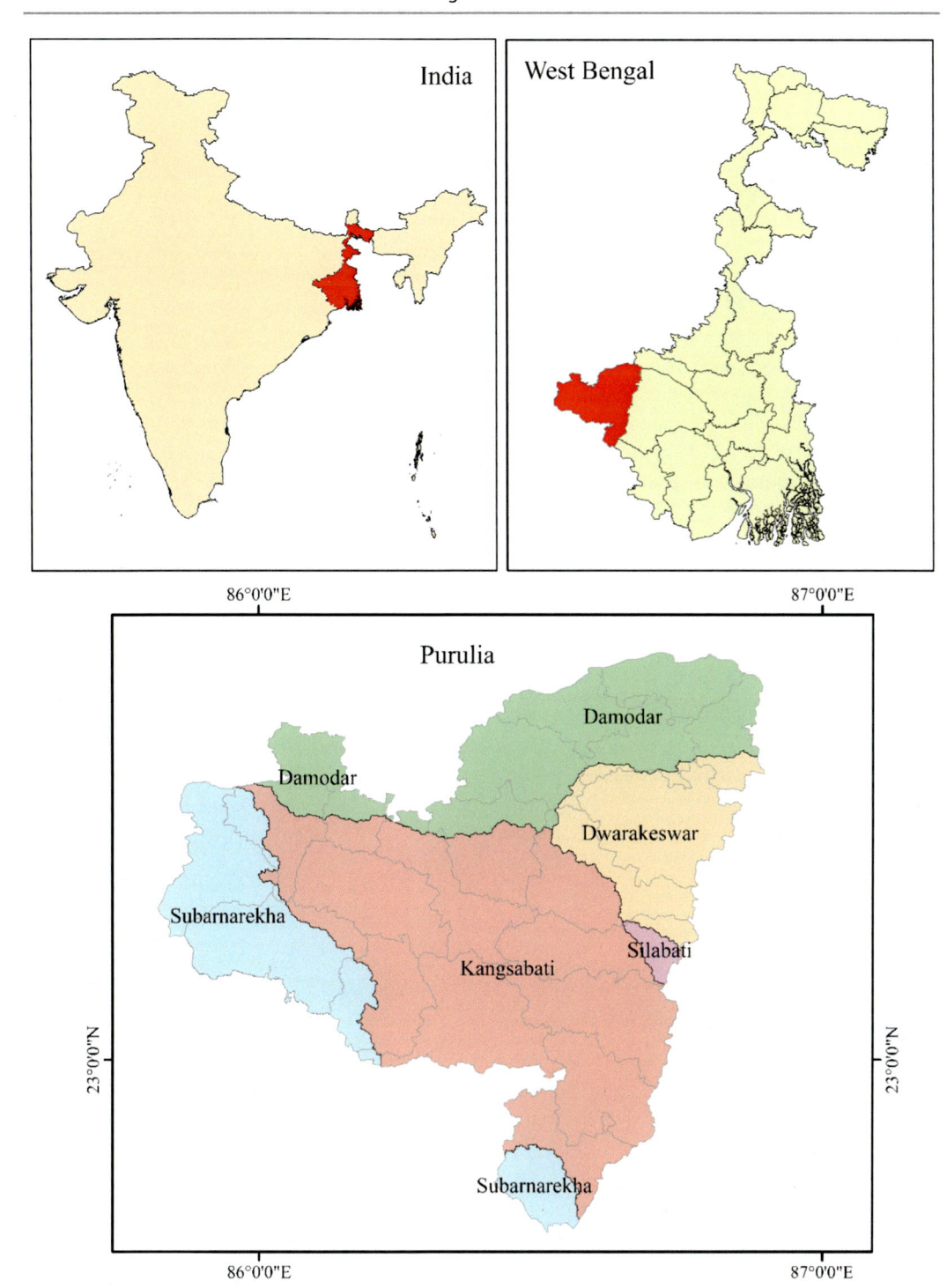

17 Assessment of Land Use and Land Cover Change in the Purulia …   333

Fig. 17.1 Location of the Purulia district and areal coverage of five watersheds

LANDSAT imageries from USGS (Table 17.1), and the prepared LULC maps were validated by using the references from Topographical maps, and Google Earth, as well as our field observations, interviews, and group discussions, were included as the primary data. The required satellite imageries for the present study are downloaded from the USGS Earth Explores. We used LANDSAT MSS image (30 m × 30 m) for 1990 and LANDSAT OLI (30 m × 30 m) for 2020. For both years two adjacent images were combined to cover the whole study area. These data have different spectral but same spatial characteristics; hence a uniform legend and scale were set before the analysis. To render all images comparable, all images were transformed to Universal Transverse Mercator (UTM) projection. Image processing and image interpretation for the development of LCLU maps were done by using the algorithm of Supervised Maximum Likelihood Classification in ERDAS Imagine 2014 software.

17.3.2 Methods

17.3.2.1 Digital Image Processing (DIP)

To enhance the image quality, DIP was manipulated by using ERDAS Imagine 2014 software. The images were geometrically corrected, calibrated, and finally subsetted. Image enhancement techniques, like histogram equalisation, were also performed on each image for improving the radiometric quality of the images.

17.3.2.2 Image Classification

The Supervised Classification was done on the pre-processed images for LULC mapping. In this classification technique, the Maximum Likelihood Algorithm will organise the pixels to a particular class based on covariance information provided by the user based on his or her knowledge of field experience and is expected a superior performance than the other classification methods (Richards 1994). The inputs were given by the user to guide the software concerning the pixels to be selected for the certain LULC types. In this study, six major LULC classes namely Waterbodies (WB), Dense forest (DF), Fallow with vegetation (FV), Fallow land (F), Settlement and built-up area (S), and Agricultural land (A) were identified (Table 17.2).

17.3.2.3 Land Use and Land Cover

The land is one of the most valuable natural resources gifted by our mother earth and never be increased its physical limit in general. The entire

Table 17.1 Data used in the study and the data details

Image (sensor)	Band	Spectral resolution	Spatial resolution (m)	Raw	Path	Date of capture
Landsat 5 (TM)	Band 1	0.45–0.52	30	139	040	11/04/1990
	Band 2	0.52–0.6	30	140	040	18/08/1990
	Band 3	0.63–0.69	30			
	Band 4	0.77–0.9	30			
	Band 5	1.55–1.75	30			
Landsat 8 (OLI)	Band 2	0.45–0.51	30	139	040	28/03/2020
	Band 3	0.53–0.59	30	140	040	25/02/2020
	Band 4	0.64–0.67	30			
	Band 5 (NIR)	0.85–0.88	30			
	Band 6 (SWIR 1)	1.57–1.65	30			
	Band 7 (SWIR 2)	2.11–2.29	30			

Table 17.2 Major Land Use Land Cover (LULC) features from visual interpretation of images

Major LULC features	Description (LULC types included in the category)
Water bodies (WB)	Rivers, lakes, reservoirs, swamp, and ponds
Dense forest (DF)	Natural forest, dense canopy contents
Fallow with vegetation (FV)	Scrubs, bushes, grassland, natural vegetation with sparse density, plantations etc
Fallow land (F)	Area without woody vegetation throughout the year, but temporarily grass in some cases, stony, rocky bare land, dry river bed etc
Settlement and built-up area (S)	Rural and urban settlements, roads, railways, industries, power stations etc
Agricultural land (A)	Dominant agriculture with patches of grass and bare land includes irrigated and unirrigated land

living world on this planet is directly or indirectly depends on land for food, energy and other needs of livelihood. Human activities have intensely changed the land cover and create imbalances between land cover and land use from the very beginning of modern civilisation. Now it is very crucial to watch the Earth from above to understand the influence of human activities on these natural resources over time. In most of the developing countries where such change is rapid and often undocumented and unrecorded, observations of the Earth through satellites provide objective information regarding land use change. At the same time, satellite images provide valuable information of the past, which were not recorded through other mediums. For the present study, LULC mapping was executed in two ways:

(a) District level LULC mapping as a whole, carried out on the LANDSAT imageries of 1990 and 2020.
(b) Watershed or basin wise LULC mapping by using the same LANDSAT imageries of 1990 and 2020.

17.3.2.4 Accuracy Assessment

The validation of the 1990 LULC image was done by using the Topographical maps and direct interviews conducted during several field visits in 2017, 2018, 2019, and 2020. The 2020 LULC results were validated by using primary data from field visits, interviewed and from Google Earth. An accuracy table (Table 17.3) was created using the observed and the classified land use data through the randomly select Ground Control Points (GCPs), and validated these points with

Table 17.3 Accuracy assessment for the LULC classes of the different periods (1990, 2020) at the watershed level

Land cover type	Kappa of each LULC class per period									
	Damodar		Dwarakeswar		Kangshabati		Silabati		Subarnarekha	
	1990	2020	1990	2020	1990	2020	1990	2020	1990	2020
WB	0.87	0.86	0.87	0.89	0.83	0.90	0.93	0.91	0.83	0.85
DF	0.91	0.90	–	–	0.81	0.83	–	–	0.87	0.90
FV	0.90	0.91	0.80	0.87	0.86	0.89	0.90	0.92	0.82	0.86
F	0.85	0.86	0.81	0.88	0.76	0.79	0.87	0.90	0.80	0.81
S	0.82	0.84	0.83	0.84	0.79	0.82	0.92	0.94	0.79	0.82
A	0.86	0.88	0.81	0.82	0.82	0.86	0.89	0.88	0.82	0.86
Kappa (overall)	0.84	0.84	0.79	0.85	0.80	0.88	0.90	0.90	0.83	0.85
Observed GCP	500	500	500	500	500	500	200	200	500	500
Correct GCP	440	435	420	435	410	440	180	180	410	430
% observed correct	88	87	84	87	82	88	90	90	82	86

the above said sources. Furthermore, both overall Kappa (accompanied by its variance) and class estimated Kappa values were calculated using the function of Accuracy Assessment in ERDAS Imagine 2014.

In the present study, 500 GCPs were randomly selected for Damodar, Dwarakeswar, Kangshabati, Subarnarekha river basin and 200 GCPs for Silabati river basin as the geographical area of this basin is only 56 km^2 (covering only 0.91% of the district's total geographical area). The accuracy assessment of the different watershed's and period's LULC maps show that land cover mapping applied in this study achieved 82–90% overall accuracy and 0.79–0.90 overall Kappa (Table 17.3). Kappa values for individual LULC classes range from 0.76 to 0.93 and these values were varying over time.

17.3.2.5 LULC Change Detection

Several techniques are applied to detect the conversion of landform one use to another, such as Dynamic of Land System (DLS) (Najmuddin et al. 2017), Simpsons-dominance index and location index (Liu and Shi 2017), Supervised Classification (Kibret et al. 2016; Hegazy and Kaloop 2015). In the present study, to make a detail discussion of the dynamics of LULC change, Transition Matrices were created both at the district and watershed level. The transition matrix is a table of symmetric rows and columns, consists of LULC classes from the initial year (period 1) on the vertical axis and the same LULC classes from the end year (period 2) on the horizontal axis. The diagonal cell of the matrix contains the surface area (in km^2) of each class of LULC that remains unaltered during the analysed period, whilst the main remaining cells contain the estimated surface area of a particular LULC class that transformed to a different class during the same period (Luenberger 1979) or transition from one class to another. Such a transition is also representing the dynamics of LULC change at different spatial scale. In this matrix, the conditional probability of LULC changes at any given time mostly depends on the present LULC, and there is no such role of previous changes (Bell and Hinojosa 1977). The model describes the result of LULC changes in aggregated ways, which are not truly spatial, but can still provide valuable information for the decision-makers (Lambin 1997). In the present study, the entire calculations of transition matrices were carried out using Land Change Modeler (LCM) which is an integrated model available in TerrSet 2020 software (v.19.0.2), developed by Clark Labs at Clark University. The LCM is a set of tools for LULC change analysis, helps users to map the changes, identify the transition between different LULC classes, model and predict the future tendency as specified by the users.

Finally, the study adopted a Markovian model to predict the future patterns of LULC change. Markov process is a random shifting from one state to another at each time step. A first-order Markov is a system in which probability distribution over the next step is assumed to only depend on the current state (Fischer and Sun 2001; Veldkamp and Lambin 2001; Pijanowski et al. 2002). These probability values calculated from the proportional area of each LULC in relation to total area (Horn 1975; Balzter 2000; Logofet and Lesnaya 2000; Lopez-Granados et al. 2001). The estimation rates of LULC change between start and end dates predict changes in type for a third date, assuming that the rate of changes is constant, and this is the main drawback of this model. The Markov model is also included in the LCM tool and makes the tool more acceptable and comprehensive to the researchers.

Rates of LULC changes were calculated by using the approach proposed by FAO (1995) described in Eq. 17.1.

$$q = \left((A2/A1)^{-1/(t2-t1)} - 1 \times 100\right) \quad (17.1)$$

where $A1$ is the surface area of the LCLU category for period 1, $A2$ is the LCLU category for period 2, $t1$ is the initial year (time 1), and $t2$ is the final year (time 2).

17.4 Results

17.4.1 Land Cover Changes at the District Level

Over the whole study period in the Purulia district, agriculture occupied the largest surface area, although this LU shows an increasing trend from 44.58 to 46.3% during the study period of 30 years (1990–2020) with an annual increase rate of 0.13%. Agricultural land (A) is distributed throughout the district, from river plains to undulating hilly sides. Most of the agricultural land is under single cropped and crops are grown mainly in the rainy season (July–October). Fallow with vegetation (FV) has consistently been the second more extensive LC in the district, with a tendency to increase its coverage from 20.69 to 23.35% with an annual rate of 0.4%. Fallow (F) is the third more extensive LC in the study area, exhibiting a remarkable decreasing trend in surface area from 20.77% in 1990 to 13.55% in 2020, with a highest annual decreasing rate of 1.41%. Similarly, dense forest (DF) is also decreasing remarkably at an annual rate of 0.79% and covers 6.82% of surface area after 2020 which was 8.64% in 1990. The most significant increase is noticed for settlement and built up (S) area with an annual rate of 2.83% and covers 7.57% after 2020, which was only 3.28% in 1990. An increase in waterbodies (W) is mainly due to non-biased erroneous image classification. In the 2020 image, a vast area of waterbodies (mainly behind the dams) is covered by water hyacinth and excluded from waterbodies after supervised classification as the spectral range is coincide with agricultural land and fallow with vegetation. A detail of gains and losses amongst the major LULC classes over the study period is included in Table 17.4.

As shown by the transition matrices (Tables 17.5 and 17.6), during the study period of 30 years, the probability of settlement remaining in the same LU is high (above 98%). The lowest probabilities of performance (i.e. higher transition probability) in the same period correspond to the fallow and dense forest respectively. Considering the transition fact, agriculture is on the top of the gainer list and received almost 409 km^2 of land from other LULC categories. This increase has taken place at the expense of fallow land, fallow with vegetation and dense forest, which have diminished respectively by 352 (27% of total fallow land), 32, and 21 km^2. And this is caused due to the increasing population, improving irrigation facilities, land conversion under MGNREGA schemes etc. Fallow with vegetation remains in the second rank and receives 398 km^2 of land and maximum from fallow land (155 km^2) and dense forest (149 km^2). Fallow is converted to fallow with vegetation due to plantation mainly (Table 17.6). The dense forest becomes fallow with vegetation due to deforestation mainly. Only 76% of dense forest having

Table 17.4 Distribution of LULC and their changes during 1990–2020

LULC	Area (km^2)		Share in %		Net change (km^2)	Change in % during 30 years	Annual change (FAO)
	1990	2020	1990	2020			
F	1300.14	848.06	20.77	13.55	− 452.08	− 34.77	− 1.41%
S	205.31	473.75	3.28	7.57	268.44	130.75	2.83%
FV	1294.71	1461.4	20.69	23.35	166.69	12.87	0.40%
A	2789.69	2897.41	44.58	46.3	107.72	3.86	0.13%
DF	540.59	426.96	8.64	6.82	− 113.63	− 21.02	− 0.79%
W	127.79	150.66	2.04	2.41	22.86	17.89	0.55%

Table 17.5 Transition matrices of change for different LULC classes during the period of 1990–2020

Transition (km²)						
Cover/km²	2020					
1990	F	S	FV	A	D	W
F	687.56 (52.85)	96.14 (7.39)	154.63 (11.89)	352.58 (27.1)	6.52 (0.51)	3.44 (0.26)
S	0 (0)	201.98 (98.38)	0 (0)	2.44 (1.19)	0 (0)	0.89 (0.43)
FV	0.98 (0.07)	8 (0.62)	1243.85 (96.07)	32.58 (2.52)	0 (0)	9.3 (0.72)
A	0 (0)	131.88 (4.74)	148.63 (5.35)	2492.48 (89.36)	0 (0)	12.92 (0.46)
D	3.57 (0.66)	5.13 (0.95)	95 (17.57)	21.33 (3.95)	412.52 (76.31)	3.04 (0.56)
W	4.77 (3.73)	1.25 (0.98)	0 (0)	0 (0)	0.14 (0.11)	121.63 (95.18)

Table 17.6 Plantation programmes and their areal coverage in Purulia district from 2009 to 2018

	Plantation type (in km²)						
Year	QGS	Namami Ganga	CAMPA	JICA (A3)	Sal	Others	Grand total
2018	0.3	1.3365	0.2597	2.24	0.20		4.34
2017	1.2	0.14		1.80	0.30		3.44
2016	2.65				0.20	Bamboo-0.40	3.25
2015	3.25					Bamboo-0.60	3.85
2014	2.7				0.25	Bambo-0.20, Fodder-0.60	3.75
2013	2.45				0.30		2.75
2012	3.2				0.30		3.50
2011	3.7		0.0648		0.40	Swing and planting-0.80, FDA-2.00	7.00
2010	5.6				0.40	Kangshabati Socoing and planting-0.80	6.80
2009	2.5				0.40	MGNREGA-0.50, RIDF-1.50	4.90

Source Divisional Forest Office, Purulia Division, Department of Forest, Govt. of West Bengal
QGS Quick generated species; *CAMPA* Compensatory Afforestation Fund Management and Planning Authority; *JICA* Japan International Cooperation Agency; *FDA* Forest Development Authority; *MGNREGA* Mahatma Gandhi National Rural Employment Guaranty Act; *RIDF* Rural Infrastructure Development Fund

the probability to stay at their same position means 24% of the dense forest has already lost or degraded. The settlement is the next important LU, receiving 242 km² of land and maximum from agricultural land (131 km²) and fallow land (96 km²) respectively (Fig. 17.2).

17.4.2 Watershed Scale Land Cover Changes

17.4.2.1 Subarnarekha Watershed

The Subarnarekha watershed is the third largest watershed of Purulia district, situated mostly

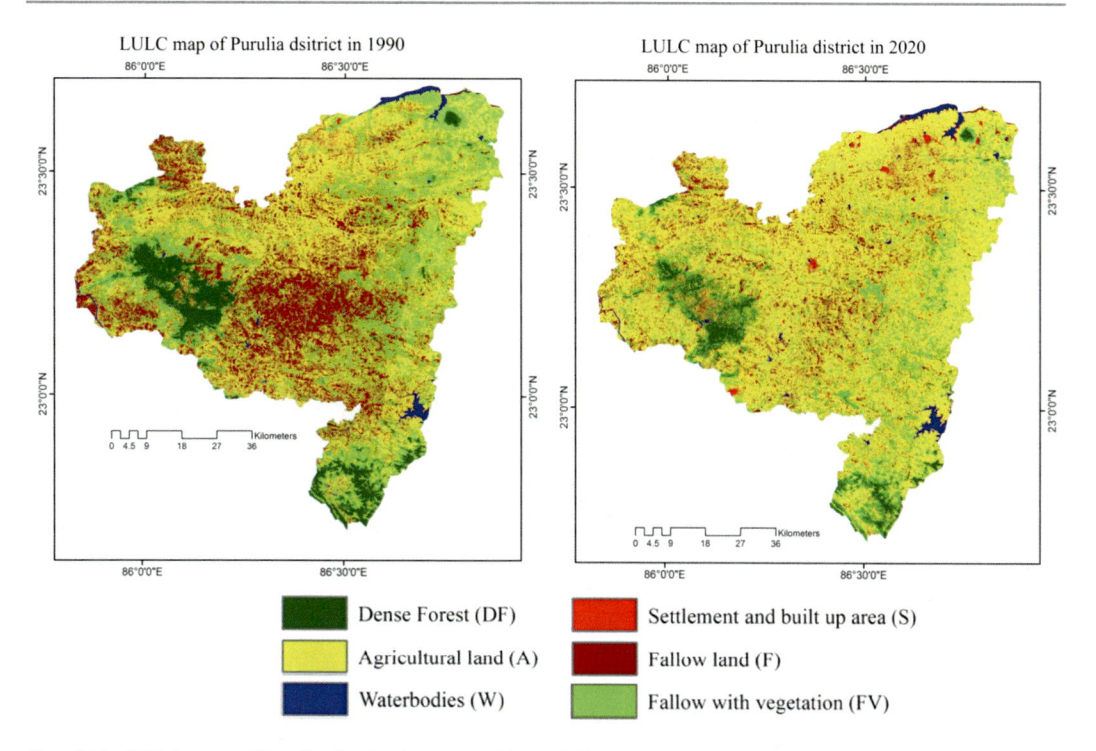

Fig. 17.2 LULC map of Purulia district for years 1990 and 2020

along the rocky and hilly tract of the southern portion of the district. The LULC analysis of the year 1990 of Subarnarekha watershed points out that agricultural land had the highest share (36.73%), followed by dense forest (31.59%), fallow land with vegetation (18.19%), barren fallow land (9.79%), settlement and built-up area (2.69%), and water bodies (1.01%) (Table 17.7). This result as usual is different in the analysis of LULC in the year 2020. Now, the order of percentage share by different landforms is similar to the result of 1990. But the amount varies. In 2020, agriculture occupies 41.40% of the land of the Subarnarekha watershed indicating an increase of about 0.4% per Annum. But the dense forest is decreasing at a rate of 1.07% per year. So the percentage share of the amount of land to total land by dense forest is 22.85% in 2020. Fallow land with vegetation (21.63%), settlement and built up area (2.86%), and water bodies (2.06%) show an increasing trend with an annual rate of 0.58%, 0.21%, and 2.38%

Table 17.7 Distribution of LULC in Subarnarekha watershed and their changes during 1990–2020

LULC classes	Area in km²		Share in %		Net change	Annual change in %
	1990	2020	1990	2020		
Waterbodies (WB)	10.48	21.23	1.01	2.06	10.75	3.42
Dense forest (DF)	326.22	236.04	31.59	22.85	− 90.19	− 0.92
Fallow with vegetation (FV)	187.89	223.42	18.19	21.63	35.53	0.63
Fallow land (F)	101.09	95.02	9.79	9.20	− 6.07	− 0.20
Settlement and built up area (S)	27.76	29.53	2.69	2.86	1.78	0.21
Agricultural land (A)	379.41	427.60	36.73	41.40	48.19	0.42

Table 17.8 Transition matrices (km² per cent) of change for different LULC classes during the period of 1990–2020 in the Subarnarekha watershed

	WB	DF	FV	F	S	A
WB	**(99.24)**	2.9 (0)	2.33 (0)	1.07 (0)	− 0.08 (0.76)	4.53 (0)
DF	− 2.9 (0.89)	**(72.35)**	− 62.11 (19.04)	− 3.57 (1.09)	− 1.75 (0.54)	− 19.87 (6.09)
FV	− 2.33 (1.24)	62.11 (0)	**(83.67)**	4.12 (0)	− 0.46 (0.24)	− 27.9 (14.85)
F	− 1.07 (1.06)	3.57 (0)	− 4.17 (4.13)	**(90.42)**	− 1.67 (1.65)	− 2.77 (2.74)
S	0.08 (0)	1.75 (0)	0.46 (0)	1.67 (0)	**(92.15)**	− 2.18 (7.85)
A	− 4.53 (1.19)	19.87 (0)	27.9 (0)	2.77 (0)	2.18 (0)	**(98.91)**

respectively. Barren fallow land (9.20%) is also squeezing.

The transition matrices (Table 17.8) show that the probability of agricultural land to remain in the same is 98.91% whilst to alter into water bodies is 1.18%. The probability of water bodies to remain in the same LC is maximum (99.24%) whilst it is minimum in the case of dense forest (72.35%) during the study period of 30 years (1990–2020). Most of the dense forest is located along the hilly tract of the Subarnarekha watershed. Most of the dense forest is altered into fallow with vegetation (19.04%) and agricultural land (6.09%) which implies a sharp deforestation process during these 30 years. Agricultural land gained most of the land from fallow land with vegetation (14.85%). The probability of barren fallow land to remain in the same is 90.42% whilst the alteration probability of this LC into fallow land with vegetation is 4.13% indicating towards some afforestation programme during 1990–2020. So the highest probability of transformation of dense forest is very common in the Subarnarekha watershed. The low lying areas of these hilly tracts are occupied by agricultural land where water is available.

17.4.2.2 Damodar Watershed

Damodar watershed is the second-largest watershed of Purulia district, situated along the northern part of the district. By analysing the supervised classification of the LANDSAT image of 1990 of this watershed, it is clear that likewise other watersheds of Purulia here also agricultural land predominates (Table 17.9). Around 50% of the land of this watershed is under agricultural activities. The percentage share of dense forest to total areal extension of this watershed is minimum (1.72% i.e. 23.94 km² out of total 1388 km²). About 40% of the land is fallow land either barren or covered by vegetation. Settlement and built-up area and water bodies share 4.87% and 3.73% of land respectively. In 2020, the percentage of agricultural land increases to 55% at a rate of 0.35% per year; settlement and built-up area increases by up to 7.67% at a rate of 1.52% per year. But the percentage of all other LULC decreases.

Table 17.9 Distribution of LULC in Damodar watershed and their changes during 1990–2020

LULC classes	Area km²		Share in %		Net change	Annual change in %
	1990	2020	1990	2020		
Waterbodies (WB)	51.84	61.44	3.73	4.43	9.60	0.617361
Dense forest (DF)	23.94	15.36	1.72	1.11	− 8.58	− 1.19433
Fallow with vegetation (FV)	349.93	259.24	25.21	18.67	− 90.69	− 0.86392
Fallow land (F)	198.08	172.96	14.27	12.46	− 25.12	− 0.42276
Settlement and built up area (S)	67.68	106.49	4.87	7.67	38.82	1.911778
Agricultural land (A)	696.86	772.83	50.19	55.67	75.98	0.363419

Table 17.10 Transition matrices (km² per cent) of change for different LULC classes during the period of 1990–2020 in the Damodar watershed

	WB	DF	FV	F	S	A
WB	**(100)**	0.14 (0)	6 (0)	2.29 (0)	0.64 (0)	0.54 (0)
DF	− 0.14 (0.6)	**(60.82)**	− 7.28 (30.41)	0.81 (0)	− 0.5 (2.08)	− 1.46 (6.09)
FV	− 6 (1.71)	7.28 (0)	**(97.03)**	28.65 (0)	− 4.41 (1.26)	35.63 (0)
F	− 2.29 (1.16)	− 0.81 (0.41)	− 28.65 (14.46)	**(10.66)**	− 16.31 (8.23)	− 128.91 (65.08)
S	− 0.64 (0.95)	0.5 (0)	4.4 (0)	16.31 (0)	**(99.05)**	18.23 (0)
A	− 0.54 (0.08)	1.46 (0)	− 35.63 (5.11)	128.91 (0)	− 18.23 (2.61)	**(92.19)**

The probability of remaining in the same LULC category is highest of water bodies (100%) followed by settlement and built-up area (99.05), fallow land with vegetation (97.03%), agricultural land (92.19%) as analysed by transition matrices (Table 17.10). But dense forest (60.82%) and barren fallow land (10.66%) show minimum probability to remain in the same category. Maximum alteration of barren fallow land into agricultural land occurred during these 30 years with a probability of 65.08% which is highest amongst all other watersheds as a result of increasing population density in this area (population data) and availability of water. As shown by the transition matrices another important deviation during this period is an alteration of dense forest to fallow land with vegetation with a probability of about 30%. The probability of extension of settlement and built-up area from other LULC is also noticeable (2.08% from dense forest, 2.61% from agricultural land, 1.26% from fallow land with vegetation, 8.23% from barren fallow land). These statistics indicate an increase in population density and establishment of new industries in this area which may result in a deforestation process and a higher probability of alteration of fallow land into agricultural land.

17.4.2.3 Kangshabati Watershed

The rivers Kangshabati and Kumari jointly drain the largest area in the middle of the Purulia district. Two amongst the three municipal towns (Jhalda & Purulia) and the maximum portion of the largest town and headquarter of the district (Purulia) and fall in this watershed. The data derived from the analysis of the supervised classification of the LANDSAT 5 image of this watershed of 1990 reveal that 42.94% of the land was under agricultural practices, 27.62% under barren fallow land, 18.10% under fallow land with vegetation, 6.33% under dense forest which is situated along the hilly tract of Ajodhya Hills, 3.26% under the settlement and built-up area, and 1.75% underwater bodies. The 2020 image shows that agricultural land is squeezing at a rate of 0.11% per year leads to a decline in percentage share to total land area (41.61%). Barren fallow land (15.76%), dense forest (5.83%), and water bodies (1.74%) are also declining. Only settlement and built-up area (10.23%) and fallow land with vegetation (24.83%) are increasing. Settlement and built-up area are expanding at a high rate of 3.9% per year (Table 17.11).

The result of transition matrices shows that Settlement and built-up area and fallow land with vegetation has a high probability to sustain in their status whilst fallow land has the lowest. Here is a probability of alteration of Barren fallow land, dense forest, agricultural land, and water bodies into Settlement and the built-up area which implies a rapid urban sprawl and establishment of new industries. Here, another important alteration faced by water bodies to fallow land (7.67%). Due to the unavailability of water, the probability of transformation of agricultural land to Barren fallow land is 7.63%, whereas an increase of population leads to alteration of later to former is 21.25% (Table 17.12). A high probability of alteration of dense forest to fallow land with vegetation (13.45%) apprises towards deforestation process near the low lying areas of Ajodhya Hills.

Table 17.11 Distribution of LULC in Kangshabati watershed and their changes during 1990–2020

LULC classes	Area (km^2)		Share in %		Net change	Annual change in %
	1990	2020	1990	2020		
Waterbodies (WB)	52.58	52.30	1.75	1.74	− 0.28	− 0.02
Dense forest (DF)	190.43	175.56	6.33	5.83	− 14.87	− 0.26
Fallow with vegetation (FV)	545.02	747.69	18.10	24.83	202.67	1.24
Fallow land (F)	831.69	474.56	27.62	15.76	− 357.13	− 1.43
Settlement and built up area (S)	98.08	307.84	3.26	10.23	209.76	7.13
Agricultural land (A)	1292.85	1252.69	42.94	41.61	− 40.15	− 0.10

Table 17.12 Transition matrices (km^2 per cent) of change for different LULC classes during the period of 1990–2020 in the Kangshabati watershed

	WB	DF	FV	F	S	A
WB	**(89.83)**	− 0.14 (0.27)	0.47 (0)	− 4.04 (7.67)	− 1.17 (2.23)	5.15 (0)
DF	0.14 (0)	**(85.04)**	− 25.61 (13.45)	5.71 (0)	− 2.88 (1.51)	7.78 (0)
FV	− 0.47 (0.09)	25.61 (0)	**(99.91)**	107.83 (0)	28.82 (0)	98.51 (0)
F	4.04 (0)	− 5.71 (0.69)	− 107.83 (13)	**(56.56)**	− 70.68 (8.5)	− 176.95 (21.25)
S	1.72 (0)	2.88 (0)	28.82 (0)	70.68 (0)	**(100)**	105.67 (0)
A	− 5.15 (0.38)	− 7.78 (0.6)	− 98.51 (7.63)	176.95 (0)	− 105.67 (8.17)	**(83.22)**

17.4.2.4 Dwarakeswar Watershed

The watershed of the Dwarakeswar River occupies a much smaller area than the watershed of river Kangshabati, Damodar, and Subarnarekha. It is situated in the easternmost part of the district. During 1990 agricultural land (51.12%), fallow land with vegetation (25.55%), barren fallow land (20.56%), water bodies (1.53), settlement and built-up area (1.24%) were the main LULC of this watershed (Table 17.13). There was no dense forest area. 2020 image shows an increase in agricultural land (53.41%) and settlement and built-up area (3.57%). Fallow land with vegetation also expanded (28.80%). In this case also barren fallow land is decreasing (12.36%).

The transition matrices show that, except barren fallow land, all other LULC has above 90% probability to lie in their categories.

Table 17.13 Distribution of LULC in Dwarakeswar watershed and their changes during 1990–2020

LULC classes	Area (km^2)		Share in %		Net change	Annual change in %
	1990	2020	1990	2020		
Waterbodies (WB)	11.80	14.28	1.53	1.85	2.48	0.70
Fallow with vegetation (FV)	196.66	221.69	25.55	28.80	25.03	0.42
Fallow land (F)	158.96	95.12	20.56	12.36	− 63.84	− 1.34
Settlement and built up area (S)	9.54	27.49	1.24	3.57	17.95	6.27
Agricultural land (A)	393.45	411.11	51.12	53.41	17.66	0.15

Table 17.14 Transition matrices (km² per cent) of change for different LULC classes during the period of 1990–2020 in the Dwarakeswar watershed

	WB	FV	F	S	A
WB	**(93.81)**	0.48 (0)	− 0.73 (6.19)	0.25 (0)	2.48 (0)
FV	− 0.48 (0.24)	**(98.25)**	13.98 (0)	− 2.96 (1.51)	14.49 (0)
F	0.73 (0)	− 13.98 (8.79)	**(59.84)**	− 7.25 (4.57)	− 42.61 (26.8)
S	− 0.25 (2.62)	2.96 (0)	7.25 (0)	**97.38**	7.98 (0)
A	− 2.48 (0.63)	− 14.49 (3.68)	42.61 (0)	− 7.98 (2.03)	**(93.66)**

A major portion of barren fallow land may have been transformed into agricultural land (26.8%). But agricultural land was also converted into fallow land with vegetation (3.68%) which was expanded during these 30 years. Settlement and the built-up area gained land from agricultural land (2.03%), fallow land with vegetation (1.51%), barren fallow land (4.57%) and enlarged its area at a rate of 3.59% per year which is the maximum rate of areal expansion within this watershed of Purulia (Table 17.14).

17.4.2.5 Silabati Watershed

Silabati watershed is the smallest basin, located on the eastern side of the Purulia district. The LULC analysis shows that here also agricultural land (63.68% in 1990 and 70.84% in 2020) occupies the maximum share of land use followed by fallow and built-up area (Table 17.15). Again fallow land and fallow with vegetation are the most distressed LC and prone to convert to other mainly to agricultural land (Table 17.16).

17.4.3 Markov Chain Analysis

The Markov chain was used to calculate the transition probability based on the period 1990–2020 for the prediction of LULC for 2050 (Table 17.17). The transition probability matrix is the cross-tabulation of two images and

Table 17.15 Distribution of LULC in Silabati watershed and their changes during 1990–2020

	Area (km²)		Share in %		Net change	Annual change in %
LULC classes	1990	2020	1990	2020		
Waterbodies (WB)	1.09	1.41	2.56	3.31	0.32	0.97
Fallow with vegetation (FV)	1.09	1.41	2.56	3.31	0.32	0.97
Fallow land (F)	11.04	7.21	25.92	16.93	− 3.86	− 0.13
Settlement and built up area (S)	2.25	2.39	5.28	5.61	0.33	0.21
Agricultural land (A)	27.12	30.17	63.68	70.84	3.05	0.74

Table 17.16 Transition matrices (km² per cent) of change for different LULC classes during the period of 1990–2020 in the Silabati watershed

	WB	FV	F	S	A
WB	**(100)**	0.02 (0)	0.08 (0)	0 (0)	0.22 (0)
FV	− 0.02 (0.13)	**(61.54)**	− 0.98 (6.44)	− 0.17 (1.12)	− 4.68 (30.77)
F	− 0.08 (0.72)	0.98 (0)	**(85.14)**	− 0.23 (2.08)	− 1.33 (12.06)
S	0 (0)	0.17 (0)	0.23 (0)	**(88.44)**	− 0.26 (11.56)
A	− 0.22 (0.8)	4.68 (0)	1.33 (0)	0.26 (0)	**(99.20)**

Table 17.17 Markov transition probability matrix of LULC changes in 2050

Subarnarekha watershed

	WB	DF	FV	F	S	A
WB	0.6794	0.0186	0.0378	0.0786	0.0462	0.1394
DF	0.0095	0.6043	0.249	0.0353	0.0107	0.0912
FV	0.0145	0.1018	0.3251	0.1089	0.041	0.4087
F	0.0187	0.0787	0.2432	0.2062	0.0416	0.4116
S	0.0145	0.0625	0.2609	0.0911	0.4975	0.0735
A	0.0158	0.026	0.1288	0.1024	0.0306	0.6964

Dwarakeswar watershed

	WB	DF	FV	F	S	A
WB	0.4156		0.0873	0.2727	0.0053	0.2191
DF						
FV	0.0077		0.4296	0.1179	0.0332	0.4117
F	0.0157		0.2349	0.2466	0.0551	0.4477
S	0.0324		0.3741	0.154	0.3476	0.0919
A	0.0129		0.2426	0.0717	0.0287	0.6441

Kangshabati watershed

	WB	DF	FV	F	S	A
WB	0.6032	0.0046	0.0235	0.1589	0.1239	0.0859
DF	0.0005	0.5743	0.2966	0.0548	0.0234	0.0505
FV	0.0031	0.0566	0.342	0.1454	0.0888	0.3641
F	0.0052	0.0194	0.2249	0.29	0.1036	0.3569
S	0.0488	0.016	0.1995	0.1575	0.4112	0.167
A	0.0075	0.0134	0.2297	0.0927	0.1129	0.5437

Damodar watershed

	WB	DF	FV	F	S	A
WB	0.803	0.0004	0.0287	0.0256	0.0618	0.0806
DF	0.0065	0.3997	0.4381	0.0113	0.0344	0.1101
FV	0.0378	0.0162	0.3535	0.0741	0.0942	0.4243
F	0.2721	0.0103	0.1238	0.0031	0.0647	0.5259
S	0.0568	0.0047	0.2105	0.0936	0.4984	0.136
A	0.0068	0.0017	0.1717	0.0791	0.0746	0.6662

Silabati watershed

	WB	DF	FV	F	S	A
WB	0.5557		0.0875	0.0826	0.0273	0.2469
DF						
FV	0.0076		0.1901	0.1858	0.035	0.5815
F	0.0152		0.167	0.2329	0.0541	0.5308
S	0.0116		0.1624	0.164	0.5456	0.1164
A	0.0181		0.1536	0.1672	0.0357	0.6254

Table 17.18 Basin wise estimation of surface and groundwater yield in million cubic meters (MCM)

River	1990			2020		
	Surface water yield (SWY)	Groundwater yield (GWY)	Total water yield (TWY)	Surface water yield (SWY)	Groundwater yield (GWY)	Total water yield (TWY)
Damodar	670.3	619.7	1290	684.4	475.6	1160
Dwarakeswar	459.2	391.7	850.9	451.57	327	778.57
Kangshabati	2101.3	1719.1	3820.4	2040.1	1360.16	3400.26
Silabati	39.3	31.2	70.5	40.92	27.28	68.2
Subarnarekha	580.84	536.16	1117	590.53	410.37	1000.9
Total	3850.94	3297.86	7148.8	3807.52	2600.41	6407.93

N.B. Estimation of surface and groundwater yield was done by using the SWAT model in the ArcGIS domain

contains the probable amount of change of any LULC class into other classes within the desired period. It is a very useful tool to monitor the rhythm, behaviour, and magnitude of LULC changes in an area.

The Markov Chain result shows that the probability of change to agricultural land (A) from any other LULC classes is remarkable in near future and the picture is more or less the same throughout the district and throughout the basins. And this will happen at the expense of transition from fallow land (F) and fallow with vegetation (FV). For example, in Damodar, 52% of fallow and 42% of fallow with vegetation will convert to agricultural land after 2050 (Table 17.16). Another disturbed LC is dense forest (DF). In the Damodar watershed, the dense forest will squeeze up to 20% after 2050 (61% in 2011 and 40% in 2050), 12% in Subarnarekha (72% in 2020 and 60% in 2050), and 28% in the Kangshabati watershed (85% in 2020 to 57% in 2050).

17.4.4 Impacts on the Basin Hydrology

Different hydrological parameters such as Evapotranspiration, surface water yield, groundwater yield etc. are the interactive outcomes of LULC, precipitation amount, slope, soil characters etc. Change in LULC is the main driving force to modify the hydrological outputs as LULC is changing at every moment. We considered surface water yield (SWY) and groundwater yield (GWY) as the measuring parameter to identify and quantify the impacts of changing LULC on basin hydrology. Soil and Water Assessment Tool (SWAT) was used in the ArcGIS domain to estimate these parameters by inputting the LULC maps of 1990 and 2020.

The SWAT output shows a decreasing trend in both SWY and GWY. TWY was decreased up to 10.4% (Table 17.18). In the case of SWY, the decreasing amount was 1.2%, whereas, it was 21% for GWY. The decreasing tendency of groundwater reveals the fact that due to deforestation the process of groundwater recharge is hampering. AS per Minor Irrigation Census (Ministry of Jal Shakti, Dept. of Water Resources, RD & GR, Government of India), the number of shallow, medium and deep tube wells, as well as the utilisation of irrigation potential, are also increasing. As per our study, settlement and built-up area are expanding due to the increase in population (Table 17.19). The whole effect of these is a declining groundwater level (Fig. 17.3).

17.5 Discussion

In this discussion, we integrated standard techniques and procedures for understanding LULC dynamics at the district and watershed level with detailed information. This integration is

Table 17.19 Decadal change of population since 1951 and projected population up to 2051

Watershed	Official census data[a]							Projected census[b]
	Area (km²)	1951	1961	1971	1981	1991	2001	2011	2051
Subarnarekha	1032.84	136,044	160,439	191,655	231,493	255,756	295,057	352,031	562,320
Damodar	1388.32	288,794	347,473	424,456	517,428	594,238	681,406	777,147	1,197,635
Kangshabati	3010.65	630,706	726,706	857,660	1,053,600	1,178,973	1,344,093	1,568,208	2,452,972
Dwarakeswar	770.41	141,292	164,768	195,546	230,314	264,957	295,482	321,865	456,439
Silabati	56.72	10,390	12,063	14,024	15,733	16,720	19,291	22,171	32,495
Total	6258.95	1,207,226	1,411,449	1,683,341	2,048,568	2,310,644	2,635,329	3,041,422	4,701,861

Source[a] District Census Handbook, Purulia (1951, 1961, 1971, 1981, 1991, 2001, 2011), Registrar General and Census Commissioner, India, Ministry of Home Affairs, GoI. Mouza wise population was summed up according to watershed level
[b] Incremental Increase Method (IIM) was applied to calculate the projected census of 2051

important for the accurate identification of driving forces of LULC dynamics, and thus provides valuable inputs to the management of the watershed. LULC dynamics are considered as the key environmental indicators to change the man–environmental set-up, but their proper evaluation has not been integrated with decision-making processes in most of the countries. The basic assumptions made in this study are that the rates of LULC change are differing with time and the final period of study can only convey information concerning the recent environmental processes that affect the study area. As a consequence, decision-making processes at a watershed level should give importance to each period, so that the drivers of change during each period should identify and integrate them with land use planning. For the present study, only two satellite images (1990, 2020) were used to cover the study period of 30 years. The assumption is here that the study area is economically and socially so backward (according to the last report of Planning Commission, Govt. of India (2010) rank one in the rural poverty rate, comes the last rank in rural monthly per capita consumption, rank 15 in per capita income amongst the 17th districts of West Bengal) that the drivers of LULC change like urbanisation, industrialisation or infrastructural developments are less active or intense in this district. The rate of LULC change is very slow and difficult to estimate from satellite images through Supervised Classification.

One of the main features of this research is to study the presence of anthropogenic impact on the natural world. For example, dense forest is shrinking remarkably (Table Five) due to deforestation and is turning into fallow with vegetation which is sparse, is indicating the overall degradation of forest health. At the same time, fallow is turning into agricultural land to support the growing needs of food, fodder etc. and into fallow with vegetation through afforestation programmes (Appendix 2). Thus, population expansion is one of the major impacts on the natural assets of the region. Another important social feature of the region is the expansion of the built-up area in terms of urbanisation, settlement, and industrialisation. Extension of

Fig. 17.3 Change in basin hydrology because of LULC change **a** TWY **b** SWY **c** GWY

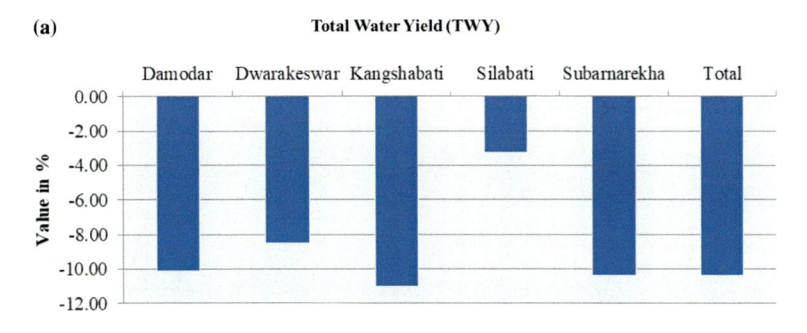

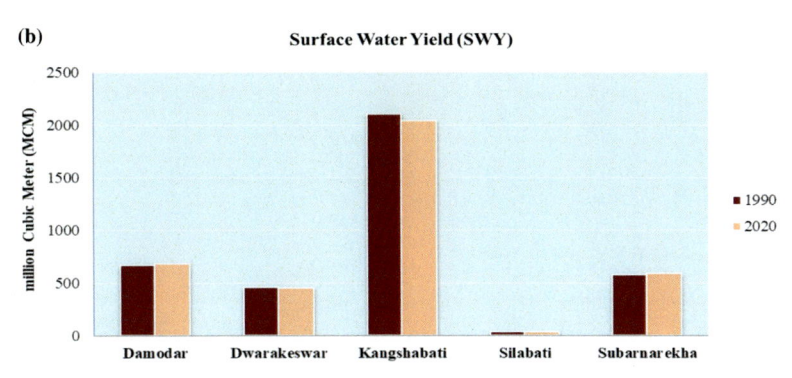

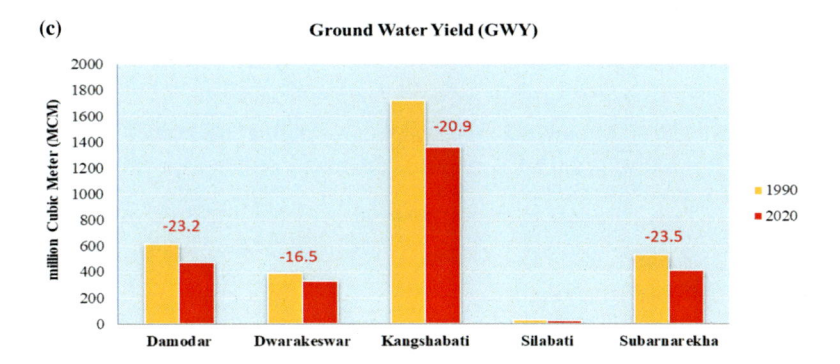

settlement area is maximum in Kangshabati watershed which is the location of Purulia town, the district headquarter and the home of 121,067 population in 2011 (92,386 in 1991), whereas, rate of industrialisation is high in Damodar watershed and maximum industries were established after 2010.

17.6 Conclusions

LULC change is a dynamic and continuous process as it ensues in any region with economic development especially in developing countries where the economic structure is shifting from primary to tertiary level. Rapid urbanisation and industrialisation are the two major pillars of altering land cover into land use and lead to overall environmental degradation.

This study relates changing LULC, detected by using Remote Sensing and GIS techniques in Purulia district during the last three decades (1990–2020), to a very interesting shifting economic settings as India is entering into the new age of liberation during this period. The use of multi-temporal satellite images combined with supervised classification and validation with real data led to improved accuracy than any conventional methods. The LULC database of two periods (1990, 2020) showed that the study area has undergone enough land cover change driven by agricultural expansion and increasing built-up area. Spatial patterns of LULC change can be linked to demographic factors, availability of fertile land, increasing irrigation facilities, distance from urban centres, improving transport facilities etc.

Appendix 1

Decadal change of population since 1951 and projected population up to 2051									
Watershed	Official census data[a]								Projected census[b]
	Area (km^2)	1951	1961	1971	1981	1991	2001	2011	2051
Subarnarekha	1032.84	136,044	160,439	191,655	231,493	255,756	295,057	352,031	562,320
Damodar	1388.32	288,794	347,473	424,456	517,428	594,238	681,406	777,147	1,197,635
Kangshabati	3010.65	630,706	726,706	857,660	1,053,600	1,178,973	1,344,093	1,568,208	2,452,972
Dwarakeswar	770.41	141,292	164,768	195,546	230,314	264,957	295,482	321,865	456,439
Silabati	56.72	10,390	12,063	14,024	15,733	16,720	19,291	22,171	32,495
Total	6258.95	1,207,226	1,411,449	1,683,341	2,048,568	2,310,644	2,635,329	3,041,422	4,701,861

Source[a] District Census Handbook, Purulia (1951, 1961, 1971, 1981, 1991, 2001, 2011), Registrar General and Census Commissioner, India, Ministry of Home Affairs, GoI. Mouza wise population was summed up according to watershed level
[b] Incremental Increase Method (IIM) was applied to calculate the projected census of 2051

Appendix 2

Irrigation schemes	Number of schemes implemented				Irrigation potential utilised (in Ha.)				Share in total irrigation
	2nd MI (1993–94)	3rd MI (2000–01)	4th MI (2006–07)	5th MI (2013–14)	2nd MI (1993–94)	3rd MI (2000–01)	4th MI (2006–07)	5th MI (2013–14)	
Surface flow scheme	19,448	21,283	13,732	12,758	65,403	59,429	63,554	42,984.01	62.4
Dugwells	17,133	13,322	3611	3048	5302	4026	2083	4362.5	6.2
Deep tube wells	NIL	NIL	NIL	275	NIL	NIL	NIL	14,838.12	21.49
Medium tube well	NIL	NIL	NIL	14	NIL	NIL	NIL	9.91	0.01
Shallow tube wells	11	2	15	28	12	3	10	26.22	0.4
Surface lift schemes	440	313	539	2016	2676	3444	8180	6517.01	9.5

Source Minor Irrigation Census (MIC) of 1993–94, 2000–01, 2006–07 and 2013–14. Ministry of Jal Shakti, Dept. of Water Resources, RD & GR, Government of India

References

Allende TC, Mendoza, ME, Granados, EML, Manilla, LMM (2009) Hydrogeographical regionalisation: an approach for evaluating the effects of land cover change in watersheds. A case study in the Cuitzeo Lake Watershed, Central Mexico. Water Resour Manag 23(12):2587–2603

Balzter H (2000) Markov chain models for vegetation dynamics. Ecol Model 126(2–3):139–154

Barlow J, Lennox GD, Ferreira J, Berenguer E, Lees AC, Mac Nally R, Thomson JR, de Barros Ferraz SF, Louzada J, Oliveira VHF, Parry L (2016) Anthropogenic disturbance in tropical forests can double biodiversity loss from deforestation. Nature 535 (7610):144–147

Basnyat B (2009) Impacts of demographic changes on forests and forestry in Asia and the Pacific. Bangkok, FAO of the United Nations Regional Office for Asia and the Pacific

Basommi LP, Guan QF, Cheng DD, Singh SK (2016) Dynamics of land use change in a mining area: a case study of Nadowli District, Ghana. J Mt Sci 13(4):633–642

Bayarsaikhan U, Boldgiv B, Kim KR, Park KA, Lee D (2009) Change detection and classification of land cover at Hustai National Park in Mongolia. Int J Appl Earth Obs Geoinf 11(4):273–280

Bhatta BR, Chalise SR, Myint AK (1999) Recent concepts, knowledge, practices, and new skills in participatory integrated watershed management: trainers' resource book

Bonan GB (2008) Forests and climate change: forcings, feedbacks, and the climate benefits of forests. Science 320(5882):1444–1449

Brink AB, Eva HD (2009) Monitoring 25 years of land cover change dynamics in Africa: a sample based remote sensing approach. Appl Geogr 29(4):501–512

Buys P (2007) At loggerheads? agricultural expansion, poverty reduction, and environment in the tropical forests. World Bank Publications

Carr DL (2004) Proximate population factors and deforestation in tropical agricultural frontiers. Popul Environ 25(6):585–612

Carr DL, Suter L, Barbieri A (2005) Population dynamics and tropical deforestation: state of the debate and conceptual challenges. Popul Environ 27(1):89–113

Chakravarty S, Ghosh SK, Suresh CP, Dey AN, Shukla G (2012) Deforestation: causes, effects and control strategies. In: Global perspectives on sustainable forest management. IntechOpen

Currit N, Easterling WE (2009) Globalization and population drivers of rural-urban land-use change in Chihuahua, Mexico. Land Use Policy 26(3):535–544

Dunjó G, Pardini G, Gispert M (2003) Land use change effects on abandoned terraced soils in a Mediterranean catchment, NE Spain. CATENA 52(1):23–37

Ewers RM (2006) Interaction effects between economic development and forest cover determine deforestation rates. Glob Environ Chang 16(2):161–169

Fischer G, Sun L (2001) Model based analysis of future land-use development in China. Agr Ecosyst Environ 85(1–3):163–176

Fuller RM, Smith GM, Devereux BJ (2003) The characterisation and measurement of land cover change through remote sensing: problems in operational applications? Int J Appl Earth Obs Geoinf 4(3):243–253

Gautam AP, Webb EL, Shivakoti GP, Zoebisch MA (2003) Land use dynamics and landscape change pattern in a mountain watershed in Nepal. Agr Ecosyst Environ 99(1–3):83–96

Geist HJ, Lambin EF (2002) Proximate causes and underlying driving forces of tropical deforestation. Bioscience 52(2):143–150

Gentry AH, Lopez-Parodi J (1980) Deforestation and increased flooding of the upper Amazon. Science 210 (4476):1354–1356

Gomez-Mendoza L, Vega-Pena E, Ramírez MI, Palacio-Prieto JL, Galicia L (2006) Projecting land-use change processes in the Sierra Norte of Oaxaca, Mexico. Appl Geogr 26(3–4):276–290

Hannah L, Carr JL, Lankerani A (1995) Human disturbance and natural habitat: a biome level analysis of a global data set. Biodivers Conserv 4(2):128–155

Hegazy IR, Kaloop MR (2015) Monitoring urban growth and land use change detection with GIS and remote sensing techniques in Daqahlia governorate Egypt. Int J Sustain Built Environ 4(1):117–124

Horn HS (1975) Markovian properties of forest succession. Ecol Evol Communities 196–211

Hosonuma N, Herold M, De Sy V, De Fries RS, Brockhaus M, Verchot L, Angelsen A, Romijn E (2012) An assessment of deforestation and forest degradation drivers in developing countries. Environ Res Lett 7(4):044009

Jansen LJ, Di Gregorio A (2002) Parametric land cover and land-use classifications as tools for environmental change detection. Agr Ecosyst Environ 91(1–3):89–100

Kibret KS, Marohn C, Cadisch G (2016) Assessment of land use and land cover change in South Central Ethiopia during four decades based on integrated analysis of multi-temporal images and geospatial vector data. Remote Sens Appl Soc Environ 3:1–19

Lal R (1996) Deforestation and land-use effects on soil degradation and rehabilitation in western Nigeria. I. Soil physical and hydrological properties. Land Degrad Dev 7(1):19–45

Lambin EF (1997) Modelling and monitoring land-cover change processes in tropical regions. Prog Phys Geogr 21(3):375–393

Lambin EF, Turner BL, Geist HJ, Agbola SB, Angelsen A, Bruce JW, Coomes OT, Dirzo R, Fischer G. Folke C, George P (2001) The causes of land-use and land-cover change: moving beyond the myths. Glob Environ Change 11(4):261–269

Liu J, Shi ZW (2017) Quantifying land-use change impacts on the dynamic evolution of flood vulnerability. Land Use Policy 65:198–210

Logofet DO, Lesnaya EV (2000) The mathematics of Markov models: what Markov chains can really predict in forest successions. Ecol Model 126(2–3):285–298

López Granados EM, Bocco G, Mendoza Cantú ME (2001) Predicción del cambio de cobertura y uso del suelo: el caso de la ciudad de Morelia. Investigaciones Geográficas 45:39–55

López E, Bocco G, Mendoza M, Duhau E (2001) Predicting land-cover and land-use change in the urban fringe: a case in Morelia city, Mexico. Landscape Urban Plann 55(4):271–285

Lopez E, Bocco G, Mendoza M, Velázquez A, Aguirre-Rivera JR (2006) Peasant emigration and land-use change at the watershed level: A GIS-based approach in Central Mexico. Agric Syst 90(1–3):62–78

Luenberger DG (1979) Introduction to dynamic systems; theory, models, and applications (No. 04; QA402, L8)

Malhi Y, Roberts JT, Betts RA, Killeen TJ, Li W, Nobre CA (2008) Climate change, deforestation, and the fate of the Amazon. Science 319(5860):169–172

Mas JF, Velázquez A, Díaz-Gallegos JR, Mayorga-Saucedo R, Alcántara C, Bocco G, Castro R, Fernández T, Pérez-Vega A (2004) Assessing land use/cover changes: a nationwide multidate spatial database for Mexico. Int J Appl Earth Obs Geoinf 5(4):249–261

Mendoza ME, Granados EL, Geneletti D, Pérez-Salicrup DR, Salinas V (2011) Analysing land cover and land use change processes at watershed level: a multitemporal study in the Lake Cuitzeo Watershed, Mexico (1975–2003). Appl Geogr 31(1):237–250

Milesi C, Hashimoto H, Running SW, Nemani RR (2005) Climate variability, vegetation productivity and people at risk. Glob Planet Change 47(2–4):221–231

Nagdeve DA (2007) Population growth and environmental degradation in India. International Institute for Population Sciences. http://paa2007.princeton.edu/papers/7192. Department of fertility studies, Govandi station road, Deonar, Mumbai 400:088

Naiman RJ, Bisson PA, Turner MG (1997) Approaches to management at the watershed scale. In: Kohm K, Franklin JF (eds) Creating a forestry for the 21st century: the science of ecosystem management. Island Press

Najmuddin O, Deng X, Siqi J (2017) Scenario analysis of land use change in Kabul River Basin–a river basin with rapid socio-economic changes in Afghanistan. Phys Chem Earth, Parts a/b/c 101:121–136

Ningal T, Hartemink AE, Bregt AK (2008) Land use change and population growth in the Morobe Province of Papua New Guinea between 1975 and 2000. J Environ Manage 87(1):117–124

Pazúr R, Bolliger J (2017) Land changes in Slovakia: past processes and future directions. Appl Geogr 85:163–175

Pijanowski BC, Brown DG, Shellito BA, Manik GA (2002) Using neural networks and GIS to forecast land use changes: a land transformation model. Comput Environ Urban Syst 26(6):553–575

Richards JA (1994) Remote sensing digital image analysis: an introduction. Second, revised and enlarged edition. Springer-Verlag, Berlin, Heidelberg, Germany

Schulz JJ, Cayuela L, Echeverria C, Salas J, Benayas JMR (2010) Monitoring land cover change of the dryland forest landscape of Central Chile (1975–2008). Appl Geogr 30(3):436–447

Shi G, Jiang N, Yao L (2018) Land use and cover change during the rapid economic growth period from 1990 to 2010: a case study of shanghai. Sustainability 10(2):426

Shriar AJ (2002) Food security and land use deforestation in northern Guatemala. Food Policy 27(4):395–414

Soltani A, Angelsen A, Eid T (2014) Poverty, forest dependence and forest degradation links: evidence from Zagros, Iran. Environ Dev Econ 19(5):607–630

Veldkamp A, Lambin EF (2001) Predicting land-use change. Ar Ecosyst Environ 85(1–3):1–6

Verburg PH, Van De Steeg J, Veldkamp A, Willemen L (2009) From land cover change to land function dynamics: a major challenge to improve land characterization. J Environ Manage 90(3):1327–1335

Wang X, Zheng D, Shen Y (2008) Land use change and its driving forces on the Tibetan Plateau during 1990–2000. CATENA 72(1):56–66

Xiao J, Shen Y, Ge J, Tateishi R, Tang C, Liang Y, Huang Z (2006) Evaluating urban expansion and land use change in Shijiazhuang, China, by using GIS and remote sensing. Landsc Urban Plan 75(1–2):69–80

Prioritization of Watershed Developmental Plan by the Identification of Soil Erosion Prone Areas Using USLE and RUSLE Methods for Sahibi Sub-Watershed of Rajasthan and Haryana State, India

Ajoy Das, Jagmohan Singh, Madan Thakur, and Asim Ratan Ghosh

Abstract

Soil erosion is a major cause for land degradation. The process is initiated when soil particles are detached from its original configuration by erosive forces such as rainfall, temperature, wind. Prior research has demonstrated that large sediment loads damages the coral reefs. Current developments in geographic information systems (GIS) make it possible to model complex spatial analysis. A GIS is used in the present work to determine soil erosion potential at watershed scale. Hydrological data has been analyzed to understand the watershed's response to the primary erosive input i.e. rainfall. The aims of the study are to develop a GIS-based soil erosion potential model of the Sahibi Sub-Watershed, located in Alwar district of Rajasthan and Rewari district of Haryana state of India and to develop a correlation between drainage density, LU/LC coverage and slope steepness. The Universal Soil Loss Equation (USLE) was used to assess soil loss in GIS environment, specifically the commercial software package ArcGIS10.0. The USLE calculates long-term average annual soil loss by multiplying six specific factors which describe the watershed characteristics such as rainfall, soil types, slope, and vegetation cover. The GIS is used to store the USLE factors as individual digital layers and multiplied together to create a soil erosion potential map and finally a priority map was made. Recent satellite imagery is used from the Google Earth and IRS LISS-III to determine the extent of vegetation cover and conservation practices. In addition to developing the GIS model, a preliminary hydrological analysis was also conducted.

Keywords

Soil erosion · USLE · Stream flow · Slope · Watershed management

A. Das (✉)
Department of Earth Science, School of Sciences, Gujarat University, Ahmedabad, Gujarat, India
e-mail: ajoydas@gujaratuniversity.ac.in

J. Singh
Cyient, Hyderabad, Telangana, India

M. Thakur
GIS Division, National Grid—HQ, Riyadh, Saudi Arabia

A. Ratan Ghosh
Department of Science and Technology and Biotechnology (DSTB), Govt. of West Bengal, Bikash Bhavan, (4th Floor) Salt Lake, Kolkata, West Bengal, India

18.1 Introduction

The surface of the earth is always modified by Endogentic and Exogenetic forces from the ancient eras. Soil erosion mentioned by Wicks and Bathurst (1996) is one major type of Exogenetic forces. Soil is one of the three most precious natural resources together with air and water, one of the fabulous products of nature and without this element there would be no imagination of life. Soil is a thin layer of material on Earth's surface which supports vegetation and life. It is a porous natural material which consists of many materials including rocks, weathered minerals, and dead decaying plants and animals. Soil formation process takes a very long time creating a thin layer of soil. About an average soil/topsoil are produced of 1 ton/year and lost 10–40 times faster than it produced. The region was known as densely forested and rich in wildlife at the time of independence. Now the ecosystem of the region has come under severe stress and continuously degrading due to excessive felling of trees to meet the increasing demand for fuel, fodder, and construction industry as well as extensive mining to meet the industrial demand for minerals. This has resulted in extensive soil erosion, loss of topsoil, silting up of river channels and reservoirs, reduced land fertility and lowering of the ground water table. Apart from this fact instead of wind, water plays dominant role in soil erosion in the Aravalli hill range. However, excessive soil erosion is responsible for surface soil loss by Symeonakis and Drake (2004) and a subsequent decrease in soil quality and agricultural production. Soil loss from cultivated and bare land is still very high elsewhere on the globe and soil erosion is the most significant and an ominous threat to food security and development prospects in many developing countries (Bekele and Drake 2003). Soil loss is greater in developing countries where the farmers are totally ignorant of soil conservation practices mentioned by Singh et al. (1992). An escalating population is also indirectly responsible for this soil loss, especially where farmers desperate to grow enough crops destroy forests and other natural areas. The fertility of soil affects the agriculture and as well as population distribution also. Agriculture is a major activity all over the world. As the recent development in technology and mechanization, agriculture has remained the world's primary industry, in which soil plays elementary role. About 66% of the globe population, comprising of farmers derives its living directly from the soil. But soil erosion reduces soil fertility. The objective of the study is to analyze the soil erosion prone areas and how it affects the land use/landcover, using RS and GIS of the Sahibi Sub-Watershed, a small river sub-basin of the Sahibi Basin is in between eastern part of Rajasthan and western part of Haryana. Apart from this study, analyze the extent of drastic impacts of soil erosion affecting the socio-economic activities, toward optimum developmental planning by Hassan and Garg (2007) of the study area with watershed approach. Attempt has also been made for designing various Check Dam parameters of the watershed by Das et al. (2012). In a nut shell, to identify the soil erosion prone areas in semi-arid region of Sahibi (Sabi) Sub Watershed and to increase food security through soil conservation planning by Garg and Sen (1994, 2001). The two goals of the present research are:

I. To develop a GIS-based soil erosion potential model of the Sahibi Sub-Watershed, located in Alwar district of Rajasthan and Rewari district of Haryana state of India.
II. To develop a correlation between drainage density, LU/LC coverage and slope steepness.

According to Morgan (1986), the rate of soil erosion of developing and developed countries is at an alarming level. In some developed countries like the United Kingdom, United States of America, and Belgium have succeeded in minimizing the soil loss to an acceptable level. In the Journal of Environmental Hydrology, Sudhanshu Sekhar Panda et al., 'Application of Geotechnology to Watershed Soil Conservation Planning at the field scale of the Kuniguda watershed of Orissa and Andhra Pradesh' has done this type of soil erosion study. Some work has been done related to this study by the DST, Govt of West Bengal, India. Water & Environmental Research Institute of the Western Pacific, Shahram et al.

(2007), worked on soil erosion and measured the soil loss areas as the name of the thesis 'A GIS based Soil Erosion Potential Model of the Ugum Watershed' also similar work done by Symeonakis and Drake (2004), Wen et al. (2007).

As a branch of earth science, it is concerned with all the elements of the environment. On-site effects are particularly important on agricultural land where the redistribution of soil within a field, the loss of soil from a field, the breakdown of soil structure, and the decline in organic matter and nutrient result in a reduction of cultivable soil depth and a decline in soil fertility. Erosion also reduces available soil moisture, resulting in more drought-prone conditions. "Soil Erosion and Conservation", Morgan (2009). A watershed is essentially a catchment, within the boundaries of which all the soil–water processes are integrated; change in any one parameter influences all the other process parameters. It is a geographical unit draining at a common point (Outlet or pour point: usually the lowest point along the watershed boundary) by a system of streams. All land everywhere is part of some watershed. Other common terms for a watershed classification (Table 18.1) are drainage basin, basin, catchment, or contributing area. The size of watershed may vary from few square kilometers to thousands of square kilometers (Khosla 1949).

Significance of soil erosion in the form of soil loss

The understanding of the soil loss behavior of watersheds is essential to the planning and implementation of integrated watershed management schemes. The major factors to be considered in the soil erosion study of watersheds are Weather records—temperature, Precipitation, Evapotranspiration, Catchment characteristics—size, shape slope, soil type, LU/LC type and Erosion characteristics, Run-off characteristics by Thakkar and Dhiman (2007)—lag time, quantity of flow and spatial and temporal distribution of flow by Das (2012), Nooka Ratnam et al. (2005). It will be difficult to change any physical parameters of a watershed such as the slope, soil, and rainfall characteristics stated by Nag and Ghosh (2012). However it may be possible to change the land use and vegetation patterns, allocation of suitable areas for housing and livestock and introduce some structural measures for soil and water conservation. Adoption of appropriate measures in a spatial framework can create a sustainable watershed ecosystem by increasing ground water recharge, moderating peak flows, enhancing low flows and improving the habitat of the plants, trees, and animals in the watershed.

The rate of erosion is affected by four main factors: Climate, which determines how much rain will occur in an area; Soil characteristics, which determine erodibility and infiltration rates; Topography or slope, which determines the velocity of runoff and the energy water will have to cause erosion; and Vegetation, which will slow runoff and prevent erosion by holding soils in place stated by Wang and Yin (1998). Remote sensing data facilitate identification of existing or potential erosion prone areas which may help in planning, reclamation or preventive measures. Methods of erosion detection and assessment

Table 18.1 Average size and size ranges for each hydrological units (Khosla in 1949)

S. No.	Category of hydrologic units	Example of code	Size range (ha)	Average size (ha)
1	Water resource region	2	27,000,000–113,000,000	55,000,000
2	Basins	A	3,000,000–30,000,000	9,500,000
3	Catchments	1	1,000,000–5,000,000	3,000,000
4	Sub-catchments	A	200,000–1,000,000	700,000
5	Watersheds	2	20,000–300,000	100,000
6	Sub-watersheds	a	5000–9000	7000
7	Micro-watersheds	2	500–1500	1000

using remote sensing techniques are based on tone, texture and physiography recognition features. Factors to be considered in water-induced erosion are drainage, precipitation, vegetation, elevation, and relief. Soil erosion features may occur in a regular sequence of types and intensities along the topographic relief in a certain area. For this study area, the soil erosion is predicted using RUSLE model. Our Indian society is an agriculturally based society where the maximum population lives in rural areas where a sustainable rural development is very much required. Watershed development is best suited here, so Govt. of India, NGOs, and other organizations like NABARD etc. have implemented watershed development in India. In terms of Mahatma Gandhi National Rural Employment Guarantee Act has implemented the watershed-based development. For proper implementation, the application of remote sensing and GIS technique is being used increasingly, for a quick as well as accurate watershed planning mentioned by Samuel (1995).

Application of Remote Sensing and GIS in Hydrological Studies

The success of planning for developmental activities depends on the quality and quantity of information available on both natural and socio-economic resources. It is, therefore, essential to develop the ways and means of organizing computerized information system. These systems must be capable of handling vast amount of data collected by modern techniques and produce up to date information. Remote Sensing technology has already demonstrated its capabilities to provide information on natural resources such as crop, land use, soils, forest etc. on regular basis. Similarly, Geographic Information Systems (GIS) are the latest tools available to store, retrieve, and analyze different types of data for management of natural resources. The space borne multispectral data enable generating timely, reliable, and cost-effective information on various natural resources, namely surface water, ground water, land use/cover, soil, forest cover and environmental hazards, namely waterlogging, salinity and alkalinity, soil erosion by water etc.

Soil Erosion and sediment yield modeling by Rompaey et al. (2005) using RUSLE in ArcGIS Environment

In this study GIS techniques have been utilized for spatial discretization of a catchment into a time-area segments to be used in numerical solutions of the governing differential equations in rainfall-runoff-erosion process. Various thematic layers such as soil, land use, slope, flow direction, DEM were generated for the study area using various tools available in GIS. These thematic layers were further utilized to generate attribute information such as, USLE 'K' and 'C' parameters for use in rainfall-runoff-soil erosion model by Yitayew et al. (1999). Based on DEM and related attribute information of the catchment, time-area map of the catchment was prepared and used for spatial discretization of the catchment.

18.2 The Study Area

Sahibi River (also known as Sabi) is an ephemeral rain-fed river, locally called SABI Nadi. It rises from Mewat hills near Jitgarh and Manoharpur close to Jaipur district in Rajasthan and after gathering volume from about a hundred tributaries, it reaches voluminous proportions, form in a broad stream around Alwar and Kotputli, it then enters the Rewari district in Haryana near the city of Rewari after which it enters Rajasthan again and then re-enters Haryana near village Jaithal.

The Rewari district, except in its Eastern part is flat and sandy and absorbs all the rainwater. The study area (Fig. 18.1) extending from 28° 03′ 45″ North latitudes to 28° 14′ 45″ North latitudes and 76° 19′ 15″ East longitudes to 76° 29′ 30″ East longitudes. The present Sahibi Sub-Watershed (Fig. 18.1) includes the north eastern portion of Alwar district of Rajasthan and some Southern portion of Rewari district of Haryana (Table 18.2). This area belongs to the river Sahibi, also locally called Sabi Nadi. Total area covered is 182.72 km^2. The major settlement areas under the different micro watersheds are Naodi, Rampura, Khundrot, Dhikwar (SW1);

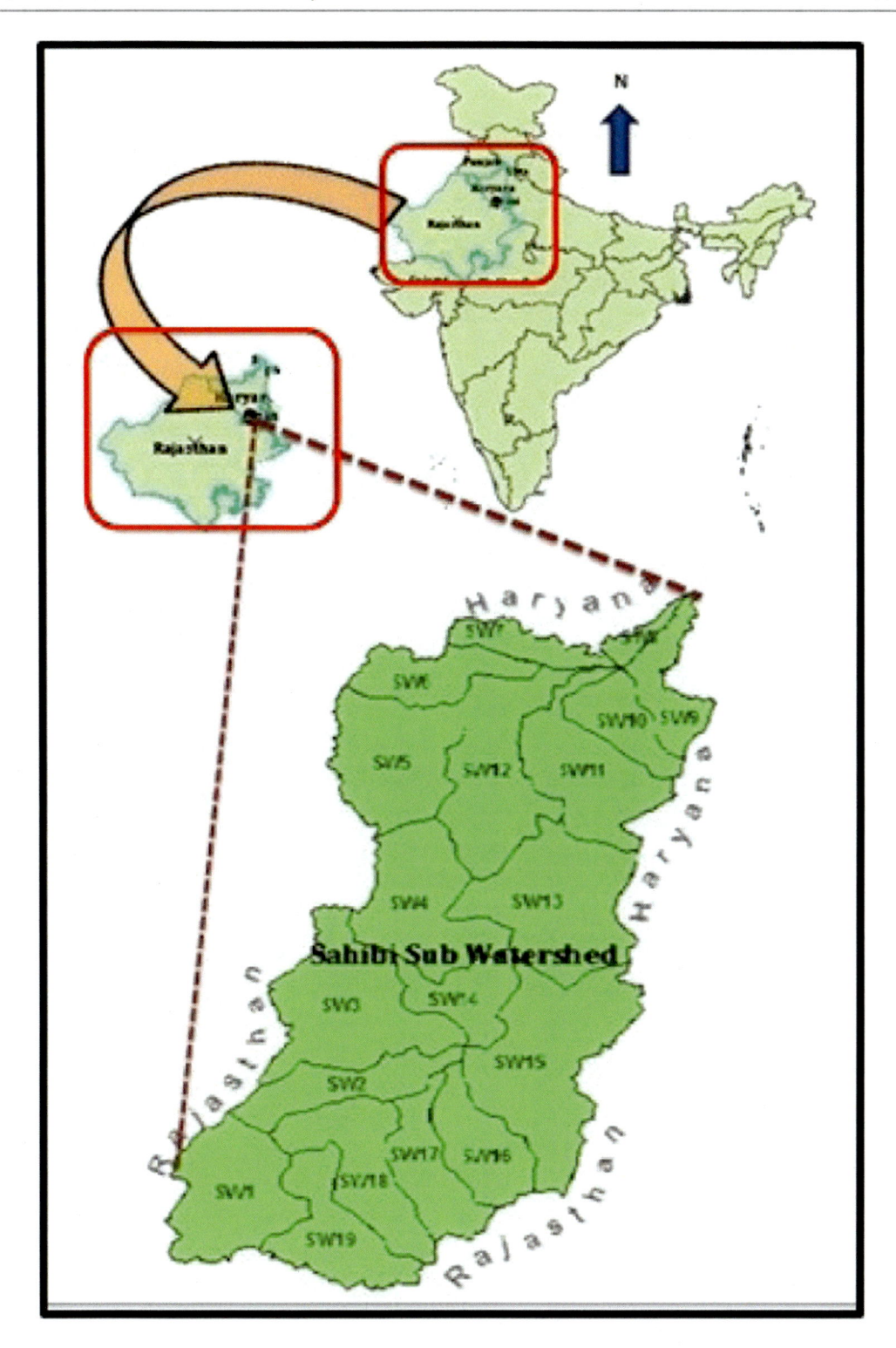

Fig. 18.1 Location map of the study area

Table 18.2 Name and location of the residential area, Sahibi Sub-Watershed

S. No.	Village name	Location in study area	District	State
1	Baturi	North East	Rewari	Haryana
2	Khaleta	North	Rewari	Haryana
3	Mayan	North	Rewari	Haryana
4	Balwari	North	Rewari	Haryana
5	Khol	North West	Rewari	Haryana
6	Nandha	North	Rewari	Haryana
7	Ahrod	North West	Rewari	Haryana
8	Mamaria Asampur	East	Rewari	Haryana
9	Mamaria Ahir	East	Rewari	Haryana
10	Mamaria Thethar	East	Rewari	Haryana
11	Gothra Tappa Khori	East	Rewari	Haryana
12	Bhalki	Central Part	Rewari	Haryana
13	Pali	East	Rewari	Haryana
14	Manethi	West	Rewari	Haryana
15	Parla	West	Rewari	Haryana
16	Kund	West	Rewari	Haryana
17	Nangli	East	Rewari	Haryana
18	Mahtawas	Central Part	Alwar	Rajasthan
19	Arind	Central Part	Alwar	Rajasthan
20	Parula	South East	Alwar	Rajasthan
21	Chela Dungra	South East	Alwar	Rajasthan
22	Giglana	South	Alwar	Rajasthan
23	Chawadi	South	Alwar	Rajasthan
24	Dabarwas	South	Alwar	Rajasthan
25	Birnwas	South	Alwar	Rajasthan
26	Naodi	South	Alwar	Rajasthan
27	Rampura	South	Alwar	Rajasthan
28	Khundrot	South West	Alwar	Rajasthan
29	Dhikwar	South West	Alwar	Rajasthan
30	Anandpur	South	Alwar	Rajasthan
31	Madhan	South	Alwar	Rajasthan
32	Nanagwas	South	Alwar	Rajasthan
33	Manglapur (Nayagaon)	South	Alwar	Rajasthan

Anandpur, Madhan (SW2); Arind, Manglapur (Nayagaon) [SW3]; Manethi, Kund (SW 4); Khol, Ahrad (SW 5); Khaleta, Mayan (SW 7); Bas, Baturi (SW 8); Manmaria Ashpur (SW 9); Manmaria Ahir, Manmaria Thethar, Nagra (SW 10); Gothra Tappa Khori, Pali (SW 11); Balwari, Nandha (SW 12); Bhalki (SW 13); Mahtawas (SW14); Nangli, Chela Dungra, Giglana (SW 15); Chawadi, Nanagwas (SW 16); Dabarwas (SW17); Birnwas (SW 19); SW6 and SW18 has no settlements.

18.2.1 Geomorphology

The district broadly forms part of Indo-Gangetic alluvial plain of Yamuna sub basin. It has vast alluvial and sandy tracts and is interspersed strike ridges which are occasionally covered with blown sand. The hill ranges are part of great Aravalli chain and contain valuable mineral deposits and natural meadows. The elevation of land in the area varies from 232 m in the north to 262 m above mean sea level in south. The master slope of the area is toward the north.

18.2.2 Soil

NBSSLUP (National Bureau of Soil Survey and Land Use Planning) has done the soil classification based on samples where minute details of the soil were studied scientifically. On the basis of soil association, chemical quality, soil series, soil horizon and sub-order association were considered to arrive at scientific soil taxonomical classification.

In this study area four types of soil were found i.e. Older Alluvial soil, Medium Brown Loam, Deep Brown sandy soil, Medium red sandy soil. These different types of soil observed through the satellite imageries and field observation (Fig. 18.2).

18.2.3 Rainfall and Climate

The climate of study area can be classified as semiarid, tropical steppe and dry with very hot summer and cold winter except during monsoon. The region experienced four seasons in a year. The winter season starts after mid of November to last week or first week of March, the hot weather season starts from mid-March to the end of June, July to September the region experienced south west monsoon. The transition period from September to October forms the post-monsoon season. The region received its maximum rainfall in monsoon season. The normal monsoon of the region is 488 mm. and annual rainfall is 550 mm. The south west monsoon contributes about 88% of annual rainfall. July and August are the wettest months. The 12% rainfall received during non-monsoon period was due to the wake of western disturbances and thunderstorms. Generally the rainfall increases from south west to north east. The minimum temperature is experienced in month of January which is about 50 °C. Maximum temperature observed in May and June about 41 °C.

18.3 Data Sources and Methodology

18.3.1 Ancillary Data

To accomplish the study SOI toposheets, hydrological and rainfall data, soil map and Satellite data were being processed properly.

- SOI toposheets, (1:50,000) viz. 54 D/8)
- IRS P6 LISS III data:
- ASTER (Advanced Spaceborne Thermal Emission and Reflection Radiometer) downloaded from Internet from the following source: http://dds.cr.usgs.gov/astr/.
- Soil Resource Map of Rajasthan and Haryana (1:250,000), by NBSS & LUP (ICAR), Dept. of Agriculture, Govt. of India, 1992.
- Available ancillary and proto-type data published information.
- Historical monthly rainfall data for eleven years (1975–1995) has been collected from Investigation and Planning Division, I&WD, Govt. of Rajasthan, for the rain gauge stations located in the watershed and has been used to determine the mean precipitation (Table 18.3).

A software package called ArcGIS version 10.0 is used for this project. Arc Map is the primary application where the data is analyzed and processed. The two spatial data types used in this project these are vector and raster (ASTER) data sets. Vector data contain features defined by a point, line, or polygon. Vector data models are useful for storing and representing discrete features such as settlements and roads, vegetation area, agricultural land, waste land etc. ArcGIS implements vector data as shape files.

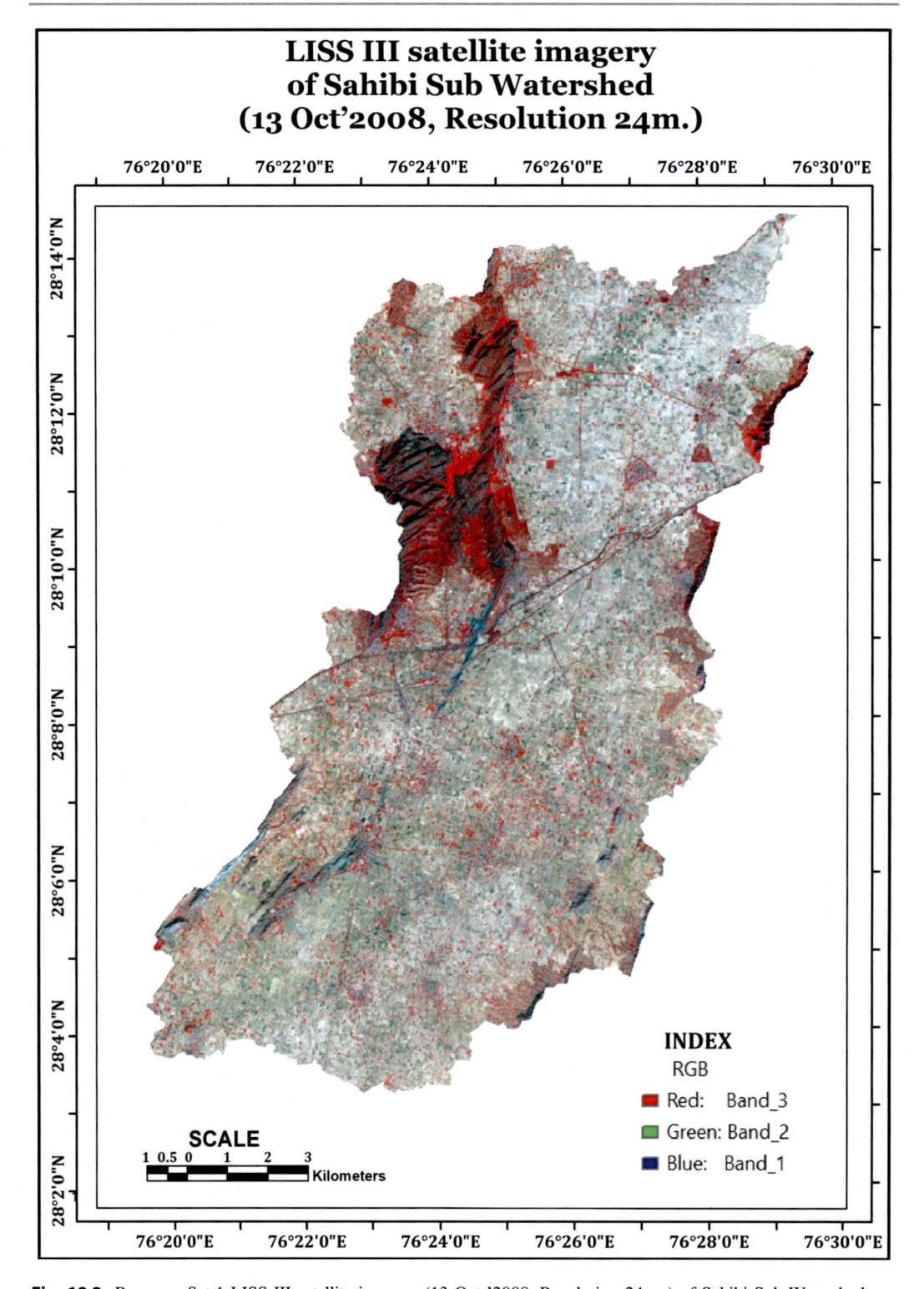

Fig. 18.2 Resource Sat-1 LISS III satellite imagery (13 Oct '2008, Resolution 24 m.) of Sahibi Sub-Watershed

Table 18.3 Satellite data information

Path–Row	Season (Date of pass)	
	Kharif	Ravi
95–051	13th October, 2008	21st January, 2008

18.3.2 Preparation and Discussion on Thematic Maps

Thematic layers help to define each element of the maps separately. We use ARCGIS 10 software for composing thematic layers. Development of Geographic Information Systems (GIS) closely follows advancements in computers. As computers can handle more data intensive operations, the use of GIS has also expanded to handle larger datasets. GIS are primarily used to process and display data having a spatial component. A flow chart is showing the work process for this study (Fig. 18.3).

18.3.2.1 Drainage, Watershed, and Surface Water Bodies

The Drainage map (Fig. 18.4) of the study area, i.e. Part of Sahibi Sub-Watershed, is prepared mainly using the Ravi season data. The streams are digitized, compared, and verified using the Kharif season data. The sub stream channel, main left bank tributaries are digitized, and name is given using the SOI toposheets in ArcGIS environment. Drainage lines are arranged according to order. Surface water bodies, like tanks, ponds are also marked (without zooming more than 1:35,000 scale) and vectorized as

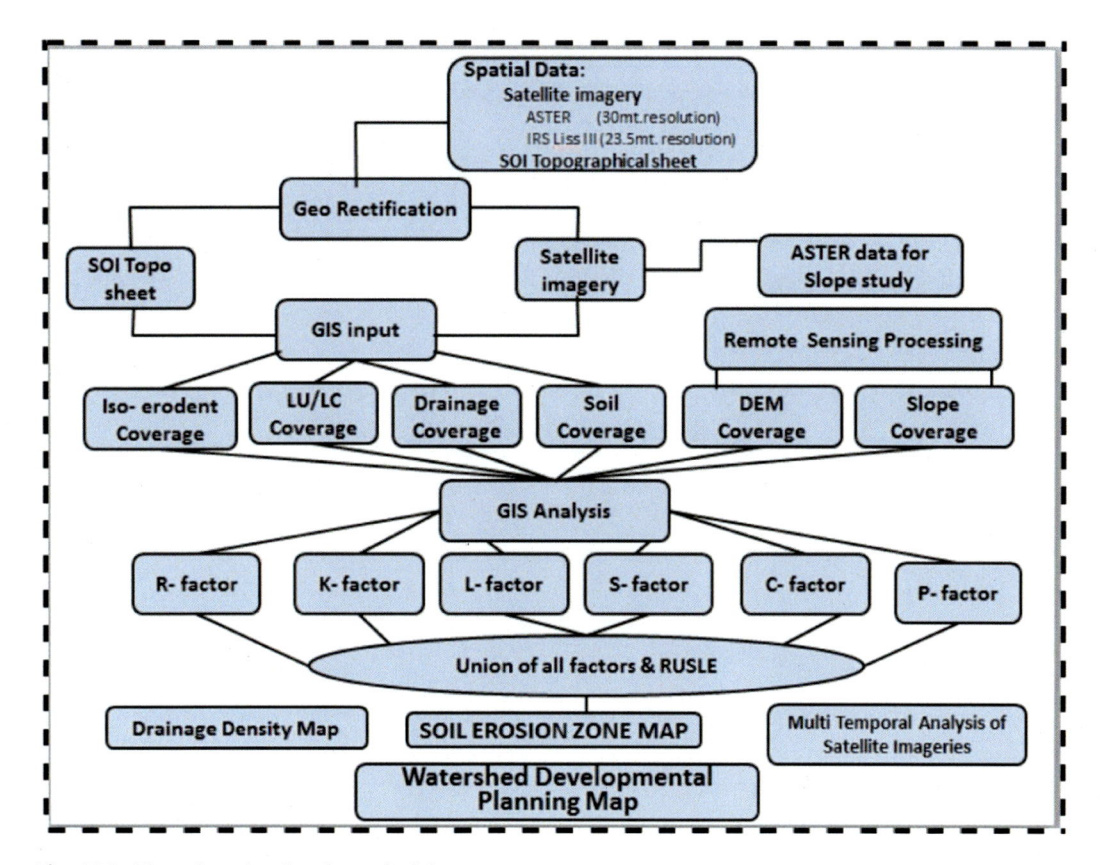

Fig. 18.3 Flow chart showing the methodology

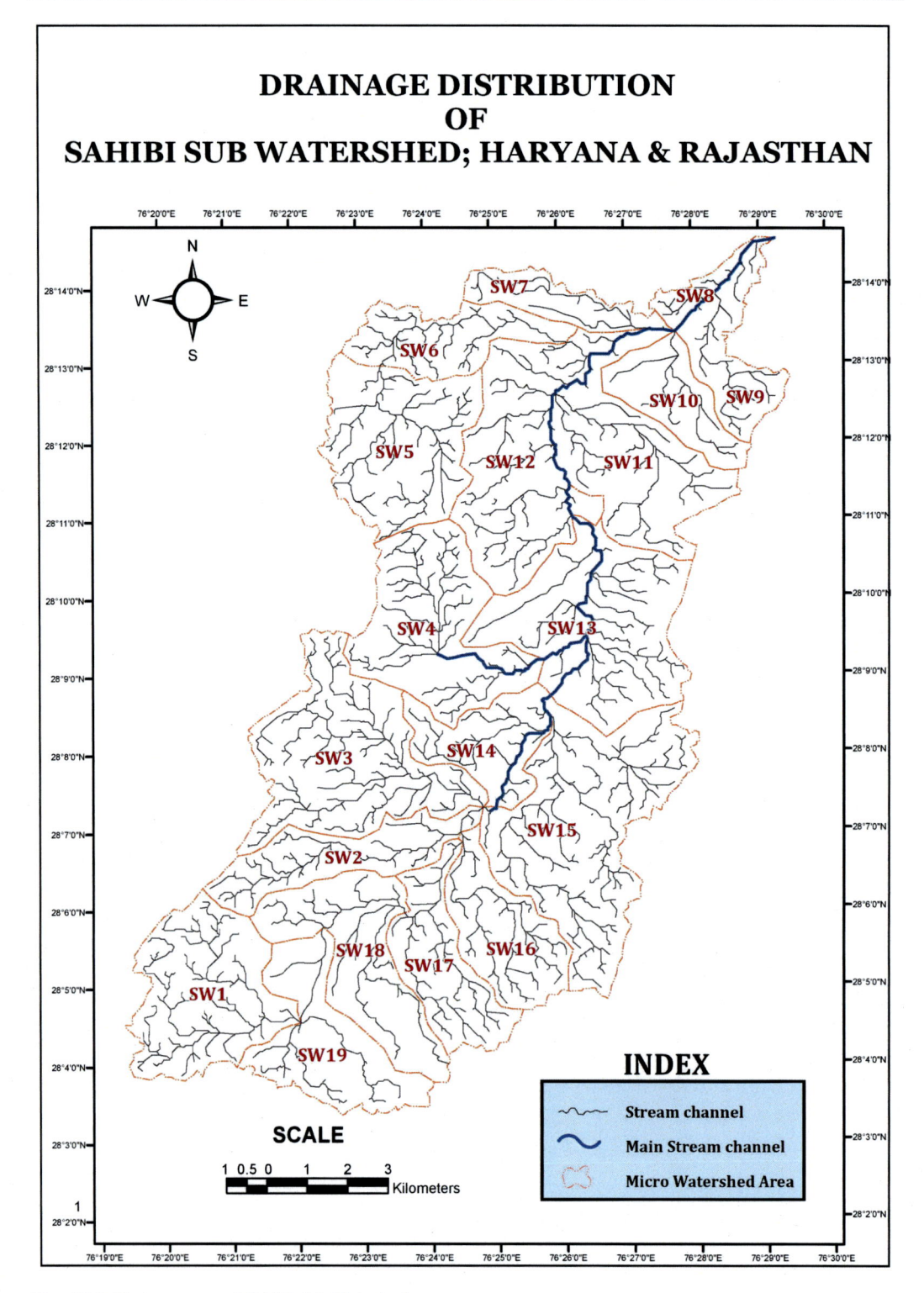

Fig. 18.4 Drainage map of Sahibi Sub-Watershed

polygons. The watershed is delineated. The ASTER data of the study area (in ASCII format) is rasterized. Then with the help of ArcGIS Spatial Analyst tools (Hydrology -fill-Flow Direction-Flow Accumulation) we extracted the drainage. Then with the help of watershed tools watershed boundary has been delineated.

The Sahibi Sub-watershed includes the Sahibi River which is an ephemeral rain-fed river, locally called SABI Nadi. It rises from Mewat hills near Jitgarh and Manoharpur close to Jaipur district in Rajasthan and after gathering volume from about a hundred tributaries, it reaches voluminous proportions, forming a broad stream around Alwar and Kotputli, it then enters the Rewari district in Haryana near the city of Rewari after which it enters Rajasthan again and then re-enters Haryana near village Jaithal. The Rewari district, except in its Eastern part is flat and sandy and absorbs all the rainwater (Dar et al. 2011). Total length of the stream channel is 509.51 km. The catchment area and length of the Sahibi Sub-Watershed have been found as 182.72 km^2. In the micro watersheds i.e. SW1, SW2, SW3, SW5, SW12, SW13, SW15, SW18 has high number of stream channels. Naodi, Rampura, Khundrot, Dhikwar villages are depends on SW1 micro watershed; Anandpur, Madhan villages are depends on SW2 micro watershed; Arind, Manglapur (Nayagaon) villages are depends on SW3 micro watershed; Manethi, Kund villages are depends on SW4 micro watershed; Khol, Ahrad SW5 micro watershed; Khaleta, Mayan SW7 micro watershed; Bas, Baturi villages are depends on SW8 micro watershed; Manmaria Ashpur villages are depends SW9 micro watershed; Manmaria Ahir, Manmaria Thethar, Nagra villages are depends on SW10 micro watershed; Gothra Tappa Khori, Pali villages are depends on SW11 micro watershed; Balwari, Nandha villages are depends on SW12 micro watershed; Bhalki village is depends on SW13; Mahtawas village is depends on SW14. Nangli, Chela Dungra, Giglana villages are depends on SW15 micro watershed; Chawadi, Nanagwas villages are depends on SW16 micro watershed; Dabarwas village is depends on SW17 micro watershed; Birnwas village is depends on SW19 micro watershed. So these settlements are dependent on the agricultural and other water need related activities on the above-mentioned micro watersheds. There are a very minimum number of surface water bodies, both natural and man-made.

18.3.2.2 Land Use/Land Cover

The land use/land cover map of the study area (Fig. 18.5 and Table 18.4) has been prepared with the help of SOI toposheets and Satellite Imageries (Pre and post monsoon data). The forest boundaries and network of roads and railway have been taken from SOI toposheets and incorporated in final map. The open forest area is identified and demarcated using the satellite imageries. Wastelands are also identified and digitized. Settlement with homestead orchards is identified from the Ravi season data. The name of the settlements is taken from SOI toposheets and added to the attribute table.

The land use pattern is a significant determinant of water availability and water use in any basin. The land in a watershed must be used for several purposes—crop and livestock production, housing roads, etc. The land can rarely be put to uses which will provide maximum or most desirable uses for watershed protection. Land use affects rates of runoff, infiltration and types and quality of vegetation cover. In response to that land use/land cover map (Fig. 18.6) is one of the most vital input in 'locale specific' land use planning procedure. The land use/landcover map is the spatial information of the physical and social cover types (i.e. both natural and cultural/anthropogenic) on the existing scenario (Andrade et al. 1988) in relation to hydro geomorphology (including rainfall/temperature), soil and socio-economic conditions of the area under consideration.

18.3.2.3 Soil/Land Capability Class

The study area, falling within eastern part of the Rajasthan and southern part of Haryana, is drawn in eye estimation from the soil map based on overlay method using Arc GIS software. Geo-referencing is done identifying sharp bends in railways, road crossing and place name with the

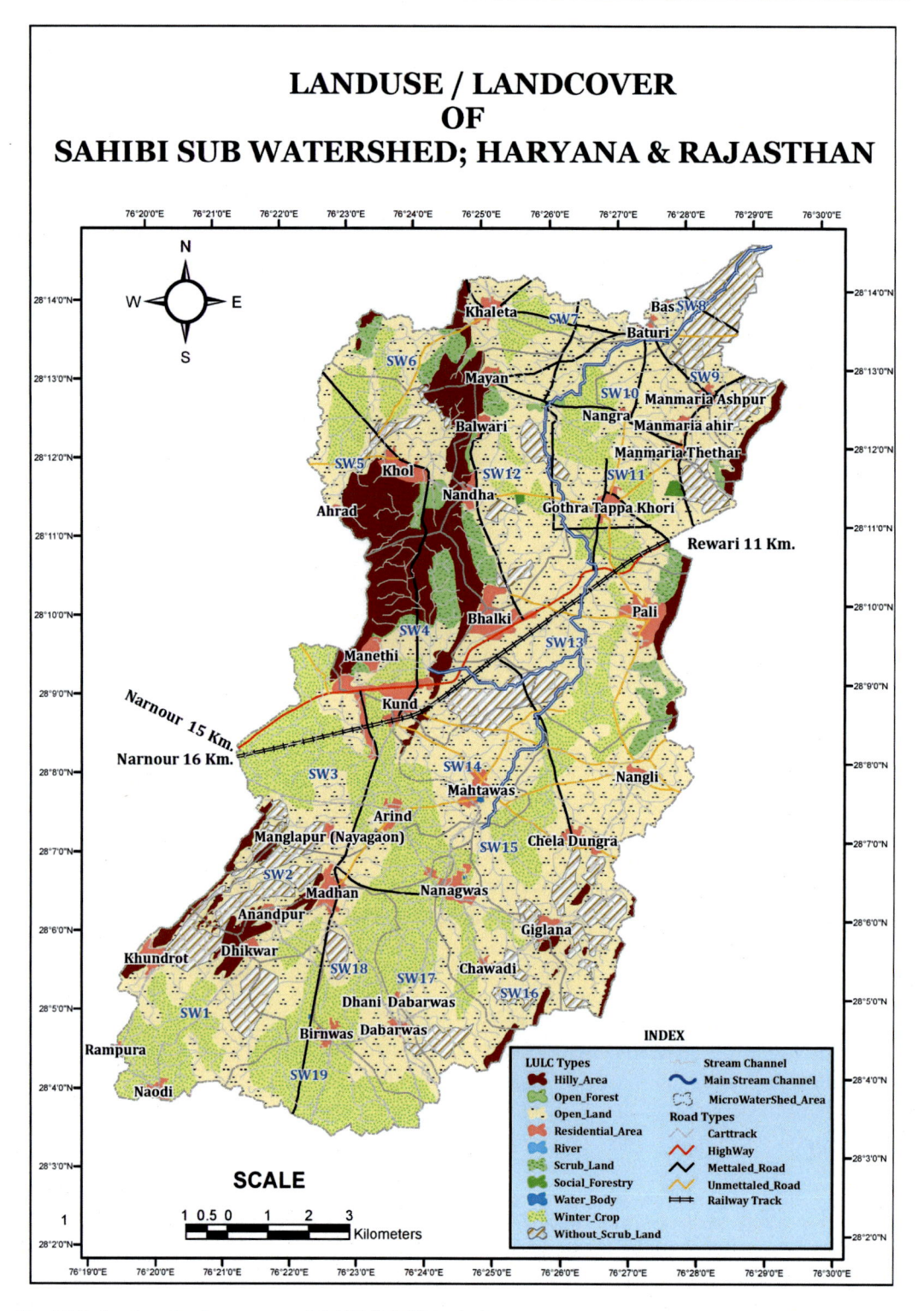

Fig. 18.5 Land use/land cover map of Sahibi Sub-Watershed

Table 18.4 Total area of different thematic layer

S. No.	Thematic layer name	Thematic layer area in km^2
1	Residential area	9.34
2	Land without scrub	16.88
3	Open forest	0.99
4	Agricultural area	48.85
5	Water body	0.06
6	Hilly area	16.36
7	Land with scrub	4.86
8	Open land	85.17
9	Social forestry	0.21
Total area		182.72

Fig. 18.6 Pie diagram showing the distribution of LU/LC coverage in percentage of Sahibi Sub-Watershed. *Source* The Soil Resource Map of Rajasthan and Haryana (1:250,000), by NBSS &LUP (ICAR)

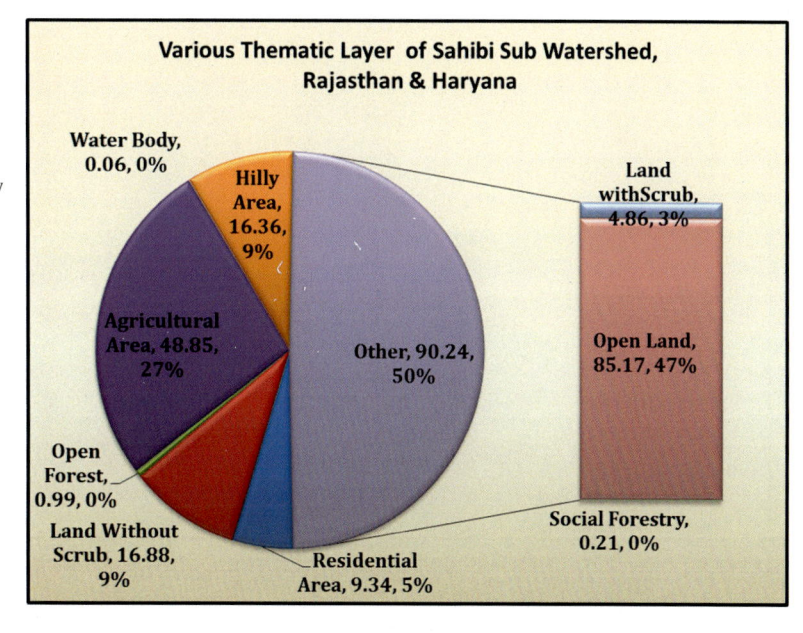

help of SOI toposheets. Unique soil polygons falling under the study area is digitized, attribute table are filled with their taxonomic name and the soil map is prepared (Fig. 18.7).

Soils of the watershed determine the amount of water which will percolate and corrective measures which will be used. The soil character also determines the amount of silt which will be washed down into water harvesting structures and the valley below. Soils are recognized as 3-dimensional bodies on the landscape whose spatial distribution and variability are mainly controlled, among other features, by the geological and geomorphological factors by Jain and Das (2010). In fact, the soil's geographic distribution patterns could be identified more reliably from the association of geomorphic environments (cf. Zinck et al. 1990). The following categories of soil and corresponding land capability classes are identified within the study area. In these study areas four types of soil are found i.e. Older Alluvial soil, Medium Brown Loam, Deep Brown sandy soil, Medium red sandy soil. Among them the Medium Brown

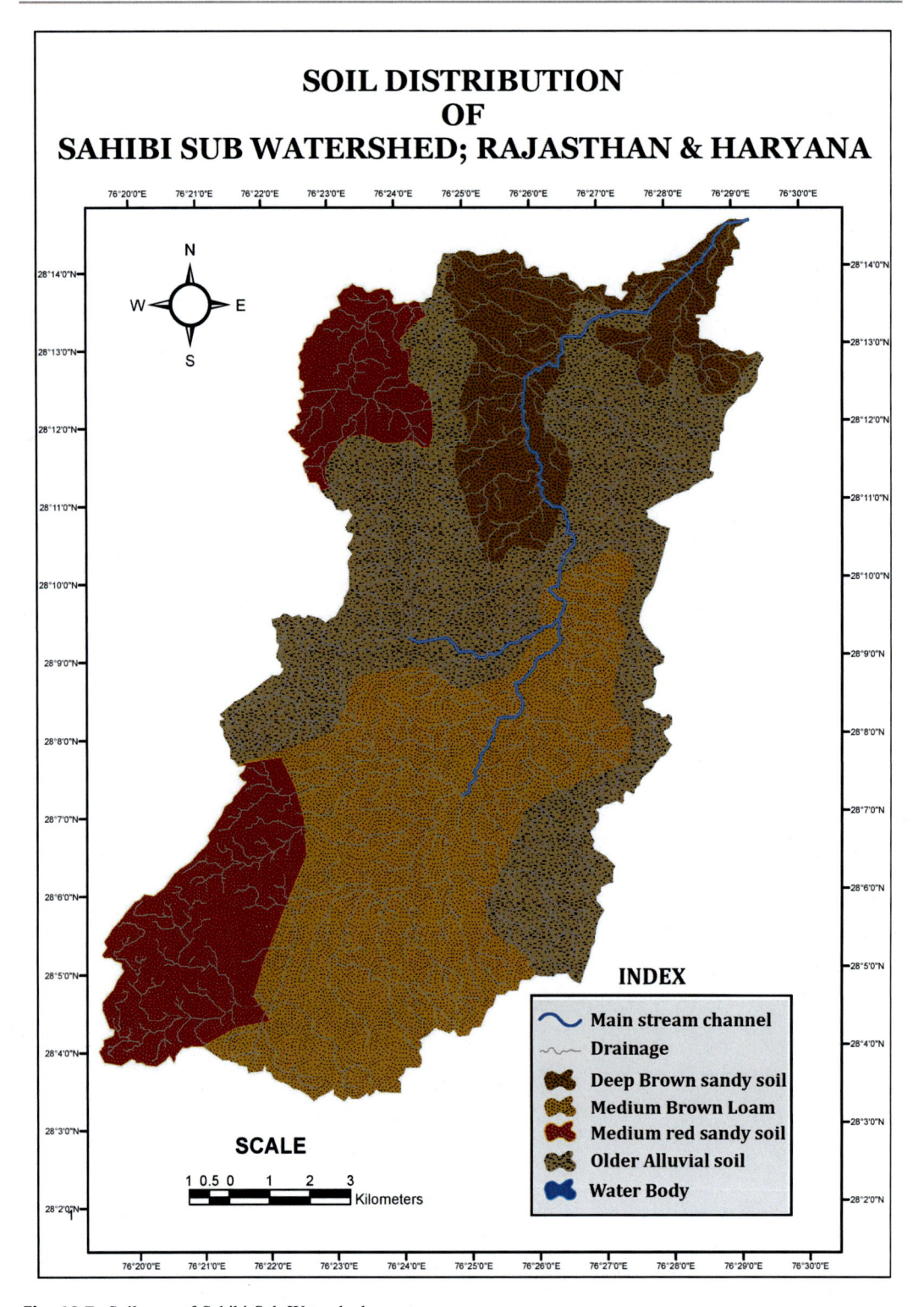

Fig. 18.7 Soil map of Sahibi Sub-Watershed

Table 18.5 Major soil categories define the 'K' value (Zinck et al. 1990)

S. No	Soil type	'K' factor
1	Older alluvial soil	0.2
2	Medium brown loam	0.23
3	Deep brown sandy soil	0.12
4	Medium red sandy soil	0.12

Table 18.6 Slope categories in study area. (NRSC, Hyderabad)

Slope categories	Percent of slope (%)
Uniform or low	0–5
Very gently sloping or moderate	5–20
Steep or high	20–35
Very steep or very high	35

Loam soil is highly fertile (Table 18.5). In this soil the agriculture is highly developed. In Older Alluvial soil, also rich in various fertility parameters as NPK (Nitrogen, Phosphorous, Potassium) and this soil area is less fertile comparatively to Medium Brown Loam soil, as a result some parts are under agriculture.

18.3.2.4 Slope Map

The ASTER (ASCII) data is rasterized. From 3D analyst tool, first we create Slope. The break values for the slope categories are given by NRSC guidelines (Table 18.6).

The slope categories identified in the study area, are:

Based on the slope categories as framed by NRSA (Table No: 5B) the slope (Fig. 18.8) of the study area could be divided into the following major groups:

  I. The undulating/rolling topography of the area, characterized by uniform very gentle slope (0.1 and 1–5%) is very clearly depicted in the study area, almost the whole region falls in this category except the north-western and south west part; south-eastern, and north-eastern part of the study area.
 II. The dissected hill (472 m) complex, shows the existence of various slope classes varying from very gentle (1–3%) to very steep (> 35%) in accordance with terrain units.
III. Scattered isolated hillocks are present in the western and eastern part of the study area.

18.3.3 Universal Soil Loss Equation (USLE)

The USLE is combined with the ArcGIS to estimate average annual soil loss (A) that is occurring in the Sahibi Watershed. Raster layers corresponding to each of the six USLE factors are created, stored, and analyzed with the ArcGIS. This combination computes the estimated soil erosion potential for the entire watershed and areas of high soil erosion potential were identified. The grid cells in each layer overlap and the USLE computation can be done by multiplying all the USLE factors. The DEM serves as the primary input for calculating the Slope Length and Slope Steepness factors (LS-factors). The R factor map is derived from the USLE standard value (Wischmeier and Smith 1978; Renard et al. 1991) which developed rainfall erosivity factor values for Sahibi Sub-Watershed. The Soil Loss Prediction is also done using Revised Universal Soil Loss Equation (RUSLE) in ArcGIS environment. The slope length and slope steepness factors were interrelated. The LS factors, R factor, K factor, C factor, and P factor were being made for performing RUSLE model (Table 18.7).

18.3.3.1 Modeling Soil Detachment with RUSLE in GIS Environment

The Universal Soil Loss Equation is an empirical equation designed for the computation of average

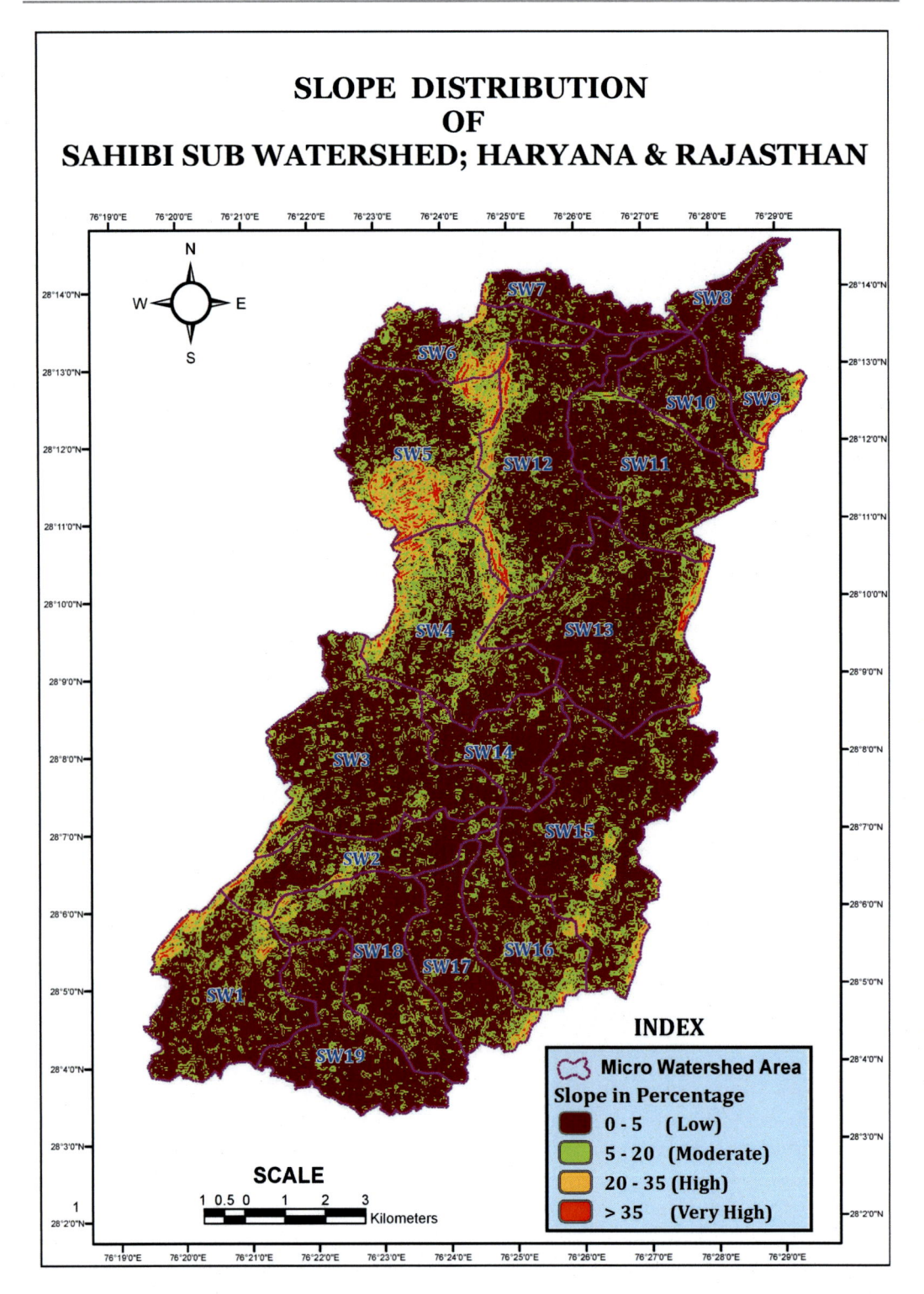

Fig. 18.8 Slope map of Sahibi Sub-Watershed

Table 18.7 USLE factor source description

USLE factor	Derived from	Source description
Slope length (L factor)	30 m Aster DEM* of Sahibi	http://dds.cr.usgs.gov/astr/
Slope steepness (S factor)	30 m DEM* of Sahibi	http://dds.cr.usgs.gov/astr/
Rainfall erosivity (R factor)	Morgan (1986)	https://doi.org/10.4236/as.2020.118043
Soil erodibility (K factor)	NBSS & LUP	http://eusoils.jrc.ec.europa.eu/esdb_archive/eudasm/asia/lists/k7_cin.htm
Vegetation cover (C factor)	Department of Agriculture, Govt. of India	http://agriharyana.nic.in/fert_rewari.htm
Prevention practices (P factor)	Reclassified DEM	NRSC, Dept. of Space, Govt. of India

soil loss in agricultural fields. Five major factors (Table 18.7) were used to calculate the soil loss for a given site. Each factor is the numerical estimate of a specific condition that affects the severity of soil erosion at a particular location. The erosion values reflected by these factors can vary considerably due to varying weather conditions. Therefore, the values obtained from the USLE more accurately represent long-term averages.

18.3.3.2 Factors Used in Erosion Equations in USLE Model

Five major factors are used to calculate the soil loss for a given site. Each factor is the numerical estimate of a specific condition that affects the severity of soil erosion at a particular location. The erosion values reflected by these factors can vary considerably due to varying weather conditions. Therefore, the values obtained from the USLE more accurately represent long-term averages.

$$\text{Universal Soil Loss Equation (USLE)}, A = R \times K \times LS \times C \times P \quad (18.1)$$

'A' represents the potential long-term average annual soil loss in tons per acre per year. This is the amount, which is compared to the 'tolerable soil loss' limits; K is the soil erodibility factor. It is the average soil loss in tons/acre per unit area for a particular soil in cultivated, continuous fallow with an arbitrarily selected slope length of 72.6 ft. and slope steepness of 9%. K is a measure of the susceptibility of soil particles to detachment and transport by rainfall and runoff. Texture is the principal factor affecting K, but structure, organic matter and permeability also contribute.

LS is the slope length-gradient factor. The LS factor represents a ratio of soil loss under given conditions to that at a site with the 'standard' slope steepness of 9% and slope length of 72.6 feet. The steeper and longer the slope, the higher is the risk for erosion. LS factor was used in USLE computation (2).

$$LS = [0.065 + 0.0456(\text{slope})] + 0.006541(\text{slope})2] \times (\text{slope\_length} \div \text{const})NN \quad (18.2)$$

where, slope = slope steepness (%); slope length = length of slope (ft.); constant = 72.5 Imperial or 22.1 metric; NN = (as per Morgan 1981).

C is the crop/vegetation and management factor. It is used to determine the relative effectiveness of soil and crop management systems in terms of preventing soil loss. The C factor is a ratio comparing the soil loss from land under a specific crop and management system to the corresponding loss from continuously fallow and tilled land. The C factor can be determined by selecting the crop type and tillage method that corresponds to the field and then multiplying these factors together. The C factor resulting from this calculation is a generalized C factor value for a specific crop that does not account for crop rotations or climate and annual rainfall distribution for the different agricultural regions of the country. This generalized C factor, however, provides relative numbers for the different cropping and tillage systems; thereby helping you weigh the merits of each system. P is the support practice factor. It reflects the effects of practices that will reduce the amount and rate of the water runoff and thus reduce the amount of erosion. The P factor represents the ratio of soil loss by a support practice to that of straight row farming up and down the slope. The most commonly used supporting cropland practices are Up and Down slope cultivation, contour farming and strip-cropping.

18.3.3.3 RUSLE

Revised USLE (RUSLE) uses the same empirical principles as USLE, however it includes numerous improvements, such as monthly factors, incorporation of the influence of profile convexity/concavity using segmentation of irregular slopes by Adediji et al. (2010), improved empirical equations for the computation of LS factor (Foster and Wischmeier1974; Renard et al. 1991).

LS factor modified for complex terrain to incorporate the impact of flow convergence, the hillslope length factor was replaced by upslope contributing area 'A' (Moore and Burch 1996; Mitasova et al. 1995, 1996; Desmet and Govers 1996). The modified equation for computation of the LS factor in GIS in finite difference form for erosion in a grid cell representing a hillslope segment was derived by Desmet and Govers (1996). A simpler, continuous form of equation for computation of the LS factor at a point $r = (x, y)$ on a hillslope, (Mitasova et al. 1996) is

$$LS(r) = (m+1)[A(r)/a0]m[\sin b(r)/b0]n$$

(18.3)

where, A[m] is upslope contributing area per unit contour width, b [deg] is the slope, m and n are parameters, and $a0 = 22.1$ m $= 72.6$ft is the length and $b0 = 0.09 = 9\% = 5.16$ deg is the slope of the standard USLE plot. According to RUSLE modifications, the upslope area better reflects the impact of concentrated flow on increased erosion. It has been shown that the values of $m = 0.6$, $n = 1.3$ give results consistent with the RUSLE LS factor for slope lengths < 100 m and slope angles < 14 deg (Moore and Wilson 1992), for slopes with negligible tangential curvature. Exponent m and n can be calibrated if the data are available for a specific prevailing type of flow and soil conditions. Both the standard and modified equations can be properly applied only to areas experiencing net erosion. Depositional areas should be excluded from the study area because the model assumes that transport capacity exceeds detachment capacity everywhere and erosion and sediment transport is detachment capacity limited. Therefore, direct application of USLE/RUSLE to complex terrain within GIS is rather restricted. The results can also be interpreted as an extreme case with maximum spatial extent of erosion possible. The direction of steepest descent from each cell center to the next closest neighboring cell center is called the FLOWDIRECTION.

18.4 Result and Discussion

18.4.1 Drainage Density

Drainage Density Map (Fig. 18.9), defined as the ratio of the total length of all streams of the catchment divided by its area, indicates the drainage efficiency of the basin. The higher the value, quicker is the runoff and lesser is the infiltration and other losses. Thus, drainage density, $Dd = Ls/A$ [where, Ls = total length of all streams in m; A = Area of the basin in km^2.]. For the study area, $Dd = 2807.06$. The catchment area and length of the Sahibi Sub-Watershed have been found as 182.72 km^2. The micro watersheds i.e. SW1, SW2, SW3, SW4, SW5, SW6, SW7, SW8, SW9, SW12, SW13, SW15, SW16, and SW17 are attributed with high and very high in drainage density. Naodi, Rampura, Khundrot, Dhikwar villages depend on SW1 micro watershed; Anandpur, Madhan villages depend on SW2 micro watershed; Arind,

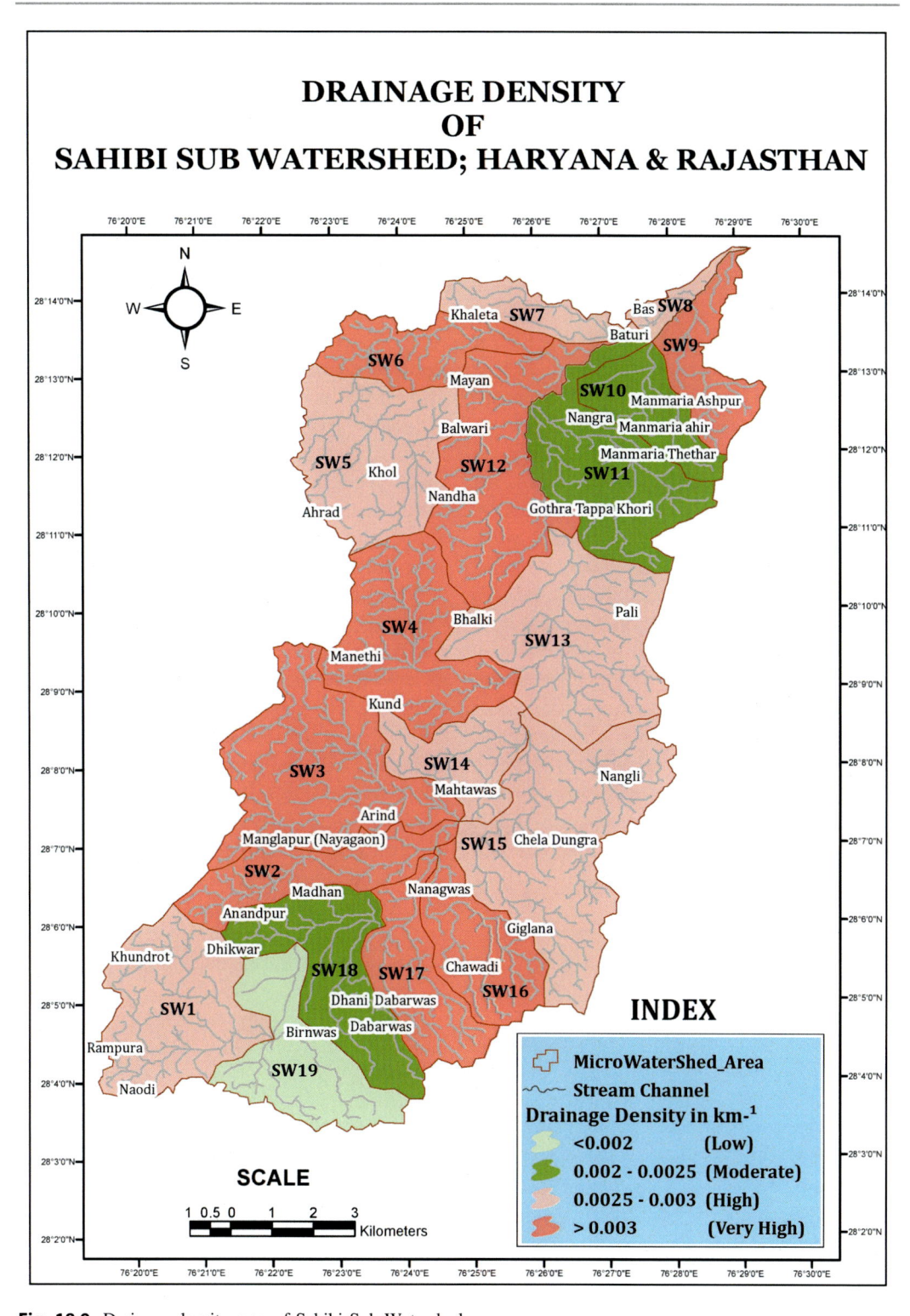

Fig. 18.9 Drainage density map of Sahibi Sub-Watershed

Manglapur (Nayagaon) villages depend on SW3 micro watershed; Manethi, Kund villages depend on SW4 micro watershed; Khol and Ahrad on SW5 micro watershed; Khaleta and Mayan on SW7 micro watershed; Bas, Baturi villages on SW8 micro watershed; Manmaria Ashpur villages depend on SW9 micro watershed; Manmaria Ahir, Manmaria Thethar, Nagra villages depend on SW10 micro watershed; Gothra Tappa Khori, Pali villages depend on SW11 micro watershed; Balwari, Nandha villages on SW12 micro watershed; Bhalki village on SW13; Mahtawas village on SW14. Nangli, Chela Dungra, Giglana villages on SW15 micro watershed; Chawadi, Nanagwas villages on SW16 micro watershed; Dabarwas village on SW17 micro watershed; Birnwas village on SW19 micro watershed. So these settlements are dependent on the agricultural and other water need related activities on the above-mentioned micro watersheds. Soil erosion basically depends on drainage density where the drainage density is high or very high there should be more erosion. Where the drainage density is low there should be less erosion. The areas have been facing more problem regarding soil erosion are Anandpur, Madhan Arind, Manglapur (Nayagaon) Manethi, Kund, Manmaria, Ashpur, Balwari, Nandha, Bhalki, Mahtawas, Dabarwas, Nangli, Chela Dungra, Giglana, Chawadi, and Nanagwas.

18.4.2 Discussion of USLE and RUSLE Method in Respect of Drainage Density

With the help of USLE and RUSLE methods, the maximum and minimum soil loss zones by Kouli et al. (2009), according to the five factor as ($R*[K]*[C]*[P]*[lsfac]$) were prioritized. All the elements of USLE model are interlinked with each other. As example increasing drainage density (Ganapuram et al. 2009) define the increasing soil erosion. On the other hand, areas covered with the vegetation experiences less soil loss viz. SW1, SW2 etc. (Table 18.12).

With the help of USLE and RUSLE method derived from the "Soil Erosion and Conservation, 3rd edition 2009" (Morgan), the study estimated soil loss Ton/ hac./year, and results were categorized in four sub classes viz. low, moderate, high, very high of the study area (Fig. 18.10). Naodi, Rampura, Khundrot, Dhikwar villages under SW1 micro watershed, Madhan villages under SW2 micro watershed, Anandpur; Arind, Manglapur (Nayagaon) villages under SW3 micro watershed, Khaleta, Mayan is under SW7 micro watershed; Bas, Baturi villages under SW8 micro watershed; Manmaria Ashpur villages under SW9 micro watershed; Manmaria Ahir, Manmaria Thethar, Nagra villages under SW10 micro watershed; Balwari, Nandha villages under SW12 micro watershed; Mahtawas village under SW14 micro watershed; Birnwas village under SW19 micro watershed falls in low to moderate level soil erosion zones.

Manethi, Kund villages under SW4 micro watershed; Khol, Ahrad under SW5 micro watershed; Gothra, Tappa, Khori, Pali villages under SW11 micro watershed; Bhalki village under SW13; Nangli, Chela Dungra, Giglana villages under SW15 micro watershed; Chawadi, Nanagwas villages under SW16 micro watershed; Dabar was village under SW17 micro watershed falls in the high and very high soil erosion category due to the high drainage density and low land use/landcover cover. Slope steepness is a very prominent factor for the soil loss. So these areas should be provided more attention for the watershed development (Fig. 18.9; Tables 18.8 and 18.9).

The relational micro watershed map has been prepared based on LU/LC categories (%) and Soil Loss Zone categories (Table 18.11; Fig. 18.11). In SW1, there is no open forest, no social forestry, no scrub land whereas agriculture is predominant. In land without scrub (13.76%) effect the upper surface of the land which caused soil erosion. The severe soil effected areas are under the SW1, SW3, SW4, SW5, SW10, SW13, and SW15 (Tables 18.9, 18.10 and 18.11).

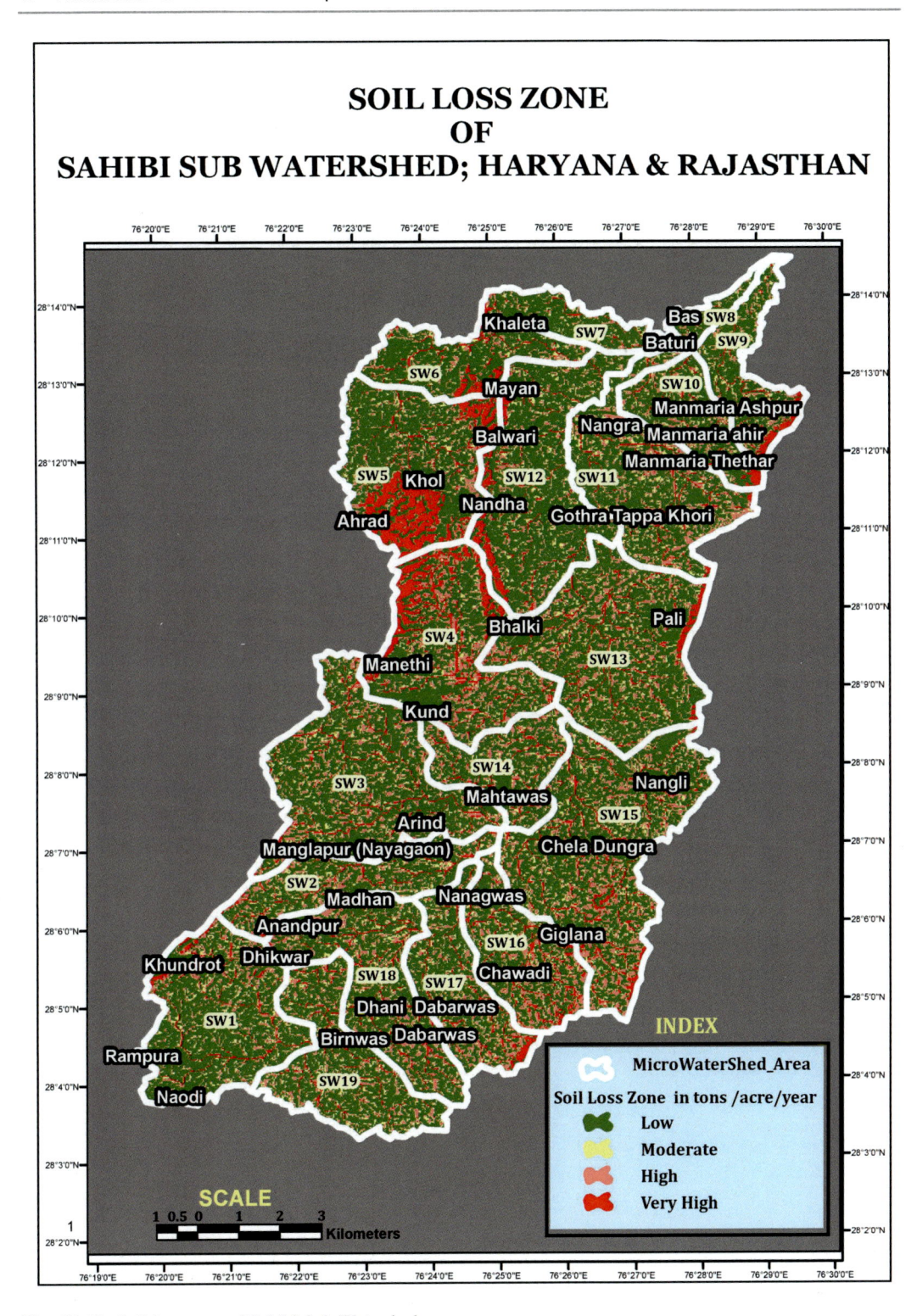

Fig. 18.10 Soil loss zone of Sahibi Sub-Watershed

Table 18.8 Micro watershed wise distribution of different soil loss zone categories related with drainage density

Watershed name	Soil loss types	Area in km^2	Total area affected in km^2	Area affected in SMW (%)	Drainage density
SW1	Low	8.06	12.04	67.0	0.0029
	Moderate	1.21		10.1	
	High	2.00		16.6	
	Very high	0.77		6.4	
SW2	Low	4.30	7.15	60.1	0.0030
	Moderate	0.71		9.9	
	High	1.66		23.2	
	Very high	0.48		6.7	
SW3	Low	9.90	15.78	62.7	0.0032
	Moderate	1.67		10.6	
	High	3.33		21.1	
	Very high	0.88		5.6	
SW4	Low	7.00	12.68	55.2	0.0030
	Moderate	0.79		6.2	
	High	2.68		21.1	
	Very high	2.22		17.5	
SW5	Low	7.24	12.36	58.6	0.0027
	Moderate	0.80		6.4	
	High	1.68		13.6	
	Very high	2.64		21.3	
SW6	Low	4.11	6.25	65.8	0.0030
	Moderate	0.54		8.6	
	High	0.88		14.1	
	Very high	0.72		11.5	
SW7	Low	2.69	3.69	72.8	0.0029
	Moderate	0.34		9.2	
	High	0.49		13.2	
	Very high	0.18		4.8	
SW8	Low	0.94	1.36	69.1	0.0024
	Moderate	0.17		12.4	
	High	0.21		15.3	
	Very high	0.04		3.3	
SW9	Low	3.25	5.18	62.8	0.0033
	Moderate	0.55		10.6	
	High	0.89		17.1	
	Very high	0.49		9.5	
SW10	Low	3.33	5.23	63.6	0.0020
	Moderate	0.51		9.7	
	High	1.03		19.7	

(continued)

Table 18.8 (continued)

Watershed name	Soil loss types	Area in km²	Total area affected in km²	Area affected in SMW (%)	Drainage density
	Very high	0.37		7.1	
SW11	Low	7.00	11.48	61.0	0.0023
	Moderate	1.30		11.3	
	High	2.46		21.5	
	Very high	0.71		6.2	
SW12	Low	9.52	14.04	67.8	0.0030
	Moderate	1.30		9.2	
	High	1.92		13.7	
	Very high	1.31		9.3	
SW13	Low	10.07	15.89	63.3	0.0027
	Moderate	1.55		9.7	
	High	3.20		20.1	
	Very high	1.08		6.8	
SW14	Low	3.71	5.99	62.0	0.0029
	Moderate	0.60		10.0	
	High	1.35		22.5	
	Very high	0.34		5.7	
SW15	Low	11.04	19.75	59.4	0.0029
	Moderate	3.30		10.0	
	High	4.35		23.4	
	Very high	1.33		7.2	
SW16	Low	4.22	7.27	58.0	0.0030
	Moderate	0.61		8.4	
	High	1.82		25.1	
	Very high	0.62		8.5	
SW17	Low	4.38	7.22	60.7	0.0031
	Moderate	0.73		10.2	
	High	1.58		21.9	
	Very high	0.53		7.3	
SW18	Low	6.45	10.34	62.3	0.0022
	Moderate	1.08		10.5	
	High	2.15		20.8	
	Very high	0.66		6.4	
SW19	Low	5.74	9.02	63.7	0.0017
	Moderate	1.14		12.6	
	High	1.71		18.9	
	Very high	0.43		4.8	
Total soil loss affected areas (km²)				182.72	

Table 18.9 Relational analysis table between drainage density, slope and soil loss zone

Micro watershed name	Drainage density in m^2	Soil loss zone in km^2	Slope weightage
SW1	0.0029	12.04	1
SW2	0.0020	7.15	1
SW3	0.0023	15.78	1
SW4	0.0030	12.68	3
SW5	0.0027	12.36	3
SW6	0.0029	6.25	2
SW7	0.0029	3.69	1
SW8	0.0030	1.36	1
SW9	0.0031	5.18	2
SW10	0.0022	5.23	2
SW11	0.0017	11.48	2
SW12	0.0030	14.04	2
SW13	0.0032	15.89	2
SW14	0.0030	5.99	1
SW15	0.0027	18.59	2
SW16	0.0030	7.27	3
SW17	0.0029	7.22	2
SW18	0.0024	10.34	1
SW19	0.0033	9.02	1

Table 18.10 Showing total soil loss zone area in km^2

Soil loss zone category	Total area in km^2
Low	112.93
Moderate	18.62
High	35.37
Very high	15.8
Total area	182.72

18.4.2.1 Prioritization of Watershed Developmental Planning

Prioritization (Biswas et al. 1999) means the evaluation of the whole matter in different categories (Table 18.12) as on priority basis. High priority means, the areas are falling in the category which is highly needed for recovery. In the study, areas having high soil erosion prone have been given more weightage and vice-versa.

Prioritization of Watershed Development planning map has been prepared for identifying the extremely needed treatment areas.

I. Find out the mostly affected villages in the Sahibi Sub-Watershed due to the soil loss.
II. Prepared a map related to LU/LC coverage and soil loss.

A Prioritization Map has been prepared for the Watershed development plan which reveals the areas that require high priority for soil conservation. In Sahibi Sub-Watershed, different weightage is given according to the settlements present in the specific micro watershed. For high density population weightage given as '3', for moderate density it is given '2' and in the low

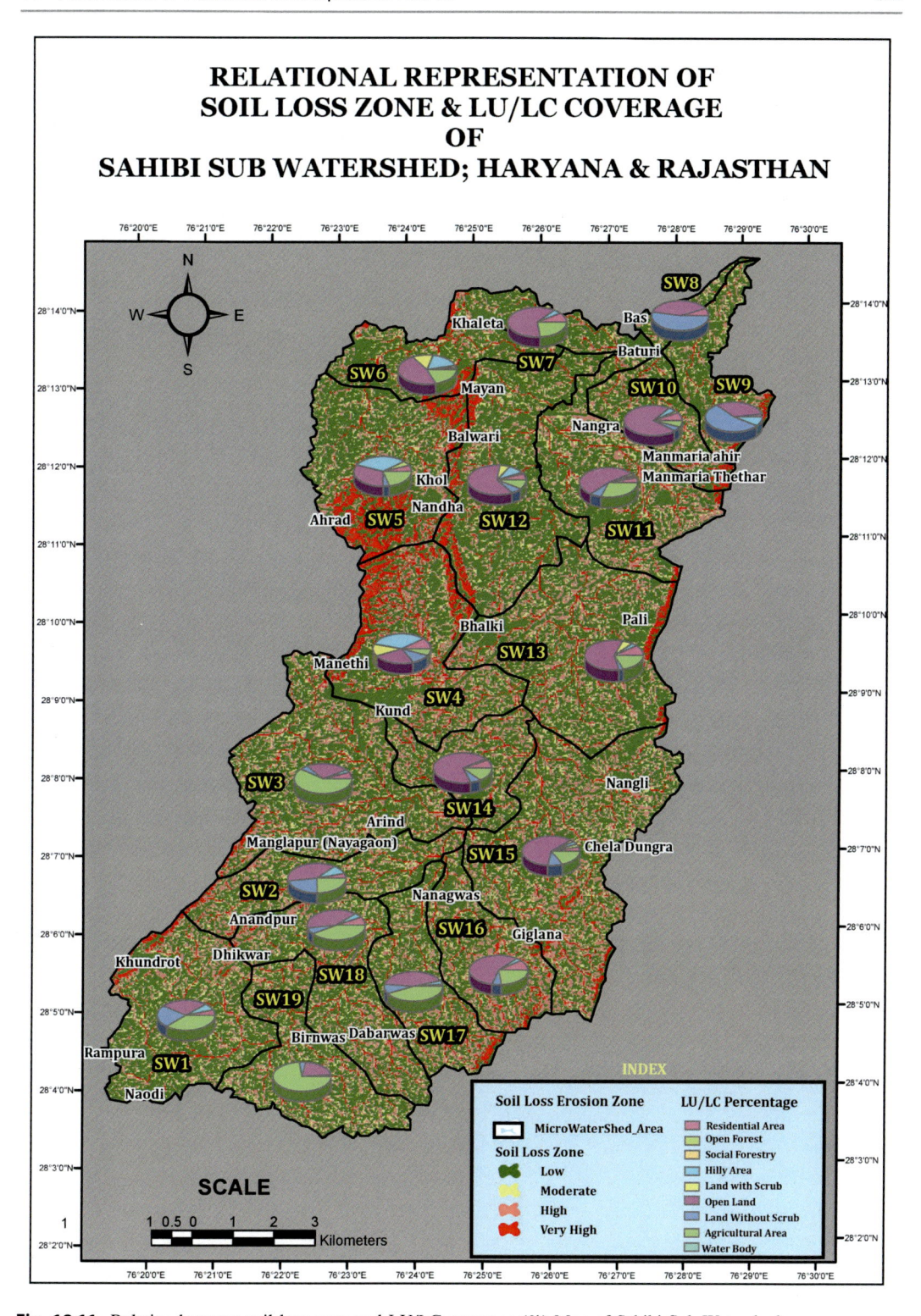

Fig. 18.11 Relation between soil loss zone and LU/LC coverage (%) Map of Sahibi Sub-Watershed

Table 18.11 Micro watershed based LU/LC distribution in percentage (%)

Watershed names	Name of the villages	SW Area in km^2	Residential area	Open forest	Social forest	Hilly area	Scrub land	Open land	Land without scrub	Agriculture	Waterbody
SW 1	Naodi, Rampura, Khundrot, Dhikwar	12.26	4.42	Nil	Nil	6.88	Nil	33.17	13.76	41.69	Nil
SW 2	Anandpur, Madhan,	7.15	3.78	Nil	Nil	8.11	Nil	40.14	23.36	24.48	Nil
SW 3	Arind, Manglapur (Nayagaon)	15.82	5.12	Nil	Nil	2.65	Nil	29.01	4.8	58.41	Nil
SW 4	Manethi, Kund,	12.7	8.98	Nil	Nil	37.32	9.76	27.64	10.55	5.67	Nil
SW 5	Khol, Ahrad	12.47	4.75	4.19	Nil	35.86	Nil	30.54	4.83	19.82	Nil
SW 6	-	6.27	1.75	1.75	Nil	16.19	15.47	45.45	0.32	18.34	Nil
SW 7	Khaleta, Mayan	3.83	8.04	Nil	Nil	4.29	Nil	64.88	Nil	22.79	Nil
SW 8	Bas, Baturi	1.39	5.76	Nil	Nil		Nil	41.01	52.52	Nil	0.72
SW 9	Manmaria Ashpur,	5.44	2.25	Nil	Nil	7.49	Nil	35.77	54.31	Nil	0.19
SW 10	Manmaria ahir, Manmaria Thethar, Nagra	5.35	6.67	2.1	Nil	4.57	Nil	77.33	4.19	5.14	Nil
SW 11	Gothra Tappa Khori, Pali	11.57	3.63	0.69	1.12	1.3	1.99	52.9	7.69	30.68	Nil
SW 12	Balwari, Nandha,	14.04	4.99	1.21	0.57	9.26	7.26	63.03	6.13	7.55	Nil
SW 13	Bhalki	15.47	8.68	Nil	Nil	3.49	6.85	62.18	3.43	18.42	Nil
SW 14	Mahtawas	5.99	6.51	1.5	Nil		Nil	72.62	6.84	12.02	0.33
SW 15	Nangli, Chela Dungra, Giglana	18.84	3.34	Nil	Nil	2.6	2.12	63.69	11.52	16.72	Nil
SW 16	Chawadi, Nanagwas	7.31	8.49	Nil	Nil	3.29	Nil	57.95	7.26	23.01	Nil
SW 17	Dabarwas	7.31	1.78	Nil	Nil	2.87	Nil	43.64	7.39	44.32	Nil
SW 18	-	10.37	6.65	Nil	Nil	4.4	Nil	40.98	5.4	42.43	Nil
SW 19	Birnwas	9.15	1.42	Nil	Nil		Nil	20.77	4.92	72.9	Nil

Table 18.12 Sahibi Sub-Watershed prioritization for management plan by weightage rank method

Micro-watershed names	Name of the villages	Weightage
SW 1	Naodi, Rampura, Khundrot, Dhikwar	3
SW 2	Anandpur, Madhan,	2
SW 3	Arind, Manglapur (Nayagaon)	2
SW 4	Manethi, Kund,	2
SW 5	Khol, Ahrad	2
SW 6	–	0
SW 7	Khaleta, Mayan	2
SW 8	Bas, Baturi	2
SW 9	Manmaria Ashpur,	2
SW 10	Manmaria ahir, Manmaria Thethar, Nagra	3
SW 11	Gothra Tappa Khori, Pali	3
SW 12	Balwari, Nandha,	2
SW 13	Bhalki	1
SW 14	Mahtawas	1
SW 15	Nangli, Chela Dungra, Giglana	2
SW 16	Chawadi, Nanagwas	2
SW 17	Dabarwas	1
SW 18	–	0
SW 19	Birnwas	1

density it is given '1' (Table 18.12). In this way, a relational map of Soil Loss Zone and LU/LC coverages for a better result of the study area i.e. Sahibi Sub-Watershed has been developed (Fig. 18.11).

18.4.2.2 Prioritization of Watershed Developmental

The study found out areas having high population density where priority should be provided for soil conservation in Sahibi Sub-Watershed. Due to surface runoff, the upper surface layer of agricultural field is washed out which affects the fertility of soil and agricultural production.

As a result indirectly this process hit the social development as well as economic growth of the area. According to our study Naodi, Rampura, Khundrot, Dhikwar, Mangalpur, Arind, Manmariya Ashpur, Manmaria Ahir, Nangli, Chela Dungra and Giglana villages will need soil Erosion treatment as soon as possible (Fig. 18.12). We should apply some soil conservation techniques for those areas such as terrace farming, contour bunding farming, rotational crop farming etc.

18.5 Conclusion

Here an attempt has been made to prioritize the Sahibi Sub-watershed for the proper development at watershed level planning using USLE and RUSLE model, RS and GIS. The correlational analysis between drainage density, LU/LC coverage and slope steepness helped to determine the high to low soil loss zones. The USLE calculates long-term average annual soil loss by multiplying six specific factors such as rainfall, soil types, slope, and vegetation cover which describes the watershed characteristics. GIS was used to store the USLE factors as individual digital layers and multiplied together to create a soil erosion potential map of Sahibi Sub-Watershed. At the final stage, the Watershed developmental priority

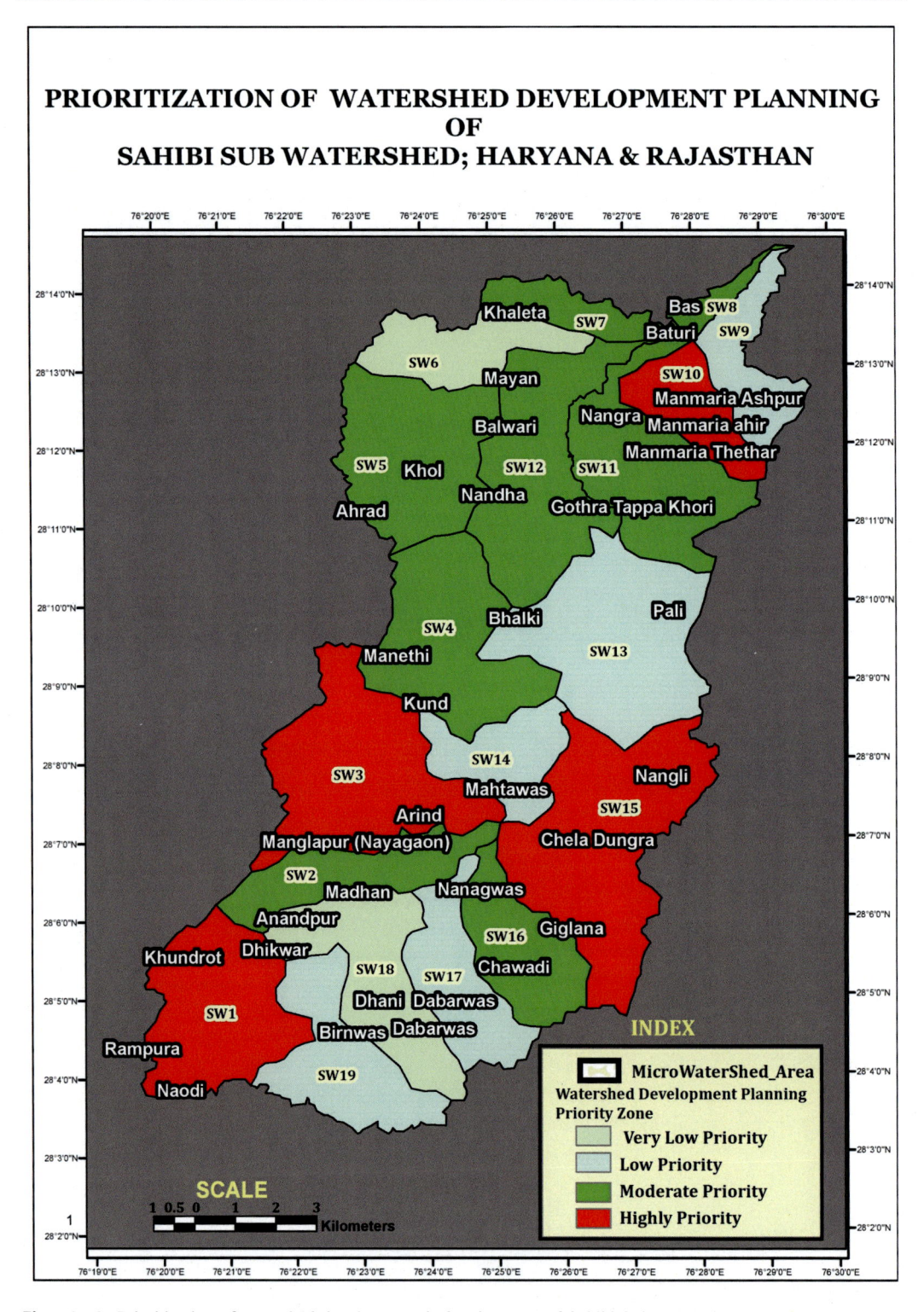

Fig. 18.12 Prioritization of watershed developmental planning map of Sahibi Sub-Watershed

map has been prepared and it revealed the most vulnerable soil erosion prone areas as well. The villages i.e. Naodi, Rampura, Khundrot, Dhikwar, Mangalpur, Arind, Manmariya Ashpur, Manmaria Ahir, Nangli, Chela Dungra and Giglana needs soil erosion treatment immediately. Some soil conservation techniques are to be applied for preserving the land resources i.e. terrace farming, contour bunding farming, rotational crop farming etc.

Acknowledgements We would like to express our special thanks to Dr. Asim Ratan Ghosh, Senior Scientist, Department of Science & Technology and Biotechnology (DSTB), Govt. of West Bengal for his expert guidance, valuable advices, suggestions, co-operation and all support throughout our project that paved our way to complete the work successfully. We are thankful to NIIT University and the all faculty members of GIS department for their continuous support during the project work, Alwar, Rajasthan for the 24 hours Lab. facility during our stay in the campus for the M.Tech. study in the year 2012. We are thankful to ISRO, India Bhuvan portal and USGS data archive, USA for accessing the required data. We are also thankful to ESRI India for their support as MTech fellowship during our study in the NIIT University.

References

Adediji A, Tukur AM, Adepoju KA (2010) Assessment of revised universal soil loss equation (RUSLE) in Kastina State of Nigeria using remote sensing and geographical information system (GIS). Iranica J Energy Environ 1(3):255–264

Bekele W, Drake L (2003) Soil and water conservation decision behavior of subsistence farmers in the eastern highlands of Ethiopia: a case study of the Hunde-Lafto area. Ecol Econ 46:437–451.https://doi.org/10.1016/S0921-8009(03)00166-6

Biswas S, Sudharakar S, Desai VR (1999) Prioritization of sub watersheds based on morphometric analysis of drainage basin: a remote sensing and GIS approach. J Indian Soc Remote Sens 27(3):155–166

Dar IA, Sankar K, Dar MA (2011) Deciphering groundwater potential zones in hard rock terrain using geospatial technology. Environ Monit Assess 173 (1–4):597–610. https://doi.org/10.1007/s10661-010-1407-6

Das A et al (2012) Analysis of drainage morphometry and watershed prioritization in Bandu Watershed, Purulia, West Bengal through remote sensing and GIS technology—a case study. Int J Geomatics Geosci 2 (4):1005–1023. https://citeseerx.ist.psu.edu/viewdoc/download?doi=10.1.1.421.4087&rep=rep1&type=pdf

Ganapuram S, Kumar G, Krishna I, Kahya E, Demirel M (2009) Mapping of groundwater potential zones in the Musi basin using remote sensing and GIS. Adv Eng Softw 40(7):506–518. https://doi.org/10.1016/j.advengsoft.2008.10.001

Garg NK, Sen D (1994) Determination of watershed features for surface runoff models. J Hydraul Eng 120:427–447

Garg NK, Sen DJ (2001) Integrated physically based rainfall-runoff model using FEM. J Hydrol Eng ASCE 6(3):179–188

Hassan Q, Garg NK (2007) Systems approach for water resources development. Global J Flexible Syst Manage 8:29–43. https://doi.org/10.1007/BF03396531

Jain MK, Das D (2010) Estimation of sediment yield and areas of soil erosion and deposition for watershed prioritization using GIS and remote sensing. Water Resour Manage 24:2091–2112. https://doi.org/10.1007/s11269-009-9540-0

Khosla AN (1949) Water Resource Regions (WPR), digital watershed Atlas of India. http://slusi.dacnet.nic.in/dwainew.html

Kouli M, Soupios P, Vallianatos F (2009) Soil erosion prediction using the Revised Universal Soil Loss Equation (RUSLE) in a GIS framework, Chania, Northwestern Crete, Greece. Environ Geol 57:483–497. https://doi.org/10.1007/s00254-008-1318-9

Morgan RPC (2009) Soil erosion and conservation. John Wiley & Sons.https://svgaos.nl/wpcontent/uploads/2017/02/Morgan_2005_Soil_Erosion_and_Conservation.pdf

Nag SK, Ghosh P (2012) Delineation of groundwater potential zone in Chhatna Block, Bankura District, West Bengal, India using remote sensing and GIS techniques. Environ Earth Sci 70(5):2115–2127. https://doi.org/10.1007/s12665-012-1713-0

Nooka Ratnam K, Srivastava YK, Venkateswara Rao V et al (2005) Check dam positioning by prioritization of micro-watersheds using SYI model and morphometric analysis—remote sensing and GIS perspective. J Indian Soc Remote Sens 33:25. https://doi.org/10.1007/BF02989988

Renard KG, Foster GR, Weesies GA, Porter JP (1991) RUSLE, revised universal soil loss equation. J Soil Water Conserv 46(1):30–33. https://www.jswconline.org/content/46/1/30.short

Samuel JC (1995) Sediment criteria for prioritizing watershed for resource development programmes. Ph.D. thesis, University of Roorkee, Roorkee, India

Singh G, Babu R, Narain P, Bhusan LS, Abrol IP (1992) Soil erosion rates in India. J Soil Water Conserv 47 (1):97–99. https://www.jswconline.org/content/47/1/97.short

Symeonakis E, Drake N (2004) Monitoring desertification and land degradation over sub-Saharan Africa. Int J Remote Sens 25(2):573–592. https://doi.org/10.1080/0143116031000095998

Thakkar AK, Dhiman SD (2007) Morphometric analysis and prioritization of mini watersheds in Mohr watershed, Gujarat using remote sensing and GIS techniques. J Indian Soc Remote Sens 35:313–321. https://doi.org/10.1007/BF02990787

Van Rompaey A, Bazzoffi P, Jones RJA, Montanarella L (2005) Modelling sediment yields in Italian catchments. Geomorphology 65:157–169. https://doi.org/10.1016/j.geomorph.2004.08.006

Wang X, Yin ZY (1998) A comparison of drainage networks derived from digital elevation models at two scales. J Hydrol 210:221–241. https://doi.org/10.1016/S0022-1694(98)00189-9

Wen Y, Khosrowpanah S, Heitz L, Park M (2007) Developing a GIS-based soil erosion potential model for the UGUM watershed. http://www.weriguam.org/docs/reports/117.pdf

Wicks JM, Bathurst JC (1996) SHESED, a physically based, distributed erosion and sediment yield component for the SHE hydrological modelling system. J Hydrol 175:213–238. https://doi.org/10.1016/S0022-1694(96)80012-6

Yitayew M, Pokrzywka SJ, Renard KG (1999) Using GIS for facilitating erosion estimation. Appl Eng Agric 15:295–301. https://elibrary.asabe.org/abstract.asp??JID=3&AID=5780&CID=aeaj1999&v=15&i=4&T=1

19. Estimation of Soil Erosion Using Revised Universal Soil Loss Equation (RUSLE) Model in Subarnarekha River Basin, India

Ujjwal Bhandari and Uttam Mukhopadhyay

Abstract

Soil erosion is utmost problem that emerges from drastic changes in land use and land cover by different degrees of anthropological activities. Soil scientists and policymakers are interested to estimate potential volume and spatial variation of soil erosion for useful planning and conservation purposes in the management of a river basin. The Subarnarekha basin (24,196 km^2: 85°08′–87°32′ E, 21°15′–23°34′ N) extends over the states of Jharkhand, Orissa and West Bengal with the maximum length and width of 297 km and 119 km, respectively. The soil erosion through the RUSLE model has been generated by multiplying all the required input thematic maps of the model in GIS platform. The potential soil erosion and the actual soil erosion maps have been derived by multiplication of the thematic maps in raster calculation through Arc GIS using the R, K, LS, C, P-factor maps, respectively. Based on the analysis using RUSLE, GIS, and remote sensing, average potential soil erosion of Subarnarekha river basin was found to be 195.11 metric tons per hectare per year (Mg or tons ha^{-1} year^{-1}). Obtained average annual soil loss distribution was subsequently grouped into five erosion intensity classes like slight, moderate, high, very high, and severe.

Keywords

Subarnarekha river basin · Soil erosion · RUSLE model · Erosion intensity zones

19.1 Introduction

Plateau fringe areas are highly susceptible to the action of flash floods and soil erosion. Soil erosion is the most vulnerable hazard in these regions. A monitoring has been conducted to reduce the adversity of soil erosion in these regions. This paper demonstrates applicability of remote sensing data with the observational field data in a multi-criteria analytical framework (MCA) to validate the soil erosion susceptibility of the sub-watersheds of the Subarnarekha basin falling in the western part of West Bengal and mainly the part of Jharkhand. In this study, MCA framework was used to define different susceptibilities zones for soil erosion. Further, the derivation from MCA, the basin is categorized into low, medium, high, and very high erosion

U. Bhandari (✉)
Department of Geography, University of Calcutta, 35 B. C. Road, Kolkata 700 019, India
e-mail: ujjwalcosmos@gmail.com

U. Mukhopadhyay
Post Graduate Department of Geography, Vidyasagar College, Block C. L., Sector-II, Bidhan Nagar, Kolkata 700 091, India

susceptibility classes. This model is suitable for the watershed management and taken into consideration as a tool to investigate soil loss in the study area.

Soil erosion is an environmental problem worldwide. Soil erosion has its process throughout the history, in recent times it has intensified (Lal and Stewart 1990; Morgan 2001, 2005). Morgan et al. predict annual soil loss with empirical model. This model has two fundamental characteristics as by the surface runoff soil particles moved and soil particles move downslope by rainfall. Morgan (2001) improved his description of differential erosion rates at different hillslopes. From agricultural land, 75 billion metric tons of soil are removed each year (Singh and Bhusan 1992; Myers 1993). This degradation causes threat to the economy (Pimentel and Kounang 1998). Saha and Pande (1993) identified the risk zones using RS and GIS databases. Bhattacharya (1997) described erosion assessment of Rakti river basin. Jong and Paracchini (1999) gave a model for Mediterranean regions named Soil Erosion Model for Mediterranean Regions. Here multi-temporal landsat TM images are used to know the properties. Topographical factors derived from DEM in GIS Platform. Finally soil moisture storage capacity and soil detachability index he calculated as an indicator of soil erosion. Several researchers examined the RUSLE model for potential soil conservation planning in GIS environment. These RS-GIS techniques are very reliable and applicable worldwide nowadays. Morgan (2001) to predict soil loss from water gave a Morgan-Morgan-Finney revised model. He includes other factors like detachment of soil particles from rain splash erosion. Jain and Varghese (2001) noted soil erosion in Himalayan watershed to understand Universal Soil Loss Equation (USLE) model. USLE gives the fluctuated rate of soil erosion. So, the Morgan model is suitable for hilly terrains (Sinha and Joshi 2012). Ouyang and Bartholic (2001) web-based RUSLE model was used to predict soil erosion. Bhusan et al. (2002) based on land resource evaluation conducted a soil erosion study. Evans (2002) described that field-based model is easier than other RS and GIS experiments. Singh et al. (2002) carried out the watershed experiment for sediment outflow in different rainfall patterns. They concluded agricultural regions are higher prone to soil erosion. Svorin (2003) used a comparative study between three models of soil erosion namely USLE, RUSLE, and SLEMSA. Amore and Santoro (2004) use Water Erosion Prediction Project (WEPP) model and volume of deposited sediments by USLE model. Li et al. (2004) carried out Revised Universal Soil Loss Equation (RUSLE) model in RS and GIS. From the spectral mixing of landsat ETM+, they developed soil erodibility maps and produces five classes very low to high. Tripathi and Singh (2004) examined Soil and Water Assessment Tool (SWAT) model to understand the relationship between runoff and sediment yield of different agricultural practices. They consider period of 18 years rainfall database to detect the susceptible soil erosion zones. Plateau fringes like Purulia and Jharkhand district face hazards like land collapse and flash floods. Bie (2005) indicates monitoring of flow surface for surveyed in the rainy seasons of Kenya. He considered soil loss is directly connected with surface soil, land cover, crop management, slope, etc. Stroosnijder (2005) reviewed that soil erosion prediction is various multi-criteria-based erosion prediction models. Chakraborty et al. (2005) described spatial modeling to predict soil erosion. They used USDA SCS curve number method for defining LULC in the platform of ARC/INFO-GIS environment. Human activities such as deforestation, land-use changes are associated with rate of soil erosion (Li and Liu 2004; Pradhan 2010a; Altaf and Romshoo 2014). Topography, landcover, amount and intensity of rainfall, and physicochemical properties of soil are the variables to predict the erosion (Christopherson 1997; Romshoo et al. 2012; Kavian et al. 2011). Watershed management involves proper utilization of land, water, forest, and soil resources (Singh and Kaur 1989; Biswas et al. 1999; Narayan and Bhusan 2002; Rashid et al. 2011; Meraj et al. 2017). Moreover, satellite-based remote sensing coupled with GIS provides a platform for quick and efficient watershed management.

The Subarnarekha basin (18,951 km^2: 85°08'–87°32' E, 21°15'–23°34' N) covers the states of

Jharkhand, Orissa, and West Bengal with the maximum length and width of 297 km and 119 km, respectively. The basin is bounded by the Chhotanagpur Plateau in the north and west, ridges spreading from the Baitarini basin in the south, the Kangsabati water divide in the east, and the Bay of Bengal in the southeast. The Subarnarekha and the Burhabalang are the major rivers of the basin. The Subarnarekha (Fig. 19.1) rises near Nagri village in the Ranchi district of Jharkhand at an elevation of 600 m and flows for 395 km before falling into the sea. Its principal tributaries joining from the right are the Rarhu, the Kanchi, the Kharkai, and the Hill but the Karru, the Sobha, the Garru and the Dulung join from the left.

Nowadays hyperspectral images are accessible and the other thing is that the previous studies were carried out in different areas but not in the flash flood areas and the last one is previous studies have lesser opportunity to validate with present condition of a study area.

19.2 Database and Methodology

This study mainly focused to enhance the use of satellite-based remote sensing data associated with the observational field data in a multi-criteria analytical (MCA) framework to estimate the soil erosion susceptibility of the watershed. For this purpose SOI (Survey of India) Toposheets, LANDSAT TM, and SRTM DEM data (30 m resolution) were used in GIS environment. SRTM data used in ARC GIS to delineate watershed, georeferencing, and mosaicing of toposheets were done in TNT Mips 2014 and ARC GIS 10.1 version software to subset the study area.

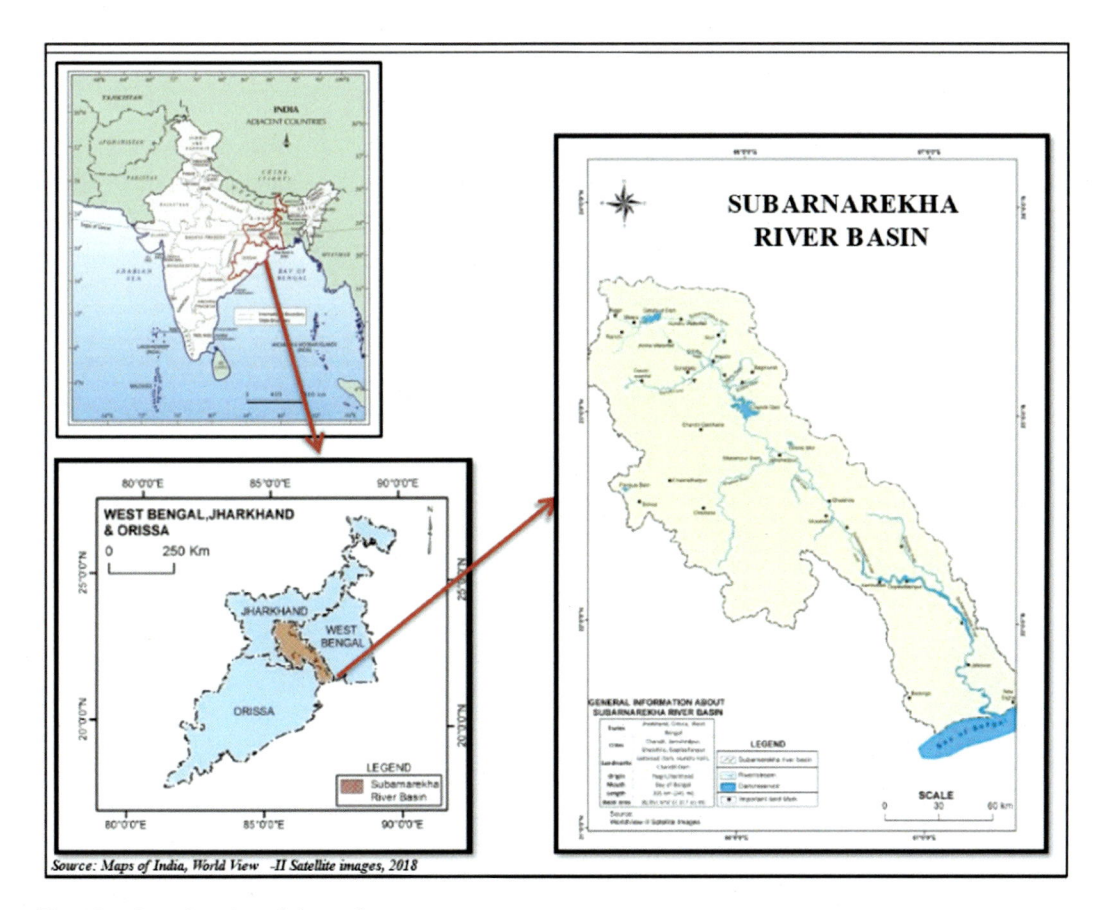

Fig. 19.1 Location map of the study area

The primary equation of Revised Universal Soil Loss Equation (RUSLE) method (Fig. 19.2) for predicting annual soil loss is:

$$A = R * K * L * S * C * P \qquad (19.1)$$

where A is the average annual soil loss per unit area (tons/ha^{-1}); R is the rainfall-runoff erosivity factor (MJ mmha^{-1} h^{-1}); K is the soil erodibility factor (ton ha h MJ^{-1} mm^{-1}); L is the slope length factor; S is the slope steepness factor; C is the cover and management factor; P is the support and conservation practices factor.

19.2.1 Rainfall Erosivity Factor (R)

The rainfall erosivity factor (R) works as the force for sheet and rill erosion. The higher rate of rainfall eroded the soil particles with high runoff flow. Heavy storm affected sheet or rill erosion. The rainfall erosivity data has been collected from the MetaData (www.indiawaterportal.org). 1901–2002, i.e., 101 years of daily rainfall data of ten rainfall gauging stations of the Subarnarekha river basin are used for the rainfall erosivity map of the study area, Renard and Freimund (1994) estimated the equation for R-factor Pandey and Mal

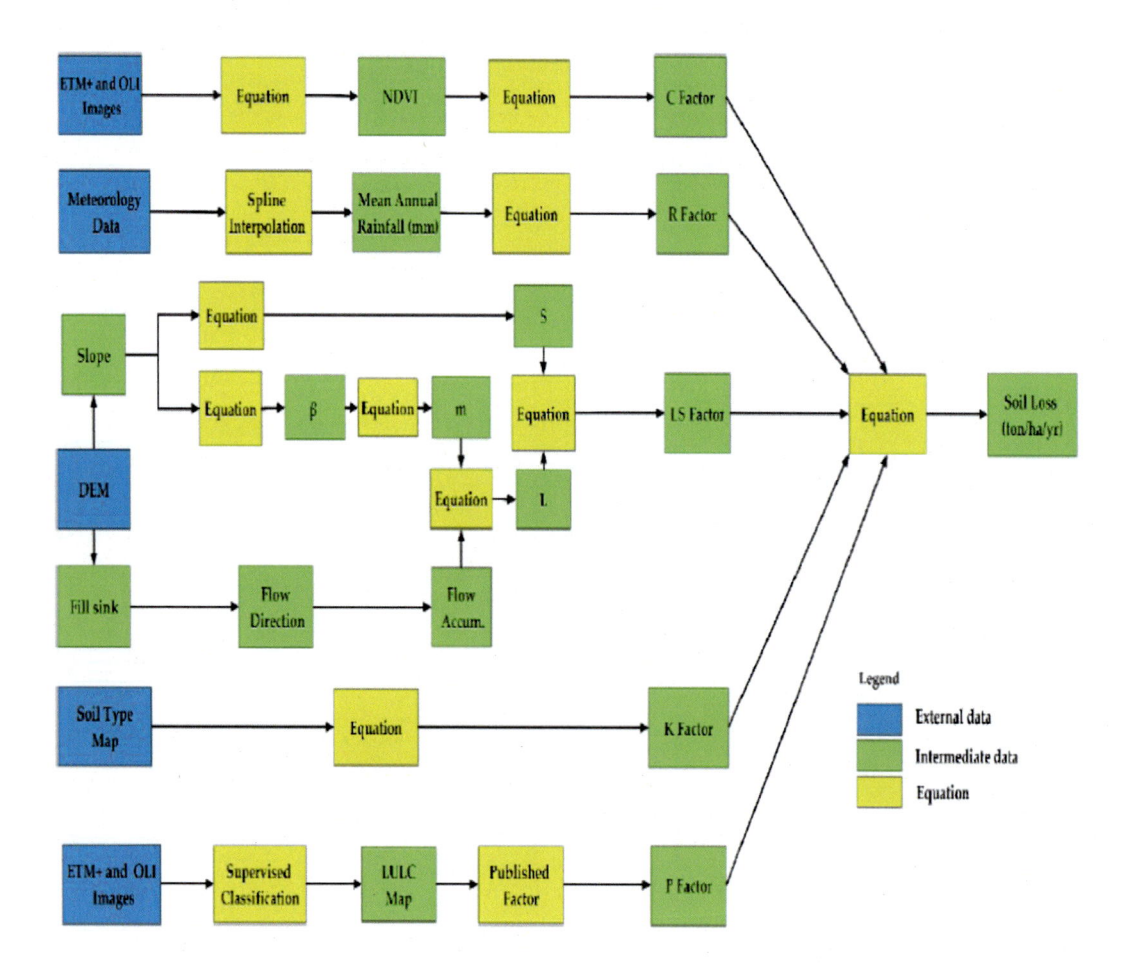

Fig. 19.2 Flow diagram of the RUSLE model developed in the Arc GIS software to estimate soil loss

(2007) also used the equation. Beskow et al. (2009) used the average monthly erosivity Eli equation (Eq. 19.2) is as follows:

$$\text{El}_i = \frac{125.92 \times \left(\frac{r^2}{P}\right)^{0.603} + 111.173 \times \left(\frac{r^2}{P}\right)^{0.691} + 68.73 \times \left(\frac{r^2}{P}\right)^{0.841}}{3}$$
(19.2)

where El$_i$ is the erosivity (MJ mm ha^{-1} h^{-1}) for the month of i (average monthly); r is the rainfall (mm) for the month of i (average monthly); and P is the mean annual precipitation (mm). The spatial distribution of R-factor has been obtained using the Kriging method in ArcGIS (version10.1) software.

19.2.2 Soil Erodibility Factor (K)

Soil erodibility (K) describes the intricint susceptibility of the soil to erosion and it depends on the attributes of the soil (Perez-Rodriguez et al. 2007). The amount of soil loss (per unit) by rainfall implicit the K factor (Brady and Weil 2007). For estimating soil erodibility factor collection of soil data and assigning the values of K factor from the field is time-consuming. Thus, the soil data have been collected from National Bureau of Soil Survey and Land-Use Planning (NBSSLUP) at 1:50,000 scale and from National Atlas and Thematic Mapping Organization (NATMO). The districts that are included in the Subarnarekha river basin have been taken into consideration for the study. The soil erodibility factor map (K) was prepared based on different soil types, textures, and organic matter of the soils. Texture of the soil has been provided by United States Department of Agriculture (USDA). K values have been assigned after Das (2008).

19.2.3 Topographic Factor (LS)

Topography has a major role in erosion. The topographic factor (LS) describes the length and steepness of the slope that exaggerated the surface runoff speed (Beskow et al. 2009). The topographic factor includes the slope length factors (L) and slope steepness factors (S). The LS-factor has been generated using the SRTM DEM 2011 (30 m resolution) that has been collected from USGS earth explorer. From the SRTM DEM, the topographic LS-factor of slope has been calculated using ArcGIS (10.1 version). McCool and Foster (1987) equation has been used to calculate slope length factor. If k is the horizontal projection of the slope length (in meter), then slope length factor (L factor) (Eq. 19.3) is given as:

$$L = \left(\frac{\lambda}{22.1}\right)^m$$
(19.3)

where L is the slope length factor; k is the contributing slope length (m); m is the variable slope length exponent changing with slope steepness. Raindrop impact is related to "m". Pandey and Mal (2007) considered slope steepness.

The values of slope length exponent carried out as 0.3 for slopes less than 3%, 4 for slope of 4%, and 5 for slope that is greater than 5%. The slope steepness factor (S factor) is evaluated using equations (Eqs. 19.4.1, 19.4.2) given by McCool and Foster (1987).

$$S = 10.8 \sin \theta + 0.03 \text{ when } s \, 9\% \quad (19.4.1)$$

$$S = 16.8 \sin \theta - 0.50 \text{ when } s \geq 9\% \quad (19.4.2)$$

where S is the slope steepness factor and h is the slope angle. The L and S factor layer has been generated by the multiplication of both L and S factors in Arc GIS.

19.2.4 Crop Management (C) and Support Practice (P) Factor

The C and P factors have vivid role in the conservation practices. The cover management factor (C) depends on the land use. Pandey and Mal (2007) show the ratio of soil loss from a cropped land in a particular condition to the soil loss in the continuous tilled fallow. The C and P-factor values are assigned based on land use and land cover of the river basin. Satellite images have been used for the classification of land use and land cover. From the satellite image, the basin area has been demarcated and classified into six land-use categories with supervised classification (Maximum Likelihood Classifier algorithm) using image processing software ERDAS IMAGINE (Joshi and Nagare 2009) The support practice factor (P) has been calculated based on the cultivation method by interlinking between terracing and slope in the field areas.

The soil erosion through the RUSLE model has been generated by multiplying all the required input of the model in GIS platform. Five input raster files namely the R-factor, K factor, LS-factor, C-factor, and P-factor maps are used for multiplication of the thematic maps in raster calculation in ArcGIS.

19.3 Result and Discussion

19.3.1 Rainfall Erosivity Factor (R)

Rainfall erosivity factor (Fig. 19.3) is difficult to determine directly. The average annual rainfall of catchment area is 1863 mm. Average R-factor value ranged from 1294.43 to 1527.05 MJ mm h^{-1} ha^{-1} year^{-1} with average of 1410.74 MJ mm h^{-1} ha^{-1} year^{-1}. Spatial distribution of R-factor value of basin area was produced in the form of a rainfall erosivity factor map (Fig. 19.3).

19.3.2 Soil Erodibility Factor (K)

Soil erodibility measures a soil resistance to the erosive powers of rainfall energy and runoff. K factor (Fig. 19.4) represents the susceptibility of a soil type to erosion, i.e., it is the reciprocal of soil opposition to erosion (Hudson 1982). It is highly related to soil's mechanical composition and organic matter content. Soil erodibility value derived from this basin varies from 0.07 to 0.93 Mg h MJ^{-1} mm^{-1} and average value of 0.50 Mg h MJ^{-1} mm^{-1} (Fig. 19.4).

19.3.3 Topographic Factor (LS)

In RUSLE, the effect of topography on soil loss is represented by L (Fig. 5a) and S (Fig. 5b) factors where S factor reflects the change in potential erosion with change in slope, and L factor reflects increasing potential erosion due to surface runoff. The upslope catchment area for each cell in a DEM was computed with number of steps. An LS-factor value in the study basin varies from 0 to 472.587.

19.3.4 Cover Management Factor (C)

Cover management (Fig. 19.6) factor (C-factor) describes conditions that can be managed most easily to reduce erosion (Renard et al. 1991, 1997). Three land cover and land-use classes were documented in the basin, predominated by bare land (71.13%) and forest (27.21%). The C-factor value in Subarnarekha basin varies from 0 to 0.45, with a relatively low mean value of 0.03. In order to improve data, further research is essential with satellite images of enhanced geometric and spatial characteristics as well as real field investigations to better correlate the C-factor with remote sensing data.

19.3.5 Support Practice (P) Factor

Conservation management (Fig. 19.7) factor (P-factor) refers to how surface conditions affect flow paths and flow hydraulics (Renard et al. 1991). The P-factor value varies from 0 to 1, and the mean value is 0.93, where 0 reflects the regions without erosion and 1 reflects the region which needs maximum care for conservation measures.

19.3.6 Spatial Distribution of Soil Erosion

Based on the analysis using RUSLE, GIS, and remote sensing, average potential soil erosion of Subarnarekha river basin was found to be

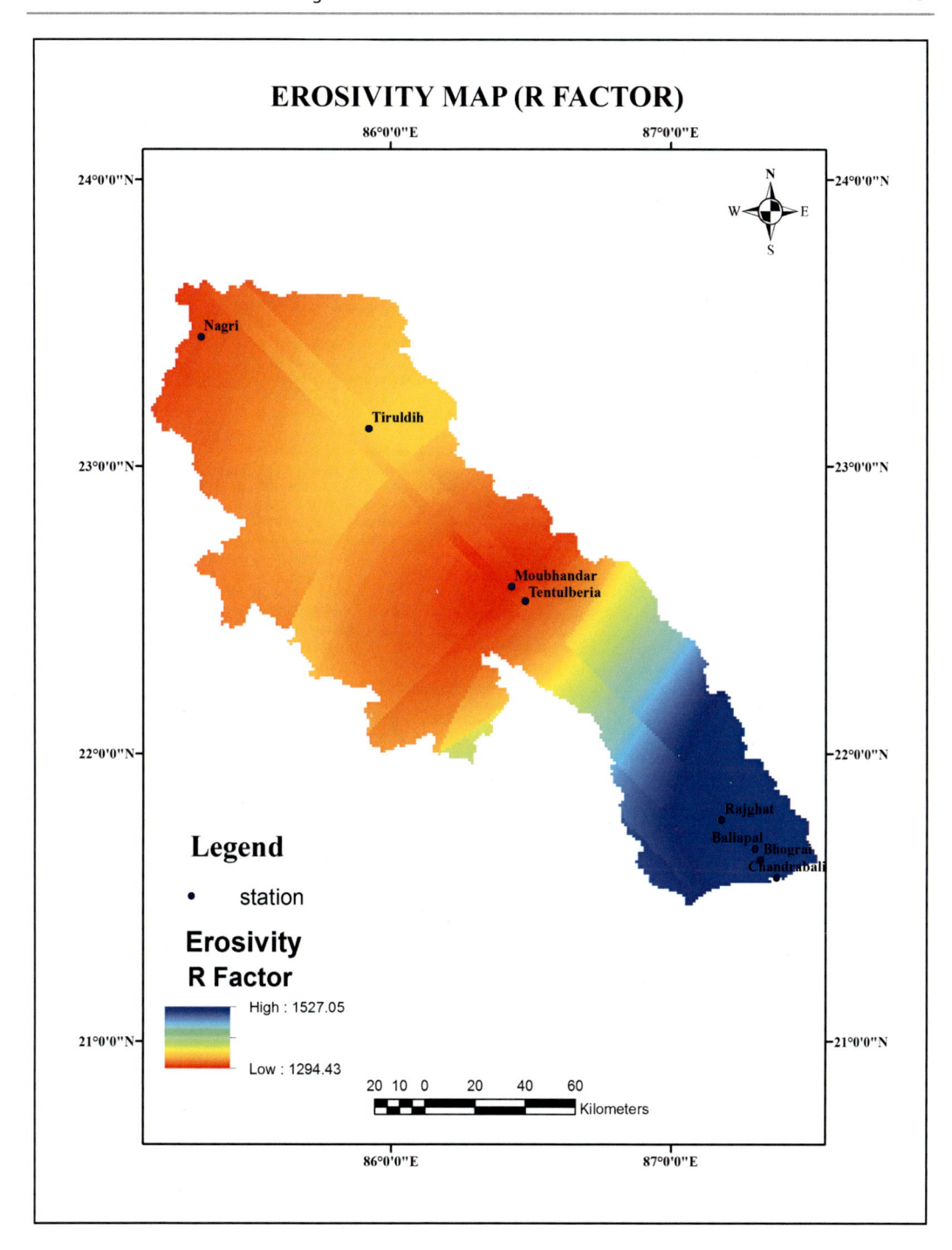

Fig. 19.3 Erosivity factor map of Subarnarekha river basin (*R*)

78.11 metric tons per hectare per year (Mg or tons ha^{-1} year^{-1}). Obtained average annual soil loss distribution was subsequently grouped into (Table 19.1) five erosion intensity classes (slight, moderate, high, very high, severe) as per guideline of Singh et al. (1992). The result from soil erosion analysis reveals that about 41% area of the basin is under category of

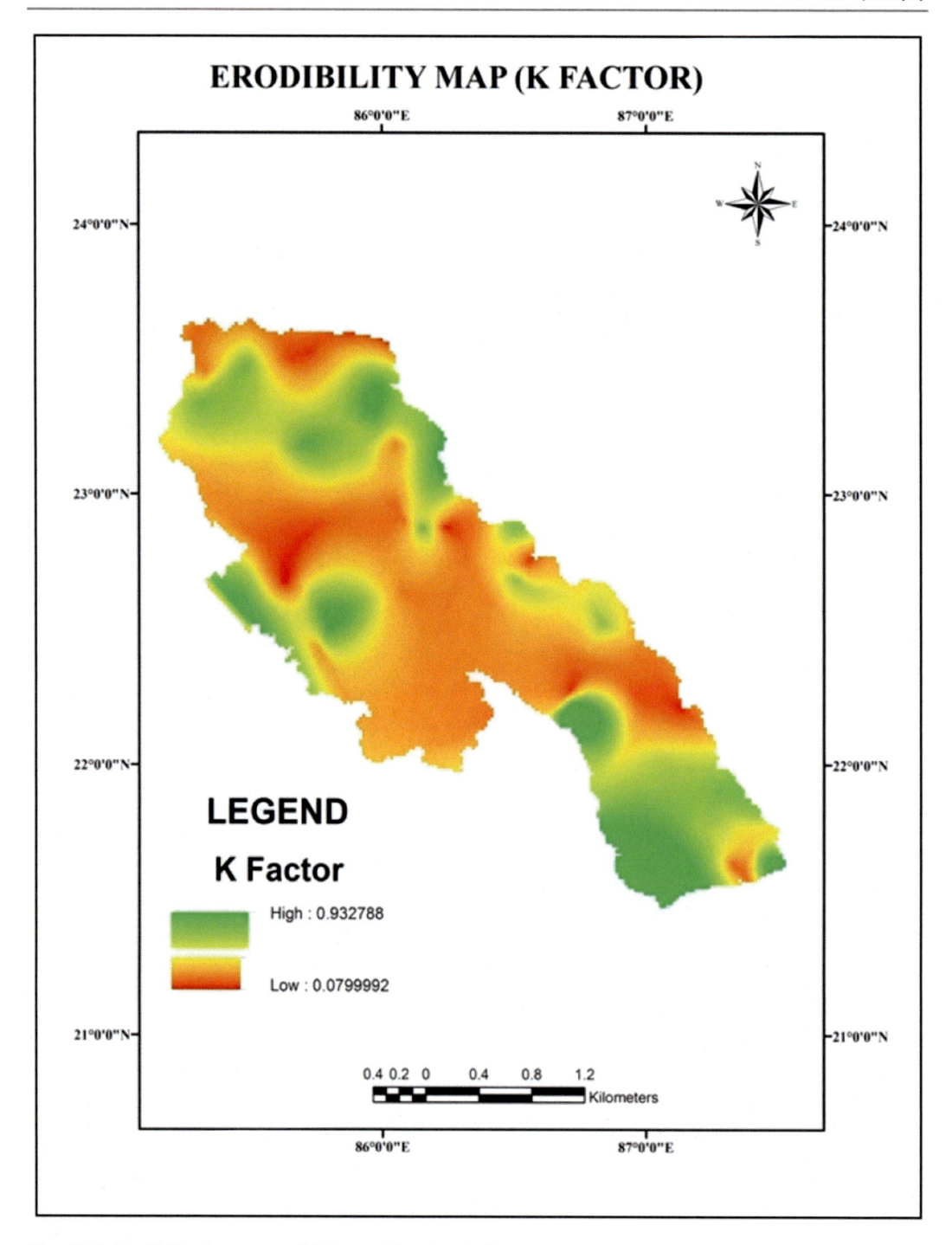

Fig. 19.4 Erodibility factor map of Subarnarekha river basin (*K*)

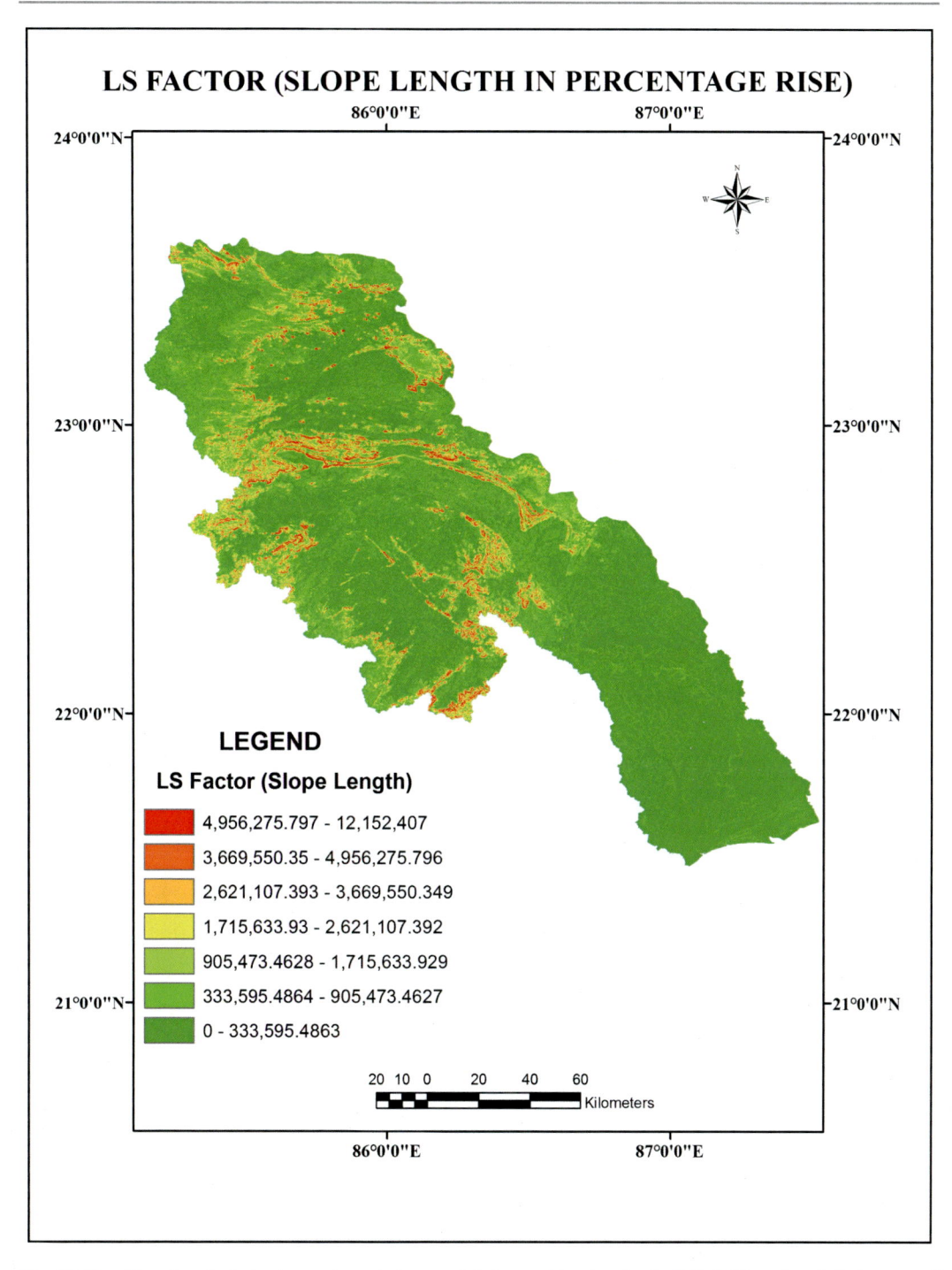

Fig. 19.5 **a** Slope length factor (percentage rise) map of Subarnarekha river basin (LS). **b** Slope length factor map of Subarnarekha river basin (LS)

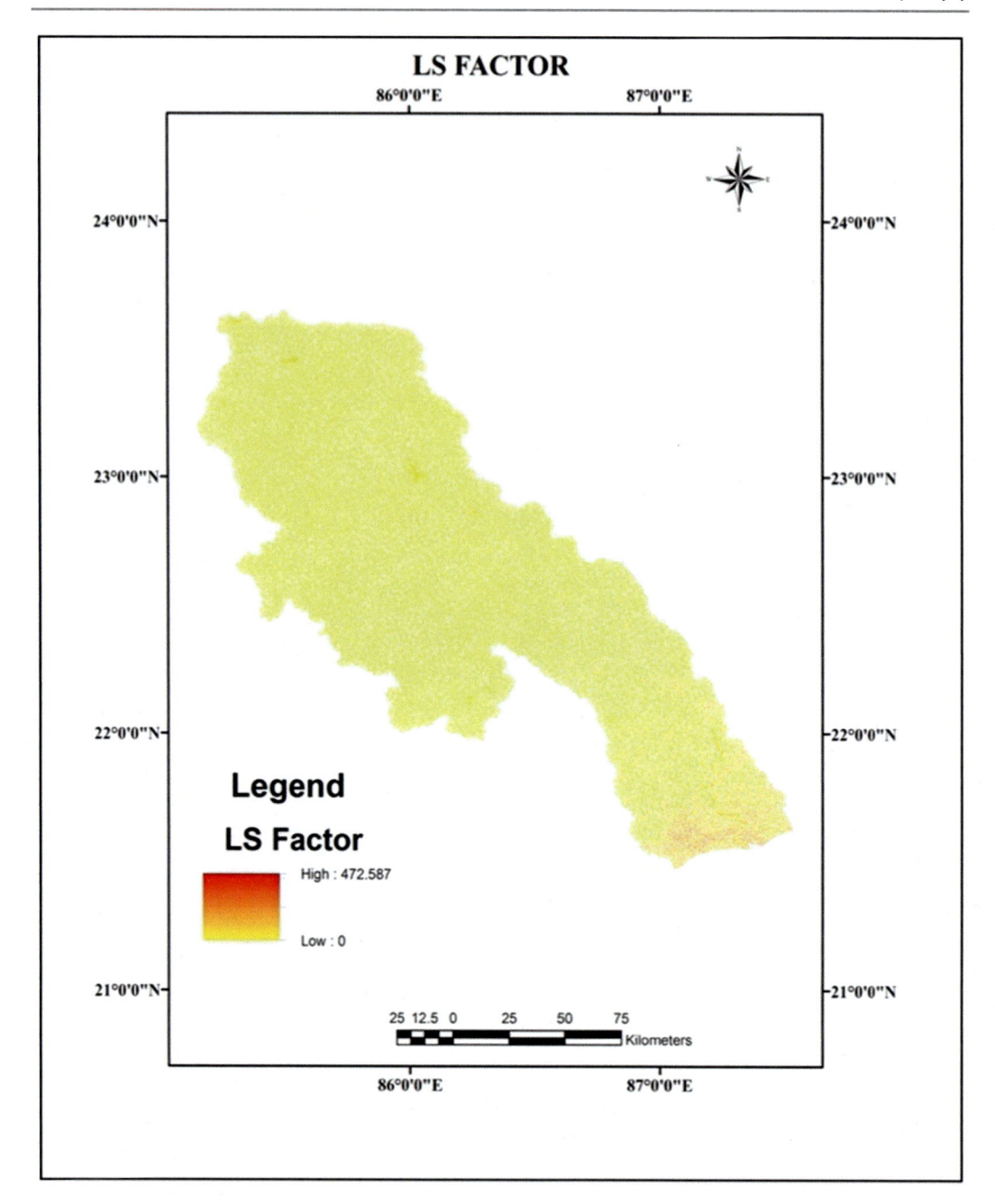

Fig. 19.5 (continued)

slight erosion and about 58% of area is under category of high to severe erosion (Fig. 19.8). The area under the range of high to severe soil erosion potential needs immediate conservation measures. Since, the spatial distribution map of soil erosion as displayed below, indicates that the soil erosion of high to severe classes is scattered all over the basin area, therefore conservation measures should be carried out in the whole basin.

The quantitative approach by the RUSLE model has shown that the average annual loss in

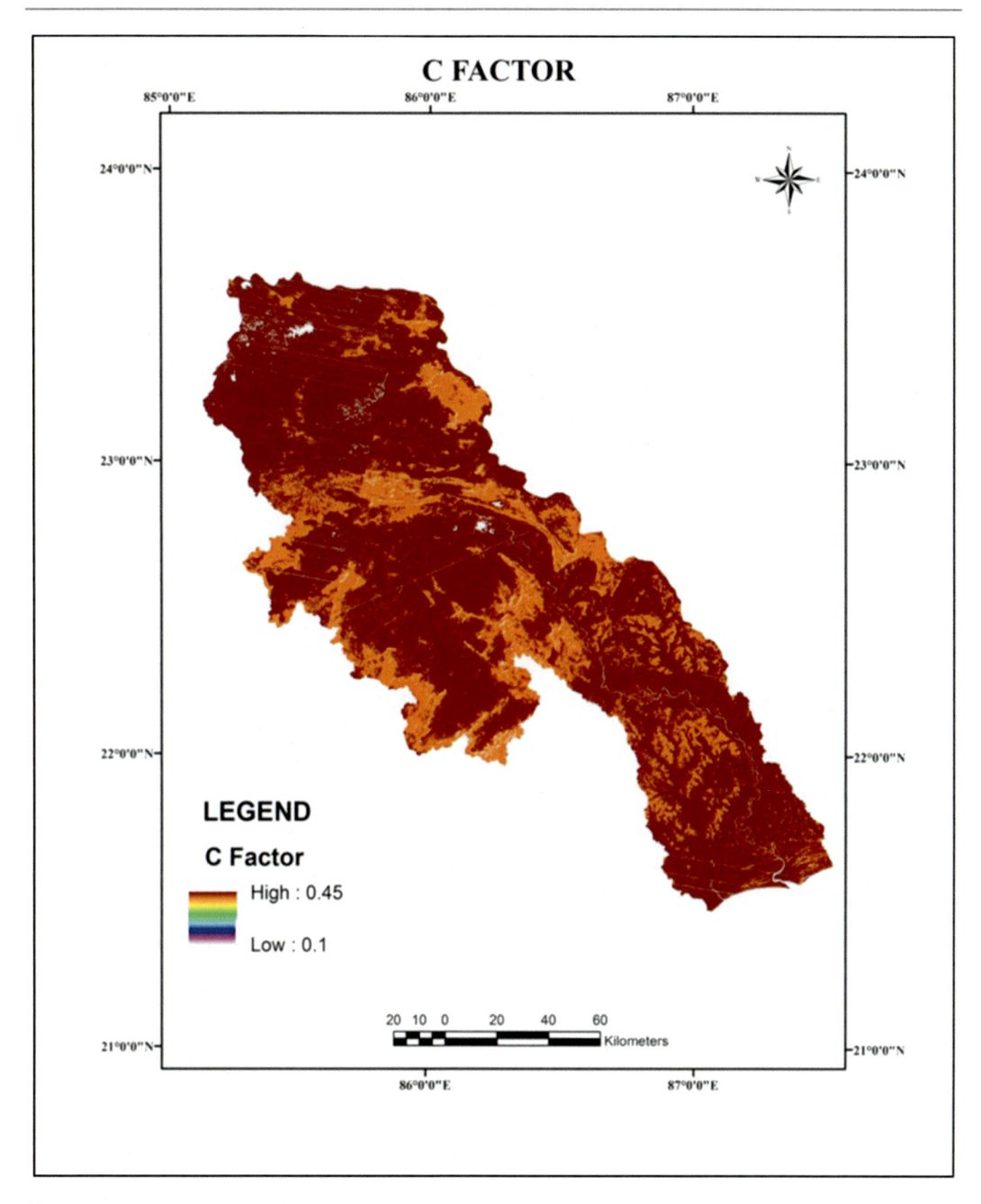

Fig. 19.6 Cover management factor map of Subarnarekha river basin (*C*)

the evaluated watershed is nearly 3500 tons ha^{-1} year^{-1} (Fig. 19.9). This soil loss quantity would be favored by the erosion factors controlling soil loss, such as rainfall and slope factor (nearly 60% of the total surface) and the other factors triggered continuously. From both the observation, it has been noticed that the main river channel, as well as confluence areas, have highest soil erosion susceptibility. The adopted method allowed us to generate soil erosion susceptibility maps that reflect the spatial distribution of gullies in the study area. The functional

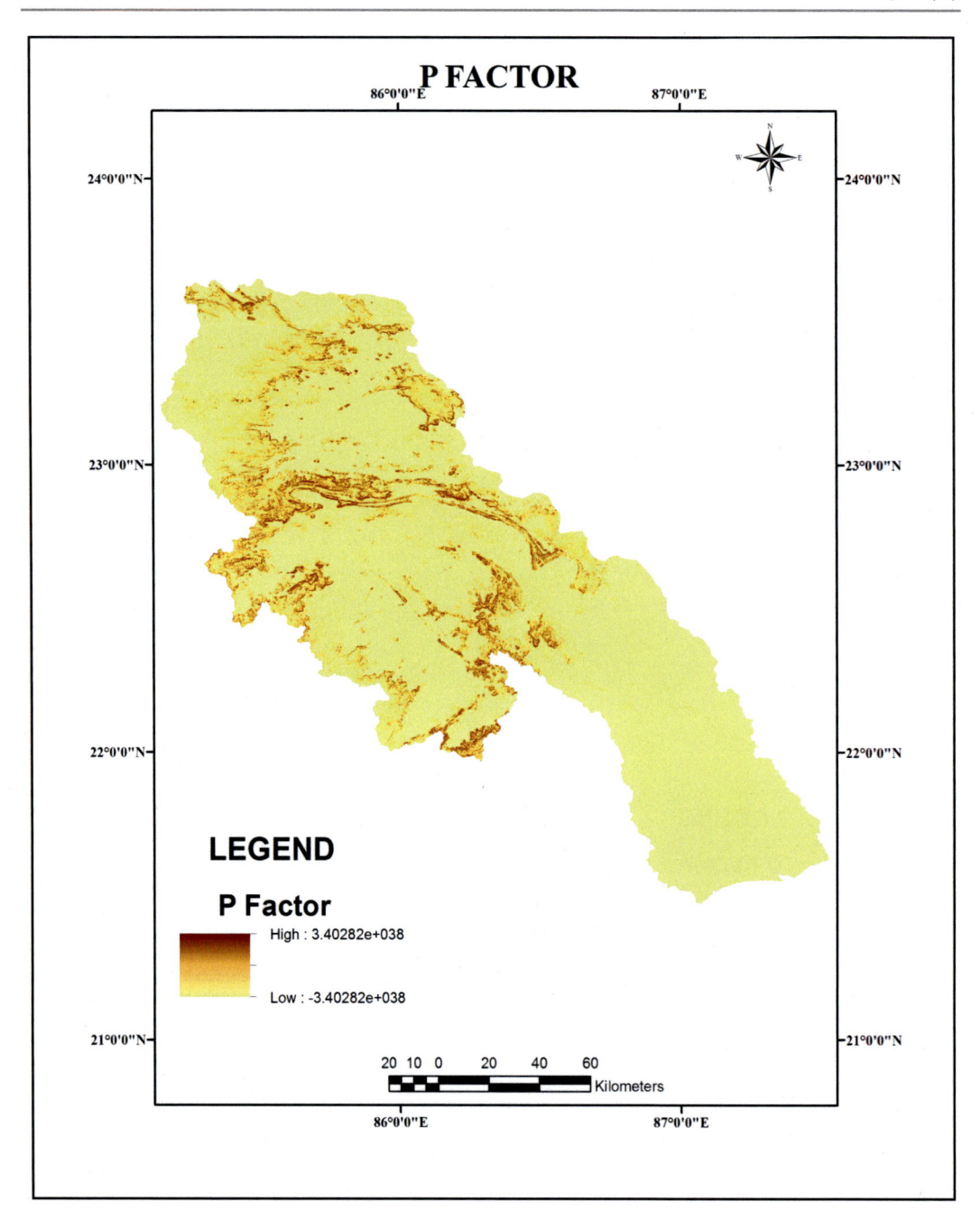

Fig. 19.7 Conservation management factor map of Subarnarekha river basin (*P*)

relationships between erosion processes and a set of environmental attributes have been assessed by means of forward stepwise ROC curve (Fig. 19.10) using three different training samples.

19.3.7 Accuracy Results

The adopted method allowed us to generate gully erosion susceptibility maps that well reflect the spatial distribution of gullies within the study

Table 19.1 Susceptible erosion zones with ranges (RUSLE)

Class	Rate of erosion (t/ha/yr)	Area (km²)	Severity index	Percentage of area
1	0–555.78	16,108.35	Slight	83.15
2	555.78–2037.85	1705.59	Moderate	9.08
3	2037.85–4538.84	568.53	High	3.78
4	4538.84–10,281.85	379.02	Very high	2.45
5	10,281.85–23,620.48	189.51	Severe	1.54

Fig. 19.8 Relation between soil loss and percentage of area

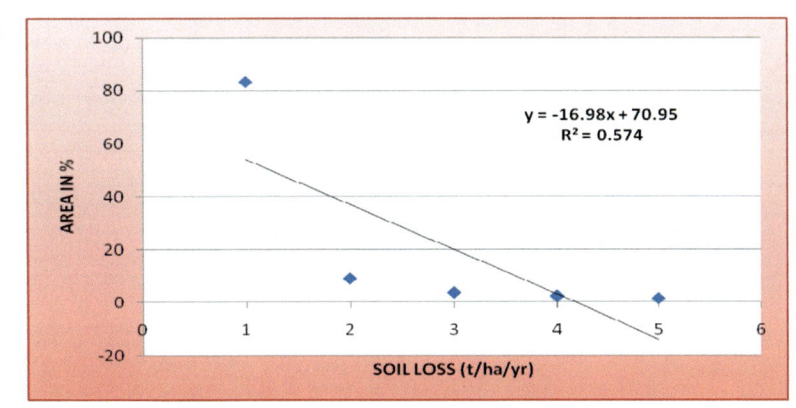

area. The functional relationships between erosion processes and a set of environmental attributes have been assessed by means of forwarding stepwise logistic regression using three different training samples of the CLUs and SLUs, and the predictive skill of the obtained susceptibility models was tested. The overall accuracy of the gully erosion susceptibility models, evaluated in terms of ROC curves and AUC values (0.830 and 0.819) is from acceptable to excellent.

This research highlights that the SLUs were adopted as sample units for analyzing gully erosion susceptibility. The use of such mapping units may overcome the intrinsic limits of purely statistical approaches to a geomorphological issue. SLUs could provide better terrain boundaries that coincide with natural limits of runoff. However, SLUs may reduce the resolution of topography with values concentrated around their mean compared to CLUs. This problem may be solved by reducing the size of the SLUs. To accurately assess erosion of the study area, the erosion model Universal soil loss equation (USLE) has been calculated to note how relevant the composite and multi-criteria-based erosion susceptibility index. For that purpose considered to be a contemporary, simple, and widely used approach to soil erosion assessment. The USLE approach is compatible with GIS environment, which has been applied for soil risk erosion assessment.

19.4 Conclusion

To accurately assess erosion in the study area, the erosion model Revised Universal Soil Loss Equation (RUSLE) has been calculated to note how much erosion is prevalent in the study area. For that purpose considered to be a contemporary, simple, and widely used approach to soil erosion assessment. The RUSLE approach is compatible with GIS environment, which has been applied for soil risk erosion assessment. The method is based on a learning process from a series of points observed in situ for which the erosion level (low, medium, or high), slope gradient, annual precipitation, land use, and drainage density shall all be known. Some of the

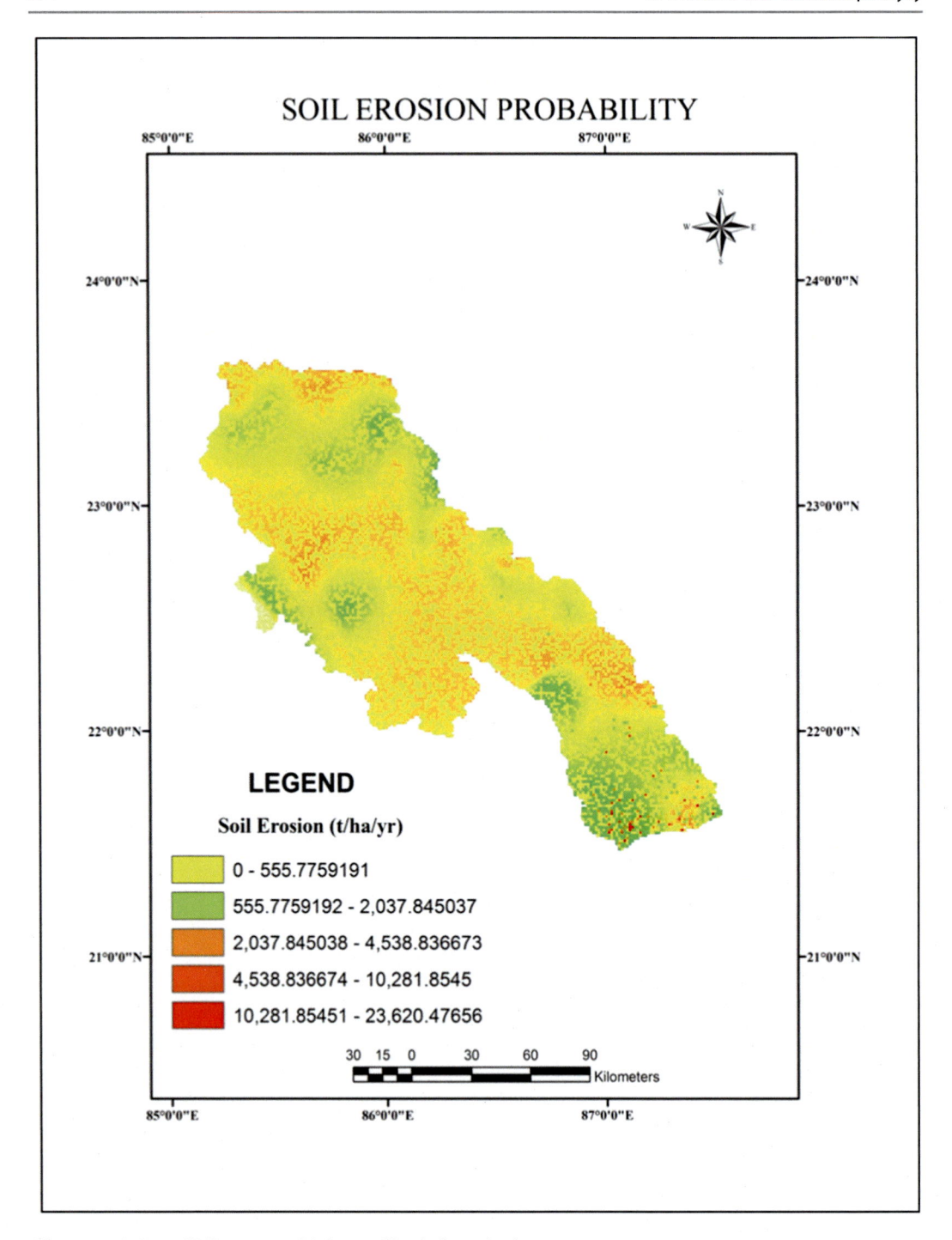

Fig. 19.9 Soil erodibility zones of Subarnarekha drainage basin

model parameters are continuous (e.g., slope and precipitation), while others are discontinuous (e.g., land use and cover).

Based on the analysis using RUSLE, GIS, and remote sensing, average potential soil erosion of Subarnarekha river basin was found to be 195.11

Fig. 19.10 Training and validation dataset by ROC curves and AUC values of the CLU-based **a** and SLU-based **b** soil erosion probability map

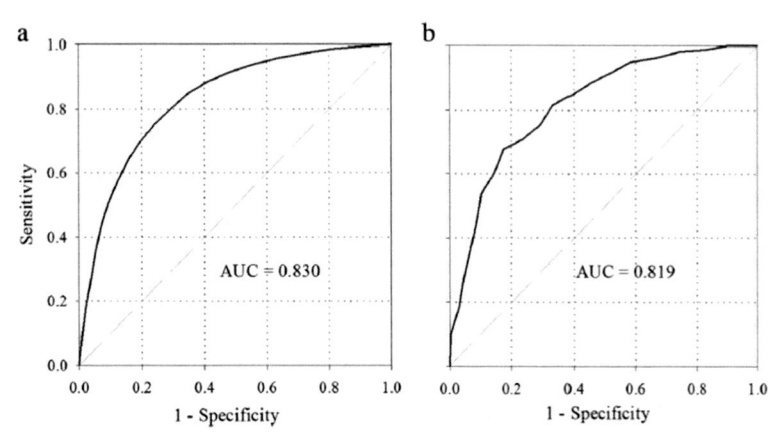

metric tons per hectare per year (Mg or tons ha^{-1} year^{-1}). Obtained average annual soil loss distribution was subsequently grouped into five erosion intensity classes (slight, moderate, high, very high, severe) as per guidelines by Singh et al. (1992). The result from soil erosion analysis reveals that about 41% area of the basin is under category of slight erosion and about 58% of area is under category of high to severe erosion. The area under the range of high to severe soil erosion potential needs immediate conservation measures. Since the spatial distribution map of soil erosion as displayed (Fig. 19.9) indicates that the soil erosion of high to severe classes is scattered all over the basin area. Therefore conservation measures should be carried out over whole basin extent.

Acknowledgements Assistance from Mrs. Mousumi Roy and Mr. Santanu Payra during fieldwork was gratefully acknowledged. Infrastructural facilities were provided by the Department of Geography, Syamaprasad College (University of Calcutta). Thanks are due to Professor Dr. Sunando Bandyopadhyay, Department of Geography, University of Calcutta, and Dr. Sandip Kr. Das. Comments from the anonymous reviewers are gratefully acknowledged.

References

Altaf S, Romshoo SA (2014) Morphometry and land cover based multi-criteria analysis for assessing the soil erosion susceptibility of the western Himalayan watershed. Environ Monit Assess 186(12):8391–8412

Amore E, Santoro VC (2004) Scale effect in USLE and WEPP application for soil erosion computation from three Sicilian basins. J Hydrol 293(1–4):100–114

Beskow S, Mello CR, Norton LD, Curi N, Viola MR, Avanzi JC (2009) Soil erosion prediction in the Grande River Basin, Brazil using distributed modeling. CATENA 79:49–59

Bhattacharya SK (1997) Erosion assessment of the Rakti river basin in the Darjeeling Himalaya. Indian J Soil Conserv 25(3):173–176

Bie De (2005) Assessment of soil erosion indicators for maize-basedagro-ecosystems in Kenya. Catena (Giessen) 59(3):231–251

Biswas S, Sudhakar S, Desai VR (1999) Prioritization of sub watershed based on morphometric analysis of drainage basin: a remote sensing and GIS approach. J Indian Soc Remote Sens 22(3):155–167

Brady NC, Weil RR (2007) The nature and properties of soils. Prentice-Hall, Upper Saddle River, New Jersey14

Chakraborty D, Dutta D, Chandrasekharan H (2005) Land use indicator of a watershed in Arid region, Western Rajasthan using RS and GIS. J Indian Soc Remote Sens 29(3)

Christopherson RW (1997) Geosystems: an introduction to physical geography. Upper Saddle River, NJ, Prentice Hall

Das DJ (2008) Identification of critical erosion prone areas for watershed prioritization using GIS and remote sensing. Dissertation, Indian Institute of Technology Roorkee

De Jong SM, Paracchini ML (1999) Regional assessment of soil erosion using the distributed model SEMMED and remotely sensed data. CATENA 37(3–4):291–308

Evans R (2002) An alternative way to assess water erosion of cultivated land field based measurements and analysis of some results. Appl Geogr 22:187–208

Hudson NW (1982) Soil conservation, research and training requirements in developing tropical countries, soil erosion and conservation in the tropics. Am Soc Agron 43:121–143

Jain SK, Varghese J (2001) Estimation of soil erosion for a Himalayan watershed using GIS technique. Water Resource Management 15:41–54

Joshi VU, Nagare V (2009) Land use change detection along the Pravara River Basin in Maharashtra using Remote Sensing and GIS Techniques. AGD Landsc Environ 3:71–86

Kavian A, Azmoodeh A, Solaimani K (2011) Modeling seasonal rainfall erosivity on a regional scale: a case study from Northeastern Iran. Int J Environ Response 5:939–950

Lal R, Stewart BA (1990) Conservation agriculture as an alternative for soil erosion control and crop production in steep slope regions cultivated by small scale farmers in Motozintla Mexico, Principles of sustainable soil management in agro ecosystems, CRC Press

Li SC, Liu Y (2004) Simulation on effects of land use change on soil erosion on Loess Plateau. J Soil Water Conserv 18:74–81

Li YB, Xie DT, Wei CF (2004) Correlation between rock desertification and variationsof soil and surface vegetation in Karst eco-system. Acta Pedologica Sinica 41(2):196–202

Meraj G, Romshoo SA, Ayoub S, Altaf S (2017) Geoinformatics based approach for estimating the sediment yield of the mountainous watersheds in Kashmir Himalaya, India. Geocarto Int. https://doi.org/10.1080/10106049.2017.1333536

Morgan RPC (2001) A simple approach to soil loss prediction: a revised morgan-morgan-finney model. Catena (44):305–322

Morgan RPC (2005) Soil erosion and conservation. Blackwell Publishing, Oxford, UK

McCool DK, Foster GR (1987) Revised slope steepness factor for the USLE. Trans ASAE 30:1387–1396

Myers N (1993) Soil erosion effects on soil productivity: a research perspective. National Soil Erosion–Soil Production Research Planning Committee. Soil Water Conserv 32:82–90

Narayan D, Bhusan LS (2002) Inter-relationship between crop canopy & erosion parameters in alluvial soils. Indian Journal of Soil Conservation 30(1)

Ouyang D, Bartholic J (2001) Web-based GIS application for soil erosion prediction, soil erosion research for the 21st century. American Society of Agricultural and Biological Engineers, St. Joseph, Michigan 260–263

Pandey A, Mal BC (2007) Identification of critical erosion prone areas in the small agricultural watershed using USLE, GIS and remote sensing. Water Resour Manag 21(4):729–746

Pimentel D, Kounang N (1998) Ecology of soil erosion in ecosystems. Ecosystems 1(5):416–426 Springer

Pradhan B (2010a) Remote sensing and GIS-based landslide hazard analysis and cross-validation using multivariate logistic regression model on three test areas in Malaysia. Adv Space Res 45:1244–1256

Rashid M, Lone MA, Romshoo SA (2011) Geospatial tools for assessing land degradation in Budgam district, Kashmir Himalaya. India J Earth Sys Sci 120(3):423-433

Renard K, Freimund JR (1994) Using monthly precipitation data to estimate R-factor in the revised USLE. J Hydrol 157:287–306

Romshoo SA, Bhat SA, Rashid I (2012) Geoinformatics for assessing the geomorphological control on the hydrological response at watershed scale in Upper Indus basin. Earth Syst Sci 121(3):659–686

Saha SK, Pande LM (1993) Integrated approach towards soil erosion inventory for environmental conservation using satellite and agro-meteorological data. Asian Pacific 5(2):21–28

Singh TV, Kaur J (1989) Studies in Himalayan ecology and development strategies. Ecology, New Dehli

Singh G, Bhusan LS (1992) Soil erosion rates in India. Soil Water Conserv 47(1):97–99

Singh RK, Aggarwal SP, Turdukulov U, Hariprasad V (2002) Prioritization of Beta River basin using Remote Sensing & GIS technique. Indian J Soil Conserv 30(3):200–205

Sinha D, Joshi VU (2012) Application of universal soil loss equation (USLE) to recently reclaimed badlands along the Adula and Mahalungi Rivers, Pravara Basin, Maharashtra. J Geol Soc India 80:341–350

Stroosnijder L (2005) Measurement of erosion: is it possible? Catena 64:162–173

Svorin J (2003) A test of three soil erosion models incorporated into a geographical information system. Hydrol Process 17(5):967–977

Tripathi MP, Singh R (2004) Hydrological modelling of a small watershed using generated rainfall in the soil and water assessment tool model. Hydrol Process 18(10):1811–1821

Perez-Rodriguez R, Marques MJ, Bienes R (2007) Spatial variability of the soil erodibility parameters and their relation with the soil map at subgroup level. Sci Total Environ 378(1–2):166–73

Renard K, Foster GR, Weesies GA, Porter JP (1991) RUSLE revised universal soil loss equation. J Soil Water Conserv 46:30–33

Renard KG, Foster GR, Weesies GA, McCool DK, Yoder DC (1997) Predicting soil loss: a guide to conservation planning with the revised soil loss equation (RUSLE); United States. Department of Agriculture, Washington D. C, USA Handbook 703

Singh G, Ram Babu, Narain P, Bhushan LS, Abrol IP (1992) Soil erosion rates in India. J Soil Water Conserv 47(1):97–99

Land Cover Changes in Green Patches and Its Impact on Carbon Sequestration in an Urban System, India

Sunanda Batabyal, Nilanjan Das, Ayan Mondal, Rituparna Banerjee, Sohini Gangopadhyay, and Sudipto Mandal

Abstract

Urbanization is an inevitable process of any city with the pace of technology and increasing population. Green patches of urban areas are the parks, urban agricultural plots, vertical gardens, wetlands, and vegetation that provide a community with different natural resources and values. Apart from the forests and agricultural lands at the fringes of any city, green patches do have importance in terms of ecosystem services. The present study aims to quantify the impact of urbanization in the Burdwan Municipality area (West Bengal, India) and loss of stored carbon in urban system. Here, remote sensing and GIS platform was applied for the period of last 15 years (2001–2015). Google Earth images were referred to carry out the change detection study. Validation of the change was done by field survey as per the change shown in Google Earth platform for the study area. The study unveiled that the vegetation cover within the Burdwan municipality decreased by 78.84% that corresponds to the total loss of carbon in the tune of $1.79E \pm 08$ kg C.

S. Batabyal · N. Das · A. Mondal · R. Banerjee · S. Gangopadhyay · S. Mandal (✉)
Ecology and Environmental Modelling Laboratory, Department of Environmental Science, The University of Burdwan, Burdwan 713104, India
e-mail: smandal@envsc.buruniv.ac.in

Keywords

Carbon sequestration · Climate change · Remote sensing · Urbanization · Vegetation loss

20.1 Introduction

A change in the land cover signifies certain continuous characteristics of the land, such as the form of vegetation, soil resources, and so on. This change in land use is a change in the way people use or manage the land for urban development, and for the developing and underdeveloped countries, it is significant (Pataki et al. 2011). It is noteworthy that this change is responsible for a number of local and global impacts, including the loss of biodiversity and its related effects on human health, and the loss of habitat and ecosystem services (Parker 2009). Natural causes may lead to a change in land cover but land use changes always require human intervention (Joshi et al. 2017). Study on land use changes helps managers to better understand how tree cover and tree populations are changing.

Global warming is happening because human activities are pumping large amounts of greenhouse gases into the atmosphere, day after day, hour after hour. Carbon or carbon dioxide (CO_2) storage, somewhere other than the atmosphere is one way to reduce emissions. There are two extremely distinct ways of carbon sequestration: biological and geological. In this chapter, biological sequestration, which is amongst the largest of the "low hanging fruits" for rapid, substantial emission cuts, will be emphasized. Many scientists are interested in the topic of carbon (C) sequestration, as urbanization and deforestation in modern times are a major concern. With ambient carbon dioxide (CO_2) rising in the recent years, there has been a growing interest in understanding the amount of carbon contained and sequestered by urban forests. Urbanization and population explosion are demanding more land to be developed at the cost of greenery.

Atmospheric CO_2 is expected to rise by approximately 2.6 billion metric tonnes (t) annually (Sedjo and Sohngen 2012). Nowak and Crane (2002) suggested that urban trees could minimize CO_2 emissions by cooling ambient air and eliminating annual heating and cooling, as well as help storing carbon directly. In addition, trees could also minimize the use of fossil fuels by transpiration, shading and wind blocking processes (Dai et al. 2013), and improve urban planning, leading to improved human and environmental health (McCarl 2001). The authors categorized CO_2 flux into four scales such as—the country, the town, the organization, and the individual. Nowak et al. (2013) proclaimed that urban cover monitoring can provide simple data useful for modelling air pollution control and CO_2 sequestration. Brack (2002) emphasized on the importance of urban forests in Canberra, Australia with particular regard to mitigating pollution. The study uses a framework of tree inventory, modelling, and decision support built to collect and use tree asset management data about trees.

In India, the ability of urban forests was studied to mitigate CO_2 in the Pune city (Howard et al. 2004). The study showed that *Dalbergia melanoxylon* and *Gliricidi sepium* are the dominant species in terms of carbon sequestration according to application of geographical information system (GIS) studies. Little attention has been paid to evaluating the impact of tree cuttings and the role of avenues in the carbon cycle during the current expansion of roads in the study. In addition, the paper discussed the effect of developmental projects and the role of pathways in estimating carbon. Another study by Mujumdar et al. (2016) onurban gardens of the same city showed that *Peltophorum pterocarpum* (Dc.) Baker, and *Acacia longifolia* Wild., are the abundant sequestering the major proportion of carbon in the garden.

The present study attempts to quantify the role of urban area in sequestering carbon. The study was conducted in the Burdwan municipality area (Fig. 20.1). It is located in the south-central part of Burdwan district, 107 km north–west to Kolkata Metropolitan Area. It is well connected to adjoining districts; Asansol-Durgapur industrial complex to north–west, Kolkata Metropolitan Area having a large consumer base in the south–east direction, Suri, Katwa, Kalna, Bankura, and Arambag (Upcoming food processing centre) in their respective districts through rail and road links.

20.2 Assessment of Land Cover Change

The land cover change study has been carried using scanning method, survey method, and geoinformatics.

20.2.1 Scanning Method

Scanning is an effective and very comprehensive way of assessing the coverage of urban vegetation and fits well with the geographic information system (Lavery et al. 2010). This approach includes social facilities and ortho-rectified images, making this technique very labour-intensive. This technique is essentially the scanning of any region's geographical area, topography, and vegetation cover. With the assistance of GIS (Geographical Information System) and remote sensing, scanning is mostly done. GIS is a technique designed to collect, store, manipulate, evaluate, handle, and present geographical or spatial data. Digitization is the most common form of data creation where, with the use of a

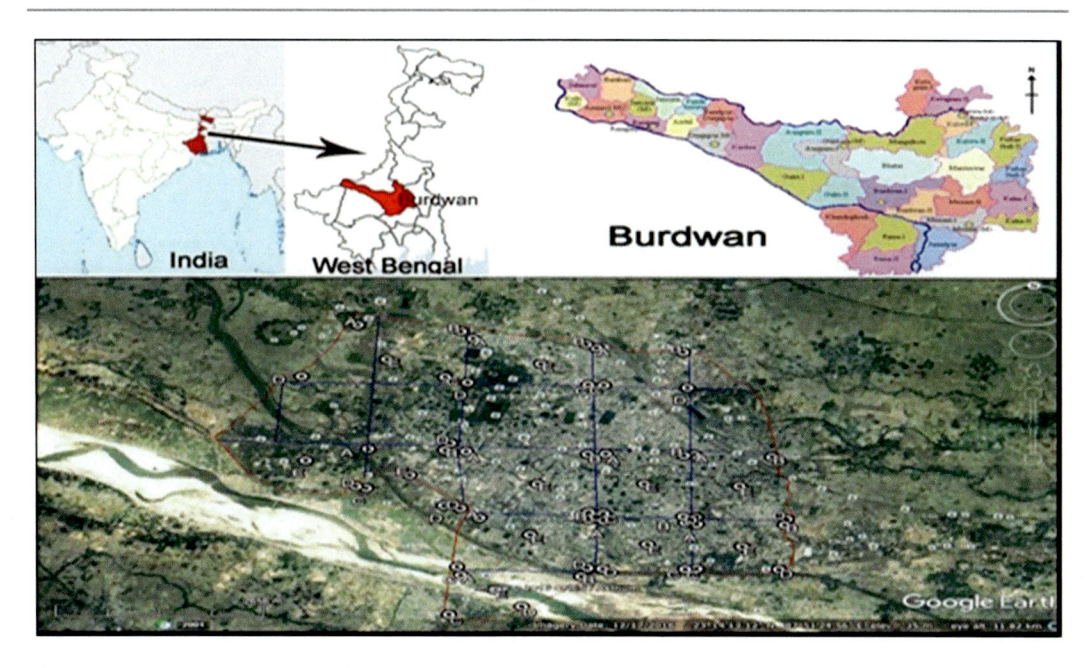

Fig. 20.1 Location map of the Burdwan municipality corporations

CAD software and geo-referencing capabilities, a hard copy map or survey plan is converted into a digital medium. Geo-coding is the method of translating a position definition, such as a pair of coordinates, an address, or a place name, to a location on the surface of the earth. Geo-coding can be achieved by entering a description of one place at a time or by presenting a table with all of them at once. The resulting locations are output as geographic features with attributes, which can be used for mapping or spatial analysis.

20.2.2 Survey Method

The process of the survey is the technique of collecting data by asking people who are considered to have desired knowledge questions. A non-disguised tactic is commonly used. Questions about their view of demographic interest are posed by the respondents. This approach was followed for verification of the ground reality. The research sites are visited and inspected by the researcher. With the current time and circumstance, the knowledge learned from the previous technique is validated.

20.2.3 Application of Geospatial Tools

Natural and human-induced environmental changes are of concern in urban environments today due to environmental degradation and human health (Jat et al. 2008). In order to better plan and use natural resources and their management, the analysis of land use land cover (LULC) changes is very important (Asselman and Middelkoop 1995). Traditional methods for gathering demographic data, censuses, and analysis of environmental samples are not adequate for multicomplex environmental studies (Maktav et al. 2005), since many problems often presented in environmental issues and great complexity of handling the multidisciplinary data set; new technologies like satellite remote sensing and Geographical Information Systems (GISs) are required. These technologies provide data to study and monitor the dynamics of natural resources for environmental management (Berlanga-Robles and Ruiz-Luna 2011). Remote sensing has become an important tool relevant to developing and understanding the global, physical processes affecting the earth (Hudak and

Wessman 1998). Recent development in the use of satellite data is to take advantage of increasing amounts of geographical data available in conjunction with GIS to assist in interpretation (Martinuzzi et al. 2007; Mandal 2018). GIS is an integrated computer hardware and software framework that can record, store, retrieve, manipulate, analyse, and view geographically referenced (spatial) information in order to help development-oriented management and decision-making processes (Annerstedt et al. 2013). Remote sensing and GIS cover a wide variety of agricultural applications, environments, and integrated eco-environment assessment (Long et al. 2008). Owing to their detrimental effects on area ecology and vegetation, some researchers have concentrated on LULC studies (El-Raey et al. 2000).

20.3 Land Use Changes in the Burdwan Municipality Area

The paired digital image analysis offers a relatively quick, easy, and cost-effective means to assess cover change, but it does have some limitations. Though Google offers high-resolution imagery in many parts of the world, paired image analysis with Google images is limited by the varying dates amongst images and varying image resolution. In urban areas many of the Google images are of sufficient resolution for accurate photo interpretation and images and varying images are continually updated. Obtaining local digital images with known and consistent dates across an area of analysis can overcome the problems associated with varying dates across a study area. Sometimes a paired city data also had different image resolution between years, but most images were 1 m or less. As image interpretation was paired image approach, information from higher resolution image could aid in interpreting the lower resolution image. After conducting the study, a decreasing trend in the vegetation cover can be seen throughout the study site in each observation years (Fig. 20.2). A brief description of research works carried out by various authors on carbon sequestration and LULC is shown in the Table 20.1.

The study site was divided into a number of discrete sites just for the ease of the study. All the sites almost followed the same trend. For the first site (R1S1, R1S2, R1S3, R1S4) in the first row of study area the vegetation covers eventually declined in the preceding years (Fig. 20.3). The data was collected from the 2001, 2006, and 2010. The data from the year 2013 and 2015 couldn't be collected, as because the maps were not visible due to clouds. For this site 3 maps were taken from the year 2001, 2006, and 2010. The other maps were not clear due to clouds. The vegetation cover in 2006 decreased 26% from 2001, whereas the vegetation cover decreased 9.47% from the 2006 in the year 2010.

Similarly, at the second site (R2S1, R2S2, R2S3) of the first row showed an overall decrease in the vegetation cover with slight changes in the values of the year 2006 (Fig. 20.4). The sudden increase in the vegetation cover value from year 2001 to 2006 was found because the crop fields were full of greeneries compared to the year 2001 which can be easily noticed from the satellite image. Then a significant fall in the value was noticed in the year 2010. But again the trend shifted to an exception showing an increase in the value during the year 2013 but eventually decreased in the year 2015. The vegetation cover increased in 2006 from the year 2001. The vegetation cover again decreased in the year 2010. The vegetation cover again increased in the year 2013. But it declined in the later year. For the second site in the first row, 5 maps were taken from the year 2001, 2006, 2010, 2013, and 2015. Here, the vegetation cover was noticed to have decreased 30.38% from year 2006 to 2010 and 29.95% from 2013 to 2015. However, it was noticed to have an increase of 6.09% from the year 2001 to 2006, and 19.17% from 2010 to 2013.

The third site (R3S1, R3S2, R3S3, R3S4) from the first row depicts the declining vegetation cover, except in the year 2006, which showed a sudden increase (Fig. 20.5). In the third site from the row, for the above mentioned years the vegetation cover has reduced 33.70,

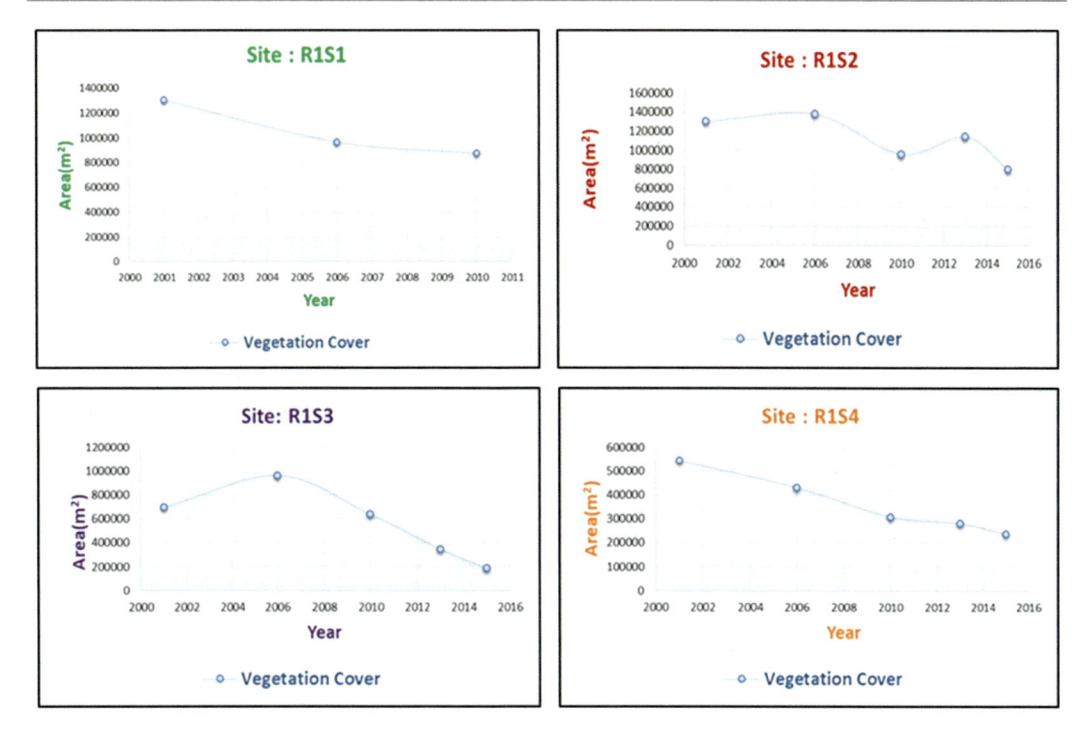

Fig. 20.2 Satellite images of study sites (R1–R4) during the study period 2000–2015

45.57, and 45.99% from the year 2006 to 2010, 2010 to 2013, and 2013 to 2015 respectively. But an increase of 38.27% of vegetation cover was found from the year 2001 to 2006. In the last site of the first row the value decreased, but a slight increase in the value of the vegetation cover was noticed in the year 2013. But the value eventually declined again in 2015.

In the fourth (R4S1, R4S2, R4S3) and the last site (R5S1) of first row, the vegetation cover decreased 20.89%, 28.19%, 8.67%, and 15.82% from 2001 to 2006, 2006 to 2010, 2010 to 2013, and 2013 to 2015 respectively (Fig. 20.6). For the first site of the second row the vegetation cover peak showed an increase in the 2006 from the year 2001. But it declined at a great rate from 2006 to 2013. But again the graph showed an increase in the year 2015. The second row was divided into 3 sites. In the first site of the second row, the vegetation cover increased 21.75% from 2001 to 2006, but it decreased 38.03% from 2006 to 2010. And again an increasing trend was found from 2010 to 2013 by 6.12%. From the year 2013 to 2015, the vegetation cover again decreased by 51.93%. The second site of the second row site showed an increase in the vegetation cover in the year 2006 and 2013. But the overall vegetation cover declined in the year 2015. The vegetation cover rate in this site increased 15.70% and 6.12% from the year 2001 to 2006 and 2010 to 2013 respectively, whereas the vegetation cover decreased 38.03% and 51.93% from the year 2006 to 2010 and 2013 to 2015 respectively.

The second row of the fifth site showed a similar trend of increase in the value of the vegetation cover in the year 2006 and 2013 (Fig. 20.7). But it eventually declined in the year 2015. In the third and last site of the second row, the vegetation cover increased 73.35% and 63.78% from the year 2001 to 2006 and 2010 to 2013 respectively, whereas the vegetation cover decreased 50.91% and 52.85% from the year 2006 to 2013 and 2013 to 2015 respectively. Continuing the declining trend in the vegetation cover, the first site of the third row showed a slight increase during the year 2006 and 2013. But overall the vegetation cover decreased in the

Table 20.1 Summary of research works done by various authors on urban vegetation loss

Authors (Year)	Place	Urban patch type	Highlights
International			
Sharpe et al. (1986)	Wisconsin, USA	Urban vegetation	Estimated the loss of urban vegetation or forest and Savanna during urbanization
Qian and Follett (2002)	USA	Turf grasses	Soil organic matter of the previous land such as agricultural or indigenous grassland is important in controlling carbon sequestration of the converted turf soils
Pataki et al. (2006)	North American cities	Urban trees	Urban areas as whole ecosystems with regard to carbon balance, including both drivers of fossil fuel emissions and carbon cycling in urban plants and soils
Ujoh and Ifatimehin (2010)	Federal Capital City (FCC), Nigeria	Urban trees	Used GIS and remote sensing technique to study the impact of sustainable urbanization in the FCC followed by remediation strategies
Liu and Li (2012)	Sejong city, South Korea	Urban vegetation	Reduction in CO_2 emission through environmentally friendly urban planning, energy efficient building design, and renewable energy use
Scherner et al. (2013)	SW Atlantic, Brazil	Coastal urbanization	Coastal urbanization is causing substantial loss of seaweed biodiversity in the SW Atlantic, and is considerably changing seaweed assemblages
Martellozzo et al. (2015)	Calgary–Edmonton Corridor, Canada	Agriculture	The average soil quality of land used for agriculture has declined in the Calgary–Edmonton corridor, confirming other studies of the food security implications of urbanization
Zhao et al. (2016)	Beijing, China	Urban vegetation	Studied the impact of urbanization on vegetation growth and heat island effect across 32 cities of China
De la Barrera and Henríquez (2017)	Macul, Chile	Urban vegetation	Studied the vegetation loss in the urban periphery due to rapid urbanization process in Macul. An overall loss of vegetation was observed in all cities has a consequence of urban expansion despite their geographical location
Zhong et al. (2019)	Shanghai, China	Urban trees	Studied the change in the impervious surface area and assessed the impacts of urbanization on vegetation greenness and gross primary productivity in Shanghai
National			
Kumar et al. (2010)	Varanasi city, India	Urban population	Loss of vegetation covers in the Varanasi city due to the development of urban sprawl. Emphasized on new policies to recover the loss
Nagendra et al. (2012)	Bangalore city, India	Urban trees	Peripheral areas are undergoing rapid urbanization, vegetation clearing, and fragmentation in Bangalore and emphasis is given on new policies to manage urban vegetation
Mahmood et al. (2014)	Rajasthan, India	Urban climate	Changes in air temperature, humidity, cloud cover, circulation, and precipitation are influenced by land cover. Such effects vary from local to regional to sub-continental and global scales
Pandey and Seto (2015)	India	Urban agriculture	A hierarchical classification technique was used to recreate land-cover transition histories using time series MODIS 250 m VI photos composited at 16-day intervals. Study demarcates the type of agricultural land loss is around the smaller cities
Das and Das (2019)	Malda city, India	Urban trees	Evaluation of the complex essence of urbanization and its effect on urban ecosystem services through LULC changes in Eastern India's Old Malda Municipal Area

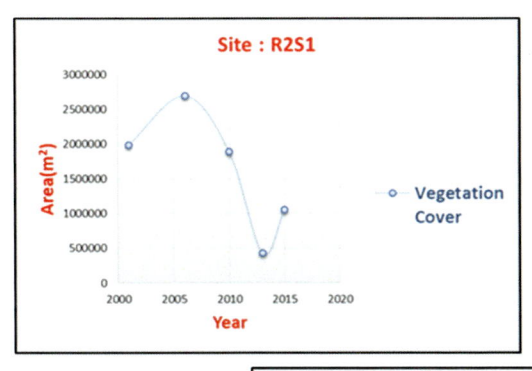

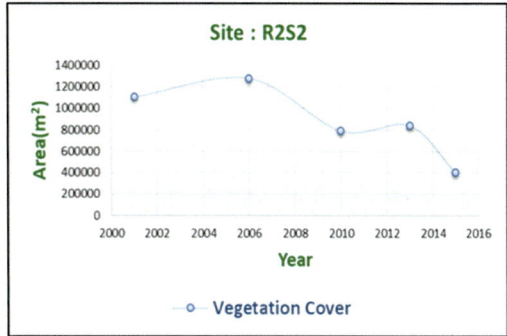

Fig. 20.3 Changes of areal vegetation cover in the first site during the study periods

year 2015. The vegetation cover in this site increased 6.07% from the area 2001 to 2006. But the vegetation cover decreased in the rate 25.37%, 9.45%, and 52.84% from the year 2006 to 2010, 2010 to 2013, and 2013 to 2015 respectively. The vegetation cover for the second site of the third row decreased from 2001 to 2010. But surprisingly the vegetation cover started to show an increasing trend from 2013. The second site in the third row showed an increase of 5.60% and 4.28% from the year 2010 to 2013 and 2013 to 2015 respectively. But the vegetation rate decreased 40.45% and 15.14% from the year 2001 to 2006 and 2006 to 2010 respectively. In the third site of the third row, the vegetation rate increased 79.62% from the year 2001 to 2006. The decreasing rate was found in the other three years like 46.74%, 28.28%, and 9.13% from the year 2006 to 2010, 2010 to 2013, and 2013 to 2015 respectively. In the fourth and the last site of the third row, decreasing rate was found in all the years viz., 5.52%, 42.62%, 29.62%, and 38.15% from the year 2001 to 2006, 2006 to 2010, 2010 to 2013, and 2013 to 2015 respectively.

In the first site of the fourth row, the vegetation cover value for the year 2015 showed decreasing trend. The vegetation cover from the year 2001 to 2006 increased 21.66%. The vegetation cover showed a decreasing rate of 37.77% and 32.54% from the year 2006 to 2010 and 2010 to 2013 respectively. The second site of the fourth row had the value decreased gradually; except in the year 2006 which showed a variation from the usual trend by showing an increase in the value. In the second site of the fourth row the vegetation rate increased 52.28% from the year 2001 to 2006. The vegetation cover decreased in the rate of 37.66%, 16.33%, and 25.39% from the year 2006 to 2010, 2010 to 2013, and 2013 to 2015 respectively. The values of the vegetation cover in the third site of the fourth row decreased from 2001 to 2013. There was slight increase during the year 2015 as it was found that some small parks were constructed during the year 2014 and small plantations were started due the efforts of the people of the residential area. The only site from the fifth row showed a massive decrease in the vegetation cover from 2001 to

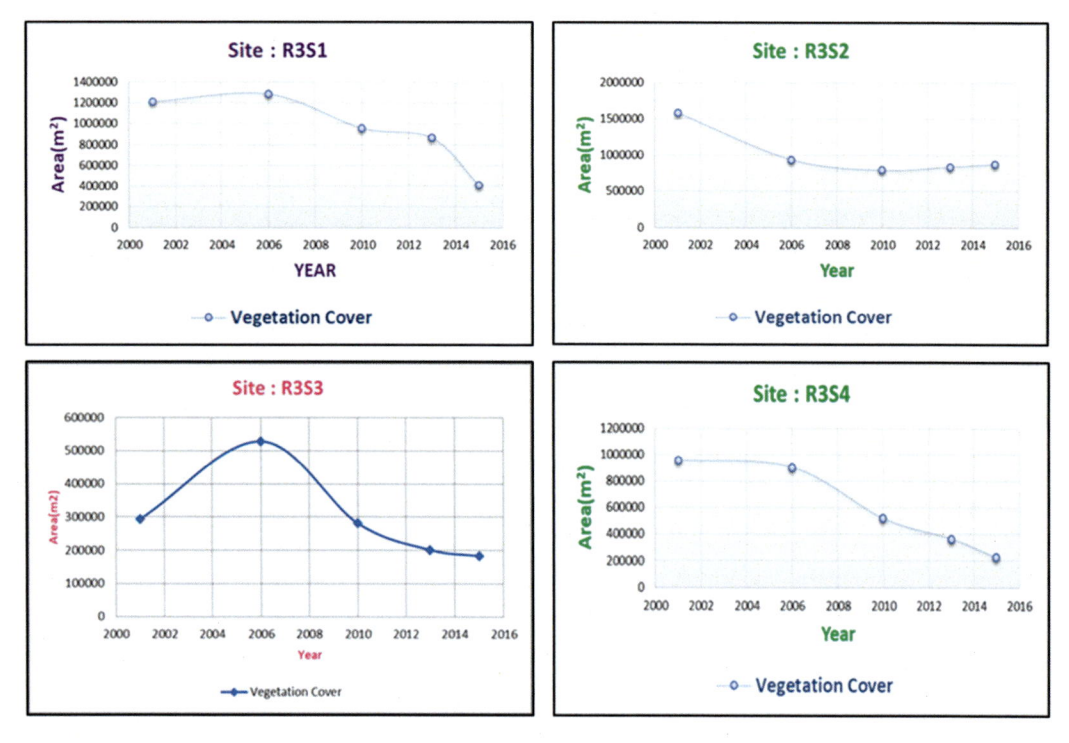

Fig. 20.4 Changes of areal vegetation cover in the second site during the study periods

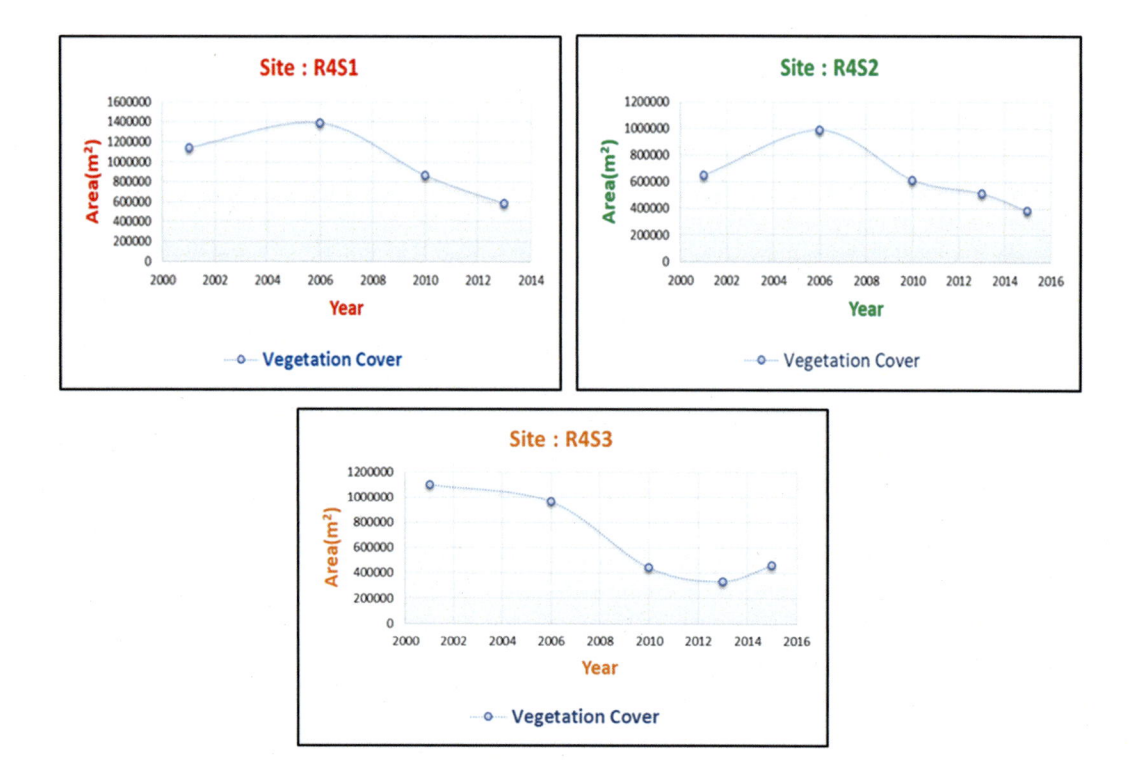

Fig. 20.5 Changes of areal vegetation cover in the third site during the study periods

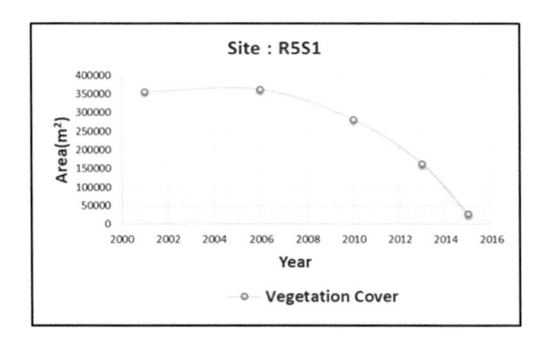

Fig. 20.6 Changes of areal vegetation cover in the fourth site during the study periods

2015 with a slight increase in the year 2006. The fifth and the last row contain only one site. In this site the vegetation cover increased by 1.77% from the year 2001 to 2006. The vegetation cover decreased in the rate of 22.34%, 43.03%, and 83.58% from the year 2006 to 2010, 2010 to 2013, and 2013 to 2015 respectively.

20.4 Prediction of Carbon Balance in the Burdwan Municipality Area

In this study, 15 patches were selected and the total gain or loss of green cover was estimated. On the basis on the present survey, a single dimension model was constructed on STELLA software considering vegetation cover as state variable (VC Total), vegetation cover gain as inflow to the vegetation pool (VC add Total), and vegetation cover loss as outflow from the vegetation pool (VC loss Total). In addition, two rate parameters gain rate (gr) and loss rate (lr) were also considered in the model as graph-time function. Here, the gain rate or loss rate corresponds to gain or loss due to anthropogenic activities only. The model considers single type of rates which includes all sorts of anthropogenic activities under a common value as it was not possible to quantify the distinct rate values of different forms of anthropogenic activities (Fig. 20.8).

The GIS map study was done for the time period of 15 years (2001–2015) and the only 5 scenarios were obtained (2001, 2006, 2010, 2013, and 2015). The model was able to calculate the gain or loss rates in between the gaps of the period of study. Here, the year 1 corresponds to 2001 and year 15 corresponds to 2015. Model shows that the vegetation cover increases slightly from 2001 to 2006 and decreases there after almost steadily (Fig. 20.9). It can be said that the anthropogenic stress increased during the period 2007–2015 may be due to removal of vegetation cover urban development like apartment construction, hospital construction etc.

It was observed that the Burdwan Municipality encounters the highest gain rate of vegetation cover during 2006 (0.021) whereas the highest loss rate was observed during 2007 with the value of 0.164. The trend of decrease rate touched down to 0.084 (lowest) during 2012 but it again increases to 0.133 during 2015 (Fig. 20.10).

Moreover, the model predicts the scenario of next 100 years of Burdwan Municipality with the study period between 2001 and 2100. The model considered the initial value of 906,247.50 (m^2) and an average gain rate (0.012) and loss rate (0.113). The model result shows that the vegetation cover after 100 years will be 37.23 m^2 if the present situation of anthropogenic stress continues within Burdwan Municipality. The model prediction for the next 100 years is shown below (Fig. 20.11). In near future, more general model can be constructed by quantifying the factors and processes associated with the anthropogenic activities.

20.5 Future Directions

Generally, that plants and vegetations are very useful for variety of way to human life. It is also useful for maintaining the environmental balance. But in modern times when the rate of urbanization is phenomenal mainly in developing countries, the adverse effect of city life is clearly noticed. Very few cities are well planned; otherwise most of urban areas are crowded, polluted, and compact. Many experts, planners recommend the green belts and green areas in cities. High levels of noise, pollution, and reduced social support could be countered by

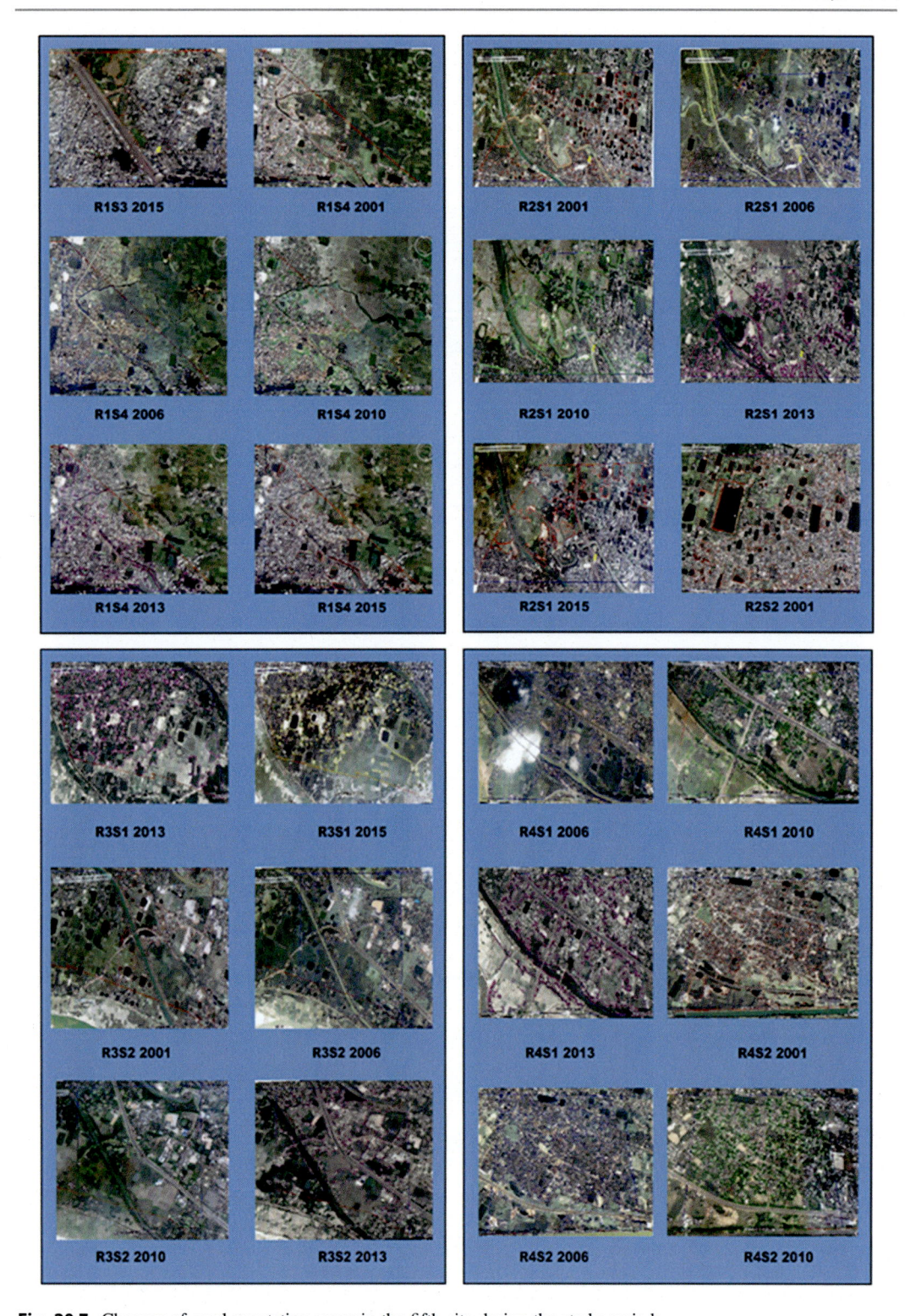

Fig. 20.7 Changes of areal vegetation cover in the fifth site during the study periods

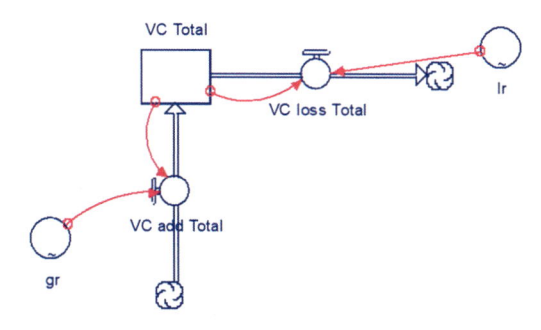

Fig. 20.8 Conceptual diagram of vegetation cover dynamics of Burdwan Municipality in STELLA 6.0 software

plants, woodland, and lawns that are reported as the most desired environment for relaxing and recovering from stress and sustain mental efforts. Long-term monitoring of urban vegetation cover can provide valuable information on how urban vegetation cover and ecosystem services are changing through time. Changes in the urban forest due to such factors as urban development, climate change, urban forest planning and management, storms, invasive plants, and insect and disease infestations will continue to affect and alter this resource in the coming years. Better

Fig. 20.9 Simulated versus observed result of vegetation cover dynamics of Burdwan Municipality

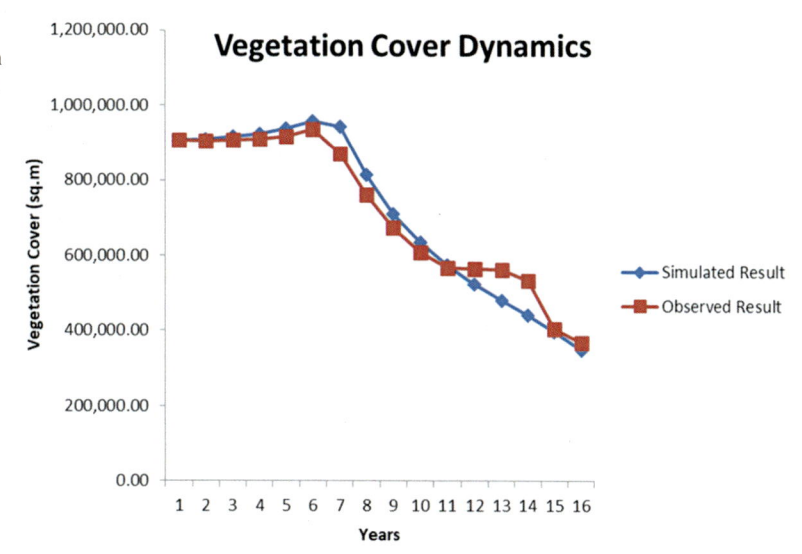

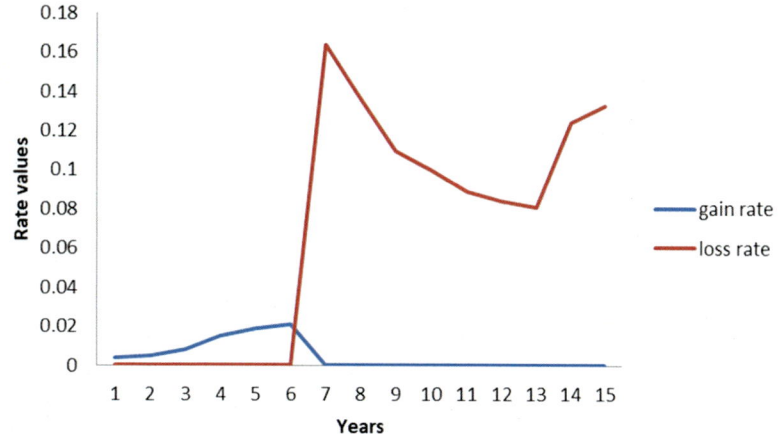

Fig. 20.10 Gain or loss rates of vegetation cover in Burdwan Municipality

Fig. 20.11 100 years prediction of vegetation cover in Burdwan Municipality

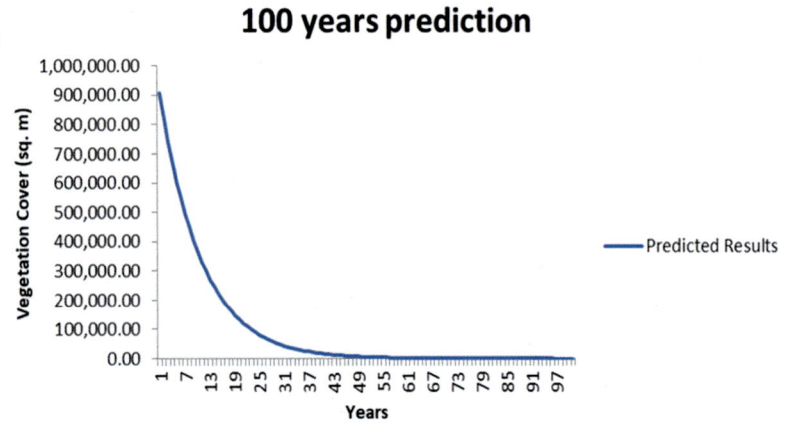

understanding of urban forests and how they are changing can facilitate better management plans to sustain ecosystem services and desired forest structure for future generations.

This decreasing vegetation cover can only be checked through plantations and construction of parks and vertical gardens. But as the population is rising day by day, the urban area is shrinking consequently. To combat the rising pollution vertical gardening and as well as terrace gardening and farming could be beautiful option. This vertical gardening can be done by the help of formation of living walls. Living walls or green walls are self-sufficient vertical gardens that are attached to the exterior or interior of a building. They differ from green façades (e.g. ivy walls) in that the plants root in a structural support which is fastened to the wall itself. The plants receive water and nutrients from within the vertical support instead of from the ground. The vertical gardens are a boon for this limited space era. The vertical and roof gardens could sequester carbon temporarily to an extent and minimize the heat generated through urban livelihoods. The last option that is left in favour of land utilization is the expansion of cities around the periphery without disturbing the large trees and grasslands along with afforestation programme. Calculation of land utilization should be done in a way that the carbon sequestration will always be greater than carbon emission. This will ensure better living and better life in an urban area.

Acknowledgements The authors are thankful to the Department of Environmental Science, The University of Burdwan for giving all sorts of laboratory facilities to conduct this research. The authors greatly acknowledge the support from SERB, DST, Govt. of India, New Delhi (Project No. EMR/2016/002618) to carry out the research work. The authors declare no conflict of interest.

References

Annerstedt M, Konijnendijk C, Busse Nielsen A, Maruthaveeran S (2013) The public health effects of urban parks—results from a systematic review. Eur J Public Health 23

Asselman NEM, Middelkoop H (1995) Floodplain sedimentation: quantities, patterns and processes. Earth Surf Proc Land 20:481–499

Berlanga-Robles CA, Ruiz-Luna A (2011) Integrating remote sensing techniques, geographical information systems (Gis), and stochastic models for monitoring land use and land cover (Lulc) changes in the northern coastal region of Nayarit, Mexico. Gisci Remote Sens 48:245–263

Brack CL (2002) Pollution mitigation and carbon sequestration by an urban forest. Environ Pollut 116:S195–S200

Dai Y, Guo S, Wu Y (2013) Study on carbon storage of urban garden vegetation—a case study of Jiangnan University. In: Proceedings of the 2013 international academic workshop on social science (Iaw-Sc-13). Atlantis Press

Das M, Das A (2019) Dynamics of urbanization and its impact on urban ecosystem services (Uess): a study of a medium size town of West Bengal, Eastern India. J Urban Manag 8(3):420–434

De La Barrera F, Henríquez C (2017) Vegetation cover change in growing urban agglomerations in Chile. Ecol Ind 1(81):265–273

El-Raey M, Fouda Y, Gal P (2000) Environ Monit Assess 60:217–233

Howard EA, Gower ST, Foley JA, Kucharik CJ (2004) Effects of logging on carbon dynamics of a jack pine forest in Saskatchewan, Canada. Glob Change Biol 10:1267–1284

Hudak AT, Wessman CA (1998) Textural analysis of historical aerial photography to characterize woody plant encroachment in South African Savanna. Remote Sens Environ 66:317–330

Jat MK, Garg PK, Khare D (2008) Monitoring and modelling of urban Sprawl using remote sensing and GIS techniques. Int J Appl Earth Obs Geoinf 10:26–43

Joshi S, Hirve S, Deshmukh A, Borse P, Desai M (2017) Detection of spam tweets by using machine learning. Int J Adv Res Comput Sci Softw Eng 7:1–4

Kumar M, Mukherjee N, Sharma GP, Raghubanshi AS (2010) Land use patterns and urbanization in the holy city of Varanasi, India: a scenario. Environ Monit Assess 167(1):417–422

Lavery TJ, Roudnew B, Gill P, Seymour J, Seuront L, Johnson G, Mitchell JG, Smetacek V (2010) Iron defecation by sperm whales stimulates carbon export in the southern ocean. Proc R Soc B: Biol Sci 277:3527–3531

Liu C, Li X (2012) Carbon storage and sequestration by urban forests in Shenyang, China. Urban for Urban Greening 11:121–128

Long H, Wu X, Wang W, Dong G (2008) Analysis of urban-rural land-use change during 1995–2006 and its policy dimensional driving forces in Chongqing, China. Sensors 8:681–699

Mahmood R, Pielke Sr RA, Hubbard KG, Niyogi D, Dirmeyer PA, Mcalpine C, Carleton AM, Hale R, Gameda S, Beltrán-Przekurat A, Baker B (2014) Land cover changes and their biogeophysical effects on climate. Int J Climatol 34(4):929–953

Maktav D, Erbek FS, Jürgens C (2005) Remote sensing of urban areas. Int J Remote Sens 26:655–659

Mandal A (2018) Changing fluvio-dynamic scenario of the Adi Ganga River, Kolkata, West Bengal, India. J Geogr Environ Earth Sci Int 17:1–14

Martellozzo F, Ramankutty N, Hall RJ, Price DT, Purdy B, Friedl MA (2015) Urbanization and the loss of prime farmland: a case study in the Calgary–Edmonton Corridor of Alberta. Reg Environ Change 15(5):881–893

Martinuzzi S, Gould WA, Ramos González OM (2007) Land development, land use, and urban sprawl in Puerto Rico integrating remote sensing and population census data. Landsc Urban Plan 79:288–297

Mccarl BA (2001) Climate change: greenhouse gas mitigation in U.S agriculture and forestry. Science 294:2481–2482

Mujumdar S, Mane S, Kumar M, More R, Salunke MS (2016) My opinion: Anrge to enrich your knowledge. Int J Adv Res 4:980–987

Nagendra H, Nagendran S, Paul S, Pareeth S (2012) Graying, greening and fragmentation in the rapidly expanding Indian City of Bangalore. Landscape Urban Plann 105(4):400–406

Nowak DJ, Crane DE (2002) Carbon storage and sequestration by urban trees in the USA. Environ Pollut 116:381–389

Nowak DJ, Greenfield EJ, Hoehn RE, Lapoint E (2013) Carbon storage and sequestration by trees in urban and community areas of the United States. Environ Pollut 178:229–236

Pandey B, Seto K (2015) Urbanization and agricultural land loss in India: comparing satellite estimates with census data. J Environ Manage 15(148):53–66

Parker DE (2009) Urban heat island effects on estimates of observed climate change. Wiley Interdisc Rev Climate Change 1:123–133

Pataki DE, Alig RJ, Fung AS, Golubiewski NE, Kennedy CA, Mcpherson EG, Nowak DJ, Pouyat RV, Romero Lankao P (2006) Urban ecosystems and the north American carbon cycle. Glob Change Biol 12:2092–2102

Pataki DE, Carreiro MM, Cherrier J, Grulke NE, Jennings V, Pincetl S, Pouyat RV, Whitlow TH, Zipperer WC (2011) Coupling biogeochemical cycles in urban environments: ecosystem services, green solutions, and misconceptions. Front Ecol Environ 9:27–36

Qian Y, Follett RF (2002) Assessing soil carbon sequestration in Turfgrass systems using long-term soil testing data. Agron J 94:930

Scherner F, Horta PA, De Oliveira EC, Simonassi JC, Hall-Spencer JM, Chow F, Nunes JMC, Pereira SMB (2013) Coastal urbanization leads to remarkable seaweed species loss and community shifts along the Sw Atlantic. Mar Pollut Bull 76(1–2):106–115

Sedjo R, Sohngen B (2012) Carbon sequestration in forests and soils. Ann Rev Resour Econ 4:127–144

Sharpe DM, Stearns F, Leitner LA, Dorney JR (1986) Fate of natural vegetation during urban development of rural landscapes in Southeastern Wisconsin. Urban Ecol 9(3–4):267–287

Ujoh F, Ifatimehin OO (2010) Understanding urban sprawl in the Federal Capital City, Abuja: towards sustainable urbanization in Nigeria. J Geogr Reg Plann 2(5):106–113

Zhao S, Liu S, Zhou D (2016) Prevalent vegetation growth enhancement in urban environment. Proc Nat Acad Sci 113(22):6313–6318

Zhong Q, Ma J, Zhao B, Wang X, Zong J, Xiao X (2019) Assessing spatial-temporal dynamics of urban expansion, vegetation greenness and photosynthesis in megacity Shanghai, China during 2000–2016. Remote Sens Environ 233:111374

Review on Sustainable Groundwater Development and Management Strategies Associated with the Largest Alluvial Multi-aquifer Systems of Indo-Gangetic Basin in India

21

Anadi Gayen

Abstract

Demand for groundwater for various uses is increasing rapidly due to accelerated growth of population, industrialization, irrigation, and urbanization. Availability of limited surface water resources is the reason for ever-increasing demand for groundwater. The diminishing quantity of replenishable groundwater resources in the unconfined, as well as confined aquifers, is facilitating the depletion of groundwater table in the Indo-Gangetic Basin (IGB). Hence, the assessment of the availability of groundwater resources is of immense importance to develop a sustainable management plan. The IGB covers an area of 1,182,689 km^2 which forms the single largest groundwater reservoir with prolific multi-aquifer systems. The basin stretches over the northern zone from west to east below the extra-peninsular region of India. The tube wells tapping deeper confined aquifers have potential yields ranging between 25 and 50 Liters per second (lps) at economic drawdown. The study reveals that a total of about 1750.92 BCM of groundwater resources are available in the alluvial area of 466,007 km^2. of IGB, of which around >91% is fresh and around 9% is brackish/saline. In 2002, arsenic in groundwater in the IGB was first detected beyond the permissible limit (>50 µg/l). But, around two decades ago arsenic menace was identified in the downstream area of IGB in the Bengal Basin and in the belt of Ganga-Meghna-Brahmaputra (GMB) River system. The severity of arsenic hazards in the Bengal Basin is considered one of the drastic health hazards. Groundwater Estimation Committee (GEC), 2015 methodology report of Central Ground Water Board (CGWB) indicates that the net annual groundwater availability of IGB is 180.38 billion cubic meters (BCM) and annual groundwater draft for domestic, industries, and irrigation uses amounts to 140.77 BCM. The stage of groundwater development is recorded as the minimum (23%) in Jharkhand state, whereas the maximum (149%) is in Punjab state. The excess exploitation and occurrence of high arsenic in shallow aquifers pose a threat to the sustainable management of groundwater resources in the IGB.

Keywords

Indo-Gangetic Basin · Replenishable groundwater assessment · Arsenic contamination · Sustainable management

A. Gayen (✉)
Rajiv Gandhi National Ground Water Training and Research Institute, Central Ground Water Board, Ministry of Jal Shakti, Government of India, Naya Raipur, Chhattisgarh, India
e-mail: anadigayen1968@gmail.com

21.1 Introduction

Indo-Gangetic basin covers an area of 1,182,689 km^2 (Indus basin: 321,289 km^2 and Ganga basin: 861,400 km^2). The basin lies within the coordinates of 27°0'0" N and 80°0'0" E. The study area encompasses 09 states and 01 Union Territory (U.T.) comprising the Haryana, Punjab, Himachal Pradesh, Rajasthan, Uttarakhand, Uttar Pradesh, Bihar, Jharkhand, and West Bengal states and Delhi U.T. (Fig. 21.1). It constitutes the single largest groundwater reservoir with a prolific multi-aquifer system stretched over the northern zone of India and is characterized by the distinctive facies associations of gravelly fan deposits, sandy and muddy interfluves deposits, and river channel deposits (Singh 1996). Longer residential period of the river channels had facilitated to build a huge pile of sand bodies in the basin. Fine sand successions have been developed in the interfluves region, which is categorized as flood plain deposits. The basin forms the significant part of ancient continent of Gondwana and is portrayed as the oldest and most geologically stable plain of the country (After Britannica.com/Gondwana). The IGB had attained its present shape during the late Quaternary period. In Indian sub-continent, the basin includes mostly northern and eastern portions of India followed by the east of Pakistan, almost entire Bangladesh, and southern plains of Nepal. It witnesses diversified geomorphological features, grain size variations, and channel pattern changes.

The fertile Indo-Gangetic alluvial plain hosts a huge quantity of freshwater resources as it has been formed due to deposition of sand and silt by numerous rivers like Ganga, Yamuna, and Brahmaputra originating from the Himalaya Mountains. The vast alluvial tract of the basin maintains a wide range in altitude from <01 m at *Sundarbans* to within 200–375 m at Uttar Pradesh and Haryana and within 150–450 m in Rajasthan. The Tertiary and Quaternary sediments in the foreland IGB overlie different older litho units in different parts of the foredeep in either the metamorphosed basement rocks, the undifferentiated un-metamorphosed Proterozoic sediments and the Gondwana Mesozoic sediments. The basement topography has a number of highs and lows traversed by several faults, which have affected the deposition of Tertiary and Quaternary sediments. In general, 2000–5000 m thick Siwalik (Tertiary) sediments of the foreland basin are underlying the alluvial plain. The thickness of these sediments increases toward north, being maximum along the northern margin, along the Himalayan foothills, and minimum in the southern margin over the Vindhyans or Bundelkhand massif (Singh 1996).

The groundwater resources of the replenishable zone have already been separately computed as dynamic groundwater resources of the basin. The present study aims at estimating the monsoon and non-monsoon recharges, replenishable groundwater resources, draft for various uses, calculation of stage of groundwater development percentage for different state/U.T. lying within the IGB along with recommendations for area specific sustainable groundwater management strategies.

21.2 Methodology Adopted for the Study

The dynamic groundwater resources of the unconfined aquifer have been estimated using the methodology that is defined by the Groundwater Estimation Committee (GEC) in 2015 by Central Ground Water Board. Adopted values of specific yield (0.6) for assessment of in-storage resources of the unconfined aquifers were considered for computing the dynamic resources of the unconfined aquifer. The aquifer system has been broadly classified into two groups viz. unconfined and confined. The extrapolation of the confined aquifer has been carried out by considering the maximum depth of exploration, used for the computation of the groundwater resources of the unconfined aquifers. Other data like the average depth of the predominant pre-monsoon water level (for the last 10 years) of unconfined aquifers of each district has been considered as

Fig. 21.1 Location map of Indo-Gangetic River Basin. *Source* GSI

top of in-storage portion of the unconfined aquifer. On other hand, the district-wise bottom of the unconfined aquifer was determined with the help of the groundwater exploration carried out by CGWB in all the states having Indo-Gangetic alluvium, and also on the basis of lithological and geophysical logs.

The average depth of the bottom of unconfined aquifer in each district and district-wise thickness of unconfined aquifers have been considered for computation of in-storage resources. Thickness of the unconfined aquifer was derived by subtracting average bottom depth of unconfined aquifer from predominant pre-monsoon groundwater levels. District-wise thickness of granular zones in the unconfined aquifers was estimated based on the interpretations of lithological logs, and electrical logs of the exploratory wells drilled by CGWB. The inter-layering of clay in unconfined and confined aquifers has been considered as single unit for computation purposes. The continuity of the various aquifer parameters such as specific yield and storativity co-efficient was taken into account while computing groundwater resources for unconfined and confined aquifers across the state borders.

The continuity of other inputs like bottom of unconfined aquifer and extension were checked and established with the neighboring districts of other states (after CGWB). Specific yield values, which have been considered for estimating dynamic groundwater resources of unconfined aquifers, by keeping in view of the reduction in the specific yield values due to increase in consolidation/compaction with depth and also the uncertainties in litho-facies variation. An attempt has been made for computing the groundwater resources for fresh and brackish/saline water separately as far as possible.

21.3 The Indo-Gangetic Plain and Aquifer System

The Indo-Gangetic Plain (IGP) is one of the most important regions of India; consists of alluvium with vast groundwater storage capacity. Based on climatic, hydrologic, and physiographic variations, the IGP consists of five homogeneous regions: (1) Trans-Gangetic plain in Pakistan; (2) Trans-Gangetic plain in India; (3) Upper-Gangetic plain; (4) Middle-Gangetic plain; and

(5) Lower-Gangetic plain. The basin is divided into four distinct physiographic zones, viz. Trans Indo-Gangetic Plains (TIGP), Upper Indo-Gangetic Plains (UIGP), Middle Indo-Gangetic Plains (MIGP), and Lower Indo-Gangetic Plains (LIGP) (Shah 2014). The strategy to develop static resources of alluvial aquifers needs a precise and sound understanding of the disposition of subsurface aquifer system and overall hydrogeological scenario. Below the zone of water table fluctuation, the groundwater, which is available in the perennially saturated zones, forms the in-storage groundwater resources of the unconfined aquifer.

A confined aquifer consists of porous and permeable geological formation, which is sandwiched in between two relatively low permeability layers. The confining layers above and below the confined aquifer are regionally extensive. The recharge and discharge areas of these aquifers may be hundreds of kilometers apart. The groundwater development in confined aquifers is much more complicated than in unconfined aquifers, due to their typical hydrogeological characteristics and hydraulic properties. In case of confined aquifer, the release of groundwater initially starts from the storage components and later on by the compression of the aquifer skeleton followed by the expansion of groundwater. This mechanism is exclusively controlled by the elastic properties of aquifer material. In the unconfined aquifer, the water releasing process is highly associated with the desaturation of the aquifer skeleton in the phreatic zone. Assessment of groundwater development potential for confined aquifers assume of crucial importance, since overexploitation of these aquifers may lead to far more adverse consequences. And, of the 1068 blocks (groundwater development units) in the three states (Punjab, Haryana, and Uttar Pradesh), 214 were characterized as dark or gray (groundwater extraction exceeds the annual recharge) in the context of groundwater depletion (Kamra 2007). The CGWB (2017) demonstrated that the entire IGB region (Fig. 21.2) depicted an abstraction ratio of >75 (Amarasinghe et al. 2016). The Projected demand for domestic and industrial uses in 2025 would be around 13.21 BCM, of which maximum demands are to be claimed by Uttar Pradesh (6.44 BCM) and minimum by Himachal Pradesh (0.07 BCM).

Central Ground Water Board (CGWB) conducted the detailed hydro-geochemical survey to understand the groundwater quality of the shallow aquifer system in the IGB. The study was appreciable enough to delineate the shallow aquifers affected by the arsenic severity (Fig. 21.3). The severity of arsenic contamination in shallow aquifers is mainly restricted in the Bengal Delta within the Holocene Younger Delta Plain and the alluvial formations, which have been approximately deposited during the period of 10,000–7000 years BP. The phenomenon may be associated with the Holocene sea-level rise and simultaneous rapid erosion in the Himalayan region (Acharyya et al. 2007).

21.3.1 Sediment Deposition History

In the broad depression and in sub-basins over the floor of the Ganga foreland basin, deposition took place during the Upper Tertiary and Quaternary periods. The depositional history in the foredeep gives an insight into the sedimentation process that occurred in different geological time scales along with subsequent digenesis. Initiation of the sedimentation mainly took place in a narrow-elongated foreland basin with the deposition of Dharmshala-Muree sediments in early Miocene period and was restricted in this narrow basin close to the Himalayan orogen. During middle Miocene to middle Pleistocene the orogen ward part of the foreland basin sediments was uplifted and thrusted toward the basin in discrete steps, while the basin expanded toward the craton (Fig. 21.4). Thus, the Siwalik sediments thickened toward foothills due to the greater load of sediments and greater concomitant sinking of the basin floor in that direction.

Thickness wise the foreland basin sediments show a strong asymmetry. Overlap of the younger Siwalik toward the south is an indication of widening of the basin over the Bundelkhand massif, whereas sediments of Middle and Upper

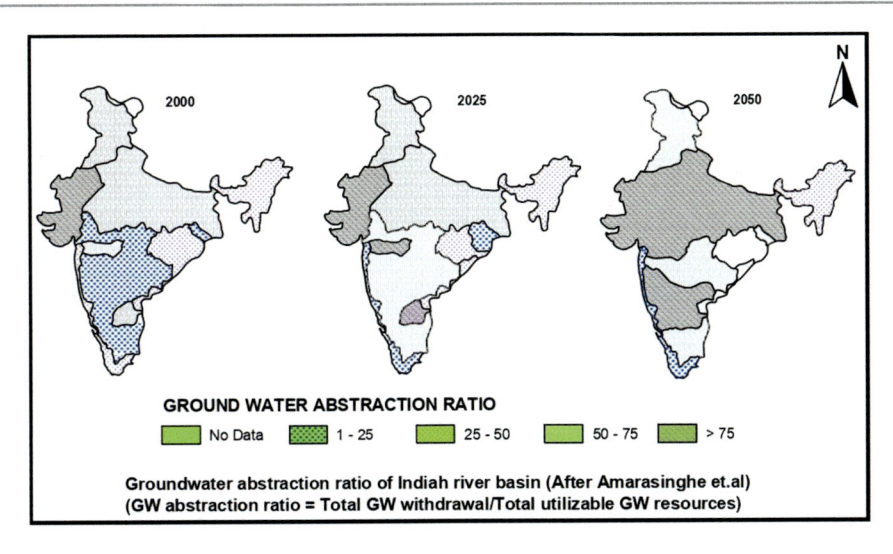

Fig. 21.2 Groundwater abstraction ratio of Indian River Basin (After Amarasinghe et al. 2016) (GW abstraction ratio = total GW withdrawals/total utilizable GW resources)

Siwalik sub-groups were laid down in the foredeep under fluvial and piedmont conditions. These sediments are coarse-grained close to Himalayas and become finer toward the craton and showed a typical coarsening upward cycle. During Upper Pleistocene, the foreland basin was subjected to upheaval due to continued subduction of Indian plate and the Siwalik was uplifted, deformed, and thrusted along the Main Boundary Thrust (MBT). This thrust gave rise to Ganga foredeep in which the Ganga alluvium has been laid down.

21.3.2 Geometric Configuration of the Multi-aquifer Systems

The alluvial plains of the Indo-Gangetic Basin (IGB) constituted 25% of the total land area of India and are composed of a thick pile of sediments belonging to the Tertiary and Quaternary ages. This vast and thick alluvial fill, exceeding 1000 m at places, constitutes the most potential and productive groundwater reservoir in India. Two distinct trends of groundwater flow direction are discernible. In the Indus basin, the elevation of water table ranges from less than 150 m *amsl* in the western part to more than 200 m *amsl* in the eastern part. The groundwater flow direction is SE to NW. In the Ganga basin, the elevation of water table ranges from about 200 m *amsl* in the western part to <20 m *amsl* in the eastern part. The general direction of groundwater flow is from NW to ESE.

The Indo-Gangetic basin is characterized by hugely productive multi-aquifer systems with long regional extensions. The deeper aquifers available in the basin provide substantial groundwater resources for further development. In the Indo-Ganga plain, wells constructed by the concerned Government organizations in the deeper aquifers have yielded within the range of 25–50 *lps* at economic drawdowns. A schematic diagram of the groundwater flow in different aquifer systems in the Indo-Gangetic basin is depicted in Fig. 21.5. The Tertiary and Quaternary sediments in the foreland of IGB overlie on different older litho units in different parts of the foredeep. The seismological study by Oil and Natural Gas Commission (ONGC) reveals that the basement topography has a number of highs and lows traversed by several faults, which have affected the deposition of Tertiary and Quaternary sediments (Sinha et al. 2005). Overall 2000–5000 m thick Siwalik (Tertiary) sediments of the foreland basin underlying the alluvial plain rest on a gentle northwardly sloping Pre-Tertiary

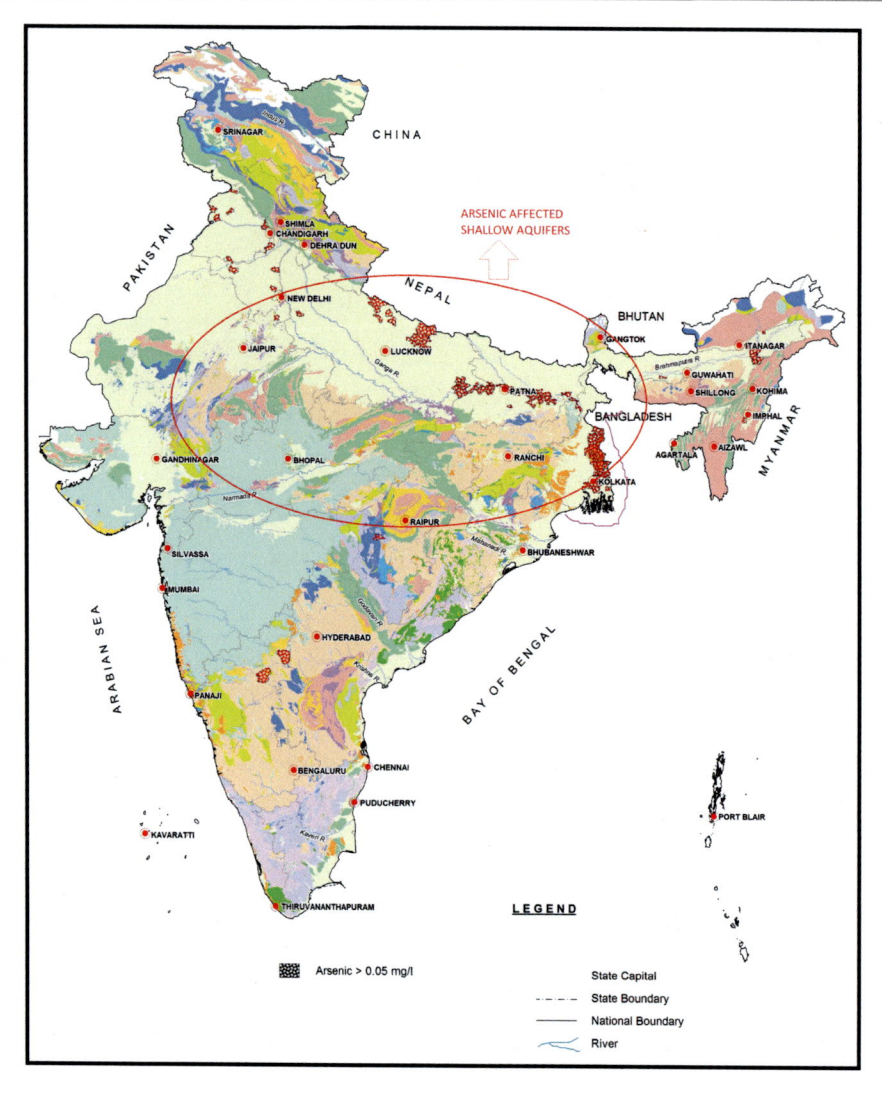

Fig. 21.3 Arsenic affected shallow aquifers in the IGB of India. *Source* CGWB

floor. The thickness of these sediments increases toward north, being maximum along the northern margin of the Himalayan foothills and minimum in the southern margin over the Vindhyans or Bundelkhand massif. Thick alluvium (590 m) and the underlying 2000–5000 m thick Siwalik formed a mammoth sequence of sedimentary deposits in the Indo-Gangetic basin.

As per Oil India Ltd., the evidences derived from the exploratory wells has conclusively established that the combined thickness of alluvium and Upper Siwalik ranges from 320 to 1500 m. This formation comprises potential aquifers with primary porosity. The prospects of suitable quality of groundwater are found only in the Alluvium of Upper Siwalik and up to certain pockets of the Middle Siwalik. The Lower Siwalik and lower parts of Middle Siwalik contain brackish/saline groundwater. The older formations are falling in the area of less groundwater development. The Gangetic alluvium deposited over the Siwalik sediments shows distinct lithological facies in different segments. In the northern part close to the Siwalik foothills, the alluvium is dominantly constituted of gravel and sand with subsidiary clays; while the inter-core area is

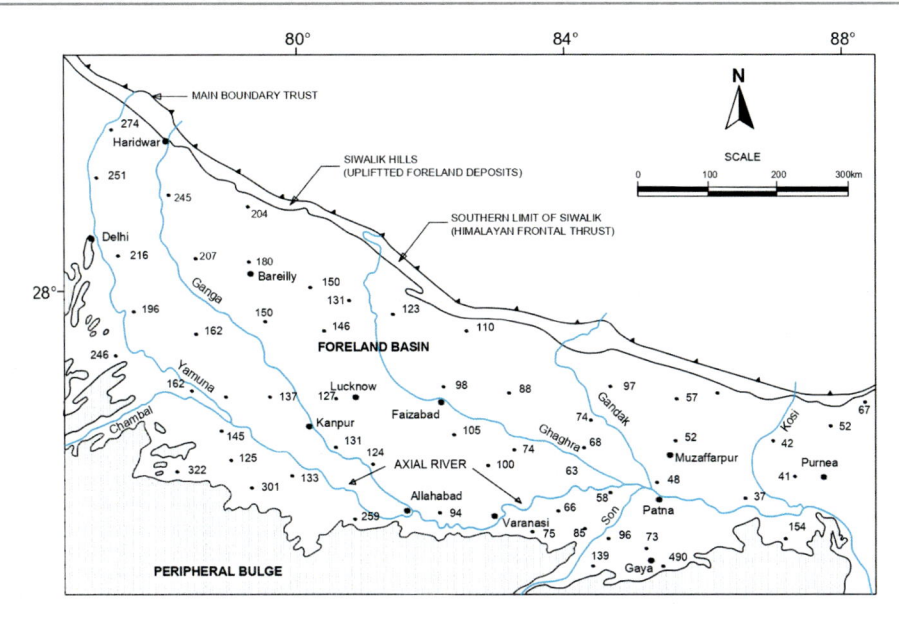

Fig. 21.4 The Ganga Plain highlighting Siwalik's point heights, slope patterns, and Foreland Basin (After Singh 1996)

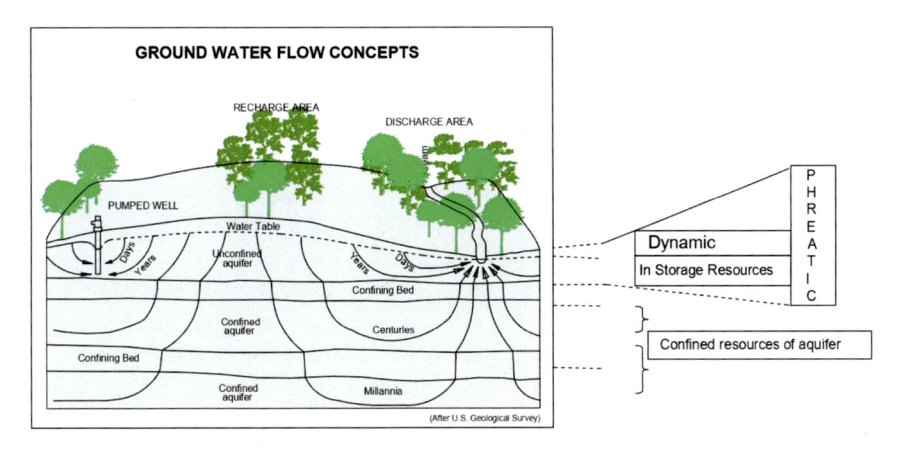

Fig. 21.5 Schematic diagram of groundwater flow system (after USGS)

enriched with clays. The Central Ganga Alluvial Plain is mainly characterized by clay, clay-kankar, and fine sand facies with occasional sand beds and sometimes comprises of gravels. The southern marginal alluvial plain shows coarse-grained sand, fine sand, and clay sequences.

21.3.3 Siwalik Aquifer System

The (I-G) alluvial plain and the underlying foreland basin are the part of (I-G) foredeep. The plain is ranging from Delhi-Aravalli ridge in the west to Rajmahal hills in the east. Its northern boundary is formed by the Siwalik Hills (Parkash et al. 1980), while the southern boundary is limited by the Bundelkhand-Vindhyan-Hazaribagh plateau. The (I-G) alluvial plain is a shallow asymmetrical depression underlain by the foreland basin having a gentle slope toward east. Northern part of the plain, adjacent to the Himalayan foothills, shows southerly and southeasterly slopes. Southern part of the plain is comparatively narrow and shows northerly to north-easterly slopes. The formation of plains is closely linked with Himalayan orogeny and the foreland basin is the resultant product of collision of Indian and Asian continental plates. Basement configuration indicates that the basin maintains slopes from west to east, and the maximum thickness of alluvium and Siwaliks is available in the Ganauli-Kadmaha section (Singh and Nambiar 1993). As studied by ONGC Ltd., in Puranpur depression, the thickness of Quaternary and Siwalik sediments varies from 1698 to 3068 m. The thickness of Quaternary and Siwalik sediments is more in Gandak depression than in Puranpur depression. A well-marked argillaceous bed is present in Tilhar and Matera areas in the Sarda depressions. Middlemiss (1900) proposed that Indo-Gangetic depression was a belt of subsidence, where the sinking of plains went on simultaneously with the rising of the mountains. It was the famous Swiss Geologist Edward Suess, who called it a foredeep of the Himalayas.

According to Rao (1973), the Indo-Gangetic depression is divided into five from east to west, i.e., (i) The Brahmaputra Basin in Assam; (ii) The Bengal Basin (Ganga-Brahmaputra) Basin in West Bengal and Bangladesh; (iii) The Ganga Basin in Uttar Pradesh and Bihar; (iv) The Punjab Basin in the Punjab, and (v) The Indus basin in Pakistan. Burrard (1915) advocated that the Indo-Gangetic trough was a great crustal crack or rift, perhaps as much as 30 km in-depth, which developed through a length of over 3200 km and got filled with alluvium. Seismic and gravity studies (Khattri 1987) suggest that the crustal thickness beneath the Indo-Gangetic plains is 28–45 km, including a 3 km top sedimentary layer. Regional continuity of freshwater aquifers and their thickening toward the north near foothills is very evident (Fig. 21.6).

21.3.4 Pleistocene and Flood Plain Deposits

Late Pleistocene is a period of severe climatic changes. Formation of different regional geomorphologic surfaces including active flood plain surfaces is related to the climatic cycles of Late Pleistocene-Holocene during 128 ka BP. These surfaces are depositional in nature and are overlain by the sediments with consecutive younger upward sequences. The Ganga plains are comprised of innumerable river networks, which have originated from parts of Himalayan Mountains, Peninsular craton, and alluvial plains formed in different periods in the late Pleistocene-Holocene. The sediments deposited in the IGB plains are originated from Himalayas undergo huge chemical weathering phenomenon resulting in the removal of Na, Ca, Sr, K, Mg and subsequent enrichment of As, Cr, Ni, and Th. In preferable hydro-chemical conditions, Smectite and Kaolinite are formed, which are partly eroded and transported to the delta region. During the Holocene period, deposition in the entranced river channels increased in response to the rising sea level (Singh 2007).

21.3.5 Geometry of Alluvial Aquifer Systems

The axis of the cusp runs in NW-SE direction passing through Saharanpur, Barielly, Sitapur, Lucknow, Basti, and Ambedkar Nagar in Uttar Pradesh, mid-central Bihar, and western portion of West Bengal states (Mathur 2003). In the states of Haryana and Punjab the basin is shallower than in Uttar Pradesh. The basin gradually becomes shallow across the central axis toward NE, where it merges with the foothills of Himalayas. The basin became comparatively shallow at the SW of the central axis, where it merges with Delhi Super Group, Bundelkhand, Vindhyans, Gondwana Super Group, and Chota Nagpur Plateau.

21.3.6 Thickness of Granular Zone in Unconfined Aquifer

The distribution of thickness of in-storage portion of unconfined aquifer in the alluvial area of the Indo-Gangetic basin is required to understand the static groundwater resources (CGWB 1997). The study on exploratory boreholes reveals that

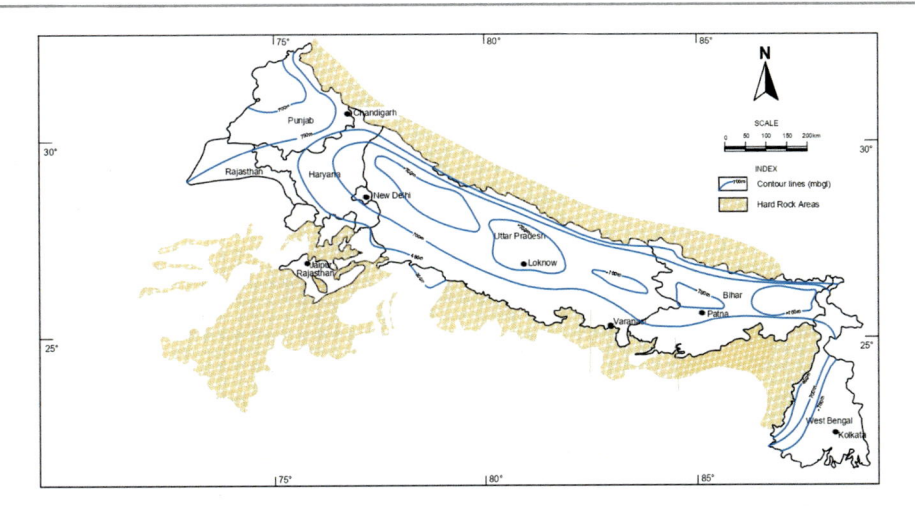

Fig. 21.6 Basin configuration map of IGB in India. *Source* CGWB

in the Indus basin the thickness of unconfined aquifer varies from less than 50 m to >100 m. The minimum thickness of the unconfined aquifer of about 50 m is observed in western part of Haryana, which increases gradually to 100 m toward northern part of the state and also in Punjab state. In the Ganga basin, central and eastern part of Uttar Pradesh is attributed to thickest aquifer material of 120 m under unconfined condition as compared to Western and North-Western part of Uttar Pradesh, where it is less than 100 m. In the state of Bihar, maximum thickness is 80 m in North-Eastern part of the state (Sinha et al. 2005). The thickness further decreases in West Bengal. The average thickness of aquifer in Bihar under unconfined conditions is about 100 m except in the regions forming basin boundary, where it is about 50 m. The state of Rajasthan indicates presence of prominent unconfined aquifer system all along the boundary with the states of Punjab, Haryana, and Uttar Pradesh and the thickness ranges between 60 and 100 m.

In the Indus basin, the cumulative thickness of granular zone ranges from 30 to 50 m in general in the states of Haryana and Punjab. However, it increases to more than 80 m in Punjab state in the bordering area of Pakistan. The cumulative thickness of granular zone in Rajasthan varies from 20 m to more than 50 m. In the Ganga basin, the central and eastern region of Uttar Pradesh., southern and eastern Bihar, northern and western parts of West Bengal State, and NE part of Rajasthan have maximum thickness of granular zones ranging between 40 and 60 m. The areas having cumulative thickness of more than 70 m of granular zone are found in eastern Bihar and western part of West Bengal. In the state of West Bengal the thickness of granular zone in the in-storage section reduces to less than 20 m in the southern part. A total of 421,112 km^2 is underlain by confined aquifers down to 450 m, in the basin.

21.3.7 Thickness of Granular Zone in the Confined Aquifer

The central region of the basin comprising the states of Uttar Pradesh, Punjab and Bihar have around 200 m thick granular material, which increases to more than 300 m at several places. The confined aquifer along the foothills of Himalayas in the state of Uttar Pradesh, Haryana, Punjab, and parts of Bihar also indicates the presence of about 100 m thick granular zone. The alluvial region of Uttar Pradesh., Bihar, Jharkhand, Haryana bordering Vindhyans, Bundelkhad, Gondwanas, and Delhi Group of formation has minimum thickness (<50 m) of granular material in the confined aquifer down to 450 m. In the state of West Bengal, maximum thickness of granular material encountered within 450 m in the confined aquifer ranges between 100 and 150 m. It is pertinent to mention here that confined aquifer does not exist in the Rajasthan part of the Indo-Gangetic basin. The distribution of cumulative thickness of granular zones occurring below the confining layer down to depth of 700 m bgl in the confined aquifer in the Indo-Gangetic basin, which needs prime attention. A total of 332,524 km^2. area is underlain by confined aquifers down to 700 m in the basin. It is observed that the thickness of granular zone is in conformity with the configuration of basin. It is of the order of 300 m to 350 m along the central axis in central and eastern parts of Uttar Pradesh (Pathak et al. 1985), 500 m in central Bihar, 250 m in Haryana, and 300 m in Punjab.

21.4 Groundwater Resources Assessment and Data Analysis

21.4.1 Water Resources of Unconfined Aquifer

The study reveals that the total replenishable groundwater resources all over the alluvial plain are 196.22 Billion Cubic Meter (BCM)/yr, which include the recharge of 144.55 BCM/yr and 51.664 BCM/yr during monsoon and non-monsoon seasons, respectively. The annual groundwater draft includes both the irrigation draft coupled with the domestic and industrial uses of groundwater. Thus, the total annual groundwater draft of the study area is 144.77 BCM/yr (Table 21.1). Among nine states and one Union Territory (U.T.) in the study area, the Uttar Pradesh state holds maximum amount of replenishable groundwater resources (76.34 BCM) and the draft (52.76 BCM), whereas the Delhi U.T. holds the least amount of replenishable groundwater resources (0.34 BCM) and the draft (0.39 BCM). Natural discharge during non-monsoon season in the study area ranges from 0.03 BCM to 4.75 BCM. The study area witnesses the stage of groundwater development within the range of 23% (Minimum) in Jharkhand state and 149% (Maximum) in Punjab state. The brackish/saline groundwater resources are available within the states of Haryana, Punjab, Rajasthan, Uttar Pradesh, and West Bengal. Net annual groundwater availability of the study area maintains the range between 0.31BCM and 71.58 BCM.

21.4.2 Groundwater Development Strategy from Confined and Unconfined Aquifers

Groundwater resources are considered by their location, their occurrence over time, their size, proportions, conditions of accessibility, the effort required to mobilize them, and therefore and their cost. All these are to be considered for planning and development. Groundwater, in Indo-Gangetic basin, occurs both under unconfined and confined conditions. The disposition of the aquifer varies from one location to another location (in space) and can be well understood using the subsurface configuration of aquifers which can be obtained from groundwater exploration data. The development plan is a function of water availability, groundwater draft, and economic viability. A comprehensive developmental plan requires the assessment of development-worthy groundwater resources, draft per tube well, and the cost to develop this resource (Prasad 1993). This helps in determining the number of tubewells feasible for construction in a particular aquifer and funds required to construct these tubewells.

Groundwater resources, in an unconfined aquifer, are classified into two categories namely the dynamic and the in-storage resources. The groundwater resource which gets replenished annually forms the dynamic resource, whereas the resource occurring below the average fluctuation zone forms the in-storage/static recourses. The dynamic groundwater resource is assessed for each district of all the states encompassed by the Indo-Gangetic basin. Block/sub-basin may be considered as the assessment unit.

Table 21.1 State-wise Groundwater Resources Availability, Utilization (in BCM), and Stage of Development (%)

S. No.	States / Union Territories (U.T.)	Annual replenishable ground water resources						Natural Discharge during non-monsoon season	Net annual ground water availability	Annual ground water draft			Projected demand for domestic and industrial uses up to 2025	Ground water availability for future irrigation use	Stage of ground water development (%)
		Monsoon season		Non-monsoon Season		Total				Irrigation	Domestic and industrial uses	Total			
		Recharge from rainfall	Recharge from other sources	Recharge from rainfall	Recharge from other sources										
1	Bihar	20.66	3.48	3.36	3.81	31.31	2.82	28.49	10.36	2.37	12.73	0.6	17.52	45
2	Delhi U.T	0.09	0.02	0.014	0.22	0.34	0.03	0.31	0.14	0.25	0.39	0.25	0.02	127
3	Haryana	3.62	3.1	1.03	3.6	11.36	1.06	10.3	13.32	0.6	13.92	0.56	−3.58	135
4	Himachal Pradesh	0.4	0.02	0.11	0.03	0.56	0.03	0.53	0.16	0.11	0.27	0.07	0.3	51
5	Jharkhand	5.61	0.06	0.73	0.16	6.56	0.57	5.99	0.63	0.72	1.35	0.17	5.19	23
6	Punjab	5.75	13.21	1.32	5.64	25.91	2.52	23.39	34.05	0.77	34.81	0.97	−11.63	149
7	Rajasthan	9.06	0.69	0.27	2.49	12.51	1.26	11.26	13.79	1.92	15.71	2.32	0.9	140
8	Uttar Pradesh	41.97	11.52	4.6	18.25	76.34	4.75	71.58	48.35	4.41	52.76	6.44	19.01	74
9	Uttarakhand	1.1	0.22	0.24	0.43	2	0.03	1.97	0.84	0.15	0.99	0.3	0.82	50
10	West Bengal	18.71	5.26	1.51	3.85	29.33	2.77	26.56	10.84	1	11.84	1.53	14.19	45
	Total (09 State & 01 U.T.)	106.97	37.58	13.184	38.48	196.22	15.84	180.38	132.48	12.3	144.77	13.21	42.74	Ranges from 23 to 149%

Source Central Ground Water Board (CGWB), India

The groundwater resources of confined aquifers have been built up over considerably long period of time and are annually not replenishable; hence they have been considered as finite in the present context. Accordingly, the development plan is based on draft from tube wells for their life period considered to be 25 years. Account of the number of wells in each state shows that Uttar Pradesh ranks first in the case of wells numbers in confined aquifers and Uttarakhand ranks last. This might be the reason for the high and low draft of groundwater in Uttar Pradesh and Uttarakhand, respectively.

The Net Ground Water Availability for future irrigation in the entire Indo-Gangetic basin, is assessed to be 180.38 BCM. To prepare the developmental plan the net available resources were divided by the unit draft/annum, of a representative tube well tapping the dynamic zone of the unconfined aquifer. This gives the number of additional tube wells, which are required to be constructed to develop the available dynamic resources. The numbers of tube wells were multiplied by the unit cost of tube well construction. The unit cost is given due to considerations of the nature of aquifer material, depth, and local factors. As per GEC 2015, state-wise groundwater resources, status of utilization, and stage of groundwater development in the IGB are depicted in Table 21.1.

21.5 Conclusion

The unconfined aquifer is covering an area of 466,007 km^2, having holding substantial groundwater resources for future irrigation development. Groundwater resources in the states of Uttarakhand, U.P., Bihar, Jharkhand, and West Bengal are available for future irrigation development. As per GEC Methodology, 2015, Net Ground Water Availability for future use in the I-G basin is maximum in the Uttar Pradesh (U.P.) state (71.58 BCM/yr.) and minimum in the Delhi U.T. (0.31 BCM/yr.). Comparative studies of two consecutive estimations indicate that total groundwater extraction in the year 2017 is 145 BCM, which was 117 BCM/yr. in the year 2009. At present stage of groundwater extraction in the I-G basin ranges from 23% (Jharkhand state) to 149% (Punjab state). Groundwater availability for future irrigation use in the year 2017 was 42.74 BCM/yr, which was 57.68 BCM/yr in the year 2009. However, it is worthwhile to point out that development potential of aquifers down to different depths should not be separately calculated, because it is ultimately one unified system on regional basis and any development carried out in the topmost aquifer will have its bearing on underlying aquifers. Demand and supply-side management to be done in a judicious manner to maintain long-term sustainable yield from the IGB aquifer systems. Conjunctive use of surface water and groundwater along with water conservation and location-specific interventions would be beneficial impact on the basin. To minimize the wastage of water in the irrigation sector, the viable options would be the precision agriculture which includes drip and sprinkler irrigation methods.

Acknowledgements The authors would like to acknowledge the support and encouragement received from Regional Director, Rajiv Gandhi National Ground Water Training and Research Institute (RGNGWTRI), Naya Raipur, Chhattisgarh during preparation of this book chapter. The author would also like to place on record his heart-felt gratitude to the Chairman of Central Ground Water Board (CGWB) for according kind permission and help in publishing this book chapter in the Springer Nature Edited Volume.

References

Acharyya SK, Shah BA (2007) Arsenic-contaminated groundwater from parts of Damodar fan-delta and west of Bhagirathi River, West Bengal, India: influence of fluvial geomorphology and quaternary morpho-stratigraphy. Environ Geol 52(3):489–501

Amarasinghe UA, Muthuwatta L, Surinaidu L, Anand S, Jain SK (2016) Reviving the Ganges water machine: potential. Hydrol Earth Syst Sci 20(3):1085–1101

Burrard S (1915) On the origin of the Indo-Gangetic trough, commonly called the Himalayan Foredeep. Proc R Soc Lond Ser A 91(628):220–238

CGWB (1997) Delineation of deep fresh water artesian aquifers in Ganga Basin by Singh, B.K., Chandra, P. C. and Srivastava, M.M.; CGWB Unpub. Report Manuscript

CGWB (2017) Report of the ground water resource estimation committee (GEC-2015)

Kamra SK (2007) Water table decline: causes and strategies for management. On-Farm Land Water Manage 103–109

Khattri KN (1987) Great earthquakes, seismicity gaps and potential for earthquake disaster along the Himalaya plate boundary. Tectonophysics 138(1):79–92

Mathur SM (2003) Physical geology of India, 2nd edn. National Book Trust, India

Middlemiss CS (1900) General report. Geological survey of India

Pathak, B.D., Sinha, B.P.C., Kidwai, A.L., Raju, T.S., Chaturvedi, P.C. and Singh, B.K. (1985) Tectonics, subsurface geology and groundwater occurrences in Ganga basin, and proposal for deep drilling with special reference to Uttar Pradesh. *Central Groundwater Board*, p.49

Parkash B, Sharma RP, Roy AK (1980) The Siwalik Group (molasse)—sediments shed by collision of continental plates. Sed Geol 25(1–2):127–159

Prasad, T. (1993) Groundwater development for economic emancipation in the lower Ganges Basin: Problems, Prospects, and Strategies. *Groundwater Irrigation and the Rural poor: Options for Development in the Gangetic Basin*, pp.139–46

Rao R, Ramachandra Rao MB(1973) The subsurface geology of the Indo-Gangetic Plains

Sinha R, Tandon SK, Gibling MR, Bhattacharjee PS, Dasgupta AS (2005) Late Quaternary geology and alluvial stratigraphy of the Ganga basin. Himalayan Geol 26(1):223–240

Shah BA (2014) Arsenic in groundwater, Quaternary sediments, and suspended river sediments from the Middle Gangetic Plain, India: distribution, field relations, and geomorphological setting. Arab J Geosci 7 (9):3525–3536

Singh IB (1996) Geological evolution of Ganga Plain—an overview. J Palaeontol Soc India 41:99–137

Singh IB (2007) The Ganga River. In: Large rivers: geomorphology and management, pp. 347–371. Wiley Chichester

Singh OP, Nambiar KV (1993) Quaternary geology and geomorphology of a part of Ganga Basin in Districts Badaun and Shahjahanpur, Uttar Pradesh; GSI Unpub. Report

https://www.britannica.com/place/gondwana-historical-region-India. Accessed 09 Apr 2020

Part III
Large Dams and River Systems

Predicting the Distribution of Farm Dams in Rural South Africa Using GIS and Remote Sensing

22

Jonathan Tsoka, Jasper Knight, and Elhadi Adam

Abstract

Farm dams are critical for their role in livestock watering and irrigation in areas that experience seasonal rainfall variability, but although cumulatively they store large volumes of water they are rarely considered as part of the water resources base or the water management system of a river basin. This study mapped farm dams and identified the factors influencing their spatial distribution in a third-order agricultural catchment of KwaZulu-Natal Province, South Africa, using GIS and remote sensing techniques. Using Landsat 8 imagery and verified using Google Earth, 864 farm dams were identified. Six physical properties of the catchment (slope, aspect, elevation, land use, soil type, and geology) were then used in a predictive model to evaluate the extent to which they influence the siting of the farm dams. Results showed that all of these factors except for aspect gave significant p-values. Slope, geology, and elevation have a greater influence on the farm dam location than soils. A multivariate logistic regression model was then created for predicting future farm dam sites. Slope, elevation, land use, and geology were fitted to the model, but aspect and soil type had no significant p-values in the model. The model was validated using 300 farm dam sites, and predicted 60% of sites correctly. This study highlights that remote sensing and GIS methods can usefully inform on strategies for more efficient water management and infrastructure planning through the locating of farm dams, and can therefore help in more efficient water use in locations that are already water-stressed.

Keywords

Catchment properties · Farms dams · Landsat imagery · Regression modelling · Water resources management

22.1 Introduction

Farm dams are privately owned and built water management structures set in agricultural landscapes that are used to intercept catchment runoff and store it for later use (Nathan and Lowe 2012; Tingey-Holyoak et al. 2013; Habets et al. 2014), most commonly for crop irrigation and livestock watering (Boardman et al. 2009). Farm dams have been constructed most commonly in locations that are water-stressed or where there is a significant dry season (Meigh 1995; Liebe et al. 2005). Therefore, farm dams can help provide reliable supplies of water all year round and reduce the need for costly groundwater extraction

J. Tsoka · J. Knight (✉) · E. Adam
School of Geography, Archaeology and Environmental Studies, University of the Witwatersrand, Johannesburg 2050, South Africa
e-mail: jasper.knight@wits.ac.za

for farms and rural communities (Arshad et al. 2013; Habets et al. 2014; Morris et al. 2019). Farm dams are particularly common—and therefore literature exists—in Australia, Brazil, South Africa, India, and New Zealand (e.g. Roohi and Webb 2012; Shao et al. 2012; Sawunyama 2013) but it is notable that they are not widely reported, and are thus under-researched, in many other parts of the world.

In South Africa, the number of farm dams has increased significantly in recent years (Hughes and Mantel 2010), with a nine-fold increase between 1944 and 1984 in some catchments (Maaren and Moolman 1986). The basic structure of a farm dam is that it is usually excavated, having a downslope bank or barrier to check streamflow, is located in a relatively shallow sloping environment, and may have water inlets and outlets that are either natural or artificial (Berg et al. 1994; Callow and Smettem 2009; Shao et al. 2012; Kollongei and Lorentz 2014). In South Africa, Section 21(b) of the National Water Act (1998) states that if more than 10,000 m^3 of water is stored or more than 50 m^3/day of water is drawn from a surface water resource on a single registered property, then this water use itself must be registered. Some farm dams may fall outside of this definition and thus may have an unregulated role in catchment water balance (Sawunyama 2013). Studies on farm dams globally have focused on their role in water management (Brainwood et al. 2004; Callow and Smettem 2009; Arshad et al. 2013; Fowler et al. 2015), sediment management (Neil and Mazari 1993; Boardman and Foster 2011; Pulley et al. 2015), and their safety concerning flood risk (Pisaniello 2009, 2010; Tingey-Holyoak et al. 2013). Although farm dams are of interest for these reasons, there is limited information on their spatial distributions or the land surface factors that favour their locations. A better understanding of these issues can help in siting dams more efficiently for agricultural water management, and to limit the negative downstream impacts such as poor water quality (Brainwood et al. 2004; Mantel et al. 2010).

Different remote sensing data types have been used to map the number and distribution of farm dams, including air photos, Landsat imagery, and Google Earth (e.g. Meigh 1995; Dare et al. 2002; Liebe et al. 2005; Sawunyama et al. 2006; Bhagat and Sonawane 2011; Roohi and Webb 2012; Shao et al. 2012). For example, in South Australia, Roohi and Webb (2012) used Landsat imagery (30 m spatial resolution) and Google Earth imagery (15 m spatial resolution) and detected 99% of all farm dams when using 4/5 band ratio and Normalized Difference Water Index (NDWI) values, which enhance the contrast between water bodies and surrounding land surfaces. In the same area, Dare et al. (2002) used Ikonos imagery (4 m multispectral resolution) and detected 92% of all farms dams. In India, Bhagat and Sonawane (2011) used Landsat 7 data with Surface Wetness Index (SWI) values, Normalized Difference Vegetation Index (NDVI) values, and land surface slope to successfully detect 92% of all farm dams. In Brazil, Rodrigues et al. (2012) used Maximum Likelihood classification of land uses using Landsat 7 imagery and achieved 96% accuracy in farm dam detection. Different approaches to spatial data analysis have also been used to address farm dam siting as well as water management more generally. For example, Singh et al. (2017) used GIS-based multi-criteria decision analysis in evaluating potential sites for rainwater harvesting in India by employing suitability criteria and GIS-based Boolean logic. Kahinda et al. (2009) used a similar approach for the suitability of rainwater harvesting sites in South Africa. Thus, different remote sensing data types and methods have been applied to the mapping of farm dams and water management systems. Although many different factors contribute to the positioning of farm dams in a landscape, such as geology, soils, relief, rainfall, agriculture type, and farmer socioeconomic characteristics (Tingey-Holyoak et al. 2013; Morris et al. 2019), these factors have not been evaluated systematically. However, understanding the role of such factors has implications for identifying strategies to increase the efficiency of catchment water management (Sawunyama 2013; Habets et al. 2014; Tingey-Holyoak 2014).

This study aims to assess the factors that influence the spatial distribution of farm dams in rural KwaZulu-Natal Province, South Africa, using remote sensing and GIS methods. In detail, this paper (1) describes the data collection and analytical methods employed in the study, (2) maps the spatial patterns of key catchment physical properties and the distribution of farm dams based on automated classification of Landsat 8 data, (3) examines the factors influencing this distribution using principal component analysis, and (4) develops a spatial predictive model for farm dam sites using multivariate logistic regression.

22.2 Study Area

The study area is the U2 sub-catchment (4439 km^2), a tertiary (third-order) catchment within the uMngeni River basin in KwaZulu-Natal Province, south-east of South Africa (Fig. 22.1). Headwater areas of this catchment fall within the Drakensburg mountain range and reach a maximum altitude of 2071 m asl; the river mouth lies at sea level where the river exits directly into the Indian Ocean. The terrain in the catchment is generally rugged and with high local relief. Mean annual rainfall in the catchment is 921 mm with peak values in the austral summer of December to February. Annual evaporation is 1214 mm and natural runoff is 152 mm (674 million m^3) (Schulze and Schütte 2019). Several large dams exist on the uMngeni River (Midmar, Albert Falls, Nagle, Inanda) with a total storage capacity of 957.5 million m^3 and these are used currently to manage river water flow and water supply, mainly for domestic but also for agricultural irrigation purposes. In this water management framework, smaller and more numerous farm dams are not considered.

The basement geology of the region comprises mainly Karoo-age sandstones and mudstones, and these form Cambisols, Ferralsols, and Luvisols (Turner 2000; Chaplot 2013). These soils are generally shallow and highly weathered, and under undisturbed conditions, the Ferralsols, in particular, are resistant to erosion. However, upon disturbance by land use change, they become highly susceptible to erosion (Dlamini et al. 2011; Makaya et al. 2019). In addition, these soils are found in different topographic contexts in the region in response to variations in geology and relief, and have different thicknesses and infiltration capacities (Chaplot 2013; Atkinson et al. 2020). As a result of the climatic and soil conditions in combination, agriculture is an important regional land use and economic activity (Roberts et al. 2003; Troskie 2013). Of the 6.5 million hectares of farming land in the province, 18% is used for crops and 82% for livestock, suggesting the need for farm dams for livestock watering (Meissner et al. 2013). KwaZulu-Natal hosts 19% of South Africa's total livestock numbers, 15% of the national dairy producers, 13% of its goats, and 10% of its pigs (Meissner et al. 2013). Ensuring sustainable water use for agricultural production is therefore an important priority.

22.3 Methods

22.3.1 Digital Data Collection and Analysis

For identification and mapping of farms dams, Landsat 8 OLI imagery (date: 14 April 2017, 0840 h) with 1 arc second resolution (30 m), was acquired from the USGS website. The image was corrected for atmospheric and radiometric errors using the FLAASH module in ENVI SAT 5.3 (Lu et al. 2004). Supervised image classification was undertaken using a Random Forest classifier. This was selected ahead of other supervised classification techniques because of its relatively higher classification accuracy, faster processing speed, and higher stability (Chan and Paelinckx 2008). Random Forest is an ensemble classifier that employs classification and regression trees to make a prediction (Breiman 2001). Trees are created by drawing a subset of training samples through the bagging approach (Belgiu and Drăguţ 2016). The in-bag samples (two-thirds of the sample) are used to train the trees whilst the out-of-bag (the remaining one-third of the sample) are used for cross-validation. Resources such as topographic maps and Google Earth were used

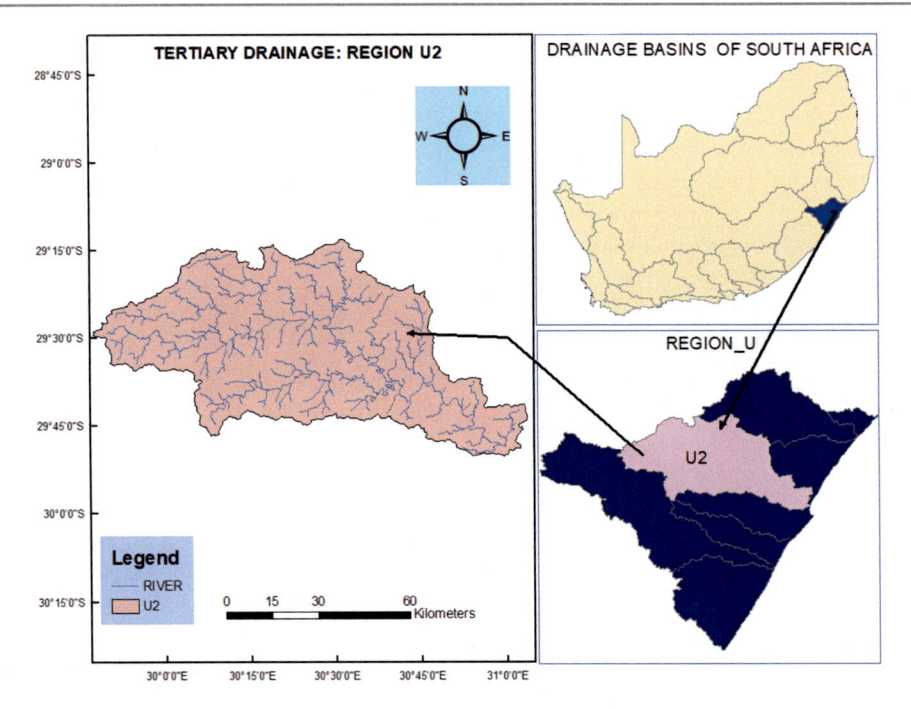

Fig. 22.1 Map of the study area U2 sub-catchment within the uMngeni River basin, South Africa

for ground-truthing. A confusion matrix was used for computing user and producer accuracies (Adam et al. 2014), and the Kappa coefficient was used to check the accuracy of the classification. Representative training and validation samples for each Land Use/Land Cover class (LULC) were obtained from the Landsat image. The six LULC classes used in this study were bare surfaces (exposed rocks and non-vegetated ground), grasslands (semi-natural grassland), plantations (forests), cropped land (agriculture), built-up areas (industrial, residential, and commercial structures), and water bodies (rivers, ponds, and reservoirs). The latter includes farm dams. A total of 220 training and validation samples per class were collected. The examination of LULC classes was supported by the calculation of the Normalized Difference Vegetation Index (NDVI) values for each pixel from the Landsat image. Low or negative NDVI values are typical of water surfaces such as large river dams, whereas bare soils have values around 0–0.1, and vegetated surfaces of greater than 0.1. NDVI values were used to help select the training sites for different LULC classes.

As classification errors can arise due to noise, spectral confusion or problems with the classification algorithm used, a quality check on the classified pixels within the image is essential. This can be done using a confusion matrix for the regions of interest to calculate the Kappa coefficient, producer accuracy, user accuracy, and overall accuracy. These items were calculated using standard equations (Richards 2012). The results of this analysis are presented below.

22.3.2 Digital Data Integration

A Digital Elevation Model (DEM) of the study area was established using Shuttle Radar Topography Mission (SRTM) data (30 m pixel resolution) obtained from the USGS Earth Explorer. These data were extracted and mosaicked in ArcMap. The hydrological network and flow directions were extracted from the DEM. The resulting stream order raster was used as input for the stream network raster, and the flow direction raster was used as the input flow direction raster. The resulting raster was a stream network, which was converted to polyline features, and stream orders 5–11 were extracted. Soil and geology shapefile data at 1:250,000-scale were acquired from the European Soil Data Centre and the City of Johannesburg, respectively. These data were projected to a UTM projection used as input for the multivariate logistic regression model.

22.3.3 Evaluation of Factors Influencing Farm Dam Spatial Distributions

A Principal Component Analysis (PCA) was carried out to determine the factors contributing to the distribution of farm dams in the study area. Sample site data were run in SAS 9.4 statistical package, producing six principal components. These components were analysed using a correlation matrix, eigenvectors, scree plots, and eigenvalues of the correlation matrix. The variables that had the highest loadings were identified and used to develop a predictive model. The assessment of the locational properties of farm dam sites was evaluated using logistic regression, and the creation of a prediction surface using multivariate logistic regression undertaken in R-Studio. Figure 22.2 illustrates the fitting process of the model, including the iterative checking and re-fitting procedure before arriving at the final multivariate logistic model.

22.3.4 Predictive Modelling of Dam Distributions

Elevation (SRTM DEM), soil, geology, LULC raster datasets, and dam site and non-site locations were prepared and added to the model in R-Studio. Six independent variables were examined (elevation, slope, aspect, soil, geology, land use). Each covariate was individually tested for association with dam site presence. In cases where the p-values were greater than 0.05, the variable was discarded. The likelihood ratio test method of testing the significance of models was also used. This assesses the difference between the error of not knowing the independent covariates, and the error when the independents are included in the model (Crema et al. 2010). This test follows a χ^2 distribution with degrees of freedom equal to the difference in the number of parameters in the model with independents compared to the model with only the intercept.

To further examine the relationship between predictor and response variables, the covariates were broken down into subgroups to increase their predictive power. These subgroups are elevation (low 0–700 m, medium 701–1600 m, high 1601–2100 m asl), aspect (southeast-facing 91°–180°, northeast-facing 0°–90°), slope (gentle 0°–5°, steep 6°–15°, very steep 16°–88°), soil (Fr20 (Ferralsol) model, Lo43 (Orthic Luvisol) model), geology (Adelaide, Karoo, Pietermaritzburg, Volksrust, which correspond to mainly sandstone and mudstone bedrock types present in the study area), and land use (grassland, agricultural land). Each of the subgroup's relationships with the dependent variable was tested to establish if they could be of predictive value. Variables at $p = 0.05$ significance level were considered for the model, and those that were not significant were discarded.

A binary multivariate logistic regression model does not show high significance levels for all covariates. Different combinations were tested to achieve the best model, evaluated based on the Akaike Information Criterion (AIC) (Crema et al. 2010). The smaller the AIC, the better the model performance. The coefficients of the logistic regression model and raster map algebra were used to generate a prediction surface. From this, the relative probability of a site existing at a particular location can be calculated.

22.4 Results

22.4.1 Landsat Image Classification

Based on LULC classification and informed by NDVI data, 54% of the study area is covered by grasslands, 20% by plantations, 13% by bare soil, 10% by agricultural land, 2% by water bodies, and 0.7% by built-up areas (Fig. 22.3).

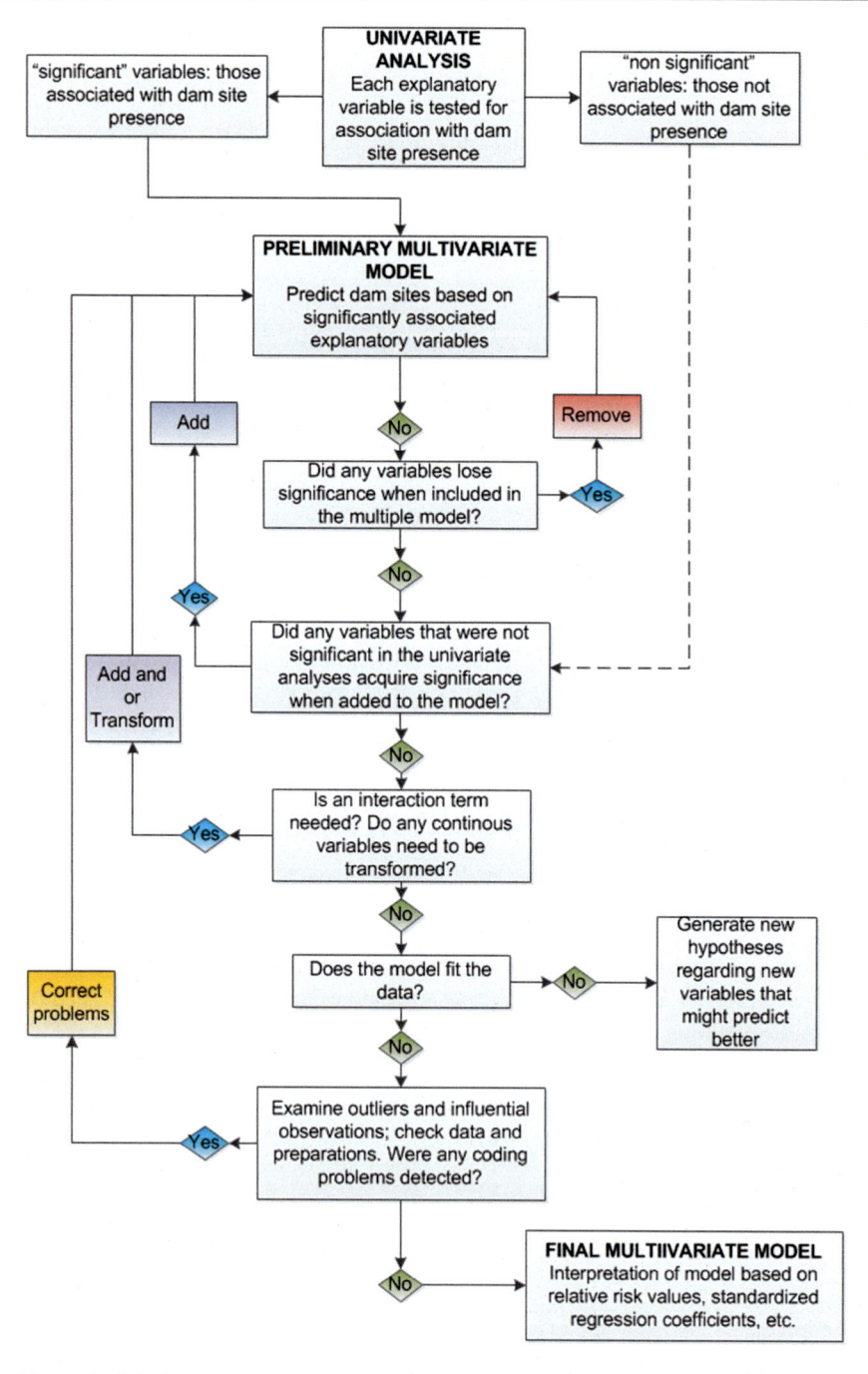

Fig. 22.2 The methodological framework employed in the multivariate logistic regression model used for dam site location prediction

Accuracy of the Landsat image classified using Random Forest was evaluated using a confusion matrix by comparison with ground control points extracted from Google Earth that were not used in the training dataset. The recommended threshold value for accuracy assessment is 85% (Anderson et al. 1976). Results show that the overall accuracy for the classified Landsat image was 96% (Table 22.1). Water bodies and built-up areas had higher user and producer accuracies compared to other land cover classes. The grass cover class had 91.3% accuracy and had the lowest user accuracy. The producer accuracy for the bare soil class was the lowest with a value of 94.3%. The reason why all classes could not attain 100% accuracy was that some pixels had mixed grass and bare soil and resulted in misclassification, leading to low producer and user accuracies.

22.4.2 Spatial Distribution of Farm Dams

In total, 864 farm dams were identified (Fig. 22.4). Complete Spatial Randomness (CSR) is often used for testing for spatial distribution point pattern analysis. It assumes that points follow a homogeneous Poisson distribution. The distribution of farm dams in the study area was tested for spatial randomness using the quadrat and nearest neighbour methods. Quadrat analysis was carried out using the quadrat count function in Spatstat (R-Studio). The density of farm dams is higher in the northwestern part of the catchment, confirmed by a χ^2 test which shows that p-values are both significant ($<\alpha = 0.05$). Therefore, the farm dam distribution pattern is spatially clustered and does not exhibit CSR. Nearest-neighbour analysis indicates a clustered distribution pattern for farm dams across a distance of 300–1200 m. Concerning farm dam size, the surface area covered ranges 0.02–22 ha. Dams between 0.022 and 6 ha surface area account for 90% of the total number of farm dams, whilst the rest account for 10%. About 90% of the largest farm dams are found in the northwest part of the catchment.

Concerning the underlying geology, 27% of the farm dams were constructed on Karoo rocks and have the highest density of 38 dams/100 km^2. Adelaide and Volksrust mudstones account for 23 and 16% of the total number of dams, with a density of 35 and 31 dams/100 km^2, respectively. These high values contrast with the Mapumulo metamorphics that account for 0.6% of all dams and 2 dams/100 km^2. Based on underlying soil type, 71% of farm dams are located on Orthic Luvisols (soil Lo43-2b) and have a density of 28 dams/100 km^2. Humic

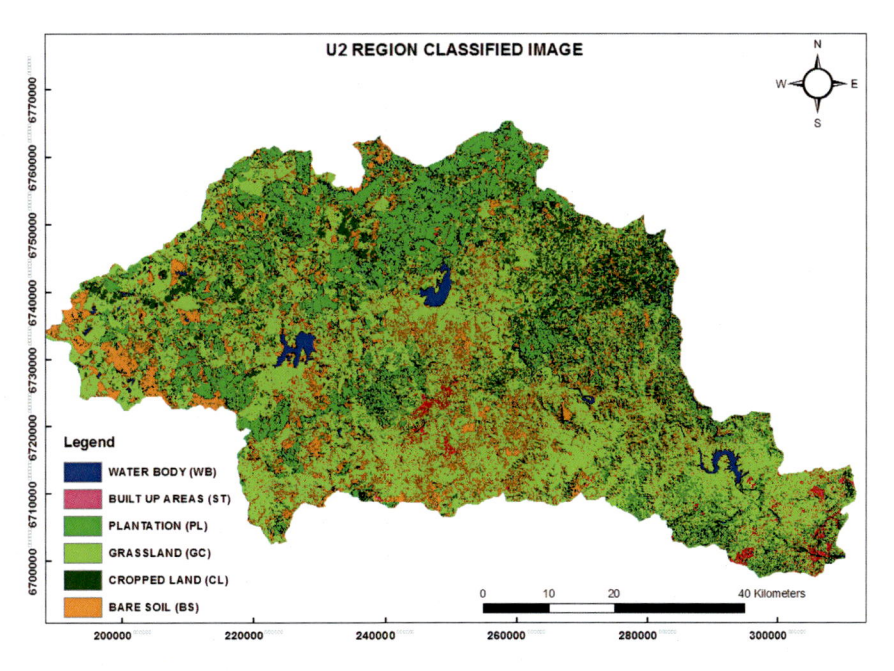

Fig. 22.3 Classified Landsat 8 image of the study catchment

Table 22.1 Confusion matrix for validation

Land cover class	Producer accuracy (%)	User accuracy (%)
Agricultural land (cropped)	97.8	94.7
Plantations	98.7	93.8
Water bodies	98.7	100
Built-up areas	100	100
Grassland	95.4	91.3
Bare soil	94.3	95.7
Kappa coefficient (K)	0.956	
Overall accuracy (%)	96.4	

Ferralsols (soil Fr20-3b) accounting for 29% of farm dams and have a density of 14 dams/100 km^2. Cambisols only account for 0.1% of dams. With respect to the elevation, 36% of farm dams are located between 1201 and 1500 m asl, 31% between 901 and 1200 m asl, and 24% between 601 and 900 m asl. Considering the slope, 50% of dams are on gentle slopes of 0°–5° and 42% on a moderate slope of 6°–10° (Fig. 22.4). Most (51%) dams are located in grasslands, 35% on cropped arable land, and 11% on forest plantations.

22.4.3 Principal Component Analysis

Results of PCA for the six physical factors affecting dam site selection are presented in Table 22.2. Most associations are weak. Geology and elevation have a negative correlation of −0.306, and geology and land use have a slight positive correlation of 0.337. Elevation and soil have negative loadings of −0.420 and −0.426. Eigenvalue loadings indicate that PC1 has a positive loading for geology (0.581) and land use (0.502) but negative loadings for soil (−0.426)

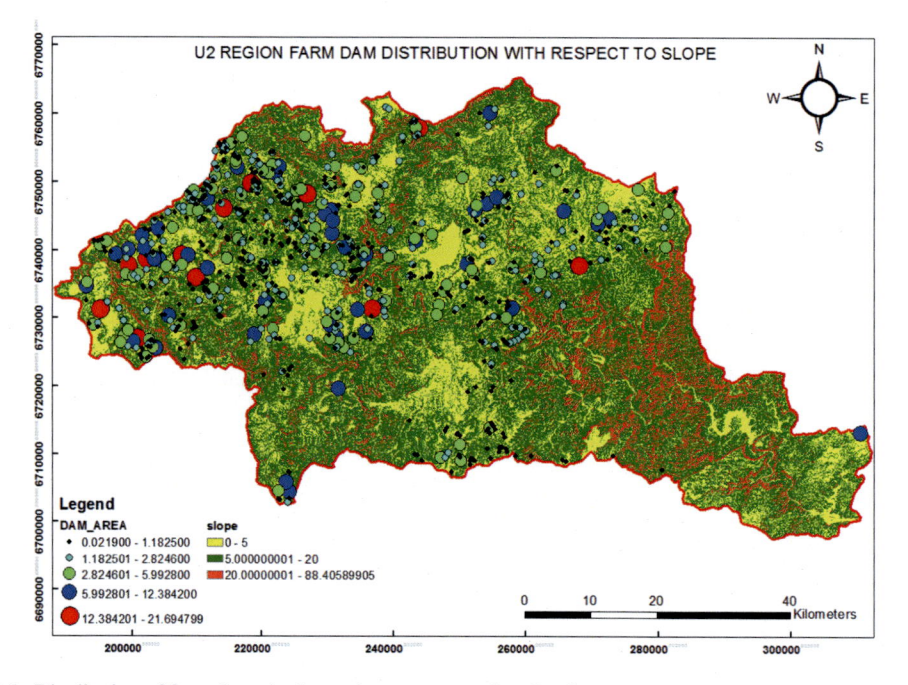

Fig. 22.4 Distribution of farm dams in the study area concerning the slope

Table 22.2 Correlation matrix of the six major landscape properties examined as variables in the dam site prediction model

	Elevation	Slope	Aspect	Geology	Soil	Land use
Elevation	1.0000	−0.0901	0.0087	−0.3061	0.0557	−0.2047
Slope	−0.0901	1.0000	0.0878	0.0637	−0.1443	0.0695
Aspect	0.0087	0.0878	1.0000	0.0161	0.0245	0.0781
Geology	−0.3061	0.0637	0.0161	1.0000	−0.3155	0.3376
Soil	0.0557	−0.1143	0.0245	−0.3155	1.0000	−0.1959
Land use	−0.2047	0.0695	0.0781	0.3376	−0.1959	1.0000

and elevation (−0.420). PC2 has high positive loadings for two terrain variables, slope (0.568) and aspect (0.790), whilst the loading of elevation is only 0.107 and geology is −0.164. In PC3 there are high positive loadings for soil (0.628) and reasonably high negative loadings for elevation (−0.499) and slope (−0.459). Eigenvalues indicate that two components describe 47% of the total variance and three components explain 63%. There is a preponderance of terrain variables over other physical variables.

the AIC. The model with the smallest AIC value was considered the best. The AIC function in R-Studio suggested that the best combination of covariates to include in the final model were a gentle slope (1°–5°), medium elevation (701–1600 m asl), land use and geology, because these variables were significant with a p-value of <0.001. The other subcategory variables had model combinations that had AIC values higher than 285.56, and so were rejected. From the covariates, the following expression was derived:

$$P(\text{location}|\text{farm dams}) = -3.0646 + (\text{gentle slope} * 2.1610) \\ + (\text{medium elevation} * 1.1779) + (\text{land use} * 0.3457) \\ + (\text{geology} * -0.1205) \quad (22.1)$$

22.4.4 Predictive Modelling of Farm Dam Site Location

The multivariate analysis involved splitting the six dependent variables into subcategories, such as three different elevational ranges, two different aspects, three different slope steepnesses, etc. These subcategories in turn also underwent univariate analysis to evaluate whether these could be used as predictors of the dependent variables. Those subcategories that were significant at $\alpha = 0.05$ level were considered for the model, with the others being discarded. These subcategory predictors were used in the multivariate regression model with their utility in the model measured by

The prediction surface was created using the raster map algebra and coefficients of the logistic regression. A total of 291 points were loaded in the model, of which 191 points were dam sites whilst 100 were non-sites. Of the dam sites, 155 were found in areas with probabilities of 0.5 or above on the prediction surface (set as the cut-off for site presence or absence), whilst 36 sites had probabilities less than 0.5. Of the 100 non-sites, 15% had probabilities of 0.5 or higher whilst 85% were less than 0.5. The model was tested using 500 other farm dam points not used in building the model for validation. This validation process found 87% of the points lay in areas with a probability greater than 0.5. Figure 22.5

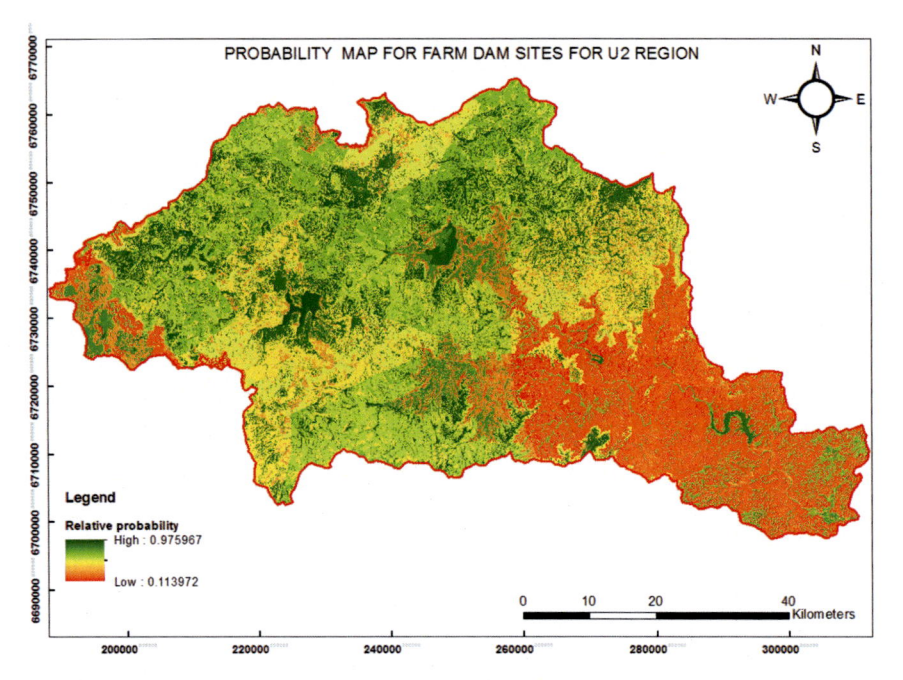

Fig. 22.5 Probability map for farm dam site locations in the study area

presents a probability map for farm dam site locations. The green colour shows areas with the highest suitability for the location of farm dams where typical presence localities have a probability of the presence of 0.5 and above. Absence may be due to factors not present in the model, such as socioeconomic conditions.

22.5 Discussion

There is a general lack of data on the number, size, and locations of farm dams, both within South Africa and globally. In the study area, a total of 864 farm dams were identified. Employing medium-resolution Landsat 8 imagery, verified by Google Earth, proved a useful methodology for mapping farm dam distributions, and compares well with previous studies using these methods (e.g. Meigh 1995; Liebe et al. 2005; Sawunyama et al. 2006). In the study area, there is a higher density of dams in the northwest and central areas where crop production and livestock farming dominate. The dominant soil types in the areas with most agricultural land are Ferralsols, which are iron-rich soils with high organic matter content, and are better suited for low input agriculture (Dowling and Fey 2007), and Orthic Luvisols, which have a clay fraction and moderately weathered, most favoured by smallholder rural farmers because of their ease of cultivation (Stocking and Murnaghan 2001). In addition, of the farm dams identified, 90% are less than 6 ha in area and 50% of all dams are found on 1°–5° slopes. This highlights the local-scale nature of farm dams, where individual farmers make use of distinctive topographic contexts within their farms when building dams. Such farm dams can therefore be seen as pragmatic responses to the need for local water management (Arshad et al. 2013; Tingey-Holyoak et al. 2013; Ogundeji et al. 2018).

The correlation matrix (Table 22.2) shows that the six major landscape properties used as variables in predictive modelling are generally weakly correlated. The strongest correlations (positive and negative) are with geology, and thus this can be considered as a first-order control on a range of other landscape properties such as soils, relief, and elevation. Of these different variables, eigenvectors were used to see which factors are of greatest importance. PC1 is correlated with increased land use and geology scores, and with decreased elevation and soil scores. This first component reflects the physical landscape attributes of farm dam sites. PC2 is correlated with increased slope and aspect and can be viewed as a measure of water flow direction across the landscape. PC3 comprises increased soil and aspect scores and decreased elevation and slope scores. Integration of the six major landscape properties resulted in the development of a spatial predictive model for the location of farm dams (Fig. 22.5). The results obtained in this study are similar to those by Roohi and Webb (2012) who plotted the spread of farm dams in western Victoria State, Australia, for the period 1973–2004 using Landsat imagery. Although based on simple mapping rather than spatial analysis, they showed that farm dam locations are closely coupled to land use, with dams concentrated in areas of dryland pasture and therefore used for cattle watering (e.g. Boardman et al. 2009). This association is also found in this study where slope and relief context, geology and land use are the key landscape factors determining farm dam location. Although the spatial model is useful and has reasonable predictive power, the model did not capture the distributions of all farm dams, and it is evident that some farm dams were constructed in suboptimal locations. Reasons potentially accounting for this may include socioeconomic factors related to farmer wealth, access to machinery or technology, and farm size; and also the distance from roads or other infrastructure. These elements could be captured in a future iteration of the predictive model.

The presence of farm dams has implications for water storage and runoff yield, especially in semi-arid areas where rainfall is seasonal or episodic (Callow and Smettem 2009; Souza da Silva et al. 2011), and can positively contribute to local biodiversity by providing new aquatic habitats (Markwell and Fellows 2008; Brainwood and Burgin 2009). However, farm dams also have negative impacts on downstream water volume and quality (Brainwood et al. 2004; Mantel et al. 2010), and sediment cascades through slope and river systems (Neil and Mazari 1993; Verstraeten and Prosser 2008; Boardman and Foster 2011). These impacts that farm dams may have upon catchment systems have not been considered as part of catchment hydrological or sediment models, and this may be an important future research theme, especially in locations such as Australia and South Africa where farm dams are common (Verstraeten and Prosser 2008; Makaya et al. 2019). Thus, understanding the distribution and properties of (often unregulated) farm dams has wider catchment-scale implications. This study shows that at least 864 farm dams exist in the mapped area, covering some 1959 ha in surface area. The potential implications of such a water body to regional hydrology and microclimate (through evaporation potential) have not been examined (e.g. Hipsey and Sivapalan 2003), but this is potentially significant in water-stressed environments (Callow and Smettem 2009; Fowler et al. 2015). Likewise, the economic role of farm dams to agricultural production in such areas has also not been examined in detail. These are important future research directions in farm dam studies.

22.6 Conclusions

Farm dams are critical for their role in livestock watering and irrigation in areas that experience seasonal rainfall variability, but although cumulatively they store large volumes of water they are rarely considered as part of the water resources of a river basin. This study showed that farm dams are common in the uMngeni River catchment in KwaZulu-Natal, South Africa, and have potential significance for wider issues of water and sediment management within the river system. Using GIS and remote sensing techniques, six key physical properties of the catchment (slope, aspect, elevation, land use, soil type, and geology) were quantified and used in a spatial prediction model to evaluate the factors contributing to the siting of farm dams within the catchment. A multivariate logistic regression model showed that slope, elevation, land use, and geology are the most important landscape factors, with the model predicting 60% of dam sites correctly. Mapping and inventorizing of farm dams can yield better estimates of area and volume of water storage within a river basin, and can contribute to more effective water management.

References

Adam E, Mutanga O, Odindi J, Abdel-Rahman EM (2014) Land-use/cover classification in a heterogeneous coastal landscape using Rapid-Eye imagery: evaluating the performance of random forest and support vector machines classifiers. Int J Remote Sens 35:3440–3458

Anderson JR, Hardy EE, Roach JT, Witmer RE (1976) A land use and land cover classification system for use with remote sensor data. USGS Prof Paper 964, Washington DC: US Government Printing Office, p 28

Arshad M, Qureshi ME, Jakeman AJ (2013) Cost-benefit analysis of farm water storage: surface storage versus managed aquifer storage. In: 20th International congress on modelling and simulation, Adelaide, Australia, 1–6 Dec 2013, pp 2931–2937

Atkinson J, de Clercq W, Rozanov A (2020) Multi-resolution soil-landscape characterisation in KwaZulu-Natal: using geomorphons to classify local soilscapes for improved digital geomorphological modelling. Geoderma Reg 22:e00291

Belgiu M, Drăguţ L (2016) Random forest in remote sensing: A review of applications and future directions. ISPRS J Photogram Remote Sens 114:24–31

Berg RR, Thompson R, Little PR, Görgens AH (1994) Evaluation of farm dam area-height-capacity relationships required for basin-scale hydrological catchment modelling. Water SA 20:265–272

Bhagat VS, Sonawane KR (2011) Use of Landsat ETM+ data for delineation of water bodies in hilly zones. J Hydroinform 13:661–671

Boardman J, Foster I, Rowntree K, Mighall T, Parsons T (2009) Small farm dams: a ticking time bomb? Water Wheel 8(4):30–35

Boardman J, Foster IDL (2011) The potential significance of the breaching of small farm dams in the Sneeuberg region, South Africa. J Soils Sed 11:1456–1465

Brainwood M, Burgin S (2009) Hotspots of biodiversity or homogeneous landscapes? Farm dams as biodiversity reserves in Australia. Biodiv Conserv 18:3043–3052

Brainwood MA, Burgin S, Maheshwari B (2004) Temporal variations in water quality of farm dams: impacts of land use and water sources. Agric Water Manag 70:151–175

Breiman L (2001) Random forests. Mach Learn 45:5–32

Callow JN, Smettem KRJ (2009) The effect of farm dams and constructed banks on hydrologic connectivity and runoff estimation in agricultural landscapes. Environ Model Softw 24:959–968

Chan JCW, Paelinckx D (2008) Evaluation of Random Forest and Adaboost tree-based ensemble classification and spectral band selection for ecotope mapping using airborne hyperspectral imagery. Remote Sens Environ 112:2999–3011

Chaplot V (2013) Impact of terrain attributes, parent material and soil types of gully erosion. Geomorphology 186:1–11

Crema ER, Bevan A, Lake MW (2010) A probabilistic framework for assessing spatio-temporal point patterns in the archaeological record. J Archaeol Sci 37:1118–1130

Dare P, Fraser C, Duthie T (2002) Application of automated remote sensing techniques to dam counting. Austral J Water Res 5:195–208

Dlamini P, Orchard C, Jewitt G, Lorentz S, Titshall L, Chaplot V (2011) Controlling factors of sheet erosion under degraded grasslands in the sloping lands of KwaZulu-Natal, South Africa. Agric Water Manag 98:1711–1718

Dowling CE, Fey MV (2007) Morphological, chemical and mineralogical properties of some manganese-rich oxisols derived from dolomite in Mpumalanga province, South Africa. Geoderma 141:23–33

Fowler K, Morden R, Lowe L, Nathan R (2015) Advances in assessing the impact of hillside farm dams on streamflow. Austral J Water Res 19:96–108

Habets F, Philippe E, Martin E, David CH, Leseur F (2014) Small farm dams: impact on river flows and sustainability in a context of climate change. Hydrol Earth Syst Sci 18:4207–4222

Hipsey MR, Sivapalan M (2003) Parameterizing the effect of a wind shelter on evaporation from small water bodies. Water Res Res 39:1339

Hughes DA, Mantel SK (2010) Estimating the uncertainty in simulating the impacts of small farm dams on streamflow regimes in South Africa. Hydrol Sci J-J Sci Hydrol 55:578–592

Kahinda JM, Taigbenu AE, Sejamoholo BBP, Lillie ESB, Boroto RJ (2009) A GIS-based decision support system for rainwater harvesting (RHADESS). Phys Chem Earth 34:767–775

Kollongei KJ, Lorentz SA (2014) Connectivity influences on nutrient and sediment migration in the Wartburg catchment, KwaZulu-Natal Province, South Africa. Phys Chem Earth 67–69:12–22

Liebe J, Van De Giesen N, Andreini M (2005) Estimation of small reservoir storage capacities in a semi-arid environment: a case study in the Upper East Region of Ghana. Phys Chem Earth 30:448–454

Lu D, Mausel P, Brondizios E, Moran E (2004) Change detection techniques. Int J Remote Sens 25:2365–2407

Maaren H, Moolman J (1986) The effects of farm dams on hydrology. In: Schulze RE (ed) Second South African National Hydrology Symposium (Pietermaritzburg, 16–18 September 1985). University of Natal, Pietermaritzburg, pp 428–441

Makaya N, Dube T, Seutloali K, Shoko C, Mutanga O, Masocha M (2019) Geospatial assessment of soil erosion vulnerability in the upper uMgeni catchment in KwaZulu Natal, South Africa. Phys Chem Earth 112:50–57

Mantel SK, Hughes DA, Muller NWJ (2010) Ecological impacts of small dams on South African rivers Part 1: drivers of change-water quantity and quality. Water SA 36:351–360

Markwell KA, Fellows CS (2008) Habitat and biodiversity of on-farm water storages: a case study in Southeast Queensland, Australia. Environ Manag 41:234–249

Meigh J (1995) The impact of small farm reservoirs on urban water supplies in Botswana. Nat Res Forum 19:71–83

Meissner HH, Scholtz MM, Palmer AR (2013) Sustainability of the South African livestock sector towards 2050 Part 1: worth and impact of the sector. South Afr J Animal Sci 43:282–297

Morris CR, Stewardson MJ, Finlayson BL, Godden LC (2019) Managing cumulative effects of farm dams in Southeastern Australia. J Water Res Plann Manag 145:05019003

Nathan R, Lowe L (2012) The hydrologic impacts of farm dams. Austral J Water Res 16:75–83

Neil DT, Mazari RK (1993) Sediment yield mapping using small dam sedimentation surveys, Southern Tablelands, New South Wales. CATENA 20:13–25

Ogundeji AA, Jordaan H, Groenewald J (2018) Economics of climate change adaptation: a case study of Ceres—South Africa. Clim Develop 10:377–384

Pisaniello JD (2009) How to manage the cumulative flood safety of catchment dams. Water SA 35:361–370

Pisaniello JD (2010) Attitudes and policy responses to Australian farm dam safety threats: comparative lessons for water resources managers. Int J Water Res Develop 26:381–402

Pulley S, Rowntree K, Foster I (2015) Conservatism of mineral magnetic signatures in farm dam sediments in the South African Karoo: the potential effects of particle size and post-depositional diagenesis. J Soils Sed 15:2387–2397

Richards JA (2012) Remote sensing digital image analysis: an introduction. Springer, Cham, p 494

Roberts VG, Adey S, Manson AD (2003) An investigation into soil fertility in two resource-poor farming communities in KwaZulu-Natal (South Africa). South Afr J Plant Soil 20:146–151

Rodrigues LN, Sano EE, Steenhuis TS, Passo DP (2012) Estimation of small reservoir storage capacities with remote sensing in the Brazilian Savannah Region. Water Res Manage 26:873–882

Roohi R, Webb J (2012) Landsat image based temporal and spatial analysis of farm dams in Western Victoria. In: Geospatial science research 2, Proceedings of the geospatial science research 2 symposium Melbourne, Australia, 10–12 Dec 2012, p 14

Sawunyama T (2013) Small farm dam capacity estimations from simple geometric relationships in support of the water use verification process in the Inkomati Water Management Area. IAHS Public 362:57–63

Sawunyama T, Senzanje A, Mhizha A (2006) Estimation of small reservoir storage capacities in Limpopo River Basin using geographical information systems (GIS) and remotely sensed surface areas: case of Mzingwane catchment. Phys Chem Earth 31:935–943

Schulze RE, Schütte S (2019) Update of potential climate change impacts on relevant water resources related issues in the uMgeni and surrounding catchments using outputs from recent global climate models as inputs to appropriate hydrological models. Centre for Water Resources Research, University of KwaZulu-Natal, for Umgeni Water, Pietermaritzburg, p 59

Shao Q, Chan C, Jin H, Barry S (2012) Statistical justification of hillside farm dam distribution in Eastern Australia. Water Res Manag 26:3139–3151

Singh LK, Jha MK, Chowdary VM (2017) Multi-criteria analysis and GIS modeling for identifying prospective water harvesting and artificial recharge sites for sustainable water supply. J Cleaner Prod 142:1436–1456

Souza Da Silva AC, Passerat De Silans AM, Souza Da Silva G, Dos Santos FA, De Queiroz PR, Almeida Neves C (2011) Small farm dams research project in the semi-arid northeastern region of Brazil. IAHS Public 347:241–246

Stocking M, Murnaghan N (2001) A handbook for the field assessment of land degradation. Earthscan, London, p 169

Tingey-Holyoak JL (2014) Water sharing risk in agriculture: perceptions of farm dam management accountability in Australia. Agric Water Manag 145:123–133

Tingey-Holyoak JL, Pisaniello JD, Burritt RL, Spassis A (2013) Incorporating on-farm water storage safety into catchment policy frameworks: international best practice policy for private dam safety accountability and assurance. Land Use Pol 33:61–70

Troskie DP (2013) Provinces and agricultural development: challenge or opportunity? Agrekon 52:1–27

Turner DP (2000) Soils of KwaZulu-Natal and Mpumalanga: recognition of natural soil bodies. Unpublished PhD thesis, University of Pretoria, p 265

Verstraeten G, Prosser IP (2008) Modelling the impact of land-use change and farm dam construction on hillslope sediment delivery to rivers at the regional scale. Geomorphology 98:199–212

23
Large Dams, Upstream Responses, and Riverbank Erosion: Experience from the Farakka Barrage Operation in India

Tanmoy Sarkar and Mukunda Mishra

Abstract

Dams are built on rivers for diverse purposes, where the most are associated with Multipurpose river valley projects. The river water is blocked to use it for domestic purposes, irrigation, industries, navigation, and hydroelectricity. Above all, most of these projects set a common "welfare" target of managing soil erosion. However, the effort of stagnating the natural flow of alluvial channels sets ample examples where adjustment of upstream channels and consequent lateral shifting exacerbates massive threats of riverbank erosion. This chapter discusses this particular phenomenon experienced in the different parts of Indian landmasses in its first part. The latter part is specific instead. The study has been carried out to examine the channel adjustment of river Ganga upstream of Farakka Barrage up to Sahebganj with the spatial extension between 24° 45′ N and 25° 30′ N latitudes and 87° 30′ E—88° 30′ E longitudes. River Ganga, the principal stream of the great Ganga basin at the Himalayan foreland, carries a substantial sedimentary load to the lower deltaic plain in Bengal. Near Rajmahal Hills, after crossing Sahebganj, the natural downstream adjustment in terms of channel slope, channel planform, and cross-section morphology of the river exhibits a drastic change in response to the modification of the channel gradient as a consequence of the human-induced changes in hydrology and sedimentary modification as well as a degree of base-level modification with the construction of the Farakka Barrage in Malda. The plateau basalt province of lower Cretaceous to the Jurassic age at the western margin and the old alluvium Barind tract in the eastern margin of the Ganga valley has a typical role. The tectonic tilting and geological settings of Rajmahal traps, which have long been a potential force of a rapid eastbound shift of the channel, are further stimulated by the Farakka Barrage. The geostrategic decision to select sites for a large Dam overrides the geomorphic and geologic realities that crave out the left bank erosion of the river Ganga in the Malda District, bringing a massive threat to human life and resources.

The content of this chapter are based on the paper entitled '*Channel Adjustment of River Ganga in Response to Human-induced Changes: Some Observations at Upstream of Farakka Barrage*', presented by the authors in Two Day International Seminar on "**Fluvial Processes and Anthropogenic Footprint**" on 11–16 March 2016, organized by the Department of Geography, University of Gour Banga, Malda in West Bengal, India.

T. Sarkar
Department of Geography, Gazole Mahavidyalay, Gazole, Malda, West Bengal, India

M. Mishra (✉)
Meghnad Saha College, Itahar, Uttar Dinajpur, West Bengal, India
e-mail: mukundamishra01@gmail.com

Keywords

Lateral shifting · Channel adjustment · Rajmahal · Geology · Meandering

23.1 Introduction

The ongoing global boom in constructing river dams and reservoirs is now at the center of perpetual debates. While the adroitness of engineering know-how contributes to modify free-flowing rivers for the benefit of society (e.g., hydropower generation, irrigation, water supplies, navigation, and flood prevention), it also triggers a range of harmful outcomes to the *locals* (e.g., directly affects stakeholder groups) (Zarfl et al. 2015), and *ecosystems* (e.g., diminishes ecological integrity terrestrial and aquatic ecosystems) (Poff et al. 1997; Poff et al. 2006; Barbarossaa et al. 2020; Siegmund-Schultze et al. 2018). However, this present study, acknowledging the importance of all these impacts, is concerned with how a river channel responds to this man-made design that traps sediments and controls its natural discharge pattern. Large dams' manifestation on flow and sediment regime subsequently control the channel morphology that has long been a concern of fluvial geomorphologists (e.g., Arroyo 1925; Gregory and Park 1974; Petts 1980; Williams and Wolman 1984; Gregory 1987; Carling 1988; Xu 1990; Graf, 1999; Phillips 2003; Petts and Gurnell 2005; Magilligan et al. 2008; Singer 2007a, b; Fu et al. 2010).

However, while there are ample shreds of literature making it evident that the downstream of the large dams are investigated frequently to examine the *alteration of flow and sediment regime* (see Magilligan and Nislow 2001; Assani et al. 2006; Lajoie et al. 2007), *bed-material grain size* (see Singer 2008, 2010), *scour development at post-dam-operation period* (see Lawson 1925; Shulits 1934; Andrews 1986; Kondolf 1997; Chen et al. 2010) and *channel aggradation* (see Malhotra 1951; Zahar et al. 2008), the *upstream response* of the dams remains undermined except few studies (e.g., Pringle 1997; Evans et al. 2007; Paola 2000; Greathouse et al. 2006) considering the same. Moreover, most of the works, embracing deterministic logic, pays attention to explain the *first-order impacts*, i.e., the *hydrologic and hydrographic alteration* (Benke 1990; Power et al. 1996a, b; Brooks and Brierley 1997; Graf 1999; Magilligan and Nislow 2001, 2005; Magilligan et al. 2003; Singer 2007a, b); *second-order impacts*, i.e., *sediment conveyance* (e.g., Andrews 1986; Topping et al. 2000; Willis and Griggs, 2003), and *channel morphology* (e.g., Gregory and Park 1974; Williams and Wolman 1984; Chien 1985; Brandt 2000) and *third-order impacts*, i.e., *ecological transformation* (e.g., Rood and Mahoney 1990; Ligon et al. 1995; Power et al. 1996a, b; Richter et al. 1996; Koel and Sparks 2002). The present study, in turn, manifests a predictive approach in a view to serve as a tool for policy prescription.

Now, glance back to India. The National Register of Large Dams (NRLD) follows the definition of *large dams* as specified by the International Commission on Large Dams (ICOLD). It states that *large dams* are those which possess a maximum height of more than 15 m from their deepest foundation to the crest. Also, a dam between 10 and 15 m in height from its deepest foundation may also be ascribed as a *large dam* provided it complies with one of the following conditions: (1) length of the crest of the dam is not less than 500 m or (2) capacity of the reservoir formed by the dam is not less than one million m^3 or (3) the maximum flood discharge dealt with by the dam is not less than 2000 m^3/s or (4) the dam has challenging foundation problems or (5) the dam is of unusual design (NRLD 2018). Figure 23.1 shows the decade-wise construction history and the distribution of large dams in India. Besides, the NRLD also defines the *Dams of National Importance*, such as dams with height of 100 m and above or gross storage capacity of 1 billion cubic m and above. As per NRLD (2018), there are 65 *completed* and 11 *under-construction* dams of national importance. Some of these dams often remain in the headlines in the national and international newspapers due to the insurgence of economic, environmental, and political debates underpinned by sensitive issues like rehabilitation of the forcibly uprooted population from the dam site, compromise with the biodiversity and forests, and, in many cases, the superimposition of the dam sites with the habitats of tribal population. India witnessed, in the recent past, the strong social movements in the form of *Narmada Bachao Andolon* (*Save Narmada River Movement*), led by Medha Patekar (see Fisher 1995; Narula 2008; Mehta 2010) and Baba Amte, and the *Anti Tehri Dam Movement*, led by the veteran environmentalist Sunderlal Bahuguna (see Kishwar 1995; Nandy 1997; Ishizaka 2006). However, post-operation reviews of large dams have multiple issues. Where the engineers monitor the dam sites, broader areas along the upstream and downstream channels seek attention from the geomorphologists and environmental scientists for long-term monitoring, which could be an effective action in policy perspectives. The task is diverse, which requires the involvement of multifarious assessment; however, the channel response envisaged by the river's planform dynamics is of deep concern that leads the discussion of this chapter.

Fig. 23.1 The distribution of large dams in Indian states along with the distribution in different periods, i.e., 1901–1950, 1951–2000, and post-2000. The term "Other States" represents the total counts in Meghalaya, Goa, Arunachal Pradesh, Assam, Manipur, Andaman & Nicobar Islands, Sikkim, Haryana, Mizoram, Nagaland, Tripura. *Source* Drawn by the authors based on NRLD 2018

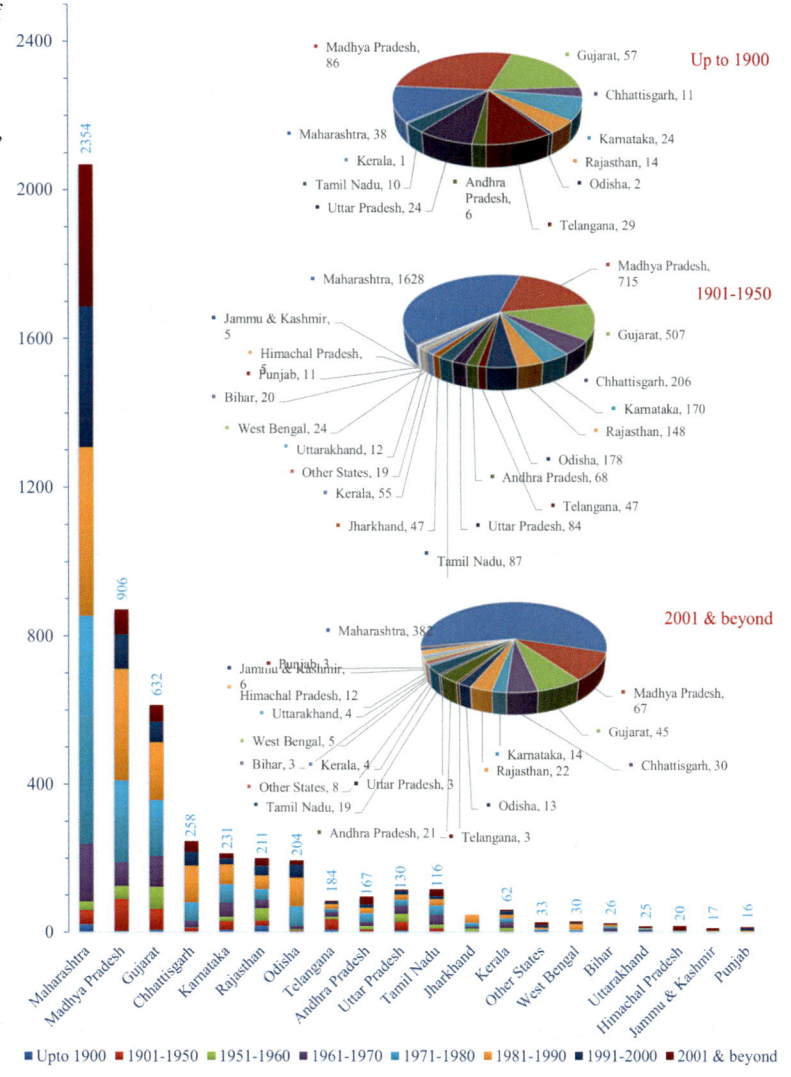

23.2 Backdrop of the Study

The Ganga river system is one of the largest river systems in India. The great river Ganga is the lifeline for millions of populations along its course. The river rises in the Gangotri Glacier, located in the western Himalaya in Uttarakhand state, and flows down south and east through the northern India Gangetic plain. This great transboundary river enters the plain of West Bengal along the Rajmahal hill on the right bank, after more than 2000 km eastward journey, and divides into two significant distributaries, Bhagirathi-Hugly and Padma, near the village called Mithipur, located in Murshidabad, West Bengal, near about 40 km downstream of Farakka (Rudra 2009). The Holocene tilt of the Gangetic delta guided the river's eastward flow, leaving the southward distributaries in decaying condition (Rudra 2018). There is several historical evidence of river course shifting of Ganga since the last three centuries, but since the Frakka barrage construction, the meander migration is the dominant character. The distributaries like Chota Bhagirathi, Kalindri, and Pagla are the geomorphic evidence of river course shifting since the late eighteenth century. The principal volume of the discharge of river Ganga is naturally carried down through the eastward flowing river Padma, which leaves India near Jalangi in Murshidabad. The Bhagirathi-Hugly is the southward flowing distributary of Ganga that enters into the Bay of Bengal after traveling around 500 km along its course on the deltaic plain of Bengal. River Ganga experiences several morphological changes in longitudinal and transverse directions.

Entering the Bengal deltaic plain, the mighty river throws off its distributaries. Kalindri, Chota Bhagirathi, and Pagla are the three moribund distributaries that indicate the old courses of Ganga. It is evident from the earlier literature, historical maps, and records that Ganga has shifted its course several times. Shifting the course across the valley with simultaneous erosion on one side and sediment deposition over another side is the natural process for meandering rivers. This natural geomorphic process of river course migration and resultant bank erosion is affected by human interventions like large dam construction and other allied activities.

The 2.62 km long Farakka barrage project was commenced in 1961 and finally commissioned on 21st May 1975 with the aim to sustain the Kolkata port by maintaining the navigability of the Bhagirathi-Hugly river system. Farakka barrage was constructed to channelize adequate water from river Ganga to the Bhagirathi-Hugly river through a 38 km long feeder canal, and it was thought that the adequate flow could manage the navigability of the river in the mouth regime. It was estimated that the transfer of 40,000 cusecs of water to the Bhagirathi-Hugly river could manage the navigability of Kolkata port by flushing the sediment into the deeper part of the estuary. However, the Farakka barrage project was taken considered only the arithmetic hydrology without proper consideration of fluvial dynamics and its multidimensional facets (Rudra 2008).

The "story" of Farakka barrage construction was started when the then "Capital Calcutta" mattered! The decision of Kolkata port establishment on an offshoot of the Bhagirathi-Hugli river could have caused the problem of siltation and future decaying of the Calcutta port. The problem of siltation in the Hugly river can be traced back to the seventeenth century, and it has been accelerated with human intervention in the form of dam construction across the river. The idea of dam construction across Ganga to divert a bulk of water in Hugly was first propped by Sir Arthor Cotton in the 19th Century, in the time of British rule in India. However, the decision of the large dam construction to save the then Calcutta was not actualized. After independence, the newly formed government had a huge responsibility to make a "New India" through its firm decisions. In 1957, Dr. W. Hensen, a British Engineer, was invited to rescue the Calcutta port (as in Chatterjee 2016). Dr. Hensen echoed the earlier ideas of Sir Cotton (1853) and suggested the construction of the Farakka dam to save Calcutta port. However, the Nehruvian penchant for large dams was opposed by Kapil

Bhattacharya, the chief superintending engineer of the Government of West Bengal, in 1961 (Singh 2019). He strongly opposed and warned that the design of the dam was bad and cautioned about the devastating consequences. The unprecedented sedimentation upstream of Farakka, the rise of the bed level, and substantial devastating floods were all predicted by Mr. Bhattacharya, the only voice against the Farakka project. The then Pakistan used the report of Bhattacharya to lodge the argument against the dam construction in Farakka, and Mr. Bhattacharya was vilified by the government and defamed through the media trial (Singh 2019).

After independence, the plan was actualized, but the aim of flushing the sediment off the mouth of the Hugly was a total failure. Actually, the role of tidal water plays a crucial role here. The diverted 40,000 cusec water to the Hugly river is freshwater, and the saline water that enters the Hugly from south to north at the time of tide is often larger in volume and makes it impossible to throw off the sediments into the sea by freshwater (Dandekar 2014). So, the earlier proposition of diverting a bulk volume of water into the Hugly river to maintain Kolkata port was overestimated, but millions of people have been suffering from the devastating impact of Frakka barrage construction.

23.3 The Research Problem

The upstream area of the Farakka barrage up to Sahibganj is taken as the focus area of this discussion. The area, which is bounded by 87.75 E—88.15 E meridians and 24.75 N—25.10 N parallels, is used for the application of different geospatial techniques (Fig. 23.2). Sahibganj is located in the northeast of Jharkhand district in India, bordering West Bengal in the East. The study area is selected to consider and limit the discussion on the river course shifting and meander migration of the great river Ganga between the

Fig. 23.2 The study area. *Source* Customized by the authors from Google earth

two nodes; the natural node at Sahibganj, near Rajmahal, and the other is the man-made node at Farakka in Murshidabad district.

Geomorphologists' speculation (e.g., Davy and Davies 1979) that meandering results from a general governing principle have evolved through scaling the width and wavelength of meandering streams over many orders of magnitude. It produces the extremal hypotheses of river meandering through the minimization of the variance of directional change in the channel (see Langbein and Leopold 1966), minimization of energy dissipation rate (see Yang, 1971), and minimization of stream power (see Chang 1988). These extremal principles invoke to constrain quantitative models of adjustments among river variables so that optimal solutions for channel geometry (*width, depth,* and *slope*) are obtained for a given set of inputs (*bankfull discharge, grain-size characteristics of the bed and banks, sediment load,* and *bank resistance*) (Rhoads 2020; see also Huang et al. 2004; Nanson and Huang 2008).

The planar curve is evolved for a meandering river by the contrasting processes of continuous elongation accelerated by local bend erosion and abrupt shortening due to the sporadic cut-off events (Camporeale et al. 2005). This natural sequence of continuous elongation and sudden shortening phases represents the long-term meander dynamics which in turn are influenced by the river flow dynamics, sediment scenario, geological processes, and anthropogenic interventions (Perona et al. 2002). In this meandering process, the river flow swings approaching the outer bank to accelerate the bank erosion process, whereas, toward the inner bank, the water spirals, slowed down by the friction force imposed by the river bed (Ahmed et al. 2009). This spiral effect toward the inside bank encourages the deposition of alluvium, and point bars are formed. Once the point bars are formed, these are continually developed with the sediment accumulation. Thus, lateral shifting of the river is influenced by these swing and spiral river processes.

Sedimentation at upstream of Farakka has had a tremendous impact on river dynamics and channel morphology of Ganga. Rudra (2008) estimated that Ganga carries an amount of 736 Million Ton silt load annually, out of which a bulk amount of 328 Million Ton per year has been trapped at upstream of Farakka. Continuous siltation and sedimentation accelerate the process of island/char formation. The continuous siltation of millions of tons of sediments shallowed the river's natural flow depth. Moreover, the water level is high as the barrage obstructs water from flowing downward. This high pond level and energy are compensated by the expansion of the river's cross-sectional area and left bank erosion. The river Ganga erodes its left bank to open a new outlet to disseminate its energy. The presence of paleochannels, ox-bow lakes, and decaying distributaries confirms the meandering character and continuous channel shifting of the mighty river Ganga throughout history. The meandering river has its own natural laws of bank erosion and bank development by deposition, but the construction of the Farakka dam entirely destabilizes the dynamic natural system. The present bank failure hazard is the response of the Ganga river system to the massive engineering structure that forced it to change the natural sediment and water flow dynamics (Rudra 2018). The forced construction accelerated the left bank erosion at an alarming rate at upstream of Ganga up to Sahibganj. Since the construction of the Farakka barrage, the impeded flow of Ganga and sediment load played havoc in left bank erosion.

The left bank is stratigraphy built up with unconsolidated sand at the base of the shelving cliff, and the post-monsoonal discharge of groundwater toward the river by seepage process is considered an important contributing factor in left bank failure (Rudra 2018). The left bank erosion is severe in the monsoonal months of June to September. The stratigraphy of the river bank depicts that the base is composed of micaceous sand overlaid by silt clay. In the late monsoon months, the high energy currents associated with the peak discharge stage easily remove the unconsolidated bank sediments from the base and the bank collapse. During the post-flood season, the discharge declines typically. However, the groundwater discharge toward the river accelerates liquefaction, which leads to the flow of basal sediments and the bank collapse. In most of the cases, linear cracks are observed along the riverbank, and mostly, the collapse has been initiated along these cracks (Rudra 2006). Rotation failure, removal of the underlying support, and collapse of the hanging upper bank are widespread forms of bank erosion in Malda, upstream of Farakka. The unconsolidated bank materials are detached and washed away by the hydraulic action and carried away by the streamflow. Even in the lean period, when the water level is relatively low, the toe erosion occurs, and the upper bank surfaces are collapsed along the shallow slip surfaces in response to the building shear stress (Mandal 2017).

The meandering character of rivers on the alluvial surface is natural. The complex linear and non-linear channel processes provide a dynamic river system. However, it is clear from the empirical observations that the process of meandering carries the benefit of the fertile soil as well as the hazard of bank erosion side by side (Fig. 23.3).

> Those who live along meandering rivers have learned that such rivers are not fixed in position but migrate over time across valley bottoms, producing renewal of fertile floodplain soils. At the same time, the changing location of the river channel can lead to loss of land and represent a hazard to infrastructure located near the river. (Rhoads 2020, p. 197)

While the human-made obstruction in the form of dams causes an undue change in the natural river dynamics, the meandering process is also magnified, and it puts havoc on the ambient human settlement as they are stressed more by the hazardous effect of bank erosion due to the shifting channels and meandering than that of, they are benefited from the renewal of fertile floodplain soils. This scenario prevails along the extensive stretch, particularly along the left bank, of the Ganga river in our study area (Fig. 23.4).

The data in Table 23.1 makes it convenient to the readers that the bank erosion along the left bank of the Ganga at the immediate upstream of Farakka is a long-run process at the post-operation period of the Farakka barrage and also indicates the lateral migration of the channel toward east passively. The situation worsens during the monsoon when the uprising water level of the river cause to wash away homes, prayer halls, schools, agricultural land—everything in its way, paralyzing the lives and livelihoods of the communities residing along the banks.[1]

23.4 Materials and Methods

The methods adopted and materials utilized for this study are discussed in accordance with the flow of the works, which will enable readers better link the methods with the sections where the result is analyzed. The present study involves two major investigations—at first, it seeks to determine the lateral shifting tendency of the river in the bounding box of the study area. Then, it analyzes regional geology and concludes.

The lateral shifting of the river is recorded with the help of geospatial techniques. Straightforwardly, the change detection techniques involve finding the changes in images of the same location taken at different times, which is of great importance in many areas, including land use and land cover change, deforestation, urban settlements, changes as part of natural calamities, and resource monitoring and management, etc. (Asokan and Anitha, 2019). Geospatial technology has proven to be an important tool to the geographers for the periodic monitoring and analysis of the changes across time. The present study incorporated the river courses for 238 years from 1776 to 2014 for detecting the change (Fig. 23.5). The Topographical Map, published by the Survey of India (Nos. $72O/7$, $72O/8, 72O/15, 72O/16, 72P/9, 72P/13$) is used as the reference map to perform georeference operations, and the LANDSAT images were used for different years between 1973 and 2014 (Table 23.2). The required Geo-

---
[1] Inquisitive readers might be interested to witness the grimed scenario of the bank erosion through the photographs by Tanmoy Bhaduri, in his photography blog 'MONGABAY' with the theme entitled 'Erosion along Ganga's riverbanks in West Bengal finds new victims' (See here: https://india.mongabay.com/2020/11/photos-erosion-along-gangas-riverbanks-in-west-bengal-finds-new-victims/).

Fig. 23.3 The meandering river causes lateral migration by eroding the valley wall along the outer bank and deposits along the inner bank (Drawn by the authors based on Marsh and Dozier 1981; see also Rhoads 2020)

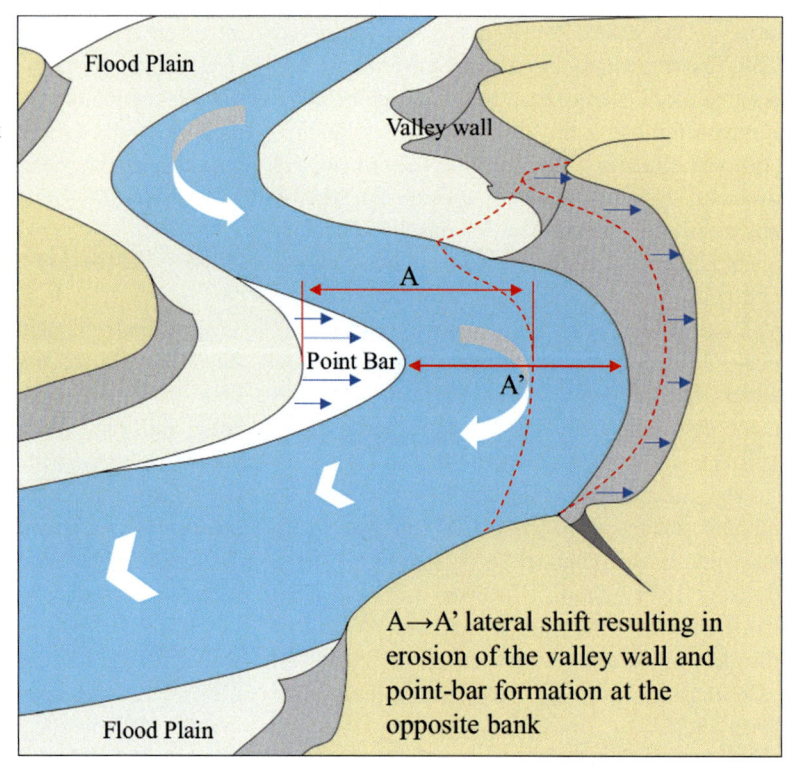

processing of the maps and pre-processing tasks for satellite imagery (e.g., atmospheric corrections, contrast enhancement, etc.) was performed on QGIS and ArcGIS Software platforms.

The last section of the study carefully analyzes the geological history of the region and stratigraphic sequences, and effort is made to find out the link between the planform geometry of the region and the regional geology, which is a least discussed matter amidst the vast literature available on the channel response, consequent by (or not consequent by) the implementation of Farakka barrage across the Ganga.

23.5 Lateral Shifting of Channels

The lateral shifting of the channel across the given time frame is assessed along with a set of ten equidistant horizontal lines (A1-A10), and 16 equidistant oblique lines (C1-C16) are used (Fig. 23.6), which cover the entire stretch of the river under present analysis. Both the horizontal and oblique lines are used to help record the shifting accurately irrespective of the different alignments and shape of the river bank at its different parts.

Crossing the river bank lines with any of the equidistant line create nodes. The values of *Easting* and *Northing* of all nodes along each line (Fig. 23.7) are recorded. It takes the following form for a given line:

$$X = \begin{bmatrix} (1776 - 1776) \\ (1955 - 1776) \\ (1971 - 1776) \\ \cdots \\ \cdots \\ \cdots \\ (2014 - 1776) \end{bmatrix}$$

$$Y_\emptyset = \begin{bmatrix} \emptyset_{1776} \\ \emptyset_{1955} \\ \emptyset_{1971} \\ \cdots \\ \cdots \\ \cdots \\ \emptyset_{2014} \end{bmatrix} \quad Y_\lambda = \begin{bmatrix} \lambda_{1776} \\ \lambda_{1955} \\ \lambda_{1971} \\ \cdots \\ \cdots \\ \cdots \\ \lambda_{2014} \end{bmatrix}$$

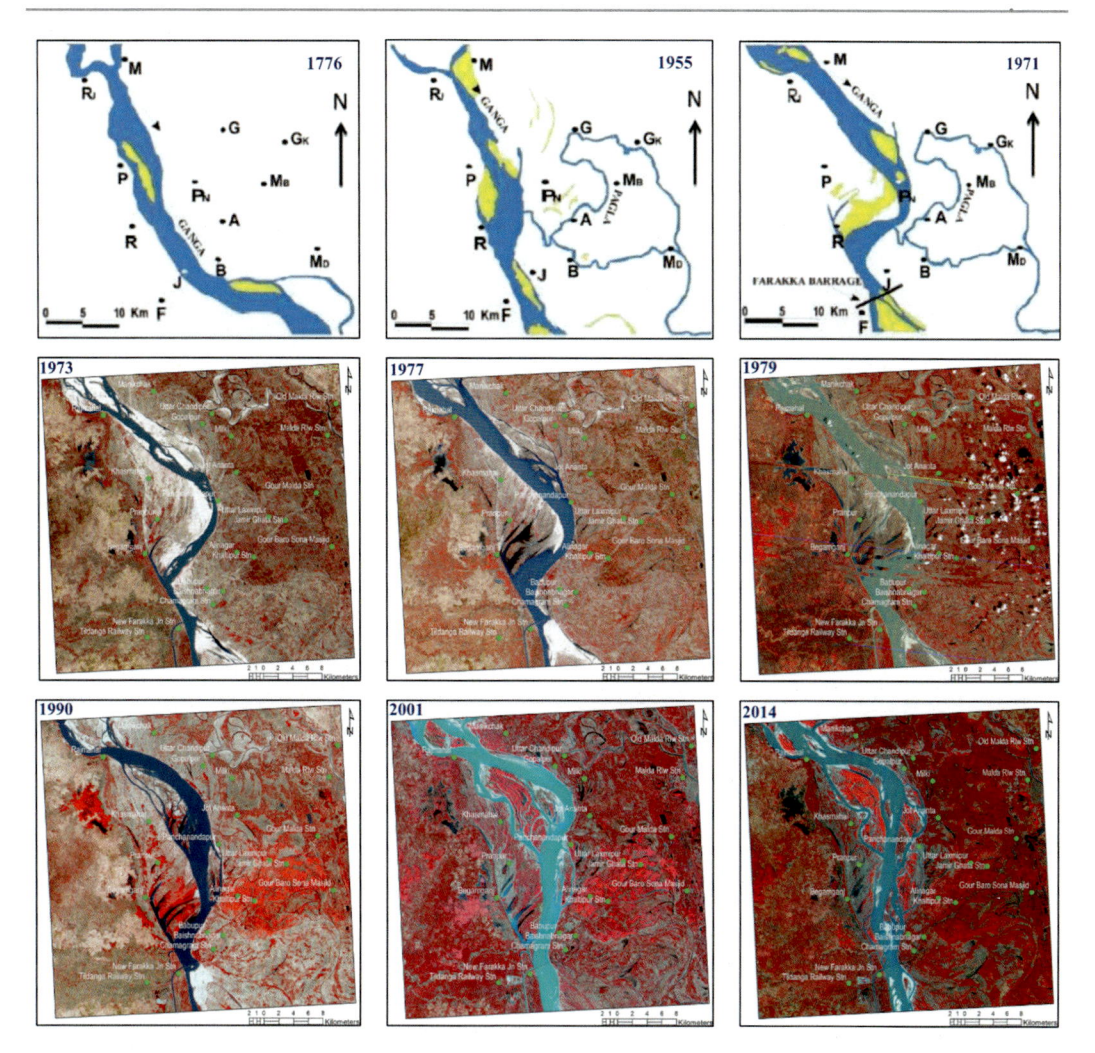

Fig. 23.4 The planform of the Ganga river in the study area during 1776, 1955, 1971, 1973, 1977, 1979, 1990, 2001, and 2014. *Source* The first three figures are adopted from Sinha and Ghosh (2012); rest are the LANDSAT images, the details of which are given in

Table 23.1 Loss and gain of land from the meandering and lateral shifting of river Ganga along its banks from the immediate upstream of the Farakka barrage to Rajmahal for a stretch of 67.09 km

Year	Bank	Estimated amount of land eroded (km^2)	Estimated amount of land deposited (km^2)	Net loss (−) or gain (+) (km^2)
1965–1980	Left	62.64	26.7	−35.94
1980–1996	Left	64.86	8.27	−56.59
1996–2017	Left	61.99	31.6	−30.39
1965–1980	Right	30.13	36.26	+6.13
1980–1996	Right	17.49	69.78	+52.29
1996–2017	Right	34.46	57.55	+23.09

Source Net loss and gain are calculated based on the estimate of Sarif et al. (2021)

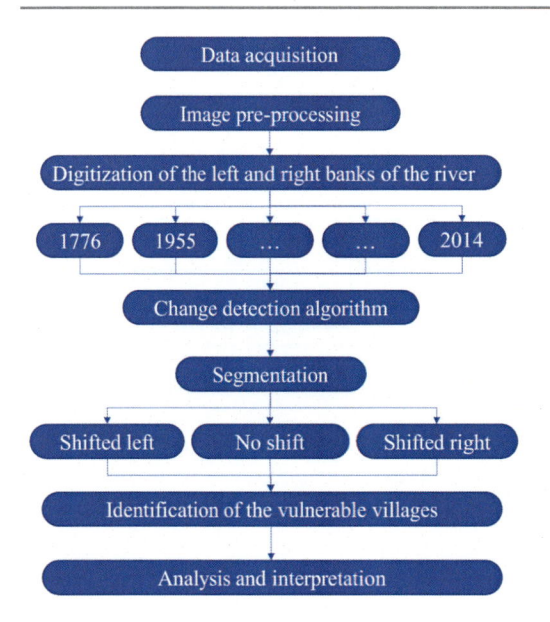

Fig. 23.5 The framework of the change detection technique applied in this study. *Source* Authors perception

where

X = Years of action
$Y_\emptyset$ = Easting after the lateral shift
Y_λ = Northing after the lateral shift.

As the present study considers the bank line of 1776 as the reference, the ($\emptyset_{1776}$, λ_{1776}) is the reference node. Hence, the departure from the reference node is calculated as per the norms of the coordinate geometry:

$$Y_0 = \sqrt[2]{(Y_\emptyset - \emptyset_{1776})^2 + (Y_\lambda - \lambda_{1776})^2} \quad (23.1)$$

Now, expressing the departure as the function of the *years of action*:

$$Y_0 = \alpha + \beta.X \quad (23.2)$$

Solving the equation with the Ordinary Least Square (OLS) method, using the recorded datasets, gives the values of α and β. The β express how rapidly or slowly the lateral shifting happens with time. The β from each line (i.e., $\beta_{A1}, \beta_{A2}, \ldots \beta_{A10}$; and, $\beta_{c1}, \beta_{c2}, \ldots \beta_{c16}$) are queued separately for the left and right banks, and the Jenks (1967) natural breaks classification method is applied, and the cluster with the highest values of β is marked. These points are considered the points of most aggressive lateral shifting, making the bank vulnerable.

Out of 49 control points, where the shifting was measured, 19 points (where 11 points are predicted with eastbound shifting and eight points are predicted with no/ westbound shifting) are used for the Ground Truth Verification (GTV) to assess the accuracy of the measurement (GTV sites are mentioned in Fig. 23.9). ROC is a popularly used method of evaluating model performance. The ROC is drawn following the method prescribed by Eng J. (2014) of Johns Hopkins University. It effectively draws ROC from small samples. The GTV sites are examined, and the real scenario is noted as *Definitely negative* (Rated as "1"), *Probably negative* ("2"), *Possibly negative* ("3"), *Possibly positive* ("4"), *Probably positive* ("5"), and *Definitely positive*

Table 23.2 List of IRS LISS-III 24 m ortho-rectified satellite images used in this study

LANDSAT	Year	WRS_Path/WRS_Row	Date of acquisition
LANDSAT 1	1973	149/043	1973-02-22
LANDSAT 2	1977	149/043	1977-10-02
LANDSAT 2	1979	149/043	1979-11-22
LANDSAT 5	1990	139/043	1990-04-11
LANDSAT 7	2001	139/043	2001-10-26
LANDSAT 8	2014	139/043	2014-11-20
LANDSAT 8	2020	139/043	2020-11-12

Source Images are downloaded from Earth Explorer, U.S. Geological Survey

Fig. 23.6 A set of equidistant horizontal and oblique lines are used to record the shift of the river banks across the time period considered for the study. *Source* Authors illustration

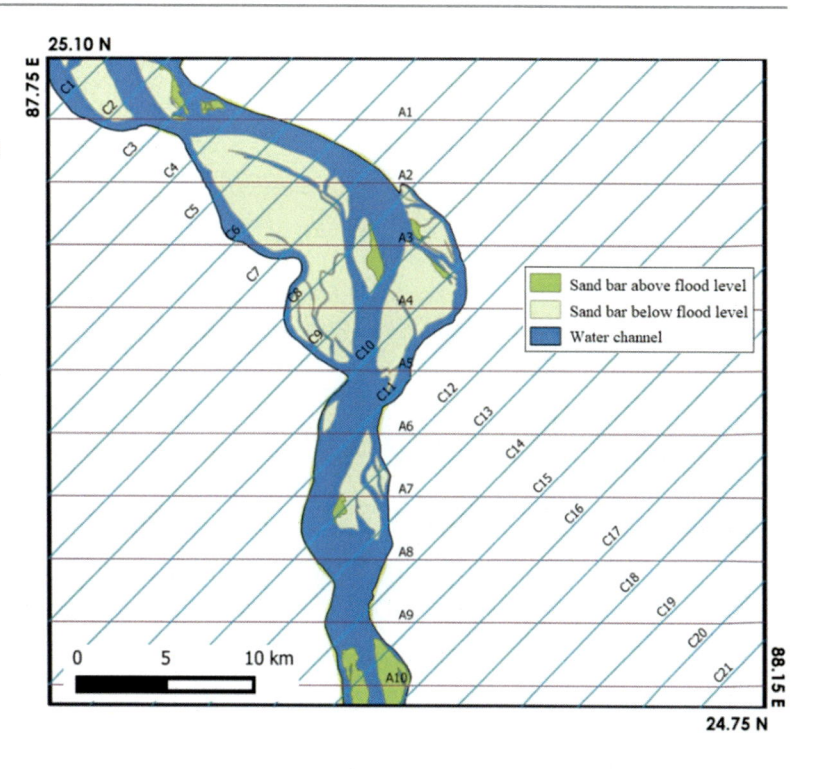

Fig. 23.7 Recording the easting and northing of the nodes created by the crossing between the line and the riverbank along a given line for different years. *Source* Authors' illustration

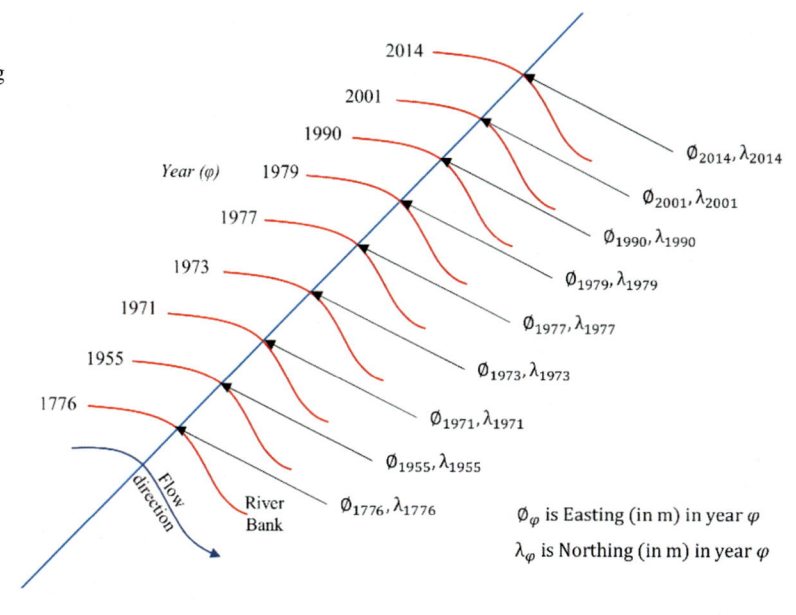

$\emptyset_\varphi$ is Easting (in m) in year φ
λ_φ is Northing (in m) in year φ

"6") in comparison to the model-predicted value as *Truly positive*, i.e., "rapid eastbound erosion exists" ("1") or *Truly negative*, i.e., "rapid eastbound erosion does not exist in reality" ("0"). From the ROC of the final model, the AUC was estimated to be "0.955" with an overall accuracy of 84.2% (Fig. 23.8). The model's performance has reasonable ground to be accepted for predicting lateral shifting of the banks with the help of historical data.

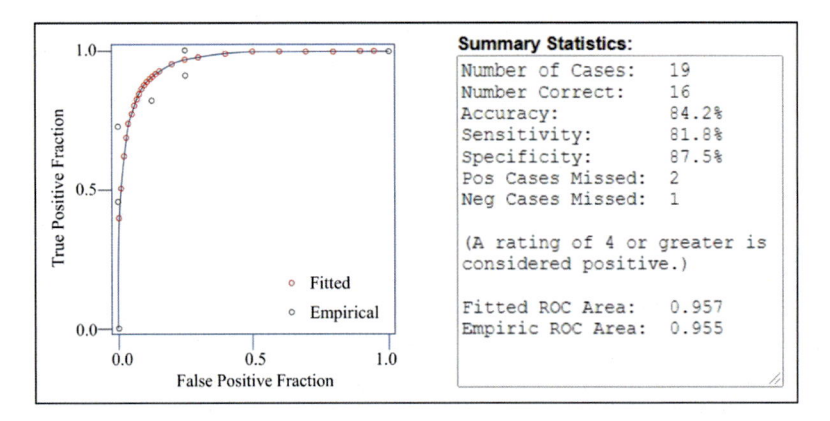

Fig. 23.8 The ROC is drawn, based on the ground truth verification

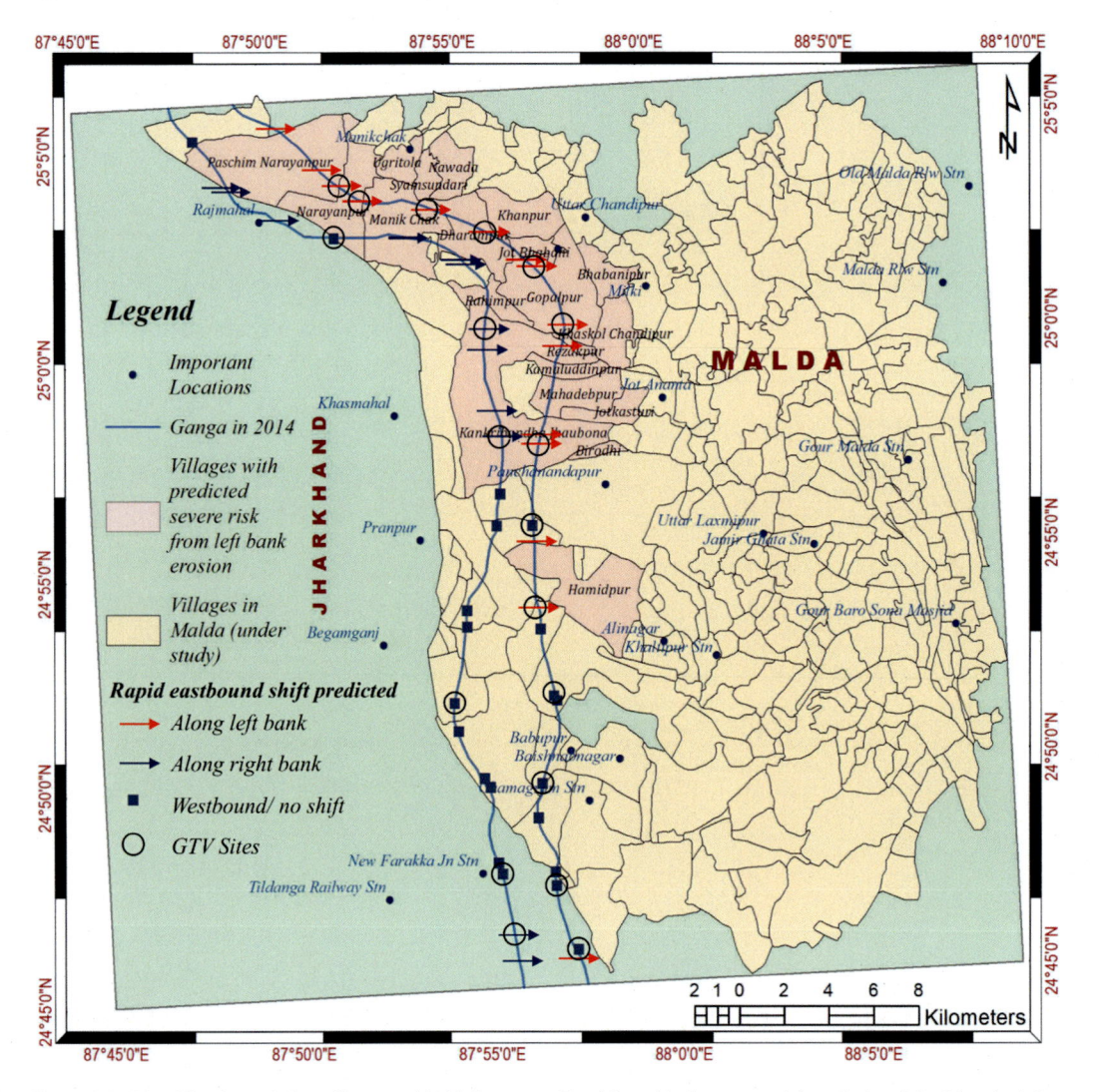

Fig. 23.9 Identification of the villages with high susceptibility to facing rapid bank erosion due to the eastbound shift of river Ganga at immediate upstream of Farakka with the geospatial analysis of the historical data for the period of 1774–2014. *Source* Computed and mapped by the authors

The gradual course shifting and meander migration of the mighty river Ganga has been analyzed between 1776 and 2014. The channel pattern and planform of the Ganga in 1776 and 1955 show a relatively less sinuous pattern in comparison to the river course after the Farakka barrage formation in 1975 (Fig. 23.4). Abrupt channel shifting along with braiding patterns for a very short period was contributed by the human intervention in the form of colossal engineering construction in Farakka across river Ganga. Since 1973, sharp eastward bending of river Ganga at a distance of 10–12 km upstream of Farakka has been observed. This human-induced sinuosity was gradually compensated in the northward part upstream of Farakka. The large part of the villages of Panchanandapur, Darijayrampur, has been gradually engulfed by the river Ganga since 1973. At the same time, the villages like Manikchak, Uttar Chandipur, Milki in the northern part of the upstream location were located far from river Ganga in 1973, the gradual eastward shifting with a rapid erosion along the left bank threatening the lives and livelihoods of the residents of these villages.

The river swallowed the land between the river and the villages of Manikchak, Uttar Chandipur, and Gopalpur at an alarming rate. The flow of water was suddenly blocked, and a bottleneck situation has been evolved for the river. The carried energy was disseminated in the upstream area, and the resultant course shifting becomes obvious. The channel at the immediate upstream of the barrage has been adjusted earlier with the human intervention. Now, this adjustment is gradually transferred to upstream further. The study has predicted the eastward course shifting of the river as during 2014 (Fig. 23.9). The prediction shows that the villages located adjacent to the left bank of Ganga (northern part of the upstream of Farakka), e.g., Paschim Narayanpur, Narayanpur, Manikchak, Dharampur, Jot Bhabani, Gopalpur, Bhabanipur, part of Milki, Khaskol Chandipur, Kamaludiinpur,

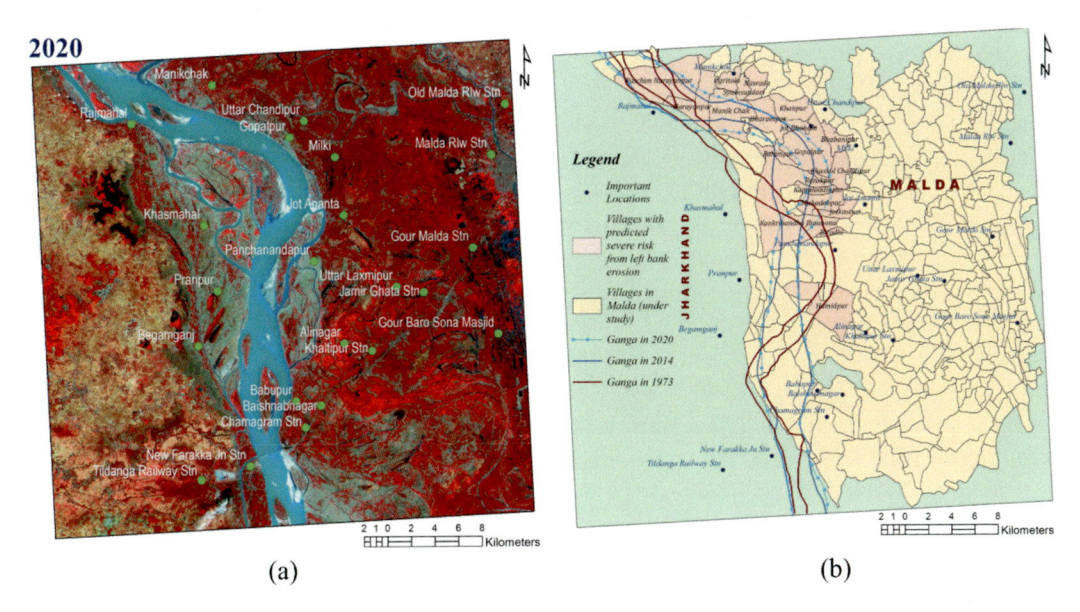

Fig. 23.10 The prediction made in 2016, based on the historical datasets from 1776–2014, was reviewed with 2020 datasets. The LANDSAT data of 2020 (**a**) shows a more updated channel position at the post prediction period, which is comparable with the channel positions in 1973 and 2014 (**b**). It is satisfactory that the prediction made in 2016 is found mostly accurate as the predicted eastbound areas in 2016 are found migrating in reality over the vulnerable villages identified (as in Fig. 23.9), causing the plight to the residing communities, particularly the situation being aggravated in 2017 West Bengal Flood. *Source* Computed and mapped by the authors

Hmidpur, the western part of Alinagar, etc. are at very high risk of left bank erosion and resultant swallowing of land by river Ganga.

23.6 Regional Geology: The Game Changer

Geology is an important aspect to be considered in the discussion of left bank erosion and eastward lateral course shifting of river Ganga between Rajmahal and Farakka. The western part of the river Ganga (from Rajmahal to Farakka) represents the eastern part of Chotanagpur Granite Gneissic Complex (CGGC), and the Rajmahal trap bears a very long and complex geological history of development. The antiquity of this part and its petrological diversity confirms its age-old complex geological history. The northern and eastern part is bounded by the quaternary deposit of Gangetic alluvium and sediments of the Bengal basin. The N-S extended Rajmahal province and Chotanagpur Granite Gneissic Complex result from past geological history. Rajmahal province indicates the NE boundary of the CGGC. The Rajmahal hills range's geological position is crucial as it is located at the juncture of the Rangpur Saddle formation of the Bengal basin in the east and the Singhbhum craton in Jharkhand to the west (Singh et al. 2016; Hossain et al. 2019). The upwelling activity of the Kerguelen plume and subsequent spreading of mantel material to the entire surface in cretaceous is responsible for the Rajmahal trap formation (Baksi et al. 1987; Roy and Chatterjee 2015; Hossain et al. 2019).

There are contradictions and controversies regarding the role of the Kerguelen hotspot in the formation of the Rajmahal basaltic trap. The Kerguelen hotspot activity triggered the breakup of the Indian plate and Australian plate and the Indian plate's northward drifting (Kent 1991; Frey et al. 2000; Bian et al. 2019). Mahoney et al. (1983) believed that the Kerguelen hotspot might provide the heat source for Rajmahal activity through its direct role in feeding the magmas for the trap formation could be overestimated. Later, Davies et al. (1989) postulated that the Kerguelen hotspot contributes to the Kerguelen Plateau, Ninetyeast Ridge, and Rajmahal flood basalt during the geological time of last 120 M years (Fig. 23.13). However, the most striking conclusion was made by Olierook et al. (2016) and Kapawar and Mamilla (2020) that the Rajmahal trap flood basalt of the Indian plate and the basalt found in Bunbury of Australian plate have preserved the geological evidence of geological connectivity that indicates their similar genesis from the Kerguelen hotspot. The similarities of cretaceous basalts of the western Australia and the rifted margins of the eastern Indian shield have been studied by several researchers (Storey et al. 1992; Baksi 1995; Kent et al. 1997) to find the link between these basaltic formations and Kerguelen hot spot activity. The Rajmahal trap basaltic formation was thought to be formed near about 117–118 $\sim$ Ma over the eastern margin of the Indian shield through the Kerguelen hotspot activity in the geological past (Courtillot 1999; Coffin et al. 2002; Kent et al. 2002; Ghatak and Basu 2013). The age of the intruded dykes in the Rajmahal hills was estimated as $\sim$ 115 Ma, and it indicates that the duration of the Rajmahal volcanic eruption was around 2–3 Million years (Kent et al. 2002). The upwelling of the mantle plume enforced a local dome formation over the crust, and it was represented by the triple point fracture made by the Tista fault, Daiki Fault, and Basin Margin Fault (N30°E–S30°W) (Roy and Chatterjee 2015; Hossain et al. 2019).

Rajmahal trap is considered one of the most important geological formations to find the paleogeographic position of the Indian plate during the early Cretaceous age. The geological shreds of evidence indicate that the Indian plate, Australian plate, and the Antarctica plate had started to separate from the juxtaposed condition during 117–130 Ma. The India-Seychelles-Madagascar block was completely detached from the Australia and Antarctica block at around 99.6 Ma (Kapawar and Mamilla 2020). The northward drifting of the Indian mass and its passage over the Reunion hotspot around $\sim$ 65 Ma could have produced the Deccan basalt formation in the western part of the Indian shield

(Aitchison et al. 2007; Kapawar and Mamilla 2020). Being separated from the eastern Gondwana, the Indian plate had drifted northward and collided with the Eurasian plate (Gaina et al. 2007). In the early cretaceous period, the northward drifting speed of the Indian plate was around 18–20 cm/year, and gradually, the speed was slowed down as it coincided with the Eurasian plate (Kapawar and Mamilla 2020), and this resulted in the close of Greater Indian Basin around ~25 Ma. Being separated from the Gondwana land, the eastern marginal part of the Indian plate experienced the tectonic formation of the Rajmahal hills. Over time, the external geomorphic processes deformed the upper surficial structure, and the present graben structure was formed (Singh et al. 2004). The Rajmahal trap formation with a thickness of around ~230 m was emplaced at ~118 Ma (Kent et al. 2002). Kapawar and Mamilla (2020) concluded that the Indian mass was in the southern hemisphere at the mid-latitude location when the Rajmahal trap was formed. The normal magnetic polarity character of the rocks formed in the Cretaceous period was preserved by most of the lava deposits in Rajmahal hills (Klootwijk 1971). Kent et al. (2002) concluded in favor of the greater possibility that the Rajmahal trap flood basalt was fed by the Kerguelen hotspot.

On the western margin of the Rajmahal trap, alluvium deposit by fluvial action by river Ganga is of the recent time. Laterite capping is also observed in many places over the traps. Rajmahal traps are predominantly composed of volcanic rocks of early cretaceous age, which are part of the Upper Gondwana Series (Sinha and Rais 2019) (Table 23.3; Fig. 23.11). The Rajmahal trap basalt is characterized by the quartz-normative tholeiitic lava of different flow sequences of around 600 m (Klootwijk 1971), whereas in the eastern part of the trap (western margin of Bengal basin) the thickness of the basalt, including alkalic lava, was found around ~332 m (Biswas 1963). There are various estimations and findings among the scientist and researchers regarding the thickness of the basaltic layer and the number of lava flows. Hobson (1929) identified at least ten distinct flow sequences, whereas Pascoe (1959) advocated for more numbers of lava flows than estimated by earlier researchers. Layers of sandstone and shale are found in the interbedded position to separate the sequences of various geological times (Tiwari and Jassal 2003). Rocks of Dubrajpur formation of Upper Triassic to Lower Jurassic age and Barakar rock formation of Permian age are overlaid by cretaceous lava flow (Sinha and Rais 2019).

Table 23.3 Stratigraphy of Rajmahal Hills

Formation/Series	Composition	Age	System	Thickness
Recent	Soil, alluvium, laterite	Quarternary		0–30 m
Rajmahal	Flows of basalt with inter-trappean sediments and doleritie dykes	Lower Jurassic	Upper Gondowana	600 m
Dubrajpur	Ferruginous Sandstone Shales and Conglomarates	Upper Triassic	Lower Gondwana	122 to 137 m
Barakar	Felsparhic Sandstone Carbonaceous Shale	Permian		Varies between 0–152 m
Talcher	Coal seams bounded by Sandstones, clays, and boulder beds	Carboniferous		55 m (approx.)
Unconformity				
Archeans	Pegmatites, quartz veins, Granite, Gneiesses with inclusion of Amphibolites, Pyroxenes and granulites	Cambrian	Archeans	

Source After Ball (1877); Ray Chowdhury (1965); Shrevastava and Shah (1966)

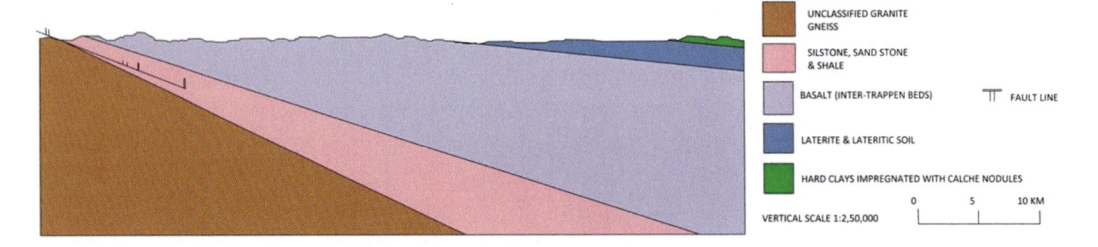

Fig. 23.11 A schematic diagram of the Stratigraphy of Rajmahal Hills

These extensive volcanic formations, along with inter-trappean rock strata, are found in the eastern part of the Eastern Indian Shield (Moiola and Weiser 1968; Madukwe 2016). Rajmahal hill is considered a fault-bounded tectonic structure located in Jharkhand, on the western margin of the Bengal basin near Malda, West Bengal. The crust thickness of the Indian plate bounding the western province of Bengal basin is nearly 38 km (Singh et al. 2015). The N-S trending Malda-Kishanganj fault lies at the north–north-west margin of the Bengal basin and Rajmahal hill (Mohanty et al. 2014; Prasad and Pundir 2017; Ghose et al. 2017). The formation of the Bengal basin as the eastward continuation of the Indo-Gangetic plain separates the northern Extra peninsular India from Peninsular India in the south (Roy 2014; Hossain et al. 2019).

The Bengal basin is bounded in the west by CGGC and Singhbhum Craton (SC), along with the Central Indian Tectonic Zone (CITZ) and the Eastern Ghats Mobile Belt (Mishra 2006). The outliers of the CGGC and Rajmahal hills in the western part of the flowing river Ganga have had an essential role in the eastward migration of the river. The flowing river with high intensity

Fig. 23.12 Geological map of the study area (Following GSI, 72O Purnia Quardangle & 72P Dumka Quardangle)

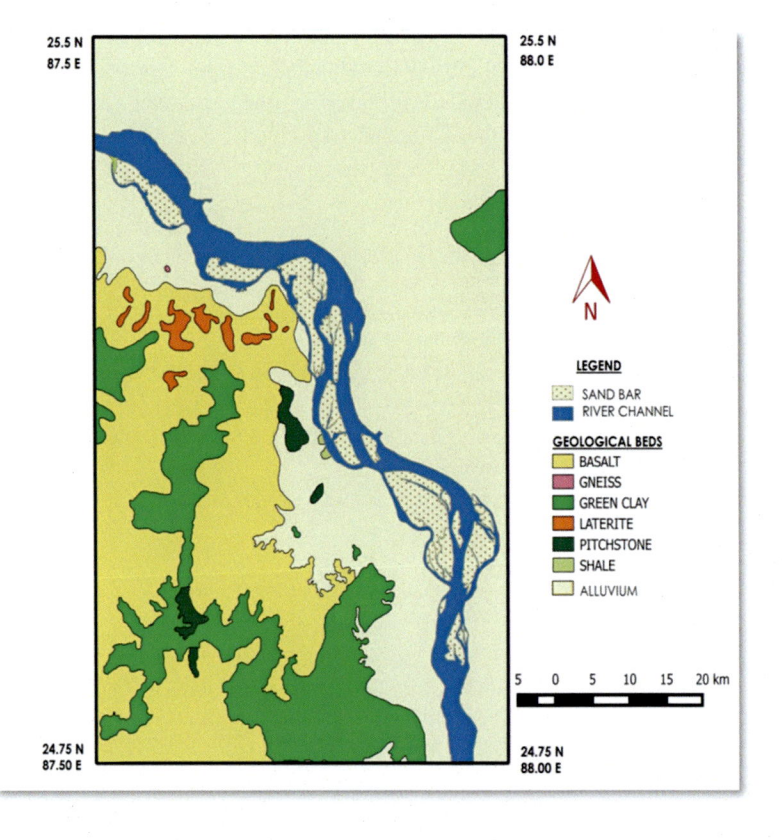

actually strikes the apparently strong geological node and deflects toward the east (Fig. 23.12). The deflected flow strikes the left bank, and the left bank has been eroded by the swinging flow. The eastward migration of river Ganga, after crossing the Rajmahal near Sahibganj, had long been observed even before the construction of the Farakka barrage in the 1960s. However, the situation has been worsened after this heavy engineering intervention. Before the construction of the Farakka barrage, only the natural node was in the form of the Rajmahal structure where the river had taken its natural bend. But the construction of the barrage has forced the Ganga to face another nodal point at Farakka, human-induced node. This artificial node completely interrupted the river flow, and the natural energy equilibrium has been devastated. After sudden obstruction, the classical energy distribution process in the youth, mature and old stages have been interrupted. Being obstructed by the heavy engineering construction and resultant sedimentation, the river was forced to find a new outlet to redistribute its energy along its cross-sectional area. However, the presence of CGGC and Rajmahal traps the mighty river Ganga migrating eastward and eroding the left bank severely.

23.7 Conclusion

Discussion under this conclusion section has two specific standpoints. The first one is particular—that is, what we have learned from this study and what could have been done; however, the second one, which runs in parallel, is vast—in the environment-development divide, is there any room for dialogues?

The study shows, or more specifically, proves that historical data could do well if handled carefully. A long-term trend has always been the center of the prediction of natural phenomena. However, building a model should ensure that it could come into the crossroads of human welfare. Prediction of the present study in 2016 has predicted which villages in Malda district would be at stake of the fury of the migrating river, and it is proved a successful model when we have compared it with the real-world scenario with 2020 datasets. The model could be extended and updated to predict the risk by utilizing modern machine learning techniques like neuro-fuzzy algorithms. The model output can further be used to risk assessment by including a range of economic and social variables. The present analysis could, in turn, to purely social analysis dealing with, on the one hand, the issues of uprooted families, and on the other hand, the extravaganza of power and politics arising out of the newly rising islands. The geomorphological understanding could be the beacon light to address all these critical issues associated with the dynamics of the river channels.

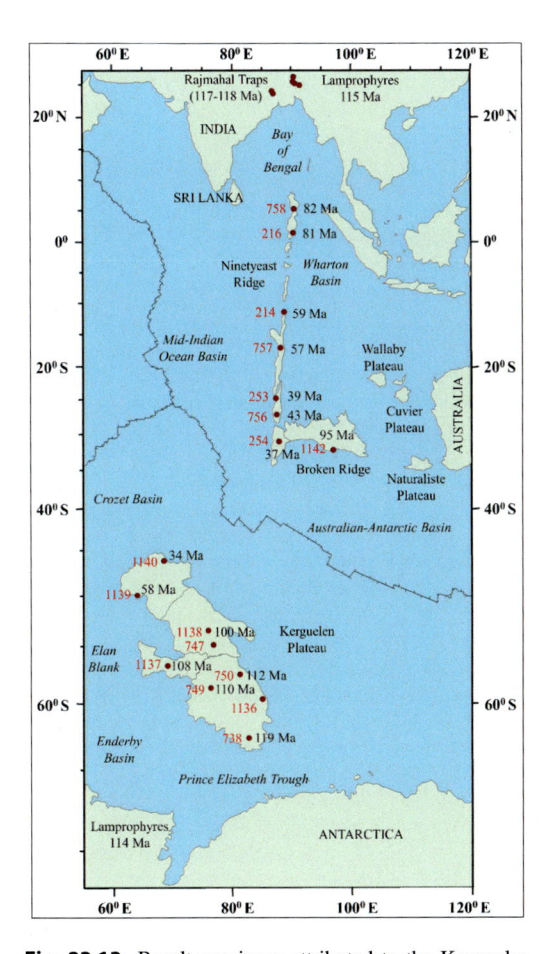

Fig. 23.13 Basalt provinces attributed to the Kerguelen plume include Kerguelen Plateau, Broken Ridge, Ninetyeast Ridge, Bunbury basalts, and Rajmahal Traps. Ghatak and Basu 2013. *Source* Drawn based on; Ingle et al. (2002), Frey et al. (2000), Yin et al. (2010)

Now, if we divert our attention to the broader issue of constructing dams, associated carnage of the biodiversity, paralyzing the age-old practiced livelihoods, and many other issues, we need to give due consideration to both the supply and the demand sides equally. The engineer's or the geomorphologist's definition of a dam does not match with those who had earned their livelihood from those forests that have been submerged into the dam site. In a similar situation what Sunderlal Bahuguna, the veteran Indian environmental activist, speculated in connection to the installation of Tahri Dam in the Tehri-Garhwal Himalaya is sensible:

> For Bahuguna, the erection of the Tehri Dam presented an ecological, social, and religious challenge. For Bahuguna, when the Ganges flow in its natural course it benefits all, irrespective of caste, creed, color, or economic circumstances. When it is dammed, it becomes the possession of the privileged and powerful who dispenses its blessings on a partisan basis. James, 2014, p. 172

It is time to accept the premise that ecological concerns are equally worthy as developmental needs, then the decision to have a mine, dam, or road needs to be weighted by its ecological detriment. The demand side should also be sensitized for setting a limit of consumption, failing which the exploitation of resources by compromising the ecological value would aggravate the delicate man-environmental relationship that would only remain confined in geographers' textbooks. The practice of maintaining world processes of productivity by replacing resources used with resources of equal value paves the way for sustainable development and man's sustainable future on the earth. Whether a proposed construction across a river to block its natural flow pre-qualifies this simple sustainability metric is to be given the most priority.

References

Ahmed AA, Fawzi A (2009) Meandering and bank erosion of the River Nile and its environmental impact on the area between Sohag and El-Minia, Egypt. Arab J Geosci. https://doi.org/10.1007/s12517-009-0048-y

JC Aitchison JR Ali AM Davis 2007 When and where did India and Asia collide? J Geophys Res 112 1 19 https://doi.org/10.1029/2006JB004706

ED Andrews 1986 Downstream effects of flaming gorge reservoir on the green river, Colorado and Utah Geol Soc Am Bull 97 1012 1023

S Arroyo 1925 Channel improvements of Rio Grande and El Paso Eng News-Rec 95 374 376

Asokan A, Anitha J (2019) Change detection techniques for remote sensing applications: a survey. Earth Sci Inform 12:143–160. https://doi.org/10.1007/s12145-019-00380-5

AA Assani E Stichelbout AG Roy F Petit 2006 Comparison of impacts of dams on the annual maximum flow characteristics in the three regulated hydrological regimes in Québec Hydrol Process 20 3485 3501

AK Baksi 1995 Petrogenesis and timing of volcanism in the Rajmahal flood basalt province, Northeastern India Chem Geol 121 73 90

AK Baksi T Barman D Paul E Farrar 1987 Widespread early cretaceous flood basalt volcanism in eastern India: geochemical data from the Rajmahal–Bengal–Sylhet Traps Chem Geol 63 133 141

Ball V (1877) Geology of the Rajmahal hills. Memories GSI 13(II):155–248

Barbarossaa V, Rafael JP, Schmittc, Mark AJH, Christiane Z, Henry K, Aafke MS (2020) PNAS 117 (7):3648. https://www.pnas.org/cgi/doi/10.1073/pnas.1912776117

A Benke 1990 Perspective on America's vanishing streams J N Am Benthol Soc 9 77 88

W Bian T Yang Y Ma J Jin F Gao S Wang W Peng S Zhang H Wu H Li L Cao Y Shi 2019 Paleomagnetic and geochronological results from the Zhela and Weimei Formations lava flows of the eastern Tethyan Himalaya: new insights into the breakup of eastern Gondwana J Geophys Res Solid Earth 124 44 64 https://doi.org/10.1029/2018JB016403.'

Biswas B (1963) Results of exploration for Petroleum in the western part of the Bengal Basin, India. In: Proceedings of the 2nd symposium on the development of petroleum resources, economic commission for Asia and far east, mineral resources development series,18. New York: United Nations, pp 241–250.

SA Brandt 2000 Classification of geomorphological effects downstream of dams CATENA 40 375 401

AP Brooks GJ Brierley 1997 Geomorphic responses of the Lower Bega River to catchment disturbance, 1851–1926 Geomorphology 1 291 304

C Camporeale P Perona A Porporato L Ridolfi 2005 On thelong-term behavior of meandering rivers Water Resour Res 41 W12403 https://doi.org/10.1029/2005WR004109

PA Carling 1988 Channel change and sediment transport in regulated U. K. Rivers. Regulated Rivers 2 369 388

HH Chang 1988 On the cause of river meandering WR White Eds International conference on River Regime Wiley New York 83 93

Chatterjee G (2016) Farakka barrage: floods and water wars imminent, but political compulsions dictate CM behaviour. Firstpost. Retrieved from https://www.firstpost.com/politics/farakka-barrage-floods-and-water-wars-imminent-but-political-compulsions-dictate-cm-behaviour-2982440.html

ZY Chen ZH Wang B Finlayson J Chen DW Yin 2010 Implications of flow control by the Three Gorges Dam on sediment and channel dynamics of the middle Yangtze (Changjiang) River China Geology 38 1043 1046

N Chien 1985 Changes in river regime after the construction of upstream reservoirs Earth Surf Proc Land 10 143 159

MF Coffin MS Pringle RA Duncan TP Gladczenko M Storey RD Muller LA Gahagan 2002 Kerguelen hotspot Magma output since 130 Ma J Petrology 43 7 1121 1137 https://doi.org/10.1093/petrology/43.7.1121

V Courtillot 1999 Evolutionary catastrophes: the science of mass extinctions Press p, Camb Uni 95

P Dandekar 2014 Lessons from Farakka as government plans more barrages on Ganga SANDRP 12 10–11 8 15

Davies HL, Sun SS, Frey FA, Gautier I, McCulloch MT, Price RC, Bassias Y, Klootwijk CT, Leclaire L (1989) Basalt basement from the Kerguelen Plateau and the trail of a DUPAL plume. Contrib Mineral Petrol 103:457–469

BW Davy TRH Davies 1979 Entropy concepts in fluvial geomorphology—re-evaluation Water Resour Res 15 1 103 106

Eng J (2014) ROC analysis: web-based calculator for ROC curves. Johns Hopkins University, Baltimore (updated 2014 Mar 19). www.jrocfit.org

JE Evans JM Huxley RK Vincent 2007 Upstream channel changes following dam construction and removal using a GIS/Remote Sensing approach J Am Water Resour Assoc 43 3. 10.1111/j.1752-1688.2007.00055.x

Fisher W (1995) Toward sustainable development? Struggling Over India's Narmada River. M. E. Sharpe. p 161. ISBN 978-1-56324-341-7

FA Frey MF Coffin PJ Wallace D Weis X Zhao SW Wise Jr V Wahnert DAH Teagle PJ Saccocia DN Reusch MS Pringle KE Nicolaysen CR Neal RD Muller CL Moore JJ Mahoney L Keszthelyi H Inokuchi RA Duncan H Delius JE Damuth D Damasceno HK Coxall MK Borre F Boehm J Barling NT Arndt M Antretter 2000 Origin and evolution of a submarine large igneous province: the Kerguelen Plateau and Broken Ridge, southern Indian Ocean Earth Planet Sci Lett 176 73 89

B Fu BF Wu YH Lu ZH Xu JH Cao D Niu GS Yang YM Zhou 2010 Three gorges project: efforts and challenges for the environment Prog Phys Geogr 34 741 754

C Gaina RD Muller B Brown T Ishihara 2007 Breakup and early seafloor spreading between India and Antarctica Geophys J Int 170 151 169

A Ghatak AR Basu 2013 Isotopic and trace element geochemistry of alkalic–maficultramafic-carbonatitic complexes and flood basalts in NE India: Origin in a heterogeneous Kerguelen Plume Geochim Cosmochim Acta 115 46 72

NC Ghose N Chatterjee BF Windley 2017 Subaqueous early eruptive phase of the Late Aptian Rajmahal volcanism, India: evidence from volcaniclastic rocks, bentonite, black shales, and oolite Geosci Front 8 809 822

WL Graf 1999 Dam nation: a geographic census of American dams and their large-scale hydrologic impacts Water Resour Res 35 1305 1311

EA Greathouse CM Pringle WH McDowell JG Holmquist 2006 Indirect upstream effects of dams: consequences of migratory consumer extirpation in Puerto Rico Ecol Appl 16 1 339 352

KJ Gregory C Park (1974) Adjustment of river channel capacity downstream from a reservoir Water Resour Res 10 870 873

KJ Gregory 1987 Environmental effects of river channel change Regulated Rivers 1 358 363

Hobson GV (1929) General report. Geol. Surv. India Records of the Geol. Surv India. LXII(1):145–146

Hossain M, Khan M, Chowdhury K, Rashed A (2019) Synthesis of the tectonic and structural elements of the Bengal Basin and its surroundings. In: Mukherjee S (ed) Tectonics and structural geology: Indian context. Springer Nature Switzerland AG 2019, pp 135–218. https://doi.org/10.1007/978-3-319-99341-6_6

H Huang HH Chang GC Nanson 2004 Minimum energy as the general form of critical flow and maximum flow efficiency and for explaining variations in river channel pattern Water Resour Res 40 4 W04502 https://doi.org/10.1029/2003wr002539

S Ingle D Weis JS Scoates FA Frey 2002 Relationship between the early Kerguelen plume and continental flood basalts of the paleo-Eastern Gondwana margins Earth Planet Sci Lett 197 35 50

S Ishizaka 2006 The Anti Tehri Dam movement as a new social movement and Gandhism J Jpn Assoc S Asian Stud 18 2006

James GA (2014) Ecology is permanent economy: the activism and environmental philosophy of Sunderlal Bahuguna. State University of New York Press, Albany. ISBN 978-1-4384-4673-8

Jenks George F (1967) The data model concept in statistical mapping. Int Yearb Cartography 7:186–190

Kapawar MR, Mamilla V (2020) Paleomagnetism and rock magnetism of early Cretaceous Rajmahal basalts, NE India: implications for paleogeography of the Indian subcontinent and migration of the Kerguelen hotspot. J Asian Earth Sci 201(2020):104517. ISSN 1367-9120. https://doi.org/10.1016/j.jseaes.2020.104517

R Kent 1991 Lithospheric uplift in eastern Gondwana: evidence for long-lived mantle plume system? Geology 19 19 23

RW Kent MS Pringle R Dietmar Müller AD Saunders NC Ghose 2002 40Ar/39Ar Geochronology of the Rajmahal Basalts, India, and Their Relationship to the

Kerguelen Plateau J Petrol 43 1141 1153 https://doi.org/10.1093/petrology/43.7.1141

RW Kent AD Saunders PD Kempton NC Ghose 1997 Rajmahal Basalts, Eastern India: mantel sources and melts distribution at a volcanic rifted margin Am Geophys Union Geophys Monogr 100 145 182

Kishwar M (1995) A Himalayan catastrophe: the controversial Tehri Dam in the Himalayas. Manushi 91 (Nov-Dec 1995), pp 5–16

CT Klootwijk 1971 Palaeomagnetism of the Upper Gondwana Rajmahal Traps, Northeast India Tectonophysics 12 449 467

TM Koel RE Sparks 2002 Historical patterns of river stage and fish communities as criteria for operations of dams on the Illinois River River Res Appl 18 3 19

GM Kondolf 1997 Hungry water: effects of dams and gravel mining on rivers channels Environ Manage 21 533 551

F Lajoie AA Assani AG Roy M Mesfioui 2007 Impacts of dams on monthly flow characteristics. The influence of watershed size and seasons J Hydrol 334 423 439

Langbein WB, Leopold LB (1966) River meanders—theory of minimum variance. U.S. Geological Survey Professional Paper 422-H. U.S. Government Printing Office, Washington, DC

Lawson JM (1925) Effect of Rio Grande storage on river erosion and deposition. Eng News Record 327–334

FK Ligon WE Dietrich WJ Trush 1995 Downstream ecological effects of dams Bioscience 45 183 192

HY Madukwe 2016 Granulometric Analysis of the Sandstone Facies of the Ise Formation, Southwestern Nigeria J Multi Eng Sci Technol (JMEST) 3 3909 3919

FJ Magilligan KH Nislow 2001 Long-term changes in regional hydrologic regime following impoundment in a humid-climate watershed J Am Water Resour Assoc 37 1551 1569

FJ Magilligan KH Nislow 2005 Changes in hydrologic regime by dams Geomorphology 71 61 78

FJ Magilligan HJ Haynie KH Nislow 2008 Channel adjustments to dams in the Connecticut River basin: implications for forested mesic watersheds Ann Assoc Am Geogr 98 267 284

FJ Magilligan KH Nislow BE Graber 2003 A scale-independent assessment of discharge reduction and riparian disconnectivity following flow regulation by dams Geology 31 569 572

JJ Mahoney JD McDougall GW Lugmair K Gopalan 1983 Kerguelen hotspot source for Rajmahal Traps and Ninetyeast Ridge? Nature 303 385 389

Malhotra SL (1951) Effects of barrages and weirs on the regime of rivers. In: Proceedings of international association of hydraulic research, 4th Meeting, pp 335–347

S Mandal 2017 Assessing the instability and shifting character of the river bank Ganga in Manikchak Diara of Malda District, West Bengal using Bank Erosion Hazard Index (BEHI), RS & GIS Eur J Geogr 8 4 6 25

W Marsh J Dozier 1981 Landscape: an introduction to physical geography Addison-Wesley Reading, MA

MG Mehta 2010 A river of no dissent: Narmada Movement and coercive Gujarati nativism South Asian Hist Cult 1 4 509 528

S Misra 2006 Precambrian chronostratigraphic growth of Singhbhum-Orissa Craton, Eastern Indian shield: an alternative model J Geol Soc India 67 356 378

WK Mohanty AK Mohapatra AK Verma KF Tiampo K Kislay 2014 Earthquake forecasting and its verification in northeast India Geomat Nat Haz Risk 7 194 214

RJ Moiola D Weiser 1968 Textural parameters: an evaluation J Sediment Petrol 38 45 53 https://doi.org/10.1306/74D718C5-2B21-11D7-8648000102C1865D

Nandy P (1992) The Old Man and the River. In: Rigzin T (ed. 1997), Fire in the heart, firewood on the back: writings on and by Himalayan Crusader Sunderlal Bahuguna, Silyara, Tehri Garhwal: Parvatiya Navjeevan Mandal, pp 9–13

GC Nanson HQ Huang 2008 Least action principle, equili brium states, iterative adjustment and the stability of alluvial channels Earth Surf Proc Land 33 6 923 942

Narula S (2008) The story of Narmada Bachao Andolan: Human Rights in the global economy and the struggle against the World Bank. New York University School of Law, p 5

NRLD (2018) National Register of Large Dams 2018, Central Water Commission, Govt of India. http://cwc.gov.in/national-register-large-dams

HKH Olierook F Jourdan RE Merle NE Timms N Kusznir J Muhling 2016 Bunbury Basalt: Gondwana breakup products or earliest vestiges of the Kerguelen mantle plume? Earth Planet Sci Lett 440 20 32 https://doi.org/10.1016/j.epsl.2016.02.008

C Paola 2000 Quantitative models of sedimentary basin filling Sedimentology 47 Suppl. 1 121 178

E Pascoe 1959 A manual of Geology of India and Burma Gov 3 485 1343

Perona P, Porporato A, Ridolfi L (2002) River dynamics after cutoff: a discussion of different approaches, In: Bousmar D, Zech Y (eds) Proceedings of the River flow 2002 international conference on fluvial hydraulics. International Association for Hydro-Environment Engineering and Research, Spain, pp 715–721

GE Petts AM Gurnell 2005 Dams and geomorphology: research progress and future directions Geomorphology 71 27 47

Petts GE (1980) Morphological changes of river channels consequent upon headwater impoundment. J Inst Water Eng Sci 34:374–382

JD Phillips 2003 Toledo Bend reservoir and geomorphic response in the lower Sabine River River Res Appl 19 137 159

NL Poff JD Allan MB Bain JR Karr BD Richter RE Sparks JC Stromberg 1997 The natural flow regime Bioscience 47 769 784

NL Poff BP Bledsoe CO Cuhaciyan 2006 Hydrologic variation with land use across the contiguous United States: geomorphic and ecological consequences for stream ecosystems Geomorphology 9 264 285

ME Power WE Dietrich JC Finlay 1996a Dams and downstream aquatic biodiversity: potential food web consequences of hydrologic and geomorphic change Environ Manage 20 887 895

ME Power WE Dietrich JC Finlay 1996b Dams and downstream biodiversity: potential food web consequences of hydrologic and geomorphic change Environ Manage 20 887 895

B Prasad BS Pundir 2017 Gondwana biostratigraphy of the Purnea Basin (Eastern Bihar, India), and its correlation with Rajmahal and Bengal Gondwana Basins J Geol Soc India 90 405 427

CM Pringle 1997 Exploring how disturbance is transmitted upstream: going against the flow J N Am Benthol Soc 16 425 438

Ray Chowdhury PE (1965) Bihar District gazetteers Santahal Pargana. Secretariat Press, Patna, pp 7-9 and p 36

Rhoads BL (2020) River dynamics: geomorphology to support management. Cambridge University Press. https://doi.org/10.1017/9781108164108

BD Richter JV Baumgartner J Powell DP Braun 1996 A method for assessing hydrologic alteration within ecosystems Conserv Biol 10 1163 1174

SB Rood JM Mahoney 1990 Collapse of riparian poplar forests downstream from dams in Western Prairies: probable causes and prospects for mitigation Envrion Manage 14 451 464

AB Roy 2014 Indian subcontinent, reference module in earth systems and environmental sciences Elsevier

AB Roy A Chatterjee 2015 Tectonic framework and evolutionary history of the Bengal Basin in the Indian subcontinent Curr Sci 109 271 279

Rudra K (2006) Shifting of the Ganga and land erosion in West Bengal/A socio-ecological viewpoint. CDEP Occasional Paper-8. Indian Institute of Management, Calcutta, pp 1–43

Rudra K (2008) Banglar nadikatha. Kolkata, Sahitya Samsad, pp 11–19

Rudra K (2009) Re-flooding the Kosi. Himal S Asian 22 (3):50–51

Rudra K (2018) The dynamic Ganga. In: Rivers of the Ganga-Brahmaputra-Meghna Delta. Geography of the physical environment. Springer, Cham. https://doi.org/10.1007/978-3-319-76544-0_4

Sarif MN, Siddiqui L, Islam MS, Parveen N, Saha M (2021) Evolution of river course and morphometric features of the River Ganga: a case study of up and downstream of Farakka Barrage. Int Soil Water Conserv Res. ISSN 2095-6339. https://doi.org/10.1016/j.iswcr.2021.01.006

Shrevastava RN, Shah SC (1966) Ginko digitata brong: from the Rajmahal hills,Santhgal Parganas (Bihar). Records GSI 96(2):309–312

Shulits S (1934) Experience with bed degradation below dams on European rivers. Eng News Record 838–839

M Siegmund-Schultze 2018 The legacy of large dams and their effects on the water-land nexus Reg Environ Change 18 1883 1888 https://doi.org/10.1007/s10113-018-1414-7

MB Singer (2007a) The influence of major dams on hydrology through the drainage network of the Sacramento Valley, California River Res Appl 23 55 72

MB Singer 2008 Downstream patterns of bed-material grain size in a large, lowland alluvial river subject to low sediment supply Water Resour Res 44 W12202 https://doi.org/10.1029/2008WR007183

MB Singer 2010 Transient response in longitudinal grain size to reduced gravel supply in a large river Geophys Res Lett 37 L18403 https://doi.org/10.1029/2010GL044381

MB Singer (2007b) The influence of major dams on hydrology through the drainage network of the Sacramento River Basin, California, River Res Applic 23 55 72

AP Singh N Kumar B Singh 2004 Magmatic underplating beneath the Rajmahal Traps: gravity signature and derived 3-D configuration J Earth Syst Sci 113 759 769

Singh A (2019) Farakka barrage: one historic stupidity that made Bihar a target of both floods and famine. Retrieved from https://tfipost.com/2019/10/farakka-barrage-one-historic-stupidity-that-made-bihar-a-target-of-both-floods-and-famine/

A Singh K Bhushan C Singh MS Steckler SH Akhter L Seeber W-Y Kim AK Tiwari R Biswas 2016 Crustal structure and tectonics of Bangladesh: new constraints from inversion of receiver functions Tectonophysics 680 99 112

A Singh C Singh BLN Kennett 2015 A review of crust and upper mantle structure beneath the Indian subcontinent Tectonophysics 644–645 1 21

A Sinha S Rais 2019 Granulometric Analysis of Rajmahal Inter-Trappen sedimentary rocks (early cretaceous), Eastern India, implications for depositional history Int J Geosci 10 238 253 https://doi.org/10.4236/ijg.2019.103015

R Sinha S Ghosh 2012 Understanding dynamics of large rivers aided by satellite remote sensing: a case study from Lower Ganga plains India Geocarto Int 27 3 207 219 https://doi.org/10.1080/10106049.2011.620180

Storey M, Kent RW, Saunders AD, Salters VJ, Hergt J, Whitechurch H, Sevigny JH, Thirlwall MF, Leat P, Ghose NC, Gifford M, Schlich R et al (1992) Lower cretaceous volcanic rocks on continental margins and their relationship to the Kerguelen Plateau, In: Proceedings of the Ocean drilling program. Sci Results 120:33–53. https://doi.org/10.2973/odp.proc.sr.120.118.1992

S Tiwari GS Jassal (2003) Origin and evolution of the Garo-Rajmahal Gap J Geol Soc India 57 389 403

DJ Topping DM Rubin JLE Vierra (2000) Colorado River sediment transport 1. Natural sediment supply limitation and the influence of Glen Canyon Dam Water Resour Res 36 515 542

Williams GP, Wolman MG (1984) Downstream effects of Dams on Alluvial Rivers. U. S. Geological Survey Professional Paper 1286. U.S. Government Printing Office, Washington, DC. 83 pp

Willis CM, Griggs GB (2003) Reductions in fluvial sediment discharge by coastal dams in California and implications for beach sustainability. J Geol 111

JX Xu 1990 An experimental study of complex response in river channel adjustment downstream from a reservoir Earth Surf Proc Land 15 43 53

CT Yang 1971 On river meanders J Hydrol 13 231 253

A Yin CS Dubey AAG Webb TK Kelty M Grove GE Gehrels WP Burgess 2010 Geologic correlation of the Himalayan orogen and Indian craton: Part 1. Structural geology, U-Pb zircon geochronology, and tectonic evolution of the Shillong Plateau and its neighboring regions in NE India Geol Soc Am Bull 122 336 359

Y Zahar A GhoRbel G Albergel 2008 Impacts of large dams on downstream flow conditions of rivers: Aggradation and reduction of the Medjerda channel capacity downstream of the Sidi Salem dam (Tunisia) J Hydrol 351 318 330

C Zarfl AE Lumsdon J Berlekamp L Tydecks K Tockner 2015 A global boom in hydropower dam construction Aquat Sci 77 1 161 170 https://doi.org/10.1007/s00027-014-0377-0

Tanmoy Sarkar is an Assistant Professor in the Department of Geography of Gazole Mahavidyalaya (College) in Malda, West Bengal, India. The college is affiliated to the University of Gour Banga. Mr. Sarkar completed his postgraduate studies in Geography and Environment Management with a specialization in Remote Sensing & GIS at Vidyasagar University. His research chiefly focuses on soil erosion modeling, geospatial tools and techniques in physical and human geography, and multi-criteria predictive models. He has been teaching Geographical Science at the undergraduate level for more than ten years. He has the working experience as investigators in different research projects, sponsored by DST and ICSSR. Mr. Sarkar has published three research papers in reputed journals, published by Springer Nature.

Mukunda Mishra is an Assistant Professor in the Department of Geography and designated Vice Principal of Dr. Meghnad Saha College, affiliated with the University of Gour Banga in West Bengal, India. Dr. Mishra completed his postgraduate studies in geography and environmental management at Vidyasagar University and holds a Ph.D. in geography from the same university. He was selected for the National Merit Scholarship by the Ministry of Human Resource Development, Government of India. His research chiefly focuses on analyzing unequal human development and creating multi-criteria predictive models. Dr. Mishra has authored/edited four books and published more than 25 research articles and book chapters. He has more than 10 years of hands-on experience in dealing with development issues at the ground level in various districts of eastern India.

Part IV

Climate Change, Geomorphic Hazards and Human Livelihood

Climate Change and Human Performance: Assessment of Physiological Strain in Male Paddy Cultivators in Hooghly, West Bengal, India

24

Ayan Chatterjee, Sandipan Chatterjee, Neepa Banerjee, and Shankarashis Mukherjee

Abstract

Climate change has a significant impact on different sectors of society, including public health. It has been reported that work performance was getting affected due to unfavorable thermal working environmental conditions existing in the working environment, particularly in outdoor occupations. Paddy cultivation is also an outdoor occupation and the food crop cultivators have to work manually throughout the year irrespective of variation in weather conditions including the change in the thermal working environmental condition. The present study has been undertaken to assess the impact of thermal working environmental conditions on cardiac response profile in terms of indices of physiological strain in male food crop cultivators. The physical and physiological parameters of participants were measured. Indicators of the thermal working conditions were measured at regular intervals during the working hours in the agricultural field. The result of the present study indicated that the thermal working environmental condition was above the recommended threshold value making the task strenuous for the human resources during the 'Aman' type of paddy cultivation.

Keywords

Paddy · Cardiac strain · Thermal working environment · Popular heat indices

A. Chatterjee (✉) · S. Chatterjee · N. Banerjee · S. Mukherjee
Human Performance Analytics and Facilitation Unit, Department of Physiology, University of Calcutta, 92, Acharya Prafulla Chandra Road, Kolkata 700009, India
e-mail: ayan4189@yahoo.com

A. Chatterjee
Guru Nanak Institutions, Ibrahimpatnam, Telangana 501506, India

S. Mukherjee
Public Health Analytics Unit, Department of Food and Nutrition, West Bengal State University, Kolkata 700126, India

24.1 Introduction

India is one of the largest developing countries with a huge population load. It is a village-oriented country and the agricultural workers of the village solely depend on cultivation. Usually, for this cultivation, agricultural workers have to depend on natural resources and climatic conditions (Adhikari et al. 2012). Moreover, it has been also reported that the impact of an increase in ambient temperature is not limited to agricultural output; it affects the work performance of a human being associated with occupational activities (Kjellstrom et al. 2016; Venugopal et al. 2016; Mukherjee 2015). Therefore, different tasks performed by agricultural workers not only demand substantial time and energy but also sources of drudgery for them (Chatterjee et al. 2020a, b, c, 2021). In this backdrop, the present study has been undertaken to assess the effect of workplace heat exposure and workload on physiological strain in terms of indices of physiological strain during 'Aman' and 'Boro' type of paddy cultivation time in male food crop cultivators primarily engaged in manual transplanting (transplanting is the most common and elaborative method of crop establishment for rice) of paddy seedlings task in the paddy field by random transplanting method (in random method, seedlings are transplanted without a definite distance or space between plants).

24.2 Materials and Methods

Initially, after getting permission from the Institutional Human Ethical Clearance Committee the study was carried out on male food crop cultivators (age range 21–30 years) occupationally engaged in different tasks during the paddy cultivation time. The study was carried out Village in Fului Gram Panchayat, Goghat II administrative Block, Arambagh Subdivision, and District Hooghly, West Bengal. Food crop cultivators having a minimum working experience of five years and regularly working for at least for a period of six to six and half hours in the agricultural field and also willingly expressing their wish for being included in the study were only considered for random selection. After obtaining initial consent from the study participants, the study necessities were explained elaborately. Data were collected from June to the middle of July (during the 'Aman' type of paddy cultivating period) and from December to January (during the 'Boro' type of paddy cultivating period). These data were presented in three spells, i.e., morning [6.15–9 a.m.] was referred to as spell 1 [S1], similarly around noon [9.30–10.00 a.m. to about 1 p.m.] was referred to as spell 2 [S2] and afternoon [2.30–4.00 p.m.] was referred to as spell 3 [S3]. It may be also mentioned that the data of individuals who were available for study during both seasons were only considered for analysis. Data were collected from 55 adult male food crop cultivators while they were taking part in the manual transplanting task during the 'Aman' and the 'Boro' type of paddy cultivation. These data were tabulated as the data from a manual threshing group (MTG-A) and (MTG-B). After obtaining the consent, the name, age (year), the ethnic background of the study participants was recorded to each individual in a pre-designed schedule. Information for the assessment of the socioeconomic status of the participants was recorded by using Kuppuswamy's socioeconomic scale [SES] (Ravikumar et al. 2013). Stature in cm and body weight (BW) in kg was measured using an anthropometric measurement set and a weighing scale, respectively. Body Mass Index (BMI) (kg m^{-2}) was calculated. Pre-work heart rate (HRPre-work) of the study participants was recorded and/or by using the Polar heart rate monitor and stopwatch before the individuals started their work and expressed in beats min^{-1}. Pre-work Systolic and diastolic blood pressure (SBPPre-work and DBPPre-work) were also recorded during the morning hours before the individuals started their work and/or by using an automated blood pressure monitor and sphygmomanometer in sitting condition and expressed in mm Hg. Heart rate was monitored using a heart rate monitor (Polar) and data were recorded at a regular interval during the activity period of paddy cultivators in three different spells and finally, the highest values of heart rates in each spell were presented as Peak heart rate (HR $_{peak}$), expressed in beats min^{-1} (Astrand and Rodhal 1986). Net cardiac cost (NCC) was obtained (Chamoux et al. 1985) as the difference between working and pre-working heart rate of the study participants and was expressed in beats min^{-1}. Peak estimated energy expenditure (EEE) of tasks was obtained (Ramanathan et al. 1967) and was expressed in kcal min^{-1}. The drudgery index and human physical drudgery index (HPDI) were calculated (Parimalam et al. 2017; Joshi et al. 2015). The 'heaviness' of work (Motamedzade and Azari 2006) was adjudged in terms of - peak heart rate (HR $_{peak}$) (beats min^{-1}), net cardiac cost (NCC) (beats min^{-1}), and peak estimated energy expenditure (EEE) (kcal min^{-1}). In the case of basic environmental parameters, Dry bulb (TDB) and Wet bulb (TWB) temperatures were measured with the help of a Hygrometer. The dry bulb temperature was recorded thereafter during the working hours in the agricultural field. Wet Bulb Globe Temperature (WBGT) index was found (Heidari et al. 2015). Corrected Effective Temperature (CET) was determined from TDB, TWB, TG, and Air velocity from specified nomograms (Brake and Bates 2002). The modified Discomfort index (MDI) (Epstein and Moran 2006) and predicted four-hour sweat rate (P$_4$SR) (lit) (McArdle et al. 1947) were determined from TDB, TWB values as follows. The collected data were tabulated, analyzed, and tested for significance with analysis of variance, as appropriate. As the thermal environmental conditions were assessed in terms of several indices, the correlation between them was found. A P-value lower than 0.05 ($P < 0.05$) was considered significant.

24.3 Result and Discussion

General characteristics in terms of age (years), ethnicity, SES, working experience (years) and working time (h day^{-1}) of the study participants has been presented in Table 24.1.

The physical and physiological variables in terms of stature (cm), body weight (kg), BMI (kg m^{-2}), HR$_{Pre-work}$ (beats min^{-1}), SBP$_{Pre-work}$ (mm Hg), and DBP$_{Pre-work}$ (mm Hg) of the study participants are presented in Table 24.2.

The mean BMI of the study participants was 20.1 kg m^{-2}, which indicated that the participants were in the 'normal weight' category as per the classification given by WHO (WHO 2000). This finding is also in tune with the findings of earlier studies, which reported that different form of recreational physical activity strategically and methodically has some advantageous role in maintaining satisfactory body composition, enhancing physical fitness, and hence facilitating maintaining a normal BMI (Mukherjee et al. 2014a, b; Chatterjee et al. 2020d). Higher values of BMI have also been found to be linked with

Table 24.1 General characteristics of study participants

Variables	Values
Age (years)	25.4 ± 1.18
Ethnicity	Bengalee
SES	Lower middle
Working experience (year)	7.8 ± 0.89
Working time (h day^{-1})	6.8 ± 0.21
Data were in AM ± SD	

Table 24.2 Physical and physiological characteristics of the study participants

Variables	Values
Stature (cm)	162.1 ± 3.11
BW (kg)	53.0 ± 4.11
BMI (kg m^{-2})	20.1 ± 1.11
HR$_{Pre-work}$ (beats min^{-1})	72.0 ± 4.19
SBP$_{Pre-work}$ (mm Hg)	116.0 ± 3.11
DBP$_{Pre-work}$ (mm Hg)	74.0 ± 7.13
Data were in AM ± SD	

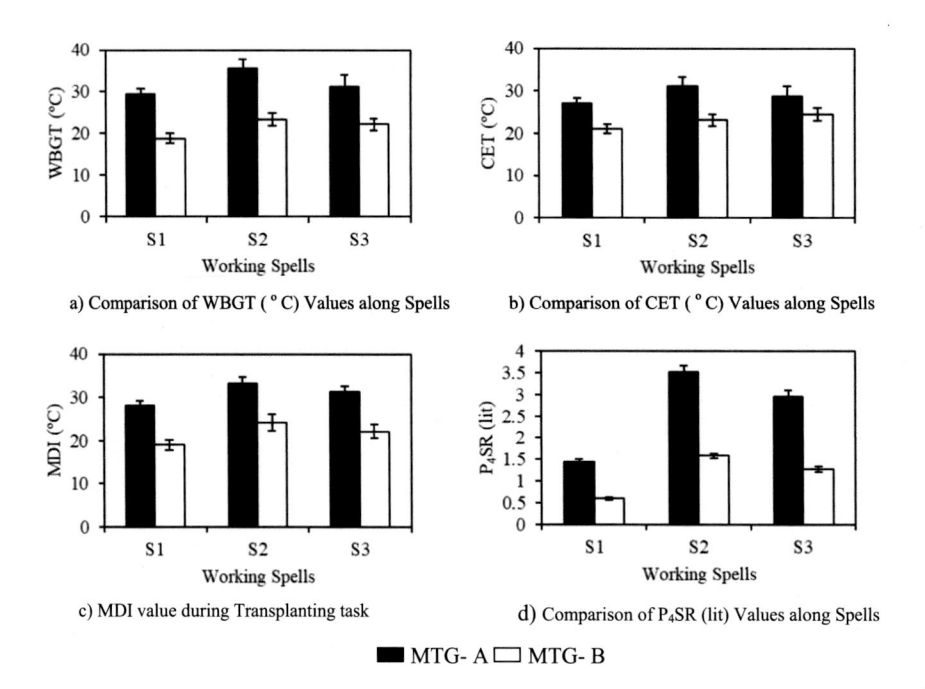

Fig. 24.1 Environmental condition in terms of four indicators of thermal environmental status- WBGT, CET, MDI, and P$_4$SR **a** Comparison of WBGT (°C) values along spells, **b** Comparison of CET (°C) values along spells, **c** MDI value during transplanting task, **d** Comparison of P$_4$SR (lit) values along spells

more chance of work-related musculoskeletal disorder among sedentary workers (Chatterjee et al. 2014, 2015d, e).

The environmental condition in terms of four indicators of thermal environmental status- WBGT, CET, MDI, and P$_4$SR are presented in Fig. 24.1.

24.3.1 In the Case of MTG-A, the Working Thermal Environmental Condition in Terms of WBGT Index

In the case of MTG-A the average values of WBGT index during 'Aman' type of paddy cultivating time in S1, S2, and S3 are 29.5 °C, 34.0 °C, and 31.3 °C, respectively. Of which, during the S2 at 34.0 °C, no work is ideally allowable as per ACGIH guidelines (ACGIH 2008; Miller and Bates 2007). In the S3 working spell at an average WBGT index value of 31.3 °C, 'light' type of work up to 50% of the time and 'moderate' type of work with the allocation of work up to 25% time each hour in the work-rest cycle is allowable. In the S1 working spell at an average WBGT index value of 29.5 °C, for 'light' type of work, there is no restriction in terms of allocation of work in the work-rest cycle; for 'moderate' type of work, up to 75% time each hour, work can be allocated in the work-rest cycle and for 'heavy' type of work, up to 50% time each hour, work can be allocated in the work-rest cycle. In the case of MTG-B - During the 'Boro' type of paddy cultivating time the average WBGT in S1, S2, and S3 are 18.9, 23.4, and 22.2 °C. During S1, S2 and S3 there is no restriction recommended against carrying out the task. In the case of MTG-A, the working thermal environmental condition in terms of CET index- The average values of CET during the 'Aman' type of paddy cultivating time along the spells are 28.0, 31.1, and 31.1 °C. In the S2 working spell, with CET value of 31.1 °C, only 'light' and in S3, with CET value of 28.8 °C, up to 'moderate' category of work could be carried out. And in the S1 working spell with an average

CET value of 28.0 °C up to 'moderate' category of work can be carried out. In case of MTG-B - During the 'Boro' type of paddy cultivating time, the average CET values are 21, 23.1, and 24.4 °C, there is no restriction recommended (WHO 1969) against carrying out of the work.

24.3.2 In the Case of MTG-A, the Working Thermal Environmental Condition in Terms of MDI Index

In the case of the 'Aman' type of paddy cultivating time the average value of MDI in the S1, S2 and S3 working spells are 28.2, 33.1, and 31.2 °C. In S1, S2, and S3 the heat load is 'severe' and the human resources engaged in physical work are at increased risk for heat illness. In the case of MTG-B - Whereas during the 'Boro' type of paddy cultivating time with the average value of MDI in the S1, S2 and S3 working spells are 19.0, 24.1, and 22.1 °C. In the S1 and S3, there are no restrictions recommended (Epstein and Moran 2006; Sohar et al. 1962) against carrying out the task whereas in the S2 working spell individual feel 'mild sensation of heat'.

24.3.3 In the Case of MTG-A, the Working Thermal Environmental Condition in Terms of P_4SR Index

In the case of P_4SR values during the 'Aman' type of paddy cultivating time in the S2 the limit in terms of P_4SR for acclimatized human resources is exceeded and for the S3, it approaches the limit (McArdle et al. 1947). In the case of MTG-B—During the 'Boro' type of paddy cultivation time, the temperature is suitable for carrying out the task.

The results of indicators of physiological strain—HR_{peak}, NCC, and EEE have been presented for both the seasons in a spell-wise manner and have been presented in Fig. 24.2.

In terms of Heart Rate $_{Peak}$ for MTG-A individuals, it is found that the values were varied from 100 to 110 beats min^{-1} in the S1 working spell, whereas during the S2 and S3 working spell of the working hours the values were varied from 115 to 125 beats min^{-1} and 115 to 129 beats min^{-1}, respectively. In terms of NCC expressed in beats min^{-1}, another important marker of physiological strain, it is found that the values were varied from 23 to 31 beats min^{-1} in the S1 working spell whereas during the S2 and S3 working spell of the working hours it varied from 40 to 51 beats min^{-1} and 32 to 44 beats min^{-1}, respectively. In terms of estimated energy expenditure (EEE) as per Ramanathan's model (1967) expressed in kcal min^{-1}, for assessment of physiological strain, it is found that the values of EEE were varied from 2.90 to 3.00 kcal min^{-1} in the S1 working spell, whereas during the S2 and S3 working spell of the working hours it varied from 3.59–3.67 kcal min^{-1} to 3.18–3.30 kcal min^{-1}, respectively.

24.3.4 For MTG-B Individuals

In terms of Heart Rate$_{Peak}$ (beats min^{-1}), it is found that the values were varied from 92 to 104 beats min^{-1} in the S1 working spell, whereas during the S2 and S3 working spell of the working hours the values were varied from 110 to 120 beats min^{-1} and 98 to 110 beats min^{-1}, respectively. In terms of NCC expressed in beats min^{-1}, it is found that the values were varied from 20 to 27 beats min^{-1} in the S1 working spell whereas during the S2 and S3 working spell of the working hours it varied from 30 to 39 beats min^{-1} and 25 to 35 beats min^{-1}, respectively. In terms of estimated energy expenditure (EEE) as per Ramanathan's model (1967) expressed in kcal min^{-1}, it is found that the values of EEE were varied from 2.93 to 3.02 kcal min^{-1} in the S1 working spell, whereas during the S2 and S3 working spell of the working hours it varied from 3.59–3.69 kcal min^{-1} to 3.20–3.30 kcal min^{-1}, respectively.

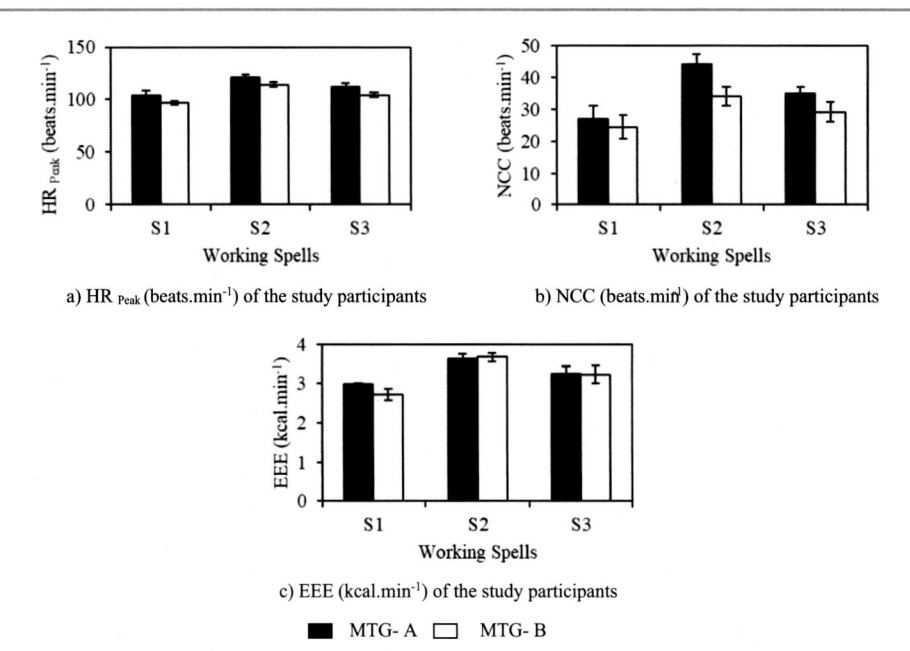

Fig. 24.2 Indicators of **a** physiological strain in terms of HR_{Peak} (beats min^{-1}), **b** NCC (beats min^{-1}), and **c** EEE (kcal min^{-1})

In terms of drudgery, is generally conceived as physical and mental strain, fatigue, monotony, and hardships experienced while doing a job. HPDI, is one of the most important indicators to assess the physical fitness profile, especially for working individuals (Parimalam et al. 2017). The value of the drudgery index and HPDI has been presented in Fig. 24.3.

24.3.5 Drudgery Index for MTG-A Individuals

The drudgery index value during the manual transplanting task varied from 34 to 44 in the S1 working spell, whereas during the S2 and S3 working spell of the working hours the value varied from 44–55 to 42–52, respectively.

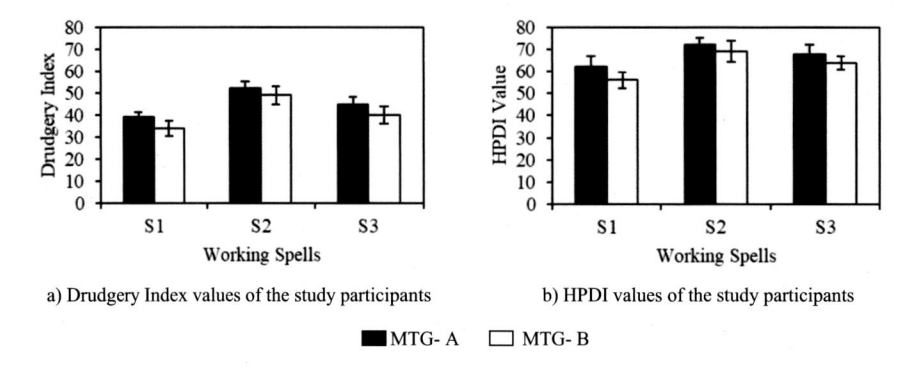

Fig. 24.3 Drudgery index and HPDI values of the study participants **a** Drudgery Index values of the study participants, **b** HPDI values of the study participants

24.3.6 Drudgery Index for MTG-B Individuals

It is found that in the paddy cultivators belonging to MTG-B individuals, the drudgery index values were varied from 30 to 41 in the S1 working spell, whereas during the S2 and S3 working spell of the working hours the values were varied from 44–56 to 36–45, respectively.

24.3.7 HPDI for MTG-A Individuals

The HPDI values during the manual transplanting task varied from 58 to 66 in the S1 working spell, whereas during the S2 and S3 working spell of the working hours the value varied from 68–76 to 64–72, respectively.

24.3.8 HPDI for MTG-B Individuals

In the case of MTG-B individuals, the HPDI values varied from 52 to 60 in the S1 working spell, whereas during the S2 and S3 working spell of the working hours the value varied from 65–73 to 60–68 kcal min^{-1}, respectively.

The heaviness of the work was carried out in terms of different indicators of physiological strain in terms of HR $_{peak}$, NCC, and EEE. The workload was categorized as 'light', 'moderate', 'heavy', etc. The heaviness of workload has been presented in Table 24.3.

The heaviness of workload for MTG-A individuals, in the S1 working spell has been adjudged as 'moderate', 'quite moderate', and 'moderate', respectively in terms of three indicators HR $_{peak}$, NCC, and EEE. In the S2 working spell the workload has been adjudged as 'heavy', 'rather heavy' and 'moderate' in terms of HR $_{peak}$, NCC, and EEE. In the S3 working spell, the workload has been adjudged as 'heavy', 'quite moderate' and 'moderate', respectively in terms of three indicators HR $_{peak}$, NCC, and EEE. For MTG-B individuals, the heaviness of workload has been adjudged as 'moderate', 'quite moderate', and 'moderate', respectively in terms of three indicators HR $_{peak}$, NCC, and EEE. In the S2 working spell the workload has been adjudged as 'heavy', 'moderate', and 'moderate' in terms of three indices of physiological strain—HR $_{peak}$, NCC, and EEE. In the S3 working spell, the workload has been adjudged as 'moderate', 'quite moderate', and 'moderate' in terms of HR $_{peak}$, NCC, and EEE. The findings of the present study regarding HR $_{peak}$ values are in agreement with the findings of earlier studies (Ojha and Kwatra 2017; Chatterjee et al. 2016a, b).

The suitability of different working environmental conditions in terms of different Indicators of Environmental Heat Indices for carrying out the specific task during paddy cultivating time has been presented in Table 24.4.

Table 24.3 Comparison of the heaviness of work in terms of indicators of physiological strain along working spells

Indicators of physiological strain	Working spells for MTG-A			Working spells for MTG-B		
	S1	S2	S3	S1	S2	S3
HR$_{peak}$ (beats min^{-1})	M	H	H	M	H	M
NCC (beats min^{-1})	QM	RH	QM	QM	M	QM
EEE (kcal min^{-1})	M	M	M	M	M	M

M Moderate, *QM* Quite Moderate, *H* Heavy

Table 24.4 Suitability of different working environmental conditions in terms of different indicators of environmental heat indices for carrying out the specific task during paddy cultivating time

Environmental heat indices	Working spells for MTG-A			Working spells for MTG-B		
	S1	S2	S3	S1	S2	S3
WBGT (°C)	R_2/L	NA	R_2/L	NR	NR	NR
CET (°C)	NR	R_L/M	R_M/M	NR	NR	NR
MDI (°C)	S	S	S	NR	L	NR

NR No restriction, *NA* Not allowed, *R* Allowed with restriction

For WBGT

Category	Suitable for intensity of work activities if work allocation in work-rest cycle per hour be			
	Less than 25%	25–50%	50–75%	75–100%
R_1	all	All except VH	L and M	Only L
R_2	Only L	NA	NA	NA

L Light, *M* Moderate, *H* Heavy, *VH* Very Heavy

For CET

Category	Intensity of work allowed
R_L	Only L
R_M	L and M

L Light, *M* Moderate

For MDI

Category	Intensity of work allowed
L	Only L
M	Only M
S	Only S

L Light, *M* Mild, *S* Severe

The present study's findings regarding energy expenditure values agreed with the finding of an earlier study carried out among paddy cultivators in Odisha (Jena and Mohanty 2014). The finding of the present study in consonance with the finding of earlier studies (Chatterjee et al. 2015a, b, c, 2017, 2018a, b, c, 2019a, b, c, d) conducted among the paddy cultivators occupationally engaged in the manual transplanting task during 'Aman' type of paddy cultivation; this is further affirmed by the finding of the present study carried out during the 'Boro' type of paddy cultivation.

24.4 Conclusions

From the present study, it may be concluded that manual transplanting is a strenuous task for the food crop cultivators. Moreover, the physiological strain was significantly higher during the 'Aman' type of paddy cultivation time, i.e., in the case of MTG-A compared to their age-matched MTG-B counterpart. Moreover, the thermal working environmental condition was also not favorable during the 'Aman' type of paddy cultivation time making the task strenuous for the food grain cultivators. Modification of work-rest schedule, i.e., early starting of the work or attempts should be made through design interventions and other means to reduce the extent of strain for ensuring better working performance.

Acknowledgements We are thankful to all the volunteers for their participation and cooperation during the study.

References

ACGIH (2008) Threshold limit values and biological exposure indices, Cincinnati, OH

Adhikari B, Bag MK, Bhowmick MK, Kundu C (2012) Rice in West Bengal-Rice Knowledge Management 88

Astrand PO, Rodhal K (1986) Threshold limit values and biological exposure indices. In: Text book of work physiology. McGraw Hill, New York

Brake R, Bates GA (2002) Valid method for comparing rational and empirical heat stress indices. Ann Occup Hyg 46:165–174

Chamoux A, Borel AM, Catilina P (1985) Pour la standardization D'unifrequence cardiaque de repos. Arch Mal Prof 46:241–250

Chatterjee A, Chatterjee S, Santra T, Mukherjee S (2014) The influence of anthropometric variables for development of musculoskeletal discomfort among computer operators in organized sectors. In: User centered design and occupational wellbeing, McGraw Hill Education, pp 499–503

Chatterjee A, Chatterjee S, Chatterjee S, Banerjee N, Santra T, Mukherjee S (2015a) Thermal comfort and HSI: a study in Bengalee male paddy cultivators. In: Quad scientific reporter, pp 148–155

Chatterjee A, Banerjee N, Chatterjee S, Santra T, Agrawal KM, Mukherjee S (2015b) Assessment of physiological strain in male paddy cultivators due to work and exposure to fluctuation in thermal conditions in working environments. Survey 55:91–98

Chatterjee A, Chatterjee S, Chatterjee S, Santra T, Bhattacharjee S, Mukherjee S (2015c) Exposure to heat from natural working environment and cardiovascular strain: a study in male agricultural workers in Southern Bengal. In: Caring for people, pp 166–171

Chatterjee A, Chatterjee S, Banerjee N, Santra T, Mondal P, Mukherjee S (2015d) Evaluation of body composition and somatic profile in male individuals: a comparison between tribal and non-tribal agricultural human resources. In: Proceedings of the national conference on agriculture and rural development issues in Eastern India, pp 25–26

Chatterjee A, Chatterjee S, Chatterjee S, Santra T, Banerjee N, Mukherjee S (2015e) Musculoskeletal discomfort in computer operators of organized sector: tracing the link with obesity status. Int Physiol 3:23–28

Chatterjee A, Chatterjee S, Banerjee N, Chatterjee S, Santra T, Mukherjee S (2016a) Seasonal distribution of thermal comfort: a study to assess physiological strain in male paddy cultivators in Southern Bengal. In: Proceedings of the international conference on humanizing work and work environment, pp 157–162

Chatterjee A, Chatterjee S, Banerjee N, Chatterjee S, Santra T, Mukherjee S (2016b) Assessment of physiological strain due to work and exposure to heat of working environments in male paddy cultivators. Adv Appl Physiol 1(1):8–11

Chatterjee A, Chatterjee S, Chatterjee S, Bhattacharjee S, Banerjee N, Mukherjee S (2017) Work place heat exposure and cardiovascular status: a study in male paddy cultivators. In: Giri B, Ghosh P (eds) Molecular physiological and nutritional responses during pathological alteration of cell function. Aaheli Publisher, pp 110–124

Chatterjee A, Chatterjee S, Chatterjee S, Banerjeee N, Mukherjee S (2018a) Diurnal variation in thermal working environment, workload and physiological strain in women workforces engaged in manual parboiling task. In: Swain KK (eds) Advance technologies in agriculture for doubling farmers' income, pp 199–210

Chatterjee A, Chatterjee S, Chatterjee S, Bhattacharjee S, Santra T, Banerjee N, Ghosh K, Mukherjee S (2018b) Assessment of physiological strain in male cultivators engaged in mechanized paddy threshing task using two different types of threshers. Sci Cult 84(5–6):199–205

Chatterjee A, Chatterjee S, Chatterjee S, Banerjee N, Mukherjee S (2018c) A comparative study on the impact of thermal working environmental factors and workload on cardiac response indicators in male food crop cultivators of two ethnic groups. Indian J Biol Sci 24:31–44

Chatterjee A, Banerjee N, Chatterjee S, Chatterjee S, Mukherjee S (2019a) A study to assess cardiac response indices in food crop cultivation task in West Bengal. Int J Innovative Knowl Concepts 7:238–243

Chatterjee A, Chatterjee S, Banerjee N, Chatterjee S, Mukherjee S (2019b) Climate change and human performance: a study in Bengalee male agricultural workers. Int J Innovative Knowl Concepts 7(1):72–80

Chatterjee A, Banerjee N, Chatterjee S, Chatterjee S, Mukherjee S (2019c) Impact of variation in working environmental condition on cardiac response profile in Bengalee male crop cultivators of a Southern district of West Bengal. J Emerg Technol Innovative Res 6:438–443

Chatterjee A., Chatterjee S., Banerjee N., Chatterjee S, Mukherjee S (2019d) Assessment of cardiac strain in male paddy cultivators using two different type of paddy thresher: a comparison. Indian Agriculturist 63:49–55. ISSN: 0019–4336

Chatterjee A, Chatterjee S, Banerjee N, Mukherjee S (2020a) Assessment of physiological strain in male food crop cultivators engaged in manual threshing task in a southern district of West Bengal. Holist Approach Environ 10(4):100–108

Chatterjee A, Chatterjee S, Banerjee N, Mukherjee S (2020b) Impact of variation in thermal working environmental condition on cardiac response indices in male human resources engaged in food crop cultivation task. J Clim Change 6(1):59–66

Chatterjee A, Chatterjee S, Banerjee N, Mukherjee S (2020c) A study to assess cardiac response profile in paddy cultivators engaged in manual paddy transplanting task in Hooghly. West Bengal. Nebio 11(1):27–34

Chatterjee A, Chatterjee S, Banerjee N, Mukherjee S (2020d) A study to assess relationship between different obesity indices and musculoskeletal discomfort score in agricultural workers in Southern Bengal, India. Open Access J Complement Altern Med 4:186–190

Chatterjee A, Chatterjee S, Chatterjee S, Santra T, Banerjee N, Mukherjee S (2021) Assessment of physiological strain in male food crop cultivators engaged in manual reaping task. In: Muzammil M, Khan AA, Hasan F (eds) Ergonomics for improved productivity. Springer Nature, Singapore

Epstein Y, Moran D (2006) Thermal comfort and the heat stress indices. Ind Health 44:388–398

Heidari H, Golbabaei F, Shamsipour A, Forushani AR, Gaeini A (2015) Evaluation of heat stress among farmers using environmental and biological monitoring: a study in North of Iran. IJOH 7:1–9

Jena D, Mohanty SK (2014) Circulo respiratory efficiency of agricultural workers in Odisha, India. Int J Sci Technol Res 3:265–269

Joshi P, Jethi R, Chandra N, Roy ML, Kharbikar HB, Atheequlla GA (2015) Ergonomics assessment of post-harvest finger millet threshing for reducing women drudgery ergonomics assessment of post-harvest finger millet threshing for reducing women drudgery. Indian Res J 15:25–30

Kjellstrom T, Briggs D, Freyberg C, Lemke B, Otto M, Hyatt O (2016) Human performance and occupational health: a key issue for the assessment of global climate change impacts. Annu Rev Public Health 37:97–112

McArdle B, Dunham W, Holling HE, Ladell WSS, Scott JW, Thomson ML, Weiner JS (1947) The prediction of effects of warm and hot environments: the P_4SR Index. Medical Research Council, London

Miller VS, Bates GP (2007) The thermal work limit is a simple reliable heat index for the protection of workers in thermally stressful environments. Ann Occup Hyg 51:553–561

Motamedzade M, Azari MR (2006) Heat stress evaluation using environmental and biological monitoring. Pakistan J Biol Sci 9:457–459

Mukherjee S (2015) Climate change: implications for human resources in informal sector of Eastern India. In: Ergonomics for rural development, pp 174–178

Mukherjee S, Banerjee N, Chatterjee S, Chatterjee S (2014a) Effect of practicing select Indian classical dance forms on body composition status of Bengalee females, an anthropometric study. Indian J Biol Sci 20:40–48

Mukherjee S, Banerjee N, Chatterjee S, Chatterjee S (2014b) Effect of Kathak dancing on obesity indices in women of sedentary avocations. Sci Cult 80:279–282

Ojha P, Kwatra S (2017) Comparative ergonomic assessment of male and female farm workers involve in rice cultivation. Int J Curr Microbiol Appl Sci 6:3439–3446

Parimalam P, Meenakshi J, Padmini DS (2017) An ergonomic analysis of drudgery prone activities in sugarcane cultivation. Int J Agric Sci Res 7:447–452

Ramanathan NL, Dutta SR, Roy BN, Chatterjee A, Mullick LN (1967) Energy cost of different muscular tests performed by Indian subjects. Indian J Occup Health 10:253–261

Ravikumar BP, Dudala SR, Rao AR (2013) Kuppuswamy's socio-economic status scale—a revision of economic parameter for 2012. Int J Res Dev Health 1:2–4

Sohar E, Tennenbaum DJ, Robinson N (1962) The thermal work limit is a simple reliable heat index for the protection of workers in thermally stressful environments biometeorology. In: Tromp SW (ed), Pergamon Press, Oxford, pp 395–400

Venugopal V, Chinnadurai JS, Lucas RAI, Kjellstrom T (2016) Occupational heat stress profiles in selected workplace in India. Int J Environ Res Public Health 89:1–13

World Health Organization (1969) Health factors involved in working under conditions of heat stress: report of a who scientific group. WHO technical report series, p 412

World Health Organization (2000) Obesity: preventing and managing the global epidemic, report of a who consultation on obesity, technical report series, No. 894, World Health Organization, Geneva, Switzerland, p 256

Study on Climate Change and Its Impact on Coastal Habitats with Special Reference to Ecosystem Vulnerability of the Odisha Coastline, India

Avijit Bakshi and Ashis Kumar Panigrahi

Abstract

Coastlines always include a diverse range of ecosystems viz., marine ecosystem, estuarine ecosystem, freshwater ecosystem, terrestrial ecosystem, and coastline interface, providing ideal habitat for a large group of organisms. Coastline habitats are very much sensitive to various consequences due to climate change such as sea-level rise, increase in the temperature of ocean water, increasing frequency and destruction potential of storms, increasing precipitation, etc. Increasing temperature of ocean water also leads to higher absorbance of carbon dioxide into the marine water causing acidification. Changing climate also has a significant role in habitat loss of the coastal inhabitants, as well as a potential societal impact on coastline communities affecting important ecological services especially regulating services. Odisha, a state of India having a coastline of about 480 km, is frequently disturbed by various natural disasters like cyclonic storms, drought, floods, heatwaves, etc. Evidence is there proving the increasing threat of climate change on the coastline ecosystems of Odisha. Management practices exist to reduce the immediate impact of climate change on the coast, but some of them are becoming unsustainable and the reason behind the coastal squeeze is real. The introduction of more natural processes is the only solution behind the sustainable development of Odisha.

Keywords

Climate change · Coastline · Habitat · Ecosystem · Odisha

25.1 Introduction

Coastlines are highly sensitive to the events associated with climate change. The risks due to climate change are in both forms, i.e., direct and indirect. Changes in temperature, unusual patterns of rainfall, and pollution can directly put some stress on the coastline habitats. Eventually, the impact of some slow but steady consequences like the melting of polar ice, sea-level rise, ocean acidification, coastal erosion, etc. can exploit the coastal habitat indirectly. The consequences of climate change not only influence the environment but also put a significant impact on society in the form of health, economic, and livelihood

A. Bakshi (✉)
Ecotoxicology, Fisheries and Aquaculture Extension Laboratory, Department of Zoology, University of Kalyani, Kalyani, Nadia, West Bengal 741235, India
e-mail: avijit1986@gmail.com

A. Kumar Panigrahi
University of Burdwan, Burdwan, West Bengal, India

issues. It is also responsible for the disruption of some important ecosystem services like coastal defenses, sequestering of carbon, etc. (Burden et al. 2020). Coastal habitats are adversely affected by various stressors that are not only environmental but also anthropogenic. The combined impact of these two factors has been evident since the industrial revolution. Climate change also threatened the coastal ecosystem reducing its resilience.

The Indian coastline is characterized by diverse landforms and ecosystems consisting of mangroves, sand beaches, salty landmasses, tablelands, and islands. It is about 7500 km long surrounded by the Bay of Bengal in the east, the Indian Ocean in the south, and the Arabian Sea in the west. Ocean, biosphere, and atmosphere interact at coastline creating opportunities for continuous exchange of mass and energy (Nayak 2017). Due to these exchanges, coastal regions become the center of huge biological diversity and productivity. Coastline productivity includes a huge variety of fishes, shellfishes, seaweeds, and other sea organisms. These products are resources of foods, various drugs, cosmetics, fertilizers, and household commodities. Wetlands of coastal areas are also responsible for nutrient storage and recycling, detoxification of pollutants, reduction of soil erosion, and also provide habitat for migratory birds (Yang et al. 2017; Nayak 2017; Saintilan et al. 2018). The presence of an ample amount of occupation opportunities makes the coastal area center of human activities. Coasts around the country are well known for their scenic beauty providing ample opportunity for the development of tourism. These products and supports directly put an impact on the socioeconomic conditions of the coastal inhabitants providing a great contribution to the Indian economy. Thus, changing climate will surely have a significant societal impact not only on the coastal communities but also on the whole population of the country.

Indian coasts are highly vulnerable to the consequences of climate change such as sea-level rise, ocean acidification, etc. mainly due to having extensively low-lying areas and dense population along with coastal areas. Rising sea levels and acidification of sea water are responsible for the degradation of cultivable land and water resources on the coasts. It is also putting a serious threat to the nursery areas of fisheries and aquaculture near the shoreline. IPCC (2013) has cleared that the global average sea-level rise over the period 1901–2010 was 0.19 m (ranging from 0.17 to 0.21 m). Studies indicate that sea-level rise along the Northern Indian Ocean has been about 1.06–1.75 mm year^{-1} during 1874–2004, but it has increased up to 3.3 mm year^{-1} during 1993–2017 (Dasgupta 2020). Increased intensity of cyclonic storms, high rate of coastal degradation, heavy rainfall, frequent flooding, and other natural hazards are proved to be potential stressors in coastal habitats. Increased frequency and intensity of cyclonic storms is highly related to the increase in sea surface temperature resulting in severe damage to the coastal biodiversity. The occurrence of floods associated with heavy rainfall has also been found to be increased in frequency affecting the biodiversity and human life in coastal cities and other areas. Besides these natural phenomena, anthropogenic activities such as coastal pollution, unplanned development, overexploitation of resources, etc. are also enhancing the threat to the coastal ecosystem. Three megacities of India (Mumbai, Chennai, and Kolkata) and many other important cities with millions of population are at a high risk of these consequences (Nayak 2017).

Being a significant resource of biological diversity, coastal ecosystems provide a variety of goods and services to the human being. Ecosystem services in coastal areas can be classified into three categories, i.e., cultural and religious opportunities, provisioning, and regulating. Cultural and religious services include recreational, educational, scientific, aesthetic, spiritual, and symbolic values of coastal ecosystems. Provisioning services of coastal ecosystem comprise the opportunity to supply food, fibers, timbers, and hydrological resources for agriculture and industries. Regulating services of coastal ecosystems are associated with regulation of hydrological flows and water cycle, climate regulation, and mitigation of different natural hazardous problems such as floods, cyclones, coastal erosion, high wind, etc. (Millennium Ecosystem Assessment 2005). The coastal ecosystem is mainly affected by two types of factors viz., terrestrial influences and marine influences (Pallewatta 2010). Terrestrial influences are primarily anthropogenic which include changes in land use pattern, nutrient loading, hydrological management, etc. However, marine influences are mostly the consequences of natural phenomena such as cyclones, storms, depressions, tsunamis, changing wave and oceanic currents, and other climatic and geomorphological changes. Degradation of the coastal ecosystem is mostly regulated by some direct (such as overexploitation, pollution, invasion of alien species, anthropogenic climate change, etc.) and some indirect (such as globalization, policy failure, market distortion, social inequality, weak developmental strategies, etc.) drivers (Daily and Matson 2008). These, in turn, prove that the consequences of climate change play key roles in coastal ecosystem vulnerability along with anthropogenic activities. The objective of the present study is to summarize the general impacts of climate change in coastal areas, with a referential case study of the Odisha coastline, in India.

25.2 Methodology

An attempt has been made here to prepare an utmost consolidated scientific review on the research topic. Extensive scrutiny has been done during the compilation and consolidation of the available scientific data to make it more inclusive and significant for future researchers. Used data has been collected, from various reputed science journals, different governmental or non-governmental published reports (particularly from national/international agencies) and published doctoral or postdoctoral theses. Precedence has been given only to the reproducible articles that are indexed in science journal databases like Web of Science, Copernicus, Scopus, Google Scholar, PubMed, etc. The articles highlighting ambiguous research methodologies are strictly avoided. Keywords, for searching the articles, have been judiciously chosen and examined based on systematic scientific approaches. The experimental findings (both laboratory and field) from our studies have been encompassed thoroughly in various parts of the article, especially to improve the essence of the chapter.

25.3 Key Vulnerabilities in Coastal Areas

Changes and establishment of coastline areas are highly influenced by various hydrological and geomorphological alterations. Thus, studies and evaluation of different hazardous changes and their impact or risk potential are very important to mitigate the vulnerabilities of coastal areas (Kantamaneni et al. 2019). The study of the coastal vulnerability index (CVI) is very useful in understanding the impact of coastal changes due to natural disasters. Gornitz (1990) has listed some parameters for developing coastal vulnerability index, which is: geomorphology of coastline (a), shoreline change rate (b), coastal slope (c), relative sea-level rise (d), mean significant wave-length (e) and mean tidal-range (f). According to Gornitz (1990), CVI can be calculated as follows (where n stands for the number of variables):

$$\text{CVI} = \sqrt{((a \cdot b \cdot c \cdot d \cdot e)/n)} \quad (25.1)$$

The above-mentioned equation is highly used for estimating the coastal vulnerability through other approaches can also be found. Rao et al. (2010) have proposed another equation using only five parameters from the above, i.e., geomorphology of coastline (a), shoreline change rate (b), coastal slope (c), mean significant wave-length (e), and mean tidal-range (f). The equation is as follows:

$$\text{CVI} = 4a + 4c + 2b + f + e \quad (25.2)$$

The coastal ecosystem supports huge diversified human activities because of its great variability. It possesses great biological diversity with different communities. Climate change is a potential stressor to the changing condition of coastal ecosystems. Consequences of climate change such as sea-level rise, coastal erosion, increased frequency of cyclones and floods, etc. put a dramatic impact on coastal productivity. An increase in atmospheric temperature, increased rainfall, warming of ocean, and pollution are also listed as the major stressors of coastal areas (Sudha Rani et al. 2015).

25.3.1 Increase in Atmospheric Temperature

According to IPCC (Intergovernmental Panel on Climate Change) special report, 2018, it has been estimated that about 0.8–1.2 °C increase in temperature may likely to be observed with a maximum of 1.5 °C between 2030 and 2052. The global mean temperature rises at a rate of 0.67 °C from 1891 to 2008 (IPCC 2013). The rising of the atmospheric temperature has been recorded to be increased significantly in the last decade. The situation becomes more serious with the increase of different greenhouse gases in the environment (IPCC 2013, 2018). It is a fact that atmospheric greenhouse gases have been gradually increased by anthropogenic activities since the industrial revolution. This confirms the increase of atmospheric temperature as the greenhouse gases can conserve heat in the atmosphere. An increase in atmospheric temperature induces the melting of continental ice and also glaciers which pour more water into the sea or ocean resulting in the rise of sea level. Studies have proved that the current energy imbalance of Earth's climate system is about 1 W/m^2 (Johnson et al. 2018; Cheng et al. 2019a, b; von Schuckmann et al. 2016). Most of the excess heat (About 90%) is absorbed by the ocean system increasing ocean heat content (OHC) which results in sea-level rise and coastal inundation (Cheng et al. 2021). As coastal areas throughout the world are the most densely populated zones because of having fertile soil, occupation opportunities for the dwellers in fisheries and shipping industries, climate change associated with the increase of atmospheric temperature (Fig. 25.1) directly influences the coastal inhabitant's life.

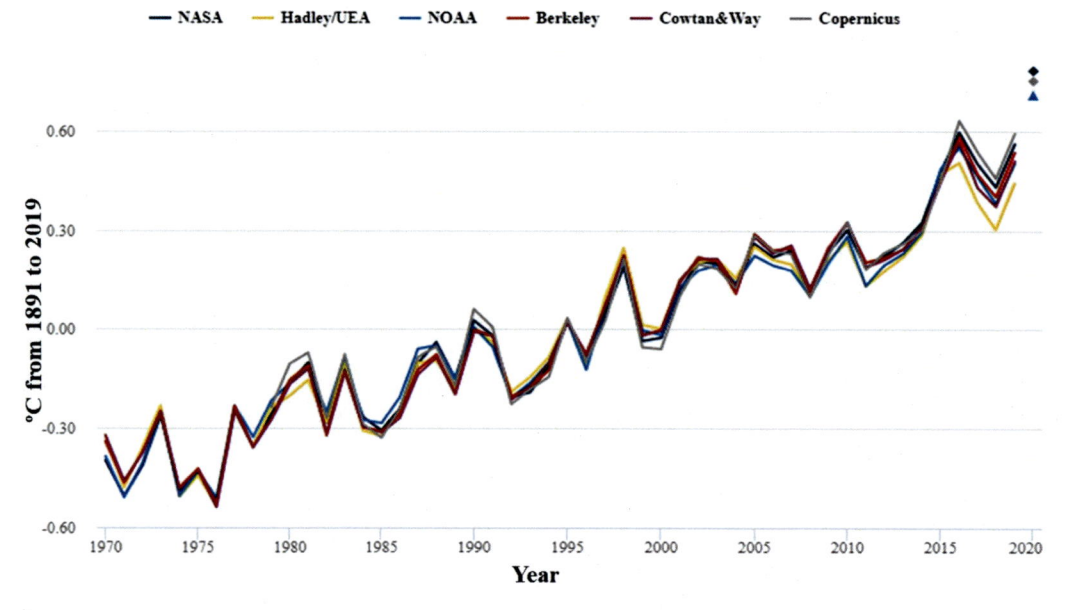

Fig. 25.1 Changes in atmospheric temperature from 1970 to 2019 (Hausfather 2020)

25.3.2 The Rise in Ocean Heat Content

With the rise of greenhouse gases in the environment, excess heat is retained in the atmosphere resulting in global warming. This excess heat is mostly absorbed by oceans leading to a rise in the ocean heat. Ocean heat in the upper 700 m is increasing very fast, influencing the slow increase of deep-sea temperature. Ocean temperature at the depth of 100 m is predicted to be increased between 0.6 and 2.0 °C by 2100 (Balmaseda et al. 2013). The heat slowly changes the deep-sea temperature altering ocean stratification and ocean current (Llovel et al. 2014; Roemmich et al. 2015). IPCC (2013) report predicts that the mean global ocean temperature would like to rise 1–4 °C by the year 2100. The warming of the ocean is directly connected with the amount of dissolved carbon dioxide in the ocean water. Due to global warming, the rate of carbon dioxide uptake by the ocean water is increased significantly which leads to acidification and deoxygenation of ocean water. Ocean warming enhances the consequences of thermal expansion of marine water along with the melting of the continental ice sheet resulting in sea-level rise. Hence many marine and coastal ecosystem is facing the threat of degeneration. Coastal areas are facing the threat of short-term or permanent inundation resulting in the loss of breeding grounds for seabirds and marine organisms. Pictorial representation of the changes in oceanic temperature in the upper layer (0–700 m) from 1970 to 2019 shows that ocean heating affected areas are continuously increasing during this period (Fig. 25.2).

25.3.3 Sea-Level Rise

Sea-level rise is now considered a major consequence of climatic changes. The problem becomes so serious mainly due to a combined effect of thermal expansion of ocean water and melting of continental glaciers. The estimated rate of mean sea level rise in the Pacific islands will be up to 0.4–0.8 m higher at the end of the century (Aucan 2018). Extreme rise in sea level will lead to inundation of huge coastal areas lasting for hours to days. The condition is further aggravated by the influence of the altered nature of waves, tides, and intensified oceanic storms. The severity and frequency of coastal flood

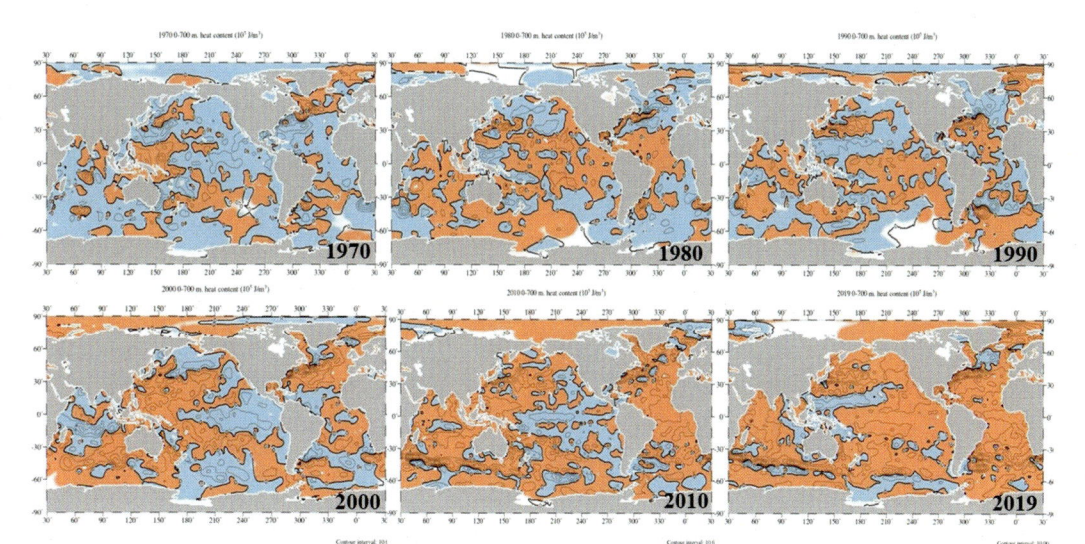

Developed from the website of National Centers for Environmental Information, https://www.ncei.noaa.gov/access/global-ocean-heat-content/bin/heatfig1.pl?action=start

Fig. 25.2 Changes in Ocean heat content from 1970 to 2019 (orange color stands for higher oceanic heat)

incidences can also be found to be altered. Measuring sea level rise can be done in two ways, i.e., estimation of "relative sea level" and "absolute sea level". Relative sea level is estimated in a particular coastline of interest, whereas absolute sea-level is measured from the center of the Earth (Fig. 25.3). Rising relative sea level enhances the chance of shoreline retreat along with beach erosion (Panigrahi and Bakshi 2018; Argo 2020). Absolute sea-level changes are observed globally by a series of highly specialized satellites which report that the rise of sea level in Pacific islands from 1993 to 2017 is about 3–6 mm/year (Aucan 2018). Coastal wetlands are highly vulnerable to the rise of sea level (Rodríguez et al. 2017). The increasing rate of inundation slows down the vegetation recovery rate in tidal marshes or coastal vegetation. Kench et al. (2018) have demanded that inundation and coastal erosion could be dangerous for many atoll islands making them uninhabitable over the next century.

25.3.4 Changes in Intensity and Frequency of Storms

Coastal areas are highly vulnerable to the current trend of increasing intensity of storms. Storm surges are also the reason behind consequent floods in low-lying areas, destroying of habitats, etc. For example, after Super-storm Sandy in 2012 large areas of New York City, New Jersey, and Long Islands have faced huge destruction of habitats. Low-lying areas of these areas have been flooded by several feet of water after the super-storm (FEMA 2008; USGCRP 2014; IPCC 2018). Sea level rise has further enhanced the impact of storm surges in coastal areas (NRC 2010). Oceanic storms generally enter into the land with heavy precipitation. An increase in cyclonic rainfall is evident at a rate of 10–15% within the 100 km of landfall. Thus increase in flooding probabilities in low-lying areas consequently occur after landfall of the storms. In the northwest Pacific basin, the latitude of the intensity of tropical cyclone shifts poleward detectably. Tropical Cyclone-induced storm surges affecting the Pearl River Delta may increase by approximately 1 m in the coming 2075–2099 period (Chen et al. 2020). The intensity of tropical cyclones in the Atlantic basin is also found to be increased in the last four decades Chen et al. 2020). Global warming further enhanced the trend of hurricane flooding by intensifying hurricanes (Mousavi et al. 2011). The projected increase in tropical cyclone intensity with a warming of the climate is about 1–10% for an increase of 2 °C global mean temperature (Knutson et al. 2019). The projection is similar to the prediction of potential intensity (PI) theory by Emanuel 1988). Bhatia et al. (2018) have also demanded that the tropical cyclone intensification rate is going to be higher at the end of this century than the recent days. Holland and Bruyère (2014) have proposed that the intensity and proportion of tropical cyclones are gradually increasing. Several researchers have proposed that the size and destructiveness of tropical cyclones are going to be higher in future which can be considered as additional characteristic to

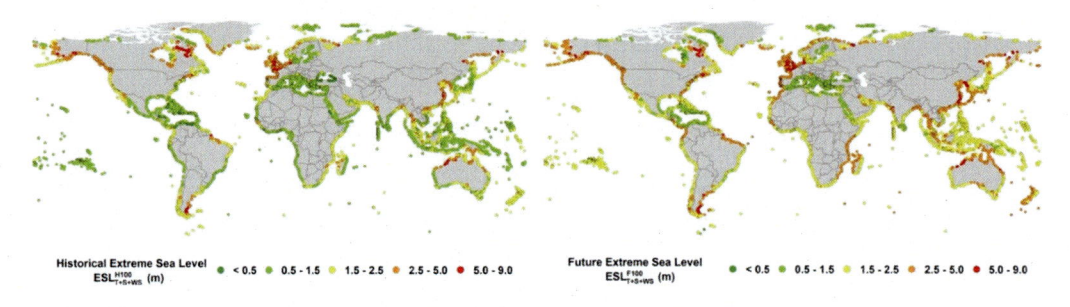

Fig. 25.3 Impact of global sea-level change (Adapted from Kirezci et al. 2020)

be faced with global warming (Sun et al. 2017; Schenkel et al. 2017).

25.3.5 Flood

The increased intensity and frequency of floods in coastal areas all over the world have been evident in the last century. Flood water penetrates the inland areas during the calamity which is highly dependent on regional topography. Ten million people all over the world have faced threats of coastal flooding after tropical cyclones. The situation is intensifying day by day. For example, the mean projection of hurricane flood in the City of Corpus Christi, Texas (U.S.) will be 0.3 m by 2030s and 0.8 m by 2080s (Mousavi et al. 2011). Coastal areas all over the world are highly populated zones, which are very much exposed to floods (Small and Nicholls 2003). An increasing rate of coastward migration indicated that more people will be directly affected by the coastal flood in future (Bijlsma et al. 1995). Some flood exposed areas are now protected by various structural and non-structural measures but most of the coastal areas are either open or with some weak resilience strategies (Balica et al. 2012). The increasing intensity of floods in coastal areas is highly connected with enhanced vulnerabilities in coastal diversity (Fig. 25.4). The coastal squeeze is a commonly used term in the UK which describes the loss of coastal habitat due to the establishment of sea defenses (Pontee 2013). These sea defenses are mainly constructed to prevent coastal flooding. But, the hard engineering interventions to prevent coastal flooding put negative impacts on biodiversity and productivity of coastal areas (Dugan and Hubbard 2006; Peterson and Bishop 2005). So, beach nourishment is another possible way to mitigate the consequences of flooding. But, beach nourishment processes also have some adverse ecological consequences to mitigate (Peterson et al. 2006; Speybroeck et al. 2006).

25.3.6 Other Natural Hazards

Coastal areas have been highly influenced by depressions witnessing heavy rainfall often. For example, depressions that originated in the Bay

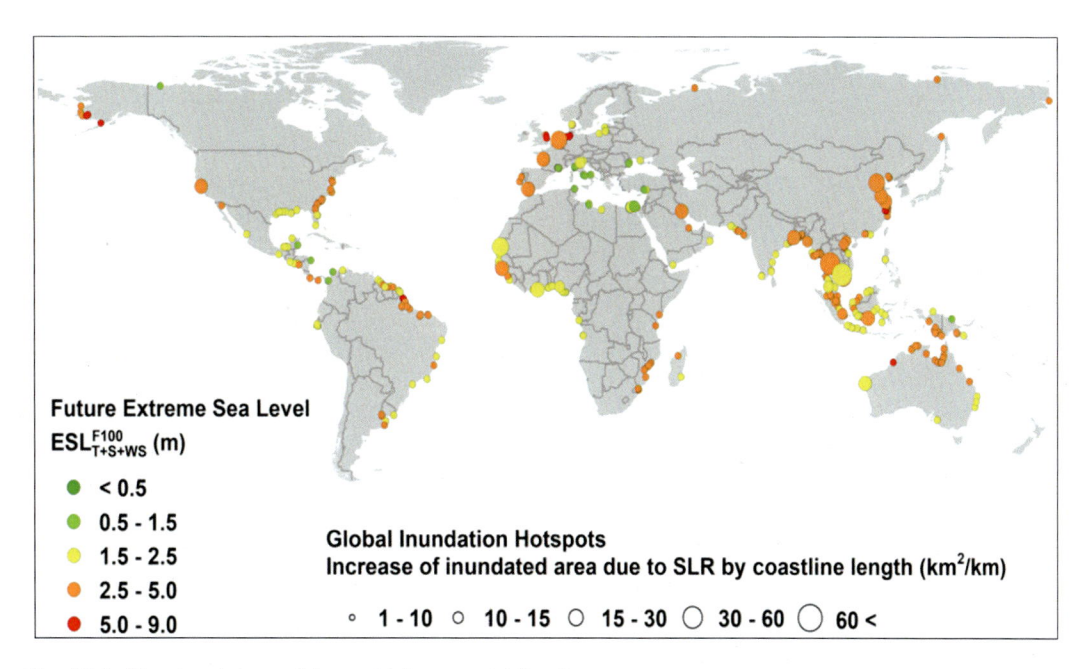

Fig. 25.4 Threatened areas of the world from coastal flooding

of Bengal and the Arabian sea are affecting the coastal states of India such as Odisha, Andhra Pradesh, Tamil Nadu (East coast), and Maharashtra, Kerala (West coast), etc. The recurrent floods in the recent past have been observed in these states due to heavy pouring.

Tsunamis, waterspouts, etc. are other natural hazards of the coastal region. A Tsunami is a series of large waves produced in oceans or seas when a large volume of water is displaced due to tectonic collision, earthquake, volcanic eruptions, underwater explosions, etc. For example, a devastating tsunami has been witnessed by the surrounding countries of the Indian Ocean in 2004. The disaster caused huge damage to the coastal ecosystems of these countries. It has uprooted many trees and has also caused potential damage to the mangroves. Initially, it has been claimed that 60–70% of the mangroves of Nicobar have been damaged during the tsunami. A recent study shows that approximately 97% of mangroves were uprooted and were wiped out completely by the high waves during the disaster. Most intertidal habitats of Nicobar Islands have been submerged permanently causing mortality of huge vegetation. Tsunamis happen frequently in the Pacific Ocean and Indonesia because of having a large number of active earthquake zones. Recently, tsunamis have occurred in the Mediterranean Sea region. Thus coastal habitats of these areas are at high-risk vulnerability due to tsunamis.

Another natural hazard of coastal areas is waterspouts, which are mostly happening in tropics and subtropics. This is an unusual disaster of seas and oceans which is made up of a large intense columnar vortex over the water that may be connected with clouds especially cumulonimbus, cumuliform, or cumulus congestus clouds. These become weaker in landmass but sometimes can suck up the small marine organisms out of the water. Thus, an uncommon phenomenon like the rain of fish has been evident in many countries in the past. Other hazards are also there like pollution, trawling, etc. which are mostly caused by anthropogenic activities, also posing a serious threat to coastal habitat.

25.4 Impact of Climatic Hazards on Coastal Ecosystem

The future risk of climatic hazards on coastal ecosystems is very high. The rate of anthropogenic climate change is very higher than the natural rate of climate change posing threats of extinction to vulnerable species. Climate changes cause a significant shift in coastal ecosystems due to the rise and magnitude of ocean heating, acidification, and sea-level rise. According to IPCC (2014), many plants and animals will not be able to adapt locally and will move fast enough to the suitable habitat during the twenty-first century. Polar ecosystem and coral reefs are highly vulnerable to the changing climate under RCP 4.5, RCP 6.0, and RCP 8.5 conditions. The extinction rate of many freshwater and marine species is increased under all RCP scenarios mainly driven by the factors like ocean warming, sea-ice loss, reduced river mouth flow, deoxygenation, and acidification of marine habitat. The fisheries sector will face the challenge of sustained provisions of productivity at lower latitudes. In the mid-twenty-first century many coastal species will be affected by species redistribution, and biodiversity reduction at low latitudes whereas, species richness will be increased at mid-high latitudes. Coastal ecosystems are highly vulnerable to the risk of natural events like tsunamis and landslides. The tsunami of 2004 has potentially devastated the coastal ecology of the Nicobar Islands. Ocean acidification and ocean heating will significantly change the species distribution and interactions (Bennett et al. 2015). Southern expansion of habitat for tropical species has been evident already. Ecosystems of large areas of the coastline of Australia have been facing the threat of extinction due to the limited capacity of range shift. Seagrass meadows of Australia is very much vulnerable to increasing water temperature and elevated sea level (Connolly 2012). Coral reefs all over the world are at high risk of extinction as increased water temperature that ultimately crosses the level for survival of symbiotic zooxanthellae leads to coral reef bleaching.

Bleached structures of reefs are highly vulnerable to storm surges and high waves. Increasing ocean temperature also has potential favor to the harmful microalgae of coral reefs. The species composition of mangroves of Sundarban, Odisha, and Australia has been to be directly impacted by sea-level rise and ocean acidification (Lovelock et al. 2012; Bakshi and Panigrahi 2015).

Coastal interface ecosystems, viz., rocky shores, headlands, sandy beaches, rocky intertidal habitats, and dune systems are widely distributed depending on ocean dynamics. Sea-level rise, tsunamis, and changes in frequency and intensity of storms may inundate these habitats affecting physical coastal processes. Erosion and landward migration of coastal habitats are likely, with the redistribution of intertidal communities. Coastal vegetation will be highly affected by atmospheric heating, drought, and a reduced amount of rainfall. It is also highly vulnerable to rising sea levels and increased extreme sea-level events like storms, tsunamis, etc. Declining vegetation health has consequences for changing coastal landform stability affecting animals, especially those who are dependent on coastal vegetation for habitat and food, including birds, reptiles, and mammals.

25.5 Assessment of Vulnerability in Odisha Coastal Area

The Odisha coastline, lined by the Bay of Bengal, India, is situated in the northeastern part of the Indian peninsula. The state contains a coastline of about 480 km long which is frequently affected by various natural disasters such as floods, drought, cyclonic storms, coastal erosion, etc. The six coastal districts (Balasore, Bhadrak, Kendrapara, Jagatsinghpur, Puri, and Ganjam) of the state are highly vulnerable to these natural calamities along with the stressors like anthropogenic exploitations. Most of the coastal part of the state contains fishing villages, tourist places, and small and middle populated towns with hotels and resorts at the seashore providing significant support to the economy of Odisha (Kumar et al. 2010). The coastal area is characterized by significant biological diversity having the second-largest mangrove in India ("*Bhitarkanika*"), the largest brackish water lagoon in Asia ("*Chilka*"), and the world's largest rookeries for Olive Ridley turtles (extreme sandy beaches of Gahirmata, Rishikulya and Debi river mouth), highly diverse estuarine and deltaic ecosystem (Kumar et al. 2010; Panigrahi and Bakshi 2018). Studies revealed that the key vulnerabilities of Odisha coasts are high variability of rainfall, flash floods during monsoon, heat waves in summer, coastal erosion, and intense impact of cyclonic storms (Fig. 25.5).

The Coastal vulnerability index (CVI) is estimated as the indication of the relative vulnerability of different segments of the Odisha coast (Table 25.1). The study has revealed that the CVI value of the Odisha coastline varied from 2.1 to 19, with a 25th and 50th percentile value of 4.75 and 9.5, respectively (Kumar et al. 2010). The vulnerability of the Odisha coastline is discussed in Table 25.1.

The coastal districts of Odisha comprise one-third population of the whole state creating a high density of population. This in turn increases the vulnerability of the coastline of the state. The present study reveals that most of the shoreline of the state is at the threat of moderate to high risk of vulnerability (Fig. 25.6). According to Kumar et al. (2010), a total of 84% of the total shoreline of the state is considered at high (22%) to medium (62%) risk of vulnerability based on eight parameters viz., shoreline change rate, significant wave height, sea-level change rate, tidal-range, Tsunami run-ups, coastal regional geomorphology, elevation, and slope (Fig. 25.7).

25.5.1 Major Climatic Risks in the Odisha Coast

The total coastline of Odisha is found to be vulnerable to several consequences of climate change such as increasing rate of drought, flood, heavy rainfall, coastal erosion, and frequent landfall of tropical cyclones.

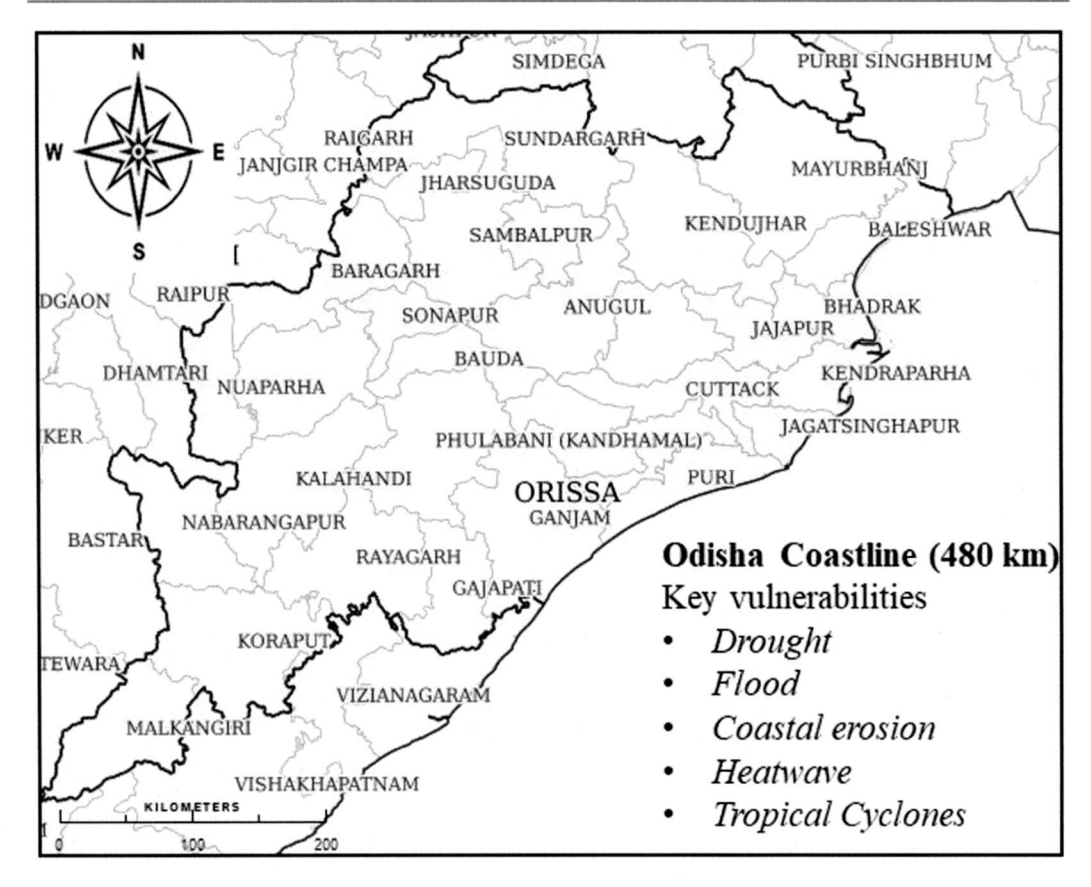

Fig. 25.5 Key vulnerabilities of Odisha coast (map source: Bhuvan portal-Odisha)

Table 25.1 Vulnerability of Odisha coastline

Vulnerability	Range of values	Areas	Length along the coastline (km)
Low	2.1–4.75	Ganjam, Chilka, southern Puri, and parts of Kendrapara	76
Moderate	>4.75 to <9.5	Northern Ganjam, parts of Chilika, central Puri, Jagatsinghpur, parts of Kendrapara, southern Bhadrak, and northern Balasore	297
High	>9.5	Northern Puri, parts of Jagatsinghpur and Kendrapara, northern and parts of southern Bhadrak and southern Balasore	107

Source Kumar et al. (2010)

i. *Drought*

Odisha has both a dry area and a coastal area with recurrent risks of drought. Our study reveals that the state can be divided into four parts, i.e., coastal, North-Central, western and southern according to the range of temperatures. We have found that the six coastal districts can be included in two different groups. Balasore, Bhadrak, Kendrapara, and the northern part of Jagatsinghpur district can be included in the north-central Odisha temperature zone where the

Jagatsingpur, Puri, and Ganjam districts have shown a range of 35–39 °C. The maximum temperature ranges between 44.5 and 48 °C in the districts of the western climatic division of Odisha. Apart from this, heat-wave from the Jharkhand plateau and deficiency of rainfall are the major reasons behind the drought in Odisha. Ghosh and Majumdar (2007) have warned about the future increasing trend of severe and extreme droughts in Odisha at the end of the twenty-first century. Though all the six coastal districts of Odisha are reported to be at low risk of severe drought (Mohanty et al. 2015).

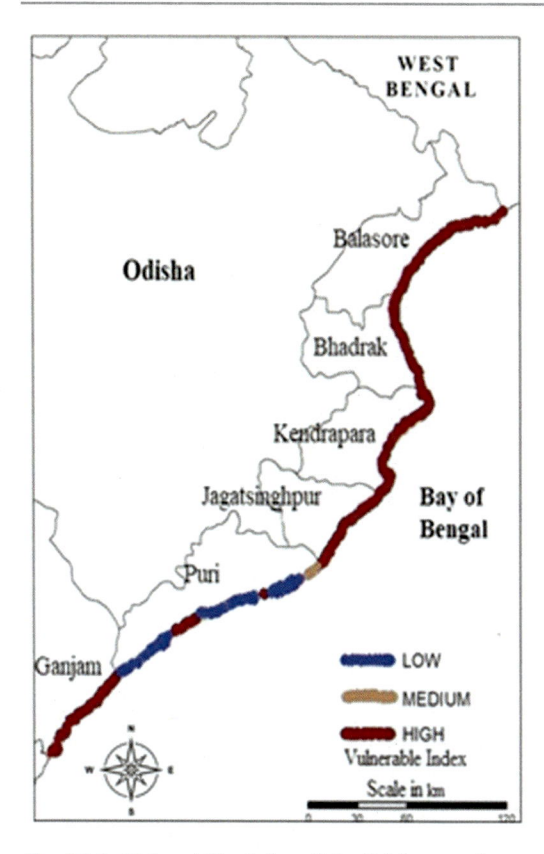

Fig. 25.6 Vulnerability index of the Odisha coastline

maximum temperature ranges from 42 to 44.5 °C. But the recorded temperatures of the past ten years revealed that the southern part of

ii. *Flood*

Odisha is a state of mostly rain-fed rivers such as Mahanadi, Dugdugi, Burha-balam, Baitarani, Brahmani, and Subarnarekha. All the rivers originated from the Chhotonagpur Plateau. Heavy rainfall in the upper catchment areas and annual heavy rainfall in Odisha are the main responsible features of coastal floods. The consequences become more severe when frequent landfalls of tropical cyclones occur along with heavy rainfall. A twenty-year analysis (1990–2009) of annual rainfall data of the Cuttack district (1649.8 ± 375.9 mm) indicates the occurrence of heavy precipitation in the coastal part of Odisha (Mohanty et al. 2015). According to a

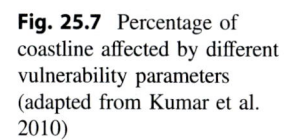

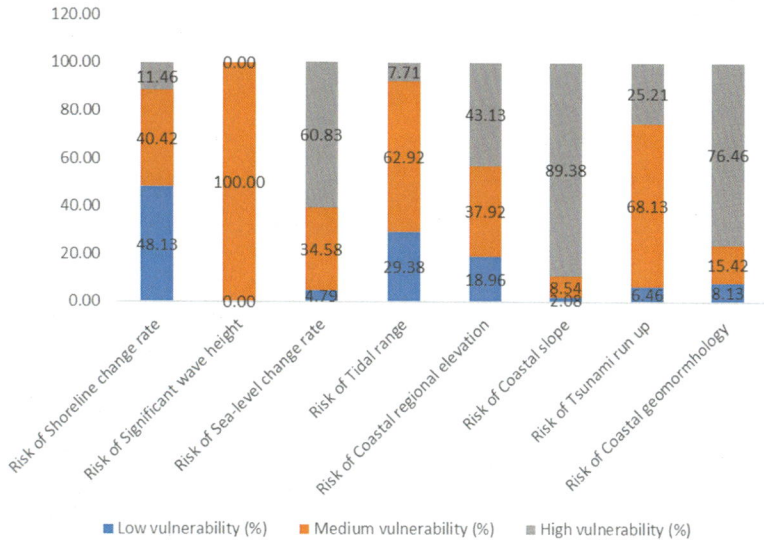

Fig. 25.7 Percentage of coastline affected by different vulnerability parameters (adapted from Kumar et al. 2010)

report by the government of Odisha (2018), rapid urbanization in low-lying areas in coastal districts enhances the risk of flooding (Source: State Action Plan on Climate Change 2018–2023). According to Barman et al. (2016), gradual decline in the water capacity of rivers Subarnarekha, Dugdugi, and Burahbalam associated with the discharge of enormous amount of sediment at the river mouth areas become the major reason behind the flooding in coastal gram panchayats of Balasore district which is further augmented by receiving a high volume of water from high magnitude cyclone events (Fig. 25.8). A report of a New Jersey-based science organization (Climate Central) has demanded that sea-level rise due to changing climate can put the entire coastline of Odisha (stretching from Balasore to Ganjam) is at potential risk of flooding deluge and coastal inundation by 2050 (Kulp and Strauss 2019).

iii. *Coastal Erosion*

Coastal erosion is considered a major problem of the coastline of Odisha. Accretion and erosion both have been recorded in the case of the Odisha coastal line. Coastline changes are highly controlled by wave characteristics, sediment dynamics, near-shore circulation, and beach form. About 220 km of the coastline has been reported to be threatened by future coastal erosion whereas, 205 km of shoreline has been facing coastal accretion (Rajawat et al. 2015). Coastal erosion is pronounced at Gopalpur port, Puri beach, Konark-Chandrabhaga-Ramchandi beach Paradip-Jatadhar Muhan, southern Dhamra port, Astaranga toward Devi river mouth, Kendrapara, etc. The Pentha village of Kendrapara district has been facing severe coastal erosion since 2004. The harmful consequences of coastal erosion are reflected mainly in natural habitats such as sand dunes, beaches, and mangroves of the shoreline (Behera et al. 2013). Basically, the southern part of breakwaters of ports and northern river mouths are termed as high erosion zone of the state. The southern part of Astaranga port of Puri district is recorded to show high erosion levels whereas, the southern part of Gopalpur (Ganjam district) and Paradip (Jagatsinghpur district) show high accretion properties. A high level of coastal erosion is observed in the northern mouths of Devi river

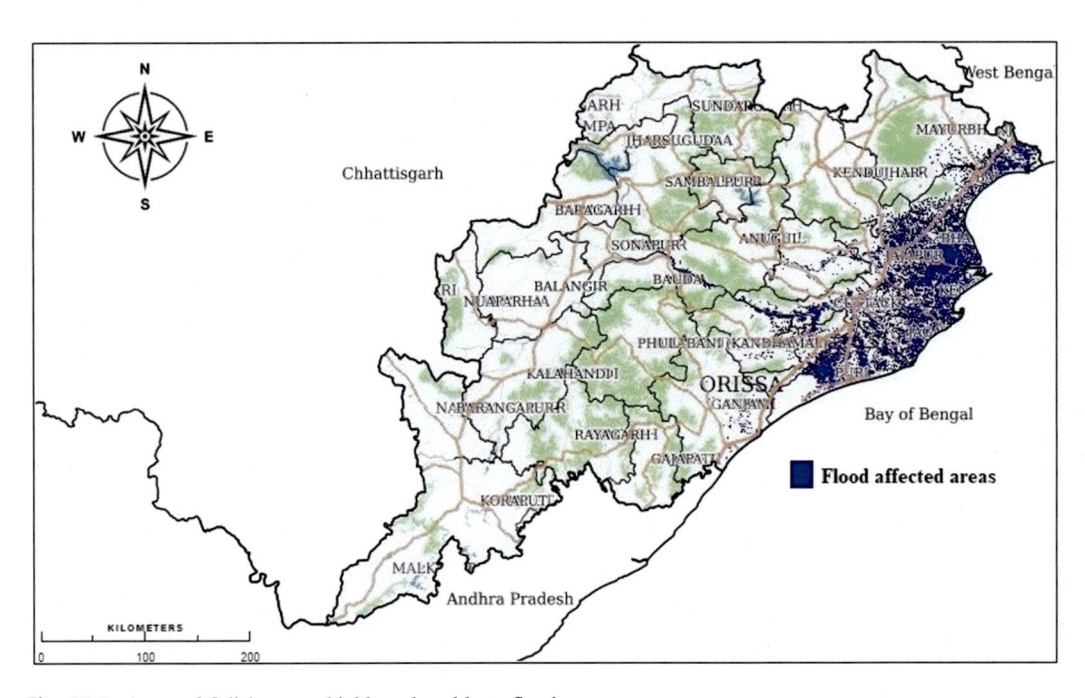

Fig. 25.8 Areas of Odisha coast highly vulnerable to flood

and Balijhori River (Jagatsinghpur district), Bansgarh River (Kendrapara district), and Subarnarekha river (Balasore district). Both the northern and southern mouths of the Bansgarh River are facing threats of a high rate of coastal erosion. Exceptionally, the southern mouth of the Bitikolia River in Jagatsingpur district is recorded to face a high degree of coastal erosion.

It has been estimated that though a large part of the Odisha coast is facing the threats of erosion (36.78%) but the major part of the coastline is either accreting (46.80%) or stable (14.37%). The shoreline of two districts, i.e., Jagatsingpur and Puri are found to be highly vulnerable to coastal erosion (Ramesh et al. 2011). It is evident that the accretion is dominant in the districts of Balasore, Bhadrak, Kendrapara, and Ganjam (Table 25.2).

Coastal engineering structures have significantly changed the shoreline movement. Seawalls and riprap revetments have been constructed on about 9.78 km of coastline to get protected them from coastal erosion (Ramesh et al. 2011). Other artificial structures like groynes, offshore breakwaters, and natural vegetation of casuarina, mangrove, and palm trees can be useful in preventing coastal erosion (Palai and Mandhaniya 2020). In our study, we have observed the coastal erosion at Pentha village of Rajnagar block in Kendrapara district. The observed reasons for the coastal erosion are mainly steep waves, sediment shifting, strong monsoon wind, and littoral drift. The construction of artificial structures with geo tubes, gabion boxes and mattresses, and geotextile filter materials can be very useful in protecting the shoreline. It is also found that the presence of dense mangrove vegetation helps in the high accretion in Balasore, Bhadrak, Kendrapara, and Ganjam districts.

Recently, it is observed that, after the construction of Gopalpur breakwaters, the corrosive nature of waves in some parts of Ganjam districts has raised significantly. Though no research data is available to prove this. A study of shoreline changes by Monalisha and Panda (2018) has indicated that coastal erosion is increasing particularly at Podampeta village. A recent study has confirmed that the villages such as Podampeta, Gokharkuda, Purunabandha, and Arjeepalli of the Ganjam district have faced the corrosive impact of coastal erosion in 2019. Erosion and further deposition are found to be natural phenomena at the coast. Generally, erosion starts during monsoon (July and August) at this coast but the process of deposition has been found to be started in September restoring the width of the coastline by December. But since the year 2007, the erosion started to continue beyond the monsoon period. Hence restoration of shoreline has been affected due to less deposition. Threats of coastal erosion have also been increased near Rishikulya rookery which is the nesting ground of endangered marine Olive Ridley turtle (Panigrahi and Bakshi 2018). It is demanded that a decrease in the width of the coast is surely affecting the mass nesting behavior of the Olive Ridley turtle. In 2016, the turtles did not prefer the coast for mass nesting though millions of them have visited the coast for mating. Proper researches by the experts in hydrology and geomorphology are very much needed for the future safety of the biodiversity of the coast.

Table 25.2 District-wise erosion and accretion of coastal districts of Odisha

District name	Coastline (km)	Erosion (km)	Accretion (km)	Artificial coast (km)	Stable coast (km)
Balasore	87.96	23.59	50.13	3.97	10.27
Bhadrak	52.61	11.48	38.32	0	2.81
Kendrapara	83.55	35.69	42.57	0	5.29
Jagatsinghpur	58.95	30.98	16.88	5.81	5.28
Puri	136.48	55.26	52	0	29.22
Ganjam	60.85	19.71	24.95	0	16.19

iv. *Heat-wave*

Heat-wave or severe heat-wave events in India are showing an increasing trend since the mid-twenteeth century. These are absolute manifestations of extreme areal temperature causing a series of acute or chronic destruction of human health, ecosystem, and economy (Singh et al. 2020). According to the Indian Meteorological Department (IMD), moderate heat-wave or heat-wave day is defined by the increase of 5–6 °C temperature where the normal temperature remains below 40 °C. If the increase of temperature occurs by +7 °C or more then the day is designated as severe heat-wave day. But, the definition is a little different in the case of places where the normal temperature is ≥ 40 °C. In this case, + (3 or 4) °C increase in temperature than normal is designated as moderate heat-wave day. Changes in temperature by +5 °C or more are designated as severe heat-wave days. If the temperature of any place continues to be 45 °C or more for consecutive two days or more, that is called a heat-wave. In the case of coastal regions, consecutive two days with a temperature ≥ 40 °C then it is termed as heat-wave condition. According to the temporal subdivision of Odisha two coastal districts, i.e., Balasore and Bhadrak are included in the north-central region with a characteristic of average temperature $\geq 42°$ C. Other four coastal districts, i.e., Ganjam, Puri, Jagatsinghpur, and Kendrapara are included in temporal coastal Odisha with a temperature range of ≥ 35 to <39 °C. Most central districts of the state are recorded to face annual heatwaves. The heat-wave condition occasionally affects the whole state at a time. An extreme heat-wave condition is found to be lasted for about two weeks in the 3rd and 4th weeks of May 2015 all over Odisha (Gouda et al. 2017). But, southern coastal districts, i.e., Ganjam, Puri, Jagatsinghpur, and Kendrapara are less vulnerable to heat-wave conditions. Thus it can be concluded that heat-wave is not a very serious threat for six coastal districts of Odisha, though occasional events may happen for shorter periods.

v. *Tropical Cyclones*

From 1891 to 2018, the number of cyclones that are formed in the Bay of Bengal and Arabian sea is 520 and 126, respectively. Odisha is the most common landfall destination for the east coast tropical cyclones that are formed in the Bay of Bengal. 34% of total tropical cyclones that are evolved in the Bay of Bengal generally show the tendency to landfalling on Odisha coast. The geography and topography of the state act as a magnetic field for the cyclones. An average of five cyclones per year has been recorded in the Bay of Bengal, which generally travels in the northwest direction toward the Indian landmass. Four of the five cyclones are formed generally during October–December, whereas, one cyclone is formed during May–June each year. Sea surface temperature and mixing of hot air coming from landmasses play a crucial role in the formation of tropical cyclones in the Bay of Bengal. So, higher sea surface temperature and hot summer on Odisha coast produce a high chance of formation of tropical cyclones in May–June. The state was highly affected by the super cyclone of 1999. After that, many major cyclones like Phailin (2013), Hudhud (2014), Titli (2018), Fani (2019), and Amphan (2020) have affected the coastal areas of the state very much (Table 25.3).

Depressions and deep depressions are very frequent in the Bay of Bengal which sometimes may transform into tropical cyclones or may cause heavy rainfall for several days in coastal districts. District-wise analysis of depressions, cyclones, and super cyclones of the Odisha coast from 1891 to 2007 show that the coastline of Balasore district is highly vulnerable to depressions and cyclones. Ganjam is least affected by depressions among the coastal districts of Odisha (Fig. 25.9). PDI or Power Dissipation Index has been widely used to measure the destructive potential of tropical cyclones. Sahoo and Bhaskaran (2018) have documented that PDI for cyclones in the present decade is with six-times higher destructive potential compared to the past.

Table 25.3 Recent tropical cyclones name, year and affected coastal districts of Odisha

Cyclone name	Year	Wind speed (km/h)	Landfall	Heavily affected coastal districts
Super cyclone	1999	276	Paradip	Balasore, Bhadrak, Kendrapara, Jagatsinghpur, Puri, Ganjam
Phailin	2013	215	Gopalpur	Ganjam, Puri
Hudhud	2014	185	Visakhapatnam (Andhra Pradesh)	Kendrapara, Jagatsinghpur, Puri, Ganjam
Titli	2018	110	Palasa (Andhra Pradesh)	Balasore, Bhadrak, Kendrapara, Jagatsinghpur, Puri, Ganjam
Fani	2019	250	Puri	Ganjam, Puri
Amphan	2020	115	Bakkhali (West Bengal)	Balasore, Bhadrak, Kendrapara, Jagatsinghpur

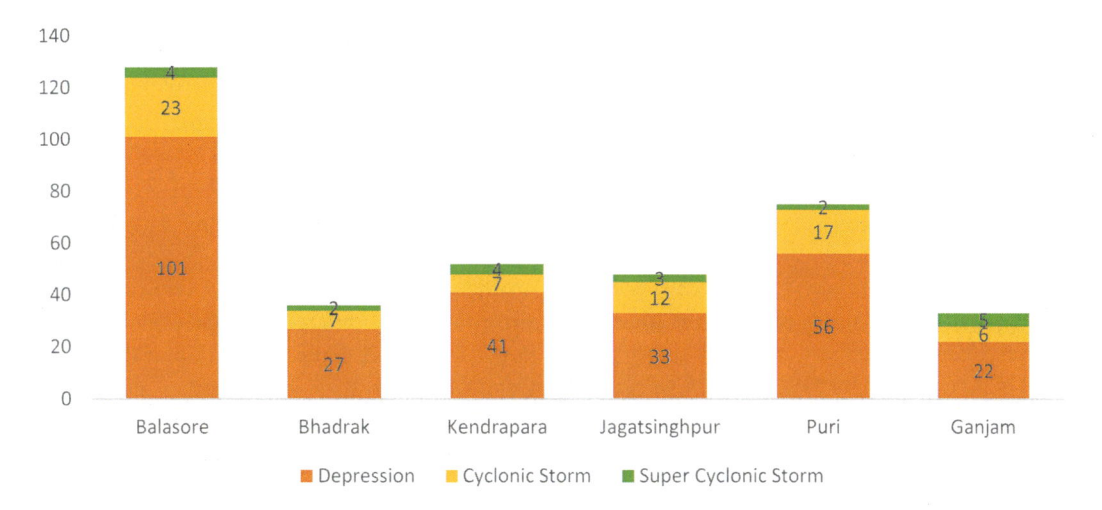

Fig. 25.9 District-wise distribution of depressions, cyclonic storms, and super cyclonic storms (1891–2007)

Thus, there is certainly a need to focus on the investigation related to coastal vulnerability and risk factors connected with an increasing rate of intense tropical cyclones highly influenced by climate change.

According to Rao et al. (2020), the coastline of Balasore, among the coastal states of Odisha, has faced the most landfalling cyclones since 1891 (till 2018). The numbers of landfalling of cyclones at the coastlines of Balasore, Bhadrak, Kendrapara, Jagatsinghpur, Puri, and Ganjam (Fig. 25.10) for the period of 127 years (from 1891 to 2018) are 28, 17, 17, 17, 6 and 13, respectively (Rao et al. 2020). About one-third of the tropical cyclones that originated in the Bay of Bengal generally show a landfall in Odisha coastal districts. Apart from this, Cyclones that cause landfall in Andhra Pradesh and West Bengal also put an adverse impact on coastal diversity. After every cyclone, coastal Odisha has faced a huge loss of biodiversity and habitat. For example, Balukhanda Wildlife Sanctuary near Puri has been found to be severely affected after the super cyclone Fani in 2019. Around 60 lakh trees including casuarina and cashew plants have been found to be completely uprooted. A large number of other plants such as neem, polanga, karanja, tamarind, etc. have been found to be broken partly. The sanctuary is a homeland for spotted deer, jungle cats, jackals, hyenas, monitor lizards, and many species of birds that are highly affected by the cyclones. Many spotted

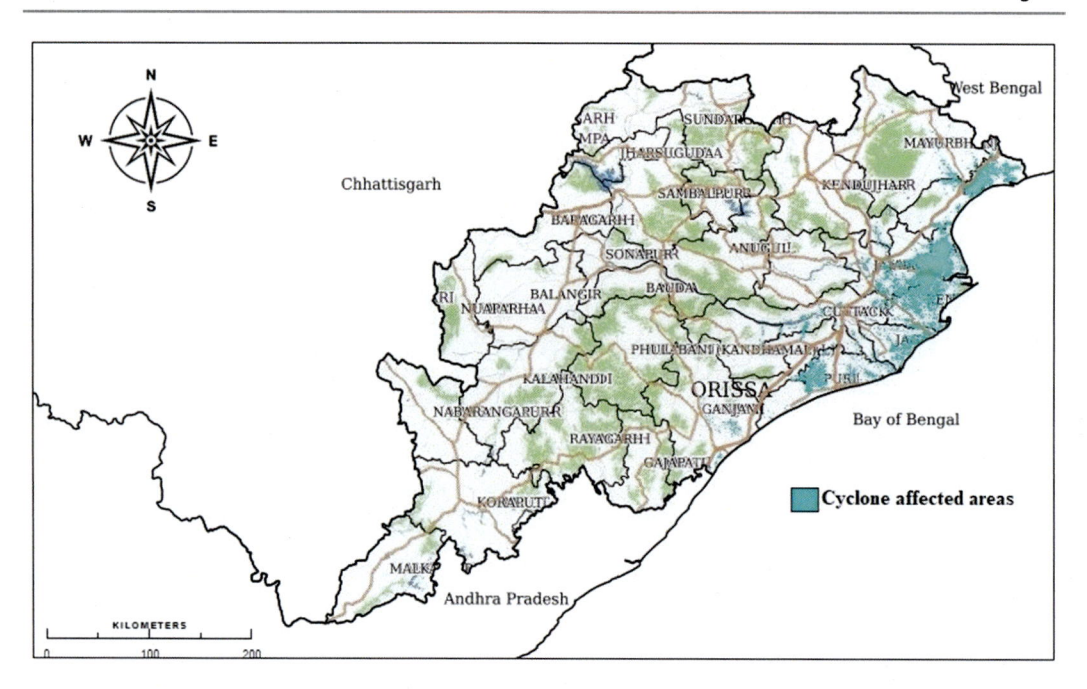

Fig. 25.10 Areas of coastal Odisha mostly affected by cyclones

deers have been found dead inside the sanctuary. An important sweet water zone for the total Puri district near the sanctuary, named Talapani, has been highly affected by the cyclone. The cyclone has severely decimated wildlife and their habitat in this region. Debata (2019) has reported that cyclone Fani has harshly affected the breeding grounds of water birds especially sandbar-nesting birds along the Mahanadi river of Odisha. As landfalling of tropical cyclones is frequently happening in coastal Odisha it may emerge as a new threat to habitat and survival of coastal fauna and flora (Debata 2019).

25.5.2 Impact of Climate Change Issues Relevant to the Odisha Coast

Odisha is a highly sensitive state on the east coast of the Indian peninsula because of having a wide range of climatic zones. The district is highly disturbed by various natural disasters like heatwaves, drought, heavy rainfall, flood, tropical cyclones, etc. There are six coastal districts in the state which are in high-risk zone due to these natural disasters. All the districts are highly sensitive to the occurrence of tropical cyclones, flood inundations, and depressions. Unlike the central Odisha districts, the coastal districts are at minimum risk of heat-wave and drought. Other climatic consequences like sea-level rise, increase rainfall, etc. have been found to affect various physical changes to the shoreline. Coastal erosion is a natural process, which is getting more dreadful due to the changing climatic situations (Table 25.4). Though most of the Odisha coastline is either stable or accreting, the shoreline shifting due to coastal erosion should be considered.

Odisha coastline has highly variable geomorphological characteristics. Due to this, the coastline is highly rich in its biodiversity. The increasing frequency of cyclones and coastal erosion are major detrimental threats to the coast. Cyclone Amphan in 2020 has highly affected the Mangrove Sanctuary of Bichitrapur (Balasore) which is an important habitat for various types of estuarine crabs (Fig. 25.11). The coastline is also affected by high anthropogenic activities at different ports and coastal cities (Fig. 25.11).

Risk due to coastal erosion is high in the districts of Kendrapara, Jagatsinghpur, and Puri.

Table 25.4 District-wise vulnerability of coastal Odisha

Coastal district	Coastline (km)	Cyclone sensitivity	Flood sensitivity	Drought sensitivity	Earthquake sensitivity
Balasore	87.96	Very high-risk zone	High risk	Minimum risk	Moderate to low risk
Bhadrak	52.61	Very high-risk zone	High risk	Minimum risk	Low risk
Kendrapara	83.55	Very high-risk zone	High risk but some areas are protected	Minimum risk	Moderate risk
Jagatsinghpur	58.95	Very high-risk zone	High risk but some areas are protected	Minimum risk	Moderate risk
Puri	136.48	Very high-risk zone	High risk but some areas are protected	Minimum risk	Moderate risk
Ganjam	60.85	High to very high-risk zone	High risk but some areas are protected	Minimum risk	Very low risk

Fig. 25.11 Photographs of different coastal regions of Odisha **a** Bichitrapur Mangrove vegetation. **b** Subarnarekha river mouth. **c** Talsari beach. **d** Chandipur coast. **e** Bhadrak coast. **f** Dhamra port of Bhadrak District. **g** densely populated coast of Puri. **h** Coastal vegetation near Chilka of Ganjam district. **i** shoreline of Chilka lake mouth

Though recently consequences of a higher rate of erosion than deposition are observed in various parts of Ganjam district (Fig. 25.12). Shifting of coastline, formation of sandbars in the inlet channels of the water bodies, littoral drift, and onshore/offshore sediment transport are also observed major risks of Odisha coast.

25.5.3 Problems of Coastal Ecosystem of Odisha

Odisha coast is characterized by a narrow shelf, though the northern part of the Odisha coast has an extended continental shelf (Mohanty et al. 2008). The 480 km long coast is bestowed with

Fig. 25.12 Photographs of some beaches on Odisha coast with coastal erosion vulnerability **a** Coastal management at Pentha village, Kendrapara. **b** Rajanagar Beach, Kendrapara. **c** Habalikhati beach near Bhitarkanika Sanctuary. **d** Eroded part of the beach at Habalikhati beach. **e** Coastal management at Siali beach, Jagatsinghpur. **f** Coastal erosion at Padampeta beach, Ganjam

the presence of huge biodiversity as the coast consists of India's second-largest mangrove forest (Bhitarkanika), the largest rookery for the Olive Ridley Turtle (Gahirmatha sandy beach), Other mass nesting grounds of Olive Ridley (Debi river mouth and Rishikulya Beach) and Asia's largest brackish water coastal lagoon (Chilka). In the past few decades, the developmental pressure of the Odisha cost puts a tremendous threat to coastal biodiversity. Various natural disasters (like coastal inundation, flood, tropical cyclones, coastal erosion, etc.) have also been considered as key risk factors for Odisha coastal ecosystems. An ecological analysis of mangroves of Odisha by Upadhyay and Mishra (2014) has reported that a total of 29 mangroves species are present in Bhitarkanika sanctuary which represents the presence of high species richness in the area. The key vulnerabilities of Odisha coastal ecosystems are sea-level rise, shifting of shoreline, formation of sandbars at waterbodies, wave velocity, and seawater current, littoral shift, siltation, onshore and offshore sediment transport, and occurrence of frequent tropical cyclones. Recurrent floods and coastal inundation have affected a large portion of coastal vegetation. The development of fisheries and aquaculture is heavily affected by tropical cyclones, floods, etc. Changes in climate along with anthropogenic activities enhance the chances of risks in coastal ecosystems.

25.5.4 Action Plans

The coastline of Odisha is very rich in its biodiversity as it includes Asia's largest brackish water lagoon, Chilka, a large area of mangroves and wetlands. Rishikulya, Debi river mouth, and Gahirmata beaches are the important nesting grounds of Olive Ridley Turtle (Panigrahi and Bakshi 2018). According to the Census report of 2011, the total population of six coastal districts of Odisha is 10,112,048 (24.1% of the total population of the state) with a population density of 462 per km^2. The coastal zone plays a significant role in the socio-economic, cultural, and environmental development of the state.

The intensity of floods and cyclones is high in coastal districts of Odisha. These are the principle threats to biodiversity conservation and livelihood security. The state government of

Odisha has taken several activities to cope with the challenges some of which are designated as key priorities. Coastal protection in high erosion zone like Kendrapara and Puri by the construction of saline embankments has been implemented as a key priority to combat the threat of coastal erosion. Pentha is a small village in Kendrapara district which is highly affected by coastal erosion but now construction of saline embankments under NCRMP (National Cyclone Risk Management Project) has been found to be beneficial.

A micro-level vulnerability assessment by the state government has revealed the importance of constructing saline embankments to protect the agricultural lands and fishing ponds from ingress of saline water during cyclones and floods. The state government has taken the initiative to construct twelve saline embankments (for a total of 57.78 km) within 2023. Restoration of Mangroves and identification of hazard mapping zones for floods and cyclones also got the attention as a key priority. The study of regional coastal processes by INCOIS (Indian National Center for Ocean Information Service) under the program of ICZM (Integrated Coastal Zone Management) has been supported by the state government as a pilot project. Initiatives have been taken to implant natural barriers such as casuarina and mangroves plants.

25.6 Future Possible Risks of Coastal Areas Related to Climate Change

As stated earlier, coastal zones are highly affected by the threats of natural phenomena such as sea-level rise, rise in ocean heat, coastal erosion, increasing intensity of storm surges, etc. Anthropogenic emissions of GHGs are connected with the gradual increase of atmospheric temperature causing continued global warming. Continuous global warming is resulting in sea-level rise affecting especially the coastal habitat and its ecosystem. Climate change models based on an estimation of thermal expansion and ice melting phenomenon have predicted that the sea-level rise will accelerate in the twenty-first century (IPCC 2013, 2014). It is likely to rise between 1 and 3 ft by the end of the century which may be furthermore because these models do not incorporate all the possible consequences of continental ice sheet melting. Representation Concentration Pathway (RCP) is the greenhouse gas concentration condition, adopted by IPCC (2014). According to report of IPCC (2014) under all emission scenario (i.e., RCP 2.6, RCP 4.5, RCP 6.0 and RCP 8.5) the projected rates of global sea level rise in 2100 are 2–7 mm/year (at RCP 2.6), 4–9 mm/year (at RCP 4.5), 5–10 mm/year (at RCP 6.0) and 8–16 mm/year (at RCP 8.5).

Global warming is highly connected with the increasing emission of carbon dioxide in the atmosphere. An increase in atmospheric carbon dioxide is also connected with the acidification of the ocean and ocean warming. At the beginning of industrialization, the atmospheric carbon dioxide concentration was 280 ppm which is elevated to approximately 400 ppm in the recent decade. This increase in carbon dioxide will affect the whole world for a long time. Even if we managed to cease our carbon dioxide emission completely, the consequences of increasing CO_2 rollover to another hundred to thousand years. This is probably due to the very slow warming property of the deep sea which will result in expanding of ocean water for the next few centuries. Another threatening fact is that the continental glaciers of Antarctica and Greenland also react very slowly to global warming. Hence continue sea-level rise is inevitable for the next few centuries which will be responsible for the permanent inundation of many coastal ecosystems.

The rise in ocean heat is another issue of concern for environmentalists. It has been predicted that the global average ocean temperature in the upper 100 m is likely to be increased between 0.6 and 2.0 °C (Balmaseda et al. 2013). The heat will slowly penetrate the deeper part of the sea and ocean putting a great impact on global oceanic circulation. (Llovel et al. 2014; Roemmich et al. 2015). It may alter the natural stratification of the ocean influencing the distribution of marine flora and fauna.

Coastal flooding and erosion due to the extreme weather condition and consequent sea-level rise is a major issue of concern to the environmentalists (Sutton-Grier et al. 2015). From the report from 2000 to 2008, it has been found that there is a rise in the level of the sea of Sundarbans. The amount of sea-level rise from the Sagar island observatory has been observed at the rate of about 12 mm/year. Between 2001 and 2009, the rate of coastal erosion in the Indian Sundarbans for the sea level rise has found to be about 5.50 km^2/year (Bakshi and Panigrahi 2015; Bakshi et al. 2020). The scenario is going to be more serious in near future.

Only a low emission scenario can change the vulnerability a little restricting the impact of climate change on coastal ecology but effects are already being exposed which are predicted to be continued for the next few centuries.

25.7 Conclusions

The coastline is highly sensitive to the changing climatic events such as increasing strength of tides, relative sea-level rise, changes in rainfall pattern, frequency and intensity of storm surges, etc. It is continuously changing depending upon some influencing factors like wave size, water depth, rate of coastal accretion and erosion, etc. Global warming and anthropogenic activities will surely be the reason behind the increasing rates of catastrophes in coastal areas. The future of the coastal areas is very much uncertain as relative sea-level rise will compel the coastal community to be displaced. Marine or coastal organisms will lose their habitat and breeding grounds. Further consequences of global population growth, increasing urbanization, settlement, and industrialization in the coastal area make this region greatly susceptible to technologies increasing the hazardous changes in climate and environment. The consequences of sea-level rise and increasing ocean heat are obvious for the next few centuries even after the restricted greenhouse gas emission. Hence coastal inundation, loss of habitat, and destruction of coastal ecosystems may be observed in different parts of the world in the upcoming time. Coastal management includes the construction of ripraps, rock armors, gabions, sea walls, revetments, bulkheads, groins, offshore breakwaters, floodgates or tidal barriers, etc. But, these artificial structures have had a potential impact on coastal ecosystems mostly affecting animal movement and breeding. The structures can interfere with processes like dynamics of sediment transport, flooding, erosion, accretion, etc. Beach filling and re-nourishment, dune building, wetland creation, and mangrove plantation may constitute examples of soft structural options for coastal defense. Research should be case-specific ensuring the increase of resilience and reducing vulnerabilities in coastal zones. Other options for coastal management such as artificial reef creation, seaweed use, natural coral growth, and replantation of mangroves can also be practiced to mitigate the consequences of climate change. Hence only natural and/or hybrid approaches (combination of natural and artificial) to infrastructure development in coastal areas can provide protection from coastal hazards. Thus more investment in and application of natural and/or hybrid infrastructural approaches and intense case-specific research are highly needed to enhance coastal resilience in future.

References

Argo (2020) Argo Float Data and Metadata from Global Data Assembly Centre (Argo GDAC). SEANOE

Aucan J (2018) Effects of climate change on sea levels and inundation relevant to the Pacific islands. Pacific marine climate change report card. Sci Rev 20188:43–49

Bakshi A, Panigrahi AK (2015) Studies on the impact of climate changes on biodiversity of a mangrove forest: case study of Sunderban delta region. J Environ Sociobiol 12(1):7–14

Bakshi A, Halder D, Panigrahi AK (2020) Impact of climate change on fisheries and aquaculture. Indian J Biol 7(1)

Balica SF, Wright NG, van der Meulen FA (2012) Flood vulnerability index for coastal cities and its use in assessing climate change impacts. Nat Hazards 64:73–105

Balmaseda MA, Trenberth KE, Källén E (2013) Distinctive climate signals in reanalysis of global ocean heat content. Geophys Res Lett 40:1754–1759

Barman NK, Chatterjee S, Paul AK (2016) Estimate the coastal vulnerability in the Balasore coast of India: a statistical approach. Model Earth Syst Environ 2:20

Behera P, Beura, D, Sarangi S (2013) Role of wind action on erosion at Puri beach of Odisha, India. Vistas Geol Res 12:116–121 (U.U. Spl. Publ. in Geology)

Bennett S, Wernberg T, Harvey ES, Santana-Garcon J, Saunders BJ (2015) Tropical herbivores provide resilience to a climate-mediated phase shift on temperate reefs. Ecol Lett 18(7):714–723

Bhatia K, Vecchi G, Murakami H, Underwood S, Kossin J (2018) Projected response of tropical cyclone intensity and intensification in a global climate model. J Climate 31:8281–8303

Bijlsma L, Ehler CN, Klein RJT, Kulshrestha SM, McLean RF, Mimura N, Nicholls RJ, Nurse LA, Perez Nieto H, Stakhiv EZ, Turner RK, Warrick RA (1995) Coastal zones and small islands. Climate change. In: Impacts, adaptations and mitigation of climate change: scientific-technical analyses, Cambridge University Press, Cambridge, pp 289–324

Burden A, Smeaton C, Angus S, Garbutt A, Jones L, Lewis HD, Rees SM (2020) Impacts of climate change on coastal habitats relevant to the coastal and marine environment around the UK. MCCIP Sci Rev 228–255

Chen J, Wang Z, Tam C-Y, Lau N-C, Dickson D-S, Mok H-Y (2020) Impacts of climate change on tropical cyclones and induced storm surges in the pearl river Delta region using pseudo-global-warming method. Sci Rep 10(1):965

Cheng LJ, Abraham J, Hausfather Z, Trenberth KE (2019a) How fast are the oceans warming? Science 363:128–129

Cheng LJ, Trenberth KE, Fasullo JT, Mayer M, Balmaseda M, Zhu J (2019b) Evolution of ocean heat content related to ENSO. J Climate 32(12):3529–3556

Cheng L, Abraham J, Trenbert KE, Fasullo J, Boyer T, Locarnini R, Zhang B, Yu F, Chen X, Song X, Liu Y, Mann ME, Reseghetti F, Simoncelli S, Gouretski V, Chen G, Mishonov A, Reagan J, Zhu J (2021) Upper ocean temperatures hit record high in 2020. Adv Atmos Sci

Connolly R (2012) Seagrass. In: Poloczanska ES, Hobday AJ, Richardson AJ (eds) A marine climate change impacts and adaptation report card for Australia 2012. Available online at https://www.nccarf.edu.au/sites/default/files/attached_files_publications/Marine_Report_Card_Australia_2012.pdf. Accessed 28 April 2016

Daily GC, Matson PA (2008) Ecosystem services: from theory to implementation. Proc Natl Acad Sci USA 105(28):9455–9456

Dasgupta S (2020) The surprisingly difficult task of measuring sea-level rise around India. Environment, the sciences. Retrieve from https://science.thewire.in/the-sciences/indian-coast-sea-level-rise-tide-gauge-satellite-altimetry-complexity

Debata S (2019) Impact of cyclone Fani on the breeding success of sandbar-nesting birds along the Mahanadi river in Odisha, India. J Threatened Taxa 11:14

Dugan JE, Hubbard DM (2006) Ecological responses to coastal armouring on exposed sandy beaches. Shore Beach 74:10–16

Emanuel KA (1988) The maximum intensity of hurricanes. J Atmos Sci 45:1143–1155

FEMA (2008) Coastal AE zone and VE zone demographics study and primary frontal dune study to support the NFIP. Federal Emergency Management Agency Technical Report, Washington, DC, p 98

Ghosh S, Majumdar PP (2007) Nonparametric methods for modelling GCM and scenario uncertainty in drought assessment. Water Resour Res 43(7)

Gornitz V (1990) Vulnerability of the east coast, USA to future sea level rise. J Coast Res 9:201–237

Gouda KC, Sahoo SK, Samantray P, Himesh S (2017) Simulation of extreme temperature over Odisha during May 2015. Weather Clim Extremes 17:17–28

Hausfather Z (2020). State of the climate: first quarter of 2020 is second warmest on record. Retrieved from https://www.carbonbrief.org/state-of-the-climate-first-quarter-of-2020-is-second-warmest-on-record

Holland GJ, Bruyère C (2014) Recent intense hurricane response to global climate change. Clim Dyn 42:617–627

IPCC (2013) Climate change 2013: the physical science basis. In: Stocker, TF, Qin D, Plattner G-K, Tignor M, Allen SK, Boschung J, Nauels A, Xia Y, Bex V, Midgley PM (eds) Contribution of working group I to the fifth assessment report of the intergovernmental panel on climate change, Cambridge University Press, Cambridge, p 1535

IPCC (2014) Climate change 2014: synthesis report. In: Pachauri RK, Meyer LA (eds) Contribution of working groups I, II and III to the fifth assessment report on the intergovernmental panel on climate change, Core writing team, Geneva, Intergovernmental Panel on Climate Change, p 151

IPCC (2018) Global warming of 1.5 °C (Summary for policymakers). In: Masson-Delmotte V, Zhai P, Pörtner H, Roberts D, Skea J, Shukla PR, Pirani A, Moufouma-Okia W, Péan C, Pidcock R, Connors S, Matthews JBR, Chen Y, ZhouX, Gomis MI, Lennoy E, Maycock T, Tignor M, Waterfield T (eds) Special report of the intergovernmental panel on climate change

Johnson G, Lyman JM, Boyer T, Domingues CM, Gilson J, Ishii M, Killick R, Monselan D, Wijffels S (2018) Ocean heat content in state of the climate in 2017. Bull Amer Meteor Soc 99:S72–S77

Kantamaneni K, Sudha Rani NNV, Rice L, Sur K, Thayaparan M, Kulatunga U, Rege R, Yenneti R, Campos LC (2019) A systematic review of coastal vulnerability assessment studies along Andhra Pradesh, India: a critical evaluation of data gathering, risk levels and mitigation strategies. Water 11:393. https://doi.org/10.3390/w11020393

Kench PS, Ford MR, Owen SD (2018) Patterns of island change and persistence offer alternate adaptation pathways for atoll nations. Nat Commun 9:605

Kirezci E, Young IR, Ranasinghe R, Muis S, Nicholls RJ, Lincke D, Hinkel J (2020) Projections of global-scale extreme sea levels and resulting episodic coastal fooding over the 21st century. Sci Rep 10:11629

Knutson T, Camargo SJ, Chan JCL, Emanuel K, Ho C-H, Kossin J, Mohapatra M, Satoh M, Sugi M, Walsh K, Wu L (2019) Tropical cyclones and climate change assessment: part I, detection and attribution. Bull Amer Meteorol Soc 100(10):1987–2007

Kulp SA, Strauss BH (2019) New elevation data triple estimates of global vulnerability to sea-level rise and coastal flooding. Nat Commun 10:4844

Kumar TS, Mahendra RS, Nayak S, Radhakrishnan K, Sahu KC (2010) Coastal vulnerability assessment for Orissa state, east coast of India. J Coast Res 26(3):2010

Llovel W, Willis JK, Landerer FW, Fukumori I (2014) Deep-ocean contribution to sea level and energy budget not detectable over the past decade. Nat Clim Chang 4:1031–1035

Lovelock CE, Skilleter G, Saintilan N (2012) Tidal wetlands. In: Poloczanska ES, Hobday AJ, Richardson AJ (eds) A marine climate change impacts and adaptation report card for Australia 2012. Available online at https://www.nccarf.edu.au/sites/default/files/attached_files_publications/Marine_Report_Card_Australia_2012.pdf. Accessed 28 April 2016

Millennium Ecosystem Assessment (2005) Ecosystems and human well-being. Island Press, Washington, DC

Mohanty PK, Panda US, Pal SR, Mishra P (2008) Monitoring and management of environmental changes along the Orissa coast. J Coast Res 24(sp2):13–27

Mohanty S, Sultan T, Panda DK, Kumar A (2015) Meteorological drought analysis based on rainfall data of coastal Odisha. In: Singh AK, Dagar JC, Arunachalam ARG, Shelat KN (eds) Climate change modelling, planning and policy for agriculture. Springer

Monalisha M, Panda GK (2018) coastal erosion and shoreline change in Ganjam coast along east coast of India. J Earth Sci Climatic Change 9(4)

Mousavi ME, Irish JL, Frey AE, Olivera F, Edge BL (2011) Global warming and hurricanes: the potential impact of hurricane intensification and sea level rise on coastal flooding. Clim Change 104:575–597

Nayak S (2017) Coastal zone management in India-present status and future needs. Geo-Spatial Inform Sci 20(2):174–183

NRC (2010) Adapting to the impacts of climate change. The National Academies Press, Washington, DC, USA, National Research Council

Palai SS, Mandhaniya P (2020) A study on coastal erosion mitigation on Odisha coastline. Int Res J Eng Technol (IRJET) 7(9):3944–3948

Pallewatta N (2010) Impacts of climate change on coastal ecosystems in the Indian Ocean region. In: Michel D, Pandya A (eds) Coastal zones and climate change. The Henry L. Stimson Center, Washington, DC

Panigrahi AK, Bakshi A (2018) Beach erosion and anthropogenic exploitation at coastal line of Odisha and its effect on olive ridley turtle and fish population. Biojournal 13(2):23–28

Peterson CH, Bishop MJ (2005) Assessing the environmental impacts of beach nourishment. Bioscience 55:887–896

Peterson CH, Bishop MJ, Johnson GA, D'Anna LM, Manning LM (2006) Exploiting beach filling as an unaffordable experiment: benthic intertidal impacts propagating upwards to shorebirds. J Exp Mar Biol Ecol 338:205–221

Pontee N (2013) Defining coastal squeeze: a discussion. Ocean Coast Manag 84:204–207

Rajawat AS, Chauhan HB, Ratheesh R, Rode S, Bhanderi RJ, Mahapatra M, Kumar M, Yadav R, Abraham SP, Singh SS, Keshri KN, Ajai, (2015) Assessment of coastal erosion along the Indian coast on 1: 25,000 scale using satellite data of 1989–1991 and 2004–2006 time frames. Curr Sci 109(2):347–353

Ramesh R., Purvaja R, Senthil VA (2011) Shoreline change assessment for Odisha coast. Online published factsheet in Collaboration of National Centre for Sustainable Coastal Management (NCSCM) Society of Integrated Coastal Management (SICOM) and Ministry of Environment and Forests (MoEF), Government of India. http://www.ncscm.org

Rao A, Babu S, Prasad K, Sadhuram MTR, Y, Mahapatra D, (2010) Investigation of the generation and propagation of low frequency internal waves: a case study for the east coast of India. Estuar Coast Shelf Sci 88:143–152

Rao AD, Upadhaya P, Ali H, Pandey S, Warrier V (2020) Coastal inundation due to tropical cyclones along the east coast of India: an influence of climate change impact. Nat Hazards 101:39–57

Rodríguez JF, Saco PM, Sandi S, Saintilan N, Riccardi G (2017) Potential increase in coastal wetland vulnerability to sea-level rise suggested by considering hydrodynamic attenuation effects. Nat Commun 8:16094

Roemmich D, Church J, Gilson J, Monselesan D, Sutton P, Wihffels S (2015) Unabated planetary warming and its ocean structure since 2006. Nat Clim Chang 5:240–245

Sahoo B, Bhaskaran PK (2018) Multi-hazard risk assessment of coastal vulnerability from tropical cyclones—A GIS based approach for the Odisha coast. J Environ Manage 206:1166–1178

Saintilan N, Rogers K, Kelleway JJ, Ens E, Sloane DR (2018) Climate change impacts on the coastal wetlands of Australia. Wetlands 39:1145–2115

Schenkel BA, Lin N, Chavas DR, Vecchi GA, Knutson TR, Oppenheimer M (2017) Will outer tropical cyclone size change due to anthropogenic warming? AGU Fall Meeting, Washington, DC

Singh S, Mall RK, Singh N (2020) Changing spatio-temporal trends of heat wave and severe heat wave events over India: an emerging health hazard. Int J Climatol

Small C, Nicholls RJ (2003) A global analysis of human settlement in coastal zones. J Coast Res 19:584–599

Speybroeck J, Bonte D, Courtens W, Gheskiere T, Grootaert P, Maelfait JP, Mathys M, Provoost S, Sabbe K, Stienen EWM, Lancker VV, Vincx M, Degraer S (2006) Beach nourishment: an ecologically sound coastal defence alternative? A review. Aquat Conserv Mar Freshw Ecosyst 16:419–435

Sudha Rani NNV, Satyanarayana ANV, Bhaskaran PK (2015) Coastal vulnerability assessment studies over India: a review. Nat Hazards 77:405–428

Sun Y, Zhong Z, Li T, Yi L, Hu Y, Wan H, Chen H, Liao Q, Ma C, Li Q (2017) Impact of ocean warming on tropical cyclone size and its destructiveness. Sci Rep 7:8154

Sutton-Grier AE, Wowk K, Bamford H (2015) Future of our coasts: the potential for natural and hybrid infrastructure to enhance the resilience of our coastal communities, economies and ecosystems. Environ Sci Policy 51:137–148

Upadhyay VP, Mishra PK (2014) An Ecological analysis of mangroves ecosystem of Odisha on the eastern coast of India. Proc Indian Natl Sci Acad 80(3):647–661

USGCRP (2014) Coastal zone development and ecosystems. In: Melillo JM, Terese TC, Richmond, Yohe GW (eds) Climate Change impacts in the United States: the third national climate assessment, Global Change Research Program, U.S. pp 579–618

von Schuckmann K, Palmer M, Trenberth K, Cazenave A, Chambers D, Champollion N, Hansen J, Josey SA, Loeb N, Mathieu P-P, Meyssignac B, Wild M (2016) An imperative to monitor earth's energy imbalance. Nat Clim Chang 6:138–144

Yang H, Ma M, Thompson JR, Flower RJ (2017) Protect coastal wetlands in China to save endangered migratory birds. PNAS 114(28):E5491–E5492

The Millennium Flood of the Upper Ganga Delta, West Bengal, India: A Remote Sensing Based Study

26

Sayantan Das and Sunando Bandyopadhyay

Abstract

The Indian part of the Upper Ganga Delta (UGD) is traversed by the Ganga River and its distributary system. As most of these distributaries, in their present condition, are unable to contain the monsoon discharge within their banks, the region is susceptible to frequent flood hazards. The topography of the UGD is mainly characterized by natural levee systems and earthen dykes alongside the rivers. Embankment-induced channel sedimentation, in some parts, has raised the elevation of riverbeds above the surroundings, causing stagnation of floodwater till it percolates down or drains out through the abandoned river courses. The levees often act as barriers and prevent the spilled water from getting back into the main channel. The largest flood in living memory occurred here at the turn of the millennium, in 2000. In September 2000, the remnant of a cyclonic depression triggered heavy downpours in the northwestern part of the UGD and the adjacent Chhotanagpur Plateau. This caused an unprecedented flood that inundated 46% of the UGD region and kept the area waterlogged for more than a month. Breaching the left bank embankment of the Bhagirathi-Hugli River, the floodwater surged southward following non-descript palaeodistributaries like the Gobra *Nala* and caused significant damages in the districts of Murshidabad and Nadia. Besides this, analysis of elevation models revealed some water movement paths that are undetectable on the ground. Following regional slope, the floodwater then moved across the Jalangi, Mathabhanga-Churni, and Ichhamati river courses to flow southeastward into the North 24 Parganas District, and eventually to Bangladesh. The waters took about a week to cover ~150 km. The pre-, syn-, and post-event satellite images related to the Millennium Flood detected only a few alterations in channel orientations and floodplain morphology. This suggests that the existing channel configuration and floodplain morphology of the UGD are the outcome of slower and lower magnitude fluvial processes. Most inhabitants, residing in the channel bars and near the rivers, have become accustomed to living with the floods during the monsoons. Construction of shelters on the higher areas detachable from the syn-flood images and elevation models can render further protection to these people.

S. Das (✉)
Department of Geography, Dum Dum Motijheel College, Kolkata 700074, India
e-mail: sayantdas@gmail.com

S. Bandyopadhyay
Department of Geography, University of Calcutta, Kolkata 700019, India

Keywords

Upper Ganga Delta · Flood · Extent of inundation · Palaeochannels

26.1 Introduction

A flood is an overflow of a body of water (rivers, lakes) that submerges land. It is basically the accumulation of water on the surface due to excess rainfall, and rise of the groundwater level above impermeable or saturated terrains (CRED 2013). Besides, floods can result from the other phenomena like coastal inundation caused by storm surges resulting from a tropical cyclone or tsunami or spring tide coinciding with higher-than-normal water levels (GA 2011). Exceptional circumstances like dam failure, triggered by an earthquake or extreme pressure from the stored water volume, may cause flooding in the downstream, even during dry weather conditions. Flooding can even occur due to melting of heavy snowpack due to sudden rise in temperature levels. Flooding is the most common environmental hazard worldwide (NDA 2013). Although it is a natural process, human activities can also trigger flooding. Floods may occur in an irregular manner and vary in magnitude, spatial extent, and duration (OQCS 2012). Normally, river floodplains and low-lying coastal areas are most susceptible to flooding. However, it is possible that floods may occur in the areas with unusually long periods of heavy rainfall. Flood can be viewed as beneficial, especially for enhancing soil fertility in the floodplains. But it is generally considered a hazard, endangering human life, property, and the environment (Godschalk 1991).

Floods affect both individuals and communities and have large-scale socio-economic and environmental consequences (OQCS 2012). These consequences vary greatly depending on the location and extent of flooding, and the vulnerability and value of the natural and constructed environments they affect. The immediate impacts of flooding include loss of human life, damage to property, destruction of crops, loss of livestock, and deterioration of health conditions owing to waterborne diseases. Loss of livelihoods, reduction in purchase power, and loss of land value in the floodplains can make communities economically vulnerable.

Geomorphic effects caused by large floods include channel widening, channel-bed erosion or deposition, and channel straightening (Osterkamp and Hedman 1981; Baker 1988; Osterkamp and Friedman 2000; Bowen and Juracek 2011). Floods cause geomorphic changes with respect to channel configuration and sediment load characteristics (Wohl 2000). These changes may occur during a single flood event or over a period of time, depending on the flood magnitude and frequency. The rate and amount of geomorphic change are influenced by flood hydraulics, resistance of riverbank materials, sediment supply, and previous occurrences of flood events. Megafloods with short duration peak flow—comparable with the discharge of more prolonged ocean currents fluxes—were common during the deglaciation that occurred in the Pleistocene (Baker 2007). Such type of large magnitude floods did not occur in the Holocene.

The geomorphic changes that take place due to a flood depend on the geomorphic effectiveness of the flood discharge, i.e., the ability of a flood event to affect the landscape (Wolman and Gerson 1978). This effectiveness is governed by the sensitivity of the landscape forming components and magnitude of the event. Factors that determine geomorphic effectiveness during floods include channel-bed and bank composition, channel morphology, channel slope, valley confinement, sediment load, flood duration, stream power, temporal ordering of floods, climate, and vegetation (Baker 1988; Osterkamp and Friedman 2000). As far as the dominant processes are concerned, erosion occurs if the velocity and shear stress are high during the flood event, while deposition occurs when the amount of transported sediment load is high. Lateral channel shifting may occur during floods in the form of meander migration, meander cutoff, and avulsion (Wohl 2000). The channel planform may alter regularly in every year with seasonal flooding or may change gradually in response to a change in the flood regime. In general, the channel configuration is shaped by high–magnitude infrequent flood events which are associated with moderate to high channel gradient, rich supply of bedload material, low bank cohesion, and relatively narrow cross-sectional geometry (Baker 1988, 1996; Wohl 2000).

Floods are the most common natural hazard of the humid tropics. The humid tropical regions of Asia, Africa, and Latin America experienced higher number of flood events due to high–magnitude storm events like tropical cyclones and extreme monsoon downpours, which account for more than 80% of the major floods occurred in these areas between 1900 and 2012 (Sanyal et al. 2013). Flood is a perpetual natural hazard in the flood plains of Asia, where most of the annual precipitation occurs during the monsoon months of June to October (Sanyal and Lu 2006; Singh 2007). With most of its area located in the tropical Asia, a large part of India is flood-prone, with extreme precipitation events like flash floods and torrential rains becoming increasingly common in the last few decades, coinciding with global warming (Agarwal and Chak 1991; Goswami et al. 2006).

26.2 The Flood-Prone Upper Ganga Delta

The problem of river flooding is of great concern for West Bengal, as a significant part of its area is nestled in the flood-prone deltaic part of the Ganga Basin. Almost 42% of West Bengal is susceptible to floods (Bandyopadhyay et al. 2014). Two of the worst floods in the state inundated 34% and 27% of its total area (88,752 km^2) in 1978 and 2000, respectively. Since the independence of India (1947), West Bengal was affected by major flood inundation on five occasions (DoIW–GoWB 2011, 2014). The portion of the Upper Ganga Delta (UGD) in West Bengal gets regularly flooded due to the capacity reduction of the channels alongside the presence of flat terrain (Bandyopadhyay et al. 2014). Force of water during high discharge events often cause embankment breaching in the plains, resulting in considerable delay in the drainage and splaying of transported particles in the floodplains (Chapman and Rudra 2007; Bandyopadhyay et al. 2014; DoIW–GoWB 2014).

The UGD is constituted by the upper part of the interfluve between the Bhagirathi-Hugli and Ganga (Padma) rivers, which expands southward from the off-take of the Bhagirathi River (24.505° N, 88.085° E). It is entirely located within the state of West Bengal, India (Fig. 26.1). The Bhagirathi-Hugli and Ganga rivers delineate the western and northern boundaries of the region, respectively. The political boundary between India and Bangladesh serves as the eastern limit of the study area, which approximately follows the Ichhamati course in its southern section. The southern margin of the UGD is demarcated by the Jamuna River—the last cross-stream between the Bhagirathi-Hugli and the Ichhamati. The southern boundary also corresponds to the southward extent of cutoff formation in the Bhagirathi-Hugli course.

The geomorphic setup of the UGD makes it continually susceptible to the flood hazard (Bandyopadhyay et al. 2014). The parent river—Ganga—usually floods the northern part of the Murshidabad District during the monsoons. Associated with the floods, the channel shifting of the Ganga is common through meander formation and braidbelt change. The districts of Murshidabad and Nadia are drained by the distributaries and palaeodistributaries of the Ganga River. Because most of these distributaries—in their present condition—are unable to contain the monsoon discharge within their banks, the region has become susceptible to flood hazards. The UGD was severely affected by floods in 1978, 1995, 1998, 2000, 2003, 2004, 2006, 2007, and 2008 (WBDMD 2014).

Among the distributaries of the Ganga, the south-flowing Bhagirathi-Hugli, which drains through the southern part of West Bengal. The off-take of the Bhagirathi-Hugli has completely dissipated now and it receives its entire headwater supply from the Farakka Barrage Project since 1975 (Das 2018). The Bhairab-Jalangi,

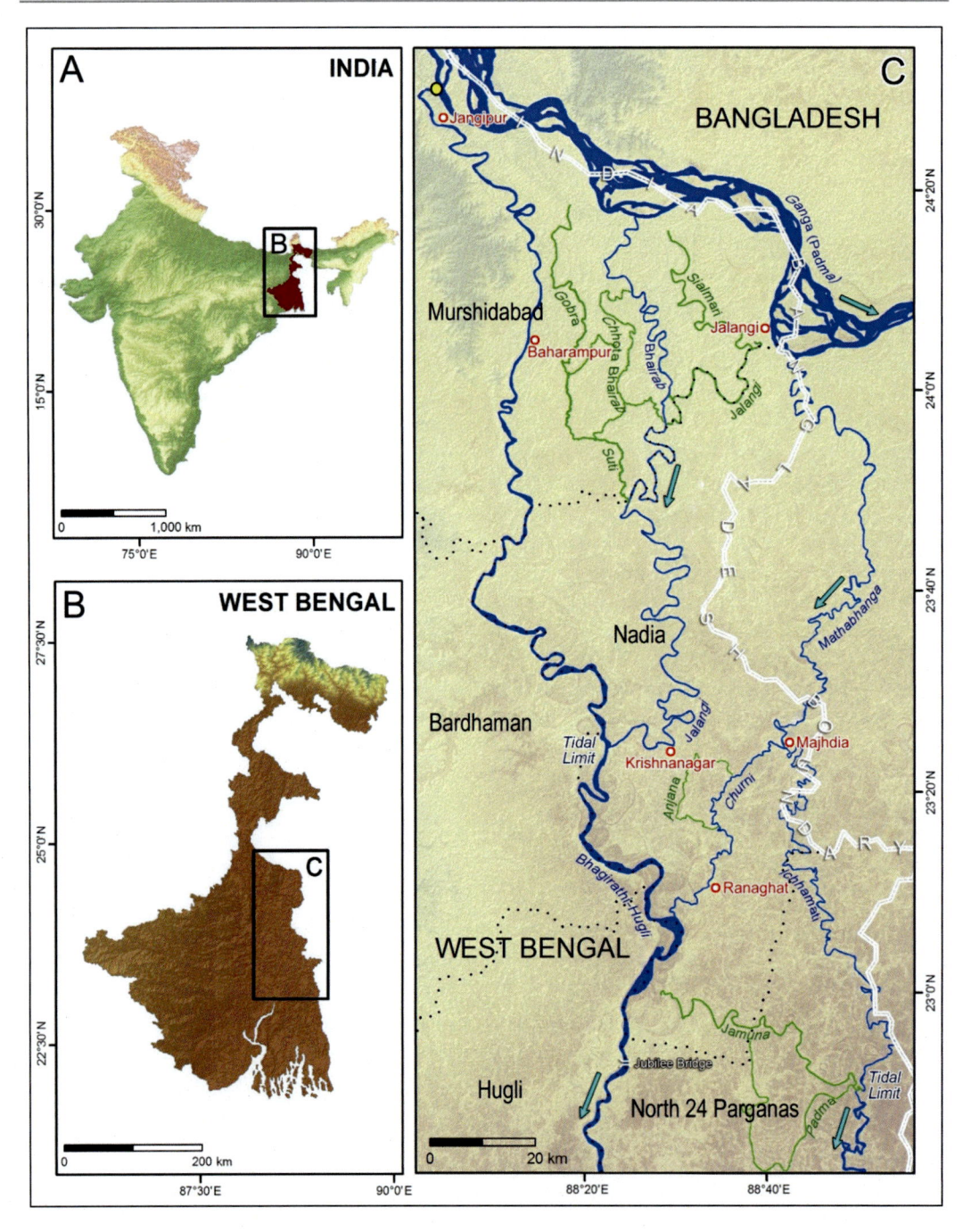

Fig. 26.1 Location map of the Upper Ganga Delta (UGD) in West Bengal, India, along with the active and non-active distributaries of the Ganga River. Blue and green watercourses in map **C** represent rivers and major palaeochannels, respectively. The yellow dot in map C locates the off-take of the Bhagirathi River, which is the northernmost point of the study area. The India–Bangladesh international boundary demarcates its eastern limit. The arrows indicate flow directions. Background image: SRTM GTOPO30 DEM (1 km), 01.12.1996

Mathabhanga-Churni, and Ichhamati are the other important distributaries of the Ganga in the Indian part of the UGD (Figs. 26.1 and 26.2). The physiography of the UGD is dominated by the presence of natural levees, floodplains, back swamps, and palaeochannels. The degeneration of the Ganga distributaries has continued in the last few centuries and many of the active or navigable channels are in various stages of degradation (Bandyopadhyay et al. 2014). A few of these streams can be identified as palaeodistributaries, viz. the Upper Jalangi, Sialmari,

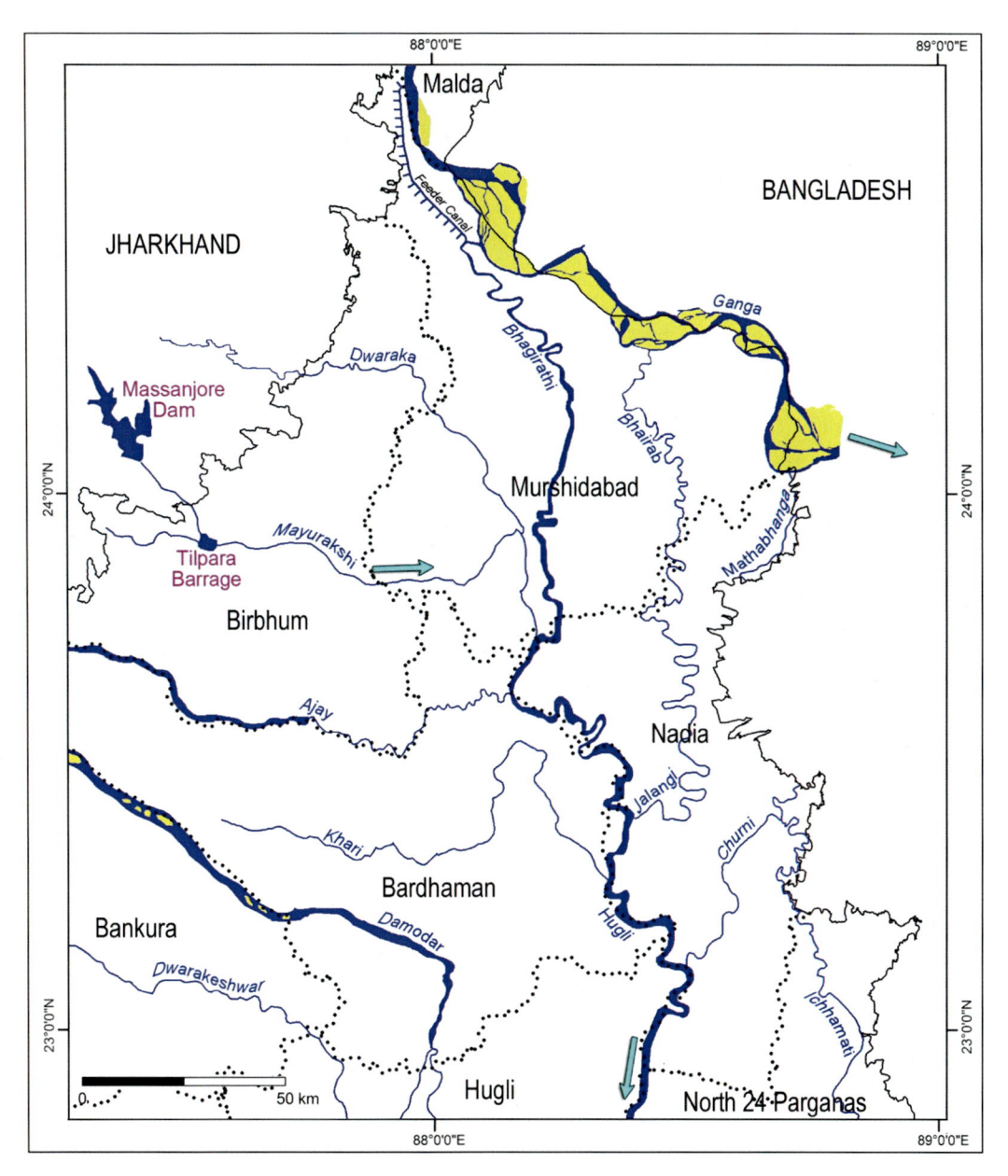

Fig. 26.2 Massanjore Dam in Jharkhand State and Tilpara Barrage in Birbhum District, along with the distributaries of the Ganga River in the UGD, and the western tributaries of the Bhagirathi-Hugli River. The arrows denote flow directions of important rivers. Some elements were incorporated from Bandyopadhyay et al. (2014)

Chhota Bhairab, Gobra Nala, Bardara, Bhandardaha, Suti, Ramdara, Anjana, Jamuna, and Padma. These palaeocourses are characterized by dry channel beds or elongated pools of stagnant water indicating previous channels.

The Jalangi and Churni rivers, near their confluence with the Bhagirathi-Hugli, are aided by the upstream-bound flood tides from the lower reaches of the later. This eventually creates a reverse hydraulic gradient that periodically prevents these rivers from discharging into the Bhagirathi-Hugli. This situation gets worsened when occasional tropical storms stray into the western plateaus or interior parts of the delta causing persistent downpours for days. The runoff, generated by the rains and brought down by the tributaries like the Mayurakshi, Dwaraka, etc. from the western Chotanagpur Plateau far exceeds the carrying capacity of the UGD, thereby causing floods (Bandyopadhyay et al. 2014). The interfluve between the Bhagirathi and Jalangi, together with the low-lying Churni Basin, forms a zone that is predisposed to flooding (Sanyal and Lu 2005). Owing to extremely low regional slope, this area is ill-drained and frequently gets waterlogged during the monsoons. Repeated flood inundations formed marshy areas in many places of the UGD. Besides, the palaeochannels and oxbow lakes retain the floodwater for a long period of time.

26.3 The Event

The Millennium Flood of 2000 that occurred in the months of September–October inundated vast areas of the UGD. This unprecedented flood in surpassed the intensity of the landmark flood of 1978 in the southern part of West Bengal (Roy 2009). The water levels in all the major distributaries of the UGD increased extraordinarily during that event. Widespread and persistent rains due to a low-pressure system over the Gangetic West Bengal and Chotanagpur Plateau areas of Jharkhand caused floods in the southern part of West Bengal (Bhan et al. 2001). The torrential rains leading to this event started on 17 September 2000 (Oxfam 2000). Murshidabad District received 1200 mm rainfall between 18 and 23 September 2000, while the Mayurakshi catchment area, upstream of Massanjore Dam, received 1008 mm between 18 and 22 September 2000 (Roy 2009). These remain the highest amounts of rainfall ever recorded in this region. All major rivers of the UGD were flowing above their danger levels by 19 September 2000 and the low-lying areas were inundated.

The accumulated water from the upstream posed severe threats to dams and barrages in Jharkhand and West Bengal. The discharge of water from the Massanjore Dam located in Jharkhand and Tilpara Barrage located in the Birbhum District of West Bengal affected a large area downstream of the Mayurakshi–Bhagirathi confluence. The water level at the confluence of the Bhagirathi and Jalangi rivers at Swarupganj in Nadia District was 9.61 m on 20 September 2000, 0.56 m above the Extreme Danger Level mark of 9.05 m (Source: Monthly Gauge Report, DoIW–GoWB), following the release of nearly 7000 cumecs of water from the Tilpara Barrage (Roy 2009). The release of additional 3400 cumecs of discharge brought the water level at the same location to 10.92 m on 24 September 2000. Close to 20,000 km^2 of land area in seven districts (Malda, Murshidabad, Nadia, North 24 Parganas, Birbhum, Bardhaman, and Hugli) were affected during the 2000 flood, with over 1200 deaths and destruction of two million houses (Oxfam 2000; Bhan et al. 2001; Rudra 2001; Chapman and Rudra 2007; WBDMD 2014). Thousands of villages were inundated by floodwater and standing crops were destroyed. 26 Community Development (CD) blocks in Murshidabad, 17 in Birbhum, eight in Nadia, and three in North 24 Parganas District were the worst hit. Bhan et al. (2001) reported that Katwa, Nabadwip, Krishnanagar, Santipur, Murshidabad, Berhampore, and Jiaganj towns were completely waterlogged and North 24 Parganas District was not flooded till 24 September 2000. Infrastructural losses were enormous, which included the destruction of many bridges, culverts, and communication lines. The total estimated damage was the highest for any single disaster that occurred in West Bengal (Roy 2009).

Table 26.1 Satellite database used in the study

Particulars	Resolution (m)	Date of imaging
Eight *Landsat-7 ETM+* images; Path 138, Rows 043, 044	30	24 April 2000 14 September 2000 30 September 2000 17 November 2000
Two *Landsat-5 TM* images; Path 138, Rows 043, 044	30	08 October 2000 24 October 2000
One *IRS-1D LISS-3* image; Path-108, Row-55	23.5	24 September 2000
Three 1 arc-second *SRTM DEM (version-3)*; Tiles n22–e88, n23–e88, n24–e88	30	February 2000

26.4 Materials and Methods

Satellite images are essential for interpreting the natural disasters occurring in the world (Bhan et al. 2001; Gianinetto et al. 2008). They provide useful information on spatial flood extent and flood damage (Bhanumurthy et al. 2010). In this study, multi-dated optical satellite datasets (Landsat-7 ETM+ and Landsat-5 TM) were procured from the United States Geological Survey (https://earthexplorer.usgs.gov) for monitoring the pre-flood, flood-time, and post-flood situations in September–October 2000 (Table 26.1). Also, Digital Elevation Models (DEM) were collated to relate the topography of the region with specific inundation levels found in the government records. The DEM datasets are essential to extract the landform characteristics of the concerned region and assess its vulnerability to the flood hazard. All these optical datasets and DEMs were reprojected into UTM on WGS-84 datum for the standardization. Subsequent analyses were done using Geomatica v.2015 and ArcGIS v.10 software.

26.5 The Millennium Flood: Occurrence, Analysis, and Discussion

26.5.1 Evidence from Satellite Images

In the pre-flood image of 14 September 2000, accumulation of cloud cover is prominent over the UGD, which resulted in exceedingly heavy downpour (Fig. 26.3). The flood inundation was at its peak between 23 and 25 September 2000 (DoIW–GoWB 2014; WBDMD 2014). The syn-flood image of 30 September 2000 reveals a clear picture of the deluge, along with the floodwater route that followed the regional slope toward south-southeast (Fig. 26.3). The Bhagirathi–Jalangi, Jalangi–Churni, and Churni–Ichhamati interfluves were severely affected due to this. The flooding around the Bhagirathi-Hugli Course occurred due to the added discharge from its major western tributaries (Dwaraka, Mayurakshi, and Ajay rivers), subsequent to the release of water from dams and barrages in the upstream. Within the UGD in West Bengal, the central and southern parts of the Murshidabad District, the northwestern, central, and southeastern parts of the Nadia District, and northeastern part of the North 24 Parganas were inundated during this flood. The satellite imagery of 8 October 2000 shows a slight decrease in inundation of the affected areas (Fig. 26.3). The dark signatures of the flooded areas signify settling down of sediments carried by the floodwater. Visual observations of 30 September and 8 October images reveal that the floods that occurred in the northern districts progressed into the northern part of North 24 Parganas. The post-flood image of 24 October 2000 shows significant numbers of inundation patches in the flood route (Fig. 26.3). These are mostly low-lying flat areas in the interfluves that remained waterlogged for months.

26.5.2 Hydraulic Routing of the Millennium Flood

In conformity with the regional slope, the flow lines were generated by using the 1 arc-second (30 m) SRTM elevation data of 2014 using ArcGIS (Fig. 26.4). The floodwater route of 2000 corresponded to these flow lines. The generated flow lines are aligned across the courses of the Jalangi and Mathabhanga-Churni, traversing toward southwest by cutting the valleys transverse to the regional slope. One of the most affected regions, the Bhagirathi–Jalangi interfluve, contains a number of palaeochannels (Fig. 26.5). The alignment of flow lines indicates definite routing of floodwater through the Gobra-Bhandardaha-Suti palaeocourse.

It is found that the 2000 flood discharge progressed from Murshidabad and Nadia districts to North 24 Parganas District, and further toward Bangladesh by following a south-southeastward route (Fig. 26.6). Three elevation profiles were drawn from the SRTM DEM of 2000. The northernmost profile, A–A′, signifies the low-lying areas between the Bhagirathi and Jalangi rivers, which act as the hydraulic conduit for the floodwater. The elevation range of the hydraulic pathways shown in the A–A′ profile is between 14 and 17 m. The B–B′ profile shows more confined hydraulic routes that correspond the Bhandardaha and Suti palaeochannels (Fig. 26.6). These constricted passages ensured spilling of the carried water in the lower part of the Bhagirathi–Jalangi interfluve. As a result, many waterlogged patches were left behind even after the recession of floodwater (Fig. 26.3). The hydraulic routes are shown in the B–B′ profile lies between 12 and 15 m. In the southernmost profile (C–C′), the hydraulic routes can be identified in the Churni–Ichhamati interfluve. However, elevation of this area is lower than the previous profiles (7–11 m). The 2000 floods passed across a number of distributaries (Jalangi, Churni, and Ichhamati) through these hydraulic pathways which corresponded to the regional slope. These cross-sections also show significant decrease in the thalweg elevation of the Bhagirathi River (14 m in profile A–A′, 10 m in B–B′, and 3 m in C–C′), indicating a southward-oriented regional slope in the western part of the UGD.

26.5.3 Extent of Inundation

Barring a few areas in the northern Murshidabad and immediate south of the Jamuna palaeochannel, the entire UGD is considered to be flood-prone (DoIW–GoWB 2014). The flooded area coverage during the Millennium Flood clearly brings out the inundation pattern in different community development (CD) blocks of the UGD (Fig. 26.7). The submerged areas during the flood were mostly identified in the central and southeastern UGD. Some 43.9% of the study area was actually flooded during the event (Table 26.2).

Considering the magnitude of the Millenium Flood, inundation was considerably less in many of the CD blocks in the north (Bhagwangola-I and -II, Raninagar-I and -II, Jalangi, Domkal, Karimpur-I and -II) and south (Ranaghat-I, Chakdaha and Haringhata). On the contrary, the inundation was widespread (covering more than 100 km^2 area) in several CD blocks, namely Berhampore, Nowda, and Beldanga-II in Murshidabad District, Tehatta-I and -II, Kaliganj, Nakashipara, Chapra, Hanskhali and Ranaghat-II in Nadia District, and Bagda, Bangaon and Gaighata in North 24 Parganas District (Table 26.2).

Seven CD blocks (Raghunathganj-II, Lalgola, Bhagwangola-I and -II, Raninagar-I and -II, and Jalangi) were flooded due to the overflowing Ganga River, whereas sporadic inundation occurred in three blocks (Domkal, Karimpur-I and -II) due to localized downpour (Fig. 26.7). Addition of huge discharge into the Bhagirathi from its western tributaries, along with incessant downpour during 18–23 September 2000 caused flooding in seven CD blocks (Murshidabad–Jiaganj, Berhampore, Beldanga-I and -II, Kaliganj, Nakashipara, and Krishnanagar-II). The south-southeastward moving floodwater submerged significant portions of 14 CD blocks (Hariharpara, Nowda, Tehatta-I and -II, Chapra, Krishnanagar-I, Krishnaganj, Hanskhali, Ranaghat-II, Haringhata, Bagda, Bangaon,

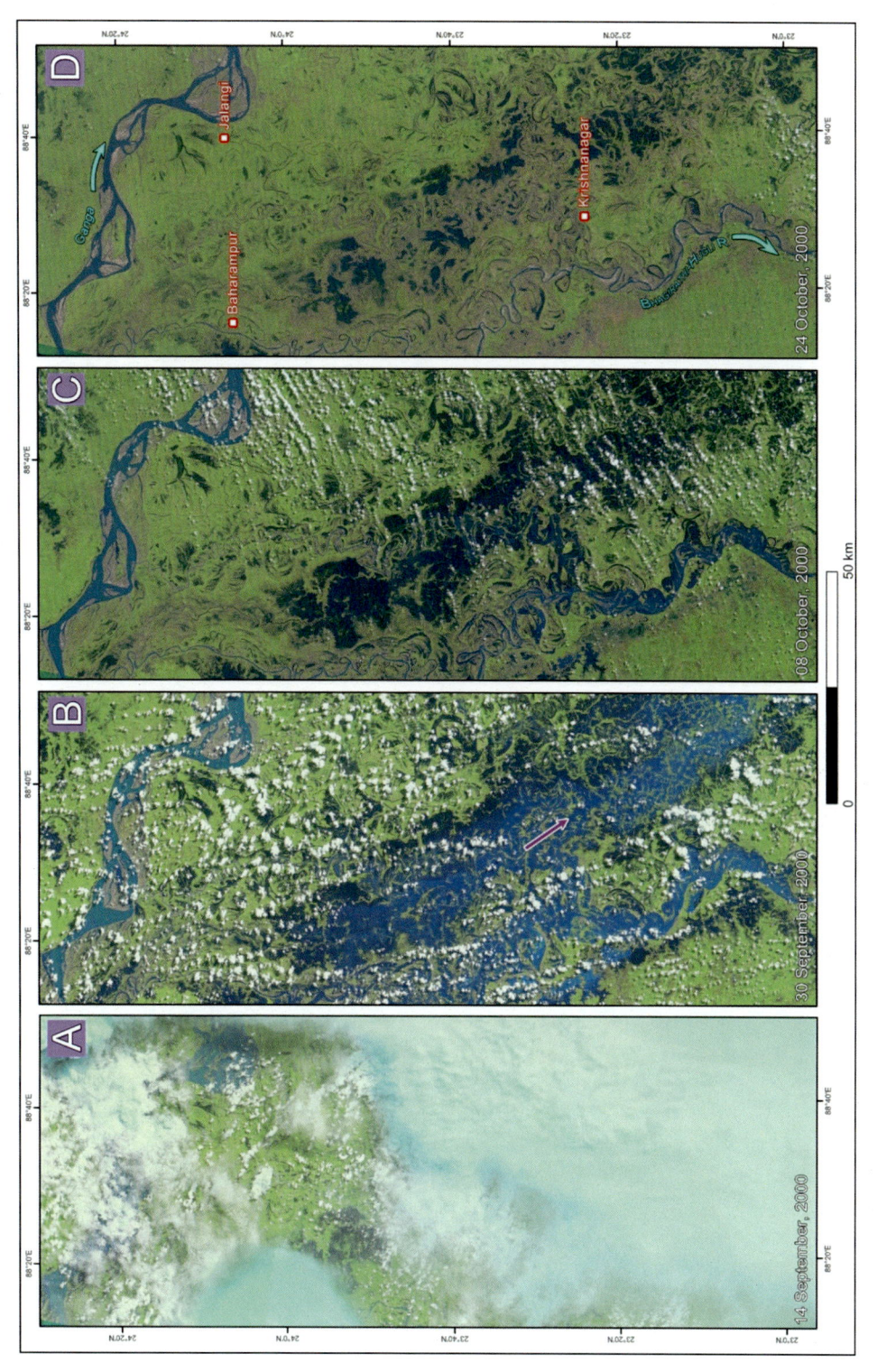

Fig. 26.3 The pre-flood (**A**), syn-flood (**B** and **C**), and post-flood (**D**) scenarios during the Millennium Flood. The purple arrow in panel **B** represents the floodwater flow direction. The flow directions of the Ganga and Bhagirathi-Hugli rivers are shown by arrows in panel **D**. Images: (**A**) Landsat-7 ETM + (Path-138, Row-43 and 44), 14.09.2000; (**A**) Landsat-7 ETM + (Path-138, Row-43 and 44), 30.09.2000; (**A**) Landsat-5 TM (Path-138, Row-43 and 44), 08.10.2000; (**D**) Landsat-5 TM (Path-138, Row-43 and 44), 24.10.2000

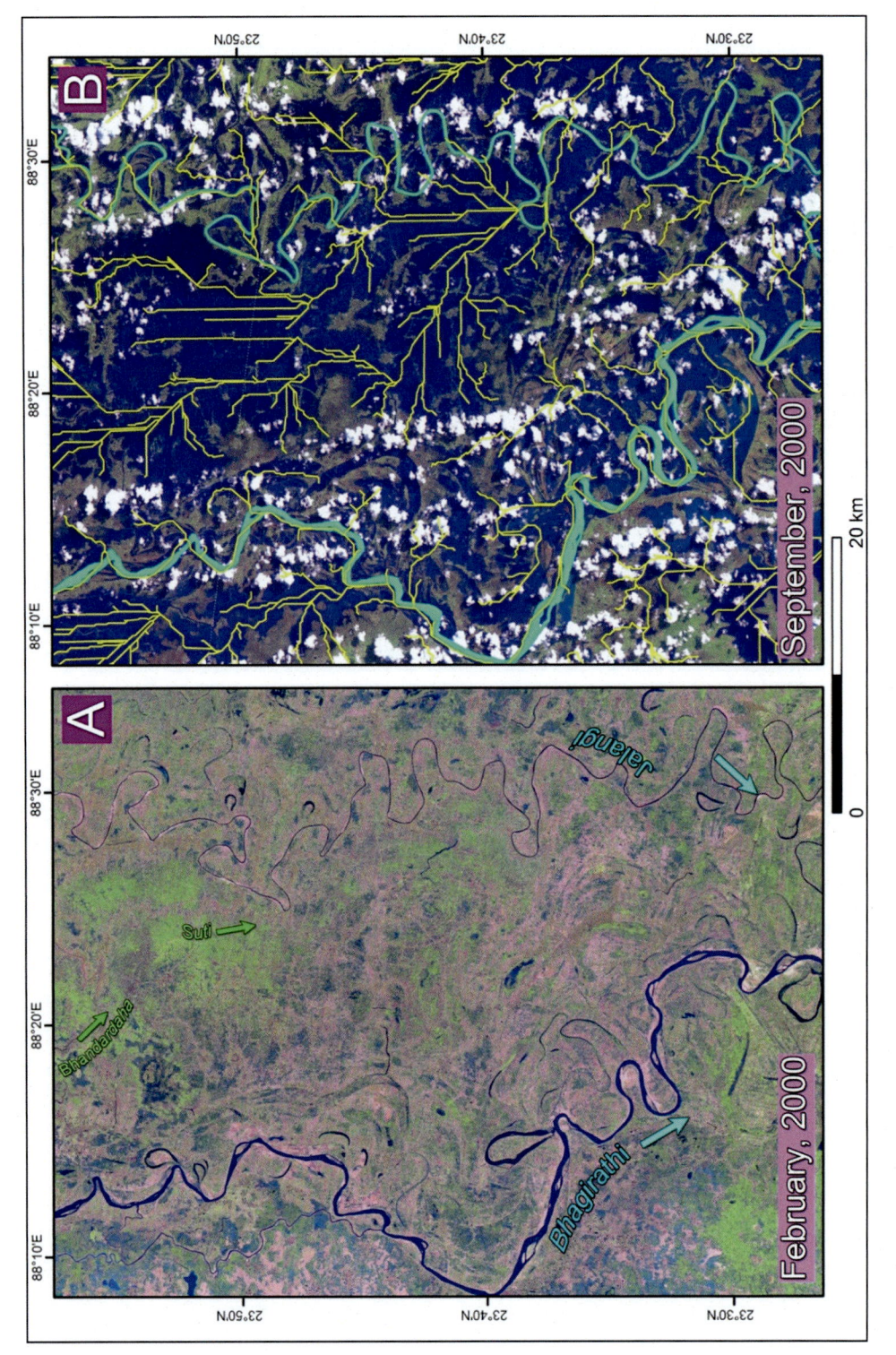

Fig. 26.4 Hydraulic routing of floodwater in the Bhagirathi–Jalangi interfluve. The green and turquoise arrows show the flow directions of the palaeochannels and distributaries, respectively. The flow lines in image B were generated by using 1 arc-second (30 m) SRTM DEM (v-3: Tile n23-e88) of February 2000. (**A**) Pre-flood image. Image: Landsat-5 TM (Path-138, Row-44), 27.02.2000. (**B**) Syn-flood image. The generated flow lines (yellow) correspond to the regional slope. Image: Landsat-7 ETM + (Path-138, Row-44), 30.09.2000

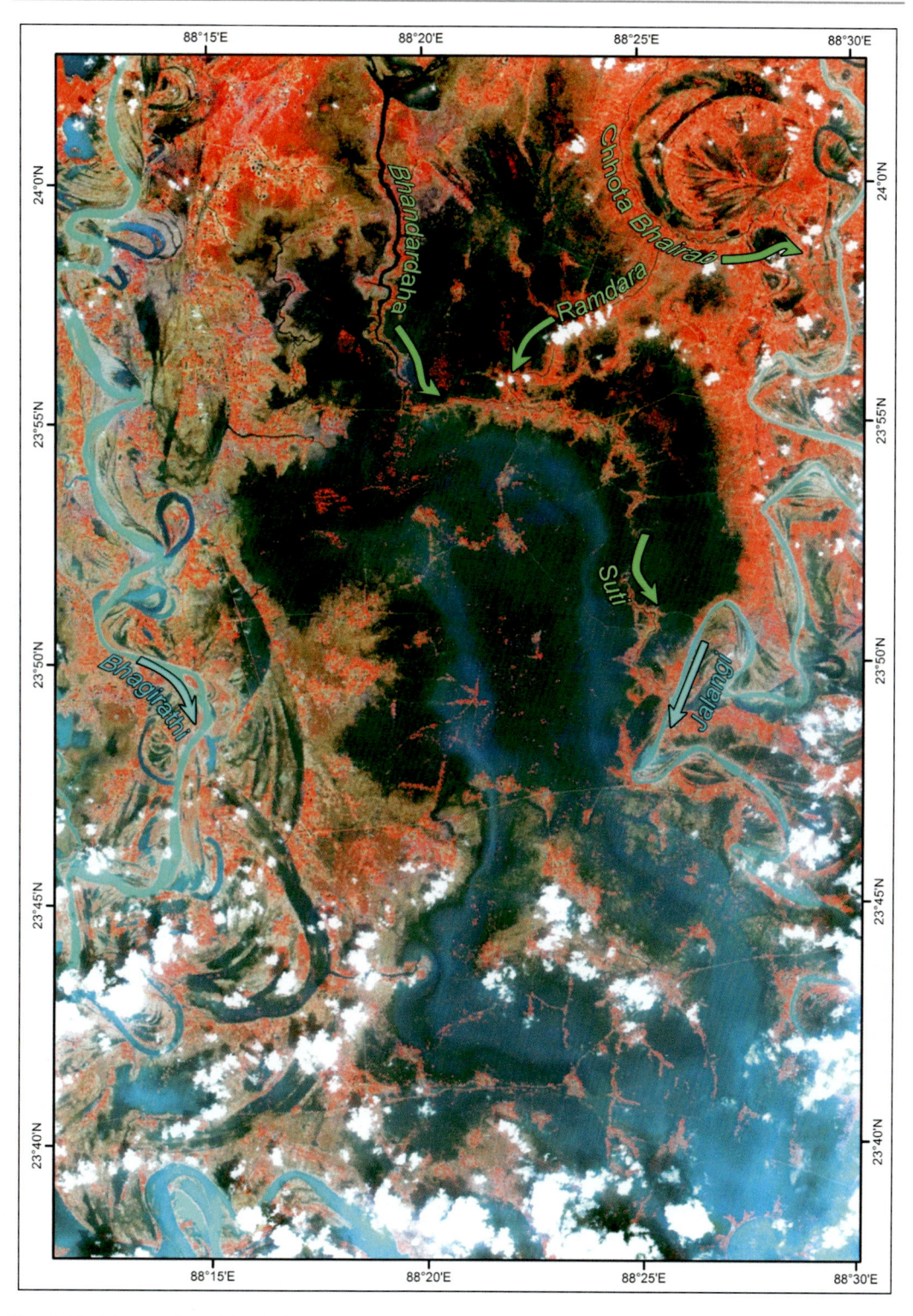

Fig. 26.5 The inundated Bhagirathi–Jalangi Interfluve during the Millennium Flood. The green arrows indicate the previous flow directions of the palaeodistributaries, while turquoise arrows indicate the flow directions of the present distributaries. Image: IRS-1D LISS-3 (Path-108, Row-55), 24.09.2000

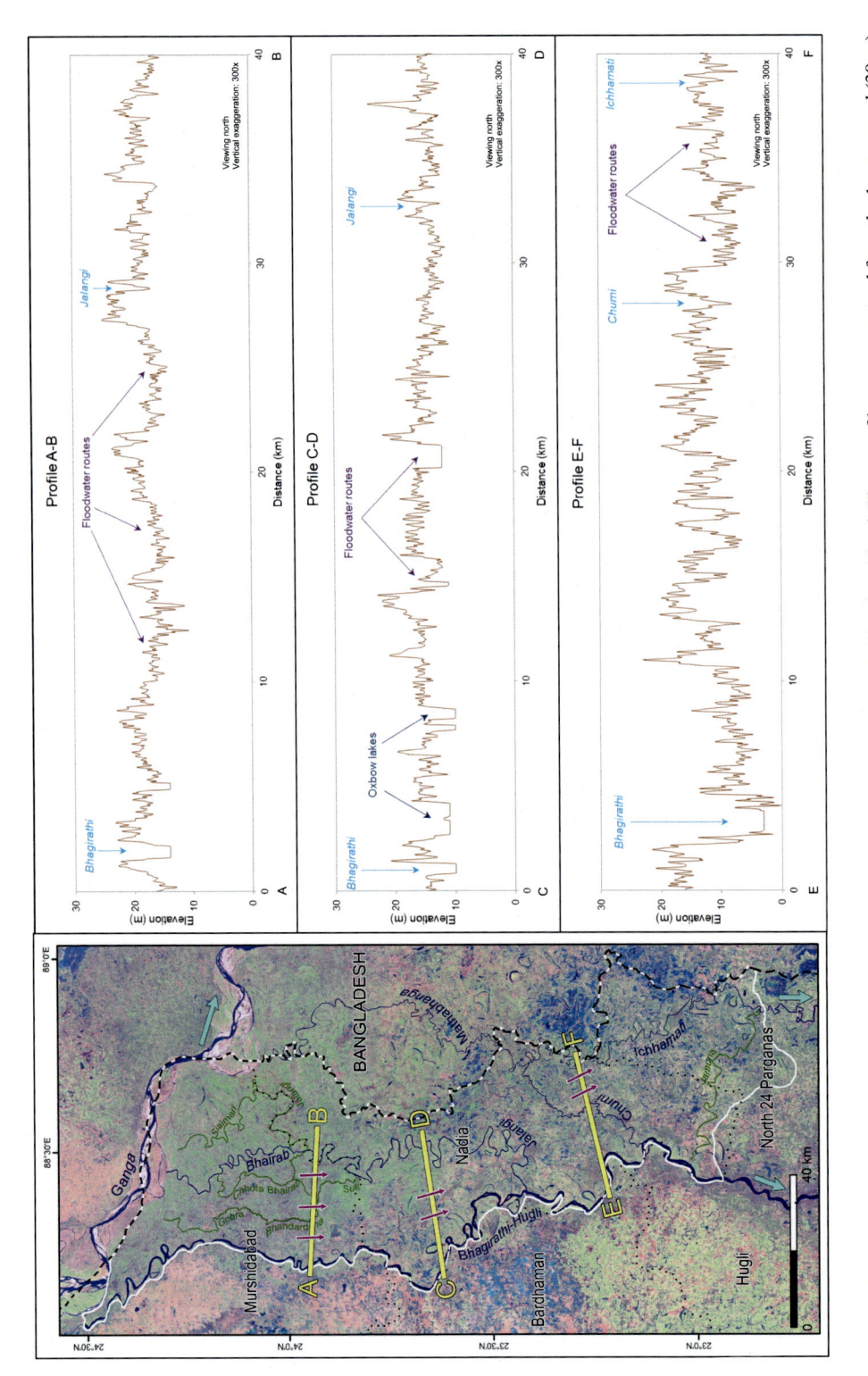

Fig. 26.6 Profiles showing the south-southeast-oriented floodwater routes during the Millennium Flood. These profiles were extracted from the 1 arc-second (30 m) SRTM DEM of February 2000. The hydraulic routes were identified by superimposing these profiles on the flood image of 30.09.2000 (Landsat-7 ETM + image, Fig. 26.3). The blue and green courses indicate rivers and important palaeocourses, respectively. The turquoise arrows show the present flow directions of major rivers, while the purple arrows indicate the progression of the floodwater during the event. The study area is delineated by the white line and the international boundary is shown by the black dashed line. Base image: Landsat-7 ETM + (Path-138, Row-43 and 44), 24.04.2000

Gaighata, and Habra-I). The remaining four-CD blocks (Nabadwip, Santipur, Ranaghat-I, and Chakdaha) were partly affected due to the overflowing Bhagirathi-Hugli River.

It is observed that inundation occurred in each of the CD blocks located in the UGD during September–October 2000 (Table 26.2; Fig. 26.7). The percentage of flooded area was maximum in Beldanga-II, Hanskhali, and Bangaon CD blocks, with more than 80% inundated area. The flooding was severe (60–80% inundation) in Nowda, Tehatta-II, Kaliganj, Nakashipara, Chapra, Krishnanagar-II, Nabadwip, Bagda, and Gaighata CD blocks. Significant proportion of flooded area (40–60% inundation) was found in Berhampore, Beldanga-I, Tehatta-I, Krishnaganj, Santipur, and Ranaghat-II CD blocks. The severity of the disaster was more in the Nadia District, which remained inundated for a longer time than the other areas (Figs. 26.3 and 26.7). The floodwater progressed into the North 24 Parganas District from north– to northwest, and subsequently inundated more than two-third area in three CD blocks (Bagda, Bangaon, and Gaighata). Within Murshidabad District, the floods inundated about 29.5% area in 14 CD blocks located in the UGD (Table 26.2). Whereas, in the 17 CD blocks of Nadia District, the floodwater had submerged 48.3% area. In four-CD blocks of the North 24 Parganas District in the UGD, the inundation affected 69.4% of the total area.

The UGD is a densely populated region. In 2001, the population of this region was 9,977,309 (CoI 2001). This corresponds to a population density of 1269 persons per km^2, which is almost four times higher than the population density of India in 2001 (324 persons per km^2). It is obvious that the Millennium Flood affected a huge number of inhabitants of the UGD.

26.5.4 Planform Changes Due to the Millennium Flood

The pre- and post-event satellite images of the 2000 disaster show very minimal alterations in the channel orientations (Figs. 26.8 and 26.9). The artificially resuscitated Bhagirathi course did not change much (Fig. 26.8), so as the combined course of the Bhairab-Jalangi (Fig. 26.9). The other distributaries and palaeochannels were flooded during the event but again went back to their usual dimensions after recession of the floodwater. In conjunction with the channel orientations, the floodplain morphology of the interfluves has not changed significantly during the Millennium Flood. Flood appears to be a major problem during the monsoon season in the UGD and a substantial proportion of land could get inundated for a long period. But even a high-intensity event like the 2000 floods did not bring about any significant change in fluvial morphology of the region. It seems that the existing channel orientations and morphological characteristics of the UGD floodplains are the outcome of relatively slower and lower magnitude fluvial processes.

26.6 Conclusions

The distributaries in the UGD—in their present condition—are unable to contain the monsoon discharge within their banks. This makes the region susceptible to floods almost every year. Following the regional slope, the floodwater moved across Jalangi, Mathabhanga-Churni, and Ichhamati courses to flow southeastward into Bangladesh. It took a week to reach North 24 Parganas District from the central part of Murshidabad District. The palaeocourses present in the Bhagirathi–Jalangi and Jalangi–Churni interfluves carried the floodwater along the regional slope. It is found that three CD blocks in Murshidabad District, ten in Nadia District, and three in North 24 Parganas District had more than 50% of their area inundated during the event.

The water level observed during this flood was significantly higher (0.5–2 m) than the Extreme Danger Level mark. In spite of the occurrence of this high–magnitude event characterized by overflowing rivers and palaeodistributaries, no major geomorphic alteration

Fig. 26.7 Flooded area of the UGD in 2000, overlayed by CD block boundaries. *Source* Flooded area in 2000 from DFO 2014

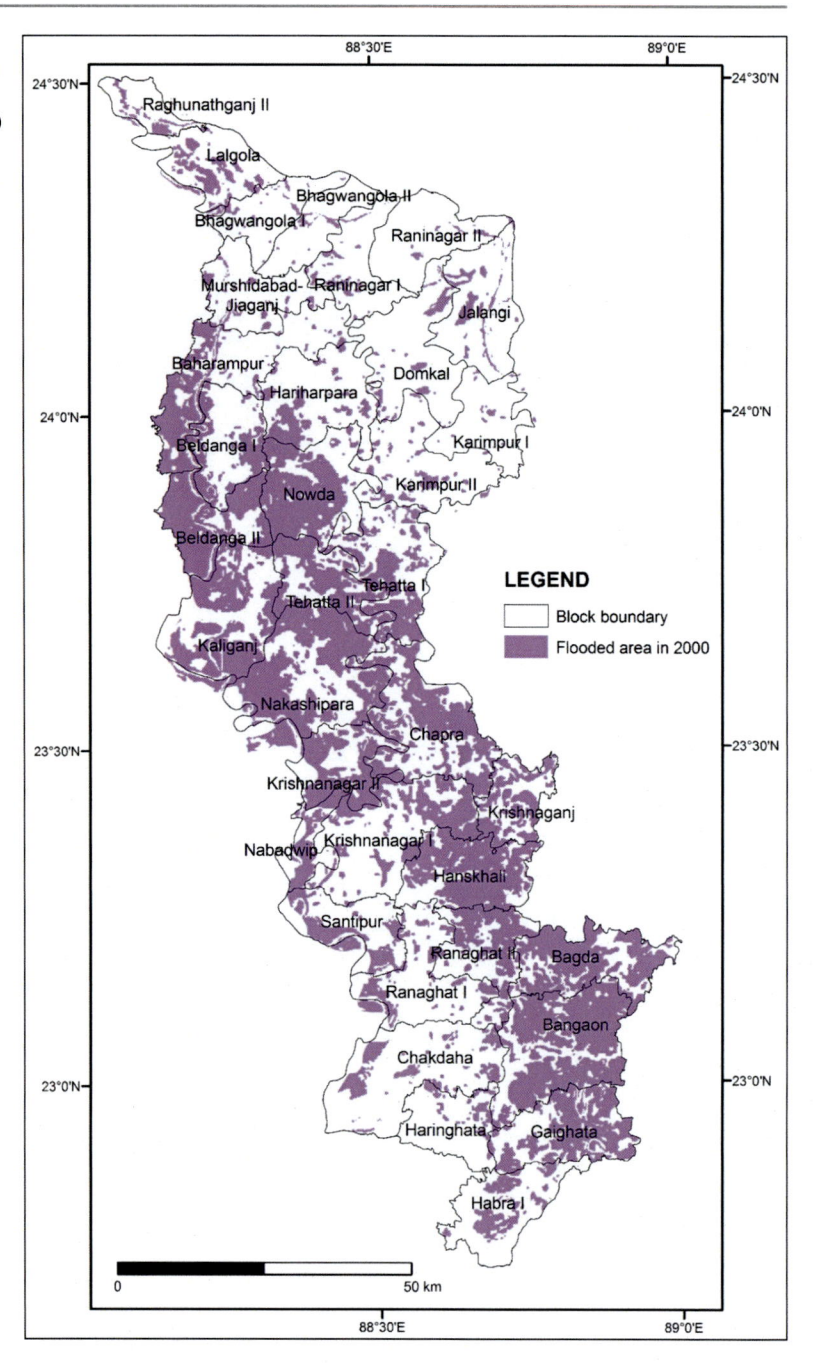

occurred in the UGD. The topography of the region is mainly characterized by natural levee systems and earthen embankments alongside the rivers. Also, quite a few palaeochannels (including oxbow lakes and meander scrolls) are found near the rivers—pointing toward shifting of the channels in the past. In some areas, due to channel siltation, cross-sectional areas of the distributaries decreased significantly and elevation of the river beds has gone higher than the surroundings. Additionally, the natural levees often act as barriers to the floodwater and prevent

Table 26.2 Inundated area coverage in the CD blocks of the UGD during the Millennium Flood

S. No.	Block name	District	Total area (km^2)	Inundated area (%)
1	Raghunathganj-II	Murshidabad	86.0	21.5
2	Lalgola		175.8	28.0
3	Bhagwangola-I		153.9	7.6
4	Bhagwangola-I		161.1	6.4
5	Raninagar-I		155.3	8.3
6	Raninagar-II		213.1	3.5
7	Jalangi		245.1	13.3
8	Domkal		364.3	9.6
9	Murshidabad–Jiaganj		219.5	11.7
10	Berhampore		294.5	47.1
11	Hariharpara		242.9	27.0
12	Nowda		269.8	74.0
13	Beldanga-I		182.2	50.9
14	Beldanga-II		208.7	85.1
15	Karimpur-I	Nadia	179.9	3.7
16	Karimpur-II		230.6	14.8
17	Tehatta-I		290.2	59.3
18	Tehatta-II		181.8	71.3
19	Kaliganj		339.2	71.2
20	Nakashipara		397.8	65.2
21	Chapra		316.2	63.4
22	Krishnaganj		139.3	57.5
23	Krishnanagar-I		260.0	33.9
24	Krishnanagar-II		129.6	72.2
25	Nabadwip		90.8	72.6
26	Hanskhali		233.6	81.8
27	Santipur		160.5	40.6
28	Ranaghat-I		235.8	22.9
29	Ranaghat-II		190.3	55.5
30	Chakdaha		334.5	18.5
31	Haringhata		166.5	14.1
32	Bagda	North 24 Parganas	241.2	75.0
33	Bangaon		366.7	81.6
34	Gaighata		241.9	68.7
35	Habra-I		164.5	34.8

Source Flooded area in 2000 from DFO 2014

it from getting back into the adjacent river channel (Fig. 26.10). These, coupled with the low regional slope, caused stagnation of floodwater at a number of places till it was percolated down or drained out by using the existing and abandoned river courses.

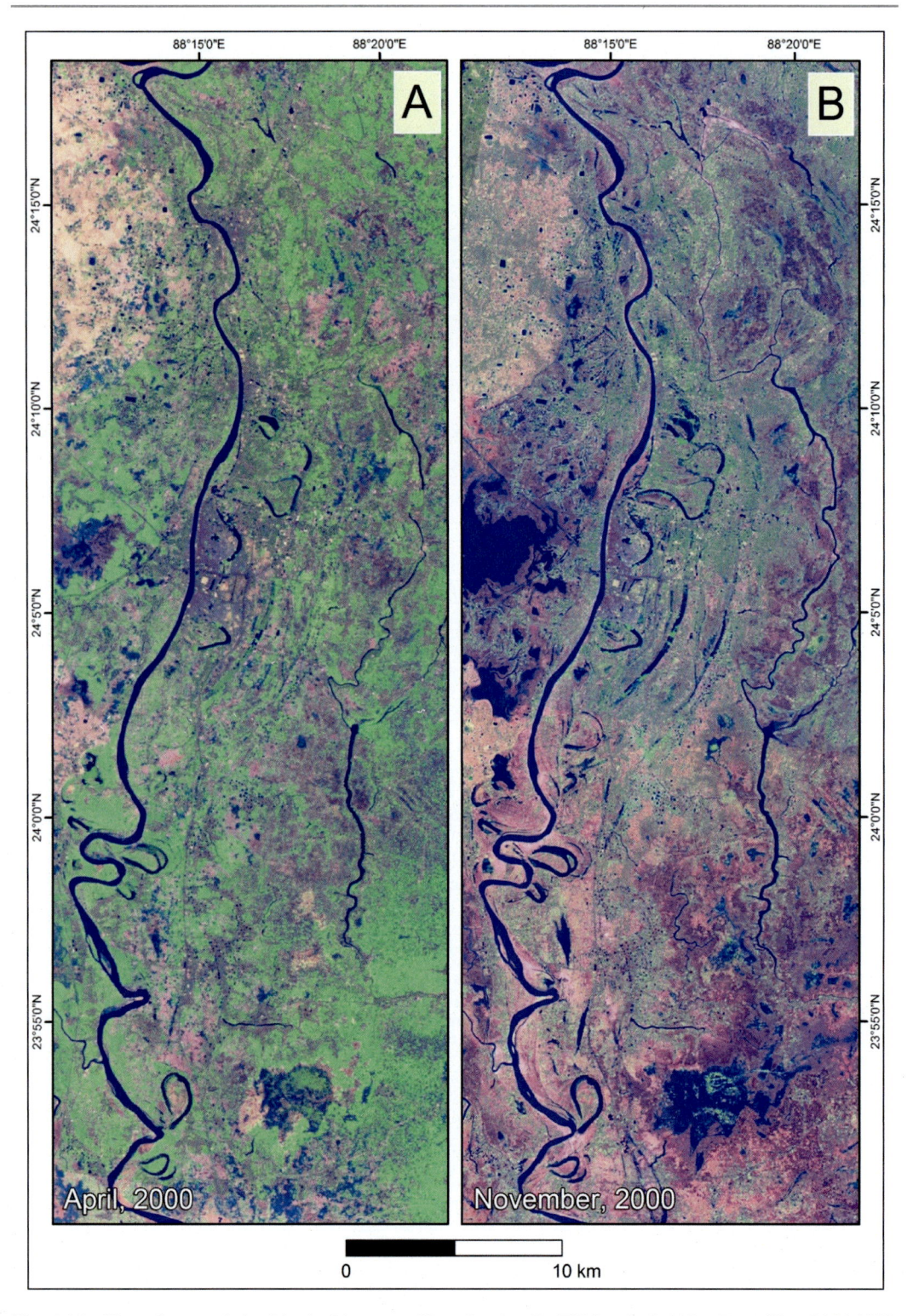

Fig. 26.8 Channels around the Bhagirathi course: No major planform change is observed by comparing the pre- and post-flood situations. (**A**) Pre-flood image. Image: Landsat-7 ETM + (Path-138, Row-43), 24.04.2000. (**B**) Post-flood image. Image: Landsat-7 ETM + (Path-138, Row-43), 17.11.2000

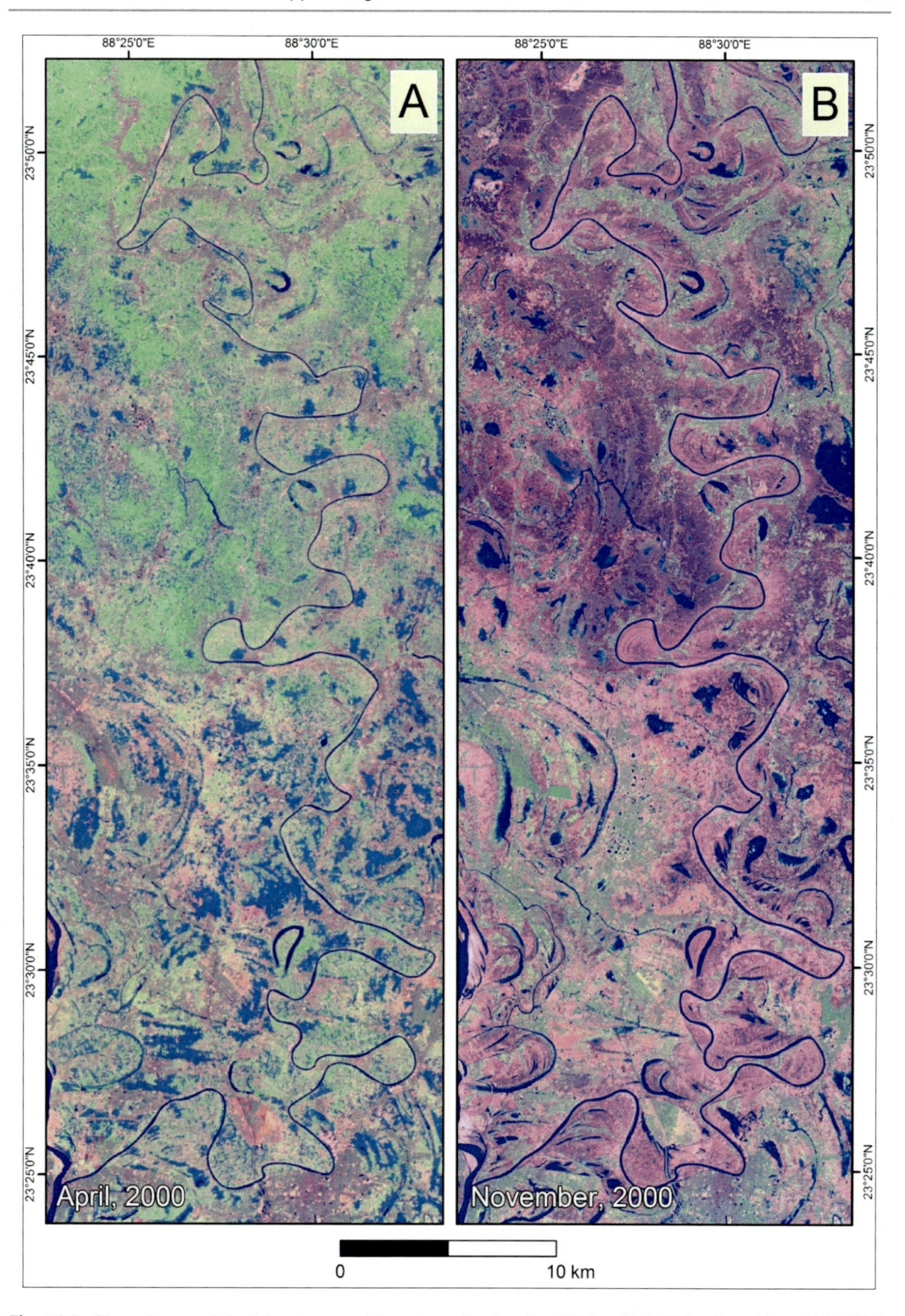

Fig. 26.9 Channels around the Jalangi course: No major planform change is noticed by comparing the pre- and post-flood situations. (**A**) Pre-flood image. Image: Landsat-7 ETM + (Path-138, Row-44), 24.04.2000. (**B**) Post-flood image. Image: Landsat-7 ETM + (Path-138, Row-44), 17.11.2000

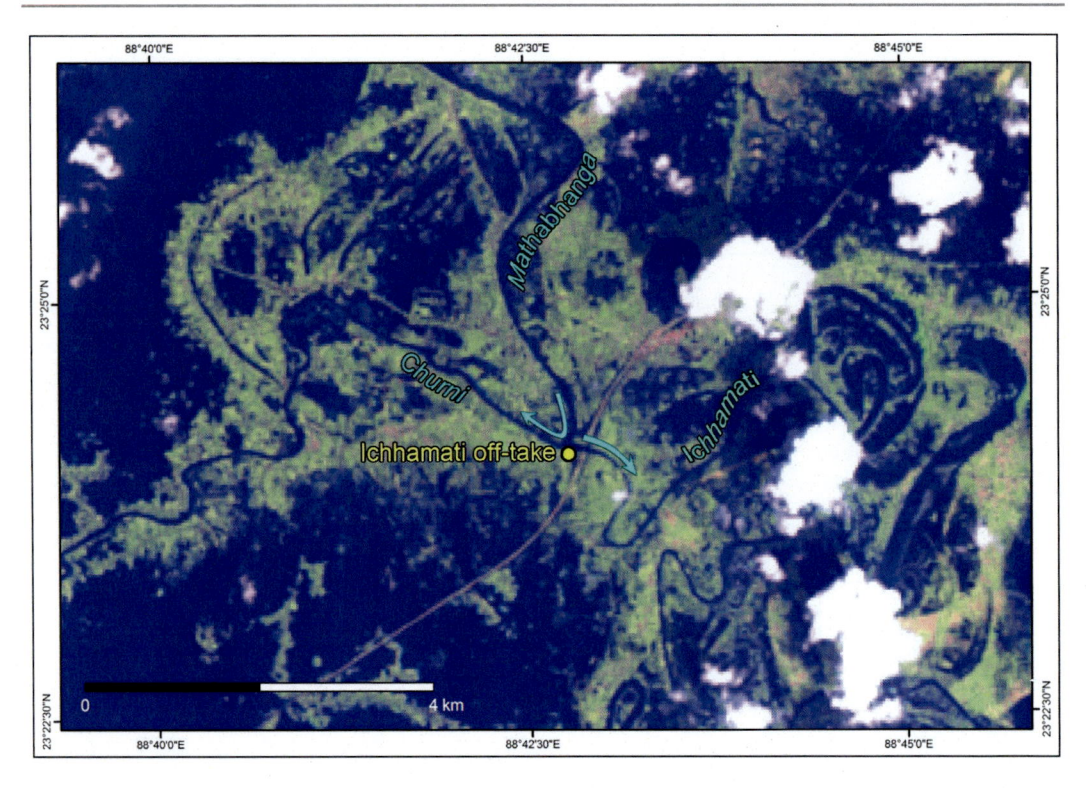

Fig. 26.10 The flooded surroundings of the Ichhamati off-take region in the Nadia District. Only the natural levees escaped inundation. Image: Landsat-7 ETM + (Path-138, Row-44), 30.09.2000

References

Agarwal A, Chak A (1991) Floods, floodplains and environmental myths. Centre for Science and Environment, New Delhi, 167 p

Baker VR (1988) Flood erosion. In: Baker VR, Kochel RC, Patton PC (eds) Flood geomorphology. Wiley, New York, pp 81–95

Baker VR (1996) Megafloods and glaciation. In: Martini IP (ed) Late glacial and post glacial environmental change. Oxford University Press, New York, pp 98–108

Baker VR (2007) Greatest floods and largest rivers. In: Gupta A (ed) Large rivers: geomorphology and management. Wiley, pp 65–74

Bandyopadhyay S, Kar NS, Das S, Sen J (2014) River systems and water resources of West Bengal: a review. In: Vaidyanadhan R (ed) Rejuvenation of surface water resources of India: potential, problems and prospects. Geological Society of India, Special Publication, vol 3, pp 63–84. https://doi.org/10.17491/cgsi/2014/62893

Bhan SK, Flood Team (2001) Study of floods in West Bengal during September, 2000 using Indian remote sensing satellite data. J Indian Soc Remote Sens 29(l-2):1–3. https://doi.org/10.1007/BF02989907

Bhanumurthy V, Manjusree P, Srinivasa Rao G (2010) Flood disaster management. In: Roy PS, Dwivedi RS, Vijayan D (eds) Remote sensing applications. National Remote Sensing Centre, Hyderabad, pp 283–302

Bowen MW, Juracek KE (2011) Assessment of the geomorphic effects of large floods using stream–gauge data: the 1951 floods in eastern Kansas, USA. Phys Geogr 32(1):52–77. https://doi.org/10.2747/0272-3646.32.1.52

Chapman GP, Rudra K (2007) Water as foe, water as friend: lessons from Bengal's millennium flood. J South Asian Dev 2(1):19–49. https://doi.org/10.1177/097317410600200102

CoI (2012) Census of India, Ministry of Home Affairs, Government of India (2001) Retrieved on 16.08.2012 from http://www.censusindia.gov.in/DigitalLibrary/Archive_home.aspx

CRED (2013) Centre for research on the epidemiology of disasters. Flood. Retrieved on 06.08.2016 from http://www.preventionweb.net/english/hazards/flood/

Das S (2018) Evolution of drainage and morphology of upper Bhagirathi Ganga interfluve region of West Bengal with special reference to Palaeochannels. Unpublished Ph.D. Thesis, University of Calcutta, Kolkata, 350 p

DFO (2014) Dartmouth Flood Observatory. Inundation image 080E030Nv3, 2 weeks ending 04.08.2014, Retrieved on 20.08.2014 from http://floodobservatory.colorado.edu/Version3/080E030Nv3.html

DoIW–GoWB (2011) Department of Irrigation and Waterways, Govt. of West Bengal. Annual Report: 2010–11, Kolkata, 60 p

DoIW–GoWB (2014) Department of Irrigation and Waterways, Govt. of West Bengal. Annual Flood Report for the Year 2013, Kolkata, 112 p

GA (2011) Geoscience Australia. What causes Floods? Retrieved on 08.06.2016 from http://www.ga.gov.au/scientific-topics/hazards/flood/basics/causes

Gianinetto M, Scaioni M, Mondino EB, Tonolo FG (2008) Sustainable approach for upgrading geographic databases based on high resolution satellite imagery. Int Arch Photogr Remote Sens Spatial Inf Sci Beijing 37(B8):1131–1137

Godschalk DR (1991) Disaster mitigation and hazard management. In: Drabek TE, Hoetmer GJ (eds) Emergency management: principles and practice for local government. International City Management Association, Washington, DC, pp 131–160

Goswami BN, Venugopal V, Sengupta D, Madhusoodanan MS, Xavier PK (2006) Increasing trend of extreme rain events over India in a warming environment. Science 314(5804):1442–1445. https://doi.org/10.1126/science.1132027

NDA (2013) Natural Disasters Association. What is Flooding? Retrieved on 06.08.2016 from http://www.n-d-a.org/flooding.php

OQCS (2012) Office of the Queensland Chief Scientist. What is a flood? Retrieved on 12.06.2013 from http://www.chiefscientist.qld.gov.au/publications/understanding-floods/what-is-a-flood

Osterkamp WR, Friedman JM (2000) The disparity between extreme rainfall events and rare floods—with emphasis on the semi-arid American west. Hydrol Process 14:2817–2829. https://doi.org/10.1002/1099-1085(200011/12)14:16/17%3c2817::AID-HYP121%3e3.0.CO;2-B

Osterkamp WR, Hedman ER (1981) Channel geometry of regulated streams in Kansas as Related to Mean Discharge, 1970–80. Kansas Water Office, Topeka, KS, USA. Technical Report No. 15

Oxfam (2000) West Bengal Floods Situation Report 27 Sept 2000. Retrieved on 16.04.2015 from http://reliefweb.int/report/india/west-bengal-floods-situation-report-27-sep-2000

Roy J (2009) The Deluge 2000. Administrative Training Institute, Kolkata, 32 p

Rudra K (2001) The Flood of September 2000. Jayasree Press, Kolkata

Sanyal J, Carbonneau P, Densmore AL (2013) Hydraulic routing of extreme floods in a large ungauged river and the estimation of associated uncertainties: a case study of the Damodar river, India. Nat Hazards 66(2):1153–1177. https://doi.org/10.1007/s11069-012-0540-7

Sanyal J, Lu XX (2005) GIS-based flood hazard mapping at different administration scales: a case study in Gangetic West Bengal, India. Singap J Trop Geogr 27:207–220. https://doi.org/10.1111/j.1467-9493.2006.00254.x

Sanyal J, Lu XX (2006) Remote sensing and GIS-based flood vulnerability assessment of human settlements: a case study of Gangetic West Bengal, India. Hydrol Process 19:3699–3716. https://doi.org/10.1002/hyp.5852

Singh IB (2007) The Ganga river. In: Gupta A (ed) Large Rivers: geomorphology and management, pp 341–371

WBDMD (2014) West Bengal Disaster Management Department. Flood. Retrieved on 13.04.2015 from http://wbdmd.gov.in/Pages/Flood2.aspx

Wohl EE (2000) Geomorphic effects of floods. In: Wohl EE (ed) Inland flood hazards: human, riparian, and aquatic communities, pp 167–193

Wolman MG, Gerson R (1978) Relative scales of time and effectiveness in watershed geomorphology. Earth Surf Process 3:189–208. https://doi.org/10.1002/esp.329003020

Tropical Cyclone: A Natural Disaster with Special Reference to *Amphan*

Biplab Biswas and Chandi Rajak

Abstract

Tropical cyclone causes huge damage to life and properties through their gusty winds, storm surge, and flooding due to excess rainfall. Tropical Cyclonic "*AMPHAN*"(*UM-PUN*), made its landfall at 14.30 h, on 20th May 2020 in between Digha in West Bengal (INDIA) and Hatiya Island (Bangladesh) with gusting winds speed more than 185 kmph, storm surge of about 12 m and some places received (24 h) rainfall which was 436% higher than monthly average. It was the first cyclone in the year 2020. Hundreds of millions of people were in its path witnessed its fury and damages. The storm took 2 human lives, and thousands of trees and houses were damaged, thousands of acres of croplands were lost with flooding, and incurred loss of billions of Indian Rupees. This paper used Geographical Information System (GIS) for mapping its origin from the Bay of Bengal; destructions caused along its path of translational movement in West Bengal (India); and socio-economic conditions along its path.

Keywords

Tropical Cyclone · AMPHAN · Depression · Super Cyclone · Storm Surge · Flood

27.1 Introduction

Tropical Cyclonic "*AMPHAN*" (Super Cyclone, a name given by Thailand *UM-PUN*), made its landfall at 14.30 h, on 20th May 2020 between Digha in West Bengal (INDIA) and Hatiya Island (Bangladesh; 21.65° N & 88.30° E) with gusting winds speed more than 185 kmph. It is the first cyclone in the year 2020. Hundreds of millions of people were in its path to witness its fury and damages. The local government has taken initiative to evacuate at least 658 thousand people from vulnerable houses to government's (and another cyclone-resistant) shelters to save their lives. However, the storm took 2 human lives; uprooted thousands of trees & houses; damaged acres of croplands with flooding and incurred loss of billions of Indian Rupees (https://www.thehindu.com/news/national/other-states/amphan-cyclone-tracker-may-202020/article31629032.ece).

According to India Meteorological Department (IMD), tropical cyclone (TCs) is a rotational low-pressure system in tropics characterized by falling central pressure by 5–6 mb from the surrounding and maximum sustained wind speed reaches 34

B. Biswas (✉) · C. Rajak
Department of Geography, The University of Burdwan, Golapbag, Burdwan, India
e-mail: bbiswas@geo.buruni.ac.in

knots (about 62 kmph). It is a vast violent whirl of 150–800 km, spiraling around a center and progressing along the surface of the sea at a rate of 300–500 km a day (Misra et al. 2020). The word cyclone has been derived from Greek word "cyclos" which means, "coiling of a snake". The word cyclone was coined by Heary Piddington who worked as a Rapporteur in Kolkata during British rule (Mathur et al. 2016).

Tropical cyclones are warm-core low-pressure systems having large vortex in the atmosphere, which is maintained by the release of latent heat by the convective clouds over warm oceans. It is a huge strong wind system, which blows around the center of intense low-pressure area. Typically, their diameter at the surface is between 100 and 1000 km and the vertical extent is about 10–15 km, with the axis tilting toward colder region (IMD, Gov. of India).

Cyclones originate in the tropical ocean region are known as tropical cyclones. TCs are also known as hurricanes, typhoons, and southern hemisphere cyclones. They are frequent phenomena and cause fatalities and major loss of life and properties. TCs have different names based on their location of origin. This is interesting to note that the TCs originate from all the oceans except the south Atlantic and south-eastern Pacific Oceans (Table 27.1 and Fig. 27.1).

TCs has two motions, one is the rotation of winds on its axis and another is the translational motion. Rotational speed is the circular motion of gusty wind along its axis. Winds blow clockwise in the Northern Hemisphere and anti-clockwise in the southern hemisphere. Based on the velocity of the gusty wind, TC is classified into different categories from Depression to Super Cyclone (Table 27.2). The translational motion is the progress of the cyclone from ocean toward the land. It is normally 15–20 km/h. When the speed of movement is 10–14 km/h, it is called as slow-moving cyclone. In the moderately moving cyclone, the speed of movement is 15–25 km/h. In a fast-moving cyclone, the speed of movement is more than 25 km/h. The gusty wind, storm surge, excessive rainfall, and associated flooding cause excessive damage to life and properties.

The tropical cyclone "*Amphan*" originated from a low-pressure area persisting a couple of hundred miles (300 km) east of Colombo, Sri Lanka, on 13th May 2020. Tracking northwestward, the disturbance organized over exceptionally warm sea surface temperatures; the Joint Typhoon Warning Center (JTWC) upgraded the

Table 27.1 Localization of tropical cyclones

Oceanic Region	Location	Name of the TCs	Timings
Tropical North Atlantic Ocean	East of the Lesser Antilles and the Caribbean, east of 70°W	Hurricane	July to October
	North of the West Indies		June to October
	Western Caribbean		June and late September to early November
Western North Pacific Ocean, including the Philippines		Typhoons	May to November
North Pacific off the West Coast of Central America			June to October
Bay of Bengal and Arabian Sea	India, Bangladesh, Myanmar	Tropical Cyclone	May to June and October to November
South Pacific Ocean, West of 140°W			December to April
South Indian Ocean	North-western Coast of Australia		December to April
	West of 90°W		November to May

Source Authors' Compilation from various sources

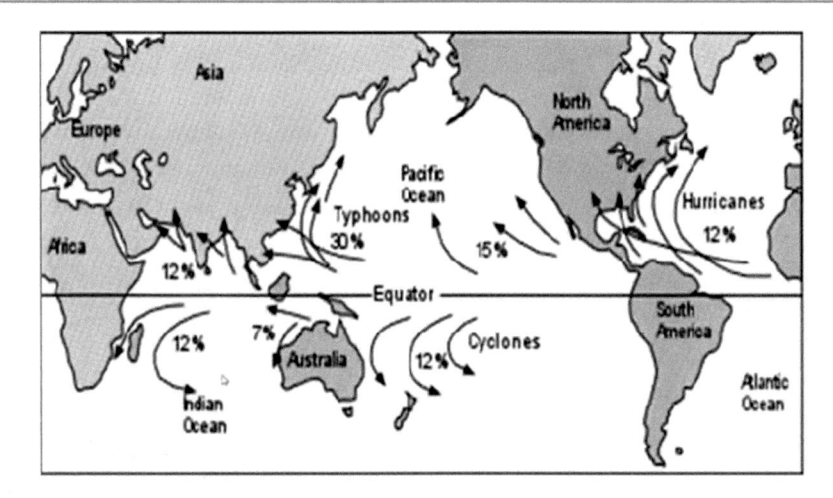

Fig. 27.1 Tropical Areas where cyclones form and typical travel paths and their name. *Source* After Abbott (1996)

Table 27.2 Classification of Tropical Disturbances and stages of *Amphan*

Category	Date (May, 2020)	Wind speed 3 min Av		Pressure change millibar (mb)	Wave height (m)	Beaufort scale
		knots	kmph			
LPA/WML	13	<17	< 31	< 4	<1.2	0–4
Dep (D)	15	17–27	31–49	04–06	0–3	04–06
Deep Dep (DD)	15	28–33	50–62	06–10	< 6	7
Cyclonic Strom (CS)	16	34–47	63–88	10–15	<9	09–11
Severe CS (SCS)	17	48–71	89–118	15–20	9–14	11–12
Very SCS (VSCS)	17	64–90	119–167	20–66	>14	>11
Extreme SCS (ESCS)	17	91–119	168–221	67–79	>14	12+
Super Cyclone (SUCS)	18	>120	>222	>80	>14	13+

LPA Low-Pressure Area; *WML* Well Marked Low; *D* Depression; *DD* Deep Depression; *CS* Cyclonic Storm; *SCS* Severe Cyclonic Storm; *VSCS* Very Severe Cyclonic Storm; *ESCS* Extreme Severe Cyclonic Storm; *SUCS* Super Cyclonic Storm
Source Compiled by Authors from various sources

system to a tropical depression on 15th May while the India Meteorological Department (IMD) followed the suit on the following day. On 17th May, it underwent rapid intensification and became a super cyclone, and made its landfall on 20th May 2020 in West Bengal (IMD, Gov. of India, press release bulletin 13th May to 22nd May 2020).

This paper would like to use Geographical Information System (GIS) to map the origin from the Bay of Bengal and destructions caused along the path of movement of the cyclone *"Amphan"* in West Bengal (India).

27.2 Origin of TCs and *Amphan*

Tropical Cyclone can be compared to a heat engine. The energy input is from warm water and humid air over tropical oceans. Release of heat happens through condensation of water vapor to water droplets/rain. Only a small percentage (3%) of this released energy is converted into Kinetic energy to maintain cyclone circulation (wind field). A mature cyclone releases energy equivalent to that of 100 hydrogen bombs.

High temperature creates very low pressure over the ocean. As the warm and moist air rises, it condenses into massive thunderclouds. Because of the Coriolis force, the wind bents inwards and then spirals upwards up to the levels of tropopause. The circulating cloud and wind can be up to 500–1000 km across. The center of the TCs is characterized by calm, cloudless area and it is known as eye–where there is no rain. The development TCs is divided into three stages—Formation and initial development; Full maturity; and Decay. The super cyclone *Amphan* has also passed through all these stages in its development process like low-pressure area (LPA) to super cyclone (SUCS) (Table 27.2 and Fig. 27.2).

27.2.1 Initial Development

Certain necessary atmospheric and oceanic conditions for the development of TCs are:

- Sea temperature in >26 °C for supply of abundant water vapors. In the case of the *Amphan*, the sea surface temperature was about 32 °C.
- High relative humidity of the atmosphere to a height of above 7000 m facilitates condensation of water vapors into water droplets and clouds; releases heat energy thereby inducing a drop in pressure;
- Small atmospheric instability which helps vertical cumulus cloud;
- The favorable region of origin of TCs is within 5° Latitudes.

According to the IMD, the cyclone *Amphan* was initiated with a low-pressure area (LPA) over southeast BoB in the early morning of that day 13th May 2020. It moved west northwestwards with favorable environmental conditions and emerged as depression (D) over central parts of south Bay of Bengal (BoB) on 15th May 2020. The wind velocity was ranging between 40 and 50 kmph. As it moved west northwestwards, it amplified into a deep depression (DD) over the adjoining southeast area of the bay in the early morning of 16th May 2020. By moving furthermore over the north Andaman Sea (BoB), *Amphan* intensified into a cyclonic storm (CS) in the late night of 16th May 2020 and developed "eye" and entered its mature stage of development.

27.2.2 Maturity

Development of "Eye of Cyclone" signifies the mature stage of the TCs. The shape of the cloud system becomes more circular. Approximately half of the cyclones of this form progress to full maturity. In the mature stage, *Amphan* turned into a severe cyclonic storm (SCS) on the evening of 17th May 2020 over west-central and southern BoB while continuing to move north north-eastwards. Moving nearly northwards, it further intensified into a very severe cyclonic storm (VSCS) in the late night (2330 IST) of 17th May 2020. It continued to move nearly northwards and it intensified into a Super Cyclone (SUCS) around noon (1200 IST) on 18th May 2020. It was associated wind speed of 220–230 km/ph. It was located about 360 km south of Paradeep Port of Odisha Coast on the evening of 19th May 2020. On 20th May 2020, between 11:30 and 14:30 (IST) *Amphan* made landfall in West Bengal.

27.2.3 Decay

TC's decay starts with landfall. The land cannot supply the required water vapor like oceans for its maintenance. The storm eye gets distorted. Wind velocity reduces to moderate. After the landfall, it moved toward Bangladesh over the landmass of West Bengal and turned into a severe cyclonic storm (SCS). It crossed West Bengal (India) border to Bangladesh in the early morning of 21st May 2020. Subsequently, it further weakened into

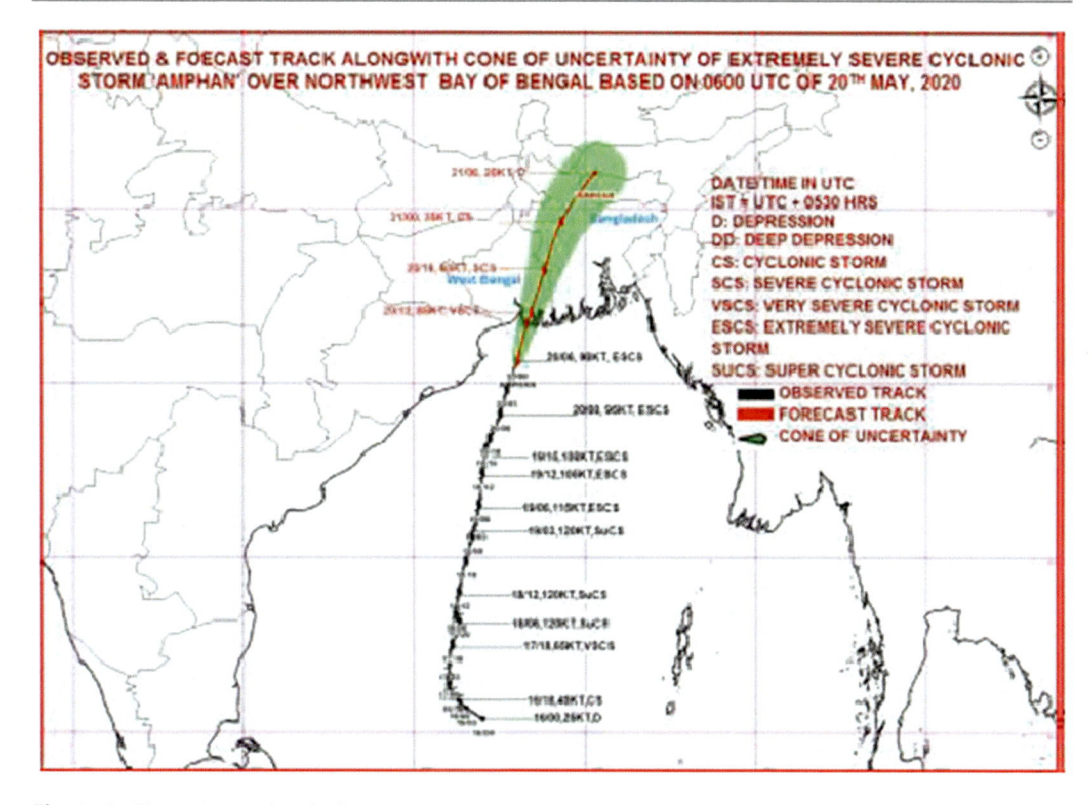

Fig. 27.2 The pathway of cyclonic storm *Amphan*. *Source* IMD New Delhi

an LPA on 22nd May 2020 morning and resulted in isolated heavy to very heavy & extremely heavy rainfall occurred over Meghalaya, and isolated heavy to very heavy rainfalls over Assam & Arunachal Pradesh.

27.2.4 Structure of TCs

A fully developed TCs has four major components.

27.2.4.1 Eye

A fully developed tropical cyclone has a central cloud-free region of calm winds, known as the "eye" of the cyclone with diameter varying from 10 to 50 km. This region is free from any weather disturbances. It is located in the center of the TCs. After the landfall, the affected region gets huge storms and other weather disturbances till the region hits by the eye of the TCs. But after the eye is over, the regions experience another severe weather disturbance.

27.2.4.2 Eyewall Cloud Region

Surrounding the eye is the "Eyewall cloud region" which is characterized by very strong winds and torrential rains, which has the width of about 10–150 km. The winds over this region rotate around the center and resemble the "coils of a snake". Major destructions of land and properties occur in this region.

27.2.4.3 Rain/Spiral Bands

Wind speed falls off gradually away from this core region, which terminates over areas of weaker winds with overcast skies and occasional squall.

27.2.4.4 Outer Storm Area

There may be one or more spiral branches in a cyclone where higher rainfall occurs. Vertical extent of the cyclone is about 15 km or the height of tropopause.

27.3 Classification and Naming of Tropical Disturbances

Tropical Cyclones start from small disturbances in the oceans. They used to grow as the Super Cyclones with increasing wind speed. Based on the wind speed and other weather parameters, the tropical disturbances are classified into eight categories ranging from LPA, i.e., Low-Pressure Area to SUCS—Super Cyclone (Table 27.2). Any "cyclonic storm" is given a name like *Amphan* or *Nisarga*. This is the convention as declared by World Meteorological Organization (WMO). Conventionally, the place of landfall of a TC was considered for its naming and record. However, it was observed that the same place might get two or more TCs in the same month. To avoid any of such complexities, the naming of the TCs started as per the guidelines of WMO in consultation with the Regional Specialized Meteorological Center (RSMC). Naming of the TCs is done by the countries located along the ocean. Normally the name is kept free from all religion, caste, and other biasness. Naming helps the scientific communities, disaster managers, media, and general masses to (i) identify each individual cyclone; (ii) create awareness of its development; (iii) remove confusion in case of simultaneous occurrence over a same region; (iv) remember TCs easily, rapidly and effectively; and (v) disseminating warning to much wider audience.

27.3.1 Path and Season of Tropical Cyclones

The paths of TCs of last 100 years in BoB are all different and they never follow any previous path. Based on the time of origin it is seen that TCs are rare in the months of January to March. The analysis of the April cyclone shows that it hits the Arakan Coast (Myanmar). While many of the pre-monsoon TCs move toward the Indian Costs, i.e., to Andhra Pradesh, Odisha, West Bengal, and Bangladesh. The post-monsoon (November) TCs generally move toward Tamil Nadu and Andhra Pradesh. Recently IMD has been able to predict the date and time and the place of landfall very precisely. This is certainly reducing the loss of human life remarkably. The predictions of the TCs path is done by seven Regional Specialized Meteorological Center (RSMC), namely (i) Miami; (ii) Honolulu; (iii) Tokyo; (iv) New Delhi; (v) La Reunion; (vi) Darwin; and (vii) Nadi (Fig. 27.3).

Since 1890, the recorded history of TCs indicates spatial variation in their landfall locations. It hits all the major administrative segments of the eastern coasts of India and there is lesser variation. However, Odisha coast gets maximum of 32.62% of the total landfall of TCs and West Bengal gets least with 20.57% of total TCs in the Bay of Bengal. Andhra Pradesh and Tamil Nadu coasts get 25.53% and 21.28% of total TCs' landfall.

From the record of TCs since 1890s, it is also very clear that the post-monsoon occurrence of TCs is more in number than the pre-monsoon. Tamil Nadu (96.67%) and Andhra Pradesh (86.11%) get their major share of TCs in the post-monsoon season. TCs affect Odisha in both the pre and post-monsoon seasons (Fig. 27.4 and Table 27.3).

The data shows that both the pre and post-monsoon origin of TCs are in slight increasing trend. Many researchers have attributed this increase as an impact of global warming and sea level rising (Sahoo and Bhaskaran 2015).

27.3.2 Monitoring of Tropical Cyclones

The IMD monitors the Tropical Cyclone very closely using five important methods:

27.3.2.1 Satellite

Satellite is one of the most important tools to monitor all the stages of development, intensification, and movement of TCs. IMD has INSAT 3D satellite with many channels or satellite products to predict the development and other aspects of TCs (Fig. 27.5).

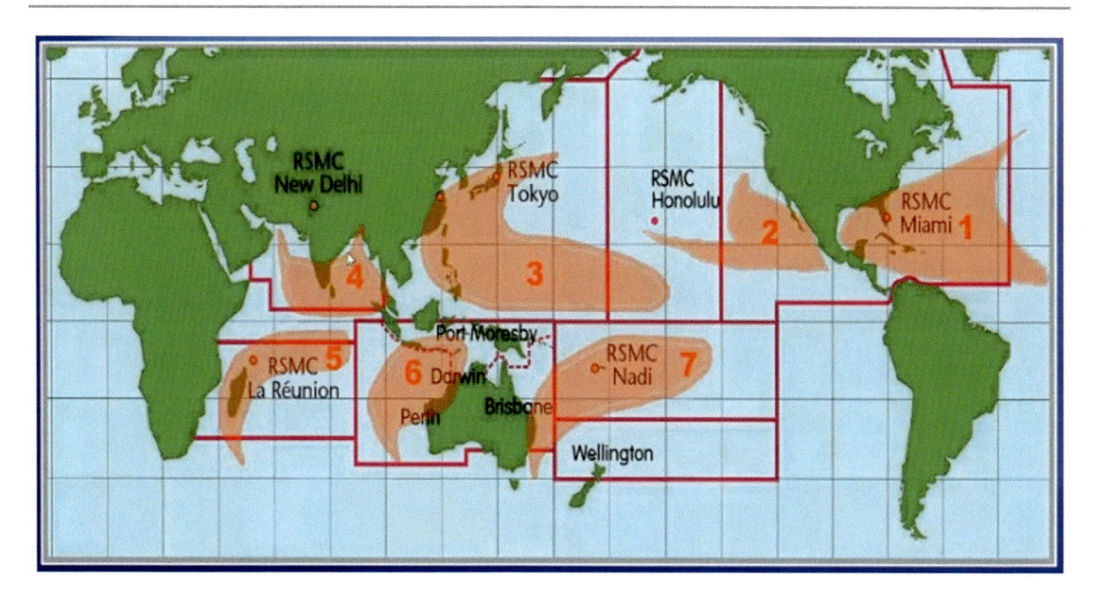

Fig. 27.3 7 Regional Specialized Meteorological Center (RSMC), namely (i) Miami; (ii) Honolulu; (iii) Tokyo; (iv) New Delhi; (v) La Reunion; (vi) Darwin; and (vii) Nadi. *Source* https://www.ducksters.com/science/earth_science/hurricanes.php

Fig. 27.4 Historical Tropical Cyclones in Bay of Bengal. *Source* Authors' Calculation

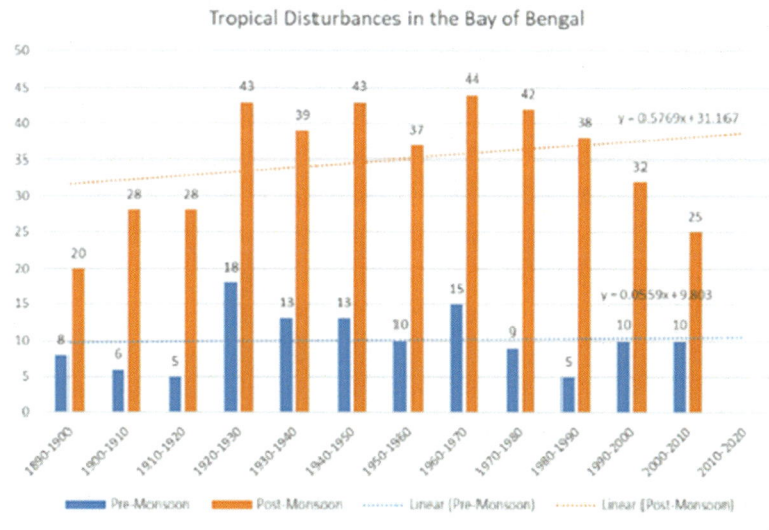

Table 27.3 Distribution of tropical disturbances in Bay of Bengal

Coastal Area	% of TCs	% in Pre-Monsoon	% in Post-Monsoon
West Bengal	20.57	31.03	68.97
Odisha	32.62	41.30	58.70
Andhra Pradesh	25.53	13.89	86.11
Tamil Nadu	21.28	3.33	96.67

Source Authors' Calculation

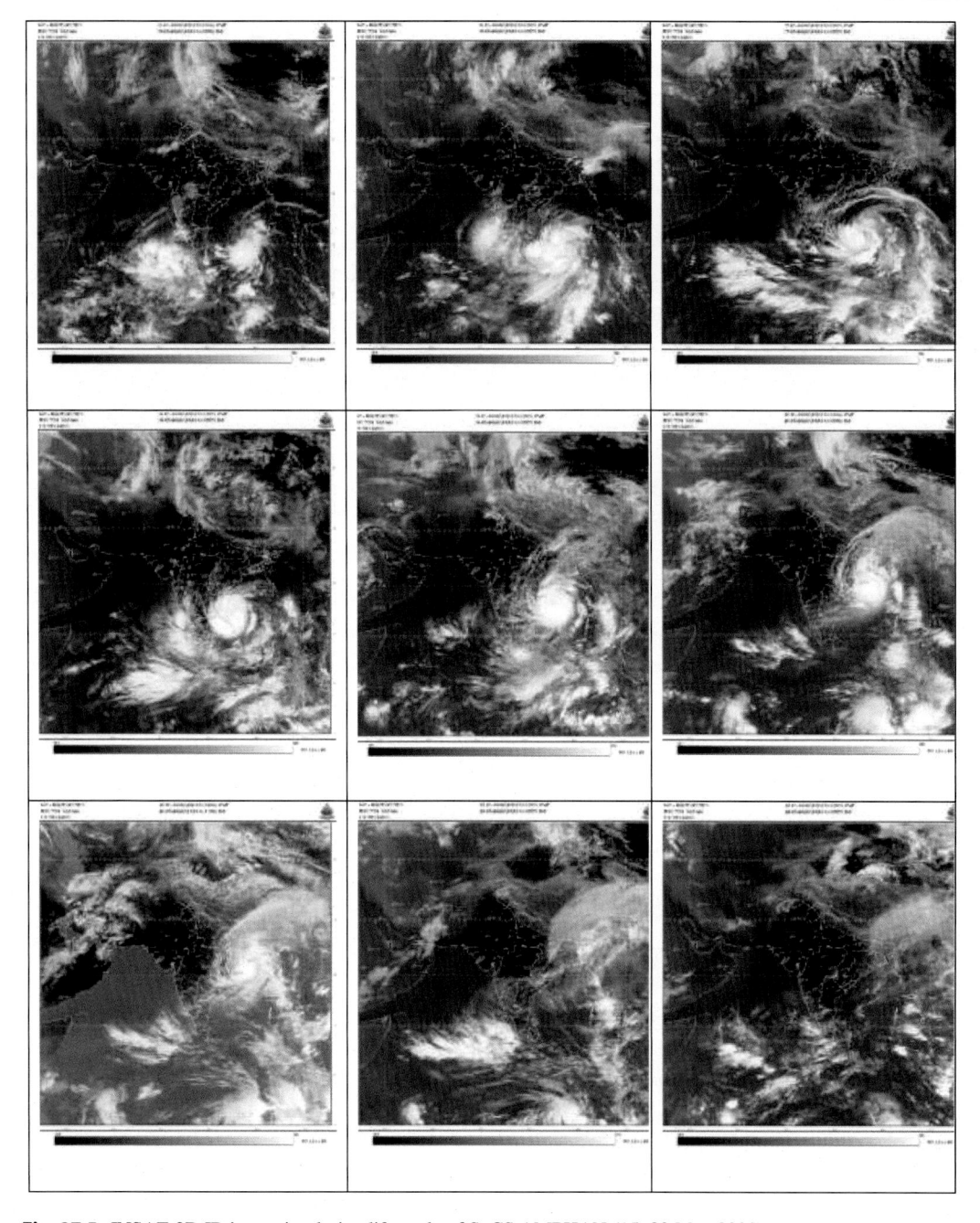

Fig. 27.5 INSAT-3D IR imageries during life cycle of SuCS AMPHAN (15–22 May 2020). *Source* IMD, New Delhi

27.3.2.2 Synoptic Chart and Bouys Data

Ministry of Forest and Environment (MoEF) has installed Bouy over different locations in Bay of Bengal and Arabian Seas. Bouys are permanent stations in oversea areas having fixed latitude and longitude. They provide weather data during cyclone. It helps to determine the intensity and possible path of movement of TCs.

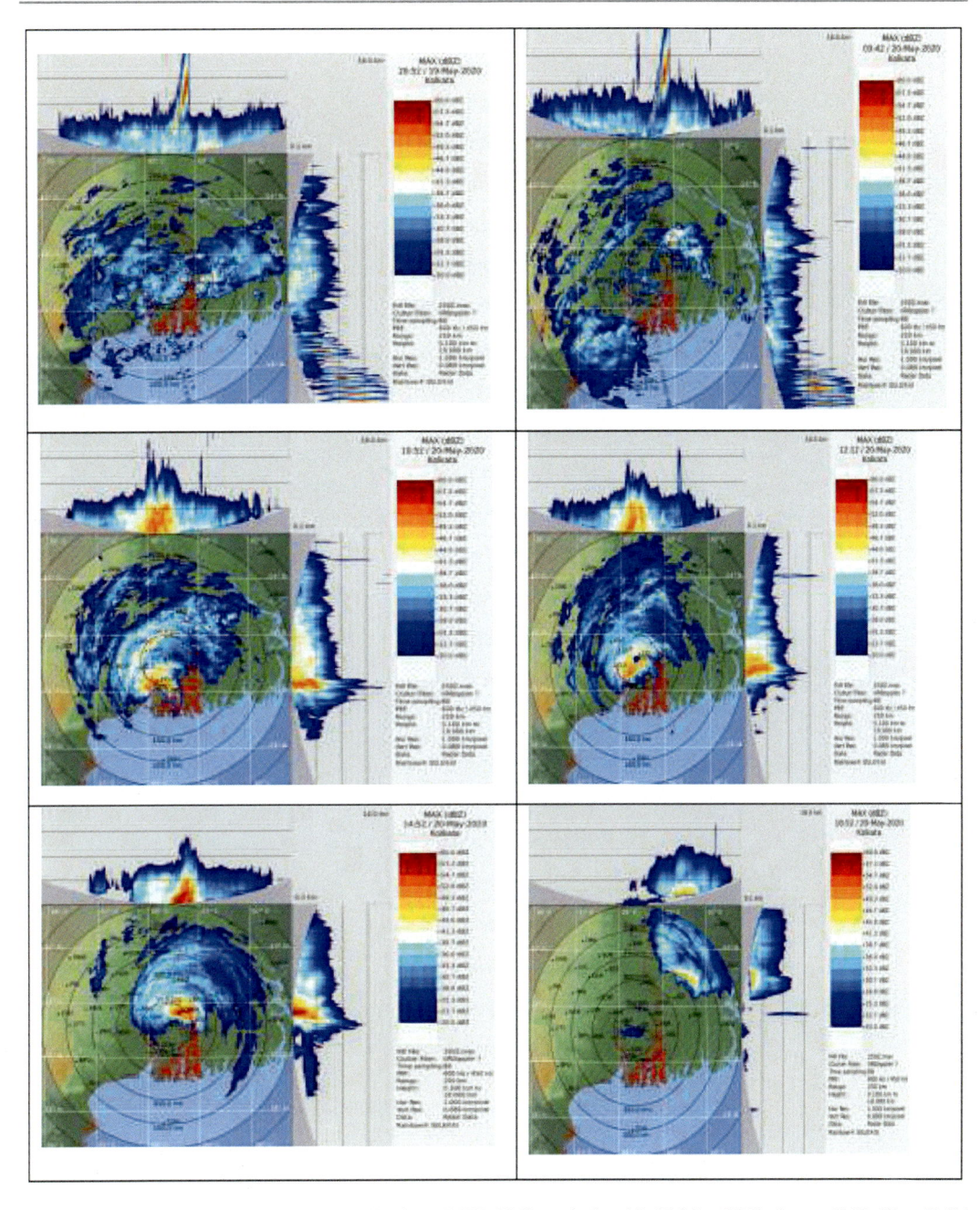

Fig. 27.6 Typical Radar Max dBZ imageries from DWR Kolkata during 19–20 May 2020. *Source* IMD, New Delhi

27.3.2.3 Radar

Radar is the most important and best tool to monitor the structure and movement of TCs near coastal regions. IMD RADARs cover all the coastal regions as well as inland of India. *Amphan* was recorded and monitored using IMD's Doppler Weather Radars. The radars are located at Visakhapatnam, Gopalpur, Paradeep, Chandipur, Kolkata, and Agartala. The RMSC, Kolkata has four RADARs, located in Kolkata, Paradeep, Gopalpur, and Patna (Fig. 27.6).

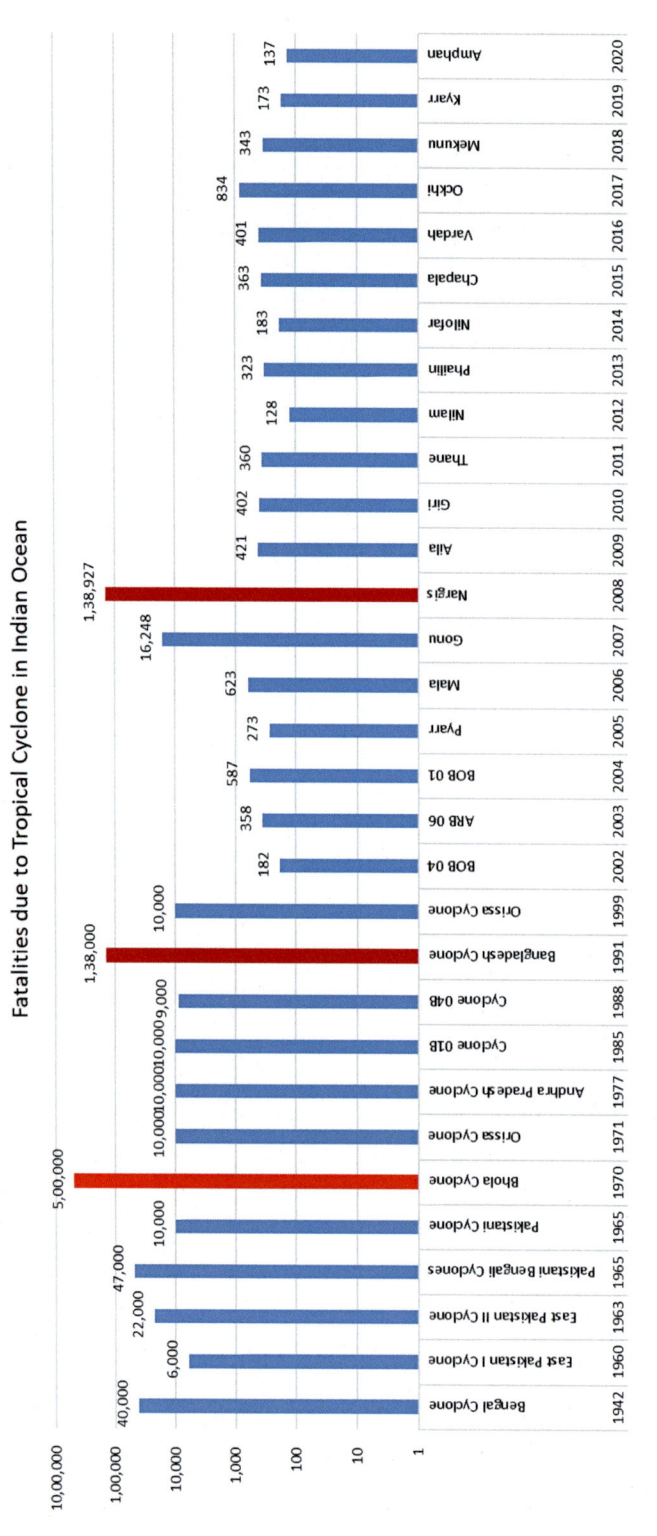

Fig. 27.7 Fatalities in recent past in the Indian Ocean Region

Table 27.4 A few tropical disturbances in Bay of Bengal

Year	Name	Fatalities
1942	Bengal Cyclone	40,000
1960	East Pakistan I Cyclone	6000
1963	East Pakistan II Cyclone	22,000
1965	Pakistani Bengali Cyclone	47,000
1965	Pakistani Cyclone	10,000
1970	Bhola Cyclone	500,000
1971	Orissa Cyclone	10,000
1977	Andhra Pradesh Cyclone	10,000
1985	Cyclone 01B	10,000
1988	Cyclone 04B	9000
1991	Bangladesh Cyclone	138,000
1999	Orissa Cyclone	10,000
2002	BOB 04	182
2003	ARB 06	358
2004	BOB 06	587
2005	Pyarr	273
2006	Mala	623
2007	Gonu	16,248
2008	Nargis	138,927
2009	Aila	421
2010	Giri	402
2011	Thane	360
2012	Nilam	128
2013	Phailin	323
2014	Nilofar	183
2015	Chapala	363
2016	Vardah	401
2017	Ockhi	834
2018	Mekunu	343
2019	Kyarr	173
2020	Amphan	137

Source Authors' Compilation from various sources

27.3.2.4 Numerical Weather Prediction (NWP)

Numerical Weather model is the state-of-the-art tool and technique for weather forecasting. It is very useful to determine the intensity and movement of TCs. IMD uses supercomputer for NWP. Presently, IMD is predicting weather up to 10-days in advance with reasonable accuracy (Fig. 27.7).

There are a few operational models for TCs tract prediction and storm surge such as T-254 model of NCMRWF, MM5 mesoscale model; Quasi-Lagrangian Limited Area Model (QLM) for track prediction; Weather Research

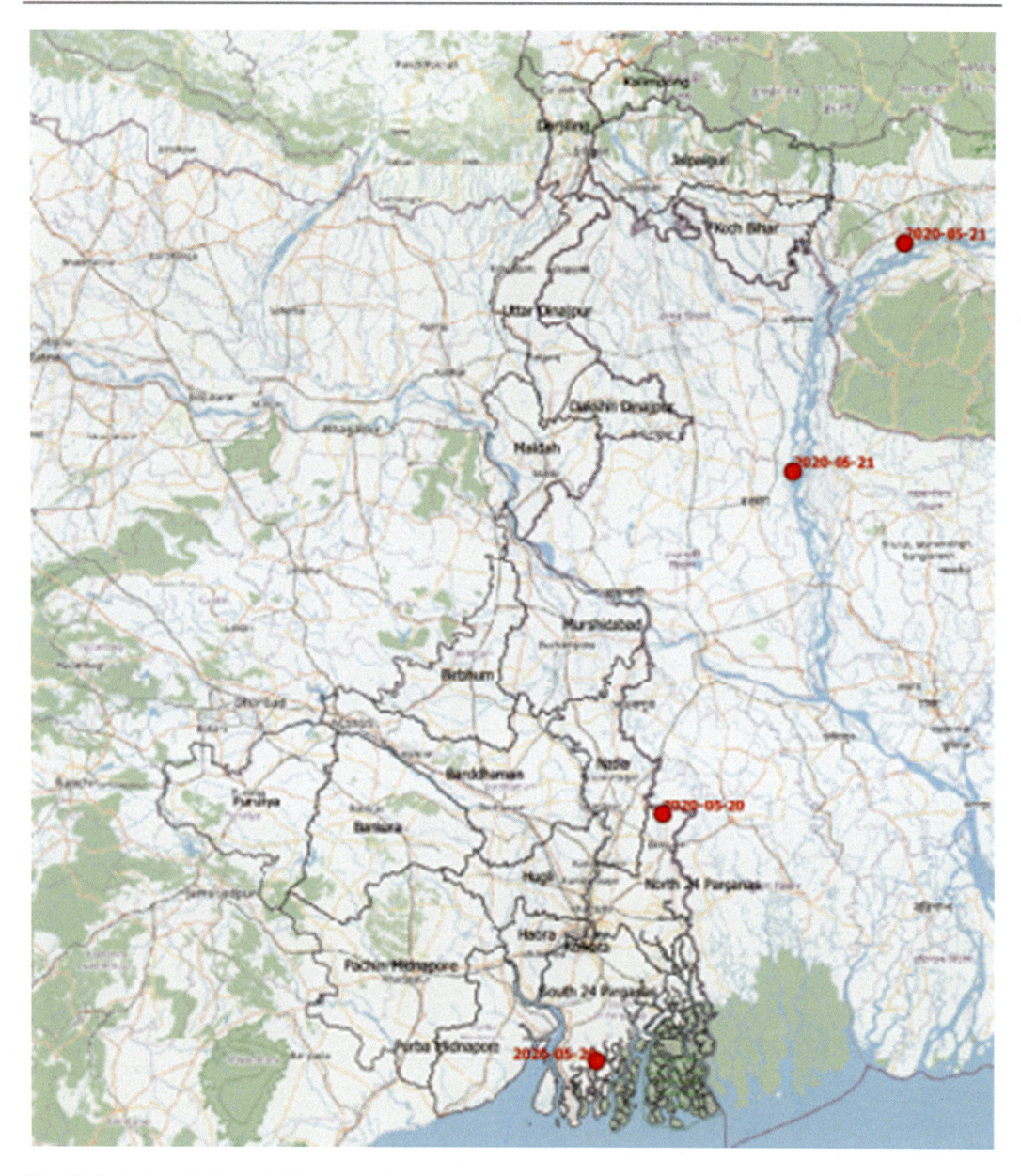

Fig. 27.8 A close look at AMPHAN path in West Bengal. *Source* Authors' Calculation

and Forecast (WRF) mesoscale model for intensification and track prediction; and Prediction Models of IIT—Delhi and NIOT Chennai. In addition to above, IMD forecasters make use of various forecasts available from international NWP models like BCBCMRF, UKMET, and COLA, etc. Probability of correct forecast decreases with increasing forecast validity period, i.e., the how many hours before the prediction is made. Forecasts are made before 12, 24, 48, and 72 h with different stages.

The pre-cyclone watch is issued by the name of the Director-General of Meteorology. It is issued irrespective of the distance from the coast.

Cyclonic Alert is issued at least 48 h before and the distance from the coasts is beyond 500 km. In every three hours, the bulletin is issued by IMD.

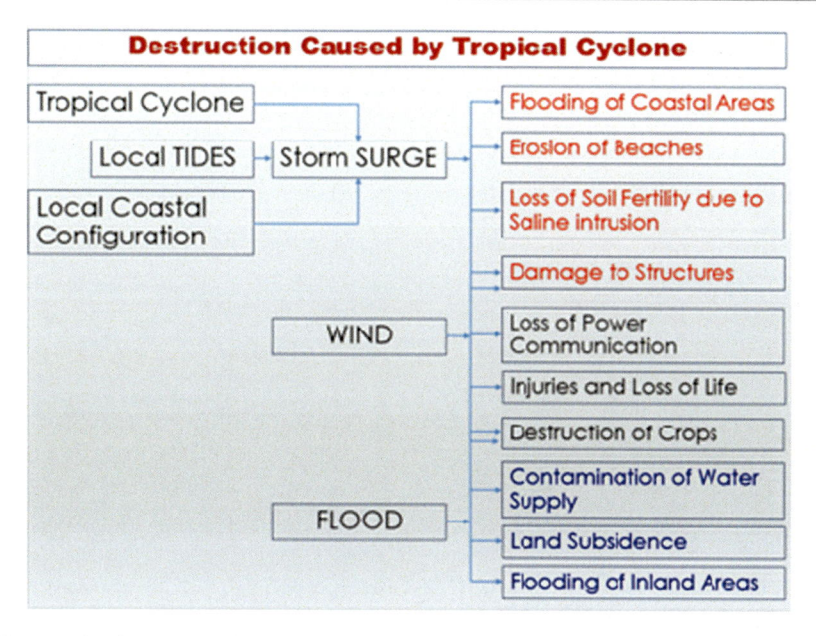

Fig. 27.9 Diagram showing nature of destructions caused due to tropical cyclone. *Source* Authors' Calculation

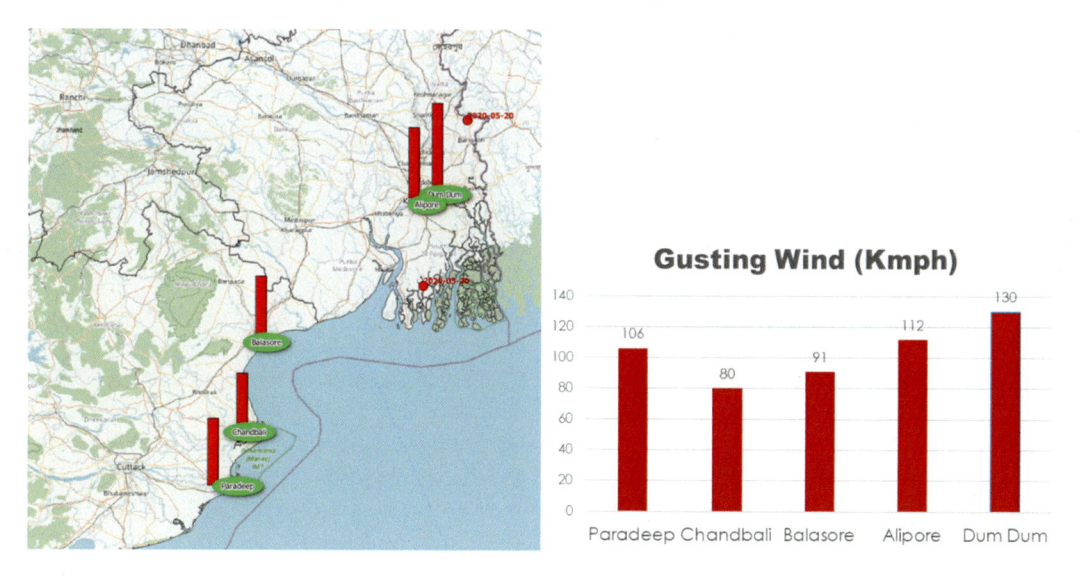

Fig. 27.10 Gusting winds as recorded. *Source* IMD, New Delhi, and Authors' Calculation

Cyclonic Warning is issued at least 24 h before the landfall and distance from the coasts is about 500 km. This bulletin includes possible landfall site.

Post landfall outlook is issued 12 h before the landfall and the cyclone is located within 200 km from the coast. This is associated with the exact landfall site and associated bad weather including storm surges.

Finally, a *De-Warning* message is issued when the Tropical Cyclone weakens into Depression stage.

Table 27.5 Rainfall received in Gangetic West Bengal region during *Amphan* on 20th May 2020

IMD Stations	% of monthly rain
Alipore	436.36
Dum Dum	322.58
Harinkhola	270.83
Digha	187.50
Mohanpur	155.56
Debgram	141.30
Kharagpur	129.63
Midnapore	129.63
Burdwan	108.70
Lalgarh	106.06
Manteswar	100.00
Bankura	83.33
Suri	79.55
Mangalkote	76.09

Source Authors' Calculation

The forecast is associated with bulletins and warnings covering every possible stakeholder in the sea area; coastal area, Indian Navy, Fisheries and Fishermen, Port, Aviation, departmental exchange, All India Radio, *Doordarshan* (TV), and press, registered designated users. The bulletins and information are disseminated using any possible media, press, TV, Facebook, WhatsApp, etc.

27.3.3 A Few Destructive Tropical Cyclones in Indian Ocean

The 1942 TCs hit India-Bangladesh border near Sunderban and caused total casualty of 40,000 people. In 1960, "*East Pakistan I Cyclone*" killed about 6000 human lives in East Pakistan (present Bangladesh). The same region was again hit by "*East Pakistan II Cyclone*" on 23rd May 1963, which resulted in 22,000 deaths. In the year 1965, "*Pakistani Bengali Cyclone*" hit twice on 11th May and 1st June resulting in 47,000 human losses. The biggest casualty in cyclone-related deaths occurred in 1970 with "*Bhola Cyclone*", which killed between 300,000 to 500,000 people in East Pakistan (Bangladesh). In 1971, "*Odisha Cyclone*" killed 10,000 people only in Cuttack. Andhra Pradesh Cyclone caused 10,000 human lives lost in 1977. Cyclone *01B*, hit Bangladesh on 25th May 1985 and resulted in 10,000 human casualties. Again in 1988, Cyclone *04B*, hit Bangladesh on 26th November. Associated storm surges, floods, and other disasters cost 9000 human lives. Bangladesh Cyclone was one of the greatest disasters in recent times. In 1991, it killed 138,000 lives. Most recently, in 1999, we witness "*Odisha Super Cyclone*", which killed no less than 10,000 human lives. Cyclone "*Sidr*" hit Bangladesh on 15th November 2007 to kill about 3500 people. Cyclone "*Nargis*" hit Myanmar in 2008 killing about 130,000 human lives (Table 27.4).

After this, the cyclone prediction system in the Indian Ocean region has improved dramatically and the New Delhi, *Mausam Bhavan* predicted the landfall site with distinct accuracy. This helped the administration with sufficient time to transfer the local resident to cyclone-resistant shelters. The *Amphan* is the recent one in this region that

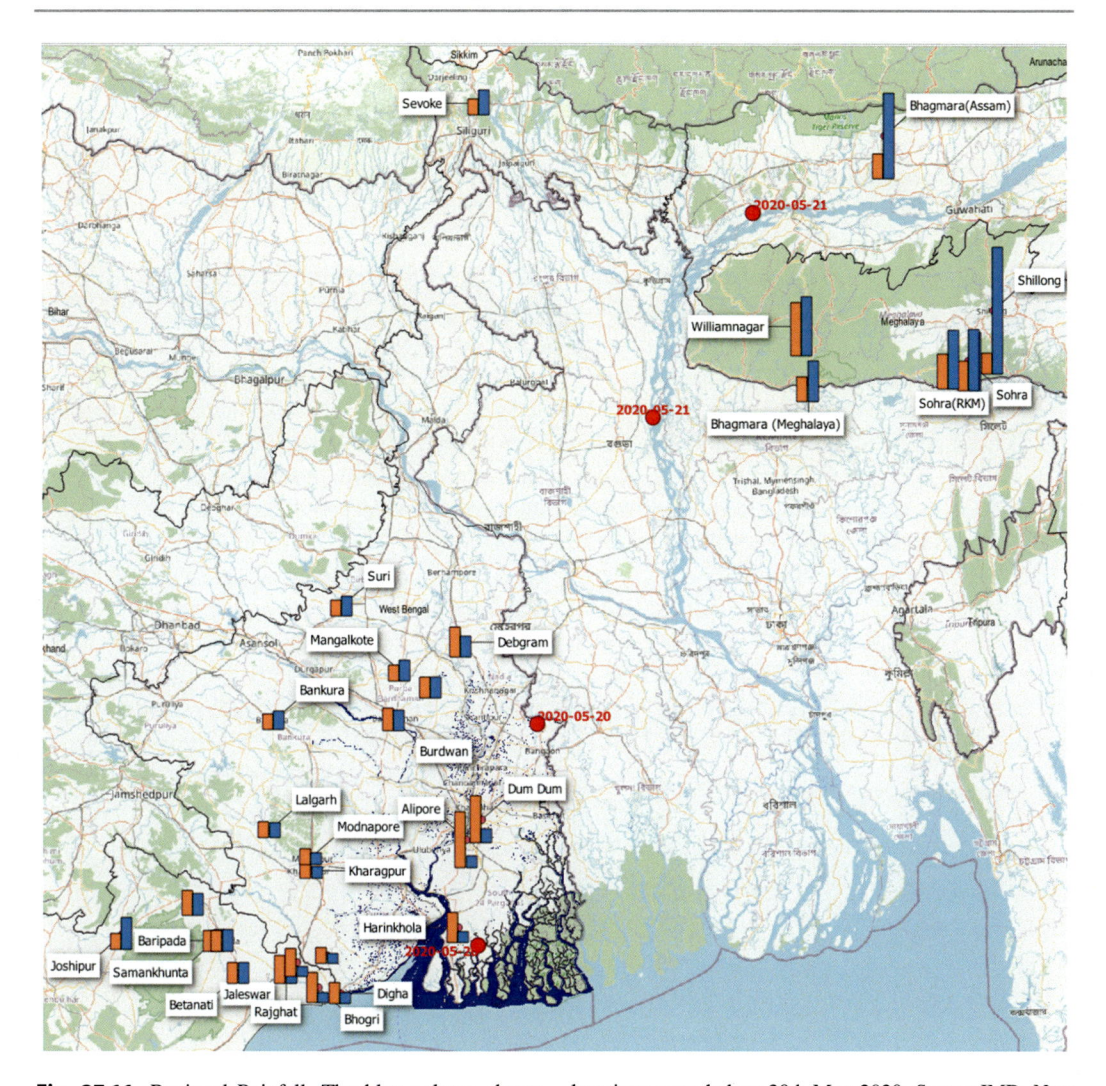

Fig. 27.11 Regional Rainfall, The blue column shows the Normal rainfall in the month of May and The saffron color shows the rainfall in the past 24 h in different locations recorded on 20th May 2020. *Source* IMD, New Delhi, and Authors' Calculation

traversed West Bengal during 20th May to 21st May 2020 (Fig. 27.8).

27.4 Hazards Associated with Tropical Cyclone *Amphan*

Tropical cyclones are associated with various kinds of hazards (Fig. 27.9). The associated hazards with TCs *Amphan* are storm surge, strong winds, high-intensity rainfall, and floods.

27.4.1 Storm Surge

TCs bring huge or abnormal rise in the sea level at the cost. It floods a large area with saline water, meaning destroying agricultural lands in the coastal region. The West Bengal coast is characterized by numerous creeks and channels. The embankments of these creeks and channels breach to destroy the life and properties of the region. *Amphan* brought a maximum storm surge of 4.6 m height. It inundated vast low-lying areas of South 24 Parganas District and eastern part of East Medinipur District of West Bengal.

27.4.2 Strong Wind

Most of the destruction cases by TCs are due to their strong gusty winds, which are in circular motion. It easily topples fences, sheds, trees, power poles, and communication systems. Loss of human lives is mainly due to collapsing of buildings.

Realized wind velocity due to *Amphan* is recorded by the meteorological stations of IMD. In West Bengal, the Dum Dum (Kolkata) station has reported maximum wind speed of 130 kmph at 1855 h IST and Alipore (Kolkata) at 112 kmph at 1752 h IST on 20th May 2020. While on the same day, in Odisha, maximum sustained wind speed in Paradip station was 106 kmph at 0630 h; in Chandbali it was 80 kmph at 0830 h.; and in Balasore it was 91 kmph 1330–1430 h. on 20th May 2020 (Fig. 27.10). Thousands of trees, electric poles, sheds, and *kuchha* houses were damaged. Many places (even in Kolkata) were in power cut condition for several days hampering normal life.

27.4.3 Rainfall and Associated Flood

TCs are always associated with intense and heavy rainfall. Heavy and prolonged rain always causes inundation in the low-lying as well as coastal areas. It is important to note that most of the rain recording stations of Gangetic West Bengal region received single-day rainfall (past 24 h on 20th May 2020) which is much more than the monthly average (Table 27.5). To name a few stations, i.e., Alipore, Dum Dum, etc. We can plot and compare the rainfall amount of 20th May 2020 with the monthly average on the map (Figs. 27.11 and 27.12). The excessive rainfall

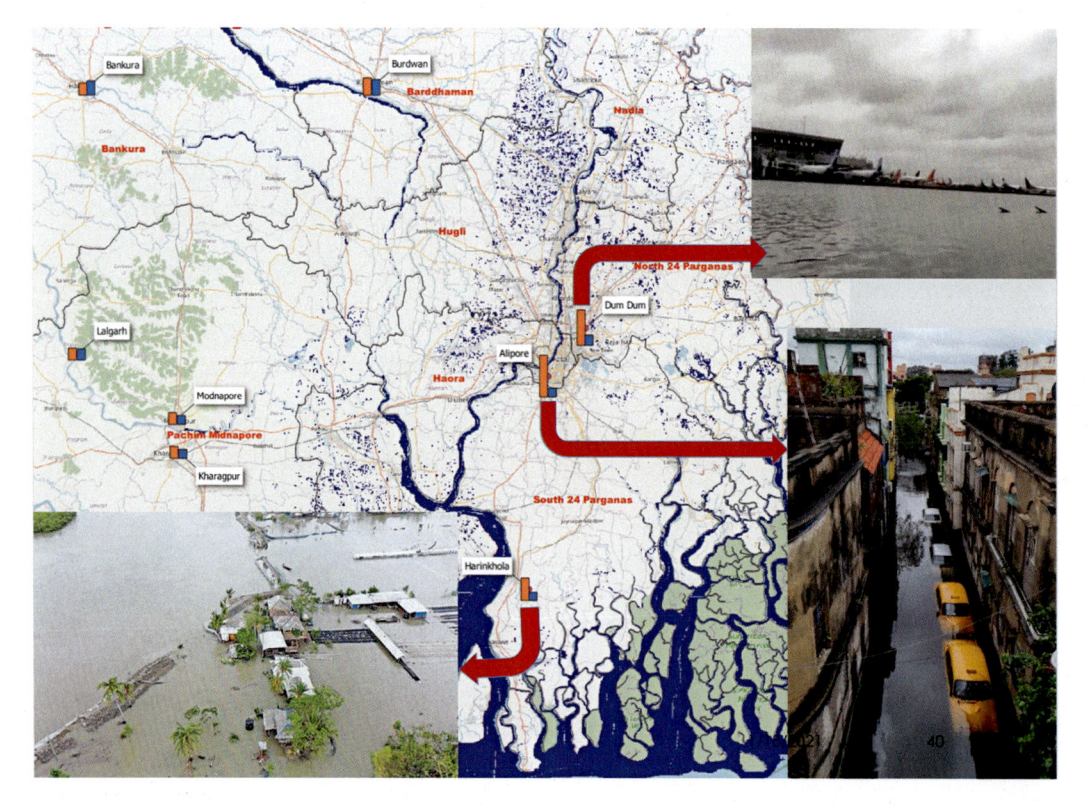

Fig. 27.12 *Amphan* related rainfall caused extensive flooding on 20th May 2020. *Source* Authors' Calculation and News Papers' Photographs

caused extensive flooding almost in the entire Gangetic West Bengal. Close look at the rainfall pattern shows that Alipur received 436.36% of monthly rainfall on 20th May 2020 and Dum Dum received 322.58%. Digha (WB Coast) received 187.50% rainfall than its average monthly total rainfall. Tropical Cyclone-*Amphan*–affected millions of population through the gusty winds as well as extensive flooding.

27.5 Conclusion

From the newspapers and other media reports, we could learn that *Amphan* damaged about 5,500 homes in North 24 Parganas. Hooghly district suffered loss of thousands of mud homes. Similarly, millions of homes were destroyed in South 24 Parganas. The storm surge breached the embankments and flooded numerous villages and along with its standing croplands. *Amphan* caused the damage to 88,000 ha of rice paddies and 200,000 ha of vegetables in West Bengal. India Meteorological Department played very important role in every step of its development. It issued notices, bulletins, and other warnings, etc. in timely manner. The state administration, state disaster management authority, national disaster management authority, and local administrative bodies. NGOs and others were well informed and prepared to deal with any eventualities. Although there were a few deaths, but a major causality has been avoided.

References

https://www.thehindu.com/news/national/other-states/amphan-cyclone-tracker-may-20-2020/article31629032.ece

Cyclone Amphan [LIVE]: Wind velocity in Kolkata clocks 112 kmph as cyclone crosses WB; 2 people dead in state", *TimesNowNews.com*. Bennett, Coleman & Company. Times Now Digital. 20 May 2020. Retrieved 22 May 2020

Subrata N, Ruma P (2020), Cyclone kills 14 in India, Bangladesh leaving trail of destruction, Reuters. Retrieved 20 May 2020

Mamata pegs cyclone Amphan damage at ₹1 lakh crore, toll rises to 86', *The Hindu*. THG Publishing. 23 May 2020. Retrieved 23 May 2020

72 killed in Amphan's march through Bengal, PM Modi to visit today", *The Times of India*. Bennett, Coleman & Company. Times News Network. 22 May 2020. Retrieved 21 May 2020

IMD, Gov. of India, press release bulletin 13th May to 22th May, 2020

Misra SP, Sethi KC, Ojha AC, Barik KK (2020) Fani, an outlier among pre-monsoon Intra-Seasonal cyclones over Bay of Bengal. Int J Emerg Technol 11(2):271–282

Mathur DK, Vora AM, Udani PM (2016) Role of remote sensing and GIS in cyclone. J Pure Appl Ind Phys 6(8):116–127

Sahoo B, Bhaskaran PK (2015) Assessment on historical cyclone tracks in the Bay of Bengal, east coast of India. Int J Climatol. Published online in Wiley Online Library, wileyonlinelibrary.com. https://doi.org/10.1002/joc.4331

Abbott PL (1996) Natural Disasters. C. Brown Publishing Co., Wm, p 438

Observed Changes in the Precipitation Regime in the Argentinean Patagonia and Their Geographical Implication

Paula B. Martin, Victoria A. Oruezabal, and María E. Castañeda

Abstract

The climate of Argentine Patagonia is characterized by extremely dry air, as a result of low rainfalls and high evaporation rates related to strong winds. Over the last 25 years, the northern Patagonian region has been affected by droughts that have threatened the operation of hydroelectric plants and the supply of drinking and irrigation water, especially in the fruit production areas of the region. However, not many research studies on rainfall variations have been carried out over Patagonia, due to the lack of continuous official records and the low density of meteorological stations in the region. This work aims to study changes in the rainfall regime and existing trends in the Patagonian zone. Monthly rainfall records of 23 meteorological stations in the Argentinean Patagonia from Argentina's national meteorological service and the Interjurisdictional Authority of the Limay, Neuquén and Negro River Basins for the period of 1980–2016. The Gamma distribution has been used to fit the observed monthly frequency distributions. Fits were tested by the non-parametric Kolmogorov–Smirnov´s test. The shape and scale parameters show seasonal and geographic variation. As the first result of this study, negative trends are shown in the west and central zone.

Keywords

Precipitation · Patagonia · Gamma distribution · Trends

P. B. Martin (✉)
Ministerio de Defensa, Servicio de Hidrografía Naval, Buenos Aires, Argentina
e-mail: martinpaulabeatriz@gmail.com

P. B. Martin
Departamento de Geografía, Facultad de Filosofía y Letras, Universidad de Buenos Aires, Buenos Aires, Argentina

P. B. Martin · M. E. Castañeda
Consejo Nacional de Investigaciones Científicas y Técnicas, Buenos Aires, Argentina

V. A. Oruezabal · M. E. Castañeda
Departamento de Ciencias de la Atmósfera y los Océanos, Facultad de Ciencias Exactas y Naturales, Universidad de Buenos Aires, Buenos Aires, Argentina

28.1 Introduction

Patagonia is a wide region in southern South America, on both sides of the Andes Cordillera. According to Prohaska (1976), the territory is set between the semi-permanent anticyclones of the Pacific and Atlantic oceans at approximately 30° S and the subpolar low-pressure belt at around 60° S. The Andes obstruct the zonal atmospheric transport of humid air from the Pacific Ocean, emphasizing the strong differences between the dry region east of the Andes and the humid side of Chile.

Patagonia is influenced by strong mid-latitude westerlies throughout the year (Garreaud et al. 2009). The prevailing westerlies throughout the year are responsible for the mean spatial precipitation distribution. The climate of Eastern Patagonian (EPAT) is described by its extreme dryness due to both low rainfall and high potential evapotranspiration caused by intense winds (Prohaska 1976). Agosta et al. (2019) studied the relationship between daily precipitation and the persistence of the east–west component of the wind measured at the weather station, along the Patagonian Atlantic coast, associated with synoptic conditions and its relationship with El Niño–Southern Oscillation. Westerly winds are characterized not only by their persistence during the year but also by their strong intensity. The seasonal meridional shift of subtropical cells and the westerly belt cause a more uniform distribution of the westerlies in winter, and a more southerly component in summer (Paruelo et al. 1998).

From the climatic perspective, Argentinean Patagonia has been an understudied area. However, Northern Patagonia is a region of special interest because of the water resources importance, such as hydropower generation to the national grid, water supply for the development of local livelihoods, the fruit production to satisfy the needs of both export and local markets and the mining activity. Therefore, the monitoring and subsequent forecasting of meteorological variables such as precipitation, are of great importance. This area has been explored by Romero and Gonzalez (2016), Gonzalez et al (2020) and Romero et al (2020).

On the Patagonian plateau, the climate is characterized by the extreme dryness of the air as a result of excess evaporation compared to precipitation. South 35° S, high rainfall over the western slope of the Andes and semi-arid conditions are observed straightway to the east leading to the temperate steppes of Argentina's Patagonia. Some authors have studied low-frequency variability of rainfall especially in Patagonia (Castañeda and Gonzalez 2008; Barros and Mattio 1978; Barros and Rodriguez Sero 1979) and observed a progressive increase in rainfall in northern Patagonia and a decrease in the mountainous west.

Climate classifications usually use the arithmetic average of the monthly, seasonal, or yearly accumulated rainfall amounts. Few of them take directly into account a characteristic that is fundamental in the rainfall of arid climates, which is the extreme variability of rainfall. The mean value is a representative parameter only of the frequency distributions of variables that have a normal or Gaussian distribution, and it is well known that one of the most salient characteristics of the frequency distributions of rainfall is their asymmetry. This is usually accentuated in the curves of rainfall recorded in arid and semi-arid regions, in which the arithmetic average, although calculated over a long period, loses all statistical significance. In these cases, it is essential to consider the incidence of rain through the analysis with theoretical frequency distributions other than Gaussian. In these cases, it is essential to consider the incidence of rainfall through the analysis with theoretical frequency distributions different from Gauss's. The two-parameter Gamma distribution (Thom 1958) is one of the most widely disseminated functions for representing precipitation frequencies. Martin and Serio (2007) studied the distribution of rainfall in the Argentine Patagonian region in the period 1961–2000, by fitting the monthly frequency distributions observed to the Gamma distribution function. The analysis of rainfall variability contributes to improve the efficiency of hydroelectric power dams and to prevent the negative impacts on productive activities derived from periods of drought or excess water. In this chapter, changes in the rainfall regime and the existing trends in the Patagonian zone have been studied.

28.2 Materials and Methods

Monthly rainfall data from 23 stations from The National Meteorological Service of Argentina and the Interjurisdictional Authority of the

Limay, Neuquén and Negro River Basins measurement networks during the period 1980–2016 were used in this study (Table 28.1).

28.2.1 Koppen-Geiger Climate Classification

To regionalize the area, the Köppen-Geiger climate classification was used. The Köppen-Geiger system classifies climate into five main classes and 30 sub-types. The classification is based on threshold values and seasonality of monthly air temperature and precipitation. The latest version of the Koppen-Geiger classification as done by Peel et al (2007) was used to divide the area within the same precipitation regime. This updated version is based on the division Köppen made in 1936, and has the philosophy to erase any non-objective decision, using a large and complete database (12,396 precipitation and 4844 temperature stations, with at least 30 observations for each month.). Data was interpolated by a simple and flexible methodology.

According to the late one classification, the Patagonia area can be regionalized into 7 sub-regions:

- BSK: Arid Desert Cold
- BWK: Arid Steppe Cold
- CFA: Temperate without a dry season and hot Summer
- CFB: Temperate without a dry season and warm summer

Table 28.1 Location and data record for the stations used to perform this study

Station	Latitude (°s)	Longitude (°w)	Height (M)	Period (Y)
Tres Arroyos	38.20	60.15	115	1980–2016
Neuquén	38.57	68.08	271	1980–2016
Bahía Blanca	38.44	62.10	83	1980–2016
Bariloche	41.09	71.10	840	1980–2016
Maquinchao	41.15	68.44	888	1980–2016
San Antonio	40.47	65.06	20	1980–2016
Viedma	40.51	63.01	7	1980–2016
Chapelco Aero	40.05	71.08	779	1991–2016
Esquel	42.56	71.09	797	1980–2016
Trelew	43.12	65.16	43	1980–2016
Comodoro Rivadavia	45.47	67.30	46	1980–2016
Gobernador Gregores	48.47	70.10	358	1980–2016
El Calafate	50.16	72.03	204	1980–2016
San Julian	49.19	67.47	62	1980–2016
Río Gallegos	51.37	69.17	19	1980–2016
Río Grande	53.48	67.45	22	1980–2016
La Angostura	40.76	71.64	900	1980–2016
Perito Moreno Aero	46.31	71.01	429	1983–2016
Paso de Indios	43.49	68.53	460	1980–2016
El Bolsón Aero	41.58	71.31	337	1980–1985 1991–2016
Rahue Estafeta	39.22	70.56	845	1980–2016
Tolhuin	54.42	67.15	105	1990–2010
Ushuaia Aero	54.48	68.19	57	1990–2016

- CFC: Temperate without a dry season and cold summer
- CSB: Temperate dry summer and warm summer
- CSC: Temperate dry summer and cold summer

But for this paper's purpose, it will be subdivided into three sub-regions following the regions above but focusing on precipitation only:
1. Areas with temperate temperature and without a dry season. It's located in the south and north Patagonia. Mostly CFA with some CFB and CFC.
2. Arid areas with, in the most of its extension, steppe and cold temperature. It is located in the centre and coastal Patagonia. Mostly BSK with two of them being BWK.
3. Areas with temperate temperature and with a dry summer. It is located in west Patagonia, near the cordillera. Mostly CSB and with one in CSC.

This regionalization is compatible with previous studies. Berman et al. (2012) divided east Patagonia according to their precipitation regime. Results indicate that the central-north areas, the southern continental region, and the southernmost islands are three independent regions of seasonal precipitation, and that each of them is associated with specific patterns of atmospheric circulation (Fig. 28.1).

28.2.2 Gamma Distribution

The Gamma Distribution which has a zero lower bound has been found to fit several such variables well. It is defined by its frequency or probability density function

$$f(x) = \frac{x^{\alpha-1} e^{-x/\beta}}{\beta^\alpha \Gamma(\alpha)} \quad (28.1)$$

where, β is a scale parameter, α is a shape parameter and $\Gamma(\alpha)$ is the Gamma Function given by:

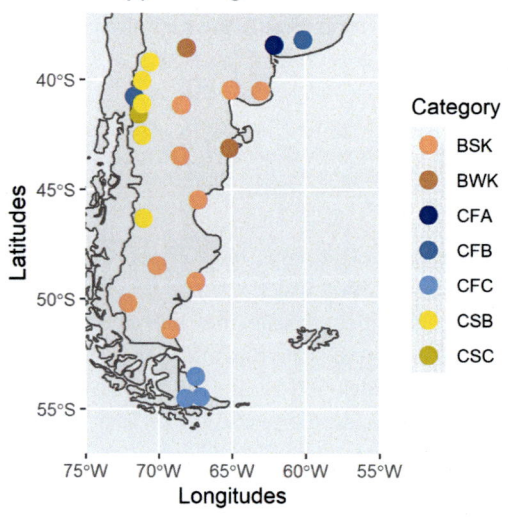

Fig. 28.1 Map of the stations used by their Koppen-Geiger classification. BSK (Río Gallegos, Paso de los Indios, San Julián, El Calafate, Gobernador Gregores, Comodoro Rivadavia, Viedma, San Antonio and Maquinchao), BWK (Trelew and Neuquén), CFA (Bahía Blanca), CFB (Tres Arroyos y La Angostura), CFC (Ushuaia, Tolhuin and Río Grande), CSB (Chapelco, Rahue Estafeta, Perito Moreno, Esquel and Bariloche) and CSC (El Bolsón)

$$\Gamma(\alpha) = \int_0^\infty x^{\alpha-1} e^{-x} dx \quad (28.2)$$

with α and β as constants to be adjusted. This distribution is advantageous over other distributions, because it adapts to skewness and is only defined for positive values of x. These characteristics make it suitable for representing precipitation.

To adjust the frequency distributions, for each month of the year and for the 23 stations, the maximum likelihood method was applied, according to the following algorithm:

$$\alpha = \frac{1 + \sqrt{1 + 4/3A}}{4A} \quad (28.3)$$

$$\beta = \frac{\bar{x}}{\alpha} \quad (28.4)$$

where, $A = \ln(\bar{x}) - \frac{\sum_0^n \ln(x)}{n} x$ is the accumulated precipitation. Since the Gamma Distribution is defined for values strictly greater than zero, the algorithm estimates the condition of replacement values by 0.1 mm, as any smaller value of precipitation is considered null joined.

The main advantages of the Gamma with respect to other distributions such as Gaussian are that it easily adapts to all kinds of asymmetries and is defined only for positive values of the variable x. This makes it suitable for representing frequencies of precipitation. The setting is carried out eliminating the months in which the precipitation was zero. The differences between theoretical and empirical distribution functions were studied by applying the test not Kolmogorov–Smirnov parametric. Then, the theoretical functions were corrected considering in each case the empirical probability of occurrence of zeros in the series of precipitation. Besides, the percentiles of the theoretical distribution functions and their seasonal variability were analysed.

28.3 Results

28.3.1 Scale and Shape Parameters of Gamma Distribution

The spatial distributions of the scale and shape parameters of the Gamma function, respectively, for January, April, July and October, where each month represents a season according to the southern hemisphere (Figs. 28.2 and 28.3). The scale parameter (Fig. 28.3) in the entire region presents the lowest values during January and has a fairly homogeneous distribution with a maximum to the north decreasing towards the south. On the other hand, in April the distribution presents a maximum in the extreme south of the Province of Buenos Aires and the lowest values are found in the south of the region. During October and July, the maximum values were

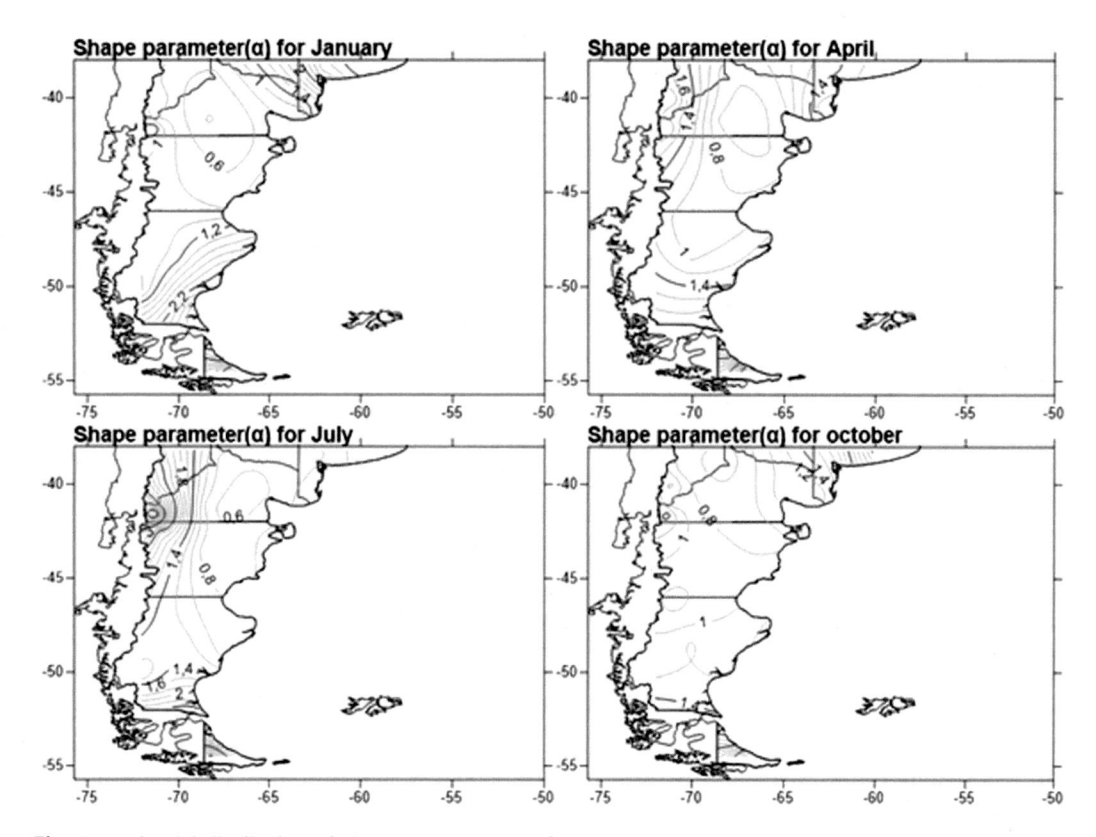

Fig. 28.2 Spatial distribution of shape parameter (α) of the Gamma function fitted to rainfall for January, April (Superior panel), July and October (Inferior Panel)

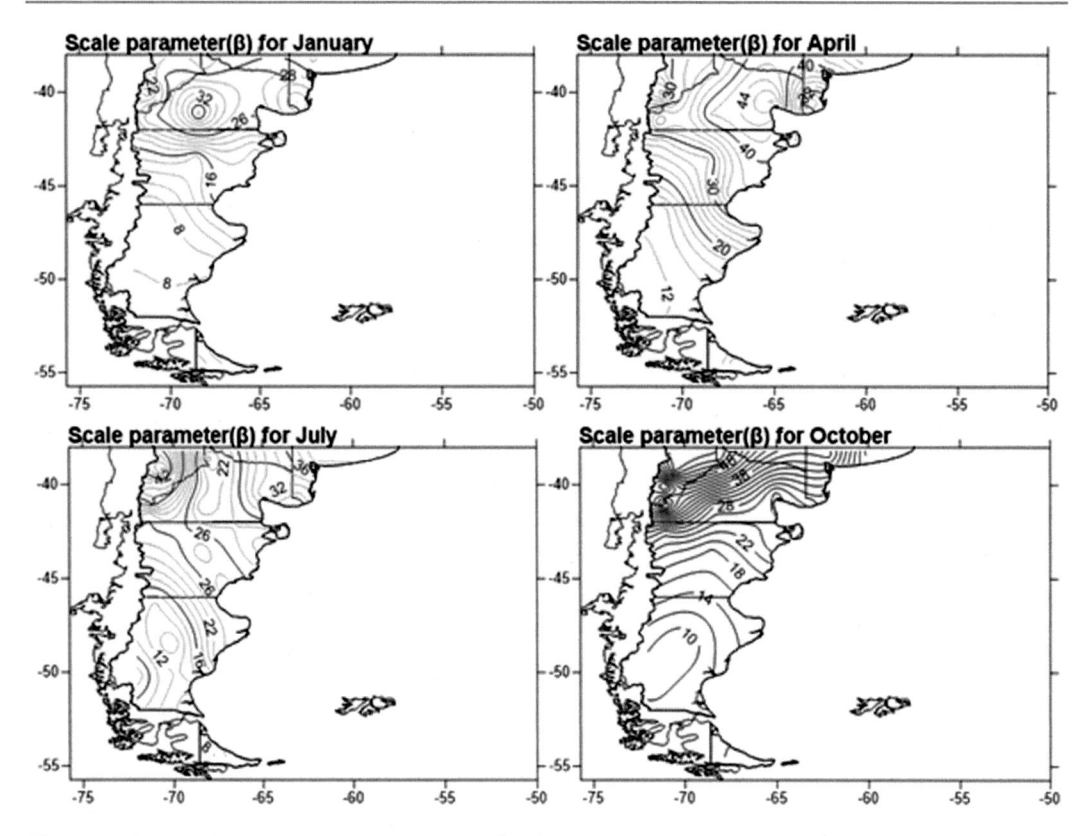

Fig. 28.3 Spatial distribution of scale parameter (β) of the Gamma function fitted to rainfall for January, April (Superior panel), July and October (Inferior Panel)

observed to the south of the Province of Neuquén. The distribution of the shape parameter (α) shows very little variability in January throughout the central-west region, and a strong gradient increasing towards the coast and the extreme south (see Fig. 28.2). In July, the behaviour is opposite; low values in the coastal region and maximum values in the west with the isolines oriented in north–south direction were detected. During the transition months (April and October) the gamma distribution is much more homogeneous, always with maximum values in the south.

The Patagonian coast maintains 10th percentile values close to zero throughout the year (Fig. 28.4). Western Patagonia shares the same configuration in the summer and reverses it towards autumn and winter, in the latter case as a result of the typical rainfall regime of this area. The northern and southern areas show a summer cycle.

All seasons show a minimum in Santa Cruz Province and rising in all directions (Fig. 28.5). North Patagonia exhibits maxima values in January, April and October and west Patagonia in July and April, probably due to the greater variability in extreme conditions.

28.3.2 Box Plot Using Gamma Distribution

The study explored percentiles 10, 25, 50 and 75 and 90% of the Gamma function for some locations over the period 1980–2016 (Fig. 28.6).

The northwest (Bariloche) and northeast (Bahía Blanca) regions show the greatest variability, although different precipitation regimes.

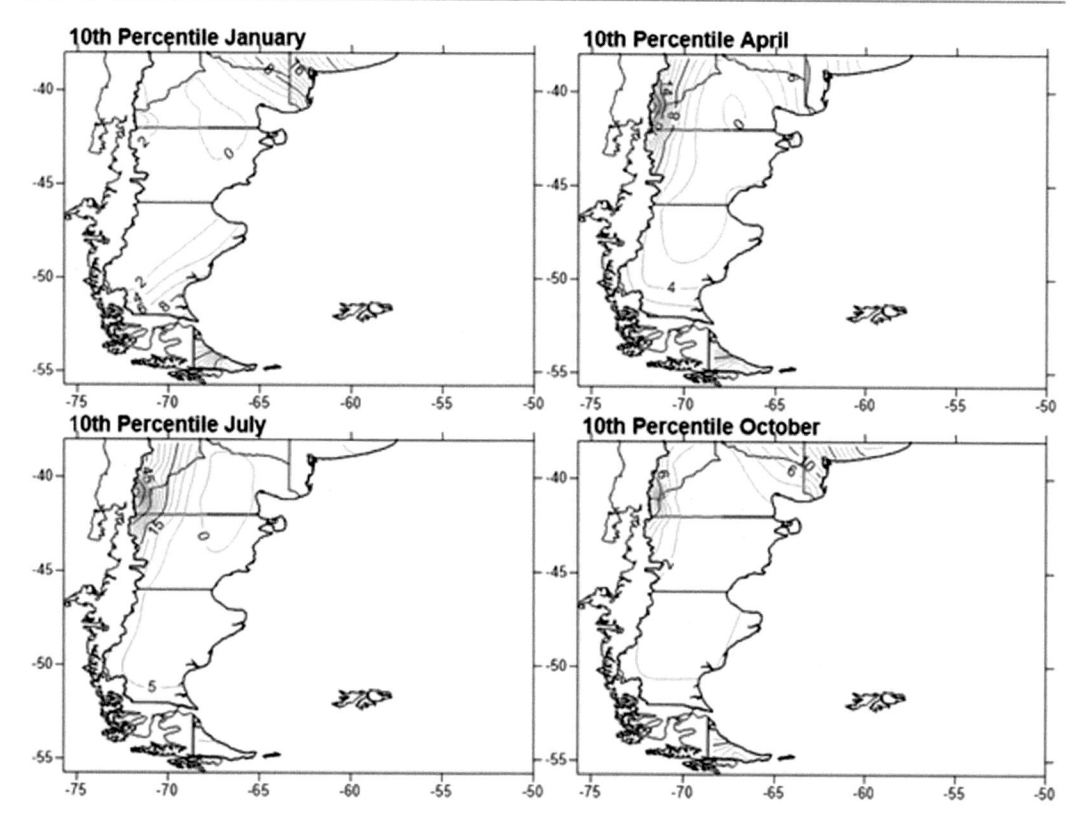

Fig. 28.4 Spatial distribution of the 10th percentile of the Gamma function fitted to rainfall for January, April (Superior panel), July and October (Inferior Panel)

At Bariloche, as well as at Chapelco Aero, El Bolsón Aero and Rahue stations (not shown here) winter rainfall predominate (Köppen-Geiger CSB/CSC), with a median monthly rainfall exceeding 100 mm. In contrast, the regimes in Bahía Blanca and Tres Arroyos (CFA/CFB) (not shown here) exhibit high percentile rainfall during the warmer semester. The transition between them is represented by Maquinchao (BSK) with a double rainy season (autumn and spring).

Apart from this transitional zone, the entire Patagonia east to the Andes has the same rainfall regime, represented by a rainy season from May to September not exceeding 20 mm. An example is Comodoro Rivadavia (BSK). The uniform rainfall distribution throughout the year of the subpolar region is represented by Ushuaia (CFC), reaching 45 and 50 mm during the autumn season. The results found in this study are similar to those reported by Martin and Serio (2007).

28.3.3 Precipitation Linear Trends

Some authors have studied the low-frequency variability of the Patagonia rainfall (Castañeda and González 2008; Castañeda and Barros 1994; Gonzalez and Herrera 2014) and observed a progressive increase in rainfall in northern Patagonia as well as a decrease in the mountainous west.

Accumulated precipitation for wet and dry seasons according to Prohaska (1976) was obtained. A transition zone is to the north, it is expressed in a double rainy season, but only of it, because of the general dryness of this region. The coast and Andes in the extreme south of Patagonia as well as the whole archipelago of Tierra del Fuego exhibit a high frequency of precipitation throughout the year, indicative of the polar character of this completely maritime precipitation regime.

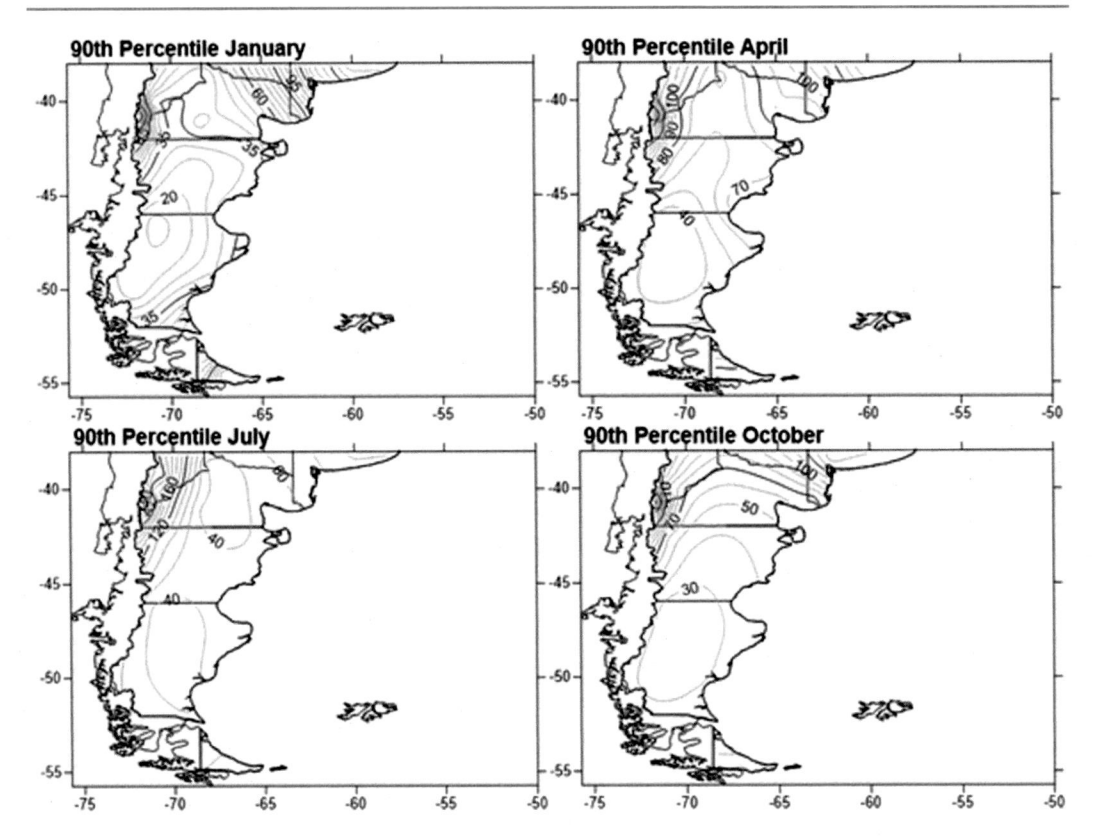

Fig. 28.5 Spatial distribution of 90th percentile of the Gamma function adjusted to rainfall from January, April (Superior panel), July and October (Inferior Panel)

According to this, the first group of the ones described above, has its wet season in summer and spring. The second one in spring and autumn. Finally, the third one has it in winter and autumn. The Pearson correlation coefficient for both seasons against time was estimated and tested with a normal distribution. The null hypothesis is that the correlation coefficient ρ is significantly different from 0. A two-sided test is used with a significance level of 90%. In this case, r is statistically significant if: $-0.27 > r$ or $r > 0.27$.

An annual precipitation linear trends field calculated with observed data registered in measurement stations (Table 28.1 and Fig. 28.7) for the dry and wet season. For the dry season, a significant decrease in precipitation is observed in the central Patagonian region, whilst a slight increase in precipitation is observed on the Patagonian coast. On the other hand, in the wet season (left panel) a centre of maximum decrease in precipitation is observed in the centre and towards the coast of the southern Patagonian region. Negative trends are shown in the west and central zone, but rather significant in the wet season in Tres Arroyos, Gobernador Gregores and La Angostura, whilst in the dry season in Neuquen, Gobernador Gregores and the Río Grande. These results are similar to those found by other authors (Castañeda and Gonzalez 2008; Gonzalez and Herrera 2014).

Therefore, we assume that between 1980 and 2016, an intensification of the dry seasons occurred in northern Patagonia. No changes between seasons have been observed in the southern region (Tierra del Fuego).

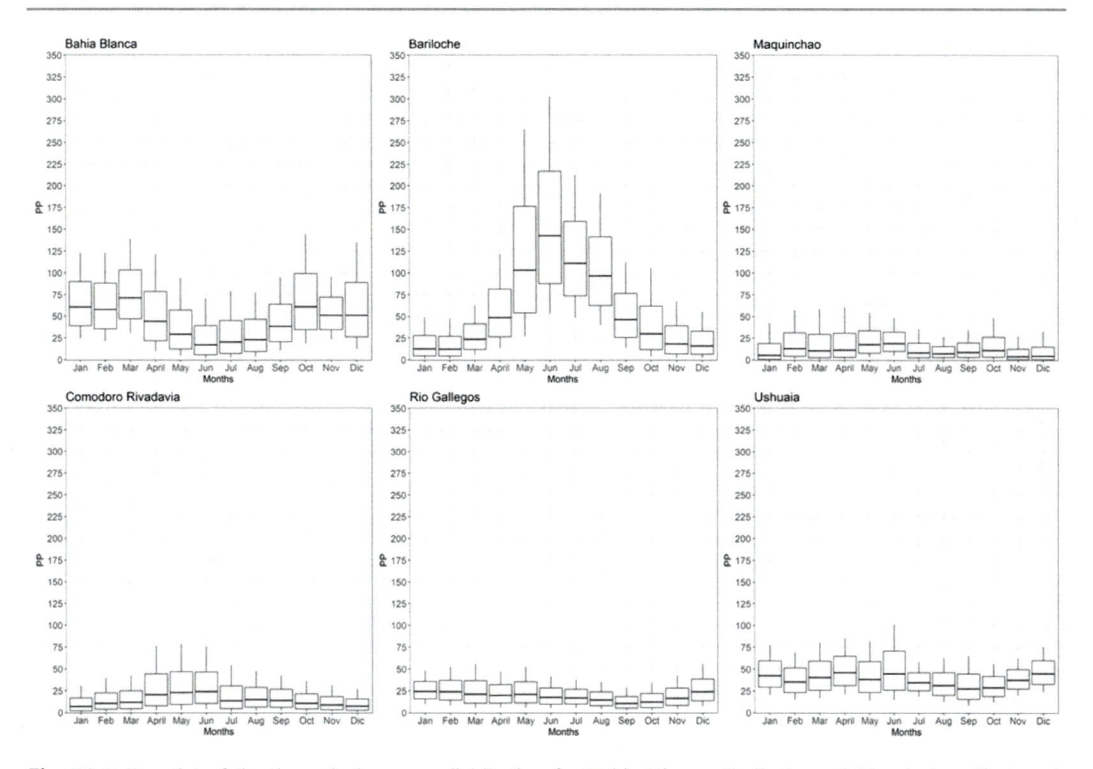

Fig. 28.6 Boxplot of the theoretical gamma distribution for Bahia Blanca, Bariloche and Maquinchao (first row), Comodoro Rivadavia, Rio Gallegos and Ushuaia (second row)

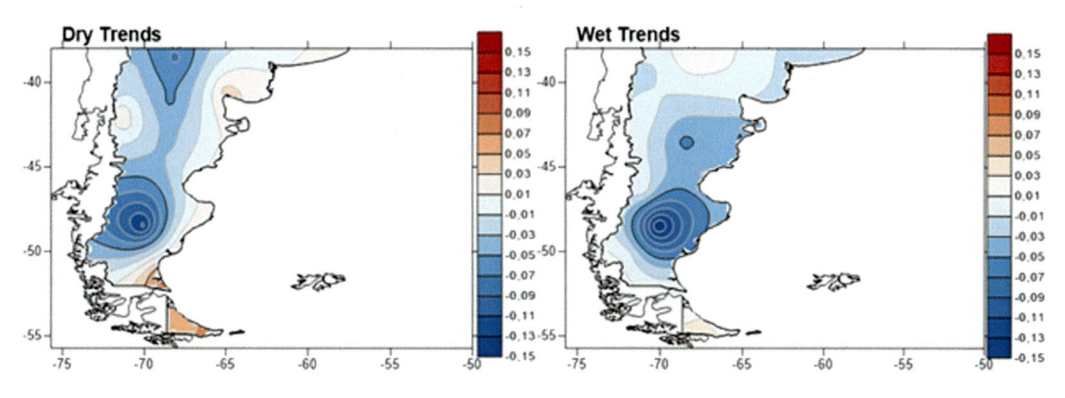

Fig. 28.7 Slope from the linear regression (Linear fit of annual precipitation tends for period 1980–2016)

28.4 Discussion and Conclusions

The parameters of the Gamma function were fitted to the observed monthly frequency distributions of precipitation of the Argentine Patagonian region. We found a good agreement between theoretical and empirical distribution functions. The spatial patterns of shape (Alpha) and scale (Beta) parameters indicated a seasonal variability in the distribution of both parameters. Prohaska's description of the climate of Patagonia provided a basis to characterize rainy and dry seasons and evaluate correlation coefficients with a significance level of 90%. As the first result of this study, negative trends are shown in the west and central zone for both the dry and wet seasons.

Northern Patagonia is a region of special interest because of the water resources importance, such as hydropower generation to the national grid and fruit production to satisfy the needs of both export and local markets. One of the most important areas when it comes to fruit production is the Alto Valle del Río Negro, located north of extra-Andean Patagonia in Argentina, on the banks of the Negro River which originates from the junction of the Limay River and Neuquén River. This area is characterized by its intensive agriculture under the irrigation of fruit trees. Therefore, from a trend point of view, there could be a higher demand for water during the dry season.

A relevant decrease in the dry season was observed in Neuquen station (Province that gathers the largest number of hydroelectric plants), which could have significant consequences for dams operation and consequently in energy supply.

Acknowledgements Rainfall data was provided by the National Meteorological Service of Argentina (SMN), the Sub-secretary of Hydric Resources (SsRH) and the Territorial Authority of Comahue basin (AIC). This research was supported by 2020–2022 UBACyT 20020190100090BA and 2018–2020 UBACyT 20620170100012BA projects.

References

Agosta Scarel E, Martin P, Serio L (2019) Persistent easterly winds leading to precipitation in the Atlantic coast of Patagonia. Int J Climatol 39(13):5063–5090

Barros V, Mattio H (1978) Tendencias y fluctuaciones de las precipitaciones en la región patagónica. Rev Meteorol VIII-IX:237–248

Barros V, Rodriguez Sero J (1979) Estudio de las fluctuaciones y tendencias de la precipitación en el Chubut utilizando funciones ortogonales empíricas. Geoacta 10(1):197–204

Berman A, Silvestri G, Compagnucci R (2012) Eastern Patagonia seasonal precipitation: influence of southern hemisphere circulation and links with subtropical South American precipitation. J Clim 25(19):6781–6795

Castañeda M, González MH (2008) Statistical analysis of the precipitation trends in the Patagonia region in Southern South America. Atmósfera 21(3):303–317

Castañeda M, Barros V (1994) Las tendencias de la precipitación en el cono sur de América al este de los Andes. Meteorológica 19(1):23–32

Garreaud R, Vuille M, Compagnucci R, Marengo J (2009) Present day South American climate. Palaeogeogr Palaeoclimatol Palaeoecol 281(3–4):180–195

Gonzalez M, Losano F, Eslamian S (2020) Rainwater harvesting reduction impact on hydro-electric in Argentina. In: Eslamian S (ed) Handbook of water harvesting and conservation. Wiley NY, USA, 1100p

Gonzalez M, Herrera N (2014) Statistical prediction of winter rainfall in Patagonia (Argentina). Horiz Earth Sci Res 11:1–21

Martin P, Serio L (2007) Estudio de la distribución de la precipitación mensual y estacional en la región patagónica argentina. Geoacta 32:145–150

Paruelo J, Beltran A, Jobbagy E, Sala O, Golluscio R (1998) The climate of Patagonia: General patterns and controls on biotic processes. Ecol Austral 8:85–101

Peel M, Finlayson B, Mcmahon T (2007) Updated World Map of the Koppen-Geiger Climate Classification. Hydrol Earth Syst Sci 11:1633–1644

Prohaska F (1976) The climate of Argentina, Paraguay and Uruguay. Climates of Central and South America, W. Schwerdtfeger

Romero P, Gonzalez M (2016) Relación entre los caudales y precipitación en algunas cuentas de la Patagonia norte. Revista de Geología Aplicaciones a la Ingeniería y Ambiente 36:7–13

Romero P, González M, Rolla A, Losano F (2020) Forecasting annual precipitation to improve the operation of dams in the Comahue region Argentina. Hydrol Sci J 65(11):1974–1983

Thom H (1958) A note on the gamma distribution. Mon Weather Rev 86:117–122

An Assessment of Severe Storms, Their Impacts and Social Vulnerability in Coastal Areas: A Case Study of General Pueyrredon, Argentina

Ignacio A. Gatti, Paula B. Martin, Elisabet C. Vargas, Mariana Gasparotto, Barbara E. Prario, Elvira E. Gentile, and Leandro G. Patané

Abstract

Severe storms are very frequent meteorological events that generate major damage in coastal areas around the world. This work aims to describe and analyze the characteristics of severe storms in relation to the spatial distribution of their impacts and the social vulnerability of the population in coastal areas. We selected the district of General Pueyrredon ($-38°00'$; $-57°33'$), Argentina, as a study case. A hydrometeorological hazard analysis was performed using data from the Estación de Observaciones Costeras (Coastal Observation Station) for the period 2013–2019. Wave height, period, and direction were computed. Besides, wind velocity and wind direction were estimated. Then a regression analysis was done for both continuous variables. Storms were selected according to three parameters: impacted neighborhoods using national and local mass media, wave height, and wind velocity using percentile and mean scores. Afterward, a storm classification was performed by establishing three different synoptic situations. Finally, the social vulnerability of the district was defined by creating an Index of Social Vulnerability to Disasters using social, economic, and housing conditions. The results between 2013 and 2019 show that 25.2% (827 events) of wave height represent values higher than 1 m, mainly from the southeast direction (42%). Strong winds (≥ 43 km h^{-1}) represented 23 cases with a higher frequency from north, northeast, and northwest (29%). Linear regression model results showed that there are probably more factors influencing wave heights. According

I. A. Gatti (✉)
Department of Sciences, Technology and Society, University School for Advanced Studies IUSS, Pavia, Italy
e-mail: ignacio.gatti@iusspavia.it

I. A. Gatti · P. B. Martin · E. C. Vargas · M. Gasparotto · E. E. Gentile · L. G. Patané
Facultad de Filosofía y Letras, Instituto de Geografía "Romualdo Ardissone", Programa de Investigaciones en Recursos Naturales y Ambiente, Universidad de Buenos Aires, Buenos Aires, Argentina

P. B. Martin · E. C. Vargas · M. Gasparotto · E. E. Gentile · L. G. Patané
Departamento de Geografía, Facultad de Filosofía y Letras, Universidad de Buenos Aires, Buenos Aires, Argentina

P. B. Martin · B. E. Prario
Servicio de Hidrografía Naval, Ministerio de Defensa, Buenos Aires, Argentina

P. B. Martin
Consejo Nacional de Investigaciones Científicas y Técnicas de, Buenos Aires, Argentina

M. Gasparotto
Universidad Nacional de Tres de Febrero (UNTREF), Sáenz Peña, Buenos Aires, Argentina

to storm selection, 45 events were analyzed. Those associated with cyclonic activity summarized 62.2% of total cases. These events produced floods, hail, and strong winds which resulted in numerous evacuees, suspension of classes, fallen trees, power outages, and infrastructure damage. A very high social vulnerability was detected in nine neighborhoods but only one on the coastline. Finally, by performing risk analysis it was found that 20 quarters are highly vulnerable and exposed to severe storms, with a small difference whether they were located in coastal (55%) or non-coastal areas (45%).

Keywords

Severe storms · Impacts · Hazard · Coastal areas · Social vulnerability · Waves

29.1 Introduction

A severe storm can produce great damage and may represent a loss of life. They are associated with high impact events like floods, coastal erosion, and hail, causing widespread damage to infrastructure, power outages, fallen trees, and the interruption of economic, educational, and productive activities. In coastal areas, these storms produce floods during storm surges and high waves, especially when combined with high tides (Seneviratne et al. 2012).

In Argentina, in the coastal region of Buenos Aires Province, severe storms are associated with extratropical cyclones (Possia et al. 2003) and storm surges from the south and southeast (southeasterlies, or "sudestadas" in Spanish). Southeasterlies can last for many days, raising the water level and causing significant differences between astronomical tide and the observed levels (Fiore et al. 2009).

Wave transformation processes from deep to shallow water include the process of refraction, bottom friction, shoaling, diffraction, and breaking (Rusu et al. 2008). These processes produce wave impacts on coastal areas, which have been studied by Sierra and Casas-Prat (2014). It was indicated that the potential changes in some parameters like wave height and wave direction will affect artificial and natural shorelines. Wind, on the other hand, has high energy potential (Ilkiliç and Aydin 2015) and can be modeled with methods like neural networks (More and Deo 2003). One of the most used models is the SWAN (Simulating Waves Nearshore) which computes random, short-crested wind-generated waves in coastal regions and inland waters (Rusu and Ivan 2010).

A storm classification, using the Storm Power Index (Dolan and Davis 1992), was utilized by Rangel-Buitrago and Afuso (2011) using wave height and storm duration as inputs. Different types of beaches were analyzed depending on the characteristics of their shorelines and slope. Thus, different storm classes were determined ranging from "weak" to "extreme" storms considering wave energy and duration. Other studies, like Zhang et al. (2001), suggested that storm tides should also be considered due to their erosion potential. However, there are limited works that deal with storm classification and impacts in urban areas.

The impacts of severe storms in coastal urban areas are different from inland zones in terms of their distribution. The media usually reflects news about the consequences of severe weather where people, goods, urban and rural infrastructure, and productive activities are affected. They are usually recorded by local, regional, and sometimes national newspapers and therefore can be adopted to determine the impact of the events (e.g., Gatti 2015). Worldwide databases examples using this type of record include the Emergency Events Database (EM-DAT) from the Center for Research on the Epidemiology of Disasters (CRED), Université catholique de Louvain (Belgium), and the Disaster Inventory System (DESINVENTAR) created by the Network of Social Studies in the Prevention of Disasters in Latin America (LA RED) and then supported by the United Nations Development Program (UNDP) and United Nation Office for Disaster Risk Reduction (UNDRR).

Another aspect to consider in the analysis is to characterize the local population through the use of vulnerability assessment studies. There is a wide range of concepts of vulnerability that could lead to defining and measuring it in a heterogeneous way (Shitangsu 2013). Vulnerability has been defined by Cardona et al. (2012) as the predisposition to being affected by adversity. Related to disasters, this includes the characteristics of a person or a group and their situation that influences the capacity to anticipate, cope, resist and recover from an adverse event. Social vulnerability refers to exploring the socio-economic characteristics of the population and how prone it is to be affected. This approach exhibits a large degree of uniformity in the index construction (Rufat et al. 2015). In the study area, some authors have characterized vulnerability using environmental quality (Celemín 2012) and socio-economic indexes (Celemín 2012; Cabral and Zulaica 2015). Others have worked on social vulnerability considering the Argentinian working crisis of 2002 (Labrunée and Gallo, 2005) and analyzing exclusively peri-urban areas (Zulaica and Celemín 2008; Zulaica and Ferraro 2010). However, no papers have correlated social vulnerability and the characteristics of severe storms and their impacts.

The present preliminary study provides an overview of the impacts of severe storms in the period 2013–2019, characterizing social vulnerability at the local scale. For this purpose, it is necessary to collect information to contribute to reducing uncertainties and bring new knowledge and tools that derive from deeper analysis of emergency response, disaster risk management, and land planning.

29.2 The Study Area

General Pueyrredon district, Buenos Aires Province, Argentina, was selected to test the methodology (Fig. 29.1). The main city and capital of the district are Mar del Plata ($-38°00'$; $-57°33'$), whose population is 618,989 considering the last National Census 2010. This city is the most important touristic place in Buenos Aires Province. During summer, tourism activities increase the population by at least 50% and the rest of the year can increase between 20 and 25% (García and Veneziano 2014).

The climate is temperate with oceanic influence or maritime subtropical regime, with an annual mean temperature of 14.1 °C. Precipitation shows an annual average of 943.6 mm with a seasonal regime, having a period of higher rainfall from the end of spring to the beginning of autumn (October–April) (Grondona 2017). Air mass interaction and alternancy from different origins (tropical, continental, polar, and maritime) generate fast weather changes, being an area with great variability of the meteorological conditions (Martos 1998). Some studies determined some variables like temperature, precipitation, and wind (García and Veneziano 2014; Albisetti 2017) for the period 1951–2010. Seasonal weather conditions and the evolution of climatic variables were characterized without any positive or negative significant trend. Although, the maximum wind speed was reduced by 7 km h^{-1} in this period.

29.3 Data and Methods

29.3.1 Hazard Assessment

A series of statistical methods were used with the data from the oceanographic station "Estación de Observaciones Costeras" (Coastal Observation Station, hereafter EOC) ($-38°\ 00'\ 02''$; $-57°\ 32'\ 18''$), located 300 m far from the coastline. EOC started to take measures in October 2012 so a 2013–2019 study period was considered. EOC obtains two values per day, including the following parameters: wave height (H), period (T), and direction (D). We define H as the altitude difference between a crest and a valley of a wave. T is the time that elapses between two consecutive peaks passing through the same point. Finally, D is the wave prevailing direction, considering 0° when the wave comes from North (D_N), 45° when it comes from Northeast (D_{NE}), 90° when it comes from East (D_E), and so on. Wave predominant direction was estimated visually (e.g., Guedes Soares 1986), so it has a certain degree of subjectivity.

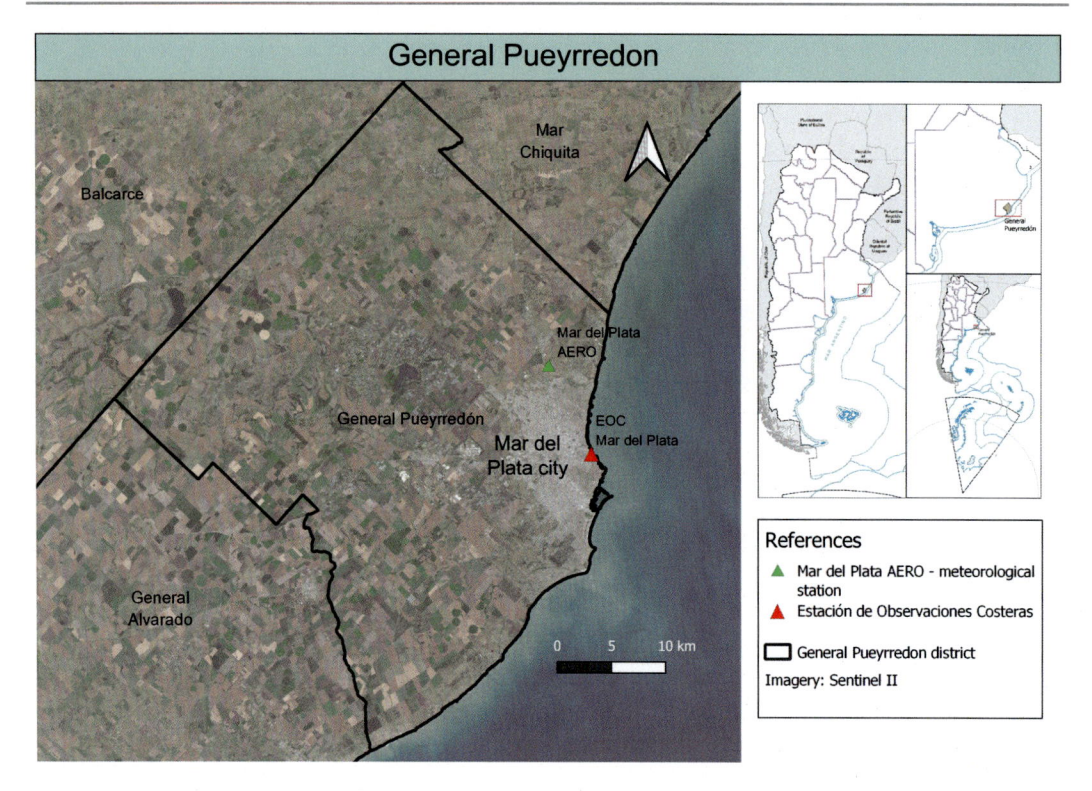

Fig. 29.1 Location of the study area and EOC and Mar del Plata AERO stations

Wind velocity (U) and wind direction (W) were obtained from Mar del Plata AERO meteorological station (−37° 55′ 48″; −57° 34′ 48″) at the same time as the wave's observations in the EOC.

We performed a regression analysis of the continuous variables H and W to describe the relationship of both parameters in coastal areas. We took W as a dependent variable (y) and H as an independent variable (x) by using a simple linear regression (Eq. 29.1):

$$y = \alpha + \beta x + \varepsilon \quad (29.1)$$

where α is the intercept, β is the slope and ε is the estimated error. By doing this, we tried to predict by fitting the model to the data. Scatterplots were produced to see the values, residuals, and linear regression. We also calculated R-squared (R^2), Residual Standard Error (RSE), and Spearman Rank Correlation to evaluate the overall quality of the fitted regression model. RSE was calculated to measure the variation of observation around the regression line. A model with a small RSE indicates that the model fits correctly. R^2 ranges from 0 to 1 and represents the proportion of information that can be explained by the model. A high value of R^2 is a good indicator of a good correlation.

Finally, a non-parametric distribution using the Spearman Rank Correlation method was selected to determine the significance and strength of the relationship. This distribution is robust to outliers (Croux and Dehon 2010), and is it estimated by (29.2):

$$p = 1 - \frac{6\Sigma d_i^2}{n(n^2 - 1)} \quad (29.2)$$

where p is the Spearman's rank correlation coefficient, d_i is the difference between ranks and n is the number of observations.

29.3.2 Severe Storms and Their Impacts

To deal with hydrometeorological hazard analysis, three dimensions were used: (I) information from national and local newspapers, social media, radio and tv shows regarding impacts effects over the urban part of the district, (II) severe storms that registered waves superior to Percentile 75 (P75) which was 1 m, (III) U over 14 km h^{-1}, which represent the local statistical mean. Between 2013 and 2019, a total of 45 events were counted. The impact on the district has been mapped on the 124 neighborhoods, obtained from the open database of General Pueyrredon district.

Additionally, we analyzed some meteorological parameters to classify storms according to their origin and development. The NCEP/NCAR Reanalysis tool from the Physical Sciences Laboratory from the National Oceanographic and Atmospheric Administration (NOAA) was used. Selected parameters were sea-level pressure, thickness 500 and 1000 mb, geopotential height 500 and 100 mb, and temperature at 850 mb. As a result, three synoptic situations were found: (1) mid-latitude front activity, (2) surface low-pressure centers over subtropical Argentina, (3) cyclonic activity: closed surface cyclone near Mar del Plata city. Furthermore, this will be related to the estimation of H and U. The duration of the storms was from one up to seven days, depending on the persistent conditions.

29.3.3 Social Vulnerability

By considering social vulnerability, a method developed by the Programa de Investigaciones en Recursos Naturales y Ambiente (PIRNA), Instituto de Geografía, Universidad de Buenos Aires was applied (Barrenechea et al. 2003; Natenzon and González 2010). Three dimensions were considered: social, housing, and economic conditions. Social conditions were divided into education, health, and population age. Housing conditions included housing characteristics and the existence of public services. Finally, economic conditions were divided into labor, education, and family.

Variables were divided into 10 different indicators: illiteracy, the density of health centers, population 0–14, population >65, house overcrowding, no sewer access, no tap water, population of unemployed, single-parent household, and education of household head. These variables were chosen because of their importance to analyze the characteristics of the population against a disaster from a natural or non-natural origin. For instance, the population between 0 and 14 and over 65 years old understands orders or strategies differently from an average adult and does not have the same strength to withhold the consequences of strong wind or flood. They probably may need assistance to be evacuated. Therefore, the availability of health centers is crucial to dealing with a primary medical assistant. Sewer and tap water access refer to the right to safe water and healthiness due to contamination. For instance, during floods, the shortage of these services can produce illness in residents due to the consumption of contaminated water. Moreover, families less educated or without a job have bad initial conditions in terms of preparation and response to withstand a catastrophe. Recovery processes are also more difficult for these citizens.

Indicators were obtained from the Argentinian 2010 National Census, by the Instituto Nacional de Estadística y Censos (INDEC). Censal radios were used as a spatial unit, that corresponds to the smaller piece of information available within the census. In urban areas, these polygons are built according to the number of households and the characteristics of the terrain. Nevertheless, in rural areas, they are selected depending on the level of accessibility and Euclidean distance of buildings.

In relation to health center analysis, some studies focus on accessibility by using a least-cost path method (Jin et al. 2015), others explored the pattern of service distribution utilizing the nearest neighborhood (Mansour 2016). However, in the present research, an approach focusing on the Euclidean distance (Ramirez

2009; Buzai 2016) between the buildings and the health centers was chosen. In a hazardous event, cars could be affected, and citizens could be forced to walk. The indicator was built by considering how much a person can walk, on average, from his home to the nearest health center. Estimating that a person walks with a mean speed of 4.4 km/h in a plain area (Gast et al. 2018), in an emergency, it was assumed that a person can walk up to 30 min to reach a health center. This corresponds to a buffer of 2500 m from his or her house.

To analyze the relationship between vulnerable and non-vulnerable groups, indicators from absolute and relative values were used. Like Roder et al. (2017) we equally weighted the components. Five categories were determined: very low, low, medium, high, and very high. In a first attempt, the Jenks natural breaks method was used to identify groups with similar values, but, finally, a manual classification was performed because histograms presented some jumps not possible to detect using the common methods (quintiles, equal interval, standard deviation, and Jenks natural breaks). Three subindexes were created directly from absolute and relative indicators, evaluating social, housing, and economic conditions and, finally, an Index for Social Vulnerability to Disasters (ISVD) was developed.

Considering the absolute values, as the units from each indicator are different, it was necessary to normalize the data (Wood et al. 2010; Jain and Bhandara 2011; Lianxiao and Morimoto 2019) using Eq. (29.3):

$$x'_{ij} = (x_{ij} - \min_{ij})/(\max x_{ij} - \min x_{ij}) \quad (29.3)$$

where, x'_{ij} = the normalized value of the attribute, x_{ij} = every value, $\min_{ij}$ = the minimum value and max x_{ij} = the maximum value.

Instead, for relative values, normalization was performed by (29.4):

$$x'_{ij} = x_{ij}/100 \quad (29.4)$$

The *IVSD* was finally built by the formula (29.5). Results between [0, 1] were obtained for absolute, relative, and synthesis values:

$$ISVD = \frac{(V_i + V_{ii} + V_{iii} + V_{iv} + V_v + V_{vi} + V_{vii} + V_{viii} + V_{ix} + V_x)}{10}$$

(29.5)

where, V_i = illiteracy; V_{ii} = density of health centers; V_{iii} = population of age 0–14; V_{iv} = population of age >65; V_v = house overcrowding; V_{vi} = no sewer access; V_{vii} = no access to tap water; V_{viii} = population of unemployed; V_{ix} = single-parent household; V_x = education of household head.

29.4 Results

29.4.1 Hazard Detection

The analysis of the wave series observed in EOC (Fig. 29.2), gave the following results: from 3199 registers of H a mean value of 0.8 m was obtained; 25.2% of H (827 events) represent values higher than 1 m. Furthermore, H was higher than 3 m 17 times, which eventually represented only 0.5% of all registers, including a single maximum event that reached 4 m on 17/9/2013. D shows that more frequent wave events come from D_{ESE} (42%), following D_{SE} (20.7%), D_E (17.6%), and D_{ENE} (15.8%), a fact that could be attributed to D_{ESE} orientation coastal shoreline.

When considering U and W, we excluded days without wind. It can be noticed that there is a wide variety of W with higher frequency from W_N, W_{NNE} and W_{NE} (around 29%) (Fig. 29.2). García (2011) and García and Veneziano (2014) consider strong winds >=43 km h^{-1} in the study area, which in this case represents 23 cases in different directions. Wind records for all observations also included sea breeze which is coastal wind caused by local temperature differences. Differently from storm surges, they represent stable surface high-pressure systems and clear skies (Weisse and Von Storch 2010).

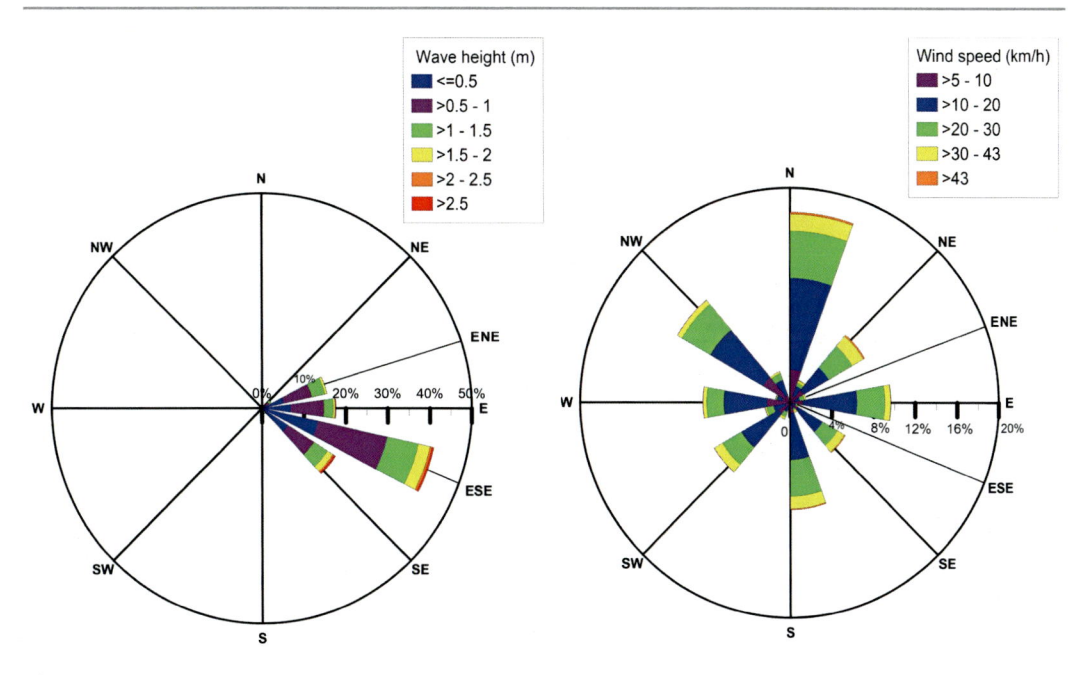

Fig. 29.2 Distribution of H, D (left) and U, W (right) observed in EOC

Besides, we also performed a linear regression model to test the relationship between H and U (Fig. 29.3). A total number of 630 observations were considered, specifically where H and U have the same direction. Results show that when H is 0.025 m higher, there is an increase of 1 km h^{-1} (significant level of 0.99). By exploring RSE and R^2 we obtained a relatively low value for the first (0.38) and a low value for the second (0.26), so there are probably more factors influencing H.

General Pueyrredon presents a coastal typology that includes sand beaches, ripraps, sea cliffs, and artificial coastal defenses like groins, seawalls, detached breakwaters, the port, and buildings. Storms cause sometimes exceptionally high storm surges and massive failure of coastal protection (Weisse and Von Storch 2010). Therefore, erosion processes and impacts on urban infrastructure are quite common.

Fig. 29.3 U-H regression analysis

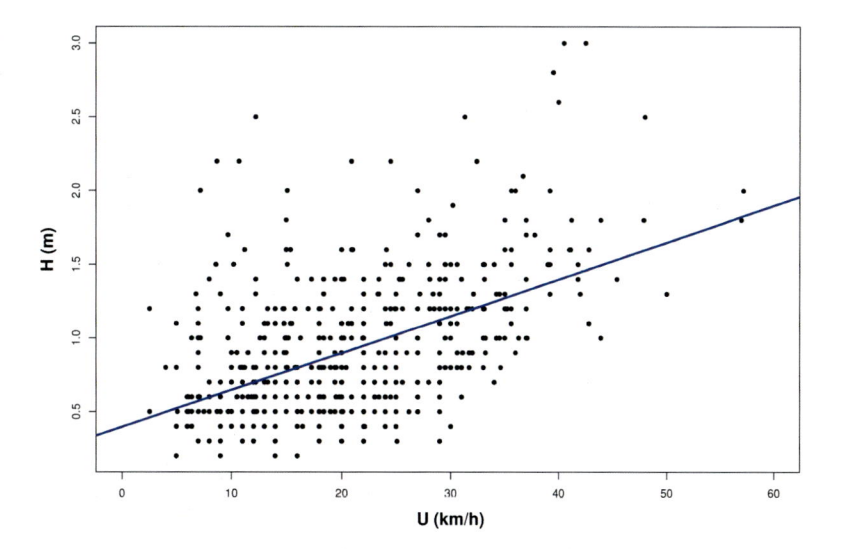

29.4.2 Storms Classification

Of the total storm events analyzed (45), 7 (15.6%) corresponded to the type number 1 (front activity), 10 (22.2%) to the type number 2 (surface low-pressure centers) and 28 (62.2%) to the type number 3 (cyclonic activity). Type 3 storms were mainly distributed in fall, winter, and spring (Fig. 29.4). Type number 2 events were not present in the fall. Furthermore, 64.4% (29 events) were related to the Percentile 95 (P95) where H is equal to 1.70 m, and 79.3% of the total events of this P95 answer to cyclonic activity, which is the 3rd type of storm (Table 29.1). Cyclonic activity associated with the event is more frequent in transitional seasons due to the thermal contrast. Some of them can be classified as southeasterlies (Garcia 2011).

29.4.3 Total Impacts

City neighborhoods represent only 32% of the district's total surface. The remaining 68% corresponds to non-urbanized/agricultural areas that were not considered in the present analysis. Figure 29.5 shows the distribution of different effects due to severe storms detected within the study area. According to news and mass media, floods affected 26 neighborhoods. Then, roof loss problems were found in 37 neighborhoods, fallen trees in 40, hail damage in 8, and power supply cuts in 56. Moreover, there were primary and secondary schools that had to suspend their classes in all quarters, on average, 2 or 3 times. By performing a summation of all types of damages, Bosque Peralta Ramos and Central Area with 14 events and The Port and Punta Mogotes with 12 events were the most affected neighborhoods (Fig. 29.6). Three of those places are located within the coastal line while the other (Bosque Peralta Ramos) is far 1.1 km considering the Euclidean distance from the nearest point.

29.4.4 Vulnerability Analysis

To determine the social vulnerability of the area, census radio polygons results were converted to the size of the neighborhoods. Indicators for relative and absolute values were classified into sub-indexes according to social, economic, and housing conditions (Fig. 29.7).

Two ISVD indexes characterized relative and absolute values. Additionally, a final synthesis index was created by using the mean of absolute and relative values (Fig. 29.8). Absolute values present higher values in the geographical center of General Pueyrredon district. On the other hand, peripheral neighborhoods in the south, west, and north have higher relative values. The Synthesis Index shows that there are nine quarters with very high vulnerability: Virgen de Luján, Santa Rosa de Lima, Dos de Abril,

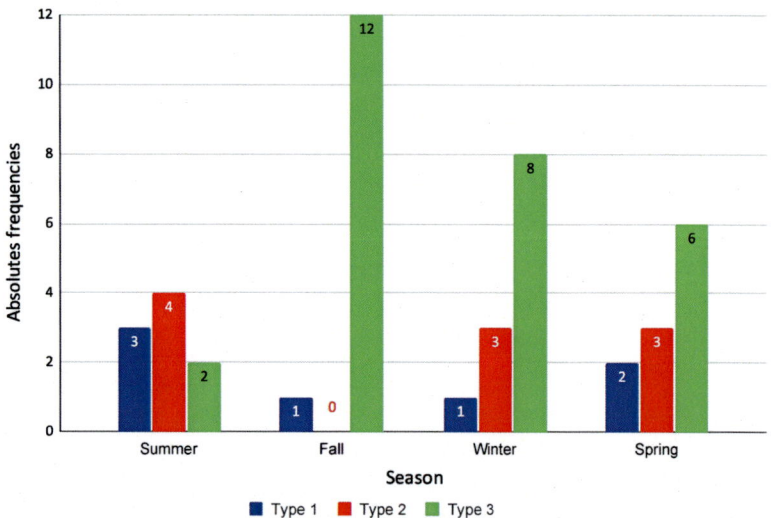

Fig. 29.4 Seasonal distribution of storms selected in the period 2013–2019

Table 29.1 Severe storms, wind, waves, rainfall, and type of storm of the 45 events that dealt with impacts from 2013 to 2019 in the study area

Date	Duration (days)	U (km h⁻¹)	W	H (m)	P (s)	D	Rainfall (mm) (total by the duration)[a]	Storm classification type	Storm type 3 with H > 1.7 (m)
24/02/2013	4	41.8	NE	2.8	10	SE	37	3	YES
3/3/2013	5	25.5	S	3.5	8	SE	77	3	YES
12/3/2013	4	19.4	S	2.2	11	ESE	1	3	YES
12/4/2013	6	25.6	SW	2.5	11	ESE	7.7	3	YES
5/8/2013	3	20.9	SW	2.5	10	ESE	0	1	NO
22/8/2013	2	17.3	SW	3.5	12	SE	12	3	YES
17/9/2013	7	43.2	S	4	10	ESE	19.6	3	YES
1/10/2013	4	42.5	ESE	3	10	ESE	85.2	3	YES
24/1/2014	1	30	SSE	2	7	ESE	31	1	NO
17/3/2014	2	24.5	SE	2.2	8	ESE	81 (4 days)	3	YES
9/4/2014	3	32.4	ENE	2.2	6	E	53	3	YES
26/8/2014	3	37	S	3	11	E	162 (7 days)	2	NO
6/9/2014	2	29.8	NE	1.3	7	E	50	1	NO
26/8/2015	3	35	NW	1.8	11	SE	3	2	NO
30/9/2015	1	27	NE	1.2	7	SE	14	3	NO
16/10/2015	1	36.7	SSE	2.5	12	SE	21.09	3	YES
23/10/2015	1	24.1	SSW	1.5	9	SE	7	3	NO
30/11/2015	1	29	E	1.4	9	ESE	3	1	NO
26/4/2016	4	32	W	2.5	10	SE	20.08	3	YES
16/5/2016	3	27.3	W	1.8	11	SE	5.2	3	YES
30/5/2016	2	57.2	SE	2	8	ESE	93	3	YES
10/6/2016	2	11	W	2	9	SE	11	3	YES
26/7/2016	2	41	SW	1.8	10	ESE	20	3	YES
12/9/2016	4	49	W	3	12	SSE	60	3	YES
6/2/2017	2	22	S	2.8	9	SE	14	3	YES
24/2/2017	1	6	ENE	0.7	10	ESE	21	2	NO
19/5/2017	1	48	E	2.6	9	E	20	3	YES
11/8/2017	1	37	E	1.7	9	ESE	44	2	NO
8/12/2017	5	20	S	0.6	9	SE	38	2	NO
23/1/2018	2	30	WSW	2.3	7	ESE	13	2	NO
21/2/2018	1	24	S	0.8	6	ESE	0	1	NO
2/3/2018	2	22	E	0.5	5	ENE	38	1	NO
7/3/2018	1	14	NNW	1	10	SE	0	3	NO
12/6/2018	1	38	SSW	1.8	7	SE	3	3	YES
19/7/2018	2	37	SE	1.9	9	ESE	26	3	YES
10/8/2018	1	16	WSW	1.8	14	ESE	0.1	3	YES

(continued)

Table 29.1 (continued)

Date	Duration (days)	U (km h^{-1})	W	H (m)	P (s)	D	Rainfall (mm) (total by the duration)[a]	Storm classification type	Storm type 3 with $H > 1.7$ (m)
22/9/2018	2	24	NW	0.5	12	ESE	44	2	NO
13/11/2018	1	14	NE	1.1	8	ESE	50 (2 days)	2	NO
14/3/2019	1	22	E	1.5	9	E	94 (2 days)	3	NO
25/3/2019	1	29	S	2	11	SE	0	3	YES
4/7/2019	2	23	WSW	2.8	12	SE	19	3	YES
2/8/2019	1	38	SW	1.2	9	SE	0.8	3	NO
29/11/2019	1	18	N	0.4	9	SSE	0.9 (2 days)	2	NO
3/12/2019	1	25	NW	1.5	10	ESE	3	1	NO
28/12/2019	2	18	E	0.6	6	ENE	33	2	NO

[a] When exceeded the duration of the event, rainfall days were marked with a ()

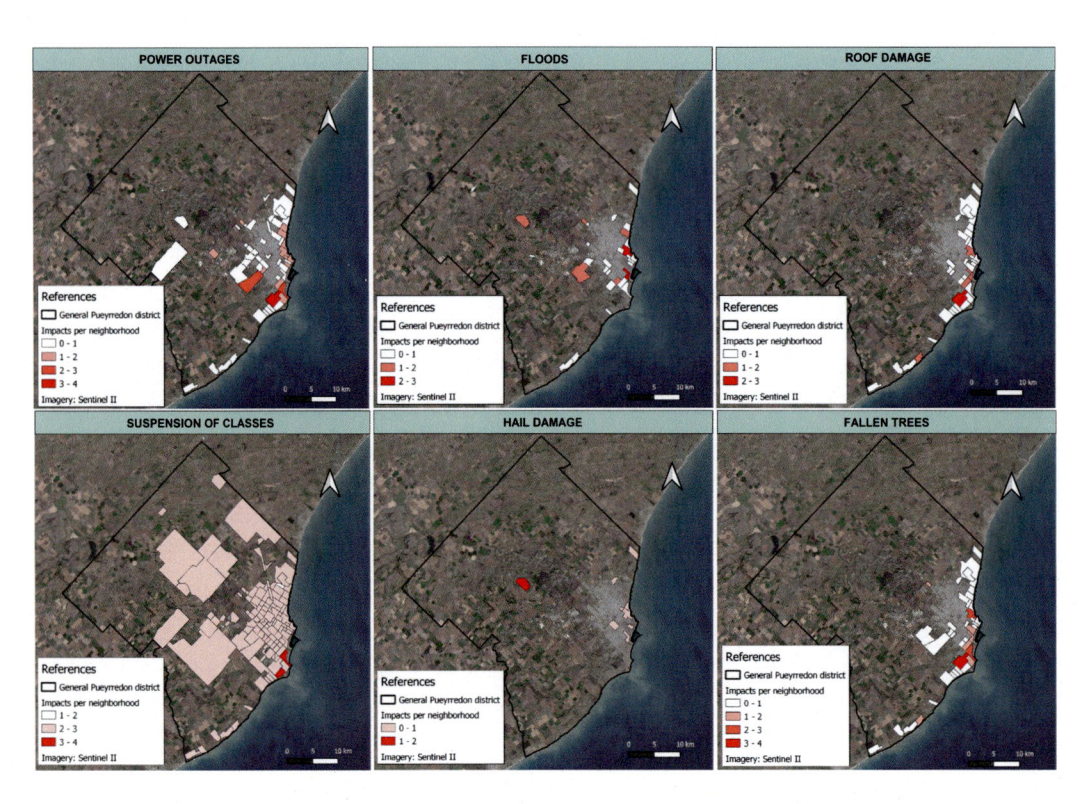

Fig. 29.5 Partial impacts per neighborhood related to several storms in General Pueyrredon

General Belgrano, Autódromo, San Jorge, Camet, La Herradura and Caribe. They are located in the west and north peripheral of Mar del Plata city.

The city center and surrounding neighborhoods show green values of ISVD indicating they are less vulnerable to impacts and probably more resilient to severe storms and floods.

Fig. 29.6 Total impacts per neighborhood related to several storms in General Pueyrredon

29.4.5 Overall Risk Analysis

To perform a classification of the impacts according to the level of vulnerability we produced a scatter plot (Fig. 29.9). To study the distribution of the population coastal and non-coastal neighborhoods were considered. It can be noticed that the ISVD synthesis histogram is a little skewed to the left, meaning that medium, high, and very high vulnerability dominate the distribution. On the other hand, neighborhood impacts have a distribution skewed to low values, and due to that a lot of quarters received three impacts. Furthermore, ellipses were created with a 0.95 confidence level. It can be observed that coastal areas are prone to impacts and many of them have a high vulnerability. In the figure, dashed lines represent the median for the y-axis and the two higher categories of ISVD for the x-axis. A total of 20 quarters are more vulnerable and exposed to storms (green square in Fig. 29.9), 11 in coastal areas (55%), and nine in non-coastal areas (45%). Among them, quarters "A" (Bosque Peralta Ramos) and "B" (La Herradura) can be classified as the zones with higher risk.

29.5 Discussion and Conclusions

Some discussion and main conclusions are presented in this part. Offshore, waves initially travel at an angle of about 70°–80° from the wind direction (Van Dorn 1994). Weisse and Von Storch (2010) then described that when the waves grow and their speed increases, this angle progressively decreases. Thus, significant surge heights are generated when the wind pushes large water masses toward or away from the coast.

Past studies like Lanfredi et al. (1992) estimated a maximum H of 2.3 m along the coastal area of Buenos Aires Province. More specifically, in General Pueyrredon, Isla (2010) stated that the highest waves come from the south with H around 1.5 m and a T of 7 s. Results from this study show that more frequent H comes from

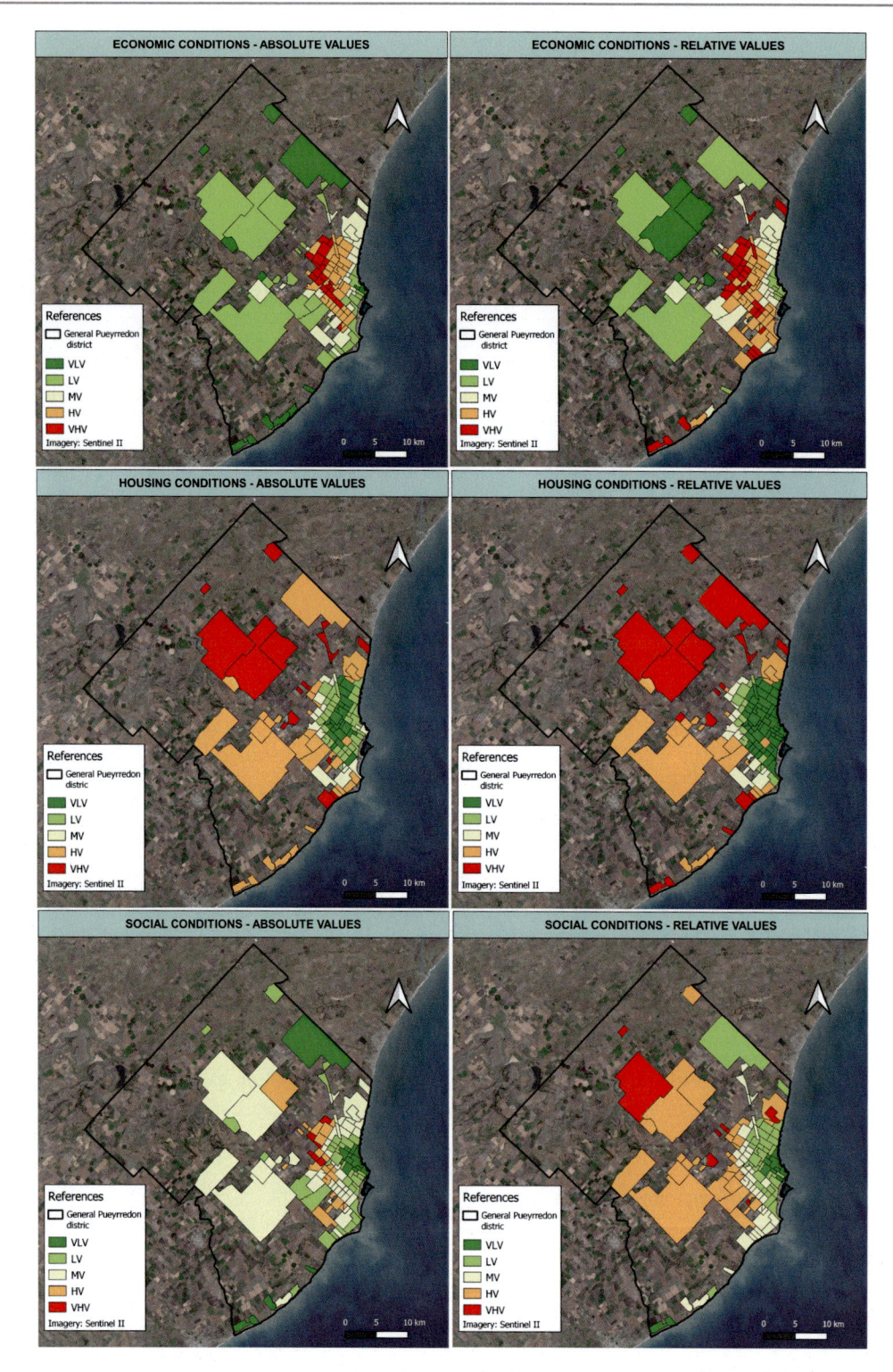

Fig. 29.7 Sub-indexes for absolute and relative values for economic (upper), housing (middle), and social conditions (lower). Categoriesare VLV (very low vulnerability), LV (low vulnerability), MV (medium vulnerability), HV (high vulnerability), VHV (very high vulnerability)

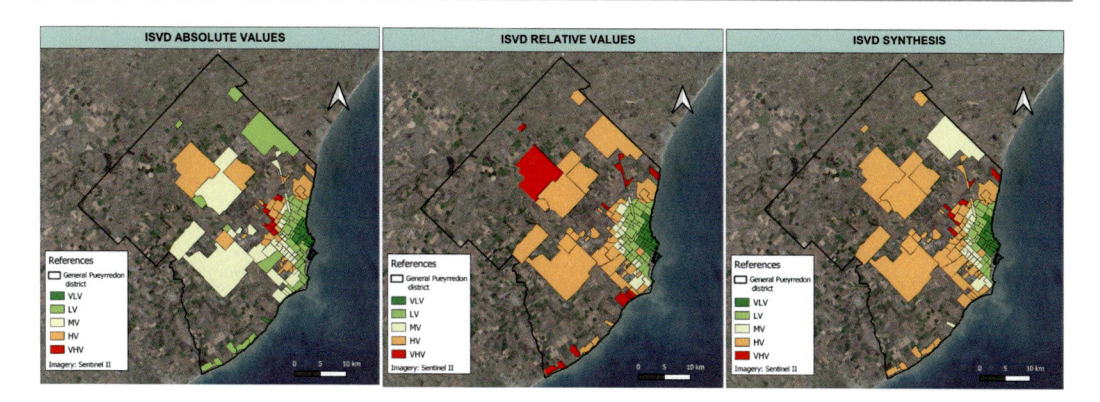

Fig. 29.8 Mosaic of ISVD according to absolute values (left), relative values (middle), and the Synthesis Index (right). Categories are VLV (very low vulnerability), LV (low vulnerability), MV (medium vulnerability), HV (high vulnerability), VHV(very high vulnerability)

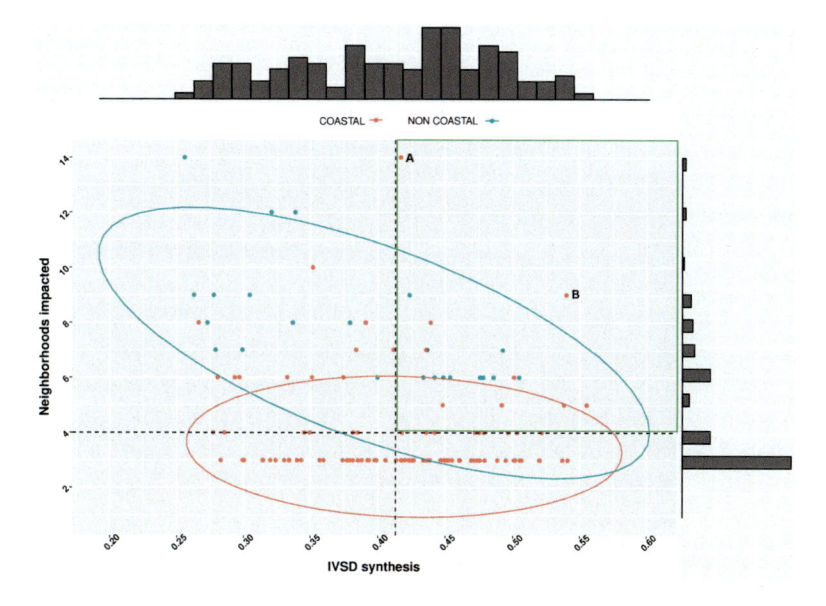

Fig. 29.9 Plot of ISVD synthesis and neighborhoods impacted. The histogram of both categories above shows the distribution of the variables. The classification was made according to the non-coastal (red dots) and coastal (light blue dots) types of the neighborhood. The green square represents the most exposed and vulnerable quarters, especially "A" and "B"

D_{ESE} (42%) and current values of T are a little higher (8.4 s). Then, for H we determined a mean equal to 0.8 m, a $P75$ of 1.1 m., a $P95$ of 1.8, and a $P99$ of 2.5 m. Finally, 17 times H was higher than 3 m which eventually represented only 0.5% of the total events, with a single situation of 4 m occurring on 17/9/2013. As stated in the regression analysis (Chap. 4.1), there are probably more factors influencing H, like local water depth related to the seabed (Battjes and Jansen 1978), type of tide, and anthropic coastal modification.

Compared with tides, surges are less predictable and regular (Weisse and Von Storch 2010). Positive storm surges can last for several days and can raise the water level producing significant differences between the observed level and the astronomical tide (Fiore et al. 2009). The mean duration of events that generated an impact was 2.2 days.

Only three cases of the impact events have $U \geq 43.5$ km h^{-1}, considered as strong wind for the area by García (2011) and García and Veneziano (2014) (see Table 29.1). However, 83.3% of the time the wind was higher than the wind average. Furthermore, D with higher frequency comes from D_N, D_{NNE} and D_{NE} (around 29%).

Sea-level rise, mainly due to the melting of ice sheets and thermal volume expansion in a Climate Change context is causing problems in many countries around the world. Even if the coastal region of Buenos Aires Province has been experiencing a sea-level rise of about 1.5 mm. per year (Fiore et al. 2008), General Pueyrredon district has not been particularly affected by this issue (Celemín and Zulaica 2012). However, during severe storms, sea-level rises more than the historical mean, therefore, wave energy increases prominently as well as erosion processes (Caldwell and Segall 2007).

In the present study case, some intense erosive processes induced by human activities started in the 70 s (García and Veneziano 2015) and last until today. Sierra and Casas-Prat (2014) indicated that potential changes in wave height will strongly affect wave transmission and longshore sediment transport. Also, inside ports, defense structures could be affected by instability, overtopping, and scouring. Rotation of the mean wave direction has to be considered as well because many beach and harbor defense structures were designed assuming a permanent directional distribution of waves (Sierra & Casas-Prat, 2014). Keeping operative the EOC station would produce a long record series of wave data to deal with storm surge characterization, erosive process, and sediment transport.

Results on impact events showed that there is a dominant type 3 storm (62.2%). Events like those on 30/5/2016 and 12/9/2016 were examples of this type. Some characteristics associated with these events were very strong winds (57.2 km h^{-1} and 49 km h^{-1}, respectively), high daily precipitation (69 and 44 mm), and a relatively high H (2 and 3 m). Impacts associated were floods, coastal erosion, widespread damage to infrastructure, power outages, fallen trees, and the interruption of classes in educational institutions.

Related to the big amount of precipitation, floods tend to occur in some local basins (Eraso 2008; Tomas 2009). Mujica (2016) pointed out that from 1969 to 2014 there was an increase in annually flooded areas from 6 km^2 to almost 19 km^2. The current infrastructure does not cover the entire city which needs larger rainwater collectors and sinks (Maya 2020). Severe storms bring sometimes hail which can cause damage like the case of events of 24/2/2013, 16/10/2015, or 22/9/2018. In coastal areas of Buenos Aires Province, they normally occur during springtime, but they could start in late winter and continue through the beginning of summer (Mezher et al. 2012). The combination of extreme rainfall events and strong winds from the W_{NE} and W_{SE} increases floods and exacerbates infrastructure damage (Albisetti 2017).

Social vulnerability is a measure of both the sensitivity of a population to natural hazards and its ability to respond to and recover from the impacts of hazards (Cutter and Finch 2008). Findings showed a city center with low and very low vulnerability, linked to the fact of a huge investment from the local and national government in infrastructure due to touristic activities. On the other hand, nine quarters have been found to have a very high vulnerability. Regarding the location, only one, Camet, is located within the shoreline.

By considering an overall risk regarding the hazard, vulnerabilities, and impacts some conclusions can be considered. According to the National Census, a total of 26,318 people live within coastal areas in the district. However, more impacted neighborhoods like Bosque Peralta Ramos, Central Area, The Port, and Punta Mogotes have a combined population of 71,126. Severe storm events have eroded natural landforms in coastal areas but also have generated problems for different neighborhoods, no matter their location.

Acknowledgements This work was supported by a grant from the FILO:CyT program (Programa de Apoyo a la Investigación de la Facultad de Filosofía y Letras - Universidad de Buenos Aires). The authors gratefully thank the Servicio Meteorológico Nacional (National Weather Service) and the Prof. Pedro Mazza for the information provided.

References

Albisetti M (2017) Riesgo Asociado a Eventos Extremos de Precipitación en Mar del Plata. Estrategias para una Gestión Sustentable, Tesis de Doctor en Geografía, Departamento de Geografía y Turismo, Universidad Nacional del Sur, Bahía Blanca, Argentina 173

Barrenechea J, Gentile E, González S, Natenzon C (2003) Una Propuesta Metodológica para el Estudio de la Vulnerabilidad Social en el Marco de la Teoría social del Riesgo. IV Jornadas de Sociología, Facultad de Ciencias Sociales, Universidad de Buenos Aires, Argentina

Battjes J, Janssen J (1978) Energy loss and set-up due to breaking in random waves. In: Proceedings of the 16th international conference on coastal engineering, American society of civil engineers, Aug 27–Sept 3, New York, pp 569–587

Buzai G (2016) Tipología de áreas de influencia de Centros de Atención Primaria de Salud en la ciudad de Luján, Provincia de Buenos Aires, Argentina. Revista Huellas 20:35–56, Instituto de Geografía, Ed UNLPam., Santa Rosa, Argentina

Cabral C, Zulaica L (2015) Análisis de la vulnerabilidad socioambiental en áreas del periurbano de Mar del Plata (Argentina) expuestas a agroquímicos. Multiciencias 15(2):172–180

Caldwell M, Craig HS (2007) No day at the beach: sea level rise, ecosystem loss, and public access along the California Coast. Ecol Law Quart 34:533–578

Cardona O, Van Aalst M, Birkmann J, Fordham M, McGregor G, Perez R, Pulwarty R, Schipper E, Sinh B (2012) Determinants of risk: exposure and vulnerability. In: Field C, Barros V, Stocker T, Qin D, Dokken D, Ebi K, Mastrandrea M, Mach K, Plattner G, Allen S, Tignor M, Midgley P (eds) Managing the risks of extreme events and disasters to advance climate change adaptation. A special report of working groups I and II of the Intergovernmental Panel on Climate Change (IPCC), pp 65–108. Cambridge University Press, New York, United Kingdom, Cambridge. Available at: https://www.researchgate.net/publication/244062037_Determinants_of_risk_exposure_and_vulnerability (01 Aug 2020)

Celemín J (2012) Asociación espacial entre fragmentación socioeconómica y ambiental en la ciudad de Mar del Plata Argentina. EURE 13(113):33–51

Celemín J, Zulaica ML (2012) Escenarios de aumento del nivel del mar y la costa de Mar del Plata. Evaluación de impactos socioeconómicos y medidas de mitigación. Lincoln Institute of Land Policy 20

Croux C, Dehon C (2010) Influence functions of the Spearman and Kendall correlation measures. Stat Methods Appl 19(4):497–515

Cutter S, Finch C (2008) Temporal and spatial changes in social vulnerability to natural hazards. Proc Natl Acad Sci 105(7):2301–2306

Dolan R, Davis R (1992) An intensity scale for Atlantic coast northeast storms. J Coastal Res 8:352–364

Eraso M (2008) Gestión del riesgo hídrico en comunidades vulnerables. Inundaciones en el arroyo El Cardalito, Mar del Plata Buenos Aires. Revista Universitaria de Geografía 17:285–307

Fiore M, D'Onofrio E, Grismeyer W, Mediavilla D (2008) El ascenso del nivel del mar en la costa de la provincia de Buenos Aires. Ciencia Hoy 18(106):18–25

Fiore M, D'Onofrio E, Pousa P, Schnack E, Bértola G (2009) Storm surges and coastal impacts at Mar del Plata Argentina. Cont Shelf Res 29(14):1643–1649

García M (2011) Escenario de riesgo climático por sudestadas y tormentas en Mar del Plata y Necochea-Quequén, provincia de Buenos Aires, Argentina. Braz Geogra J Geosci Humanit Res Medium 2(2):286–304

García M, Veneziano M (2014) Comportamiento temporal y tendencias climáticas en la ciudad de Mar del Plata período (1971–2010). Actas Congreso Internacional de Geografía 75° Semana de Geografía 77–93

García M, Veneziano M (2015) Análisis sobre rompeolas y playas regeneradas en el sur de General Pueyrredon R. Argentina. Contribuciones Científicas gæa 27:93–108

Gast K, Kram R, Riemer R (2018) Preferred walking speed on rough terrain; is it all about energetics? J Exp Biol 24

Gatti I (2015) Precipitaciones, sudestadas y su relación con el riesgo de inundación. Entre la gestión del riesgo y adaptación al Cambio Climático. Caso del barrio de Belgrano, Ciudad de Buenos Aires. Período 1981–2012, Tesis de Lic. en Geografía, Facultad de Filosofía y Letras, UBA, p 310

Grondona S (2017) El clima de Mar del Plata de los últimos 40 años, in Alejandra Merlotto et al. Proyecto WATERCLIMA LAC 2015–2017: compilación de informes técnicos producidos en el Área Piloto Mar del Plata. 1st ed., Universidad Nacional de Mar del Plata, Mar del Plata, pp 73–75

Guedes Soares G (1986) Calibration of visual observation of wave period. Ocean Eng 13(6):539–547

Ilkiliç C, Aydin H (2015) Wind power potential and usage in the coastal regions of Turkey. Renew Sustain Energy Rev 44:78–86

Isla F (2010) Natural and Artificial Reefs at Mar del Plata, Argentina. J Integr Coastal Zone Manage 10(1):81–93

INDEC Instituto Nacional de Estadísticas y Censo, Ministerio de Economía (2010) Censos Nacionales de Población y Vivienda, Buenos Aires

Jain YK, Bhandara SK (2011) Min max normalization-based data perturbation method for privacy protection. Int J Commun Comput Technol 2:45–50

Jin C, Cheng J, Lu Y, Huang Z, Cao F (2015) Spatial inequity in access to healthcare facilities at a county level in a developing country: a case study of Deqing County, Zhejiang, China. Int J Equity Health 14:67

Labrunée M, Gallo M (2005) Vulnerabilidad social: el camino hacia la exclusión. En: Lanari ME (ed) Trabajo decente: diagnóstico y aportes para la medición del mercado laboral local. Mar del Plata 1996–2002, Mar del Plata: Suárez, pp 133–154

Lanfredi N, Pousa L, Mazio CA, Dragani W (1992) Wave-power potential along the coast of the Province of Buenos Aires Argentina. Energy 17(11):997–1006

Lianxiao Morimoto T (2019) Spatial analysis of social vulnerability to floods based on the MOVE Framework and information entropy method: case study of Katsushika Ward Tokyo. Sustainability 11:529

Mansour S (2016) Spatial analysis of public health facilities in Riyadh Governorate, Saudi Arabia: a GIS-based study to assess geographic variations of service provision and accessibility. Geo-Spatial Inf Sci 19(1):26–38

Martos P (1998) Características climáticas en el Río, Evaluación de Impacto Ambiental. 2da etapa. Estación Depuradora de Efluentes Cloacales de Mar del Plata, UNMDP, Mar del Plata

Maya M (2020) Flood hazard estimation: actors and social responses to the risk scenario in the southern neighborhoods of Mar del Plata, Buenos Aires (Argentina). Estudios Socioterritoriales. Revista De Geografía 27(040):15

Mezher R, Doyle M, Barros V (2012) Climatology of hail in Argentina. Atmos Res 114–115:170–182

More A, Deo M (2003) Forecasting wind with neural networks. Mar Struct 16(1):35–49

Mujica C (2016). Servicios Ambientales. Regulación de Inundaciones en Mar Del Plata (Partido De General Pueyrredón) durante el período 1969/2015. Tesis de grado, Universidad Nacional del Centro de la provincia de Buenos Aires, Tandil, 92

Natenzon C, González S (2010) Risk, social vulnerability and indicators development. Samples from Argentina., en: Argentina y Brasil, Posibilidades y Obstáculos en el Proceso de Integración Territorial, Universidad de Buenos Aires 195–217

Possia N, Cerne S, Campetella C (2003) A diagnostic analysis of the Río de la Plata Superstorm, May 2000. Meteorological Appl 10:1–13

Ramirez, L (2009) Spatial patterns of the accessibility of the population to health centres in the Metropolitan Area of Gran Resistencia-Chaco (Argentina). In: Proceedings of the 24th international cartographic conference, Chile, p 15

Rangel-Buitrago N, Afuso G (2011) Coastal storm characterization and morphological impacts on sandy coasts. Earth Surface Process Land 36:1997–2010

Roder G, Sofia G, Wu Z, Tarolli P (2017) Assessment of Social Vulnerability to floods in the floodplain of northern Italy. Weather Clim Soc 9(4):717–737

Rufat S, Tate E, Burton C, Maroof A (2015) Social vulnerability to floods: review of case studies and implications for measurement. Int J Disaster Risk Reduction 14(4):470–486

Rusu L, Ivan A (2010) Modelling wind waves in the Romanian coastal environment. Environ Eng Manage J 9(4):547–552

Rusu L, Pilar P, Guedes Soares C (2008) Hindcast of the wave conditions along the west Iberian coast. Coast Eng 55:906–919

Seneviratne, SI, Nicholls, N, Easterling, D, Goodess, CM, Kanae, S, Kossin, J, Luo, Y, Marengo, J, McInnes, K, Rahimi, M, Reichstein, M, Sorteberg, A, Vera, C, Zhang, X (2012) Changes in climate extremes and their impacts on the natural physical environment. In: Field CB, Barros V, Stocker TF, Qin D, Dokken DJ, Ebi KL, Mastrandrea MD, Mach KJ, Plattner G-K, Allen SK, Tignor M, Midgley PM (eds) Managing the risks of extreme events and disasters to advance climate change adaptation. A special report of working groups I and II of the intergovernmental panel on climate change (IPCC). Cambridge University Press, Cambridge, UK, and New York, NY, USA, 109–230

Shitangsu P (2013) Vulnerability concepts and its applications in various fields: a review on geographical perspective. J Life Earth Sci 8:63–81

Sierra J, Casas-Prat M (2014) Analysis of potential impacts on coastal areas due to changes in wave conditions. Clim Change 124:861–876

Tomas M (2009) Expansión Urbana y Riesgo de Inundación. El Caso de Estudio de la Cuenca del Arroyo del Barco, Partido de General Pueyrredon. Tesis de Maestría, Facultad de Arquitectura, Urbanismo y Diseño, Universidad Nacional de Mar del Plata, p 129

Van Dorn W (1994) Oceanography and seamanship, 2nd edn. Cornell Maritime Press, p 440

Veneziano, MF, García, MC (2014) Protección costera y regeneración de playas en el sur del municipio de Gral. Pueyrredon. II Jornadas Nacionales de Ambiente, Tandil 19–21

Weisse R, Von Storch H (2010) Marine climate and climate change. Storms, wind waves and storm surges. Springer, Praxis Publishing, Chichester, p 243

Wood N, Burton C, Cutter S (2010) Community variations in social vulnerability to Cascadia-related tsunamis in the U.S Pacific Northwest. Nat Hazards 52:369–389

Zhang K, Douglas B, Leatherman S (2001) Beach erosion potential for severe nor'easters. J Coastal Res 17(2):309–321

Zulaica L, Ferraro R (2010) Vulnerabilidad Socioambiental y Dimensiones de la Sustentabilidad en el Sector Periurbano Marplatense, Estudios Socioterritoriales. Revista de Geografía 8:197–219

Zulaica L, Celemín J (2008) Análisis territorial de las condiciones de habitabilidad en el periurbano de la ciudad de Mar del Plata (Argentina), a partir de la construcción de un índice y de la aplicación de métodos de asociación espacial. Revista de Geografía Norte Grande 41:129–146

Modelling and Mapping Landslide Susceptibility of Darjeeling Himalaya Using Geospatial Technology

Biplab Mandal, Subrata Mondal, and Sujit Mandal

Abstract

Landslide susceptibility (LS) mapping is essential for planning and development activities, as well as disaster management in mountainous terrains. The present research deals with the application of bivariate statistical models (LNRF and SI) to identify the spatial distribution of LS of Darjeeling Himalaya, West Bengal, India. To establish this models, the spatial database of 2079 landslides location and 15 landslide affecting factors i.e. elevation, slope angle, slope aspect, slope curvature, geology, soil, drainage density, distance to drainage, lineament density, distance to lineament, SPI, TWI, NDVI, LULC and rainfall were considered. All the thematic data layers were prepared incorporating Google Earth images, SRTM Digital Elevation Model, satellite images, topo-sheets and various recognized sources using ArcGIS (ver. 10.1) environments. The relative importance of each factor was calculated using LNRF and SI methods. Finally, by overlapping all landslide triggering data layers, landslide susceptibility zonation (LSZ) maps were prepared. The AUC values of the ROC curve indicated a reasonable prediction accuracy of 76.50% and 77.90% with an area ratio of 0.7650 and 0.7790 for LNRF and SI models respectively. Very low, low, moderate, high and very high LS zones are associated with LFR of 0.20, 0.50, 0.73, 1.31, 1.67 for the LNRF model and 0.20, 0.29, 0.55, 0.85, 2.29 for the SI model.

Keywords

Darjeeling Himalaya · LSZ · LNRF and SI · Validation · RS and GIS

30.1 Introduction

Landslides are one of the most destructive natural hazards and are caused mostly by the interaction of various geomorphological, geotectonical, geological, hydrological, environmental and anthropogenic factors especially in the mountainous regions (Cubito et al. 2005; Moreiras 2005). This is one of the most threatened, widespread and frequent hazards among various

B. Mandal
Department of Geography, The University of Burdwan, Golapbag, Bardhaman, West Bengal 713104, India

S. Mondal
Department of Geography, A.B.N. Seal College, Cooch Behar, West Bengal 736101, India

S. Mandal (✉)
Department of Geography, Diamond Harbour Women's University, Sarisha, West Bengal 743368, India
e-mail: mandalsujit2009@gmail.com

natural hazards in hilly regions causing human and economic losses worldwide (Mohammady et al. 2012). All over the world, every year around 1000 human lives losses and approximately 4 billion US properties are being damaged due to landslides (Lee and Pradhan 2007). According to the International disaster database, it accounts for about 4.9% of all disasters and most of the landslides occurred (54%) in Asia (Guha-Sapir et al. 2018). According to National Disaster Management Authority (NDMA), approximately fifteen percent (15%) area is landslide hazard-prone and the most vulnerable landslides regions are commonly situated in the Himalayan mountain terrains, the North-eastern hill ranges, the Western Ghats, the Nilgiris, the Vindhya and the Eastern Ghats areas in India.

Many times the Darjeeling Himalaya faced destructive landslides, which took away several lives and brought damages to properties (Starkel and Basu 2000; Sarkar 2010). It is, therefore, necessary to prepare LSZ maps of Darjeeling Himalaya to minimize losses and controls landslides. In the past, Basu and Sarkar (1987), Sengupta (1995), Kanungo et al. (2006), Mandal and Maiti (2014) investigated landslide and correlated damages in Darjeeling Himalaya. Different models and techniques were used to prepare landslide susceptibility zonation maps of the small fragments of Darjeeling Himalaya using geomorphological, lithological, hydrological and environmental factors (Sarkar and Kanungo 2004; Ghosh et al. 2011; Mandal and Maiti 2014; Mandal and Mandal 2016, 2017; Mondal and Mandal 2017; Basu and Pal 2018). Worldwide, many scientific models and techniques have been applied for landslide susceptibility zonation mapping in the last few decades (van Westen et al. 2003; Guzzetti et al. 2005; Lee et al. 2013). For this purpose, bivariate statistical models like frequency ratio (Pradhan and Lee 2010; Yalcin et al. 2011; Mondal and Maiti 2013; Mandal and Mandal 2016, Mondal and Mandal 2017), information value (Sarkar et al. 2006; Pereira et al. 2012; Chen et al. 2014; Sharma et al. 2015), modified information value (Wang et al. 2015a, b; Mandal and Mandal 2017; Mandal and Mondal 2019), landslide nominal risk factor (Gupta and Joshi 1990; Saha et al. 2005; Torkashvand et al. 2014), weights of evidence (Mohammady et al. 2012; Pourghasemi et al. 2013a), fuzzy logic (Guettouche 2013; Sharma et al. 2013), relative effect (Naveen Raj et al. 2011; Ramesh and Anbazhagan 2015) and statistical index (Pradhan et al. 2012; Raman and Punia 2012; Pourghasemi et al. 2013b; Bourenane et al. 2015; Mandal and Mandal 2017) were frequently used. Multivariate techniques like logistic regression (Akgun et al. 2012; Ghosh et al. 2011), discriminant analysis (Davis 2002; Venables and Ripley 2002) and probabilistic models like Bayesian probability (Lee et al. 2002; Sujatha et al. 2014) and certainty factor approach (Pradhan and Kim 2014; Wang et al. 2015a, b) were also used in different parts of the world for landslide susceptibility mapping. Many machine learning algorithms, like Bayes-based algorithms, Naïve Bayes (Bui et al. 2012; Pourghasemi and Rahmati 2018) and Bayes Networks (Pham et al. 2016); functional algorithms such as Artificial Neural Networks (Park et al. 2013; Tsangaratos and Benardos 2014), Support Vector Machine (Peng et al. 2014; Hong et al. 2015), Decision Tree algorithms (Pradhan 2013; Naghibi et al. 2016) were frequently used by researchers for landslide modelling. The main focus of the present study is to prepare LSZ maps of the Darjeeling Himalaya based on landslide nominal risk factor (LNRF) model and Statistical Index (SI) model, considering 15 landslide causative geomorphological, geological, hydrological and environmental factors.

30.2 The Study Area

The present study focuses on the Darjeeling Himalaya area, which is covering of about 2320.31 km^2 which is located between Nepal and Bhutan Himalaya and the north-western side of the state of West Bengal in India which packed the Tethys geosynclinals residue and created different thrust planes, bedding plane, foliation plane, fractures, various lineaments, active faults etc. and lastly distorted the whole lithology. It's enclosed between the latitudes and longitudes of 26° 45′ 44″ N to 27° 13′ 51″ N and 87° 58′ 44″ E to 88° 53′ 10″ E. The main rivers of Darjeeling Himalaya are Tista, Great Rangit, Little Rangit, Mahanadi, Balason and Mechi which increased the steepness of the valley side slope, sharpened the interfluves area and made the slope more susceptible to landslides through surface erosion and mass movement processes. The region slope decreases from north to south, ranging from 42.55° to 0° (mean = 16.41°, standard deviation = 7.58°). The highest and lowest elevation and mean annual temperature ranges from 3600 to 130 m (mean = 1212.46 m, standard deviation = 718.90 m) and 24–12 °C (plain-ridge). During winter month the temperature drops at 5–6 °C on the ridge and summer month reaches 16–17 °C. Generally, Darjeeling Himalaya comprises the lithology of Proterozoic, Upper Palaeozoic-Mesozoic and Tertiary area and the Darjeeling gneiss, the Buxa, the Gondwana, the Daling series, the Damuda series, Nahan group and alluvium. The soil of Darjeeling Himalaya is mixed with an adequate amount of organic matter and humus and this fragile nature of soil helps in the initiation of slope failure. The maximum average rainfall ranges from 2000 and 5000 mm, which occurred during the monsoon period mainly in June, July and August. The geomorphic threshold of the mountain slope materials are influenced by the geomorphic, geotectonic and geohydrologic characteristics, which promote landslides. The hill ranges of Darjeeling Himalaya are structurally controlled, highly rugged and constantly under the active denudational processes. Main boundary thrust (MBT) and main central thrust (MCT) traverse the southern parts of this area which helps in the chemical and mechanical decomposition of rocks rapidly and mass movement processes. The high precipitation, the rapid expansion of settlements and the establishment of multi-storied buildings without proper planning over steep slope increased the chance of landslide occurrences (Fig. 30.1).

30.3 Materials and Methods

30.3.1 Preparation of Landslide Inventory and Landslide Conditioning Thematic Data Layers

The landslide events mapping of past through landslide inventory as well as distribution map is the primary stage in LSZ studies (Yalcin et al. 2011), which was prepared combining Google Earth imagery, Arc map high-resolution world imagery, high-resolution Sentinel-2 imagery, SOI topo-sheets and GPS data. The total number of 2079 polygon coverage with different types (shallow translational debris slides, shallow translational rockslides and deep-seated rockslides) of landslide locations were identified covering an area of about 6.47 km^2. 1455 landslide locations (70%) were used for training the model whereas the remaining 624 landslide locations (30%) were used for validating the model.

In the present study, a total of fifteen geomorphological, lithological/geological, hydrological and environmental landslide causative factors were taken into consideration. Elevation, slope angle, slope aspect, slope curvature, SPI and TWI data layers were directly prepared from downloaded (earthexplorer.usgs.gov) Shuttle Radar Topography Mission Digital Elevation Model (SRTM DEM with 30 m spatial resolution) in ArcGIS (10.1) software environment. The SPI and TWI maps were prepared using the following equations (30.1 & 30.2):

$$SPI = Ln\{(A_s + 0.001) \\ \times (\tan_\beta \text{ in percent\_rise}/100) + 0.001)\} \quad (30.1)$$

$$TWI = Ln\{(A_s + 0.001)/ \\ (\tan_\beta \text{ in percent\_rise}/100) + 0.001)\} \quad (30.2)$$

where, Ln is the natural log, A_s is the flow accumulation, $\tan_\beta$ is the slope and percent_rise is the slope in %.

District soil and geological maps are collected from NBSS (National Bureau of Soil Science)

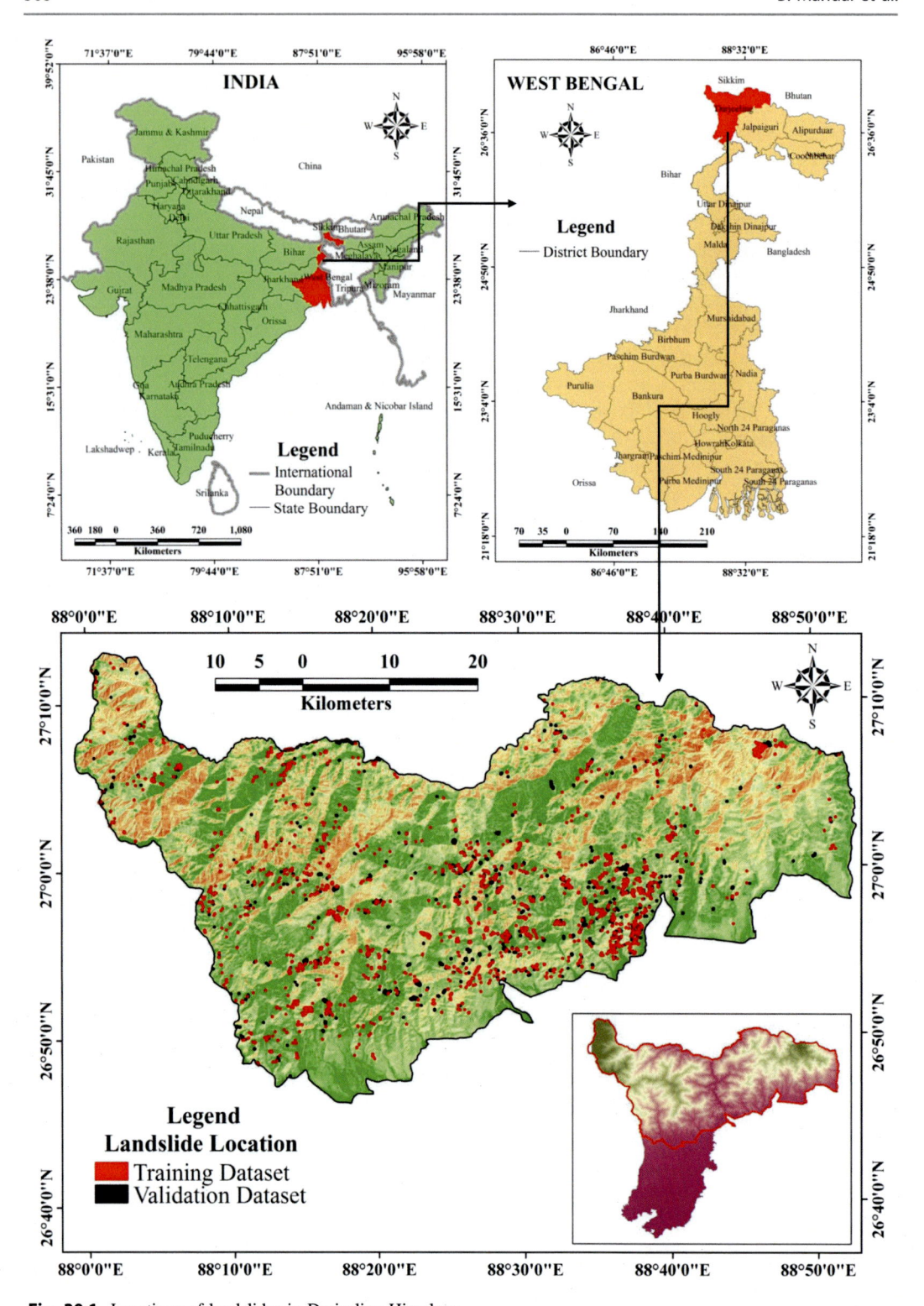

Fig. 30.1 Locations of landslides in Darjeeling Himalaya

and LUP (Land Use Planning) Regional Centre (Kolkata) and Geological Survey of India (GSI, Kolkata) and collected maps were georeferenced and digitised and overlaid on study area in ArcGIS (10.1) software tools to extract the soil and geology of the Darjeeling Himalaya. A drainage map was prepared combining topo-sheets and disaster mitigation map of Darjeeling district (http://darjeeling.gov.in/gismaps.html) and updated on Google Earth Imagery. The drainage density and distance to drainage maps were prepared from the drainage map using line density and buffer analysis in ArcGIS (10.1) software environment. NDVI map was extracted from downloaded (glovis.usgs.gov) Landsat Thematic Mapper (TM) data (February 2009, 30 m spatial resolution). The NDVI was assessed by using the formula of (IR − R)/(IR + R), where IR is the infrared portion and R is the Red portion of the electromagnetic spectrum. Lineament density and distance to lineament maps were derived from www.wbphed.gov.in. The whole basin was divided into a 1×1 km vector grid with centroid points. Then each point was assigned the length of lineament within the vector grid. The lineament density and distance to lineament maps were prepared using IDW interpolation and buffer analysis. The land use/land cover map was prepared from www.banglarbhumi.gov.in and lastly, the rainfall distribution map was derived from downloaded world climate data (www.worldclim.org). All the factors map were prepared from primary and various secondary sources and classified into 10 classes except geology (6 types), soil (8 types) and LULC (9 types) using ArcGIS (10.1) software environments (Table 30.1 and Figs. 30.2, 30.3, 30.4 and 30.5).

30.3.2 Application of LNRF and SI Models for Landslide Susceptibility Mapping

30.3.2.1 Landslide Nominal/Normal/ Numerical Risk Factor (LNRF)

Landslide nominal risk factor attempts to reduce each map layer thoughts to be important to a single metric value and adds up the scores to produce an overall index. LNRF is one of the most important bivariate statistical models by Gupta and Joshi (1990), Saha et al. (2005), Torkashvand et al. (2014) and Mondal and Mandal (2019) to produce the landslide susceptibility map applying the following equations (Eqs. 30.3 and 30.4).

$$\text{LNRF}_{ij} = \frac{\text{Npi}x(S_j)}{(\sum_{i=1}^{n} \text{Npi}x(S_j))/n} \quad (30.3)$$

where, LNRF_{ij} is the Landslide Nominal Risk Factor of the jth class of the ith landslide conditioning factor, Npi$x$ (S_i) is the number of pixels containing landslides in the jth class, n is the number of classes present in the particular conditioning factor.

$$\text{LSI}_{\text{LNRF}} = \sum_{i=1}^{n} \text{LNRF} \quad (30.4)$$

where, n is the number of conditioning factors.

30.3.2.2 Statistical Index (SI) Method

The statistical index method is associated with the determination of weighted values of a factor's class which is defined as the natural algorithm of the landside density of a particular class divided by the landslide density of the study area after Pourghasemi et al. (2013b), Tay et al. (2014), Bourenane et al. (2015) and Mandal and Mandal (2017) (Eq. 30.5).

$$\begin{aligned} wi &= \ln\left(\frac{\text{Densclass}}{\text{Densmap}}\right) \\ &= \ln\left[\left(\frac{\text{Npi}x(\text{Si})}{\text{Npi}x(\text{Ni})}\right) \div \left(\frac{\sum \text{Npi}x(\text{Si})}{\sum \text{Npi}x(\text{Ni})}\right)\right] \end{aligned} \quad (30.5)$$

where, wi is the weight given to the parameter class, ln is the natural logarithm, Densclass is the landslide density within the parameter class, Densmap is the landslide density within the area, Npi$x$ (Si) is the number of landslide pixels in parameter class ii, Npi$x$(Ni) is the number of pixels in parameter class i, $\sum \text{Npi}x$(Si) is the total number of landslide pixels in the area,

Table 30.1 Sources of different data layers used

Data layers	Data employed	Sources
Administrative boundary of India and West Bengal	Map of the census of India, 2011	www.censusindia.gov.in/2011census/maps/atlas/00part1.pdf
Landslide inventory/distribution map	Google Earth imagery, toposheets and GPS field survey data (2016–2017)	Google Earth imagery (about 15 m per pixel); high-resolution Sentinel-2 imagery (resolution ranging from 10 to 60 m); Arc map high-resolution world imagery (resolution 30 cm to 15 m); Survey of India (SOI with scale 1:50,000), Kolkata, toposheets (78A/4, 78A/8, 78A/12, 78A/16, 78B/1, 78B/2, 78B/5, 78B/6, 78B/9 and 78B/13) and field studies
Elevation, aspect, slope, curvature, SPI and TWI	SRTM DEM (30 m spatial resolution)	earthexplorer.usgs.gov
Geology	Geological map (1:250,000)	Geological Survey of India, Kolkata
Soil	Soil map (1:250,000)	NBSS and LUP Regional Centre, Kolkata
Lineament density and distance to lineament	Drinking water prospects map (1:50,000)	www.wbphed.gov.in
Drainage density and distance to drainage	Disaster mitigation map of Darjeeling district (1:100,000)	darjeeling.gov.in/gismaps.html
Rainfall	Rainfall data (1950–2010) (1 km × 1 km)	www.worldclim.org
NDVI	Landsat TM image, Feb. 2009 (30 m spatial resolution)	glovis.usgs.gov
LULC data layer	Land use map (1:50,000)	www.banglarbhumi.gov.in

$\sum Npix(Ni)$ is the total number of pixels in the area.

Finally, the landslide susceptibility index (LSI) map was prepared by summing the weighted parameters, following Eq. (30.6):

$$LSI = \sum_{j=1}^{n} Wij \qquad (30.6)$$

30.4 Result and Discussion

30.4.1 Relationship Between Landslide Causative Factors and Landslide Occurrences

All the district towns i.e. Darjeeling, Kalimpong, Kurseong and Mirik towns experienced destructive landslides in the past. Weak lithological units along with the presence of steep slopes, moderate to high drainage density, closely spaced lineaments, the permeability of surface soil and moderate to the high intensity of rainfall have made the Darjeeling Himalaya more prone to landslides. In the present day anthropogenic activities such as the construction of roads and buildings, expansion of tea garden area and development of tourism are putting enormous pressure on fragile geological units of Darjeeling Himalaya and making mountain slope landslide-prone. The present study classified all the causative factors into four i.e. geomorphological factors group, lithological factors group, hydrological factors group and triggering and other factors group.

30.4.1.1 Geomorphological Factors and Landslides

Geomorphological factors of elevation, slope angle, slope aspect and slope curvature played an important role in driving landslides in Darjeeling

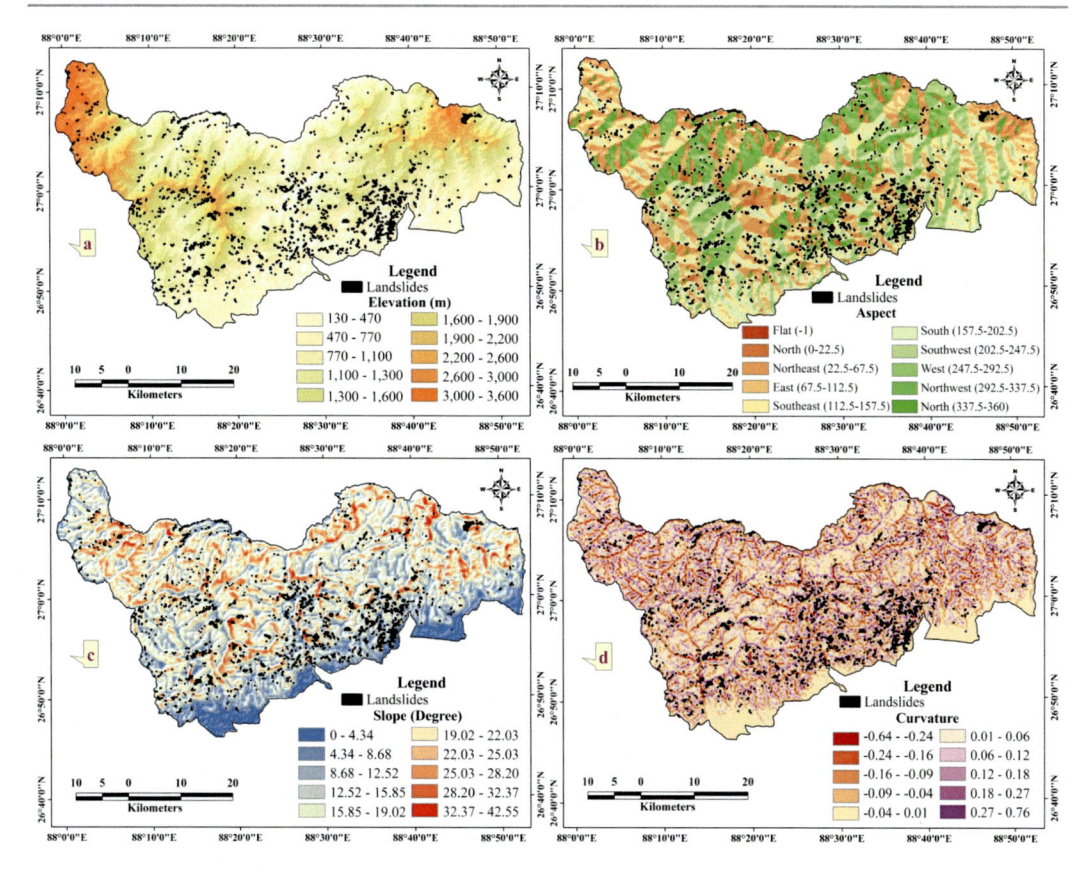

Fig. 30.2 Landslide triggering geomorphological factors; **a** elevation, **b** aspect, **c** slope and **d** curvature of Darjeeling Himalaya

Himalaya. All the major urban centres of Darjeeling Himalaya such as Kalimpong, Darjeeling, Kurseong and Mirik are situated at higher elevated areas of the mountains steep slope. Darjeeling Himalaya was classified into 10 elevation zones out of which two zones of 470–770 m and 770–1100 m are affected mostly by landslides where LNRF values are 3.11 and 2.57 respectively. The SI values of these two zones are positive i.e. 0.608 and 0.418. The remaining eight elevation classes are registered with an LNRF value of less than unity and negative SI values where landslide probability is low. The slope angle is one of the most causative factors of landslides in Darjeeling Himalaya. Landslide probability is very high at places having more than 15° slopes. Slope classes of 12.52–15.85, 15.85–19.02, 19.02–22.03 and 22.03–25.03 are having maximum landslide affected pixels as well as high LNRF values. Moderate slope areas of Darjeeling Himalaya provide a suitable environment for the percolation of rainwater and slope saturation over the space. Positive SI values are found above 22° slopes. Very low negative SI values are being observed in the slope classes of 12.52–15.85, 15.85–19.02, and 19.02–22.03 where large numbers of pixels are being affected by landslides. The south, south-east, and south-west facing slopes of Darjeeling Himalaya receive a high intensity of orographic rainfall every year. All these slope facets are attributed with maximum run-off, soil erosion and slope materials movement downslope. LNRF values of more than unity are found in the south, south-east, south-west and east facing slopes where SI values are also positive. Other slope facets i.e. north, west and north-west slope facets are not experienced with active and re-activated

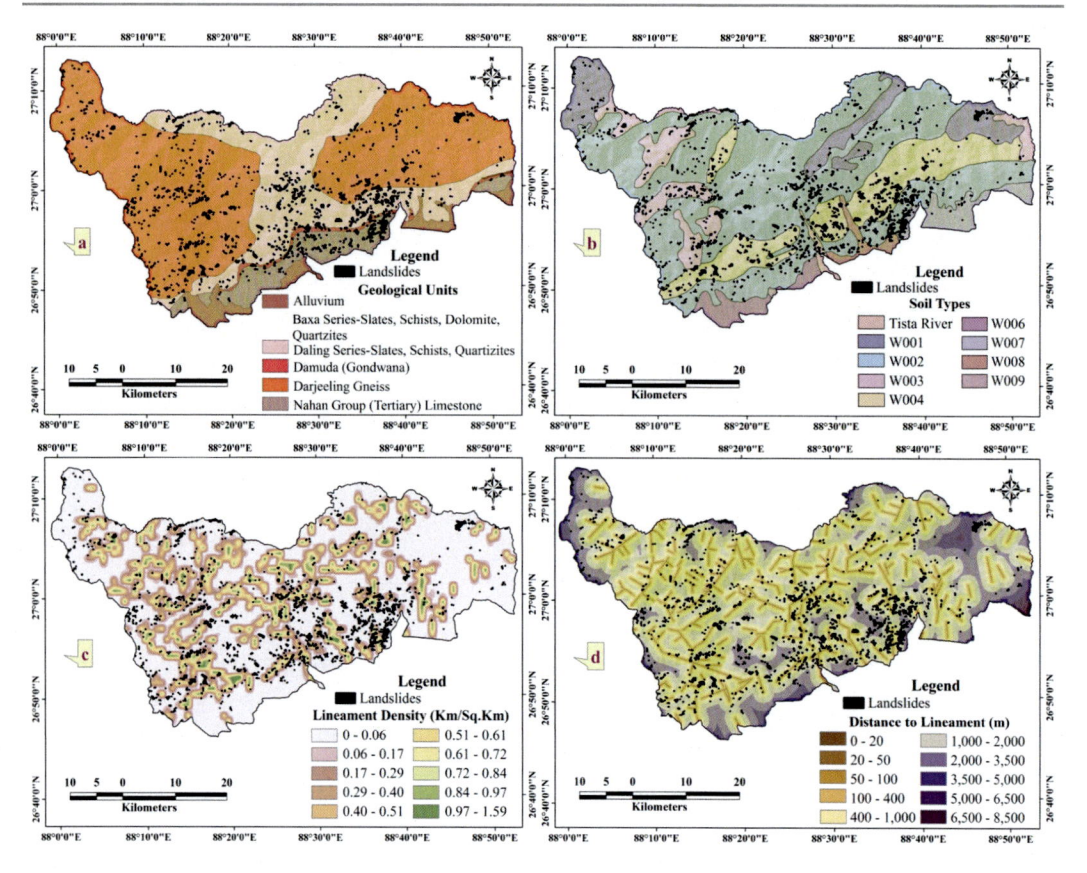

Fig. 30.3 Landslide triggering lithological factors; **a** geology, **b** soil, **c** lineament density and **d** distance to lineament of Darjeeling Himalaya

landslide phenomena and consequently having low LNRF values and negative SI values as well as low probability of landslide occurrences. High positive and negative slope curvature are very much prone to slope materials saturation and slope movement. In Darjeeling Himalaya, maximum numbers of the landslide-affected pixels are being found in negative slope curvature or concave slope segments where LNRF values are more than unity. The foothills of the mountain slope is characterized by concave slope where various sizes of slope materials are being deposited. During monsoons, materials get saturated and make the concave slope more vulnerable to landslides. The positive slope curvature is located in the upper segment of the mountain slope with less rock-soil debris. The positive slope curvature areas are attributed to less number of landslide-affected pixels and low LNRF values and negative SI values (Fig. 30.2 and Table 30.2).

30.4.1.2 Lithological Factors and Landslides

Geology, soil, lineament density and distance to lineaments are included in the lithological factors group. Geologically, Darjeeling Himalaya is divided into six major units i.e. alluvium; Baxa Series-Slates, Schists, Dolomite, Quartzites; Daling Series-Slates, Schists, Quartzites; Damuda (Gondwana); Darjeeling Gneiss; and Nahan Group (Tertiary) Limestone with LNRF values of 0.033, 0.005, 2.427, 0.143, 3.235 and 0.157 respectively. The lithological units of the Darjeeling Gneiss and Daling series exhibit maximum LNRF values with a large number of landslides affected pixels. These two units also depicted SI values of −0.092 and 0.381. The

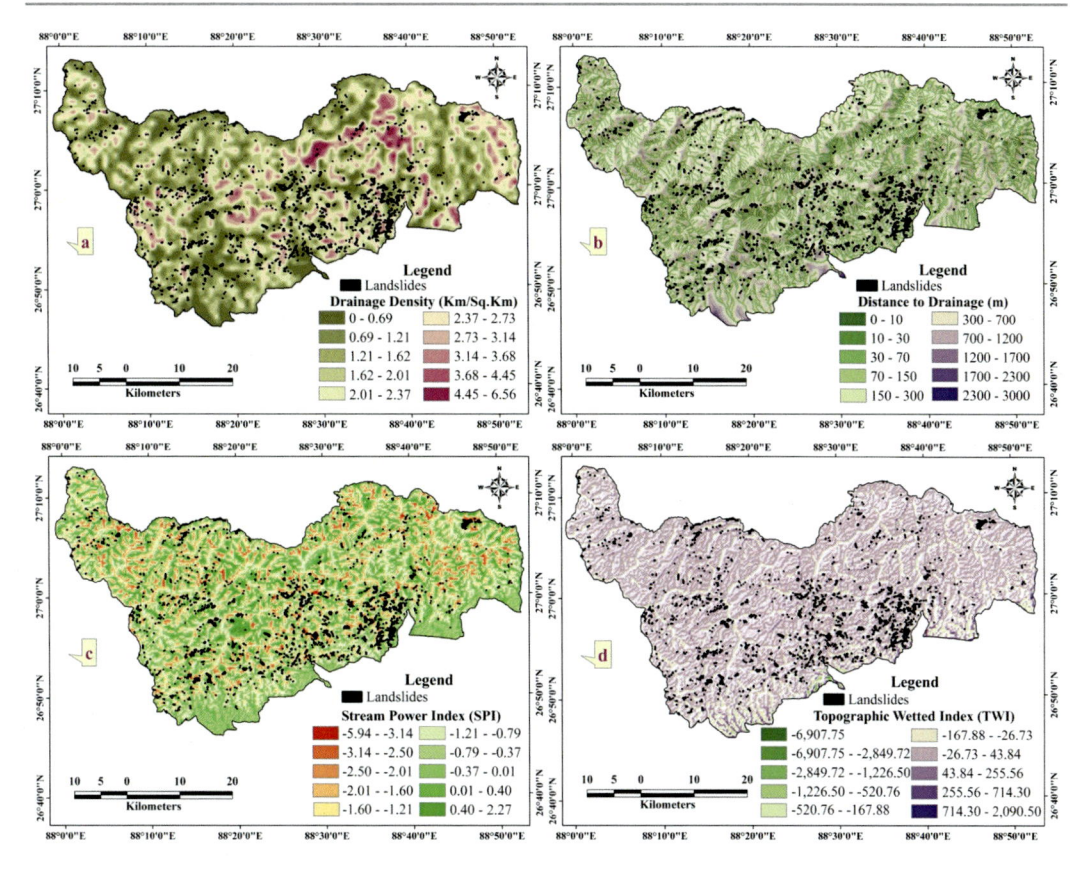

Fig. 30.4 Landslide triggering hydrological factors; **a** drainage density, **b** distance to drainage, **c** SPI and **d** TWI of Darjeeling Himalaya

presence of various weaknesses planes such as thrust planes, bedding planes, lineaments, joints and cracks caused due to heavy compressive forces created Darjeeling Himalaya as one of the fragile lithologies of eastern Himalaya. A thick layer of soil caused due to disintegration and decomposition of rocks in Darjeeling Himalaya has made the area more susceptible to landslides. W002 and W004 group of soils having a large number of landslide affected pixels and LNRF values of 4.024 and 3.626. W006, W007, W009 and W008 group of soils exhibit very few landslide-affected pixels with low LNRF values. All the soil groups are with negative SI values except W004. Lineament density and distance to lineaments are two important lithological factors of slope instability in Darjeeling Himalaya. High lineament density is being found in the geological units of the Darjeeling gneiss and Daling series. The middle-most section of Darjeeling Himalaya is registered with more lineaments and high lineament density. Most of the landslide-affected pixels are being observed below the lineament density of 0.61 km km^{-2} where the LNRF value ranges from 0.690 to 4.760. High and low lineament density zones are attributed to negative SI values. The probability of landslide decreases with increasing distance from lineaments. Distance from lineament classes of 100–400, 400–1000, 1000–2000 and 2000–3500 having LNRF values of 1.686, 2.929, 2.966 and 1.786 respectively. Negative SI values are being observed beyond the distance of 3500 m from lineaments (Fig. 30.3 and Table 30.2).

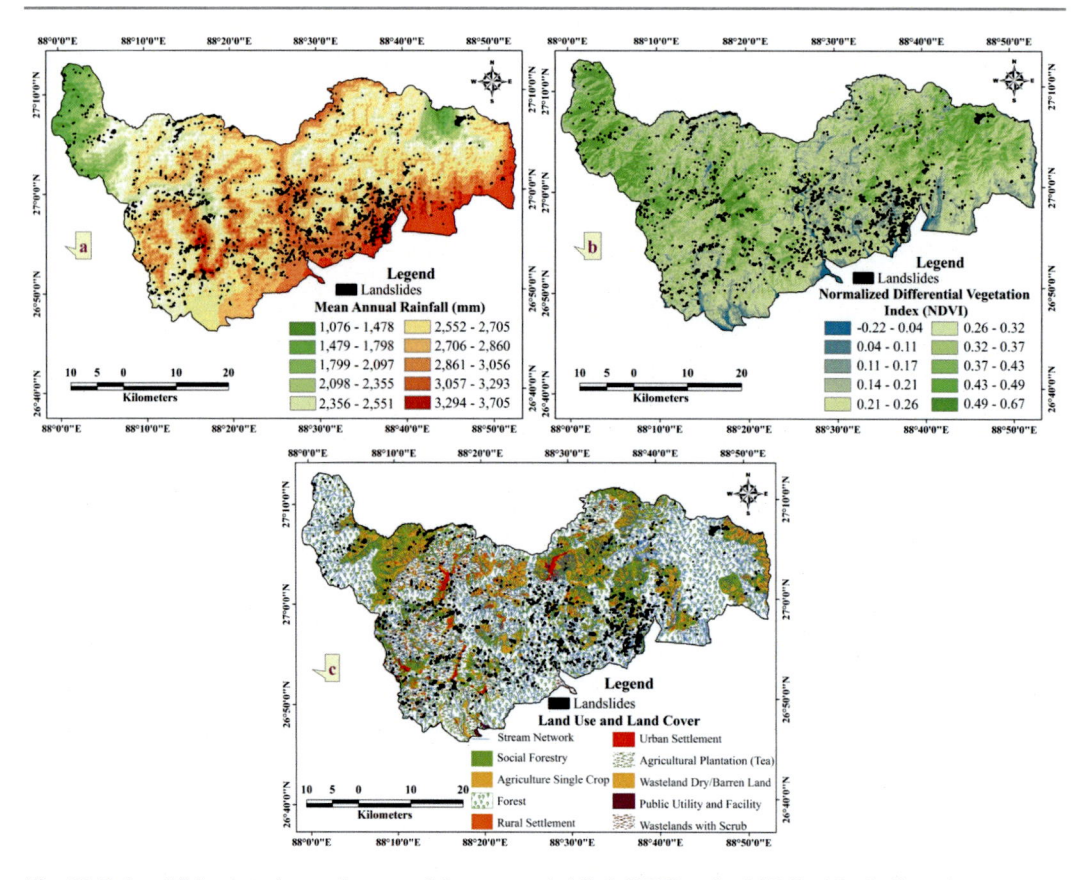

Fig. 30.5 Landslide triggering environmental factors; **a** rainfall, **b** NDVI and **c** LULC of Darjeeling Himalaya

30.4.1.3 Hydrological Factors and Landslides

The hydrological factors group comprises drainage density, distance to drainage, stream power index (SPI) and topographic wetness index (TWI). All these hydrological parameters are very much significant in landslide occurrences. High drainage density makes the slope more susceptible to soil erosion and slope materials movement downslope. Moderate to high drainage density is found in the east–west direction of Darjeeling Himalaya. In the middle and lower most section of Darjeeling Himalaya with moderate drainage density, more landslide activities are depicted. The maximum area of Darjeeling Himalaya exists below the drainage density of 3.14 km km^{-2}. Most of the active and reactivated landslide phenomena took place below the drainage density of 3.14 km km^{-2} where LNRF values are high. Beyond the drainage density of 3.14 km km^{-2}, minimum numbers of pixels are being affected by landslides where SI values are negative. Drainage density classes of 1.21–1.62, 1.62–2.01 and 2.01–2.37 depicted a large number of landslide-affected pixels with LNRF values of 1.800, 2.460 and 1.542 respectively. All these three classes have SI values of −0.010, 0.240, and 0.010. Places away from the drainage have a low probability of soil erosion and slope materials movement. Within 700 m from drainage lines in Darjeeling Himalaya, the occurrences of landslide events are high and consequently have more landslide affected pixels and high LNRF values. Areas within the distance of 300 m registered with positive SI values and beyond 300 m have negative SI values. More landslide-affected pixels were being observed at the places of having high stream power index values. The study depicted a positive relationship between stream power index and LNRF values.

Table 30.2 Weights of LNRF and SI models in different landslide triggering factors

Landslide triggering factors	Sub-classes	Number of pixels in domain	Number of landslide pixels in domain	LNRF	SI
Elevation (m asl)	130–470	409,786	465	0.952	−0.512
	470–770	437,151	1520	3.113	0.608
	770–1100	437,589	1259	2.579	0.418
	1100–1300	246,682	445	0.912	−0.049
	1300–1600	342,538	465	0.952	−0.333
	1600–1900	260,070	178	0.365	−1.018
	1900–2200	171,257	268	0.549	−0.191
	2200–2600	142,581	207	0.424	−0.266
	2600–3000	80,867	69	0.141	−0.797
	3000–3600	49,605	6	0.012	−2.751
Aspect	Flat (−1)	231	0	0.000	−4.086
	North (0–22.5)	136,124	90	0.184	−1.052
	Northeast (22.5–67.5)	296,730	706	1.446	0.228
	East (67.5–112.5)	395,973	890	1.823	0.171
	Southeast (112.5–157.5)	413,694	893	1.829	0.131
	South (157.5–202.5)	386,071	777	1.592	0.061
	Southwest (202.5–247.5)	299,310	844	1.729	0.398
	West (247.5–292.5)	266,605	396	0.811	−0.243
	Northwest (292.5–337.5)	250,427	252	0.516	−0.632
	North (337.5–360)	132,961	34	0.070	−2.002
Slope (degree)	0–4.34	187,019	134	0.274	−0.972
	4.34–8.68	255,476	403	0.825	−0.183
	8.68–12.52	343,632	590	1.209	−0.098
	12.52–15.85	374,691	668	1.368	−0.060
	15.85–19.02	427,123	746	1.528	−0.081
	19.02–22.03	382,230	704	1.442	−0.028
	22.03–25.03	278,609	573	1.174	0.083
	25.03–28.20	189,918	459	0.940	0.244
	28.20–32.37	107,393	501	1.026	0.902
	32.37–42.55	32,035	104	0.213	0.539
Curvature	−0.64 to −0.24	48,330	159	0.326	0.552
	−0.24 to −0.16	168,473	765	1.567	0.875
	−0.16 to −0.10	286,017	855	1.751	0.457
	−0.10 to −0.04	390,224	859	1.760	0.151
	−0.04 to 0.02	577,689	672	1.376	−0.487
	0.02–0.06	406,708	548	1.122	−0.340
	0.06–0.12	330,604	435	0.891	−0.364
	0.12–0.19	234,280	330	0.676	−0.296
	0.19–0.27	108,702	192	0.393	−0.070
	0.27–0.76	27,099	67	0.137	0.267

(continued)

Table 30.2 (continued)

Landslide triggering factors	Sub-classes	Number of pixels in domain	Number of landslide pixels in domain	LNRF	SI
Geology	Alluvium	104,604	27	0.033	−1.993
	Baxa Series-Slates, Schists, Dolomite, Quartzites	2344	4	0.005	−0.104
	Daling Series-Slates, Schists, Quartizites	712,203	1975	2.427	0.381
	Damuda (Gondwana)	35,988	116	0.143	0.532
	Darjeeling Gneiss	1,523,339	2632	3.235	−0.092
	Nahan Group (Tertiary) Limestone	199,648	128	0.157	−1.083
Soil	Tista River	14,861	0	0.000	−4.086
	W001	261,470	492	0.907	−0.006
	W002	1,434,693	2183	4.024	−0.219
	W003	221,855	183	0.337	−0.831
	W004	446,612	1967	3.626	0.844
	W006	75,842	1	0.002	−4.967
	W007	85,141	17	0.031	−2.250
	W008	35,338	39	0.072	−0.540
	W009	2314	0	0.000	−4.086
Lineament density (km km^{-2})	0–0.06	1,309,367	2324	4.760	−0.065
	0.06–0.17	300,996	482	0.987	−0.168
	0.17–0.29	230,519	577	1.182	0.279
	0.29–0.40	192,603	337	0.690	−0.079
	0.40–0.51	162,361	342	0.701	0.106
	0.51–0.61	141,605	390	0.799	0.375
	0.61–0.72	102,252	248	0.508	0.247
	0.72–0.84	73,150	118	0.242	−0.160
	0.84–0.97	48,815	53	0.109	−0.556
	0.97–1.59	16,458	11	0.023	−1.041
Distance from lineament (metre)	0–20	23,691	42	0.086	−0.066
	20–50	36,643	61	0.125	−0.129
	50–100	63,438	119	0.244	−0.009
	100–400	406,582	823	1.686	0.067
	400–1000	787,207	1426	2.921	−0.044
	1000–2000	781,676	1448	2.966	−0.022
	2000–3500	369,169	872	1.786	0.221
	3500–5000	95,146	91	0.186	−0.683
	5000–6500	10,670	0	0.000	−4.086
	6500–8500	3904	0	0.000	−4.086

(continued)

Table 30.2 (continued)

Landslide triggering factors	Sub-classes	Number of pixels in domain	Number of landslide pixels in domain	LNRF	SI
Drainage density (km km^{-2})	0–0.69	243,293	266	0.545	−0.549
	0.69–1.21	409,926	732	1.499	−0.059
	1.21–1.62	469,031	879	1.800	−0.010
	1.62–2.01	498,755	1201	2.460	0.240
	2.01–2.37	393,587	753	1.542	0.010
	2.37–2.73	272,462	665	1.362	0.254
	2.73–3.14	166,766	279	0.571	−0.124
	3.14–3.68	76,489	83	0.170	−0.557
	3.68–4.45	34,741	20	0.041	−1.191
	4.45–6.56	13,076	4	0.008	−1.823
Distance to drainage (metre)	0–10	92,856	210	0.430	0.178
	10–30	184,399	412	0.844	0.165
	30–70	356,416	754	1.544	0.111
	70–150	628,425	1291	2.644	0.081
	150–300	763,840	1600	3.277	0.101
	300–700	498,808	600	1.229	−0.454
	700–1200	45,871	11	0.023	−2.066
	1200–1700	5659	4	0.008	−0.985
	1700–2300	1593	0	0.000	−4.086
	2300–3000	259	0	0.000	−4.086
SPI	−5.939 to −3.140	14,266	20	0.041	−0.301
	−3.140 to −2.496	57,082	92	0.188	−0.161
	−2.496 to −2.013	132,095	221	0.453	−0.124
	−2.013 to −1.595	223,338	379	0.776	−0.110
	−1.595 to −1.209	296,386	482	0.987	−0.152
	−1.209 to −0.791	361,250	612	1.254	−0.111
	−0.791 to −0.373	379,786	691	1.415	−0.040
	−0.373 to 0.014	571,397	1040	2.130	−0.040
	0.014–0.400	391,010	752	1.540	0.016
	0.400–2.266	151,516	593	1.215	0.726
TWI/CTI	−6907.755	231	0	0.000	−4.086
	−6907.755 to −2849.717	848	0	0.000	−4.086
	−2849.717 to −1226.502	2601	0	0.000	−4.086
	−1226.502 to −520.757	6953	0	0.000	−4.086
	−520.757 to −167.884	36,026	15	0.031	−1.515
	−167.884 to −26.735	454,856	639	1.309	−0.299
	−26.735 to 43.840	1,987,529	4151	8.503	0.098
	43.840–255.564	76,829	67	0.137	−0.775
	255.564–714.298	10,964	10	0.020	−0.731
	714.298–2090.502	1289	0	0.000	−4.086

(continued)

Table 30.2 (continued)

Landslide triggering factors	Sub-classes	Number of pixels in domain	Number of landslide pixels in domain	LNRF	SI
Mean annual rainfall (mm)	1076–1478	54,493	40	0.082	−0.948
	1478–1798	70,178	43	0.088	−1.128
	1798–2097	73,731	82	0.168	−0.532
	2097–2355	88,710	287	0.588	0.536
	2355–2551	268,934	199	0.408	−0.940
	2551–2705	551,234	909	1.862	−0.138
	2705–2860	608,234	1342	2.749	0.153
	2860–3056	423,558	1165	2.386	0.373
	3056–3293	277,977	628	1.286	0.177
	3293–3705	161,077	187	0.383	−0.489
NDVI	−0.200 to 0.037	34,132	324	0.664	1.612
	0.037–0.110	77,506	482	0.987	1.189
	0.110–0.166	167,231	520	1.065	0.496
	0.166–0.214	273,220	683	1.399	0.278
	0.214–0.263	381,618	758	1.553	0.048
	0.263–0.315	451,330	753	1.542	−0.127
	0.315–0.368	428,060	587	1.202	−0.323
	0.368–0.427	364,538	425	0.871	−0.485
	0.427–0.493	257,616	245	0.502	−0.689
	0.493–0.667	142,875	105	0.215	−0.947
LULC	Social forestry	309,800	453	0.835	−0.259
	Agricultural single crop	331,856	537	0.990	−0.157
	Forest	1,430,826	2978	5.490	0.095
	Rural settlement	126,193	150	0.277	−0.466
	Urban settlement	16,672	18	0.033	−0.562
	Agricultural plantation (tea)	337,588	514	0.948	−0.218
	Wasteland dry/barren land	399	7	0.013	2.226
	Public utility and facility	3052	0	0.000	−4.086
	Wastelands with scrub	21,740	225	0.415	1.698

The locations lying between −1.209 and 2.266 stream power index values are attributed with SPI values of more than unity. SI values are positive at the places having high SPI values. It is assumed that the stream network and its activities are crucial for slope instability in Darjeeling Himalaya. Landslide susceptibility is high to very high where SPI is high. Topographic wetness index classes of −167.884 to −26.735 and −26.735 to 43.840 are dominated by landslide-affected pixels with LNRF values of 1.309 and 8.503. A positive SI value of 0.098 is observed where TWI ranges from −26.735 to 43.840. In Darjeeling Himalaya, very close to the main river slope is very gentle with maximum soil saturation. Very high topographic wetness indexed areas are not associated with landslides where LNRF value is 0 and SI value is high negative (Fig. 30.4 and Table 30.2).

30.4.1.4 Environmental Factors and Landslides

Rainfall is the triggering factors of landslides in Darjeeling Himalaya. Historical landslide studies revealed that almost all the massive landslides caused as a result of high intensity of rainfall which took several lives and damaged properties a lot. Most of the landslides in Darjeeling Himalaya took places in moderate to high mean annual rainfall zones. Except for the extreme north-western and eastern part, almost every places received moderate to high rainfall. The maximum numbers of landslide pixels were found in mean annual rainfall zones of 2551–2705, 2705–2860, and 2860–3056 with LNRF values of 1.862, 2.749 and 2.386. These three classes also depicted SI values of −0.138, 0.153 and 0.177. It was observed that maximum mean annual rainfall occurred in the higher elevated zones with dense vegetation cover, minimum soil depth, the minimum length of overland flow and low drainage density where the probability of landslide occurrences is low. In Darjeeling Himalaya, very high-intensity rainfall classes and very low-intensity rainfall classes demonstrated a minimum number of landslide affected pixels and minimum LNRF values. Minimum mean rainfall classes are associated with minimum landslide affected pixels and negative SI values. LNRF values are more than unity in the NDVI classes of 0.110–0.166, 0.166–0.214, 0.214–0.263, 0.263–0.315, and 0.315–0.368. Higher NDVI values denote negative SI values and minimum LNRF values. All the places have NDVI values ranging from −0.200 to 0.263 attributed with positive SI values. Forest, tea plantation and social forestry are the dominating LULC in Darjeeling Himalaya. Forest area of Darjeeling Himalaya experienced maximum landslide affected pixels followed by agricultural single crop, agricultural plantation (Tea), social forestry etc. Settlement area of both rural and urban comprised 168 landslide-affected pixels. LNRF values of forestry, social forestry, plantation, rural settlement and urban settlement are 5.490, 0.835, 0.948, 0.277 and 0.033 respectively. Positive SI values were found in LULC classes of forest, wasteland dry/barren land and wasteland with scrub. It has been found that the tea plantation area and forest area of Darjeeling Himalaya faced many destructive landslide events. Besides settlement area over steep slope became more susceptible to landslides (Fig. 30.5 and Table 30.2).

30.4.2 Models Development and LSZ Mappings

The landslide susceptibility index (LSI) maps ware prepared applying LNRF and SI models, summing up the contrast value of each factor (Eq. 30.7).

$$\begin{aligned}\text{LSI}_{\text{LNRF/SI}} = \{&(\text{elevation\_LNRF/SI}) + (\text{aspect\_LNRF/SI}) \\&+ (\text{slope\_LNRF/SI}) + (\text{curvature\_LNRF/SI}) \\&+ (\text{geology\_LNRF/SI}) + (\text{soil\_LNRF/SI}) \\&+ (\text{lineament density\_LNRF/SI}) \\&+ (\text{distance to lineamnet\_LNRF/SI}) \\&+ (\text{drainage density\_LNRF/SI}) \\&+ (\text{distance to drainage\_LNRF/SI}) \\&+ (\text{SPI\_LNRF/SI}) + (\text{TWI\_LNRF/SI}) \\&+ (\text{rainfall\_LNRF/SI}) + (\text{NDVI\_LNRF/SI}) \\&+ (\text{LULC\_LNRF/SI})\}\end{aligned}$$

(30.7)

The prepared landslide susceptibility index value ranges from 10.04 to 48.60 for the LNRF model and −18.33 to 6.28 for the SI model, where a greater LSI value indicates more susceptible for landslides and a lower value suggests less susceptibility to slope failure. Both LNRF and SI models and prepared landslide susceptibility maps were classified into five different zones using Jenks natural breaks classification method i.e. very low (10.04–24.85 and −18.33 to −8.10), low (24.85–30.15 and −8.10 to 4.62), moderate (30.15–34.54 and −4.62 to −2.31), high (34.54–39.07 and −2.31 to −0.09) and very high (39.07–48.60 and −0.09 to 6.28) (Fig. 30.6). Most of the areas of the central, south-eastern and south-western parts of the Darjeeling Himalaya registered with very high landslide susceptibility which covers a landslide area of 31.82% under LNRF and 50.83% under SI of the total landslide area followed by high (38.04 and 30.64%), moderate (20.05 and 14.27%), low (8.70 and 3.44%) and very low (1.39 and 0.81%). More than 48 and 58% of the area pixels were found in high to very high landslide susceptibility zones under LNRF and SI models of total area and the chance of landslide probability was very high due to slope saturation and reduction of cohesion of the slope materials. The 51.92% area under LNRF and 41.75% area under the SI model were situated under very low to moderate susceptibility zones. These zones were located mainly along the southern fringe area, north-eastern part, north-western part and some middle portion of the Darjeeling Himalaya (Table 30.3). The landslide density or landslide frequency ratio

Fig. 30.6 LSZ maps produced by LNRF and SI models

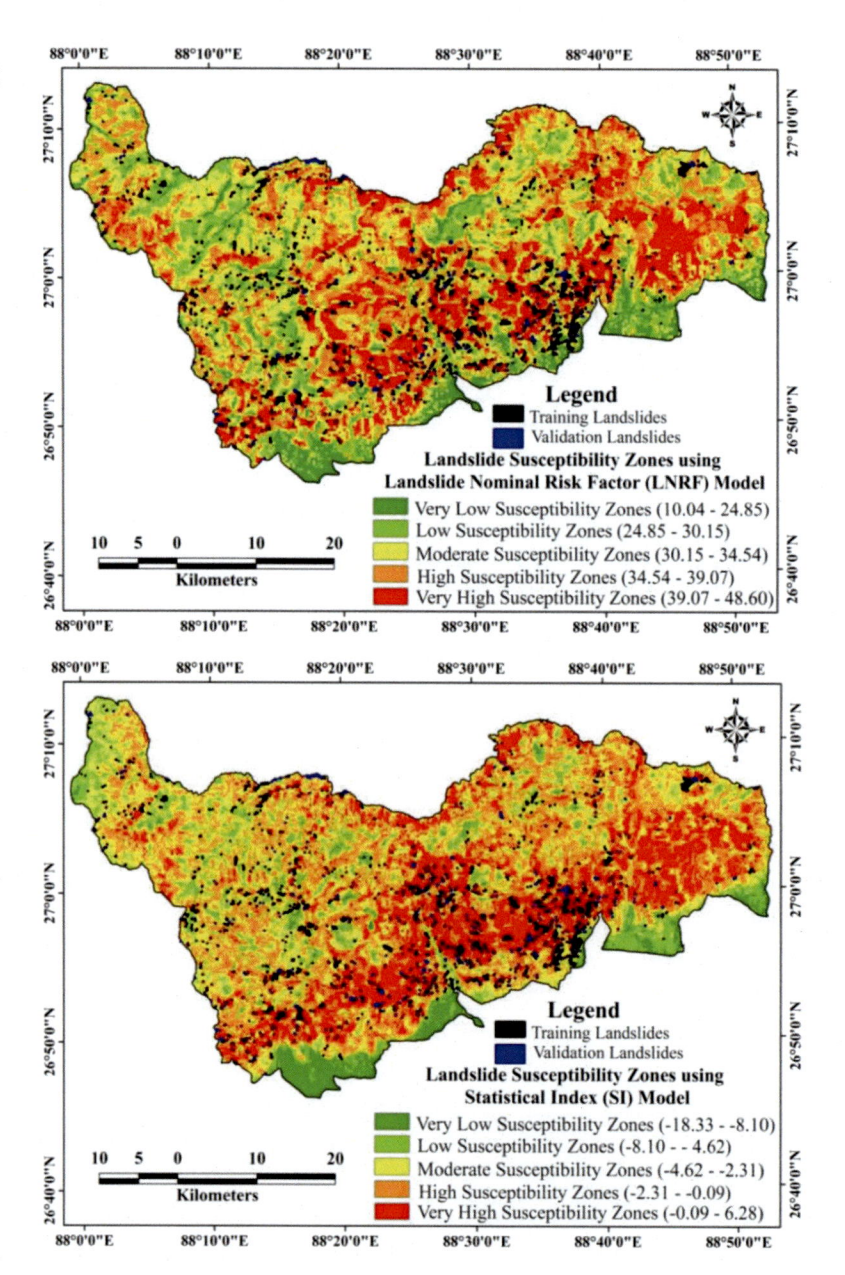

Table 30.3 Frequency ratio value of LNRF and SI models in five different landslide susceptibility zones

Landslide susceptibility zones (LSZs)	% of total pixels		% of total landslide pixels		Frequency ratio	
	LNRF model	SI model	LNRF model	SI model	LNRF model	SI model
Very low	7.00	4.00	1.39	0.81	0.20	0.20
Low	17.45	11.98	8.70	3.44	0.50	0.29
Moderate	27.47	25.77	20.05	14.27	0.73	0.55
High	29.01	36.01	38.04	30.64	1.31	0.85
Very high	19.06	22.24	31.82	50.83	1.67	2.29

value increases from very low to very high susceptibility zones for both models.

30.4.3 Validation and Accuracy Assessments of the LSZ Maps

An important aspect of every projected model is to validate the reality that can offer expressive results. Worldwide different validations as well as accuracy assessments techniques, have been repeatedly applied for the LSZ mapping, out of which frequency ratio (FR) value and receiver operating characteristics (ROC) curve plots are the remarkable ones and these two validation techniques have developed for assessing the accuracy level of the produced landslide susceptibility zonation maps of Darjeeling Himalaya by LNRF and SI models.

30.4.3.1 Validation of LSZ Maps Using ROC Curve

The ROC exploration is a method in the assessment classification of the predictive rule and here the susceptibility map is compared with a dataset reporting the presence or positive/absence or negative of landslide occurrences in the study area where its area under curve (AUC) values represents the prediction accuracy. The values close to 1 indicate a very good fit for the LSZ map. ROC curve deals with the graphical representation of the sensitivity and 1-specificity of landslide susceptibility models (Mohammady et al. 2012). Akgun et al. (2012), Bui et al. (2012) and Ozdemir and Altural (2013) used the ROC curve for validation and accuracy assessment of LSZ map in different areas. In the present study out of 100% (2079 landslides location), 30% (624 landslides location) training dataset were randomly used for validating the models. The result of the area under curve (AUC) value of ROC curve indicated 76.50% and 77.90% prediction accuracy with an area ratio of 0.7650 and 0.7790 for LNRF and SI models which is reasonable and accepted (Fig. 30.7).

30.4.3.2 Validation of LSZ Maps Using FR Value

The FR value is increasing continuously from very low to very high LS zones (Pourghasemi et al. 2013a; Tsangaratos and Ilia 2016; Mandal and Mandal 2016, 2017). The FR values of each LS zones were calculated from 'zonal statistics as a table' tools in ArcGIS (10.1) software

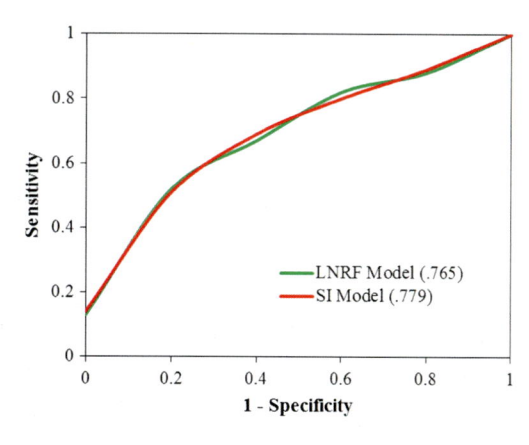

Fig. 30.7 ROC curve representing the prediction accuracy of the LSI maps achieved by LNRF and SI models

Fig. 30.8 FR plots in different LSZs of the LNRF and SI models

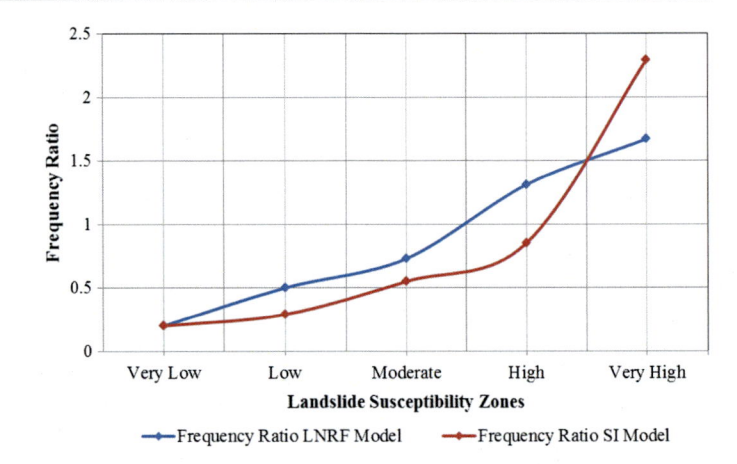

environment overlapping landslide inventory on the LS map (Mandal and Mandal 2018). The calculated and plotted FR value of five LS zones gradually increased from very low to very high LS zones. The study depicted that there is a positive relationship between LS and FR value. The results showed very low to very high LS zones recorded with the FR values of 0.20, 0.50, 0.73, 1.31, 1.67 for LNRF and 0.20, 0.29, 0.55, 0.85, 2.29 for SI models respectively. The prepared LSZ maps of Darjeeling Himalaya depicted most of the landslide affected areas are being found in very high LS zones (Fig. 30.8 and Table 30.3).

30.5 Conclusions

Landslides represent the biggest challenges for the development of human activities in mountainous areas. In fact, such destructive natural hazards cause massive losses of lives and properties every year, especially during the rainy season in Darjeeling Himalaya. The main drivers of landslides are geotechnical factors, environmental factors, triggering factors and sustaining factors. These factors act differently from place to place and differ from one landslide type to another one. The monsoon rainfall, dynamics of land use pattern, steep slopes, very high relative relief, complex geological structure and unscientific construction, cultivation approach, the pressure of population and deforestation has made Darjeeling Himalaya susceptible to slope failure. To minimize landslides damage, identification of hazard-prone areas and preparation of landslide potential zones is paramount for planning and development. Many scientific approaches have been proposed for landslide susceptibility and hazard zonation mapping. The bivariate statistical approaches are most suited for the regional-scale LSZ, which incorporates remote sensing and GIS techniques because these are very useful in data acquisition, processing, analysis and management and to assess and predict landslides susceptible areas incorporating geomorphic, geologic, hydrologic, triggering, protective and anthropogenic factors. In the present study, bivariate statistical methods (LNSF and SI) were applied and LSZ maps were made for Darjeeling Himalaya on ArcGIS (10.1) software environment. The result of the AUC values of the ROC curve indicated that 76.50% prediction accuracy with an area ratio of 0.7650 against the LNRF model, and 77.90% prediction accuracy with an area ratio of 0.7790 against the SI model. Both models provided satisfactory results but the SI model depicted a high accuracy level. Very low, low, moderate, high and very high LS zones recorded with the frequency ratio values of 0.20, 0.50, 0.73, 1.31 and 1.67 in the case of LNRF. In the case of the SI model, FR values of 0.20, 0.29, 0.55, 0.85 and 2.29 were found in very low, low, moderate, high and very high LS zones. The prepared LSZ maps of Darjeeling Himalaya may help in the decision-making process and policy makers may use them for site-specific landslide hazard management and mitigation, construction of roads, train lines, water pipe lines, establishment of settlement and tea garden area and others developmental activities.

Acknowledgements The authors would like to express their sincere thanks and gratitude to the Geological Survey of India (GSI, Kolkata), Survey of India (SOI, Kolkata), USGS (The United States Geological Survey), National Bureau of Soil Survey and Land Use Planning (NBSS and LUP Regional Centre, Kolkata), www.banglarbhumi.gov.in, darjeeling.gov.in/gismaps.html, www.wbphed.gov.in and www.worldclim.org for providing necessary data, facilities and support during the present work.

Funding There is no funding agency against this research.

Conflict of Interest Statement The authors declare that there is no conflict of interest.

Ethical Statement This research work is originally in nature and it is based on geoinformation technology and it is not published anywhere.

References

Akgun A, Sezer EA, Nefeslioglu HA, Gokceoglu C, Pradhan B (2012) An easy-to-use MATLAB program (MamLand) for the assessment of landslide susceptibility using a Mamdani fuzzy algorithm. Comput Geosci 38(1):23–34

Basu T, Pal S (2018) Identification of landslide susceptibility zones in Gish River basin, West Bengal, India. Georisk Assess Manag Risk Eng Syst Geohazards 12(1):14–28

Basu SR, Sarkar, S (1987) Ecosystem vis-a-vis landslide: a case study in Darjeeling Himalaya. In: Proceedings of impact of development on environment, vol 2. Geographical Society of India, pp 45–53

Bui DT, Pradhan B, Lofman O, Revhaug I, Dick OB (2012) Spatial prediction of landslide hazards in Hoa Binh province (Vietnam): a comparative assessment of the efficacy of evidential belief functions and fuzzy logic models. Catena 96:28–40

Bourenane H, Bouhadad Y, Guettouche MS, Braham M (2015) GIS-based landslide susceptibility zonation using bivariate statistical and expert approaches in the city of Constantine (Northeast Algeria). Bull Eng Geol Env 74(2):337–355

Chen W, Li W, Hou E, Zhao Z, Deng N, Bai H, Wang D (2014) Landslide susceptibility mapping based on GIS and information value model for the Chencang District of Baoji, China. Arab J Geosci 7(11):4499–4511

Cubito A, Ferrara V, Pappalardo G (2005) Landslide hazard in the Nebrodi mountains (Northeastern Sicily). Geomorphology 66(1–4):359–372

Davis JC (2002) Statistics and data analysis in geology, 3rd edn. Wiley, p 638

Ghosh S, Carranza EJM, van Westen CJ, Jetten VG, Bhattacharya DN (2011) Selecting and weighting spatial predictors for empirical modeling of landslide susceptibility in the Darjeeling Himalayas (India). Geomorphology 131(1–2):35–56

Guettouche, MS (2013) Modeling and risk assessment of landslides using fuzzy logic. Application on the slopes of the Algerian Tell (Algeria). Arab J Geosci 6(9):3163–3173

Guha-Sapir D, Below R, Hoyois PH (2018) EM-DAT: international disaster database. Université Catholique de Louvain, Brussels, Belgium. http://www.emdat.be. Last access 19 Feb 2018

Gupta RP, Joshi BC (1990) Landslide hazard zoning using the GIS approach-a case study from the Ramganga catchment, Himalayas. Eng Geol 28(1–2):119–131

Guzzetti F, Reichenbach P, Cardinali M, Galli M, Ardizzone F (2005) Probabilistic landslide hazard assessment at the basin scale. Geomorphology 72(1–4):272–299

Hong H, Pradhan B, Xu C, Bui DT (2015) Spatial prediction of landslide hazard atthe Yihuang area (China) using two-class kernel logistic regression, alternating decision tree and support vector machines. Catena 133:266–281

Kanungo DP, Arora MK, Sarkar S, Gupta RP (2006) A comparative study of conventional, ANN black box, fuzzy and combined neural and fuzzy weighting procedures for landslide susceptibility zonation in Darjeeling Himalayas. Eng Geol 85(3–4):347–366

Lee S, Pradhan B (2007) Landslide hazard mapping at Selangor, Malaysia using frequency ratio and logistic regression models. Landslides 4(1):33–41

Lee S, Choi J, Min K (2002) Landslide susceptibility analysis and verification using the Bayesian probability model. Environ Geol 43(1–2):120–131

Lee S, Hwang J, Park I (2013) Application of data-driven evidential belief functions to landslide susceptibility mapping in Jinbu, Korea. Catena 100:15–30

Mandal S, Maiti R (2014) Role of lithological composition and lineaments in landsliding: a case study of Shivkhola watershed, Darjeeling Himalaya. Int J Geol Earth Environ Sci 4(1):126–132

Mandal B, Mandal S (2016) Assessment of mountain slope instability in the Lish River basin of Eastern Darjeeling Himalaya using frequency ratio model (FRM). Model Earth Syst Environ 2(3):121

Mandal B, Mandal S (2017) Landslide susceptibility mapping using modified information value model in the Lish river basin of Darjiling Himalaya. Spat Inf Res 25(2):205–218

Mandal S, Mandal K (2018) Bivariate statistical index for landslide susceptibility mapping in the Rorachu river basin of eastern Sikkim Himalaya, India. Spat Inf Res 26(1):59–75

Mandal S, Mondal S (2019) Statistical approaches for landslide susceptibility assessment and prediction. Springer International Publishing AG, part of Springer Nature

Mohammady M, Pourghasemi HR, Pradhan B (2012) Landslide susceptibility mapping at Golestan Province, Iran: a comparison between frequency ratio, Dempster-Shafer, and weights-of-evidence models. J Asian Earth Sci 61:221–236

Mondal S, Maiti R (2013) Integrating the analytical hierarchy process (AHP) and the frequency ratio (FR) model in landslide susceptibility mapping of Shiv-khola watershed, Darjeeling Himalaya. Int J Disaster Risk Sci 4(4):200–212

Mondal S, Mandal S (2017) Application of frequency ratio (FR) model in spatial prediction of landslides in the Balason river basin, Darjeeling Himalaya. Spat Inf Res Inf Res 25:337–350

Moreiras SM (2005) Landslide susceptibility zonation in the Rio Mendoza valley, Argentina. Geomorphology 66(1–4):345–357

Naghibi SA, Pourghasemi HR, Dixon B (2016) GIS-based groundwater potential mapping using boosted regression tree, classification and regression tree, and random forest machine learning models in Iran. Environ Monit Assess 188(1):44

Naveen Raj T, Ram Mohan V, Backiaraj S, Muthusamy S (2011) Landslide hazard zonation using the relative effect method in south eastern part of Nilgiris, Tamilnadu, India. Int J Eng Sci Technol 3(4):3260–3266

Ozdemir A, Altural T (2013) A comparative study of frequency ratio, weights of evidence and logistic regression methods for landslide susceptibility mapping: Sultan Mountains, SW Turkey. J Asian Earth Sci 64:180–197

Park S, Choi C, Kim B, Kim J (2013) Landslide susceptibility mapping using frequency ratio, analytic hierarchy process, logistic regression, and artificial neural network methods at the Inje area, Korea. Environ Earth Sci 68(5):1443–1464

Peng L, Niu R, Huang B, Wu X, Zhao Y, Ye R (2014) Landslide susceptibility mapping based on rough set theory and support vector machines: a case of the Three Gorges area, China. Geomorphology 204:287–301

Pereira S, Zezere JLGMDS, Bateira C (2012) Assessing predictive capacity and conditional independence of landslide predisposing factors for shallow landslide susceptibility models. Nat Hazards Earth Syst Sci 12:979–988

Pham BT, Tien Bui D, Pham HV (2016) Spatial prediction of rainfall induced landslides using Bayesian network at Luc Yen District, Yen Bai Province (Viet Nam). In: International conference on environmental issues in mining and natural resources development (EMNR 2016). Hanoi University of Mining and Geology (HUMG), Viet Nam, pp 1–10

Pourghasemi HR, Rahmati O (2018) Prediction of the landslide susceptibility: Which algorithm, which precision? Catena 162:177–192

Pourghasemi HR, Pradhan B, Gokceoglu C, Mohammadi M, Moradi HR (2013a) Application of weights-of-evidence and certainty factor models and their comparison in landslide susceptibility mapping at Haraz watershed, Iran. Arab J Geosci 6(7):2351–2365

Pourghasemi HR, Moradi HR, Aghda SF (2013b) Landslide susceptibility mapping by binary logistic regression, analytical hierarchy process, and statistical index models and assessment of their performances. Nat Hazards 69(1):749–779

Pradhan B (2013) A comparative study on the predictive ability of the decision tree, support vector machine and neuro-fuzzy models in landslide susceptibility mapping using GIS. Comput Geosci 51:350–365

Pradhan AMS, Kim YT (2014) Relative effect method of landslide susceptibility zonation in weathered granite soil: a case study in Deokjeok-ri Creek, South Korea. Nat Hazards 72(2):1189–1217

Pradhan B, Lee S (2010) Landslide susceptibility assessment and factor effect analysis: backpropagation artificial neural networks and their comparison with frequency ratio and bivariate logistic regression modelling. Env Model Soft 25(6):747–759

Pradhan AMS, Dawadi A, Kim YT (2012) Use of different bivariate statistical landslide susceptibility methods: a case study of Khulekhani watershed, Nepal. J Nepal Geolog Soc 44:1–12

Raman R, Punia M (2012) The application of GIS-based bivariate statistical methods for landslide hazards assessment in the upper Tons river valley, Western Himalaya, India. Georisk Assess Manag Risk Eng Syst Geohazards 6(3):145–161

Ramesh V, Anbazhagan S (2015) Landslide susceptibility mapping along Kolli hills Ghat road section (India) using frequency ratio, relative effect and fuzzy logic models. Environ Earth Sci 73(12):8009–8021

Saha AK, Gupta RP, Sarkar I, Arora MK, Csaplovics E (2005) An approach for GIS-based statistical landslide susceptibility zonation-with a case study in the Himalayas. Landslides 2(1):61–69

Sarkar S (2010) Geo-hazards in sub Himalayan North Bengal. Department of Geography and Applied Geography. University of North Bengal

Sarkar S, Kanungo DP (2004) An integrated approach for landslide susceptibility mapping using remote sensing and GIS. Photogramm Eng Remote Sens 70(5):617–625

Sarkar S, Kanungo D, Patra A, Kumar P (2006) Disaster mitigation of debris flows, slope failures and landslides: GIS based landslide susceptibility mapping case study in Indian Himalaya. Universal Academy Press, Tokyo, Japan, pp 617–624

Sengupta CK (1995) Detailed study of geofactors in selected hazard prone stretches along the surface communication routes in parts of Darjeeling and Sikkim Himalaya, phase I, part-I (Rongtong-Kurseong road section). Progress report (FS1993–94). Geological Survey of India

Sharma LP, Patel N, Ghose MK, Debnath P (2013) Synergistic application of fuzzy logic and geoinformatics for landslide vulnerability zonation—a case study in Sikkim Himalayas, India. Appl Geomat 5(4):271–284

Sharma LP, Patel N, Ghose MK, Debnath P (2015) Development and application of Shannon's entropy integrated information value model for landslide susceptibility assessment and zonation in Sikkim Himalayas in India. Nat Hazards 75(2):1555–1576

Starkel L, Basu S (eds) (2000) Rains, landslides, and floods in the Darjeeling Himalaya. Indian National Science Academy

Sujatha ER, Kumaravel P, Rajamanickam GV (2014) Assessing landslide susceptibility using Bayesian probability-based weight of evidence model. Bull Eng Geol Env 73(1):147–161

Tay LT, Lateh H, Hossain MK, Kamil AA (2014) Landslide hazard mapping using a Poisson distribution: a case study in Penang Island, Malaysia. In: Landslide science for a safer geoenvironment. Springer, Cham, Switzerland, pp 521–525

Torkashvand AM, Irani A, Sorur J (2014) The preparation of landslide map by landslide numerical risk factor (LNRF) model and geographic information system (GIS). Egypt J Remote Sens Space Sci 17(2):159–170

Tsangaratos P, Benardos A (2014) Estimating landslide susceptibility through a artificial neural network classifier. Nat Hazards 74(3):1489–1516

Tsangaratos P, Ilia I (2016) Comparison of a logistic regression and Naïve Bayes classifier in landslide susceptibility assessments: the influence of models complexity and training dataset size. Catena 145:164–179

Van Westen CJ, Rengers N, Soeters R (2003) Use of geomorphological information in indirect landslide susceptibility assessment. Nat Hazards 30(3):399–419

Venables WN, Ripley BD (2002) Modern applied statistics with S, 4th edn. Springer, Berlin, Germany, p 495

Wang Q, Wang D, Huang Y, Wang Z, Zhang L, Guo Q, Chen W, Chen W, Sang M (2015a) Landslide susceptibility mapping based on selected optimal combination of landslide predisposing factors in a large catchment. Sustainability 7(12):16653–16669

Wang Q, Li W, Chen W, Bai H (2015b) GIS-based assessment of landslide susceptibility using certainty factor and index of entropy models for the Qianyang County of Baoji city, China. J Earth Syst Sci 124(7):1399–1415

Yalcin A, Reis S, Aydinoglu AC, Yomralioglu T (2011) A GIS-based comparative study of frequency ratio, analytical hierarchy process, bivariate statistics and logistics regression methods for landslide susceptibility mapping in Trabzon, NE Turkey. Catena 85(3):274–287

Climate Change Induced Coastal Hazards and Community Vulnerability in Indian Sundarban

31

Biraj Kanti Mondal

Abstract

Climate change does not have any boundary and the coastal parts of the world are getting affected largely by the climate-induced hazards frequently. The deltaic parts of India are most vulnerable and Sundarban is severely affected by such events over the last few decades. The Sundarban is an active coastal delta spreading over two countries, i.e. India and Bangladesh, and known for its unique physical, ecological and climatic features. Indian Sundarban has been victimized by diverse etiquettes of climate-induced hazards and consequently enormous human population displaces, increasing stress of economic demands, enhance the rate of vulnerability and risk in a diverse manner across marginalized groups and thereby making the region socio-economically assailable. The inhabitants of Sundarban especially the islanders are facing enormous difficulties and challenges to cope up with any kind of extreme events due to various drawbacks that are operating in the area and thus they are often bound to live with the edge. Hence, an in-depth analysis of the adaptability and livelihood in order to identify the challenges and uncertainties in the extreme vulnerable conditions due to hazards, climate change and other adverse environmental conditions. Indian Sundarban comprises 102 islands out of which 54 are inhabited by almost 4.4 million people in 4493.60 km^2 of landmass with 19 blocks of North and South 24 Parganas districts of West Bengal. The application of geoinformatics was carried out with the combination of primary and secondary data to record extensive mapping and analysis to tease out the effect of hazards, vulnerability of Sundarban at even micro-level by reaffirmation method in the present submission.

Keywords

Climate change · Hazards · Vulnerability · Sundarban · Adaptation · Geoinformatics

31.1 Introduction

Climate change is known global phenomena and climate-induced hazards affect the coastal parts of the world very adversely and Asian countries are very much prone to such phenomena and thus the last few decades provoked the attention of researchers and consequently encouraged to ascertain the issues at varied spatial to local scale (Frich et al. 2002; Hansen and Indeje 2004). The

B. K. Mondal (✉)
Department of Geography, School of Sciences, Netaji Subhas Open University, Kolkata, India
e-mail: birajmondal@wbnsou.ac.in

scholars have assessed the increasing frequency of climate-induced hazards and other extreme weather events and its variability and impacts. Asia is considered as a most vulnerable continent to climate-induced hazards as the impact of climate change is observed distinctly in Asian countries including India (Hijioka et al. 2014). Various literature reveals that India secures the sixth rank among various countries in terms of facing extreme weather events (Kreft et al. 2014) and the deltaic parts are more vulnerable and thus few attempts were made to assess the climatic variability in the deltaic ecosystem. Several scholars used meteorological variables and accurate meteorological data from India Meteorological Department to analyze the climate change related issues in India and Bangladesh Sundarban (Nandy and Bandopadhyay 2011; Pramanik 2015; Kusche et al. 2016; Ghosh 2018). It becomes evident that climate change and variability and coastal disasters largely impacted on socio-economic structure (Roy and Guha 2017).

The IPCC fifth assessment report states that one metre sea level rise towards the end of the twentieth century (Mengel et al. 2016) which is higher than last two millennium and in such paradigm the issues of climate change induced hazards and its impacts in Sundarban needs to be assessed. Climate change-induced hazards and extreme environmental events have larger persuade on those sections of the society or population, especially in a developing country like India, that are most reliant on natural resources for livelihood. This weaker or marginalized section of the population has the least capacity to respond to natural hazards, such as super cyclone, flood, drought, etc. and Sundarban is not an exception. Various socio-economic and environmental aspects were considered for assessing the vulnerability and adaptations with the effects of climate change-induced hazards. The inhabitants of this Sundarban Biosphere Reserve (SBR) are solely dependent on primary economic activities like agriculture, fishing, and furthermore, and they are inclined towards the mangrove forest resources for their livelihood. As the challenges of adaptations, displacement of habitat, or the economy is largely affected by such events, thus this is considered as the vital one to portray the risk, vulnerability status due to changing climate related issues and hazards in the SBR. The poverty ratio of the region was also taken into consideration, as most of the households are facing tremendous challenges to cope up with any kind of hazardous circumstances. To assess these aspects minutely micro-level study was conducted in the Gosaba block to portray the ground level scenario of vulnerability. The present study explores climate change-induced hazards, vulnerability and adaptation strategies for livelihood with the help of geoinformatics. To derive the data field surveys, interviews, and unobtrusive observation were being made to tease out posers in areas of climate change, hazards, vulnerability, adaptations of the inhabitants of Sundarban.

31.2 The Study Area

The Indian Sundarban, is contoured by numerous rivers, complex network of tidal waterways, mudflats and mangrove forest (Das 2006) and attacked by cyclones frequently as a consequence of climate change. The study area is distributed across North and South 24 Parganas districts of West Bengal (Fig. 31.1) and the northern of the Indian Sundarban is demarcated by the Dampier-Hodges line and the southern portion are open to the Bay of Bengal. This active coastal delta has a significant amount of mangrove forests and known as World Heritage Site (1987) and Ramsar Site (2019). This unique region has a total of 102 islands out of which 54 islands are inhabited (Das 2006) by population. The 4493.60 km^2 landmass of Indian Sundarban is inhabited by more than 4.4 million people, which at present close to 5 million (Census of India, 2011). The area comprises 19 blocks; 6 blocks of North 24 Parganas and 13 blocks of South 24 Parganas districts of West Bengal respectively. The area extends from 21° 00′ North to 22° 30′ North and 88° 00′ East to 88°29′ East latitude and longitude respectively. To assess the vulnerability and adaptation strategies of the region, one of the most vulnerable area

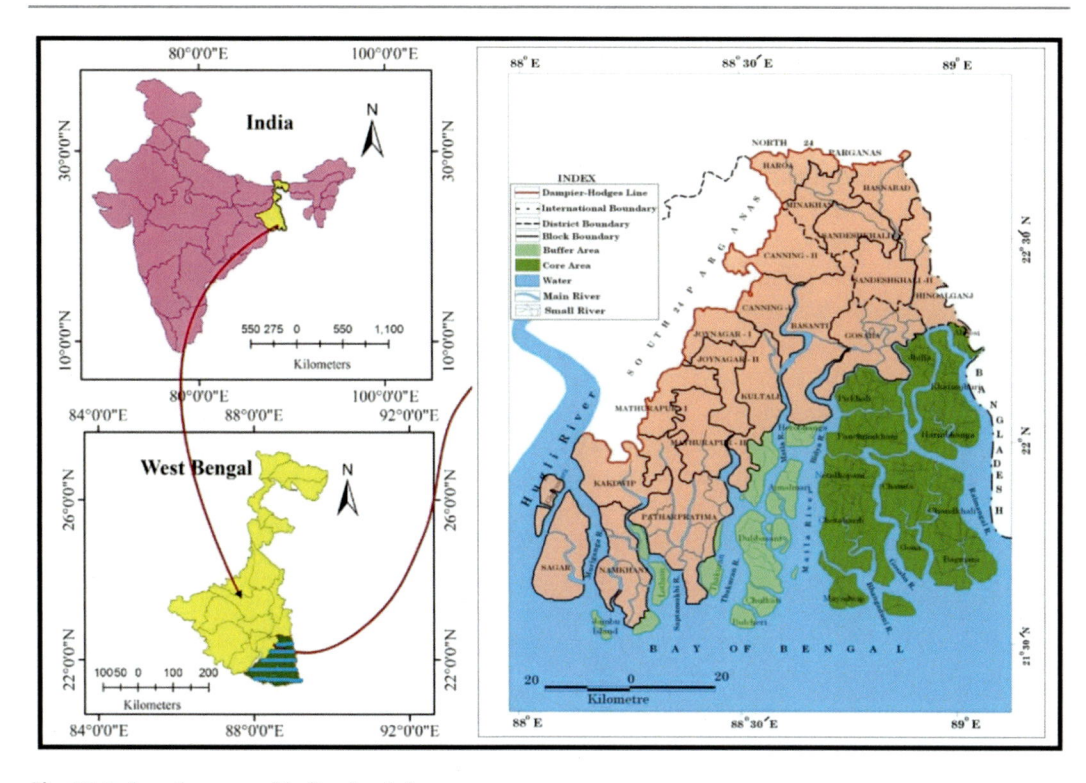

Fig. 31.1 Location map of Indian Sundarban

i.e. Gosaba block has been considered for the in-depth investigation.

31.3 Database and Methodology

The present study is a combination of primary and secondary data. The primary data was collected through proper questionnaire survey along with the minute observations and interviews. The secondary data has been collected from Census of India, District Census Handbook, District Statistical Handbook, District Human Development Report, Report of State Disaster Management Authority, and a few other published reports. The primary data was collected through a primary survey, filed visit, observation, household survey, direct interview and discussion with households and individuals were conducted following two stage cluster purposive random sampling method. At the initial stage, three *Mouzas* (administrative unit) were chosen randomly (for micro-level investigation) from altered Gram Panchayats (located in a diverse island) of Gosaba block; while in the second stage, households within these *mouzas* were selected randomly for the survey. A direct interview was carried out during the survey choosing 40 households from each *mouzas* covering a total of 120 households in the year 2019. The questionnaire scheduled was designed using questions in regard to households, livelihood and individual level features i.e. house types, occupation, duration of spending time in economic activities, land use land cover change etc. Some other structured questions were asked concerning the climate change, climate-induced events and hazards, suffering status, types of losses, level of impact on life, livelihood, environment and health, nature of the vulnerability, challenges and adaptation strategies. Geoinformatics was applied at this juncture to analyze all the data by using statistical techniques and quantitative methods to address the aimed issues of the study.

31.4 Results and Analysis

31.4.1 The Issues of Climate Change, Hazard, Risk and Vulnerability

Indian Sundarban region is geo-physically and ecologically susceptible (Bakshi and Panigrahi 2015; Sahana et al. 2020a) and often devastated by climate-induced super-cyclones and floods. The Sundarban region has been victimized by diverse etiquettes of climate change (WWF 2010; Raha et al. 2012; Mukherjee et al. 2019) due to which invites displacement of population and increases stress of economic demands in turn making the region socio-economically assailable. What is even more alarming is the fact that as a consequence of such events of climate change, this area is facing tremendous pressure due to sea level rise. The situation is culminating into loss of habitat and cultivable land, compelling people to dislocate. Risk and vulnerability to natural hazards mostly due to climate change and extreme weather events like severe cyclone attack etc. are integrally related to prevailing backwardness and poor socio-economic, as well as environmental conditions in Sundarban. It is evident from some studies that the islands of Sundarban become warmer (Mandal et al. 2013; Danda 2010) and the seasonal variability of temperature and rainfall is reported (Nandargi and Barman 2018). Furthermore, one metre rise in the sea level would lead to a land loss of 30,000 km^2 in Bangladesh and 600 km^2 in India, displacing nearly 15 million people in Bangladesh and seven million in India. The occurrence of various climate change events and effects on various segment accumulating various literature (Hazra et al. 2002; Mitra et al. 2009a, b; Banerjee 2013; Raha 2014; Chakraborty 2015; Pramanik 2015; Trivedi et al. 2016; Sahana et al. 2016; Nandargi and Barman 2018). The study area faced climate change issues and its immediate consequences (Table 31.1).

Climate change-induced hazards, sea level rise, salinization and deterioration of mangroves have significantly affected the coastal ecosystem of Sundarban (Mahadevia and Vikas 2012; Hajra et al. 2017; Sahana and Sajjad 2019). Most of the studies have found that an increase in temperature and its consequent cyclone intensity is linked to storm surge (Raha 2014; Sahana and Sajjad 2019). The hazards, environmental degradation due to climate change and related issues are acting as the diverse force that poses a challenge in the restoration of the mangrove ecosystem and human progress.

The geo-spatial tools were applied to prepare the risk zonation map of Sundarbans (Figs. 31.2 and Fig. 31.3) that was made using storm surge height, coastal flooding, damages of earthen embankments and cyclonic effects. The Sundarban blocks are categorized into three zones (Fig. 31.2), namely very high risk (surge height <2 m), high risk (surge height 2–2.5 m) and moderate risk (surge height >2.5 m) zone based on the inundation area due to storm surge height, and risk exposure. A large part of the coastal and fringing blocks are belonging to very high risk zones (the Sagar block from the south to Hingalganj block in the north-eastern part of Sundarban). A large portion of Patharpratima, Namkhana, Kakdwip, Mathurapur-II, Kultali, Gosaba, Minakhan, Sandeshkhali blocks fall in the high risk zone while the remaining parts mostly belong to the moderate risk zone. Furthermore, the embankments of Sundaraban are enhancing the level of vulnerability and the most susceptible embankments are associated largely with Gosaba, Namkhana, Sagar, Patharpratima, Kultali, Mathurapur-II, Kakdwip, Hingalganj, Sandeshkhai-II blocks. It signifies that the increasing distance from the Dampier and Hodges line (northern limit of Sundarban) mean the closing to the forest area, mounting risk, and vulnerability due to embankment collapse and damages (Ghosh and Mistri 2020).

The climatic vulnerability of Sundarban region is extremely high due to the severe attack of cyclonic events for decades (Fig. 31.3). Most of the coastal and fringing blocks are at high risk in terms of cyclones (Chakraborty 2015; Mukherjee et al. 2019) and other atmospheric hazards (Hazra et al. 2002; Ghosh 2017;

Table 31.1 Issue and consequences of climate change in Sundarban

Climate change	Facts	Impact/vulnerability	Facts
Temperature	Increased 0.5 °C	River flow	Changing
Rainfall amount	Decreases	River bank erosion	Increased
Cyclone intensity	Increased	Flood hazard	Increased
Super cyclone	Greater than before	Embankment damaged by cyclone	Increased
Overall weather	Changed	Soil erosion	Increased
Frequency of cyclone	Increased	Saline water inclusion	Increased
Intensity of cyclone	Changed	Salinity of soil	Increased
Damages by cyclone	Increased	Deforestation	Increased
Rainy days	Fewer	Mangrove forest	Decreased
Rainy season	Shorter	Agricultural area	Decreased
Winter	Late coming and warmer	Inundation of land	Increased
Summer	Coming early and longer	Land degradation	Increased
Sea level rise	Increased	Agricultural potentialities	Decreased
Fresh water	Less availability	Crop production	Decreased
Ecosystem services	Diminishes	Fish production	Decreased
Surface water	Decreased	Honey collection	Decreased
Ground water	Decreased	Damages of house	Damaged
Coastal areas	Destroyed	Inhabitants	Displaced

Source Tabulated by the author

Fig. 31.2 Risk zonation map of Sundarban

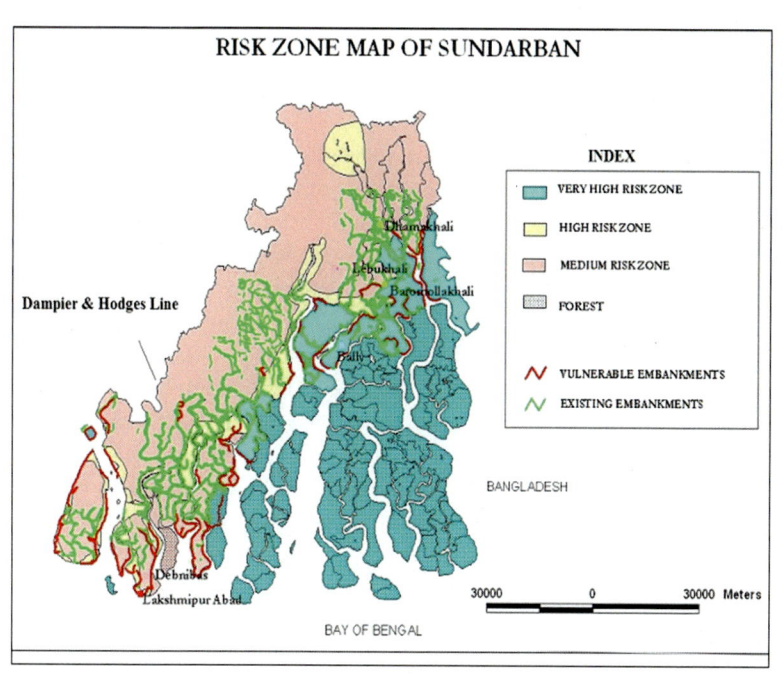

Fig. 31.3 Climatic vulnerability of Sundarban

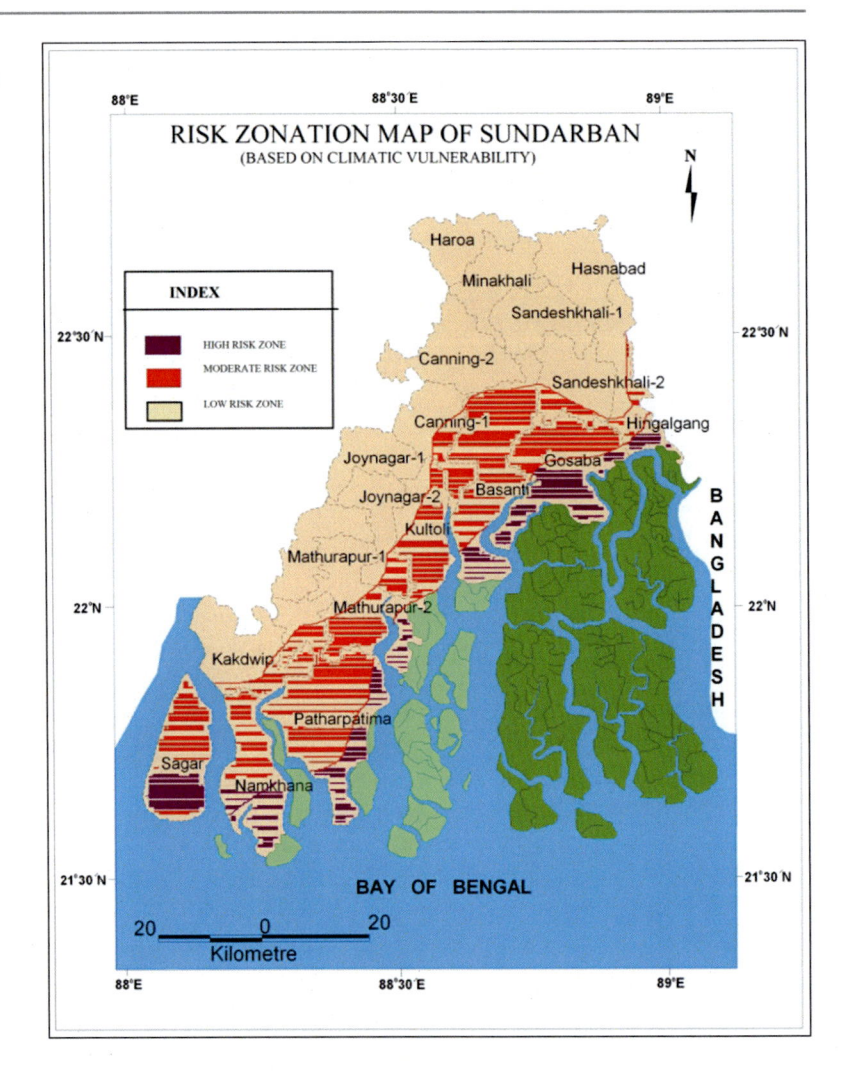

Bandyopadhyay 2018; Karmakar and Roy 2019) and therefore, natural calamities become a part and parcel of human livelihood. The quantity and intensity of cyclonic events, especially at the juncture of the twentieth and twenty-first century make us think about the future of such a unique mangrove forest area along with its huge inhabitants to a large extent. The number of cyclonic events observed from 1851 up to 2010 in Sundarban (Table 31.2) and its states variability but the number is good enough to destroy the ecosystem and thereby the residents are bound to cope up with these calamitous.

The increasing frequency, and intensity of cyclonic events, and related hazards damages and destroys the embankments (Ghosh and Mistri 2020) and associated loss of mangroves and biodiversity rendered the region more susceptible and thus such recurrent damage leads to property. Some of the recent remarkable cyclonic catastrophes which have an adverse distress to the Indian Sundarban region (Table 31.3).

The hitting of cyclone in the year 1999 started to affect a large portion of Sundarban (ten blocks, namely Kultali, Joynagar-II, Canning-I, Gosaba, Basanti, Kakdwip, Sagar, Mathurapur-II, Namkhana, Patharpratima) adversely and probably this was the beginning. In 2009, when cyclone Aila hit Sundarban on 25th May, the inhabitants totally get distorted and perhaps feeling most helplessness after a long back. This incident affected all the blocks adversely and about

Table 31.2 Natural calamities occurred in Sundarban

Year	Before 1700	1701–1750	1751–1800	1801–1850	1851–1900	1901–1950	1951–2000	After 2000
Frequency	1	4	3	13	49	36	20	15

Source State Disaster Management Authority (2019)

Table 31.3 Recent history of cyclone in Sundarban

Type of hazard	Year of occurrence	Blocks affected	Impact on livestock
Cyclone	1999	About ten blocks of the south part of Sundarban get severely affected and rest blocks are affected slightly	Affected the life and livelihood to a great extent in the affected blocks and they have to adapt to this adverse situation after a long time as this type of hazards make them vulnerable
Cyclone (Aila)	2009	All the blocks are very severely affected	Affected to a great extent of the entire inhabitants in most of the blocks
Cyclone (Bulbul)	2019	Almost all the blocks are moderately affected	Affected to some extent of livelihood in most of the blocks
Cyclone (Amphan)	2020	Almost all the blocks are severely affected	Affected to great extent mostly costal and fringing livelihood

Data source Report of State Disaster Management Authority (2019) and Newspaper (2020)

400 km elongated embankment was damaged (Kundu 2014; Sarkar et al. 2016; Das et al. 2012; Ghosh and Mistri 2020) and the saline water encroached land before earthen embankment by crossing limits (Ghosh and Mukhopadhyay 2016; Dhara and Paul 2016; Mondal 2015). As a consequence, damages of agricultural crops, shortage of drinking water and related risky situation arrested more than 2 million inhabitants who were stuck in such water-logged situation for several days. Furthermore, the majority of the thatched houses were smashed making the population helpless. Even the people don't have the land for crop production as most of the agricultural land became salt ridden and remained as non-productive for about a couple of years ahead after the event.

In the year 2019, again the cyclone Bulbul threatened the people and invited lot of damages. The very recent cyclone Amphan (May 2020) attacked the entire Sundarban region very rudely and caused severe damages. As per the government report, about 28% of the Sundarban has been damaged. The affected blocks were totally devastated and almost all the resources, houses, trees were vanished due to its very high magnitue (>120 km/h), and the consequences of such are still continuing. Some of the academicians reported about the overall scale that the villages had suffered largely by this cyclone and several parts of the regions were inundated with the salt water inclusion due to embankment breakage and damages. The damaged embankments were not recovered properly and the people are sustaining in such a distressing situation which states their degree of helplessness in front of natural disasters and related phenomena.

31.4.2 The Demographic Attributes

31.4.2.1 Population Growth and Density

Sundarbans is one of the most densely populated parts of West Bengal with a high population growth rate. The average population density of Sundarban (Fig. 31.4) in 2001 is about 929 persons/km^2 which has increased to 1082 persons/km^2 in 2011 (Census of India 1991, 2001, 2011).

Fig. 31.4 Population density of Sundarban

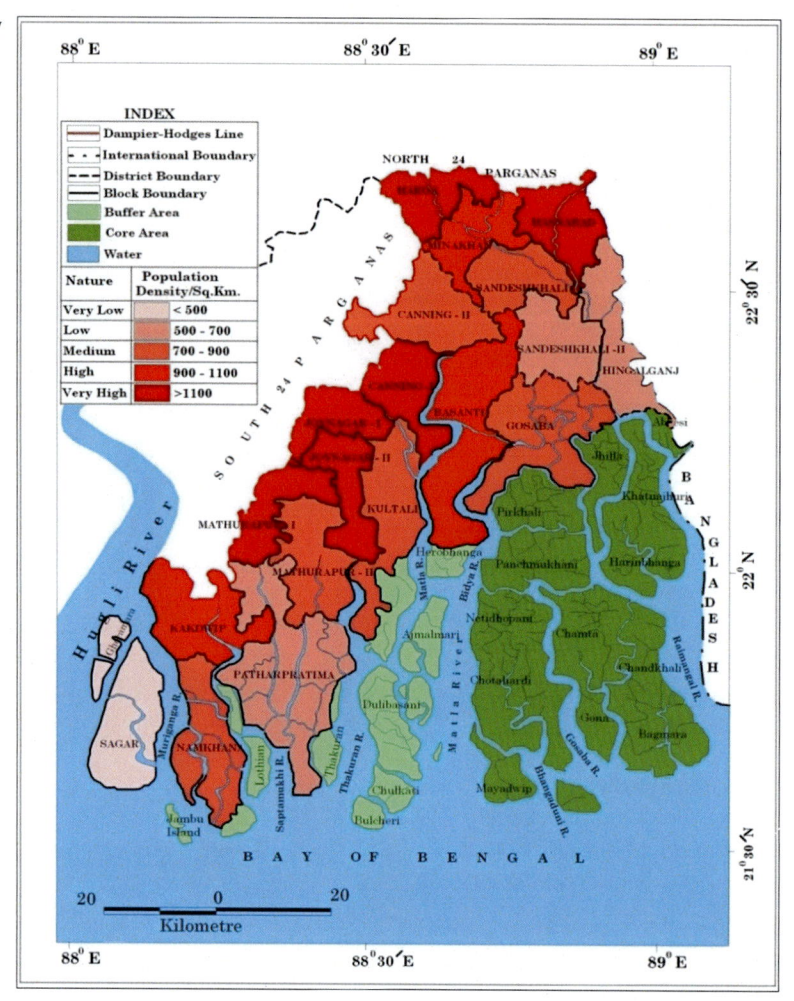

The block-wise variation of population growth rate (Fig. 31.5) signifies that the maximum growth rate is found in Canning-II block, followed by Canning-I, Basanti blocks. Almost all other blocks have also a growth rate of above 10 and only a few blocks have a higher growth rate in 2011 than the year 2001. The overall population increase of the region is quite high mostly due to natural increase and migration in the region due to environment-induced immigration.

31.4.2.2 Poverty and Concentration of BPL Populace in Sundarban

In Sundarban, about 44% population is breathing below the poverty line and the poverty ratio of all the blocks was quite high (Fig. 31.6). The highest poverty ratio is observed in Basanti block, followed by Canning-II block and these high rates of poverty are aggravating the demographic and socio-economic vulnerability of the inhabitants of Sundarban. Moreover, the exchange of occupation and breaks from the over-dependence of the biotic pressure of the Sundarbans, howsoever, is a far-away daydream. The BPL households of Sundarban have shown high rates of concentration in all the respective 19 blocks of Sundarban of the two corresponding districts.

The maximum number of BPL (Below Poverty Line) population concentrates in Gosaba, Basanti, Patharpratima, Namkhana, and

Fig. 31.5 Status of block-wise population growth rate

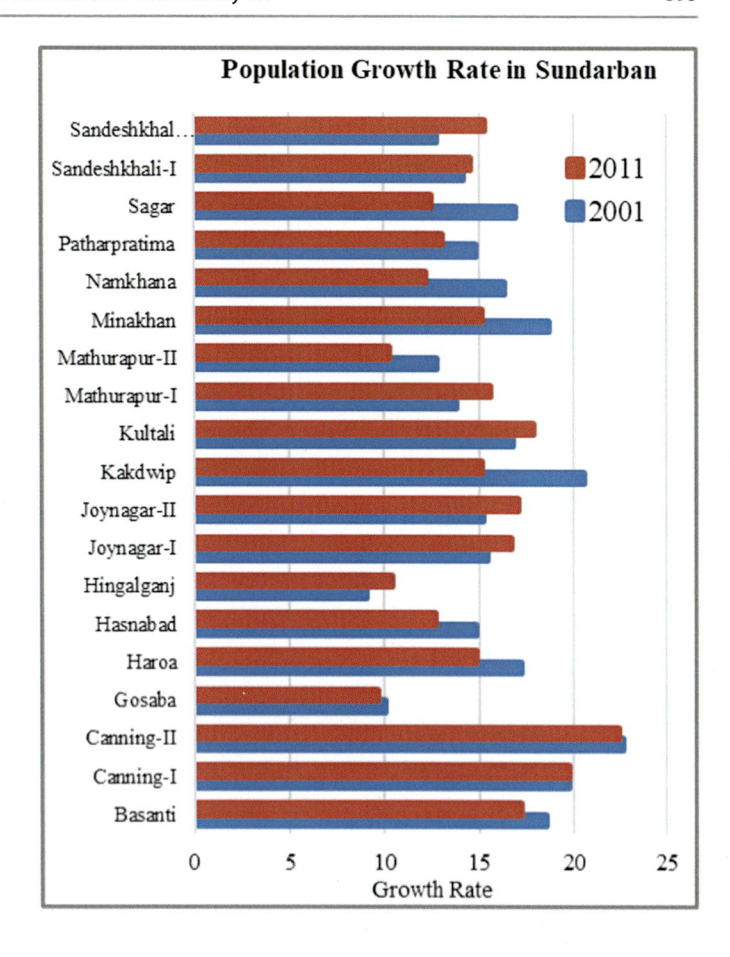

Sandeshkhali-II since the percentage of BPL households to a total number of households in these blocks is above 50. It is due to the over-dependency on the forest and fishing-based economic structure and insufficient productive land for compound agriculture. Furthermore, the frequent attack of natural hazards like cyclone, flood, and breakage of river embankments, etc. are the factors which make them helpless. Looking at Sundarban, Basanti block has the majority BPL households i.e. more than 40,000 in 2007 followed by Patharpratima where the BPL household figure stood at 35,000 in 2007 (Fig. 31.7). In Haroa, Hasnabad, Minakhan, Hingalganj, Mathurapur-I and II, Kultali, Kakdwip, Namkhana, Canning-I and II and Gosaba blocks BPL household range from 8000 to 15,000. The sketch of poverty in Sundarban therefore reveals that most of the inhabitants of Sundarban indicate a high attentiveness towards vulnerability in response to climate change and related hazardous consequences.

31.4.3 Living with the Edge in Climate Change-Induced Hazardous Situation

The Sundarban is a geo-physically feeble region (Pound et al. 2018; Ghosh et al. 2018; Sahana et al. 2020b) where most of the inhabiting people depend on natural resource especially on mangrove forest and primary activities like agriculture and fishing. Any sort of hazards, catastrophic events and climate change gets deeply rooted in the region and it enhances the environmental degradation, vulnerability (Bakshi and Panigrahi 2015; Karmakar and Roy 2019) which increases poverty and inequality in various

Fig. 31.6 Status of block-wise poverty ratio

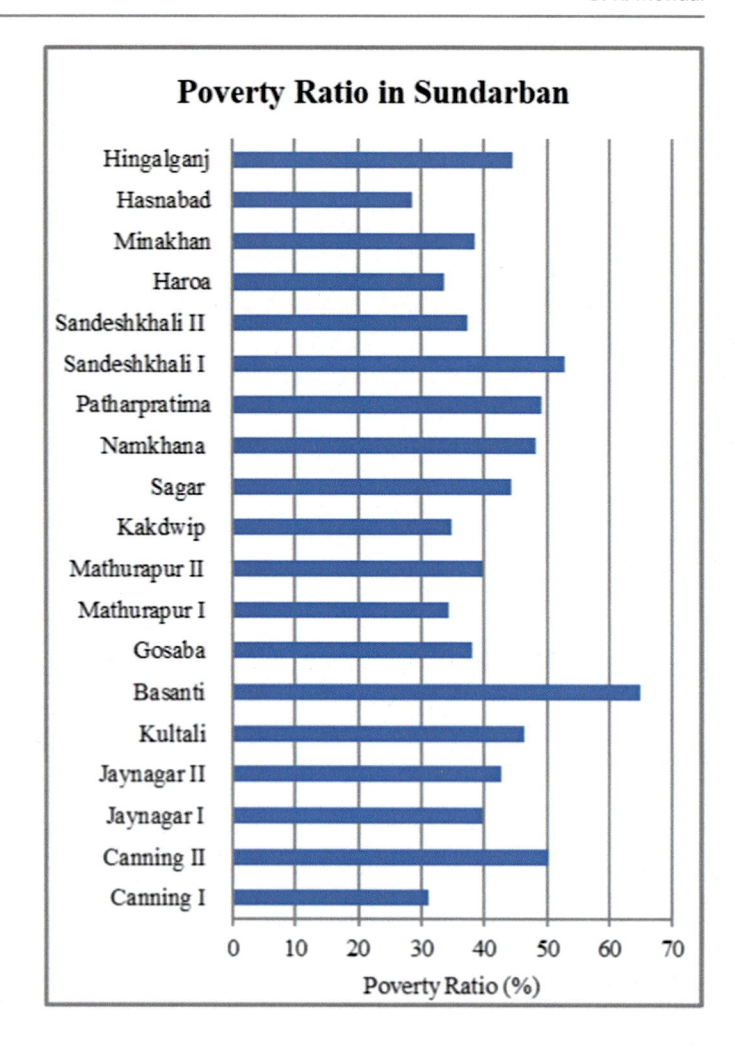

socio-economic sectors. On the other hand poverty and deprivation throw challenges in the way of disaster preparedness often exacerbating the humanitarian crisis. In this backdrop, it is, therefore, crucial to understand how the issues of hazards amplify vulnerability, diminishes the capacity of women and increases the propensity of environment-induced displacement (Samling et al. 2015; Ghosh et al. 2018) and uncertainty by reducing their adaptability in this region.

The vulnerability perspective has become a significant constituent of hazard and disaster due to climate (Ciurean et al. 2013; Orencio 2014; Ahsan and Warner 2014; Brooks 2003; Slettebak 2013) change by judging the degree of vulnerability (Sahana et al. 2019b) in various components (Table 31.4). The physical vulnerability enhances demographic, economic, social sectors (Dumenu and Obeng 2016; Hahn et al. 2009) which in turn alters the availability of health, education, infrastructure as well as food and drinking water. The consequences of these maximizes the vulnerability which encompasses inhabitant's helplessness.

The environmental degradation enhances the ecological vulnerability and people are getting more risk prone and thereafter their capacity in response to the vulnerable situation has been increased (Table 31.5) (Ghosh et al. 2015; Mitra et al. 2009a, b; Mukhopadhyay 2009; Mukherjee et al. 2012; Chand et al. 2012; Kar and Bandopadhyay 2015; Sahana and Sajjad 2019). This is judged by the concentration of poor households, women headed households, homelessness

Fig. 31.7 The concentration of BPL household

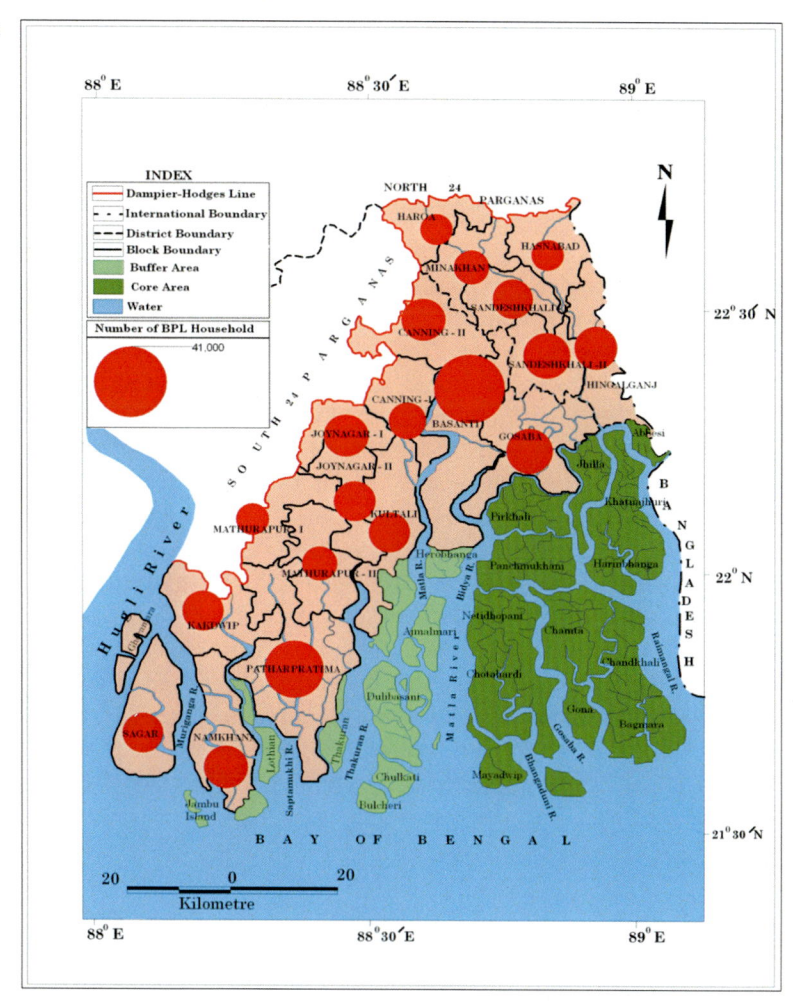

Table 31.4 Components of vulnerability

Components	Physical	Demographic	Economic	Social	Health	Education	Food and water	Infrastructural
Vulnerability level	High to very high	Moderate to high	High to very high	Moderate to high	High to very high	Moderate to high	High to very high	Moderate to high

and household near the embankment and those who are mostly dependent on primary economic activities. Physical vulnerability boosts up demographic and socio-economic vulnerability and make people more sensitive and reduce their adaptive capacity (Sahana et al. 2019a, b).

The people who breathe in the Sundarban are in a very helpless situation since most of that low-lying area of it regularly submerged by the tidal activity and flow (Mukherjee et al. 2019; Hazra et al. 2002; Jahan 2018) where any alteration in sea level means a direct peril to life,

Table 31.5 Capacity to respond during climatic events and vulnerable situation

Vulnerability aspects	Physical	Demographic	Economic	Social	Exposure	Sensitivity	Adaptive capacity
Level	High to very high	Moderate to high	High to very high	Moderate to high	High to very high	Moderate to high	Low to moderate

livelihood and assets. World Bank Report articulates that one metre rise of sea level will demolish the entire Sundarban. The mangrove forest-dependent 'ecosystem people' harvesting food and other resources from the forest (Dutta et al. 2019; Chowdhury et al. 2008; Hajra and Ghosh 2018) or exercise their own labour to survive by cultivating (Bandyopadhyay and Basu 2015; Das 2016; Mondal et al. 2015; Jahan 2018; Dutta et al. 2019; Laha 2019) or largely dependent on fishing activities for their livelihood and women holds a greater part of that. These consequences of climate change like loss of coastal wetlands would decrease breeding ground for numerous estuarine fish, which would decrease their population thereby hampering the livelihood of the inhabitants and increasing their proclivity towards poverty.

31.4.4 Climate Change-Induced Community Vulnerability: A Micro-level Study

To understand the degree of vulnerability in respect to climate change-induced hazards, an in-depth micro-level study was conducted to demonstrate the issue of livelihood and adaptation challenges that are being faced by the inhabitants in vulnerable block of Gosaba (Fig. 31.8), which is located close to the mangrove forest in Sundarban. Gosaba block belongs to a vulnerable or very high risk zone with respect to any climatic events; even though it is often considered as one of the least accessible blocks of Sundarban in terms of connectivity and remotely located island position. The Gosaba block (District Human Development Report 2009; District Statistical Handbook 2009) encompasses 15 g Panchayat, 51 *mouzas*, moderate population density (825 people/km^2), moderate decadal population growth rate (8.83%), overly dependent on primary economic activities like agriculture (70%) and fishing (20%); high concentration of poor households. Most of the households of the study area are thatched and kutcha (80%) by nature and about 2% households do not have any electricity and telecommunication facilities. Consequently, here the lives are in isolation as it inter-linked by crisscross drainage network (Fig. 31.8) and livelihood opportunities are very much restricted; thereafter the condition of women is extremely susceptible (Kar and Bandopadhyay 2016) not only in terms of hazards but also socio-economic position and empowerment (Bandyopadhyay and Basu 2015).

In terms of infrastructural setting, Gosaba block has been facing tremendous troubles as there is no formal motorized public transportation. Most of the roads are non-metalled and the condition of such roads and jetties is getting worse during the rainy season. As the block is entirely surrounded by rivers and earthen embankments (Ghosh and Mistri 2020), thus the coastal exposure is very high that invites challenges during hazards. Many of such embankments and roads are destroyed during cyclone Aila (Kar and Bandopadhyay 2016; Mondal 2018) leading to prolonged flooding, saline water stagnation, loss of agricultural land along with the existing issue of man–animal conflicts due to close location to the forest.

The relevant information regarding the inhabitant's occupation, working status along with the land use land cover changes of the block as a whole and demographic attributes such as house types, occupational engagement and working staus are exposed to climatic variability,

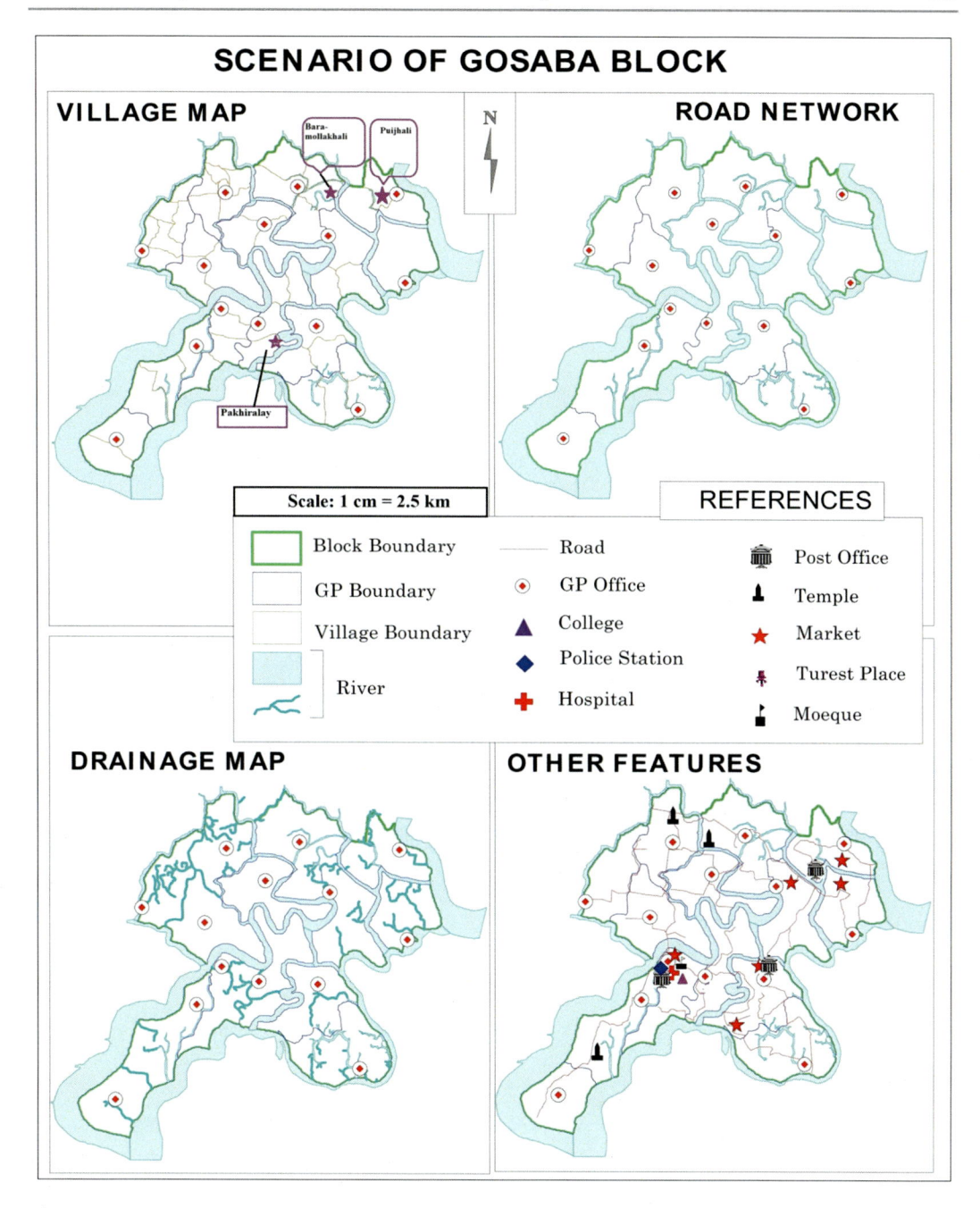

Fig. 31.8 Physical and infrastructural setting of Gosaba block of Sundarban

climate-induced hazards were derived from perception studies using well structured questionaries. In the backdrop of high exposure to climate change-induced vulnerability and poor infrastructural conditions, the inhabitants of the block are being isolated which have changed the working status of the common people (Fig. 31.9).

It reveals that a large number of people is under non-working group and engaged in some marginal and main workers while the other major parts are engaged as cultivators and agricultural

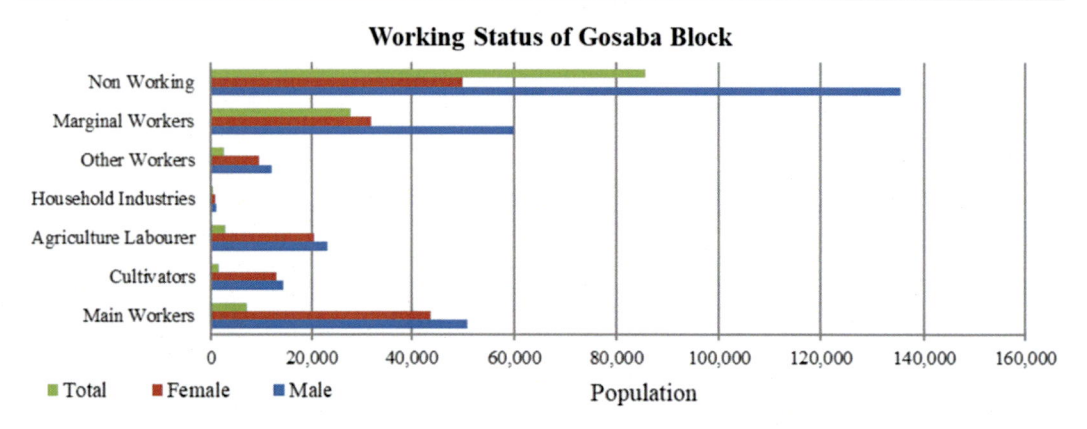

Fig. 31.9 Working status of Gosaba Block. *Data source* Census of India (2011)

labourers and other works. On the other hand, household industrial workers are very insignificant.

The total number of households, male and female is higher in Puijhali than the other two *mouzas* (Fig. 31.10). Thatched house made up of mud, semi-pucca houses with mud wall, and straw/asbestos/tin shaded roof are very common in all the three *mouzas* while the pucca houses of concrete structure are less in number (Fig. 31.11). The majority of the inhabitants in the area are associated with agricultural land, fishing, forest and some unpaid household activities. It reveals from the primary survey of women that above 30% belongs to agricultural activities, followed by fishing and prawn seed collection (above 20%). A less proportion engaged in daily wage, business and job including tourism (Fig. 31.12).

Based on the National Sample Survey Office's (NSSO) categorization of economic activities, there are three groups namely, *prime* (work throughout the year), *subordinate* (work foremost part of the year) and *both* (work in one action for a longer period with an alternate action for a shorter phase). These categorization of all respondents is based on the duration of spending time in economic activity throughout the year. It reveals that most of the respondents are engaged in both the activities (approximately half), followed by prime activity at all *mouzas* (Fig. 31.13). This can be assumed that, as most of the households are frequently affected by the climate-induced hazards almost in each and

Fig. 31.10 Number of households and male–female in the study villages *Data source* Census of India (2011)

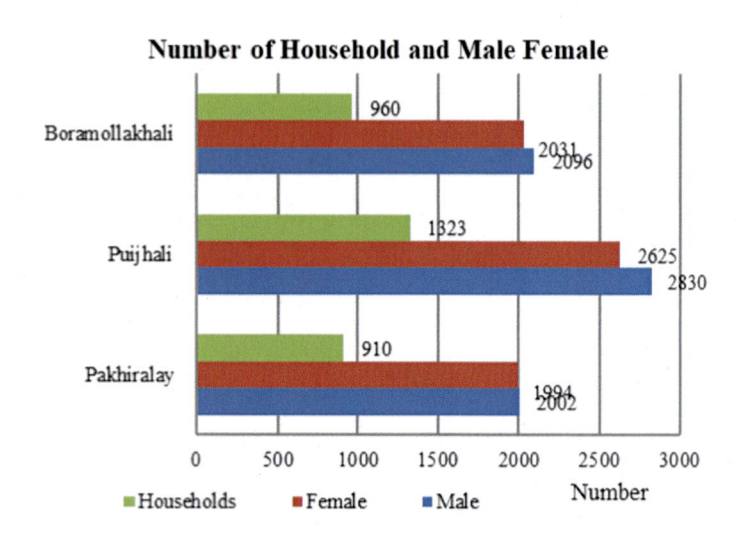

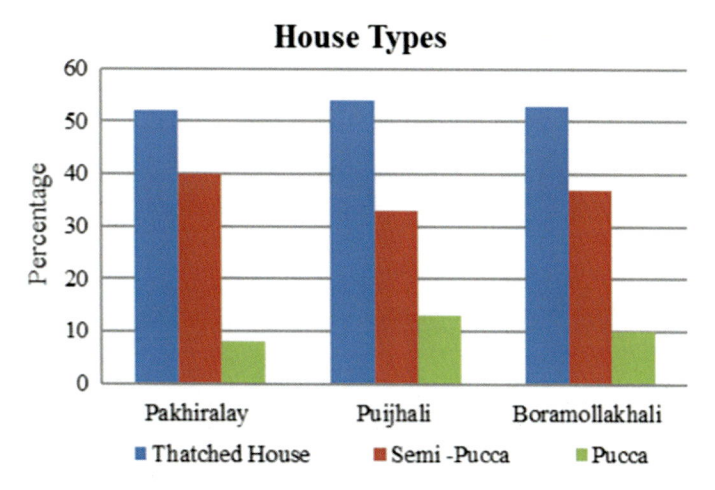

Fig. 31.11 House types in the study villages *Data source* Primary Survey (2019)

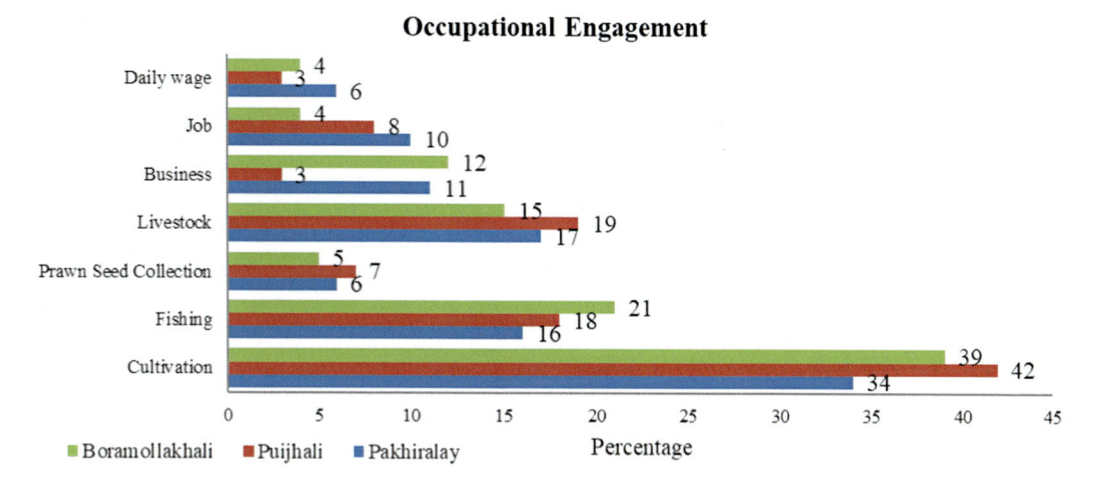

Fig. 31.12 Occupational engagement in the study villages *Data source* Primary Survey (2019)

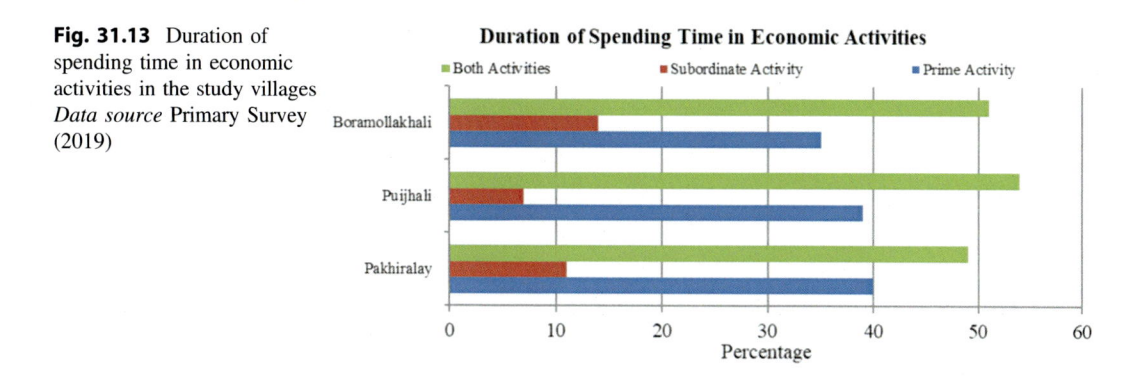

Fig. 31.13 Duration of spending time in economic activities in the study villages *Data source* Primary Survey (2019)

Fig. 31.14 Land use land cover changes of Gosaba block. *Data source* Debnath (2018)

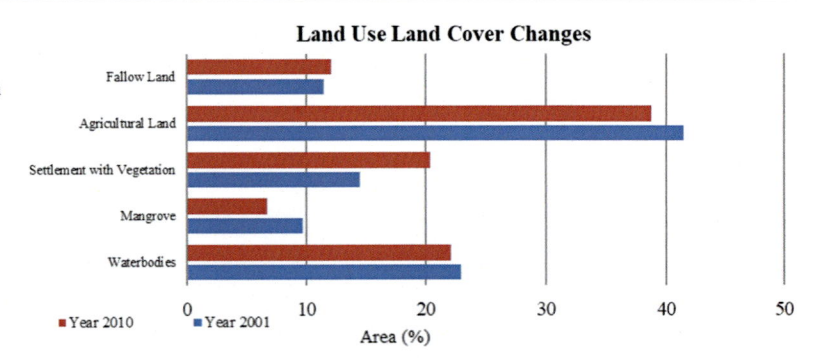

every year, therefore, their prime economic sectors are being hampered and thereafter people have to change their economic activities to sustain and adaptation.

Sundarban and its several parts has a dynamic land use and land cover (LULC) and it has a strong connection with climate change and induced hazards. Several research works were completed on the LULC changes (Debnath 2018) due to the attack of cyclone Aila. It has been observed that most of the islands of diverse blocks of the entire Sundarban area were affected and noticable LULC changes happened (Fig. 31.14). All the LULC types were being transformed and the foremost alterations was initiated in settlement area, vegetation and fallow land. The reduction in the areas of mangrove forest took place significantly which is followed by agricultural land and water bodies.

The study on diverse etiquette of climate change related various issues occurrences that more than 40% respondents strongly agreed and about 20% respondents agrees the early coming of summer and their longer stayed, late and warmer winter, shorter rainy season and fewer rainy days along with decrease of rainfall and increase of temperature are pertaining in the area and all these have more dynamic in the present day (Fig. 31.15).

Morethan 60% repondents are strongly agreed with overall weather change, frequency and intensity of cyclone changes, damages by the cyclone, soil salinity and embankment damages. Around 70% respondents agreed with the reduction of surface water area (Fig. 31.16).

The status of suffering from climate change-induced hazards and vulnerability reveals that more or less 60% of the respondents are severely suffered, about 20% or more are moderately suffered and rest portions are least suffered by the occurrence of flood hazard, river bank erosion, changes in river flow, embankment collapse, reduction of mangrove vegetation, decrease in fish and honey collection, lessening of crop production, salinization of agricultural land, land inundation and degradation, soil erosion etc. (Fig. 31.17).

31.5 Discussion

The Sundarban has always been a land of natural resources with physical toughness and livelihood is always been attached largely with primary economic activities like agriculture, fishing, forest product collection, etc. (Chowdhury et al. 2008) without deeply of impression to industrial improvement. The islanders of Sundarban constantly have been tolerating both detectable and not-so-visible uncertainties to uphold human livelihood based on primary activities and therefore people are often bound to work in a turbulent situation. The climatic hazards and related effects like sea level rise, breaking of earthen embankments, inundation of households and agricultural lands and salinity incursion in land and water always invites such uncertainties at household level. Even the tiger attack during fishing and forest product collection (Datta, Chattopadhyay and Deb 2011) made the women

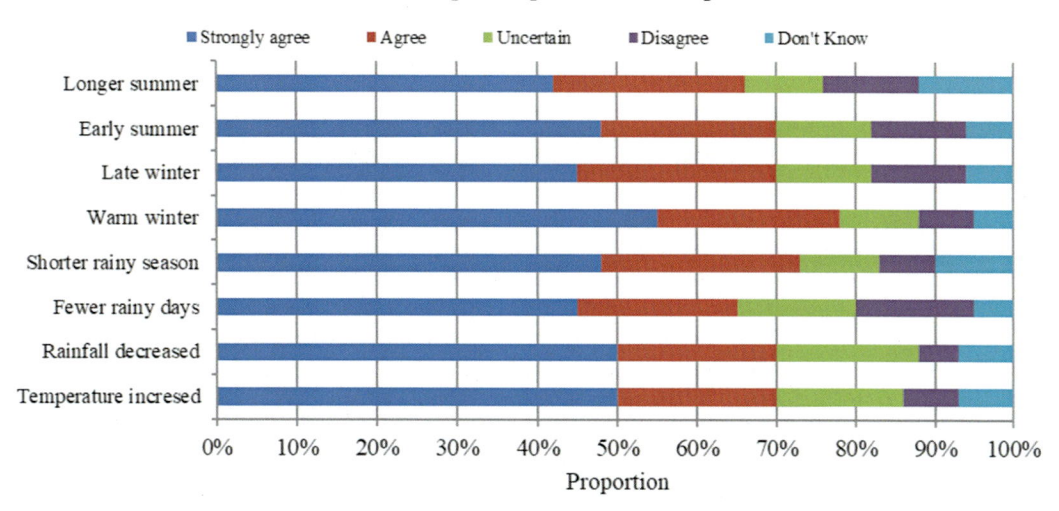

Fig. 31.15 Respondent's perception about diverse issues of climate change. *Data source* Primary Survey (2019)

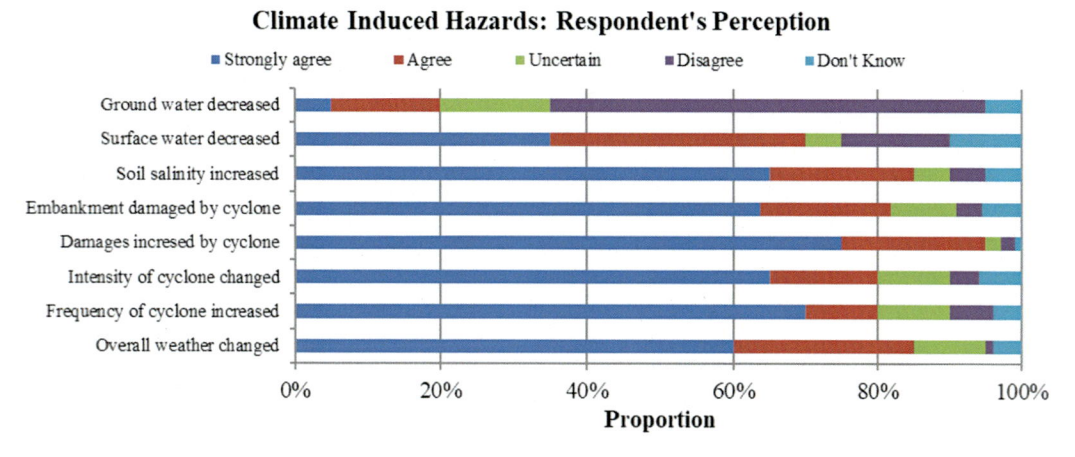

Fig. 31.16 Respondent's perception of climate change and induced Hazards. *Data source* Primary Survey (2019)

more vulnerable and such a situation persist in a village of Gosaba island and probably many more where all the widows are living and thus the village named as 'widow village' ('*bidhoba gram/para*'). Moreover, significant diminution of agricultural land and production, reduction of fish-catch make them economically more vulnerable, and due to such economic instability and food crisis results in migration of men (Danda 2019). Thus, women-headed households are emerging where they have to arrange supplemental livelihood, and child care along with household chores. In this condition, women are frequently addicted to '*meendhar*' (tiger prawn seed collection) or risky crab catching, other fishing and sometimes become a part of the marginal workforce as unskilled labour in handicraft activities. Subsequently, the people living in Rangabelia G.P. (Pakhiralay Mouza), Amtali G.P. (Puijhali Mouza) and Radhanagar Taranagar G.P. (Boramollakhali Mouza) of Gosaba block of South 24 Parganas are under problematic situation and the nature of which tabulated below (Table 31.6).

The perception of the community of the Gosaba block on various issues of climate

Status of suffering by various hazards

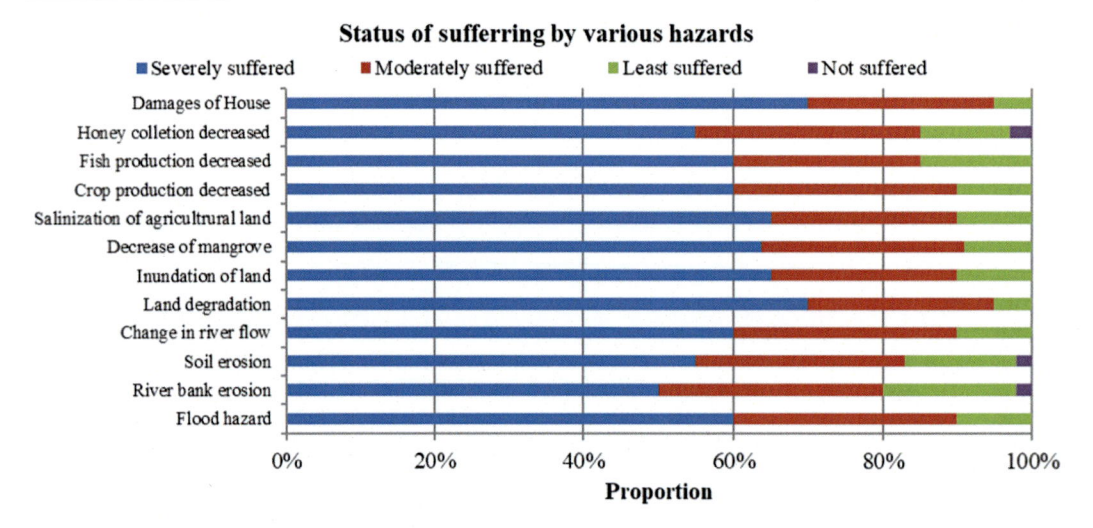

Chart showing proportions of Severely suffered, Moderately suffered, Least suffered, and Not suffered across categories: Damages of House, Honey collection decreased, Fish production decreased, Crop production decreased, Salinization of agricultrural land, Decrease of mangrove, Inundation of land, Land degradation, Change in river flow, Soil erosion, River bank erosion, Flood hazard.

Fig. 31.17 Status of suffering by climate-induced hazards. *Data source* Primary Survey (2019)

Table 31.6 Diverse aspects, nature of hazards, vulnerability and adaptations

Aspects	Nature	Climate change and related hazards
Environment	Impact	Coastal erosion, loss of mangroves, breaching and damages of the earthen embankment, coastal erosion
	Level of impact	Very severe and long term
	Coping style/adaptation Strategies	Building new embankment, repairing embankments by bamboo, sandbags, brick pitching, etc., taking government compensations, more inclined to the mangrove forest area
Livelihood	Impact	Damages of house, roads, displacement, the intrusion of saline water, life risk, crop loss, loss of fish collection, shortage of drinking water problems, increased poverty
	Level of impact	Very severe to severe, long term effect
	Coping style/adaptation strategies	Migrated, selling household assets, borrowing money from a money lender, non-farm activities, wage labour, cultivate salinity resistant crops, planting different crops/vegetables, changing planting dates, livestock farming
Health	Impact	Waterborne disease, women and child affected much, fever, cough like disease, malnutrition
	Level of impact	Very severe to severe, increasing helplessness
	Coping style/adaptation strategies	Dependent of quack doctors or local medical practitioners, rarely treated in hospitals, primary health centres, sub-centres, private clinic, migrated to cities

Source Perception study during Primary Survey (2019)

change and induced hazards along with their effects on livelihood, socio-economic, and ecological systems is very essential to identify the vulnerability groups and to introduce planning and development of the backward region of Sundarbans (Table 31.7).

It revealed that almost all the livelihood dependent issues operating in Gosaba block like

Table 31.7 Livelihood issues affected by climate change induced condition

Livelihood issues	Cyclone	Flood	River bank erosion	Embankment damages	Salinity	Sea level rise	Soil erosion
Agriculture	Yes	Yes	Yes	Yes	Yes	Yes	Yes
Livestock	Yes	Yes	Yes	Yes	Yes	*No*	Yes
Fishing	Yes	Yes	Yes	Yes	Yes	Yes	Yes
Shrimp culture	Yes	Yes	Yes	Yes	Yes	Yes	*No*
Honey production	Yes	Yes	Yes	Yes	*No*	Yes	*No*

Source Perception study during Primary Survey (2019)

agriculture, livestock, fishing, shrimp culture, honey collection, etc. are affected largely by the cyclone, flood, river bank erosion, sea level rise, embankment damages, salinity, and soil erosion. The shrimp culture and honey collection are not much affected by soil erosion and salinity issues. Moreover, it can be assumed that all these climate change related issues are strongly pertaining in the area, and the inhabitants of the block are often bound to cope up with such vulnerable situations.

The overall assessment of the vulnerability, adaptability of the inhabitants are scrutinized minutely in the current study and sorted out the probable prime causes of such helplessness at the end and these may be pointed as:

1. *Geographical:* (i) Natural hazards like, cyclone, flood, earthquake; (ii) Unequal distribution of resources; (iii) Relief and climatic harshness; (iv) Existence of huge riverine division; saline water; (v) saline soil, etc.
2. *Economical:* (i) Unemployment; (ii) Unequal economic development; (iii) Huge dependency on agriculture, fishing; (iv) Less opportunity of other income generation; (v) Very less industrial development; (vi) Low capital investment; (vii) Improper economic planning; (viii) Price hike of basic commodities, etc.
3. *Social:* (i) Massive illiteracy; (ii) Superstitions; (iii) Traditionalism; (iv) Caste system; (v) Malnutrition; (vi) Poor health; (vii) Fewer women empowerment (viii) Less involvement of women in planning; (ix) Male–female disparity, etc.
4. *Political:* (i) Inefficient political system; (ii) Corruption; (iii) Central and state government conflict; (iv) Disparity in the decision; (v) Improper government planning etc.
5. *Demographic:* (i) Increasing population; (ii) Immigration; (iii) High dependency ratio; (iv) Large family size etc.

The multidimensional analysis of the various livelihood and adaptive strategies of the inhabitants of Gosaba block of South 24 Parganas of West Bengal exposed the consistency of climate change and their hazardous effects and severity. This situation persist in almost all other coastal, islandic, and fringing areas of Sundarban which necessitates a holistic approach involving the process of hazard identification, community awareness and participation considering ecological, climatic, economic and social aspects of utmost importance for planning and development.

31.6 Conclusions

The environmental and socio-economic vulnerability necessitates instantaneous attention and effective strategies for the holistic progress of the Sundarban Biosphere Reserve (SBR) and strengthening the public system. Thus the present attempts to record extensive scrutiny on climate change related shocks and stresses are putting pressure to existing livelihoods of marginalized people in the Sundarban region. The diverse etiquettes of climatic hazards enhance the vulnerability to the inhabitants who are already suffering by the poverty, massive illiteracy, health hazards and in such situation, marginalized and principally dependent people on natural resources mostly on mangrove, often get displaced and migrated for better wages, and so on and so forth. The increasing economic stress due to the physical vulnerability, which is strong enough to boost other sort of vulnerability is totalling the adverse state of affairs and creating pressure to the marginalized, coastal, fringing, and remotely positioned populace in the Sundarban region.

The in-depth micro-level analysis of climate change and vulnerability of Sundarban has portrayed some social and demographic issues that have been empirically analyzed and reaffirmed across the gender groups. The primary surveyed block i.e. Gosaba is frequently exposed to all sorts of climate-induced hazards and the entire area remains vulnerable and risky as it is experiencing cyclone, sea level rise, coastal erosion, changing of the river course, embankment collapse, saline water inclusion and many more. Furthermore, as the physical vulnerability due to climatic hazards enhances demographic, economic and social susceptibility resulting environmental degradation which not only enhanced the ecological vulnerability and but also making people more risk prone and thereafter their capacity in response to vulnerable situation decreases by losing their adaptive capacity and thus the sensitized people are getting more exposed to risk. Henceforth, implementation of any kinds of policies and developmental schemes are very critical for the Gosaba block mostly due to physical isolation and unawareness of people about different related issues which plays a vital role in this regard. Therefore, appropriate adaptation strategies along with the participatory appraisal by empowering communities with education, employment, technological development and awareness enhancement can be the best way to address the issue and cope up with such vulnerable situations.

Acknowledgements The author would like to thank Shri Subrata Sarkar, without whom the field work in the Gosaba block of Sundarban would have been a hundred times difficult. This current effort is a tribute to the inhabitants of Sundarban for their spontaneous response during interviews and discussions.

Funding A special thanks to Netaji Subhas Open University for providing the supportive research funding (No. AC/140/2021-22, Dated 01/11/2021) and necessary support.

Conflict of Interest The author declares that there has been no conflict of interest.

Consent to Participate All the respondents were informed about the nature of the study before proceeding with the interview and survey and thereafter participated in the research was voluntary. The study was also assured to participants that the data would be used for academic purposes only and would be handled properly to guarantee their confidentiality and safety.

References

Ahsan MN, Warner J (2014) The socioeconomic vulnerability index: a pragmatic approach for assessing climate change led risks—a case study in the southwestern coastal Bangladesh. Int J Disaster Risk Reduct 8:32–49

Bakshi A, Panigrahi AK (2015) Studies on the impact of climate changes on biodiversity of mangrove forest of Sundarban delta region. J Environ Sociobiol 12(1):7–14. Available via: https://www.researchgate.net/publication/320244331

Bandyopadhyay S (2018) Long-term Island area alterations in the Indian and Bangladeshi Sundarban: an assessment using cartographic and remote sensing sources. Available via: https://www.sundarbansonline.org/wp-content/uploads/2020/03/Paper-Long-term-Island-Area-Alterations-in-the-Indian-and-Bangladeshi-Sundarbans.pdf. Accessed 10 Mar 2020

Bandyopadhyay M, Basu R (2015) A peep into spatial variation in the level of development in South 24 Parganas District, West Bengal. Int J Sci Res 6:2319–7064. Available via: https://www.ijsr.net

Banerjee K (2013) Decadal change in the surface water salinity profile of Indian Sundarbans: a potential indicator of climate change. J Mar Sci Res Dev Available via: https://doi.org/10.4172/2155-9910.S11-002

Brooks N (2003) Vulnerability, risk and adaptation: a conceptual framework. Working paper 38. Tyndall

Centre for Climate Change Research, pp 1–16. Available via: https://www.researchgate.net/publication/200032746_Vulnerability_Risk_and_Adaptation_A_Conceptual_Framework. Accessed 20 Mar 2020

Census of India (1991, 2001, 2011) Provisional population totals. Registrar General and Census Commissioner of India, Ministry of Home Affairs, New Delhi, India. Available via: https://censusindia.gov.in/2011census/dchb/DCHB_A/19/1917_PART_A_DCHB_SOUTH%20TWENTY%20FOUR%20PARGANAS.pdf. Accessed 02 Mar 2020

Chakraborty S (2015) Investigating the impact of severe cyclone Aila and the role of disaster management department—a study of Kultali block of Sundarban. Am J Theoret Appl Bus 1(1):6–13

Chand BK, Trivedi RK, Dubey SK, Beg MM (2012) Aquaculture in changing climate of Sundarban: survey report on climate change vulnerabilities, aquaculture practices and coping measures in sagar and basanti blocks of Indian Sundarban. West Bengal University of Animal and Fishery Sciences, Kolkata. Available via: http://www.nicra-icar.in/nicrarevised/images/Books/Aquaculture%20in%20Changing%20Climate%20of%20Sundarban.pdf

Chowdhury AN, Mondal R, Bramha A, Biswas MK (2008) Eco-psychiatry and environmental conservation: study from Sundarban Delta, India. Environ Health Insights 2:61–76. https://doi.org/10.4137/EHI.S935

Ciurean RL, Schroter D, Glade T (2013) Conceptual frameworks of vulnerability assessments for natural disasters reduction. In: Tiefenbacher J (ed) Approaches to disaster management—examining the implications of hazards, emergencies and disasters. Intech Open, London, UK, pp 3–31. https://doi.org/10.5772/55538

Danda A (2010) Sundarbans: future imperfect climate adaptation report. World Wide Fund for Nature, New Delhi. Available via: http://awsassets.wwfindia.org/downloads/sundarbans_future_imperfect__climate_adaptation_report_1.pdf

Danda A (2019) Environmental security in the Sundarban in the current climate change era: strenghening India-Bangladesh Cooperation. Observer Research Foundation, New Delhi. https://www.orfonline.org/research/environmental-security-in-the-sundarban-in-the-current-climate-change-era-strengthening-india-bangladesh-cooperation-57191/. Accessed 02 Feb 2020

Das GK (2006) The Sundarban. Sarat Book Distributor, Kolkata

Das K (2016) Livelihood dynamics as a response to natural Hazards: a case study of selected places of Basanti and Gosaba Blocks, West Bengal, Earth sciences. Sci Publ Group 5(1):13. https://doi.org/10.11648/j.earth.20160501.12

Das M, Das TK, Maity A (2012) Managing embankment breaching in North-East Sundarban. ACB Publications, Kolkata, pp 8–17. ISBN: 81-87500-59-X

Datta D, Chattopadhyay RN, Deb S (2011) Prospective livelihood opportunities from the mangroves of the Sunderbans, India. Res J Enviorn Sci Alert 5(6):536–543. https://doi.org/10.3923/rjes.2011.536.543

Debnath A (2018) Land use and land cover change detection of Gosaba Island of the Indian Sundarban region by using multitemporal satellite image. Int J Humanit Soc Sci 7(1):209–217. https://www.thecho.in/files/26_uf3v344k.-Dr.-Ajay-Debnath.pdf. Accessed 02 Feb 2020

Dhara S, Paul D (2016) Impact of cyclone and flood on social vulnerability—a case study at Kakdwip Block, South 24 Parganas, West Bengal. IJISET-Int J Innov Sci Eng Technol 3. Available via: http://www.ijiset.com

District Human Development Report of South and North 24 Parganas (2009) Development and Planning Department, Government of West Bengal, pp 1–20. Available via: http://www.wbpspm.gov.in/publications/District%20Human%20Development%20Report. Accessed 12 Mar 2020

District Statistical Handbook of South and North 24 Parganas (2009) Development and Planning Department, Government of West Bengal, pp 1–20. Available via: http://www.wbpspm.gov.in/publications/District%20Statistical%20Handbook. Accessed 12 Mar 2020

Dumenu WK, Obeng EA (2016) Climate change and rural communities in Ghana: social vulnerability, impacts, adaptations and policy implications. Environ Sci Policy 55:208–217

Dutta S, Maiti S, Garai S, Bhakat M, Mandal S (2019) Socio economic scenario of the farming community living in climate sensitive Indian Sundarbans. Int J Curr Microbiol Appl Sci 8(02):3156–3164. https://doi.org/10.20546/ijcmas.2019.802.369

Frich P, Alexander LV, Della-Marta PM, Gleason B, Haylock M, Tank AK (2002) Observed coherent changes in climatic extremes during the second half of the twentieth century. Climate Res 19(3):193–212

Ghosh A (2017) Quantitative approach on erosion hazard, vulnerability and risk assessment: case study of Muriganga-Saptamukhi interfluve, Sundarban, India. Nat Hazards 87(3):1709–1729. https://doi.org/10.1007/s11069-017-2844-0

Ghosh KG (2018) Analysis of rainfall trends and its spatial patterns during the last century over the Gangetic West Bengal, Eastern India. J Geovisual Spat Anal 2(2):15. https://doi.org/10.1007/s41651-018-0022-x

Ghosh A, Mukhopadhyay S (2016) Bank erosion and its management: case study in Muriganga-Saptamukhi Interfluves Sundarban, India. Geogr Rev India 78(2):146–161. https://doi.org/10.1007/s40808-016-0130-x

Ghosh S, Mistri B (2020) Geo-historical appraisal of embankment breaching and its management on active tidal land of Sundarban: a case study in Gosaba Island, South 24 Parganas, West Bengal. Space Cult India 7(4):166–180. https://doi.org/10.20896/SACI.V7I4.587

Ghosh A, Schmidt S, Fickert T, Nüsser M (2015) The Indian Sundarban mangrove forests: history, utilization, conservation strategies and local perception. Diversity 7(2):1–13. https://doi.org/10.3390/d7020149

Ghosh S, Chakraborty D, Dash P, Patra S, Nandy P, Mondal PP (2018) Climate risks adaptation strategies for Indian Sundarbans. https://doi.org/10.7287/peerj.preprints.26963v2

Hahn MB, Riederer AM, Foster SO (2009) The livelihood vulnerability index: a pragmatic approach to assessing risks from climate variability and change—a case study in Mozambique. Glob Environ Chang 19(1):74–88. https://doi.org/10.1016/j.gloenvcha.2008.11.002

Hajra R, and Ghosh T (2018) Agricultural productivity, household poverty and migration in the Indian Sundarban Delta, Elementa. University of California Press, p 6. https://doi.org/10.1525/elementa.196

Hajra R, Szabo S, Tessler Z, Ghosh T, Matthews Z, Foufoula-Georgiou E (2017) Unravelling the association between the impact of natural hazards and household poverty: evidence from the Indian Sundarban delta. Sustain Sci 12(3):453–464

Hansen JW, Indeje M (2004) Linking dynamic seasonal climate forecasts with crop simulation for maize yield prediction in semi-arid Kenya. Agric Meteorol 125(1–2):143–157

Hazra S, Ghosh T, DasGupta R, Sen G (2002) Sea level and associated changes in the Sundarbans. Sci Cult 68(9/12):309–321

Hijioka Y, Lin E, Pereira JJ, Corlett RT, Cui X, Insarov GE, Lasco RD, Lindgren E, Surjan A (2014) Asia. In: Climate change 2014: impacts, adaptation, and vulnerability, part B: regional aspects. Contribution of working group II to the fifth assessment report of the intergovernmental panel on climate change. Cambridge University Press, Cambridge

Jahan A (2018) The effect of salinity in the flora and fauna of the Sundarbans and the impacts on local livelihood. Master thesis in sustainable development 2018/33. Published at Department of Earth Sciences, Uppasala University. Available via: http://uu.diva-portal.org/smash/get/diva2:1261398/FULLTEXT01.pdf. Accessed 07 Mar 2020

Kar NS, Bandyopadhyay S (2015) Tropical storm Aila in Gosaba Block of Indian Sundarban: remote sensing based assessment of impact and recovery. Geogr Rev India 77:40–54

Kar NS, Bandyopadhyay S (2016) Tropical storm Aila in Gosaba Block of Indian Sundarban: remote sensing based assessment of impact and recovery. Available at: https://www.academia.edu/27512407/. Accessed 10 Mar 2020

Karmakar M, Roy M (2019) Rise of sea level and the sinking Islands of Sundarban Region: a study of Mousuni Island In India, UGC-CARE listed. J Group D 6(01). ISSN Online 2455–2445

Kreft S, Eckstein D, Junghans L, Kerestan C, Hagen U (2014) Global climate risk index 2015. Who suffers most from extreme weather events?, pp 1–31. Available via: https://germanwatch.org/sites/germanwatch.org/files/publication/10333.pdf

Kundu AK (2014) Embankment in Sundarban and reconstruction of damaged embankment in Aila. Sechpatra, Waterways and Irrigation Department, Government of West Bengal, pp19–25. Available via: https://wbiwd.gov.in/index.php/applications/aila

Kusche J, Uebbing B, Rietbroek R, Shum CK, Khan ZH (2016) Sea level budget in the Bay of Bengal (2002–2014) from GRACE and altimetry. J Geophys Res Oceans 121(2):1194–1217

Laha A (2019) Mitigating climate change in Sundarbans role of social and solidarity economy in mangrove conservation and livelihood generation implementing the sustainable development goals: what role for social and solidarity economy? Available via: http://unsse.org/wp-content/uploads/2019/07/258_Laha_Mitigating-Climate-Change-in-Sundarbans_En.pdf. Accessed on 10 Mar 2020

Mahadevia K, Vikas M (2012) Climate change—impact on the Sundarbans: a case study. Int J Environ Sci 2(1):7–15. Available via: https://www.researchgate.net/publication/311607858

Mandal S, Choudhury BU, Mondal M, Bej S (2013) Trend analysis of weather variables in Sagar Island, West Bengal, India: a long-term perspective (1982–2010). Curr Sci 105(7):947–953

Mengel M, Levermann A, Frieler K, Robinson A, Marzeion B, Winkelmann R (2016) Future sea level rise constrained by observations and long-term commitment. Proc Natl Acad Sci 113:2597–2602

Mitra A, Banerjee K, Sengupta K, Gangopadhyay A (2009a) Pulse of climate change in Indian Sundarbans: a myth or reality? Natl Acad Sci Lett (india) 32(1):19

Mitra A, Gangopadhyay A, Dube A, Schmidt AC, Banerjee K (2009b) Observed changes in water mass properties in the Indian Sundarbans (northwestern Bay of Bengal) during 1980–2007. Curr Sci 97:1445–1452

Mondal BK (2015) Nature of Propensity of Indian Sundarban. Int J Appl Res Stud 4(1):1–17

Mondal BK (2018) Assessment of effects of global warming and climate change on the vulnerability of Indian Sundarban. In: Shukla PS (ed) Sustainable development: dynamic perspective. Anjan Publisher, Kolkata, pp 63–74

Mondal I, Bandopadhyay J, Chakrabarti P, Santra D (2015) Morphodynamic change of Fraserganj and Bakkhali coastal stretch of Indian Sundarban, South 24 Parganas, West Bengal, India. Int J Rem Sens Appl 5:1–10. https://doi.org/10.14355/ijrsa.2015.05.001

Mukherjee S, Chaudhuri A, Sen S, Homechaudhuri S (2012) Effect of cyclone Aila on estuarine fish assemblages in the Matla River of the Indian Sundarbans. J Trop Ecol, 405–415

Mukherjee N, Siddique G, Basak A, Roy A, Mandal MH (2019) Climate change and livelihood vulnerability of the local population on Sagar Island, India. Chin Geogr Sci 29(3):417–436. https://doi.org/10.1007/s11769-019-1042-2

Mukhopadhyay A (2009) Cyclone Aila and the Sundarbans: an enquiry into the disaster and politics of aid and relief. Mahanirban Calcutta Research Group, Kolkata, pp 18–20

Nandargi SS, Barman K (2018) Evaluation of climate change impact on rainfall variation in West Bengal. Acta Sci Agricult 2(7):74–82. https://www.actascientific.com/ASAG/pdf/ASAG-02-0125.pdf

Nandy S, Bandyopadhyay S (2011) Trend of sea level change in the Hugli estuary, India. IJMS 40(6). Available via: http://nopr.niscair.res.in/handle/123456789/13266

Orencio PM (2014) Developing and applying composite indicators for assessing and characterizing vulnerability and resilience of coastal communities to environmental and social change. https://doi.org/10.14943/doctoral.k11531

Pound B, Lamboll R, Croxton S, Gupta N (2018) Climate-resilient agriculture in South Asia: an analytical framework and insights from practice. Available via: https://reliefweb.int/sites/reliefweb.int/files/resources/OPM_Agriculture_Pr2Final_WEB.pdf

Pramanik MK (2015) Assessment of the impacts of sea level rise on mangrove dynamics in the Indian part of Sundarbans using geospatial techniques. J Biodivers Bioprospect Dev 3:155. https://doi.org/10.4172/2376-0214.1000155

Raha AK (2014) Sea level rise and submergence of Sundarban Islands: a time series study of estuarine dynamics. J. Ecol Environ Sci. ISSN 0976-9900. Available via: https://www.researchgate.net/publication/268813865

Raha A, Das S, Banerjee K, Mitra A (2012) Climate change impacts on Indian Sunderbans: a time series analysis (1924–2008). Biodivers Conserv 21(5):1289–1307. https://doi.org/10.1007/s10531-012-0260-z

Roy C, Guha I (2017) Economics of climate change in the Indian Sundarbans. Glob Bus Rev 18(2):493–508. https://doi.org/10.1177/0972150916668683

Sahana M, Sajjad H (2019) Vulnerability to storm surge flood using remote sensing and GIS techniques: a study on Sundarban Biosphere Reserve, India. Rem Sens Appl Soc Env 13:106–120. https://doi.org/10.1016/j.rsase.2018.10.008

Sahana M, Ahmed R, Sajjad H (2016) Analyzing land surface temperature distribution in response to land use/land cover change using split window algorithm and spectral radiance model in Sundarban Biosphere Reserve, India. Model Earth Syst Environ 2(2):81. https://doi.org/10.1007/s40808-016-0135-5

Sahana M, Hong H, Ahmed R, Patel PP, Bhakat P, Sajjad H (2019a) Assessing coastal island vulnerability in the Sundarban Biosphere Reserve, India, using geospatial technology. Environ Earth Sci 78(10):1–22. https://doi.org/10.1007/s12665-019-8293-1

Sahana M, Rehman S, Paul AK, Sajjad H (2019b) Assessing socio-economic vulnerability to climate change-induced disasters: evidence from Sundarban Biosphere Reserve, India. Geol Ecol Landsc 00(00):1–13. https://doi.org/10.1080/24749508.2019.1700670

Sahana M, Rehman S, Ahmed R, Sajjad H (2020a) Analyzing climate variability and its effects in Sundarban Biosphere Reserve, India: reaffirmation from local communities. Environ Dev Sustain. https://doi.org/10.1007/s10668-020-00682-5.

Sahana M, Rehman S, Sajjad H, Hong H (2020b) Exploring effectiveness of frequency ratio and support vector machine models in storm surge flood susceptibility assessment: a study of Sundarban Biosphere Reserve, India. Catena 189:104450. https://doi.org/10.1016/j.catena.2019.104450.

Samling CL, Das S, Hazra S (2015) Migration in the Indian Bengal Delta and the Mahanadi Delta: a review of the literature, Migration in the Indian Bengal Delta and the Mahanadi Delta: a review of the literature. DECCMA working paper, deltas, vulnerability and climate change: migration and adaptation. IDRC Project Number 107642. Available via: http://www.deccma.com

Sarkar H, Roy A, Siddique G (2016) Impact of embankment breaching and rural livelihood: a case study in Ghoramara Island of the Sundarbans Delta in South 24 Parganas. J Bengal Geogr 5(4):97–117

Slettebak RT (2013) Climate change, natural disasters, and post-disaster unrest in India. India Rev 12(4):260–279. https://www.tandfonline.com/doi/abs/10.1080/14736489.2013.846786

Trivedi S, Zaman S, Chaudhuri TR, Pramanick P, Fazli P, Amin G, Mitra A (2016) Inter-annual variation of salinity in Indian Sundarbans. NISCAIR Online Periodicals Repository. http://nopr.niscair.res.in/handle/123456789/35043. Accessed 20 Mar 2020

32

Sea-Level Changes Along Bangladesh Coast: How Much Do We Know About It?

M. Shahidul Islam

Abstract

Bangladesh is considered to be one of worst victims due to climate change-induced sea-level rise, which would force millions of people to leave their homesteads and become climate victims. However, based on limited and poor quality information, it is very difficult to make such a straightforward statement, without knowing the critical and complex local situation of the coast. The quantity and quality of data are not adequate enough to make any precise conclusion about the rate of sea-level changes in the past, present, and future. However, in this chapter, it has been attempted to make a critical overview of the existing knowledge to depict the sea-level scenarios along the coastal belt of Bangladesh. It shows that during the Holocene period sea level was oscillating with an average rising rate of 1.75 mm/year, and shows at least five transgressive–regressive episodes. Based on tidal records and satellite altimeters, the assessment of current trends of sea-level rise made by different authors are not in agreement and varies from 2.1 to 25 mm/year (mostly around 8 mm/year). The projection by year 2100 is also dubious, which varies from 0.85 to 4.5 m (mostly around 1.4 m). This synthesis shows that in addition to climate changed induced global contributions, the major components for sea-level movements of Bangladesh are regionally and locally driven. Reduction of sediment supply from upstream and stoppage of regular sediment influxes to the coastal floodplain in the downstream are major anthropogenic causes to increase the sea-level height along the coastal belt. Practically, what we currently observe in the coastal belt is not the secular sea-level rise; rather, it is the amplified local tide-level rise, which is mostly related to humanly induced non-climatic factors.

Keywords

Climate change · Sea-level rise · Anthropogenic processes · Bangladesh coast

32.1 Introduction

Based on agro-ecological pattern, soil type, surface topography, depth of flood inundation and climatic condition, Brammer (2012) has divided Bangladesh into 14 broad physiographic units, of which coastal belt includes four units; these are Ganges Tidal Floodplain, Lower Meghna River Floodplain, Meghna Estuarine Floodplain, and

M. S. Islam (✉)
Department of Geography and Environment,
University of Dhaka, Dhaka, Bangladesh
e-mail: shahidul.geoenv@du.ac.bd

Chittagong coastal plain. These units have further been subdivided, each having distinctive geomorphic characteristics and different responses to sea-level rise (Brammer 2014).

The average elevation of the coastal belt of Bangladesh is 1–2 m, which is slightly higher on the southeast coast (4–7 m). It covers an area of about 47,201 km^2, which lies within 3 m of the mean sea level, occupies 32% of the national territories, and includes 28% of total population of the country (Islam 2004). It is widely assumed that due to climate change-induced sea-level rise by year 2100, 16% landmass of the country will be inundated, and will adversely affect the water resources, agriculture, mangrove ecosystem, food security, livelihood, and overall coastal environment (see Sarwar and Khan 2007; Khan and Awal 2009). Climate change-induced sea-level rise is, thus, a great concern to the academia, policymakers, and all relevant stakeholders, including general public. However, there remains a dearth of understanding regarding the process, scale, and rates of sea-level movements along Bangladesh coast.

The general understanding, even by many academics, is that there is a positive link between the current temperature rise due to global warming and global sea-level rise. Due to addition of ice-melting water, there would be a sea-level rise throughout the world, and low-lying countries, like Bangladesh and the Maldives, would be badly affected. Besides the melting water, ocean thermal expansion is another important challenge for future global sea-level sceneries. There are some miss-concepts too, particularly at local and regional-scale sea-level assessment. The major wrong concept is that if a certain volume of ice-melted water is added to the ocean system, there will not be a uniform change (rise) of sea level throughout the planet, due to differential effects of astronomical force, geoidal impact, and isostatic readjustment, although the oceans are interconnected. Greenhouse effect is not the only factor but contributes only a small fraction to sea-level rise, and in many cases, it leads to relative sea-level fall by excessive evaporation (Morner 2007). Moreover, the impacts of regional and local causes are not properly understood and addressed to assess local-relative sea-level movement. However, adequate attention is given to IPCCassessments (IPCC 1990, 1996, 2001) in national policy and programs, without doing any further investigation on local sea-level scenarios. In case of Bangladesh, the knowledge-base and understanding of the characteristics of local sea-level trends and scenarios, both past, present, and future are inadequate, imperfect, noisy, and sporadic. In this chapter, it has been attempted to synthesize the existing knowledge and explore the challenges for better understanding of the sea-level movements along Bangladesh coast.

32.2 Causes of Sea-Level Change

32.2.1 Global Causes

The important component of global causes is the glaciation and degraciation. The global process has an esutatic (worldwide) impact, but not at equal scale, and varies considerably over time and space. For example, due to melting of glaciers and ice sheets in the mid- and high latitude, the sea level was rising very fast worldwide, which continued unlit the mid-Holocene. The rate was much higher in the high latitude than in the tropics. Tooley (1978) has shown that at 7800 years BP, the rate of sea-level rise was 34–44 mm/year in North-East England, which was only less than 3 mm/year in Bengal Delta (Islam 2001). However, the global sea-level rise was slow-downed and established by mid-Holocene and since then there has been less than 0.5 m eustatic sea-level rise due to ice-melting (Clark et al. 1978).

Since the industrial revolution, humanly induced climate change (Fig. 32.1), with additional concentration of greenhouse gasses in the atmosphere, has been treated as an important ingredient for global (eustatic) mean sea-level rise (IPCC 1990, 1996, 2001, 2007, 2014). However, the rates and direction of this climate change-induced sea-level rise are not equal, uniform, and identical across the world, Rather, there are many places where sea-level would fall due to global warming (Morner 2007). The concept of eustatc sea-level rise is thus no more valid and is replaced by the concept of local-relative sea-level change. The global processes, off-course, have a share of local sea-level movement, along with other regional and local factors.

Three main variables derived from climate change to contribute to global mean sea-level rise are; changes in ocean water volume, thermal expansion of ocean water, and redistribution of water masses over the globe (Morner 2011). Even, no ice-melting water is further added to global ocean system, only due to redistribution of ocean water masses can change sea-level significantly. Morner (2016) has argued that at Solar Maxima, due to slower earth's rotation, heat is transmitted North-Westward and as consequence, there is a rise of sea level in the high latitude and fall in low latitude. During Solar Minima, opposite is the case.

32.2.2 Regional Causes

Crustal deformation due to tectonic activities may result in subsidence or uplift covering a wider geographical area and can change the vertical relationship between land and ocean. Due to glacio-isostatic rebound, areas once depressed by glaciations started to emerge to its original position. As consequence of such glacio-isostatic readjustment, the sea level will fall (up to 69 ± 9 m/ka) in the Near-field region (e.g. Antarctica, Greenland, Canada, Sweden, and Scotland); will rise fast (up to 10 ± 1 m/ka) in the Intermediate-filed regions (e.g. mid-Atlantic and Pacific coasts of the USA, Netherlands, Southern France); and will fluctuate between 1 and 6 m in the Far-field region (e.g. South America, Africa, Asia, and Oceania region) (Khan et al. 2015).

Oceanographic and atmospheric processes at regional-scale also have tremendous effect on regional sea-level movement, both seasonally and in long-term trends. Wind stress, atmospheric pressure, ocean current, sea surface temperature, salinity and tidal force collectively may changes sea level annually. Because of general geostrophic balance, there would be a variation in sea level between the central and peripheral parts of an ocean gyre. In countries around the Bay of Bengal, particularly in Bangladesh, the seasonal changes in oceanographic processes have its impact on local sea-level variations. Along Bangladesh coast, the sea level during the monsoon period is about 50 cm higher than that of the normal.

32.2.3 Local Process

At local scale, two important driving forces are local geological causes, particularly local sedimentation, subsidence and uplift, and anthropogenic impacts. These two forces also have wider regional significance. Sedimentation and subsidence of floodplain, river beds, coastal belt, and off-shore areas have remarkable effects on the relative sea-level movement of deltaic countries, like Bangladesh. Land subsidence due to compaction and/or consolidation is controlled by differential sedimentation rate, sediment composition, void ratio, over burden sediment layers, and tectonic activities, Moreover, large scale human interventions to regular river flows, flood inundations, and sediment dispersal, both at the upstream and downstream, have their long-lasted impacts on land–ocean interface. Withdrawal of surface water in the upstream and construction of embankments along the tidal rivers in the coastal belt escalate tidal range and more area suffers from saline water intrusion, as is the case in Bangladesh. Moreover, extraction of ground water and hydrocarbon accelerates the local subsidence and invites coastal inundation. Ericson et al. (2006) based on their studies on 40 major deltas across the world have shown that anthropogenic effects on accelerated sea-level rise of deltaic coast are more prominent, with a relatively less importance on the role of eustatic cause.

The relative sea-level changes of any location are driven by combine effects of global, regional, and local factors. Individual role of each of these factors is site dependents, and thus, no single

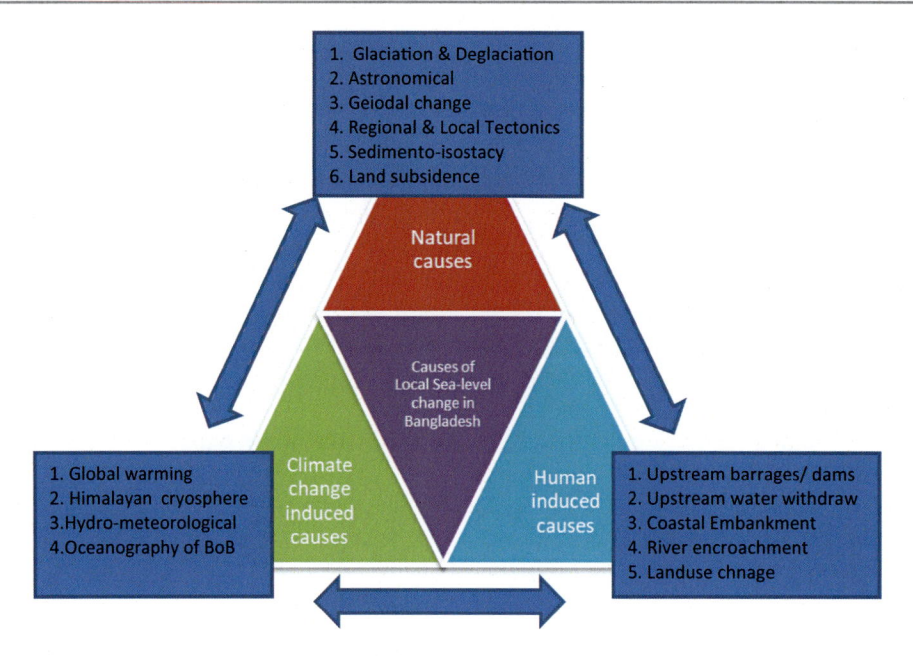

Fig. 32.1 The model shows different driving forces of sea-level movement along Bangladesh coast and their interlinks (shown by arrows)

cause should be ignored to assess and measure the past, present, and future sea-level scenarios of any particular coastal location.

32.3 Global Sea-Level Change

Study of tidal data is the common and simplest way to measure the secular trends of sea level globally, regionally, and locally. There are about 2300 tidal stations across the world of which NOAA has selected 240 representative stations from global tidal network to measure the global mean sea-level changes. On the other hand, PSMSL maintains global tidal database for 170 stations having records of more than 60 years (Morner 2016).

Based on available long-term tidal records, it has been suggested by many authors that the global sea level has been rising continuously since the early twentieth century. Douglas (2001) considered 25 best representative stations from stable land regions having tidal records of more than 70 years and has shown global mean sea-level rise of 1.8 mm/year. Based on evidence from altimeter data and tidal records from 177 stations covering 13 geographical regions, Holgate and Woodworth (2004) measured the global mean sea-level rise of 1.7 ± 0.4 mm/year between 1948 and 2002. They have found that during the decade 1950–1960, there was a significant sea-level rise, than any other decades in the past.

Using TOPEX and Jaso-1 satellite data, Leuliette et al. (2004) have observed a steady rise in global sea level at the rate of 2.8 mm/year and a large change of 15 cm during the 1997–1998 El-Nino event. From satellite altimetry since 1993 and tidal records, Cazenave et al. (2008) have measured a secular rise of global sea level at the rate of 2.5 mm/year and have argued that ocean thermal expansion contributed 50% of the rate of rising between 1993 and 2003. Since 2003, the contribution of ocean thermal expansion has significantly reduced or stopped.

Allison et al. (2009) have shown global sea-level rise of 3.4.mm/year since the satellite altimeter data is available and this rate is 80% faster that the best estimation of 1.9 mm/year by IPCC thirds Assessment Report (IPCC 2001) of same period. Using the continuous series of satellite altimeter data from T/P (launched 1992), Jason-1 (launched 2001), and Jason-2 (launched 2008) Nerem et al. (2010) have calculated that the sea level has risen at the rate of 3.4 ± 0.4 mm/year over the period of 17 years (1993–2009). However, Nerem et al. (2018) in another research, using 25 years of satellite altimeter data, have estimated a reduced rate (2.9 mm/year) of global sea-level rise, coupled with global climate change. The measurement of such accelerated rate of global sea-level rise, based on satellite altimetry in the past 20 years, has raised questions about the reliability of satellite-based interpretation of such short-duration data (Brammer 2014). Based on sea-level records collected from different sources (i.e., tide gauges, coastal morphology, and satellite altimetry), Morner (2016) has given the argument that the current trend of mean global sea-level rise remains within the range of ±0.0 to +1.0 mm/year, and by the year 2100, the rise will remain within the limit of 5 ± 15 cm.

32.4 Regional Sea-Level Changes

Using TOPEX/Poseidon satellite altimeter data and available tidal data over the period of 1950–2000, Church et al. (2004) have suggested a rise of sea level closer to 2.0 mm/year in the north Indian Ocean, except a rapid rise of more than 4 mm/year[1] in the north-eastern Bay of Bengal. Unnikrishnan et al. (2006) used the tidal data till 1996 of four selected stations (Mumbai, Kochi, Chennai, and Vishakhapatnam) from east coast of India and have shown an increase of sea level at the rate of 1 mm/year along the coast, except a slight fall at Chennai. Unnikrishnan and Shankar (2007) further have studied the tidal records of >40 years period of 16 tidal stations in the north Indian Ocean of which 10 stations are in coastal belt of the Bay of Bengal. The tidal records of those stations (Chennai, Vishakhapatnam, Paradip, Sagar, Gangra, Diamond Harbor, Hiron Point, Cox's Bazaar, Yangon, and Ko Taphao Noi) show the trend of secular sea-level rise at the rate of 1–2 mm/year (regional average of 1.29 mm/year), for stations in the Bay of Bengal, except for the northeast coast (Diamond Harbor) with a rise of 5.74 mm/year, since 1940s. They have the argument that, except for the north-eastern Bay of Bengal (e.g. Diamond Harbor), the sea-level trends along the Bay are within the range of globally estimated range of 1–2 mm/year (Douglas 2001; Church et al. 2004; Holgate and Woodworth 2004). Unnikrishnan and Shankar (2007) have also shown that in the Bay of Bengal, there was a drop in sea level at Vishakhapatnam, Sagor, and Ko Taphao Noi, but not at Diamond Harbor after 1960. This drop in sea level is related to intense rainfall in India.

On the other hand, using altimetry records since 1993, Cazenave et al. (2008) have measured a rising sea-level trend of 2–5 mm/year in the northern part of the Bay of Bengal. Jana et al. (2014) have studied the monthly and annual mean sea-level data of Haldia, Paradip, and Gangra over a period of 1972–2006, and have shown a secular increasing trend of 5.32 mm/year until 2000 and a sharp decreasing trend (7.9 mm/year) subsequently. However, Morner (2007, 2010, 2015, 2016) from a series of studies in and around north Indian Ocean has concluded that in the last 40–50 years sea level has remained virtually stable in this region.

32.5 Sea-Level Changes in Bangladesh

32.5.1 Past Sea-Level Change

No written documents of past sea-level changes in Bangladesh are available and it can only be reconstructed from proxy records. Tooley (1981) has demonstrated a series of techniques to reconstruct the past sea-level changes, of which some widely used proxies are lithological signature, microfossils (pollen, diatom, and foraminifera), and C^{14} dating. Using the above proxies, only limited attempts have so far been undertaken to unveil the Holocene sea-level movements of Bangladesh. Based on sediment signature and biomarkers from a 60 m long core at Dawlatpur (Khulna), Umitsu (1987) has proposed the first Holocene sea-level curve of

Bangladesh, showing a continuous rise, but a relative fall between 12,000–10,000 years BP. The maximum rise was 7.27 mm/year during the early Holocene period.

Based on pollen, diatom, C^{14} dates, and lithostratigraphic signature, Islam (2001) made a comprehensive study on the characteristics of the Holocene sea-level changes in Bangladesh. He has proposed an oscillation sea-level curve showing five phases of marine transgressions. During transgression-I (9150–6770 years BP) the level was 7.5 m below the present, with an average rise of 1.33 mm/year (maximum 2.25 mm/year). During the transgression-II (6615–6415 years BP), transgression-III (6315–5915 years BP), transgression-IV (4665–3715 years BP) and transgression-V (2175–1765 years BP), the sea-level was 6.8, 6.2, 5.2, and 4.2 m below the present, with an average rise of 1.29 mm/year (max. 2.68 mm/year), 2.17 mm/year (max. 3.65 mm/year), 0.75 mm/year (max. 1.00 mm/year) and 0.80 mm/year (max. 1.52 mm/year), respectively. His reconstruction shows that during the Transgression-I there was a slow rise of sea level, which was due to excessive siltation from the Himalayan ice-melting and rapid delta progradation.

Based on lithostratigraphic signature, biomarker (pollen), and C^{14} dates from Kolkata in India (closed to south-western border of Bangladesh), Banerjee and Sen (1987) proposed a Holocene sea-level curve in the Bengal Basin, with a continuous rise until 5000 years BP. None of these three curves (Umitsu 1987; Banerjee and Sen 1987; and Islam 2001) shows any evidence of higher sea level than that of the present. However, Rashid et al. (2013) have reconstructed the Holocene sea-level curve of central Bangladesh, and have recognized that at about 6000 years BP there was a transgressive episode and sea level was at least 5 m above the modern sea-level (also Monsur and Kamal 1994).

32.5.2 Current Sea-Level Trends

Recent trends of sea-level changes can be measured on the basis of instrumental records, such as tidal data and satellite information. In Bangladesh, BIWTA (Bangladesh Inland Water Transport Authority) is primarily responsible for collection, compilation, processing, storage, and publication of tidal data from a network of 47 tide gauge stations along the coastal belt and offshore islands, of which tides are forecasted in the form of tide table for 17 stations using computer modulation since 1987. BWDP (Bangladesh Water Development Board) also collects water-level data at 188 stations, of which many are located in the coastal area. Moreover, Chittagong and Mongla port authorities collect and store tide data at regular basis. The oldest tide data available in Bangladesh is at Chittagong port station since 1937.

Few attempts have been made in the past to measure the secular rise of sea level along Bangladesh coast using available tidal records (Fig. 32.2). The researcher commonly prefers to use the tidal records mostly available from a selected number of stations covering the wide geographical areas along the coastal belt, which are Hiron Point (available from 1967), Mongla (1975), Char Changa (1965), Khepupara (1976), Khal-10 (1982), Sadarghat (1976), and Cox's bazaar (1976), Das (1992) made the first attempt to measure the changes of sea-level at Chittagong site and found a rise of 25 cm between 1944 and 1964 (a rate of 1.25 cm/year). Mahmood (1992) also attempted to measure the sea-level changes at Chittagong and Mongla sites, and demonstrated a fall at the rate of 2.24 mm/year at Chittagong and a rise at the rate of 5.18 mm/year at Mongla port stations.

A comprehensive picture of sea-level trends has been portrayed in a study conducted by SAARC Meteorological Research Center (SMRC) in 2003. This study shows a sea-level rise of 4 mm/year at Hiron Point, representing the west coast; 6 mm/year at Char Changa, representing the central coast and 7.8 mm/year at Cox's Bazar, representing the east coast (SMRC 2003). According to this estimate, the rate of sea-level rise in the eastern coastal belt is nearly double than that of the western coast. Wahid et al. (2007) estimated an increase of about 0.75 m sea-level rise at Mongla and Nalianala

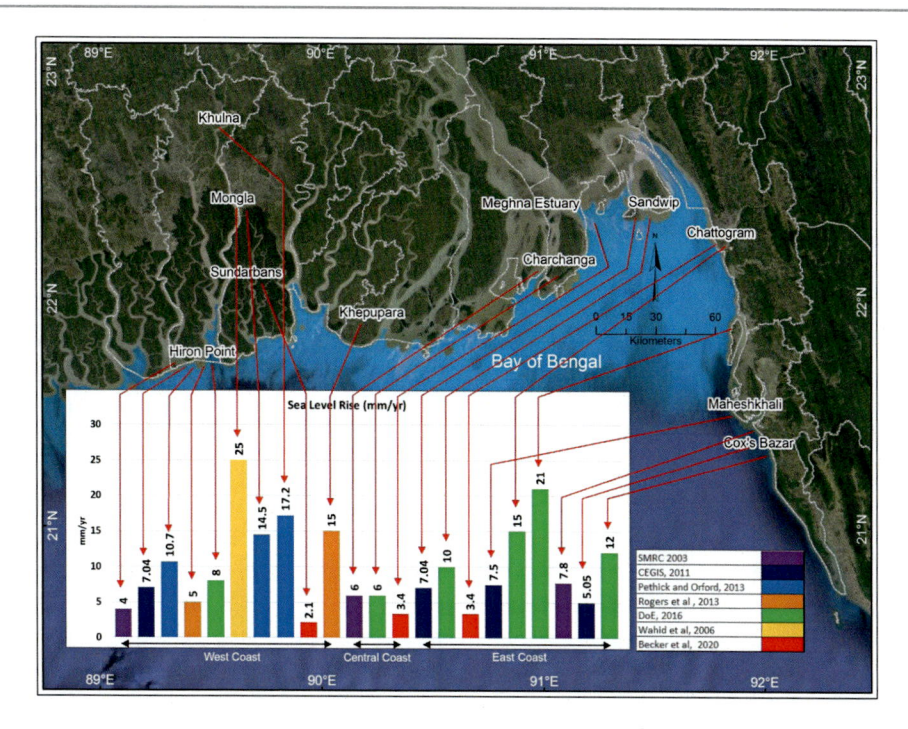

Fig. 32.2 The bar diagrams show the rates of sea-level rise calculated by different individual researchers and their geographical locations. Sources are shown inside the box

(Sundarban area) over a period of 30 years, with an increasing rate of 25 mm/year, which is faster than any other estimation. CEGIS report (2011) under the SNC (Second National Communication) project shows an estimation of 5.5 mm/year sea-level rise at Hiron Point, which is 7.04 mm/year at Sandwip, 7.5 mm/year at Moheskhali, and 5.05 mm/year at Cox's Bazar (see DoE 2016). The estimate shows that along Bangladesh coast, the mean sea-level rise is 5.05–7.5 mm/year. Based on tidal records from three stations along Pussur estuary in the southwest Bangladesh (Sundarban region) Pethick and Orford (2013) have shown an accelerated relative sea-level rise, which is 10.7 mm/year at Hiron Point (mouth of the estuary), 14.5 mm/year at Mongla and 17.2 mm/year at Khulna. Using the tidal records of 21 years duration, Rogers et al. (2013) have estimated mean sea-level rise of 5.0 mm/year at Hiron Point, 15.0 mm/year at Khepupara, and 5.0 mm/year at Hatia.

An important document on secular trend analysis was published by the Govt. of Bangladesh in 2016 (DoE 2016). In this consultancy research it has been suggested that trends of sea-level rise at Hiron Point (1981–2013), Char Changa (1980–2012), Sandwip (1977–2012), Khal No 10, (1983–2012), Cox's Bazar (1980–2012) and Lemiskhali (1981–2012) are at the rate of 8 mm/year, 6 mm/year, 10 mm/year, 15 mm/year, 11 mm/year and 21 mm/year, respectively. The report has summarized that in the western coastal region (covering the Ganges Tidal Plain) the trend of sea-level rise is 5–7 mm/year, which is 9–10 mm/year in the central coastal belt (covering the Meghna Deltaic Plain) and 14–23 mm/year in the eastern coastal belt (covering the Chittagong coastal plain) (DoE 2016). The rate is almost three times higher in the eastern coastal zone than that of the western coast and shows nearly a similar trend to that of the SMRC study (SRMC 2003).

Using tidal gauge data from 1968 to 2012 and satellite altimetry during 1993 and 2012, Becker et al. (2020) have shown that the sea-level movement at Ganges Tidal Floodplain (Sundarbans) is 2.7 mm/year (altimetry result 2.1 mm/year), which is 3.6 mm/year (3.2 mm/year) at Ganges Tidal Floodplain (central), 3.0 mm/year (3.4 mm/year) at Meghna Floodplain (Meghna Estuary) and 1.3 mm/year (3.4 mm/year) at Chittagong coastal plain. This recent research shows that along the coastal belt of Bangladesh the rate of sea-level rise is much lower than previously estimated. The mid-term report (on-going and unpublished) of DoE based on satellite altimetry, it has been proposed that there is an increasing trend of sea-level rise ranging from 2 mm/year to 12 mm/year, with an average of 3.6 mm/year, in the northern Bay of Bengal (Islam, personal contact). Along Bangladesh coast, the estimated mean annual rising trend is 6 mm/year, which is 3.8 mm/year, while seasonality is considered.

32.5.3 Future Sea-Level Projections

Future sea-level scenarios can be predicted through computer simulation based on proxy records, documentary records, and/or instrumental records. It is projected that due to global warming the average temperature of the earth would rise 1.4–5.8 °C, SST would rise 0.2–2.5 °C and global mean sea level would rise 98 mm by 2100 (IPCC 2014). Using TOPEX/Poseidon, Jason-1, Jason-2, and Jason-3 satellite altimeter data, Nerem et al. (2018) project global mean sea-level rise of 65 ± 12 cm by 2100, which is slightly lower than the projection made by AR5 (IPCC 2014). Bangladesh has not yet projected any future sea-level scenarios. The country has accepted the IPCC scenarios (IPCC 1990, 1996, 2001) of secular uniform sea-level rise for all its policy and planning measures (NAPA 2005; BCCSAP 2009). However, the coastal belt of Bangladesh shows spatial variations in terms of its tectonic activities, sedimentation rate, subsidence rates, and shoreline geometry (Brammer 2014), so as its sea-level scenarios.

Sea-level rise does not always fit with the trends of temperature rise and SST as predicted by model studies and there are always risks for uncertainties. IPCC reports (1990, 1996, 2001, 2007, 2014) show that thermal expansion and ice-melting are two important causes of future global sea-level rise. However, their projections until 2100 show different future scenarios, such as 66 cm (rate 6 mm/year), 50 cm (rate 4.45 mm/year), 48 cm (rate 4 mm/year), and 28–48 cm (rate 2.76–4.1 mm/year) in the First (1990), Second (1995), Thirds (2001) and Fourth (2007) assessment reports, respectively (IPCC 1990, 1996, 2001, 2007). Fifth Assessment Report (IPCC 2014) has some regional projections, including the projection of 20 cm to 1 m mean sea level in the Bay of Bengal, under low and high emission scenarios by 2100. However, at high emissions, this report predicts a global mean sea-level rise of 52–98 cm by 2100. In all scenarios, only the climate change-induced parameters are considered, and thermal expansion is identified as the main contributor (30–55%), followed by glacier and ice-melt water.

Milliman et al. (1989) have argued that by the year 2100, sea level along Bangladesh will rise up to 4.5 m higher than the present. Pethick (2012) has extrapolated the tidal records of Mongla and Khulna stations and has predicted a mean sea-level rise of 2.31 m and 3.71 m, respectively by the year 2100. He also suggested that if there were no eustatic component of sea-level rise due to global warming, the southwestern coastal belt of Bangladesh would also experience a sea-level rise of 2 m in next 100 years. Recently, using satellite altimetry Becker et al. (2020) have projected a sea-level rise of 85–140 cm in Bangladesh by 2100 under the RCP 4.5 scenarios.

32.6 Issues and Challenges

32.6.1 Reliability on Proxy Data

The Holocene sea-level reconstruction and curve presented by different authors vary significantly in their methodology, data presented, and interpretation. Curves, presented by Umitsu (1987), and Banerjee and Sen (1987) are the offshoot of their missing study the sedimentary sequences of

Bengal delta, and reconstruction of palaeo-vegetational history of Bengal, respectively. Unlike Umitsu (1987), Banerjee and Sen (1987) have suggested that the early Holocene sea-level rise was fairly rapid. However, in both researches. C^{14} dates were used to infer the past sea level and both have shown that the shoreline was far inland between 7000 and 6000 years BP. Islam (2001) primarily focused on the reconstruction of Holocene sea-level changes and their implications, based on multi-proxy records (pollen and diatom) and C^{14} dates. This research shows sea-level fluctuation during the Holocene.

The reconstruction by Rashid et al. (2013) is quite opposite to those of previous studies by Umitsu (1987), Banerjee and Sen (1987) and Islam (2001). Mid-Holocene high-stand of at least 5 m above the present sea level is questionable. A closer look at these four published Holocene curves, except Rashid et al. (2013) other three, despite their minor variations, are in agreement in terms of their rates and directions of Holocene sea-level movements in the Bengal Basin (Fig. 32.3). Rashid et al. (2013) dealt with the sea-level index points derived from a tectonically uplifted Madhupur tract (Morgan and McIntire 1959; Khandoker 1987; Islam 1998).

Chowdhury et al. (2009) based on beach rock evidence at St. Martine's island, have argued that sea-level rise at this location took place at about 10,000 years BP and continued till 6000 years BP, and after ca. 5000 years BP, there was a lowering of sea level. During the last 6000 years BP, there were fluctuations in sea level of 1 m or more over a thousand years. However, without any absolute dating result (C^{14}), such projection and time-frame of sea-level oscillations near St. Martins Island are questionable. The proxy records available from the Bengal delta are not adequate, accurate, and comprehensive, to reconstruct the past sea-level history with confidence. It is mostly because of high cost of C^{14} dating (the most reliable tool), and lack of appropriate laboratory supports.

However, despite all such limitations, it can be suggested that the sea-level movement along Bangladesh coast was primarily contributed by the early Holocene ice-melting water (as eustatic component), which continued until the mid-Holocene, with few minor functions. During the last 5000 years, the regional and local geological factors played the vital roles over global causes, and the coastal belt of Bangladesh experienced at least four transgressions, followed by a regression, of about a thousand years cycles. During this period, there was balance between sediment influx, natural subsidence, and relative sea-level movements in the Bengal delta (Islam 2001).

32.6.2 Accuracy of Tidal Data

The major uncertainty and criticism that remains with the measurement of present-day secular sea-level trend are the quality of tidal data. Measurement of sea level in tidal gauge stations simply shows the changes in respect to the instrument itself, without reflecting any eustatic component, unless the station is located in a stable location without any local crustal movements. In sea-level research, no stable landmass is available on the earth's surface (Morner 2016) and each location is distorted vertically either by subsidence or uplift.

Very high precision long-term data is required for accurate measurement of sea-level changes. For meaningful assessment, the length of tidal data must be at least >18.6 years (the lunar-tidal cycle) and preferable >60 years (Morner 2016). In Bangladesh, the oldest tidal data is available at Chittagong port station since 1937 and in other most cases, the records are not continuous. Tidal data in Bangladesh is not procured for sea-level studies, and are not authentic, stable, and consistent. They have calibration problems between manually collected and auto-gauge collected data. For sea-level research tidal data of millimeter accuracy is required, but BIWTA collects data at centimeter accuracy, mostly for navigational purposes. Moreover, in most cases, the data is available only for short duration (since

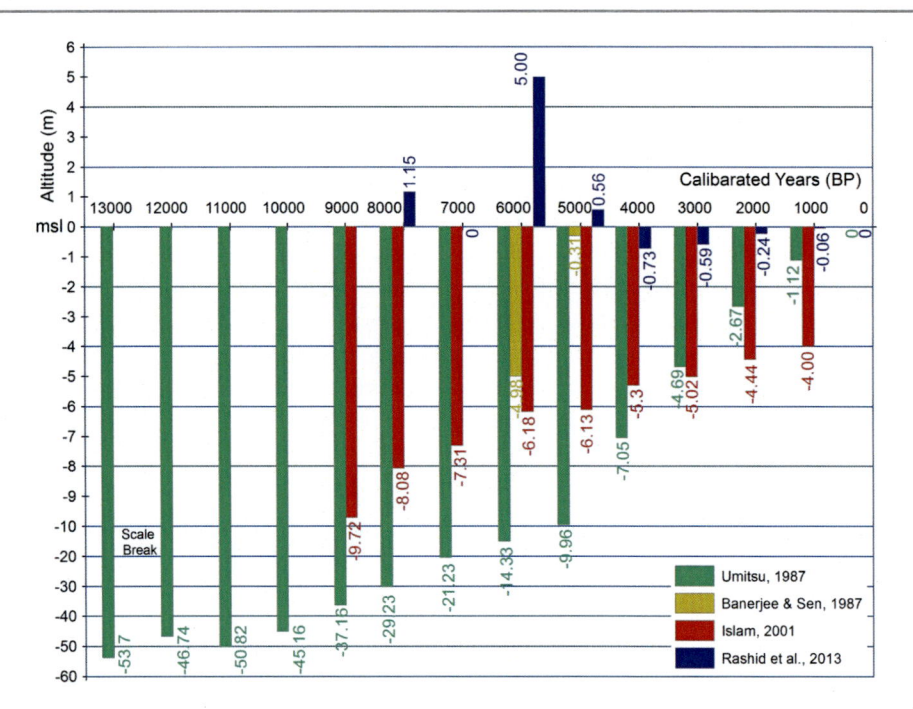

Fig. 32.3 Holocene sea-level position at thousand years interval since 12,000 to 1000 years BP, as suggested by four different studies

1977), and in some cases, data are discontinuous and show vertical errors due to subsidence and bank erosion. Nahmias and Islam (2014) have examined the quality of tidal data of five stations across the coastal belt (Cox's Bazar, Teknaf, Char Changa, Dosmonia, and Dhulia) of Bangladesh, and have found that the available tidal data from all those stations are not suitable for sea-level research. It is because of erroneous characteristics of each station due to shifting of tide gauge, datum change, conversion of scale, automation, and mechanical problems.

Until today, nearly half a dozen attempts have been made to measure site-specific trends of sea-level movement (which is truly the tide-level movement) using the available tidal records, mostly collected by BIWTA and Port Authorities. The results are confusing, inconsistent (even for the same station), and non-reliable. This is because of the use of erroneous tidal data, and different approaches to investigation by different scientists. The use of such assessments in national policies and programs is unsafe and risky and must be avoided.

32.6.3 Measurement of Subsidence

Subsidence is an important consideration, particularly in the present era, to measure the secular trends of relative sea-level movement, projecting the future and assessing the coastal vulnerabilities of a deltaic country, like Bangladesh. Available records from Bangladesh and West Bengal (India) show that in the Ganges–Brahmaputra-Meghna deltaic region subsidence level is very high, not uniform, and shows considerable regional and locational variations. Meem et al. (2016) have assessed the subsidence data available from 120 locations across the coast, and have found that the average subsidence rate of the coast belt is 3.6 mm/year. The geographical distribution of available subsidence data is very skewed, 105 records are from southwest coastal zone and only 10 from central coastal zone. No published subsidence data is available from the south-eastern zone. Their assessment shows that high subsiding areas (>20 mm/year) are at Magura and Hatia Trough; medium subsiding areas (5–20 mm/year) are at

Patuakhali, Barisal and Barguna; and low subsiding areas (<5 mm/year) are at Noakhali, Bagerhat, Khulna, Satkhira and West Bengal, suggesting that the south-central part of the coast is subsiding at a faster rate than that of the southwestern coast (Fig. 32.4).

There are remarkable methological differences to calculate the subsidence rate by different scientists. Based on geological records (lithostratigraphy, pollen evidences, and C^{14} dates) the subsidence rates of 0.6–0.9 mm/year were calculated by Vishnu-Mittre and Gupta (1972) and Umitsu (1987); 1.73 mm/year by Islam (1998); and 5 mm/year by Stanley and Hait (2000) for the Sundarban region of Bangladesh and India. Using information from a well in the southern Bangladesh, Hoque and Alam (1997) have calculated the subsidence rate of 2.0 cm/year, the ever highest estimation from Bengal delta. Based on stratigraphic signature and C^{14} dates Goodbred and Kuehl (2000) have measured the subsidence rate of 2–4 mm/year for the central coastal belt. They have argued that unlike many other deltas of the world, the subsidence of Ganges–Brahmaputra delta is not related to compaction, but is tectonically controlled. Using the GPS data, Steckler et al. (2008) have measured a subsidence rate of 7–9 mm/year in Khulna region, and this is due to compaction of organic content of the deltaic sediment (cited in Pethick and Orford 2013).

Sarker et al. (2013) studied the level of the plinth of a mosque of fifteenth century at Bagerhat, a temple of 400 year at old in the Sundarban forest, and another temple of 200 years old at Khepupara, and have calculated the subsidence rate of 1–2.5 mm/year in the southwestern part of Bangladesh. Hanebuth et al. (2013) observed the bases of a number of salt kilns at Kotka (in Sundarbans) at about 155 cm below the modern surface level. The C^{14} results of charcoal samples (found in the kilns) show that these kilns were last fired about 300 years ago. Based on these kilns records, they have

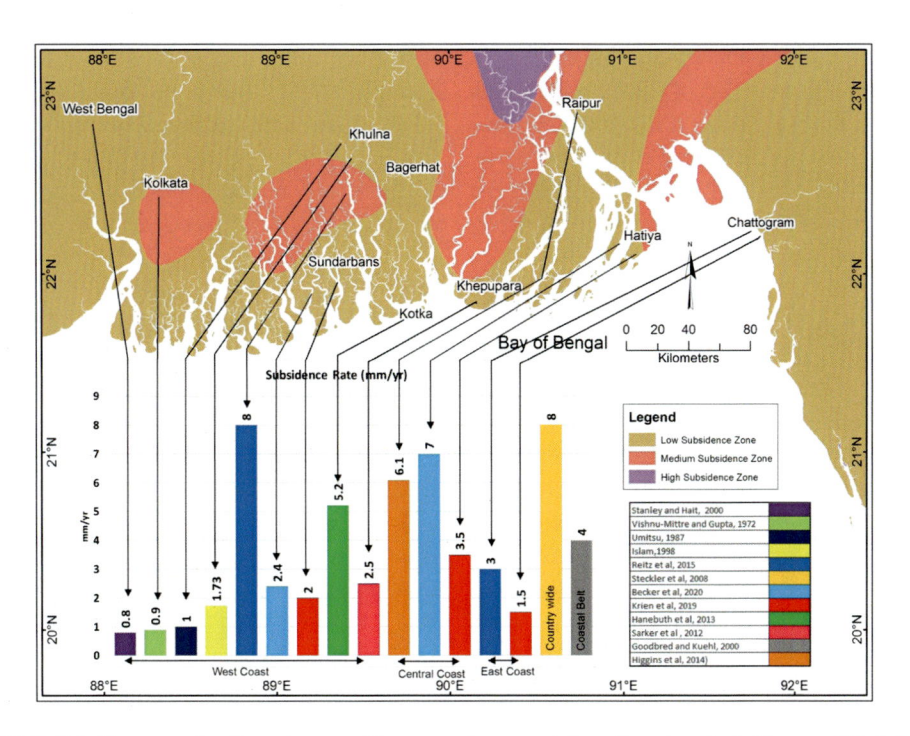

Fig. 32.4 Subsidence of the lower Ganges–Brahmaputra-Meghna delta. The map in the upper part of the diagram shows the regional extent of subsidence as compiled by Meem et al. (2016). Bar diagrams in the lower part show the rates of subsidence and their locations as proposed by different authors. The sources are stated inside the figure box

measured the subsidence rate of 5.2 ± 1.2 mm/year, Based on InSAR information for a period of 2007–2011, Higgins et al. (2014) have shown that subsidence rate in the south-central part (Raipur) of Bangladesh is 6.1 mm/year. From site-specific continuous GPS observations between 2003 and 2013, Reitz et al. (2015) have measured the subsidence rates in Bangladesh ranging from 8 mm/year at Khulna to less than 3 mm/year at Chittagong regions.

Krien et al. (2019) used 3D numerical model to measure the deformation of the earth's surface of the delta due to sediment weight carried by the major rivers, and have measured subsidence rate of 2–3 mm/year. The highest subsidence rate is at 3.5 mm/year at Hatiya Trough and in Chittagong coastal belt it is less than 1.5 mm/year. Using satellite altimetry, in a recent study, Becker et al. (2020) have shown that the land subsidence at Ganges Tidal Floodplain (Sundarbans) is 2.4 mm/year, which is 7.0 mm/year at Ganges Tidal Floodplain (central), and 5.2 mm/year at Meghna Floodplain (Meghna Estuary); for Chittagong coastal plain no data is available.

The above discussion shows that there is a wide range (1–8 mm/year) of subsidence rates in the coastal belt of Bangladesh, which is site-specific and shows considerable variations even for a short distance (Fig. 32.4). There also exist different opinions regarding the causes of subsidence at local scale. Subsidence may be due to either tectonic activities (Goodbred and Kuehl 2000; Reitz et al. 2015), crustal deformation under the weight of sediment load, compaction and drying (Hanebuth et al. 2013), and/or extraction of excessive groundwater, withdrawing water in the upstream and floodplain engineering in the downstream (Syvitski et al. 2009).

Local-relative sea-level scenarios in Bangladesh significantly depend on the rate of local subsidence. Until today, at least a dozen of attempts have been made to measure site-specific subsidence rate on the coast, and the results are confusing, and inconsistent (even for the same station). However, unless an accurate assessment of local land subsidence is available, and such vales are taken into consideration, the measurement of local-relative sea-level rate remains questionable.

32.6.4 Measurement of Sedimentation

The Ganges–Brahmaputra river system together carries about 1.6 billion tons of sediment annually (Broadus 1993). Other studies (Goodbred and Kuehl 1999; Islam et al. 1999) also show that the annual sediment influx through this river system is approximately 1.00 billion tons, of which 51% is deposited in the coastal area, 28% in the flood plain and remaining 21% within the river channel (Islam et al. 1999). Goodbred and Kuehl (1999) however, have suggested that sediments are deposited roughly in equal proportion at the floodplains, shelf zone, and deep-sea areas. A small portion of these sediments is drifted toward the west by wave and tidal action and finally deposited in the Sundarbans areas, at the rate of 1 cm/year, which is equivalent to the relative sea-level rise of Bangladesh coast (Rogers et al. 2013).

Measurement of sedimentation is a complicated task and requires a wide variety of inputs, such as geographical location, climatic condition, source of sediment, human disturbance, and absolute dating. Based on geological records, Islam (2001) has calculated a sediment rate of 0.82 mm/year during the Holocene and 3.12 mm/year during the last one thousand years in the Ganges–Brahmaputra deltaic region. Using ^{137}Cs geochronology of sediment cores Allison et al. (1998) have suggested that the sedimentation rate varies from >4 cm/year (on the natural levees) to <1 cm/year in the basin areas. The geological data shows that the sedimentation rate in the Ganges–Brahmaputra delta has kept equilibrium with land subsidence and relative sea-level moments (Islam 2001).

In Bangladesh, sediment accumulation in the river beds is a widely known event. CEGIS (2010) has shown that due to sediment accumulation, the Jamuna and Ganges river bed will rise 11–26 cm and Padma river bed will rise 33–52 cm in next 100 years, which would have obvious impact on sea-level scenarios of Bangladesh. The topography and landscape of greater part of Bangladesh, particularly in the coastal area, are built, shaped, and re-shaped by the continuous natural supply of sediment from the upstream. The major rivers of the country, particularly the Ganges–Brahmaputra river system plays the vital role in natural supply of sediment to the downstream floodplain, which is very vital for the existence of the country. The supply of large volume of sediment in the deltaic system is the compensation for frequent river avulsion, land subsidence, and local sea-level rise. At present, the sediment budget in the low-lying deltaic area is disrupted in two ways: the reduction of sediment supply from the upstream, particularly due to the construction of Farakka barrage; and interruption of natural sediment distribution to the coastal floodplain, particularly due to the construction of embankments along the tidal rivers. Khalequzzaman (2016) has argued that the lower part of the delta receives less sediment supply than the previous, and if such reduction in sediment supply continues, then the growth of the delta, both vertically and horizontally, will be seriously curtailed and the coastal belt of Bangladesh would be more vulnerable to future sea-level rise any disruption to natural supply of sediment in the deltaic landmass by human interventions is an important cause for local level subsidence and a threat to relative sea-level rise.

32.6.5 Regional Tectonic

The Ganges is the first river to develop in the Bengal Basin, which during the Pleistocene upheaval used to flow to the Bay of Bengal through the Bhagirathi-Hooghly system (Khan 1991, Niyogi 1975). Islam (2001) has argued that during the last glacial maximum, the head of the "Swatch of No Ground" was in the same alignment with the Gorai-Madhumati river, which was possibly the original course of the Ganges. The eastward shifting of the Ganges river and its distributaries from Bhagirathi-Hooghly to its present course since the 1600AD reflects the continuous tilting of the delta toward the east (Akhter et al. 2005). Many of the active rivers of the delta overlie the active plate boundaries and other tectonic features, which continuously modifies its landscape and the rivers are forced to migrate due to tectonic tilting (Reitz et al. 2015).

The rapid subsidence in the Ganges–Brahmaputra-Meghna deltaic landmass is the combined effect of crustal deformation due to enormous sediment load and tectonic activities (Stanley and Hait 2000; Goodbred and Kuehl 2000; Reitz et al. 2015). The delta experiences vertical elastic deformation of up to 6 cm due to annual loading and unloading of water during the monsoon (Steckler et al. 2010). Krien et al. (2019) based on 3D numerical model have also noticed the deformation of the surface of the delta due to sediment weight carried by the major rivers, and associated land subsidence. Indian plate is slowly moving north-east direction at the rate of 6 mm/year and subducting under the Eurasian plate and Burmese plate at the rate of 45 mm/year and 35 mm/year, respectively. From GPS instrumental observation it shows that due to convergence motion of Indian and Burmese plates, the Bengal delta is shortening by 13 mm/year (Steckler et al. 2008). Because of this horizontal shortening, the uplift and subsidence of the delta are tectonically attributed. Tectonic subsidence and variable channel dynamics are strong enough for vertical land deformation and relative sea-level movements along the coastal belt of Bangladesh.

Bangladesh had experienced a number of mega earthquakes in the past, which had brought significant changes in the topography. Catastrophic events are also responsible for crustal deformation in this region. The records of subsidence in Noakhali and Chittagong regions due to 1762 earthquake is an evidence of tectonic crustal deformation of the delta (Cummins 2007). Based on field evidences, Akhter et al. (2015) have shown that this 1762 earthquake brought changes in the landscape in the Chittagong region. The Teknaf peninsula and St. Martins Island were uplifted by about 2–2.5 m due to this catastrophic event.

32.6.6 Himalayan Cryospheric Impact

Cryosphere is an important part of earth surface system and has its links to global and regional climate change. The climate change impact on cryosphere is strong and contributes significantly to regional-scale sea-level moments. The impact of climate change on Himalayan glaciers is well noticed (see Mukherji et al. 2019). Kulkarni et al. (2011) have shown that between 1962 and 2002, the Himalayan cryosphere has reduced by 16% due to climate change. Khan et al. (2017) have shown that the size of the Gangotri glacial, which constitutes the main source of water in the regional river system, is receding very fast.

Himalayan cryosphere, which covers an area of about 33,000 km^2 and delivers about 8.6×10^6 m^3 of water annually, is an important element of regional hydrological budget. Ganges and Brahmaputra rivers are largely influenced by Himalayan ice-met water, which finally discharges into the Bay of Bengal through Bangladesh. The retreat of Himalayan glaciers and volume of stream run-off from ice-melt water is influenced by global warming on the Himalayan cryosphere. Glacial meltwater significantly regulates the river discharge in Bangladesh, which has high seasonal variability.

Seasonal flow of ice-melt fresh water into the northern Bay of Bengal has its effect on SST, salinity, tidal range, and water density, which have their individual and collective impact on seasonal and inter-annual sea-level movement along Bangladesh coast. The mean sea level along the east coast of India is 30 cm higher than that of west coast, due to distinct seasonal variation of ocean dynamics and salinity (Shankar and Shetye 2001), and has its connectivity with the Himalayan cryospheric impact. The major rivers of Bangladesh are fed by both rainfalls in the catchment and ice-melting water from the Himalayas. Due to lack of information on their proportionate input, the contribution of Himalayan cryosphere to the ocean dynamic of the Bay and its impact on local sea-level movements has not yet been assessed. Cryosphere component has a long-term impact on regional sea-level moments. It is, thus, necessary to monitor and evaluate the impact of climate change on the Himalayan cryosphere for better understanding of the sea-level sceneries of Bangladesh.

32.6.7 Hydro-Meteorological Impact

Relative sea-level movement not only depends on astronomical force (as regular tide) but also on local meteorological and hydrological causes. Rivers originating from the Himalayan regime and flowing through Bangladesh carry large amount of fresh water to the Bay of Bengal during the monsoon and raise the local sea level up to 25 cm higher than in the winter (Antony et al. 2005). The deviation of tidal range due to non-tidal (as residual tide) contributions (known as residual sea-level elevation), is well noticed along the coast of the Bay of Bengal. Inter-annual variation of sea level along Bangladesh coast is largely affected by global atmospheric disturbances. El-Nino is known to create climatic anomalies at global and regional scales and has a link to SST, pressure, and rainfall variations in the northern Bay of Bengal. Singh et al. (2001) have shown that El-Nino and La-Nino events are important components of the sea-level variations in the estuarine zone of Bangladesh. Based on tidal records at Char Changa (1979–1998) Singh (2002a) has shown that during the La-Nina phase, the sea level along Meghna estuary is 20–30 cm higher than that of the El-Nino phase, and the coastal belt of Bangladesh had experienced worst floods during the La-Nina episode of 1998. There is a close relationship between SOI (southern oscillation index) and mean tide level in the Meghna estuary.

The sea-level deviation due to local causes is also well documented by Pethick and Orford (2013) in the south-west Bangladesh, and have shown that the increasing tide limit at Khulna (122 km inland) is 6.7 mm/year higher (the rate is 17.2 mm/year) compare to that of the mouth of estuary (at Hiron Point) (rate is 10.7 mm/year). Wahid et al. (2007) have suggested that such an amplification of tidal limit could be due to reduction of fresh water in this area, backwater effect from the south, and funnel effect of the estuary. Such residual sea-level rise, which has no link with eustatic component, is primarily contributed by the hydro-meteorological impacts and human interventions.

32.6.8 Ocean Dynamics

Oceanographic conditions, such as ocean bathymetry, water mass, and coastal topography largely influence the semidiurnal sea-level oscillation along Bangladesh coast. Atmospheric pressure anomaly, wind force, ocean current, and ocean bathymetry are dominant factors to deviate artificially the tidal pick much higher or lower than a normal tide level. The fall of barometric pressure increase the sea surface elevation. It has been observed that for every 1-mbar fall of atmospheric pressure, there is a sea-level rise of 1 cm. Such inverted barometer effect of sea-level rise is higher in the shallow water (Antony et al. 2005), as is the ideal case in the shelf zone of Bangladesh, which is very sensitive to seasonal atmospheric pressure differences.

The relative importance of non-tidal impact on local sea-level variation is seasonal. The distinct seasonality and annual cycle of coastal water circulation in the northern Bay of Bengal has its effects on salinity, water density, and water-masses movement in the Bay. The strong south-west monsoon wind stimulates ocean water-mass to move toward the north-east corner of the Bay and artificially increases the elevation of sea level in the Meghna estuary. Singh (2001) has shown that half of the contribution for annual tide-level trends (3 mm/year) is derived from monsoon-induced components. Bangladesh coast being located in the northern hemisphere, the Coriolis force acting on the ocean current further escalates this rise. Wahid et al. (2007) have shown that due to seasonal variation and monsoon effect there may a 15 cm change in sea level for a difference of 15 mbar air pressure, 5–10 cm for wind-set, and 20–30 cm due to density difference between saline and freshwater, showing an increasing effect in the northern Bay of Bengal. Until now, the seasonal and annual contributions of oceanographic components to local scale sea-level movement along Bangladesh coast are not well studied and understood. However, a very general conclusion can be made that because of funnel effect, the seasonal rise of sea level (during the monsoon) due to hydro-meteorological and oceanographic causes is higher in the Meghna estuary than in the south-west and south-east coastal regions of Bangladesh.

32.6.9 Human Interventions

Local and regional-scale human interventions, particularly mega projects of floodplain engineering, have tremendous impact on local sea-level movements, particularly in deltaic environments. Woodworth et al. (1991) have demonstrated the impact of anthropogenic effects on the amplification of tidal range along the UK coast. In Bangladesh, the accelerated local sea-level rise at Khulna is due to artificial amplification of tidal limit, of which anthropogenic causes, such as polder effect, contribute about 55% (Pethick and Orford 2013). While artificial barrier contributes 0 mm/year sea-level rise at Hiron Point (no human intervention), it is 2.2 mm/year higher at Mongla and 6.1 mm/year at Khulna (polder area).

Among all human interventions, the construction of Farakka barrage in the upstream and coastal embankments in the downstream, are two important driving forces to escalate the tidal limit and engulf more area under coastal inundation. The rivers of the south-western Bangladesh, particularly the Gorai-Madhumati system are primarily fed by water available from the Ganges flow, which has reduced drastically after the operation of Farakka barrage in India in 1975. The long-term effect of such freshwater withdraw the upstream is the main cause of secular rise of water level at an accelerated rate in the south-western coastal belt of Bangladesh (Sherin et al. 2020). It has no link with eustatic components and global warming. The observation of recent rapidly rising inundation level and increasing salinity intrusion, by local people and researchers, in the coastal area is simply the reflection of amplified backwater effect of the tide, which has its relationship with the reduced freshwater flow in the region (Sherin et al. 2020). The problem has further been escalated by groundwater extraction for irrigation and construction of embankments along the tidal rivers.

Since 1960s, under the Coastal Embankment Project, about 5100 km of embankments have been constructed under 123 polders in the south-western coastal area of Bangladesh, with the goal to protect the tidal floodplain from tidal inundation and to foster agricultural output. However, these polders not only have stopped tidal inundation but also have ceased sediment to flow into the floodplains. The sediments carried from upstream and also through tidal reworking are forced to settle on the river beds. Sediment starvation in the floodplain has accelerated land subsidence, on one hand, and rapid siltation in the river channel, on the other hand, has raised the riverbeds, which are not only the main cause of permanent waterlogging but also artificial escalation of tidal height in this region. Because of the confinement of river flow within the embankments, the environmental condition and tidal inundation of this coastal belt vary significantly with different physiographic units (Brammer 2014). Auerbach et al. (2015) have found that due to polderization the south-western coastal plainland has lost its elevation by 1–1.5 m or at the rate of 2–3 cm/year, compared to neighboring Sundarbans forest area. This lowering of land elevation due to interventions of natural influx of sediment in the floodplain attributes greatly to local-relative sea-level rise.

Besides massive interventions, small-scale local engineering projects are also playing vital role to accelerate local tidal range. For example, with the aim to increase crop production by preventing saline water intrusion and storing rainwater for irrigation, the Muhuri irrigation project was implemented in 1986, by constructing a 3.4 km long cross-dam and a regulator of 40 sluice gates across the Feni river. Islam and Paul (2004) have shown that this project, despite its agricultural benefits, has tremendously affected the hydro-dynamics at the Feni river estuary. The tidal range has increased significantly at the river mouth, down to the cross-dam, which has led the areas more vulnerable to tidal inundation.

Groundwater extraction is another challenge to land subsidence, salinity intrusion, and coastal inundation. In Bangladesh, particularly in the coastal belt, the main source of dry season irrigation is the use of groundwater. In a study, BADC (2011) has shown that due to groundwater extraction the problem of salinity intrusion is very high in the south and south-western coastal belt. In this belt, the hydrogeology and aquifer system is very complex and varies considerably within a short distance (Zahid et al. 2018). Because of restricted lateral movement of groundwater, the presence of brackish and saline water in the coastal aquifer does not follow any regular trend. Salinity concentration in groundwater is coupled with the supply of freshwater flooding and salinity intensity due to accelerated tidal effects. Water taken from groundwater sources for domestic, industrial or irrigation purposes is, thus, an attribute to local-relative sea-level rise.

32.6.10 Impact of Climate Change

From most of the existing literature, there is a general assumption that due to global warming and climate change-driven causes, there will be an accelerated sea-level rise along the coastal belt of Bangladesh, which would inundate 16% of total landmass by 2100 (WARPO 2006). This will be due to addition of ice-melting water from the polar regions. However, there is widespread misconception that ice-melted water is equally distributed in the global ocean system, and raises the sea level equally worldwide. The distribution and redistribution of ocean water largely depend on astronomical forces, geoidal influence, and regional oceanographic conditions. Along Bangladesh coast, except IPCC assumptions (IPCC 1996, 2001, 2007), the contribution of ice-melt water from the polar regions, primarily due to global warming, has not yet been studied. On the other hand, Morner (2016) has argued that impact of climate change in the northern Indian region is negligible and there was no sea-level rise during the last 50 years. Morner (2007) mentioned that in the Indian Ocean, due to strong regional evaporation, as derived from global warming, there is a lowering of sea level in this region.

Most importantly, the direct discharge of ice-melting water, primarily due to climate change, from the Himalayan glacier into the Ganges–Brahmaputra river system, has its immediate impact on relative sea-level movements in the Bay of Bengal, particularly along Bangladesh coast. Such an important climate changed induced regional component has not yet been studied and measured to assess the sea-level movement in the northern part of the Bay. Without scientific evidences, the assumption of contour by contour rise of sea-level due to climate change, along Bangladesh coast, and its vulnerability assessment, however, is not desirable. It requires thorough investigation to prove how and to what extent global climate change has its links with local sea-level movement of Bangladesh. The observation of elevated tidal range, inundation of more area under tidal water, and saline water intrusion along the coastal belt are all justified, but difficult to prove how these are related to climate change-induced ice-meting water. Rather there are ample field level evidence and documentation to prove how due to massive human interventions by floodplain engineering, both upstream and downstream, the tidal range has artificially been amplified many folds to inundate the coastal plainland and are forced more coastal people to become vulnerable to the sea-level rise. Such an accelerated sea-level rise is nothing, but the humanly enforced accelerated tide-level rise at local scale. It may have a link to global climate change, but not significant.

32.7 Discussion

Assessment of sea-level trends along a deltaic coast in general, and Bangladesh coast, in particular, is a complicated task and requires an analysis of multivariate good quality data. These are proxy data to retrieve the past sea-level scenarios, instrumental data to assess the current trends of change, and computer-based model analysis of geological and instrumental data to project the future scenarios. In case of Bangladesh, the quantity and quality of such data and their interpretation are poor and are not reliable enough for policy and planning purposes.

The available long-term geological records on sea-level movement are inadequate. However, based on available proxy records it can be assumed that in the geological past, particularly during the Holocene, the sea level in Bengal delta was not continuously rising, rather it was oscillating with rises and falls following the phases of transgression and regression, respectively, with an average rise of 1.75 mm/year (Islam 2001). However, the debate yet remains on the issue of mid-Holocene high-stands. Such transgressive–regressive episodes in the past had tremendous impacts on shoreline migration, river shifting, changes of mangrove belts and human occupancy in the Bengal.

The information available on current trends of sea-level changes along Bangladesh coast is dubious. The major challenge is the quality of tidal data. Tradition approach to measure current tends of sea level relies on tidal records, although satellite altimeter data is available since 1992. In Bangladesh, most of the assessments are based on analysis of available tidal data, but the results are inconsistent and doubtful. Among all assessments, the report published by the Department of Environment (DoE 2016), as an official publication of the Govt. of Bangladesh, is a mess and has created lots of confusion about the trends of sea-level rise along Bangladesh coast. It shows that in the south-western coastal region the trend of sea-level rise is 7–8 mm/year, which is 11–21 mm/year, in the south-eastern coastal belt. Such an unusual difference (nearly 3–4 times higher in the eastern belt) only within less than 350 km ocean distance can neither be justified geologically, oceanographically, nor in the light of climate changes induced causes. Singh (2002b) also made nearly similar conclusions and tried to justify that such an accelerated rate in the eastern coastal belt is due to high rate of subsidence. However, the available published data (see Fig. 32.4) show that the subsidence rate along Chittagong coast is less than 3 mm/year, which is much lower than that of the south-western coastal belt. On the other hand, Chowdhury et al. (2009) and Akhter et al. (2015) have given the evidence of tectonic upliftment of the Chittagong Coastal belt.

Due to inter-annual seasonal variations, atmospheric disturbances, Himalayan cryospheric effect, and oceanographic causes, there might have significant regional deviations of sea-level trends along Bangladesh coast, but in such case, the higher sea-level rise would be in the funnel-shaped Meghna estuary of the central coastal belt, off-course not in the eastern coast. If the DoE (2016) assessment is taken as the real case, the 135 years old Chittagong port or any other old establishments on Chittagong coast would have been underwater by this time.

Besides the DoE (2016) assessment, a number of efforts were made by individual scientists, and they all, except Morner (2010), are in agreement that the sea level along Bangladesh coast is rising, although there remain differences regarding the rates they had measured and the causes they had identified. Among them, Pethick and Orford (2013) have shown the most excessarated rates (10.7–17.2 mm/year) for south-west coastal belt. At these rates of sea-level rise, the present-day Sundarbans forest did not survive and by this time would have been submerged under seawater. However, the first satellite altimeter based assessment of sea-level changes along Bangladesh coast is available from Becker et al. (2020), which unlike most previous works, shows a relatively slower rate of sea-level rise; 2.1 mm/year at Sundarban areas, 3.4 mm/year at Meghna estuary and 3.4 mm/year at Chittagong coastal plain.

A complete picture was drawn by Morner (2010). Based on the morphological and stratigraphical evidence from Sundarbans areas, he has given the argument that there is no sign of global component of sea-level rise in this coastal belt; local sea level is stabilized and attributes a minor regression of the sea. The predicted sea-level scenario of Bangladesh is also uncertain. Using TOPEX/Poseidon, Jason-1, Jason-2, and Jason-3 satellite altimeter data, the projection made by Nerem et al. (2018) shows global sea-level rise of 65 ± 12 cm by 2100, which is very closed to the projection (52–98 cm) by IPCC5th Assessment Report (IPCC 2014). From the synthesis of available information in Bangladesh, it can be assumed that under the present situation of local and regional geographical context, and human interventions-both at upstream and downstream, the non-climatic driving forces, such as artificial tidal amplification, local subsidence, and/or tectonic crustal deformation are more prominent than the climate change-driven global contributions to project the future sea-level scenario of Bangladesh. However, in addition to global contribution, the cumulative impacts of all local inputs on the projected sea-level rise of Bangladesh would not cross 50–60% more than that of the global scenarios. It would, thus, be unrealistic to accept that the sea level in Bangladesh will rise up to 4.5 m, or 3.71 m by year 2100, as projected by Milliman et al, (1989) and Pethic (2012), respectively. The impression given to date indicates that one-third of the country would be inundated and would go under the sea, and the country will be helpless and unable to resist against this rising sea level. Such assumptions, descriptions, and explanations are incorrect (Brammer 2014). Without any site-specific need and study, not necessary to upgrade all 4800 km of existing embankment, build additional 4000 km new embankment, and spend billions of US dollars, as suggested by Huq et al. (1995), to combat the future sea-level threat.

At this stage, what Bangladesh needs to consider is to give priority attention to ensuring river and tidal water and sediment flow without any interruptions, and sediments to distribute naturally to the floodplains. However, the recent water-resource-related development plan of the Govt. indicates that we are not moving in the right direction. The formulation and implementation of Delta Plan-2100 is a typical example. The Delta Plan-2100 is a visionary plan of the Government to face the long-term challenges of climate change, and to achieve SDGs in due course (see http://www.bangladeshdeltaplan2100.org/). However, the Delta Plan-2100 includes massive scale alteration to the landscape of the country, which is truly the extension of the "Cordon" approach of water resource management, as has been implemented in Bangladesh since 1960s, which has already created massive adverse consequences, including waterlogging and land subsidence, in the south-western coastal belt (Islam 2018). Khalequzzaman (2019) has suggested that the implementation of the Delta Plan-2100 in its present shape will heavily interrupt the water and sediment budget, and due to sediment curtail, the coastal belt will face saline water intrusion, long-lasted waterlogging, and rapid subsidence. Considering the future sustainability of the coastal belt of Bangladesh and to avoid humanly induced accelerated tidal inundation, the relevant components of Delta Plan-2100 need to be readjusted.

32.8 Conclusions

Bangladesh is considered to be one of the most vulnerable counties in the world to be affected by global climate change. The main impact will be on the water resources, particularly due to sea-level rise. The coastal belt will be more exposed to saline water intrusion, increase in water and soil salinity, decrease in agricultural production, expansion of prolonged waterlogging, and considerable damage to the Sundarbans forest. However, based on this critical overview, it can be suggested that such an assumption of climate change-induced sea-level rise and associated adverse impacts on the environment and ecosystem of the coastal belt is not simple and straightforward, and requires special attention. In addition to the global climate change-induced components, the sea-level scenario of Bangladesh should be investigated closely under the lenses of regionally and locally induced factors, a few of which are climatic and many of which are non-climatic ingredients. Regionally active climatic components, such as Himalayan cryospheric inputs and hydro-metrological behaviors of the Bay of Bengal, as contribution to the sea-level movements of Bangladesh, have not yet been properly understood. Regional and local scale non-climatic components, particularly massive human interventions to prevent, diver, or reduce natural flow of water and sediment into the floodplain, are the major input to accelerating local level tidal height along the coast. A great share of amplified tidal range along Bangladesh coast lies on the reduction of water and sediment discharge at Farakka point Moreover, once the Indian inter-river linking project is implemented, due to diversion of water from Himalayan regime to the south, it is not only the Bangladesh part of the lower Ganges delta, but also the West Bengal part of the delta will experience an amplified tidal limit, and it has nothing to link with climate change. Similarly, once the Delta Plan-2100 is implemented, it will inundate more coastal land due to local scale rise of tide level and has nothing to do with climate change. The secular sea-level rising trend, currently being observed along the coastal belt of Bangladesh is a misleading; practically what is currently being detected is the amplified local tidal level, which is mostly related to non-climatic factors. The coastal policy (MoWR 2005) of Bangladesh has not given any attention to such critical coastal issues. The future coastal geomorphology, landscape, environmental setting, vulnerability, and sustainability, due to amplified tidal height, largely depend on; (a) how Bangladesh could manage to ensure adequate sediment supply from the upstream neighboring countries, and (b) how local level interruption by polders, embankments, cross-dams, and river encroachments can be stopped, and let the river and tide flow naturally. Otherwise, the existence of the coastal belt and survival of its people in future will be under threat. Bangladesh should combat such a critical situation not emotionally, but based on comprehensive research-based knowledge.

Acknowledgements Financial and technical support available from Bangladesh Center for Coastal and Ocean Studies (BACCOS) is duly acknowledged. Sincere thanks to Premanondo Debnath for his wholehearted support and to assist the work.

References

Akhter S, Seeber L, Steckler M, Armbruster J, Kogan M, Small C, Ho E (2005) Subduction and accretion across the Ganges-Brahmaputra delta: is it seismogenic? AGU fall meeting abstracts

Akhter SM, Seeber L, Steckler MS (2015) The Northern rupture of the 1762 Arakan Meghathrust earthquake and other potential earthquake sources in Bangladesh. In: American geophysical union, fall meeting 2015, abstract id T41B-2887

Allison MA, Kuehl SA, Martin TA, Hassan A (1998) Importance of flood-plain sedimentation for river sediment budgets and Terrigenous input to the oceans: insights from the Brahmaputra-Jamuna River. Geology 26(2):175–178

Allison I et al (2009) The Copenhagen diagnosis: climate science report. University of New South Wales Climate Change Research Centre, Sydney, Australia

Antony J, Prabhudesai RG, Kumar V, Mehra P, Nagvekar (2005) Meteorologically Induced Modulation in Sea Level off Tikkavanipalem Coast—Central East Coast of India. J Coast Res 21(5):880–886

Auerbach LW, Goodbred SL Jr, Mondal DR, Wilson CA, Ahmed KR, Roy K, Steckler MS, Small C, Gilligan JM, Ackerly BA (2015) Flood risk of natural and embanked landscapes on the Ganges-Brahmaputra Tidal delta plain. Nat Clim Change 5:53–157

BADC (2011) Forecasting saline water intrusion, irrigation water quality and waterlogging program in Southern Area. Ministry of Agriculture, Government of Bangladesh

Banerjee M, Sen PK (1987) Palaeobiology in understanding the changes of sea-level and coastline in Bengal Basin during Holocene period. Indian J Earth Sci 14(3–4):307–320

BCCSAP (2009) Bangladesh climate change strategy and action plan 2009. Ministry of Environment and Forest, Govt. of Bangladesh

Becker M, Papab F, Karpytcheva M, Delebecqueb C, Kriend Y, Khan JU, Valérie Ballua V, Durandb F, Cozannetf GL, Islam AKMS, Calmant S, Shumg CK (2020) Water level changes, subsidence, and sea level rise in the Ganges–Brahmaputra–Meghna delta. PNAS 117(4):1867–1876

Broadus JM (1993) Possible impacts of, and adjustment to, sea level rise: the cases of Bangladesh and Egypt. In: Warrick RA, Barrow EM, Wighley ML (eds) Climate and sea level change: observation, projection and implication. Projection and Implication, Cambridge University Press, Cambridge, pp 263–275

Brammer H (2012) The physical geography of Bangladesh. University Press Ltd., Dhaka

Brammer H (2014) Bangladesh's dynamic coastal regions and sea-level rise. Clim Risk Manag 1:51–62

Cazenave A, Lombard A, Llovel W (2008) Present-day sea level rise: a synthesis. C R Geosci 340:761–770

CEGIS (2010) Impacts of climate change on the morphological processes of the main rivers and Meghna Estuary of Bangladesh. Technical report

CEGIS (2011) Final report on programmes containing measures to facilitate adaptation to climate change of the second national communication project of Bangladesh. Department of Environment. Govt. of Bangladesh

Chowdhury SQ, Haq ATMF, Hasan K (2009) Beachrock in the St. Martins Island: implication of sea level changes in beachrock cementation. Mar Geodesy 20(1):89–104

Church JA, White N, Coleman R, Lambeck K, Mitrovica JX (2004) Estimates of the regional distribution of sea level rise over the 1950–2000 period. J Clim 17:2609–2625

Clark JA, Farrell WE, Peltier WR (1978) Global change of postglacial sea-level: a numerical calculation. Quatern Res 9:265–287

Cummins PR (2007) The potential for giant Tsunamigenic earthquakes in the Northern Bay of Bengal. Nature 449:75–78

Das SC (1992) Physical oceanography of the Bay of Bengal. In: Elahi KM, Sharif AHMR, Kalam AKMA (eds) Bangladesh: geography, environment and development. Bangladesh National Geographical Association, Dhaka, pp 36–52

DoE (2016) Assessment of sea level rise on Bangladesh coast through trend analysis, climate change cell. Department of Environment, Bangladesh

Douglas BC (2001) Sea level change in the era of the recording tide gauge, pp 37–64

Ericson JP, Vörösmarty CJ, Dingman SL, Ward LG, Michel Meybeck M (2006) Effective sea-level rise and deltas: causes of change and human dimension implications. Global Planet Change 50:63–82

Goodbred SL Jr, Kuehl SA (1999) Holocene and modern sediment budgets for the Ganges-Brahmaputra river system: evidence for highstand dispersal to floodplain, shelf, and deep-sea depocenters. Geology 27(6):559–562

Goodbred SL Jr, Kuehl SA (2000) The significance of large sediment supply, active tectonism, and eustasy on margin sequence development: late quaternary stratigraphy and evolution of the Ganges-Brahmaputra delta. Sed Geol 133:227–248

Hanebuth TJJ, Kudrass HR, Jörg Linstädter J, Islam B, Zander AM (2013) Rapid coastal subsidence in the central Ganges-Brahmaputra delta (Bangladesh) since the 17th century deduced from submerged salt-producing kilns. Geology 41(9):987–990

Higgins SA, Overeem I, Steckler MS, James PM, Syvitski JPM, Seeber L, Akhter SH (2014) InSAR measurements of compaction and subsidence in the Ganges-Brahmaputra delta, Bangladesh. J Geophys Res Earth Surf 119:1768–1781

Holgate SJ, Woodworth PL (2004) Evidence for enhanced coastal sea level rise during the 1990s. Geophys Res Lett 31(L07305):1–4

Hoque M, Alam M (1997) Subsidence in the lower deltaic area of Bangladesh. Mar Geodesy 20(1):105–120

Huq S, Ali SI, Rahman A (1995) Sea-level rise and Bangladesh: a preliminary analysis. J Coast Res 14:44–53

IPCC (1990) Climate change: the IPCC first scientific assessment report. In: Contribution of working group I to the first assessment report of the intergovernmental panel on climate change

IPCC (1996) Climate change 1995: the science of climate change. In: Contribution of working group I to the second assessment report of the intergovernmental panel on climate change

IPCC (2001) Climate change 2001: the scientific basis. In: Contribution of working group I to the third assessment report of the intergovernmental panel on climate change

IPCC (2007) Climate change 2007: the physical science basis. In: Contribution of working group I to the fourth assessment report of the intergovernmental panel on climate change

IPCC (2014) Climate change 2014: synthesis report. In: Contribution of working groups I, II and III to the fifth assessment report of the intergovernmental panel on climate change

Islam MS (1998) Implication of relative sea-level movements on vertical land displacement and coastal dynamics: example from the Bengal Basin. Indian J Earth Sci 25(1–4):42–57

Islam MS (2001) Sea-level changes in Bangladesh: the last ten thousand years. Asiatic Society Bangladesh

Islam MR (2004) Where land meets the sea: a profile of the coastal belt of Bangladesh. University Press Ltd., Dhaka

Islam SN (2018) Bangladesh delta plan-2100: a review. Eastern Academic, Dhaka

Islam S (personal communication) Estimation of sea-level rise (SLR) in Bangladesh using satellite altimetry data. Mid-term report, Department of Environment, Govt. of Bangladesh

Islam MS, Paul A (2004) Muhuri irrigation project at the Feni River Mouth in the Coastal Zone of Bangladesh: a geoenvironmental study. J NOAMI 21(1):63–84

Islam MR, Begum SF, Yamaguchi Y, Katsuro Ogawa K (1999) The Ganges and Brahmaputra rivers in Bangladesh: Basin Denudation and Sedimentation. Hydrol Process 13:2907–2923

Jana A, Biswas A, Maiti S, Bhattacharya AK (2014) Shoreline changes in response to sea level rise along Digha coast, Eastern India: an analytical approach of remote sensing, GIS and statistical techniques. J Coast Conserv 18:145–155

Khalequzzaman M (2016) Coastal zone management in the context of sustainable development goal. In: Islam MS, Khalequzzaman M (eds) Coastal and marine environment of Bangladesh. BAPA-BEN, pp 45–66

Khalequzzaman M (2019) Assessment of the Bangladesh delta plan 2100 and recommendation to enhance its implementation strategies. In: Islam MS, Khalequzzaman M (eds) Delta plan 2100 and sustainable development in Bangladesh, BAPA-BEN, pp 20–52

Khan FH (1991) Geology of Bangladesh, University Press Ltd. Dhaka

Khan MAH, Awal MA (2009) Global warming and sea level rising: impact on bangladesh agriculture and food security. Final report. National Food Policy Capacity Strengthening Programme. Govt. of Bangladesh

Khan NS, Ashe E, Shaw TA, Vacchi M, Walker J, Peltier WR, Koop RE, Horton BP (2015) Holocene relative sea-level changes from near-, intermediate-, and far-field locations. Curr Climatol Change Rep 1:247–262

Khan AA, Pant NC, Sarkar A, Tandon SK, Thamban M, Mahalinganathan K (2017) The Himalayan cryosphere: a critical assessment and evaluation of glacial melt fraction in the Bhagirathi Basin. Geosci Front 8:107–115

Khandoker RA (1987) Origin of elevated Barind-Madhupur areas, Bengal Basin: result of neotectonic activities. Bangladesh J Geol 6:1–9

Krien Y, Karpytchev M, Ballu V, Becker M, Grall C, Goodbred S, Calmant S, Shum CK, Khan Z (2019) Present-day subsidence in the Ganges-Brahmaputra-Meghna delta: Eastern amplification of the Holocene sediment loading contribution. Geophys Res Lett 46 (19):10764–10772

Kulkarni AV, Rathore BP, Singh SK, Bahuguna M (2011) Understanding changes in the Himalayan cryosphere using remote sensing techniques. Int J Remote Sens 32(3):601–615

Leuliette EW, Nerem RS, Mitchum GT (2004) Calibration of TOPEX/Poseidon and Jason altimeter data to construct a continuous record of mean sea level change. Mar Geodesy 27(1–2):79–94

Mahmood N (1992) Sea level rise solution in Bangladesh: problem identification, policy implication, and research need. In: Proceedings of workshop on coastal zone management in Bangladesh

Meem MS, Islam MS, Debnath P (2016) Subsidence in the coastal belt of Bangladesh and its implication to the coastal dynamics and ecosystem. In: Islam MS, Khalequzzaman M (eds) Coastal and marine environment of Bangladesh. BAPA-BEN, pp 138–151

Milliman JD, Broadus JM, Gable F (1989) Environmental and economic implication of rising sea level and subsiding deltas: the Nile and Bengal examples. Ambio 18(6):340–345

Monsur MH, Kamal ASMM (1994) Holocene sea level changes along the Moheskhali-Cox's Bazar-Teknaf coast of the bay of Bengal. J NOAMI 11(1):15–21

Morgan JP, McIntire WG (1959) Quaternary geology of the Bengal Basin; East Pakistan and India. Geol Soc Am Bull 70:319–342

Morner N (2007) Sea level changes and Tsunamis, environmental stress and migration overseas: the case of the Maldives and Sri Lanka. Int Asienforum 38(3–4):353–374

Morner N (2010) Sea level changes in Bangladesh: new observational facts. Energy Environ 21(3):235–249

Morner NA (2011) Setting the frames of expected future sea level changes by exploring past geological sea level records. In: Easterbrook D (ed) Evidenced-basted climate science, Elsevier, pp 185–196

Morner N (2015) Natural science is ruled by observational facts, not ephemeral model outputs. Glob J Res Anal 4 (11):193–194

Morner N (2016) Sea level changes in the real world: models vs observational facts. In: Mörner N (ed) London climate change conference, pp 69–72

MoWR (2005) Coastal zone policy-2005. Ministry of Water Resources, Govt. of Bangladesh

Mukherji A, Sinisalo A, Marcus Nüsser M, Garrard R, Eriksson M (2019) Contributions of the cryosphere to mountain communities in the Hindu Kush Himalaya: a review. Reg Environ Change 19:1311–1326

Nahmias BMA, Islam MS (2014) Available tidal records in Bangladesh: how reliable are they for sea-level research? Orient Geogr 55(1&2):13–36

NAPA (2005) National Adaptation Programme of Action (NAPA). Final report. Ministry of Environment, Govt. of Bangladesh

Nerem RS, Chambers DP, Choe C, Mitchum GT (2010) Estimating mean sea level change from the TOPEX and Jason altimeter missions. Mar Geodesy 33 (S1):435–446

Nerem RS, Beckley BD, Fasullo JT, Hamlington BD, Masters D, Mitchum GT (2018) Climate change driven accelerated sea-level rise detected in the altimeter era. PNAS 115(9):2022–2025

Niyogi D (1975) Quaternary geology of the coastal plain of West Bengal and Orissa. Indian J Earth Sci 2 (1):51–61

Pethick J (2012) Assessing changes in landform and geomorphology due to sea level rise in the Bangladesh Sundarbans. In: Climate change adaptation, biodiversity conservation and socio-economic development of the Bangladesh Sundarbans Bangladesh: a non-lending technical assistance, World Bank

Pethick J, Orford JD (2013) Rapid rise in effective sea-level in Southwest Bangladesh: its causes and contemporary rates. Glob Planet Change 111:237–245

Rashid T, Suzuki S, Sato H, Monsur MH, Saha SK (2013) Relative sea-level changes during the Holocene in Bangladesh. J Asian Earth Sci 64:136–150

Reitz MD, Pickering JL, Goodbred SL, Paola C, Steckler MS, Seeber L, Akhter SH (2015) Effects of tectonic deformation and sea level on river path selection: theory and application to the Ganges-Brahmaputra-Meghna River delta. J Geophys Res Earth Surf 120:671–689

Rogers KG, Goodbred SL Jr, Mondal DR (2013) Monsoon sedimentation on the 'abandoned' tide-influenced ganges Brahmaputra delta plain. Estuar Coast Shelf Sci 131:297–309

Sarker MH, Choudhury GI, Akter J, Hore SK (2013) Bengal delta not subsiding at a very high rate. CEGIS report, Available from: www.cegisbd.com/pdf/bengaldelta.pdf

Sarwar GM, Khan MH (2007) Sea level rise: a threat to the coastal of Bangladesh. Int Asienforum 38(3–4):375–397

Shankar D, Shetye SR (2001) Why is mean sea level along the indian coast higher in the bay of Bengal than in the Arabian sea? Geophys Res Lett 28(4):563–565

Sherin VR, Durand F, Papa F, Islam AKMS, Gopalakrishna VV, Khaki M, Suneel V (2020) Recent salinity intrusion in the Bengal delta: observations and possible causes. Cont Shelf Res 202:104142

Singh OP (2001) Cause-effect relationships between sea surface temperature, precipitation and sea level along the Bangladesh coast. Theoret Appl Climatol 68:233–243

Singh OP (2002a) Predictability of sea level in the Meghna estuary of Bangladesh. Global Planet Change 32:245–251

Singh OP (2002b) Spatial variation of sea level trend along the Bangladesh coast. Mar Geodesy 25(3):205–212

Singh OP, Khan TMA, Murty TS, Rahman MS (2001) Sea level changes along the Bangladesh coast in relation to the Southern oscillation phenomenon. Mar Geodesy 24:65–72

SMRC (2003) The vulnerability assessment of the SAARC coastal region due to sea-level rise: Bangladesh case study. SAARC Meteorological Research Centre, Dhaka

Stanley DJ, Hait AK (2000) Holocene depositional patterns, neotectonics and Sundarban Mangroves in the Western Ganges-Brahmaputra delta. J Coastal Res 16:26–39

Steckler MS, Akter SH, Seeber L (2008) Collision of the Ganges-Brahmaputra delta with the Burma arc: implications for earthquake hazard. Earth Planet Sci Lett 273:367–378

Steckler MS, Nooner SL, Akhter SH, Chowdhury SK, Bettadpur S, Seeber L, Kogan MG (2010) Modelling earth deformation from monsoonal flooding in Bangladesh using hydrographic, GPS, and gravity recovery and climate experiment (GRACE) data. J Geophys Res 115:B08407

Syvitski JPM, Kettner AJ, Overeem I, Hutton EWH, Hannon MT, Brakenridge GR, Day J, Vörösmarty C, Saito Y, Giosan L, Nicholls RJ (2009) Sinking deltas due to human activities. Nat Geosci 2:681–686

Tooley MJ (1978) Sea-level changes: Northwest England during the Flandrian stage. Clarendon Press, Oxford

Tooley MJ (1981) Methods of reconstruction. In: Simmons IG, Tooley MJ (eds) The environment in British prehistory. Duckworth, London, pp 1–48

Umitsu M (1987) Late quaternary sedimentary environment and landform evolution in the Bengal Lowland. Geogr Rev Jpn Ser B 60(2):164–178

Unnikrishnan AS, Shankar D (2007) Are sea level rise trends along the coasts of the North Indian Ocean consistent with global estimates? Global Planet Change 57:301–307

Unnikrishnan AS, Kumar KR, Fernandes SE, Michael GS, Patwardhan SK (2006) Sea level changes along the Indian coast: observations and projections. Curr Sci 90(3):362–368

Vishnu-Mittre, Gupta HP (1972) Pollen analytical study of quaternary deposits in the Bengal Basin. Palaeobotanist 19(3):297–306

Wahid SM, Babel MS, Bhuiyan AR (2007) Hydrologic monitoring and analysis in the Sundarbans Mangrove ecosystem, Bangladesh. J Hydrol 332:381–395

WARPO (2006) Coastal developmental strategy. Water Resources Planning Organization. Ministry of Water Resources, Government of the People's Republic of Bangladesh

Woodworth P, Shaw S, Blackman D (1991) Secular trends in mean tidal range around the British Isles and along the adjacent European coast. Geophys J Int 104:593–609

Zahid A, Hossain AFMA, Ali MH, Islam K, Abbassi SU (2018) Monitoring the coastal groundwater of Bangladesh. In: Mukherjee A (ed) Groundwater of South Asia. Springer Hydrogeology, pp 341–351

Assessing Channel Migration, Bank Erosion Vulnerability and Suitable Human Habitation Sites in the Torsa River Basin of Eastern India Using AHP Model and Geospatial Technology

Sourav Dey and Sujit Mandal

Abstract

Human settlements are under threat on both sides of the Torsa River due to channel migration and associated bank erosion. The present study focused on the identification of suitable sites for human habitation in the lower segment of the Torsa River basin, Eastern India. *Analytic Hierarchy Process* (AHP) model was applied to integrate slope, altitude, distance from rivers and fault lines, drainage density, land use/land cover, distance from *channel migration zone* (CMZ), and bank erosion vulnerability zones as well as to explore suitable habitation sites. *Bank erosion hazard index* (BEHI) and *near bank stress* (NBS) techniques were used to prepare bank erosion vulnerability zonation map. The *CMZ* map was prepared by superimposing the Torsa River courses for the year 1913, 1943, 1955, 1972, 1977, 1990, 2001, 2010, 2013, and 2018. A total of 182 cross sections were selected to derive data from 364 locations to assess stability character of the river bank applying BEHI and NBS techniques. The study of channel migration since 1913–2018 revealed that a total 37 villages out of 109 are severely vulnerable to bank erosion processes. The prepared site suitability map comprising the area of 413.33 km^2 showed that 23.33 km^2, 68.61 km^2, 122.46 km^2, and 198.96 km^2 are characterized by good, fair, poor, and not suitable for human habitation respectively. About 5.65% of the total area of 413.33 km^2 of the study reach is very much suitable for human habitation.

Keywords

Bank erosion · BEHI-NBS · AHP · Habitat suitability · Remote sensing · GIS

33.1 Introduction

The economy of the Himalayan Foothill region largely depends on cultivation and fishing. The ideal climatic condition and lack of job opportunities may lead to expansion of the agricultural and fishing activities. The river dwellers have been facing challenges of various natural hazards or disasters since very long in the riverine environment of Torsa river basin. The human settlements are being damaged frequently on both sides of the Torsa River due to channel migration, avulsion, flooding as well as associated

S. Dey
Department of Geography, Darjeeling Government College, Darjeeling, West Bengal, India

S. Mandal (✉)
Department of Geography, Diamond Harbour Women's University, South 24 Parganas, Diamond Harbour, West Bengal, India
e-mail: mandalsujit2009@gmail.com

riverbank erosion. Moreover, riverbank erosion is the major problem in the Torsa basin. Consequently, the villagers lost their farming lands, homesteads, houses, livestock, and other economic properties. The land losses have been changing villager's economic activities from cultivator to marginal worker or daily labour in the study area. Identification of suitable sites for human settlement or habitation is necessary for reducing the damages of settlement and economic properties and improving the wellbeing of the inhabitants of the Torsa River. Identification of extremely suitable sites for human habitation is of immense importance for reducing the probability of property losses as well as growing the socio-economic development of the community.

Several methods have been used for evaluating the site suitability in GIS platform, such as analytic network process (Saaty 2007), index of regional sustainability spatial decision support system (Graymore et al. 2009), analytic hierarchy process with GIS (Mohit and Ali 2006; Shukla et al. 2017), etc. The AHP is a frequently used decision-making method in various fields of studies, such as agriculture, planning, habitat, environment (Gumusay et al. 2016). The Analytic Hierarchy Process (AHP) was developed by Saaty (1980), for introducing a hierarchical model on the basis of various criteria (Roig-Tierno et al. 2013). The AHP method has been carried out to explore the suitable sites for groundwater recharge with reclaimed water in Tunisia (Gdoura et al. 2015), to assess the land suitability for tobacco production in Shandong, China (Zhang et al. 2015), to analyze the sustainability analysis of corn production (Houshyar et al. 2014), to identify suitable sites for agricultural land use of Darjeeling district (Pramanik 2016), to assess the site suitability for marina construction in Istanbul, Turkey, (Gumusay et al. 2016), to identify the suitable sites for organic farming in Uttarakhand (Mishra et al. 2015), etc. The AHP model has also been used to land suitability analysis for urban planning in Srinagar and Jammu urban centers, India (Parry et al. 2018), as well as to analyze the habitat suitability for waterbirds in the West Songnen Plain, China (Dong et al. 2013). Moreover, the Analytical Hierarchical Process (AHP) has been applied as a decision-making tool for the identification of suitable human habitation sites. The Geospatial analysis and AHP method as well as GPS survey are essential tools for multi-criterion decision-making analysis of land suitability (Shukla et al. 2017; Parry et al. 2018). The AHP has been taken into account to find out the dependable and objective weights for the suitability factors (Dong et al. 2013). The present study focused to explore suitable sites for human habitation in the lower segment of the Torsa River basin, Eastern India.

33.2 The Study Area

The Torsa River is a lower catchment tributary of the Brahmaputra River covering the countries of Tibet, Bhutan, India, and Bangladesh and it is a well-known name in the river scenario of the Duars and Tal region, West Bengal, India. The catchment of the Torsa River is one of the largest river systems in the Eastern Himalaya and its foothills region. The Torsa River basin covers a geographical area of 7486.31 km^2 and it is located between 27° 56′ 42.5″ N to 25° 54′ 16.5′ N latitude and 88° 44′ 22.5″ E to 89° 50′ 35.5′ E longitude (Fig. 1.1). The catchment area is a part of Eastern Himalaya (Tibet and Bhutan), the Duars (India) and Tal (India and Bangladesh) region and lies between the catchments of Jaldhaka of the west and the River Sankosh of the east. The Torsa River originates from the *Chumbi valley* in the Southern *Tibet* at the height of 6996 m above mean sea level, and known as *Proma Chhu*, in Tibet, *Amo Chu* in Bhutan, *Torsa River* in India and *Dudhkumar River* in Bangladesh. The north–south elongated basin having 295 km long river of which 99 km lies in West Bengal, India. This river is also known as 'Toyrosa' (meaning 'sorrow of river') in Tibet. The River Torsa is fed by the melting water of glaciers near the *Chumbi valley* as well as a maximum portion of its course is fed by rainwater. The entire course of the Torsa River flows from *Greater-Lesser-Shiwalik Himalayas* through the steep mountain gorges to *alluvial plain region* and this course, ultimately meets as an important right bank tributary of *Brahmaputra* in the Rangpur district of Bangladesh. Torsa River and its tributaries contain the *Brahmaputra river system*. The upper course of the Torsa River is linked with many perennial lakes; consequently, the water level of this river is retained during non-monsoon.

The northern part of the Torsa River Basin covers hills and mountains which belongs to the Himalayan range marked by the above 300 m contour and consists of a number of landforms like *ridges, gorges, low hills, steep V-shaped valleys*, toe cutting of *hill slopes, uplifted terraces* and *scarps*. In the south, the *Northern Plain* region is homogeneous and featureless that spread up to the Jamuna (Brahmaputra) River in the Bangladesh. This region has led to the formation of numerous wetlands (bills), oxbow lakes, marshy tracts, cutoffs, abandoned river tracks and wide valley floors. Innumerable tributaries of the Torsa River branch out in many channels giving birth to river meandering and leading to the formation of bills, swamps, and oxbow lakes in the southern part of the *Northern Plain,* and this lowland region also known as *Tal* (*Tal* means wetland or a bill or lake) region (Fig. 33.1). The Torsa River Basin is experienced by high intensity rainfall. The lower catchment of the Torsa River basin is characterized by recent alluvium mainly very deep, moderately well drained, and coarse loamy soils.

33.3 Materials and Methods

33.3.1 Suitability Factors for Human Habitation

The selection of site suitability factors is the most important aspect in habitat suitability analysis (Dong et al. 2013). The availability of road tends to have direct effects on the human habitat or settlement. Some geophysical factors can strongly influence on the human habitation and the livelihood for river dwellers during monsoon, such as steep slope, low altitude below flood level, high drainage density, existence of fault line, etc. (Figs. 33.2, 33.3, 33.4, 33.5 and 33.6). The site suitability for human habitation in the riparian areas requires gentle slope, medium altitude, and availability of drainage system. Whereas, the study area having more vulnerable to river bank erosion as well as avulsion and flooding (Plate 33.1), so the fault line, channel migration zone, bank erosion vulnerable sites, and land use/land cover are also the decision-making processes to identify suitable sites for human settlement or habitation (Figs. 33.7, 33.8, 33.9 and 33.10). The human habitation sites can be secured if the settlements are being established beyond the vulnerability zones of channel migration and riverbank erosion. The channel migration zone and bank erosion vulnerability sites are not suitable for human habitation. These suitability factors are good indicators of housing for the river dwellers in the Torsa basin. All the criteria were taken into consideration during the analysis to identify suitable human habitation site.

33.3.2 Generation of Suitability Maps

The suitability factors or geophysical factors for human habitation were being analyzed using geospatial techniques with the intensive field survey (Plate 33.1). Slope and altitude map were derived using ASTER DEM of 30 m resolution acquired from the USGS. The fault line maps were generated using a geological quadrangle map of GSI. Drainage density and system maps were developed using the topographical maps of SOI and Landsat-8 image of 30 m spatial resolution obtained from the USGS. The land use/land cover map was derived from the Landsat-8 satellite image of 30 m spatial resolution which was acquired from the USGS.

The bank erosion vulnerability site maps were prepared using the field generated data. Two models have been used to determine the bank erosion vulnerability of the Torsa River, i.e., *Bank Erosion Hazard Index* (BEHI) and *Near Bank Stress* (NBS). Different kinds of data were derived by field survey for the assessment of

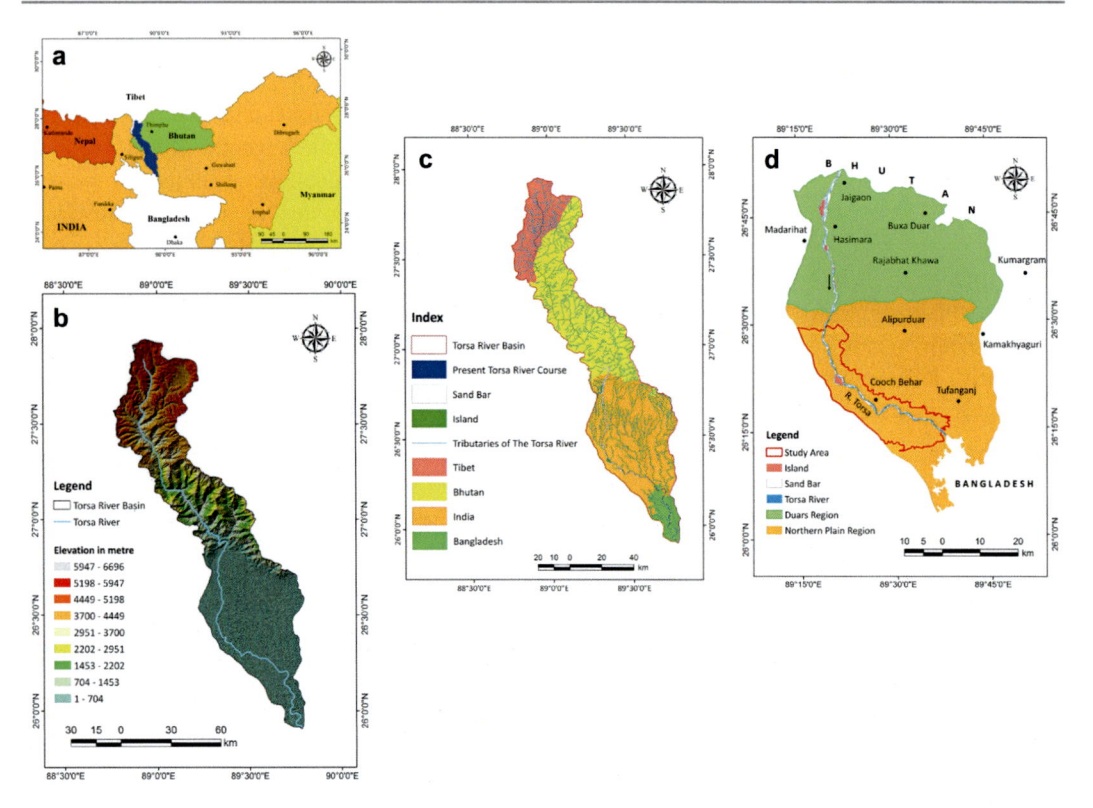

Fig. 33.1 Location map of the study area; **a** location of the Torsa River basin in the South Asia, **b** Torsa River basin along with altitude map, **c** Torsa River basin covers a geopolitical area of Tibet, Bhutan, India, and Bangladesh, and **d** study area in the Northern Plain region

BEHI and NBS parameters. The data were derived from the field with the help of various measuring instruments such as G.P.S. receiver, levelling stuff, eco-sounder, digital water current meter, clinometer and dumpy level. The riverbank *instability* has been assessed on the basis of different parameters of BEHI and NBS. The BEHI ratings were assigned considering the seven parameters, such as the *ratio of bank height* and *bank-full height, ratio of riparian root depth and bank height, root density, bank angle, surface protection, bank material adjustment,* and *stratification adjustment* (Rosgen 2006; Starr 2009; Ghosh et al. 2016; Dey and Mandal 2019). The *NBS* is determined on the basis of the ratio of *near bank shear stress* to *bank-full shear stress* (Rosgen 2006). The 182 cross-sections were selected to derive data from 364 bank sites for determining the bank erosion vulnerability along the Torsa River in the Northern Plain regions.

The *Channel migration zone* (CMZ) map was prepared using historical maps and field generated data. The *Channel migration zone* model is widely recognized as more scientific and proper classificatory techniques to delineate the CMZ. Delineation of the CMZ is the cumulative product of historical analysis and rigorous field experimentation, which has been formulated as (Eqs. 33.1, 33.2).

$$\begin{aligned}\text{Channel Migration Zone (CMZ)} \\ = [\{\text{Historical Migration Zone(HMZ)} \\ + \text{Avulsion Hazard Zone(AHZ)} \\ + \text{Erosion Hazard Area(EHA)}\} \\ - \text{Disconnected Migration Area(DMA)}]\end{aligned}$$

(33.1)

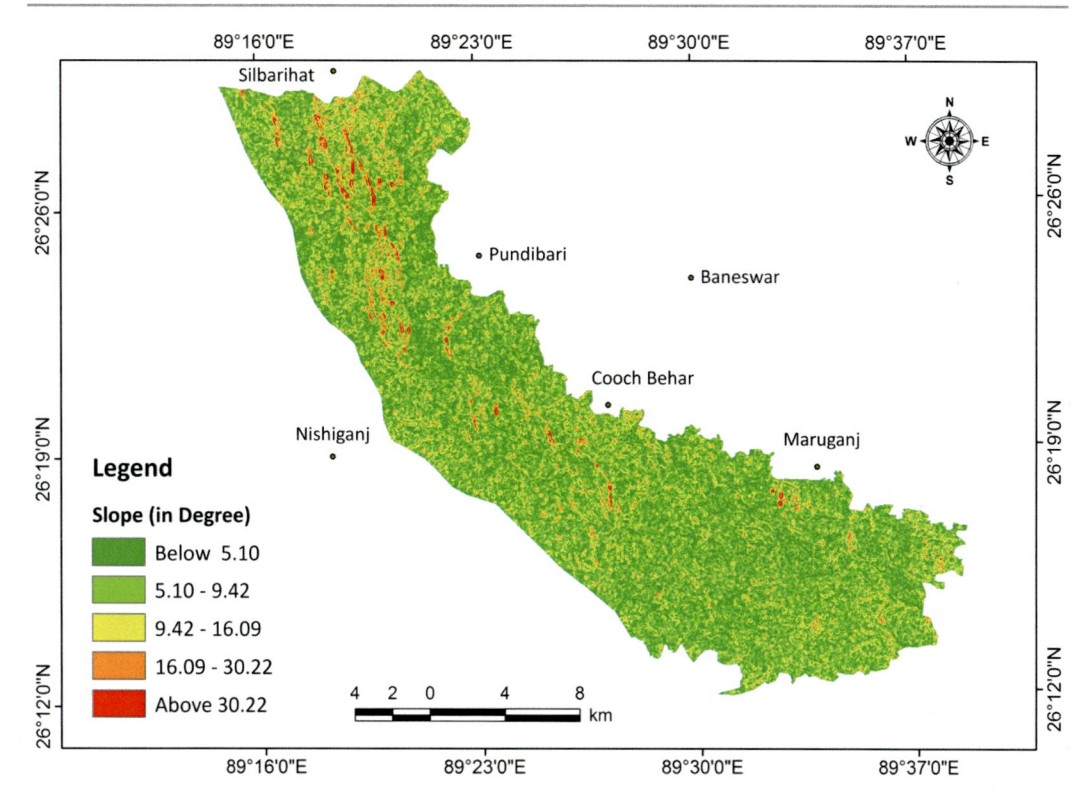

Fig. 33.2 Slope map of the study area

$$[EHA = \text{Erosion Setback}(ES) + \text{Geotechnical Setback}(GS)] \quad (33.2)$$

33.3.3 Analysis of Suitability Factors

The slope map clearly shows that the maximum part of the study area having gentle to moderate slope ranges from 0° to 9.42°. However, the slope is more near the river banks which is up to 90° (Fig. 33.2). The altitude of the selected mouzas varies from 11 to 80 m. The high altitude is found in the mouzas like Khagribari, Chhat Singimari, Singimari Paschimpar, etc. However, the southeastern part of the study area experiences low altitude (Fig. 33.3). The buffer map of the fault line has been considered because the maximum chance of avulsion hazard can happen along the fault line (Fig. 33.4). The drainage density of the selected mouzas ranges from 0.03 to 0.82 km/km^2, which were divided into five classes (Fig. 33.5). The buffer map of the River is also important for suitability analysis for human habitation because the locations which are very near to the river indicate most vulnerability to bank erosion as well as flooding and avulsion (Plate 33.1). Therefore, the vulnerability decreases with increasing distance from the river, which was divided into five classes (Fig. 33.6).

The Land use/Land cover map shows that the 28.34% area is under current fallow which is followed by agricultural field (28.08%), grassland (17.82%), vegetation (9.79%), built-up, and other areas (5.78%), etc. (Fig. 33.7). The 100 m buffer map of the bank erosion vulnerability sites has been taken into account because the maximum risk of bank erosion is observed near the high and moderate BEHI and NBS ratings (Figs. 33.8, 33.9). The buffer map of the Channel Migration Zone (CMZ) has been considered because the maximum risk of channel migration as well as avulsion takes place within 1 km buffer of CMZ (Fig. 33.10).

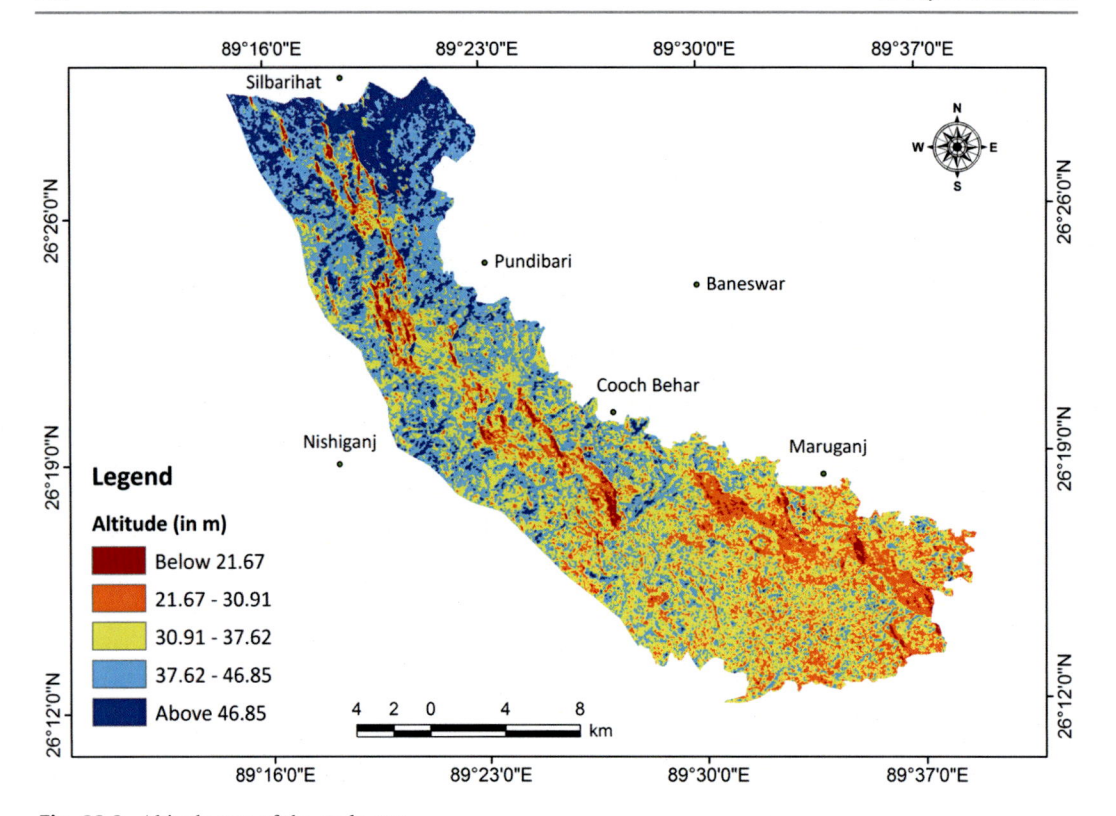

Fig. 33.3 Altitude map of the study area

33.3.4 Standardization of Criteria Maps

To carry out the weighted overlay method, the selected criteria or suitability factor maps having different units need to be changed in the same units and therefore required standardization. The procedures of standardization make the measurement to uniform units; moreover the outcome score loses their dimension for all criteria along with their measurement unit (Effat and Hassan 2013). The vector layers of the criteria maps were converted to a raster layer. After that, the layers are reclassified using the reclassify tool in spatial analyst of ArcGIS for the input data to the weighted overlay. Finally, the layers of weighted criteria maps have been combined and superimposed to create the site suitability map for human habitation (Fig. 33.12). The values of all selected criteria were standardized for comparative significance. The sub-criteria were categorized into different ranking of the scale, where the maximum rank having highest significance and the minimum rank having least significance.

33.3.5 Criteria Wise Weight Calculation

Analytical Hierarchical Process (AHP) is most important multi criteria decision-making method for suitability analysis. The AHP is applied to a group of criteria or sub-criteria to set up a hierarchical structure by assigning the weightage of every criterion in the entire decision-making process (Mishra et al. 2015). The values of weight reflect the relative significance of each criterion and hence are to be chosen cautiously. Analytic hierarchy process can be used to make pairwise strong comparisons between the criteria and factors and thus reduces the difficulty of the decision-making process (Miller et al. 1998; Saaty 1977).

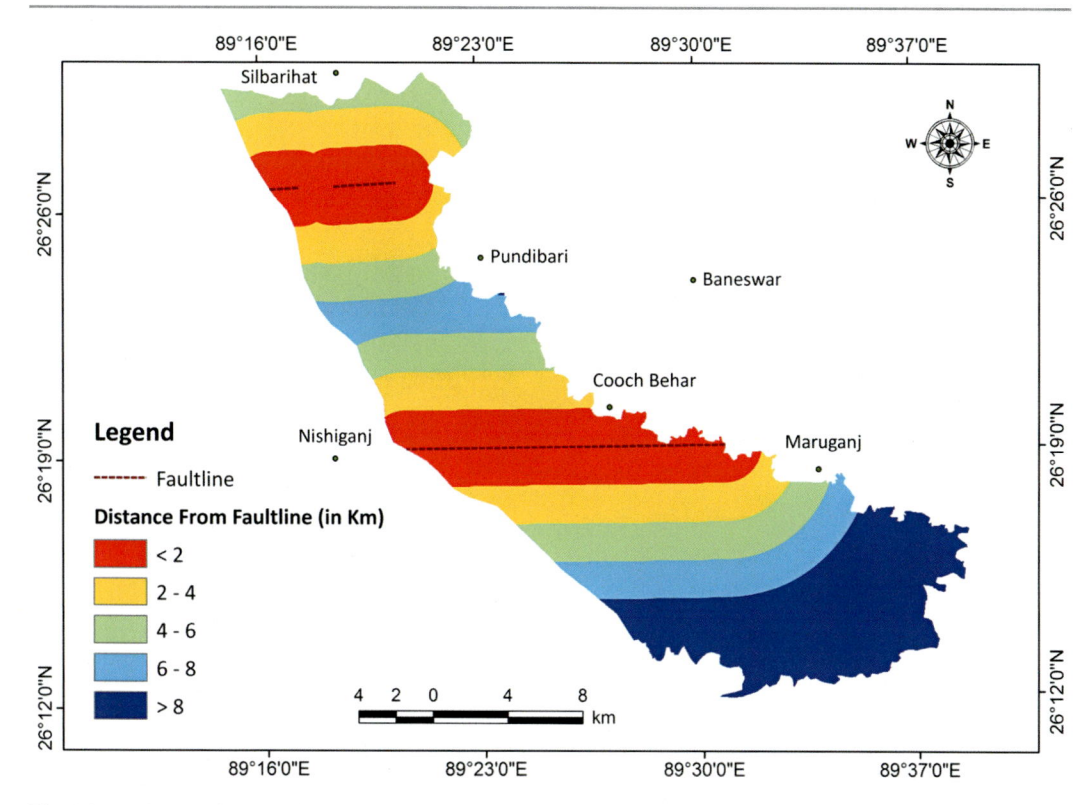

Fig. 33.4 Distance from the fault line map of the study area

The pairwise comparison matrix was calculated using Saaty's (1980) pairwise comparison scale (Table 33.1). The pairwise comparison scale ranges from 1 to 9, where 1 having equal significance and 9 having extremely significant in between the criterion shown in Table 33.1 (Saaty 1980; Malczewski 1999). The comparison matrix normally has the property of reciprocity, which is mathematically expressed as follows: $n(n − 1)/2$, where n is the number of elements in the pairwise comparison matrix (Saaty 1980; Akinci et al. 2013). The pairwise comparison matrix indicates the significance of criteria between each other (Table 33.2). After the computation of the pairwise comparison matrix, the normalized matrix is developed by dividing each number of the pairwise comparison matrix to the sum of each column value of the pairwise comparison matrix (Gumusay et al. 2016). The normalized matrix has been calculated (Table 33.3) using the Eq. 33.3,

$$\bar{p}_{ij} = \frac{p_{ij}}{\sum_{i=1}^{n} p_{ij}} \qquad (33.3)$$

where, p_{ij} indicates each member of the pairwise comparison matrix, and $\sum_{i=1}^{n} p_{ij}$ indicates the sum of each column value of the pairwise comparison matrix.

Once the normalized matrix is completed, the criteria weight is computed using the normalized matrix (Saaty 1980; Gumusay et al. 2016). Weights for each criterion have been calculated applying Saaty's (1980) AHP method (Table 33.4). The criteria weight is calculated through the averaging of raw values of the normalized matrix using the Eq. 33.4.

$$W_i = \frac{\sum_{j=1}^{n} P_{nom_{ij}}}{n} \qquad (33.4)$$

where, $\sum_{j=1}^{n} P_{nom_{ij}}$ indicates the sum of each row values of the normalized matrix (Table 33.5), and n indicates the number of compared factors.

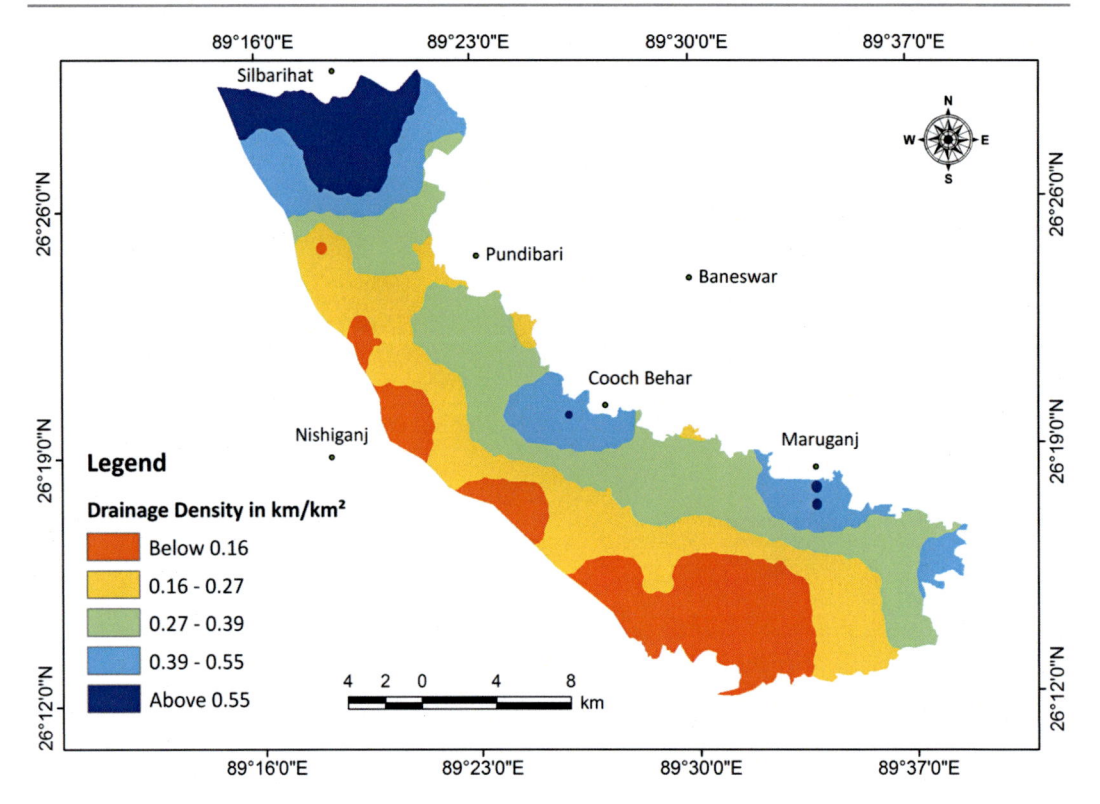

Fig. 33.5 Drainage density map of the study area

However, the AHP identifies and takes into account the inconsistencies of decision-makers, which is one of the important characteristics (Saaty 1980; Garcia et al. 2014). The efficiency criteria of AHP were measured by Consistency Relationship (CR) using the following Eq. 33.5,

$$CR = \frac{CI}{RI} \quad (33.5)$$

where, CI is the consistency index, and RI is the random matrix.

The CR is calculated through the λ calculation, and the determination of the consistency index (CI) and random index (RI). The λ and CI are measured applying the Eqs. 33.6 and 33.7, respectively.

$$\lambda_{max} = \frac{\sum_{i=1}^{n}(w/p)_i}{n} \quad (33.6)$$

where, w indicates the calculated weighted sum, p indicates the priority, and n is the number of compared factors.

$$CI = \frac{\lambda_{max} - n}{n - 1} \quad (33.7)$$

Equation 33.7 indicates the CI where n represents the number of compared factors.

The calculated values of λ and CI are about 13.151 and 0.105 applying the Eqs. (33.6) and (33.7), respectively.

The RI is the average value of the consistency index depending on the order of computed matrix prearranged by Saaty (1977) (Table 33.6). The CR has been computed to recognize whether the AHP is consistent or inconsistent.

If the calculated CR value is more than 0.10, then the pairwise comparison matrix is inconsistent, and may not continue the decision-

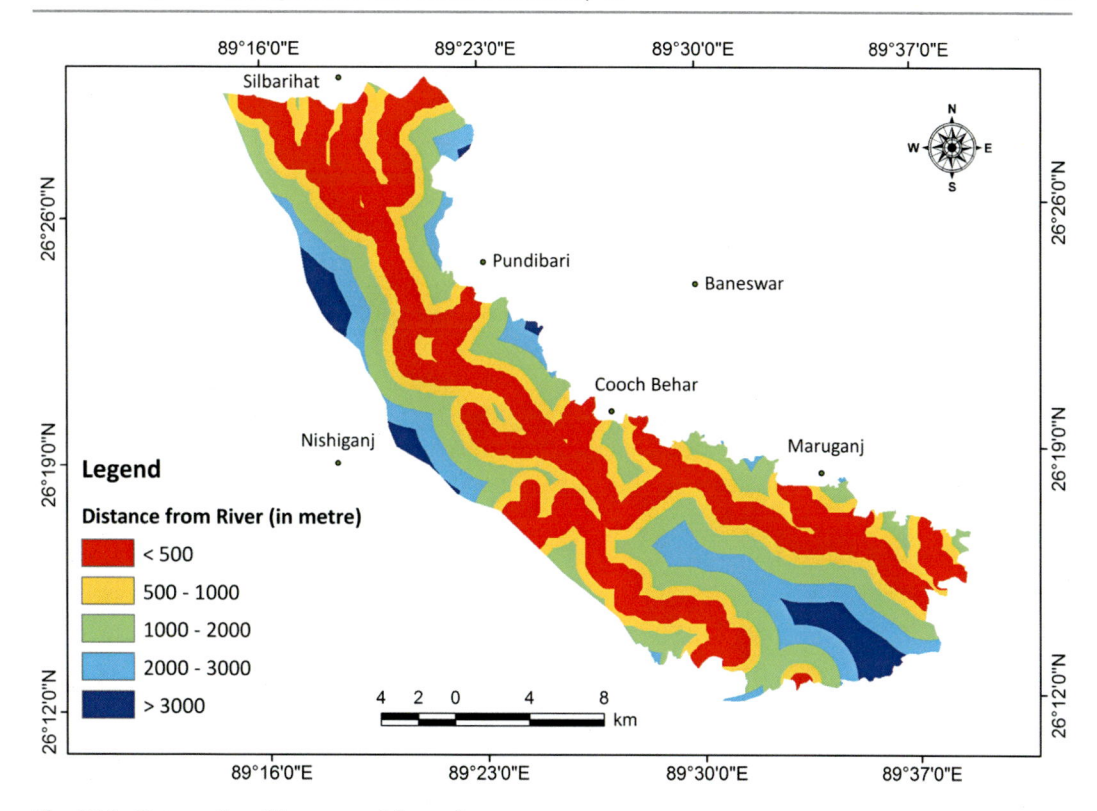

Fig. 33.6 Distance from River map of the study area

Plate 33.1 a Lost of lands and houses due to bank erosion of the Torsa River during flood 2017 at municipal ward-18 and **b** during flood 2018 at Damodarpur mouza

making process applying AHP (Saaty 1988). In the present study, the calculated CR was 0.055 which proves that the pairwise comparison matrix is consistent or under acceptable limits, and we may continue the decision-making process using AHP.

The Analytic Hierarchy Process (AHP) was applied to integrate slope, altitude, distance from fault lines, drainage density, distance from river, land use/land cover, bank erosion vulnerability sites based on BEHI and NBS, and distance from channel migration zone (Fig. 33.12), and finally

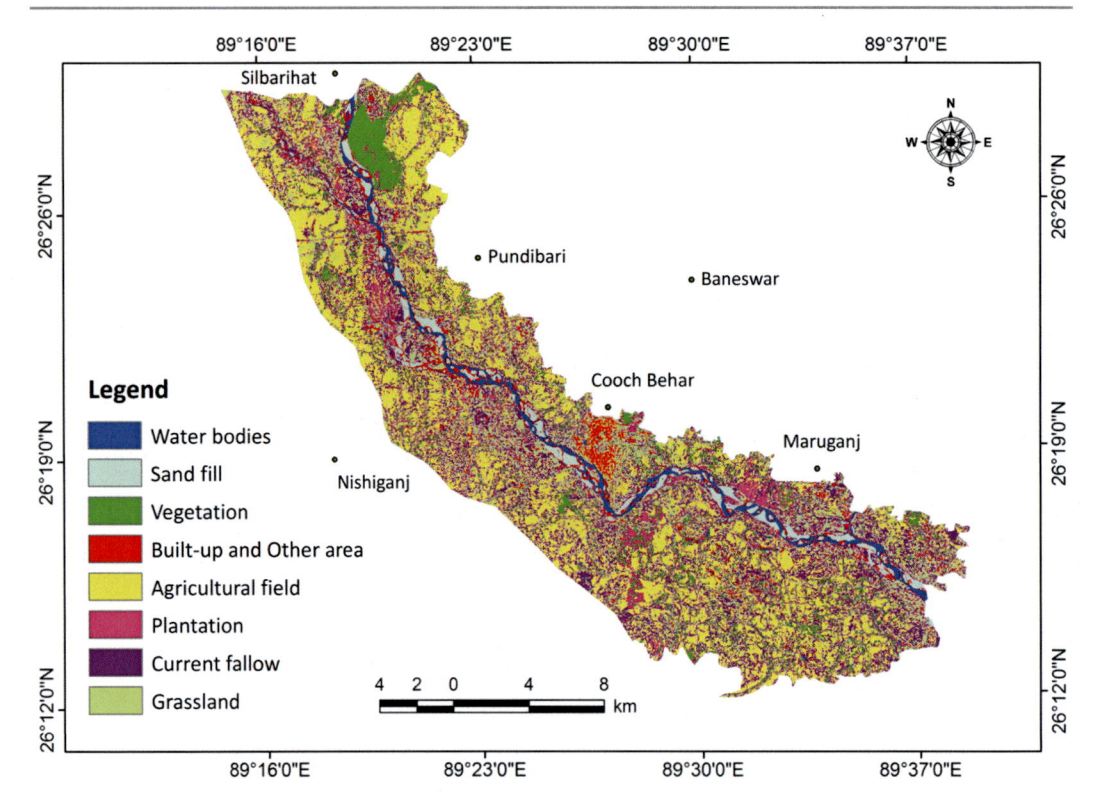

Fig. 33.7 Land use/land cover map of the study area

explored the site suitability map for human habitation in the Torsa River basin (Fig. 33.13).

33.4 Result and Discussion

The pairwise comparison matrix is calculated to find out the weights of each criterion applying the AHP model (Table 33.4). Results of AHP show that the maximum weighted criteria for the decision-making process is the bank erosion vulnerability based on the BEHI model (25.98%), and it is followed by distance from the channel migration zone (21.01), bank erosion vulnerability based on the NBS model (17.12%), distance from river (10.27%), etc. Whereas, the minimum weighted criteria are altitude, land use/land cover, distance from fault line, and slope (Table 33.4). The chosen scores of the sub-criteria were applied in the weighted overlay analysis to create the site suitability map for human settlement or habitation in the selected mouzas (Fig. 33.12).

The site suitability map for human habitation is classified into four categories, such as good, fair, poor, and not suitable (Table 33.7, Fig. 33.13). The spatial distribution of habitat suitability of each category demonstrates significant spatial dissimilarities. According to the generated site suitability map, it was found that 23.33 km^2 areas (5.65% area of the total area under study) are good for human habitation (Table 33.7, Fig. 33.13). The good category human habitation sites covered the smallest area of the total area under study. Most of the good and fair sites were located far away from the riverbank erosion vulnerability sites, expected bank lines, avulsion hazard zone, and channel migration zone. It was also observed that the not suitable sites for human habitation covered the maximum area i.e. 198.93 km^2 (48.13% area of the total study area) (Table 33.7, Fig. 33.13). Alternatively, 29.63% (122.46 km^2) area was poor for human habitation in the study area (Table 33.7, Fig. 33.13). The present study indicates that the

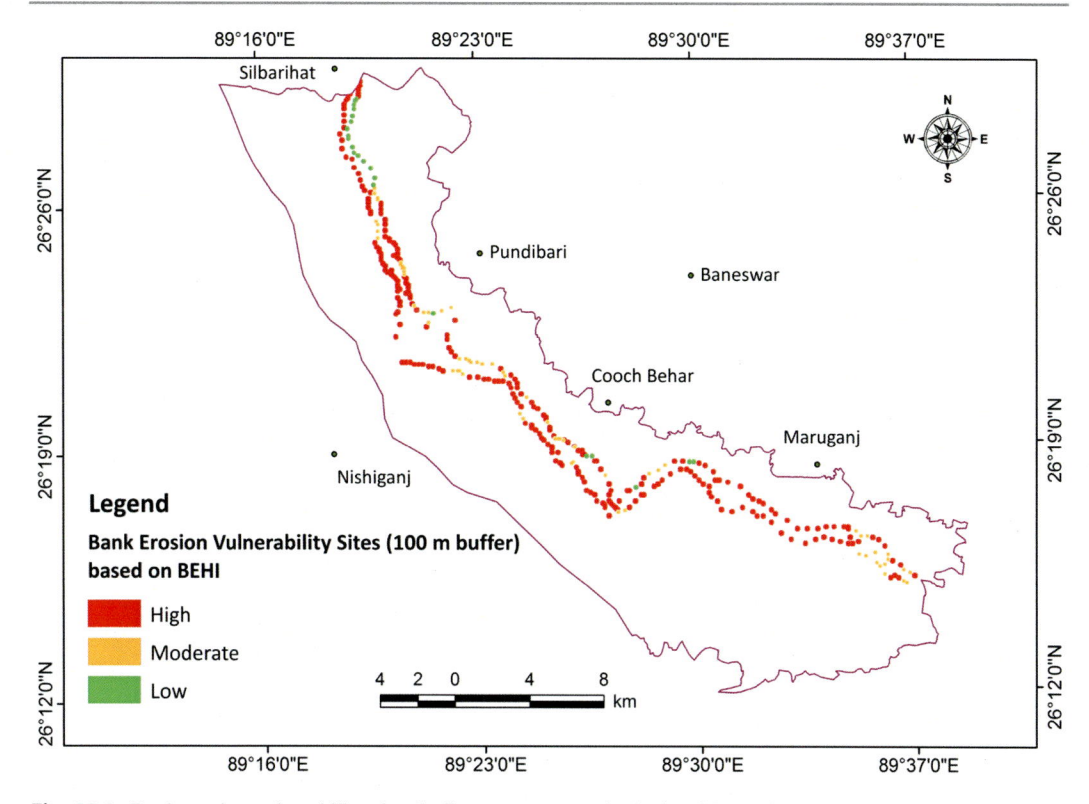

Fig. 33.8 Bank erosion vulnerability sites buffer zone map on the basis of BEHI in the study area

large part of the study area is poorly suitable for human habitation because this zone is located beyond the channel migration zone and also protected by embankments, but most of this area coincides with the built-up area which is unsuitable for resettlement. The southern part of the poorly suitable areas for human habitation may be regularly damaged by flooding or inundation during monsoon due to low altitude. A total of 16.60% (68.61 km^2) of the study area is fair for suitable human habitation sites which coincide with agricultural fields and grassland with current fallow (Table 33.7, Fig. 33.13).

The good and fair sites are situated beside the protected forest in the Patlakhawa region, and along the side of embankments in Ghokshadanga and Balarampur regions (Fig. 33.13). The poor sites are located adjacent to the not suitable areas (Fig. 33.13). Compared with the good to fair sites, the poor and not suitable sites have been primarily located near the vulnerable river bank and lowland regions.

The very recent geological formation consists of the alternate layers of coarse to fine sand, silt, and clay with higher intensity of riverbank erosion as well as avulsion, resulting the least suitable human habitation in the selected mouzas. Similarly, the adjacent area of the Torsa River is not suitable for human habitation due to low altitude, and more vulnerability of riverbank erosion as well as the existence of estimated CMZ (Fig. 33.12). The Patlakhawa forest area is also not suitable for settlement because of the protected forest areas rules and regulations. Recently, the existence of vegetation cover and embankments play a significant role to reduce the vulnerability of the bank erosion and channel migration of the Torsa River, which may lead to the increase of human habitation site in the study area.

A total of 91 locations in the selected mouzas were surveyed through GPS to validate the accuracy of the site suitability for human habitation. Out of 91 locations, 22 locations are under

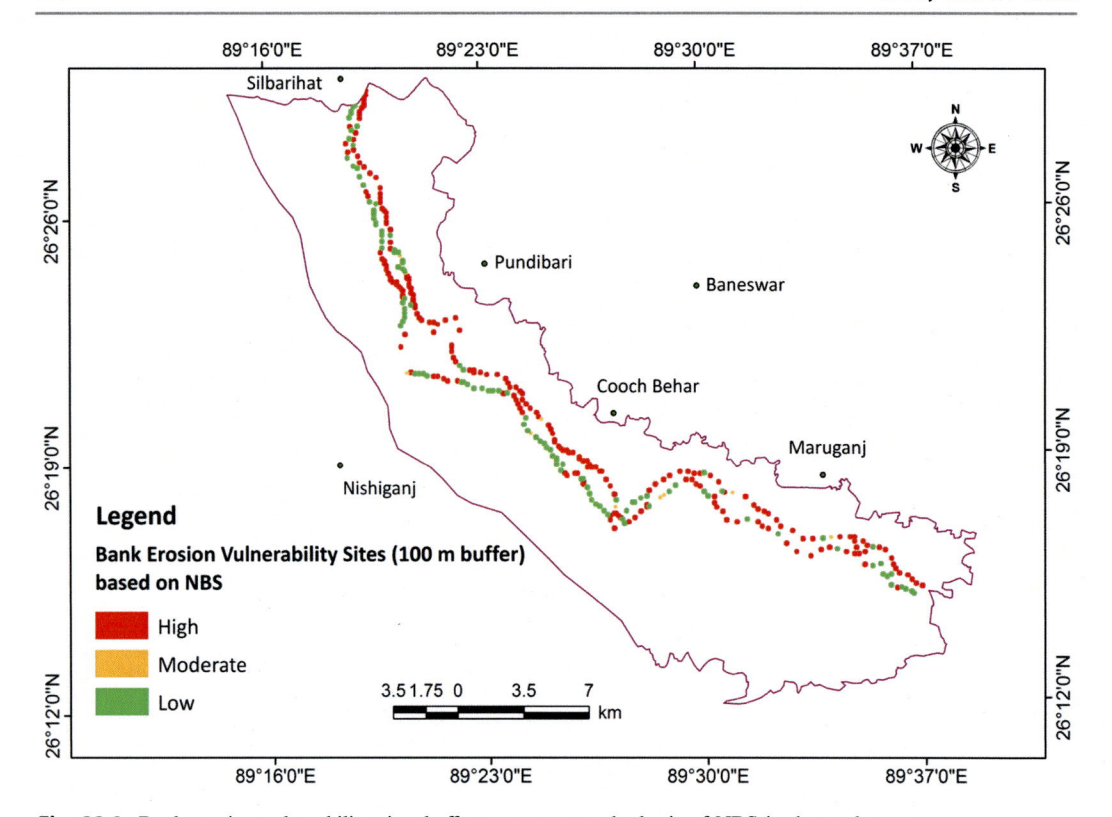

Fig. 33.9 Bank erosion vulnerability sites buffer zone map on the basis of NBS in the study area

good category with 100% producer accuracy and 95.65% user accuracy and 27 locations are under fair category out of 28 locations with 100% producer accuracy and 96.43% user accuracy (Table 33.8). The overall accuracy and kappa coefficient values for the suitability map are about 95.60% and 0.94 respectively (Table 33.8). The Kappa coefficient value indicates that the suitability classes are having almost perfect match between the classified and the referenced (field) data. As a result, it can be considered as a valid database for habitat suitability analysis.

33.5 Conclusion

The combination of analytic hierarchy process, geospatial techniques and object-oriented segmentation recognized an effective way for identification of the suitable sites for human habitation. The results of the present study demonstrate that 5.65% (23.33 km^2) of the total area of the selected mouzas is good for human habitation. So, the river dwellers should be remembered to be resettled in their houses either

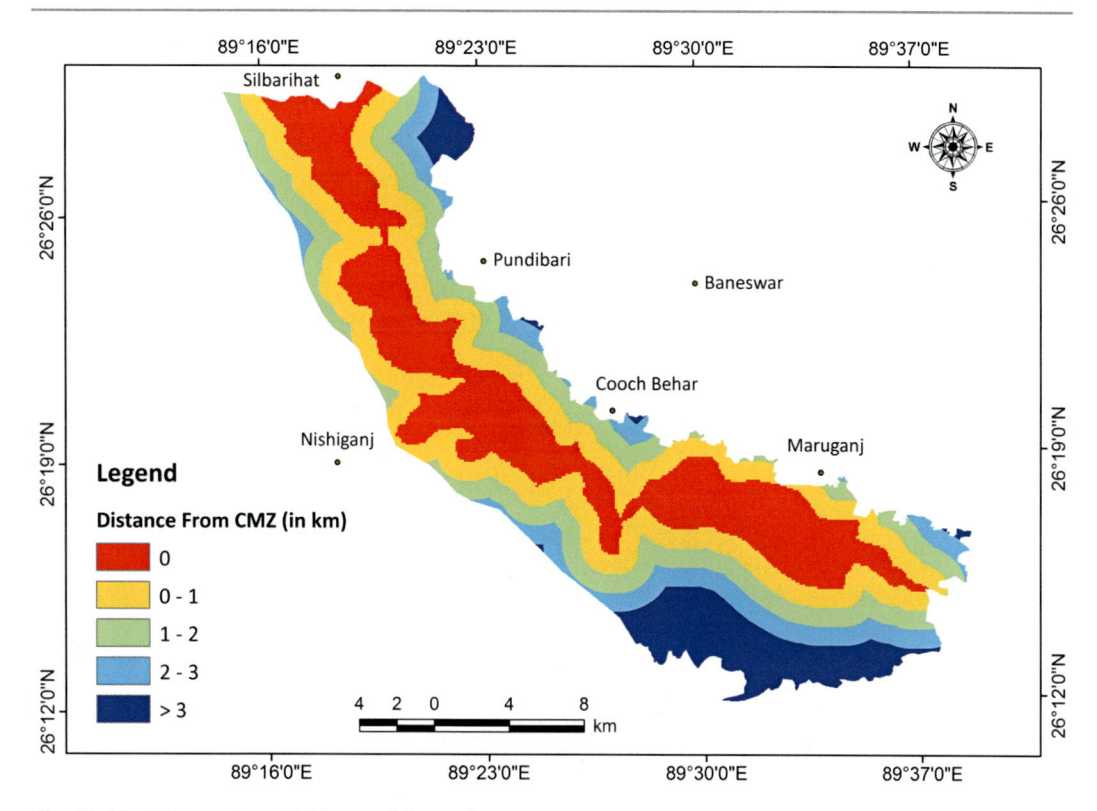

Fig. 33.10 Distance from CMZ map of the study area

in the fair to good habitation sites or beyond the CMZ and bank erosion vulnerability sites. The poor and not suitable sites are more vulnerable to river bank erosion as well as avulsion, so, the river dwellers of not suitable areas, should be shifted to other suitable habitation sites for reducing future risk. The site suitability map may be used for expansion of human settlement, construction of roads, expansion of agricultural land and other developmental activities for regional planning and development of Himalayan foothills environment.

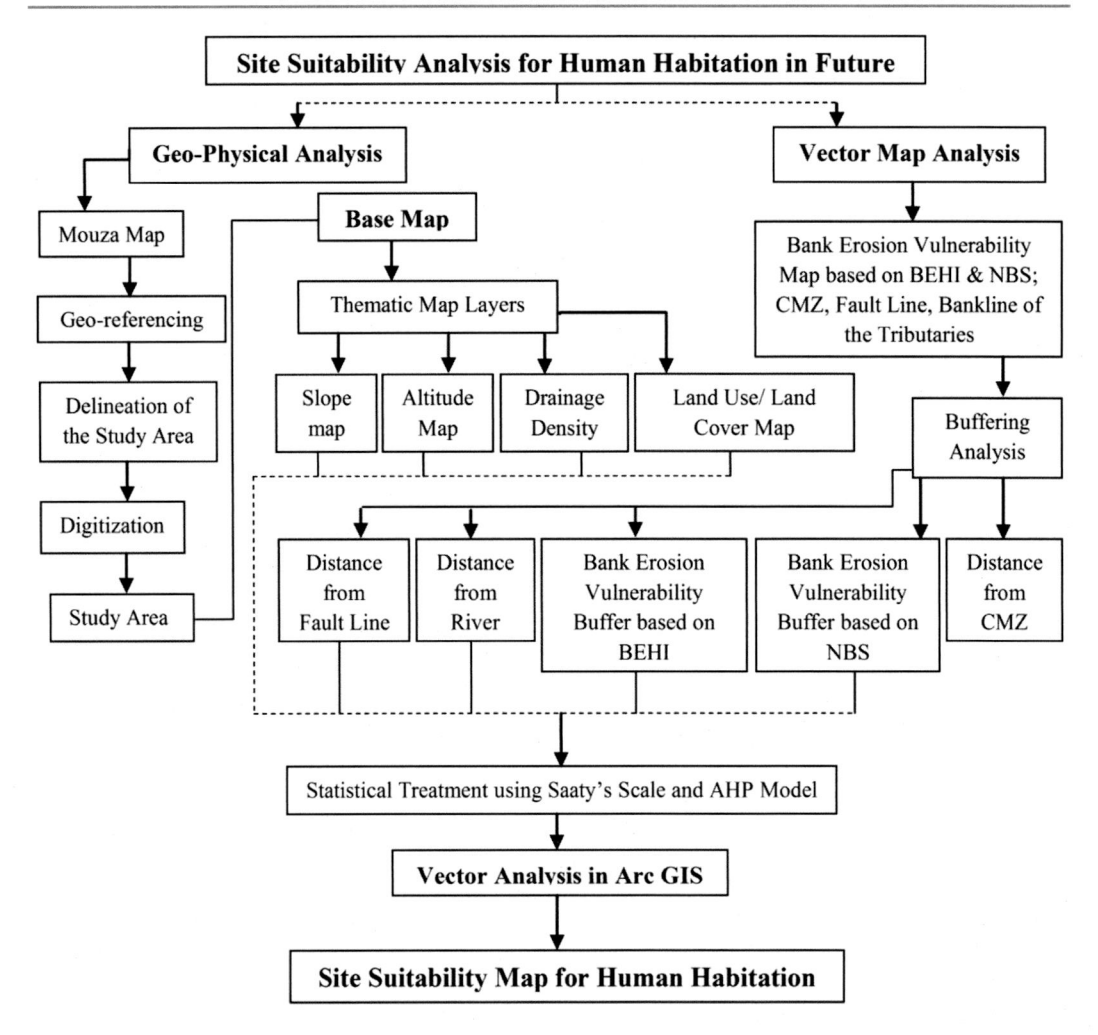

Fig. 33.12 Flow chart of methodology

Table 33.1 Saaty's pairwise comparison scale

Absolute scale	Degree of importance	Explanation
1	Equal	Two activities contribute equally or uniformly to the objective
3	Moderate	Experience and judgement moderately favour one activity over another
5	Strong or essential	Experience and judgement essentially or strongly favour one activity over another
7	Very strong	An activity is very strongly favoured and its dominance is showed in practice
9	Extreme	The evidence of favouring one activity over another is of the utmost possible order of an affirmation
2, 4, 6 and 8	Intermediate values	Applied to represent compromises between the weights 1, 3, 5, 7 and 9
Reciprocals	Opposite relation	Applied for inverse comparison

Source Saaty (1987)

Table 33.2 Criteria pairwise comparison matrix

Criteria	Slope	Altitude	Distance from fault line	Drainage density	Distance from river	Land use land cover	Bank erosion vulnerability sites (BEHI)	Bank erosion vulnerability sites (NBS)	Distance from CMZ
Slope	1.000	1.200	1.000	1.000	0.333	2.100	0.250	0.286	0.250
Altitude	0.833	1.000	0.500	1.000	0.333	1.000	0.143	0.200	0.167
Distance from fault line	1.000	2.000	1.000	1.000	1.000	0.333	0.143	0.238	0.143
Drainage density	1.000	1.000	1.000	1.000	1.000	3.500	0.286	0.313	0.500
Distance from river	3.000	3.000	1.000	1.000	1.000	3.000	0.286	1.000	0.500
Land use land cover	0.476	1.000	3.000	0.286	0.333	1.000	0.167	0.200	0.167
Bank erosion vulnerability sites (BEHI)	4.000	7.000	7.000	3.500	3.500	6.000	1.000	2.000	1.000
Bank erosion vulnerability sites (NBS)	3.500	5.000	4.200	3.200	1.000	5.000	0.500	1.000	1.000
Distance from CMZ	4.000	6.000	7.000	2.000	2.000	6.000	1.000	1.000	1.000

Table 33.3 Normalized matrix

Criteria	Slope	Altitude	Distance from fault line	Drainage density	Distance from river	Land use land cover	Bank erosion vulnerability sites (BEHI)	Bank erosion vulnerability sites (NBS)	Distance from CMZ
Slope	0.053	0.044	0.039	0.072	0.032	0.075	0.066	0.046	0.053
Altitude	0.044	0.037	0.019	0.072	0.032	0.036	0.038	0.032	0.035
Distance from fault line	0.053	0.074	0.039	0.072	0.095	0.012	0.038	0.038	0.030
Drainage density	0.053	0.037	0.039	0.072	0.095	0.125	0.076	0.050	0.106
Distance from river	0.159	0.110	0.039	0.072	0.095	0.107	0.076	0.160	0.106
Land use land cover	0.025	0.037	0.117	0.020	0.032	0.036	0.044	0.032	0.035
Bank erosion vulnerability sites (BEHI)	0.213	0.257	0.272	0.250	0.333	0.215	0.265	0.321	0.212
Bank erosion vulnerability sites (NBS)	0.186	0.184	0.163	0.229	0.095	0.179	0.132	0.160	0.212
Distance from CMZ	0.213	0.221	0.272	0.143	0.190	0.215	0.265	0.160	0.212

Table 33.4 Weights of each criterion

Criteria	Weights (%)
Bank erosion vulnerability sites (BEHI)	25.98
Distance from channel migration zone	21.01
Bank erosion vulnerability sites (NBS)	17.12
Distance from river	10.27
Drainage density	7.25
Slope	5.33
Distance from fault line	5.01
Land use land cover	4.20
Altitude	3.83

Table 33.5 Calculations of weighted columns

Criteria	Slope	Altitude	Distance from fault line	Drainage density	Distance from river	Land use land cover	Bank erosion vulnerability sites (BEHI)	Bank erosion vulnerability sites (NBS)	Distance from CMZ	Calculation of weighted sum
Slope	0.053	0.046	0.050	0.072	0.034	0.088	0.065	0.049	0.053	0.511
Altitude	0.044	0.038	0.025	0.072	0.034	0.042	0.037	0.034	0.035	0.363
Distance from fault line	0.053	0.077	0.050	0.072	0.103	0.014	0.037	0.041	0.030	0.477
Drainage density	0.053	0.038	0.050	0.072	0.103	0.147	0.074	0.053	0.105	0.697
Distance from river	0.160	0.115	0.050	0.072	0.103	0.126	0.074	0.171	0.105	0.977
Land use land cover	0.025	0.038	0.150	0.021	0.034	0.042	0.043	0.034	0.035	0.423
Bank erosion vulnerability sites (BEHI)	0.213	0.268	0.350	0.254	0.360	0.252	0.260	0.342	0.210	2.510
Bank erosion vulnerability sites (NBS)	0.187	0.192	0.210	0.232	0.103	0.210	0.130	0.171	0.210	1.644
Distance from CMZ	0.213	0.230	0.350	0.145	0.205	0.252	0.260	0.171	0.210	2.037

Table 33.6 Random index (RI)

N	1	2	3	4	5	6	7	8	9
RI	0	0	0.58	0.90	1.12	1.24	1.32	1.41	1.45

Source Saaty (1977)

Table 33.7 Human habitation suitability final score classes

Suitability classes	Area (in km^2)	Area (%)
not suitable	198.93	48.13
poor	122.46	29.63
fair	68.61	16.60
good	23.33	5.65

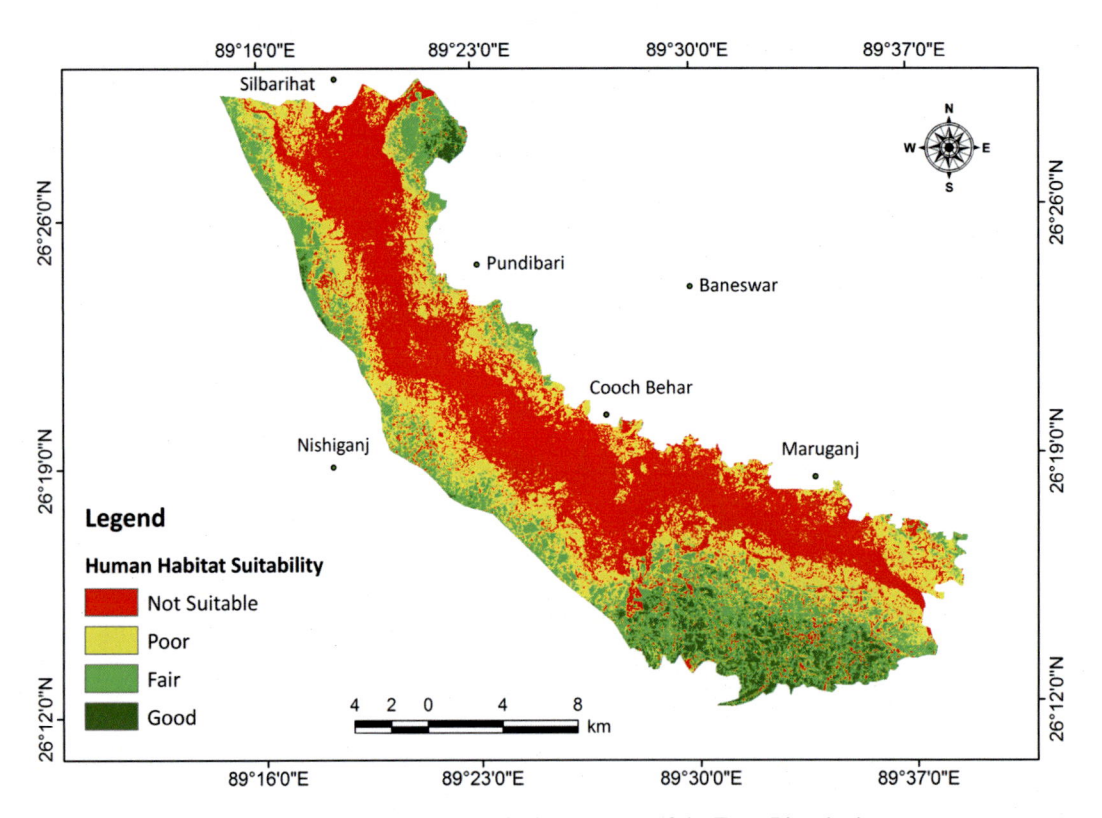

Fig. 33.13 Suitability map for human habitation in the lower course of the Torsa River basin

Table 33.8 Confusion matrix table of site suitability classes for human habitation

Suitability classes		User's accuracy				User accuracy calculation	Producer accuracy calculation
		Good	Fair	Poor	Not suitable		
Producer's accuracy	Good	22	0	1	0	95.65	100.00
	Fair	0	27	0	1	96.43	100.00
	Poor	0	0	27	1	96.43	93.10
	Not Suitable	0	0	1	11	91.67	84.62
Kappa coefficient = 0.94						Overall accuracy = 95.60	

Acknowledgements I would like to express my deepest thanks to Mrs. Moumita Dutta, Research Scholar, Department of Geography, Diamond Harbour Women's University and my beloved students Prakash Dhar, Pankaj Das for their support in my field work. I am grateful to Survey of India, for proving me Topographical Maps; Geological Survey of India, for providing Geological map; USGS for providing satellite imageries; and Block Land and Land Reforms Officer, Coochbehar-II for providing me the Mouza Maps of the study area.

References

Akıncı H, Ozalp AY, Turgut B (2013) Agricultural land use suitability analysis using GIS and AHP technique. Comput Electron Agric 97:71–82. https://doi.org/10.1016/j.compag.2013.07.006

Dey S, Mandal S (2019) Fluvial processes and channel stability of the Torsa River, West Bengal (India). J Geogr Stud 2(2):62–78. https://doi.org/10.21523/gcj5.18020202

Dong Z, Wang Z, Liu D, Li L, Ren C, Tang X, Jia M, Liu C (2013) Assessment of habitat suitability for waterbirds in the West Songnen Plain, China, using remote sensing and GIS. Ecol Eng 55:94–100. https://doi.org/10.1016/j.ecoleng.2013.02.006

Effat HA, Hassan OA (2013) Designing and evaluation of three alternatives highway routes using the analytical hierarchy process and the least-cost path analysis, application in Sinai Peninsula, Egypt. Egypt J Remote Sens Space Sci 16(2):141–151. https://doi.org/10.1016/j.ejrs.2013.08.001

García JL, Alvarado A, Blanco J, Jiménez E, Maldonado AA, Cortés G (2014) Multi-attribute evaluation and selection of sites for agricultural product warehouses based on an analytic hierarchy process. Comput Electron Agric 100:60–69. https://doi.org/10.1016/j.compag.2013.10.009

Gdoura K, Anane M, Jellali S (2015) Geospatial and AHP-multicriteria analyses to locate and rank suitable sites for groundwater recharge with reclaimed water. Resour Conserv Recycl 104(Part A):19–30. https://doi.org/10.1016/j.resconrec.2015.09.003

Ghosh KG, Pal S, Mukhopadhyay S (2016) Validation of BANCS model for assessing stream bank erosion hazard potential (SBEHP) in Bakreshwar River of Rarh region Eastern India. Model Earth Syst Environ 2 (95):1–15. https://doi.org/10.1007/s40808-016-0172-0

Graymore MLM, Wallis AM, Richards AJ (2009) An index of regional sustainability: a GIS-based multiple criteria analysis decision support system for progressing sustainability. Ecol Complex 6(4):453–462. https://doi.org/10.1016/j.ecocom.2009.08.006

Gumusay MU, Koseoglu G, Bakirman T (2016) An assessment of site suitability for marina construction in Istanbul, Turkey, using GIS and AHP multicriteria decision analysis. Environ Monit Assess 188(12):677. https://doi.org/10.1007/s10661-016-5677-5

Houshyar E, SheikhDavoodi MJ, Almassi M, Bahrami H, Azadi H, Omidi M, Sayyad G, Witlox F (2014) Silage corn production in conventional and conservation tillage systems. Part I: sustainability analysis using combina-tion of GIS/AHP and multi-fuzzy modeling. EcologicalIndicators 39:102–114. https://doi.org/10.1016/j.ecolind.2013.12.002

Malczewski J (1999) GIS and multicriteria decision analysis. JohnWiley & Sons

Miller W, Collins MG, Steiner FR, Cook E (1998) An approach for greenway suitability analysis landscape and urban planning. Int J Geogr Inform Sci 42(2–4):91–105. https://doi.org/10.1016/S0169-2046(98)00080-2

Mishra AK, Deep S, Choudhary A (2015) Identification of suitable sites for organic farming using AHP & GIS. Egypt J Remote Sens Space Sci 18(2):181–193. https://doi.org/10.1016/j.ejrs.2015.06.005

Mohit A, Ali M (2006) Integrating GIS and AHP for land suitabiltiy analysis for urban development in a secondary city of Bangladesh. J Alam Bina Jilid 8(1)

Parry JA, Ganaie SA, Bhat MS (2018) GIS based land suitability analysis using AHP model for urban services planning in Srinagar and Jammu urban centers of J&K, India. J Urban Manage 7(2):46–56. https://doi.org/10.1016/j.jum.2018.05.002

Pramanik MK (2016) Site suitability analysis for agricultural land use of Darjeeling district using AHP and GIS techniques. Model Earth Syst Environ 2(56). https://doi.org/10.1007/s40808-016-0116-8

Roig-Tierno N, Baviera-Puig A, Buitrago-Vera J, Mas-Verdu F (2013) The retail site location decision process using GIS and the analytical hierarchy process. Appl Geogr 40:191–198. https://doi.org/10.1016/j.apgeog.2013.03.005

Rosgen DL (2006) A Watershed assessment for river stability and sediment supply (WARSSS). Wildl and Hydrology Books, Fort Collins, Colorado. 23 (1–4). http://www.epa.gov/warsss/Accessed 20 June 2010

Saaty RW (1987) The analytic hierarchy process-what it is and how it is used. Mat/d Model 9(3–5):161–176. https://doi.org/10.1016/0270-0255(87)90473-8

Saaty TL (1977) A scaling method for priorities in hierarchical structures. J Math Psychol 15:234–281. https://doi.org/10.1016/0022-2496(77)90033-5

Saaty TL (1980) The analytic hierarchy process. McGraw-Hill, NewYork

Saaty TL (1988) The analytic hierarchy process. Typesetters Ltd., Becceles, Suffolk. https://doi.org/10.1007/978-3-642-83555-1_5

Saaty TL (2007) Time dependent decision-making; dynamic priorities in the AHP/ANP: generalizing from points to functions and from real to complex variables. Math Comput Model 46:860–891. https://doi.org/10.1016/j.mcm.2007.03.028

Shukla A, Kumar V, Jain K (2017) Site suitability evaluation for urban development using remote sensing, GIS and analytic hierarchy process (AHP). Springer, Singapore, pp 377–388.https://doi.org/10.1007/978-981-10-2107-7_34

Starr RR (2009) Stream assessment protocol Anne Arundel County, Maryland. Stream habitat assessment and restoration program. U.S. Fish and Wildlife Service. Chesapeake Bay Field Office

Zhang J, Su Y, Wu J, Liang H (2015) GIS based land suitability assessment for tobacco production using AHP and fuzzy set in Shandong province of China. Comput Electron Agric 114:202–211. https://doi.org/10.1016/j.compag.2015.04.004

Spatiotemporal Assessment of Drought Intensity and Trend Along with Change Point: A Study on Bankura District, West Bengal, India

Shrinwantu Raha, Suman Kumar Dey, Madhumita Mondal, and Shasanka Kumar Gayen

Abstract

This research was designed for spatiotemporal assessment of drought of Bankura District, West Bengal using Standardized Precipitation Index (SPI) during 1979–2013 along with change point. Peak intensity, average drought intensity, and trend, were considered as the drought evaluation parameters. Application of Discrete Wavelet Transform (DWT) to SPI in different time steps indicated the fact that the time frame 1979–2013 experienced several surpluses and deficit phases. Execution of Mann-Whittney and Kruskal-Walis test in 12 months time step, confirmed that the region has a change-point phase from 1989 to 1995–96 but the most probable and convenient change point is considered as 1995–96. Overall, peak and average drought were intensified in north-western portions of Bankura and as the time step started to increase, peak and average drought intensity started to decrease by a significant proportion. With advancing time steps, from three months to six months, the trend had decreased but at 12 months' time step trend had increased again. It is very interesting to note that the intensity of drought had decreased but the tendency of drought had increased after the change point. The north-western portions were noticed with a negative change in peak and average drought intensity and southern portions were noticed with a positive change in the peak intensity. Average drought intensity had decreased slightly at a slower pace in the north-eastern and south-eastern patches of Bankura. Northern and Southern portions of the study region were observed, respectively with negative and positive changes. For the MK test statistic, also a similar pattern of change was noticed. Combining all stations, peak and average drought were observed with decreasing nature and MK and Sen's slopes were observed with increasing rate. In the recent decade, Bankura District had faced drought several times but the drought phenomena in this region were far from the proper conclusive statement. In such circumstances, the spatiotemporal assessment of drought intensity and trend along with change point is a very noble attempt.

Keywords

Drought · Changepoint · Mann-Whittney test · Trend · Intensity

---

S. Raha (✉) · S. K. Dey · S. K. Gayen
Department of Geography, Cooch Behar Panchanan Barma University, Cooch Behar, West Bengal 736101, India
e-mail: shrinwanturaha1@gmail.com

M. Mondal
Department of Geography, Bhairab Ganguly College, Belgharia, West Bengal 700056, India

34.1 Introduction

The climate of a region is determined by the long-term average and extremes of several climatological variables (Patel et al. 2007). Drought is a periodic phenomenon that exerts multifaceted negative impacts on a wide range of water-related sectors (Spinoni et al. 2013). It is generally considered a prolonged period with significantly lower precipitation relative to normal levels (Shadeed 2012). Meteorological drought is identified with precipitation shortages which cause diminishes in water supplies for residential and other purposes (Gupta et al. 2011). In other words, meteorological drought can be defined as the prolonged period of shortage of water (Chhajer et al. 2015). Almost every part of the world is affected by drought (Kwon et al. 2019). Drought is rather different from other water-related hazards in terms of spatiotemporal characteristics resulting in structured spatial coverage with varying durations (Thomas et al. 2015). Overall drought evaluation parameters intensity and trend are the most important aspects as these parameters exert the most resourceful imprints on drought assessment and monitoring (Kulkarni et al. 2020). Drought events are reported and expected to be strengthening in this era of global warming (IPCC 2007). Thus drought monitoring and early warning systems on global and local scales have emerged as powerful platforms for mitigating and preventing negative impacts of drought (Kwon et al. 2019).

Nowadays, researchers and climatologists have recommended detecting the change in the hydro-meteorological time series by change point analysis (Lee et al. 2019). Spatiotemporal analysis of drought should be assessed concerning different time scales and change points (Okal et al. 2020). Change points should be assessed in the long-term time frame and meaningful statistical characteristics of drought before and after change points should be addressed (IPCC 2013). Unfortunately, considering both aspects together, there is a very limited number of works done on drought assessment and monitoring. Thus this study was focused on the following aspects:

(1) Spatiotemporal assessment of drought using intensity and trend in several time steps (here 3, months. six months, and 12 months' time step); (2) Estimation of change point in the long-term time frame (here 12 months) and assessment of change of those before and after the change point.

The study area is surrounded by Purba Bardhaman district and Paschim Bardhaman district in the north, Purulia District in the west, Jhargram and Paschim Medinipur district in the south, and some part of Hugli district in the east. Bankura is such an agrarian tract that is very much sensitive to drought. Monsoon rainfall in this tract is essential not only for agriculture during monsoon months, but also for recharging the groundwater for irrigation. Above 70% of the total population of Bankura depends on rainfed agriculture. Thus the assessment of drought phenomena is so much important for this region which was not highlighted by previous research papers focused on this region.

While the region receives about 1600 mm of rainfall, almost all rainfall is confined to the rainy season. Also, the rainfall is not evenly distributed over the study region. According to the last Census Report (2011), the overall total population of the district was 3,596,674 (District Census Handbook 2011) and the population growth rate was 12.64% which appeared higher than the national average.

34.2 Materials and Methods

34.2.1 Data Sources

Station-wise rainfall data on daily basis was downloaded from the Climate Forecast System Reanalysis (CFSR) project which started in January 1979 and ended in July 2014. CFSR is one of the widely used, trusted, and flexible datasets that can easily be used in drought identification and monitoring (Sommerlot 2017). The reanalysis datasets are converted into the station data which is available in the SWAT format at globalweatherdata.tamu.edu. Our study extracted

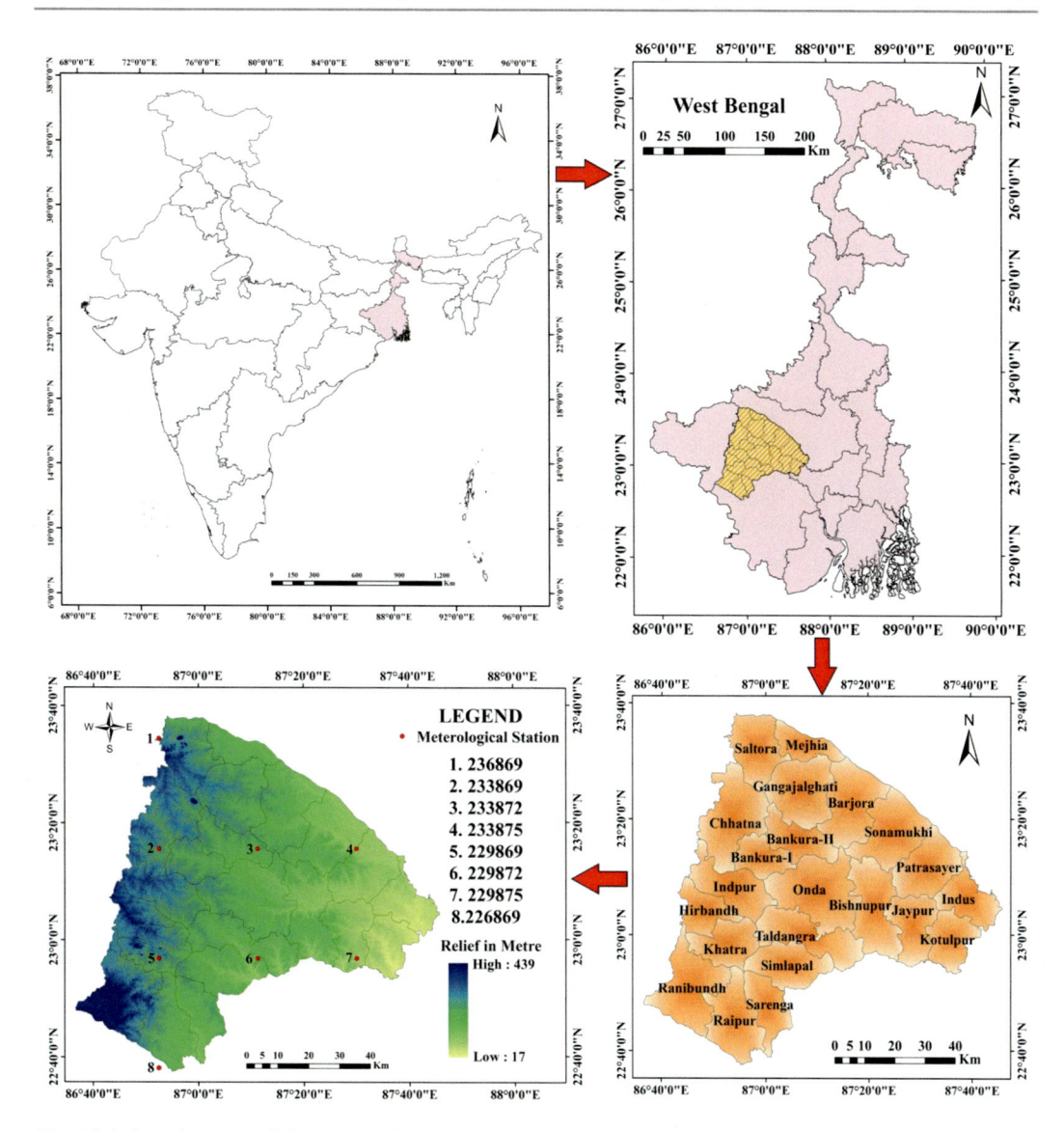

Fig. 34.1 Location map of the study region

those daily basis station data and converted them into monthly values. Table 34.1 determines the location of meteorological stations with the mean and standard deviation of rainfall.

34.2.2 Standardized Precipitation Index (SPI)

Mckee et al. (1993) developed SPI to estimate drought. SPI is found most suitable and reliable and can be applied in different parts of the world (Turkes and Tatli 2009) SPI is simple, flexible can be applicable in shorter and longer time steps (Dogan et al. 2012). To identify drought effectively it should be estimated in shorter and longer time steps. Here, the classification of SPI is based on three months, six months, and 12 months' time steps. Three months and six months are considered a short-term, 12-month time frame is considered a long-term time frame. The monthly precipitation

Table 34.1 List of meteorological stations with latitude, longitude, elevation (m)

Id of stations	Longitude	Latitude	Elevation (m)	Mean rainfall (mm)	Standard Deviation (SD) of rainfall
229,869	86.875	22.9488	133	251.48	279.805
229,872	87.1875	22.9488	61	164.35	217.435
229,875	87.5	22.9488	34	158.45	211.01
233,869	86.875	23.261	127	269.08	228.14
233,872	87.1875	23.261	95	157.22	257.34
233,875	87.5	23.261	46	157.89	212.49
236,869	86.875	23.5733	234	120.45	210.33
226,869	86.875	22.6366	176	278.67	245.66

series is modeled here, using a gamma distribution. Gamma distribution as the best fit for the precipitation time series. That method has been applied here as follows:

$$g(x) = \frac{1}{\beta^\alpha \lambda(\alpha)} x^{\alpha-1} e^{x/\beta} \quad (34.1)$$

where, $\alpha > 0$ is a shape parameter, $\beta > 0$ is a scale parameter, $x > 0$ is the amount of precipitation, $\lambda(\alpha)$ defines gamma distribution. α and β are estimated for each station at each time step. According to Edwards and Mckee (1997) those two parameters can be estimated using the approximation of Thom (1958) by maximum likelihood:

$$\alpha^\circ = \frac{1}{4A}\left(1 + \sqrt{1 + \frac{4A}{3}}\right) \quad (34.2)$$

$$\beta^\circ = \frac{\bar{x}}{\alpha^\circ} \quad (34.3)$$

where,

$$A = In(\bar{x}) - \frac{\sum In(x)}{n} \quad (34.4)$$

n = no. of precipitation observations.

Integrating the probability density function with respect to x and inserting the estimates of α and β yields the expression of cumulative probability distribution $G(x)$ of an observed amount of precipitation occurring for a given month at a particular time step

$$G(x) = \int_0^x g(x)dx = \frac{1}{\beta^\alpha \lambda(\alpha)} \int_0^x x^{\alpha-1} e^{-x/\beta} \quad (34.5)$$

The cumulative probability becomes as further

$$H(x) = q + (1-q)G(x) \quad (34.6)$$

Based on Mckee et al. (1993) drought severity classes are stretched from −2 to +2. −2 indicates extreme drought and +2 indicates the extreme wet condition. The SPI class −0.99 to +0.99 denotes the normal condition.

34.2.2.1 Phase and Periodicity Estimation by Discrete Wavelet Transform (DWT)

Wavelet transform is a mathematical tool can be effectively used to detect the rapid and varied changes of drought in a slow time pace. DWT sequentially able to develop different periods of drought very meticulously (Nason and Sachs 1999):

$$W_f(j,k) = \int_{-\infty}^{+\infty} f(t)\varphi_{j,k}*(t)dt \text{ with } \varphi_{j,k}(t) = a_0^{\frac{j}{2}} \varphi(a_0^{-j} t - b_0^k) \quad (34.7)$$

Where, $\varphi(t)$ is the mother wavelet, $f(t)$ is the series analysed, and t is an index representing time, a_0 and b_0 are constants, integer j is the decomposition level and k is the translation factor, $\varphi*(t)$ is the complex conjugate.

34.2.2.2 Peak (PI$_D$) and Average Drought Intensity (MI$_D$)

In a particular time frame, the lowest value of SPI denotes peak drought (Mckee et al. 1993). On the other hand, the average drought event refers to the magnitude of drought divided by the duration of drought. In this research MI$_D$ is:

$$MI_D = \sum_{i=1}^{n} SPI_{ij}\Big/n - WI_p \quad (34.8)$$

where SPI$_{ij}$ are the SPI values of drought and wet event in j-th time and n is the number of months. MI$_D$ is the mean of drought period intensity and WI$_p$ is the mean of wet period intensity.

34.2.2.3 Mann Kendal Trend Test (MK Test)

The Mann Kendall test (Kendall 1975) is recommended by the World Meteorological organization and this method is used here as follows:

$$S = \sum_{i=1}^{n-1} \sum_{j=i+1}^{n} sgn(x_j - x_i) \quad (34.9)$$

where, n is the number of data points, xi and xj are the data values of the separate time series i and j ($j > i$), respectively, and $sgn(x_j - x_i)$ is the sign function

$$sgn(x_j - x_i) = \begin{cases} +1 \; if \; x_j - x_i > 0 \\ 0 \; if \; x_j - x_i = 0 \\ -1 \; if \; x_j - x_i = 0 \end{cases} \quad (34.10)$$

The variance is computed as:

$$Var(S) = \frac{n(n-1)(2n+5) - \sum_{i=1}^{p} t_i(t_i - 1)(2t_i + 5)}{18} \quad (34.11)$$

where n is the number of data points, P is the number of tied groups, the summation sign indicates that the summation over all tied groups. In the case of a sample size greater than 30 the standard normal format of Zs can be computed as follows

$$Z = \begin{cases} \frac{S-1}{\sqrt{var(S)}} \; if \; s > 0 \\ 0 \; if \; s = 0 \\ \frac{S+1}{\sqrt{var(S)}} \; if \; s < 0 \end{cases} \quad (34.12)$$

Positive trends of the whole series Z value indicate a positive trend and negative Zs value of the whole series indicates a negative trend of drought. Testing trends is performed by the specific α significance level. When the null hypothesis is rejected it indicates that there is a significant trend exists in the data series.

For the visualization of spatial maps, the Inverse Distance Weightage Method (IDW) has been applied in ArcGIS 10.2.2 environment.

34.2.2.4 Sen's Slope Estimator

The magnitude of the trend in the SPI time series was estimated in the following way (Theil 1950; Sen 1968):

$$\beta = median\left(\frac{X_i - X_j}{i - j}\right), \forall j < i \quad (34.13)$$

In the above Eq. (34.11), $1 < j < i < n$. According to Tabari et al. (2011), β is the median overall combination of all recorded pairs, not affected by the extreme values in the observations.

34.2.2.5 Analysis of the Homogeneity Using Kruskal-Walis Test

The Kruskal-Walis Chi-square test is applied to determine whether any significant difference exists before and after the change point. The procedure is as follows (Kruskal and Walis 1952):

(a) Order the observations from least to the greatest and give them corresponding rank
(b) Then Kruskal-Walis Chi-square test is then given in the following way:
  i. If the value of the observations has no ties then the value of H is obtained as follows:

$$H = \frac{12}{n(n+1)} \sum_{i=1}^{k} \frac{R_i^2}{n_i} - 3(n+1)$$

$$(34.14)$$

where, n_i is the number of observations in the i-th group, $\sum_{i=1}^{k} n_i$ and R_i = sum of the ranks in the i-th group.

ii. If the number of identical observations exceeds 25% of the number of the samples the H can be modified as follows:

$$H = \frac{\frac{12}{n(n+1)} \sum_{i=1}^{k} \frac{R_i^2}{n_i} - 3(n+1)}{1 - \frac{\sum_{j=1}^{c}(t_j^3 - t_j)}{n^3 - n}} \quad (34.15)$$

where c is the number of tied groups and t_j is the size of tied group j. When n_i exceeds five, the distribution of H is approximately the χ_2 distribution.

34.2.2.6 Estimation of Change of Drought as a Percentage of Mean

Percentage change of different drought evaluation parameters before and after change point can be expressed as (Yue and Hasino 2003):

$$\%Change = \frac{D_{p(a)} - D_{p(b)}}{D_{(a)}} \times 100 \quad (34.16)$$

where, $D_{p(a)}$ is the drought evaluation parameter after change point, and $D_{p(b)}$ is the drought evaluation parameter before change point, $D_{(a)}$ is the mean of drought evaluation parameter before and after the change point.

34.3 Results and Discussion

Figure 34.2 describes the periodicity of the drought in Bankura District. Based on phase estimated by Discrete Wavelet Transform (DWT) the drought of the last 32 years were roughly categorized into consecutive surplus and deficit phases (Fig. 2a–c): (a) *acute deficit phase:* October 1982–December 1983; (b) *short surplus phase:* July 1984–April 1987; (c) *short drought phase:* May 1987–March 1990; (d) *longest surplus phase*: April 1991–August 1996; (e) *short drought phase:* September 1997–January 1998; (f) *short surplus phase:* January 1999–July 2000; (g) *long Drought Phase:* August 2000–December 2002; (h) *surplus Phase*: January 2003–February 2005; (i) *short Drought phase:* March 2005– October 2007; (j) *long Surplus Phase:* November 2008–December 2013.

In three months' time step, drought was at its' peak (PI$_D$ value −3.22) in station 236,869 in December 1990 (Table 34.2). The other five stations experienced the highest drought intensity with below −2 PI$_D$ value (stations 229,869, 229,872, 229,875, 233,869, and 233,872, respectively). The drought was observed at its' highest intensity in the post-monsoon period. Average drought intensity (SPI ≤ −1.0) was also highest at station 236,869 with a −1.590 SPI value (Table 34.2). Station 236,869 (−1.95 SPI value) was observed with the highest average drought intensity and on the contrary, stations 226,869 and 233,869 (−1.445 SPI value) were the two with the highest and lowest average drought intensity, respectively (Table 34.2). Except the stations, 229,869 and 229875 all the stations were noticed with significant MK value and in case of Sen's slope all the stations are noticed with the significant positive value (Table 34.3). At this time step, PI$_D$ and MI$_D$ dominated in the north-western and south-western portions of the study region (Figs. 3a and 4a). At the three months' time step, Sen's slope was highest in north-western portions of the study region whereas it is low in the eastern and north-western portions of the study region (Fig. 5a). MK trend was noticed as the lowest in the south-western and north-western portions of the study region (Fig. 6a). By considering all stations, MK and Sen's slopes were positive and they were significant at a 95% confidence interval. So, it was expected that at three months' time step, for Bankura, drought had a significant positive trend.

However, at six months' time step, the scenario changed a bit. At six months' time step, drought was at its' peak in January 1980 at station 233,872 with PI$_D$ value of −5.061 (Table 34.2). At this time step, all other stations were characterized with below − PI$_D$ value except station 226,869. At this time step, almost all the meteorological stations faced the highest

Fig. 34.2 Periodicity of drought using Discrete Wavelet Transform (DWT) (Here all stations are combined)

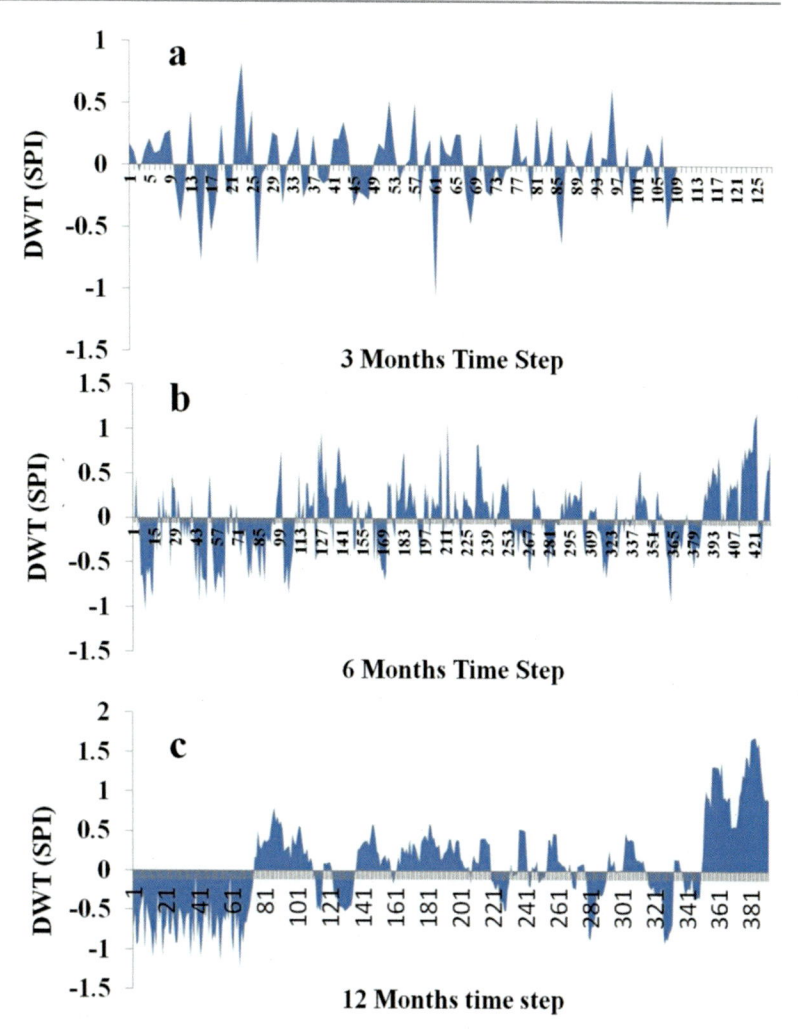

Table 34.2 List of meteorological stations with peak and average drought intensity

Id of Stations associated Bankura	Peak drought intensity (three months)	Average drought intensity (three months)	Peak drought intensity (six months)	Average drought intensity (six months)	Peak drought intensity (12 months)	Average drought intensity (12 months)
229,869	−3.12	−1.556	−3.121	−1.408	−0.982	−0.972
229,872	−3	−1.473	−2.583	−1.404	−2.232	−1.438
229,875	−2.86	−1.506	−2.412	−1.426	−2.241	−1.410
233,869	−2.79	−1.453	−2.621	−1.425	−2.161	−1.536
233,872	−3.1	−1.525	−5.061	−1.550	−5.291	−1.962
233,875	−3.01	−1.526	−2.291	−1.463	−2.345	−1.429
236,869	−3.22	−1.590	−3.56	−1.590	−2.03	−1.559
226,869	−1.97	−1.453	−1.97	−1.253	−1.97	−1.424
Average	−2.85	−1.510	−2.92	−1.43988	−2.61	−1.46625

Table 34.3 List of meteorological stations with MK test and Sen's slope at different time steps

Id of Stations associated Bankura	Mann Kendal Test (MK test) (three months)	Sen's Slope (three months)	Mann Kendal Test (MK) (six months)	Sen's (S) Slope (six months)	Mann Kendal test (12 months)	Sen's slope
229,869	−1.4136	0.00042*	2.6053*	0.00057*	5.8328*	0.006*
229,872	2.43*	0.01*	3.965*	0.0016*	7.264*	0.0023*
229,875	2.7344	0.0011*	4.2787*	0.0017*	7.1729*	0.0027*
233,869	3.1835*	0.0013	4.7772*	0.0019*	7.4266*	0.002*
233,872	3.5523*	0.0014*	4.2787*	0.0017*	7.471*	0.00211*
233,875	3.574*	0.00123*	5.0538*	0.0021*	7.471*	0.00285*
236,869	1.4136*	0.003*	4.324*	0.0020*	5.7616*	0.0023*
226,869	3.516*	0.004*	2.600*	0.0004*	7.3226*	0.00226*
Average	2.650*	0.0028*	4.06175*	0.0015625*	6.9653125*	0.0028575*

* At 95% confidence interval

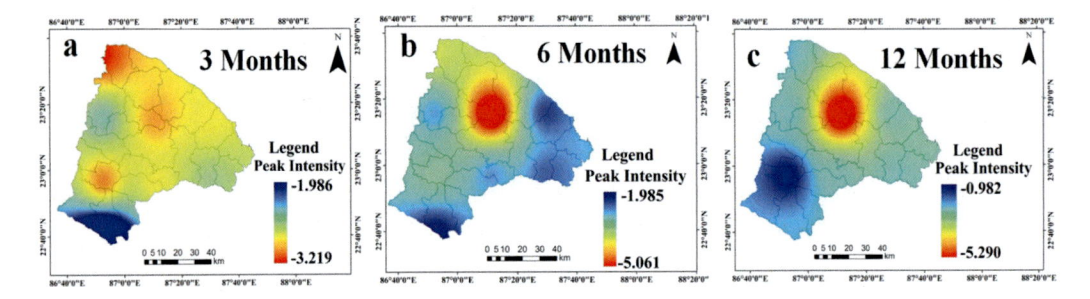

Fig. 34.3 Spatial Assessment of Peak Intensity of drought at **a** three months **b** six months and **c** 12 months' time step

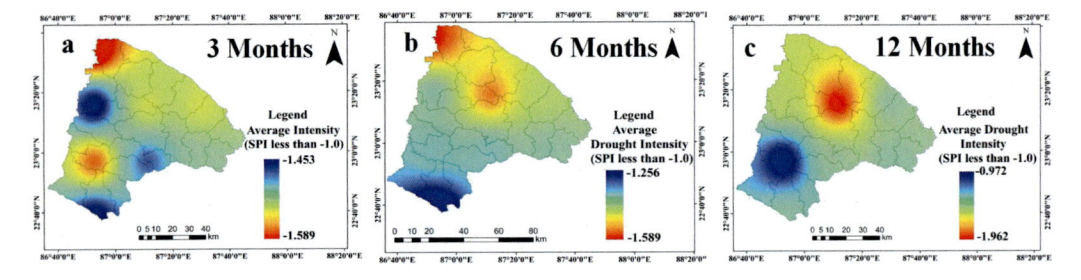

Fig. 34.4 Spatial assessment of the average intensity of drought **a** three months **b** six months and **c** 12 months' time step

drought intensity in the post-monsoon phase. At this time step, the average drought was intensified at 236,869. Significant (at 95% significance level) positive MK and Sen's slope value was noticed at all the stations. At six months' time step also, PI_D and MI_D dominated in the north-western and south-western portions of the study region (Figs. 3b and 4b). At this time step, Sen's slope (Fig. 5b) was lowest (intensity high) in the south-western portions of the study region and MK trend (Fig. 6b) was also lowest (intensity high) in the south-western portions of the study region. Overall at six months' time step, drought had a significant positive trend. So overall, drought was expected to be intensified at this time step.

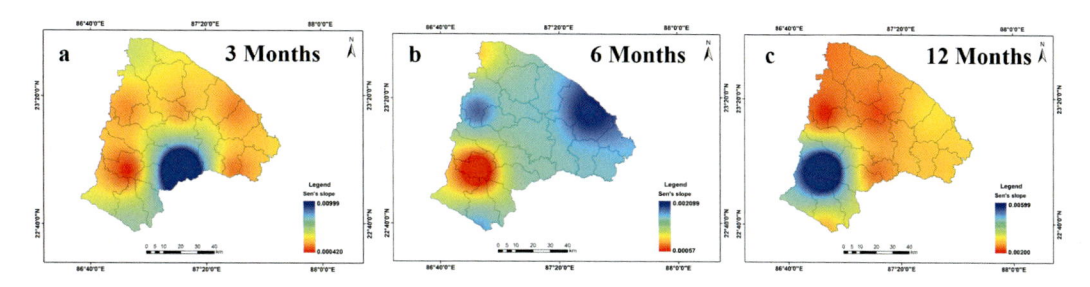

Fig. 34.5 Sen's slope of the study region **a** three months **b** six months **c** 12 months' time step

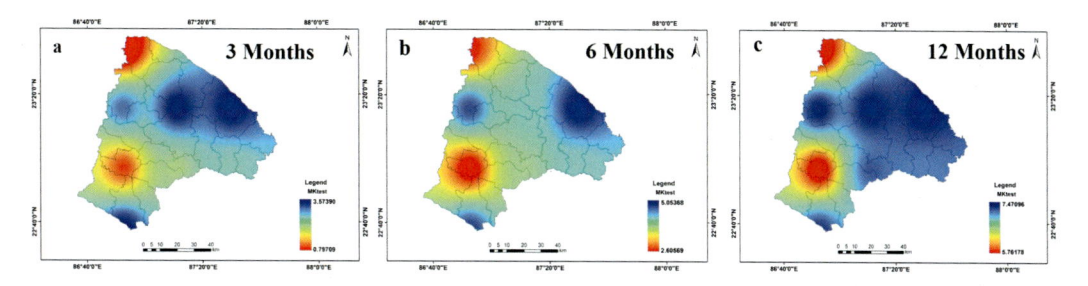

Fig. 34.6 MK test of the study region **a** three months **b** six months **c** 12 months' time step

At the 12 months' time step, PI_D and MI_D were highest at station 233,872 (Table 34.2). PI_D and MI_D were observed at the lowest level at station 229,869. All the stations are noticed with significant positive trend. Rest stations were observed with non-significant positive MK and Sen's slope value. At this time step, peak intensity and average drought intensity were high (Figs. 3c and 4c) in the north-western portions of the study area. Trend of the drought is high at the south-western portions and relatively low at the remaining portions (Figs. 5c and 6c).

Mann-Whitney (MW) test was utilized in 12 months' time step, to estimate the change point. All stations experienced a change point phase in 1995–1996. Kruskal-Walis (KW) test statistic (H) was further utilized to judge the homogeneity of the series and it was clear that stations 229,869, 233,869, 236,869, and 226,869 were heterogeneous. Rest stations were found with homogeneity (Table 34.4). PI_D was remarkably decreased at station 233,872. As a result, PI_D decreased at a rapid pace (negative change) in the north-western portions. Only station 229,869 was observed with a slight increase in peak intensity and as a result, southern portions were noticed with a positive change in the peak intensity (Fig. 7a). The amount of MI_D decreased remarkably for station 233,872 as a result the north-western portions were noticed with the highest negative change with respect to MI_D. A negative change in MI_D was prominent in the station 233,875, 229,872, and 229,875. As a result, MI_D decreased slightly at a slower pace in the north-eastern and south-eastern portions of Bankura (Fig. 7b). A negative change in the MK test statistic is observed at all stations. A negative change in the Sen's slope is also observed in all stations except station 229,869. North-eastern portions of the study region were observed with negative change (−133%) in Sen's slope whereas north-western portions were noticed with positive change (+42.2628%) in the Sen's slope (Fig. 7c). For the MK test statistic, also a similar pattern of change was noticed (Fig. 7d). Combining all stations, PI_D and MI_D were observed, respectively at 35% and 24% decreasing rate. These features conclude that even though after change point intensity was observed at a decreasing rate but the tendency of the drought

Table 34.4 Change point of homogeneity of time series of SPI of Bankura (12-month)

Station Id	Change point (Mann Whitney Test)	Homogeneity (Kruskal-Walis Test)					
		Pre change point period	Post change point period	Chi-square test statistic (H)	Degree of freedom	Probability > Chi-Square	Series
229,869	1995–96	1979–1994	1995–2013	22.445	7.7	0.00517	S_1
229,872	1995–96	1979–1994	1997–2013	24.635	119	0.06216	S_0
229,875	1995–96	1979–1994	1995–2013	34.799	137	0.25034	S_0
233,869	1995–96	1979–1994	1995–2013	23.662	27	9.31E−20	S_1
233,872	1995–96	1979–1994	1995–2013	24.712	119	0.06216	S_0
233,875	1995–96	1979–1994	1995–2013	26.169	147	0.26332	S_0
236,869	1995–96	1979–1994	1995–2013	34.868	20	1.09E-17	S_1
226,869	1995–96	1979–1994	1995–2013	34.871	21	8.10E-24	S_1

All values are significant at 95% confidence interval, S_0 means homogeneous, S_1 means heterogeneous

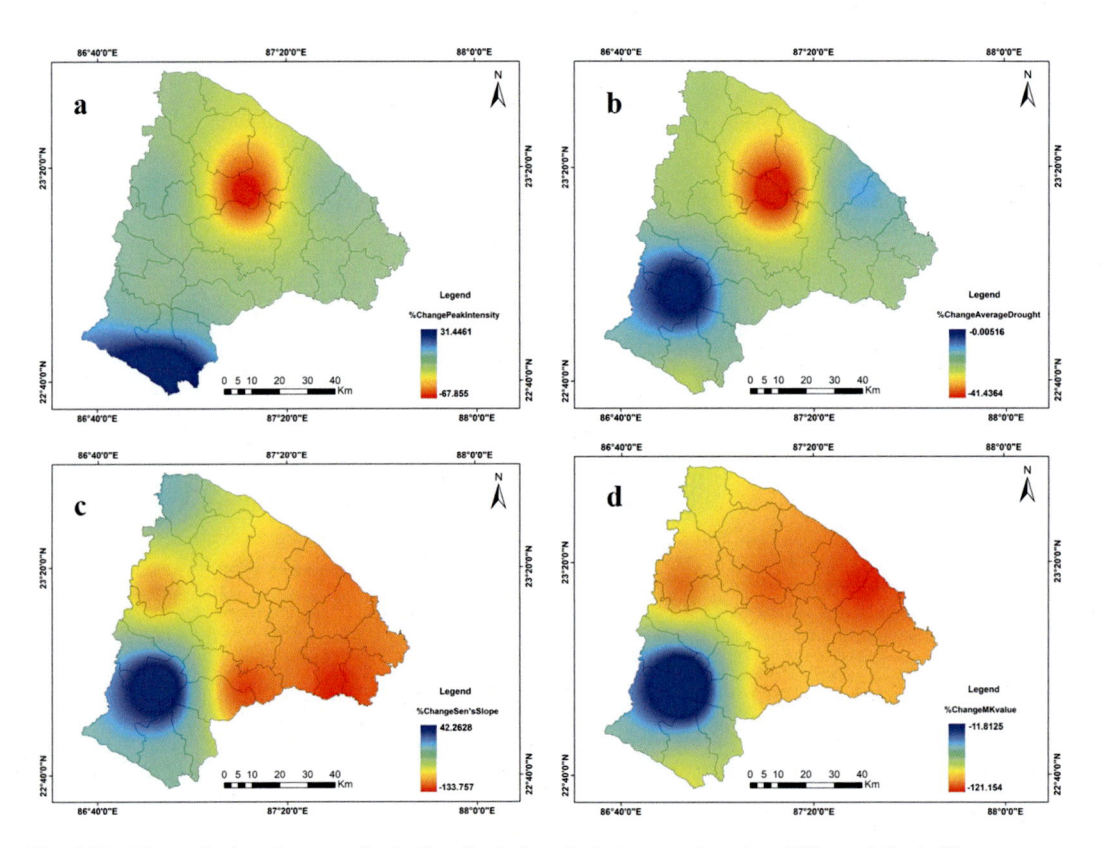

Fig. 34.7 Change in drought properties in %: **a** Peak drought **b** Average drought **c** MK test **d** Sen's Slope

was found at the increasing mode. The drought had started to shift from the western portions to the southern portions of the study area after the change point. With increasing time steps, a completely reversed picture was noticed for the trend.

34.4 Conclusions

In this study, drought was assessed spatiotemporally at three months, six months and 12 month time step along with the change point. Intensity and trend of drought were portrayed using visual interpretative maps and statistical assessment. Application of Discrete Wavelet Transform (DWT) method over SPI revealed that the period 1979–2013 faced several deficit and surplus phases and it could be pointed out there existed some inconsistent periodicity of drought within 2–4 years which was observed in accordance of the recent accordance with the recent weakening relationship between Indian Summer Monsoon Rainfall and El Nino Southern Oscillation. The year 1995–96 was confirmed as the change point by the Mann-Whittney U test Kruskal-Walis test was further applied to reveal peak and average drought were intensified in north-western portions of the study area at every time step. Even with increasing time steps, peak and average drought were intensified in these portions. Combining all stations, average drought had decreased (MI_D 4.64% decrease) from three months to six months' time steps but from six months to 12 months' time steps average drought had increased again (MI_D 1.83% increase). On the other hand, from three months to six months' time step, PI_D had increased by about 2.74% and it had decreased by about 49% from six months to 12 months' time step. As the time step had increased from three months to six months trend had decreased but at 12 months' time step trend had increased again and drought shifted from northern to southern portions of the study region.

After the change point, considering all stations, peak and average drought intensity had decreased at 35% and 24% rate, respectively. Southern portions were noticed with a positive change in peak intensity. The average drought had decreased (slightly at a slower pace) in the north-eastern and south-eastern patches of the study region. Northern portions were observed with negative change (less intensity) in Sen's slope and southern portions were observed with positive change (more intensity) in the Sen's slope. Thus, from the overall assessment, it could be concluded that drought intensity was decreasing but the tendency (trend) of the drought was increasing after the change point.

Assessment of trend and intensity of drought in different time steps and also before and after change point helps to understand the true nature of drought for Bankura. This estimation demonstrates a potential risk to agrarian agricultural practices which is prevalent in the present study area. Therefore, this assessment can help in early warning responses, local-scale planning, and food security management. This study is an essential step to enhancing drought risk management strategy in the study area.

References

Chhajer V, Prabhakar S, Prasad PRC (2015) Development of index to assess drought conditions using geospatial data a case study of Jaisalmer District, Rajasthan India. Geoinformatica Polonica 14(1):29–39

District Census Handbook (2011) Accessed: 30th May 2018

Dogan S, Berktay A, Singh VP (2012) Comparison of multi-monthly rainfall-based drought severity indices, with application to semi-arid Konya closed basin, Turkey. J Hydrol 470–471:255–268

Edwards DC, McKee TB (1997) Characteristics of 20th century drought in the United States at multiple time scales, Climatology report no. 97–2, 155. Department of Atmospheric Science, Colorado State University, Fort Collins. Available at: https://apps.dtic.mil/sti/pdfs/ADA325595.pdf

Gupta AK, Tyagi P, Sehgal VK (2011) Drought disaster challenges and mitigation in India: strategic appraisal. Curr Sci 100(12):1795–1806

IPCC (2007) Climate change The physical science basis. In: Solomon S, Quin D, Manning M, Chen X, Marquis M, Averyt KB, Tignor HL, Miller M (eds) Contribution of working group I to the fourth assessment report of the intergovernmental panel on climate change, pp 1–996. Cambridge University Press, Cambridge

IPCC Intergovernmental Panel on Climate Change (2013) Detection and attribution of climate change: from global to regional. Climate change—the physical science basis, pp 867–952

Kendall MG (1975) The advanced theory of statistics: design and analysis, and time-series 3. Charles Griffin and Company Limited, London

Kruskal WH, Wallis WA (1952) Use of ranks in one-criterion variance analysis. J Am Stat Assoc 47(260):583–621

Kulkarni SS, Wardlow BD, Bayissa YA, Tadesse T, Svoboda MD, Gedam SS (2020) Developing a remote sensing-based combined drought indicator approach for agricultural drought monitoring over Marathwada India. Remote Sens 12(13):2091

Kwon M, Kwon H, Han D (2019) Spatio-temporal drought patterns of multiple drought indices based on precipitation and soil moisture: A case study in South Korea. Int J Climatol 39(12):4669–4687

Lee M-H, Kim E-S, Bae D-H (2019) A comparative assessment of climate change impacts on drought over Korea based on multiple climate projections and multiple drought indices. Clim Dyn 53:389–404

Mckee TB, Doesken NJ, Kleist J (1993) The relationship of drought frequency and duration to time scales. In: AMS 8th conference on applied climatology (17–22 January). American Meteorological Society, Anaheim, CA, pp 179–184. Available at: https://climate.colostate.edu/pdfs/relationshipofdroughtfrequency.pdf

Nason GP, Sachs RV (1999) Wavelets in time-series analysis. Philos Trans Royal Soc Math Phys Eng Sci 357(1760):2511–2526

Okal H, Ngetich F, Okeyo J (2020) Spatio-temporal characterisation of droughts using selected indices in Upper Tana River Watershed, Kenya. Sci Afr e00275

Patel NR, Chopra P, Dadhwal K (2007) Analyzing spatial patterns of meteorological drought using standardized precipitation index. Meteorol Appl 14(4):329–336

Sen PK (1968) Estimates of the regression coefficient based on Kendall's Tau. J Am Stat Assoc 63(324):1379

Shadeed S (2012) Spatio-temporal drought analysis in arid and semi-arid regions: a case study from Palestine. Arab J Sci Eng 38(9):2303–2313

Sommerlot AR (2017) Coupling physical and machine learning models with high resolution information transfer and rapid update frameworks for environmental applications (Doctoral dissertation, Virginia Tech). Accessed 15 April 2020

Spinoni J, Naumann G, Carrao H, Barbosa P, Vogt J (2013) World drought frequency, duration, and severity for 1951–2010. Int J Climatol 34(8):2792–2804

Tabari H, Somee BS, Zadeh MR (2011) Testing for long-term trends in climatic variables in Iran. Atmos Res 100(1):132–140

Theil H (1950) A rank-invariant method of linear and polynomial regression analysis, 3; confidence regions for the parameters of polynomial regression equations. Indag Math 1(2):467–482

Thom HC (1958) A note on the gamma distribution. Mon Weather Rev 86(4):117–122

Thomas T, Nayak PC, Ghosh NC (2015) Spatiotemporal analysis of drought characteristics in the bundelkhand region of central India using the standardized precipitation index. J Hydrol Eng 20(11):05015004

Turkes M, Tatli H (2009) Use of the standardized precipitation index (SPI) and a modified SPI for shaping the drought probabilities over turkey. Int J Climatol 29:2270–2282

Yue S, Hashino M (2003) Long term trends of annual and monthly precipitation In Japan. JAWRA J Am Water Resour Assoc 39(3):587–596

Landslide Susceptibility Assessment and Management Using Advanced Hybrid Machine Learning Algorithms in Darjeeling Himalaya, India

Anik Saha and Sunil Saha

Abstract

The present study evaluated the landslide (LS) susceptibility using RBF net, Naïve-Bayes Tree, Random subspace, and Rotational forest advanced hybrid machine learning (HML) algorithm in landslide hazard-prone area of Kurseong, Darjeeling Himalaya, India. The locations of landslides were detected by field surveys. 352 LS coordinates have been derived, displayed as an LS inventory map to calibrate LS susceptibility models, and used to authenticate the models. 16 LCFs (landslide conditioning factors) were utilized to prepare LS susceptibility maps. The developed landslide models were validated using two statistical methods, i.e., the mean absolute error (MSE) and the root mean square error (RMSE) as well as the receiver operating characteristics (ROC), efficiency, and accuracy. The results of the accuracy measures (area under curve for RBF net = 85.76%, NB tree = 86.54%, Random sub = 87.29%; and Rotational for = 84.81%) revealed that all models have good potentiality to forecast the landslide susceptibility in the Kurseong region of Darjeeling Himalaya. The Random subset model achieved higher accuracy (ROC = 87.29%; MSE = 0.145; and RMSE = 0.118) than other used models. The research revealed that in the fellow land, plantation areas, sides of highways, and structural hills where elevation is above 1150 m and slope ranges from 26° to 69° the susceptibility to landslide is very high. The prepared landslide susceptibility maps can be helpful in introducing location-specific proper management strategies for reducing landslide hazards in Kurseong region of Darjeeling Himalaya.

Keywords

Hybrid machine learning models · RBF net · Naïve-Bayes Tree · Random subspace · Rotational forest · Kurseong

35.1 Introduction

Landslide (LS) is a downslope movement of landmass, triggered mainly by slope instability, heavy rainfall, and earthquake (Wooten et al. 2016; Saha et al. 2020a). It is the reason not only for the collapse of thousands of houses and some other infrastructure assets every year but also loss of lives and environmental resources (Alexander 2005; Grima et al. 2020). It inevitably deteriorates the natural system which causes major damage and overwhelms people's capacity to cope and prevent it. This type of natural calamity

A. Saha · S. Saha (✉)
Department of Geography, University of Gour Banga, Malda 732103, India
e-mail: sunilgeo.88@gmail.com

mostly occurs in rocky terrain. India is a country where 0.42 million sq. km of its land area excluding snow-covered areas is highly prone to land sliding. It includes Himalayan region, Western and Eastern Ghats (gsi.gov.in/webcenter/portal). These landslides prone areas in Himalayan terrain mainly fall under the high to very high earthquake-prone zone (GSI, 2014). The slope instability is initiated mostly by the human intervention specially quarrying, illegal hill-cutting, road and house constructions, etc. (Pandey et al. 2020a; b; Saha and Saha 2020a). According to Haque et al. (2016) 1370, death and more than 4.75 billion euros were ruined for LS in 27 European countries from 1995 to 2014. Another research paper highlighted that 163,658 deaths and 11,689 personal injuries since 1995 to 2014 were caused by the landslides in the world (Haque et al. 2019).

Risk estimation relies on the chances of occurrence of landslide (LS) in a region, so a particular measure to predict the future probability of the LS needs to apply. LS risk management includes a combination of concentrated efforts, but LS susceptibility modeling is the most effective way of reducing its consequences by identifying the area susceptibility to LS (Arabameri et al. 2020). A number of parameters, including topographic, hydrological, climate conditioning, types of landslides, failure mechanisms, coverage of affected areas, frequency, and intensity are normally arranged for LS susceptibility mapping (Yalcin et al. 2011). Depending on the local environment and affecting variables, the landslide susceptibility maps present the distribution of the spatial probabilities of LS occurrence in a given study region (Huang et al. 2020). A sound LS susceptibility can offer as important tool for managing the landslide effects in the landslide-prone area (Pham et al. 2017a). In this context, several studies focused on landslide susceptibility mapping in the area facing high rate of landslide frequency (Arabameri et al. 2020; Saha et al. 2021). Physical models using detailed geotechnical and geological data to create susceptibility maps were early quantitative landslide susceptibility models. Statistical models were developed to examine associations between causative variables and the frequency of landslides due to the complexities of physical model implementation, such as data criteria, costs, and infeasibility for large regions (Basu et al. 2020; Saha et al. 2020a, b).

Different types of models, such as analytic hierarchy process (Saha et al. 2020a), frequency ratios (Mandal and Maiti 2015), the weight of evidence (Saha et al. 2020a), fuzzy logic (Roy and Saha 2019), logistic regression (Mandal and Mandal 2018), and ensemble models (Saha et al. 2021) have been implemented to map LS susceptibility. Several standard models have become obsolete, and more efficient models have been built based on ensemble techniques and algorithms of machine learning techniques (Saha et al. 2020b). It has been found that the ensemble of several statistical models could produce better results than a single model (Saha et al. 2020a). In LS susceptibility modeling, machine learning algorithms are presently being applied for the better prediction performance and especially for the effectiveness of these models to cope with different predictor variables (Merghadi et al. 2020). The evaluation of various machine learning strategies is important because of the complexity of landslide conditioning variables and their heterogeneity across research areas to obtain the best results for a given set of environmental characteristics (Dou et al. 2020; Fang et al. 2020). Recent across-validation approach-based landslide susceptibility mapping was performed in Garhwal Himalaya, Uttarakhand showed that novel ensemble conditional probability and boosted regression tree performed better than the individual conditional probability and boosted regression tree methods (Saha et al. 2021). In some experiments, composite configurations of several models in hybrid or ensemble frameworks have proved to be more effective (Nguyen et al. 2017; Abedini et al. 2019). It is now important to assess the variables that impact landslide occurrence in quantifying the danger and predict the LS susceptibility to minimize the losses with practical mitigation techniques.

The Kurseong Himalaya is characterized by complex geomorphologic, geological and seismotectonic setup. The hill ranges of Kurseong area are highly rugged, structurally controlled, and are constantly under the highly dynamic and active denudation processes (Saha and Saha 2020b). Geologically, the mountainous regions of the Darjeeling and Sikkim Himalayas are the part of active Himalayan Fold-Thrust Belt (FTB). All these complexities trigger slope instability. This region is one of the most LSs occurring regions of Darjeeling Himalayas in terms of frequency (GSI report 2016). The main purpose of this research is to analyze the LS susceptibility for this region, applying some rare advanced hybrid machine learning (HML) approaches, that is RBF network, Naïve-Bayes Tree, Random subspace, and Rotational forest. Such hybrid methods are exceptional and have not been used for LS analysis in the Indian Himalayan region. The changing climatic pattern along with various geological, geomorphologic, meteorological, tectonic conditions, and unplanned urbanization due to population growth, is aggravating the possibility of landslide occurrence. Hence the studies on landslides have been emphasized so, that the human-induced triggering factors may be controlled and on the other hand, the magnitude of damage can be reduced by identifying the landslide susceptibility areas in advance.

35.2 The Study Area

The study area Kurseong block is located between longitudes 26°46′30″ N to 26°57′47″ N and latitudes 88°08′25″ E to 88°27′54″ E (Fig. 35.1) having an area of 380.34 km^2. According to the Kurseong municipality weather record from 1980 to 2016 the mean annual rainfall is about 325 cm. and mean annual temperature is 23.69 °C in the study area. Geologically, the study area is occupied by igneous, metamorphic, and sedimentary rocks belonging to the Gneiss (32.46%), States & schists (19.52%), Damuda (9.24%), Limestone (19.02%), and alluvial complex (19.76%). The study area is frequently affected by the landslides. Most of the landslides occurred in

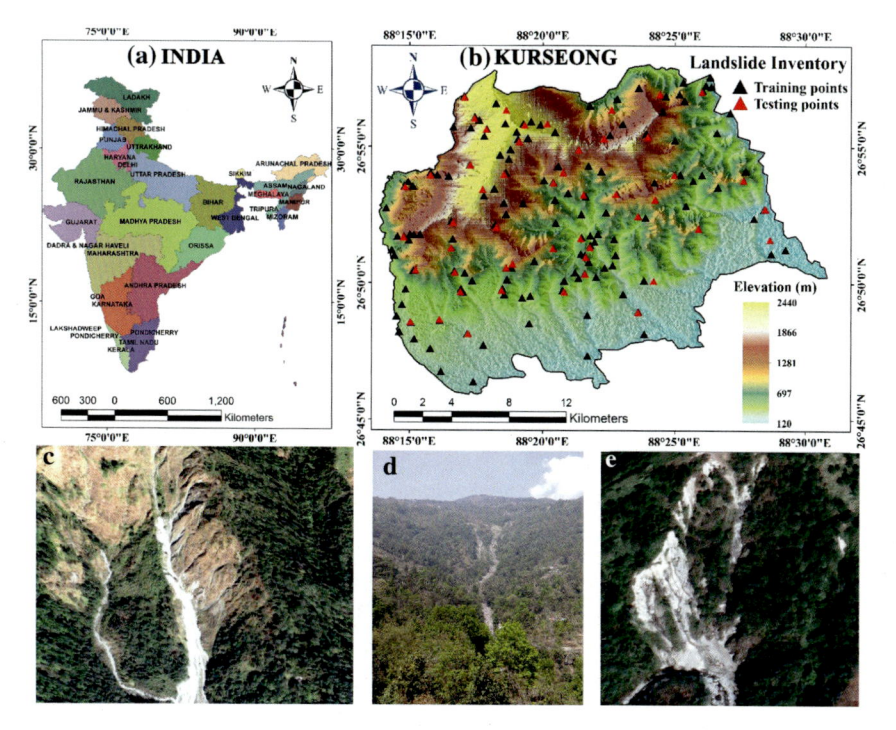

Fig. 35.1 Location of study area map with landslide inventory map **a** India, **b** Kurseong, some landslide satellite imagery **c** 26°53′09″ N; 88°27′07″ E **d** 26°53′09″ N; 88°19′28″ E; **e** 26°52′23″ N; 88°21′ E of study area

Monsoon season in the study area. Besides, the heavy rainfall, steep slope changes in the land-use pattern for cultivation of tea on terraces, and other developmental activities increased the landslide occurrences in the area. Narrow valleys and steep hillslopes are some of the main factors causing landslides, besides heavy rains and anthropogenic activity (Saha and Saha 2020a). It is essential to prepare the landslide susceptibility map that can be used as tool for managing and formulating the strategies to reduce the effect the landslides.

35.3 Materials and Methods

35.3.1 Data Acquisition and Their Source

In the present study, required datasets were gathered from various government organizations, websites, and some filed surveys. Topographical map (sheet no. 78 B/5) was derived from Survey of India (SOI) and DEM was downloaded from the website of USGS (United States Geological Survey) and acquired in September 2014 with 30 m spatial resolution. Land Use/Land Cover (LULC) map was prepared from the satellite imagery, i.e., Landsat 8 OLI/TIRS (Operational Land Imager/Thermal Infrared Sensor) acquired on 4th February 2019 (Path/row—139/41). Geology, Geomorphology, and soil types were obtained from GSI (Geological Survey of India) and NBSSLUP, India (National Bureau of Soil Survey and Land-Use Planning) and resembled to match the spatial resolution of DEM (Fig. 35.2).

35.3.2 Inventory Map of Landslides (IML)

Inventory map of Landside (IML) is essential to provide baseline information about the pattern of distribution, characteristics of LSs, and specific LS processes involved in relation to the

Table 35.1 Multicollinearity analyses of landslide conditioning factors

Landslide conditional factors	Multicollinearity	
	Tolerance	VIP
Rainfall	0.685	1.647
Drainage density	0.898	1.114
TWI	0.924	1.083
SPI	0.659	1.653
Slope	0.919	1.048
Elevation	0.861	1.162
Aspect	0.835	1.195
Curvature	0.980	1.021
Soil types	0.764	1.124
Soil depth	0.887	1.184
Geology	0.869	1.309
Geomorphology	0.798	1.254
Relative relief	0.889	1.465
Land use/land cover (LU/LC)	0.863	1.158
NDVI	0.916	1.093
Distance from roads	0.977	1.284

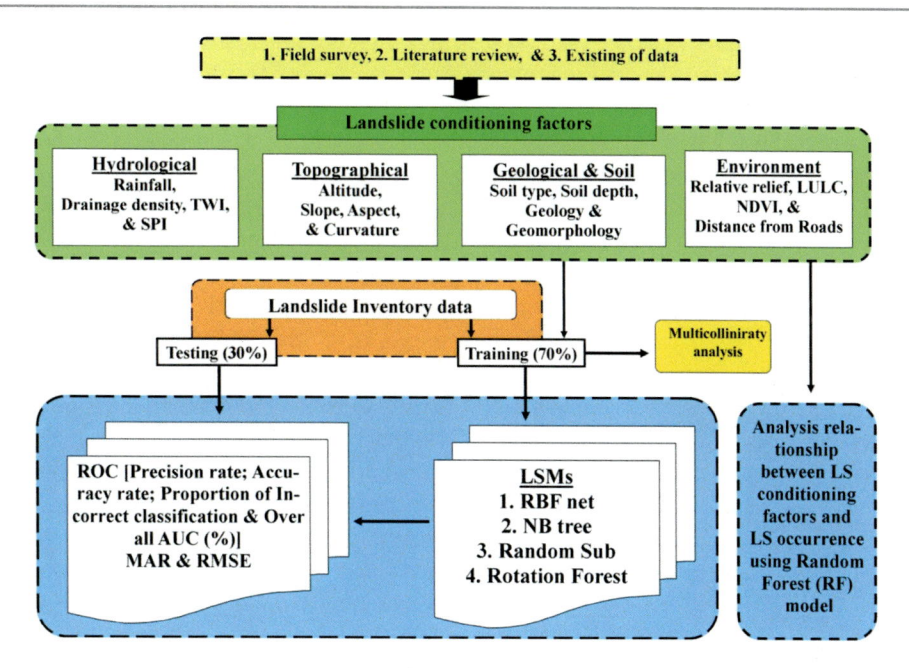

Fig. 35.2 Flowchart of this work

geophysical components present in the affected areas. In our earlier study (Saha and Saha 2020a), we were mapped a total of 273 LSs in the study area, among those, some were mapped using filed study, and others were demarcated using Google Earth images. In the current study, to specific the IML, an extensive field survey was carried out for the second time using GPS and a total of 352LSs were identified in the LS inventory map in polygon (Fig. 35.1b). Finally, all the LSs were overlapped on a high-resolution google earth satellite image to verify the LS locations. In most of the research literature related to the LS susceptibility mapping, the IML was classified into the 70%:30% ratio (Kalantar et al. 2018; Roy et al. 2019). Regarding these researches, the current study followed the same process, and LIM were classified into two parts such as training and validation datasets. The 70% of previous landslides were applied as training data to simulate these models even as 30% were used for models validation as validate data (Fig. 35.1b).

35.3.3 Selecting the LS Condition Factors (LSCFs)

For evaluating the LS susceptibility hazard in any region selection of effective factors is essential. The selection of effective factors is very complicated task because there are no standard criteria. In this study firstly, we have preferred the factors for modeling the LS susceptibility assessment considering the previous literature. Afterward, to select the effective landslide conditioning factors (LCFs) multi-collinearity test and Random forest method were used. Finally, 16 geo-environmental LS susceptibility conditioning factors, i.e., rainfall, drainage density, stream power index (SPI), topographical wetness index (TWI), slope, elevation, aspect, curvature, geology, geomorphology, soil texture, soil depth, relative relief (RR) land use/land cover (LULC), normalized differential vegetation index (NDVI), and distance from roads were selected for modeling the LS susceptibility assessment in the present study.

35.3.3.1 Hydrological Factors

Torrential rainfall over a span of several hours contributes greatly to huge landslides (Kirschbaum and Stanley 2018; Dikshit et al. 2020). To create a rainfall map, we used a dataset from http://worldclim.org/version2. There are five categories of the rainfall map (Fig. 35.3a). Approximately 70% of the LS pixels are in the first three groups of rainfall. The drainage density layer has five categories and, as calculated, the percentage of LS pixels is highest in the maximum drainage density level. Topographic Wetness Index (TWI) is a measure of wetness condition of an area that is dependent on the ratio of the area to the angle of slope. It gives an indicator of soil moisture and can be positively associated with the frequency of LSs (Wang et al. 2020). The Stream Power Index (SPI) is a parameter related to the discharge of a particular area within a stream and measures the amount of the erosive power of the flowing water. The higher the value of the SPI, the more likely the proficiency of the LS is to be (Ozdemir 2009).

35.3.3.2 Topographical Factors

In particular, steeper slopes seem to be more vulnerable to LSs than gentle slopes (Dai and Lee 2002). We have divided the slope angle into five groups. Over 94% of the LS pixels are in the three classes (18.20–69.28°). There is an opposite relationship between altitude & landslides: as altitude rises, the number of landslides and streamflow increases (Sheng and Shen 2017). We created five altitude ranges, with an altitude over 750 m, contributing to 82.25% of the landslides. Aspect influences moisture levels, evaporation and transpiration, absorption, daylight, and local atmosphere. Although aspect

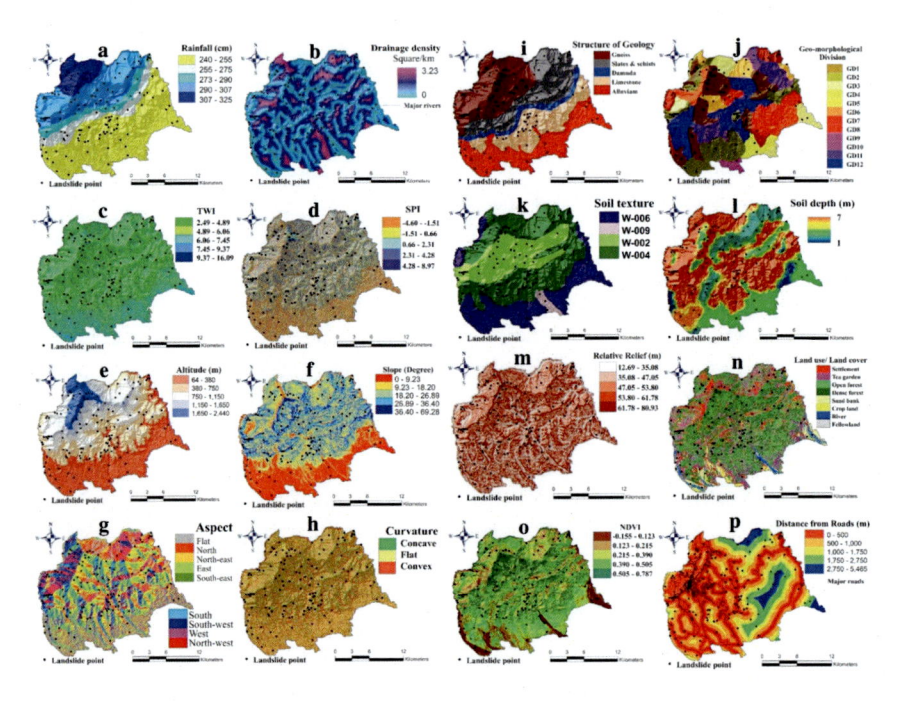

Fig. 35.3 Landslide conditioning factors: **a** rainfall, **b** drainage density, **c** TWI, **d** SPI, **e** slope, **f** elevation, **g** aspect, **h** curvature, **i** geology, **j** geomorphology (GD1 = deeply entrance denudation valley; GD2 = moderately dissected valley; GD3 = fault related faceted slope; GD4 = alluvial deposition fan; GD5 = dissected screenland slope; GD6 = lowly intermountain valley; GD7 = old alluvial deposition terrace; GD8 = alluvial debris intermountain fan; GD9 = deposition alluvial flood plain; GD10 = fluvial denudation mountain valley; GD11 = intermountain moderate slope structural plateau; and GD12 = alluvial debris deposition new plain), **k** soil texture, **l** soil depth, **m** relative relief, **n** land use/land cover, **o** NDVI, **p** distance from road

influences LS merely indirectly, the majority of researchers mentioned LS as being one of the predictor variables for LS susceptibility mapping (Achour et al. 2017; Sun et al. 2020). Aspect has been mapped and positioned in nine divisions corresponding to the cardinal directions. LSs pixels are spread reasonably uniformly among all nine groups. Curvature is a slope form that corresponds to the diversity of the surface runoff (Constantin et al. 2011). A significant curvature value is a convex slope, and a negative value is a concave slope, whereas a relatively planar surface is indicated by ranging between 0 and 1 (SahaSaha et al. 2020a). We classify the curvature of the slope into three groups.

35.3.3.3 Geomorphologic and Soil-Based Factors

The present geological map (1:50,000) was reproduced to the resolution of the 30 m. Geological map of the study region was obtained from the Survey of India. The geological divisions identified in this study region are Gneiss, Slate & Schist, Damuda, Limestone, and Alluvial (Fig. 35.3i). The research region's geomorphologic map (Fig. 35.3j) was extracted from Bhuvan (https://bhuvan.nrsc.gov.in) produced by ISRO (Indian Space Research Organization) and Ministry of Mines together with 15 collaborator institutions. The area of research is fragmented into separate medium and narrower spurs; ridges, broad valleys "V" shaped.

The soil properties seem to be a very influential element to measure the vulnerability of LSs (Gökceoglu and Aksoy 1996; Kitutu et al. 2009). In this field of study, four soil textures, i.e., WOO2 (Loamy skeletal, Typical Udorthents, Loamy skeletal, Typical Dystrochrepts), WOO4 (Loamy-skeletal, Typical Udorthents, Loamy skeletal, Typical Haplumb brepts) and WOO6 (Course loamy, Umbric Dystrochrets) were identified, depending on the taxonomy. Two factors related to soil namely, soil textures (Fig. 35.3k) and soil depth (Fig. 35.3l) were prepared. Soil category datasets were obtained from the NBSS & LUP. To produce the soil depth map, a field study was done.

35.3.3.4 Environment Based Factors

This study revealed that the frequency of landslides is highly affected by elevation. The land-use map provided by maximum likelihood method using Landsat eight image has eight categories. The classes with larger areas are dense forest areas (26.84%), open forests (23.58%), settlements (7.45%), and tea gardens (8.10%). Most LSs pixels are in settlement areas (18.10%), croplands (19.6%), near tea gardens (14.4%), and open forests (18.7%). In the context of the NDVI, the vulnerability increases rapidly with the decrease in the NDVI value. The roads, which are built on a very steep slope, undermine the safety of the ground above them. Increasing the slope of the area raises the risk of topographic loss. The distance from the roads in this study is divided into five categories.

35.3.4 Multi-collinearity Analysis

Collinearity specifies a linear function between one independent parameter and other independent parameters pointing toward a strong correlation among the independent parameters which sinks the accurateness of the operational model (Arabameri et al. 2020). VIF (Variance Inflation Factor) and tolerance are considered as suitable and efficient indicators for diagnosing multi-collinearity between the parameters (Pandey et al. 2020a, b). A VIF of ≥ 5 and tolerance of ≤ 0.20 denotes the multicollinearity problem (Saha and Saha 2020a). This analysis was prepared by importing the data into SPSS.

35.3.5 Hybrid Machine Learning Algorithm

35.3.5.1 Radial Basis Function Neural Network (RBF Net)

The RBF net (radial basis function neural network) is a kind of (non-linear) neural network. The RBF Neural Network is consist of three-layers layer, i.e., neural network comprising input layers, hidden layer, and an output layer.

Each unit in the hidden layer receives the elements of the input vector from the input layer. Following that, each unit in the hidden layer generates activation depending on the RBF. After that, the output layer computes a linear combination of the hidden units' activations (Yavari et al. 2019).

35.3.5.2 Naïve-Bayes Tree (NB Tree)

Kohavi (1996) suggested the Naïve Bayes Tree (NBTree) model, which incorporates two classifiers: the ID3 decision tree, which is responsible for the classification process and tree splitting, and Nave Bayes. It outperforms other machine learning models in terms of its ability to (1) reflect information, (2) control uncertainty, (3) pick candidate concepts, (4) process small datasets, and (5) eliminate noise in training datasets (Luger 2005). Also, a limited volume of data can be used in the simulation and classification methods (Pham et al. 2017a, b).

35.3.5.3 Random Subspace (Ran Sub)

A key hybrid ensemble and parallel learning algorithm is the random subspace classifier. Random subspace was implemented by Ho (1998). The optimization of the subset was used to integrate several classifier decisions for this algorithm. These subsets of function space are selected at random from the training classifiers. In comparison, the random subspace ensemble solution is distinct from the others by an ensemble algorithm since it uses several sample numbers (Pham et al. 2020). In the first step, q dimensional training of subsets L was used to classify the original function space. For both of these subsets, the RBFnet was used as the basis classifier in this research. Finally, the weighted majority vote was derived using the base classifier's integration.

35.3.5.4 Rotational Forest (RoTF)

Rotation Forest is a relatively recent technique focused on bootstrap sampling that produces a classifier ensemble using principal component analysis (PCA) as a feature extraction technique (Rodriguez et al. 2006). The training dataset F is randomly divided into K subsets (K is the algorithm's parameter), and a rotation sparse matrix is built by conducting extraction of features on each subset with bootstrapped samples corresponding to 75% of the original training samples. The final result is obtained by integrating the contribution of the multiple classifiers, which are based on the features repeatedly predicted by the matrix multiple times. It works for almost every base classifier, and the feature extraction for each classifier keeps all of the features that promote heterogeneity.

35.3.6 Validation Methods

Validation is essential when assessing the reliability of the model (Guillard and Zezere 2012; Saha et al. 2020b). In this analysis, three threshold-dependent statistics [receiver operating characteristics (ROC) area under curve (AUC), wrongly categorized percentage, accuracy, and precision] and two statistical approaches [i.e., mean absolute error (MSE) and root-mean-square error (RMSE)] were used to determine the reliability of the models.

35.3.6.1 Threshold Dependent Method ROC

The methods used to determine vulnerability areas are not accurate unless they are tested (Saha and Saha 2020b). Here, more than one validation process was used to evaluate predictability, and it is more appropriate for everyone to support models. The literature has assured that the ROC (AUC-ROC) curve area (AUC) is a useful method for model validation and comparison (Pham et al. 2017a, b; Aditian et al. 2018). The AUC was calculated using Eq. 35.1.

$$\text{AUC} = \frac{(\sum \text{TP} + \sum \text{TN})}{(P + N)} \quad (35.1)$$

where TP is a valid positive, TN is a valid negative, FP is a false positive, and FN a false negative. P is the amount of LSs overall, and N represents the number of non-LSs overall.

$$\text{PPV or ``Precision''} = \frac{\text{TP}}{(\text{TP} + \text{FP})} \quad (35.2)$$

$$\text{Accuracy} = \frac{(\text{TP} + \text{TN})}{(\text{TP} + \text{FP} + \text{FN} + \text{TN})} \quad (35.3)$$

The AUC values between 0 and 1 and the value close to 1.0 imply a model's higher efficiency (Tien Bui et al. 2012; Saha et al. 2020b). Gorsevski et al. (2006) classified the AUC scores into groups to calculate the performance of the prediction: bad (50–60%), fair (60–70%t), decent (70–80%), really nice (80–90%), and excellent (90–100%). The reliability of forecasting was also tested using inaccurate classifications of time, consistency, and proportion. Higher values of both accuracy and precision suggest that the model has greater predictability. In case of inaccurate classifications, lower proportions of values suggest greater precision.

35.3.6.2 Statistical Techniques MSE and RMSE

In this analysis, MSE and RMSE were used to test the simulations. For a database, MSE is defined as specified as the amount of the discrepancies between predicted values and the actual values. RMSE is represented by the square root of MSE. The calculation of the MSE and RMSE was carried out using Eqs. 35.4 and 35.5:

$$\text{MSE} = \frac{i}{n} \sum_{i=1}^{n} |x_{\text{predicted}} - x_{\text{actual}}| \quad (35.4)$$

$$\text{RMSE} = \sqrt{\frac{i}{n} \sum_{i=1}^{n} |x_{\text{predicted}} - x_{\text{actual}}|} \quad (35.5)$$

Here, n is the sample size of the training or testing dataset; X_{actual} is considered to be the testing data set values. The $X_{\text{predicted}}$ is reflecting the expected values produced for the LSMs. Willmott et al. (2005) used this strategy and established a limit value of 0.5. Values above 0.5 suggest poor results, whereas values below 0.5 imply strong success in the experiment. Such two tests were used to determine the efficacy of models of vulnerability, also including landslide (Roy et al. 2019; Saha et al. 2020a), and gully erosion (Pourghasemi et al. 2020).

35.4 Results and Discussion

35.4.1 Multi-collinearity Analysis

The TOL and VIF indicate that there are no complications with multi-collinearity (MCoA) among LS conditioning variables (Bai et al. 2010; Lee et al. 2018). The outcomes (Table 35.1) revealed that the lowest TOL is 0.318 for rainfall. Other hands the highest TOL is 0.914 for soil type. In this study, the minimum and maximum VIF values are 0.982, and 2.869 (Table 35.1). From this review, it revealed that the chosen 17 LS conditioning factors in this study field are ideal for modeling LS susceptibility.

35.4.2 Landslide Susceptibility Maps

The four LS susceptibility maps were produced with the 70% of total landslides datasets using RBF net, NB tree, Random-sub, and Rotation-forest models in GIS environment (Fig. 35.4). The areal distributions of different models are given in Table 35.2. The simulations of 4 HML models reveal that the areas of very low LS susceptibility class in RBF net, NB tree, Random-sub and Rotation-forest models are 92.09 km^2 (24.08%), 88.15 km^2 (23.05%), 88.16 km^2 (23.08%) and 61.11 km^2 (16.24%) and moderate LS susceptibility class are 102.99 km^2 (26.93%); 82.57 km (21.59%); 80.45 (22.87%) and 95.88 km^2 (25.07%), respectively. Comparatively, very high LS susceptibility zone covers 63.60 km^2 (16.63%) in RBF net, 60.64 km^2 (15.91%) in NB tree; 60.64 (15.43%) in Random subset and 41.07 km^2 (10.74%) in Rotational forest, respectively (Fig. 35.4). According to the generated landslide susceptibility models, the middle and western portions of the study region are highly susceptible to landslides.

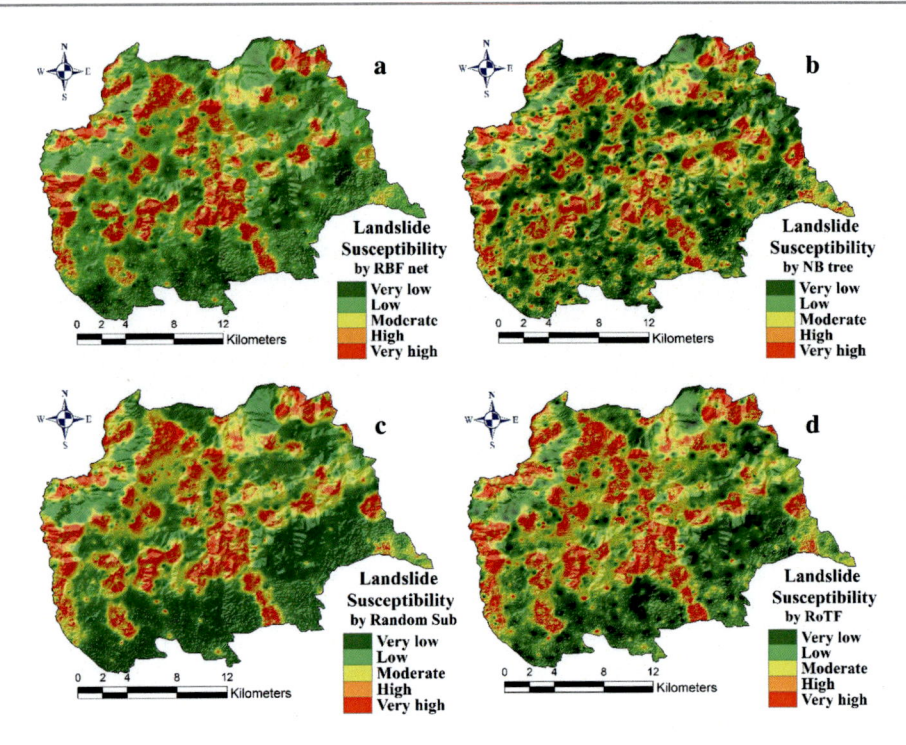

Fig. 35.4 Landslide susceptibility map using hybrid machine learning models **a** RBF net, **b** NB tree, **c** random subset, and **d** rotational forest

Table 35.2 Areal distribution of RBF net, NB tree, Random sub, and Rotation forest approach-based landslide susceptibility maps (LSMs)

Landslide susceptibility classes	RBF net		NB tree		Random subset		Rotational forest	
	area in km^2	% of area	area in km^2	% of area	area in km^2	% of area	area in km^2	% of area
Very low	92.09	24.08	88.15	23.05	88.15	23.08	62.11	16.24
Low	72.09	18.85	88.42	23.12	82.57	19.94	101.46	26.53
Moderate	102.99	26.93	82.57	21.59	80.45	22.87	95.88	25.07
High	51.67	13.51	62.45	16.33	72.42	18.68	81.91	21.42
Very high	63.6	16.63	60.84	15.91	60.64	15.43	41.07	10.74

35.4.3 Models Performance Validation

The ROC, accuracy, precision, and proportion are incorrectly classified. MSE and RMSE were used to evaluate the LS susceptible mapping of the Kurseong area. For this analysis, the results of the 4 HML-based models applied for landslide susceptibility were validated, integrating fieldwork datasets as indicated above. In addition, the AUC was determined to compare the effects of certain simulations (Fig. 35.5). The findings of the ROC curve showed that all the models often have enormous potential for predicting susceptibility. The AUC-ROC values of the RBF net, NB tree, Random subset, and ROTF models are 85.76%, 86.54%, 87.29%, and 84.81% for validation dataset (Table 35.3).

Table 35.3 Values of ROC, precision, accuracy, proportion incorrectly classified, MAE and RMSE methods

Hybrid machine learning models	Precision	Accuracy	Proportion incorrectly classified	Overall AUC (%)	MAE	RMSE
RBF net	0.853	0.843	0.144	85.76	0.208	0.121
NB tree	0.856	0.868	0.138	86.54	0.217	0.118
Random subset	0.893	0.883	0.123	87.29	0.145	0.095
Rotational forest	0.828	0.826	0.203	84.81	0.294	0.146

Therefore, the results indicated considering the testing dataset the HML Random-subset (AUC = 87.29%) model performed better than ROTF, RBF net, and NB tree in mapping landslide susceptibility in the Kurseong range (Fig. 35.5). The Random subset HML method achieved the highest accuracy and precision followed Rotational forest, RBF net, and NB tree, respectively (Table 35.3). Contrarily, the Random-subset HML model achieved the lowest proportion of incorrectly classified, MSE and RMSE values followed by the NB tree, RBF net, and ROTF, respectively. The Random-subset technique proved to be the best method for landslide susceptibility simulation with present confounding factors.

35.5 Discussion

35.5.1 Comparison of Models' Predictive Performance

Many approaches have been used to assess the vulnerabilities of the environmental hazards. Modeling mechanisms and methods, however, are varied and have produced precise outcomes and predictive performance. GIS-based geographical forecasts for different modeling methods are important tools for environmental and spatial-environmental studies that can help in the management and efficient development (Giordan et al. 2018). Accessibility to the range of

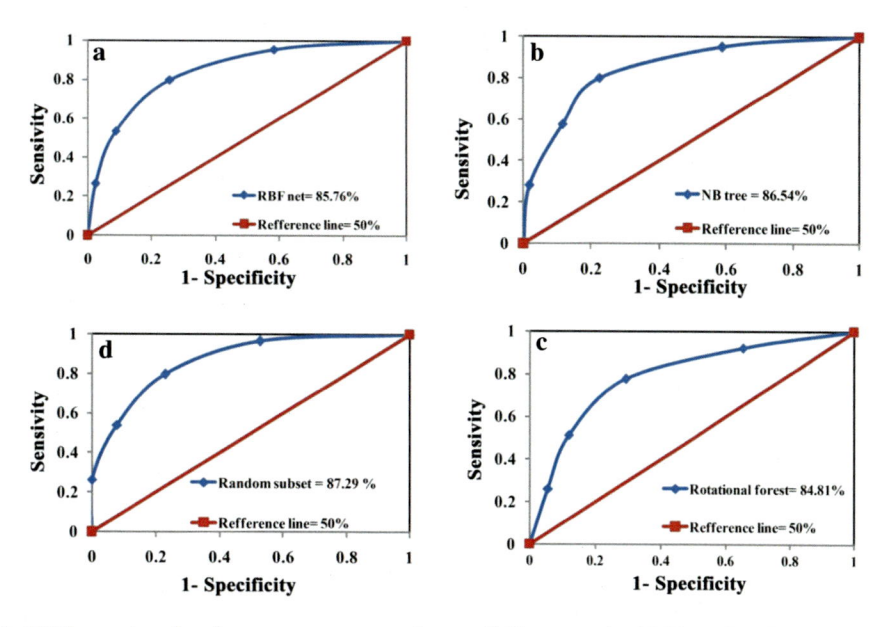

Fig. 35.5 ROC curve based performance assessment of susceptibility map using RBF net (85.76%), NB tree (86.54%), random subset (87.29%), and rotational forest (84.81%) model

solutions has considerably improved decision-makers' willingness to ensure continuity in environmental development (Giordan et al. 2018). Frattini et al. (2010) cautioned that the application of modeling methods would produce quite different outcomes and very different outputs in many other domains or with other applications. With this purpose, simulation results evaluations are critical for model efficiency and accuracy assessments (Pisello et al. 2012; Chen et al. 2018).

The predictive efficiency of the models for predicting LS susceptibility was compared in this analysis using AUC-ROC, precision, accuracy, proportion incorrectly classified, MSE and RMSE. The four hybrid models (RBF net, NB tree, Random sub, and Rotational for) have a varying performance. All models showed incredible performance and the Random sub (AUC = 87.29%; MSE = 0.145; and RMSE = 0.118) model had the highest prediction accuracy. Specific independent variables were used to forecast LS susceptibility involving (drainage density, SPI, TWI), topography (slope, elevation, aspect, and curvature), geomorphic (geology structure and geomorphology), soil (texture and depth), landform, and anthropogenic behavior (LULC, NDVI, and distance to road). The main indicator of the LSMs in this analysis is slope, rainfall, geomorphology, and elevation.

In addition, the AUC-ROC values >0.8 already indicate excellent results. The areal distribution of the four HML models is not really the same (Fig. 35.5). For this study, we found that Random subset output is outstanding while the Rotational forest model's consistency is just quite fine. The other methods of validation (precision, accuracy, MSE, and RMSE) had given same results. Across all efficiency measures, the Random subset had the best and Rotational forest alone the lowest precision. Previous research (Nguyen et al. 2019; Pham et al. 2018) deduced that the LS forecasting of HML models would enhance in comparison with traditional methods. After all, these approaches would be different based on the location and the variables utilized.

35.5.2 Contribution of LSCFs Analyzed by RF

For the management of environmental resources, the LS conditioning factors' significance assessment is important (Lai et al. 2018). Arabameri et al. (2017) proposed that a quantitative assessment of the associations between LCFs and LS would enable planners and stakeholders to forecast the impacts of LCFs. The conditioning variables which work well in one simulation can, therefore, be negligible in other simulations. For this current research, the substantial contribution of LCFs in a wide range of earlier research was evaluated using the RF model (Chen et al. 2018; Saha et al. 2020b). The RF model's results showed that the maximum weight was acquired by the slope and the minimum weight was obtained by the soil depth (Table 35.4).

Table 35.4 Values of mean decrease accuracy of Gini using Random Forest model

Selected factors	Mean decrease Gini	Selected factors	Mean decrease Gini
Rainfall	112.325	Geomorphology	131.251
Drainage density	114.203	Geology	100.284
TWI	72.263	Soil depth	29.480
SPI	59.345	Soil type	42.622
Slope	174.309	Relative relief	42.385
Altitude	131.094	LULC	77.240
Aspect	84.351	Road density	103.968
Curvature	62.055	NDVI	89.250

35.6 Conclusion

Reliable LS susceptibility maps offer a strategies for handling LS threats to land-use planners and government officials. This paper introduced some rare hybrid machine learning (HML) algorithms like RBF net, NB Tree, Random sub, and Rotational and used it to generate an effective LS susceptibility map. In our study region, Kurseong, the most significant LS modeling factors among these 16 LSCFs are slope, elevation, rainfall, geomorphic division, and road distance. Random Forest Model was effectively used to identify suitable conditioning factors and thus systematically face the challenge of choosing an acceptable testing dataset for simulation. The incorporated HML model faced the challenges of data heterogeneity through parameter tuning and the description of suitable network architectures. Used model indicates that high mountainous slopes with heavy rainfall in the sample region are very susceptible to LS. The conceptual framework can be expanded and improved by integrating hydraulic and topographic data and landslide risk mitigation models, especially in Darjeeling Himalaya, where there are no remote sensing systems for constructing a real-time landslide warning system. The rare novel HML model used and evaluated in this study can be further used in hazard susceptibility mapping, LS protection and prevention, and emergency recovery in LS-prone regions.

Acknowledgements Authors would like to thanks the inhabitants of study area because they have helped a lot during our field visit. At last, authors would like to acknowledge all of the agencies and individuals specially, Survey of India (SOI), Geological Survey of India (GSI) and USGS for providing the maps and data required for the study.

Conflict of Interest There is no conflict of interest.

Funding No funding was received for this work.

Credit Author Statement
Anik Saha: Methodology, format analysis, investigation, original draft preparation, software;
Sunil Saha: Methodology, format analysis, original draft preparation, review and editing.

References

Abedini M, Ghasemian B, Shirzadi A, Shahabi H, Chapi K, Pham BT, Tien Bui D (2019) A novel hybrid approach of Bayesian logistic regression and its ensembles for landslide susceptibility assessment. Geocarto Int 34(13):1427–1457

Achour Y, Boumezbeur A, Hadji R, Chouabbi A, Cavaleiro V, Bendaoud EA (2017) Landslide susceptibility mapping using analytic hierarchy process and information value methods along a highway road section in Constantine, Algeria. Arab J Geosci 10(8): 194

Aditian A, Kubota T, Shinohara Y (2018) Comparison of GIS-based landslide susceptibility models using frequency ratio, logistic regression, and artificial neural network in a tertiary region of Ambon, Indonesia. Geomorphology 318:101–111

Alexander D (2005) Vulnerability to landslides. Landslide hazard and risk, pp 175–198

Arabameri A, Pourghasemi HR, Yamani M (2017) Applying different scenarios for landslide spatial modeling using computational intelligence methods. Environ Earth Sci 76(24):1–20

Arabameri A, Saha S, Roy J, Chen W, Blaschke T, Tien Bui D (2020) Landslide susceptibility evaluation and management using different machine learning methods in the Gallicash River Watershed, Iran. Remote Sens 12(3):475

Bai SB, Wang J, Lü GN, Zhou PG, Hou SS, Xu SN (2010) GIS-based logistic regression for landslide susceptibility mapping of the Zhongxian segment in the Three Gorges area, China. Geomorphology 115(1–2): 23–31

Basu T, Das A, Pal S (2020) Application of geographically weighted principal component analysis and fuzzy approach for unsupervised landslide susceptibility mapping on Gish River Basin, India. Geocarto Int 1–24

Chen W, Zhang S, Li R, Shahabi H (2018) Performance evaluation of the GIS-based data mining techniques of best-first decision tree, random forest, and naïve Bayes tree for landslide susceptibility modeling. Sci Total Environ 644:1006–1018

Constantin M, Bednarik M, Jurchescu MC, Vlaicu M (2011) Landslide susceptibility assessment using the bivariate statistical analysis and the index of entropy in the Sibiciu Basin (Romania). Environ Earth Sci 63(2): 397–406

Dai FC, Lee CF (2002) Landslide characteristics and slope instability modeling using GIS, Lantau Island, Hong Kong. Geomorphology 42(3–4):213–228

Dikshit A, Sarkar R, Pradhan B, Segoni S, Alamri AM (2020) Rainfall induced landslide studies in Indian Himalayan region: a critical review. Appl Sci 10(7): 2466

Dou J, Yunus AP, Merghadi A, Shirzadi A, Nguyen H, Hussain Y, Yamagishi H (2020) Different sampling strategies for predicting landslide susceptibilities are deemed less consequential with deep learning. Sci Total Environ 720:137320

Fang Z, Wang Y, Peng L, Hong H (2020) A comparative study of heterogeneous ensemble-learning techniques for landslide susceptibility mapping. Int J Geogr Inf Sci 1–27

Frattini P, Crosta G, Carrara A (2010) Techniques for evaluating the performance of landslide susceptibility models. Eng Geol 111(1–4):62–72

Giordan D, Cignetti M, Wrzesniak A, Allasia P, Bertolo D (2018) Operative monographies: development of a new tool for the effective management of landslide risks. Geosciences 8(12):485

Gökceoglu C, Aksoy HÜSEYİN (1996) Landslide susceptibility mapping of the slopes in the residual soils of the Mengen region (Turkey) by deterministic stability analyses and image processing techniques. Eng Geol 44(1–4):147–161

Gorsevski PV, Gessler PE, Foltz RB, Elliot WJ (2006) Spatial prediction of landslide hazard using logistic regression and ROC analysis. Trans GIS 10(3):395–415

Grima N, Edwards D, Edwards F, Petley D, Fisher B (2020) Landslides in the Andes: forests can provide cost-effective landslide regulation services. Sci Total Environ 745:141128

Guillard C, Zezere J (2012) Landslide susceptibility assessment and validation in the framework of municipal planning in Portugal: the case of Loures Municipality. Environ Manag 50(4):721–735

Haque U, Blum P, Da Silva PF, Andersen P, Pilz J, Chalov SR, Malet JP, Auflič MJ, Andres N, Poyiadji E, Lamas PC (2016) Fatal landslides in Europe. Landslides 13(6):1545–1554

Haque U, Da Silva PF, Devoli G, Pilz J, Zhao B, Khaloua A, Wilopo W, Andersen P, Lu P, Lee J, Yamamoto T (2019) The human cost of global warming: deadly landslides and their triggers (1995–2014). Sci Total Environ 682:673–684

Huang F, Cao Z, Guo J, Jiang SH, Li S, Guo Z (2020) Comparisons of heuristic, general statistical and machine learning models for landslide susceptibility prediction and mapping. CATENA 191:104580

Kalantar B, Pradhan B, Naghibi SA, Motevalli A, Mansor S (2018) Assessment of the effects of training data selection on the landslide susceptibility mapping: a comparison between support vector machine (SVM), logistic regression (LR) and artificial neural networks (ANN). Geomat Nat Hazard Risk 9(1):49–69

Kirschbaum D, Stanley T (2018) Satellite-based assessment of rainfall-triggered landslide hazard for situational awareness. Earth's Future 6(3):505–523

Kitutu MG, Muwanga A, Poesen J, Deckers JA (2009) Influence of soil properties on landslide occurrences in Bududa district, Eastern Uganda. Afr J Agric Res 4(7):611–620

Kohavi R (1996) Scaling up the accuracy of naive-Bayes classifiers: a decision-tree hybrid. In: Kdd, vol 96, pp 202–207

Lai C, Chen X, Wang Z, Xu CY, Yang B (2018) Rainfall-induced landslide susceptibility assessment using random forest weight at basin scale. Hydrol Res 49(5):1363–1378

Lee JH, Sameen MI, Pradhan B, Park HJ (2018) Modeling landslide susceptibility in data-scarce environments using optimized data mining and statistical methods. Geomorphology 303:284–298

Luger GF (2005) Artificial intelligence: structures and strategies for complex problem solving. Pearson Education

Mandal S, Maiti R (2015) Application of analytical hierarchy process (AHP) and frequency ratio (FR) model in assessing landslide susceptibility and risk. In: Semi-quantitative approaches for landslide assessment and prediction. Springer, Singapore, pp 191–226

Mandal S, Mandal K (2018) Modeling and mapping landslide susceptibility zones using GIS based multivariate binary logistic regression (LR) model in the Rorachu river basin of eastern Sikkim Himalaya, India. Model Earth Syst Environ 4(1):69–88

Merghadi A, Yunus AP, Dou J, Whiteley J, ThaiPham B, Bui DT, Avtar R, Abderrahmane B (2020) Machine learning methods for landslide susceptibility studies: a comparative overview of algorithm performance. Earth-Sci Rev 103225

Nguyen QK, Tien Bui D, Hoang ND, Trinh PT, Nguyen VH, Yilmaz I (2017) A novel hybrid approach based on instance based learning classifier and rotation forest ensemble for spatial prediction of rainfall-induced shallow landslides using GIS. Sustainability 9(5):813

Nguyen VV, Pham BT, Vu BT, Prakash I, Jha S, Shahabi H, Shirzadi A, Ba DN, Kumar R, Chatterjee JM, Tien Bui D (2019) Hybrid machine learning approaches for landslide susceptibility modeling. Forests10(2):157

Ozdemir A (2009) Landslide susceptibility mapping of vicinity of Yaka Landslide (Gelendost, Turkey) using conditional probability approach in GIS. Environ Geol 57(7):1675–1686

Pandey BW, Anand S, Negi VS, Pathak U, Prasad AS (2020a) Ecological challenges and vulnerability assessment for exploring the adaptation-development nexus for sustainability in Alaknanda River Basin, Uttarakhand, India. In: Geoecology of landscape dynamics. Springer, Singapore, pp 359–377

Pandey VK, Pourghasemi HR, Sharma MC (2020b) Landslide susceptibility mapping using maximum entropy and support vector machine models along the Highway Corridor, Garhwal Himalaya. Geocarto Int 35(2):168–187

Pham BT, Bui DT, Pourghasemi HR, Indra P, Dholakia MB (2017a) Landslide susceptibility assessment in the Uttarakhand area (India) using GIS: a comparison study of prediction capability of naïve Bayes, multi-layer perceptron neural networks, and functional trees methods. Theor Appl Climatol 128(1–2):255–327

Pham BT, Bui DT, Prakash I, Nguyen LH, Dholakia MB (2017b) A comparative study of sequential minimal optimization-based support vector machines, vote feature intervals, and logistic regression in landslide susceptibility assessment using GIS. Environ Earth Sci 76(10):371

Pham BT, Prakash I, Bui DT (2018) Spatial prediction of landslides using a hybrid machine learning approach based on random subspace and classification and regression trees. Geomorphology 303:256–270

Pham BT, Phong TV, Nguyen-Thoi T, Parial K, Singh SK, Ly HB, Nguyen KT, Ho LS, Le HV, Prakash I (2020) Ensemble modeling of landslide susceptibility using random subspace learner and different decision tree classifiers. Geocarto Int 1–23. https://doi.org/10.1080/10106049.2020.1737972

Pisello AL, Taylor JE, Xu X, Cotana F (2012) Inter-building effect: simulating the impact of a network of buildings on the accuracy of building energy performance predictions. Build Environ 58:37–45

Pourghasemi HR, Sadhasivam N, Kariminejad N, Collins AL (2020) Gully erosion spatial modelling: role of machine learning algorithms in selection of the best controlling factors and modelling process. Geosci Front 11(6):2207–2219

Rodriguez JJ, Kuncheva LI, Alonso CJ (2006) Rotation forest: a new classifier ensemble method. IEEE Trans Pattern Anal Mach Intell 28:1619–1630

Roy J, Saha S (2019) Landslide susceptibility mapping using knowledge driven statistical models in Darjeeling District, West Bengal, India. Geoenviron Disasters 6(1):1–18

Roy J, Saha S, Arabameri A, Blaschke T, Bui DT (2019) A novel ensemble approach for landslide susceptibility mapping (LSM) in Darjeeling and Kalimpong districts, West Bengal, India. Remote Sens 11(23):2866

Saha A, Saha S (2020a) Comparing the efficiency of weight of evidence, support vector machine and their ensemble approaches in landslide susceptibility modelling: a study on Kurseong region of Darjeeling Himalaya, India. Remote Sens Appl: Soc Environ 19:100323

Saha A, Saha S (2020b) Application of statistical probabilistic methods in landslide susceptibility assessment in Kurseong and its surrounding area of Darjeeling Himalayan, India: RS-GIS approach. Environ Dev Sustain 1–31

Saha A, Mandal S, Saha S (2020a) Geo-spatial approach-based landslide susceptibility mapping using analytical hierarchical process, frequency ratio, logistic regression and their ensemble methods. SN Appl Sci 2(10):1–21

Saha S, Arabameri A, Saha A, Blaschke T, Ngo PTT, Nhu VH, Band SS (2021) Prediction of landslide susceptibility in Rudraprayag, India using novel ensemble of conditional probability and boosted regression tree-based on cross-validation method. Sci Total Environ 142928

Saha S, Roy J, Pradhan B, Hembram TK (2021) Hybrid ensemble machine learning approaches for landslide susceptibility mapping using different sampling ratios at East Sikkim Himalayan, India. Adv Space Res 68(7):2819–2840

Saha S, Saha A, Hembram TK, Pradhan B, Alamri AM (2020b) Evaluating the performance of individual and novel ensemble of machine learning and statistical models for landslide susceptibility assessment at Rudraprayag District of Garhwal Himalaya. Appl Sci 10(11):3772

Sheng T, Chen Q (2017) An altitude based landslide and debris flow detection method for a single mountain remote sensing image. In: International conference on image and graphics. Springer, Cham, pp 601–610

Sun X, Chen J, Han X, Bao Y, Zhan J, Peng W (2020) Application of a GIS-based slope unit method for landslide susceptibility mapping along the rapidly uplifting section of the upper Jinsha River, South-Western China. Bull Eng Geol Environ 79(1):533–549

Tien Bui D, Pradhan B, Lofman O, Revhaug I (2012) Landslide susceptibility assessment in Vietnam using support vector machines, decision tree, and Naive Bayes models. Math Probl Eng 2012

Wang S, Zhang K, van Beek LP, Tian X, Bogaard TA (2020) Physically-based landslide prediction over a large region: scaling low-resolution hydrological model results for high-resolution slope stability assessment. Environ Model Softw 124:104607

Willmott CJ, Matsuura K (2005) Advantages of the mean absolute error (MAE) over the root mean square error (RMSE) in assessing average model performance. Clim Res, 30(1):79–82

Wooten RM, Witt AC, Miniat CF, Hales TC, Aldred JL (2016) Frequency and magnitude of selected historical landslide events in the southern Appalachian Highlands of North Carolina and Virginia: relationships to rainfall, geological and ecohydrological controls, and effects. In: Natural disturbances and historic range of variation. Springer, Cham, pp 203–262

Yalcin A, Reis S, Aydinoglu AC, Yomralioglu T (2011) A GIS-based comparative study of frequency ratio, analytical hierarchy process, bivariate statistics and logistics regression methods for landslide susceptibility mapping in Trabzon, NE Turkey. CATENA 85(3):274–287

Yavari H, Pahlavani P, Bigdeli B (2019) Landslide hazard mapping using a radial basis function neural network model: a case study in Semirom, Isfahan, Iran. Int Arch Photogr Remote Sens Spat Inf Sci

Predicting the Landslide Susceptibility in Eastern Sikkim Himalayan Region, India Using Boosted Regression Tree and REPTree Machine Learning Techniques

Kanu Mandal, Sunil Saha, and Sujit Mandal

Abstract

In the mountainous parts of the world, landslides are considered as the most dangerous to people and property. The number and the amount of damage caused by landslides have been steadily growing globally. As a result, slope instability management is critical in the hilly region's ecological and socioeconomic dynamics. The Rorachu river basin of Sikkim Himalaya has been selected for the present study which is Sikkim's most landslide-prone area. The key intention of the research is to use computer-based machine learning techniques to create landslide susceptibility maps (LSMs) and compare the models' efficiency. Nineteen variables, including stimulating and environmental factors, were chosen to better explain the current spatial relationship with the landslide. Two popular machine learning techniques, i.e., Reduced Error Pruning Tree (REPTree) and Boosted Regression Tree (BRTree) have been incorporated to prepare LSMs. Randomly selected landslide and non-landslide points were used to build two different databases: training data and evaluation data. During the collection of both training and testing sites, a 70:30 ratio was retained. Tolerance (TOL) and Variance Inflation Factor (VIF) was used to estimate multicollinearity, and Information Gain Ratio (IGR) was taken to evaluate the importance of the variables. The findings show that multicollinearity is minimum in the landslide causing factors, and rainfall is perhaps the most crucial component in landslide occurrence. Receiver Operating Characteristics curve (ROC) coupled with statistical techniques like Root Mean Square Error (RMSE), and Mean Absolute Error (MAE) were exercised to portray the models' accuracy for both training and testing datasets. The result enlightens that higher landslide susceptibility classes contain most of the slide area. In both the training and testing datasets, the BRTree model has the highest Area Under Curve (AUC) value. The BRTree model's AUC values for training and testing datasets were found to be 0.896 and 0.9, respectively. The result of RMSE and MAE also hails the superior representation of the REPTree model. The RMSE and MAE value of the REPTree model was noticed 0.120, 0.189 (for training dataset), and 0.129, 0.192 (for validation dataset), while the value was obtained 0.150, 0.235 (for training dataset) and 0.139, 0.21 (for validation dataset) in BRTree model

K. Mandal · S. Saha (✉)
Department of Geography, University of Gour Banga, Malda, West Bengal 732103, India
e-mail: sunilgeo.88@gmail.com

S. Mandal
Department of Geography, Diamond Harbour Woman's University, Diamond Harbour, West Bengal, India

respectively. As a result, both models performed admirably, but the BRTree model performed better than the REPTree model.

Keywords

Reduced Error Pruning Tree (REPTree) · Boosted Regression Tree (BRTree) · Landslide susceptibility · ROC curve · Rorachu river basin

36.1 Introduction

The quasi-natural hazard landslide consequences the downward movement or lateral spread of slope-building components like soil, rock, or a composition of both. The involvement of natural and anthropogenic triggering and causative factors (rainfall, earthquake, volcanic eruption, marine erosion, construction of road, and agricultural field) has made landslide a quasi-natural hazard (Froude and Petley 2018). Rainfall has been identified among the most effective landslide stimuli. The National Aeronautics and Space Administration (NASA) data depicts that globally 10,804 landslides occurred between 2007 and 2017 due to rainfall (NASA 2019). Landslide is gaining attention as a dangerous hazard because of the frequency of occurrence, loss of lives, and livelihoods. Generally, when supernumerary pressure on slope mass crosses the shear strength limit, a landslide occurs (Mandal and Mandal 2018). Landslide leaves an impression on the landscape by causing river course modification, massive deforestation, human settlement wreckage, flash flood, etc. (Saha et al. 2020). It is a known fact that triggering factors play a leading role in landslide occurrence, and sometimes triggering factors like seismicity alone can make a slope unstable. Alongside a combined effect of causative factors also have a great potentiality in landslide initiation. Hence a precise understanding of the relation between landslide and various triggering and causative factors is essential (Saha et al. 2020). The World Bank has estimated that nearly 3.7 million square kilometers of land are severely threatened by landslides globally, which may risk some 300 million lives (Dilley 2005). In a study, Froude and Petley (2018) discovered that between 2004 and 2016, a record of 4862 landslides erupted all around the globe with 55,997 people killed (Petley 2012). It is easy to imagine that the globe encounters tremendous economic losses every year as a consequence of landslides. In terms of loss, the landslide has become the earth's seventh most severe hazard (Nadim et al. 2006). The countries of the world confront a massive loss of US$ 20 billion due to landslides (Klose et al. 2016). In India, a region of 25% is accompanied by numerous landslides in hill slopes and rugged mountains (Dubey et al. 2005). The study of Raman and Punia (2012) depicts that India's 15% of the land area is already under a serious landslide threat zone. Landslide causes an annual loss of US$ 500 million to India (Dubey et al. 2005). According to NASA data (2019), 6779 people were killed in 958 slope failures in India between 2008 and 2015, with Uttarakhand taking the lead with 5226 casualties. Himalayan regions of India and China have witnessed the highest occurrence of the landslide; countries like Laos, Bangladesh, Myanmar, Indonesia, and the Philippines are in the next position in landslide occurrence (Pham et al. 2019). In past years India recorded several destructive landslides. For instance, the Amboori landslide in Kerala in 2001, the Kedarnath landslide in 2014 in Uttarakhand, the Pune landslide in Maharastra (Sharma et al. 2014; NASA. Global Landslide Catalog, 2019). It is true that even if the magnitude of the landslide or the amount of damage can be reduced, it is not possible to stop this catastrophe completely (Pham et al. 2019). Hence a proper scientific mitigation technique and policy for slope instability is necessary to check the magnitude of risk (Carrara et al. 1991). Toward landslide management, the first should be the delineation of landslide-prone zones (Hong et al. 2016a). In this regard, a landslide susceptibility map may play a vast role as it expresses the places with the various probability of future landslide (Bui et al. 2016a). Knowledge-based qualitative and data-driven quantitative methods are two important segments of landslide susceptibility analysis.

Qualitative methods widely applied in the previous decades, but due to its cost-effective and time-consuming nature, recently quantitative methods are getting more attention (Pham et al. 2019). Data collection in highly inaccessible and data sparsed regions is challenging, which is another significant disadvantage of the qualitative method. The quantitative methods easily overcome the problem of data collection by using various Remote Sensing (RS) and Geographical Information System (GIS) methodologies. In the case of quantitative analysis, the quality of the data and employed methodology determines the overall precision of the study (Saha et al. 2020). In many quantitative studies of landslide susceptibility analysis, it was found that remote sensing data has very high reliability (Chen et al. 2018; Saha et al. 2020). The literature of landslide susceptibility study strongly supports the fact that quantitative methods were abruptly applied in the recent past. Traditional double variable and multivariable statistical methodologies like Statistical Index (SI) (Mandal and Mandal 2018), Weight-of-Evidence (WoE) (Chen et al. 2019), Logistic Regression (LR) (Chen et al. 2019) successfully employed to build maps of landslide susceptibility throughout the globe. But in landslide modeling, recently, the use of conventional statistical models decreases due to low precision (Bui et al. 2016a), the presence of pre-calculation conjuncture in the model (Benediktsson et al. 1990).

On the other hand, due to the robustness, very high predictive capacity, and more accurate result, machine learning techniques are being comprehensively used in the spatial analysis of slope instability and landslides (Chen et al. 2018; Pham et al. 2019, Saha et al. 2020; Bui et al. 2020; Ngo et al. 2021). More efficient and reliable approaches are being implemented using data mining techniques in landslide hazard modeling work (Chen et al. 2018). Several advantages of machine learning techniques include no need for statistical assumptions before the calculation; in the case of a Geographical Information System (GIS) based database, no conversion of categorical and continuous data is required (Chen et al. 2017). On the contrary, ensemble-based machine learning techniques can easily solve the non-linearity problems in the data (Pham et al. 2020; Saha et al. 2022). Several types of landslide analysis have also been completed successfully using machine learning and deep learning algorithms such as CNN (Saha et al. 2022; Pham et al. 2020; Ngo et al. 2020), decision tree algorithms viz. Functional Tree (FTree) (Zhao and Chen 2019), Random Forest method (RF) (Zhang et al. 2017), Alternating Decision Tree (ADTree) (Chen et al. 2017), classification and regression tree (Chen et al. 2017), and data mining functional algorithms like Artificial Neural Network method (ANN) (Sameen et al. 2020; Nhu et al. 2020), Support Vector Machine method (SVM) (Pandey et al. 2020; Pham et al. 2020). In order to determine the location of future landslides or to make predictions of landslides, it is essential to have a comprehensive knowledge of landslides and use them in spatial studies of landslides (Pham et al. 2019). The accuracy of the model is a very sensitive issue when it comes to spatial modeling of landslides. Various studies on landslide modeling have found that the multiple model ensemble gives more accuracy than using one model (Saha et al. 2020). Considering this aspect, the research paper used different models individually as well as the ensemble of the models. The basic aim of the study is to formulate landslide susceptibility models for the Rorachu river basin by using machine learning algorithm-based Boosted Regression Tree (BRTree) and Reduced Error Pruning Tree (REPTree) model for understanding which model is better in landslide susceptibility modeling.

36.2 Study Area

The research field Rorachu river basin is found in the East Sikkim district's middle and northern regions. The basin is characterized by the areal coverage of 71.73 km^2, and spatial extension ranges between 27°17′19″ to 27°23′52″ N and 88°35′37″ to 88°43′17″ E (Fig. 36.1). The multiphase metamorphism, polycyclic deformation, highly rugged and dissected topography with moderate to very high relief portrays the geological and topographic character of the basin. Tectono-stratigraphic characteristics of the Rorachu river basin mainly depend on the inner zone (metabasic rocks, lingtse gneiss, and pelitic rock of the Daling group) and central crystalline zone (Chungthang formation and Kanchenjunga gneiss). Dissection of topography due to fluvial and glacial erosion largely controls the geomorphic character of the Rorachu river basin. Major geomorphological divisions of the study area consist of structural origin highly dissected hills and valleys, structural origin moderately dissected hills and valleys, and snow-covered area. The entire northern part of the Rorachu river basin belongs to the highly dissected hills; on the contrary, moderately dissected hills were mainly found in the basin's southern and western reaches. Relief of the basin encourages such kind of geomorphological divisions because the highest altitude of 4213 m (Near Kyongnosla) was found in the northern periphery of the basin, where the lowest relief (834 m) was recorded in the southwestern boundary (near Ranipool) of the basin. Topographic dissection mainly occurs due to the heavy downcutting of the slope by umpteen number of streams. With light summer and parky winter, the study area is under the subtropical highland climate (Cwb) as per Koppen's climatic division scheme. The weather condition of the Rorachu river basin is highly similar to the weather of the spring season. The temperature data show that average annual highest and lowest registered temperatures are 22 °C and 4 °C, respectively, and that mild to low temperatures prevail across the year. The state capital, Gangtok, is situated on the southwestern outskirts of the Rorachu river basin.

36.3 Material and Methods

36.3.1 Landslide Inventory Map (LIM)

The LIM can be termed as a landslide distribution map (Fig. 36.1). For the creation of LIM, the subsisted landslide of a particular area should be mapped either with field investigation or Remote Sensing (RS) and GIS techniques. LIM is treated as an integral component both in the case of RS and GIS-based traditional statistical modeling or computer algorithm-based highly advanced machine learning modeling of the landslide. Various literatures (Mandal and Mandal 2018; Yao et al. 2020) hails the same fact that spatial modeling of landslide is incomplete without LIM. For the establishment and proper understanding of the spatial relationship between landslide and its causative factors, a LIM is the only way to fulfill the objective. Hence in spatial modeling of landslide or landslide susceptibility study, an accurate LIM is highly essential. In the present study, both rigorous field study and standard RS and GIS techniques have been carefully applied to construct the LIM of the Rorachu river basin. Varnes's (1978) classification was taken to identify various types of landslides with proper characteristics during the field study. The Rorachu river basin is characterized by a large number of debris flows followed by rotational slides, translational slides, and rockfalls (Fig. 36.2). To create the final LIM, 80 landslides with different traits have been demarcated and plotted after a thorough analysis of landslides in the study region. All the landslides possess an area of 0.85 km^2 of the study area.

36.3.2 Database and Preparation of Landslide Causative Factor

Spatial modeling of the landslide (Fig. 36.3), including remote sensing and GIS-based conventional statistical modeling or advanced computer algorithm-based machine learning modeling, requires a concrete database. In spatial modeling of landslide susceptibility, probable landslide occurrence zones are identified based on subsisted landslides and the existing relation between landslide and its causative factors. Hence it is clear that the formulation of an optimal landslide susceptibility model is dependent on the construction of the spatial database to a great extent. Apart from preparing the dataset, selecting the proper and most responsible factors

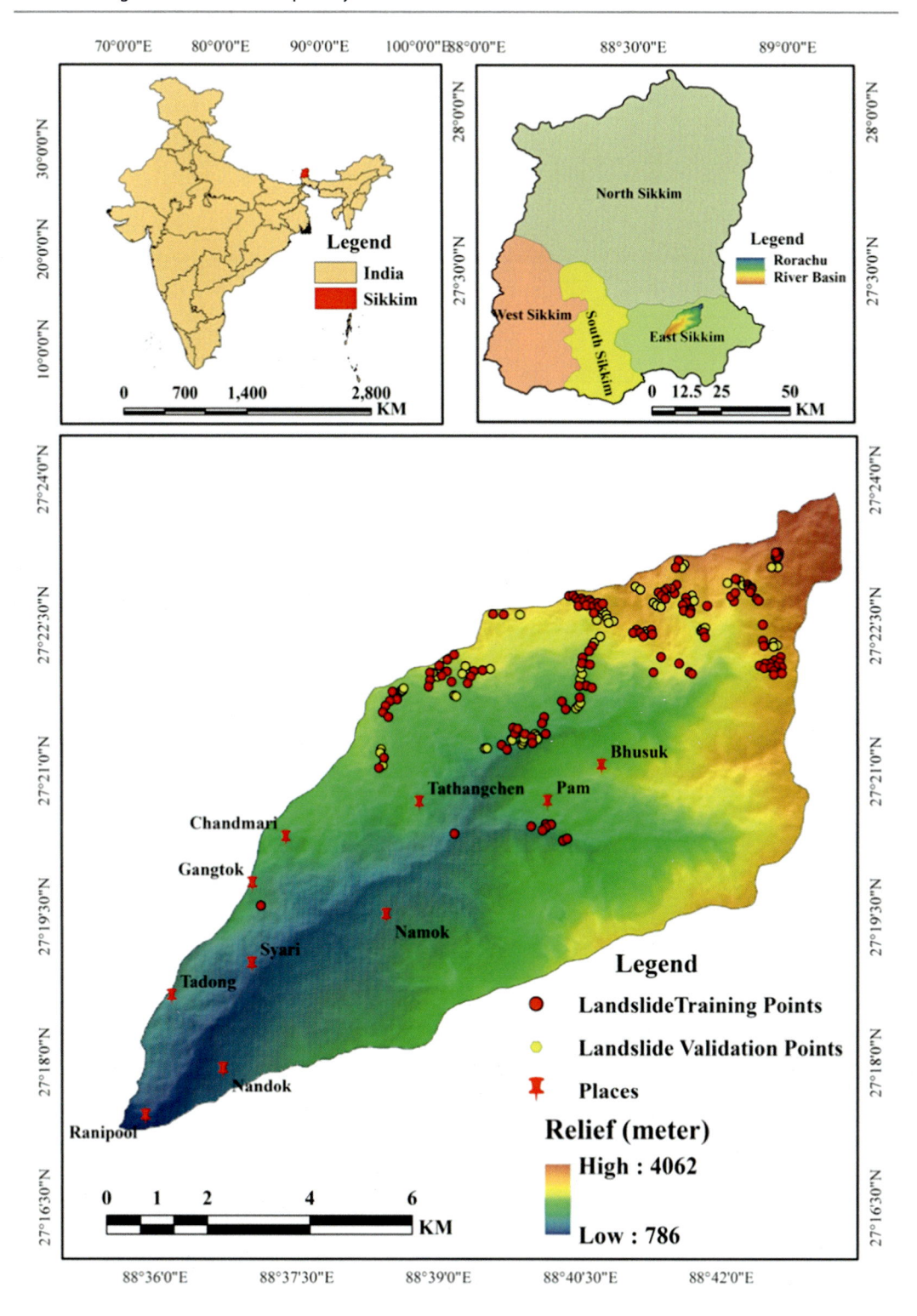

Fig. 36.1 Location of the study area **a** India, **b** Sikkim and **c** Rorachu river basin

Fig. 36.2 Ground photograph of various landslide types of the Rorachu river basin: **a** rock slide near Ranipool (27°17′30.17″ N, 88°35′49.09″ E), translational slide near 13th Mile Bridge (27°22′54.75″ N, 88°41′30.79″ E), debris flow near 7th Mile (27°21′56.22″ N, 88°39′8.31″ E), crown of a rotational slide near Namok (27°19′13.18″ N, 88°38′14.70″ E)

for the occurrence of landslide over a region is crucial. Previous researches have shown that a wide range of geological, topographical, hydrological, and environmental factors are responsible for the commencement of landslides (Roy et al. 2019; Arabameri et al. 2020; Chen et al. 2009, 2019, 2020; Mandal et al. 2021). However, it is also true that the effect of the factors varies by location. In the present study, special attention has been given to this matter of landslide conditioning factor selection. Before selecting prime factors for the initiation of the landslide in the Rorachu river basin, several rigorous field survey has been done to identify various types of slope movements and their probable cause of occurrence. Previous research works coupled with several official articles of the Geological Survey of India (GSI) have helped to develop a proper knowledge about the landslide causative factors of the Rorachu river basin. After a thorough study and field investigation, 19 landslide conditioning factors, including atmospheric, physiographic, geological, hydrological, and environmental factors, have been taken to prepare the landslide susceptibility model (Table 36.1 and Fig. 36.4). After the selection of relevant factors, another important matter in landslide modeling is the quality of the data because the forecasting or prediction precision of the model is largely dependent on the quality of the data. All the data layers, including the landslide inventory, were prepared from authentic data sources like the Survey of India (SOI) toposheet, district resource map of GSI, report of the ministry of environments and forest, satellite images from the United States Geological Survey (USGS), and thematic maps from Bhuvan (https://bhuvan.nrsc.gov.in/bhuvan_links.php#).

The rainfall map (Fig. 36.4a) was produced using the rainfall data of the "worldclim" website

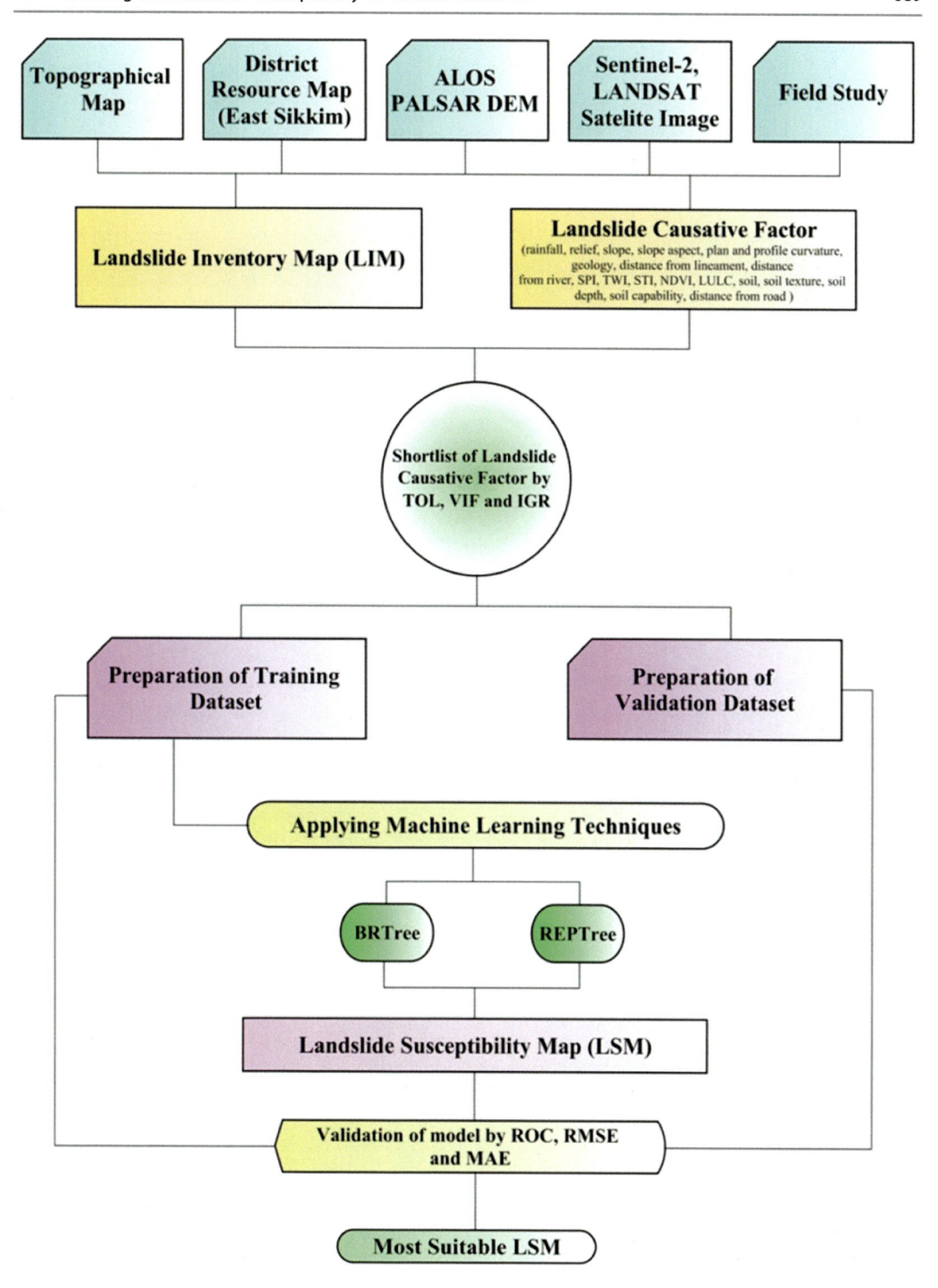

Fig. 36.3 Methodological flow chart of the study

Table 36.1 Reason of selection of landslide causative factors

Data layer	Source	Reason behind the picking up of landslide conditioning factor with reference
Rainfall	https://www.worldclim.org/	Rainfall triggers landslide by reducing soil and rock mass strength (Sharma et al. 2014)
Relief, slope, slope aspect, plan curvature, profile curvature	ALOS-*PALSAR* DEM	Morphometric factors have aggressively been applied in landslide modeling due to their role in the commencement of landslide (Shit et al. 2016; Mandal and Mandal 2018)
Geology, distance from lineament	District resource map of East Sikkim district, thematic data layer of lineament (https://bhuvan.nrsc.gov.in/bhuvan_links.php#)	The nature of the slope failure largely depends on the attitude of geological properties like joint, foliation, etc. (Mandal and Mandal 2018; Siddique and Khan 2019)
Distance from river	SOI Toposheet No. 78A/11, ALOS PALSAR DEM	The presence of drainage encourages the erosion and reduction of slope building material's strength (Roy et al. 2019)
Stream power index (SPI)	ALOS PALSAR DEM	SPI estimates the erosive capacity of running water, and it is treated as a severe factor of landslide occurrence (Regmi et al. 2014)
Topographic wetness index (TWI)	ALOS PALSAR DEM	Topographic wetness favors a slope to become unstable (Zêzere et al. 2017)
Sediment transportation index (STI)	ALOS PALSAR DEM	STI measures the amount of overland flow erosion and is considered a critical factor for mass movement and landslide (Lay et al. 2019)
Normalized difference vegetation index (NDVI), land use and land cover (LULC)	USGS Sentinel 2 satellite image	The concentration of vegetation inversely affects the incidents of the landslide (Mandal and Mandal 2018). Anthropogenic activities like road construction, agricultural expansion make the slope more vulnerable by the direct modification of slope properties (Lee and Kim 2020)
Soil, soil texture, soil depth, soil capability	Forest environment and wildlife management department, Government of Sikkim (http://www.sikkimforest.gov.in/ccstbs.html?reload)	Landslide resistance capacity of the soil depends on the fundamental properties like texture, soil capability, etc. (Roy et al. 2019)
Distance from road	SOI Toposheet No. 78A/11, Google Earth satellite image (https://www.earth.google.com)	The modification of slope properties due to the construction of new roads and the expansion of old ones is one of the prime causes of slope instability (Ghosh et al. 2020)

by the IDW tool of ArcMap 10.3 software. The spatial analyst function in ArcMap 10.3 was used to obtain morphometric factors such as relief (Fig. 36.4b), slope (Fig. 36.4c), slope aspect (Fig. 36.4d), curvature i.e. plan (Fig. 36.4e) and profile (Fig. 36.4f) from an ALOS PALSAR DEM with an image resolution of 12.5 m. The geological framework (Fig. 36.4g) was produced

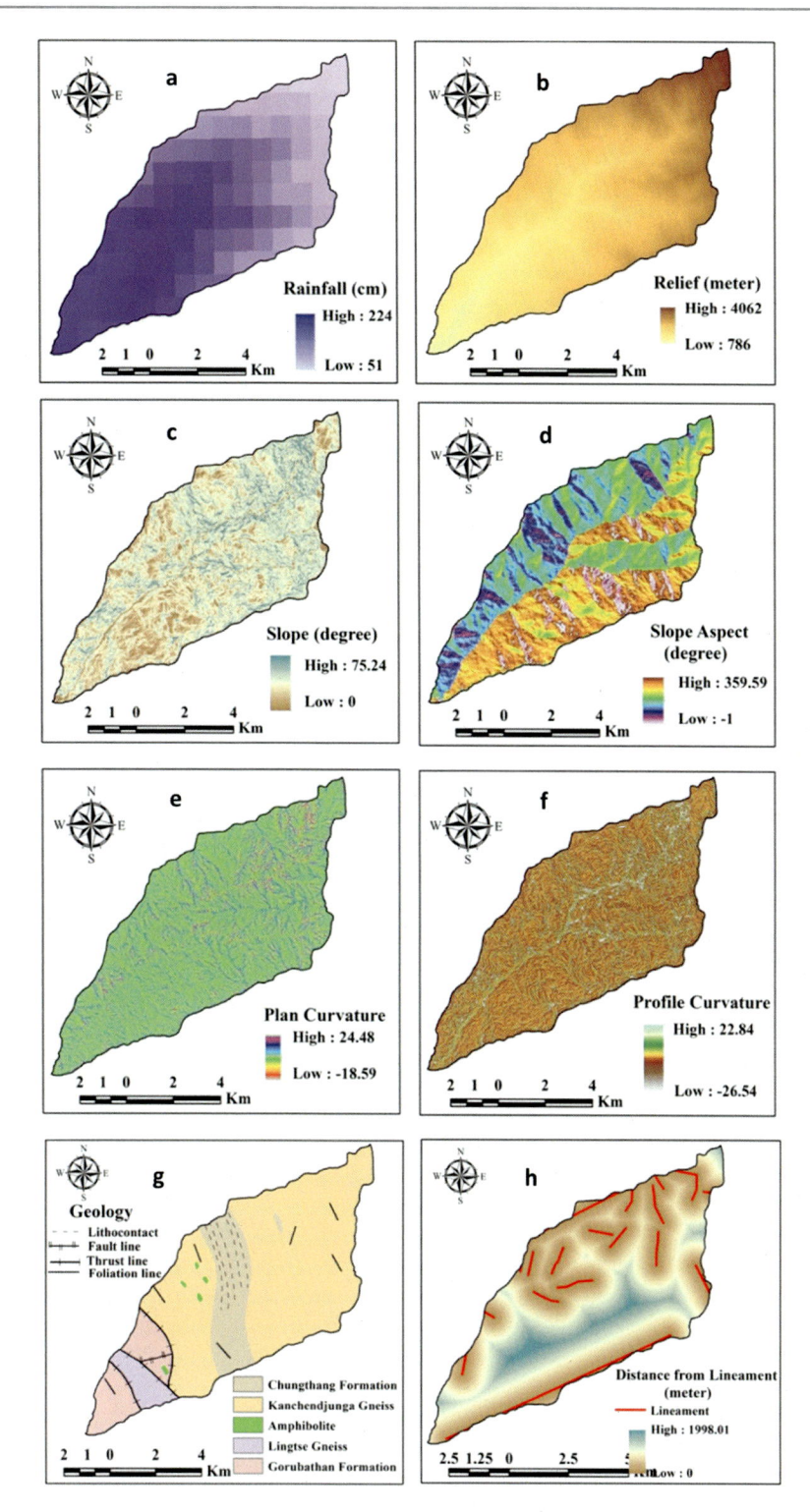

Fig. 36.4 Landslide conditioning factors: **a** rainfall, **b** relief, **c** slope, **d** slope aspect, **e** plan curvature, **f** profile curvature, **g** geology, **h** distance from lineament, **i** distance from drainage, **j** SPI, **k** TWI, **l** STI, **m** distance from road, **n** NDVI, **o** LULC, **p** soil, **q** soil texture, **r** soil depth, **s** soil capability

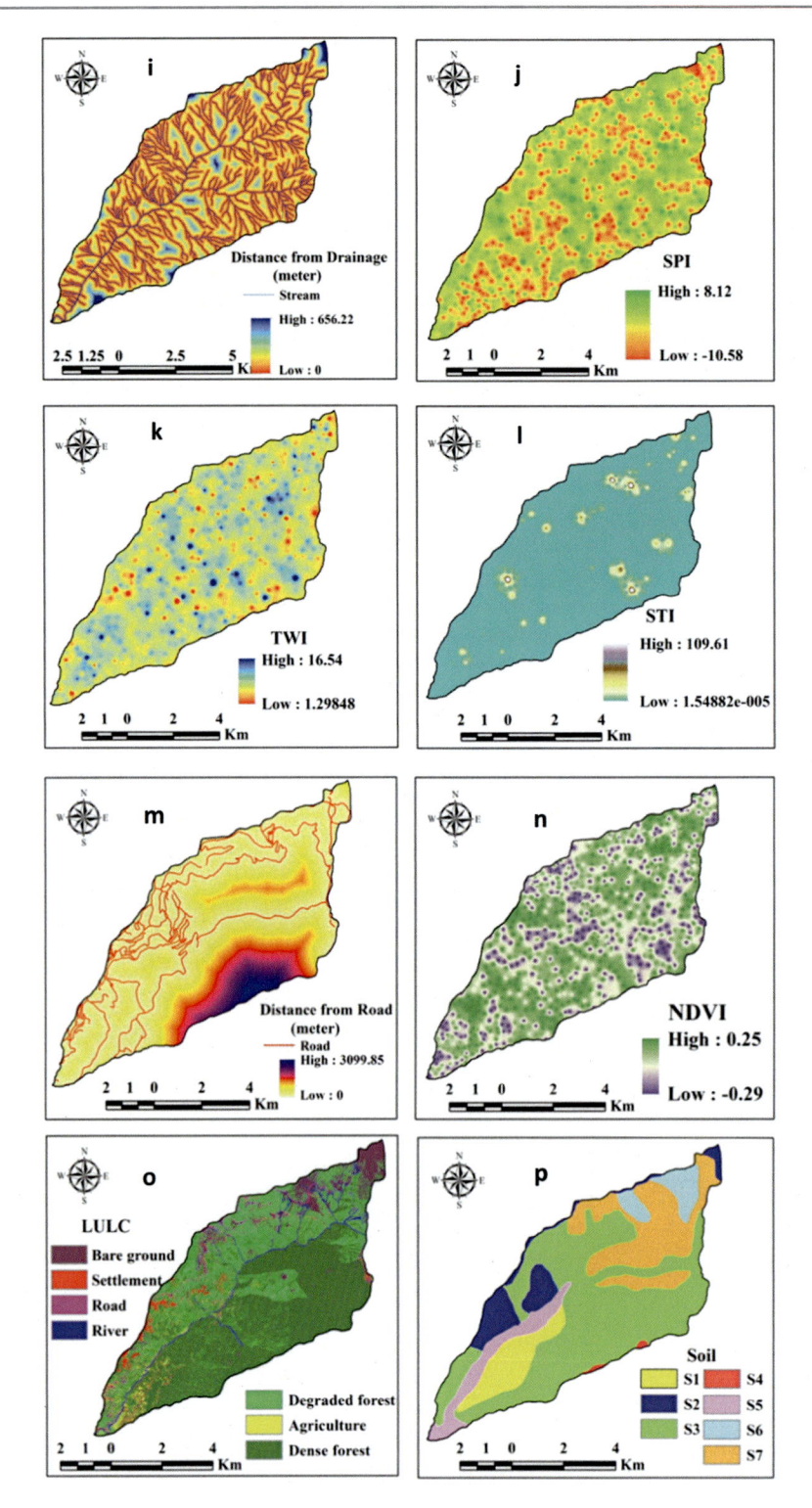

Fig. 36.4 (continued)

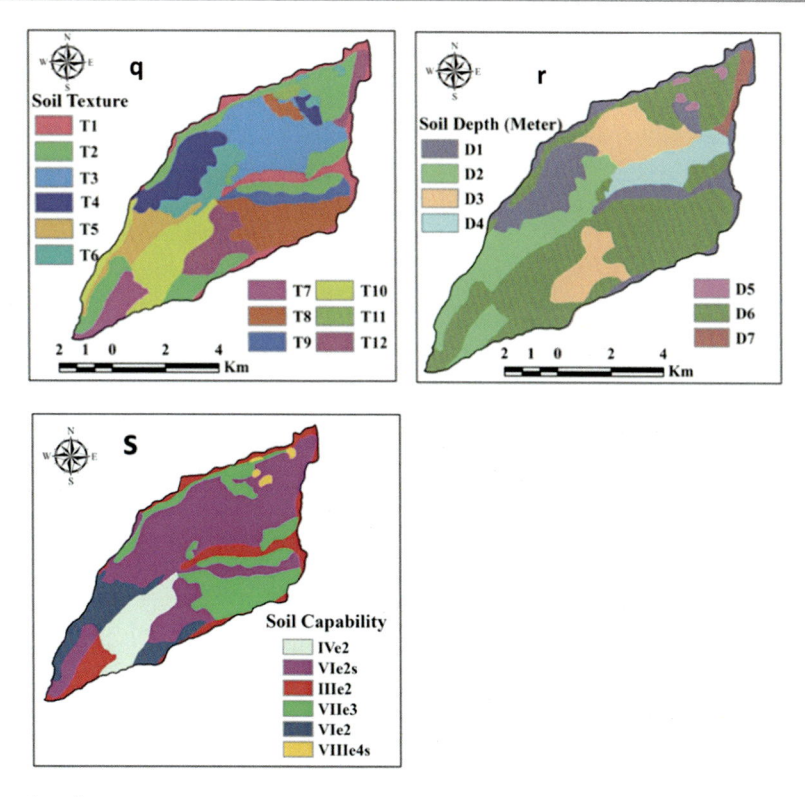

Fig. 36.4 (continued)

by vectorizing the East Sikkim district resource map and then converting it to raster format. Lineaments were also digitized from the same map and compared with the thematic layer of Bhuvan. Finally, a distance from the lineament map (Fig. 36.4h) was constructed using the Euclidean distance tool of ArcMap 10.3. Drainage segments were vectorized from the toposheet and extracted from DEM using the flow accumulation tool of ArcMap 10.3. The Euclidian distance tool of ArcMap 10.3 was utilized for the preparation of distance from the river (Fig. 36.4i) data layer. Flow accumulation raster was used to measure the Stream Power Index (SPI) (Fig. 36.4j), Topographic Wetness Index (TWI) (Fig. 36.4k) and Sediment Transportation Index (STI) (Fig. 36.4l). The flowing water's erosive capacity is measured by stream power index (SPI) (Park and Kim 2019). Moore et al. (1991), considered SPI to be a significant factor in a region's stability. SPI is mostly dependent on the characteristics of the terrain and the amount of run-off of any site (Zhang et al. 2017). In order to evaluate the accumulation trend of water in any given catchment area, the TWI index was established by Beven and Kirkby (1979) (Zhang et al. 2017; Park and Kim 2019). STI is another important index to measure surface erosion rate (Moore and Burch 1986). The following formula was applied to the SPI (Eq. 36.1), TWI SPI (Eq. 36.2), and STI SPI (Eq. 36.3):

$$\text{SPI} = A_s \times \tan \beta \tag{36.1}$$

$$\text{TWI} = \ln\left(\frac{A_s}{\tan \beta}\right) \tag{36.2}$$

$$\text{STI} = \left(\frac{A_s}{22.13}\right)^{0.6} \left(\frac{\sin \beta}{0.0896}\right)^{1.3} \tag{36.3}$$

where A_s and β denotes a particular catchment area and slope (in degree), the sentinel-2 satellite image with 10 m × 10 m ground resolution has been utilized to enumerate the Normalized Difference Vegetation Index (NDVI) (Fig. 36.4n) and land use-land cover (LULC) (Fig. 36.4o). The current study's NDVI index was calculated with the Erdas imagine 9.2 program, and an NDVI map was created with the raster calculator of ArcMap 10.3 (Eq. 36.4).

$$NDVI = \frac{(NIR - R)}{(NIR + R)} \qquad (36.4)$$

NDVI denotes the final value (ranges between −1 and +1), near-infrared expressed by NIR, and R indicates red. A supervised satellite image classification technique was incorporated to portray the actual scenario of land use and land cover, and the Kappa coefficient further examined the precision of the map. The map of soil (Fig. 36.4p) and various soil properties like soil texture (Fig. 36.4q), soil depth (Fig. 36.4r), and soil capability (Fig. 36.4s) were vectorized from a published report of the forest environment and wildlife management department, the Government of Sikkim (https://www.sikkimforest.gov.in/ccstbs.html?reload) and converted into the raster format. The present study's road network was digitized from various authentic sources like the Survey of India Toposheet No. 78A/11 and Google Earth satellite image (https://www.earth.google.com). Finally, a distance from the road raster layer (Fig. 36.4m) was generated using the ArcMap 10.3 software package's Euclidean distance tool.

36.3.3 Selection of Landslide Causing Factors (LCFs)

36.3.3.1 Multicollinearity Analysis

Worldwide research on the mechanism and causes of the landslide has already established that landslide is a complex geoenvironmental phenomenon where many factors work together. In statistics-based spatial modeling of landslide susceptibility, the spatial database has enormous significance. The data of landslide and causative factors establish the relationship and, based on which the potential areas of future landslide are identified. So the proper establishment of the relationship between the event and causative factors has immense importance. Apart from the relation between landslide and conditioning factors, some connection between the causative factors is also noticed. This type of statistical relationship among the causative factors is termed multicollinearity. A large amount of multicollinearity among the predictor variables affects the overall precision of the model. Hence proper assessment of multicollinearity is necessary. The tolerance (TOL) and variance inflation factor (VIF) are highly effective methods to justify multicollinearity between the various independent variables (Chen et al. 2019; Arabameri et al. 2020). The lower TOL value expresses the higher presence of multicollinearity. Multicollinearity is severe when the TOL value lies less than 0.1 (Menard 1995). On the contrary, multicollinearity among the independent variables is considerable only when the VIF value crosses 10 (Bui et al. 2011).

36.3.3.2 Information Gain Ratio (IGR)

In multi-factor spatial modeling like landslide susceptibility modeling, selecting a proper predictor variable is a higher priority task. Though many factors play a role in a landslide occurrence, there is a sharp difference in participation. Worldwide research on landslide (Dou et al. 2020; Yao et al. 2020) also established that the predictor variable's role varies from place to place. Quantitative estimation of the part of landslide causative factors is essential in this regard. Because the factors with the lesser significant role may reduce the overall precision of the model (Bui et al. 2015). Therefore toward the construction of an optimal landslide susceptibility model, apart from the inclusion of most relevant factors, it is crucial to exclude the less important factors (Bui et al. 2015; Shirzadi et al. 2019). IGR is a well-recommended method to fulfill the objective of critical factors selection for a high-quality spatial model (Quinlan 1993; Nhu et al. 2020; Yao et al. 2020). IGR was introduced by Quinlan (1993) to volition features (Zhou et al. 2018). In selecting prime factors for landslide occurrence, IGR works in how an element with a higher IGR value plays a more significant role in a landslide event. IGR (Eq. 36.5) can be expressed for a given landslide causative factor "*relief*"(R), where S is the training data with n number of total inputs, $n(L_i S)$ is the samples in the training data S belong to L_i (landslide and non-landslide points) (Bui et al. 2015; Shirzadi et al. 2019).

$$\text{IGR}(S, R) = \frac{\text{Entropy}(S) - \text{Entropy}(S, R)}{\text{SplitEntropy}(S, R)} \tag{36.5}$$

$$\text{Entropy}(S) = -\sum_{i=1}^{2} \frac{n(L_i, R)}{|S|} \log_2 \frac{n(L_i, R)}{|S|} \tag{36.6}$$

$$\text{Entropy}(S, R) = \sum_{j=1}^{m} \frac{S_j}{|S|} \text{Entropy}(S) \tag{36.7}$$

$$\text{SplitEntropy}(S, R) = -\sum_{j=1}^{m} \frac{|S_j|}{|S|} \log_2 \frac{|S_j|}{|S|} \tag{36.8}$$

36.3.4 Models Used for Modeling the Landslide Susceptibility

36.3.4.1 Boosted Regression Tree (BRTree)

In the field of artificial intelligence-based advanced machine learning programs, the non-parametric BRTree algorithm is considered a well-justified and felicitated methodology. The amalgamation of statistics and machine learning programs leads to the formation of this model (Park and Kim 2019). Statistical regression coupled with machine learning-based "boosting" algorithm is the central architecture of the BRTree model. "Boosting" in the BRTree model mainly enhances the precision of the model's prediction capacity by decreasing variance (Aertsen et al. 2010; Rahmati et al. 2018). The "boosting" algorithm mainly works based on the assumption that getting aggregate value from a bunch of decision trees (DT) is more fruitful than one precise DT (Elith et al. 2008). On the contrary, the application and union of various algorithms to better the performance is another vital characteristic of this model (Schapire 2003; Elith et al. 2008). Another significant advantage of the BRTree model is that pre-calculation conjuncture about the relationship between the event and predictor is absent (Saha et al. 2020). Apart from the classification of the data, the BRTree model also examines the role of predictor variables that leads to eradicating the forecasting problem of the model (Aertsen et al. 2010). The "boosting" process also eliminates the necessity of converting, modifying, or processing the data (Ebrahimy et al. 2020a, b). Hence BRTree model easily handles continuous, categorical, and missing data (Roy et al. 2020). The construction of a BRTree model that yields accurate forecasting is a very crucial task (Ebrahimy et al. 2020a, b). The splitting of the dependent variable using a correct predictor (divisor) is a preliminary task toward the formulation of the model (Saha et al. 2020). Control over the splitting process (tree complexity), assessing the role of DT in the model (learning rate) are also essential steps for a precise BRTree model.

36.3.4.2 Reduced Error Pruning Tree (REPTree)

The REPTree model was suggested by Quinlan (1987). The architecture of this algorithm follows the mechanism of information gain (IG) of entropy and reduction of error in the variance (Quinlan 1987; Devasena 2014). REPTree is an advanced machine learning algorithm (MLA) with the benefit of rapid computational capacity (Pham et al. 2019). Union of reduced error pruning (REP) function with the decision tree (DT) algorithm prepares the design of the REPTree model (Quinlan 1987). The algorithm has been constructed and applied to eliminate several well-known statistical problems like backward overfitting of the data, large output of the data, etc. Finding a solution for backward overfitting is the prime task of the pruning function (Saha et al. 2020). On the contrary, using the training data, DT clears the issues associated with large output. The DT function of the REPTree model ensues the anatomy of general regression tree (Saha et al. 2020). The selection of DT in the REPTreemodel depends on the tree, as the regression tree generates alternatives from which the best tree is taken for the calculation (Devasena 2014). Apart from the solution of the backward overfitting problem, the pruning function is also very significant for the selection of DT by post-pruning technique (Chen et al. 2009).

36.3.5 Model Validation Methods

36.3.5.1 Receiver Operating Characteristics (ROC) Curve

The ROC curve has proven to be a reliable tool for determining the precision of a probabilistic model all over the world (Park and Kim 2019). Specificity and sensitivity are two significant components of the ROC curve for measuring the performance of a statistical model (Mandal and Mandal 2018). Training data of a model construct the ROC curve is termed as a Success Rate Curve (SRC). On the other hand, the ROC curve, which uses the validation dataset, is known as the Prediction Rate Curve (PRC) (Park and Kim 2019). The SRC and PRC assess the suitability of the data for a model and the forecasting capacity of that particular model, respectively. The area under the ROC curve is referred to as the Area Under Curve (AUC), and it is used to assess the model's overall precision. The AUC value varies from 0 to 1, with the higher value indicating higher prediction performance of the model. On the contrary, the AUC value of less than 0.5 indicates the poor performance of the model (Mandal and Mandal 2018; Saha et al. 2020).

36.3.5.2 RMSE and MAE

The RMSE and MAE are considered a well-tested statistical index for the proper assessment of the machine learning-based hybrid model's precision. Plenty of researchers across the globe (Pham et al. 2019; Chen and Li 2020) have been successfully applied these statistical methodologies to examine the accuracy of spatial modeling of the landslide. The main architecture of RMSE depends on the four outcomes, i.e., true positive (TP), true negative (TN), and false positive (FP), false negative (FN). Based on these outcomes, RMSE forecasts a model's error, and the lower value of the error indicates greater precision of the model (Gorum et al. 2008; Shirzadi et al. 2017a). For the calculation of RMSE in the present study, the formula has been incorporated (Eq. 36.9) (Nguyen et al. 2019).

$$\text{RMSE} = \sqrt{\frac{\sum_{i=1}^{n}(X_{\text{ac}} - X_{\text{ob}})^2}{n}} \quad (36.9)$$

where X_{ac} denotes the actual value, X_{ob} indicates the output value, and n is the total points taken to calculate RMSE. Among the most well-tested statistical methods for checking a spatial model's accuracy, MAE is one of the most reliable methods. Many research scholars (Chen and Li 2020; Lei et al. 2020) have extensively used this method across the globe. MAE measures the error of a model using the difference between actual and forecasted variables (Hong et al. 2019). The accuracy and forecasting ability of a model also depends on the value of MAE. A smaller MAE value indicates greater precision of the model and vice-versa. Based on the following equation (Eq. 36.10), MAE was calculated in the present study.

$$\text{MAE} = \frac{(p_1 - a_1) + (p_2 - a_2) + \cdots + (p_n - a_n)}{n} \quad (36.10)$$

The estimated values are represented by p, the real values are denoted by a, and the total number of observations is denoted by n (Bui et al. 2014).

36.4 Results

36.4.1 Analyzing the Multicollinearity

The statistical modeling of landslide susceptibility with multiple triggering and causative

Table 36.2 Result of multicollinearity assessment

Factor	Tolerance	VIF
Rainfall	0.250	3.873
Relief	0.271	4.044
Slope	0.890	1.111
Aspect	0.821	1.132
Plan curvature	0.756	1.258
Profile curvature	0.787	1.255
Geology	0.861	1.170
Distance from lineament	0.865	1.152
Drainage density	0.845	1.172
Distance from drainage	0.814	1.139
SPI	0.881	1.138
TWI	0.883	1.131
STI	0.980	1.016
NDVI	0.975	1.015
LULC	0.632	1.610
Soil	0.821	1.233
Soil texture	0.450	2.193
Soil depth	0.771	1.295
Soil capability	0.805	1.171
Distance from road	0.924	1.089

factors requires special attention before selecting elements. Apart from assessing the relationship with the landslide, proper estimation of the relationship among various factors is essential. Because sometimes strong inter-relationship among the dependent variables leads to the decrease of the model's ultimate precision. TOL and VIF is a well-applied (Arabameri et al. 2020; Pourghasemi et al. 2020) methodology to solve the multicollinearity problem that has been successfully applied in the present study. The result of the multicollinearity assessment portrays that multicollinearity is absent among the selected landslide conditioning factors. The lowest TOL value (0.250) was obtained in rainfall, which is much higher than the threshold value of severe multicollinearity (Table 36.2). On the contrary, the relief yields the highest VIF value (4.044) is much lower than 10. Hence all the selected causative factors were included both in the BRTree and REPTree models.

36.4.2 Selecting Significant Landslide Causing Factors (LCFs) by IGR

For landslide susceptibility evaluation of the Rorachu river basin, the current study includes one triggering factor, namely rainfall, and 18 causative factors. Various research and scientific analysis of landslide throughout the globe have successfully established that triggering factors like rainfall and seismicity alone can make favorable conditions for the landslide initiation. In India, rainfall triggered many destructive landslides in several hilly states like Uttarakhand, Sikkim, etc. For instance, Sikkim faced plenty of times rainfall-triggered severe landslides, among which the landslide of 1968 and 28th August 2010 are some of the most destructive ones (Sharma et al. 2014). The IGR was used as a standard method in this research to evaluate the major contributing factors for slope instability in

Table 36.3 Identification of important factors by IGR

Factors	IGR
Rainfall	0.915
Relief	0.698
Slope	0.641
Aspect	0.512
Plan curvature	0.048
Profile curvature	0.022
Geology	0.256
Distance from lineament	0.235
Drainage density	0.073
Distance from drainage	0.495
SPI	0.007
TWI	0.052
STI	0.742
NDVI	0.094
LULC	0.789
Soil	0.432
Soil texture	0.224
Soil depth	0.274
Soil capacity	0.312
Distance from road	0.374

the research region. The result of IGR (Table 36.3) displays that with the value of 0.915, triggering factor rainfall is the most important landslide conditioning factor followed by LULC (0.789), STI (0.742), relief (0.698), slope (0.641), slope aspect (0.512), distance from drainage (0.495), etc. Conversely, SPI is the least significant factor for the landslide with the lowest IGR value of 0.007 (Table 36.3).

36.4.3 Analysis of Landslide Susceptibility Map (LSM)

The LSM is an important tool to build proper knowledge about the probable location of future landslide occurrence (Park and Kim 2019; Saha et al. 2020). It is also significant for selecting appropriate places for planning and development in hilly areas. In this analysis, BRTree and REPTree, two common machine learning methods, were used to create landslide susceptibility maps for the Rorachu river basin in eastern Sikkim Himalaya. Values of causative factors were extracted using the training data of 135 landslide and 780 non-landslide points. After the calculation, the probability values of a total of 915 points (135 landslide and 780 non-landslide points) were plotted against the points, and the inverse distance weighted (IDW) interpolation technique was exercised to formulate landslide susceptibility index (LSIs). The whole work has been successfully done by the ArcMap 10.3 software. LSI of both BRTree and REPTree models were classified into very low (VL), low (L), moderate (M), high (H), and very high (VH) landslide susceptibility classes to produce landslide susceptibility maps (LSMs) (Fig. 36.5 a, b). The quantile reclassification technique was engaged for the classification of LSIs. Plenty of previous works have prescribed the quantile reclassification scheme as the most suitable LSI classification technique (Pham et al. 2019). Landslide susceptibility class-wise distribution of

Fig. 36.5 Landslide susceptibility map: **a** BRTree model, **b** REPTree model

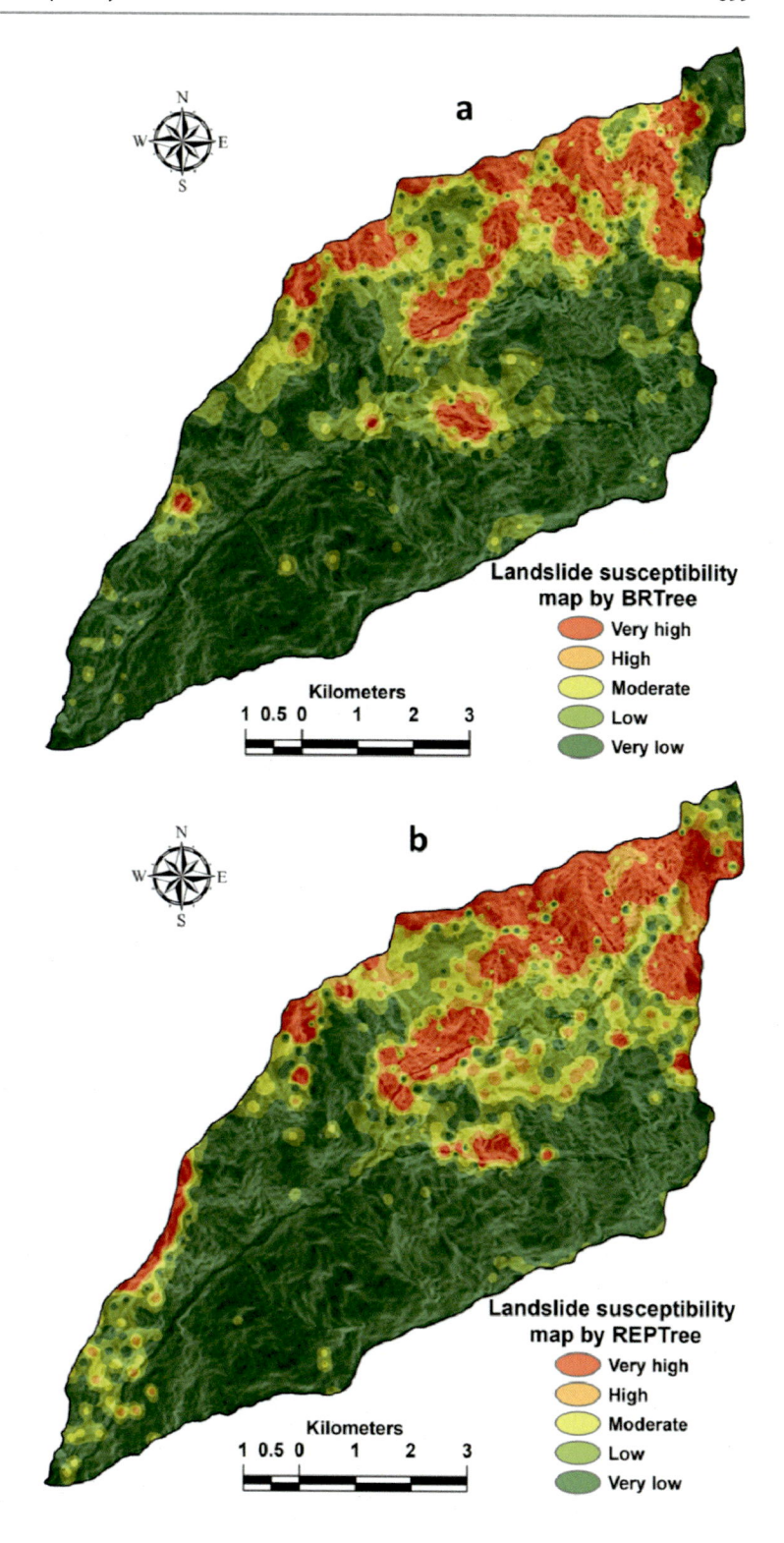

Table 36.4 LSI wise distribution of total area and total landslide area of the Rorachu river basin

LSI	LSI class wise total area distribution				LSI class wise landslide area distribution			
	BRTree model		REPTree model		BRTree model		REPTree model	
	Number of pixel	Area (%)	Number of pixel	Area (%)	Number of pixel	Area (%)	Number of pixel	Area (%)
VL	267,566	60.39905	239,905	54.15499	12	0.201072	156	2.613941
L	72,640	16.3974	69,206	15.62223	31	0.519437	198	3.317694
M	32,640	7.367996	44,437	10.03099	138	2.312332	802	13.43834
H	21,526	4.859175	28,227	6.371826	425	7.121314	744	12.46649
VH	48,625	10.97637	61,222	13.81996	5362	89.84584	4068	68.16354

area shows that the very low (VL) landslide susceptibility class acquires the maximum area of 60.399% and 54.154% in the BRTree and REPTree model respectively (Table 36.4). On the other hand, the high (H) landslide susceptibility class grabbed the lowest area in the model (Table 36.4). Landslide susceptibility class-wise distribution of landslide area depicts more or less similar results for both the models. Very high landslide susceptibility classes (VH) have covered a maximum landslide area of 89.845% and 68.163% for the BRTree and REPTree models, respectively. Conversely, very low (VL) landslide susceptibility classes have acquired landslide areas of 0.201% for BRTree and 2.613% for the REPTree model (Table 36.4). Table 36.4 further shows that high (H) and very high (VH) landslide susceptibility classes together possess 99.631% and 96.797% of landslide area in the BRTree and REPTree model.

36.4.4 Model Validation

36.4.4.1 Receiver Operating Characteristics (ROC) Curve

The ROC curves of the training (Fig. 36.6a) and validation (Fig. 36.6b) datasets were used to determine the overall precision of both models. The training data consists of 135 landslide and 780 non-landslide points, and validation data contains 59 landslide and 336 non-landslide points. The ROC curve result shows that the

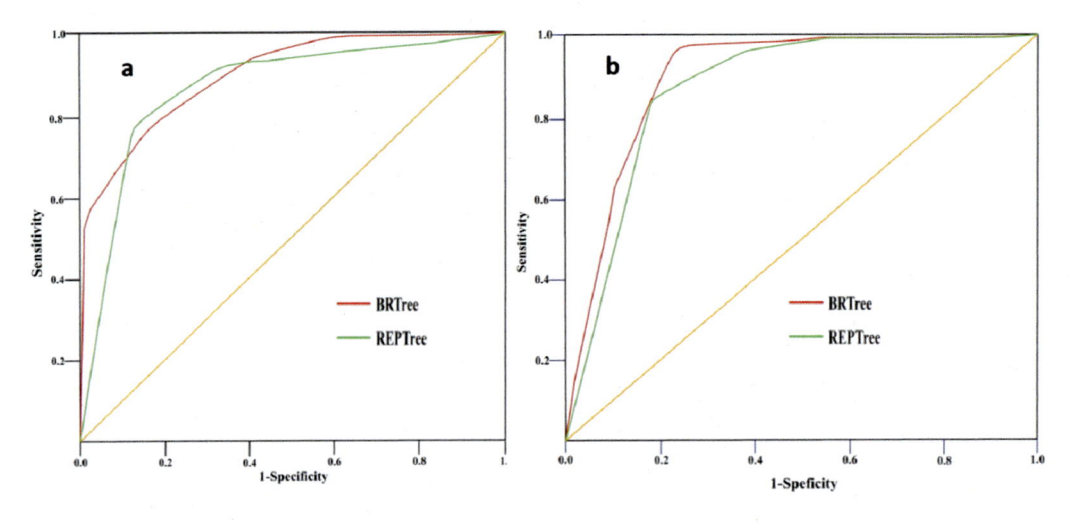

Fig. 36.6 ROC curves for the landslide susceptibility models using **a** training data and **b** validation dataset

Table 36.5 Accuracy assessment by ROC curve

Models	Area	Std. error	Asymptotic Sig.	Asymptotic 95% confidence interval	
				Lower bound	Upper bound
Area under the curve (training data)					
BRTree model	0.896	0.014	0.000	0.868	0.924
REPTree model	0.868	0.017	0.000	0.836	0.901
Area under the curve (validation data)					
BRTree model	0.900	0.017	0.000	0.867	0.932
REPTree model	0.873	0.020	0.000	0.835	0.912

AUC value of the BRTree model is more than the REPTree model in both training and validation data (Table 36.5). The AUC value of the BRTree model was obtained by 0.896 and 0.900 for the training and validation dataset, respectively, where the training and validation datasets of the REPTree model produced an AUC value of 0.868 and 0.873.

36.4.4.2 Analyzing the Results of RMSE and MAE

The accuracy of the present work was further estimated using two useful statistical methodologies, viz. RMSE and MAE. The smaller value of both the statistical techniques indicates the better performance of the model. RMSE value of BRTree model for training and validation dataset was obtained 0.112 and 0.079 where the value was 0.126 and 0.129 for REPTree model (Table 36.6). In the case of the MAE, the BRTree model has attained 0.162 and 0.178 for training and validation datasets. The training and validation datasets have yield 0.189 and 0.203 MAE values in the REPTree model (Table 36.6).

36.5 Discussion

Landslide is one of the most dangerous geographical hazards, posing a significant threat to the environmental and socioeconomic resources of mountainous areas (Ghasemain et al. 2020). Landslide can be stunned hilly lives and livelihoods for a few hours to few weeks by damaging roads, collapsing houses, and destructing bridges, for instance, the Chandmari landslide of Sikkim (1997). This quasi-natural hazard is also a severe threat to India's hilly states with international boundaries like Sikkim, Uttarakhand, etc. Because roads play a pivotal role in national security and landslide, sometimes border areas remain disconnected for few days, e.g., 15 Mile landslide, en route to Tsomgo Lake and Nathu la, Sikkim (2012). Hence a proper management strategy is essential to check landslide-induced risk in the mountains. Landslide susceptibility map (LSM) is a very useful tool in this regard. Many researchers (Nhu et al. 2020; Chen and Li 2020) and scientists have already successfully

Table 36.6 Understanding the precision of models by RMSE and MAE

Matrices	BRTree model	REPTree model
Training data		
RMSE	0.112	0.126
MAE	0.162	0.189
Validation data		
RMSE	0.079	0.129
MAE	0.178	0.203

established the usefulness and applicability of LSM in landslide risk reduction. An LSM not only reflects the periphery of landslide-prone areas but also highlights the safe places for developmental activities (Mandal and Mandal 2018). Consequently, the preparation of an accurate LSM is the key matter in the landslide susceptibility analysis (Mandal et al. 2021). Researchers and geoscientists (Lei et al. 2020; Fang et al. 2020) are continuously working on precise LSM making techniques globally. Still, the task isn't easy—successive research on landslide susceptibility yields many computer-based hybrid machine learning techniques which have already produced LSM with excellent exactitude (Chen et al. 2018; Mandal et al. 2021). In search of the most precise landslide susceptibility analysis technique, the present study has compared two well-applied machine learning algorithms, i.e., BRTree (Ebrahimy et al. 2020a, b; Roy et al. 2020), REPTree (Pham et al. 2019; Saha et al. 2020) in the Sikkim Himalayan context. Plenty of research on landslide susceptibility analysis has been successfully carried out in the Sikkim Himalayan region, but modern machine learning techniques aren't thoroughly applied. The present study aims to fulfill this research gap with hybrid machine learning algorithms.

Multi-factor geohazard landslide consists of two sets of factors, i.e., triggering and causative. Triggering factors alone (like seismicity and intensive downpour) can create massive slope instability in the mountains, whereas causative factors make an ideal landslide situation. Side by side, both triggering and causative factors' combined role destruct slope stability and lead to a massive landslide. Hence the proper study of landslide susceptibility requires specific knowledge about the role of both the factors in a particular place because the role of factors in the landslide occurrence may differ in different sites (Mandal et al. 2021). Apart from this, the landslide conditioning factor selection is also important for the precision of the landslide susceptibility model. Factors with less significance or minimum role in the landslide event create noise in the model and increase error.

Formulation of a precise landslide susceptibility model depends on selecting factors and estimating the role of those factors in the landslide occurrence. However, there are no fixed methods or techniques for picking landslide conditioning factors (Chen et al. 2018; Mandal et al. 2021). Based on the previous literature and field study, 19 landslide conditioning factors were selected. Before using those factors in the model, inter-relationship among the factors was estimated using multicollinearity assessment. Multicollinearity test outcome allows all landslide conditioning components to be included in the model. After that, the role of landslide conditioning factors was statistically estimated by the IGR method. IGR method has perfectly reflected the fundamental role of the landslide conditioning factors. Triggering factor rainfall dominates the other factors in terms of role in the landslide occurrence (Table 36.3). Increasing anthropogenic activities like the construction of roads and settlements, deforestation, etc., is one of the major causes of slope failure in the Rorachu river basin (Mandal and Mandal 2018). IGR value of LULC also indicates a similar fact (Table 36.3). The high IGR value of STI directly highlights the active downcutting of lower-order streams, making the slope unstable. In the Rorachu river basin, most of the landslides were noticed along the roads and lower-order streams in the northern, northeastern, and northwestern parts, where relief and slope are high. These factors possess the highest value of IGR, which indicates the proper evaluation of landslide condition factors' role by the IGR method. In the present research, the outcome of IGR is similar to those of studies of Dou et al. (2020), Ghasemain et al. (2020), and Yao et al. (2020). The present study employed two advanced machine learning techniques, and a comparative study has been successfully carried out. Behind the selection of methods, the prime factor was the accuracy of the model. The BRTree method is suitable for analyzing the non-linear relationship of the physical process (Youssef et al. 2016). Side by side, the "boosting" mechanism of the BRTree model enhances the model's precision, which is important to attain the goal of the study (Youssef et al. 2016). The REPTree model, on the contrary, has a high footprint for reducing the problem of overfitting (Chen and Li 2020). REPTree model successfully deals with the problem of noise in the model. The "pruning" feature mainly removes the noise, which leads to the increment of precision of the model (Phong et al. 2019; Chen and Li 2020).

The landslide susceptibility model highlights the areas amenable to landslide based on the subsisted landslide data (Mandal et al. 2021). A landslide susceptibility map is a useful tool because it contains both the existing landslide area and the areas with a greater tendency to become landslide-prone. Table 36.4 shows that only 8.946% area of the basin possesses 84.84% of the total landslide area in the BRTree model. On the other hand, in the REPTree model, 76.941% landslide area is found under 19.27% basin area. This condition indicates the alarming situation of the areas under higher landslide susceptibility classes in the Rorachu river basin. Landslide density is unfolding the similar fact that landslide concentration increases sharply with increasing landslide susceptibility classes. LSM's of both BRTree and REPTree model shows that northern, northeastern parts, and southwestern periphery of the basin are more vulnerable to landslide. Both the LSM's produced more or less similar results except the distribution of very low and low landslide susceptibility classes. In the BRTree model, southern and most southwestern parts of the Rorachu river basin belong to very low and low landslide susceptibility classes (Fig. 36.5a). On the other hand, a very low landslide susceptibility class is absent in the REPTree model, and low landslide susceptibility class captures most of the southern parts of the basin (Fig. 36.5b). Several research works (Mandal and Mandal 2018; Mandal et al. 2021) on the Rorachu river basin have reflected the similar result that southern part of the basin comparatively stable and less prone to landslide. The present study depicts that several places of Gangtokviz. Deorali, Tathangchen, Samdur, etc., are in severe threat of landslide. These places are one of the major habitable areas of Sikkim's capital city which belong to very high landslide susceptibility zone. Apart from these places, 15th Mile (near Kyangnosla), 7th Mile along J.N. Road, Rongyek in the northern stretch, Hanumantok, upper Chandmari in the northwestern stretch, Namok in central, and Bhusuk in the eastern stretch of the Rorachu river basin also belongs to very high and high landslide susceptibility classes. In the REPTree model, some scattered distribution of high landslide susceptibility zone is found in the vicinity of Nandok. Rorachu river basin is an important area for both Sikkim and India. Because Gangtok, the capital of Sikkim, is located in the western-southwestern periphery of the basin and the communication with Nathu La, the Indo-China border is largely maintained by J.N. Road, which is extended through the Rorachu river basin. LSMs show that very high and high landslide susceptibility cover more than 30% of the total area of such an important basin (Fig. 36.5a, b) (Table 36.4).

Accuracy assessment and validation of the model using proper methods are considered as key works of landslide susceptibility modeling. A model is useful only when it satisfies a level of very high precision. Several statistical indices, such as the ROC curve, RMSE, and MAE, have been integrated in the current study to support the purposes of accuracy evaluation, model validation, and model comparison. The AUC value of both training and validation datasets in the ROC curve reflects that both models have secured enough precision. Still, BRTree has exceeded the REPTree model in terms of accuracy (Table 36.5). In RMSE and MAE test (Table 36.6), the BRTree model acquires better accuracy by securing the lowest value than the REPTree model. So it can be said that in the present study, the BRTree model has outperformed the REPTree model in the Indian Himalayan context. In landslide modeling, plenty of literature has established the BRTree model better than other machine learning techniques. But a comparative study of landslide susceptibility analysis using BRTree and REPTree model in the Indian Himalayan context has not been done yet. Ebrahimy et al. (2020a, b), in the study of land subsidence susceptibility analysis, have found the better performance of the BRTree model. A similar result was noticed in the research works of Youssef et al. (2016), Kim et al. (2018), and Park and Kim (2019). Arabameri et al. (2019) have shown that even in the ensemble model, BRTree performs better than the other ensemble models. The present study has successfully identified the most precise and suitable machine learning technique among two very well-applied decision tree algorithms. This study is also important because it fulfilled the research gap of comparative study among "boosting" based BRTree and "pruning" based REPTreemodels.

36.6 Conclusion

Quasi-natural geohazard landslide is a serious threat to the hilly mountainous terrain's natural and socio-economic environment. Roads are treated as the artillery of the mountains, which is the most landslide-affected anthropogenic component. Due to rapid population growth and expansion of tourism during the last few decades, the slope modification process by anthropogenic activities (like construction of houses, roads, etc.) sharply increased. As a result, the problem of slope instability, as well as landslide, has also energized. The present study has employed two machine learning algorithms, i.e., BRTree and REPTree, to fulfill the goal of the preparation of precise LSM. GIS is widely used mainly to prepare the entire database of the study and integrate various data layers. Twenty data layers (including 19 causative factors and LIM) have been prepared using Arc Map 10.3 GIS software package. Landslide susceptibility distribution maps of the Rorachu river basin enlighten that 30.666% and 38.93% of the basin's total area belong to very high and high landslide susceptibility classes in BRTree and REPTree model, respectively. Validation of landslide susceptibility models was validated using AUC of ROC curve, RMSE, and MAE. These statistical indices show that the models have secured very good results, but the BRTree model performs better than the REPTree model in terms of accuracy. Despite the good result, it is admissible that the present study has some limitations too. The study has encountered some major limitations like lack of large-scale soil and geological map, scarcity of geophysical data in high altitudes like 15th Mile near Kyangnosla, and the absence of structural geology data in the model. Finally, it can be concluded that this research work has successfully overcome the limitations and yields two high-precision landslide susceptibility maps. These maps can be a useful tool for government authorities and planners to identify stable sites for future development.

Conflict of Interest None.

Funding None.

References

Aertsen W, Kint V, Van Orshoven J, Özkan K, Muys B (2010) Comparison and ranking of different modelling techniques for prediction of site index in Mediterranean mountain forests. Ecol Model 221(8):1119–1130

Arabameri A, Pradhan B, Rezaei K, Sohrabi M, Kalantari Z (2019) GIS-based landslide susceptibility mapping using numerical risk factor bivariate model and its ensemble with linear multivariate regression and boosted regression tree algorithms. J Mt Sci 16 (3):595–618

Arabameri A, Saha S, Roy J, Chen W, Blaschke T, Tien Bui D (2020) Landslide susceptibility evaluation and management using different machine learning methods in the Gallicash River Watershed, Iran. Remote Sens 12(3):475

Beven KJ, Kirkby MJ (1979) Un modèle à base physique de zone d'appel variable de l'hydrologie du bassin versant (A physically based, variable contributing area model of basin hydrology). Hydrol Sci J 24(1):43–69

Bui DT, Ho TC, Pradhan B, Pham BT, Nhu VH, Revhaug I (2016a) GIS-based modeling of rainfall-induced landslides using data mining-based functional trees classifier with AdaBoost, Bagging, and MultiBoost ensemble frameworks. Environ Earth Sci 75 (14):1–22

Bui DT, Ho TC, Revhaug I, Pradhan B, Nguyen DB (2014) Landslide susceptibility mapping along the national road 32 of Vietnam using GIS-based J48 decision tree classifier and its ensembles. In: Cartography from pole to pole. Springer, Berlin, Heidelberg, pp 303–317

Bui DT, Lofman O, Revhaug I, Dick O (2011) Landslide susceptibility analysis in the HoaBinh province of

Vietnam using statistical index and logistic regression. Nat Hazards 59(3):1413

Bui DT, Pradhan B, Revhaug I, Nguyen DB, Pham HV, Bui QN (2015) A novel hybrid evidential belief function-based fuzzy logic model in spatial prediction of rainfall-induced shallow landslides in the Lang Son city area (Vietnam). Geomat Nat Hazards Risk 6(3):243–271

Chen CH, Ke CC, Wang CL (2009) A back-propagation network for the assessment of susceptibility to rock slope failure in the eastern portion of the Southern Cross-Island Highway in Taiwan. Environ Geol 57(4):723–733

Chen W, Li Y (2020) GIS-based evaluation of landslide susceptibility using hybrid computational intelligence models. CATENA 195:104777

Chen W, Pourghasemi HR, Kornejady A, Xie X. (2019) GIS-based landslide susceptibility evaluation using certainty factor and index of entropy ensembled with alternating decision tree models. In: Natural hazards GIS-based spatial modeling using data mining techniques. Springer, Cham, pp 225–251

Chen W, Xie X, Peng J, Wang J, Duan Z, Hong H (2017) GIS-based landslide susceptibility modelling: a comparative assessment of kernel logistic regression, Naïve-Bayes tree, and alternating decision tree models. Geomatics Nat Hazards Risk 8(2):950–973

Chen W, Zhang S, Li R, Shahabi H (2018) Performance evaluation of the GIS-based data mining techniques of best-first decision tree, random forest, and naïve Bayes tree for landslide susceptibility modeling. Sci Total Environ 644:1006–1018

Devasena CL (2014) Comparative analysis of random forest, REP tree and J48 classifiers for credit risk prediction. Int J Comput Appl 0975–8887

Dilley M (2005) Natural disaster hotspots: a global risk analysis, vol 5. World Bank Publications

Dou J, Yunus AP, Bui DT, Merghadi A, Sahana M, Zhu Z, Chen CW, Han Z, Pham BT (2020) Improved landslide assessment using support vector machine with bagging, boosting, and stacking ensemble machine learning framework in a mountainous watershed, Japan.Landslides 17, 641–658

Dubey CS, Chaudhry M, Sharma BK, Pandey AC, Singh B (2005) Visualization of 3-D digital elevation model for landslide assessment and prediction in mountainous terrain: a case study of Chandmari landslide, Sikkim, eastern Himalaya. Geosci J 9(4):363–373

Ebrahimy H, Feizizadeh B, Salmani S, Azadi H (2020a) A comparative study of land subsidence susceptibility mapping of Tasuj plane, Iran, using boosted regression tree, random forest and classification and regression tree methods. Environ Earth Sci 79(223):223

Ebrahimy H, Feizizadeh B, Salmani S, Azadi H (2020b) A comparative study of land subsidence susceptibility mapping of Tasuj plane, Iran, using boosted regression tree, random forest and classification and regression tree methods. Environ Earth Sci 79:1–12

Elith J, Leathwick JR, Hastie T (2008) A working guide to boosted regression trees. J Anim Ecol 77(4):802–813

Fang Z, Wang Y, Peng L, Hong H (2020) Integration of convolutional neural network and conventional machine learning classifiers for landslide susceptibility mapping. Comput Geosci 139:104470

Froude MJ, Petley DN (2018) Global fatal landslide occurrence from 2004 to 2016. Nat Hazards Earth Syst Sci 18(8):2161–2181

Ghasemain B, Asl DT, Pham BT, Avand M, Nguyen HD, Janizadeh, SJVJOES (2020) Shallow landslide susceptibility mapping: A comparison between classification and regression tree and reduced error pruning tree algorithms. Vietnam J Earth Sci 42(3):208–227

Ghosh T, Bhowmik S, Jaiswal P, Ghosh S, Kumar D (2020) Generating substantially complete landslide inventory using multiple data sources: a case study in Northwest Himalayas, India. J Geol Soc India 95(1):45–58

Gorum T, Gonencgil B, Gokceoglu C, Nefeslioglu HA (2008) Implementation of reconstructed geomorphologic units in landslide susceptibility mapping: the Melen Gorge (NW Turkey). Nat Hazards 46(3):323–351

Hong H, Naghibi SA, Pourghasemi HR, Pradhan B (2016a) GIS-based landslide spatial modeling in Ganzhou City, China. Arab J Geosci 9(2):1–26

Hong H, Liu J, Zhu AX (2019) Landslide susceptibility evaluating using artificial intelligence method in the Youfang district (China). Environ Earth Sci 78(15):488

Kim JC, Lee S, Jung HS, Lee S (2018) Landslide susceptibility mapping using random forest and boosted tree models in Pyeong-Chang, Korea. Geocarto Int 33(9):1000–1015

Klose M, Maurischat P, Damm B (2016) Landslide impacts in Germany: a historical and socioeconomic perspective. Landslides 13(1):183–199

Lay US, Pradhan B, Yusoff ZBM, Abdallah AFB, Aryal J, Park HJ (2019) Data mining and statistical approaches in debris-flow susceptibility modelling using airborne LiDAR data. Sensors 19(16):3451

Lee SR, Kim YT (2020) Spatial probability assessment of landslide considering increases in pore-water pressure during rainfall and earthquakes: case studies at Atsuma and Mt. Umyeon. CATENA 187:104317

Lei X, Chen W, Pham BT (2020) Performance evaluation of GIS-based artificial intelligence approaches for landslide susceptibility modeling and spatial patterns analysis. ISPRS Int J Geo Inf 9(7):443

Mandal S, Mandal K (2018) Bivariate statistical index for landslide susceptibility mapping in the Rorachu river basin of eastern Sikkim Himalaya, India. Spat Inf Res 26(1):59–75

Mandal K, Saha S, Mandal S (2021) Applying deep learning and benchmark machine learning algorithms for landslide susceptibility modelling in Rorachu river basin of Sikkim Himalaya, India. Geosci Front 12(5):101203

Menard S (1995) An introduction to logistic regression diagnostics. Appl Logist Regress Anal 58–79

Moore ID, Burch GJ (1986) Sediment transport capacity of sheet and rill flow: application of unit stream power theory. Water Resour Res 22(8):1350–1360

Moore ID, Grayson RB, Ladson AR (1991) Digital terrain modelling: a review of hydrological, geomorphological, and biological applications. Hydrol Process 5(1):3–30

Nadim F, Kjekstad O, Peduzzi P, Herold C, Jaedicke C (2006) Global landslide and avalanche hotspots. Landslides 3(2):159–173

NASA. Global Landslide Catalog. Available online: https://data.nasa.gov/EarthScience/Global-Landslide-Catalog/h9d8-neg4#About (accessed on 30 March 2019)

Ngo PTT, Panahi M, Khosravi K, Ghorbanzadeh O, Kariminejad N, Cerda A, Lee S (2021) Evaluation of deep learning algorithms for national scale landslide susceptibility mapping of Iran. Geosci Front 12(2):505–519

Nguyen VV, Pham BT, Vu BT, Prakash I, Jha S, Shahabi H, Shirzadi A, Ba DN, Kumar R, Chatterjee JM, Tien Bui D (2019) Hybrid machine learning approaches for landslide susceptibility modeling. Forests 10(2):157

Nhu VH, Shirzadi A, Shahabi H, Chen W, Clague JJ, Geertsema M, Jaafari A, Avand M, Miraki S, Asl DT, Pham BT (2020) Shallow landslide susceptibility mapping by random forest base classifier and its ensembles in a semi-arid region of Iran. Forests 11(4):421

Pandey VK, Pourghasemi HR, Sharma MC (2020) Landslide susceptibility mapping using maximum entropy and support vector machine models along the Highway Corridor, Garhwal Himalaya. Geocarto Int 35(2):168–187

Park S, Kim J (2019) Landslide susceptibility mapping based on random forest and boosted regression tree models, and a comparison of their performance. Appl Sci 9(5):942

Petley D (2012) Global patterns of loss of life from landslides. Geology 40(10):927–930

Pham BT, Prakash I, Singh SK, Shirzadi A, Shahabi H, Bui DT (2019) Landslide susceptibility modeling using Reduced Error Pruning Trees and different ensemble techniques: hybrid machine learning approaches. CATENA 175:203–218

Pham BT, Prakash I, Dou J, Singh SK, Trinh PT, Tran HT, Le TM, Van Phong T, Khoi DK, Shirzadi A, Bui, DT (2020) A novel hybrid approach of landslide susceptibility modelling using rotation forest ensemble and different base classifiers. Geocarto Int 35(12):1267–1292

Phong TV, Phan TT, Prakash I, Singh SK, Shirzadi A, Chapi K, Ly HB, Ho LS, Quoc NK, Pham BT (2019) Landslide susceptibility modeling using different artificial intelligence methods: a case study at Muong Lay district, Vietnam. Geocarto Int 1–24

Pourghasemi HR, Kornejady A, Kerle N, Shabani F (2020) Investigating the effects of different landslide positioning techniques, landslide partitioning approaches, and presence-absence balances on landslide susceptibility mapping. CATENA 187:104364

Quinlan JR (1987) Simplifying decision trees. Int J Man Mach Stud 27(3):221–234

Quinlan JT (1993) C4.5: programs for machine learning. Morgan Kaufmann Publishers, San Francisco, CA

Rahmati O, Naghibi SA, Shahabi H, Bui DT, Pradhan B, Azareh A, Rafiei-Sardooi E, Samani AN, Melesse AM (2018). Groundwater spring potential modelling: comprising the capability and robustness of three different modeling approaches. J Hydrol 565:248–261

Regmi AD, Devkota KC, Yoshida K, Pradhan B, Pourghasemi HR, Kumamoto T, Akgun A (2014) Application of frequency ratio, statistical index, and weights-of-evidence models and their comparison in landslide susceptibility mapping in Central Nepal Himalaya. Arab J Geosci 7(2):725–742

Roy J, Saha S, Arabameri A, Blaschke T, Bui DT (2019) A novel ensemble approach for landslide susceptibility mapping (LSM) in Darjeeling and Kalimpong Districts, West Bengal, India. Remote Sens 11(23):2866

Roy P, Chandra Pal S, Arabameri A, Chakrabortty R, Pradhan B, Chowdhuri I, Lee S, Tien Bui D (2020) Novel ensemble of multivariate adaptive regression spline with spatial logistic regression and boosted regression tree for gully erosion susceptibility. Remote Sens 12(20):3284

Saha S, Saha A, Hembram TK, Mandal K, Sarkar R, Bhardwaj D (2022) Prediction of spatial landslide susceptibility applying the novel ensembles of CNN, GLM and random forest in the Indian Himalayan region. Stochast Environ Res Risk Assess 1–20

Saha S, Saha M, Mukherjee K, Arabameri A, Ngo PTT, Paul GC (2020) Predicting the deforestation probability using the binary logistic regression, random forest, ensemble rotational forest and REPTree: a case study at the Gumani River Basin, India. Sci Total Environ 139–197

Sameen MI, Pradhan B, Lee S (2020) Application of convolutional neural networks featuring Bayesian optimization for landslide susceptibility assessment. Catena 186:104249

Schapire RE (2003) The boosting approach to machine learning: an overview. In: Nonlinear estimation and classification. Springer, New York, NY, pp 149–171

Sharma LP, Patel N, Ghose MK, Debnath P (2014) Application of frequency ratio and likelihood ratio model for geo-spatial modelling of landslide hazard vulnerability assessment and zonation: a case study from the Sikkim Himalayas in India. Geocarto Int 29(2):128–146

Shirzadi A, Bui DT, Pham BT, Solaimani K, Chapi K, Kavian A, Shahabi H, Revhaug I (2017a) Shallow landslide susceptibility assessment using a novel hybrid intelligence approach. Environ Earth Sci 76(2):60

Shirzadi A, Solaimani K, Roshan MH, Kavian A, Chapi K, Shahabi H, Keesstrad S, Ahmade BB, Bui DT (2019). Uncertainties of prediction accuracy in shallow landslide modeling: sample size and raster resolution. CATENA 178:172–188

Shit PK, Bhunia GS, Maiti R (2016) Potential landslide susceptibility mapping using weighted overlay model (WOM). Model Earth Syst Environ 2(1):21

Siddique T, Khan EA (2019) Stability appraisal of road cut slopes along a strategic transportation route in the Himalayas, Uttarakhand, India. SN Appl Sci 1(5):409

Varnes DJ (1978) Slope movement types and processes. Special Rep 176:11–33

Yao J, Qin S, Qiao S, Che W, Chen Y, Su G, Miao Q (2020) Assessment of landslide susceptibility combining deep learning with semi-supervised learning in Jiaohe County, Jilin Province, China. Appl Sci 10(16):5640

Youssef AM, Pourghasemi HR, Pourtaghi ZS, Al-Katheeri MM (2016) Landslide susceptibility mapping using random forest, boosted regression tree, classification and regression tree, and general linear models and comparison of their performance at Wadi Tayyah Basin, Asir Region, Saudi Arabia. Landslides 13(5):839–856

Zêzere JL, Pereira S, Melo R, Oliveira SC, Garcia RA (2017) Mapping landslide susceptibility using data-driven methods. Sci Total Environ 589:250–267

Zhang K, Wu X, Niu R, Yang K, Zhao L (2017) The assessment of landslide susceptibility mapping using random forest and decision tree methods in the Three Gorges Reservoir area, China. Environ Earth Sci 76(11):1–20

Zhou C, Yin K, Cao Y, Ahmed B, Li Y, Catani F, Pourghasemi HR (2018) Landslide susceptibility modelling applying machine learning methods: a case study from Longju in the Three Gorges Reservoir area, China. Comput Geosci 112:23–37

An Exploratory Analysis of Mountaineering Risk Estimation Among the Mountaineers in the Indian Himalaya

Chinmoy Biswas, Koyel Roy, Rupan Dutta, and Shasanka Kumar Gayen

Abstract

Mountaineering is likely a significant risk than any other type of adventure tourism. While mountaineers seek adventure on high mountain peaks, they are inevitably exposed to dangerous and deadliest factors such as avalanches, steep slopes, bad weather, rock fall, and other natural diasters, as well as to the physical risk of climbing. This study particularly emphasizes on the mountaineering risks estimation and heterogeneity of mountaineering risks in the Indian Himalaya. Information was derived from the Indian mountaineer journal and the expedition reports of the IMF (Indian Mountaineering Foundation). Our findings of mountaineering risks are almost uniformly coherent among the four significant mountain states (Sikkim Himachal Pradesh, Uttarakhand, Jammu Kashmir & Ladakh). Another risk component of mountaineering has been predicted; however, it differs from technically easy to the high elevation Himalayan peak. Even though, mountaineering deaths are few they are closely linked to mountaineering tourism. Every expedition member (including Sherpa) who lost their lives exhibits all mechanisms of death. The main goal of frequency distribution analysis on Indian mountaineers is to derive from the unsuccessful mountain summit factors. The ward's hierarchical cluster approach has been used to classify mountaineering tourism risks into five groups to understand better which risk cluster is most or least vulnerable to mountain climbing on the high mountain. Only 12 mountaineering risk components have been identified from 125 unsuccessful Indian expeditions, with bad weather and ration shortage being the two most and least vulnerable risk components, respectively.

Keywords

Risk estimation · State-wise mountaineering risk · Risk classification · Mountaineering deaths

37.1 Introduction

High-risk adventure tourism is a rapidly growing global industry (Bott 2009). Mountaineering is also one of the adventure tourism that is growing rapidly all over the world and is a risk-taking sport. Maurice Herzog and Louis Lachenal became the first mountaineers to reach the top of an 8000 m peak (Annapurna, 8091 m) in 1950. Since then pioneering achievement, thousands of mountain climbers have ventured to Nepal,

---

C. Biswas (✉) · K. Roy · R. Dutta · S. Kumar Gayen
Department of Geography, Coochbehar Panchanan Barma University, Cooch Behar, West Bengal 736101, India
e-mail: earthsciencegeo@gmail.com

Pakistan, Tibet, India, and China to climb the 14 peaks above 8000 m from sea level. But, it is challenging to summit a mountain peak for a climber in an uncertain situation. Generally, the dangers of mountain climbing such as falling, avalanches, adverse or destructive weather, exposure, high altitude sickness, cardiac arrest, rock fall, ice fall, snow blindness, crevasses, injury, death, frost bite, mismanagement, and navigational errors are called the mountaineering risk. But, it is very tough to determine the risks. Priest and Gass (1997) demonstrate that risk is the potential to lose something. Therefore the risk is a potential and more or less foreseeable danger (Seigneur 2006). Risk can be defined as a part of the manifest attraction of certain activities, so the voluntary nature of risk engagement in leisure activity is essential to grasp (Cater 2006). Mountaineering risk is vital for every mountain explorer or adventure lover. In every step of climbing, mountaineers and mountain guides expose their lives in adverse situations, so any kind of altitudinal risk can be revealed during their climbing time and can lead to death. So risk classification and agglomeration is an important work to emphasize awareness of high-altitude expedition members and guides.

Huey et al. (2020), reported that extreme events are always associated with mountaineering at high altitudes. It brings challenges of extreme risks of hypoxia and weather as well as the usual hazards of climbing. Fa-Hien (whose dates of birth and death are not known) produced what was possibly the first description of high-altitude mountain sickness and pulmonary oedema. While he was crossing the "Little Snowy Mountains," probably near the Safed Khirs Pass in northeastern Afghanistan, Fa-Hien's companion (HWUY-KING) became ill, "a white froth came from his mouth," and he died. The description is so striking and typical that it seems likely to be the first account of high-altitude mountain sickness and pulmonary oedema (West 1998). Williamson (1985) has given importance to three events (a. fall, slip on a rock, b. falling rock, and c. avalanche) in a review of climbing accidents report of the American Alpine Club Safety Committee that should indicate that the levels of risks contribute to most accidents. The first quantitative risk estimates for high altitude mountaineering were published in the late seventies/early eighties of the last century (Wilson et al. 1978; Ridden 1983). In the same contemporary periods, the number of publications on quantitative risk estimates for high-altitude mountaineering has increased significantly (Pollard and Clarke 1988; Christensen and Lacsina 1999), but data are mainly restricted to the Nepalese Himalayas and Denali in Alaska. Notably, it is the fact that the Himalayan mountaineering tragedies are observable yet (Krakauer 2009). According to the Himalayan database report, the average death rates are 1.97 at several altitudes of the high Himalayan Mountain. Between 5000–5999 m, 6000–6999 m, 7000–7999, and 8000–8850 m, the death rates are 0.00, 0.59, 1.32, and 5.26, respectively, it gradually increased with altitude from 1950 to 2018 (Himalayan data base). Some psychological and sociological phenomena also directly affect a climber's evaluations of risks and hazards in a specific situation (Helms 2015). Undoubtedly, high-altitude mountaineering (as attempting a mountain with an altitude above 5000 m) is inherently associated with high risks due to the harsh environmental conditions, low temperatures, strong winds, and steep & difficult terrain (Nordby and Weinbruch 2013). An excellent example of the death while Albert Frederick Mummery and two Gorkha porters, Raghubir, and Goman Singh, in 1895 at Nanga Parbat, is during the first serious attempt on an 8000 m peak (Sale and Clare 2000). North America's highest peak, Denali, poses a unique combination of hazards related to high altitude, arctic climate, treacherous terrain, and remote wilderness (Clare 1993). Some specific areas on the mountain are notorious among climbers for environmental hazards (Clare 1993). He also claimed that avalanches and crevasses are significant hazards in the high mountain. An injury that occurred on snow was more likely to be fatal and may have been due in part to the increased appearance of hazards that are not immediately under the climber's control, such as avalanches, icefall, and bad weather, and apart from these, ice climbing is frequently done in remote areas, often at high altitude and in severe weather conditions" (Addis and Baker 1989).

Smith (2006) reported that mountaineering's specific injury and death patterns are challenging to determine, and the climbing member is at risk for various injuries and illness. Perhaps, the risks and hazards involved with climbing make it the challenging and rewarding sport it is (Helms 2015). Windsor et al. (2009) also claim that mountaineering deaths will inevitably occur when the mountain landscape and its environment are inevitable. Firth et al. (2008), Windsor et al. (2009), and Westensee et al. (2013) are reported that deaths can occurr during mountaineering activities, and a common cause of death was bad weather, altitude sickness, trauma, disappearance, heart failure and avalanche while they have studied on mountaineering fatalities. Climbing above 8000 m is a severe test of human performance (West 2000). Massive snow and avalanches cause the most fatalities, and hurricane-force winds, hypothermia and the notorious Khumbu ice fall take many lives on Everest (Bott 2009).

Most of the research on mountaineering risk has been done in the Nepal Himalaya, Andes, Alps and the Rocky Mountains. But, in the Indian Himalaya (Fig. 37.1), no study has been done about mountaineering risks. West (2000), Smith (2006), Firth et al. (2008), and Windsor et al. (2009) clarified the mountaineering risk during their study but explicitly avoided risk classification and risk agglomeration of mountaineering. The main purpose of this study was to create an idea of what kind of hazards are concentrated during mountaineering and find out the mountaineering risks classification.

37.2 Database and Methodology

37.2.1 Sample Selection

The information is extracted and gathered from the Indian Mountaineer Journal, Himalayan data base, and Indian expedition reports and finally compiled systematically. Repetitive samples for different peaks were considered as a sample collection process. Figure 37.1 depicts the sample collection sites. It is a long process to make the unsuccessful factors (components) of Indian mountaineering in the Indian Himalaya. In this regard, the Indian Mountaineering Foundation (New Delhi) had no specific data risk events. Firstly, each unsuccessful expedition and its factor was separated, recorded, and compiled according to the needs of the study. On account of this, four famous Himalayan states (Sikkim = 8, Himachal Pradesh = 50, Uttarakhand = 60, and JK & Ladakh = 7) were selected. A total of 125 samples were considered for the unsuccessful expedition of Indian climbers. Based on reports of the unsuccessful expedition, the uncertain factors (i.e. bad weather, death, avalanche, steep slope, mountain sickness, frostbite/snow blind, crevasse, fall, technical difficulties, lack of geographical area knowledge, alpine style, cardiac arrest, drowned into the river and shortage of ration/mismanagement) were selected. Secondly, 250 unsuccessful factors have been considered for risk classification. The fatality records and their causes have been made based on the 125 unsuccessful expeditions from IMF data. The death rate has been calculated by dividing the number of deaths by the total number of individuals who have climbed above the base camp and multiplying by 100 on the basis Himalayan database (as of 05.12.2020) between 1950 and 2018.

37.2.2 Diagnosis of Risk Agglomeration Using Hierarchical Clustering Technique

The hierarchical cluster analysis technique is very important for understanding risks agglomeration. For this, Ward's Linkage method has been followed. Ward's method is known as Ward's linkage method which relies on the minimum variance of the dataset (Gong and Richman 1995). At each stage, the method generates clusters that integrate minimum distances of two-factor weights (Güçdemir and Selim 2015). In this research, the method follows the

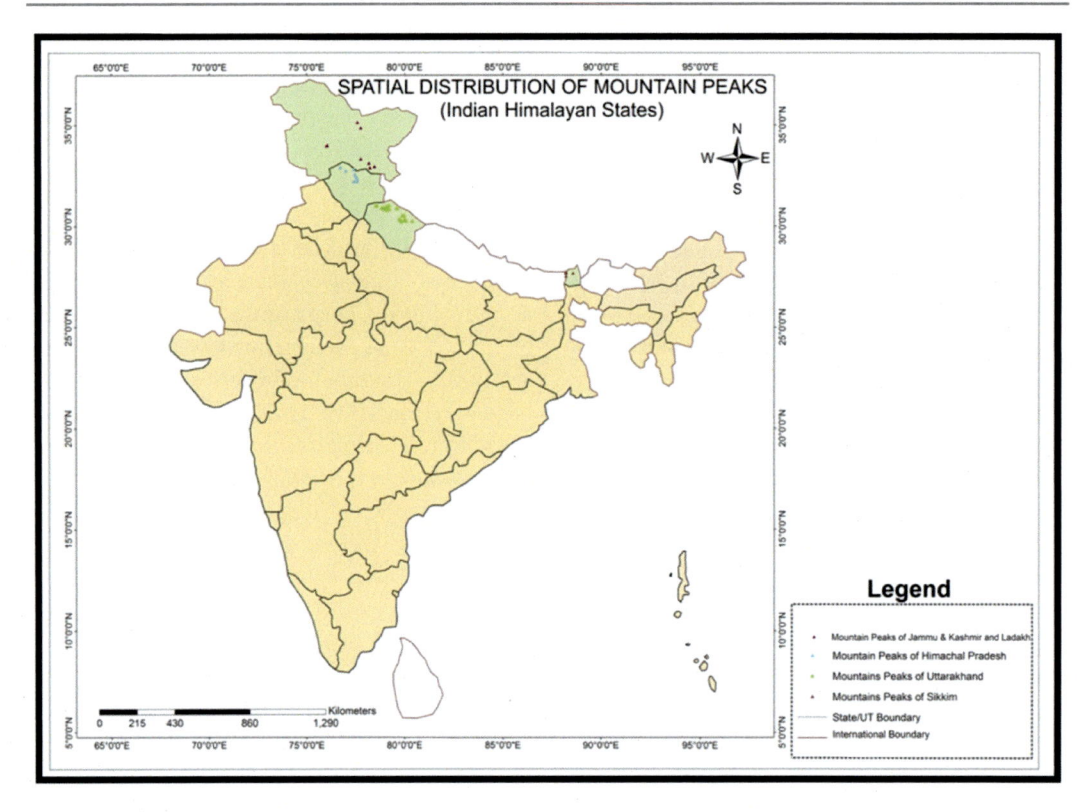

Fig. 37.1 The sample location in four states of India. *Source* Prepared by author

technique of weighted Group Sum of Squares (WGSS). The increase of WGSS while merging two groups namely k and m is as follows.

$$W_{km} = \frac{N_K N_m}{N_m + N_K} \left(\overline{X}_k - \overline{X}_m\right)^T \left(\overline{X}_k - \overline{X}_m\right) \quad (37.1)$$

where, $\overline{X}_k$ and $\overline{X}_m$ denote the centroid of clusters k and m.

37.2.3 Frequency Distribution, ANOVA, and Post-hoc Tests

Frequency distribution and percentage calculation were used to estimate the risk events of mountaineering. Statistical Package Social Software (SPSS) and Microsoft Excel software were applied as a tool to assist in data analysis. A line diagram has been used for showing the trend line of fatalities and their related risk factors. ANOVA analysis was made to identify the variance of several factors to compare the mean between Sikkim, Himachal Pradesh, Uttarakhand, and JK & Ladakh. Further, Post-hoc Tukey test was applied to denote where the significant difference of variances lie within the data set.

37.2.4 Risk Estimation of Mountaineering Tourism

Mountaineering is generally a very high-risk and individualistic activity in which participation is rarely recorded, and there are no standardized teams, national league competitions, or player rankings (Malcolm 2001). In 1993, Clare Lattimore published a research article on quantifying the risks of climbing Denali. Foray (1978, 1980), Hackett and Rennie (1979), and Mahajan (1981) describe that most of the researchers have not dealt primarily with climbing accidents but rather with specific medical problems associated with climbing, such as altitude problems. In particular, mountaineering risks agglomerations have never been classified, and it is challenging to predict the vulnerable factors. It was also very challenging to estimate the mountaineering risk events at high altitude. It is also difficult to identify a suitable denominator for the calculation of rates of climbing-related mortality or morbidity (Malcolm 2001). Despite the risk factors, data are available in other mountains such as the Alps, Aconcagua, Andes, and Nepal Himalaya. But, the Indian Himalaya has been neglected in this regard for a long time. Probably, it is the first-ever attempt at mountaineering risk classification and summarized risk estimation concerning the Indian mountaineer's perspective.

37.3 Results

37.3.1 Descriptive Statistics of Several Risks in Indian Himalaya

The bad weather (66.4%) dominates the mountain expedition in the Indian Himalaya (Table 37.1). Although other parameters are very small in percentage, they are indiscriminately involved in mountaineering risks' contributing factors. The risks are not evenly distributed over a given mountain and during an expedition (Nordby and Weinbruch 2013). Physical hazards such as avalanches, rock, and ice falls are present in ice climbing, and physiological altitude-induced adaptations must also be factored into the climbs (Schoffl et al. 2010). Collinson et al. (2018) also express their experience in their high-altitude mountaineering study; the weather is an integral part of high-altitude mountaineering, requiring expedition members to pay close attention to and monitor weather conditions at every moment. Acute mountain sickness, frostbite, mountain sickness, and cardiac arrests are some of the considerable concrete challenges for those who venture to these heights. 12 major mountaineering risk events were identified,: troubled mountaineers and Sherpa guides for climbing in the snow-capped mountain. Nordby and Weinbruch (2013) reported that, human and environmental factors confound risk estimates. It clarifies that except bad weather (66.4%), steep slope (8.8%), avalanche (5.6%), technical cause (4%), mountain sickness (2.4%), rock fall (1.6%), fall from mountain slope (1.6%), ration shortage (1.6%), lack of area knowledge (0.8%) and alpine style of mountain climbing (0.8%) are the other important alternative factors which might create a problematic situation behind of successful summit bid in the high mountain. The trend of mountaineering risks demonstrates and teaches us that these are more common and influential factors in high altitude-mountaineering.

Due to the lack of equal samples from different four states, the present study reviewed the variance of dangers of mountaineering in the four mountainous states of India. A closer look at the dangers of mountaineering reveals how the bad weather, as it has in the past, has affected various states, meaning that weather is a hazardous component in mountaineering, and in the case of four Himalayan states, only bad weather is at the forefront of the various risks. For example, out of 8 unsuccessful expeditions of Sikkim, 5 expeditions were abandoned due to bad weather, 1 expedition by crevasse, 1 expedition by fall from the mountain slope, and 1 expedition postponed due to alpine style.

Of the 50 unsuccessful expeditions of Himachal Pradesh, 36 expeditions were abandoned due to bad weather. 1 expedition abandoned due to death of all participants. Another 2 expeditions abandoned due to avalanche. 2 expeditions abandoned due to steepness of the mountain slope. Another 2 expeditions were unsuccessful due to the steep mountain slope and mountain sickness respectively. 2 expeditions were postponed due to huge rock fall, and another 2 expeditions abandoned due to crevasse. 4 expeditions abandoned due to technical causes. In the state of Uttarakhand, of the 60 unsuccessful expeditions, 38 expeditions were abandoned due to bad weather, 4 expeditions were unsuccessful due to death, 4 expeditions were

Table 37.1 Unsuccessful expedition and related risk factors

Sl. No.	Risk factors	Number of risk factors $N = 125$	% of risk factors
1	Bad weather	83	66.4
2	Death	6	4.8
3	Avalanche	7	5.6
4	Steep slope	11	8.8
5	Mountain sickness	3	2.4
6	Rock fall	2	1.6
7	Crevasse	2	1.6
8	Fall	2	1.6
9	Technical Issue	5	4.0
10	Lack of area knowledge	1	0.8
11	Alpine style	1	0.8
12	Ration shortage	2	1.6

Source IMF (New Delhi, 2019), data compiled by author

unsuccessful due to the avalanche, 9 expeditions were unsuccessful due to the steep slope, 2 expeditions becomes unsuccessful due to the mountain sickness, and 1 expedition was unsuccessful due to the technical causes. In Jammu Kashmir & Ladakh, out of 7 unsuccessful expeditions, 4 expeditions were abandoned due to the bad weather, 1 was due to death, 1 expedition was abandoned by the avalanche, and 1 expedition was abandoned due to shortages of ration (Table 37.2).

37.3.2 ANOVA and Post Hoc Tukey Test

According to the characteristics of risk factor in different states, almost all factors are homogeneous and there are no significant difference in the risks in the four states ($p = 0.346$) of Indian Himalaya. However, the Post hoc Tukey test (significance level, $p = 0.05$) indicated the risk factors were not significantly different between Sikkim and Himachal Pradesh ($p = 0.448$), Sikkim and Uttarakhand ($p = 0.345$), Sikkim and JK & Ladakh ($p = 0.954$), Himachal Pradesh and Uttarakhand ($p = 0.989$), Himachal Pradesh and JK & Ladakh ($p = 0.882$), Uttarakhand and JK & Ladakh ($p = 0.806$).

37.3.3 Risk Factors Agglomeration

Classification of mountaineering risks is not too much easy like other risks. On the basis of the limited number of data, the study classified the mountaineering risk into five clusters resembling events or components are more or less risky along with various factors (Table 37.3).

37.3.4 Classification of Mountaineering Risks

Five types of mountaineering risks have been classified through the hierarchical cluster method, i.e. (i) Very high vulnerable risk factors, (ii) High vulnerable risk factors, (iii) Moderately vulnerable risk factors, (iv) Low vulnerable risk factors, and (v) Very low vulnerable risk factors. The only bad weather event has fallen under the very high vulnerable factors, and 141 expeditions (56.4%) have been postponed due to this component. Bad weather is one of the most extreme and challenging environments for high-altitude mountaineering (Collinson et al., 2018). Mountain sickness (17), steep slope (22), rock fall (3), and frost bite (1) are in the high vulnerable risk group. The prevalence of acute mountain sickness (AMS) is correlated with altitude

Table 37.2 State wise unsuccessful expeditions and causes

Sl. No.	Unsuccessful factors	Sikkim N = 8	Himachal Pradesh N = 50	Uttarakhand N = 60	Jammu Kashmir and Ladakh N = 7
1	Bad weather	5	36	38	4
2	Death	0	1	4	1
3	Avalanche	0	2	4	1
4	Steep slope	0	2	9	0
5	Mountain sickness	0	1	2	0
6	Rock fall	0	2	0	0
7	Crevasse	1	1	0	0
8	Fall	1	1	0	0
9	Technical Issue	0	4	1	0
10	Lack of area knowledge	0	0	1	0
11	Alpine style	1	0	0	0
12	Ration shortage	0	0	1	1

Source IMF, New Delhi [compiled by the researcher]

Table 37.3 Unsuccessful expedition and risk agglomerations for mountaineering

Group	Cluster name	Agglomeration of risk factors	Number of cases	% of risks
Cluster 1	Very high vulnerable risk factors	Bad weather	141	56.4
Cluster 2	High vulnerable risk factors	Mountain sickness, steep slope, rock fall, frost bite	43	17.2
Cluster 3	Moderately vulnerable risk factors	Avalanche, death	28	11.2
Cluster 4	Low vulnerable risk factors	Crevasse, fall, technical cause	21	8.4
Cluster 5	Very low vulnerable risk factors	Alpine style, lack of area knowledge, ration shortage, drowned into the river, cardiac arrest	17	6.8

Source IMF (New Delhi), data compiled and risk classification has made by the researcher using hierarchical cluster method

(Maggiorini et al., 1990; Mairer et al., 2009). In this group, 43 expeditions (17.2%) have been postponed due to these 4 components in the Indian Himalaya. Avalanches (17) and deaths (11) are under the moderate vulnerable risk group. Avalanche is relatively common as the immediate cause of death (Boyd et al., 2009; McClung, 2016). In cluster number 3, 28 expeditions (11.2%) were postponed due to both risk components. Three components are crevasse (4), fall (4), technical causes, or technical issues (13) have fallen in the low vulnerable group. These three risk components are responsible for 21 unsuccessful expeditions. Alpine style (5), lack of area knowledge (1), ration shortage (4), drowning in the river (1), and cardiac arrest (6) have fallen in the low vulnerable risks groups. These five risk components are responsible for 17 unsuccessful expeditions (6.8%) several times (Table 37.3) of the expedition.

37.4 Discussions

Indeed, mountaineering and death are coherently related from a few centuries ago. According to climbing history while the British climber Albert Mummery led the first expedition in 1895 to make a serious attempt to climb one of the 14 mountains in the world that are more than 8000 m in height (Boyce and Bischak 2010). But unfortunately, during the expedition of Nanga Parbat (now in Pakistan), Mummery and two companions died in this expedition's attempt. Deaths are more common in mountaineering adventures. Pollard and Clarke (1988) identified some factors involved in deaths and fatalities in high mountains; they have studied the deaths on British expeditions to peaks above 7000 m in Greater Himalaya. In 2001, Murray Malcolm estimated the risk of death associated with mountaineering in the Mt Cook National Park and described some fatal events' characteristics. The American Alpine Club (AAC) reported three times as many injured climbers and nearly twice as many climbing-related deaths per year despite significant advances in safety equipment and widely available technical instruction (Williamson 1985). The summit bid is generally the most dangerous part of an expedition for members, whereas most high-altitude porters die during route preparation (Nordby and Weinbruch 2013). The Canadian Alpine Club (CAC) recorded 958 accidents and then analyzed ice climbing accidents over 30 years to reveal that 92 mountaineers were injured while ice climbing and 30 were fatal among them. According to the UIAA medical commission meeting report (2008), fatalities are technically possible, but there is no danger in indoor climbing. Fatalities, and falls are not very dangerous; risk is calculable for sports climbing, low elevation, and technically easy peaks. Falls, fatalities, and deaths are more frequent and difficult to calculate in traditional and high Himalayan (7000–8000 m) or challenging peaks (Schöffl et al. 2012). But the fatality rate from mountaineering accidents varies from one region to another. Lack et al. (2012a, b) reported a mortality rate of 5.5%, Ferris (1963) found a fatality rate of 41% in the USA, Forrester et al. (2018) reported a mortality rate of <1% in the USA, Bowie et al. (1988) found 6% in Yosemite National Park (YNP) and Schussman and Lutz (1982) 20% in Grand Teton National Park (GTNP). Addis and Baker (1989) classified mountain climbing injuries into two types, fatal and non-fatal and 11 significant causes responsible for both types of injuries. Risk factors are generally unpredictable during the climbing time, and no one knows what can be happened ahead of climbing members. Addis and Baker (1989) claimed that avalanche or icefall caused 9% of all injuries, but 28% of deaths occurred, while they were investigated in the entire National Park Service (NPA) of the USA. Shussman and Lutz (1982) claimed mountaineering risks and accidents in Grand Teton National Park from 1970–1980, with 144 accidents and 30 deaths. Out of 30 deaths, 16 (53%) involved ascending rather than descending, 10 (30%) occurred while progressing on the rock, and 16 of the victims (53%) were leading personnel, but they did not clarify the specific causes of deaths. They identified 8 major causes of fatalities in GTNP (Grand Teton National Park), which were (a) fall from rock, (b) avalanche, (c) slipping on snow, (d) failed rappel and fall, (e) glissading, (f) rock pulled loose and caused fall, (g) pulled from a belay and (h) stranded and hypothermia.

During various climbing expeditions, Indian mountaineers have been involved in numerous accidents. At a high altitude, several dreadful events can occur at any time in any stage, and fatalities or deaths may occur as a response to these causes. We have identified eight significant causes of Indian mountaineers' deaths while mountaineering in the Indian Himalaya. After evaluating the casualty variables, it was observed that avalanches were responsible for many casualties. Total 13 (50%) fatalities were occurred due to avalanches, 2 (7.7%) were by mountain sickness, 1 (3.8%) were by fall from the steep slope during ascent or descent time. Only 3 (11.5%) were died by technical causes, 2 (7.7%) were by unknown, 2 (7.7%) were by exhaustion, 2 (7.7%) were disappeared in the mountain. For example, perhaps George Mallory disappeared in 1920 with another climber into the mists near the summit of Everest and never returned (Thompson 1997). Only 1 climber (3.8%) died by drowning into the strong glacial river while mountaineering in the Indian Himalaya (Table 37.4).

Altitude is another measure to estimate the risk and deaths in the high Himalayan region. Death rates are gradually increasing with altitude (Table 37.5). It can imagine that altitude is another prime factor in the deaths of climbing and other members. The probability of dying from falling (rock or ice fall) increased with altitude (Huey et al. 2020). According to the Himalayan database, no casualties or deaths happened between 5000 and 5999 m, only 1 member has died between 6000 and 6999 m, 3 members have died in between 7000 and 7999 m, and 6 members have died between 8000 and 8850 m in height. There were no deaths of female climbers, though their participation is very poor in number. Drastically all casualties were made by male members only.

Hence with increasing height, the probability of death will also increase. Cold wind, thin air, low oxygen, exhaustion, and other co-factors are responsible for that. Mountaineers can live only for short periods at the extreme heights where countless challenges are existed such as weather, open and hidden crevasses, falling into the avalanche, rock and icefall, fall from the mountain slope, ration shortage, etc. make the high altitude form of mountaineering an extremely dangerous realization (Fig. 37.2). "Difficulties asserted by weather and atmospheric condition,

Table 37.4 Deaths and factors in mountaineering

Causes of deaths	No. of deaths of Indian mountaineers	% of deaths
Avalanche	13	50.00
Mountain sickness	2	7.7
Fall	1	3.8
Technical cause	3	11.5
unknown cause	2	7.7
Exhaustion	2	7.7
Disappeared	2	7.7
Drowned into river	1	3.8
Total	26	100

Source IMF (2019, New Delhi) [data compiled by author]

Table 37.5 The death rate for mountaineers climbing between 5000 and 8850 m in the Indian Himalaya between 1950 and 2018

Peak altitude range (in metre)	Individuals above base camp	Total deaths above base camp	Death rate	Male		Female	
				Number	Rate	Number	Rate
5000–5999	0	0	0.00	0	0.00	0	0.00
6000–6999	201	1	0.59	1	0.65	0	0.00
7000–7999	244	3	1.32	3	1.33	0	0.00
8000–8850	173	6	5.26	6	5.50	0	0.00

Source Himalayan Data Base as of 5/12/2020

Fig. 37.2 The number of deaths and its factors. *Source* Figure prepared by author

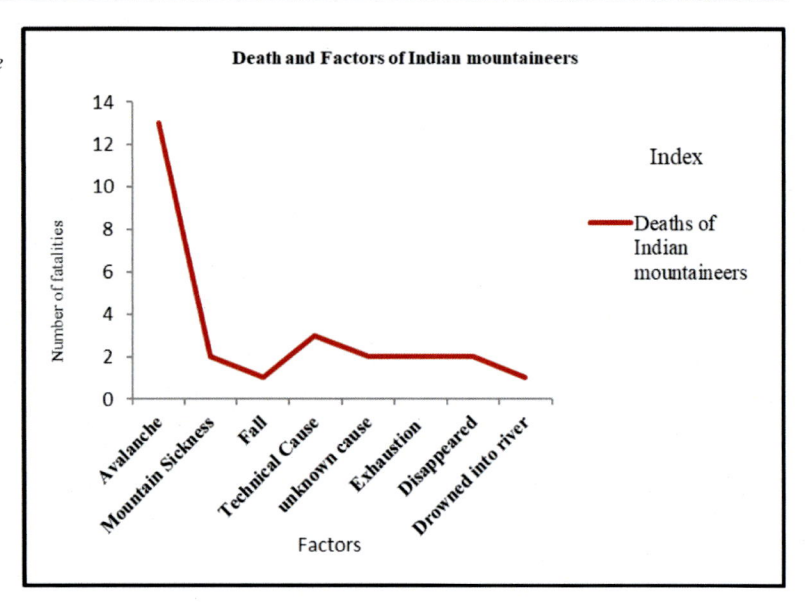

[Graph: Death and Factors of Indian mountaineers — Number of fatalities vs Factors (Avalanche, Mountain Sickness, Fall, Technical Cause, unknown cause, Exhaustion, Disappeared, Drowned into river); Index: Deaths of Indian mountaineers]

the concrete risks come from reduced oxygen at high altitudes, and extremely low and cold temperatures are severe (Collinson et al. 2018)."

According to the Himalayan Database, after a successful ascent or summit, of the 10 deaths, 6 climbers have died after a successful ascent of a mountain peak. Among them, 2 climbers have died between 7000–7999 m and 4 climbers between 8000–8850 m. It also proves that altitude is a major factor in the death of climbing members. The death rates are 6.66 persons/100 climbers between 7000–7999 m and 28.57 persons/100 climbers between 8000–8850 m altitude ranges. This (8000–8850 m) range is called the death zone (Table 37.6). Anyone can be dying within this altitude range at any time.

37.5 Conclusion

The study concludes that weather plays a vital role in a high-altitude mountain expedition, risks vary from one state to another, deaths can be produced at any time by the different factors during the climbing hours, and death rates are varied in different altitudes, and death rates are increasing with increasing altitude. There is more than one risk factor that influences mountaineers to climb the mountain. Some factors might have more influenced to climb than the other factors. Bad weather is the most vulnerable component among other risk events. Every component can be harmful at any time during high-altitude

Table 37.6 Death of Indian climbing members after the ascent of a mountain peak (including Sherpa), 1950–2018

Height in meter	Total deaths	Rate	Male		Female	
			Number	Rate	Number	Rate
5000–5999	0	0.00	0	0.00	0	0.00
6000–6999	0	0.00	0	0.00	0	0.00
7000–7999	2	6.66	2	6.66	0	0.00
8000–8850	4	28.57	4	28.57	0	0.00

Source Himalayan Data Base as of 5/12/2020

climbing. Acclimatization is a better process for the prevention of mountain sickness. Mountaineers can be well aware of the weather report from time to time with the help of modern instruments. Every mountaineer should be well equipped and trained from a reputed institute. They should not be over-ambitious in summiting the mountain, and while they will face a problem of life risks at high altitude, at that moment, they will be getting down from there. Mountaineering risk classification can be helpful to the new climbers to take the decisions before the start of their expedition. Due to the limited data future research can be needed to make a relatively easy way to the mountaineers.

References

Addis DG, Baker SP (1989) Mountaineering and rock-climbing injuries in US national parks, pp 125–129
Bott E (2009) Big mountain, big name: globalised relations of risk in Himalayan mountaineering. J Tour Cult Chang 7(4):287–301
Bowie WS, Hunt TK, Allen HA (1988) Jr. Rock climbing injuries in Yosemite National Park. West J Med 149:172–177
Boyce RJ, Bischak PD (2010) Learning by doing, knowledge spillovers, and technological and organizational change in high-altitude mountaineering. J Sports Econ 11(5):496–532
Boyd J, Haegali P, Abulaban RB, Shuster M, Butt JC (2009) Patterns of death among avalanche fatalities: a 21-year review. CMAJ 180(5):507–512
Canadian Alpine Club (n.d.) Accidents in North American mountaineering. Canadian Alpine Club
Cater CI (2006) Playing with risk? Participant perceptions of risk and management implications in adventure tourism. Tour Manag 27(2):317–325
Christensen ED, Lacsina EQ (1999) Mountaineering fatalities on Mount Rainier, Washington, 1977–1997: autopsy and investigative findings. Am J Forens Med Pathol 20(2):173–179
Clare LM (1993) Mountaineering emergencies on Denali. J Wildern Med 4:358–362
Collinson JA, Crust L, Swann C (2018) Embodiment in high-altitude mountaineering: sensing and working with the weather. SAGE 25(1):90–115
Ferris BG (1963) Mountain-climbing accidents in the United States. N Engl J Med 268:430–431
Firth PG, Zheng H, Windsor JS, Sutherland AI, Imray CH, Moore GWK, Semple JL, Roach RC, Salisbury RA (2008) Mortality on Mount Everest, 1921–2006: descriptive study. Bmj 337
Foray JJF (1978) Les gelures de montagne, vol 74, pp 352–355
Foray JCC (1980) Less accidents de haute montagne, pp 375–389
Forrester JD, Tran K, Tennakoon L, Staudenmayer K (2018) Climbing-related injury among adults in the United States: 5 year analysis of the national emergency department sample. Wildern Environ Med 29:425–430
Gong X, Richman MB (1995) On the application of cluster analysis to growing season precipitation data in North America east of the rockies. J Clim 8(4):897–931
Güçdemir H, Selim H (2015) Integrating multi-criteria decision making and clustering for business customer segmentation. Ind Manag Data Syst 115(6):1022–1040
Hackett PH, Rennie D (1979) Rales, peripheral edema, retinal hemorrhage and acute mountain sickness. Am J Med 67:214–218
Helms M (2015) Factors affecting evaluations of risks and hazards in mountaineering. North Carolina Sate University
Huey RB, Carroll C, Salisbury R, Wang J-L (2020) Mountaineers on Mount Everest: effects of age, sex, experience, and crowding on rates of success and death. PLoS ONE 15(8):1–16
Krakauer J (2009) Into thin air: a personal account of the Mount Everest disaster. Anchor
Lack DA, Sheets AL, Entin JM, Christenson DC (2012a) Rock Climbing rescues: causes, injuries, and trends in Boulder County, Colorado. Wildern Environ Med 23:223–230
Lack DA, Sheets AL, Entin JM, Christenson DC (2012b) Rock climbing rescues: causes, injuries and trends in Boulder country, Colorado. Wildern Environ Med 23:223–230
Maggiorini M, Bühler B, Walter M, Oelz O (1990) Prevalence of acute mountain sickness in the Swiss Alps. BMJ 301(853):853–855
Mahajan SL, Myers TJ, Baldini MG (1981) Disseminated intravascular coagulation during rewarming following hypothermia. JAMA 245:2517–2518
Mairer K, Wille M, Bucher T, Burtscher M (2009) Prevalence of acute mountain sickness in the Eastern Alps. High Alt Med Biol 10(3):239–245
Malcolm M (2001) Mountaineering fatalities in Mt Cook National Park. N Z Med J 114(1127):78–80
Mandelli G, Angriman A (n.d.) Scales of difficulty in climbing, pp 1–26
McCLUNG DM (2016) Avalanche character and fatalities in the high mountains of Asia. Ann Glaciol 57(71):114–118
Nordby K-C, Weinbruch S (2013) Fatalities in high altitude mountaineering: a review of quantitative risk estimates. High Alt Med Biol 14(4):346–359
Pollard A, Clarke C (1988) Deaths during mountaineering at extreme. UIAA Mountain Medicine Data Centre, The Lancet

Priest S, Gass MA (1997) Effective leadership in adventure programming. Human Kinetics. Inc., Champaign

Ridden JMC (1983) An estimate of the fatal accident frequency rate from mountaineering fatalities in Peru. Accid Anal Prev 15:309–312

Sale R, Clare J (2000) To the top of the world. Climbing the world's 14 highest mountains

Schöffl V, Morrison A, Schöffl I, Küpper T (2012) Epidemiology of injury in adventure and extreme sports. Med Sports Sci 58:17–43

Schoffl V, Audry M, Ulrich S, Schoffl I, Thomas K (2010). Evaluation of injury and fatality risk in rock and ice climbing. Sports Med 40(8):657–679

Schussman LC, Lutz LJ (1982) Mountaineering and rock-climbing accidents. Phys Sportsmed 10(6):52–61

Seigneur V (2006) The problems of the defining the risk: the case of mountaineering. Hist Soc Res/Historische Sozialforschung 245–256

Smith LO (2006) Alpine climbing: injuries and illness. Phys Med Rehab Clin 17(3):633–644

Thompson M (1997) Thick resistance: death and the cultural construction of agency in Himalayan mountaineering. JSTOR 59:135–162

West JB (1998) High life a history of high-altitude physiology and medicine. Springer, New York

West JB (2000). Human limits for hypoxia: the physiological challenge of climbing Mt. Everest. Ann New York Acad Sci 899(1):15–27

Westensee J, Rogé I, Van Roo JD, Pesce C, Batzli S, Courtney DM, Lazio MP (2013) Mountaineering fatalities on Aconcagua. High Alt Med Biol 14(3):298–303

Williamson JE (1985) Accidents in North American mountaineering

Wilson R, Mills WJ Jr, Rogers DR, Propst MT (1978) Death on Denali: fatalities among climbers in Mount McKinley National Park from 1903 to 1976-analysis of injuries, illnesses and rescues in 1976. West J Med 128(6):471

Windsor JS, Firth PG, Grocott MP, Rodway GW, Montgomery HE (2009) Mountain mortality: a review of deaths that occur during recreational activities in the mountains. Postgrad Med J 85(1004):316–321